T0398733

GRAPHENE SCIENCE HANDBOOK

Nanostructure and Atomic Arrangement

GRAPHENE SCIENCE HANDBOOK

Nanostructure and Atomic Arrangement

EDITED BY

Mahmood Aliofkhazraei • Nasar Ali
William I. Milne • Cengiz S. Ozkan
Stanislaw Mitura • Juana L. Gervasoni

CRC Press
Taylor & Francis Group
Boca Raton London New York

CRC Press is an imprint of the
Taylor & Francis Group, an **informa** business

CRC Press
Taylor & Francis Group
6000 Broken Sound Parkway NW, Suite 300
Boca Raton, FL 33487-2742

© 2016 by Taylor & Francis Group, LLC
CRC Press is an imprint of Taylor & Francis Group, an Informa business

No claim to original U.S. Government works

Printed on acid-free paper
Version Date: 20151124

International Standard Book Number-13: 978-1-4665-9137-0 (Hardback)

This book contains information obtained from authentic and highly regarded sources. Reasonable efforts have been made to publish reliable data and information, but the author and publisher cannot assume responsibility for the validity of all materials or the consequences of their use. The authors and publishers have attempted to trace the copyright holders of all material reproduced in this publication and apologize to copyright holders if permission to publish in this form has not been obtained. If any copyright material has not been acknowledged please write and let us know so we may rectify in any future reprint.

Except as permitted under U.S. Copyright Law, no part of this book may be reprinted, reproduced, transmitted, or utilized in any form by any electronic, mechanical, or other means, now known or hereafter invented, including photocopying, microfilming, and recording, or in any information storage or retrieval system, without written permission from the publishers.

For permission to photocopy or use material electronically from this work, please access www.copyright.com (http://www.copyright.com/) or contact the Copyright Clearance Center, Inc. (CCC), 222 Rosewood Drive, Danvers, MA 01923, 978-750-8400. CCC is a not-for-profit organization that provides licenses and registration for a variety of users. For organizations that have been granted a photocopy license by the CCC, a separate system of payment has been arranged.

Trademark Notice: Product or corporate names may be trademarks or registered trademarks, and are used only for identification and explanation without intent to infringe.

Library of Congress Cataloging-in-Publication Data

Names: Aliofkhazraei, Mahmood, editor. | Ali, Nasar, editor. | Milne, W. I. (William I.), editor. | Ozkan, Cengiz S., editor. | Mitura,
    Stanislaw, 1951- editor. | Gervasoni, Juana L., editor.
Title: Graphene science handbook. Nanostructure and atomic arrangement / edited by Mahmood Aliofkhazraei, Nasar Ali, William I.
    Milne, Cengiz S. Ozkan, Stanislaw Mitura, and Juana L. Gervasoni.
Other titles: Nanostructure and atomic arrangement
Description: Boca Raton, FL : CRC Press, Taylor & Francis Group, 2016. | "2016 | Includes bibliographical references and index.
Identifiers: LCCN 2015042989| ISBN 9781466591370 (hardcover ; alk. paper) |
ISBN 1466591374 (hardcover ; alk. paper)
Subjects: LCSH: Graphene--Handbooks, manuals, etc. | Nanostructured materials--Handbooks, manuals, etc.
Classification: LCC QD341.H9 G687 2016 | DDC 547/.61--dc23
LC record available at http://lccn.loc.gov/2015042989

Visit the Taylor & Francis Web site at
http://www.taylorandfrancis.com

and the CRC Press Web site at
http://www.crcpress.com

Contents

Preface...ix
Editors...xi
Contributors...xiii

SECTION I Atomic Arrangement and Defects

Chapter 1   Graphene Heterostructures...3

               Zheng Liu and Hong Wang

Chapter 2   Atomic-Scale Defects and Impurities in Graphene 21

               Rocco Martinazzo

Chapter 3   Atomic Arrangement and Its Effects on Electronic Structures of Graphene from Tight-Binding Description 39

               Sirichok Jungthawan and Sukit Limpijumnong

Chapter 4   Graphene Plasmonics: Light–Matter Interactions at the Atomic Scale 63

               Pai-Yen Chen and Mohamed Farhat

Chapter 5   Graphene/Polymer Nanocomposites: Crystal Structure, Mechanical and Thermal Properties 77

               Fabiola Navarro-Pardo, Ana Laura Martínez-Hernández, and Carlos Velasco-Santos

Chapter 6   Graphene-Like Structures as Cages for Doxorubicin 99

               Iva Blazkova, Pavel Kopel, Marketa Vaculovicova, Vojtech Adam, and Rene Kizek

Chapter 7   Mathematical Modeling for Hydrogen Storage Inside Graphene-Based Materials...........111

               Yue Chan

Chapter 8   Morphology of Cylindrical Carbon Nanostructures Grown by Catalytic Chemical Vapor Deposition Method.............123

               S. Ray, M. Jana, and A. Sil

Chapter 9   sp^2 to sp^3 Phase Transformation in Graphene-Like Nanofilms.......................................147

               Long Yuan, Zhenyu Li, and Jinlong Yang

Chapter 10   Symmetry and Topology of Graphenes...159

               A. R. Ashrafi, F. Koorepazan-Moftakhar, and O. Ori

SECTION II Modified Graphene

Chapter 11   N-Doped Graphene for Supercapacitors 167

               Dingsheng Yuan and Worong Lin

Chapter 12 Electrical and Optical Properties and Applications of Doped Graphene Sheets 179

　　　　　　Ki Chang Kwon and Soo Young Kim

Chapter 13 Chemical Modifications of Graphene via Covalent Bonding 207

　　　　　　Liang Cui, Dongjiang Yang, and Jingquan Liu

Chapter 14 Functionalization and Vacancy Effects on Hydrogen Binding in Graphene 221

　　　　　　A. Tapia, C. Cab, and G. Canto

Chapter 15 Modifications of Electronic Properties of Graphene by Boron (B) and Nitrogen (N) Substitution 231

　　　　　　Debnarayan Jana, Palash Nath, and Dirtha Sanyal

SECTION III Characterization

Chapter 16 Electronic Structure and Topological Disorder in sp^2 Phases of Carbon 249

　　　　　　Y. Li and D. A. Drabold

Chapter 17 3D Macroscopic Graphene Assemblies 263

　　　　　　Marcus A. Worsley, Juergen Biener, Michael Stadermann, and Theodore F. Baumann

Chapter 18 3D-AFM-Hyperfine Imaging of Graphene Monolayers Deposit on YBCO-Superconducting Surface 277

　　　　　　Khaled M. Elsabawy

Chapter 19 Phonon Spectrum and Vibrational Thermodynamic Characteristics of Graphene Nanofilms 289

　　　　　　Alexander Feher, Sergey Feodosyev, Igor Gospodarev, Eugen Syrkin, and Vladimir Grishaev

Chapter 20 Tuning Atomic and Electronic Properties of Graphene by Selective Doping 305

　　　　　　Cecile Malardier-Jugroot, Michael N. Groves, and Manish Jugroot

Chapter 21 Scanning Electron Microscopy of Graphene 319

　　　　　　Yoshikazu Homma, Katsuhiro Takahashi, Yuta Momiuchi, Junro Takahashi, and Hiroki Kato

Chapter 22 Tunneling Current of the Contact of the Curved Graphene Nanoribbon with Metal and Quantum Dots 327

　　　　　　Mikhail B. Belonenko, Natalia N. Konobeeva, Alexander V. Zhukov, and Roland Bouffanais

Chapter 23 Using Few-Layer Graphene Sheets as Ultimate Reference of Quantitative Transmission Electron Microscopy 341

　　　　　　Wang-Feng Ding, Bo Zhao, and Fengqi Song

SECTION IV Recent Advances

Chapter 24 Computational Modeling of Graphene and Carbon Nanotube Structures in the Terahertz, Near-Infrared, and Optical Regimes 359

　　　　　　M. F. Pantoja, D. Mateos Romero, H. Lin, S. G. Garcia, and D. H. Werner

Chapter 25 Design and Properties of Graphene-Based Three Dimensional Architectures 375

Chunfang Feng, Ludovic F. Dumée, Li He, Zhifeng Yi, Zheng Peng, and Lingxue Kong

Chapter 26 Electronic Structure of Graphene-Based Materials and Their Carrier Transport Properties 401

Wen Huang, Argo Nurbawono, Minggang Zeng, Gaurav Gupta, and Gengchiau Liang

Chapter 27 Graphene-Enabled Heterostructures: Role in Future-Generation Carbon Electronics 423

Nikhil Jain and Bin Yu

Chapter 28 Recent Progresses and Understanding of Lithium Storage Behavior of Graphene Nanosheet Anode for Lithium Ion Batteries 435

Xifei Li and Xueliang Sun

Chapter 29 Study of Transmission, Transport, and Electronic Structure Properties of Periodic and Aperiodic Graphene-Based Structures 453

Heraclio García-Cervantes, Rogelio Rodríguez-González, José Alberto Briones-Torres, Juan Carlos Martínez-Orozco, Jesús Madrigal-Melchor, and Isaac Rodríguez-Vargas

Chapter 30 Benefits of Few-Layer Graphene Structures for Various Applications 479

I. V. Antonova and V. Ya. Prinz

Chapter 31 Designing Carbon-Based Thin Films from Graphene-Like Nanostructures 497

Cecilia Goyenola and Gueorgui K. Gueorguiev

Chapter 32 Graphene-Based Hybrid Composites 517

Antonio F. Ávila, Diego T. L. da Cruz, Hermano Nascimento Jr., and Flávio A. C. Vidal

Chapter 33 Graphene Dispersion by Polymers and Hybridization with Nanoparticles 529

Po-Ta Shih, Kuo-Chuan Ho, and Jiang-Jen Lin

Chapter 34 Magnetocaloric Effect of Graphenes 541

M. S. Reis and L. S. Paixão

Chapter 35 Mode-Locked of Fiber Laser Employing Graphene-Based Saturable Absorber 555

Pi Ling Huang, Chao-Yung Yeh, Jiang-Jen Lin, Lain-Jong Li, and Wood-Hi Cheng

Index 573

Preface

The theory behind "graphene" was first explored by the physicist Philip Wallace in 1947. However, the name "graphene" was not actually coined until 40 years later, where it was used to describe single sheets of graphite. Ultimately, Professor Geim's group in Manchester (UK) was able to manufacture and see individual atomic layers of graphene in 2004. Since then, much more research has been carried out on the material, and scientists have found that graphene has unique and extraordinary properties. Some say that it will literally change our lives in the twenty-first century. Not only is graphene the thinnest possible material, but it is also about 200 times stronger than steel and conducts electricity better than any other material at room temperature. This material has created huge interest in the electronics industry, and Konstantin Novoselov and Andre Geim were awarded the 2010 Nobel Prize in Physics for their groundbreaking experiments on graphene.

Graphene and its derivatives (such as graphene oxide) have the potential to be produced and used on a commercial scale, and research has shown that corporate interest in the discovery and exploitation of graphene has grown dramatically in the leading countries in recent decades. In order to understand how this activity is unfolding in the graphene domain, publication counts have been plotted in Figure P.1. Research and commercialization of graphene are both still at early stages, but policy in the United States as well as in other key countries is trying to foster the concurrent processes of research and commercialization in the nanotechnology domain.

Graphene can be produced in a multitude of ways. Initially, Novoselov and Geim employed mechanical exfoliation by using a Scotch tape technique to produce monolayers of the material. Liquid-phase exfoliation has also been utilized. Several bottom-up or synthesis techniques developed for graphene include chemical vapor deposition, molecular beam epitaxy, arc discharge, sublimation of silicon carbide, and epitaxy on silicon carbide.

The *first volume* of this handbook concerns the fabrication methods of graphene. It is divided into four sections: (1) fabrication methods and strategies, (2) chemical-based methods, (3) nonchemical methods, and (4) advances of fabrication methods.

Carbon is the sixth most abundant element in nature and is an essential element of human life. It has different structures called carbon allotropes. The most common crystalline forms of carbon are graphite and diamond. Graphite is a three-dimensional allotrope of carbon with a layered structure in which tetravalent atoms of carbon are connected to three other carbon atoms by three covalent bonds and form a hexagonal network structure. Each one of these aforementioned layers is called a graphene layer or sheet. Each sheet is placed in parallel on other sheets. Hence, the fourth valence electron connects the sheets to each other via van der Waals bonding. The covalent bond length is 0.142 nm. The bonds that are formed by carbon atoms between layers are weak; therefore, the sheets can slide easily over each other. The distance between layers is 0.335 nm. Due to its unique structure and geometry, graphene possesses remarkable physical–chemical properties, including a high Young's modulus, high fracture strength, excellent electrical and thermal conductivity, high charge carrier mobility, large specific surface area, and biocompatibility.

These properties enable graphene to be considered as an ideal material for a broad range of applications, ranging from quantum physics, nanoelectronics, energy research, catalysis, and engineering of nanocomposites and biomaterials. In this context, graphene and its composites have emerged as a new biomaterial, which provides exciting opportunities for the development of a broad range of applications, such as nanocarriers for drug delivery. The building block of graphene is completely different from other graphite materials and three-dimensional geometric shapes of carbon, such as zero-dimensional spherical fullerenes and one-dimensional carbon nanotubes.

The *second volume* of this handbook is predominantly about the nanostructure and atomic arrangement of graphene. The chapters in this volume focus on atomic arrangement and defects, modified graphene, characterization of graphene and its nanostructure, and also recent advances in graphene nanostructures. The planar structure of graphene provides an excellent opportunity to immobilize a large number of substances, including biomolecules and metals. Therefore, it is not surprising that graphene has generated great interest for its nanosheets, which nowadays can serve as an excellent platform for antibacterial applications, cell culture, tissue engineering, and drug delivery.

It is possible to produce composites reinforced with graphene on a commercial scale and low cost. In these composites, the existence of graphene leads to an increase in conductivity and strength of various three-dimensional materials. In addition, it is possible to use cheaply manufactured graphene in these composites. For example, exfoliation of graphite is one of the cheapest graphene production techniques. The behavior of many two-dimensional materials and their equivalent three-dimensional forms are completely different. The origin of the aforementioned differences in the behavior of these materials is associated with the weak forces that hold a large number of single layers together to create a bulk material. Graphene can be used in nanocomposites. Currently, researchers have been able to produce several tough and light materials by adding small amounts of graphene to metals, polymers, and ceramics. The composite materials usually show better electrical conductivity characteristics compared with pure bulk materials, and they are also more resistant against heat.

The *third volume* describes graphene's electrical and optical properties and also focuses on nanocomposites and their applications. The *fourth volume* relates to the mechanical

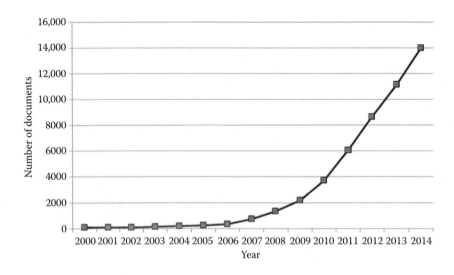

FIGURE P.1 Number of documents published around graphene during recent years, extracted fromScopus search engine by searching "graphene" in title + keywords + abstract.

and chemical properties of graphene and cites recent developments. The *fifth volume* presents other topics, such as size effects in graphene, characterization, and applications based on size-affected properties. In recent years, scientists have produced advanced composites using graphene, which are excellent from the point of view of mechanical and thermal properties. However, in some of these composites, high electrical conductivity only is desirable. For example, the Institute of Metal Research, Chinese Academy of Sciences (IMR, CAS) has created a polymer matrix composite reinforced with graphene, which has a high electrical conductivity. In this composite, a flexible network of graphene has been added to a polydimethylsiloxane matrix (of the silicon family).

Investigation of early corporate trajectories for graphene has led to three major observations. First, the discovery-to-application cycle for graphene seems to be accelerated, for example, compared to fullerene. Even though the discovery of graphene is relatively new, large and small firms have contributed to an upsurge in early corporate activities. Second, a rapid globalization has occurred by companies in the United States, Europe, Japan, South Korea, and other developed economies, which were involved in early graphene activities. Chinese companies are currently starting to enter the graphene domain, resulting in the expansion of research capability of nanotechnology. Nevertheless, science alone does not guarantee commercial exploitation. To clarify the issue, the level of corporate patenting in the United Kingdom, which is a pioneer in graphene research, is slightly ahead of Canada and Germany; however, it is dramatically lower than in the United States, Japan, and South Korea. Third, the potential applications of graphene are rapidly expanding. Corporate patenting trends are indicative of their enthusiasm to utilize the features of graphene in various areas, including transistors, electronic memory and circuits, capacitors, displays,

solar cells, batteries, coatings, advanced materials, sensors, and biomedical devices. Although graphene was initially proposed as an alternative to silicon, its initial applications have been in electronic inks and additives to resins and coatings. We have identified six areas of emerging applications for graphene, including displays/screens, memory chips, biomedical devices, batteries/fuel cells, coatings, inks, and materials. In the investigation of the corporate engagement in graphene, we sought to understand early corporate activity patterns related to broader research and invention trends. In traditional innovation models, a lag between research publication and patenting is consistent with the linear model. However, more recent innovation models are stressing concurrent launch, open innovation, and strategic property management.

The *sixth volume* of this handbook is about the application and industrialization of graphene, starting with chapters about biomaterials and continues onto nanocomposites, electrical/sensor devices, and also new and novel applications.

The editorial team would like to thank all contributors for their excellent chapters contributed to the creation of this handbook and for their hard work and patience during its preparation and production. We sincerely hope that the publication of this handbook will help people, especially those working with graphene, and benefit them from the knowledge contained in the published chapters.

Winter 2016
Mahmood Aliofkhazraei
Nasar Ali
William I. Milne
Cengiz S. Ozkan
Stanislaw Mitura
Juana L. Gervasoni

Editors

Mahmood Aliofkhazraei is an assistant professor in the Materials Engineering Department at Tarbiat Modares University. Dr. Aliofkhazraei's research interests include nanotechnology and its use in surface and corrosion science. One of his main interests is plasma electrolysis, and he has published more than 40 papers and a book in this area. Overall he has published more than 12 books and 90 journal articles. He has delivered invited talks, including keynote addresses in several countries. Aliofkhazraei has received numerous awards, including the Khwarizmi award, IMES medal, INIC award, best-thesis award (multiple times), best-book award (multiple times), and the best young nanotechnologist award of Iran (twice). He is on the advisory editorial board of several materials science and nanotechnology journals.

Nasar Ali is a visiting professor at Meliksah University in Turkey. Earlier he held the post of chief scientific officer at CNC Coatings Company based in Rochdale, UK. Prior to this Dr. Ali was a faculty member (assistant professor) at the University of Aveiro in Portugal where he founded and led the Surface Engineering and Nanotechnology group. Dr. Ali has extensive research experience in hard carbon-coating materials, including nanosized diamond coatings and CNTs deposited using CVD methods. He has over 120 international refereed research publications, including a number of book chapters. Dr. Ali serves on a number of committees for international conferences based on nanomaterials, thin films, and emerging technologies (nanotechnology), and he chairs the highly successful NANOSMAT congress. He served as the fellow of the Institute of Nanotechnology for 2 years on invitation. Dr. Ali has authored and edited several books on surface coatings, thin films, and nanotechnology for leading publishers, and he was also the founder of the *Journal of Nano Research*. Dr. Ali was the recipient of the Bunshah prize for presenting his work on time-modulated CVD at the ICMCTF-2002 Conference in San Diego, California.

William I. Milne, FREng, FIET, FIMMM, was head of electrical engineering at Cambridge University from 1999 until 2014 and has been director of the Centre for Advanced Photonics and Electronics (CAPE) since 2004. He earned a BSc at St. Andrews University in Scotland in 1970 and later earned a PhD in electronic materials at the Imperial College London. In 2003 he was awarded a DEng (honoris causa) by the University of Waterloo, Canada, and he was elected as Fellow of the Royal Academy of Engineering in 2006. He received the JJ Thomson medal from the Institution of Engineering and Technology in 2008 for achievement in electronics and the NANOSMAT prize in 2010. He is a distinguished visiting professor at Tokyo Institute of Technology, Japan, and a distinguished visiting professor at Southeast University in Nanjing, China, and at Shizuoka University, Japan. He is also a distinguished visiting scholar at KyungHee University,

Seoul and a high-end foreign expert for the Changchun University of Science and Technology in China. In 2015, he was elected to an Erskine Fellowship to visit the University of Canterbury, New Zealand. His research interests include large area silicon and carbon-based electronics, thin film materials, and, most recently, MEMS and carbon nanotubes, graphene, and other 1-D and 2-D structures for electronic applications, especially for field emission. He has published/presented approximately 800 papers, of which around 200 were invited/keynote/plenary talks—his "h" index is currently 57 (Web of Science).

Cengiz S. Ozkan has been a professor of mechanical engineering and materials science at the University of California, Riverside, since 2009. He was an associate professor from 2006 to 2009 and an assistant professor from 2001 to 2006. Between 2000 and 2001 he was a consulting professor at Stanford University. He earned a PhD in materials science and engineering at Stanford University in 1997. Dr. Ozkan's areas of expertise include nanomaterials for energy storage; synthesis/processing including graphene, III–V, and II–VI materials; novel battery and supercapacitor architectures; nanoelectronics; biochemical sensors; and nanopatterning for beyond CMOS (complementary metal-oxides semiconductor). He organized and chaired 20 scientific and international conferences. He has written more than 200 technical publications, including journal papers, conference proceedings, and book chapters. He holds over 50 patent disclosures, has given more than 100 presentations worldwide, and is the recipient of more than 30 honors and awards. His important contributions include growth of hierarchical three-dimensional graphene nanostructures; development of a unique high-throughput metrology method for large-area CVD-grown graphene sheets; doping and functionalization of CVD-grown and pristine graphene layers; study of digital data transmission in graphene and InSb materials; memory devices based on inorganic/organic nanocomposites, novel lithium-ion batteries based on nano-silicon from beach sand and silicon dioxide nanotubes; fast-charging lithium-ion batteries based on silicon-decorated three-dimensional nano-carbon architectures; and high-performance supercapacitors based on three-dimensional graphene foam architectures.

Stanislaw Mitura has been a professor in biomedical engineering at Koszalin University of Technology from 2011. He is a visiting professor at the Technical University (TU) of Liberec and was awarded a doctor honoris causa from TU Liberec. He was professor of materials science at Lodz University of Technology from 2001 to 2014. He earned an MSc in physics at the University of Lodz in 1974, a PhD in mechanical engineering at the Lodz University of Technology (1985), and a DSc in materials science at the Warsaw University of Technology in 1993. Professor Mitura's most prominent cognitive achievements comprise the following: from the concept of nucleation of diamond powder particles to the synthesis of nanocrystalline diamond coatings (NDC); discovery of diamond bioactivity; a concept of the gradient transition from carbide forming metal to diamond film; and technology development of nanocrystalline diamond coatings for medical purposes. Professor Mitura has published over 200 peer-reviewed articles, communications, and proceedings, over 50 invited talks, and contributed to 7 books and proceedings, including *Nanotechnology for Materials Science* (Pergamon, Elsevier, 2000) and *Nanodiam* (PWN, 2006). He organized and co-organized several conferences focused on materials science and engineering, especially diamond synthesis under reduced pressure. He is an elected member of the Academy of Engineering in Poland, guest editor in few international journals, including *Journal of Nanoscience and Nanotechnology*, *Journal of Superhards Materials* and also a member of the editorial boards of several journals and an elected Fellow of various foreign scientific societies.

Juana L. Gervasoni earned her doctorate in physics at the Instituto Balseiro, Bariloche, Argentina, in 1992. She has been head of the Department of Metal Materials and Nano-structured, Applied Research of Centro Atomico Bariloche (CAB), National Atomic Energy Commission (CNEA), since 2012. She has been member of the Coordinating Committee of the CNEA Controlled Fusion Program since 2013. Her area of scientific research involves the interactions of atomic particles of matter, electronic excitations in solids, surfaces, and nano-systems, the absorption of hydrogen in metals, and study of new materials under irradiation. Gervasoni is a researcher at the National Atomic Energy Commission of Argentina and the National Council of Scientific and Technological Research (CONICET, Argentina). She teaches at the Instituto Balseiro and is involved in directing graduate students and postdoctorates. She has published over 100 articles in international journals, some of which have a high impact factor, and she has attended many international conferences. Gervasoni has been a member of the Executive Committee and/or the International Scientific Advisory Board of the International Conference on Surfaces Coatings and Nanostructured Materials (Nanosmat) since 2010, Latin American Conference on Hydrogen and Sustainable Energy Sources (Hyfusen), and the International Conference on Clean Energy (International Conference on Clean Energy, ICCE-2010) and guest editor of the *International Journal of Hydrogen Energy* (Elsevier). Recently she has focused her research on the study of hydrogen storage in carbon nanotubes. Along with her academic and research work, Gervasoni is heavily involved in gender issues in the scientific community, especially in Argentina and Latin America. She is a member of the Third World Organization for Women in Science (TWOWS), branch of the Third World Academy of Science (TWAS), Trieste, Italy, since 2010, as well as of Women in Nuclear (WiN) since 2013.

Contributors

Vojtech Adam
Department of Chemistry and Biochemistry
Mendel University in Brno
and
Central European Institute of Technology
Brno University of Technology
Brno, Czech Republic

I. V. Antonova
Rzhanov Institute of Semiconductor Physics SB RAS
Novosibirsk, Russia

A. R. Ashrafi
Department of Nanocomputing
Institute of Nanoscience and Nanotechnology
University of Kashan
Kashan, Iran

Antonio F. Ávila
Department of Mechanical Engineering
Universidade Federal de Minas Gerais
Belo Horizonte, Minas Gerais, Brazil

Theodore F. Baumann
Physical and Life Sciences Directorate
Lawrence Livermore National Laboratory
Livermore, California

Mikhail B. Belonenko
Volgograd Institute of Business
and
Volgograd State University
Volgograd, Russia

Juergen Biener
Physical and Life Sciences Directorate
Lawrence Livermore National Laboratory
Livermore, California

Iva Blazkova
Department of Chemistry and Biochemistry
Mendel University in Brno
Brno, Czech Republic

Roland Bouffanais
Singapore University of Technology and Design
Singapore

José Alberto Briones-Torres
Unidad Académica de Física
Universidad Autónoma de Zacatecas
Zacatecas, México

C. Cab
Facultad de Ingeniería
Universidad Autónoma de Yucatán
Yucatán, México

G. Canto
Centro de Investigación en Corrosión
Universidad Autónoma de Campeche
Campeche, México

Yue Chan
School of Mathematical Sciences
University of Nottingham, Ningbo China
Ningbo, China

Pai-Yen Chen
Department of Electrical and Computer Engineering
Wayne State University
Detroit, Michigan

Wood-Hi Cheng
Graduate Institute of Optoelectronic Engineering
National Chung Hsing University
Taichung, Taiwan

Liang Cui
Chemical and Environmental Engineering
Qingdao University
Qingdao, China

Diego T. L. da Cruz
Graduate Studies Program on Mechanical Engineering
Universidade Federal de Minas Gerais
Belo Horizonte, Brazil

Wang-Feng Ding
Department of Physics
Hangzhou Normal University
Hangzhou, China

D. A. Drabold
Department of Physics and Astronomy
Ohio University
Athens, Ohio

Ludovic F. Dumée
Institute for Frontier Materials
Deakin University
and
Institute for Sustainability and Innovation
Victoria University
Victoria, Australia

Khaled M. Elsabawy
Materials Science Unit, Chemistry Department
Tanta University
Tanta, Egypt

and

Department of Chemistry
Taif University
Taif, Saudi Arabia

Mohamed Farhat
Division of Computer, Electrical, and Mathematical
    Sciences and Engineering
King Abdullah University of Science and Technology
Thuwal, Saudi Arabia

Alexander Feher
Institute of Physics
P. J. Šafárik University
Košice, Slovakia

Chunfang Feng
Institute for Frontier Materials
Deakin University
Victoria, Australia

Sergey Feodosyev
B. I. Verkin Institute for Low Temperature Physics
    and Engineering NASU
Kharkov, Ukraine

S. G. Garcia
Department of Electromagnetics
University of Granada
Granada, Spain

Heraclio García-Cervantes
Unidad Académica de Física
Universidad Autónoma de Zacatecas
Zacatecas, México

Igor Gospodarev
B. I. Verkin Institute for Low Temperature Physics
    and Engineering NASU
Kharkov, Ukraine

Cecilia Goyenola
Department of Physics, Chemistry, and Biology (IFM)
Linköping University
Linköping, Sweden

Vladimir Grishaev
B. I. Verkin Institute for Low Temperature Physics
    and Engineering NASU
Kharkov, Ukraine

Michael N. Groves
Department of Chemistry and Chemical Engineering
Royal Military College of Canada
Kingston, Ontario, Canada

Gueorgui K. Gueorguiev
Department of Physics, Chemistry, and Biology (IFM)
Linköping University
Linköping, Sweden

Gaurav Gupta
Department of Electrical and Computer Engineering
National University of Singapore
Singapore

Li He
Institute for Frontier Materials
Deakin University
Victoria, Australia

Kuo-Chuan Ho
Department of Chemical Engineering
National Taiwan University
Taipei, Taiwan

Yoshikazu Homma
Department of Physics
Tokyo University of Science
Tokyo, Japan

Pi Ling Huang
Department of Photonics
National Sun Yat-sen University
Kaohsiung, Taiwan

Wen Huang
Department of Electrical and Computer Engineering
National University of Singapore
Singapore

Nikhil Jain
College of Nanoscale Science and Engineering
State University of New York
Albany, New York

Debnarayan Jana
Department of Physics
University of Calcutta
West Bengal, India

M. Jana
Advanced Materials and Process Technology Centre
    (AMPTC)
Crompton Greaves Limited
Global R&D Centre
Mumbai, Maharashtra, India

Manish Jugroot
Advanced Materials and Process Technology Centre
(AMPTC)
Crompton Greaves Limited
Global R&D Centre
Mumbai, Maharashtra, India

Sirichok Jungthawan
School of Physics, Institute of Science and
NANOTEC-SUT Center of Excellence on
Advanced Functional Nanomaterials
Suranaree University of Technology
Nakhon Ratchasima, Thailand

Hiroki Kato
Department of Physics
Tokyo University of Science
Tokyo, Japan

Soo Young Kim
School of Chemical Engineering and
Materials Science
Chung-Ang University
Seoul, Republic of Korea

Rene Kizek
Department of Chemistry and Biochemistry
Mendel University in Brno
and
Central European Institute of Technology
Brno University of Technology
Brno, Czech Republic

Lingxue Kong
Institute for Frontier Materials
Deakin University
Victoria, Australia

Natalia N. Konobeeva
Volgograd State University
Volgograd, Russia

F. Koorepazan-Moftakhar
Department of Nanocomputing
Institute of Nanoscience and Nanotechnology
University of Kashan
Kashan, Iran

Pavel Kopel
Department of Chemistry and Biochemistry
Mendel University in Brno
and
Central European Institute of Technology
Brno University of Technology
Brno, Czech Republic

Ki Chang Kwon
School of Chemical Engineering and Materials Science
Chung-Ang University
Seoul, Republic of Korea

Lain-Jong Li
Institute of Atomic and Molecular Science
Taipei, Taiwan

Xifei Li
Department of Mechanical and Materials Engineering
University of Western Ontario
London, Ontario, Canada

Y. Li
Department of Physics and Astronomy
Ohio University
Athens, Ohio

Zhenyu Li
Hefei National Laboratory for Physical Sciences
at Microscale
University of Science and Technology of China
Hefei, China

Gengchiau Liang
Department of Electrical and Computer Engineering
National University of Singapore
Singapore

Sukit Limpijumnong
School of Physics, Institute of Science
Suranaree University of Technology
Nakhon Ratchasima, Thailand

and

Thailand Center of Excellence in Physics
Commission on Higher Education
Ministry of Education
Bangkok, Thailand

H. Lin
Central China Normal University
Wu Han, China

Jiang-Jen Lin
Institute of Polymer Science and Engineering
National Taiwan University
Taipei, Taiwan

Worong Lin
Department of Chemistry and Institute of Nanochemistry
Jinan University
Guangzhou, China

Jingquan Liu
Chemical and Environmental Engineering
Qingdao University
Qingdao, China

Zheng Liu
Center for Programmable Materials
School of Materials Science and Engineering
and
NOVITAS
Nanoelectronics Centre of Excellence
School of Electrical and Electronic Engineering
and
CINTRA CNRS/NTU/THALES
Nanyang Technological University
Singapore

Jesús Madrigal-Melchor
Unidad Académica de Física
Universidad Autónoma de Zacatecas
Zacatecas, México

Cecile Malardier-Jugroot
Department of Chemistry and Chemical Engineering
Royal Military College of Canada
Kingston, Ontario, Canada

Rocco Martinazzo
Dipartimento di Chimica
Università degli Studi di Milano
and
Istituto di Scienze e Tecnologie Molecolari
Consiglio Nazionale delle Ricerche
Milan, Italy

Ana Laura Martínez-Hernández
División de Estudios de Posgrado e Investigación
Instituto Tecnológico de Querétaro
and
Centro de Física Aplicada y Tecnología Avanzada
Universidad Nacional Autónoma de México
Santiago de Querétaro, Querétaro, México

Juan Carlos Martínez-Orozco
Unidad Académica de Física
Universidad Autónoma de Zacatecas
Zacatecas, México

D. Mateos Romero
Department of Electromagnetics
University of Granada
Granada, Spain

Yuta Momiuchi
Department of Physics
Tokyo University of Science
Tokyo, Japan

Hermano Nascimento Jr.
FIAT Automobile Inc.
Betim, Brazil

Palash Nath
Department of Physics
University of Calcutta
West Bengal, India

Fabiola Navarro-Pardo
División de Estudios de Posgrado e Investigación
Instituto Tecnológico de Querétaro
and
Énergie Matériaux Télécommunications
Institut national de la recherche scientifique
Varennes, Quebec, Canada

Argo Nurbawono
Department of Electrical and Computer Engineering
National University of Singapore
Singapore

O. Ori
Actinium Chemical Research
Rome, Italy

L. S. Paixão
Instituto de Física
Universidade Federal Fluminense
Rio de Janeiro, Brazil

M. F. Pantoja
Department of Electromagnetics
University of Granada
Granada, Spain

Zheng Peng
Institute for Frontier Materials
Deakin University
Victoria, Australia
and
Agricultural Product Processing Research Institute
Chinese Academy of Tropical Agricultural Sciences
Guangdong, China

V. Ya. Prinz
Rzhanov Institute of Semiconductor Physics SB RAS
Novosibirsk, Russia

S. Ray
School of Engineering
Indian Institute of Technology Mandi
Himachal Pradesh, India

M. S. Reis
Instituto de Física
Universidade Federal Fluminense
Rio de Janeiro, Brazil

Rogelio Rodríguez-González
Unidad Académica de Física
Universidad Autónoma de Zacatecas
Zacatecas, México

Isaac Rodríguez-Vargas
Unidad Académica de Física
Universidad Autónoma de Zacatecas
Zacatecas, México

Dirtha Sanyal
Variable Energy Cyclotron Centre
Bidhannagar
West Bengal, India

Po-Ta Shih
Institute of Polymer Science and Engineering
National Taiwan University
Taipei, Taiwan

A. Sil
Department of Metallurgical and Materials
    Engineering
Indian Institute of Technology Roorkee
Uttarakhand, India

Fengqi Song
College of Physics
Nanjing University
Nanjing, China

Michael Stadermann
Physical and Life Sciences Directorate
Lawrence Livermore National Laboratory
Livermore, California

Xueliang Sun
Department of Mechanical and Materials
    Engineering
University of Western Ontario
London, Ontario, Canada

Eugen Syrkin
B. I. Verkin Institute for Low Temperature Physics
    and Engineering NASU
Kharkov, Ukraine

Junro Takahashi
Department of Physics
Tokyo University of Science
Tokyo, Japan

Katsuhiro Takahashi
Department of Physics
Tokyo University of Science
Tokyo, Japan

A. Tapia
Facultad de Ingeniería
Universidad Autónoma de Yucatán
Yucatán, México

Marketa Vaculovicova
Department of Chemistry and Biochemistry
Mendel University in Brno
and
Central European Institute of Technology
Brno University of Technology
Brno, Czech Republic

Carlos Velasco-Santos
División de Estudios de Posgrado e Investigación
Instituto Tecnológico de Querétaro
and
Centro de Física Aplicada y Tecnología
    Avanzada
Universidad Nacional Autónoma de México
Santiago de Querétaro, Querétaro, México

Flávio A. C. Vidal
FIAT Automobile Inc.
Betim, Brazil

Hong Wang
School of Materials Science and
    Engineering
Nanyang Technological University
Singapore

D. H. Werner
Pennsylvania State University
University Park, Pennsylvania

Marcus A. Worsley
Physical and Life Sciences Directorate
Lawrence Livermore National Laboratory
Livermore, California

Dongjiang Yang
College of Chemistry
Chemical and Environmental Engineering
Laboratory of Fiber Materials and Modern
    Textile
Growth Base for State Key Laboratory
Qingdao University
Qingdao, China

Jinlong Yang
Hefei National Laboratory for Physical Sciences
    at Microscale
University of Science and Technology of China
Hefei, China

Chao-Yung Yeh
Metal Industries Research and Development Center
Kaohsiung, Taiwan

Zhifeng Yi
Institute for Frontier Materials
Deakin University
Victoria, Australia

Bin Yu
College of Nanoscale Science and Engineering
State University of New York
Albany, New York

Dingsheng Yuan
Department of Chemistry and Institute of Nanochemistry
Jinan University
Guangzhou, China

Long Yuan
Hefei National Laboratory for Physicial Sciences at
   Microscale
University of Science and Technology of China
Hefei, China

Minggang Zeng
Department of Electrical and Computer Engineering
National University of Singapore
Singapore

Bo Zhao
College of Physics
Nanjing University
Nanjing, China

Alexander V. Zhukov
Singapore University of Technology and Design
Singapore

Section I

Atomic Arrangement and Defects

1 Graphene Heterostructures

Zheng Liu and Hong Wang

CONTENTS

Abstract 3
1.1 Introduction 3
1.2 Lateral Graphene Heterostructures 3
    1.2.1 Lateral Graphene/h-BN Heterostructures with Random Domains 5
    1.2.2 Lateral Graphene/h-BN Heterostructures with Patterned Domains 8
1.3 Vertical Graphene Heterostructures 8
1.4 All-Carbon G/CNT Vertical Heterostructure 15
1.5 Conclusions 16
References 16

ABSTRACT

It is well known that graphene has ultrahigh carrier mobility. However, the zero bandgap of graphene will lead to a large leaking current in graphene-based transistors and, therefore, limits its application in high-performance semiconducting electronics. Creation of graphene heterostructures will not only result in graphene-based artificial architectures that open the bandgap of graphene, but also pave a promising way to the landscape of full-integrated and multifunctional graphene electronics.

By integrating graphene with its analogs such as hexagonal boron nitride (h-BN)[1] and dichogenides (e.g., MoS_2),[2–5] one can tailor the graphene transport in terms of mobility, ON/OFF ratio,[6] radio frequency,[7] and etc. Generally, one can divide the graphene heterostructures into two classes: planar layers[2,7] and vertical stack, which can be approached by various ways. With controlled number of layers, ordering of layers, and manipulation of the positions of the layers, such heterostructures will lead to vast applications such as in high-performance transistors, radio frequency devices, and in THz rectifiers. Besides the two-dimensional graphene heterostructures, the three-dimensional heterostructure of graphene and carbon nanotubes (CNTs) can be realized via bonding CNTs with graphene for energy applications.[8,9]

1.1 INTRODUCTION

The low resistivity and extremely high mobility of graphene make it a promising material for electronic and optoelectronic devices such as transparent conducting electrodes,[10] ultrafast transistors,[11,12] frequency multipliers,[13] broadband radio mixers, and so on.[14] However, due to the large leaking current from the zero bandgap, the ON/OFF ratio of graphene-based field-effect transistors (FETs) are typically less than 10.[15] These values are far less than the silicon-based transistors ($>10^6$). As such, opening the bandgap of graphene becomes critical for the electronic applications of graphene. Creation of the graphene heterostructure is one of the most promising approaches to open the bandgap of graphene. Furthermore, by integrating graphene with other low-dimensional materials such as hexagonal boron nitride (h-BN),[16–19] transition metal dichalcogenides (TMDs) family (MX_2, M: Mo, W, and X: S, Se, Te),[4,20–28] and CNTs, various graphene heterostructures can be prepared, providing versatile applications such as in high ON/OFF FET transistors,[2,6,29] spin filters,[30] atomically thin circuitry,[31] one-dimensional (1D) transport channel,[32] solar cells,[33] resonators,[34] hydrogen evolution,[35] and in harvesting and conversion of energy (Table 1.1).[36]

In this chapter, we review the recent progress on the synthesis, fabrication, characterization, electric transport, and potential applications of graphene heterostructures, particularly, graphene/h-BN heterostructures. The graphene heterostructures with other two-dimensional (2D) and 1D material will be also briefly discussed. This chapter mainly consists of three parts. Planar graphene heterostructures will be discussed in the first part which involved preparation and characterization of atomic layers of hybridized h-BN graphene random domains, shape engineering of planar graphene/h-BN (G/h-BN), and doped graphene. In the following part we focus on the vertical graphene heterostructures with other 2D crystals (h-BN, TMDs) in the form of stacks and superlattice, namely van der Waals (vdW) solid. The first two parts show the vast possibilities of graphene heterostructures with other 2D materials. In the final part, three-dimensional (3D) graphene/carbon nanotube (G/CNT) will be discussed briefly.

1.2 LATERAL GRAPHENE HETEROSTRUCTURES

Lateral graphene heterostructures are layered materials consisting of graphene and other materials in the same plane. Generally, it could be classified into two classes: (I) hybridizations of graphene with other materials by doping and substitutions (Figure 1.1a and b) and (II) shape engineering

TABLE 1.1

Fabrication Techniques and Application Fields of Various Graphene Heterostructures

Heterostructure Type	Lateral Heterostructures	Vertical Heterostructures		G/CNTs Heterostructures
	G/h-BN	G/h-BN	G/TMDs	
Fabrication technique	CVD[29,31,34]	Transfer,[12,37,38] CVD,[39–46] liquid exfoliation[47]	Transfer,[6,36,48,49] CVD[50–52]	CVD[8,53]
Application	Atomically thin circuitry[31] Resonator[34]	High-mobility FET[12] Tunneling FET (G/h-BN/G[6])	Tunneling FET (G/ MoS$_2$/G,[6] G/WS$_2$/G[2]) Memory cells[48,49] Optoelectronic devices[36]	Supercapacitors[8,53] Field emission[8]

of graphene by patterning graphene and other 2D crystals (Figure 1.1c and d) via fabrications.

For group I, graphene is laterally hybridized other elementals/compound such as nitrogen, fluorine, h-BN, etc. The feature sizes of the domains range from single atoms to the order of microns.[1,29,3454–59] (Figure 1.1a). By controlling the foreign materials, for example, composition and concentration, one can tune the physical properties. Random domains of graphene and other 2D materials can be experimentally prepared, such as the atomic layers of hybridized h-BN and graphene domains consisting of hybridized, randomly

distributed domains of h-BN and C phases and the bandgap of h-BNC films can be tuned by the atomic concentration of h-BN (Figure 1.1a).[1,60] A more challenging goal will be the synthesis of crystals with various stoichiometric ratios of carbon and foreign atoms such as the B$_x$N$_y$C$_z$ system. It is predicted that the B$_x$N$_y$C$_z$ crystals can possess metallic or semiconducting behaviors depending on the stoichiometry.[61] The maximum challenge for creating such heterostructures will be the phase separations.[62]

In the case of group II, graphene may retain its nature, such as its highly electrical conductivity. Here, the geometries may

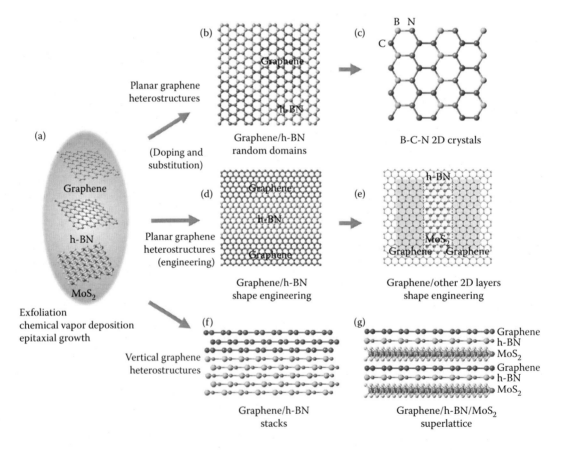

FIGURE 1.1 Approaches to graphene heterostructures. (a) Graphene and graphene analogs (h-BN, TMDs, etc.) can be prepared by three methods: mechanical/liquid exfoliation, CVD, and epitaxial growth. (b)–(e) Landscape of lateral graphene heterostructures. (b) Heterostructures with random domains via doping and substitution. (c) Homogeneously doped graphene heterostructures. (d) and (e) Shape engineering/patterning of graphene/2D materials. (f) and (g) Vertical graphene heterostructures containing h-BN and MoS$_2$.

play a more important role and dominate the nature of the samples. Full-integrated and multifunctional devices are proposed for versatile applications. Multistep growth and fabrications are usually adopted for such graphene heterostructures. Graphene/h-BN patterns have been demonstrated in the laboratory using a two-step growth protocol.[34,55] A further challenge includes understanding the interfaces of graphene and other 2D materials and seeking approaches to how to incorporate more 2D materials into a single film.

1.2.1 LATERAL GRAPHENE/h-BN HETEROSTRUCTURES WITH RANDOM DOMAINS

As a representative architecture of lateral graphene heterostructures, planar graphene/h-BN films have attracted much attention in recent years. Geometrically, the small lattice mismatch between graphene and h-BN (~1.8%) makes it possible to stitch both materials in an individual film.[31,34,63,64] Moreover, both graphene and h-BN can be grown via chemical vapor deposition (CVD) on copper, nickel, and other metals such as Co, Pt, Rh, Ru, Ag, etc.[16,18,65–73] with a similar protocol, which makes the growth of graphene/h-BN heterostructures possible. More important, h-BN is a wide-bandgap insulator (~5.9 eV) while graphene is a highly electronic conductor. The integrated BN-doped graphene (h-BNC) film will have a tunable bandgap.[55,60]

Recently a systematic approach is proposed for the large-scale synthesis of planar h-BCN heterostructures, containing random BN and C nanodomains, via a CVD method using ammonia borane and methane as the BN and C sources. A typical setup for the CVD growth of h-BNC films contains a heater for the BN precursors and a furnace (Figure 1.2a).[29,74] The as-grown samples can be transferred onto silicon substrates. The optical images show that the film is quite uniform over a large

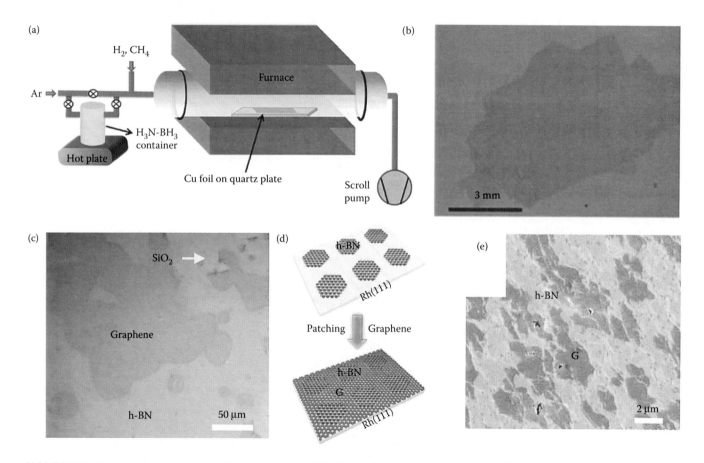

FIGURE 1.2 Synthesis and characterizations of planar h-BNC films with random composition. (a) Schematic of the setup for the growth of random h-BCN films via CVD.[29] Various solid precursors such as ammonia boron (BH₃NH₃),[16] borazine (B₃N₃H₆)[18] have been used as the sources of B and N. (b) Optical image of hybrid h-BNC film. The size of graphene and h-BN domains is at the scale of a few nanometers. The thickness of the h-BNC films can be down to a few layers. (c) Graphene/h-BN film with domains of up to hundred microns.[34] (d) Schematic of epitaxial growth of planar graphene/h-BN on Rh (111) surface under ultrahigh vacuum. The vaporized ammonia borane and ethylene are used as the source of B, N, and C, respectively.[63] (e) Optical image of the h-BCN film by epitaxial growth. (Reprinted with permission from (a) Chang, C.-K. et al. Band gap engineering of chemical vapor deposited graphene by in situ BN doping. *ACS Nano* **7**, 1333–1341. Copyright 2012 American Chemical Society; (b) Ci, L. et al. Atomic layers of hybridized boron nitride and graphene domains. *Nature Materials* **9**, 430–435. Copyright 2010 American Chemical Society; (c) Reprinted by permission from Macmillan Publishers Ltd. *Nature Nanotechnology*, Liu, Z. et al. In-plane heterostructures of graphene and hexagonal boron nitride with controlled domain sizes. **8**, 119–124, copyright 2013; (d, e) Reprinted with permission from Gao, G., Gao, W. and Cannuccia, E. Artificially stacked atomic layers: Toward new van der Waals solids. *Nano Letters* **12**, 3518–3525. Copyright 2012 American Chemical Society.)

area (Figure 1.2b).[1] Optimizing the growth condition will make the graphene and h-BN domain size increase to the order of a few hundreds of microns (Figure 1.2c).[34] The darker regions are graphene, while lighter area is h-BN. The SiO₂ substrate is indicated by the arrows.[34] The hybrid h-BNC films can also be prepared on a single crystalline Rh(111) substrate via a two-step epitaxial growth (graphene first, h-BN second) approach under UHV (ultrahigh vacuum) conditions.[63] Figure 1.2d and e is a schematic of the h-BNC patching growth and SEM (scanning electron microscope) images of the resulting samples, respectively. The domain size is around a few microns. The stoichiometric ratio of BN and C can be tuned by changing the growing parameters such as the amount/flow rate of sources, growing temperature, etc.

Although planar h-BNC heterostructures with nanodomains have been demonstrated by different approaches, the synthesis of homogeneous h-BNC or h-BNC crystal remains a problem such as the well-known phase separation in ternary materials.[62] The positive interface energy between graphene and BN implies that for the same amount of BN area fewer number of BN islands will be favored over multiple smaller islands with longer total interface length.[75–79]

The atomic structures of planar hybridized h-BNC films have been characterized by transmission electron microscopy (TEM) and scanning tunneling microscopy (STM). The typical thickness of CVD and epitaxy grown h-BCN films are from monolayer to few layers (Figure 1.3a). High-resolution TEM (HRTEM) image (Figure 1.3b) shows various Moiré patterns. In their electron energy loss spectroscopy (EELS, Figure 1.3c), there are three visible K-shell edges of B, C, and N, starting at 185, 276, and 393 eV, respectively,[75,80] Boron is very sensitive even in very small amounts in the sample while the signal from nitrogen is relatively weak. Employing energy-filtered imaging technique, the distribution of each element can be mapped (Figure 1.2c).

The interface of graphene and h-BN plays an important role in the transport behavior of h-BNC films. However, because of the two-step growth of materials and unavoidable contamination during the TEM sample preparation, it is a great challenge to obtain a clean sample to study their interfaces. In this case, STM is a powerful tool to study the epitaxial G/h-BN films grown under high vacuum, complementary to TEM characterization. A two-step growth method was used to prepare hybrid h-BN and graphene film on Rh(111): h-BN

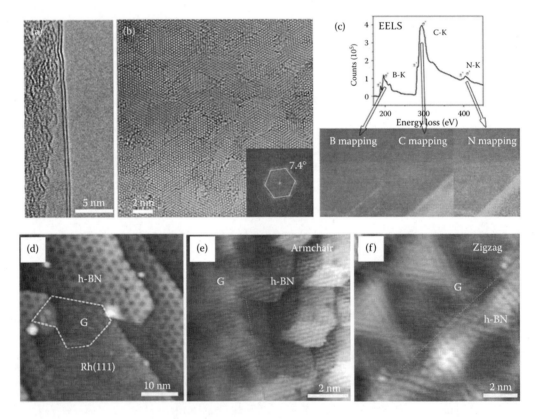

FIGURE 1.3 TEM and STM characterization of atomic structures of h-BNC films. (a) Transmission electron microscope (TEM) image of a typical edge of double-layered h-BCN film prepared by CVD on Cu foils, in which methane and ammonia boron were supplied at the same time to provide C, B, and N. (b) Atomic TEM image of h-BNC film containing BN and graphene patches. (c) EELS of h-BNC film and chemical mapping. (d–f) STM images of BNC hybrid grown on Rh(111). (d) STM image showing the preferred linking of graphene to preexisting h-BN domains with two typical hybrids: graphene-embracing h-BN (h-BN at G) and h-BN-embracing graphene (G at h-BN). (e) Atomically resolved STM images with an armchair linking edge, and (f) with a zigzag linking edge. (Reprinted with permission from (a–c) Ci, L. et al. Atomic layers of hybridized boron nitride and graphene domains. *Nature Materials* **9**, 430–435. Copyright 2010 American Chemical Society; (d–f) Gao, Y. et al. Towards single-layer uniform hexagonal boron nitride–graphene patchworks with zigzag linking edges. *Nano Letters* **13**, 3439–3443. Copyright 2013 American Chemical Society.)

domains were first grown with ammonia borane as B and N sources, then ethylene was introduced for graphene growth on uncovered Rh surface. It is evident that graphene preferably grows from the edge of h-BN domains (Figure 1.3d). Moreover, the graphene patches are atomically perfect, as shown in the close-up of the linking boundaries (Figure 1.3e and f). The large amount of dangling bonds at the h-BN edges may be a reason for the epitaxial growth of graphene, mediated by the underlying Rh(111) lattice.

Optical technologies such as Raman and UV–visible absorption spectrum are utilized to estimate the quality of the h-BCN films and their optical gap. Figure 1.4a shows the typical Raman spectra of pristine graphene and atomic-layered h-BCN film. The h-BNC film has a border D band and a shoulder at the G band, an indication of increasing defects in graphene caused by the BN. A similar behavior is also found after oxidizing graphene at a temperature above 300°C. The 2D band is suppressed, probably, by the photoluminescence background. The optical gap of h-BNC film can be estimated by the UV–vis absorption. Tauc's plot is depicted in Figure 1.4b. However, the value of 2.36 eV is related to the resonant excitonic effects due to the electron–hole interaction in the π and π* bands at the M point[81] and should not be considered the optical bandgap of graphene. It is observed that h-BCN films show high transparency at long wavelengths with two absorption peaks (Figure 1.4c and d).[1,82] For h-BNC with 2% BN (Figure 1.4c), one absorption edge corresponds to an optical gap of 4.48 eV, which may come from the h-BN domains in the film but smaller than that of pure h-BN. In fact, this value can be used to track the concentration of BN. It has been found that for a high concentration of BN, this value will become close to the bulk one.[29] Another absorption edge corresponding to an optical band ranging from 1 to 2 eV is from graphene nano-domains.[83] The bandgap evolution of graphene nanodomains in h-BNC films has been insensitively investigated by theoretical calculation.[32] In the h-BNC system, the domains may be circular or small disks that have been seen experimentally most recently due to the positive interface energy between graphene layers during their growth.

The electronic structure and properties of h-BCN vary depending on the stoichiometry of carbon and BN.[84–86] The resistivity of h-BCN ranges from 10^{-4} Ω cm (less than ~2% BN) to 10^{-3} Ω cm (~6% BN) to an insulating temperature (>80% BN). The h-BCN with random domains exhibits an ambipolar semiconducting behavior, an indication that such h-BCN films have an atomic structure consisting of hybridized h-BN and graphene domains, since theoretical and experimental investigations have indicated that B, C, N mixed phases are typically p-type semiconductors. Furthermore, this h-BCN film shows a clear insulator to metal transition at ~10 K,[87] The author suggests that the observed insulator to metal transition in h-BCN is as a result of the coexistence between two distinct mechanisms, namely, percolation through metallic graphene networks and hopping conduction between edge states on randomly distributed insulating h-BN domains. Besides, a small gap could be opened in graphene due to quantum confinement and/or spin polarization at a specific C-BN boundary.[1,88,89] This is further confirmed by electric measurement from an *in situ* BN-doped graphene film. A significant bandgap of ~600 meV is observed for low BN concentrations and is attributed to the opening of the π–π* bandgap of graphene due to isoelectronic BN doping.

In addition to graphene doping with h-BN, planar graphene heterostructures can be also prepared by doping graphene with other elements, such as nitrogen,[90–99] sulfur,[90,100] boron,[101] and fluorine.[102–104] They may dramatically open the bandgap of graphene. Doped graphene also provides us a good opportunity to know how the geometry of materials alters its properties in mechanics. There are three main approaches for doping graphene: growing graphene using a carbon source containing or mixing the doping elements such as PMMA (poly(methyl methacrylate)), melamine as carbon sources, or mixing *n*-hexane and sulfur as carbon sources, etc.;[96] annealing graphene with doping gas such as NH_3;[96] interaction of doping elements, especially for metals such as Li, Xe, etc.[105,106]

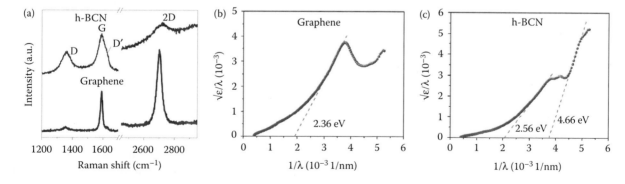

FIGURE 1.4 Raman spectrum and Tauc's plot for pristine graphene and BNG. (a) Raman spectra of pristine CVD-grown graphene (bottom). Compared with graphene, for the h-BNC film, the D peak is stronger and 2D peak is suppressed. (b) and (c) Tauc's plot ($\sqrt{\varepsilon}/\lambda$ versus $1/\lambda$, data is acquired from the UV–via absorption) of (b) pristine graphene and (c) h-BNC sample with 2% BN.[29] (Reprinted with permission from (a) Ci, L. et al. Atomic layers of hybridized boron nitride and graphene domains. *Nature Materials* **9**, 430–435. Copyright 2010 American Chemical Society; (b, c) Chang, C.-K. et al. Band gap engineering of chemical vapor deposited graphene by in situ BN doping. *ACS Nano* **7**, 1333–1341. Copyright 2012 American Chemical Society.)

1.2.2 LATERAL GRAPHENE/h-BN HETEROSTRUCTURES WITH PATTERNED DOMAINS

In the section above, we have discussed the synthesis, characterization, and transport behaviors of the lateral h-BNC heterostructure with random domains. However, in order to fully utilize the unique nature of graphene, a sophisticated shape engineering or patterning using combining graphene and other 2D material is pursued by the community, which will bring more opportunities to keep the merits of both graphene and the incorporated materials.[107] Dry-etching-based technologies are usually used toward the precisely controlled lateral heterostructures containing graphene and h-BN.[31,34] The synthesis of shape-engineered graphene/h-BN heterostructures can start from the growth of h-BN atomic layers on metal foils, following an etching process of h-BN and regrowth of graphene at the etched regions (Figure 1.5a).[34] Owing to the high transparency of h-BN (>98% for few-layered h-BN), the h-BN film covers the copper foils but is invisible (Figure 1.5b). Graphene is grown on the patterned areas and is seen in gray. Like the conventional polymer-assisted transfer, this G/h-BN films can be transferred by PMMA (Figure 1.5c)[108] to arbitrary substrates (Figure 1.5d on SiO₂).

An inverse routine has also been demonstrated by growing and patterning graphene first and subsequent growth of h-BN (Figure 1.5e).[31] These works imply the possibility of epitaxial growth of graphene on h-BN edges or the inverse way. The complementary dry-etching technologies provide for versatile approaches to graphene lateral heterostructures among intrinsic graphene, substitutional doped graphene, insulating h-BN, and TMDs (MoS₂, WSe₂, etc.).[20,68,109–119] Focus ion beam (FIB) can fabricate samples with nanoscale feature size, while a combination of photo/e-beam lithography and reaction ion etching (RIE) can be used for the fabrication of large-scale samples. Up to date, the minimal feature size of G/h-BN can be less than 100 nm using FIB etching, while millimeter-sized graphene/doped graphene and graphene/h-BN planar patterns were produced by photolithography.[34] In addition, both methods produced mechanically continuous sheets across the heterojunctions, that is, the whole film containing heterostructure domains remain mechanically intact during the transfer from their growth substrates.[31,34] Such heterostructures can be transferred to rigid substrates such as SiO₂ or flexible substrates such as PMDS (polydimethylsiloxane).[31,34] Their electronic transport has been also carefully examined. For the graphene/doped graphene junctions, there is no considerable change in the mobility and junction resistance.[31] In contrast, the geometries of the patterns will dominate the electrical transport of graphene/h-BN films and may result in an anisotropy behavior. Such as, for graphene/h-BN alterative stripes, along the graphene stripes, the mobility of the device is close to pristine graphene. However, no gating effect was observed if the current flows perpendicular to the stripes. No available resistance can be measured.[34] In terms of applications, a prototype of graphene/h-BN split closed-loop resonators has been demonstrated with a resonating frequency of ~1.95 GHz. This value is comparable to that of copper microstripes.[34] More promising applications have been predicted by theoretical calculation such as the 1D transport channels,[32] resonant tunneling diodes,[120] and rectifiers.[121]

1.3 VERTICAL GRAPHENE HETEROSTRUCTURES

Beyond lateral graphene heterostructures, various 2D layers could be artificially stacked due to the weak vdW interaction, forming vertical heterostructures or the so-called vdW solids (Figure 1.1f and g).[119,122] Graphene vertical heterostructures including G/h-BN stack, G/MoS₂ stack, G/h-BN superlattice, and other hybrid architectures have been investigated in both experiments and theory and show great potential for use in applications such as electronics, optoelectronics, catalysts, etc.[123–126]

There are quite a few methods that are developed to prepare vertical graphene heterostructures, including manipulation of individual layers, direct CVD growth,[96,108,129] liquid exfoliation, molecular beam epitaxy,[50,130,131] etc. toward a great landscape of vertical architectures. Among them, mechanical manipulation of exfoliated 2D layers is a widely used method to fabricate graphene heterostructures that have been demonstrated in the success of the G/h-BN stack (Figure 1.6a). Polymers such as PMMA are usually used to assist the mechanical transfer process. It was reported by Dean et al.[12] for the first time, paving a way to many subsequent work on vdW reassembly of vertical graphene heterostructures. The multilayer graphene heterostructures, h-BN/G/h-BN, are first demonstrated by Ponomarenko et al.[132] by this method. A sophisticated h-BN/G superlattice (Figure 1.6b) was made by Britnell et al. Fabrication of this devices requires multiple times of dry transfers, alignments, electron-beam lithography, plasma etching, and metal depositions.[6,127] However, these two-step or multistep mechanical transfer may break graphene and other 2D materials. Moreover, the size of mechanical exfoliated layers is limited. These will lead to a low sample output.[2] Transfer of other 2D materials to large-sized CVD graphene can increase the output, as the SEM image shown in Figure 1.6c.[128] The direct CVD growth of G/h-BN stacks paves a promising way to the large-scale synthesis of graphene heterostructures (Figure 1.6d).[129,133] Further, the ordering of stack can be controlled by direct transfer of CVD films. A G/h-BN/G sandwich structure (Figure 1.6e) is realized by PMMA-assisted transfer. In addition, liquid exfoliation can produce a large amount of vertical heterostructures with random ordering among various layers (Figure 1.6f). Molecular beam epitaxy provides an alternative way for high-quality samples (h-BN on graphene as seen in Figure 1.6g).[130] Furthermore, other graphene heterostructures beyond G/h-BN have been demonstrated such as graphene/MoS₂ (Figure 1.6h)[51] and h-BN/G/MoS₂ (Figure 1.6i and k).[2] Figure 1.6i is a false-colored SEM image of h-BN/G/MoS₂ structures. The HAADF-STEM imaging and chemical mapping (Figure 1.6j and k) further confirm this heterostructure containing graphene, nitride, and TMDs.[2]

Among many vertical graphene heterostructures, graphene/h-BN stacks have seized a lot of attention. The main reason is

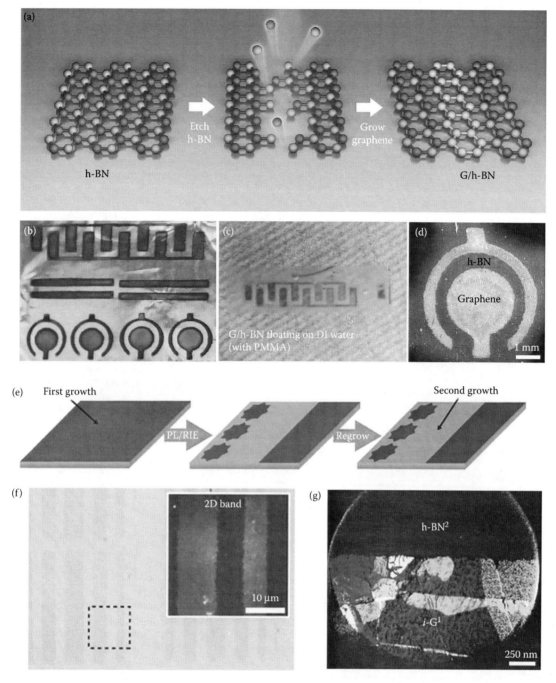

FIGURE 1.5 Synthesis and characterizations of shape engineered graphene/h-BN heterostructures. (a)–(d) Graphene/h-BN heterostructures by growth h-BN first, etching h-BN, and re-grow graphene.[34] (a) Illustration of the fabrication procedure for in-plane graphene/h-BN heterostructures with three steps: preparation of h-BN films using the CVD method; partial etching of h-BN by argon ions to give predesigned patterns and subsequent CVD growth of graphene on the etched regions. (b) Photo of the graphene/h-BN patterns (shaped as combs, bars, and rings) on a copper foil. Light yellow regions are covered by h-BN and dark yellow areas are covered by graphene. (c) A graphene/h-BN/PMMA film floating on the DI water after etching the copper foil. The transparent and gray regions are PMMA covered h-BN and graphene, respectively. (d) SEM image of an h-BN ring surrounded by graphene.[34] (e)–(g) Fabrication of lateral graphene/h-BN heterostructure via an inverse way.[31] (e) Schematic of the fabricating process. The graphene is grown first, following a pattern and etching process to remove unwanted graphene. h-BN was grown on the etched areas. (f) Optical image of a planar graphene (darker areas)/h-BN (lighter areas) film on a Si/SiO$_2$ substrate. Inset, Raman graphene 2D band (area indicated by the dotted box) showing a stark contrast between the regions. (g) False-color DF-TEM image of a suspended graphene/h-BN sheet with the junction region clearly starting at the sharp line that ends the graphene area. (Reprinted by permission from Macmillan Publishers Ltd. (a–d) *Nature Nanotechnology*, Liu, Z. et al. In-plane heterostructures of graphene and hexagonal boron nitride with controlled domain sizes. **8**, 119–124, copyright 2013; (e–g), *Nature, Levendorf*, M.P. et al. Graphene and boron nitride lateral heterostructures for atomically thin circuitry. **488**, 627–632, copyright 2012.)

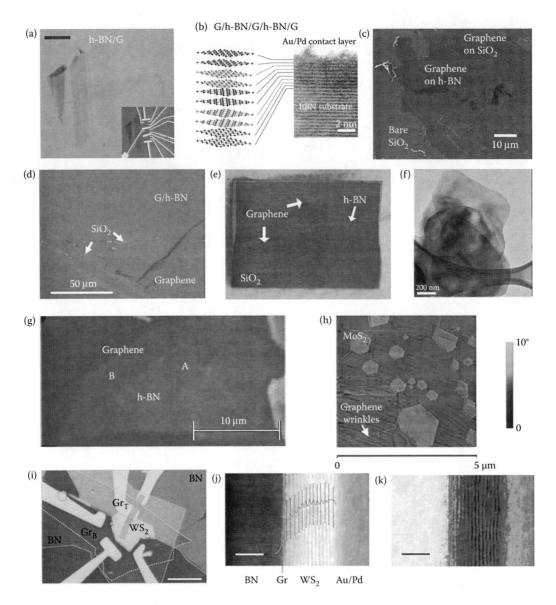

FIGURE 1.6 Landscape of vertical graphene heterostructure. (a–c) High-quality vertical graphene/h-BN stacks from mechanical exfoliated graphene and h-BN. (a) Optical image of a typical graphene/h-BN stack with graphene atop h-BN. Scale bar, 10 μm. Inset: electrical contacts patterned by e-beam lithography.[12] (b) TEM image of h-BN/G/h-BN/G/h-BN multilayer stacks with patterned contacts.[6,127] (c) SEM image of CVD graphene on an exfoliated h-BN flake.[128] (d)–(h) Bulk vertical graphene/h-BN stacks from chemical deposited and isolated graphene and h-BN. (d) Optical images of large-area few-layered G/h-BN via direct deposition of h-BN on graphene.[129] The films are uniform and continuous on the substrate. (e) h-BN/G/h-BN sandwich from PPMA-assisted transfer of CVD-grown graphene, h-BN, and graphene, respectively. The size of the wafer is ~1 cm × 0.8 cm. (f) TEM image of liquid exfoliated h-BN/G.[47] (g) Optical image of epitaxy h-BN on graphene. (h–k) Vertical graphene heterostructure with TMDs.[136] (h) AFM phase imaging of MoS_2 flakes grown on graphene.[51] (i)–(k) Complex graphene vertical heterostructure consisting of three difference 2D materials.[2] (i) Optical image of graphene/WS_2/h-BN devices. Scale bar, 10 μm. (j) Cross-section high-resolution high-angle annular dark-field scanning transmission electron microscopy (HAADF-STEM) imaging of the sample. Scale bar, 5 nm. (k) False-colored bright-field STEM image. Scale bar, 5 nm. ((a) Reprinted by permission from Macmillan Publishers Ltd. Nature Nanotechnology, Dean, C.R. et al. Boron nitride substrates for high-quality graphene electronics. **5**, 722–726, copyright 2010; (b) Reprinted by permission from Macmillan Publishers Ltd. *Nature Materials*, Haigh, S.J. et al. Cross-sectional imaging of individual layers and buried interfaces of graphene-based heterostructures and superlattices. **11**, 764–767, copyright 2012; (c) Reprinted with permission from Gannett, W. et al. Boron nitride substrates for high mobility chemical vapor deposited graphene. *Applied Physics Letters* **98**, 242105. Copyright 2011, American Institute of Physics; (d) Reprinted with permission from Liu, Z. et al. Direct growth of graphene/hexagonal boron nitride stacked layers. *Nano Letters* **11**, 2032–2037. Copyright 2011 American Chemical Society; (f) Reprinted with permission from Gao, G., Gao, W. and Cannuccia, E. Artificially stacked atomic layers: Toward new van der Waals solids. *Nano Letters* **12**, 3518–3525. Copyright 2012 American Chemical Society; (h) Reprinted with permission from Shi, Y. et al. van der Waals epitaxy of MoS2 layers using graphene as growth templates. *Nano Letters* **12**, 2784–2791. Copyright 2012 American Chemical Society; (i–k) Reprinted by permission from Macmillan Publishers Ltd. *Nature Nanotechnology*, Georgiou, T. et al. Vertical field-effect transistor based on graphene-WS2 heterostructures for flexible and transparent electronics. **8**, 100–103, copyright 2013.)

that the transport properties of graphene can be considerably modified by placing h-BN beneath, atop, or by sandwiching graphene. As we mentioned, a hurdle has been the absence of an energy gap for graphene as an alternative material to silicon. The zero bandgap in graphene results in a high power dissipation in the OFF state (typically, ON/OFF ratio < 10). Bokdam et al.[134] predicted that graphene could be electrostatically doped through ultrathin hexagonal boron nitride films. Britnell et al.[6] report that sandwiching graphene in h-BN will obtain an ON/OFF ratio of up to 10^2. More interestingly, h-BN will enhance the mobility of graphene as a substrate,[12,128,135] due to its atomically smooth surface and close in-plane lattice match to graphene.[128,7,108,136]

h-BN has an atomic-level smooth surface. The root mean square (rms) roughness is ~50 pm. As a comparison, the rms roughness of SiO_2 is ~250 pm (Figure 1.7a).[7,39] the height distributions of graphene on SiO_2 (squares) and graphene on h-BN (triangles) are clearly shown in Figure 1.7b.[39] This ultra-flat surface can suppress rippling in graphene, which has been shown to mechanically conform to both corrugated and flat substrates.[137,138] Moiré patterns can be found in G/h-BN stack, arising from a lattice mismatch between the graphene layer and the underlying BN surface (Figure 1.7c, the rotation angle is 4°).[37] The DOS mapping of 100 nm area of graphene on both h-BN and SiO_2 is found in Figure 1.7d and e. The Dirac points of graphene were determined by the minimum voltage of the tip. The distribution of energy of the Dirac point is shown in Figure 1.7f, which is used to evaluate the disorder in graphene. With h-BN as a substrate, the full width at half maximum (FWHM) of the energy of the Dirac point is ~20 meV, only ~1/5 of that on SiO_2 (>100 meV). This suggests that the h-BN substrates may improve the electronic properties of graphene. Complementary to DOS mapping, a significant reduction in local microscopic charge inhomogeneity was observed as well at the same system. Topography and charge density of G/h-BN and G/SiO_2 are mapped in Figure 1.7g and h. The roughness of graphene charge density is 8.2×10^{10} e cm^{-2} rms (right pane in Figure 1.7h) when it sits on SiO_2, more than three times higher than on h-BN (~2.3×10^{10} e cm^{-2} rms, left panel in Figure 1.7h).[37]

G/h-BN stacks have an enhanced mobility due to fact that the extremely smooth surface will dramatically reduce the scattering effect of carriers in graphene. High-quality graphene/h-BN stacks can be prepared using mechanical exfoliated graphene and h-BN following a transfer protocol with the help of glass slides and PMMA,[2,7,128,139-141] as we mentioned. The mobility of graphene was reported up to 270,000 cm^2/Vs.[12,128,135,139,141] In addition to the high mobility, G/h-BN rapidly rises as an excellent system for the studies of correlated states of Dirac fermions in graphene under strong magnetic fields, namely fractional quantum Hall effect (FQHE).[142] The FQHE in graphene has been explored from the aspect of interplay between man–body correlations and the SU(4) symmetry of quasiparticles.[142] However, the FQHE requires high-quality, extremely clean graphene, and less influence from the substrates such as impurities and charge traps. That is why it is difficult to observe the high mobility

and FQHE for the graphene on SiO_2. Nevertheless, FQHE is still observed by Bolotin et al.[142] in the suspended exfoliated graphene (Figure 1.8a and b) at a low temperature (6 K) and high field (14 T). They also found that at low carrier density graphene becomes an insulator with a magnetic-field-tunable energy gap.

h-BN is believed as a perfect substrate for graphene for the measurement of FQHE. The 1.8% mismatches of their lattice will lead to the formation of Moiré patterns with a much larger wavelength than the lattice constant (Figure 1.8a). The periodic patterns form a smooth superlattice potential and tune the local asymmetry between the graphene sublattices induced by the potential difference between boron and nitrogen atoms in the h-BN.[141] Most recently, Hunt, Dean, and Ponomarenko independently reported the FQHE from G/h-BN-stacked devices. They transferred graphene flakes onto h-BN substrates via polymer-assisted mechanical transfer. With an increasing perpendicular magnetic field, extra peaks (namely, neutrality points) appear in the plot of longitudinal and Hall resistivities (ρ_{xx} and ρ_{xy}) versus magnetic field (Figure 1.8b and c), arising from the superlattice potential induced by h-BN substrates.[140] Moreover, as shown in Figure 1.8d, the peak of ρ_{xx} becomes broader with increasing perpendicular magnetic field and then splits into multiple maxima. The numbers (the superlattice filling factors are labeled from −6 to 22; $\nu_S = n\phi_0/B = \pm2, \pm6, \pm10, \ldots$ where ϕ_0 is the flux quantum) indicate the QHE states extending from the main Dirac point. The superlattice quantum states (evolved from $\nu_S = \pm2$) are marked by arrows. There is a little bit shifting to make the white stripes clear.[140] This deep minimum reveals that that electron-like cyclotron trajectories in graphene's valence band persist under quantized magnetic field.[140] Such behavior can somehow be characterized by the Hofstadter butterfly spectrum (Figure 1.8e). The emergence of states with integer-quantized conductance at noninteger filling of a single Landau level, severing the canonical relationship between quantized conductance and filling fraction, is the signature of the Hofstadter butterfly (Figure 1.8e).[141]

We have discussed the surface morphologies and transport behaviors of G/h-BN vertical heterostructures, in particular, the h-BN stacks, prepared by mechanical exfoliation. This method leads to high-quality samples yet limited output. The multistep growth is proposed for scalable synthesis of graphene/h-BN vertical heterostructures on various substrates. Liu et al. reported the direct growth of h-BN on graphene. Graphene is first grown on Cu substrates using liquid carbon sources. Subsequently, the as-grown graphene/Cu has been transferred to another CVD system for the growth of h-BN on the graphene using ammonia boron as the source of B and N (Figure 1.9a). Figure 1.9b shows an optical image of as-grown G/h-BN stacked landscape transferred on silica. HR TEM in Figure 1.9c reveals that the stacked structure consists of two to few layers. The ordering of graphene and h-BN is further confirmed by XPS (x-ray photoelectron spectroscopy) depth profile. They observed that the signal from carbon increased first and then decreased for the graphene/h-BN films because graphene grew beneath the h-BN layer. With

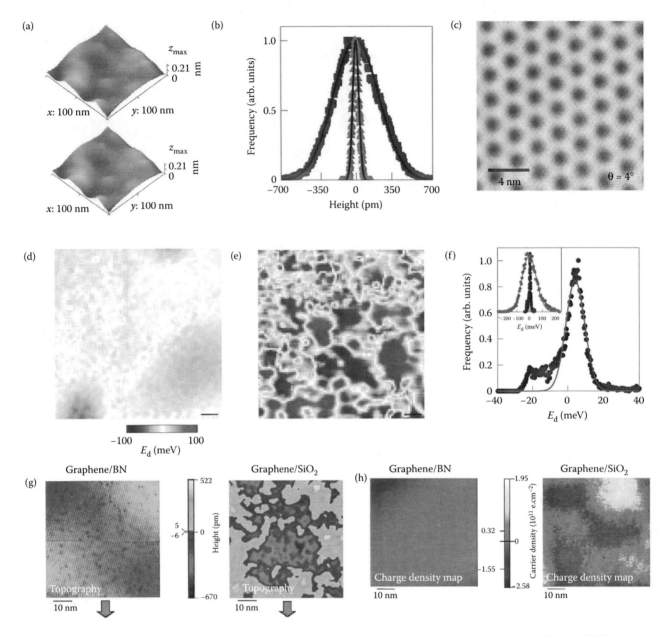

FIGURE 1.7 Comparison of the surface roughness of graphene and h-BN and impact on the spatial density of state (DOS) and charge density distribution of graphene. (a) AFM images showing the surface roughness of h-BN surface (top) and 285 nm SiO$_2$ from thermal oxidation.[38] (b) Height distribution from the topographies of h-BN and SiO$_2$ surface.[38] (c) STM imaging of typical Moiré pattern from graphene/h-BN stack with a lattice rotation angle of ~4° between graphene and h-BN. Scale bar, 4 nm.[37] (d) and (e) spatial DOS mapping of graphene on h-BN and SiO$_2$, respectively. Scale bar, 10 nm.[38] (f) Distribution of energy of the Dirac point of graphene on h-BN (inset: graphene on SiO$_2$).[37] (g) and (h) Topography and charge density mapping for graphene on h-BN and SiO$_2$, respectively. Scale bar, 10 nm.[37] ((a, b, d, e) Reprinted by permission from Macmillan Publishers Ltd. *Nature Materials*, Xue, J.M. et al. Scanning tunneling microscopy and spectroscopy of ultra-flat graphene on hexagonal boron nitride. **10**, 282–285, copyright 2011; (c, f–h) Reprinted with permission from Decker, R.G. et al. Local electronic properties of graphene on a BN substrate via scanning tunneling microscopy. *Nano Letters* **11**, 2291–2295, Copyright 2011 American Chemical Society.)

the Ar ion etching, the top h-BN layer becomes thinner and the buried graphene layer appears resulting in the maximum value of carbon concentration in XPS data. Then the value decreases because graphene also gets removed by the continuous ion etching.[129]

Complementary to the growth of h-BN on graphene, the CVD growth of graphene on h-BN is also demonstrated via

a general transfer-free method. The growth was proved to be applicable for large areas of uniform bilayer graphene growth on h-BN and other insulating substrates, from solid carbon sources such as films of poly(2-phenylpropyl)methysiloxane (PPMS), PMMA, polystyrene (PS), and poly(acrylonitrile-co-butadiene-co-styrene).[128] The carbon feedstock was deposited on the insulating substrates and then caped with a

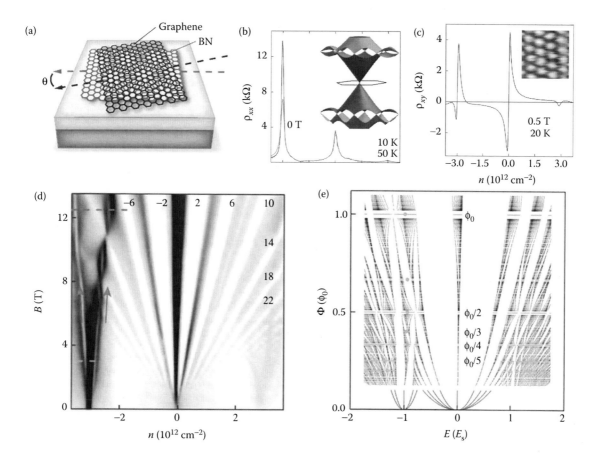

FIGURE 1.8 FQHE and Hofstadter butterfly spectrum in G/h-BN stacks. (a) Schematic of a Moiré pattern from graphene /h-BN stack. The Moiré wavelength depends on the mismatch angle, θ.[139] (b) Longitudinal resistivity ρ_{xx} versus carrier density n. Inset plots a possible band structure of graphene on h-BN only for the first and second.[140] (c) Hall resistivity ρ_{xy} versus carrier density n. The secondary Dirac point is found. Inset shows the AFM image of the Moiré patters of the sample.[140] (d) Longitudinal resistivity dependency on the magnetic field from a graphene/h-BN stacks with a rotation angle of ~1.2°, measured at 20 K. Grayscale: white, 0 kΩ; black, 8.5 kΩ. Grayscale: white, 0 kΩ; black, 1.1 kΩ. Numbers denote n for the QHE states extending from the main Dirac point. The arrows mark the superlattice quantum states that evolve along $\nu_s = \pm 2$.[140] (e) Hofstadter-like butterfly for the graphene/h-BN superlattice. The energies of states are shown for simple fractions.[139] (Reprinted by permission from Macmillan Publishers Ltd. *Nature, Dean*, C.R. et al. Hofstadter's butterfly and the fractal quantum Hall effect in Moire superlattices. **497**, 598–602, copyright 2013; Reprinted by permission from Macmillan Publishers Ltd. *Nature*, Ponomarenko, L.A. et al. Cloning of Dirac fermions in graphene superlattices. **497**, 594–597, copyright 2013.)

layer of nickel. At 1000°C, under low pressure and a reducing atmosphere, the carbon source could be transformed into a bilayer graphene film on the insulating substrates. The Ni layer was removed by dissolution, affording the bilayer graphene directly on the insulator such as h-BN with no traces of polymer left from a transfer step. A homogeneous bilayer of graphene was synthesized between the h-BN and the Ni film. Marble's reagent was used to dissolve the Ni layer. The end result was that bilayer graphene was directly grown on the h-BN surface.[143]

In addition to h-BN, graphene could be grown by CVD method on different substrates, including transition metals such as nickel (Ni),[56,144–151] copper (Cu),[56,148,152–157] cobalt (Co),[148,158,159] ruthenium (Ru),[160–162] iridium (Ir),[163,164] palladium (Pd),[165] gold (Au)[166] and their alloys (Cu–Ni, Cu–Zn),[148,167,168] etc., insulating substrates such as silicon oxide (SiO₂),[143,169,170] magnesium oxide (MgO),[171] silicon nitride (Si₃N₄),[143] aluminum oxide (Al₂O₃),[143,170] and diamond (C),[172]

and other substrates such as silicon (Si)[173–176] and silicon carbide (SiC)[177–182] as well. These works provide an exciting potential for graphene-based applications in composite/functional material, batteries, electronics, and transparent electrodes.[183–186]

Graphene could be obtained by reducing SiC at high temperatures (>1100°C). The sample size depends on the size of SiC wafer. SiC is the first substrate used to grow graphene with good quality in 2004.[180] Since then, significant advances have been made in this area. The quality of graphene is considerably improved.[187–190] The possibility of large integrated electronics on SiC-epitaxial graphene was also first proposed in 2004. Hundreds of transistors on a single chip[183] are made by Massachusetts Institute of Technology and in 2009, very high frequency transistors were produced on monolayer graphene on SiC.[185] Most recently, researchers at IBM reported graphene transistors with an on and off rate of 100 GHz, exceeding the speed of silicon transistors with an equal gate length.[184]

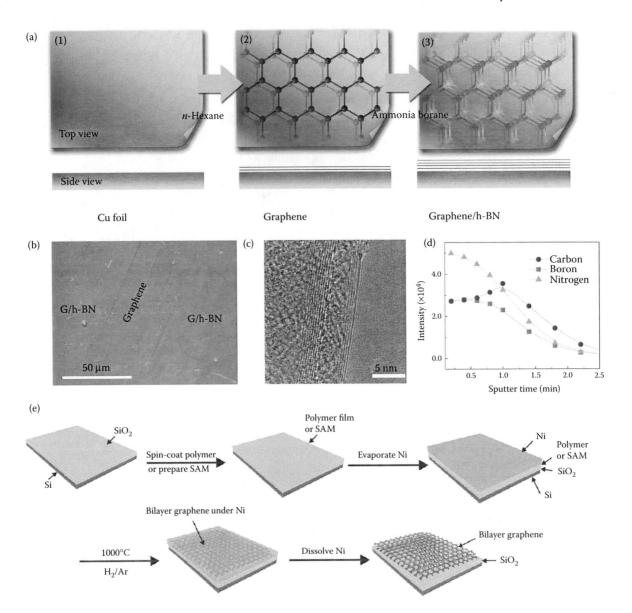

FIGURE 1.9 Direct growth of graphene/h-BN stacks by chemical vapor deposition. (a–d) CVD growth of h-BN on graphene.[129] (a) Schematic showing the preparation of G/BN stacked film. (1) The Cu foil cleaned and prepared as the growth substrate; (2) high-quality and large-area graphene films grown based on CVD of *n*-hexane at a temperature of 950°C; (3) some graphene transferred onto SiO_2 for further characterization and the rest loaded into another furnace for the growth of h-BN film on top. (b) Optical microscopy image of graphene/h-BN stacked atomic layers supported by the SiO_2 substrate. The wide stripe is graphene, located at the center. Some tiny SiO_2 regions are noted. The rest of the area is covered by G/h-BN stack. (c) TEM image of the film edge show the number of layers and thickness of the G/h-BN film. (d) XPS depth profile show the evolution of count/second for B, N, and C 1s core levels during etching with Ar ion sputtering. The signal from B and N becomes weak during etching. (e) Synthetic protocol and spectroscopic analysis of bilayer graphene.[143] (Reprinted with permission from (a–d) Liu, Z. et al. Direct growth of graphene/hexagonal boron nitride stacked layers. *Nano Letters* 11, 2032–2037. Copyright 2011 American Chemical Society; (e) Yan, Z. et al. Growth of bilayer graphene on insulating substrates. *Acs Nano* 5, 8187–8192. Copyright 2011 American Chemical Society.)

A variety of transition metals are demonstrated for the growth of graphene via simple thermal decomposition of hydrocarbons on the surface or surface segregation of carbon on cooling from a metastable carbon–metal solid solution such as the CVD method. The carbon solubility in the metal and the growth conditions determine the deposition mechanism which ultimately also defines the morphology and thickness of the graphene films.[191] Especially, Ni and Cu are found to be excellent substrates for the growth of large-scale and high-quality graphene. Among the substrates, Cu and Ni structures are the most common material used for growing graphene. It is also demonstrated the direct CVD of a large-size and homogeneous single- or few-layer graphene film on dielectric surfaces, including SiO_2, Si_3N_4, and Al_2O_3, via sacrificial metal films (Cu and Ni) as we mentioned above; the metal layer could be removed by dissolution, affording the graphene directly on the insulator with no traces of polymer left from a transfer step.

1.4 ALL-CARBON G/CNT VERTICAL HETEROSTRUCTURE

CNTs, as warped graphene, have unique electric transport properties. The electrons in CNTs move only along the confined tube's axis (typically less than 2 nm for single-walled carbon nanotubes, SWNTs) and the electron transport involves quantum effects. Subsequently, CNTs are frequently referred to as 1D conductors. It is well-known that the maximum electrical conductance of an SWNT is two times of quantum conductance ($2G_0$, ~7.7 × 10^{-5}/Ω). Theoretical work suggested that a covalently bonded graphene/SWNT hybrid material would extend those properties to three dimensions, and be useful in energy storage and nanoelectronic technologies.[8,9,54,192–194]

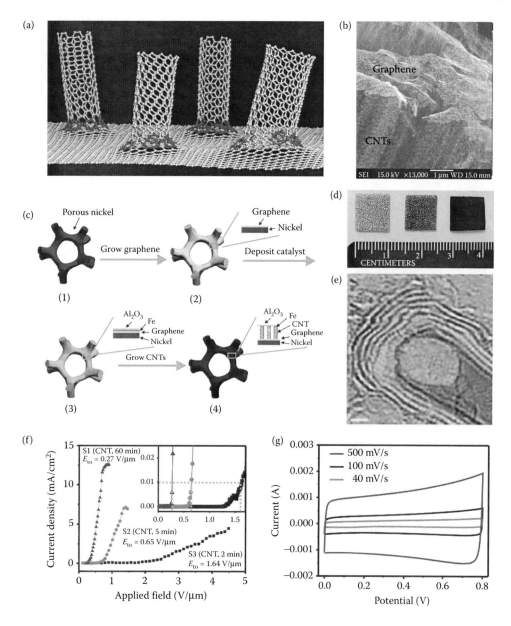

FIGURE 1.10 Synthesis, characterization, and applications of graphene/CNTs heterostructures. (a) Schematic of graphene/CNTs heterostructure connected by seven-atom rings at the transition from graphene to nanotube. This hybrid may be the best electrode interface material possible for many energy storage and electronics applications.[53] (b) SEM image of graphene/CNTs heterostructures. Scale bar, 1 μm. (c) Scheme for the synthesis of CNT forests on graphene porous nickel. (1) The porous nickel substrate; (2) few-layer graphene is formed on the porous nickel by a CVD method; (3) Fe and Al_2O_3 are sequentially deposited on the graphene using e-beam evaporation; (4) a CNT forest is directly grown from the graphene surface while lifting the Fe/Al_2O_3 catalyst layer.[8] (d) Photographs of porous nickel, graphene porous nickel, and CNT graphene porous nickel (from left to right).[8] (e) High-resolution BF-STEM images of the root of CNTs. Scale bar, 1 nm.[53] (f) Variation of the emission current density as a function of the applied field for S1 (CNT, 60 min), S2 (CNT, 5 min), and S3 (CNT, 2 min). The inset is the enlarged data, from which the turn-on fields are determined at the current density of 0.01 mA/cm^2.[9] (Reprinted by permission from Macmillan Publishers Ltd. *Nature Communications*, Zhu, Y. et al. A seamless three-dimensional carbon nanotube graphene hybrid material. **3**, 1225, copyright 2012 and Reprinted with permission from Yan, Z. et al. Three-dimensional metal–graphene–nanotube multifunctional hybrid materials. *ACS Nano* **7**, 58–64. Copyright 2012 American Chemical Society.)

Figure 1.10a illustrates a graphene/CNT heterostructure with covalent bonds. The 5–7 rings may play an important role that provides the C–C bonding between graphene and CNTs.[53] Figure 1.10b shows a typical graphene/CNT by the CVD method. Figure 1.10c is a schematic of the growth protocol. First, graphene was grown on porous nickel. Alumina and iron were deposited on graphene as catalyst for the subsequent growth of CNTs. The photography of as-grown graphene/CNTs on Ni foams is shown in Figure 1.10d. From left to right are porous nickel, graphene porous nickel, and CNT graphene porous nickel, respectively. Owing to the 3D architectures, it becomes a great challenge to characterize the interfaces between graphene and CNTs (Figure 1.10e). The work by Zhu et al.[53] implies the interfaces consist of 5–7 rings.

The graphene/CNT heterostructure has a very large surface area (~2000 m^2/g), making it a promising material for energy storage.[53] In addition, graphene/CNT hybrid is a good candidate for high-performance field emission. Compared with previous reports, the current density at a field of 0.87 V/μm is ~13 mA/cm^2 (Figure 1.10f), one of the highest reported values, to date.[8] The capacitive properties are also examined. The rectangular and symmetric shape of the cyclic voltammetry curves (CVs) were observed at high scan rates of 500 mV/s, supporting the suggestion of low contact resistance for the CNT graphene porous nickel interface (Figure 1.10g).[8] An almost constant specific capacitance as the scan rate up to 1.00 A/g is confirmed, which indicates the high diffusion conductivity of the active material and the electrolyte. Lin et al. reported that the graphene/CNTs based micro-supercapacitors can deliver a high volumetric energy density of up to 2.42 mW h/cm^3 in the ionic liquid, more than 2 orders of magnitude higher than that of aluminum electrolytic capacitors. The ultrahigh rate capability of 400 V/s enables a maximum power density of 115 W/cm^3 in aqueous electrolyte using the microdevices.[9] These results show the great potential of graphene/CNT heterostructures for their promising applications in electronic and energy devices.

1.5   CONCLUSIONS

Graphene heterostructures have recently emerged as a promising area of research beyond graphene. Graphene on h-BN architecture, as the first graphene heterostructure, was introduced in 2010 and has been recognized as an approach to realize high-quality graphene as that of suspended graphene. Owing to the lack of a bandgap, graphene FET shows very high mobility yet poor "ON/OFF" ratio, which is a main obstacle to the application of graphene in electronic industry. Graphene heterostructures with particular architectures were expected to open a bandgap in graphene, which has been confirmed by recent experiments. Graphene heterostructures can also find many applications such as in high "ON/OFF" ratio FETs, photodetectors, solar cells, etc. It is expected that an increase in discoveries and continuing progress will emerge in this research area in the near future.

In this chapter, we first demonstrated the synthesis, characterizations, and applications of planar graphene heterostructures including the phase-separated h-BNC materials and doped graphene. Experimentally, it has been proved that the bandgap of hybridized BNC films can be tuned in a large range by changing the concentrations of BN and graphene. In the next part, we summarized the progress on various vertical graphene heterostructures such as graphene-h BN, graphene-TMDs stacks, and superlattice. The fabrication technique is of critical importance in the research and applications of graphene heterostructures. We have demonstrated that CVD can be used as a most effective way for achieving all of these lateral and vertical graphene heterostructures. Finally, we introduced the 3D G/CNT as an example of other types of graphene heterostructures, which show great promise in applications such as energy harvest and field emission.

REFERENCES

1. Ci, L. et al. Atomic layers of hybridized boron nitride and graphene domains. *Nature Materials* **9**, 430–435, 2010.
2. Georgiou, T. et al. Vertical field-effect transistor based on graphene-WS2 heterostructures for flexible and transparent electronics. *Nature Nanotechnology* **8**, 100–103, 2013.
3. Li, Y. et al. MoS$_2$ nanoparticles grown on graphene: An advanced catalyst for the hydrogen evolution reaction. *Journal of the American Chemical Society* **133**, 7296–7299, 2011.
4. Najmaei, S. et al. Vapour phase growth and grain boundary structure of molybdenum disulphide atomic layers. *Nature Materials* **12**, 754–759, 2013.
5. Xu, M., Liang, T., Shi, M., and Chen, H. Graphene-like two-dimensional materials. *Chemical Reviews* **113**, 3766–3798, 2013.
6. Britnell, L. et al. Field-effect tunneling transistor based on vertical graphene heterostructures. *Science* **335**, 947–950, 2012.
7. Wang, H. et al. BN/graphene/BN transistors for RF applications. *Electron Device Letters, IEEE* **32**, 1209, 2011.
8. Yan, Z. et al. Three-dimensional metal–graphene–nanotube multifunctional hybrid materials. *ACS Nano* **7**, 58–64, 2012.
9. Lin, J. et al. 3-Dimensional graphene carbon nanotube carpet-based microsupercapacitors with high electrochemical performance. *Nano Letters* **13**, 72–78, 2012.
10. Kim, K.S. et al. Large-scale pattern growth of graphene films for stretchable transparent electrodes. *Nature* **457**, 706–710, 2009.
11. Bolotin, K.I. et al. Ultrahigh electron mobility in suspended graphene. *Solid State Communications* **146**, 351–355, 2008.
12. Dean, C.R. et al. Boron nitride substrates for high-quality graphene electronics. *Nature Nanotechnology* **5**, 722–726, 2010.
13. Lin, Y.-M. et al. Wafer-scale graphene integrated circuit. *Science* **332**, 1294–1297, 2011.
14. Liu, M. et al. A graphene-based broadband optical modulator. *Nature* **474**, 64–67, 2011.
15. Novoselov, K.S. et al. Electric field effect in atomically thin carbon films. *Science* **306**, 666–669, 2004.
16. Song, L. et al. Large scale growth and characterization of atomic hexagonal boron nitride layers. *Nano Letters* **10**, 3209–3215, 2010.
17. Sutter, P., Lahiri, J., Zahl, P., Wang, B., and Sutter, E. Scalable synthesis of uniform few-layer hexagonal boron nitride dielectric films. *Nano Letters* **13**, 276–281, 2012.
18. Shi, Y. et al. Synthesis of few-layer hexagonal boron nitride thin film by chemical vapor deposition. *Nano Letters* **10**, 4134–4139, 2010.

19. Lee, K.H. et al. Large-scale synthesis of high-quality hexagonal boron nitride nanosheets for large-area graphene electronics. *Nano Letters* **12**, 714–718, 2012.

20. Lee, Y.-H. et al. Synthesis of large-area MoS_2 atomic layers with chemical vapor deposition. *Advanced Materials* **24**, 2320–2325, 2012.

21. Osada, M. and Sasaki, T. Two-dimensional dielectric nanosheets: Novel nanoelectronics from nanocrystal building blocks. *Advanced Materials* **24**, 210–228, 2012.

22. van der Zande, A.M. et al. Grains and grain boundaries in highly crystalline monolayer molybdenum disulphide. *Nature Materials* **12**, 554–561, 2013.

23. Liu, K.-K. et al. Growth of large-area and highly crystalline MoS_2 thin layers on insulating substrates. *Nano Letters* **12**, 1538–1544, 2012.

24. Shi, Y. et al. Highly ordered mesoporous crystalline $MoSe_2$ material with efficient visible-light-driven photocatalytic activity and enhanced lithium storage performance. *Advanced Functional Materials* **23**, 1832–1838, 2012.

25. Zhang, Y., Liu, Z., Najmaei, S., Ajayan, P.M., and Lou, J. Large-area vapor-phase growth and characterization of MoS(2) atomic layers on a SiO(2) substrate. *Small* **8**, 966–971, 2012.

26. Zhang, Y., Ye, J., Matsuhashi, Y., and Iwasa, Y. Ambipolar MoS_2 thin flake transistors. *Nano Letters* **12**, 1136–1140, 2012.

27. Fang, H. et al. High-performance single layered WSe_2 p-FETs with chemically doped contacts. *Nano Letters* **12**, 3788–3792, 2012.

28. Yu, Y. et al. Controlled scalable synthesis of uniform, high-quality monolayer and few-layer MoS_2 films. *Scientific Reports* **3**, 1866, 2013.

29. Chang, C.-K. et al. Band gap engineering of chemical vapor deposited graphene by *in situ* BN doping. *ACS Nano* **7**, 1333–1341, 2012.

30. Dubois, S.M.-M., Declerck, X., Charlier, J.C., and Payne, M.C. Spin filtering and magneto-resistive effect at the graphene/h-BN ribbon interface. *ACS Nano* **7**, 4578–4585, 2013.

31. Levendorf, M.P. et al. Graphene and boron nitride lateral heterostructures for atomically thin circuitry. *Nature* **488**, 627–632, 2012.

32. Jung, J., Qiao, Z., Niu, Q., and MacDonald, A.H. Transport properties of graphene nanoroads in boron nitride sheets. *Nano Letters* **12**, 2936–2940, 2012.

33. Bernardi, M., Palummo, M., and Grossman, J.C. Semiconducting monolayer materials as a tunable platform for excitonic solar cells. *ACS Nano* **6**, 10082–10089, 2012.

34. Liu, Z. et al. In-plane heterostructures of graphene and hexagonal boron nitride with controlled domain sizes. *Nature Nanotechnology* **8**, 119–124, 2013.

35. Li, Y. et al. MoS_2 Nanoparticles grown on graphene: An advanced catalyst for hydrogen evolution reaction. *Journal of the American Chemical Society* **133**, 7296–7299, 2011.

36. Bernardi, M., Palummo, M., and Grossman, J.C. Extraordinary sunlight absorption and one nanometer thick photovoltaics using two-dimensional monolayer materials. *Nano Letters* **13**, 3664–3670, 2013.

37. Decker, R.G. et al. Local electronic properties of graphene on a BN substrate via scanning tunneling microscopy. *Nano Letters* **11**, 2291–2295, 2011.

38. Xue, J.M. et al. Scanning tunnelling microscopy and spectroscopy of ultra-flat graphene on hexagonal boron nitride. *Nature Materials* **10**, 282–285, 2011.

39. Ding, X., Ding, G., Xie, X., Huang, F., and Jiang, M. Direct growth of few layer graphene on hexagonal boron nitride by chemical vapor deposition. *Carbon* **49**, 2522–2525, 2011.

40. Yang, W. et al. Epitaxial growth of single-domain graphene on hexagonal boron nitride. *Nature Materials* **12**, 792–797, 2013.

41. Wang, M. et al. A platform for large-scale graphene electronics—CVD growth of single-layer graphene on CVD-grown hexagonal boron nitride. *Advanced Materials* **25**, 2746–2752, 2013.

42. Tang, S. et al. Precisely aligned graphene grown on hexagonal boron nitride by catalyst free chemical vapor deposition. *Scientific Reports* **3**, 2666, 2013.

43. Son, M., Lim, H., Hong, M., and Choi, H.C. Direct growth of graphene pad on exfoliated hexagonal boron nitride surface. *Nanoscale* **3**, 3089–3093, 2011.

44. Shujie, T. et al. Nucleation and growth of single crystal graphene on hexagonal boron nitride. *Carbon* **50**, 329–31, 2012.

45. Roth, S., Matsui, F., Greber, T., and Osterwalder, J. Chemical vapor deposition and characterization of aligned and incommensurate graphene/hexagonal boron nitride heterostack on Cu(111). *Nano Letters* **13**, 2668–2675, 2013.

46. Kim, S.M. et al. Synthesis of patched or stacked graphene and hBN flakes: A route to hybrid structure discovery. *Nano Letters* **13**, 933–941, 2013.

47. Gao, G., Gao, W., and Cannuccia, E. Artificially stacked atomic layers: Toward new van der Waals solids. *Nano Letters* **12**, 3518–3525, 2012.

48. Sup Choi, M. et al. Controlled charge trapping by molybdenum disulphide and graphene in ultrathin heterostructured memory devices. *Nature Communications* **4**, 1624, 2013.

49. Bertolazzi, S., Krasnozhon, D., and Kis, A. Nonvolatile memory cells based on MoS_2/graphene heterostructures. *ACS Nano* **7**, 3246–3252, 2013.

50. Shi, Y. et al. van der Waals epitaxy of MoS_2 layers using graphene as growth templates. *Nano Letters* **12**, 2784–2791, 2012.

51. Lin, Y.-C. et al. Direct synthesis of van der Waals solids. *ACS Nano* **8**, 3715–3723, 2014.

52. Ling, X. et al. Role of the seeding promoter in MoS_2 growth by chemical vapor deposition. *Nano Letters* **14**, 464–472, 2014.

53. Zhu, Y. et al. A seamless three-dimensional carbon nanotube graphene hybrid material. *Nature Communications* **3**, 1225, 2012.

54. Zhao, L. et al. Visualizing individual nitrogen dopants in monolayer graphene. *Science* **333**, 999–1003, 2011.

55. Fiori, G., Betti, A., Bruzzone, S., and Iannaccone, G. Lateral graphene-hBCN heterostructures as a platform for fully two-dimensional transistors. *ACS Nano* **6**, 2642–2648, 2012.

56. Ao, Z.M., Jiang, Q., Zhang, R.Q., Tan, T.T., and Li, S. Al doped graphene: A promising material for hydrogen storage at room temperature. *Journal of Applied Physics* **105**, 074307, 2009.

57. Liu, H.T., Liu, Y.Q., and Zhu, D.B. Chemical doping of graphene. *Journal of Materials Chemistry* **21**, 3335–3345, 2011.

58. Jariwala, D., Srivastava, A., and Ajayan, P.M. Graphene synthesis and band gap opening. *Journal of Nanoscience and Nanotechnology* **11**, 6621–6641, 2011.

59. Kim, K.K. et al. Enhancing the conductivity of transparent graphene films via doping. *Nanotechnology* **21**, 285205, 2010.

60. Muchharla, B. et al. Tunable electronics in large-area atomic layers of boron–nitrogen–carbon. *Nano Letters* **13**, 3476–3481, 2013.

61. Kumar, N. et al. Borocarbonitrides, BxCyNz. *Journal of Materials Chemistry A* **1**, 5806–5821, 2013.

62. Kaner, R.B., Kouvetakis, J., Warble, C.E., Sattler, M.L., and Bartlett, N. Boron–carbon–nitrogen materials of graphite-like structure. *Materials Research Bulletin* **22**, 399–404, 1987.

63. Gao, Y. et al. Toward single-layer uniform hexagonal boron nitride–graphene patchworks with zigzag linking edges. *Nano Letters* **13**, 3439–3443, 2013.

64. Sutter, P., Cortes, R., Lahiri, J., and Sutter, E. Interface formation in monolayer graphene-boron nitride heterostructures. *Nano Letters* **12**, 4869–4874, 2012.

65. Kim, K.K. et al. Synthesis and characterization of hexagonal boron nitride film as a dielectric layer for graphene devices. *ACS Nano* **6**, 8583–8590, 2012.

66. Sutter, P., Lahiri, J., Albrecht, P., and Sutter, E. Chemical vapor deposition and etching of high-quality monolayer hexagonal boron nitride films. *ACS Nano* **5**, 7303–7309, 2011.

67. Orofeo, C., Suzuki, S., Kageshima, H., and Hibino, H. Growth and low-energy electron microscopy characterization of monolayer hexagonal boron nitride on epitaxial cobalt. *Nano Research* **6**, 335–347, 2013.

68. Lee, Y.-H. et al. Growth selectivity of hexagonal-boron nitride layers on Ni with various crystal orientations. *RSC Advances* **2**, 111–115, 2012.

69. Ćavar, E. et al. A single h-BN layer on Pt(111). *Surface Science* **602**, 1722–1726, 2008.

70. Preobrajenski, A.B. et al. Influence of chemical interaction at the lattice-mismatched h-BN/Rh(111) and h-BN/Pt(111) interfaces on the overlayer morphology. *Physical Review B* **75**, 245412, 2007.

71. Dong, G., Fourré, E.B., Tabak, F.C., and Frenken, J.W.M. How boron nitride forms a regular nanomesh on Rh(111). *Physical Review Letters* **104**, 096102, 2010.

72. Müller, F., Hüfner, S., and Sachdev, H. Epitaxial growth of boron nitride on a Rh(111) multilayer system: Formation and fine tuning of a BN-nanomesh. *Surface Science* **603**, 425–432, 2009.

73. Müller, F. et al. Epitaxial growth of hexagonal boron nitride on Ag(111). *Physical Review B* **82**, 113406, 2010.

74. Terrones, M., Souza Filho, A.G., and Rao, A.M. Doped carbon nanotubes: Synthesis, characterization and applications. In: Jorio, A., Dresselhaus, G., and Dresselhaus, M. (eds.), *Carbon Nanotubes*, 531–566, Springer, Berlin, Heidelberg, 2008.

75. Martins, T.B., Miwa, R.H., da Silva, A.J.R., and Fazzio, A. Electronic and transport properties of boron-doped graphene nanoribbons. *Physical Review Letters* **98**, 196803, 2007.

76. Watanabe, M.O., Itoh, S., Sasaki, T., and Mizushima, K. Visible-light-emitting layered BC2N semiconductor. *Physical Review Letters* **77**, 187–189, 1996.

77. Stephan, O. et al. Doping graphitic and carbon nanotube structures with boron and nitrogen. *Science* **266**, 1683–1685, 1994.

78. Suenaga, K. et al. Synthesis of nanoparticles and nanotubes with well-separated layers of boron nitride and carbon. *Science* **278**, 653–655, 1997.

79. Kouvetakis, J. et al. Novel aspects of graphite-intercalation by fluorine and fluorides and new B/C, C/N and B/C/N materials based on the graphite network. *Synthetic Metals* **34**, 1–7, 1990.

80. Golberg, D., Bando, Y., Dorozhkin, P., and Dong, Z.C. Synthesis, analysis, and electrical property measurements of compound nanotubes in the B-C-N ceramic system. *MRS Bulletin* **29**, 38–42, 2004.

81. Yang, L., Deslippe, J., Park, C.-H., Cohen, M.L., and Louie, S.G. Excitonic effects on the optical response of graphene and bilayer graphene. *Physical Review Letters* **103**, 186802, 2009.

82. Han, W.-Q., Mickelson, W., Cumings, J., and Zettl, A. Transformation of $B_xC_yN_z$ nanotubes to pure BN nanotubes. *Applied Physics Letters* **81**, 1110–1112, 2002.

83. Warner, J.H., Rümmeli, M.H., Gemming, T., Büchner, B., and Briggs, G.A.D. Direct imaging of rotational stacking faults in few layer graphene. *Nano Letters* **9**, 102–106, 2008.

84. Moulder, J.F. and Chastain, J. *Handbook of X-ray Photoelectron Spectroscopy: A Reference Book of Standard Spectra for Identification and Interpretation of XPS Data.* Perkin-Elmer, Eden Prairie, MN, 1992.

85. Park, K.S., Lee, D.Y., Kim, K.J., and Moon, D.W. Observation of a hexagonal BN surface layer on the cubic BN film grown by dual ion beam sputter deposition. *Applied Physics Letters* **70**, 315–317, 1997.

86. Zunger, A., Katzir, A., and Halperin, A. Optical properties of hexagonal boron nitride. *Physical Review B* **13**, 5560–5573, 1976.

87. Blase, X., Rubio, A., Louie, S.G., and Cohen, M.L. Quasi-particle band structure of bulk hexagonal boron nitride and related systems. *Physical Review B* **51**, 6868–6875, 1995.

88. Rubio, A., Corkill, J.L., and Cohen, M.L. Theory of graphitic boron nitride nanotubes. *Physical Review B* **49**, 5081–5084, 1994.

89. Miyamoto, Y., Rubio, A., Cohen, M.L., and Louie, S.G. Chiral tubules of hexagonal BC2N. *Physical Review B* **50**, 4976–4979, 1994.

90. Dai, J., Yuan, J., and Giannozzi, P. Gas adsorption on graphene doped with B, N, Al, and S: A theoretical study. *Applied Physics Letters* **95**, 232105, 2009.

91. Palnitkar, U.A. et al. Remarkably low turn-on field emission in undoped, nitrogen-doped, and boron-doped graphene. *Applied Physics Letters* **97**, 063102, 2010.

92. Guo, B.D. et al. Controllable N-doping of graphene. *Nano Letters* **10**, 4975–4980, 2010.

93. Jin, Z., Yao, J., Kittrell, C., and Tour, J.M. Large-scale growth and characterizations of nitrogen-doped monolayer graphene sheets. *ACS Nano* **5**, 4112–4117, 2011.

94. Qian, W., Cui, X., Hao, R., Hou, Y.L., and Zhang, Z.Y. Facile preparation of nitrogen-doped few-layer graphene via supercritical reaction. *ACS Applied Materials and Interfaces* **3**, 2259–2264, 2011.

95. Reddy, A.L.M. et al. Synthesis of nitrogen-doped graphene films for lithium battery application. *ACS Nano* **4**, 6337–6342, 2010.

96. Sun, Z. et al. Growth of graphene from solid carbon sources. *Nature* **468**, 549–552, 2010.

97. Wang, X.R. et al. N-doping of graphene through electrothermal reactions with ammonia. *Science* **324**, 768–771, 2009.

98. Wei, D.C. et al. Synthesis of N-doped graphene by chemical vapor deposition and its electrical properties. *Nano Letters* **9**, 1752–1758, 2009.

99. Yu, S.S., Zheng, W.T., Wang, C., and Jiang, Q. Nitrogen/boron doping position dependence of the electronic properties of a triangular graphene. *ACS Nano* **4**, 7619–7629, 2010.

100. Denis, P.A., Faccio, R., and Mombru, A.W. Is it possible to dope single-walled carbon nanotubes and graphene with sulfur? *ChemPhysChem* **10**, 715–722, 2009.

101. Chen, A.Q., Shao, Q.Y., Wang, L., and Deng, F. Electronic structure and optical property of boron doped semiconducting graphene nanoribbons. *Science China-Physics Mechanics and Astronomy* **54**, 1438–1442, 2011.

102. Nair, R.R. et al. Fluorinated graphene: Fluorographene: A two-dimensional counterpart of teflon (Small 24/2010). *Small* **6**, 2773–2773, 2010.

103. Robinson, J.T. et al. Properties of fluorinated graphene films. *Nano Letters* **10**, 3001–3005, 2010.

104. Markevich, A., Jones, R., and Briddon, P.R. Doping of fluorographene by surface adsorbates. *Physical Review B* **84**, 115439, 2011.

105. Denis, P.A. Chemical reactivity of lithium doped monolayer and bilayer graphene. *Journal of Physical Chemistry C* **115**, 13392–13398, 2011.

106. Lu, D., Yan, X.H., Xiao, Y., and Yang, Y.R. Charge distributions of Li-doped few-layer graphenes on C-terminated SiC surfaces. *Physica B-Condensed Matter* **406**, 4296–4299, 2011.

107. Wengsieh, Z. et al. Synthesis of BxCyNz nanotubules. *Physical Review B* **51**, 11229–11232, 1995.

108. Li, X. et al. Large-area synthesis of high-quality and uniform graphene films on copper foils. *Science* **324**, 1312–1314, 2009.

109. Bertolazzi, S., Brivio, J., and Kis, A. Stretching and breaking of ultrathin MoS_2. *ACS Nano* **5**, 9703–9709, 2011.

110. Matte, H.S.S.R. et al. MoS_2 and WS_2 analogues of graphene. *Angewandte Chemie* **122**, 4153–4156, 2010.

111. Matte, H.S.S.R., Plowman, B., Datta, R., and Rao, C.N.R. Graphene analogues of layered metal selenides. *Dalton Transactions* **40**, 10322–10325, 2011.

112. Abdallah, W.E.A. and Nelson, A.E. Characterization of $MoSe_2(0001)$ and ion-sputtered $MoSe_2$ by XPS. *Journal of Materials Science* **40**, 2679–2681, 2005.

113. Peng, Y. et al. Hydrothermal synthesis of MoS_2 and its pressure-related crystallization. *Journal of Solid State Chemistry* **159**, 170–173, 2001.

114. Addou, R., Dahal, A., and Batzill, M. Growth of a two-dimensional dielectric monolayer on quasi-freestanding graphene. *Nature Nanotechnology* **8**, 41–45, 2013.

115. Splendiani, A. et al. Emerging photoluminescence in monolayer MoS_2. *Nano Letters* **10**, 1271–1275, 2010.

116. Balendhran, S. et al. Atomically thin layers of MoS_2 via a two step thermal evaporation-exfoliation method. *Nanoscale* **4**, 461–466, 2012.

117. Radisavljevic, B. et al. Single-layer MoS_2 transistors. *Nature Nanotechnology* **6**, 147–150, 2011.

118. Tonndorf, P. et al. Photoluminescence emission and Raman response of monolayer MoS_2, $MoSe_2$, and WSe_2. *Optics Express* **21**, 4908–4916, 2013.

119. Coleman, J.N. et al. Two-dimensional nanosheets produced by liquid exfoliation of layered materials. *Science* **331**, 568–571, 2011.

120. Nguyen, V.H., Mazzamuto, F., Bournel, A., and Dollfus, P. Resonant tunnelling diodes based on graphene/h-BN heterostructure. *Journal of Physics D: Applied Physics* **45**, 325104, 2012.

121. Modarresi, M., Roknabadi, M.R., and Shahtahmassebi, N. Rectifying behavior of graphene/h-boron-nitride heterostructure. *Physica B: Condensed Matter* **415**, 62–66, 2013.

122. Gao, G. et al. Artificially stacked atomic layers: Toward new van der Waals solids. *Nano Letters* **12**, 3518–3525, 2012.

123. Liu, W., Wang, Z.F., Shi, Q.W., Yang, J., and Liu, F. Band-gap scaling of graphene nanohole superlattices. *Physical Review B* **80**, 233405, 2009.

124. Bai, J., Zhong, X., Jiang, S., Huang, Y., and Duan, X. Graphene nanomesh. *Nature Nanotechnology* **5**, 190–194, 2010.

125. Balog, R. et al. Bandgap opening in graphene induced by patterned hydrogen adsorption. *Nature Materials* **9**, 315–319, 2010.

126. Qiao, Z., Jung, J., Niu, Q., and MacDonald, A.H. Electronic highways in bilayer graphene. *Nano Letters* **11**, 3453–3459, 2011.

127. Haigh, S.J. et al. Cross-sectional imaging of individual layers and buried interfaces of graphene-based heterostructures and superlattices. *Nature Materials* **11**, 764–767, 2012.

128. Gannett, W. et al. Boron nitride substrates for high mobility chemical vapor deposited graphene. *Applied Physics Letters* **98**, 242105, 2011.

129. Liu, Z. et al. Direct growth of graphene/hexagonal boron nitride stacked layers. *Nano Letters* **11**, 2032–2037, 2011.

130. Koma, A. Van der Waals epitaxy—A new epitaxial growth method for a highly lattice-mismatched system. *Thin Solid Films* **216**, 72–76, 1992.

131. Garcia, J.M. et al. Graphene growth on h-BN by molecular beam epitaxy. *Solid State Communications* **152**, 975–978, 2012.

132. Ponomarenko, L.A. et al. Tunable metal-insulator transition in double-layer graphene heterostructures. *Nature Physics* **7**, 958–961, 2011.

133. Song, L. et al. Binary and ternary atomic layers built from carbon, boron, and nitrogen. *Advanced Materials* **24**, 4878–4895, 2012.

134. Bokdam, M., Khomyakov, P.A., Brocks, G., Zhong, Z.C., and Kelly, P.J. Electrostatic doping of graphene through ultrathin hexagonal boron nitride films. *Nano Letters* **11**, 4631–4635, 2011.

135. Zomer, P.J., Dash, S.P., Tombros, N., and van Wees, B.J. A transfer technique for high mobility graphene devices on commercially available hexagonal boron nitride. *Applied Physics Letters* **99**, 232104, 2011.

136. Song, L., Balicas, L., Mowbray, D.J., and Capaz, R.B. Anomalous insulator-metal transition in boron nitride-graphene hybrid atomic layers. *Physical Review B* **86**, 075429, 2012.

137. Ishigami, M., Chen, J.H., Cullen, W.G., Fuhrer, M.S., and Williams, E.D. Atomic structure of graphene on SiO_2. *Nano Letters* **7**, 1643–1648, 2007.

138. Lui, C.H., Liu, L., Mak, K.F., Flynn, G.W., and Heinz, T.F. Ultraflat graphene. *Nature* **462**, 339–341, 2009.

139. Dean, C.R. et al. Hofstadter's butterfly and the fractal quantum Hall effect in moire superlattices. *Nature* **497**, 598–602, 2013.

140. Ponomarenko, L.A. et al. Cloning of Dirac fermions in graphene superlattices. *Nature* **497**, 594–597, 2013.

141. Hunt, B. et al. Massive Dirac fermions and Hofstadter butterfly in a van der Waals heterostructure. *Science* **340**, 1427–1430, 2013.

142. Bolotin, K.I., Ghahari, F., Shulman, M.D., Stormer, H.L., and Kim, P. Observation of the fractional quantum Hall effect in graphene. *Nature* **462**, 196–199, 2009.

143. Yan, Z. et al. Growth of bilayer graphene on insulating substrates. *ACS Nano* **5**, 8187–8192, 2011.

144. Xu, M.S., Fujita, D., Sagisaka, K., Watanabe, E., and Hanagata, N. Production of extended single-layer graphene. *ACS Nano* **5**, 1522–1528, 2011.

145. Chae, S.J. et al. Synthesis of large-area graphene layers on poly-nickel substrate by chemical vapor deposition: Wrinkle formation. *Advanced Materials* **21**, 2328–2333, 2009.

146. Park, H.J., Meyer, J., Roth, S., and Skakalova, V. Growth and properties of few-layer graphene prepared by chemical vapor deposition. *Carbon* **48**, 1088–1094, 2010.

147. Iwasaki, T. et al. Long-range ordered single-crystal graphene on high-quality heteroepitaxial Ni thin films grown on MgO(111). *Nano Letters* **11**, 79–84, 2011.

148. Liu, N. et al. Universal segregation growth approach to wafer-size graphene from non-noble metals. *Nano Letters* **11**, 297–303, 2011.

149. Wang, R., Hao, Y.F., Wang, Z.Q., Gong, H., and Thong, J.T.L. Large-diameter graphene nanotubes synthesized using Ni nanowire templates. *Nano Letters* **10**, 4844–4850, 2010.

150. Dedkov, Y.S., Fonin, M., Rudiger, U., and Laubschat, C. Graphene-protected iron layer on Ni(111). *Applied Physics Letters* **93**, 022509, 2008.

151. Amini, S., Garay, J., Liu, G., Balandin, A.A., and Abbaschian, R. growth of large-area graphene films from metal-carbon melts. *Journal of Applied Physics* **108**, 094321, 2010.

152. Srivastava, A. et al. Novel liquid precursor-based facile synthesis of large-area continuous, single, and few-layer graphene films. *Chemistry of Materials* **22**, 3457–3461, 2010.

153. Gao, L., Guest, J.R., and Guisinger, N.P. Epitaxial graphene on Cu(111). *Nano Letters* **10**, 3512–3516, 2010.

154. Rasool, H.I. et al. Continuity of graphene on polycrystalline copper. *Nano Letters* **11**, 251–256, 2011.

155. Robertson, A.W. and Warner, J.H. Hexagonal single crystal domains of few-layer graphene on copper foils. *Nano Letters* **11**, 1182–1189, 2011.

156. Meyer, J.C., Chuvilin, A., Algara-Siller, G., Biskupek, J., and Kaiser, U. Selective sputtering and atomic resolution imaging of atomically thin boron nitride membranes. *Nano Letters* **9**, 2683–2689, 2009.

157. Reddy, K.M., Gledhill, A.D., Chen, C.-H., Drexler, J.M., and Padture, N.P. High quality, transferrable graphene grown on single crystal Cu(111) thin films on basal-plane sapphire. *Applied Physics Letters* **98**, 113117, 2011.

158. Ago, H. et al. Epitaxial chemical vapor deposition growth of single-layer graphene over cobalt film crystallized on sapphire. *ACS Nano* **4**, 7407–7414, 2010.

159. Ramirez-Caballero, G.E., Burgos, J.C., and Balbuena, P.B. Growth of carbon structures on stepped (211)Co surfaces. *Journal of Physical Chemistry C* **113**, 15658–15666, 2009.

160. Zhang, H., Fu, Q., Cui, Y., Tan, D.L., and Bao, X.H. Growth mechanism of graphene on Ru(0001) and O(2) adsorption on the graphene/Ru(0001) surface. *Journal of Physical Chemistry C* **113**, 8296–8301, 2009.

161. Sutter, P.W., Flege, J.I., and Sutter, E.A. Epitaxial graphene on ruthenium. *Nature Materials* **7**, 406–411, 2008.

162. Sutter, E., Albrecht, P., and Sutter, P. Graphene growth on polycrystalline Ru thin films. *Applied Physics Letters* **95**, 133109, 2009.

163. Coraux, J. et al. Growth of graphene on Ir (111). *New Journal of Physics* **11**, 023006, 2009.

164. Coraux, J., Plasa, T.N., Busse, C., and Michely, T. Structure of epitaxial graphene on Ir (111). *New Journal of Physics* **10**, 043033, 2008.

165. Murata, Y. et al. Orientation-dependent work function of graphene on Pd (111). *Applied Physics Letters* **97**, 143114, 2010.

166. Oznuluer, T. et al. Synthesis of graphene on gold. *Applied Physics Letters* **98**, 183101, 2011.

167. Chen, S.S. et al. Synthesis and characterization of large-area graphene and graphite films on commercial Cu–Ni alloy foils. *Nano Letters* **11**, 3519–3525, 2011.

168. Liu, X. et al. Segregation growth of graphene on Cu–Ni alloy for precise layer control. *Journal of Physical Chemistry C* **115**, 11976–11982, 2011.

169. Peng, Z., Yan, Z., Sun, Z., and Tour, J.M. Direct growth of bilayer graphene on SiO_2 substrates by carbon diffusion through nickel. *ACS Nano* **5**, 8241–8247, 2011.

170. Ismach, A. et al. Direct chemical vapor deposition of graphene on dielectric surfaces. *Nano Letters* **10**, 1542–1548, 2010.

171. Rummeli, M.H. et al. Direct low-temperature nanographene CVD synthesis over a dielectric insulator. *ACS Nano* **4**, 4206–4210, 2010.

172. Lee, J.-K. et al. The growth of AA graphite on (111) diamond. *The Journal of Chemical Physics* **129**, 234709, 2008.

173. Li, L.M. et al. Epitaxial growth of multi-layer graphene on the substrate of Si(111). *Journal of Inorganic Materials* **26**, 472–476, 2011.

174. Hackley, J., Ali, D., DiPasquale, J., Demaree, J., and Richardson, C. Graphitic carbon growth on Si (111) using solid source molecular beam epitaxy. *Applied Physics Letters* **95**, 133114, 2009.

175. Suemitsu, M. and Fukidome, H. Epitaxial graphene on silicon substrates. *Journal of Physics D: Applied Physics* **43**, 374012, 2010.

176. Ouerghi, A. et al. Epitaxial graphene on cubic SiC (111)/Si (111) substrate. *Applied Physics Letters* **96**, 191910, 2010.

177. Emtsev, K.V. et al. Towards wafer-size graphene layers by atmospheric pressure graphitization of silicon carbide. *Nature Materials* **8**, 203–207, 2009.

178. Huang, H., Chen, W., Chen, S., and Wee, A.T.S. Bottom-up growth of epitaxial graphene on 6H-SiC(0001). *ACS Nano* **2**, 2513–2518, 2008.

179. Sprinkle, M. et al. Scalable templated growth of graphene nanoribbons on SiC. *Nature Nanotechnology* **5**, 727–731, 2010.

180. Berger, C. et al. Ultrathin epitaxial graphite: 2D electron gas properties and a route toward graphene-based nanoelectronics. *The Journal of Physical Chemistry B* **108**, 19912–19916, 2004.

181. Park, J. et al. Epitaxial graphene growth by carbon molecular beam epitaxy (CMBE). *Advanced Materials* **22**, 4140–4145, 2010.

182. Ferralis, N., Maboudian, R., and Carraro, C. Evidence of structural strain in epitaxial graphene layers on 6H-SiC (0001). *Physical Review Letters* **101**, 156801, 2008.

183. Kedzierski, J. et al. Epitaxial graphene transistors on SiC substrates. *IEEE Transactions on Electron Devices* **55**, 2078–2085, 2008.

184. Lin, Y.-M. et al. 100-GHz Transistors from wafer-scale epitaxial graphene. *Science* **327**, 662, 2010.

185. Moon, J.S. et al. Epitaxial-graphene RF field-effect transistors on Si-face 6H-SiC substrates. *IEEE Electron Device Letters* **30**, 650–652, 2009.

186. Bae, S. et al. Roll-to-roll production of 30-inch graphene films for transparent electrodes. *Nature Nanotechnology* **5**, 574–578, 2010.

187. Emtsev, K., Speck, F., Seyller, T., Ley, L., and Riley, J. Interaction, growth, and ordering of epitaxial graphene on SiC {0001} surfaces: A comparative photoelectron spectroscopy study. *Physical Review B* **77**, 155303, 2008.

188. Hass, J., De Heer, W., and Conrad, E. The growth and morphology of epitaxial multilayer graphene. *Journal of Physics: Condensed Matter* **20**, 323202, 2008.

189. Nakatsuji, K. et al. Shape, width, and replicas of π bands of single-layer graphene grown on Si-terminated vicinal SiC (0001). *Physical Review B* **82**, 045428, 2010.

190. de Heer, W.A. et al. Epitaxial graphene. *Solid State Communications* **143**, 92–100, 2007.

191. Mattevi, C., Kim, H., and Chhowalla, M. A review of chemical vapour deposition of graphene on copper. *Journal of Materials Chemistry* **21**, 3324–3334, 2011.

192. Pei, T. et al. Electronic transport in single-walled carbon nanotube/graphene junction. *Applied Physics Letters* **99**, 113102–3, 2011.

193. Rout, C.S. et al. Synthesis of chemically bonded CNT-graphene heterostructure arrays. *RSC Advances* **2**, 8250–8253 2012.

194. Youn-Su, K., Kitu, K., Frank, T.F., and Eui-Hyeok, Y. Out-of-plane growth of CNTs on graphene for supercapacitor applications. *Nanotechnology* **23**, 015301, 2012.

2 Atomic-Scale Defects and Impurities in Graphene

Rocco Martinazzo

CONTENTS

Abstract 21
2.1 Introduction 21
2.2 The π-Electron Cloud 22
2.3 Defect Formation and Characterization 23
2.4 Midgap States 24
2.5 Charge Transport Properties 28
2.6 Magnetic Properties 31
2.7 Chemical Properties 33
2.8 Conclusions 34
References 35

ABSTRACT

Atomic-scale defects such as carbon atom vacancies and neutral impurities such as adatoms play an important role in graphene because of the huge impact they have on its low-energy electronic structure. This chapter summarizes the current understanding of the effects they have on the structural, electronic, magnetic, chemical, and transport properties of this material. The emphasis is on a few theoretical aspects that help to critically examine available experimental data.

2.1 INTRODUCTION

In spite of its extraordinary and fascinating properties, graphene is not immune to defects. Native defects such as missing carbon atoms, topological imperfections, and grain boundaries appear spontaneously in graphene and can also be easily introduced in the lattice by electron or heavy-atom bombardment. Foreign species such as adsorbed hydrogen atoms, small hydrocarbon, and hydroxyl groups, are unavoidable contaminants due to, for example, the same hydrocarbon sources used in the growing process (when using chemical vapor deposition) or the "raw" precursor material employed (as it happens for graphene oxide). Treatments with relatively "high energy" sources are routinely used to introduce atoms forming stable (chemical) bonds with the substrate, while simple exposure to several gases (from inert molecules to metals) leave weaker bound ad-species on the surface, which are commonly employed for charge doping.

Obviously, because of its one-atom thickness, graphene is extremely sensitive to any lattice defect or adsorbed atom present on its "surface," with negative effects for device applications. For many reasons though, this sensitivity is better considered as an added feature, which prospectively could also be employed to tune its properties.

In this chapter we review the basic features of commonly found atomic-scale defects such as carbon atom vacancies and (monovalent) chemically bound ad-species, for which a common (and relatively simple) picture of their physics and chemistry has emerged in the last years. These defects have in common the property that they strongly affect a single lattice site at a time, and essentially remove a p_z orbital from the π-electron cloud. In turn, they act as strong, short-range ("resonant") scatterers for charge carriers, leave an unpaired electron nearby (a quasi-localized spin-1/2 magnetic moment) and bias chemical reactivity toward specific neighboring lattice positions. Some other defects (multiple vacancies, point, and extended topological defects) have more complex structures and more specific effects on graphene properties that a proper account would require a much ampler discussion, and we defer the interested reader to the excellent review articles available in the literature (Banhart et al. 2011, Kotakoski and Krasheninnikov 2011, Araujo et al. 2012, Terrones et al. 2012; Foa Torres et al. 2014). Defects such as physisorbed molecules or alkali metals have in many respects simpler effects (they act mainly as charge dopant) and will not be discussed here.

This chapter is organized as follows. After a brief introduction (Section 2.2) of the main properties determining the peculiar electronic structure of graphene (aiming mainly at establishing notation), and of the experimental realization and characterization of defective substrates (Section 2.3), in Section 2.4, we focus on isolated p_z vacancies and discuss in detail the formation of the so-called midgap states. In Section 2.5, we discuss their role in charge transport, while Section 2.6 addresses their magnetic properties. Section 2.7 deals with chemical properties, and Section 2.8 summarizes and concludes.

2.2 THE π-ELECTRON CLOUD

As is well known, carbon atoms in graphene are arranged in a honeycomb lattice by strong and directional σ bonds between sp^2 orbitals, which form occupied σ bands at energies well below the Fermi level and are responsible for the extraordinary mechanical properties of graphene. The remaining valence electrons (one for each carbon atom) populate a π band and spread over the lattice. The band is completely filled at zero temperature and charge neutrality, and comes with an empty, "antibonding" π* band and no energy gap between the two (differently from what happens in many conjugated polymers). Such simple π/π* band system governs the low-energy (up to ~1 eV) behavior of charge carriers in graphene and is responsible for the remarkable electronic and optical properties of this material. It was first described by Wallace more than 60 years ago with a model tight-binding Hamiltonian employing one p_Z orbital per site, which describes the dynamics of conduction electrons in the effective potential created by the σ network (Wallace 1947). In modern (second quantized) language it reads as

$$H = -t \sum_{<i,j>} a_i^\dagger b_j - t \sum_{<i,j>} b_i^\dagger a_j - t' \sum_{\ll i,j \gg} a_i^\dagger a_j - t' \sum_{\ll i,j \gg} b_i^\dagger b_j$$

where $a_i^\dagger$, $b_i^\dagger$ (a_i, b_i) describe creation (destruction) of an electron on the i-th lattice site of the A, B sublattices of which graphene is made, the first two sums run over the sites that are nearest neighbors (t is the corresponding hopping energy), the second ones over the sites that are nearest neighbors in each sublattice (with hopping $|t'| \ll t$) and the on-site energies (the energy of p_Z orbitals in carbon atoms) have been set equal to zero. In the absence of magnetic fields, the hopping integrals can be chosen real, and the accepted value for t is ~2.7 eV while t' depends on the parameterization used. Neglecting overlap between orbitals on different C atoms, the usual anticommutation rules hold and introducing the Fourier-transformed creation (annihilation) operators $a_k^\dagger$, $b_k^\dagger$ (a_k, b_k) the Hamiltonian takes a block diagonal form. In the matrix notation of the sublattice components $c_k^\dagger = [a_k^\dagger \quad b_k^\dagger]$ it reads as

$$H = \sum_k c_k^\dagger H(k) c_k$$

where the sum runs over k points in the first Brillouin zone, and the matrix elements of the 2×2 k-Hamiltonians contain "structure factors" for the nearest- and next-nearest neighbors,

$$H(k) = -\begin{bmatrix} t'g(k) & tf(k) \\ tf^*(k) & t'g(k) \end{bmatrix}$$

$$f(k) = \sum_{i=1,3} e^{ik\delta_i}, \quad |f(k)|^2 = 3 + g(k)$$

(δ_i are the vectors joining an A site to its neighboring B sites). Diagonalization gives the energy bands $\varepsilon(k)$,

$$\varepsilon(k) = -t'g(k) \pm t \,|f(k)|$$

where the minus (plus) sign solution correspond to the π(π*) band (see, e.g., Wallace 1947, Castro Neto et al. 2008, Katsnelson 2013). Close to the K point (the one at $2\pi/3(-2/\sqrt{3},0)d_{CC}$ if the x axis is set along the *zig-zag* direction, d_{CC} being the carbon–carbon bond length ~1.42 Å) we have $f(K + q) \sim 3d_{CC}/2(q_x - iq_y)$ and the dispersion is linear,

$$H(K + q) \approx \hbar v_F \sigma q$$

thereby giving rise to the well-known Dirac cones where the low-energy carrier dynamics takes place. Here, $v_F = (3/2) td_{CC}/\hbar$, $q = (q_x, q_y)$ is the momentum vector with origin at K, and $\sigma = (\sigma_x, \sigma_y)$ are Pauli matrices describing a "pseudo-spin" (the sublattice degree of freedom). This latter Hamiltonian describes massless, relativistic fermions in $(2 + 1)$D, and forms the basis for the continuum approximation of the carrier dynamics in graphene, that is, for the (2-component) envelope wavefunctions at energies close to the Fermi level of the neutral system. Similarly, around K', whose Hamiltonian is related to the one given above by time-reversal symmetry, $H(K' + q) = H^*(K - q)$. Consequently, the density-of-states (DOS) is linearly vanishing at zero energy (it reads as $\rho(\varepsilon) = |\varepsilon|/\pi(\hbar v_F)^2$, per unit area per spin), one of the fingerprints of massless Dirac electrons. This behavior challenges one's intuition since experiments find a finite, nonzero minimum conductivity at this energy.

Albeit simple, this tight-binding model is accurate enough in describing much of the behavior of conduction electrons in graphene. If only hopping between nearest neighbors is allowed the two sublattices form two disjoint sets where A-type sites connect to B-type sites only and vice versa. The Hamiltonian describes a *bipartite* system and displays an interesting symmetry (Coulson and Rushbrooke 1940): for each nonzero energy level $\varepsilon(k)$ and eigenfunction $|\psi_k\rangle = c_A|A\rangle + c_B|B\rangle$ (where $|A\rangle/|B\rangle$ is nonzero on A/B lattice sites only), there exists a "conjugate" level with energy $-\varepsilon(k)$ and wavefunction $|\psi'_k\rangle = c_A|A\rangle - c_B|B\rangle$. This is called *electron–hole* (e–h) symmetry since at *half-filling* (which is the case of graphene with one electron per site), the Fermi level lies at zero energy, and the above symmetry relates electrons and holes. For a proof, let P be the projector on the A subspace and $Q = 1 - P$ that on the B subspace. The operator $C = P - Q$ is self-adjoint, null-potent ($C^2 = 1$), and satisfies $CHC = -H$ if only nearest-neighbors terms are retained in H. Then $HC|\psi_k\rangle = -\varepsilon(k)C|\psi_k\rangle$ and $|\psi'_k\rangle = C|\psi_k\rangle$. Notice that in the continuum (long wavelength) approximation C reduces to the z component of the pseudo-spin in spinor notation (σ_z), and describes *chirality* of graphene Dirac fermions.

As we shall see in the following, e–h symmetry plays an important role in graphene, even if it holds only approximately. Here, we just notice that, because of such symmetry, the gapless nature of graphene would be accidental without a specific reason for having energy levels exactly at zero. The specific reason is provided by the *spatial* symmetry of the substrate. Graphene has the same point symmetry of benzene, with sixfold rotation axis perpendicular to the lattice plane and mirror planes between them, in addition to the mirror plane of the lattice itself and the ensuing inversion center. It belongs to the so-called D_{6h} point symmetry group, which is the point group appropriate for symmetry operations in real space. For a Bloch electronic state with k-vector k, symmetry is reduced to that subgroup of D_{6h} which either leaves k invariant or transform it into one its reciprocal lattice images, $k \rightarrow k + G$, where G is a reciprocal lattice vector. This is the k-group at k and determines the possible symmetry of the electronic states. For instance, the k-group at K is D_{3h} since only threefold rotation axes and three of the mirror planes transform the K images into themselves (the remaining symmetry elements determine the so-called *star* of the given k point, i.e., the set of points generated by applying these symmetry elements to k. Such a set of physically distinct points in k space are degenerate in energy; this is the case, for instance, of K and K', which belong to the star of each other).

The point here is that graphene is symmetric enough that allows high-symmetry k-groups (i.e., with rotation axis of order >2) supporting two-dimensional (real) irreducible representations (E irreps), namely D_{6h} at Γ (obviously) and D_{3h} at K, K'. Since spatial symmetry is almost compatible with e–h symmetry (i.e., it does not mix one- and two-dimensional irreps, even though it may transform a specific representation into a different one of the same dimension), a single level at zero energy may only appear if the electronic wavefunctions span a *two-dimensional* irreducible representation. This is exactly the case of the $K(K')$ point, where Bloch functions built with p_z orbitals of the A and B sublattice span the E irrep of the above D_{3h} k-group. Furthermore, this symmetry argument is enough to explain the conical dispersion of the energy at the $K(K')$ point which makes graphene so attractive; without an inversion center degeneracy is lifted already at first order in $k \cdot p$ perturbation theory when moving away from the corners of the Brillouin zone.

While spatial symmetry is exact, e–h symmetry holds in the nearest-neighbor approximation only. Nevertheless, since inclusion of higher-order hopping terms does not modify the level ordering (i.e., the minimum of the π^* band lies always above the top of the π band) the Fermi level at charge neutrality exactly matches the energy where the E irrep is found.

The tight-binding Hamiltonian above describes single-electron properties in graphene quite accurately but is necessarily inadequate for investigating many-body phenomena such as magnetism. Improvements can be achieved by including short-range electron–electron interactions in the form of a repulsive on-site Coulomb term, as originally proposed by Hubbard in the physical literature (Hubbard 1963) and Pariser–Parr–Pople in the chemical literature (Pariser and Parr 1953a,b, Pople 1953). As is shown below, the resulting Hubbard–Pariser–Parr–Pople (HPPP) model Hamiltonian, though not simple enough to be analytically solvable, offers some exact results on the behavior of the electrons in graphene, which are very useful in practice, at least as long as the system is charge neutral. Numerically, the model can be solved

exactly for few electrons only (it remains a correlated problem) and thus its mean-field approximation is often used as a surrogate. In this form it was long used to investigate (very) large systems, that is, of dimensions far outside the range of applicability of *first principles* density-functional theory (DFT) methods applied to the exact Hamiltonian. Nowadays, however, computational power and software development have been making the pseudo-potential DFT approach the method of choice for solving (at least in principle) the *exact* electronic problem for systems of ever-increasing size (thousands of atoms), and the number of realistic DFT studies of graphene-related systems has increased enormously.

Closely related to the above HPPP model, and very useful in providing *simple* ways of looking at graphene electronic properties, is the resonating valence bond (RVB) model proposed by Pauling soon after Wallace introduced his tight-binding picture. The model relies on the traditional picture of chemical bonds in terms of Lewis structures—properly modified to describe the "chemical resonance" phenomenon—and account for both localization and band-like behavior, much like the above HPPP model (see, e.g., Cooper et al. (1986) for an application to benzene, the basic building block of graphene). In fact, it is not hard to show, for instance, that the Hubbard model for the H_2 molecule can be obtained from a simple (the so-called Coulson–Fisher) VB *ansatz* to the two-electron wavefunction. In the following we will make a qualitative use of this chemical picture, as it provides insights into the formation of defects and their properties; see though Wassmann et al. (2010) who successfully applied the VB concept of Clar's sextets to investigate the influence of the edge termination on the structure of graphene nanoribbons.

2.3 DEFECT FORMATION AND CHARACTERIZATION

In this section, we briefly describe how defects such as adatoms and carbon atom vacancies are typically introduced in graphene and eventually imaged or identified. As we shall see in the following, even though the above defects have similar effects on the electronic structure, they are introduced in rather different ways on the lattice, that is, low-energy beams for adsorbing neutrals versus high-energy electrons or ions for creating vacancies. Consequently, ad-species are produced under *kinetic control* and tend to cluster on the surface, while vacancies form random patterns and typically come with a number of defects of similar formation energy.

Irradiation by high-energy particles is a main tool for creating defects in graphene, and it has long been employed on many carbon structures (graphite, fullerenes, carbon nanotubes, etc.). In contrast to bulk materials, when irradiating graphene energetic particles typically pass through the lattice without dissipating all their energy into the system. In addition, whether displacing a carbon atom, this is definitely ejected from the target and hardly enters a collision cascade or gets trapped somewhere else.

In general, the outcome of the irradiation process depends on how much energy is transferred to the lattice in individual collisions. Whether this energy is higher than the displacement energy (~22 eV for carbon atoms in graphene, Kotakoski et al. 2010) one or more carbon atoms can be removed from the lattice, thereby leaving a single or a multiple vacancy aside (Banhart et al. 2011, Kotakoski and Krasheninnikov 2011, Terrones et al. 2012). Knock-on displacements are readily achievable with a transmission electron microscope (TEM) where electrons can acquire energies up to several hundred kiloelectron volts. This gives TEM a central role in this context, since the very same experimental setup is used to create the defects and to image the surface. Controlled formation can be achieved at low energy (~100 keV), right above the threshold for the electron-induced carbon displacement (Meyer et al. 2010), and indeed aberration-corrected high-resolution (HR) TEM operating at 80 keV has been used to create defects of various kinds, image them with atomic scale resolution and follow their evolution on a second time scale (Meyer et al. 2008); even imaging the atomic-level random walk of single point defects has been shown possible in this way (Kotakoski et al. 2014). Generally speaking, the defects introduced in the lattice are Thrower–Stone–Wells defects (the $\pi/2$ bond rotation which forms a heptagon–pentagon pair), single and double carbon atom vacancies, and pentagon–heptagon and pentagon–octagon clusters. Carbon atoms tend to keep their threefold coordination and rearrange their bonding pattern to accommodate the energy transferred to the lattice; in addition, at low irradiation energy, only combination of five-, seven-, and eight-membered rings that avoid dislocations and keep the surface planar appear.

Ion irradiation allows a much larger momentum transfer to the surface and is typically used at lower energy than electrons. If the ions used in the irradiation process are carefully selected, the technique also allows some chemical modification of the substrate, that is, replacement of one or more carbon atoms with foreign species (Kotakoski and Krasheninnikov 2010, Terrones et al. 2012).

Neutral atomic or molecular beams are more difficult to handle, and adsorption experiments are typically performed with thermal beams. In the case of graphene, adsorption (chemisorption) is an activated process, which requires hyperthermal beams to dissociate a precursor molecule and create atomic/molecular species that are able to bind to the substrate. For instance, chemisorption of hydrogen atoms has a relatively high activation barrier of ~0.2 eV that hot hydrogen atoms produced by dissociating H_2 molecules at ~2000 K are needed to observe sticking and, because of that, for a long time graphite was considered to be unable to adsorb (chemisorb) hydrogen atoms (Martinazzo et al. 2013). Furthermore, as we shall discuss later in this chapter, hydrogen atoms have been shown to be immobile on the surface (Hornekær et al. 2006) and, in contrast to the defects discussed above, tend to cluster at all but the smallest coverages (<1%). The case of hydrogen atoms has long been studied because of its relevance in many fields but most of the results found in this context hold for similar monovalent species that bind to the substrate through a covalent bond.

Characterization of defects is possible with several techniques. As already mentioned above, aberration-corrected HR-TEM allows one to image individual carbon atoms and defects in the lattice and to follow their dynamics on the time scale needed to acquire images (seconds). Similar resolution can be achieved with the scanning tunneling microscope, even though in this case one rather images electronic states close to the Fermi level. Raman spectroscopy has been shown to be particularly useful in this context, though the kind of information obtained refers to the whole area illuminated by the laser spot. A distinct, defect-induced peak (D band) in the Raman spectrum appears in the presence of atomic-scale defects at a frequency well below the G peak thanks to inter- and intravalley scattering processes, along with a shoulder slightly above the G peak (D′ band) (Ferrari 2007). The intensity ratio D/D' has been shown to be characteristic of the type of defect introduced: ~13 for sp^3 defects such as the adsorbed H atoms mentioned above, ~7 for vacancies, and ~3.5 for boundaries (Eckmann et al. 2012). In addition, a number of surface science techniques have long been used to characterize adatoms in graphite, for example, temperature programmed desorption and high-resolution electron energy loss spectroscopy (Zecho et al. 2002) and, when possible, information readily translated to graphene; see Martinazzo et al. (2013) for a recent summary of available experimental and theoretical results concerning hydrogen species.

2.4 MIDGAP STATES

As mentioned in the introduction, the defects described in this chapter have in common the removal of a p_Z orbital from the π cloud. This occurs either because one carbon atom is removed from the lattice or because its p_Z orbital gets engaged in a strong chemical bond with an adsorbed species.

The latter situation is best exemplified by the adsorption of H atoms, which has long been studied on *graphite* because of its importance for the chemistry of the interstellar medium (see, e.g., Casolo et al. 2013). Jeloaica and Sidis (1999) and Sha and Jackson (2002) first showed that chemisorption on graphite is possible if the substrate is allowed to relax and that, among the four possible adsorption sites, the *hollow* and *bridge* are not binding, while the two *atop* sites (with or without a carbon atom on the second layer) give essentially the same behavior, in line with the expectation that only strong and directional CH bonds are plausible. Given that graphene chemistry is hardly affected by the presence of underlying weakly bound layers (in graphite the first one lies 3.4 Å below) quantitatively similar results are expected (and found) for free-standing graphene (see, e.g., Boukhvalov et al. 2008, Casolo et al. 2009).

Importantly, adsorption on the top site induces a surface reconstruction ("puckering")—an outward motion of the carbon atom beneath the adsorbed hydrogen—which occurs as a consequence of the sp^2–sp^3 rehybridization of the carbon valence orbitals needed to form the chemical bond. Such rehybridization induces a change in the local geometry of the substrate from planar (sp^2) to tetrahedral (sp^3), which actually extends several Angstroms away from the defect (Ivanovskaya et al. 2010). It is the combined effect of the strong hybridization between C and H orbitals and the reduced overlaps with p_z orbitals of neighboring sites that decouples the CH electrons from the π bands. In addition, such lattice distortion greatly enhances the otherwise negligible spin–orbit coupling in graphene (Castro Neto and Guinea 2009, Zhou et al. 2010) and makes the realization of the spin Hall effect possible (Balakrishnan et al. 2013).

In this context, it is worth noticing that the energy required for the surface reconstruction (defined as the energy difference between the relaxed and the puckered configuration) is substantial (~0.8 eV, to be compared with the net binding energy of ~0.8–0.9 eV) and this has important consequences for the hydrogen dynamics. First, as mentioned above, adsorption is an activated process with a barrier ~0.2 eV high, which prevents hydrogenation of (clean) graphene samples with room temperature gases; second, the diffusion barrier matches the desorption energy (the binding energy *plus* the adsorption barrier), what makes H species *immobile* on the surface (they desorb rather than diffusing) (Hornekær et al. 2006).

From an electronic point of view a p_z vacancy in graphene, being due to either a missing carbon atom or an adsorbed species, creates an *imbalance* between the number of sites in the two sublattices, thereby strongly affecting the electronic states at *low* energy (i.e., at the Fermi level). This is already evident by counting the states and remembering the e–h symmetry: upon removal of a single site the odd-numbered system necessarily has a zero energy level ("mode"). This zero-energy state is a singly occupied molecular "orbital" dubbed *midgap* state even though a gap is not really present. When relaxing the nearest-neighbor approximation such state moves away from the Fermi level, but remains very close to it for reasonable values of the hoppings.

More generally, the appearance of midgap states is a common feature of bipartite systems, which has long been investigated and rediscovered several times in different contexts. Here, for bipartite system we mean a system whose Hamiltonian takes an off-block diagonal form, $H = H_{AB} + H_{BA}$, where A, B identifies two complementary subspaces of the Hilbert state space and (call P_A and P_B the corresponding projectors) $H_{AB} = P_A H P_B$, etc.; in graphene with only nearest-neighboring hoppings A and B obviously identify with the space spanned by the p_z orbitals of the two sublattices, respectively, and thus in the following we refer to A, B "sites" to mean A, B "states." A simple result can then be stated as follows (Inui et al. 1994, Yazyev 2010): in any bipartite lattice in which the numbers of sites in the two sublattices (call them N_A and N_B) are not equal to each other, there exist at least $\eta = |N_A - N_B|$ linearly independent eigenfunctions of the Hamiltonian at zero energy, all with *null amplitudes* on the minority sublattice sites. The proof is simple: for let $N_A > N_B$ and $|\psi\rangle = \Sigma_i \alpha_i |\alpha_i\rangle$ be a trial solution at zero energy, with $\{|\alpha_i\rangle\}$ a complete set in A; the coefficients α_i have to satisfy $\Sigma_i \langle\beta_j|H|\alpha_i\rangle\alpha_i = 0$ for $j = 1,\dots N_B$,

which is a set of N_B equations for the $N_A > N_B$ coefficients α_i, that is, having *at least* η linearly independent solutions. This also shows that the zero-energy modes localize on the A lattice sites.

We have thus proved the *imbalance rule*, a very useful result in graphenic systems, which gives the *minimum* number of zero energy modes to be expected. More generally, the concept of nonadjacent sites in an N-site bipartite system helps counting the number of such states (Longuet-Higgins 1950, Fajtlowicz et al. 2005, Yazyev 2010), as the simple argument below shows.

We say that two sites are nonadjacent if they are not bound (connected by a transfer integral) to each other; for instance, two sites on the same sublattice are nonadjacent. Clearly, there exists a maximal set of nonadjacent sites and we call α the sites in this set, and β the remaining ones (N_α, $N_\beta = N - N_\alpha$ in number, respectively). Each site α binds at least one site β, otherwise it would be completely isolated and could be removed at the outset. Arranging one electron per site, however, we can form at most N_β bonds at a time, and therefore we are left with $\eta = N_\alpha - N_\beta$ unpaired electrons. Equivalently, we end up with $\eta = 2N_\alpha - N$ midgap states *localized on the maximal set of nonadjacent sites*, and, eventually, restate this result by defining η as the number of unpaired electrons in the Lewis structure(s) with the *maximum* number of bonds; for π bonds in graphenic systems the latter are known as *principal resonance structures*.

Rigorously speaking (Fajtlowicz et al. 2005), for a generic (bipartite) system this result gives yet a lower bound only to the number of zero energy states, $\eta \geq 2N_\alpha - N$, though greater or equal than the above one based on the lattice imbalance (the reader may easily recognize that the latter is just a special result of this counting rule). In fact, there may exist further states at zero energy for *specific* values of the hoppings (H matrix elements), which are known as *supernumerary* modes (Longuet-Higgins 1950). However, they cannot occur in hexagonal (benzenoid) systems for which $\eta \equiv 2N_\alpha - N$ strictly holds.

The above results become largely relevant to "real" graphene because e–h symmetry (bipartitism) holds to a large extent. Numerical calculations, both at the tight binding and at higher levels of theory (e.g., density functional theory), confirm the appearance of "zero-energy" modes associated with p_z vacancies. Interestingly, in graphene (at low defect density) it has been shown (Pereira et al. 2006, 2008) that these modes partially localize around each defect and decay with a $1/r$ power law (corresponding yet to a nonnormalizable state), a result which has been nicely confirmed by experiments (Ugeda et al. 2010). See also Peres et al. (2006) and Pereira et al. (2008), who performed a comprehensive analysis of the effect that low-density defects have on the graphene density of states by means of analytic results and numerical tight-binding calculations for millions of lattice sites. Analogous results have been found in DFT studies of vacancies and adsorbed species (H, F, OH, CH_3, etc.), though at much higher defect densities, because of the periodic setting used in the

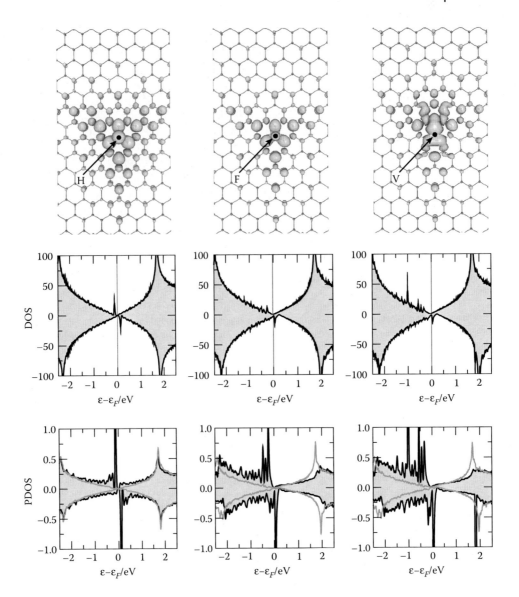

FIGURE 2.1 Midgap states from periodic, *first principles* DFT calculations on large simulation cells (see text for details). From left to right, the case of H and F adatoms, and of a carbon atom vacancy. Top row: isosurfaces of the spin density in the relaxed structures (± 0.004 $\mu_B/\text{Å}^3$); middle and bottom rows: the total DOS and PDOS, respectively. The PDOS was summed over the three sites that are nearest neighbors of the defect, and compared with the same quantity summed over three sites far away from it (gray lines).

calculation. *First principles* results for vacancies and simple adatoms (H and F), for instance, are reported in Figure 2.1 for a rather large simulation cell, a rectangular unit cell of dimensions 4.7×3.4 nm² containing more than 600 atoms.* The spin-polarized density of states (DOS) for each of the above species is seen to have two peaks, one slightly below the Fermi level and one slightly above it, thus describing a

singly occupied level. This is shown in the middle and the bottom rows of Figure 2.1, which display, respectively, the total DOS and its projection (PDOS) onto the three sites that are nearest neighbors of the defect. Such singly occupied state mainly localizes on the sites of the hexagonal sublattice opposite to the defective one, as is evident from the maps of the spin density reported in the top row of Figure 2.1.

A complementary, easy-to-use picture of the appearance of midgap states can be obtained by applying the resonant valence bond picture mentioned in Section 2.2, and it has long been taught in organic chemistry classes when discussing electrophilic aromatic substitution reactions. Considering H adsorption as a specific example, it is not hard to realize that binding of a H atom breaks the aromatic network and leaves one unpaired electron on the lattice (see Figure 2.2). The latter is *free* to move by *bond switching*, as can be easily seen by

* In these calculations the Perdew-Burke-Ernzerhof functional was used to handle exchange and correlation effects, core electrons were described by separable, norm-conserving pseudopotentials, and the single particle wavefunctions were expanded on a set of numerical atomic orbitals with compact support of double-ζ *plus* polarization quality, as implemented in SIESTA (Soler et al. 2002). Fine *k*-meshes were employed to sample the Brillouin zone (4×5 in the self-consistent calculations and 16×22 for computing the density of states) and geometry optimizations were performed till the maximum component of each atomic force fell below 0.005 eV/Å (Martinazzo, 2014 [unpublished]).

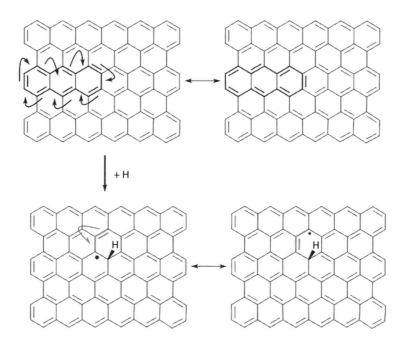

FIGURE 2.2 Resonating valence bond model of the itinerant electron in the zero-energy mode, as applied to the case of hydrogen adsorption on graphene. Bond switching (arrows) in graphene (top) and in hydrogenated graphene (bottom) connects different chemical resonance structures.

drawing the corresponding resonance (Kekulé) structures of similar energies: spin recoupling with a neighboring double bond creates an unpaired electron in one of every two lattice sites, namely on the same majority set predicted by the counting rules given above (see Figure 2.2, bottom). In fact, such itinerant electron is equivalently described by a singly occupied molecular orbital delocalized on such a set. As we shall see in the following, despite its simplicity, such picture is powerful in describing spin and/or charge density localization in conjugated molecules in general and graphene in particular, and thus very useful for predicting chemical reactivity.

Finally, one further complementary way to look at the zero-energy modes in graphene is provided by the (noninteracting) Anderson model, which we discuss here in some detail because it is instrumental for the following section. In this model the ad-species is described by a single level at energy ε_d (e.g., the $1s$ level of a H atom), which is let to hybridize with a carbon atom of the lattice, say of A type at the origin, by adding the term

$$H' = \varepsilon_d d^\dagger d + V(a_0^\dagger d + d^\dagger a_0)$$

to the lattice Hamiltonian (here V is the hybridization energy). The problem can be solved exactly by means of the Green's function or, equivalently, of the T-matrix. To this end, we first introduce the two complementary subspaces that describe the "lattice" and the "impurity" (call P and Q the corresponding projectors) and, upon partitioning, write the exact Green's function of the lattice (G_{PP}) in terms of an effective, energy-dependent lattice Hamiltonian, that is, $G_{PP}(\varepsilon) = (\varepsilon - H_P^{\text{eff}})^{-1}$ with

$$H_P^{\text{eff}}(\varepsilon) = H + H_{PQ}g_{QQ}H_{QP} = H + \frac{V^2}{\varepsilon - \varepsilon_d}a_0^\dagger a_0 = H + v_{\text{eff}}a_0^\dagger a_0$$

where H is the lattice Hamiltonian and g_{QQ} (the uncoupled impurity Green's function) introduces a resonant, on-site potential v_{eff} on the defective lattice site. Similarly, $G_{QQ}(\varepsilon) = (\varepsilon - H_Q^{\text{eff}})^{-1}$ and

$$H_Q^{\text{eff}}(\varepsilon) = \varepsilon_d d^\dagger d + H_{QP}g_{PP}H_{PQ} = (\varepsilon_d + V^2 g_0(\varepsilon))d^\dagger d$$

where $g_0(\varepsilon)$ is the on-site Green's function of the defect-free lattice. The latter reads as $g_0(\varepsilon) = -\varepsilon/D^2 \ln[(D/\varepsilon)^2 - 1] - i\pi|\varepsilon|/D^2$ for $\varepsilon < D$ and $g_0(\varepsilon) = -\varepsilon/D^2 \ln[1 - (D/\varepsilon)^2]$ otherwise, if the Dirac-cone approximation is employed; here D is the bandwidth and can be chosen to give the correct low-energy behavior of the DOS, namely $D^2 = \sqrt{3}\pi t^2$. Next, we solve the lattice problem with the scattering potential $V_{\text{eff}} = v_{\text{eff}}a_0^\dagger a_0$ by solving the Lippmann–Schwinger equation for the T-matrix, $T = (1 - V_{\text{eff}}G_0)^{-1}V_{\text{eff}}$ (here G_0 is the bare Green's function of H_P^{eff}, i.e., of the lattice). This is possible for any strength of the coupling because V_{eff} has a separable form, and gives the T-matrix in the form $T(\varepsilon) = t(\varepsilon)a_0^\dagger a_0$ where

$$t(\varepsilon) = \frac{v_{\text{eff}}}{1 - v_{\text{eff}}g_0(\varepsilon)} = \frac{V^2}{\varepsilon - \varepsilon_d - V^2 g_0(\varepsilon)} = V^2 g_d(\varepsilon)$$

with $g_d(\varepsilon) = \langle d|G_{QQ}(\varepsilon)|d\rangle = \text{Tr}(G_{QQ}(\varepsilon))$. In turn, T determines the Green's function of the H_P^{eff} problem,

$$G_{PP}(\varepsilon) = G_0(\varepsilon) + G_0(\varepsilon)T(\varepsilon)G_0(\varepsilon)$$

and the DOS can be computed from the imaginary part of its trace,

$$\mathrm{Tr}(G_{PP}(\varepsilon)) = \mathrm{Tr}(G_0(\varepsilon)) + t(\varepsilon)\langle 0 \mid G_0(\varepsilon)^2 \mid 0 \rangle$$

$$= \mathrm{Tr}(G_0(\varepsilon)) - t(\varepsilon)\frac{dg_0(\varepsilon)}{d\varepsilon}$$

namely,

$$\rho_{\mathrm{latt}}(\varepsilon) = \rho_0(\varepsilon) + \frac{1}{\pi}\,\mathrm{Im}\!\left(t(\varepsilon)\frac{dg_0(\varepsilon)}{d\varepsilon} \right)$$

where $\rho_0(\varepsilon) = -(1/\pi)\mathrm{Im}(\mathrm{Tr}(G_0(\varepsilon))) = -(N/\pi)\mathrm{Im}(g_0(\varepsilon)) = N|\varepsilon|/D^2$ is the DOS of the bare lattice with N sites. Adding the impurity level DOS $\rho_d(\varepsilon) = -(1/\pi)\mathrm{Im}(g_d(\varepsilon))$ and rearranging, it follows:

$$\rho(\varepsilon) = \rho_0(\varepsilon) + \frac{1}{\pi}\,\mathrm{Im}\!\left(\frac{d\ln t(\varepsilon)}{d\varepsilon} \right)$$

Of interest is the strong-coupling, "unitary" limit (V large) where the impurity level strongly hybridizes with the given lattice site and effectively removes it from the π band system. In this limit a bonding–antibonding pair of orbitals forms between the impurity and the neighboring lattice site and it draws the whole spectral density of the impurity level *outside* the lattice band, at energies $\sim\pm V$. Indeed, upon computing $\rho_d(\varepsilon)$ for large energies in this limit

$$\rho_d(\varepsilon) = \delta(\varepsilon - \varepsilon_d - V^2 g_0(\varepsilon)) \approx \frac{1}{2}\delta(\varepsilon - V) + \frac{1}{2}\delta(\varepsilon + V)$$

A resonance appears close to the neutrality point, irrespective of the original position of the impurity level ($t(\varepsilon) \sim -1/g_0(\varepsilon)$ in this limit), which however has negligible weight on the impurity ($g_d(\varepsilon) = t(\varepsilon)/V^2$): it corresponds to a resonant level close to the neutrality point, which localizes *on the lattice*, that is, the above-mentioned midgap state housing an itinerant electron.

Even though common adsorbates are not in the true unitary limit and not necessarily develop the bonding–antibonding pair, the appearance of a low-energy resonance is a common feature of strongly interacting impurity levels, which can be easily traced back to the behavior of the graphene DOS. Notice though that a *dip* appears at low energies in $\Delta\rho(\varepsilon) = \rho(\varepsilon) - \rho_0(\varepsilon)$ (i.e., $\Delta\rho(\varepsilon) < 0$) that makes results physically meaningful in the true thermodynamic limit $N \to \infty$ only: this means that the above T-matrix approach, when applied at the neutrality point, actually works for infinitely diluted systems only.

2.5 CHARGE TRANSPORT PROPERTIES

As seen in the previous section, p_z vacancies strongly affect the electronic states in the vicinity of the Fermi level and therefore act as important scattering centers for charge carriers in graphene, both at zero and finite electron density (Peres et al. 2006, Robinson et al. 2008, Ferreira et al. 2011, Peres 2010, Wehling et al. 2010).

Charge transport at the neutrality point is in the quantum regime and cannot be addressed with semiclassical means, but it is obvious that the enhancement of the DOS due to the appearance of zero energy modes may provide a mean for sustaining transport at neutrality, where a nonzero conductivity of the order of the conductance quantum $\sim e^2/h$ is experimentally found under different conditions (Castro Neto et al. 2008, Mucciolo and Lewenkopf 2010, Peres 2010). Although at such low energies (long wavelengths) real samples are no longer homogeneous and e–h puddles develop (Martin et al. 2008) that likely dominate the behavior of charge carriers in realistic situations (Polini et al. 2008, Rossi and Das Sarma 2008), theory predicts a conductivity minimum of $4e^2/\pi h$ for a wide range (several orders of magnitude) of neutral impurity concentrations (Peres et al. 2006, Stauber et al. 2008), which is also found by a number of numerical Green–Kubo quantum calculations with model resonant scatterers. Numerical calculations further show that at increasing defect density a *plateau* develops in the conductivity around the neutrality point (Ferreira et al. 2011, Wehling et al. 2010, Yuan et al. 2010), in correspondence of the appearance of an impurity band, and increasing further the impurity concentration leads to conventional Anderson localization of the electronic wavefunctions (Cresti et al. 2013, Jayasingha et al. 2013, Lherbier et al. 2013). Indeed, p_z vacancies act as strong, short-range scatterers and introduce inter-valley and, in general, backscattering processes. In the absence of inter-valley scattering mechanisms, graphene undergoes weak *anti*-localization, and shows a conductivity that *increases* with increasing system size and/or disorder (see, e.g., Das Sarma et al. 2011).

At finite electron densities, for some time, scattering by neutral impurities (including short-range scatterers of the type discussed here) was considered marginal only, because results obtained within the first Born approximation were clearly at odds with the linear (or sublinear) dependence of the conductivity on the doping charge density (Castro Neto et al. 2008). Experiments indeed clearly show that mobility of graphene samples is essentially independent on the doping charge n_e and, according to standard Drude–Boltzmann theory,

$$\sigma_{dc} = \frac{2e^2}{h}k_F l_e = \frac{2e^2}{h}k_F v_F \tau_e$$

(where l_e is the transport mean free path and τ_e the corresponding relaxation time; the factor of 2 comes from the valley degeneracy) this requires a relaxation time linear in k_F to explain the observed dependence on $k_F^2 = \pi n_e$. In this context, while short-range scatterers in the widely used Born approximation give rise to $\tau_e \sim 1/k_F$, Coulomb scatterers (supposedly present on the surface of the insulating layer where graphene is often supported) give the required $\tau_e \sim k_F$ dependence (Castro Neto et al. 2008). Scattering of 2D Dirac fermions off Coulomb impurities has been analyzed in detail (Nomura and MacDonald 2007) and quantitative agreement with experiments has been found when correctly taking into account screening and the wave vector and temperature dependence of the graphene static dielectric function (Hwang and Das Sarma 2007, 2008, Das Sarma et al. 2011). There are though a number of experimental facts that are hardly accounted for if charges on the supporting substrates were the primary sources of scattering (see the discussion on this point in the review article by Peres 2010), and some measurements do indicate that scattering off neutral, resonant impurities may explain the observed mobilities at more reasonable impurity concentrations (Ni et al. 2010).

Here, without attempting to answer the question of whether the Coulomb or the short-range impurities are the primary source of scattering (which would require a careful consideration of the specific experimental set-up used for transport measurements, e.g., the graphene preparation method, the device geometry and its composition) we explore the consequences of the presence of the above-mentioned neutral impurities on the transport properties of graphene.

Strong scattering potentials of the kind introduced by the impurities here considered cannot be handled with perturbation theory and require different approaches (Ferreira et al. 2011). Among these, we focus on the T-matrix approach for the Anderson model described in the previous section, an approach pursued by several authors (Peres et al. 2006, Robinson et al. 2008, Peres 2010, Wehling et al. 2010), but equivalent results can be obtained with the phase-shift method for similar physical situations (scattering of Dirac fermions by hard-disk potentials, see Ferreira et al. (2011), Peres (2010), and references therein).

According to scattering theory, the T-matrix determines the scattering amplitude, and thus the scattering cross-section. The actual relationship for Dirac fermions in 2D is given by

$$\delta\sigma(\boldsymbol{k'},\boldsymbol{k}) = (2\pi)^3 \frac{k_F}{(\hbar v_F)^2} |\tilde{T}_{\boldsymbol{k'k}}|^2 = \frac{A^2}{2\pi} \frac{k_F}{(\hbar v_F)^2} |T_{\boldsymbol{k'k}}|^2$$

where in the first term T-matrix elements are taken between $\boldsymbol{k}$-states normalized on the momentum scale and in writing the second equality a large area A containing N graphene cells has been introduced, and T-matrix elements taken between states normalized on such area. For the T-matrix obtained in the previous section and graphene $\boldsymbol{k}$-states

$$|T_{\boldsymbol{k'k}}|^2 = \frac{|t(\varepsilon)|^2}{N^2 n_c^2}$$

(where $n_c = 2$ is the number of carbon atoms per cell) scattering is isotropic and the transport cross-sections is just the total cross-section σ

$$\sigma = 2 \frac{A_C^2}{(\hbar v_F)^2} k_F |t(\varepsilon)|^2$$

Here A_c is the area per C atom (half the area of the unit cell) and the factor of 2 stems from summing over both valleys. This expression can also be arranged in a form involving the DOS per unit area (and per spin) $\rho_a(\varepsilon)$, namely,

$$\sigma = 2\pi\rho_a(\varepsilon) \frac{A_c^2}{\hbar v_F} |t(\varepsilon)|^2$$

which gives the relaxation time in a Fermi's golden-rule-like form

$$\tau_\varepsilon^{-1} = \frac{v_F}{l_e} = v_F n\sigma = \frac{2\pi}{\hbar} \upsilon\rho_C(\varepsilon) |t(\varepsilon)|^2 \left(\equiv -\frac{2}{\hbar} \upsilon \, \text{Im}(t(\varepsilon))\right)$$

where n and υ are the number of impurities per unit area and per carbon atom, respectively, and $\rho_C(\varepsilon)$ is the DOS per C atom. It is easy to see from the last expression how the observed behavior of the relaxation time may arise: in the unitary limit $t(\varepsilon) \sim -1/g_0(\varepsilon)$ and, at low energies, $\text{Im}(-(1/g_0(\varepsilon))) \sim -(\pi D^2/4)/[\varepsilon \ln^2(D/\varepsilon)]$, from which it follows:

$$\upsilon\tau_e \sim \frac{2\hbar}{\pi D^2} |\varepsilon| \ln^2\left[\frac{|\varepsilon|}{D}\right]$$

which is linear in ε to within a logarithmic correction. In this limit, the d.c. conductivity reads as

$$\sigma_{dc} = \frac{2e^2}{h} \frac{1}{\upsilon} \frac{2|\varepsilon|^2}{\pi D^2} \ln^2\left[\frac{|\varepsilon|}{D}\right]$$

(where $D^2 = \sqrt{3}\pi t^2$) or, equivalently, introducing the length scale $R = \sqrt{3\sqrt{3}}/2 d_{CC} \approx d_{CC}$ ($R^2 \equiv A_C$, which is the area per C atom) and the density of charge carriers n_e

$$\sigma_{dc} = \frac{2e^2}{h} \frac{1}{\upsilon} \frac{n_e R^2}{2\pi} \ln^2[n_e R^2]$$

which explicitly shows the dependence of the conductivity on the number of carriers and impurities per C atom, namely $n_e R^2$ and υ, respectively.

Similar results can be obtained in the continuum limit for hard-disk potentials with the phase-shift method (Ferreira et al. 2011). In the limit of a δ-like potential $V(\boldsymbol{r}) = v_0\delta(\boldsymbol{r})$ the problem can be solved exactly, similarly as shown above since the potential becomes separable in that limit. In that case, the T-matrix is diagonal in pseudo-spin space and reads as $T = \bar{t}(\varepsilon)\delta(\boldsymbol{r})$ with

$$\bar{t}(\varepsilon) = \frac{v_0}{(1 - v_0 G_0(\varepsilon))}$$

where $G_0(\varepsilon)$ relates to the Green function of the 2D Dirac equation at the scatterer position. The appropriate Green function is a 2×2 matrix, which in $\boldsymbol{k}$-space takes the simple form

$$G(\boldsymbol{q}|\varepsilon) = \frac{\varepsilon + \hbar v_F \boldsymbol{\sigma q}}{\varepsilon^2 - (\hbar v_F)^2 q^2}$$

and thus $G(x, x'|\varepsilon)$ (depending on $x - x'$ only) follows readily from this expression upon Fourier transforming in real space. For $x = x'$ the function is diagonal in pseudo-spin space, $G(x, x|\varepsilon) = G_0(\varepsilon)\mathbf{1}$, and

$$G_0(\varepsilon) \approx \frac{k_F}{2\pi\hbar v_F} \ln(k_F R_c)$$

Here $k_F = |\varepsilon|/\hbar v_F$ has been used and a momentum cutoff $1/R_c$ has been introduced to regularize the divergent Green's function at the origin (this is reasonable since the continuum approximation does not hold for arbitrary momenta. A reasonable limit for q is provided by the lattice constant of the reciprocal lattice and thus R_c has to be of the order of d_{CC}). Accordingly, the differential cross-section follows as

$$\delta\sigma(k', k) = (2\pi)^3 \frac{k_F}{(\hbar v_F)^2} |T_{k'k}|^2 = \frac{k_F}{(\hbar v_F)^2} |\bar{t}(\varepsilon)|^2 \frac{1+\cos\theta}{4\pi}$$

where the last term on the right-hand side involves the scattering angle θ and arises from the incoming and outgoing spinors (note how backscattering is here suppressed). As different from before, scattering is no longer anisotropic and the transport cross section reads as

$$\sigma = \frac{k_F}{4(\hbar v_F)^2} |\bar{t}(\varepsilon)|^2$$

which is similar to the previous expression

$$\sigma = \frac{A_c^2}{2(\hbar v_F)^2} k_F |t(\varepsilon)|^2$$

now written in terms of the unit cell area A_c. Plugging in the expression for the T-matrix, the d.c. conductivity follows similarly as above (Ferreira et al. 2011):

$$\sigma_{dc} = \frac{2e^2}{h} \frac{4}{n} \left[\frac{\hbar v_F}{v_0} - \frac{k_F}{2\pi} \ln(k_F R_c) \right]^2 \approx \frac{2e^2}{h} \frac{1}{v} \frac{n_e R^2}{4\pi} \ln^2[\pi n_e R_c^2]$$

where the second equality holds in the unitary limit and vR have been defined above ($\pi R_c^2 \equiv R^2$ if the same cutoff is used in both approaches).

The above results hold, strictly speaking, for vacancies only, since realistic adsorbates are not in the true unitary limit, as *first principles* results show: on fitting DFT-derived energy bands to the above Anderson model, for instance, Wehling et al. (2010) found $V \sim 2t$ and $|\varepsilon_d| \sim 0.1$ eV for several neutral species, including H, OH, and alkyl groups. Under such conditions, the complete T-matrix has to be used in the expression of the cross section and the d.c. conductivity computed correspondingly. After elementary algebra, the result can be written in the form

$$\sigma_{dc} = \frac{2e^2}{h} \frac{1}{v} \frac{1}{2\pi} \left[\left(\eta + q\sqrt{(n_e R^2)} \ln(n_e R^2) \right)^2 + n_e R^2 \right]$$

where the dimensionless quantity η is given by $\eta = D(\varepsilon - \varepsilon_d)/V^2$ and $q = 1$ for holes and -1 for electrons. Importantly, this expression introduces an e–h asymmetry, which is related to the position of the impurity level, as seen in Figure 2.3 where the DOS, the cross sections, and the conductivities obtained with the T-matrix approach are given for representative values of the parameters of the Anderson model and for a concentration $v = 0.1\%$ of impurities.

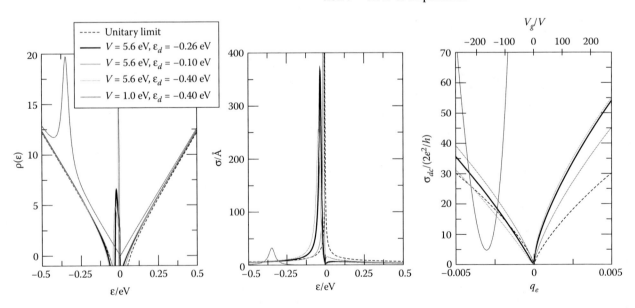

FIGURE 2.3 From left to right: DOS, cross sections, and conductivities for the noninteracting Anderson model described in the main text, with representative values of the parameters given in the legend, and for an impurity concentration of 0.1% defects per carbon atom. The upper scale on the rightmost panel gives the voltages for a hypothetical 300 nm-thick SiO₂ insulating layer between graphene and the gate electrode.

2.6 MAGNETIC PROPERTIES

We have seen in the previous sections that a p_z defect introduces a midgap state in its neighborhoods which, at charge neutrality, is singly occupied. Thus, in dilute systems where hybridization does not occur each defect-induced level may act as a spin-half paramagnetic center. The question arises then whether this π moment can be quenched, at least partially, by interacting with conduction electrons—a common issue in metallic systems related to the appearance of the Kondo effect. In fact, charge transport measurements at low temperature on graphene-irradiated samples showed a logarithmic increase of the resistivity, which may well be explained as a spin-half Kondo effect (Chen et al. 2011), though alternative explanations involving electron–electron interactions in the disordered system are possible (Chen et al. 2012, Jobst et al. 2012). There are actually notable differences between the conventional Kondo problem involving a spin-half local moment and the possible one related to the π midgap states: the zero energy modes extend over large regions, are built from the same π states responsible for conduction and no hopping exists with the latter (Haase et al. 2011).

The problem of how π-midgap states interact with conduction electrons has been considered theoretically by Haase and coworkers (2011), who combined dynamical mean-field theory with quantum Monte Carlo simulations to solve a model for a resonant scatterer including locally the electron–electron interactions, that is, with a Hubbard on-site repulsive term. The results of such calculations show that the magnetic susceptibility retains Curie-law behavior down to the lowest temperatures, thereby suggesting a ferromagnetic coupling between the π moment and the conduction electrons. Nair et al. (2012) have performed superconductive quantum-interference magnetometry experiments on carefully controlled fluorinated and irradiated graphene samples and measured a paramagnetic response, which could only be due to spin-half moments, thereby supporting the picture that π moments are not quenched.

At a closer look, however, a carbon atom vacancy is expected to behave differently from an adsorbate. This is because in the first case, in addition to the π electron, three additional dangling bonds are left on the σ network upon vacancy formation, and possibly couple with the π one. Even though a lattice reconstruction occurs that reduces the number of dangling bonds, *two* unpaired electrons are left on the reconstructed vacancy, apparently at odds with a spin-half paramagnetism.

Lattice reconstruction upon vacancy formation is due to a structural instability of the Jahn–Teller type. An isolated, unrelaxed vacancy has D_{3h} symmetry, and the dangling σ bonds span the $a_1' + e'$ irreducible representations of this group, whereas the π state belongs to the a_2'' symmetry species. Among these symmetry adapted states, a_1' is lowest in energy since it contains a purely bonding combination of σ orbitals; hence, the lowest energy many-body electronic states arise from configurations of the type $\ldots (a_1')^2 (e')^1 (a_2'')^1$ and are of E'' symmetry for both the parallel and antiparallel

alignment. It follows that the ground state is doubly degenerate for both spin alignments and undergoes (proper or pseudo) Jahn–Teller distortion. In the specific case of interest here, such distortion occurs because coupling of the electronics with in-plane e' vibrations ($[E'']^2 = [E']^2 = A' + E''$) breaks the above symmetry and leaves a pentagonal ring and an "apical" carbon atom opposite to it (threefold degenerate) behind, see Figure 2.4. This is confirmed by several theoretical investigations (Lethinen et al. 2004, Yazyev and Helm 2007, Dharmawardana and Zgierski 2008, Palacios and Yndurain 2012) and HR-TEM analysis (Meyer et al. 2008). Such distortion actually induces a long-ranged strain field—recently observed in graphene on a metal surface (Blanc et al. 2013)—that impacts on the chemical reactivity several Angstroms away from the defect position (Krasheninnikov and Nieminen 2011); see Figure 2.4 for a map of such strain field as obtained by the *first principles* calculations mentioned in Figure 2.1.

As for the electronic structure, coupling between the two unpaired electrons is mainly governed by the *vertical* position of the apical carbon (Casartelli et al. 2013): the ground state is a *triplet* with a planar equilibrium geometry and lies 0.2 eV lower in energy than the open-shell *singlet* with one spin flipped; the latter is a *bistable* system with two equivalent equilibrium lattice configurations (for the apical C atom above or below the lattice plane) and a barrier 0.1 eV high separating them. Accordingly, the exchange (Hund) coupling

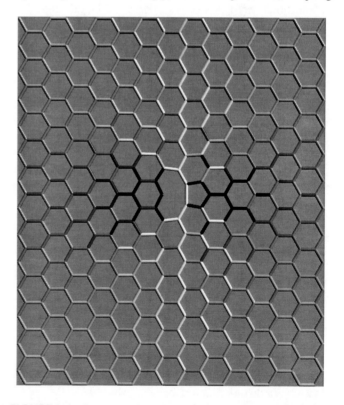

FIGURE 2.4 Structure of a carbon atom vacancy in graphene with the accompanying strain field, as obtained by the *first principles* calculations of Figure 2.1. Colors vary from black to white for increasing bond lengths, in the range [0.995,1.005] times the equilibrium CC distance in graphene. See also Krasheninnikov and Nieminen (2011).

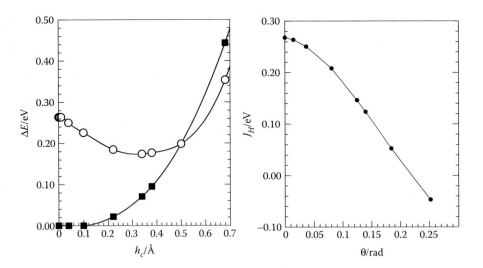

FIGURE 2.5   Left: Energetics of a carbon atom vacancy as a function of the height of the apical carbon above the surface (filled and open symbols for the triplet and the singlet state, respectively). Right: Corresponding Hund's coupling constant, as a function of the angle that the apical atom forms with the surface plane. (Adapted from Casartelli, M. et al. 2013, *Phys. Rev. B* 88, 195424.)

decreases as the carbon atom moves out of the plane, starting from ~0.27 eV at planar geometry, up to reverting sign for geometry close to the singlet minimum, as reported in Figure 2.5. Its behavior essentially reflects the dependence of the hybridization energy between s and p states on the angle that the apical carbon atom forms with the lattice plane.

The value of the computed Hund coupling is clearly too large to give a decoupled response of the two electrons to magnetic fields (spin-half behavior), thereby suggesting that a carbon atom vacancy should behave as a spin-one paramagnetic species, a result actually found in some experiments (Ney et al. 2011). However, depending on the environment, interaction with a substrate may well affect the energetics and stabilize the system with an out-of-plane apical carbon, as indeed found, for instance, when graphene lies on a metallic substrate (Ugeda et al. 2011). Furthermore, charge doping or binding with adsorbates (vacancies are obviously highly reactive) may have the same effect. For instance, thermodynamic and structural analysis based on *first principles* results have shown that the triply hydrogenated state is the most stable under a wide range of temperature and hydrogen partial pressure, and this is a spin-half paramagnetic species (Casartelli et al. 2014). Also, rippling has been proved to have a large effect on the vacancy-induced moment (Santos et al. 2012), thereby suggesting that strain could be used to tune magnetism in the graphene sheet.

In fact, recent experiments have shown that the paramagnetic response of irradiated graphene, free from any other defect, has a dual origin: a σ moment, which is unaffected by charge doping (thanks to the large Coulomb repulsion in the localized σ state); and a π moment, which can be quenched upon shifting the Fermi level (Nair et al. 2013). Though Nair et al. used molecular doping in their experiment, the obtained results open the way for controlling the magnetic response by means of the electric field effect. The question remains as

to how the two electrons decouple from each other in their samples.

Having discussed some of the magnetic properties of diluted defects, we can now focus on the possibility that interactions between moments lead to ferromagnetic order, a long-standing issue for the search of s–p magnetism. Ferromagnetism in graphite and later in graphene has been reported (Esquinazi et al. 2002, 2003, Barzola-Quiquia et al. 2007, Wang et al. 2009) but later questioned in the light of the ubiquitous presence of magnetic contaminants. Recent measurements under carefully controlled conditions have indeed shown that graphene, like graphite, is strongly diamagnetic and has only the weak paramagnetic contribution described above due to adatoms and/or carbon atom vacancies (Sepioni et al. 2010).

Coupling between defects may in principle give rise to magnetically ordered structures if it favored the parallel alignment of π moments of the defect-induced midgap states, that is, if a sort of Hund's rule held. From a theoretical point of view, energy ordering in such situations is entirely determined by electron correlation, and requires that Coulomb repulsion is taken into account. If this is done at the level of on-site repulsion—that is, in the framework of the repulsive Hubbard model—a theorem due to Lieb gives a definite answer to this question (Lieb 1989): Lieb showed that for any bipartite system at half-filling the ground-state spin S is given by the sublattice imbalance only, $S \equiv |n_A - n_B|/2$, or, in other words, that the coupling between π moments is ferromagnetic for defects in the same sublattice and antiferromagnetic otherwise. Similar results follow by analyzing the Ruderman–Kittel–Kasuya–Yosida interactions between local magnetic moments (Kogan 2011, Sherafati and Satpathy 2011).

Lieb's result sets important constraints for building up macroscopic magnetic moments: ordered domains can only be obtained if defects are *unevenly* distributed between the two sublattices. This also holds for the long discussed issue of the edge-related magnetism in zig-zag graphene nanoribbons: a *singlet* ground state always arises between opposite *zig-zag* edges ($n_A = n_B$ overall) and this excludes the possibility of having a nonvanishing spin density (at least as long as the edges are at a distance smaller than the coherence length): spin density is *everywhere zero* in a singlet state, as a consequence of the vector character of this quantity.

Lieb's result also shows that parallel alignment of electrons in midgap states does occur, but only when such states originate from a sublattice imbalance: if midgap states appear *without* a sublattice imbalance, antiparallel alignment must be favored. This is a subtle effect of electron correlation, which would lead to an energetically unfavorable spin polarization of the remaining occupied orbitals if the above Hund's rule were followed in the absence of a sublattice imbalance.

2.7 CHEMICAL PROPERTIES

The presence of singly occupied energy states that localize on specific lattice positions has also obvious consequences for the chemical reactivity of defective graphene. This occurs because, as mentioned above, chemisorption of typical ad-species discussed here (e.g., H, F, alkyl groups) is an activated process, and thus any change in the height of the energy barrier to sticking reflects exponentially on the kinetics of the adsorption process. These changes are brought about by the ad-species themselves, thereby making adsorption on graphene quite unique. The situation is best explained by focusing again on the hydrogenation process, for which a wealth of experimental and theoretical data have been collected in the last few years (Martinazzo et al. 2013).

When adsorbing hydrogen atoms on graphene (or graphite) under kinetic control (i.e., at energies comparable to the barrier height, ~0.2 eV) STM images clearly show the formation of dimers and clusters (Hornekær et al. 2006, Balog et al. 2009). Since H atoms are immobile on the surface, such clustering must be due to a *preferential sticking* mechanism (Hornekær et al. 2006). The preference for forming specific dimers, rather than randomly adsorbate structures, is a direct

effect of the defect-induced midgap states (and their spatial properties) that are introduced in the substrate at earlier times of the exposure.

The overall picture is best understood with the help of the resonating valence bond model (Casolo et al. 2009): when a first H atom adsorbs on an A-type site, the unpaired electron left on the B sublattice may readily (i.e., with a small or even vanishing activation barrier) couple to the electron of an incoming H: bond formation is an easy process on these sites, an "AB" dimer is formed and a singlet ground-state results in which the aromaticity of the substrate is partially restored. Conversely, if adsorbtion occurred on the same sublattice to form an "A_2" dimer, the incoming H atom would *not* take advantage of the available unpaired electron density (spin density), and adsorption of a second atom would be as difficult as the first one.

Figure 2.6 shows the results of DFT calculations on a number of dimers as function of the *site-integrated* magnetization, which is a rough measure of the average number of unpaired electrons left in each site as a consequence of the first adsorption event. Binding (barrier) energies increase (decrease) linearly by increasing the local magnetization, thereby showing how the spatial distribution of the midgap state determines the thermodynamic stability and the chemical reactivity (Casolo et al. 2009). The result is a genuine electronic effect (in principle tunable by charge doping, Huang et al. 2011): substrate relaxation effects, though substantial, are site-independent for all but the so-called *ortho* dimer (with two H on neighboring sites), since surface puckering upon adsorption involves—to a first approximation—nearest-neighboring C atoms only. In this respect, the *ortho* dimer represents an exception (rightmost data point in the graphs), due to the too close proximity of the two hydrogen atoms. Notice further that a linear relationship between barrier and binding energies results, a rather common tendency in activated chemical reactions (Bronsted–Evans–Polanyi rule).

The preferential sticking mechanism outlined above works only for *dimers*. Once an AB dimer is formed there are no further unpaired electrons available, and no bias on the

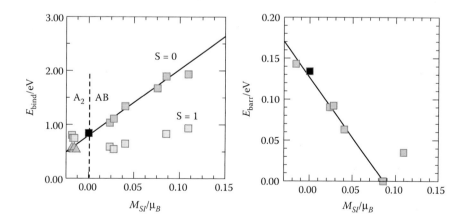

FIGURE 2.6 Binding (left) and barrier (right) energies for adsorbing a second hydrogen atom on graphene, as functions of the site-integrated spin-density (see Figure 2.1). Binding energies are shown for both the triplet (light grey) and the singlet (dark grey) states, while barriers refer to singlet states only. Black symbols are data for the first hydrogen adsorption. (Adapted from Casolo, S. et al. 2009, *J. Chem. Phys.* 130, 054704.)

Electronic structure of graphene	Wallace (1947), Castro Neto et al. (2008), Katsnelson (2013), Foa Torres et al. (2014).
Defect formation and characterization	Zecho et al. (2002), Hornekær et al. (2006), Ferrari (2007), Meyer et al. (2008), Balog et al. (2009), Kotakoski et al. (2010), Kotakoski and Krasheninnikov (2010), Jani Kotakoski and Krasheninnikov (2011), Banhart et al. (2011), Terrones et al. (2012), Eckmann et al. (2012), Meyer et al. (2010), Martinazzo et al. (2013), Kotakoski et al. (2014).
Electronic and structural properties of atomic-scale defects	Inui et al. (1994), Jeloaica and Sidis (1999), Sha and Jackson (2002), Lethinen et al. (2004), Fajtlowicz et al. (2005), Peres et al. (2006), Pereira et al. (2006), Yazyev and Helm (2007), Pereira et al. (2008), Boukhvalov et al. (2008), Dharma-wardana and Zgierski (2008), Casolo et al. (2009), Castro Neto and Guinea (2009), Yazyev (2010), Ugeda et al. (2010), Zhou et al. (2010), Ivanovskaya et al. (2010), Ugeda et al. (2011), Palacios and Yndurain (2012), Balakrishnan et al. (2013), Martinazzo et al. (2013), Casartelli et al. (2013).
Charge transport properties	Peres et al. (2006), Nomura and MacDonald (2007), Hwang and Das Sarma (2007), Hwang and Das Sarma (2008), Robinson et al. (2008), Stauber et al. (2008), Martin et al. (2008), Rossi and Das Sarma 2008; Polini et al. (2008), Castro Neto et al. (2008), Wehling et al. (2010), Ni et al. (2010), Mucciolo and Lewenkopf (2010), Wehling et al. (2010), Yuan et al. (2010), Peres (2010), Das Sarma et al. (2011), Ferreira et al. (2011), Cresti et al. (2013), Lherbier et al. (2013), Jayasingha et al. (2013), Foa Torres et al. (2014).
Magnetic properties	Lieb (1989), Esquinazi et al. (2002), Esquinazi et al. (2003), Lethinen et al. (2004), Yazyev and Helm (2007), Barzola-Quiquia et al. (2007), Dharma-wardana and Zgierski (2008), Wang et al. (2009), Castro Neto and Guinea (2009), Sepioni et al. (2010), Zhou et al. (2010), Sherafati and Satpathy (2011), Kogan (2011), Chen et al. (2011), Haase et al. (2011), Ney et al. (2011), Jobst et al. (2012), Chen et al. (2012), Nair et al. (2012), Palacios and Yndurain (2012), Santos et al. (2012), Nair et al. (2013), Blanc et al. (2013), Casartelli et al. (2013), Balakrishnan et al. (2013), Casartelli et al. (2014).
Chemical properties	Hornekær et al. (2006), Casolo et al. 2009, Balog et al. (2009), Balog et al. (2010), Huang et al. (2011), Krasheninnikov and Nieminen (2011), Casartelli et al. (2014).

adsorption of additional H atoms due to the midgap states. This is confirmed by DFT calculations, which show that adsorption of a third hydrogen atom to a stable AB dimer is quantitatively similar to the first H adsorption event (Casolo et al. 2009). Energy barriers for further adsorption follow a similar trend: barriers for sticking a third H atom compare rather well with that for single H atom adsorption for the processes AB → A_2B and A_2 → A_3, and are considerably smaller for A_2 → A_2B ones.

Few exceptions to this simple picture are found for compact clusters, in which substrate relaxation does play some role: some structures are more favored than others by structural effects, namely because of the substrate *softening* which occurs upon adsorption. This effect helps explaining the clustering of adatoms: experiments at intense H atom flux do not find a random distribution of *dimers*, as expected on the basis of electronic effects only, rather clusters made up of a number of atoms. When stable (balanced) structures on the surface are created and electronic effects are turned off, it is the lattice that provides the driving force for clustering.

It follows that when exposing graphene to large doses of hydrogen atoms an amorphous product is likely to form, with little (if any) crystalline order. Annealing helps in relaxing strain, but forming crystalline hydrogenated form of graphene in this way appears unlikely. *Graphane*, for instance, the fully hydrogenated form of graphene with H atoms alternating above *and* below the surface plane, would require a perfect correlation between the sublattice position and the surface face, at odds with the above-discussed tendency to form "balanced" structures, irrespective of the surface face (only small differences arises between the so-called *syn-* and *anti*-configurations).

Different results can be obtained if graphene interacts more or less strongly with a supporting substrate, which can then act

as a template. For instance, graphene on Iridium (111) forms a commensurate (Moiré) structure, whose unit cell is 10×10 graphene and 9×9 Ir and contain several different binding "environments" for adatoms. Binding of graphene to Ir is "quasi" covalent only, nevertheless H adsorption occurs preferentially close to the C atoms which lie on top of an Ir atom, as if covalent C-Ir bonds were present that leave unpaired electrons in neighboring sites. The result is an ordered hydrogenated structure with a sizeable bandgap (Balog et al. 2010).

2.8 CONCLUSIONS

In this chapter, we summarized the basic effects that simple atomic-scale defects, collectively referred here as "p_Z vacancies," have on some of the graphene properties. This kind of defects, for example carbon atom vacancies and (monovalent) chemically bound ad-species, are commonly found in realistic graphene samples and have the distinguishing feature of essentially removing a p_Z orbital from the π-electron cloud. Consequently, they introduce sharp resonances in the lattice electronic structure at the neutrality point, act as strong, short-range (resonant) scatterers for charge carriers, behave as semilocalized spin-1/2 magnetic moment, and bias chemical reactivity toward specific neighboring lattice positions.

The impact of these defects on the graphene properties can be hardly overemphasized, and can be traced back to the *enhancement* of the DOS at the Fermi level which they give rise to, along with the peculiar spatial features of the associated "midgap" states. Actually, even though not discussed here, it can be shown that the effect of such defects is so huge that, if they were properly arranged, a sizable gap in the bandstructure—of the kind needed for digital applications—could be opened already at very small impurity concentrations (Martinazzo et al. 2010).

From a different perspective, the defects discussed in this chapter highlight the unique position that graphene has in condensed matter, since it lies exactly at the border between metals and insulators, and most often display features of both.

REFERENCES

Araujo, P. T., M. Terrones, and M. S. Dresselhaus, 2012. Defects and impurities in graphene-like materials, *Mater. Today* 15, 98.

Balakrishnan, J., G. K. W. Koon, M. Jaiswal, A. H. Castro Neto, and B. Özyilmaz, 2013. Colossal enhancement of spin–orbit coupling in weakly hydrogenated graphene, *Nat. Phys.* 9, 284.

Balog, R., B. Jørgensen, J. Wells, E. Lægsgaard, P. Hofmann, F. Besenbacher, L. Hornekaer, 2009. Atomic hydrogen adsorbate structures on graphene, *J. Am. Chem. Soc.* 131, 8744.

Balog, R., B. Jørgensen, L. Nilsson, M. Andersen, E. Rienks, M. Bianchi, M. Fanetti et al. 2010. Bandgap opening in graphene induced by patterned hydrogen adsorption, *Nat. Mater.* 9, 315.

Banhart, F., J. Kotakoski, and A. V. Krasheninnikov, 2011. Structural defects in graphene, *ACS Nano* 5(1), 26.

Barzola-Quiquia, J., P. Esquinazi, M. Rothermel, D. Spemann, T. Butz, and N. Garcia, 2007. Experimental evidence for two-dimensional magnetic order in proton bombarded graphite, *Phys. Rev. B* 76, 161403.

Blanc, N., F. Jean, A.V. Krasheninnikov, G. Renaud, and J. Coraux, 2013. Strains induced by point defects in graphene on a metal, *Phys. Rev. Lett.* 111, 085501.

Boukhvalov, D. W., M. I. Katsnelson, and A. I. Lichtenstein, 2008. Hydrogen on graphene: Electronic structure, total energy, structural distortions and magnetism from first-principles calculations, *Phys. Rev. B* 77, 035427.

Casartelli, M., S. Casolo, G. F. Tantardini, and R. Martinazzo, 2013. Spin coupling around a carbon atom vacancy in graphene, *Phys. Rev. B* 88, 195424.

Casartelli, M., S. Casolo, G. F. Tantardini, and R. Martinazzo, 2014. Structure and stability of hydrogenated of carbon atom vacancies in graphene, *Carbon*, 77, 165.

Casolo, S., O. M. Løvvik, R. Martinazzo, and G. F. Tantardini, 2009. Understanding adsorption of hydrogen atoms on graphene, *J. Chem. Phys.* 130, 054704.

Casolo, S., G. F. Tantardini, and R. Martinazzo, 2013. Insights into H_2 formation in space from ab initio molecular dynamics, *Proc Natl Acad Sci* 110, 667.

Castro Neto, A. H. and F. Guinea, 2009. Impurity-induced spin-orbit coupling in graphene, *Phys. Rev. Lett.* 103, 026804.

Castro Neto, A. H., F. Guinea, N. M. R. Peres, K. S. Novoselov, and A. K. Geim, 2008. The electronic properties of graphene, *Rev. Mod. Phys.* 81, 109.

Chen, J.-H., L. Li, W. G. Cullen, E. D. Williams, and M. S. Fuhrer, 2011. Tunable Kondo effect in graphene with defects, *Nature Phys.* 7, 535.

Chen, J.-H., L. Li, W. G. Cullen, E. D. Williams, and M. S. Fuhrer, 2012. Reply to "Origin of logarithmic resistance correction in graphene," *Nature Phys.* 8, 353.

Cooper, D. L., J. Gerratt, and M. Raimondi, 1986. The electronic structure of the benzene molecule, *Nature* 329, 492.

Coulson, C. A. and Rushbrooke, G. S. 1940. Note on the method of molecular orbitals, *Proc. Cambridge Phil. Soc.* 36, 193.

Cresti, A., F. Ortmann, T. Louvet, D. Van Tuan, and S. Roche, 2013. Broken symmetries, zero-energy modes, and quantum transport in disordered graphene: From supermetallic to insulating regimes, *Phys. Rev. Lett.* 110, 196601.

Das Sarma, S., S. Adam, E. H. Hwang, and E. Rossi, 2011. Electronic transport in two-dimensional graphene, *Rev. Mod. Phys.* 83, 407.

Dharma-Wardana, M. and M. Z. Zgierski, 2008. Magnetism and structure at vacant lattice sites in graphene, *Physica E* 41, 80.

Eckmann, A., A. Felten, A. Mishchenko, L. Britnell, R. Krupke, K. S. Novoselov, and C. Casiraghi, 2012. Probing the nature of defects in graphene by Raman spectroscopy, *Nano Letters* 12(8), 3925.

Esquinazi, P., A. Setzer, R. Höhne, C. Semmelhack, Y. Kopelevich, D. Spemann, T. Butz, B. Kohlstrunk, and M. Lösche, 2002. Ferromagnetism in oriented graphite samples, *Phys. Rev. B* 66, 024429.

Esquinazi, P., D. Spemann, R. Höhne, A. Setzer, K.-H. Han, and T. Butz, 2003. Induced magnetic ordering by proton irradiation in graphite, *Phys. Rev. Lett.* 91, 227201.

Fajtlowicz, S., P. E. Johni, and H. Sachs, 2005. On maximum matchings and eigenvalues of benzenoid graphs, *Croat. Chem. Acta* 78, 195.

Ferrari, A., 2007. Raman spectroscopy of graphene and graphite: Disorder, electron–phonon coupling, doping and nonadiabatic effects, *Solid State Commun.* 143, 47.

Ferreira, A., J. Viana-Gomes, J. Nilsson, E. R. Mucciolo, N. M. R. Peres, and A. H. Castro Neto, 2011. Unified description of the dc conductivity of monolayer and bilayer graphene at finite densities based on resonant scatterers, *Phys. Rev. B* 83, 165402.

Foa Torres, L. E. F., S. Roche, J.-C. Charlier. 2014. *Introduction to Graphene-based Nanomaterials–From Electronic Structure to Quantum Transport.* Cambridge University Press, Cambridge.

Haase, P., S. Fuchs, T. Pruschke, H. Ochoa, and F. Guinea, 2011. Magnetic moments and Kondo effect near vacancies and resonant scatterers in graphene, *Phys. Rev. B* 83, 241408(R).

Hornekær, L., E. Rauls, W. Xu, Z. Sljivancanin, R. Otero, I. Stensgaard, E. Lægsgaard, B. Hammer, and F. Besenbacher, 2006. Clustering of chemisorbed H(D) atoms on the graphite (0001) surface due to preferential sticking, *Phys. Rev. Lett.* 97, 186102.

Huang, L. F., M. Y. Ni, G. R. Zhang, W. H. Zhou, Y. G. Li, X. H. Zheng, and Z. Zeng, 2011. Modulation of the thermodynamic, kinetic, and magnetic properties of the hydrogen monomer on graphene by charge doping, *J. Chem. Phys.* 135, 064705.

Hubbard, J., 1963. Electron correlations in narrow energy bands, *Proc. R. Soc. London, Ser. A* 276, 238.

Hwang, E. H. and S. Das Sarma, 2007. Dielectric function, screening, and plasmons in two-dimensional graphene, *Phys. Rev. B* 75, 205418.

Hwang, E. H. and S. Das Sarma, 2008. Single-particle relaxation time versus transport scattering time in a two-dimensional graphene layer, *Phys. Rev. B* 77, 195412.

Inui, M., S. A. Trugman, and E. Abrahams, 1994. Unusual properties of midband states in systems with off-diagonal disorder, *Phys. Rev. B* 49, 3190.

Ivanovskaya, V. V., A. Zobelli, D. Teillet-Billy, N. Rogeau, V. Sidis, and P. R. Briddon, 2010. Hydrogen adsorption on graphene: A first principles study, *Eur. Phys. J. B* 76, 481.

Jayasingha, R., A. Sherehiy, S.-Y. Wu, and G. U. Sumanasekera, 2013. In situ study of hydrogenation of graphene and new phases of localization between metal–insulator transitions, *Nano Lett.* 13, 5098.

Jeloaica, L. and V. Sidis, 1999. DFT investigation of the adsorption of atomic hydrogen on a cluster-model graphite surface, *Chem. Phys. Lett.* 300, 157.

Jobst, J., D. Waldmann, I. V. Gornyi, A. D. Mirlin, and H. B. Weber, 2012. Electron-electron interaction in the magnetoresistance of graphene, *Phys. Rev. Lett.* 108, 106601.

Katsnelson, M. I., 2013, *Graphene: Carbon in Two Dimensions,* Cambridge University Press.

Kogan, E., 2011. RKKY interaction in graphene, *Phys. Rev. B* 84, 115119.

Kotakoski, J., C. H. Hin, O. Lehtinen, K. Suenaga, and A. V. Krasheninnikov, 2010. Electron knock-on damage in hexagonal boron nitride monolayers, *Phys. Rev. B* 82, 113404.

Kotakoski, J. and Krasheninnikov, A. V., Native and irradiation-induced defects in Graphene: What can we learn from atomistic simulations? In *Computational Nanoscience*, 2011, Ed. E. Bichoutskaia, RSC Publishing, Cambridge, p. 334.

Kotakoski, J., Mangler C., and Meyer, J.C., 2014. Imaging atomic-level random walk of a point defect in graphene, *Nat. Commun.* 5, 3991.

Krasheninnikov, A. V. and R.M. Nieminen, 2011. Attractive interaction between transition-metal atom impurities and vacancies in graphene: A first-principles study, *Theo Chem Acc*, 129, 625.

Lethinen, P. O., A. S. Foster, Y. Ma, A. V. Krasheninnikov, and R. M. Nieminen, 2004. Irradiation-induced magnetism in graphite: A density functional study, *Phys. Rev. Lett.* 93, 187202.

Lherbier, A., S. Roche, O. A. Restrepo, Y.-M. Niquet, A. Delcorte, and J.-C. Charlier, 2013. Highly defective graphene: A key prototype of two-dimensional Anderson insulators, *Nano Res.* 6(5), 326.

Lieb, E. H., 1989. Two theorems on the Hubbard model, *Phys. Rev. Lett.* 62, 1201.

Longuet-Higgins, H. C., 1950. Some studies in molecular orbital Theory I. Resonance structures and molecular orbitals in unsaturated hydrocarbons, *J. Chem. Phys.* 18, 265.

Martin, J., N. Akerman, G. Ulbricht, T. Lohmann, K. von Klitzing, J. H. Smet, and A. Yacoby, 2008. Observation of electron–hole puddles in graphene using a scanning single-electron transistor, *Nat. Phys.* 4, 144.

Martinazzo, R., S. Casolo, and L. Hornekaer, 2013, Dynamics of Gas Surface Interactions (Springer), Chapter 7. Hydrogen recombination on graphitic surfaces.

Martinazzo, R. S. Casolo and G. F. Tantardini, 2010. Symmetry-induced band-gap opening in graphene superlattices, *Phys. Rev. B* 81, 245420.

Meyer, J. C., F. Eder, S. Kurasch, V. Skakalova, J. Kotakoski, H. J. Park, S. Roth et al. 2010. Accurate measurement of electron beam induced displacement cross sections for single-layer graphene, *Phys. Rev. Lett.* 108, 196102.

Meyer, J. C., C. Kisielowski, R. Erni, M. D. Rossell, M. F. Crommie, and A. Zettl, 2008. Direct imaging of lattice atoms and topological defects in graphene membranes, *Nano Letters* 8(11), 3582.

Mucciolo, E. R. and Lewenkopf, C. H. 2010. Disorder and electronic transport in graphene. *J. Phys.: Condens. Matter*, 22, 273201.

Nair, R. R., M. Sepioni, I.-L. Tsai, O. Lehtinen, J. Keinonen, A. V. Krasheninnikov, T. Thomson, A. K. Geim, and I. V. Grigorieva, 2012. Spin-half paramagnetism in graphene induced by point defects, *Nat. Phys.* 8, 199.

Nair, R. R., I.-L. Tsai, M. Sepioni, O. Lehtinen, J. Keinonen, A. V. Krasheninnikov, A. H. C. Neto, A. K. Geim, and I. V. Grigorieva, 2013. Dual origin of defect magnetism in graphene and its reversible switching by molecular doping, *Nat. Commun.* 4, 2010.

Ney, A., P. Papakonstantinou, A. Kumar, N.-G. Shang, and N. Peng 2011. Irradiation enhanced paramagnetism on graphene nanoflakes, *Appl. Phys. Lett.* 99, 102504.

Ni, Z. H., L. A. Ponomarenko, R. R. Nair, R. Yang, S. Anissimova, I. V. Grigorieva, F. Schedin et al. 2010. On resonant scatterers as a factor limiting carrier mobility in graphene, *Nano Letters* 10(10), 3868.

Nomura, K. and A. H. MacDonald, 2007. Quantum Transport of Massless Dirac Fermions, *Phys. Rev. Lett.* 98, 076602.

Palacios, J. J. and F. Yndurain. Critical analysis of vacancy-induced magnetism in monolayer and bilayer graphene, 2012, *Phys. Rev. B* 85, 245443.

Pariser, R. and R. G. Parr, 1953a. A semi-empirical theory of the electronic spectra and electronic structure of complex unsaturated molecules. I., *J. Chem. Phys.* 21, 466.

Pariser, R. and R. G. Parr, 1953b. A semi-empirical theory of the electronic spectra and electronic structure of complex unsaturated molecules. II, *J. Chem. Phys.* 21, 767.

Pereira, V. M., F. Guinea, J. Lopes dos Santos, N. Peres, and A. Castro Neto, 2006. Disorder induced localized states in graphene, *Phys. Rev. Lett.* 96, 036801.

Pereira, V. M., J. M. B. Lopes dos Santos, and A. H. Castro Neto, 2008. Modeling disorder in graphene, *Phys. Rev. B* 77, 115109.

Peres, N. M. R., 2010. *Colloquium*: The transport properties of graphene: An introduction, *Rev. Mod. Phys.* 82, 2673.

Peres, N. M. R., F. Guinea, and A. H. Castro Neto, 2006. Localized description of band structure effects on Li atom interaction with graphene, *Phys. Rev. B* 73, 125411.

Polini, M., A. Tomadini, R. Asgari, and A. H. MacDonald, 2008. Density functional theory of graphene sheets, *Phys. Rev. B* 78, 115426.

Pople, J. A., 1953. Electron interaction in unsaturated hydrocarbons., *Trans. Faraday Soc.* 49, 1375.

Robinson, J. P., H. Schomerus, L. Oroszlány, and V. I. Fal'ko, 2008. Adsorbate-limited conductivity of graphene, *Phys. Rev. Lett.* 101, 196803.

Rossi, E. and S. Das Sarma, 2008. Ground state of graphene in the presence of random charged impurities, *Phys. Rev. Lett.* 101, 166803.

Santos, E. J. G., S. Riikonen, D. Sànchez-Portal, and A. Ayuela, 2012. Magnetism of single vacancies in rippled graphene, *J. Phys. Chem. C* 116, 7602.

Sepioni, M., R. R. Nair, S. Rablen, J. Narayanan, F. Tuna, R. Winpenny, A. K. Geim, and I. V. Grigorieva, 2010. Limits on intrinsic magnetism in graphene, *Phys. Rev. Lett.* 105, 207205.

Sha, X. and B. Jackson, 2002. First-principles study of the structural and energetic properties of H atoms on a graphite (0001) surface, *Surf. Sci.* 496, 318.

Sherafati, M. and S. Satpathy 2011. RKKY interaction in graphene from the lattice Green's function, *Phys. Rev. B* 83, 165425.

Soler, J. M., E. Artacho, G. J. D., A. García, J. Junquera, P. Ordejón, and D. Sánchez-Portal, 2002. The SIESTA method for ab initio order-N materials simulation, *J. Phys.: Condens. Matter* 14, 2745.

Stauber, T., N. M. R. Peres, and A. H. Castro Neto, 2008. Conductivity of suspended and non-suspended graphene at finite gate voltage, *Phys. Rev. B* 78, 085418.

Terrones, H., R. Lv, M. Terrones, and M. S. Dresselhaus, 2012. The role of defects and doping in 2D graphene sheets and 1D nanoribbons, *Rep. Prog. Phys.* 75, 062501.

Ugeda, M. M., I. Brihuega, F. Guinea, and J. M. Gómez-Rodríguez, 2010. Missing atom as a source of carbon magnetism, *Phys. Rev. Lett.* 104(9), 096804.

Ugeda, M.M., D. Fernandez-Torre, I. Brihuenga, P. Pou, A.J. Martinez-Galera, R. Perez and J. M. Gómez-Rodríguez, 2011. Point defects on graphene on metals, *Phys. Rev. Lett.* 107, 116803.

Wallace, P. R., 1947. The band theory of graphite, *Phys. Rev.* 71, 622.

Wang, Y., Y. Huang, Y. Song, X. Zhang, Y. Ma, J. Liang, and Y. Chen, 2009. Room-temperature ferromagnetism of graphene, *Nano Lett.* 9, 220.

Wassmann, T., A. P. Seitsonen, A. M. Saitta, M. Lazzeri, and F. Mauri, 2010. Clar's theory, π-electron distribution, and geometry of graphene nanoribbons, *J. Am. Chem. Soc.* 132(10), 3440.

Wehling, T. O., S. Yuan, A. I. Lichtenstein, A. K. Geim, and M. I. Katsnelson, 2010. Resonant scattering by realistic impurities in graphene, *Phys. Rev. Lett.* 105, 056802.

Yazyev, O. V., 2010. Emergence of magnetism in graphene materials and nanostructures, *Rep. Prog. Phys.* 73, 056501.

Yazyev, O. V. and L. Helm, 2007. Defect-induced magnetism in graphene, *Phys. Rev. B* 75, 125408.

Yuan, S., H. De Raedt, and M. I. Katsnelson, 2010. Modeling electronic structure and transport properties of graphene with resonant scattering centers, *Phys. Rev. B* 82, 115448.

Zecho, T., Güttler, A., Sha, X., Jackson, B., and Küppers, J., 2002. Adsorption of hydrogen and deuterium atoms on the (0001) graphite surface, *J. Chem. Phys.* 117, 8486.

Zhou, J., Q. F. Liang, and J. M. Dong, 2010. Enhanced spin–orbit coupling in hydrogenated and fluorinated graphene, *Carbon* 48, 1405.

3 Atomic Arrangement and Its Effects on Electronic Structures of Graphene from Tight-Binding Description

Sirichok Jungthawan and Sukit Limpijumnong

CONTENTS

Abstract...39
3.1 Tight-Binding Method...39
3.2 Electronic Structures of Pristine Graphene and Honeycomb Lattice..44
3.3 Electronic Structures of Graphene via Brick-Type Lattice..48
3.4 Electronic Structures of Graphene Nanoribbons...50
3.5 Atomic Arrangement and Its Effects on Electronic Structures of Graphene: A Case of Graphene/Boron Nitride Sheet Superlattices...55
3.6 Enumeration Method to Generate Graphene–Alloy Configurations: A Case of Graphene/Boron Nitride Alloys...........58
Acknowledgments...59
References...59

ABSTRACT

An exhaustive enumeration method to generate graphene–alloy configurations and the method to calculate or estimate the electronic structures of those configurations will be discussed. The electronic structures of pristine graphene can be qualitatively described by tight-binding method. Tight-binding model is a simple method to understand the contributions of each atomic state. The method is helpful to investigate how chemical bonding, atomic arrangement, and structural symmetry reflect to the electronic structure of a system. Structural stability of monolayer graphene with dopants or impurities is able to be systematically investigated by means of first-principles calculations. However, the method is more expensive than simple tight-binding approximation. Tight-binding method provides an insightful of how the interactions between the constituents influence on the characteristic of electronic structure, which is sensitive to the detailed arrangement of the constituents. Tight-binding calculations of several representative ordering patterns including ribbon, superlattice (SL) or stripe, and scattering arrangements are given to illustrate an idea of how to construct Hamiltonian matrix for such systems. These matrix elements are considered as parameters, which are fitted to reproduce certain properties from experimental data or first-principles calculations. The properties of nanoribbons and superlattices along armchair and zigzag direction have been discussed in the context of the tight-binding approximation, as they provide an informative trend of the electronic properties related to edge modification and inversion symmetry of structure.

3.1 TIGHT-BINDING METHOD

During the past decade, graphene (Castro Neto et al. 2009; Geim and Novoselov 2007; Novoselov et al. 2004, 2005), a single layer of graphite with a planar honeycomb structure as illustrated in Figure 3.1, has been extensively studied because of its astonishing properties that are mostly attributed to quantum phenomena from 2D confinement effects (Novoselov et al. 2004, 2005; Zhou et al. 2006). In a perfect graphene, each carbon atom forms σ-bonds with its three nearest neighbors (sp^2 hybridization). The electronic states near the Fermi energy are dominated by the π and π* bands, which are derived from the weakly interacting p_z orbitals. The most important characteristic of the electronic structure of graphene is the degenerate states π and π* at the K point of graphene (hexagonal) Brillouin zone (BZ) (inset of Figure 3.2), making graphene a zero-bandgap semi-metal (Reich et al. 2002; Wallace 1947).

The tight-binding method is an approximation assuming that the wave functions tightly bound to the atoms, so-called "tight-binding," such that the atomic wave functions can be used as a basis for expanding the crystal wave functions. The method is helpful to investigate how chemical bonding, atomic arrangement, and structural symmetry reflect to the characteristic of electronic structure. This method is the simplest method for computationally calculating band structures. Since each carbon atom forms bonds with its three nearest neighbors, it is assumed that the wave functions of s, p_x, and p_y orbitals are tightly bound to the atoms. The method has been widely used and proven to be efficient for studying this class of materials as summarized in Table 3.1. Generally, suppose

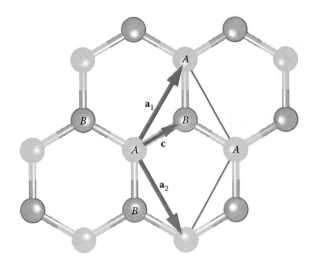

FIGURE 3.1 Crystal structure of honeycomb lattice showing the two sublattices, A and B, in the unit cell and the basis vector **c**. The primitive unit cell is defined by the primitive lattice vectors **a₁** and **a₂**.

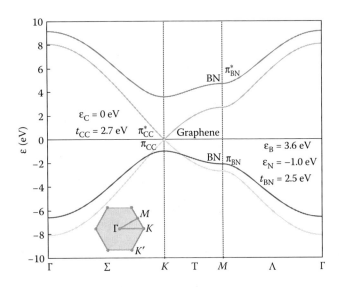

FIGURE 3.2 Band structures of graphene and boron nitride with the first nearest-neighbor interactions. The values of the parameters used in tight-binding calculations are shown. The horizontal line at 0 eV is considered as the Fermi level.

that we start with an atomic wave function, for example, p_z orbital, on a sublattice A (Figure 3.1), which is centered at coordinate $\mathbf{R}_A$,

$$\langle \mathbf{r} \mid \mathbf{R}_A \rangle = \phi(\mathbf{r} - \mathbf{R}_A). \tag{3.1}$$

$\phi(\mathbf{r} - \mathbf{R}_A)$ is an atomic wave function associated with this atom. It is assumed that this state interacts with an atomic wave function on a sublattice B (Figure 3.1), which is centered at site $\mathbf{R}_B$,

$$\langle \mathbf{r} \mid \mathbf{R}_B \rangle = \phi(\mathbf{r} - \mathbf{R}_B). \tag{3.2}$$

These two orbitals dominantly attribute to π and π^* bands near the Fermi level. As long as we consider about the

electronic structures near the Fermi level, we can form electronic states that can be used as the basis functions for the crystal wave functions assuming that there are no other orbitals, for example, s, p_x, and p_y orbitals can mix into the states near the Fermi level since the bands that correspond to the dispersion of bonding and antibonding molecular orbital are π and π^* bands. Given that there are two atoms in the unit cell at sublattices A and B, we can construct two Bloch states as

$$\mid \mathbf{k}_i \rangle = \frac{1}{\sqrt{N}} \sum_{\mathbf{R}} e^{i\mathbf{k} \cdot \mathbf{R}} \mid \mathbf{R}_i + \mathbf{R} \rangle, \tag{3.3}$$

where $i = A, B$, with the summation running over all the N unit cells in the crystal (the vectors $\mathbf{R}$). It can be verified that these states obey Bloch's theorem. Let $\mathbf{R}'$ be another lattice vector,

$$\langle \mathbf{r} \mid \mathbf{k}_i \rangle = \frac{1}{\sqrt{N}} \sum_{\mathbf{R}'} e^{i\mathbf{k} \cdot \mathbf{R}'} \phi(\mathbf{r} - \mathbf{R}_i - \mathbf{R}'). \tag{3.4}$$

Then, this state at position $\mathbf{r} + \mathbf{R}$ is given by

$$
\begin{aligned}
\langle \mathbf{r} + \mathbf{R} \mid \mathbf{k}_i \rangle &= \frac{1}{\sqrt{N}} \sum_{\mathbf{R}'} e^{i\mathbf{k} \cdot \mathbf{R}'} \phi((\mathbf{r} + \mathbf{R}) - \mathbf{R}_i - \mathbf{R}') \\
&= \frac{1}{\sqrt{N}} \sum_{\mathbf{R}'} e^{i\mathbf{k} \cdot (\mathbf{R}' - \mathbf{R})} e^{i\mathbf{k} \cdot \mathbf{R}} \phi(\mathbf{r} - \mathbf{R}_i - (\mathbf{R}' - \mathbf{R})) \\
&= \frac{1}{\sqrt{N}} e^{i\mathbf{k} \cdot \mathbf{R}} \sum_{\mathbf{R}'} e^{i\mathbf{k} \cdot (\mathbf{R}' - \mathbf{R})} \phi(\mathbf{r} - \mathbf{R}_i - (\mathbf{R}' - \mathbf{R})),
\end{aligned}
\tag{3.5}
$$

where $\mathbf{R}' - \mathbf{R}$ is another lattice vector. This leads to

$$\langle \mathbf{r} + \mathbf{R} \mid \mathbf{k}_i \rangle = e^{i\mathbf{k} \cdot \mathbf{R}} \langle \mathbf{r} \mid \mathbf{k}_i \rangle \tag{3.6}$$

that satisfies Bloch's theorem. The crystal eigenstates can be expanded in these two basis functions, $\mid \mathbf{k}_A \rangle$ and $\mid \mathbf{k}_B \rangle$. We can construct the approximate crystal eigenstates as

$$\mid \mathbf{k} \rangle = \sum_i c_i \mid \mathbf{k}_i \rangle = c_A \mid \mathbf{k}_A \rangle + c_B \mid \mathbf{k}_B \rangle, \tag{3.7}$$

where the expansion coefficients c_A and c_B are to be determined. The eigenstates $\mid \mathbf{k} \rangle$ are normalized by $|c_A|^2 + |c_B|^2 = 1$. We want to find the eigenvalues $\varepsilon(\mathbf{k})$ that are exactly the number of bands that we can expect at each $\mathbf{k}$-point of a Hamiltonian operator $\hat{H}$ such that

$$\hat{H} \mid \mathbf{k} \rangle = \varepsilon(\mathbf{k}) \mid \mathbf{k} \rangle. \tag{3.8}$$

Inserting the unit operator $\hat{I} = \sum_j \mid \mathbf{k}_j \rangle \langle \mathbf{k}_j \mid = \mid \mathbf{k}_A \rangle \langle \mathbf{k}_A \mid + \mid \mathbf{k}_B \rangle \langle \mathbf{k}_B \mid$ in front of $\mid \mathbf{k} \rangle$ on both sides of the equation and

TABLE 3.1

Example of Tight-Binding Method Used to Study Electronic Structures of Graphene and Its Derivatives

System	Main Result	Journal	Author
Artificial graphene	Experimental investigation comparison with tight-binding method	*Phys. Rev. B* 88, 115437 (2013)	M. Bellec, U. Kuhl, G. Montambaux, and F. Mortessagne (Bellec et al. 2013a)
Bent graphene	Modeling graphene bending	*Phys. Rev. B* 89, 155437 (2014)	I. Nikiforov, E. Dontsova, R. D. James, and T. Dumitrică (Nikiforov et al. 2014)
Biased bilayer Graphene	Gap tunable by electric Field	*Phys. Rev. Lett.* 99, 216802 (2007)	Eduardo V. Castro, K. S. Novoselov, S. V. Morozov, N. M. R. Peres, J. M. B. Lopes dos Santos, Johan Nilsson, F. Guinea, A. K. Geim, and A. H. Castro Neto (Castro et al. 2007)
Bilayer graphene	Electronic transmission and conductance	*Phys. Rev. B* 79, 155402 (2009)	Michaël Barbier, P. Vasilopoulos, F. M. Peeters, and J. Milton Pereira, Jr. (Barbier et al. 2009)
Bilayer graphene	Optical properties	*Phys. Rev. B* 89, 045419 (2014)	Faris Kadi and Ermin Malic (Kadi and Malic 2014)
Bilayer graphene	Potential difference between the layers and band gap	*Phys. Rev. B* 74, 161403(R) (2006)	Edward McCann (McCann 2006)
Bilayer graphene	Spin–orbit coupling	*Phys. Rev. B* 85, 115423 (2012)	S. Konschuh, M. Gmitra, D. Kochan, and J. Fabian (Konschuh et al. 2012)
Bilayer graphene	Theoretical model	*Phys. Rev. B* 89, 035405 (2014)	Jeil Jung and Allan H. MacDonald (Jung and MacDonald 2014)
Boron-doped graphene field-effect transistors	Transistor characteristics	*ACS Nano* 6, 7942 (2012)	P. Marconcini, A. Cresti, F. Triozon, G. Fiori, B. Biel, Y. M. Niquet, M. Macucci, and S. Roche (Marconcini et al. 2012)
Bottom-gated bilayer graphene	Tight-binding parameters and gate-voltage-dependent bandgap	*Phys. Rev. B* 80, 165406 (2009)	A. B. Kuzmenko, I. Crassee, D. van der Marel, P. Blake, and K. S. Novoselov (Kuzmenko et al. 2009)
Deformed GNRs	Electronic structure	*J. Chem. Phys.* 129, 074704 (2008)	L. Sun, Q. Li, H. Ren, H. Su, Q. W. Shi, and J. Yang (Sun et al. 2008)
Disordered graphene	Experimental investigation comparison with tight-binding method	*Phys. Rev. B* 87, 035101 (2013)	S. Barkhofen, M. Bellec, U. Kuhl, and F. Mortessagne (Barkhofen et al. 2013)
Few-layer graphene	Quasiparticle dispersion	*Phys. Rev. B* 78, 205425 (2008)	A. Grüneis, C. Attaccalite, L. Wirtz, H. Shiozawa, R. Saito, T. Pichler, and A. Rubio (Grüneis et al. 2008)
Graphene	Band structure and density of states	*Phys. Rev. B* 83, 115404 (2011)	C. Bena and L. Simon (Bena and Simon 2011)
Graphene	Intrinsic and Rashba spin-orbit interaction	*Phys. Rev. B* 74, 165310 (2006)	Hongki Min, J. E. Hill, N. A. Sinitsyn, B. R. Sahu, Leonard Kleinman, and A. H. MacDonald (Min et al. 2006)
Graphene	Phonon-limited electron mobility	*J. Appl. Phys.* 112, 053702 (2012)	N. Sule and I. Knezevic (Sule and Knezevic 2012)
Graphene	Review article	*Rev. Mod. Phys.* 81, 109 (2009)	A. H. Castro Neto, F. Guinea, N. M. R. Peres, K. S. Novoselov, and A. K. Geim (Castro Neto et al. 2009)
Graphene	Spin–orbit coupling	*Phys. Rev. B* 82, 245412 (2010)	S. Konschuh, M. Gmitra, and J. Fabian (Konschuh et al. 2010)
Graphene	Vacancy defects	*Phys. Rev. B* 74, 245411 (2006)	Gun-Do Lee, C. Z. Wang, Euijoon Yoon, Nong-Moon Hwang, and K. M. Ho (Lee et al. 2006)
Graphene	Vacancy defects	*Phys. Rev. B* 96, 036801 (2006)	Vitor M. Pereira, F. Guinea, J. M. B. Lopes dos Santos, N. M. R. Peres, and A. H. Castro Neto (Pereira et al. 2006)
Graphene SL (antidot)	Electronic states	*Phys. Rev. B* 80, 045410 (2009)	M. Vanević, V. M. Stojanović, and M. Kindermann (Vanević et al. 2009)
Graphene SL (line defect)	Electronic transmission	*Phys. Rev. B* 86, 045410 (2012)	Lü Xiao-Ling, Liu Zhe, Yao Hai-Bo, Jiang Li-Wei, Gao Wen-Zhu, and Zheng Yi-Song (Xiao-Ling et al. 2012)
Graphene and carbon nanotubes	Analytic expression for the tight-binding dispersion	*Phys. Rev. B* 66, 035412 (2002)	S. Reich, J. Maultzsch, C. Thomsen, and P. Ordejón (Reich et al. 2002)
Graphene and graphite	Electronic structure	*Phys. Rev.* 71, 622 (1947)	P. R. Wallace (Wallace 1947)

(*Continued*)

TABLE 3.1 (*Continued*)
Example of Tight-Binding Method Used to Study Electronic Structures of Graphene and Its Derivatives

System	Main Result	Journal	Author
Graphene multilayers	Electronic structure	*Phys. Rev. B* 75, 193402 (2007)	B. Partoens and F. M. Peeters (Partoens and Peeters 2007)
GNRs	Electronic states	*Phys. Rev. B* 73, 235411 (2006)	L. Brey and H. A. Fertig (Brey and Fertig 2006)
GNRs	Quantum thermal transport properties	*Phys. Rev. B* 79, 115401 (2009)	J. Lan, J.-S. Wang, C. K. Gan, and S. K. Chin (Lan et al. 2009)
GNRs	Scaling rules for band gap	*Phys. Rev. B* 97, 216803 (2006)	Y.-W. Son, M. L. Cohen, and S. G. Louie (Son et al. 2006)
GNRs	Transport model	*Phys. Rev. B* 81, 245402 (2010)	Y. Hancock, A. Uppstu, K. Saloriutta, A. Harju, and M. J. Puska (Hancock et al. 2010)
Graphene nanorods	Transport properties	*Nano Lett.* 12, 2936 (2012)	J. Jung, Z. Qiao, Q. Niu, and A. H. MacDonald (Jung et al. 2012)
Graphene nanostrips	Effective mass, electron–phonon coupling constant	*Phys. Rev. B* 77, 115116 (2008)	D. Gunlycke and C. T. White (Gunlycke and White 2008)
Graphene over pillars	Electronic structure	*Phys. Rev. B* 86, 041405(R) (2012)	M. Neek-Amal, L. Covaci, and F. M. Peeters (Neek-Amal et al. 2012)
Graphene quantum rings	Electronic states	*Phys. Rev. B* 89, 075418 (2014)	D. R. da Costa, Andrey Chaves, M. Zarenia, J. M. Pereira, Jr., G. A. Farias, and F. M. Peeters (da Costa et al. 2014)
Graphene ribbons	Localized edge state	*Phys. Rev. B* 54, 17954 (1996)	K. Nakada, M. Fujita, G. Dresselhaus, and M. S. Dresselhaus (Nakada et al. 1996)
Graphene/BN superlattices	Electronic structure	*Phys. Rev. B* 84, 235424 (2011)	S. Jungthawan, S. Limpijumnong, and J.-L. Kuo (Jungthawan et al. 2011)
Graphene-like lattice	Localized edge state	*Phys. Rev. Lett.* 110, 033902 (2013)	M. Bellec, U. Kuhl, G. Montambaux, and F. Mortessagne (Bellec et al. 2013b)
Graphite	Electronic structure evolution	*Phys. Rev. B* 74, 075404 (2006)	B. Partoens and F. M. Peeters (Partoens and Peeters 2006)
Graphite ribbons	Electronic states of armchair and zigzag ribbons	*J. Phys. Soc. Jpn.* 65, 1920 (1996)	M. Fujita, K. Wakabayashi, K. Nakada, and K. Kusakabe (Fujita et al. 1996)
Nanographite ribbons	Electronic and magnetic properties	*Phys. Rev. B* 59, 8271 (1999)	K. Wakabayashi, M. Fujita, H. Ajiki, and M. Sigrist (Wakabayashi et al. 1999)
Nitrogen-doped graphene	Electronic properties	*Phys. Rev. B* 86, 045448 (2012)	Ph. Lambin, H. Amara, F. Ducastelle, and L. Henrard (Lambin et al. 2012)
Strained bilayer graphene	Electronic structure	*Phys. Rev. B* 85, 125403 (2012)	B. Verberck, B. Partoens, F. M. Peeters, and B. Trauzettel (Verberck et al. 2012)
Strained graphene	Band structure	*New J. Phys.* 11, 115002 (2009)	R. M. Ribeiro, Vitor M. Pereira, N. M. R. Peres, P. R. Briddon, and A. H. Castro Neto (Ribeiro et al. 2009)
Trilayer graphene	Quasiparticle band structure	*Phys. Rev. B* 89, 035431 (2014)	Marcos G. Menezes, Rodrigo B. Capaz, and Steven G. Louie (Menezes et al. 2014)
Twisted bilayer graphene	Electronic structure	*Phys. Rev. B* 82, 121407(R) (2010)	E. Suárez Morell, J. D. Correa, P. Vargas, M. Pacheco, and Z. Barticevic (Suárez Morell et al. 2010)
Twisted bilayer graphene	Electronic structure and quantum Hall effect	*Phys. Rev. B* 85, 195458 (2012)	Pilkyung Moon and Mikito Koshino (Moon and Koshino 2012)
Twisted bilayer graphene	Optical absorption	*Phys. Rev. B* 87, 205404 (2013)	Pilkyung Moon and Mikito Koshino (Moon and Koshino 2013)
Twisted bilayer graphene	van Hove singularities	*Phys. Rev. Lett.* 109, 196802 (2012)	I. Brihuega, P. Mallet, H. González-Herrero, G. Trambly de Laissardière, M. M. Ugeda, L. Magaud, J. M. Gómez-Rodríguez, F. Ynduráin, and J.-Y. Veuillen (Brihuega et al. 2012)
Twisted graphene flakes	Electronic structure	*Phys. Rev. B* 87, 075433 (2013)	W. Landgraf, S. Shallcross, K. Türschmann, D. Weckbecker, and O. Pankratov (Landgraf et al. 2013)
Twisted trilayer graphene	Electronic properties	*Phys. Rev. B* 87, 125414 (2013)	E. Suárez Morell, M. Pacheco, L. Chico, and L. Brey (Suárez Morell et al. 2013)
Uniaxial strain in graphene	Electronic structure	*Phys. Rev. B* 80, 045401 (2009)	Vitor M. Pereira, A. H. Castro Neto, and N. M. R. Peres (Pereira et al. 2009)

multiplying by $\langle \mathbf{k}_i |$, where $i = A, B$, we can write this equation in the form

$$\langle \mathbf{k}_i \,|\, \hat{H} \left(\sum_j |\mathbf{k}_j\rangle\langle \mathbf{k}_j| \right) |\,\mathbf{k}\rangle = \varepsilon(\mathbf{k})\langle \mathbf{k}_i \,| \left(\sum_j |\mathbf{k}_j\rangle\langle \mathbf{k}_j| \right) |\,\mathbf{k}\rangle,$$

$$\sum_j \left(\langle \mathbf{k}_i \,|\, \hat{H} \,|\, \mathbf{k}_j\rangle\langle \mathbf{k}_j \,|\, \mathbf{k}\rangle \right) = \varepsilon(\mathbf{k}) \sum_j \left(\langle \mathbf{k}_i \,|\, \mathbf{k}_j\rangle\langle \mathbf{k}_j \,|\, \mathbf{k}\rangle \right),$$

$$\sum_j \left[\langle \mathbf{k}_i \,|\, \hat{H} \,|\, \mathbf{k}_j\rangle - \varepsilon(\mathbf{k})\langle \mathbf{k}_i \,|\, \mathbf{k}_j\rangle \right]\langle \mathbf{k}_j \,|\, \mathbf{k}\rangle = 0, \qquad (3.9)$$

$$\sum_j \left[H_{ij} - \varepsilon(\mathbf{k})\langle \mathbf{k}_i \,|\, \mathbf{k}_j\rangle \right] c_j = 0.$$

In order to solve this system of linear equations, we need to evaluate $\langle \mathbf{k}_i | \mathbf{k}_j \rangle$

$$\langle \mathbf{k}_i \,|\, \mathbf{k}_j \rangle = \frac{1}{N} \sum_{\mathbf{R}',\mathbf{R}''} e^{i\mathbf{k}\cdot(\mathbf{R}''-\mathbf{R}')} \langle \mathbf{R}_i + \mathbf{R}' \,|\, \mathbf{R}_j + \mathbf{R}'' \rangle. \quad (3.10)$$

It is obvious that $\mathbf{R}'' - \mathbf{R}'$ is another lattice vector. We can define $\mathbf{R}'' - \mathbf{R}' = \mathbf{R}$ given that

$$\langle \mathbf{k}_i \,|\, \mathbf{k}_j \rangle = \frac{1}{N} \sum_{\mathbf{R}',\mathbf{R}} e^{i\mathbf{k}\cdot\mathbf{R}} \langle \mathbf{R}_i + \mathbf{R}' \,|\, \mathbf{R}_j + \mathbf{R} + \mathbf{R}' \rangle. \quad (3.11)$$

The integral $\langle \mathbf{R}_i + \mathbf{R}' | \mathbf{R}_j + \mathbf{R} + \mathbf{R}' \rangle = \langle \mathbf{R}_i | \mathbf{R}_j + \mathbf{R} \rangle$ is independent to the lattice vectors $\mathbf{R}'$ because the relative position between $\mathbf{R}_i$ and $\mathbf{R}_j + \mathbf{R}$ is unchanged under the same translation vector $\mathbf{R}'$. The summation over the lattice vectors $\mathbf{R}'$ will cancel with the factor $1/N$ because there is no explicit dependence on $\mathbf{R}'$ so that

$$\langle \mathbf{k}_i \,|\, \mathbf{k}_j \rangle = \frac{1}{N} \sum_{\mathbf{R}',\mathbf{R}} e^{i\mathbf{k}\cdot\mathbf{R}} \langle \mathbf{R}_i \,|\, \mathbf{R}_j + \mathbf{R} \rangle$$

$$= \sum_{\mathbf{R}} e^{i\mathbf{k}\cdot\mathbf{R}} \langle \mathbf{R}_i \,|\, \mathbf{R}_j + \mathbf{R} \rangle. \qquad (3.12)$$

In the framework of the tight-binding approximation, the overlap integral $\langle \mathbf{R}_i | \mathbf{R}_j + \mathbf{R} \rangle$ is nonzero only for the same orbital on the same atom, that is, only for $i = j$ and $\mathbf{R} = 0$, therefore,

$$\langle \mathbf{k}_i \,|\, \mathbf{k}_j \rangle = \langle \mathbf{R}_i \,|\, \mathbf{R}_j \rangle = \delta_{ij}. \qquad (3.13)$$

The two basis functions, $|\mathbf{k}_A\rangle$ and $|\mathbf{k}_B\rangle$, are orthogonal. With this approximation, Equation 3.9 can be written as

$$\sum_j \left[H_{ij} - \varepsilon(\mathbf{k})\delta_{ij} \right] c_j = 0 \qquad (3.14)$$

with

$$H_{ij} = \langle \mathbf{k}_i \,|\, \hat{H} \,|\, \mathbf{k}_j \rangle. \qquad (3.15)$$

This system of equations can have nonzero solutions only if its determinant vanishes,

$$\det[H_{ij} - \varepsilon(\mathbf{k})\delta_{ij}] = 0. \qquad (3.16)$$

This is known as the secular or characteristic equation. The Hamiltonian matrix $H_{ij} = \langle \mathbf{k}_i | \hat{H} | \mathbf{k}_j \rangle$ is $N \times N$ matrix, where N is the number of basis functions used for expanding the crystal eigenstates. Alternatively, the eigenvalues can be obtained by directly diagonalizing the Hamiltonian matrix with all the known values of elements H_{ij}. The eigenvalues of a symmetric matrix are real. The solutions of this equation yield N eigenvalues $\varepsilon_1(\mathbf{k})$, $\varepsilon_2(\mathbf{k})$, ..., $\varepsilon_N(\mathbf{k})$, which are exactly the number of bands that we can expect for each value of $\mathbf{k}$.

The Hamiltonian matrix elements H_{ij} between atomic states can be obtained by

$$\langle \mathbf{k}_i \,|\, \hat{H} \,|\, \mathbf{k}_j \rangle = \frac{1}{N} \sum_{\mathbf{R}',\mathbf{R}''} e^{i\mathbf{k}\cdot(\mathbf{R}''-\mathbf{R}')} \langle \mathbf{R}_i + \mathbf{R}' \,|\, \hat{H} \,|\, \mathbf{R}_j + \mathbf{R}'' \rangle$$

$$= \frac{1}{N} \sum_{\mathbf{R}',\mathbf{R}} e^{i\mathbf{k}\cdot\mathbf{R}} \langle \mathbf{R}_i + \mathbf{R}' \,|\, \hat{H} \,|\, \mathbf{R}_j + \mathbf{R} + \mathbf{R}' \rangle$$

$$\langle \mathbf{k}_i \,|\, \hat{H} \,|\, \mathbf{k}_j \rangle = \sum_{\mathbf{R}} e^{i\mathbf{k}\cdot\mathbf{R}} \langle \mathbf{R}_i \,|\, \hat{H} \,|\, \mathbf{R}_j + \mathbf{R} \rangle. \qquad (3.17)$$

At this point, an important approximation in the framework of the tight-binding method is introduced by taking the Hamiltonian matrix elements to be nonzero only if: (i) the orbital is on the same atom, that is, for $i = j$ and $\mathbf{R} = 0$,

$$\langle \mathbf{k}_i \,|\, \hat{H} \,|\, \mathbf{k}_j \rangle = \langle \mathbf{R}_i \,|\, \hat{H} \,|\, \mathbf{R}_j \rangle = \varepsilon_i \delta_{ij}, \qquad (3.18)$$

where the diagonal elements ε_i are referred to as the onsite energies or the energy offset for an atom at site i, or (ii) the orbital is on atoms at nearest-neighbor sites located at $\mathbf{c}_{NN}$,

$$\langle \mathbf{k}_i \,|\, \hat{H} \,|\, \mathbf{k}_j \rangle = \sum_{\mathbf{R}} e^{i\mathbf{k}\cdot\mathbf{R}} \langle \mathbf{R}_i \,|\, \hat{H} \,|\, \mathbf{R}_j + \mathbf{R} \rangle$$

$$= \sum_{\mathbf{R}} e^{i\mathbf{k}\cdot\mathbf{R}} t_{ij} \langle \mathbf{R}_i \,|\, \mathbf{R}_j + \mathbf{R} + \mathbf{c}_{NN} \rangle, \qquad (3.19)$$

that is, $\mathbf{R}_i = \mathbf{R}_j + \mathbf{R} + \mathbf{c}_{NN}$. The summation running over all $\mathbf{R}$ will be nonzero when the nearest neighbors are in the same unit cell ($\mathbf{R} = 0$) or when they are across unit cells ($\mathbf{R}$ is a combination of the primitive lattice vectors). The t_{ij} is referred to as the hopping integral between atoms at sites i and j. A range of nearest-neighbor sites can be included into hopping elements (Equation 3.19) for better description of Hamiltonian matrix but more hopping integrals t_{ij} have to be included to evaluate the matrix elements. In practice, we can consider ε_i and t_{ij} as parameters or we can even consider all the matrix elements as parameters as well. In both cases, these parameters are fitted to reproduce certain properties from experimental data or first-principles calculations. Then these parameters can be used to calculate other properties. Alternatively, these parameters can be used as variables to investigate a change to the characteristic of electronic structure under influence of that parameter, that is, ε_i, t_{ij}.

The Hamiltonian matrix elements H_{ij} in Equation 3.15 can be obtained by another form of Hamiltonian that is defined as

$$\hat{H} = \sum_i \varepsilon_i a_i^\dagger a_i - \sum_{i,j} t_{ij}(a_i^\dagger a_j + \text{h.c.}), \qquad (3.20)$$

where ε_i is the site energy for an atom at site i, t_{ij} is the hopping integral between atoms at sites i and j, and $a_i^\dagger$ and a_i are the creation and annihilation operators, respectively, of π-electron at site i. By this definition, $a_i^\dagger$ is equivalent to an atomic state $|\mathbf{R}_i\rangle$, and a_i is equivalent to an atomic state $\langle\mathbf{R}_i|$. j represents the index of summation that is taken over only the nearest neighbors of interest, typically first neighbors, with truncation of farther neighbors. The interactions with farther neighbors, higher-order correction, are ignored for simplicity in most qualitative interpretations. The abbreviation "h.c." stands for Hermitian conjugate of the term $a_i^\dagger a_j$. Tight-binding method provides an insightful of how the interactions between the constituents influence on the electronic structures.

3.2 ELECTRONIC STRUCTURES OF PRISTINE GRAPHENE AND HONEYCOMB LATTICE

The lattice structure of graphene contains two sublattices per primitive unit cell as shown in Figure 3.1. The primitive unit cell is defined by the primitive lattice vectors,

$$\mathbf{a}_1 = \frac{1}{2}a\hat{x} + \frac{\sqrt{3}}{2}a\hat{y} \quad \text{and} \quad \mathbf{a}_2 = \frac{1}{2}a\hat{x} - \frac{\sqrt{3}}{2}a\hat{y}, \qquad (3.21)$$

where a is a lattice constant of graphene. With this choice of lattice vectors, the reciprocal primitive lattice vectors are given by

$$\mathbf{b}_1 = \frac{2\pi}{a}\left(\hat{x} + \frac{1}{\sqrt{3}}\hat{y}\right) \quad \text{and} \quad \mathbf{b}_2 = \frac{2\pi}{a}\left(\hat{x} - \frac{1}{\sqrt{3}}\hat{y}\right). \qquad (3.22)$$

The lattice vectors $\mathbf{R}$ are formed by all possible combinations of lattice vectors

$$\mathbf{R} = n_1\mathbf{a}_1 + n_2\mathbf{a}_2, \qquad (3.23)$$

where n_1 and n_2 are integers. The lattice vectors connect all equivalent points in space. The two sublattices are usually referred to as sublattice "A" and "B" which are indicated by different colors in Figure 3.1. Given that there are two atoms that are located at the two sublattices in the primitive unit cell. Suppose that we choose the position of sublattice A to be at lattice point $\mathbf{R}_A$, then sublattice B will be located at $\mathbf{R}_B = \mathbf{R}_A + \mathbf{c}$ with the basis vector

$$\mathbf{c} = \frac{2}{3}\mathbf{a}_1 + \frac{1}{3}\mathbf{a}_2. \qquad (3.24)$$

The electronic structures of pristine graphene can be qualitatively described by tight-binding model. The Hamiltonian is defined as

$$\hat{H} = \sum_i \varepsilon_i a_i^\dagger a_i - \sum_{i,j} t_{ij}(a_i^\dagger a_j + \text{h.c.}). \qquad (3.25)$$

For simplicity, let us start by considering only the hopping integral or the interactions from first nearest neighbors. The summation is taken over only the first nearest neighbors. The interactions with farther neighbors are ignored for a moment. In the unit cell as shown in Figure 3.1, sublattice A is surrounded by three neighbors which are located at $\mathbf{R}_B \equiv \mathbf{R}_A + \mathbf{c}$, $\mathbf{R}_B - \mathbf{a}_1$, and $\mathbf{R}_B - \mathbf{a}_1 - \mathbf{a}_2$. Likewise, sublattice B is surrounded by three neighbors, which are located at $\mathbf{R}_A$, $\mathbf{R}_A + \mathbf{a}_1$, and $\mathbf{R}_A + \mathbf{a}_1 + \mathbf{a}_2$. Therefore, the Hamiltonian in Equation 3.25 can be explicitly written as

$$\hat{H} = \sum_{\mathbf{R}} (\varepsilon_A |\mathbf{R}_A\rangle\langle\mathbf{R}_A| + \varepsilon_B |\mathbf{R}_B\rangle\langle\mathbf{R}_B|)$$

$$- t_{AB} \sum_{\mathbf{R}} (|\mathbf{R}_A\rangle\langle\mathbf{R}_B| + |\mathbf{R}_A\rangle\langle\mathbf{R}_B - \mathbf{a}_1| + |\mathbf{R}_A\rangle\langle\mathbf{R}_B - \mathbf{a}_1 - \mathbf{a}_2|)$$

$$- t_{AB} \sum_{\mathbf{R}} (|\mathbf{R}_B\rangle\langle\mathbf{R}_A| + |\mathbf{R}_B\rangle\langle\mathbf{R}_A + \mathbf{a}_1| + |\mathbf{R}_B\rangle\langle\mathbf{R}_A + \mathbf{a}_1 + \mathbf{a}_2|),$$

$$(3.26)$$

where $|\mathbf{R}_A\rangle$ and $|\mathbf{R}_B\rangle$ are the π atomic states associated with the atoms at sublattices A and B, respectively. ε_A and ε_B are the site energies for an atom at sites A and B, respectively. t_{AB} is the hopping integral between atoms at sites A and B. To evaluate the Hamiltonian matrix elements in Equation 3.15, we need to use two bases for expansion of the crystal wave functions which obey Bloch's theorem. These bases are given by Equation 3.3,

$$|\mathbf{k}_A\rangle = \frac{1}{\sqrt{N}} \sum_{\mathbf{R}} e^{i\mathbf{k}\cdot\mathbf{R}} |\mathbf{R}_A + \mathbf{R}\rangle \quad \text{and}$$

$$|\mathbf{k}_B\rangle$$

$$= \frac{1}{\sqrt{N}} \sum_{\mathbf{R}} e^{i\mathbf{k}\cdot\mathbf{R}} |\mathbf{R}_B + \mathbf{R}\rangle, \qquad (3.27)$$

with the summation running over all the N unit cells in the crystal (the vectors $\mathbf{R}$), since the Hamiltonian matrix as a function of $\mathbf{k}$ is defined by

$$\hat{H}(\mathbf{k}) = \begin{bmatrix} \langle \mathbf{k}_A | \hat{H} | \mathbf{k}_A \rangle & \langle \mathbf{k}_A | \hat{H} | \mathbf{k}_B \rangle \\ \langle \mathbf{k}_B | \hat{H} | \mathbf{k}_A \rangle & \langle \mathbf{k}_B | \hat{H} | \mathbf{k}_B \rangle \end{bmatrix} \quad (3.28)$$

or

$$\hat{H}(\mathbf{k}) = \begin{bmatrix} H_{AA} & H_{AB} \\ H_{BA} & H_{BB} \end{bmatrix}. \quad (3.29)$$

The Hamiltonian matrix elements can be evaluated by using,

$$\langle \mathbf{k}_A | = \frac{1}{\sqrt{N}} \sum_{\mathbf{R}} e^{-i\mathbf{k}\cdot\mathbf{R}} \langle \mathbf{R}_A + \mathbf{R} | \quad \text{and}$$

$$\langle \mathbf{k}_B | = \frac{1}{\sqrt{N}} \sum_{\mathbf{R}} e^{-i\mathbf{k}\cdot\mathbf{R}} \langle \mathbf{R}_B + \mathbf{R} |. \quad (3.30)$$

The first diagonal matrix element $H_{AA} = \langle \mathbf{k}_A | \hat{H} | \mathbf{k}_A \rangle$ is given by

$$\hat{H} | \mathbf{k}_A \rangle = \sum_{\mathbf{R}} \left(\varepsilon_A | \mathbf{R}_A \rangle \frac{1}{\sqrt{N}} \langle \mathbf{R}_A | \mathbf{R}_A \rangle \right)$$

$$- t_{BA} \sum_{\mathbf{R}} \left(| \mathbf{R}_B \rangle \frac{1}{\sqrt{N}} \langle \mathbf{R}_A | \mathbf{R}_A \rangle \right)$$

$$- t_{BA} \sum_{\mathbf{R}} \left(| \mathbf{R}_B \rangle \frac{1}{\sqrt{N}} e^{i\mathbf{k}\cdot\mathbf{a}_1} \langle \mathbf{R}_A + \mathbf{a}_1 | \mathbf{R}_A + \mathbf{a}_1 \rangle \right)$$

$$- t_{BA} \sum_{\mathbf{R}} \left(| \mathbf{R}_B \rangle \frac{1}{\sqrt{N}} e^{i\mathbf{k}\cdot(\mathbf{a}_1 + \mathbf{a}_2)} \langle \mathbf{R}_A + \mathbf{a}_1 + \mathbf{a}_2 | \mathbf{R}_A \right.$$

$$\left. + \mathbf{a}_1 + \mathbf{a}_2 \rangle \right). \quad (3.31)$$

The overlap integral $\langle \mathbf{R}_i | \mathbf{R}_j \rangle = \delta_{ij}$ is nonzero only for the same orbital on the same atom, that is, only for $i = j$; hence

$$\hat{H} | \mathbf{k}_A \rangle = \sum_{\mathbf{R}} \left(\varepsilon_A | \mathbf{R}_A \rangle \frac{1}{\sqrt{N}} \right) - t_{BA} \sum_{\mathbf{R}} \left(| \mathbf{R}_B \rangle \frac{1}{\sqrt{N}} \right)$$

$$- t_{BA} \sum_{\mathbf{R}} \left(| \mathbf{R}_B \rangle \frac{1}{\sqrt{N}} e^{i\mathbf{k}\cdot\mathbf{a}_1} \right)$$

$$- t_{BA} \sum_{\mathbf{R}} \left(| \mathbf{R}_B \rangle \frac{1}{\sqrt{N}} e^{i\mathbf{k}\cdot(\mathbf{a}_1 + \mathbf{a}_2)} \right). \quad (3.32)$$

The element $H_{AA} = \langle \mathbf{k}_A | \hat{H} | \mathbf{k}_A \rangle$ is

$$\langle \mathbf{k}_A | \hat{H} | \mathbf{k}_A \rangle = \sum_{\mathbf{R}} \left(\varepsilon_A \frac{1}{N} \langle \mathbf{R}_A | \mathbf{R}_A \rangle \right) = \varepsilon_A \sum_{\mathbf{R}} \left(\frac{1}{N} \right) = \varepsilon_A. \quad (3.33)$$

By using Equations 3.30 and 3.32, the off-diagonal element $H_{BA} = \langle \mathbf{k}_B | \hat{H} | \mathbf{k}_A \rangle$ is given by

$$\langle \mathbf{k}_B | \hat{H} | \mathbf{k}_A \rangle = -t_{BA} \sum_{\mathbf{R}} \left(\frac{1}{N} \langle \mathbf{R}_B | \mathbf{R}_B \rangle + \frac{1}{N} e^{i\mathbf{k}\cdot\mathbf{a}_1} \langle \mathbf{R}_B | \mathbf{R}_B \rangle \right.$$

$$\left. + \frac{1}{N} e^{i\mathbf{k}\cdot(\mathbf{a}_1 + \mathbf{a}_2)} \langle \mathbf{R}_B | \mathbf{R}_B \rangle \right),$$

$$\langle \mathbf{k}_B | \hat{H} | \mathbf{k}_A \rangle = -t_{BA} \frac{1}{N} \sum_{\mathbf{R}} (1 + e^{i\mathbf{k}\cdot\mathbf{a}_1} + e^{i\mathbf{k}\cdot(\mathbf{a}_1 + \mathbf{a}_2)})$$

$$= -t_{AB}(1 + e^{i\mathbf{k}\cdot\mathbf{a}_1} + e^{i\mathbf{k}\cdot(\mathbf{a}_1 + \mathbf{a}_2)}). \quad (3.34)$$

Similarly, the second diagonal matrix element $H_{BB} = \langle \mathbf{k}_B | \hat{H} | \mathbf{k}_B \rangle$ can be evaluated, by first considering

$$\hat{H} | \mathbf{k}_B \rangle = \sum_{\mathbf{R}} \left(\varepsilon_B | \mathbf{R}_B \rangle \frac{1}{\sqrt{N}} \langle \mathbf{R}_B | \mathbf{R}_B \rangle \right)$$

$$- t_{AB} \sum_{\mathbf{R}} \left(| \mathbf{R}_A \rangle \frac{1}{\sqrt{N}} \langle \mathbf{R}_B | \mathbf{R}_B \rangle \right)$$

$$- t_{AB} \sum_{\mathbf{R}} \left(| \mathbf{R}_A \rangle \frac{1}{\sqrt{N}} e^{-i\mathbf{k}\cdot\mathbf{a}_1} \langle \mathbf{R}_B - \mathbf{a}_1 | \mathbf{R}_B - \mathbf{a}_1 \rangle \right)$$

$$- t_{AB} \sum_{\mathbf{R}} \left(| \mathbf{R}_A \rangle \frac{1}{\sqrt{N}} e^{-i\mathbf{k}\cdot(\mathbf{a}_1 + \mathbf{a}_2)} \langle \mathbf{R}_B - \mathbf{a}_1 - \mathbf{a}_2 | \mathbf{R}_B \right.$$

$$\left. - \mathbf{a}_1 - \mathbf{a}_2 \rangle \right),$$

$$\hat{H} | \mathbf{k}_B \rangle = \sum_{\mathbf{R}} \left(\varepsilon_B | \mathbf{R}_B \rangle \frac{1}{\sqrt{N}} \right) - t_{AB} \sum_{\mathbf{R}} \left(| \mathbf{R}_A \rangle \frac{1}{\sqrt{N}} \right)$$

$$- t_{AB} \sum_{\mathbf{R}} \left(| \mathbf{R}_A \rangle \frac{1}{\sqrt{N}} e^{-i\mathbf{k}\cdot\mathbf{a}_1} \right)$$

$$- t_{AB} \sum_{\mathbf{R}} \left(| \mathbf{R}_A \rangle \frac{1}{\sqrt{N}} e^{-i\mathbf{k}\cdot(\mathbf{a}_1 + \mathbf{a}_2)} \right). \quad (3.35)$$

The element $H_{BB} = \langle \mathbf{k}_B | \hat{H} | \mathbf{k}_B \rangle$ is

$$\langle \mathbf{k}_B | \hat{H} | \mathbf{k}_B \rangle = \sum_{\mathbf{R}} \left(\varepsilon_B \frac{1}{N} \langle \mathbf{R}_B | \mathbf{R}_B \rangle \right) = \varepsilon_B \sum_{\mathbf{R}} \left(\frac{1}{N} \right) = \varepsilon_B. \quad (3.36)$$

By using Equations 3.30 and 3.35, the off-diagonal element $H_{AB} = \langle \mathbf{k}_A \mid \hat{H} \mid \mathbf{k}_B \rangle$ is given by

$$\langle \mathbf{k}_A \mid \hat{H} \mid \mathbf{k}_B \rangle = -t_{AB} \sum_{\mathbf{R}} \left(\frac{1}{N} \langle \mathbf{R}_A \mid \mathbf{R}_A \rangle + \frac{1}{N} e^{-i\mathbf{k} \cdot \mathbf{a}_1} \langle \mathbf{R}_A \mid \mathbf{R}_A \rangle \right.$$

$$\left. + \frac{1}{N} e^{-i\mathbf{k} \cdot (\mathbf{a}_1 + \mathbf{a}_2)} \langle \mathbf{R}_A \mid \mathbf{R}_A \rangle \right),$$

$$\langle \mathbf{k}_A \mid H \mid \mathbf{k}_B \rangle = -t_{AB} \frac{1}{N} \sum_{\mathbf{R}} \left(1 + e^{-i\mathbf{k} \cdot \mathbf{a}_1} + e^{-i\mathbf{k} \cdot (\mathbf{a}_1 + \mathbf{a}_2)} \right)$$

$$= -t_{AB} \left(1 + e^{-i\mathbf{k} \cdot \mathbf{a}_1} + e^{-i\mathbf{k} \cdot (\mathbf{a}_1 + \mathbf{a}_2)} \right). \quad (3.37)$$

Therefore, the Hamiltonian matrix is

$$\hat{H}(\mathbf{k}) = \begin{bmatrix} \varepsilon_A & -t_{AB} \left(1 + e^{-i\mathbf{k} \cdot \mathbf{a}_1} + e^{-i\mathbf{k} \cdot (\mathbf{a}_1 + \mathbf{a}_2)} \right) \\ -t_{AB} \left(1 + e^{i\mathbf{k} \cdot \mathbf{a}_1} + e^{i\mathbf{k} \cdot (\mathbf{a}_1 + \mathbf{a}_2)} \right) & \varepsilon_B \end{bmatrix}.$$

$$(3.38)$$

In this matrix, one can readily understand that the diagonal elements H_{AA} and H_{BB} are the energies at the atomic site called "onsite energy" or the energy offset for an atom at sites A and B. The H_{AB} is the interaction from atom at B sublattices to the atom at sublattice A, or vice versa for H_{BA}. Notice that, the Hamiltonian matrix is a Hermitian matrix that has two real eigenvalues for each value of $\mathbf{k}$. The solutions to this Hamiltonian can be obtained by diagonalizing the Hamiltonian or from the secular equation

$$\det \left[\hat{H}(\mathbf{k}) - \varepsilon(\mathbf{k})\mathbf{I} \right] = 0,$$

$$\begin{vmatrix} H_{AA} - \varepsilon(\mathbf{k}) & H_{AB} \\ H_{BA} & H_{BB} - \varepsilon(\mathbf{k}) \end{vmatrix} = 0, \quad (3.39)$$

which is

$$\left[H_{AA} - \varepsilon(\mathbf{k}) \right]\left[H_{BB} - \varepsilon(\mathbf{k}) \right] - H_{AB} H_{BA} = 0 \quad (3.40)$$

having solutions

$$\varepsilon(\mathbf{k}) = \frac{1}{2} \left[(H_{AA} + H_{BB}) \pm \sqrt{(H_{AA} - H_{BB})^2 + 4 H_{AB} H_{BA}} \right] \quad (3.41)$$

or

$$\varepsilon(\mathbf{k}) = \frac{1}{2} \left[\varepsilon_A + \varepsilon_B \pm \sqrt{\begin{array}{c}(\varepsilon_A - \varepsilon_B)^2 + 4 t_{AB}^2 \left(1 + e^{-i\mathbf{k} \cdot \mathbf{a}_1} + e^{-i\mathbf{k} \cdot (\mathbf{a}_1 + \mathbf{a}_2)} \right) \\ \left(1 + e^{i\mathbf{k} \cdot \mathbf{a}_1} + e^{i\mathbf{k} \cdot (\mathbf{a}_1 + \mathbf{a}_2)} \right)\end{array}} \right],$$

$$(3.42)$$

where the multiplication

$$\left(1 + e^{-i\mathbf{k} \cdot \mathbf{a}_1} + e^{-i\mathbf{k} \cdot (\mathbf{a}_1 + \mathbf{a}_2)} \right)\left(1 + e^{i\mathbf{k} \cdot \mathbf{a}_1} + e^{i\mathbf{k} \cdot (\mathbf{a}_1 + \mathbf{a}_2)} \right) = 3 + 2\cos(\mathbf{k} \cdot \mathbf{a}_1)$$

$$+ 2\cos(\mathbf{k} \cdot \mathbf{a}_2)$$

$$+ 2\cos \mathbf{k} \cdot (\mathbf{a}_1 + \mathbf{a}_2).$$

$$(3.43)$$

With $\mathbf{k} = k_x \hat{x} + k_y \hat{y}$ and the lattice vectors defined by Equation 3.21, we can evaluate product of

$$\mathbf{k} \cdot \mathbf{a}_1 = \frac{1}{2} k_x a + \frac{\sqrt{3}}{2} k_y a,$$

$$\mathbf{k} \cdot \mathbf{a}_2 = \frac{1}{2} k_x a - \frac{\sqrt{3}}{2} k_y a, \quad (3.44)$$

$$\mathbf{k} \cdot (\mathbf{a}_1 + \mathbf{a}_2) = k_x a.$$

For simplicity, we can define

$$f(\mathbf{k}) = 2\cos(\mathbf{k} \cdot \mathbf{a}_1) + 2\cos(\mathbf{k} \cdot \mathbf{a}_2) + 2\cos \mathbf{k} \cdot (\mathbf{a}_1 + \mathbf{a}_2). \quad (3.45)$$

For a simple nearest-neighbor tight-binding model, one obtains the dispersion relation (Reich et al. 2002; Wallace 1947):

$$\varepsilon(\mathbf{k}) = \frac{1}{2} \left[\varepsilon_A + \varepsilon_B \pm \sqrt{(\varepsilon_A - \varepsilon_B)^2 + 4 t_{AB}^2 \left[3 + f(\mathbf{k}) \right]} \right]. \quad (3.46)$$

If the atom at sublattice A is identical to the atom at sublattice B ($\varepsilon_A = \varepsilon_B \equiv \varepsilon$ and $t_{AB} \equiv t$), then the above equation will get reduced to

$$\varepsilon(\mathbf{k}) = \varepsilon \pm t\sqrt{3 + f(\mathbf{k})}. \quad (3.47)$$

Here, the atoms at sites A and B are carbon atoms; we can choose the energy offset $\varepsilon = 0$. The values of ε and t are found by fitting experimental data or first-principles calculations. In practice, ε and t are adjusted to reproduce a good description of π bands at the K point. Based on the first-principles calculations, a typical value of $\varepsilon = 0$ eV and $t = 2.7$ eV are used (Reich et al. 2002). With this choice of reciprocal lattice vectors in Equation 3.22, we can plot the electronic structures of graphene, Equation 3.47, as illustrated in Figure 3.2 within the first BZ as shown in the inset. The band structure of graphene is plotted along $\Gamma - K - M - \Gamma$ directions. The lines connecting the high symmetry points, $\Gamma - K - M - \Gamma$, are defined by

$$\Sigma = \left[\frac{2k}{3}, 0 \right],$$

$$T = \left[\frac{2}{3}(1-k) + \frac{k}{2}, \frac{k}{2\sqrt{3}} \right], \quad (3.48)$$

$$\Lambda = \left[\frac{1}{2}(1-k), \frac{1}{2\sqrt{3}}(1-k) \right].$$

where $0 \leq k \leq 1$ and the unit along these paths is in the unit of $2\pi/a$. In the case that the atom at sublattice A is equivalent to the atom at sublattice B, there will be doubly degenerate energy levels at the K point or at the six corners of the hexagonal BZ, so-called Dirac point (Zhou et al. 2006). Among the six Dirac points, only two are independent denoted by K and K' in the inset of Figure 3.2, where the others are related to K and K' by a combination of the reciprocal lattice vectors in Equation 3.22. This characteristic is strongly related to the inversion symmetry of crystal structure, where the inversion center is located between the two sublattices, such that the doubly degenerate levels at the K point will always be presented in other atomic species as well, for example, silicene (Şahin et al. 2009). If the atom at sublattice A is different from the atom at sublattice B, the degeneracy is lifted as in the case of *hexagonal* form of boron nitride (h-BN) (Pease 1952). The band structures of h-BN can be calculated from Equation 3.46 with the onsite energies of boron $\varepsilon_B = 3.6$ eV and nitrogen $\varepsilon_N = -1.0$ eV, and the hopping integral $t_{BN} = 2.5$ eV, where the values of all parameters are found by fitting first-principles calculations (Jungthawan et al. 2011). The difference between the two sublattices creates a direct bandgap instead of doubly degenerate levels at K point in h-BN. The bandgap of graphene can be created by several approaches (Peng and Ahuja 2008), one of which is to break the equivalence of A and B sublattices. This simple model clearly shows that the equivalence of A and B sublattices is crucial to electronic structures of honeycomb structure.

A higher-order correction can be made by including the interactions from second nearest neighbors. The Hamiltonian in Equation 3.25 can be written as the Hamiltonian with the interactions from first nearest neighbors $\hat{H}_{1NN}$, given by Equation 3.26, plus the Hamiltonian with the interactions from second nearest neighbors $\hat{H}_{2NN}$. From the unit cell in Figure 3.1, sublattice A is surrounded by six neighbors which are located at $\mathbf{R}_A + \mathbf{a}_1$, $\mathbf{R}_A + \mathbf{a}_1 + \mathbf{a}_2$, $\mathbf{R}_A + \mathbf{a}_2$, $\mathbf{R}_A - \mathbf{a}_1$, $\mathbf{R}_A - \mathbf{a}_1 - \mathbf{a}_2$, and $\mathbf{R}_A - \mathbf{a}_2$. Likewise, the second neighbors of sublattice B are located at $\mathbf{R}_B + \mathbf{a}_1$, $\mathbf{R}_B + \mathbf{a}_1 + \mathbf{a}_2$, $\mathbf{R}_B + \mathbf{a}_2$, $\mathbf{R}_B - \mathbf{a}_1$, $\mathbf{R}_B - \mathbf{a}_1 - \mathbf{a}_2$, and $\mathbf{R}_B - \mathbf{a}_2$. Therefore, the Hamiltonian up to the second neighbor interactions $\hat{H}_{2NN}$ can be explicitly written as

$$
\hat{H}_{2NN} = -t'_A \sum_{\mathbf{R}} \begin{pmatrix} |\mathbf{R}_A\rangle\langle\mathbf{R}_A + \mathbf{a}_1| + |\mathbf{R}_A\rangle\langle\mathbf{R}_A + \mathbf{a}_1 + \mathbf{a}_2| \\ + |\mathbf{R}_A\rangle\langle\mathbf{R}_A + \mathbf{a}_2| + |\mathbf{R}_A\rangle\langle\mathbf{R}_A - \mathbf{a}_1| \\ + |\mathbf{R}_A\rangle\langle\mathbf{R}_A - \mathbf{a}_1 - \mathbf{a}_2| + |\mathbf{R}_A\rangle\langle\mathbf{R}_A - \mathbf{a}_2| \end{pmatrix}
$$

$$
-t'_B \sum_{\mathbf{R}} \begin{pmatrix} |\mathbf{R}_B\rangle\langle\mathbf{R}_B + \mathbf{a}_1| + |\mathbf{R}_B\rangle\langle\mathbf{R}_B + \mathbf{a}_1 + \mathbf{a}_2| \\ + |\mathbf{R}_B\rangle\langle\mathbf{R}_B + \mathbf{a}_2| + |\mathbf{R}_B\rangle\langle\mathbf{R}_B - \mathbf{a}_1| \\ + |\mathbf{R}_B\rangle\langle\mathbf{R}_B - \mathbf{a}_1 - \mathbf{a}_2| + |\mathbf{R}_B\rangle\langle\mathbf{R}_B - \mathbf{a}_2| \end{pmatrix}.
$$

$$(3.49)$$

t'_A and t'_B are the hopping integrals between corresponding second neighbors and atoms at sites A and B, respectively. The second neighbors Hamiltonian is corresponding to the interactions between the same atomic species with a

translation with lattice vector $\mathbf{R}$. This correction will affect only on the diagonal elements, that is, H_{AA} and H_{BB}. The off-diagonal elements, that is, H_{AB} and H_{BA}, will be the same as the matrix in Equation 3.38. In this case, the first diagonal matrix element $H_{AA} = \langle\mathbf{k}_A | \hat{H} | \mathbf{k}_A\rangle$ is given by

$$
\begin{aligned}
\langle\mathbf{k}_A | \hat{H} | \mathbf{k}_A\rangle &= \varepsilon_A - t'_A \frac{1}{N} \sum_{\mathbf{R}} \left(e^{i\mathbf{k}\cdot\mathbf{a}_1} + e^{i\mathbf{k}\cdot(\mathbf{a}_1+\mathbf{a}_2)} + e^{i\mathbf{k}\cdot\mathbf{a}_2} + e^{-i\mathbf{k}\cdot\mathbf{a}_1} \right. \\
&\qquad \left. + e^{-i\mathbf{k}\cdot(\mathbf{a}_1+\mathbf{a}_2)} + e^{-i\mathbf{k}\cdot\mathbf{a}_2} \right) \\
&= \varepsilon_A - t'_A \left(e^{i\mathbf{k}\cdot\mathbf{a}_1} + e^{-i\mathbf{k}\cdot\mathbf{a}_1} + e^{i\mathbf{k}\cdot\mathbf{a}_2} + e^{-i\mathbf{k}\cdot\mathbf{a}_2} \right. \\
&\qquad \left. + e^{i\mathbf{k}\cdot(\mathbf{a}_1+\mathbf{a}_2)} + e^{-i\mathbf{k}\cdot(\mathbf{a}_1+\mathbf{a}_2)} \right) \\
&= \varepsilon_A - 2t'_A \left[\cos(\mathbf{k}\cdot\mathbf{a}_1) + \cos(\mathbf{k}\cdot\mathbf{a}_2) \right. \\
&\qquad \left. + \cos\mathbf{k}\cdot(\mathbf{a}_1+\mathbf{a}_2) \right].
\end{aligned}
$$

$$(3.50)$$

Similarly, the second diagonal matrix element $H_{BB} = \langle\mathbf{k}_B | \hat{H} | \mathbf{k}_B\rangle$ is given by

$$
\begin{aligned}
\langle\mathbf{k}_B | \hat{H} | \mathbf{k}_B\rangle &= \varepsilon_B - t'_B \frac{1}{N} \sum_{\mathbf{R}} \left(e^{i\mathbf{k}\cdot\mathbf{a}_1} + e^{i\mathbf{k}\cdot(\mathbf{a}_1+\mathbf{a}_2)} + e^{i\mathbf{k}\cdot\mathbf{a}_2} + e^{-i\mathbf{k}\cdot\mathbf{a}_1} \right. \\
&\qquad \left. + e^{-i\mathbf{k}\cdot(\mathbf{a}_1+\mathbf{a}_2)} + e^{-i\mathbf{k}\cdot\mathbf{a}_2} \right) \\
&= \varepsilon_B - t'_B \left(e^{i\mathbf{k}\cdot\mathbf{a}_1} + e^{-i\mathbf{k}\cdot\mathbf{a}_1} + e^{i\mathbf{k}\cdot\mathbf{a}_2} + e^{-i\mathbf{k}\cdot\mathbf{a}_2} \right. \\
&\qquad \left. + e^{i\mathbf{k}\cdot(\mathbf{a}_1+\mathbf{a}_2)} + e^{-i\mathbf{k}\cdot(\mathbf{a}_1+\mathbf{a}_2)} \right) \\
&= \varepsilon_B - 2t'_B \left[\cos(\mathbf{k}\cdot\mathbf{a}_1) + \cos(\mathbf{k}\cdot\mathbf{a}_2) \right. \\
&\qquad \left. + \cos\mathbf{k}\cdot(\mathbf{a}_1+\mathbf{a}_2) \right].
\end{aligned}
$$

$$(3.51)$$

With these forms of the diagonal elements, we can construct the Hamiltonian matrix, Equation 3.29, for each value of $\mathbf{k}$,

$$
\begin{aligned}
H_{AA} &= \varepsilon_A - 2t'_A \left[\cos(\mathbf{k}\cdot\mathbf{a}_1) + \cos(\mathbf{k}\cdot\mathbf{a}_2) + \cos\mathbf{k}\cdot(\mathbf{a}_1+\mathbf{a}_2) \right], \\
H_{AB} &= -t_{AB} \left(1 + e^{-i\mathbf{k}\cdot\mathbf{a}_1} + e^{-i\mathbf{k}\cdot(\mathbf{a}_1+\mathbf{a}_2)} \right), \\
H_{BA} &= -t_{AB} \left(1 + e^{i\mathbf{k}\cdot\mathbf{a}_1} + e^{i\mathbf{k}\cdot(\mathbf{a}_1+\mathbf{a}_2)} \right), \\
H_{BB} &= \varepsilon_B - 2t'_B \left[\cos(\mathbf{k}\cdot\mathbf{a}_1) + \cos(\mathbf{k}\cdot\mathbf{a}_2) + \cos\mathbf{k}\cdot(\mathbf{a}_1+\mathbf{a}_2) \right],
\end{aligned}
$$

$$(3.52)$$

and obtain the solutions by Equation 3.41. If the atom at sublattice A is identical to the atom at sublattice B ($t'_A = t'_B \equiv t'$), then the solution is

$$
\varepsilon(\mathbf{k}) = \varepsilon_{1NN}(\mathbf{k}) - 2t' \left[\cos(\mathbf{k}\cdot\mathbf{a}_1) + \cos(\mathbf{k}\cdot\mathbf{a}_2) + \cos\mathbf{k}\cdot(\mathbf{a}_1+\mathbf{a}_2) \right]
$$

$$(3.53)$$

or

$$\varepsilon(\mathbf{k}) = \varepsilon \pm t\sqrt{3 + f(\mathbf{k})} - t'f(\mathbf{k}), \qquad (3.54)$$

where $\varepsilon_{1NN}(\mathbf{k})$ is the dispersion relation from first neighbors interactions in Equation 3.47. $f(\mathbf{k})$ is a function of $\mathbf{k}$ defined by Equation 3.45. According to Reich et al. (2002), the value of t' for graphene is in the range of $-0.2t \leq t' \leq -0.02t$ depending on the parameterization of ε and t. This second neighbor interaction $t' \neq 0$ is responsible for asymmetric feature of the π and π^* bands as shown in Figure 3.3.

The Hamiltonian with third nearest-neighbor interactions $\hat{H}_{3NN}$ can be made by considering the unit cell in Figure 3.1. Sublattice A is surrounded by three neighbors which are located at $\mathbf{R}_B - \mathbf{a}_2$, $\mathbf{R}_B + \mathbf{a}_2$, and $\mathbf{R}_B - 2\mathbf{a}_1 - \mathbf{a}_2$. Likewise, the third neighbors of sublattice B are located at $\mathbf{R}_A - \mathbf{a}_2$, $\mathbf{R}_A + \mathbf{a}_2$, and $\mathbf{R}_A + 2\mathbf{a}_1 + \mathbf{a}_2$. Therefore, the Hamiltonian up to the third neighbors $\hat{H}_{3NN}$ is given by

$$\begin{aligned}
H_{3NN} = &-t''_A \sum_{\mathbf{R}} (|\mathbf{R}_A\rangle\langle\mathbf{R}_B - \mathbf{a}_2| + |\mathbf{R}_A\rangle\langle\mathbf{R}_B + \mathbf{a}_2| + |\mathbf{R}_A\rangle \\
&\langle\mathbf{R}_B - 2\mathbf{a}_1 - \mathbf{a}_2|) \\
&-t''_B \sum_{\mathbf{R}} (|\mathbf{R}_B\rangle\langle\mathbf{R}_A - \mathbf{a}_2| + |\mathbf{R}_B\rangle\langle\mathbf{R}_A + \mathbf{a}_2| + |\mathbf{R}_B\rangle \\
&\langle\mathbf{R}_A + 2\mathbf{a}_1 + \mathbf{a}_2|).
\end{aligned} \qquad (3.55)$$

t''_A and t''_B are the hopping integrals between corresponding third neighbors and atoms at sites A and B, respectively. The third neighbors Hamiltonian is corresponding to the interactions between different atomic species. This correction will affect the off-diagonal elements, that is, H_{AB} and H_{BA}, in

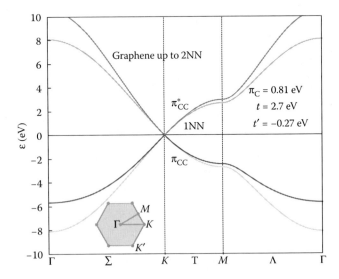

FIGURE 3.3 Band structures of graphene in comparison with the first and second nearest-neighbor interactions. The values of the parameters used in tight-binding calculations are shown. The horizontal line at 0 eV is considered as the Fermi level.

Equation 3.52. The correction to the third nearest-neighbor interactions is given by the product of $\langle\mathbf{k}_A | \hat{H}_{3NN} | \mathbf{k}_B\rangle$ and $\langle\mathbf{k}_B | \hat{H}_{3NN} | \mathbf{k}_A\rangle$, which are

$$\langle\mathbf{k}_A | \hat{H}_{3NN} | \mathbf{k}_B\rangle = -t''_A\left(e^{-i\mathbf{k}\cdot\mathbf{a}_2} + e^{i\mathbf{k}\cdot\mathbf{a}_2} + e^{-i\mathbf{k}\cdot(2\mathbf{a}_1+\mathbf{a}_2)}\right) \quad (3.56)$$

and

$$\langle\mathbf{k}_B | \hat{H}_{3NN} | \mathbf{k}_A\rangle = -t''_B\left(e^{-i\mathbf{k}\cdot\mathbf{a}_2} + e^{i\mathbf{k}\cdot\mathbf{a}_2} + e^{i\mathbf{k}\cdot(2\mathbf{a}_1+\mathbf{a}_2)}\right). \quad (3.57)$$

With these corrections to the Hamiltonian matrix elements in Equation 3.52, all the elements in Equation 3.29 are given by

$$\begin{aligned}
H_{AA} = &\varepsilon_A - 2t'_A\left[\cos(\mathbf{k}\cdot\mathbf{a}_1) + \cos(\mathbf{k}\cdot\mathbf{a}_2) + \cos\mathbf{k}\cdot(\mathbf{a}_1+\mathbf{a}_2)\right], \\
H_{AB} = &-t_{AB}\left(1 + e^{-i\mathbf{k}\cdot\mathbf{a}_1} + e^{-i\mathbf{k}\cdot(\mathbf{a}_1+\mathbf{a}_2)}\right) \\
&-t''_A\left(e^{-i\mathbf{k}\cdot\mathbf{a}_2} + e^{i\mathbf{k}\cdot\mathbf{a}_2} + e^{-i\mathbf{k}\cdot(2\mathbf{a}_1+\mathbf{a}_2)}\right), \\
H_{BA} = &-t_{AB}\left(1 + e^{i\mathbf{k}\cdot\mathbf{a}_1} + e^{i\mathbf{k}\cdot(\mathbf{a}_1+\mathbf{a}_2)}\right) \\
&-t''_B\left(e^{-i\mathbf{k}\cdot\mathbf{a}_2} + e^{i\mathbf{k}\cdot\mathbf{a}_2} + e^{i\mathbf{k}\cdot(2\mathbf{a}_1+\mathbf{a}_2)}\right), \\
H_{BB} = &\varepsilon_B - 2t'_B\left[\cos(\mathbf{k}\cdot\mathbf{a}_1) + \cos(\mathbf{k}\cdot\mathbf{a}_2) + \cos\mathbf{k}\cdot(\mathbf{a}_1+\mathbf{a}_2)\right].
\end{aligned}$$

$$(3.58)$$

If the atom at sublattice A is identical to the atom at sublattice B, we can let $t''_A = t''_B \equiv t''$. The eigenvalues of this matrix can be obtained by Equation 3.41 for each value of $\mathbf{k}$. Usually, the second and third neighbor interactions are not significant and generally ignored. However, the change in these interactions can substantially affect the band structure. The value of t' and t'' can change the number of Dirac points, their position, and their properties in graphene and graphene-related materials (Bena and Simon 2011). It is interesting to note that the simple first nearest-neighbors tight-binding results give a reasonable feature at the band edge near K point. Hence, for the larger system, first nearest-neighbor interactions will be our main focus.

3.3 ELECTRONIC STRUCTURES OF GRAPHENE VIA BRICK-TYPE LATTICE

The honeycomb lattice structure can be simplified by performing a lattice transformation to the brick-type lattice structure (Wakabayashi et al. 1999) as shown in Figure 3.4. In this model, the direction of each bond is aligned into two perpendicular axes. This transformation can be considered as a distorted honeycomb lattice under the influence of compression in $\hat{y}$ direction together with extension in $\hat{x}$ direction. The topology of brick-type lattice is still similar to honeycomb lattice but with different lattice vectors. This transformation

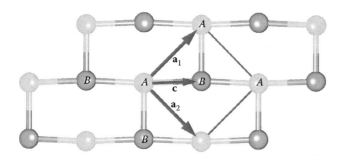

FIGURE 3.4 Crystal structure of brick-type lattice that is a transformation from honeycomb lattice showing sublattices A and B in the unit cell and the basis vector $\mathbf{c}$. The primitive unit cell is defined by the primitive lattice vectors $\mathbf{a}_1$ and $\mathbf{a}_2$.

reduces complexity in the investigation of electronic states near the Fermi energy when considering the overlap between each orbital, that is, s, p_x, p_y, and p_z. The unit cell of brick-type lattice contains two sublattices per primitive unit cell as shown in Figure 3.4. The primitive unit cell is defined by the primitive lattice vector

$$\mathbf{a}_1 = \frac{1}{\sqrt{2}} a\hat{x} + \frac{1}{\sqrt{2}} a\hat{y} \quad \text{and} \quad \mathbf{b}_2 = \frac{2\pi}{a}\left(\frac{1}{\sqrt{2}}\hat{x} - \frac{1}{\sqrt{2}}\hat{y}\right), \quad (3.59)$$

where a is a lattice constant. With this choice of lattice vectors, the reciprocal primitive lattice vectors are given by

$$\mathbf{b}_1 = \frac{2\pi}{a}\left(\frac{1}{\sqrt{2}}\hat{x} + \frac{1}{\sqrt{2}}\hat{y}\right) \quad \text{and} \quad \mathbf{b}_2 = \frac{2\pi}{a}\left(\frac{1}{\sqrt{2}}\hat{x} - \frac{1}{\sqrt{2}}\hat{y}\right).$$

$$(3.60)$$

The Hamiltonian based on the tight-binding model (with second nearest-neighbor interactions) is defined as

$$\hat{H} = \sum_{\mathbf{R}} (\varepsilon_A |\mathbf{R}_A\rangle\langle\mathbf{R}_A| + \varepsilon_B |\mathbf{R}_B\rangle\langle\mathbf{R}_B|)$$

$$- t_{AB} \sum_{\mathbf{R}} \begin{pmatrix} |\mathbf{R}_A\rangle\langle\mathbf{R}_B| + |\mathbf{R}_A\rangle\langle\mathbf{R}_B - \mathbf{a}_1| + |\mathbf{R}_A\rangle \\ \langle\mathbf{R}_B - \mathbf{a}_1 - \mathbf{a}_2| + |\mathbf{R}_B\rangle\langle\mathbf{R}_A| + |\mathbf{R}_B\rangle \\ \langle\mathbf{R}_A + \mathbf{a}_1| + |\mathbf{R}_B\rangle\langle\mathbf{R}_A + \mathbf{a}_1 + \mathbf{a}_2| \end{pmatrix}$$

$$- t'_A \sum_{\mathbf{R}} \begin{pmatrix} |\mathbf{R}_A\rangle\langle\mathbf{R}_A + \mathbf{a}_1| + |\mathbf{R}_A\rangle\langle\mathbf{R}_A + \mathbf{a}_1 + \mathbf{a}_2| + |\mathbf{R}_A\rangle \\ \langle\mathbf{R}_A + \mathbf{a}_2| + |\mathbf{R}_A\rangle\langle\mathbf{R}_A - \mathbf{a}_1| + |\mathbf{R}_A\rangle \\ \langle\mathbf{R}_A - \mathbf{a}_1 - \mathbf{a}_2| + |\mathbf{R}_A\rangle\langle\mathbf{R}_A - \mathbf{a}_2| \end{pmatrix}$$

$$- t'_B \sum_{\mathbf{R}} \begin{pmatrix} |\mathbf{R}_B\rangle\langle\mathbf{R}_B + \mathbf{a}_1| + |\mathbf{R}_B\rangle\langle\mathbf{R}_B + \mathbf{a}_1 + \mathbf{a}_2| + |\mathbf{R}_B\rangle \\ \langle\mathbf{R}_B + \mathbf{a}_2| + |\mathbf{R}_B\rangle\langle\mathbf{R}_B - \mathbf{a}_1| + |\mathbf{R}_B\rangle \\ \langle\mathbf{R}_B - \mathbf{a}_1 - \mathbf{a}_2| + |\mathbf{R}_B\rangle\langle\mathbf{R}_B - \mathbf{a}_2| \end{pmatrix},$$

$$(3.61)$$

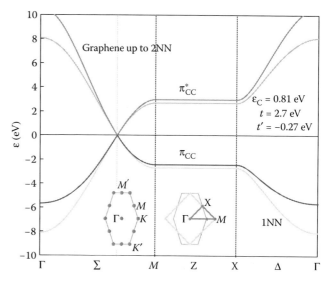

FIGURE 3.5 Band structures of graphene in brick-type lattice in comparison with the first and second nearest-neighbor interactions. The values of the parameters used in tight-binding calculations are shown. The horizontal line at 0 eV is considered as the Fermi level.

where $|\mathbf{R}_A\rangle$ and $|\mathbf{R}_B\rangle$ are the π atomic states associated with the atoms at A and B sublattices, respectively. We can use the two bases in Equation 3.27 for expansion of the crystal wave functions yielding the similar result to Equation 3.54 if the atom at sublattice A is identical to the atom at sublattice B ($t'_A = t'_B$). With the reciprocal lattice vectors in Equation 3.60, we can plot the band structures of this system along high symmetry lines connecting the $\Gamma - M - X - \Gamma$ points as shown in Figure 3.5 within the first BZ as illustrated in the inset of Figure 3.5. These paths are defined by

$$\Sigma = \left[\frac{k}{\sqrt{2}}, 0\right],$$

$$Z = \left[\frac{1}{\sqrt{2}}(1-k) + \frac{k}{2\sqrt{2}}, \frac{k}{2\sqrt{2}}\right], \quad (3.62)$$

$$\Delta = \left[\frac{1}{2\sqrt{2}}(1-k), \frac{1}{2\sqrt{2}}(1-k)\right],$$

with the unit of $2\pi/a$ and $0 \le k \le 1$. The same set of parameters used in Figure 3.3 is applied to calculate the dispersion relation $\varepsilon(\mathbf{k})$ obtained from Equation 3.54 giving the band structures of graphene in brick-type lattice as shown in Figure 3.5. The first BZ of the reciprocal lattices in Equation 3.60 is a square shape as shown in the inset of Figure 3.5. This square BZ transforms the dispersion relation $\varepsilon(\mathbf{k})$ of hexagonal BZ to a distorted hexagonal BZ (inset of Figure 3.5). The original K point of hexagonal BZ is located at $(\sqrt{2}/3, 0)$ on the Σ path. The idea of brick-type lattice transformation can be used to study a distorted honeycomb structure under the influence of external loading that lifts threefold symmetry of the bonding between the atoms at sublattices A and B. In Figure 3.4, the bonding along $\hat{y}$ axis is the difference from the other two which can be formulated by introducing another hopping integral for this bond. We can define different hopping integral t_y for the bonding along $\hat{y}$ axis and obtain the Hamiltonian as

$$
\begin{aligned}
\hat{H} = &\sum_{\mathbf{R}} \left(\varepsilon_A \mid \mathbf{R}_A \rangle\langle \mathbf{R}_A \mid + \varepsilon_B \mid \mathbf{R}_B \rangle\langle \mathbf{R}_B \mid \right) \\
&- t_{AB} \sum_{\mathbf{R}} \left(\mid \mathbf{R}_A \rangle\langle \mathbf{R}_B \mid + \mid \mathbf{R}_A \rangle\langle \mathbf{R}_B - \mathbf{a}_1 - \mathbf{a}_2 \mid + \mid \mathbf{R}_B \rangle \right. \\
&\left. \langle \mathbf{R}_A \mid + \mid \mathbf{R}_B \rangle\langle \mathbf{R}_A + \mathbf{a}_1 + \mathbf{a}_2 \mid \right) \\
&- t_y \sum_{\mathbf{R}} \left(\mid \mathbf{R}_A \rangle\langle \mathbf{R}_B - \mathbf{a}_1 \mid + \mid \mathbf{R}_B \rangle\langle \mathbf{R}_A + \mathbf{a}_1 \mid \right).
\end{aligned} \quad (3.63)
$$

The Hamiltonian matrix elements for each value of $\mathbf{k}$ are given by

$$
\begin{aligned}
H_{AA} &= \varepsilon_A, \\
H_{AB} &= -t_{AB} \left(1 + e^{-i\mathbf{k}\cdot(\mathbf{a}_1 + \mathbf{a}_2)} \right) - t_y e^{-i\mathbf{k}\cdot\mathbf{a}_1}, \\
H_{BA} &= -t_{AB} \left(1 + e^{i\mathbf{k}\cdot(\mathbf{a}_1 + \mathbf{a}_2)} \right) - t_y e^{i\mathbf{k}\cdot\mathbf{a}_1}, \\
H_{BB} &= \varepsilon_B.
\end{aligned} \quad (3.64)
$$

The dispersion relation is given by

$$
\varepsilon(\mathbf{k}) = \frac{1}{2} \left[(H_{AA} + H_{BB}) \pm \sqrt{(H_{AA} - H_{BB})^2 + 4 H_{AB} H_{BA}} \right], \quad (3.65)
$$

where

$$
\begin{aligned}
H_{AB} H_{BA} = &2 t_{AB}^2 + t_y^2 + 2 t_{AB} t_y \left[\cos(\mathbf{k}\cdot\mathbf{a}_1) + \cos(\mathbf{k}\cdot\mathbf{a}_2) \right] \\
&+ 2 t_{AB}^2 \cos \mathbf{k}\cdot(\mathbf{a}_1 + \mathbf{a}_2)
\end{aligned} \quad (3.66)
$$

and

$$
\begin{aligned}
\mathbf{k}\cdot\mathbf{a}_1 &= \frac{1}{\sqrt{2}} k_x a + \frac{1}{\sqrt{2}} k_y a, \\
\mathbf{k}\cdot\mathbf{a}_2 &= \frac{1}{\sqrt{2}} k_x a - \frac{1}{\sqrt{2}} k_y a, \\
\mathbf{k}\cdot(\mathbf{a}_1 + \mathbf{a}_2) &= \sqrt{2} k_x a.
\end{aligned} \quad (3.67)
$$

With additional parameter t_y, the band structure of distorted honeycomb structure can be investigated. The difference between t_y and t_{AB} depends on the amount of distortion.

For distorted graphene, the atoms at sites A and B are carbon atoms ($\varepsilon_A = \varepsilon_B \equiv \varepsilon$ and $t_{AB} \equiv t$). For qualitative analysis, the value of ε and t can be taken from the case of pristine graphene. t_y can be varied to see how this parameter affects a particular region or electronic level in the band structures. Alternatively, these parameters can be parameterized by fitting to empirical or first-principles data according to interaction type and distance for quantitative analysis. The brick-type lattice is also useful for evaluating the bandgap of graphene nanostructures such as nanoribbons or superlattices.

3.4   ELECTRONIC STRUCTURES OF GRAPHENE NANORIBBONS

To utilize graphene as a semiconductor, several approaches to lift the degeneracy and open up the bandgap have been introduced. One approach is to confine or reduce the dimension of graphene. Graphene can be cut to graphene nanoribbons (GNRs). The hexagonal lattice structure can be cut with three different kinds of shape edges, that is, armchair, zigzag, and chiral shape edges. Previous studies show that GNRs can exhibit both metallic and semiconducting properties depending on the geometric and the width of GNRs (Fujita et al. 1996; Nakada et al. 1996; Son et al. 2006; Wakabayashi et al. 1999). Here, we consider two types of GNRs with armchair and zigzag edges, as shown in Figure 3.6, which are the two most relevant ribbon orientations with different ribbon widths. The width of ribbon N is defined by the number of dimer lines for armchair graphene nanoribbons (AGNR) and by the number of zigzag molecular chains for zigzag graphene nanoribbons (ZGNR). The smallest width of AGNR and ZGNR is $N = 3$ containing 6 atoms per unit cell and $N = 2$ containing 4 atoms per unit cell, respectively. AGNR and ZGNR give different ribbon width with the same number of N. The lattice structure of each ribbon can be defined by rectangle with the relevant lattice vector, as shown in Figure 3.6,

$$
\mathbf{a}_1 = a\hat{x} \quad \text{and} \quad \mathbf{a}_2 = a\sqrt{3}\hat{y} \quad (3.68)
$$

for ZGNR and AGNR, respectively, where a is a lattice constant. With this choice of lattice vector, the reciprocal primitive lattice vector for ZGNR and AGNR is given, respectively, by

$$
\mathbf{b}_1 = \frac{2\pi}{a}\hat{x} \quad \text{and} \quad \mathbf{b}_2 = \frac{2\pi}{a\sqrt{3}}\hat{y}. \quad (3.69)
$$

The electronic structures of GNRs can be calculated from the Hamiltonian matrix which can be obtained in the same manner as the previous section. The number of bands is corresponding to the number of states included in the model. If we consider only π orbital from each atom, the number of bands is equal to the number of atom n which is the dimension of the Hamiltonian matrix,

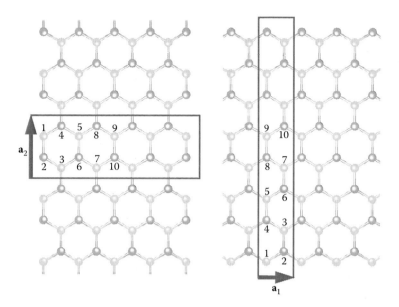

FIGURE 3.6 Structure of GNRs with armchair and zigzag shape edges. The ribbon width is defined by the number of atoms in the unit cell. The number in figure denotes the site of the atom in the unit cell for forming the Hamiltonian matrix. Sublattices A and B are drawn by different colors. $\mathbf{a}_1$ and $\mathbf{a}_2$ are the lattice vectors corresponding to zigzag and armchair nanoribbons, respectively.

$$\hat{H}(\mathbf{k}) = \begin{bmatrix} H_{11} & H_{12} & \cdots & H_{1n} \\ H_{21} & H_{22} & \cdots & H_{2n} \\ \vdots & \vdots & \ddots & \vdots \\ H_{n1} & H_{n2} & \cdots & H_{nn} \end{bmatrix}. \tag{3.70}$$

The matrix elements H_{ij} of Hamiltonian matrix are basically derived by investigating the interactions on each atom in the unit cell to its neighbors which are located in *the same unit cell* as well as *the neighbor cells*. Suppose that the interaction between atoms at sites i and j is t_{ij}. If atom at site i interacts with interacting atom at site j that is *within the same unit cell*, the product $\langle \mathbf{k}_i | \hat{H} | \mathbf{k}_j \rangle$ or matrix element H_{ij} is equal to t_{ij} where $\mathbf{k}_i$ and $\mathbf{k}_j$ are bases for expansion of the crystal wave functions given by Equation 3.3 which obey Bloch's theorem and have orthogonality given by Equation 3.13. If atom at site i interacts with interacting atom at site j that is *within the different unit cell*, the matrix element H_{ij} is equal to t_{ij} multiplied by the factor $\exp(-i\mathbf{k}\cdot\mathbf{R})$ where the exponent $\mathbf{R}$ is the lattice vector that translates the position of interacting atom at site j to the corresponding site within the same unit cell of atom at site i. For instance, the Hamiltonian matrix of AGNR with $N = 3$ is given by

$$\hat{H}(\mathbf{k}) = \begin{bmatrix} \varepsilon & t & 0 & t & 0 & 0 \\ t & \varepsilon & t & 0 & 0 & 0 \\ 0 & t & \varepsilon & te^{-i\mathbf{k}\cdot\mathbf{a}_2} & 0 & t \\ t & 0 & te^{i\mathbf{k}\cdot\mathbf{a}_2} & \varepsilon & t & 0 \\ 0 & 0 & 0 & t & \varepsilon & t \\ 0 & 0 & t & 0 & t & \varepsilon \end{bmatrix} \tag{3.71}$$

The Hamiltonian matrix of ZGNR with $N = 2$ is given by

$$\hat{H}(\mathbf{k}) = \begin{bmatrix} \varepsilon & t(1+e^{-i\mathbf{k}\cdot\mathbf{a}_1}) & 0 & 0 \\ t(1+e^{i\mathbf{k}\cdot\mathbf{a}_1}) & \varepsilon & t & 0 \\ 0 & t & \varepsilon & t(1+e^{i\mathbf{k}\cdot\mathbf{a}_1}) \\ 0 & 0 & t(1+e^{-i\mathbf{k}\cdot\mathbf{a}_1}) & \varepsilon \end{bmatrix}. \tag{3.72}$$

Here we apply typical value of tight-binding parameters ($\varepsilon = 0$ eV and $t = -2.7$ eV) of pristine graphene to GNR system. Nevertheless, the form of analytical solutions to the Hamiltonian matrix is quite complicate. An appropriate approach is to use the numerical method to solve for the eigenvalues corresponding with the Hamiltonian matrix for each value of $\mathbf{k}$. The plots of electronic structures of AGNR with the number of dimer lines $N = 3$, 4, 5 along relevant direction ($\mathbf{k} = k\hat{y}$ where $-\pi/a < k < \pi/a$) in the reciprocal space are shown in Figure 3.7. Figure 3.8 shows the electronic structures of ZGNR with the number of zigzag molecular chains $N = 2$, 3, 4 along $\mathbf{k} = k\hat{x}$ where $-\pi/a < k < \pi/a$.

With these simple models, AGNRs are either metallic or semiconducting depending on their widths, and ZGNRs are always metallic regardless of the width (Nakada et al. 1996). The bandgap of semiconducting AGNRs decreases as a function of increasing widths as shown in Figure 3.9. The plot of calculated bandgap of AGNRs versus N exhibits three distinct family behaviors that can be separated into three different categories as $N = 3p$, $3p + 1$, and $3p + 2$ (where p is positive integer). AGNR could be metallic if $N = 3p + 2$ and be semiconducting for $N = 3p$ or $N = 3p + 1$. From these simple models, the bandgap of ZGNR and AGNR with $N = 3p + 2$

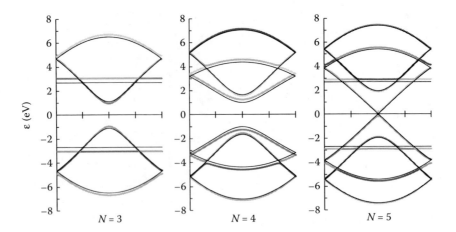

FIGURE 3.7 Band structures of armchair GNRs (in the range of $-\pi/a < k < \pi/a$) for the number of dimer lines $N = 3, 4, 5$ without the correction $\delta = 0$ and with the correction $\delta = 0.12$ to the hopping parameter t are indicated by black and gray solid curves, respectively. The onsite energy $\varepsilon = 0$ and the hopping integral $t = -2.7$ eV are used in tight-binding calculations. The horizontal line at 0 eV could be considered as the Fermi level.

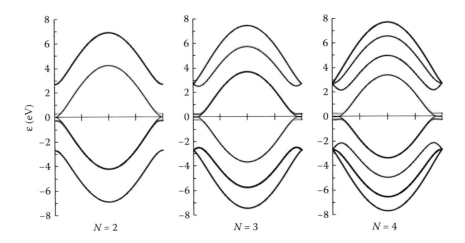

FIGURE 3.8 Band structures of zigzag GNRs (in the range of $-\pi/a < k < \pi/a$) for the number of zigzag molecular chains $N = 2, 3, 4$ without the correction $\varepsilon_\alpha = \varepsilon_\beta = 0$ and with the correction $\varepsilon_\alpha = -\varepsilon_\beta = 0.25$ eV to the onsite energies of atoms at the edges are indicated by black and gray solid curves, respectively. The onsite energy $\varepsilon = 0$ and the hopping integral $t = -2.7$ eV are used for other terms in tight-binding calculations. The horizontal line at 0 eV could be considered as the Fermi level.

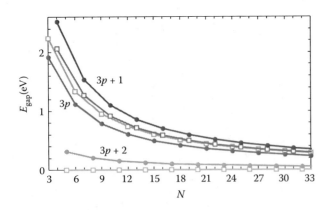

FIGURE 3.9 Bandgap of armchair GNRs as a function of width calculated from simple tight-binding models with the hopping integral $t = -2.7$ eV (square symbol) and with the correction $\delta = 0.12$ to the hopping parameter t (circle symbol).

are always zero. The bandgap of AGNR with $N = 3p$ and $N = 3p + 1$ is comparable for the same p. However, the results from these simple tight-binding models are different from the first-principles calculations or higher-level calculation method and there are no metallic GNRs. The discrepancies are due mainly to the different behaviors of chemical bonding at the edges of ribbons which cannot be described by a single hopping integral t. To be more realistic, the edge atoms of GNRs are passivated by some atoms or molecules so that the hopping integrals and the onsite energies of atoms at the edges would be different from the atoms in the middle of the GNRs. Such effects would reduce the bonding distances between atoms at the edges and induce approximately 12% increase in the hopping integral between π-orbitals (Son et al. 2006). The correction to tight-binding models could be done by replacing the hopping integral of atoms at the edges with $(1 + \delta)t$ where $\delta \approx 0.12$. The onsite energy of atoms at the edges could be defined by an additional parameter ε_δ. However, from previous study, there is no contribution from the variation in the onsite energies (ε_δ) at the edges to the first order (Son et al. 2006), so we can use $\varepsilon_\delta = \varepsilon = 0$. With these corrections, the Hamiltonian matrix of AGNR with $N = 3$ is given by

$$\hat{H}(\mathbf{k}) = \begin{bmatrix} \varepsilon_\delta & (1+\delta)t & 0 & t & 0 & 0 \\ (1+\delta)t & \varepsilon_\delta & t & 0 & 0 & 0 \\ 0 & t & \varepsilon & te^{-i\mathbf{k}\cdot\mathbf{a}_2} & 0 & t \\ t & 0 & te^{i\mathbf{k}\cdot\mathbf{a}_2} & \varepsilon & t & 0 \\ 0 & 0 & 0 & t & \varepsilon_\delta & (1+\delta)t \\ 0 & 0 & t & 0 & (1+\delta)t & \varepsilon_\delta \end{bmatrix}.$$

(3.73)

Similarly, the Hamiltonian matrix of AGNR for larger N ($N = 4$ and so on) could be obtained in the same manner. For example, the Hamiltonian matrix of AGNR with $N = 4$ is given by

$$\hat{H}(\mathbf{k}) = \begin{bmatrix} \varepsilon_\delta & (1+\delta)t & 0 & t & 0 & 0 & 0 & 0 \\ (1+\delta)t & \varepsilon_\delta & t & 0 & 0 & 0 & 0 & 0 \\ 0 & t & \varepsilon & te^{-i\mathbf{k}\cdot\mathbf{a}_2} & 0 & t & 0 & 0 \\ t & 0 & te^{i\mathbf{k}\cdot\mathbf{a}_2} & \varepsilon & t & 0 & 0 & 0 \\ 0 & 0 & 0 & t & \varepsilon & t & 0 & t \\ 0 & 0 & t & 0 & t & \varepsilon & t & 0 \\ 0 & 0 & 0 & 0 & 0 & t & \varepsilon_\delta & (1+\delta)te^{-i\mathbf{k}\cdot\mathbf{a}_2} \\ 0 & 0 & 0 & 0 & t & 0 & (1+\delta)te^{i\mathbf{k}\cdot\mathbf{a}_2} & \varepsilon_\delta \end{bmatrix}.$$

(3.74)

The band structures of AGNR with $N = 3, 4, 5$ are shown in Figure 3.7. It clearly shows how the electronic structures change with an additional parameter δ. The correction to hopping integral of atoms at the edges is strongly correlated to the lift of the doubly degenerate states at the zone edge ($-\pi/a$ or π/a) for $N = 3, 5$ and at the crossing bands for $N = 4, 5$. This correction creates bandgap for AGNR with $N = 5$ or group of AGNRs with $N = 3p + 2$. The variation of bandgaps of AGNRs with correction to the hopping integral of atoms at the edges is shown in Figure 3.9. All AGNRs are now semiconductors with bandgaps that decrease as a function of ribbon width.

With the correction to atoms at the edges, the Hamiltonian matrix of ZGNR with $N = 2$ is given by

$$\hat{H}(\mathbf{k}) = \begin{bmatrix} \varepsilon_\alpha & t(1+e^{-i\mathbf{k}\cdot\mathbf{a}_1}) & 0 & 0 \\ t(1+e^{i\mathbf{k}\cdot\mathbf{a}_1}) & \varepsilon & t & 0 \\ 0 & t & \varepsilon & t(1+e^{i\mathbf{k}\cdot\mathbf{a}_1}) \\ 0 & 0 & t(1+e^{-i\mathbf{k}\cdot\mathbf{a}_1}) & \varepsilon_\beta \end{bmatrix}.$$

(3.75)

Since the atoms on opposite edge of ZGNRs belong to different sublattices, the onsite energy of atoms at the edges could be defined separately as ε_α and ε_β for sublattices A and B, respectively. Similarly, the Hamiltonian matrix of ZGNR for larger N ($N = 3$ and so on) could be obtained in the same manner. For example, the Hamiltonian matrix of ZGNR with $N = 3$ is given by

$$\hat{H}(\mathbf{k}) = \begin{bmatrix} \varepsilon_\alpha & t(1+e^{-i\mathbf{k}\cdot\mathbf{a}_1}) & 0 & 0 & 0 & 0 \\ t(1+e^{i\mathbf{k}\cdot\mathbf{a}_1}) & \varepsilon & t & 0 & 0 & 0 \\ 0 & t & \varepsilon & t(1+e^{-i\mathbf{k}\cdot\mathbf{a}_1}) & 0 & 0 \\ 0 & 0 & t(1+e^{i\mathbf{k}\cdot\mathbf{a}_1}) & \varepsilon & 0 & 0 \\ 0 & 0 & 0 & \varepsilon & t & 0 \\ 0 & 0 & t & \varepsilon & t(1+e^{-i\mathbf{k}\cdot\mathbf{a}_1}) & \varepsilon_\beta \end{bmatrix}.$$

(3.76)

If the $\varepsilon_\alpha = \varepsilon_\beta = \varepsilon$, ZGNRs are always metallic as shown in Figure 3.8. The first-order correction can be made by using different values for ε_α and ε_β. From first-principles calculations based on density functional theory (DFT) (Fujita et al. 1996), it has been shown that ZGNRs are semiconductors, and they present spin-polarized edges. This opens up bandgap of ZGNR, which can be investigated by setting value of ε_α and ε_β. This correction lifts the doubly degenerate states at the zone edge ($-\pi/a$ or π/a) resulting in band gaps for ZGNRs. We can clearly see that the electronic structures around the Fermi level and the zone edges are sensitive to the modification of the edges of GNRs. Especially, for ZGNRs, the onsite energies of atoms at the edges affect specific states around the Fermi level as clearly shown in Figure 3.8. These states mainly belong to the orbitals of atoms at the edges, so-called edge states. The edge states around the Fermi level form flat bands that attribute to a very large density of states at the Fermi level and play a crucial role in the magnetic properties of GNRs (Fujita et al. 1996).

It is interesting to see what would happen if an atom at sublattice A is different from an atom at sublattice B, which lifts the inversion symmetry of nanoribbons. Suppose that boron and nitrogen atoms are at sublattices A and B, respectively. The tight-binding parameters taken from h-BN (Jungthawan et al. 2011), that is, the onsite energy of boron $\varepsilon_B = \varepsilon_A = 3.6$ eV, the onsite energy of nitrogen $\varepsilon_N = \varepsilon_B = -1$ eV, and the hopping integral $t = -2.5$ eV, could be used for atoms that are not on the edges of ribbon. Similar to AGNRs, the correction to the hopping integral of atoms at the edges could be done by replacing t with $(1 + \delta)t$. The onsite energies of atoms at sublattices A and B at the edges could be defined by an additional

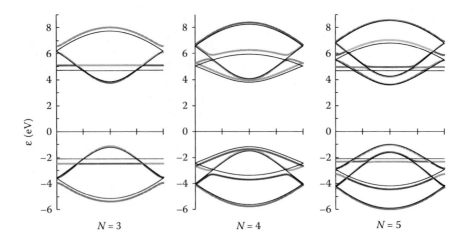

$N = 3$ $N = 4$ $N = 5$

FIGURE 3.10 Band structures of armchair boron nitride nanoribbons (in the range of $-\pi/a < k < \pi/a$) for the number of dimer lines $N = 3$, 4, 5 without the correction $\delta = 0$ and with the correction $\delta = 0.2$ to the hopping parameter t are indicated by black and gray solid curves, respectively. Suppose that boron and nitrogen atoms are at sublattices A and B, respectively. The onsite energies $\varepsilon_A = 3.6$ eV, $\varepsilon_B = -1$ eV and the hopping integral $t = -2.5$ eV are used in tight-binding calculations. The horizontal line at 0 eV could be considered as the Fermi level.

parameter ε_α and ε_β, respectively. For armchair boron nitride nanoribbon (ABNR) with $N = 3$, the Hamiltonian matrix is given by

$$
\hat{H}(\mathbf{k}) = \begin{bmatrix}
\varepsilon_\alpha & (1+\delta)t & 0 & t & 0 & 0 \\
(1+\delta)t & \varepsilon_\beta & t & 0 & 0 & 0 \\
0 & t & \varepsilon_A & te^{-i\mathbf{k}\cdot\mathbf{a}_2} & 0 & t \\
t & 0 & te^{i\mathbf{k}\cdot\mathbf{a}_2} & \varepsilon_B & t & 0 \\
0 & 0 & 0 & t & \varepsilon_\alpha & (1+\delta)t \\
0 & 0 & t & 0 & (1+\delta)t & \varepsilon_\beta
\end{bmatrix}.
$$

$$(3.77)$$

Similarly, the Hamiltonian matrix of ABNR for larger N ($N = 4$ and so on) could be obtained in the same manner. For example, the Hamiltonian matrix of ABNR with $N = 4$ is given by

$$
\hat{H}(\mathbf{k}) = \begin{bmatrix}
\varepsilon_\alpha & (1+\delta)t & 0 & t \\
(1+\delta)t & \varepsilon_\beta & t & 0 \\
0 & t & \varepsilon_A & te^{-i\mathbf{k}\cdot\mathbf{a}_2} \\
t & 0 & te^{i\mathbf{k}\cdot\mathbf{a}_2} & \varepsilon_B \\
0 & 0 & 0 & t \\
0 & 0 & t & 0 \\
0 & 0 & 0 & 0 \\
0 & 0 & 0 & 0
\end{bmatrix}
$$

$$
\begin{bmatrix}
0 & 0 & 0 & 0 \\
0 & 0 & 0 & 0 \\
0 & t & 0 & 0 \\
t & 0 & 0 & 0 \\
\varepsilon_A & t & 0 & t \\
t & \varepsilon_B & t & 0 \\
0 & t & \varepsilon_\alpha & (1+\delta)te^{-i\mathbf{k}\cdot\mathbf{a}_2} \\
t & 0 & (1+\delta)te^{i\mathbf{k}\cdot\mathbf{a}_2} & \varepsilon_\beta
\end{bmatrix}.
$$

$$(3.78)$$

The band structures of ABNR with $N = 3$, 4, 5 are shown in Figure 3.10. The splitting of doubly degenerate states and crossing bands caused by parameter δ is similar to the case of AGNRs. For zigzag boron nitride nanoribbons (ZBNR), the Hamiltonian matrix for ZBNR with $N = 2$ is given by

$$
\hat{H}(\mathbf{k}) = \begin{bmatrix}
\varepsilon_\alpha & t(1+e^{-i\mathbf{k}\cdot\mathbf{a}_1}) & 0 & 0 \\
t(1+e^{i\mathbf{k}\cdot\mathbf{a}_1}) & \varepsilon_B & t & 0 \\
0 & t & \varepsilon_A & t(1+e^{i\mathbf{k}\cdot\mathbf{a}_1}) \\
0 & 0 & t(1+e^{-i\mathbf{k}\cdot\mathbf{a}_1}) & \varepsilon_\beta
\end{bmatrix}.
$$

$$(3.79)$$

Similarly, the Hamiltonian matrix of ZBNR for larger N ($N = 3$ and so on) could be obtained in the same manner. For example, the Hamiltonian matrix of ZBNR with $N = 3$ is given by

$$
\hat{H}(\mathbf{k}) = \begin{bmatrix}
\varepsilon_\alpha & t(1+e^{-i\mathbf{k}\cdot\mathbf{a}_1}) & 0 \\
t(1+e^{i\mathbf{k}\cdot\mathbf{a}_1}) & \varepsilon_B & t \\
0 & t & \varepsilon_A \\
0 & 0 & t(1+e^{-i\mathbf{k}\cdot\mathbf{a}_1}) \\
0 & 0 & 0 \\
0 & 0 & 0 \\
\\
0 & 0 & 0 \\
0 & 0 & 0 \\
t(1+e^{i\mathbf{k}\cdot\mathbf{a}_1}) & 0 & 0 \\
\varepsilon_B & t & 0 \\
t & \varepsilon_A & t(1+e^{-i\mathbf{k}\cdot\mathbf{a}_1}) \\
0 & t(1+e^{i\mathbf{k}\cdot\mathbf{a}_1}) & \varepsilon_\beta
\end{bmatrix}.
$$

$$(3.80)$$

The band structures of ZBNR with $N = 2$, 3, 4 are shown in Figure 3.11. This type of nanoribbons shows interesting

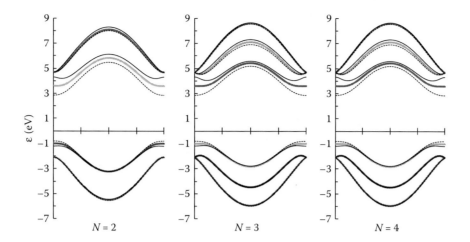

FIGURE 3.11 Band structures of zigzag boron nitride nanoribbons (in the range of $-\pi/a < k < \pi/a$) for the number of zigzag molecular chains $N = 2, 3, 4$ without the correction $\varepsilon_\alpha = \varepsilon_\beta = 0$ (gray solid curves), and with the correction to the onsite energies of atoms at the edges if the onsite energies of atoms at the edges are increased by 20% (black solid curves) or decreased by 20% (dashed curves). Suppose that boron and nitrogen atoms are at sublattices A and B, respectively. The onsite energies $\varepsilon_A = 3.6$ eV, $\varepsilon_B = -1$ eV, and the hopping integral $t = -2.5$ eV are used in tight-binding calculations. The horizontal line at 0 eV could be considered as the Fermi level.

features of the electronic structures that depend on onsite energies at the edges (ε_α or ε_β). If the onsite energy is stronger (more positive or more negative) than the onsite energy of middle atom, the extrema of valence band or conduction band (depending on the sign of onsite energy) are slightly shifted away from the zone edge ($-\pi/a$ or π/a). For example, if the onsite energies of atoms at the edges are increased by 20% that are $\varepsilon_\alpha = 1.2\varepsilon_A$ and $\varepsilon_\beta = 1.2\varepsilon_B$, the valence band maximum and the conduction band minimum are shifted about $\pi/4a$ from the zone edge as shown by black solid line in Figure 3.11. If the onsite energies of atoms at the edges are decreased by 20% that are $\varepsilon_\alpha = 0.8\varepsilon_A$ and $\varepsilon_\beta = 0.8\varepsilon_B$, the flat bands of the valence band and the conduction band near the zone edges are changed. The flat bands at the zone edges are significantly changed due to edge modification which implies that these states would be highly localized on atoms at the edges. Therefore, the modification of nanoribbon edges by some foreign atoms or molecules could manipulate the edge states and the properties of nanoribbons. The boundary regions play an important role so the edge effects influence strongly the π electron states near the Fermi level.

3.5 ATOMIC ARRANGEMENT AND ITS EFFECTS ON ELECTRONIC STRUCTURES OF GRAPHENE: A CASE OF GRAPHENE/ BORON NITRIDE SHEET SUPERLATTICES

An approach for modification of GNR edges and manipulation of the properties of GNRs is forming an SL structure between graphene and boron nitride (BN). A SL structure is formed by mixing graphene and BN with different stripe widths (number of molecular chains). The simple models of the two most relevant orientations of the SLs, that is, zigzag (Z) or armchair (A) molecular chains, are shown in

Figures 3.12 and 3.13, respectively. The zigzag and armchair SLs with one alternating chain are referred as 1Z and 1A, respectively. A corresponding SL structure (for 1Z or 1A) with inversion symmetry could be modeled by doubling the supercell and swap the atomic types in one of the stripes as illustrated in Figures 3.12 (1Zi) and 3.13 (1Ai). Other SL structures of graphene and BN with different number of stripe widths could be modeled in the same manner. Based on nearest-neighbor tight-binding model, there are three possible onsite energy values ε_C, ε_B, and ε_N. In addition, there are four hopping integrals t_{CC}, t_{CB}, t_{CN}, and t_{BN}. These seven parameters (three onsite energies and four hopping integrals) could be adjusted or fitted to reproduce certain properties from first-principles calculations. One of the advantage of tight-binding model is that the contribution from each parameter could be examined by adjusting its value and

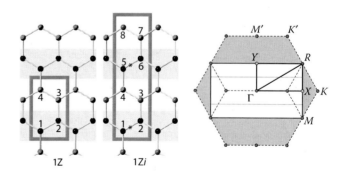

FIGURE 3.12 Zigzag superlattices with one alternating molecular chain, 1Z and 1Zi. The gray, dark gray, and black spheres represent B, N, and C atoms, with light gray highlighting the graphene stripe. The first BZ of 1Z structure is shown together with the primitive hexagonal BZ. The gray line shows the first BZ of 1Zi structure. The inversion center is marked by asterisks.

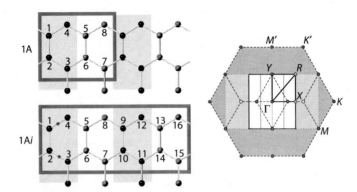

FIGURE 3.13 Armchair superlattices with one alternating molecular chain, 1A and 1A*i*. The gray, dark gray, and black spheres represent B, N, and C atoms, with light gray highlighting the graphene stripe. The first BZ of 1A structure is shown together with the primitive hexagonal BZ. The gray line shows the first BZ of 1A*i* structure. The inversion center is marked by asterisks.

inspecting the change of electronic structures. The general form of the Hamiltonian matrix is

$$\hat{H}(\mathbf{k}) = \begin{bmatrix} H_{11}(\mathbf{k}) & H_{12}(\mathbf{k}) & \cdots & H_{1N}(\mathbf{k}) \\ H_{21}(\mathbf{k}) & H_{22}(\mathbf{k}) & \cdots & H_{2N}(\mathbf{k}) \\ \vdots & \vdots & \ddots & \vdots \\ H_{N1}(\mathbf{k}) & H_{N2}(\mathbf{k}) & \cdots & H_{NN}(\mathbf{k}) \end{bmatrix} \quad (3.81)$$

The Hamiltonian matrix is Hermitian matrix, so we can explicitly write down only the upper triangular elements. The Hamiltonian matrix with nearest-neighbor interactions for 1Z SL is given by

$$\hat{H}(\mathbf{k})$$
$$= \begin{bmatrix} \varepsilon_C & t_{CC}(1+e^{-i\mathbf{k}\cdot\mathbf{a}_1}) & 0 & t_{CN}e^{-i\mathbf{k}\cdot\mathbf{a}_2} \\ t_{CC}(1+e^{i\mathbf{k}\cdot\mathbf{a}_1}) & \varepsilon_C & t_{CB} & 0 \\ 0 & t_{CB} & \varepsilon_B & t_{BN}(1+e^{i\mathbf{k}\cdot\mathbf{a}_1}) \\ 0 & 0 & t_{BN}(1+e^{-i\mathbf{k}\cdot\mathbf{a}_1}) & \varepsilon_N \end{bmatrix},$$
$$(3.82)$$

where the upper triangular elements are

$$H_{11} = \varepsilon_C, H_{12} = t_{CC}(1+e^{-i\mathbf{k}\cdot\mathbf{a}_1}), H_{14} = t_{CN}e^{-i\mathbf{k}\cdot\mathbf{a}_2},$$
$$H_{22} = \varepsilon_C, H_{23} = t_{CB},$$
$$H_{33} = \varepsilon_B, H_{34} = t_{BN}(1+e^{i\mathbf{k}\cdot\mathbf{a}_1}),$$
$$H_{44} = \varepsilon_N. \quad (3.83)$$

The Hamiltonian matrix for 1Z*i* SL is given by

$$H_{11} = \varepsilon_C, H_{12} = t_{CC}(1+e^{-i\mathbf{k}\cdot\mathbf{a}_1}), H_{18} = t_{CN}e^{-i\mathbf{k}\cdot\mathbf{a}_2},$$
$$H_{22} = \varepsilon_C, H_{23} = t_{CN},$$
$$H_{33} = \varepsilon_N, H_{34} = t_{BN}(1+e^{i\mathbf{k}\cdot\mathbf{a}_1}),$$
$$H_{44} = \varepsilon_B, H_{45} = t_{CB},$$
$$H_{55} = \varepsilon_C, H_{56} = t_{CC}(1+e^{-i\mathbf{k}\cdot\mathbf{a}_1}),$$
$$H_{66} = \varepsilon_C, H_{67} = t_{CB},$$
$$H_{77} = \varepsilon_B, H_{78} = t_{BN}(1+e^{i\mathbf{k}\cdot\mathbf{a}_1}),$$
$$H_{88} = \varepsilon_N. \quad (3.84)$$

The Hamiltonian matrix for 1A SL is given by

$$H_{11} = \varepsilon_C, H_{12} = H_{14} = t_{CC}, H_{18} = t_{CN}e^{-i\mathbf{k}\cdot\mathbf{a}_1},$$
$$H_{22} = \varepsilon_C, H_{23} = t_{CC}, H_{27} = t_{CB}e^{-i\mathbf{k}\cdot\mathbf{a}_1},$$
$$H_{33} = \varepsilon_C, H_{34} = t_{CC}e^{-i\mathbf{k}\cdot\mathbf{a}_2}, H_{36} = t_{CN},$$
$$H_{44} = \varepsilon_C, H_{45} = t_{CB},$$
$$H_{55} = \varepsilon_B, H_{56} = H_{58} = t_{BN},$$
$$H_{66} = \varepsilon_N, H_{67} = t_{BN},$$
$$H_{77} = \varepsilon_B, H_{78} = t_{BN}e^{-i\mathbf{k}\cdot\mathbf{a}_2},$$
$$H_{88} = \varepsilon_N. \quad (3.85)$$

The Hamiltonian matrix for 1A*i* SL is given by

$$H_{11} = \varepsilon_C, H_{12} = H_{14} = t_{CC}, H_{1,16} = t_{CN}e^{-i\mathbf{k}\cdot\mathbf{a}_1},$$
$$H_{22} = \varepsilon_C, H_{23} = t_{CC}, H_{2,15} = t_{CB}e^{-i\mathbf{k}\cdot\mathbf{a}_1},$$
$$H_{33} = \varepsilon_C, H_{34} = t_{CC}e^{-i\mathbf{k}\cdot\mathbf{a}_2}, H_{36} = t_{CB},$$
$$H_{44} = \varepsilon_C, H_{45} = t_{CN},$$
$$H_{55} = \varepsilon_N, H_{56} = H_{58} = t_{BN},$$
$$H_{66} = \varepsilon_B, H_{67} = t_{BN},$$
$$H_{77} = \varepsilon_N, H_{78} = t_{BN}e^{-i\mathbf{k}\cdot\mathbf{a}_2}, H_{7,10} = t_{CN},$$
$$H_{88} = \varepsilon_B, H_{89} = t_{CB},$$
$$H_{99} = \varepsilon_C, H_{9,10} = H_{9,12} = t_{CC},$$
$$H_{10,10} = \varepsilon_C, H_{10,11} = t_{CC},$$
$$H_{11,11} = \varepsilon_C, H_{11,12} = t_{CC}e^{-i\mathbf{k}\cdot\mathbf{a}_2}, H_{11,14} = t_{CN},$$
$$H_{12,12} = \varepsilon_C, H_{12,13} = t_{CB},$$
$$H_{13,13} = \varepsilon_B, H_{13,14} = H_{13,15} = t_{BN},$$
$$H_{14,14} = \varepsilon_N, H_{14,15} = t_{BN},$$
$$H_{15,15} = \varepsilon_B, H_{15,16} = t_{BN}e^{-i\mathbf{k}\cdot\mathbf{a}_2},$$
$$H_{16,16} = \varepsilon_N. \quad (3.86)$$

To start with the fitting, one would optimize all three onsite energies (ε_C, ε_B, and ε_N) and two of the hopping integrals (t_{CC} and t_{BN}) to reproduce the π band and bandgap of graphene and BN obtained from experimental data or first-principles calculations. Then these five parameters would be kept fixed while the remaining two hopping integrals (t_{CB} and t_{CN}) are optimized, to reproduce the π band near the Fermi level, or adjusted, to gain qualitative interpretation, for each SL. First, ε_C is set to zero and t_{CC} is optimized (optimized value is $t_{CC} = -2.7$ eV) to reproduce the π_{CC} band of graphene (Figure 3.2). Second, ε_B, ε_N, and t_{BN} are optimized (optimized values are $\varepsilon_B = 3.6$ eV, $\varepsilon_N = -1$ eV, and $t_{BN} = -2.5$ eV) to reproduce the π_{BN} band and bandgap of BN (Figure 3.2) (Jungthawan et al. 2011). In order to investigate the effects of the bonding at the edges, t_{CB} and t_{CN} are set to zero, that is, turning off the interactions between C–B and C–N at the edges. By diagonalizing the Hamiltonian matrix (Equations 3.83 through 3.86), one obtains the dispersion relation for the case of non-interacting edges as shown in Figures 3.14b and 3.15b for zigzag and armchair SLs, respectively.

The electronic structure of 1Z and 1Zi SLs can be viewed as a combination of electronic structures of the corresponding width ZGNR and ZBNR. In Figure 3.14b, the bands of graphene domain have closed bandgap and the bands of BN domain have wide bandgaps. Because the ZBNR has a wide bandgap, the electronic structure from the ZGNR plays the dominant roles near the band edges. To gain qualitative interpretation, we can turn on and adjust the interactions between C–B and C–N at the edges. When the hopping integrals t_{CB} and t_{CN} are turned on, we can optimize the bands to fit the first-principles calculations or adjust the bands to investigate the effect of each interaction to the electronic structures. The hopping integrals t_{CB} and t_{CN} lift the degeneracy of the ZGNR edge states at X where t_{CB} (or t_{CN}) allows a coupling of energy levels between C and B (or N) at the edges. The coupled π-states are split into C–B (or C–N) bonding state (π_{CB} or π_{CN}) and antibonding state (π^*_{CB} or π^*_{CN}), as shown in

Figure 3.14a for 1Z SL and Figure 3.14c for 1Zi SL. The π_{CB} and π^*_{CB} states (coupled π_{CC} and π^*_{BN} states) are located above π_{CN} and π^*_{CN} (coupled π_{BN} and π^*_{CC} states), respectively. As a result, the π_{CB} and π^*_{CN} states become valence band maximum and conduction band minimum states, respectively. For 1Z SL, the difference between the two graphene edges breaks the symmetry, leading to bandgap opening. In contrast, for 1Zi SL, the two edges of any graphene stripe are identical because of the inversion symmetry. As a result, C–B and C–N edge states can cross, leading to a closed bandgap. Note that the optimized values of t_{CB} and t_{CN} to fit the first-principles calculations for 1Z and 1Zi are $t_{CB} = -2.1$ eV and $t_{CN} = -2.3$ eV, and $t_{CB} = -1.8$ eV and $t_{CN} = -1.0$ eV, respectively. The optimized values are quite different, indicating that they are not very transferable.

In the previous section, we have demonstrated that within nearest-neighbor tight-binding model, AGNRs are either metallic or semiconducting depending on their widths, which are defined by the number of dimer lines N. The bandgap of AGNRs is categorized into three distinct groups as $N = 3p$, $3p + 1$, and $3p + 2$ (where p is integer). AGNR could be metallic if $N = 3p + 2$ and be semiconducting for $N = 3p$ or $N = 3p + 1$. From these simple models, the bandgap of ZGNR and AGNR with $N = 3p + 2$ are always zero. The bandgap of AGNR with $N = 3p$ and $N = 3p + 1$ is comparable for the same p, so that 1A and 1Ai SLs have a graphene domain width equal to the closed bandgap AGNRs. On the other hand, ABNRs always have wide bandgap. The electronic structures of 1A and 1Ai SLs can be considered as a combination of the electronic structures of the corresponding width AGNR and ABNR as shown in Figure 3.15b. Since the ABNR has a wide bandgap, the electronic structure from the graphene stripe plays a dominant role near the band edges. When the hopping integrals t_{CB} and t_{CN} are turned on, the interactions

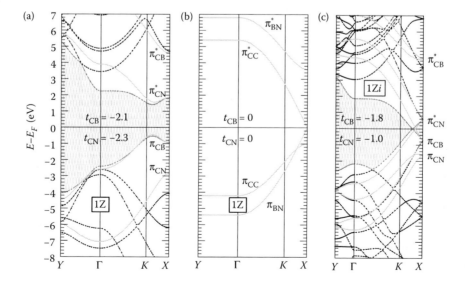

FIGURE 3.14 Calculated π and π^* bands of 1Z (a) and 1Zi (c) based on a tight-binding method (gray solid curves) in comparison with the band structures obtained from first-principles calculations (black dashed curves) (Jungthawan et al. 2011). The tight-binding bands calculated without interactions between graphene and BN stripes (t_{CB} and t_{CN} are set to zero) for 1Z (b) are shown. The band edge states are shown using gray and dark gray dashed curves. The unit of tight-binding parameters is eV.

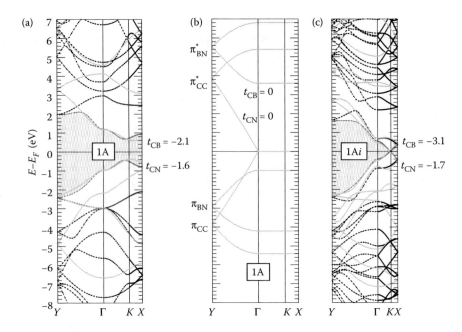

FIGURE 3.15 The calculated π and π^* bands of 1A (a) and 1Ai (c) based on a tight-binding method (gray solid curves) in comparison with the band structures obtained from first-principles calculations (black dashed curves) (Jungthawan et al. 2011). The tight-binding bands calculated without interactions between graphene and BN stripes (t_{CB} and t_{CN} are set to zero) for 1A (b) are shown. The band edge states are shown using gray and dark gray dashed curves. The unit of tight-binding parameters is eV.

at the ribbon edges open up the bandgap, as shown in Figure 3.15a for 1A SL, by allowing couplings between AGNR and ABNR at the edges that lift the degeneracy between π_{CC} and π_{CC}^* state. In armchair SLs, each edge contains an equal number of C–B and C–N bonds. This leads to a mixed C–B and C–N character at each band edge, with energy levels comparable to those of the graphene states (π_{CC} and π_{CC}^*) causing the splitting of AGNR state. The bandgap opening effect is reduced by the inversion symmetry as can be seen in Figure 3.15c for 1Ai SL, where the splitting of AGNR state is less than the reduction by the inversion symmetry. As a result, the crossing by the inversion symmetry is observed. Note that the optimized values of t_{CB} and t_{CN} to fit the first-principles calculations for 1A and 1Ai are $t_{CB} = -2.1$ eV and $t_{CN} = -1.6$ eV, and $t_{CB} = -3.1$ eV and $t_{CN} = -1.7$ eV, respectively. The orientation and ordering play crucial roles in the electronic structure of SLs. The electronic structures reveal particular states that are sensitive to the detailed arrangement of constituents.

3.6 ENUMERATION METHOD TO GENERATE GRAPHENE–ALLOY CONFIGURATIONS: A CASE OF GRAPHENE/BORON NITRIDE ALLOYS

A systematic enumeration method to generate graphene–alloy configurations can be performed by the algorithm for generating superstructures based on Hart and Forcade (2008). The tight-binding method can be used to calculate or estimate the electronic structures of those configurations. Under nonequilibrium growth condition, whereas some atomic configuration can be controlled, the properties of graphene can be modified

by forming nanostructure or changing atomic arrangement. In order to understand the properties of graphene, the tight-binding method is a simple and an effective approach to understand the underlying mechanism of the change of properties due to atomic arrangement. The configurations of graphene–alloy compounds could be generated by Alloy Theoretic Automated Toolkit (ATAT) (van de Walle 2009; van de Walle et al. 2002). As a representative case, we will discuss about the atomic arrangement of single-sheet graphene/BN alloys with the BC_2N stoichiometric ratio ($1 - BC_2N$). The number of configurations

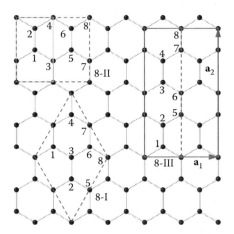

FIGURE 3.16 Unique unit cell shapes found by ATAT. The 4/8 atoms cell is indicated by gray solid line/dark gray dashed line, respectively. The smallest supercell (16 atoms) containing all structures is indicated by dark gray solid line. The numerical values in the unit cells indicate atomic position corresponding to the bases listed in Table 3.2.

and cell shapes generated by ATAT for $1 - BC_2N$ with up to 8 atoms in a unit cell are $6 + 236$ configurations (6 and 236 configurations for 4 and 8 atoms in a unit cell, respectively) and $2 + 5$ cell shapes. There are a number of equivalent configurations (some of the configurations can be represented by more than one unit cell). Previous theoretical works showed that $1 - BC_2N$ compounds disfavor B–B and N–N neighboring (Blase 2000; Blase et al. 1999; Liu et al. 1989), and hence the configurations having such disfavored bonds can be discarded.

TABLE 3.2

Lattice Vectors, Bases, and Number of Configurations for Each Unique Unit Cell Containing 4 and 8 Atoms

Cell Shape	Lattice Vectors (in Unit of a)	Bases (in Unit of a)	Configuration Number (n_{10})	Number of Configuration
4	$(1, 0)$ $(0, \sqrt{3})$	$(1/2, \sqrt{3}/2)$ $(1/2, 5/\sqrt{12})$ $(1, 1/\sqrt{3})$ $(1, \sqrt{3})$	5, 21, <u>57</u>	3
8-I	$(1, \sqrt{3})$ $(1, -\sqrt{3})$	$(1/2, 1/\sqrt{12})$ $(1, -1/\sqrt{3})$ $(1, 0)$ $(1, 2/\sqrt{3})$ $(3/2, -\sqrt{3}/2)$ $(3/2, 1/\sqrt{12})$ $(3/2, \sqrt{3}/2)$ $(2, 0)$	318, 396, 1746, 1800	4
8-II	$(2, 0)$ $(0, \sqrt{3})$	$(1/2, \sqrt{3}/2)$ $(1/2, 5/\sqrt{12})$ $(1, 1/\sqrt{3})$ $(1, \sqrt{3})$ $(3/2, \sqrt{3}/2)$ $(3/2, 5/\sqrt{12})$ $(2, 1/\sqrt{3})$ $(2, \sqrt{3})$	412, 996, 4622, 4628, 4636, 4638, 4672, 4780, 5348, 5364, 5400, <u>5508</u>	12
8-III	$(1, 0)$ $(0, \sqrt{12})$	$(1/2, 1/\sqrt{12})$ $(1/2, \sqrt{3}/2)$ $(1/2, 7/\sqrt{12})$ $(1/2, 9/\sqrt{12})$ $(1, 2/\sqrt{3})$ $(1, \sqrt{3})$ $(1, 5/\sqrt{3})$ $(1, \sqrt{12})$	68, 500, 502, 550, 552, 578, 630, 902, 1226, 1230, <u>1278</u>, 1472, 1474, 1954, 1956, 2034, <u>2684</u>, 2688, 2736, 3660, 3816, 4140, 4872, 4950, 5118, 5274, 5598	27

Note: The atomic ordering for each unit cell is indicated by base-10 number. The actual position is obtained by converting configuration number to base-3 number (see text). The lowest four energetic configurations are underlined.

After exclusion of the redundant configurations that are the configurations with disfavor bonds or repetitive configurations (i.e., some of configurations can be represented by more than one unit cell), the total number of unique configurations is $3 + 43$ configurations from $1 + 3$ unique unit cells.

The four unique unit cells are illustrated in Figure 3.16. For each unit cell, the lattice vectors and bases are tabulated in Table 3.2. The smallest supercell containing all these structures is 16 atoms cell with lattice vectors of $\mathbf{a}_1 = 2a\hat{x}$ and $\mathbf{a}_2 = 2\sqrt{3}a\hat{y}$ in Cartesian coordinate. This supercell contains roughly 3^{16} structures including 900, 900 structures of $1 - BC_2N$. A unit cell with 8 atoms contains roughly 3^8 structures (420 structures are $1 - BC_2N$). It is clearly shown that most of the structures are equivalent and only 46 representative structures belong to the four unique unit cells. The direct enumeration approach to generate alloy configurations is vastly efficient. The method provides a practical number of configurations (irreducible representation), which are computationally feasible for comprehensive study by first-principles calculations or for qualitative study by tight-binding method. In Table 3.2, the atomic position of each configuration can be obtained by converting configuration number from base-10 (n_{10}) to base-3 (n_3) number. The digits 0, 1, and 2 correspond to C, B, and N. For instance, the configuration 57_{10} (2010_3) is a 4-atoms configuration containing two C at $(1/2, \sqrt{3}/2)$ and $(1, 1/\sqrt{3})$, one B at $(1/2, 5/\sqrt{12})$, and one N at $(1, \sqrt{3})$. The calculation by tight-binding method can be performed similarly to the previous section.

ACKNOWLEDGMENTS

This work has been supported by the Thailand Research Fund (TRF) under the TRF Senior Research Scholar (Grant No. RTA5680008) and the National Nanotechnology Center (NANOTEC), NSTDA, Ministry of Science and Technology, Thailand, through its program of Center of Excellence Network. S. Jungthawan acknowledges financial support from the TRF and Suranaree University of Technology (Grant No. TRG5680092).

REFERENCES

Barbier, Michaël, P. Vasilopoulos, F. M. Peeters, and J. Milton Pereira. 2009. Bilayer graphene with single and multiple electrostatic barriers: Band structure and transmission. *Physical Review B* 79 (15):155402.

Barkhofen, S., M. Bellec, U. Kuhl, and F. Mortessagne. 2013. Disordered graphene and boron nitride in a microwave tight-binding analog. *Physical Review B* 87 (3):035101.

Bellec, M., U. Kuhl, G. Montambaux, and F. Mortessagne. 2013a. Tight-binding couplings in microwave artificial graphene. *Physical Review B* 88 (11):115437.

Bellec, M., U. Kuhl, G. Montambaux, and F. Mortessagne. 2013b. Topological transition of Dirac points in a microwave experiment. *Physical Review Letters* 110 (3):033902.

Bena, C. and L. Simon. 2011. Dirac point metamorphosis from third-neighbor couplings in graphene and related materials. *Physical Review B* 83 (11):115404.

Blase, X. 2000. Properties of composite BxCyNz nanotubes and related heterojunctions. *Computational Materials Science* 17 (2–4):107–114.

Blase, X., J. C. Charlier, A. De Vita, and R. Car. 1999. Structural and electronic properties of composite BxCyNz nanotubes and heterojunctions. *Applied Physics A: Materials Science and Processing* 68 (3):293–300.

Brey, L. and H. A. Fertig. 2006. Electronic states of graphene nanoribbons studied with the Dirac equation. *Physical Review B* 73 (23):235411.

Brihuega, I., P. Mallet, H. González-Herrero, G. Trambly de Laissardière, M. M. Ugeda, L. Magaud, J. M. Gómez-Rodríguez, F. Ynduráin, and J. Y. Veuillen. 2012. Unraveling the intrinsic and robust nature of van Hove singularities in twisted bilayer graphene by scanning tunneling microscopy and theoretical analysis. *Physical Review Letters* 109 (19):196802.

Castro, Eduardo V., K. S. Novoselov, S. V. Morozov, N. M. R. Peres, J. M. B. Lopes dos Santos, Johan Nilsson, F. Guinea, A. K. Geim, and A. H. Castro Neto. 2007. Biased bilayer graphene: Semiconductor with a gap tunable by the electric field effect. *Physical Review Letters* 99 (21):216802.

Castro Neto, A. H., F. Guinea, N. M. R. Peres, K. S. Novoselov, and A. K. Geim. 2009. The electronic properties of graphene. *Reviews of Modern Physics* 81 (1):109.

da Costa, D. R., Andrey Chaves, M. Zarenia, J. M. Pereira, Jr., G. A. Farias, and F. M. Peeters. 2014. Geometry and edge effects on the energy levels of graphene quantum rings: A comparison between tight-binding and simplified Dirac models. *Physical Review B* 89 (7):075418.

Fujita, M., K. Wakabayashi, K. Nakada, and K. Kusakabe. 1996. Peculiar localized state at zigzag graphite edge. *Journal of the Physical Society of Japan* 65 (Copyright (C) 1996 The Physical Society of Japan):1920.

Geim, A. K. and K. S. Novoselov. 2007. The rise of graphene. *Nature Materials* 6 (3):183–191.

Grüneis, A., C. Attaccalite, L. Wirtz, H. Shiozawa, R. Saito, T. Pichler, and A. Rubio. 2008. Tight-binding description of the quasiparticle dispersion of graphite and few-layer graphene. *Physical Review B* 78 (20):205425.

Gunlycke, D. and C. T. White. 2008. Tight-binding energy dispersions of armchair-edge graphene nanostrips. *Physical Review B* 77 (11):115116.

Hancock, Y., A. Uppstu, K. Saloriutta, A. Harju, and M. J. Puska. 2010. Generalized tight-binding transport model for graphene nanoribbon-based systems. *Physical Review B* 81 (24):245402.

Hart, Gus L. W., and Rodney W. Forcade. 2008. Algorithm for generating derivative structures. *Physical Review B* 77 (22):224115.

Jung, J. and A. H. MacDonald. 2014. Accurate tight-binding models for the π bands of bilayer graphene. *Physical Review B* 89 (3):035405.

Jung, J., Z. Qiao, Q. Niu, and A. H. MacDonald. 2012. Transport properties of graphene nanoroads in boron nitride sheets. *Nano Letters* 12 (6):2936–2940.

Jungthawan, S., S. Limpijumnong, and J.-L. Kuo. 2011. Electronic structures of graphene/boron nitride sheet superlattices. *Physical Review B* 84 (23):235424.

Kadi, F. and E. Malic. 2014. Optical properties of Bernal-stacked bilayer graphene: A theoretical study. *Physical Review B* 89 (4):045419.

Konschuh, S., M. Gmitra, D. Kochan, and J. Fabian. 2012. Theory of spin–orbit coupling in bilayer graphene. *Physical Review B* 85 (11):115423.

Konschuh, S., M. Gmitra, and J. Fabian. 2010. Tight-binding theory of the spin–orbit coupling in graphene. *Physical Review B* 82 (24):245412.

Kuzmenko, A. B., I. Crassee, D. van der Marel, P. Blake, and K. S. Novoselov. 2009. Determination of the gate-tunable band gap and tight-binding parameters in bilayer graphene using infra-red spectroscopy. *Physical Review B* 80 (16):165406.

Lambin, Ph, H. Amara, F. Ducastelle, and L. Henrard. 2012. Long-range interactions between substitutional nitrogen dopants in graphene: Electronic properties calculations. *Physical Review B* 86 (4):045448.

Lan, J., J.-S. Wang, C. K. Gan, and S. K. Chin. 2009. Edge effects on quantum thermal transport in graphene nanoribbons: Tight-binding calculations. *Physical Review B* 79 (11):115401.

Landgraf, W., S. Shallcross, K. Türschmann, D. Weckbecker, and O. Pankratov. 2013. Electronic structure of twisted graphene flakes. *Physical Review B* 87 (7):075433.

Lee, G.-D, C. Z. Wang, E. Yoon, N-M. Hwang, and K. M. Ho. 2006. Vacancy defects and the formation of local haeckelite structures in graphene from tight-binding molecular dynamics. *Physical Review B* 74 (24):245411.

Liu, A. Y., R. M. Wentzcovitch, and M. L. Cohen. 1989. Atomic arrangement and electronic structure of BC2N. *Physical Review B* 39 (3):1760.

Lü, X.-L., Z. Liu, H.-B. Yao, L.-W. Jiang, W.-Z. Gao, and Y.-S. Zheng. 2012. Valley polarized electronic transmission through a line defect superlattice of graphene. *Physical Review B* 86 (4):045410.

Marconcini, P., A. Cresti, F. Triozon, G. Fiori, B. Biel, Y.-M. Niquet, M. Macucci, and S. Roche. 2012. Atomistic boron-doped graphene field-effect transistors: A route toward unipolar characteristics. *ACS Nano* 6 (9):7942–7947.

McCann, E. 2006. Asymmetry gap in the electronic band structure of bilayer graphene. *Physical Review B* 74 (16):161403.

Menezes, M. G., R. B. Capaz, and S. G. Louie. 2014. *Ab initio* quasiparticle band structure of ABA and ABC-stacked graphene trilayers. *Physical Review B* 89 (3):035431.

Min, H., J. E. Hill, N. A. Sinitsyn, B. R. Sahu, L. Kleinman, and A. H. MacDonald. 2006. Intrinsic and Rashba spin–orbit interactions in graphene sheets. *Physical Review B* 74 (16):165310.

Moon, P. and M. Koshino. 2012. Energy spectrum and quantum Hall effect in twisted bilayer graphene. *Physical Review B* 85 (19):195458.

Moon, P. and M. Koshino. 2013. Optical absorption in twisted bilayer graphene. *Physical Review B* 87 (20):205404.

Nakada, K., M. Fujita, G. Dresselhaus, and M. S. Dresselhaus. 1996. Edge state in graphene ribbons: Nanometer size effect and edge shape dependence. *Physical Review B* 54 (24):17954–17961.

Neek-Amal, M., L. Covaci, and F. M. Peeters. 2012. Nanoengineered nonuniform strain in graphene using nanopillars. *Physical Review B* 86 (4):041405.

Nikiforov, I., E. Dontsova, R. D. James, and T. Dumitrică. 2014. Tight-binding theory of graphene bending. *Physical Review B* 89 (15):155437.

Novoselov, K. S., A. K. Geim, S. V. Morozov, D. Jiang, Y. Zhang, S. V. Dubonos, I. V. Grigorieva, and A. A. Firsov. 2004. Electric field effect in atomically thin carbon films. *Science* 306 (5696):666–669.

Novoselov, K. S., A. K. Geim, S. V. Morozov, D. Jiang, M. I. Katsnelson, I. V. Grigorieva, S. V. Dubonos, and A. A. Firsov. 2005. Two-dimensional gas of massless Dirac fermions in graphene. *Nature* 438 (7065):197–200.

Partoens, B. and F. M. Peeters. 2006. From graphene to graphite: Electronic structure around the K point. *Physical Review B* 74 (7):075404.

Partoens, B. and F. M. Peeters. 2007. Normal and Dirac fermions in graphene multilayers: Tight-binding description of the electronic structure. *Physical Review B* 75 (19):193402.

Pease, R. 1952. An X-ray study of boron nitride. *Acta Crystallographica* 5 (3):356–361.

Peng, Xiangyang, and R. Ahuja. 2008. Symmetry breaking induced bandgap in epitaxial graphene layers on SiC. *Nano Letters* 8 (12):4464–4468.

Pereira, Vitor M., A. H. Castro Neto, and N. M. R. Peres. 2009. Tight-binding approach to uniaxial strain in graphene. *Physical Review B* 80 (4):045401.

Pereira, Vitor M., F. Guinea, J. M. B. Lopes dos Santos, N. M. R. Peres, and A. H. Castro Neto. 2006. Disorder induced localized states in graphene. *Physical Review Letters* 96 (3):036801.

Reich, S., J. Maultzsch, C. Thomsen, and P. Ordejón. 2002. Tight-binding description of graphene. *Physical Review B* 66 (3):035412.

Ribeiro, R. M., M. Pereira Vitor, N. M. R. Peres, P. R. Briddon, and A. H. Castro Neto. 2009. Strained graphene: Tight-binding and density functional calculations. *New Journal of Physics* 11 (11):115002.

Şahin, H., S. Cahangirov, M. Topsakal, E. Bekaroglu, E. Akturk, R. T. Senger, and S. Ciraci. 2009. Monolayer honeycomb structures of group-IV elements and III–V binary compounds: First-principles calculations. *Physical Review B* 80 (15):155453.

Son, Y.-W., M. L. Cohen, and S. G. Louie. 2006. Energy gaps in graphene nanoribbons. *Physical Review Letters* 97 (21):216803.

Suárez Morell, E., J. D. Correa, P. Vargas, M. Pacheco, and Z. Barticevic. 2010. Flat bands in slightly twisted bilayer graphene: Tight-binding calculations. *Physical Review B* 82 (12):121407.

Suárez Morell, E., M. Pacheco, L. Chico, and L. Brey. 2013. Electronic properties of twisted trilayer graphene. *Physical Review B* 87 (12):125414.

Sule, N. and I. Knezevic. 2012. Phonon-limited electron mobility in graphene calculated using tight-binding Bloch waves. *Journal of Applied Physics* 112 (5):053702.

Sun, L., Q. Li, H. Ren, H. Su, Q. W. Shi, and J. Yang. 2008. Strain effect on electronic structures of graphene nanoribbons: A first-principles study. *The Journal of Chemical Physics* 129 (7): 074704.

van de Walle, A., M. Asta, and G. Ceder. 2002. The alloy theoretic automated toolkit: A user guide. *Calphad – Computer Coupling of Phase Diagrams and Thermochemistry* 26 (4):539–553.

van de Walle, A. 2009. Multicomponent multisublattice alloys, nonconfigurational entropy and other additions to the Alloy Theoretic Automated Toolkit. *Calphad – Computer Coupling of Phase Diagrams and Thermochemistry* 33 (2):266–278.

Vanević, M., V. M. Stojanović, and M. Kindermann. 2009. Character of electronic states in graphene antidot lattices: Flat bands and spatial localization. *Physical Review B* 80 (4):045410.

Verberck, B., B. Partoens, F. M. Peeters, and B. Trauzettel. 2012. Strain-induced band gaps in bilayer graphene. *Physical Review B* 85 (12):125403.

Wakabayashi, K., M. Fujita, H. Ajiki, and M. Sigrist. 1999. Electronic and magnetic properties of nanographite ribbons. *Physical Review B* 59 (12):8271–8282.

Wallace, P. R. 1947. The band theory of graphite. *Physical Review* 71 (9):622.

Zhou, S. Y., G. H. Gweon, J. Graf, A. V. Fedorov, C. D. Spataru, R. D. Diehl, Y. Kopelevich, D. H. Lee, Steven G. Louie, and A. Lanzara. 2006. First direct observation of Dirac fermions in graphite. *Nature Physics* 2 (9):595–599.

4 Graphene Plasmonics
Light–Matter Interactions at the Atomic Scale

Pai-Yen Chen and Mohamed Farhat

CONTENTS

Abstract..63
4.1 Introduction 63
4.2 Graphene Surface Conductivity...64
4.3 Terahertz Wave Scattering, Propagation, and Guidance by Graphene...................................65
    4.3.1 Reflection, Transmission, and Scattering Properties of Graphene.................................66
    4.3.2 Surface Wave Excitation of Graphene Monolayer..69
    4.3.3 Graphene Parallel-Plate Waveguide Interconnect..70
4.4 Graphene-Based THz Plasmonic Nanoantenna 72
4.5 Graphene-Based THz Devices and Metamaterials 72
Conclusion 74
References..74

ABSTRACT

Graphene plasmonics, which studies the collective oscillation of massless Dirac fermions inside graphene, merges two vibrant fields of study: graphene physics and plasmonics. The propagation of surface plasmon polaritons (SPP) waves in the one-atom-thick graphene can be largely controlled by graphene's tunable surface conductivity via chemical doping or electrostatic gating. Graphene is a well-known material whose plasma frequency can range broadly from direct current (DC) to infrared, sensibly depending on the carrier density or Fermi level. The intriguing plasmonic properties of graphene open tremendous new possibilities in tunable and switchable novel terahertz (THz) and infrared optoelectronic devices, with features of compact size, ultrahigh speed, and low power consumption. In this chapter, we will review the theory and recent findings on graphene plasmonics. We present current advances and future applications of graphene plasmonics in the extreme manipulation of light at the nanoscale.

4.1 INTRODUCTION

Graphene is a single layer of sp²-bonded carbon atoms arranged in a hexagonal structure, which is effectively a 2-D version of the 3-D crystalline graphite. Ever since graphene has been discovered in 2004 [1], it has attracted wide attention due to its exotic electronic, thermal, mechanical, chemical, and optical properties, such as high carrier mobility, exceeding 20,000 cm²/Vs and Fermi velocity (υ_F) of 10^8 cm/s at low temperatures, quantum Hall effect, unexpectedly high opacity, and mechanical flexibility, among others [1–5]. In graphene, the linear energy–momentum dispersion relation is obtained over a wide range of energies such that electrons in graphene

behave as massless relativistic particles (Dirac fermions) with an energy-independent velocity. As expected, a high-quality graphene monolayer may exhibit a ballistic transport over at least submicron distances, as has been demonstrated in its two-dimensional (2-D) version: cylindrical carbon nanotube (wrapped graphene monolayer [6]). Graphene's exotic band structure and its finite density of state associated with its extreme thinness results in a pronounced ambipolar electric field effect and quantum-capacitance-limited nanodevices [1–5], of which the charge carriers can be tuned continuously between electrons and holes in concentrations as high as 10^{13} cm^{-2} [2] by the chemical doping or the electrostatic gating. Recently, it became possible to fabricate large-area mono- or few-layer graphene by the chemical vapor deposition (CVD) at low temperatures [7–9]. Graphene is considered to be one of the most promising material candidates for post-silicon nanoelectronic devices and interconnects [1–5], and as transparent electrodes for solar cells and flexible flat-panel displays [9].

Recently, the discovery of graphene's inherent plasmonic-like properties in the THz and infrared (IR) spectrum has further enriched this multifunctional nanomaterial platform [10–26]. The surface plasmon bounded to one-atom-thick graphene sheet shows numerous favorable properties, making graphene a promising alternative to typical plasmonic materials. Typically, plasmonics exploits the mass inertia of conduction band free electrons to create propagating charge density waves at the metal (semiconductor)–dielectric interface in the infrared and visible region [27–29], being particularly useful for overcoming the diffraction limit and applications in intra-chip signal transmission and information processing, ultra-sensitive optical biosensing, optical metamaterials, and ultimately, a synergistic bridge between nanoelectronics and nanophotonics [27–30]. However, the plasma frequency of most plasmonic materials, such as noble metals, typically falls into the visible and ultraviolet (UV) spectrum, which is also hardly tunable. Besides, noble metals have large Ohmic losses, which in turn reduce the lifetime of SPP excitation and the maximum field confinement. On the other hand, due to its finite free carrier density, graphene's plasma frequency can fall into the THz and IR spectrum. The extreme light–matter interaction at the atomic scale is followed by a significant wavelength shortening and dramatically enhanced local fields. Most interestingly, the dynamic tuning of the plasmon spectrum is possible through the control of carrier concentration or Fermi energy of graphene by the electrical, magnetic, or chemical means [10–18]. Besides, the large-area, low-defect, highly crystalline graphene is expected to raise the surface plasmon lifetime of graphene ideally up to hundreds of optical cycles, circumventing a major challenge of noble-metal plasmonics [17,20]. Last but not least, this atomically thin carbon layer has excellent electronic and optical properties, unprecedented mechanical strength, and thermal stability [1–3]. As a result, *graphene plasmonics* [10–26] has become a rapidly emerging field, particularly because of the gate-tunable plasmon resonance of massless Dirac fermions followed by controllable electromagnetic wave transitions. Graphene has been experimentally demonstrated to support the surface wave in the IR spectrum, with moderate loss and very strong field localization [24], which is envisioned to realize flatland transformation optics and metamaterials in the same spectrum [17,31].

Graphene plasmonics has become the subject of intensive research, with extremely promising applications for tunable and frequency-configurable optoelectronic devices [32,33], metamaterials [31,34–36], polarizers [19], invisibility cloaks [18,37,38], nanoantennas [22,39–43], phase shifters [39], photomixers [40], and oscillators [14], spanning the frequency from THz, far-infrared, to mid-infrared. In general, for those applications the ultrafast, low-power, and broadband modulations for amplitude and phase are potentially possible. In this chapter, we will briefly review the recent developments of theory and practical applications of graphene plasmonics and enhanced light–matter interactions in the THz and infrared spectrum.

4.2 GRAPHENE SURFACE CONDUCTIVITY

Typically, the one-atom-thick graphene is represented by a complex-valued surface conductivity $\sigma_s(\omega) = \sigma'_s + j\sigma''_s$ [10–13] derived from a microscopic quantum-dynamical model. We assume in the following an $e^{j\omega t}$ time dependence. In this scenario, the classical Maxwell's equations are solved exactly for an arbitrary electrical current. In infrared frequency range and below, the surface conductivity of graphene monolayer is sensitively dependent on Fermi energy (chemical potential), largely tuned either by the chemical doping [44] or external static electric field [24] (providing an isotropic scalar surface conductivity) or external static magnetic field via Hall effect [10–13] (providing anisotropic and tensor surface conductivity). In the absence of external magnetic field, graphene's surface conductivity, which includes both semi-classical

intraband conductivity σ_{intra} and quantum-dynamical interband conductivity σ_{inter} can be expressed as

$$\sigma_s(\omega, E_F, \tau, T) = \sigma_{intra} + \sigma_{inter}$$

$$= \frac{jq^2(\omega - j\tau^{-1})}{\pi\hbar^2}\left[\int_{-\infty}^{+\infty}\frac{|\varepsilon|}{(\omega - j\tau^{-1})^2}\frac{\partial f_d(\varepsilon)}{\partial \varepsilon}d\varepsilon\right.$$

$$\left.-\int_0^{+\infty}\frac{\partial f_d(-\varepsilon) - \partial f_d(\varepsilon)}{(\omega - j\tau^{-1})^2 - 4(\varepsilon/\hbar)^2}d\varepsilon\right], \quad (4.1)$$

where $f_d = 1/(1 + \exp[(\varepsilon - E_F)/(k_B T)])$ is the Fermi–Dirac distribution, ε is the energy, E_F is the Fermi energy, T is the temperature, q is the electron charge, $\hbar$ is the reduced Planck's constant, and τ is the momentum relaxation time associated with plasmon loss due to impurities and defects. The first and second terms in Equation 4.1 account for the intraband and interband contributions, respectively. In the THz region, σ_{intra} dominates over σ_{inter} and, therefore, the surface conductivity of graphene can be explicitly expressed as

$$\sigma_{intra} = -j\frac{q^2 K_B T}{\pi\hbar^2(\omega - j\tau^{-1})}\left[\frac{E_F}{K_B T} + 2\ln\left(1 + \exp\left(-\frac{E_F}{K_B T}\right)\right)\right].$$

$$(4.2)$$

For moderately doped graphene, that is, $|E_F| \gg k_B T$ and for frequencies far below the interband transition threshold, $\hbar\omega < 2|E_F|$, Equation 4.2 reduces to a Drude-type dispersion:

$$\sigma_{intra} \approx -j\frac{q^2 E_F}{\pi\hbar^2(\omega - j\tau^{-1})}. \quad (4.3)$$

The interband conductivity is on the order of $q^2/\hbar$. In general, σ_{inter} must be evaluated numerically, however, for $\hbar\omega, |E_F| \gg k_B T$, it can be approximated as [11–13]

$$\sigma_{inter} = -j\frac{q^2}{4\pi\hbar}\ln\left[\frac{2|E_F| - \hbar(\omega - j\tau^{-1})}{2|E_F| + \hbar(\omega - j\tau^{-1})}\right]. \quad (4.4)$$

At room temperatures and for frequencies below the THz regime, the interband conductivity is almost negligible when compared with the intraband contributions (4.2). The unique features of ballistic transport and ultrahigh electron mobility (in excess of 20,000 cm^2 V^{-1} s^{-1}) [2] may provide a large value of $|\sigma''/\sigma'|$. This implies that graphene can be considered as a purely reactive sheet at high frequencies (THz and far-infrared), playing a role analogous to a moderately lossy reactive frequency selective surface or metasurface at lower frequencies, without even the need of patterning the surface.

Figure 4.1 reports the complex-valued conductivity of graphene with different values of Fermi energy. It is seen from Figure 4.1 that the imaginary part of the sheet conductivity is

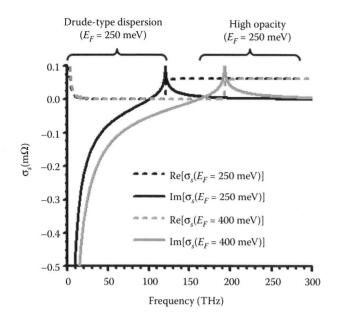

FIGURE 4.1 Frequency dispersion of sheet conductivity for a monolayer graphene, varying the Fermi energy. The results are calculated using Equation 4.1.

tunable with the Fermi energy, implying that the sheet reactance associated with the near-field stored energy is largely tunable. In the near-infrared and visible region ($\hbar\omega \gg 2|E_F|$), graphene's sheet conductivity obtained from Equation 4.1 is a real constant $q^2/4\hbar$ and a high opacity about 97.7% is obtained. For an N-layer graphene, one may assume $\sigma_{N\text{-}layer} = N\sigma_{mono}$ in the classical model, which is approximately valid for at least up to $N = 10$, as observed in experiments [45].

In the local Hall-effect regime, with a magnetic bias and possibly electrostatic bias, the surface conductivity is a tensor that can be explicitly expressed as [10,11,13]

$$\overline{\overline{\sigma}} = \begin{bmatrix} \sigma_{xx} & \sigma_{xy} \\ \sigma_{yx} & \sigma_{yy} \end{bmatrix} = \begin{bmatrix} \sigma_d(\mu_c(E_0), B_0) & \sigma_o(\mu_c(E_0), B_0) \\ -\sigma_o(\mu_c(E_0), B_0) & \sigma_d(\mu_c(E_0), B_0) \end{bmatrix},$$

(4.5)

where

$$\sigma_d(E_F(E_0), B_0) = -\frac{q^2 v_F^2 |eB_0|(\omega - j\tau^{-1})\hbar}{j\pi}$$

$$\sum_{n=0}^{\infty} \left\{ \frac{f_d(M_n) - f_d(M_{n+1}) + f_d(-M_{n+1}) - f_d(-M_n)}{(M_{n+1} - M_n)^2 - (\omega - j\tau^{-1})^2 \hbar^2} \right.$$

$$\times \left(1 - \frac{\Delta^2}{M_n M_{n+1}}\right) \frac{1}{M_{n+1} - M_n}$$

$$+ \frac{(f_d(-M_n) - f_d(M_{n+1}) + f_d(-M_{n-1}) - f_d(M_n))}{(M_{n+1} + M_n)^2 - (\omega - j\tau^{-1})^2 \hbar^2})$$

$$\left. \times \left(1 + \frac{\Delta^2}{M_n M_{n+1}}\right) \frac{1}{M_{n+1} + M_n} \right\},$$

(4.6)

$$\sigma_o(E_F(E_0), B_0) = -\frac{q^2 v_F^2 eB_0}{\pi}$$

$$\sum_{n=0}^{\infty} \left\{ [f_d(M_n) - f_d(M_{n+1}) - f_d(-M_{n+1}) + f_d(-M_n)] \times \right.$$

$$\times \left[\left(1 - \frac{\Delta^2}{M_n M_{n+1}}\right) \frac{1}{(M_{n+1} - M_n)^2 - (\omega - j\tau^{-1})^2 \hbar^2} + \right.$$

$$\left. \left. + \left(1 + \frac{\Delta^2}{M_n M_{n+1}}\right) \frac{1}{(M_{n+1} + M_n)^2 - (\omega - j\tau^{-1})^2 \hbar^2} \right] \right\},$$

(4.7)

and $M_n = \sqrt{\Delta^2 + 2n v_F^2 |qB_0| \hbar}$, and E_0 and B_0 are the external static electric and magnetic fields, respectively. Table 4.1 summarizes the graphene's conductivity at different operating frequencies and biasing conditions.

Typically, the Fermi energy of graphene can be tuned from −1 eV to 1 eV by an externally applied bias. Due to the electron–hole symmetry in the graphene band structure, both negative and positive signs of Fermi energy provide the same complex surface conductivity. The Fermi energy in a graphene monolayer is determined by the carrier density n_s as [6]

$$n_s = \frac{2}{\pi(\hbar v_F)^2} \int_0^{\infty} \varepsilon[f_d(\varepsilon - E_F) - f_d(\varepsilon + E_F)] d\varepsilon. \quad (4.8)$$

This implies that heavily doped polysilicon or metal gate positioned behind the insulating oxide may tune the Fermi energy and, therefore, the associated conductivity by applying a gate voltage. In this scenario, the 2-D surface charge density can be controlled by the displacement current on a charged surface $en_s = C_{ox}V_g$, where $C_{ox} = \varepsilon_{ox}/d_{ox}$ is the gate capacitance and d_{ox} is the oxide thickness.

4.3 TERAHERTZ WAVE SCATTERING, PROPAGATION, AND GUIDANCE BY GRAPHENE

This section starts with the terahertz wave reflection and transmission of electromagnetic waves at the graphene–dielectric interface, as well as the scattering of electromagnetic radiation by graphene-covered objects. It will follow with discussions on the surface wave excitation on the graphene sheet and guidance properties of the graphene parallel-plate waveguide, a graphene–dielectric–graphene structure. For simplicity, unless otherwise mentioned, $\tau = 1$ ps is assumed. This value is consistent with the ballistic transport features of graphene, whose mean free path was measured to be up to 500 nm at room temperature and larger than 4 μm at low temperatures [2]. Here, we assume that there is no external magnetic field and the conductivity is, therefore, isotropic without the Hall conductivity.

TABLE 4.1

Summary of Graphene's Conductivity at Different Operating Frequency and Bias Conditions

Frequency	Biasing Field	Conductivity
All frequency range	Electrostatic	$\sigma_s(\omega, E_F, \tau, T) = \sigma_{intra} + \sigma_{inter}$

$$= \frac{jq^2(\omega - j\tau^{-1})}{\pi\hbar^2}\left[\int_{-\infty}^{+\infty}\frac{|\varepsilon|}{(\omega - j\tau^{-1})^2}\frac{\partial f_d(\varepsilon)}{\partial\varepsilon}d\varepsilon - \int_{0}^{+\infty}\frac{\partial f_d(-\varepsilon) - \partial f_d(\varepsilon)}{(\omega - j\tau^{-1})^2 - 4(\varepsilon/\hbar)^2}d\varepsilon\right]$$

$\hbar\omega \ll E_F$	Electrostatic	$\sigma_s \simeq \sigma_{intra} \simeq -j\dfrac{q^2 K_B T}{\pi\hbar^2(\omega - j\tau^{-1})}\left[\dfrac{E_F}{K_B T} + 2\ln\left(1 + \exp\left(-\dfrac{E_F}{K_B T}\right)\right)\right]$

$$\approx -j\frac{q^2 E_F}{\pi\hbar^2(\omega - j\tau^{-1})}$$

$\hbar\omega > 2|E_F|$	Electrostatic	$\sigma_s \simeq \sigma_{inter} \simeq -j\dfrac{q^2}{4\pi\hbar}\ln\left[\dfrac{2|E_F| - \hbar(\omega - j\tau^{-1})}{2|E_F| + \hbar(\omega - j\tau^{-1})}\right].$
All frequency range	Magnetic/electrostatic	$\overline{\sigma} = \begin{bmatrix} \sigma_{xx} & \sigma_{xy} \\ \sigma_{yx} & \sigma_{yy} \end{bmatrix} = \begin{bmatrix} \sigma_d(\mu_c(E_0), B_0) & \sigma_o(\mu_c(E_0), B_0) \\ -\sigma_o(\mu_c(E_0), B_0) & \sigma_d(\mu_c(E_0), B_0) \end{bmatrix}$

$$\sigma_d(E_F(E_0), B_0) = -\frac{q^2 v_F^2 |eB_0|(\omega - j\tau^{-1})\hbar}{j\pi}$$

$$\sum_{n=0}^{\infty}\left\{\frac{f_d(M_n) - f_d(M_{n+1}) + f_d(-M_{n+1}) - f_d(-M_n)}{(M_{n+1} - M_n)^2 - (\omega - j\tau^{-1})^2\hbar^2}\times\left(1 - \frac{\Delta^2}{M_n M_{n+1}}\right)\frac{1}{M_{n+1} - M_n} + \right.$$

$$\left. + \frac{(f_d(-M_n) - f_d(M_{n+1}) + f_d(-M_{n-1}) - f_d(M_n))}{(M_{n+1} + M_n)^2 - (\omega - j\tau^{-1})^2\hbar^2}\times\left(1 + \frac{\Delta^2}{M_n M_{n+1}}\right)\frac{1}{M_{n+1} + M_n}\right\}$$

$$\sigma_o(E_F(E_0), B_0) = -\frac{q^2 v_F^2 eB_0}{\pi}\sum_{n=0}^{\infty}\left\{[f_d(M_n) - f_d(M_{n+1}) - f_d(-M_{n+1}) + f_d(-M_n)]\times\right.$$

$$\times\left[\left(1 - \frac{\Delta^2}{M_n M_{n+1}}\right)\frac{1}{(M_{n+1} - M_n)^2 - (\omega - j\tau^{-1})^2\hbar^2} + \right.$$

$$\left.\left. + \left(1 + \frac{\Delta^2}{M_n M_{n+1}}\right)\frac{1}{(M_{n+1} + M_n)^2 - (\omega - j\tau^{-1})^2\hbar^2}\right]\right\}$$

$$M_n = \sqrt{\Delta^2 + 2nv_F^2|qB_0|\hbar},$$

4.3.1 REFLECTION, TRANSMISSION, AND SCATTERING PROPERTIES OF GRAPHENE

The surface conductivity of graphene, which corresponds to a surface impedance $Z_s = 1/\sigma_s$, behaves like an inductive surface with moderate losses at THz and infrared frequencies, because of the small real part and negative imaginary part of its complex conductivity: $\text{Re}[\sigma_s] > 0$, $\text{Im}[\sigma_s] < 0$ (or $\text{Re}[Z_s] > 0$, $\text{Im}[Z_s] > 0$). Since a graphene monolayer can be described with the frequency-dependent complex sheet conductivity, the plane-wave reflection R and transmission T of a large-area graphene monolayer can be obtained by applying the two-sided impedance boundary conditions at the graphene–dielectric interface [25]. Figure 4.2a reports the reflectance $|R|^2$ and transmittance $|T|^2$ for a suspended, large-area graphene monolayer (much larger than the electron mean free path) on a silicon dioxide (SiO_2) substrate with permittivity $\varepsilon_{SiO2} = 4\varepsilon_0$, varying the Fermi energy of graphene. It is seen that a graphene monolayer is highly reflective on the air–SiO_2 interface at low-THz frequencies, in some sense similar to inductive metallic partially reflective surface (PRS) (i.e., an inductive metallic mesh-grid array at RF and microwaves [46–48]). A graphene sheet behaves like a "high-pass filter" at millimeter-wave and low-THz frequencies (see Figure 4.2a). Notice that the transparency of graphene (including both amplitude and phase of R and T) depends on its shiftable Fermi energy [24]. With this intriguing property, we may be able to make dynamically tunable graphene filters, cavities, or bandgap structures at THz and infrared frequencies.

Next, we consider a Fabry–Perot resonant cavity (see the inset of Figure 4.2b), which is made of a thin SiO_2 slab and two graphene sheets. The transmittance, reflectance, and absorption of this multilayer structure can be readily analyzed using the transmission-line circuit model and the transfer-matrix method [39,25,47], where a graphene sheet can be modeled as a shunt admittance $Y_s = \sigma_s$. Figure 4.2b presents the transmittance of this graphene–dielectric–graphene cavity; here, the Fermi energy of graphene $E_F = 0.4$ eV (gray solid line) and the thickness of SiO_2 layer $d = 25$ μm. The transmittance of the SiO_2 substrate without the graphene sheets (gray dash-dot line) is also calculated for comparison. It is observed in Figure 4.2b that the transmission resonance appears at lower frequencies compared to the typical Fabry–Perot resonance of dielectric slab. The graphene sheet behaves like a shunt loading inductive element, which introduces addition phase shift for reflected waves and, thus, reduces the electrical length of cavity. Figure 4.2b presents also a stacked graphene Fabry–Perot cavity with N dielectric films (here $N = 4$) and $N + 1$ graphene sheets (black solid line); here, the Fermi energy of graphene and the thickness of SiO_2 are fixed. It can be seen that when the number of stacked layers increases, N peaks are obtained for an N-identical-layers graphene cavity. Those transmission peaks at low frequencies ($\omega/2\pi < 1/\tau$) have a smaller magnitude, due to the serious plasmon loss. Notice that all transmission peaks are located in the passband region followed by a stop-band region. This can be understood from the physics in relation to the electromagnetic bandgap of a transmission line periodically loaded with shunt-connected reactances (Chapter 8 of Reference 49). In addition, since the surface reactance of graphene sensitively depends on Fermi energy, the resonant peaks are tunable by shifting graphene's Fermi level, resulting in a tunable and switchable THz Fabry–Perot cavity.

Recently, specific interest has been devoted to novel material platforms (plasmonic noble metals at infrared frequencies and visible, or graphene at THz and infrared frequencies), which may scatter and reradiate light in anomalous and exotic ways, providing new phenomena and applications. One of the most exciting applications of plasmonic materials is to achieve exotic transparency effects, or "hiding" an opaque object and making it effectively invisible to the electromagnetic radiation [18,36,37,50–54]. This passive camouflaging technique has the iconic term of *invisibility cloaking*. In addition to camouflaging and mirage applications, the cloaking technique has been proposed to improve near-field sensors and detectors, such that their presence may cause less disturbance to the environment in complex sensing networks and systems [50–54]. This "cloaking a sensor" concept may be useful in many setups, for example, near-field scanning optical microscopes (NSOM) or receiving antennas in the future sub-THz and THz sensing and communication networks [51–53]. Since most sub-diffractive measurements are usually performed in the very near-field of the details to be imaged, their accuracy is intrinsically limited by the disturbance introduced by the close proximity of the sensing instrument, that is, a sensing probe that may perturb the near-field distributions and influence the measurement [51–53]. Hence, a "cloaked" sensor with much reduced scattering may significantly improve sensing, detection, near-field measurements, and even communication technologies requiring low levels of interference and noise [51–53].

Plasmonic cloaking is a scattering cancellation technique based on a homogeneous layer with low- or negative-permittivity and/or permeability of plasmonic materials or

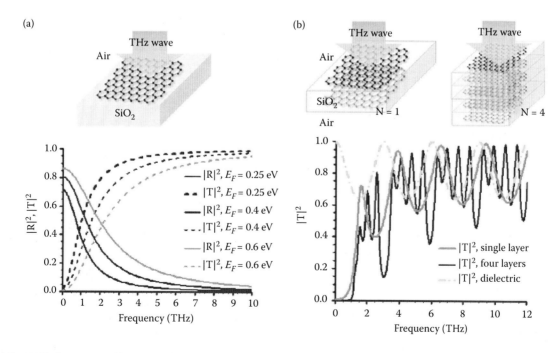

FIGURE 4.2 (a) Reflectance and transmittance versus frequency for a suspended graphene monolayer laid on a SiO_2 substrate (permittivity $\varepsilon_{SiO_2} = 4\varepsilon_0$), varying the Fermi energy. (b) The transmittance for a graphene–SiO_2–graphene Fabry–Perot cavity (gray solid line), cavity with four identical layers (black solid line), and a SiO_2 dielectric slab with thickness $d = 25$ μm (gray dashed dot line).

metamaterials, which may produce a local polarization vector that is in "*anti-phase*" with respect to that of the object to be cloaked. Therefore, a cancellation between the scattering of the object and of a properly designed plasmonic cover may restore the incident wavefront in the near- and far-field, and the cloaking effect is, in general, independent of the polarization and incidence angle of the impinging wave, as well as the position of the observer. A single homogeneous plasmonic layer, such as noble metal thin-film was shown to significantly suppress the scattering from a few multipolar scattering orders, allowing effectively cloaking objects on the scale of the wavelength of operation.

In the THz and far-infrared spectrums, the relevant plasmonic properties of graphene would be best suited to manipulate the scattering responses of graphene-covered objects. The physical principle behind such graphene cloak resides in a similar scattering cancellation effect of bulk plasmonic covers, of which the induced *surface current* on graphene surface is properly tailored to radiate "*anti-phase*" scattered fields [18,37]. The tunable sheet reactance of graphene, however, offers the unique advantage of being frequency-reconfigurable.

The inset of Figure 4.3a shows a graphene microtube (or single-walled carbon microtube [55]) realized by wrapping a graphene monolayer around a seamless cylinder [2], being designed for cloaking a dielectric cylinder, illuminated by a normally incident transverse magnetic (TM) plane wave. Here, we employ the Lorentz–Mie scattering theory and apply impedance boundary conditions for the graphene conductivity, forcing a discontinuity on the tangential magnetic field distribution on the graphene surface [18], which is proportional to the induced averaged surface current:

$$\mathbf{H}_{\tan}\big|_{r=a^+} - \mathbf{H}_{\tan}\big|_{r=a^-} = \frac{\hat{r} \times \mathbf{E}_{\tan}\big|_{r=a}}{\mathbf{Z}_s}. \qquad (4.9)$$

The scattering coefficients may be obtained by matching the tangential field components at the different boundaries and expressed as [53]

$$c_n^{TM} = -\frac{P_n^{TM}}{P_n^{TM} - jQ_n^{TM}}. \qquad (4.10)$$

The expressions for the nth order P_n^{TM} and Q_n^{TM} scattering coefficients may be found in Reference 53, as a function of the geometry and the sheet conductivity (impedance) of graphene. For an isotropic surface with negligible cross-polarization coupling, such as graphene, the total scattering width (SW), as a quantitative measure of the overall visibility of the object at the frequency of interest, is given by [53]

$$\sigma_s = \frac{4}{k_0} \sum_{n=-\infty}^{n=\infty} \left| c_n^{TM} \right|^2. \qquad (4.11)$$

In the quasi-static limit, the closed-form cloaking condition for a dielectric cylinder under the TM-polarized illumination can be derived as $X_{diel} = 2/[\omega a \varepsilon_0 (\varepsilon_r - 1)]$, where a is the radius of the cylinder. Obviously, an inductive surface will be required for effectively cloaking a moderate-size dielectric object.

Figure 4.3a presents the total scattering width for an infinite SiO$_2$ cylinder with diameter $2a = 40$ μm, covered by a graphene-wrapped microtube with different values of Fermi energy; here, an uncloaked SiO$_2$ cylinder is also presented for a fair comparison (gray dash line). From Figure 4.3a, we see how significant scattering suppression is achieved at specific frequencies in the THz spectrum, using a conformal, atomically thin graphene cloak. It is verified how the cloaking frequency may be widely tuned by varying the Fermi energy, realizing a tunable and switchable cloaking device. At specific frequencies, it is possible to vary the total scattering width by

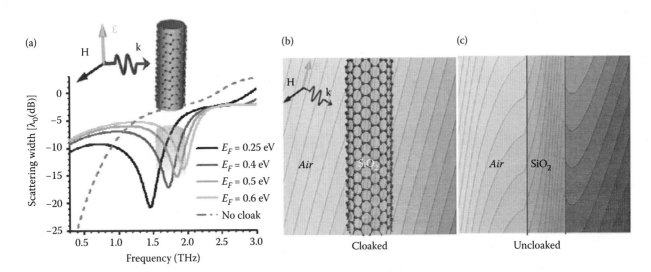

FIGURE 4.3 (a) Total scattering width of a SiO$_2$ cylinder with diameter $2a = 40$ μm, without and with THz cloak made of a graphene-wrapped microtube. Here the Fermi energy of graphene is varied to show its great tunability on design frequency. The phase of magnetic fields on the E plane (b) with and (c) without the graphene cloak with Fermi energy $E_F = 0.25$ eV at the design frequency $f_0 = 1.45$ THz.

over two orders of magnitude, simply through the variation of Fermi energy. Figure 4.3b and c presents, respectively, the near-field phase contours of the magnetic field on the E-plane for a cloaked ($E_F = 0.25$ eV) and uncloaked SiO_2 cylinder at the operating frequency $f_0 = 1.5$ THz. In Figure 4.3b and c, we consider an incident angle $\alpha = 60°$ (the angle between the incident wave direction and the cylinder axis; see the inset of Figure 4.3a). In both panels, the plane wave excites the geometry from the left side and contours are plotted on the same color scale. It is found that for all positions around the cylinder, the significant scattering reduction and restoration of original phase fronts are obtained, when compared with a pure dielectric cylinder. The graphene cloak shows the remarkable property of drastically suppressing the scattering even in the very near-field of the object. It is also worth noting that, similar to those plasmonic cloaking technique [51–53], the THz wave can penetrate the graphene cloak and object, thereby enabling applications in low-interference cloaked sensing and non-invasive probing in the THz and infrared spectrum. The multiband operation and cloaking larger objects may also be possible by using multiple graphene layers, which offer more degrees-of-freedom and allow suppressing a larger number of scattering orders. A theoretical discussion on the possibility of cloaking a dielectric sphere with multiband operation can be found in Reference 37. In addition, microstructured graphene (i.e., graphene patches or strips [38]) may tailor the surface reactance over a wide range capacitive or inductive value, depending on the geometry and dimensions of graphene microstructures. This may allow the effective cloaking of dielectric or conducting objects [38].

4.3.2 Surface Wave Excitation of Graphene Monolayer

The extremely confined surface wave sustained by graphene sheet may be of interest for passively guiding and actively gate-tuning/-modulating THz and infrared signals (with a graphene–insulator–gate configuration), dynamically tuning the dispersion relation of propagating surface plasmon waves with a voltage-controlled guiding wavelength. By matching the two-sided impedance boundary, the dispersion equation for transverse magnetic (TM) surface waves on a planar graphene layer in Figure 4.4a can be expressed as

$$\frac{\varepsilon_2}{\varepsilon_1}\sqrt{\beta^2 - k_1^2} + \sqrt{\beta^2 - k_2^2} + \frac{\sigma_s}{j\omega\varepsilon_1}\sqrt{\beta^2 - k_1^2}\sqrt{\beta^2 - k_2^2} = 0, \quad (4.12)$$

whereas for transverse electric (TE), the dispersion relation is

$$\frac{\mu_2}{\mu_1}\sqrt{\beta^2 - k_1^2} + \sqrt{\beta^2 - k_2^2} + j\omega\mu_2\sigma_s = 0, \quad (4.13)$$

where β is the complex phase constant of surface wave, k_i, ε_i, and μ_i are the wave number, permittivity, and permeability of the ith medium, respectively. The excitation of surface wave modes depends on both signs and value of imaginary part of graphene's conductivity. In general, only modes on the proper Riemann sheet may provide meaningful physical wave phenomena, and leaky modes on the improper sheet are used to approximate parts of the spectrum in restricted spatial regions and to explain certain radiation phenomena [11,13,35]. Consider a free-standing graphene in vacuum ($\varepsilon_1 = \varepsilon_2 = \varepsilon_0$, and $\mu_1 = \mu_2 = \mu_0$), the eigenmodal solutions for Equation 4.12 can be explicitly expressed as

$$\beta = \sqrt{k_0^2 - \left(\frac{2\omega\varepsilon_0}{\sigma_s^2}\right)^2}, \quad (4.14)$$

whereas the similar explicit solution for Equation 4.13 is

$$\beta = \sqrt{k_0^2 - \left(\frac{\omega\mu_0\sigma_s}{2}\right)^2}. \quad (4.15)$$

From Equations 4.14 and 4.15, we notice that if $Re[\sigma_s]$ is small and $Re[\sigma_s] < 0$, a TM-mode slow surface-wave exists at THz and far-infrared frequencies, at which the intraband conductivity

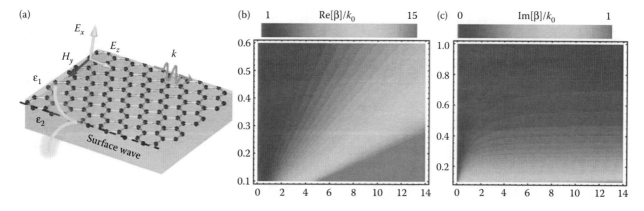

FIGURE 4.4 (a) Illustrative schematic of TM-mode surface plasmon polariton wave propagating along the graphene surface. Normalized (b) real and (c) imaginary parts of complex wave number for a TM-mode surface wave propagating on a free-standing graphene in vacuum.

dominates over interband contribution for a moderately doped graphene (see Figure 4.1). In this case, the strongly confined TM mode arises from the imaginary conductivity (see Figure 4.1). However, if Im[σ_s] > 0, where the interband transition dominates, a leaky mode on the improper Riemann sheet [11,13,35] is obtained for the TM-mode surface wave. On the other hand, if Im[σ_s] > 0, the TE-mode surface wave can exist in the mid-infrared region (see Figure 4.1). Unlike strongly localized TM-mode surface wave, TE-mode surface wave is, however, only weakly localized at the surface. Also, the plasmon losses and optical phonon scattering in the mid-infrared spectrum may limit the propagation length of TE-mode surface wave.

Figure 4.4b and c presents, respectively, contours of real and imaginary phase constant (normalized by the free space wave number) for TM surface wave, varying the operating frequency and Fermi energy of graphene. In the low frequency region, where $\omega \ll \tau^{-1}$ (i.e., microwaves), the TM surface wave is poorly confined to the graphene surface and relatively fast ($\beta \cong k_9$), where τ^{-1} is the phenomenological scattering rate. In the THz and far-infrared region, the surface wave is strongly confined and becomes slow ($\beta \gg k_0$), as its energy is concentrated in the near-field of graphene surface. We note that around the interband transition threshold ($\hbar\omega = 2|E_F|$), a moderate change in Fermi energy can dramatically change the sign of imaginary conductivity from negative to positive, thereby supporting the TE surface wave. In the range of $\tau^{-1} \ll \omega < 2|E_F|/\hbar$ (i.e., THz and far-infrared region), only a strongly confined TM surface wave can propagate. This TM mode is of particular interest, since it is dispersive with the Fermi energy in the frequency range of interest. This allows one to tune and modulate the dispersion of surface waves by electrical gating or chemical doping. In contrast, the TE mode is poorly confined to the graphene sheet and it is essentially non-dispersive. Since the TE and TM modes do not coexist, this allows one to make a broadband graphene polarizer, which supports the TE surface wave propagation for frequencies above the interband transition threshold [19].

4.3.3 GRAPHENE PARALLEL-PLATE WAVEGUIDE INTERCONNECT

Typically, a subwavelength parallel-plate plasmonic waveguide (i.e., metal–insulator–metal (MIM) heterostructure) may be used to further confine and control the surface plasmon modes. A low-loss, low-signal-dispersion THz interconnect will soon be required by the insatiable demand for high speed devices and wideband communication. Figure 4.5a illustrates a graphene-based THz interconnect, constructed with two graphene sheets with width much larger than separation distance d (a parallel-plate waveguide configuration with the propagation axis oriented along the z-axis). In many senses, the propagation properties of this graphene waveguide are analogous to the ones of the MIM optical waveguide supporting similarly confined surface modes, but suitable for sub-THz, THz, and infrared frequencies. For TM_z waves with magnetic field $H_x(y)\exp[j(\omega t - \beta z)]$, the complex eigenmodal phase constant β can be evaluated by solving the dispersion equation:

$$\tanh\left[\sqrt{\beta^2 - \omega^2\mu_0\varepsilon_2}\,\frac{d}{2}\right]\frac{\sqrt{\beta^2 - \omega^2\mu_0\varepsilon_2}}{\sqrt{\beta^2 - \omega^2\mu_0\varepsilon_1}}$$
$$= -\frac{\varepsilon_2}{\varepsilon_1 - j\sqrt{\beta^2 - \omega^2\mu_0\varepsilon_1}\,\dfrac{\sigma_s}{\omega}}, \qquad (4.16)$$

where ε_2 and ε_1 are the material permittivity inside and outside the waveguide, respectively. When the separation between the graphene monolayers is reduced to a deep-subwavelength scale $d \ll \lambda_0$, such graphene waveguide supports a quasi-TEM mode [12–39]. Under the long-wavelength approximation $d \ll \min(2\pi/|\omega\sqrt{\varepsilon_2\mu_0}|, 2\pi/|\omega\sqrt{\varepsilon_1\mu_0}|)$ and $\sigma_s\sqrt{\beta^2 - \omega^2\mu_0\varepsilon_2}/\omega\varepsilon_2$, $\sigma_s\sqrt{\beta^2 - \omega^2\mu_0\varepsilon_{out}}/\omega\varepsilon_{out} \gg 1$, a simple explicit dispersion relation can be obtained as

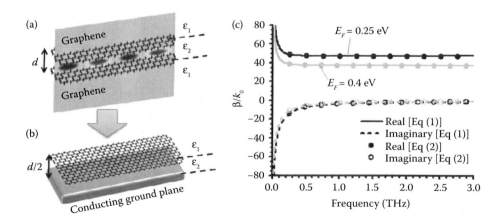

FIGURE 4.5 Schematics of (a) a graphene parallel-plate waveguide and (b) a simplified design inspired by the field symmetry of quasi-TEM mode. (c) The complex phase constant of a graphene waveguide in (b) with thickness $d/2 = 50$, supporting a quasi-TEM mode. Lines and symbols represent the eigenmodal solutions calculated by Equations 4.16 and 417, respectively.

$$\beta/k_0 \cong \sqrt{\frac{\varepsilon_2}{\varepsilon_0}\left(1 - j\frac{2}{\mu_0\omega\sigma_s d}\right)}. \qquad (4.17)$$

As expected, for sufficiently large conductivity, the permittivity of outer cladding layer has a negligible effect on the complex phase constant β, since the mode is tightly confined between the graphene layers. For the quasi-TEM mode that, in principle, has no cutoff frequency, the longitudinal field E_z is nonzero but very small compared to the uniform transverse field E_x, provided that the waveguide dimension (d) is very small. Due to the nature of field symmetry of the quasi-TEM mode, the graphene parallel-plate waveguide in Figure 4.5a can be replaced by a half-size waveguide in Figure 4.5b, of which a single graphene sheet is separated from a conducting ground plane by a half-thickness dielectric layer. Figure 4.5c presents the normalized phase constant β/k_1 for a graphene waveguide in Figure 4.5b, filled with a 50 nm-thick SiO_2 ($\varepsilon_2 = 4\varepsilon_0$ and $d/2 = 50$ nm) in a background of air ($\varepsilon_1 = \varepsilon_0$). It can be seen that in the low-THz region, the quasi-TEM mode is relatively non-dispersive, supporting a slow-wave propagation with strongly confined THz waves inside the waveguide, and a guiding wavelength much smaller than the free-space wavelength $\lambda_g \ll \lambda_0$, thanks to the large plasmonic properties (kinetic inductance) of graphene. In the microwave and millimeter-wave regions ($\omega \ll 1/\tau$), the value of the attenuation constant corresponding to Im[β] is quite high and the real phase constant is dispersive, due to the high plasmon losses. As a result, the group velocity $\partial\omega/\partial\beta$ is frequency-dependent, which usually causes the undesired signal dispersion. In the low frequency range, the graphene waveguide is basically a lossy RC transmission line (where the R and C are mainly contributed from the intrinsic plasmon loss of graphene and electrostatic capacitance yielded by the geometry, respectively). In some recent experiments, graphene-based transmission line has been demonstrated to be useful for making attenuators at RF and microwave frequencies [56]. For frequencies above sub-THz, the attenuation constant becomes small and the group velocity is almost constant without dispersion. In this case, a low-loss LC transmission line with minimized signal

attenuation and dispersion is obtained, where the inductance is mainly contributed by graphene's large kinetic inductance resulted from the collective electron-inertia effect), [6]. From Figure 4.5c, we note that the phase constant of graphene waveguide is tunable by shifting the Fermi level, enabling an adaptively tunable transmission line at sub-THz and THz frequencies.

The transmission line model and transfer matrix method can be used to evaluate the propagation characteristics of integrated graphene-based waveguide circuits, components, and systems. The characteristic impedance of the graphene waveguide shown in Figure 4.5 can be defined as $Z_c = -E_y/H_x = \beta/\omega\varepsilon_2$. It is found that the phase constant and characteristic impedance of this transmission line may be largely tuned from relatively high (with a low E_F) to low (with a high E_F). By combining the graphene waveguide in Figure 4.5a with two gates (or the structure in Figure 4.5b with a back gate), graphene's Fermi energy can be controlled by the applied gate voltage to dynamically tune the propagation constant, phase velocity, and local impedance. This is arguably the most significant advantage of graphene over conventional noble-metal plasmonic waveguides, providing an exciting venue to realize electronically programmable THz tunable transmission lines, which is an analog to a microwave tunable transmission line realized by mounting varactors, diodes, or transistors [49].

Figure 4.6 shows an integrally gated, tunable transmission line, based on the quasi-TEM graphene parallel-plate waveguide and three double gates. As an example to illustrate the potential of such tunable THz transmission line, the device in Figure 4.6 can be used as a loaded-line phase shifter designed for phased arrays [39], with low insertion loss, low return loss, and small phase error. In practical designs, the return loss is inherently present, due to the reflection at the mismatched loaded-line. The design in Figure 4.6 consists of a 3-bit phase shifter with 8 phase shift states: 0/45/90/135/225/270/315° [39]. When gate voltages are applied, the length of each loaded section must be a multiple of a half guided wavelength, in order to minimize the impedance mismatch and return loss. In order to create impedance matching for all binary states, the ith transmission line segment must satisfy the conditions: $\beta_{bias}l_i = \pi/2$

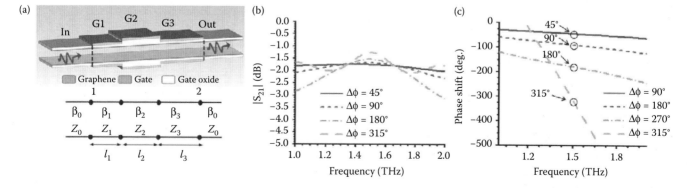

FIGURE 4.6 (a) Schematic diagram (top) and equivalent circuit (bottom) of a simplified digital (3-bit) loaded-line phase shifter [39]. (b) Magnitude transmission coefficients related to insertion loss and (c) phase shift between the biased and unbiased conditions for the load-line phase shifter shown in (a).

and $\beta_{unbias}l_i - \beta_{bias}l_i = \Delta\phi_i$, where l_i and $\Delta\phi_i$ are defined as the gate length and relative phase shift between the unbiased and biased conditions for the ith section, respectively. In this scenario, the total phase shift is $\Delta\phi = \Delta\phi_1 + \Delta\phi_2 + \Delta\phi_3$. In order to satisfy these conditions, the required Fermi energy and length for each bit need to be properly designed. For instance, if we want to produce phase shifts of 45° (bit 0), 90° (bit 1), and 180° (bit 2), the propagation constants must satisfy the relationships $\beta_1 = 4/5\beta_0$, $\beta_2 = 2/3\beta_0$, and $\beta_3 = 1/2\beta_0$, and the associated length of each line must satisfy the condition $\beta_{bias}l_i = \pi/2$. Figure 4.6b and c shows numerical results for the magnitude of S_{21} and phase shifts of 45°, 90°, 180°, and 315°, respectively. Those phase angles are obtained by applying the proper voltages to gate G1, G2, G3, and all, respectively. It is seen that good input return losses are obtained, with desired relative phase shifts at the design frequency $f_0 = 1.5$ THz. This graphene phase shifter may pave the way towards practical realization of sub-THz, THz, and infrared high-gain, steerable phased-array antennas with digital beam-forming and beam-steering functions.

The gated graphene waveguides, with tunable and strongly confined mode, may enable a number of chip-scale passive and active THz nanocircuit components, including switching, matching, coupling, power dividing/combining, and filtering devices. It can be envisioned that the integrally gated graphene transmission line may serve as a practical paradigm and platform for sub-THz and THz nano-communication, biosensors, and information processing systems on wafer, with *real-time* electro-optically tuning, phasing, and modulating functionalities [57–59]. The integration of several active graphene-based THz nanocircuit components, including a graphene nanoantenna discussed below, into a single entity will present a fundamental step towards design architectures and protocols for innovative THz communications, biomedicine, sensing, and actuation [57–59].

4.4 GRAPHENE-BASED THz PLASMONIC NANOANTENNA

A graphene nanoantenna is proposed to conduct high-speed data transformation with data rate up to 100 terabits per second. The idea of using graphene nanoantenna for THz communication was first proposed by Jornet and Akyildiz at the Georgia Institute of Technology [59]. The strongly confined surface wave on graphene sheet, with a wavelength much shorter than free space wavelength, may realize an ultracompact submicron graphene-strip antenna, which allows the oscillation of surface waves and broadcast radiation in the THz near-field communication systems [57]. As expected, the gate-tunable complex phase constant and local impedance of propagating waves on graphene may enable the active graphene-patch antennas or traveling-wave antennas, with dynamic beamforming and beamsteering ability [22,41–43]. Moreover, we may be able to build a graphene-based THz transmit or reflect array by properly modulating the phase shift of transmitted or reflected THz waves and the spatial phase distributions. This is achievable with arrays of gate-tuning graphene scatters. An active graphene transmit or reflect array may perform dynamic beamforming and beamsteering with high antenna directivity and narrow beamwidth.

4.5 GRAPHENE-BASED THz DEVICES AND METAMATERIALS

Metamaterials are artificially engineered composite materials which have electromagnetic properties that can not be found in nature, such as negative refractive index and magnetism above the THz frequency. Metamaterials usually gain their properties from their deep-subwavelength constituent inclusions designed to tailor the electric, magnetic, or both responses [60,61]. Plasmonic metamaterials, typically made of noble-metal nanostructures, further exploit surface plasmons to achieve optical properties not seen in nature, such as sub-diffraction imaging [62] and transformation optics devices [63]. THz plasmonic (graphene) metamaterials are a class of artificial electromagnetic materials that may realize exotic and anomalous electromagnetic properties at THz frequencies, where natural materials typically have weak responses. Potential applications include efficient generation of THz waves, sub-diffraction lenses, switches, modulators for phase and amplitude, beam-steering devices, and sensors. The THz metasurfaces are 2-D planarized version of metamaterials, composed of subwavelength inclusions over an otherwise homogeneous host surface. In this section, we combine the ideal properties of graphene and its tunability to realize tunable metamaterials and metasurfaces, and transformation optics operating in the THz spectrum, which may have potential impact in a variety of applications of interest.

Here, we present a THz magnetic metamaterial or metaferrite as an example to illustrate the great potential of graphene metamaterials. The idea of microwave metaferrites is realized using a high-impedance metasurface placed on top of a dielectric substrate backed by a conducting plane [64]. The electromagnetic properties of this grounded artificial slab are equivalent to a grounded magnetic slab, providing an interesting alternative to split-ring and other magnetic resonators at microwaves [60]. Figure 4.7a shows a graphene magnetic metamaterial or metaferrite designed to realize tunable low or negative effective permeability at THz frequencies. This metaferrite is composed of a graphene nano-patch metasurface placed on top of a dielectric substrate and backed by a metallic gate. We notice that according to the semi-empirical results [6], the bandgap of a graphene nanoribbon with width $a - g > 500$ nm is not open yet. Thus, the previously derived surface conductivity for a graphene monolayer is still valid. For an atomically thin graphene metasurfaces made of a nanostrips array (Figure 4.7a), its surface impedance $Z_{sheet} = R_{sheet} + jX_{sheet}$, ignoring the higher-order Floquet harmonics, has a simple, yet accurate expression [46–48]

$$Z_{sheet} = \frac{1}{\sigma_s}\frac{1}{1-g/a} + \frac{1}{j\omega\left(\dfrac{2\varepsilon_{avg}a}{\pi c\eta_0}\right)\ln[\csc(\pi g/2a)]f(\alpha)},$$

(4.18)

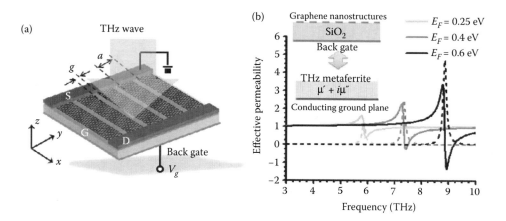

FIGURE 4.7 (a) THz metaferrite composed of parallel graphene nanostrips on top of a SiO₂ slab, backed by a conducting metal gate, which can be equivalent to a ferrite with relative permeability $\mu = \mu' - j\mu''$ backed by a conducting plane, as shown in the inset of (b). (b) Frequency dispersion of the effective permeability for a graphene metaferrite shown in (a) with different biased Fermi energy (solid line: real, dash line: imaginary). The thickness of SiO₂ is 1 µm, and the dimensions of nano-patch are $a = 4$ µm and $g = 0.5$ µm. (From P. Y. Chen, C. Argyropoulos, and A. Alù, *IEEE Trans. Antennas Propagat.* vol. 61, 1528–1537, © 2013 IEEE.)

where a and g are, respectively, the period and gap, c is the speed of light, η_0 is the intrinsic impedance of free space, $\varepsilon_{avg} = (\varepsilon_{substrate} + \varepsilon_{hostmedium})/2$ is the averaged permittivity of the upper and lower media, α is the incident angle, and $f_{TM}(\alpha) = 1$, $f_{TE}(\theta) = 1 - \sin^2\alpha/(2\varepsilon_{avg})$ are angular correlation functions corresponding to TM- and TE-incident waves, respectively. If we assume that the incident electric field is y-polarized, the incident planes for TM and TE incidence, respectively, correspond to the x–z and y–z planes. Consider now a thin grounded slab with thickness t and effective permeability $\mu = \mu' - j\mu''$, the input impedance, looking into the slab interface, is

$$Z_{in,metaferrite} = \eta_0 \sqrt{\mu' - j\mu''} \tanh\left(j\sqrt{\mu' - j\mu''}k_0 t\right). \quad (4.19)$$

By equating Equation 4.19 to the input impedance $Z_{in} = R_{in} + jX_{in}$ at the graphene–dielectric interface and taking the quasi-static limit (i.e., $\tanh(x) \approx x$), the effective real and imaginary parts of permeability for a graphene metaferrite slab in the inset of Figure 4.7b can be expressed as

$$\mu'(\omega) \approx \frac{X_{in}}{\eta_0 k_0 t},$$
$$\mu''(\omega) \approx \frac{R_{in}}{\eta_0 k_0 t}. \quad (4.20)$$

We may be now able to synthesize the desired values and frequency dispersion of the effective permeability of such grounded metaferrite slab by either the period and gap of nanostrips (which determines the electrostatic capacitance), materials and thickness of the thin dielectric slab (which determines the magnetic inductance), or the Fermi energy of graphene (which varies the surface resistance and kinetic inductance). Figure 4.7b shows the frequency dependency of effective permeability, retrieved using Equation 4.19, for

the metaferrite in Figure 4.7a under normal incidence. The solid and dashed lines represent the real and imaginary parts of effective permeability, respectively. The geometry of nanostrips are $g = 0.5$ µm and $a = 4$ µm, and the dielectric slab is made of SiO₂ with thickness $t = 1$ µm. Here, we note two main features of this graphene-based THz metamaterial: (i) the large kinetic inductance yielded by the extreme subwavelength confinement of surface plasmon along the structured graphene may help scaling the inclusion size to deep subwavelength; (ii) the largely tunable surface impedance controlled by the carrier concentration may be achieved with the back gate biasing. From Figure 4.7b, it is seen that the Lorenzian resonance and effective permeability can be tuned over a very broad range of values (from low or negative to positive). This tunable and switchable magnetic response is of interest for a variety of tunable THz and far-infrared magnetic metamaterial applications, such as the artificial magnetic conductor [65], ultracompact Salisbury absorbing films [66,67], polarizing filters [68], and selective thermal emitters or perfect absorber [68,69].

Recent progress in the growth and lithographic patterning of large-area epitaxial graphene [33–35] presents great opportunities for the commercialization of THz and infrared metamaterials, readily integrated with silicon-based plasmonics and photonics systems. Figure 4.8a shows a graphene nanopatch metamaterial recently released by IBM Corporation [35]. A wafer-scale, large-area graphene metamaterial have been successfully fabricated and its properties have been characterized. Figure 4.8b shows the measured rejection ratio spectrum. It is seen that by changing the dimensions of fabricated graphene patch and lattice constant, the resonant peak is tunable. Furthermore, the gate-tuning of surface plasmon waves on graphene ribbon has been experimentally demonstrated using the scatter-type scanning near-field microscopy [24]. The measurement is based on the feedback scheme taken from atomic force microscopy (AFM), of which the nano-manipulated nanotip positioned and scanned in the

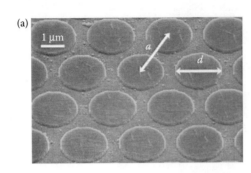

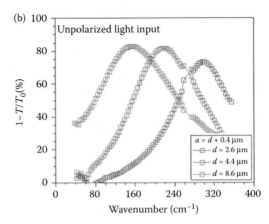

FIGURE 4.8 (a) Scanning electron microscopy (SEM) image for nanofabricated infrared metamaterial made of multilayer circular graphene patches [55]. (b) Measured electromagnetic wave rejection ratio of the graphene metamaterial in the mid-infrared and far-infrared spectrum. (From H. Yan et al., *Nat. Nanotech.* vol. 7, 330–334, 2012.)

immediate vicinity of graphene surface. The advancement of nanotechnology and nanofabrication will benefit the rapidly expanding field of graphene metamaterials and plasmonic devices. The strong local field confinement and gate-tuning standing-wave field patterns due to the interference in a finite-length graphene ribbon have been successfully characterized in Reference 24. The tunable electromagnetic response is one of the most exciting areas in current metamaterial research [70–80], since they may add a large degree of control and flexibility to the exotic right-/left-handed properties of metamaterials. In this case, instead of tuning wave properties using nonlinear loading, such as diodes and transistors at RF and microwaves, or nonlinear optical materials at optical frequencies, the gate-tuning of graphene plasmons may allow significantly larger tunability. In addition, extending these effects to the THz spectrum may provide new tools to push metamaterials in a frequency range that is more difficult to work in, compared to microwave or optical regimes.

Vakil and Engheta have theoretically shown that by designing and manipulating spatially inhomogeneous and nonuniform conductivity patterns across a graphene sheet (for instance by using a corrugated gate) [17], this material can be used as a flatland platform for THz and infrared transformation optics devices. In Reference 17, Vakil and Engheta present several numerical examples demonstrating a variety of infrared modulation and transformation functions. In Reference 31, a Fourier optics lens, transforming a broadcasting surface wave from source into a perfect planar wavefront on a single sheet of graphene, has been numerically studied in the mid-infrared region [17].

CONCLUSION

This chapter has addressed an emerging field of graphene plasmonics, which shows great potential to merge high-frequency electronics and photonics in the THz and infrared regions. We have presented some of the most recent and most representative innovations in this area. First, we have introduced the electrically/magnetically tunable conductivity and plasmonic resonances of graphene sheets at THz and infrared frequencies. We have presented the exotic light reflection, transmission, and scattering over an one-atom-thick graphene sheets in the THz and infrared regions. Then, we have overviewed the propagation of surface plasmon polaritons waves in graphene and graphene-based waveguides with a strong mode confinement and slow-wave propagation, which are possibly controlled by the electrostatic gating. In particular, we presented novel graphene-based THz/infrared optoelectronic devices, with adaptively tunable functions of phase-shifting, modulation, switching, radiation, and frequency and polarization filtering. In this chapter, we also presented some graphene-based metamaterials and transformation optical devices for future THz and infrared systems.

REFERENCES

1. K. S. Novoselov, A. K. Geim, S. V. Morozov, D. Jiang, Y. Zhang, S. V. Dubonos, I. V. Grigorieva, and A. A. Firsov, Electric field effect in atomically thin carbon films, *Science* vol. 306, 666–669, 2004.

2. A. K. Geim and K. S. Novoselov, The rise of graphene, *Nat. Mater.* vol. 6, 183–191, 2007.

3. A. H. C. Neto, F. Guinea, N. M. R. Peres, K. S. Novoselov, and A. K. Geim, The electronic properties of graphene, *Rev. Mod. Phys.* vol. 81, 109–162, 2009.

4. I. Meric, M. Y. Han, A. F. Young, B. Ozyilmaz, P. Kim, and K. L. Shepard, Current saturation in zero-bandgap, top-gated graphene field-effect transistors, *Nat. Nanotechnol.* vol. 3, 654–659, 2008.

5. Y. M. Lin, C. Dimitrakopoulos, K. A. Jenkins, D. B. Farmer, H. Y. Chiu, A. Grill, and Ph. Avouris, 100-GHz transistors from wafer-scale epitaxial graphene, *Science* vol. 327, 662, 2010.

6. H. S. P. Wong and D. Akinwande, *Carbon nanotube and graphene device physics*, Cambridge University Press, London, 2009.

7. S. Stankovicha, D. A. Dikina, R. D. Pinera, K. A. Kohlhaasa, A. Kleinhammesc, Y. Jiac, Y. Wuc, S. T. Nguyenb, and R. S. Ruoff, Synthesis of graphene-based nanosheets via chemical reduction of exfoliated graphite oxide, *Carbon* vol. 45, 1558–1565, 2007.

8. X. Li, W. Cai, J. An, S. Kim, J. Nah, D. Yang, R. Piner, A. et al. Large-area synthesis of high-quality and uniform graphene films on copper foils, *Science* vol. 324, 1312–1314, 2009.

9. K. S. Kim, Y. Zhao, H. Jang, S. Y. Lee, J. M. Kim, K. S. Kim, J. H. Ahn, P. Ki, J. Y. Choi, and B. H. Hong, Large-scale pattern growth of graphene films for stretchable transparent electrodes, *Nature* vol. 457, 706–710, 2009.

10. V. P. Gusynin, S. G. Sharapov, J. P. Carbotte, Magneto-optical conductivity in graphene, *J. Phys.Cond. Matter.* vol. 19, 026222, 2007.

11. G. W. Hanson, Dyadic Green's functions and guided surface waves for a surface conductivity model of graphene, *J. Appl. Phys.* vol. 103, 064302, 2006.

12. G. W. Hanson, Quasi-transverse electromagnetic modes supported by a graphene parallel-plate waveguide, *J. Appl. Phys.* vol. 104, 084314, 2008.

13. G. W. Hanson, Dyadic Green's functions for an anisotropic, non-local model of biased graphene, *IEEE Trans. Antenna Propagat.* vol. 56, 747–757, 2008.

14. L. A. Falkovsky and S. S. Pershoguba, Optical far-infrared properties of a graphene monolayer and multilayer, *Phys. Rev. B* vol.76, 153410, 2007.

15. F. Rana, Graphene terahertz plasmon oscillators, *IEEE Trans. Nanotech.* vol. 7, 91–99, 2008.

16. M. Jablan, H. Buljan, and M. Soljacic, Plasmonics in graphene at infrared, *Phys. Rev. B* vol. 80, 245435, 2009.

17. A. Vakil and N. Engheta, Transformation optics using graphene, *Science* vol. 332, 1291–1294, 2011.

18. P. Y. Chen and A. Alù, Atomically thin surface cloak using graphene monolayers *ACS Nano*, vol. 5, 5855–5863, 2011.

19. Q. Bao, H. Zhang, B. Wang, Z. Ni, C. Haley, Y. X. Lim, Y. Wang, D. Y. Tang, K. P. Loh, Broadband graphene polarizer, *Nat. Photonics*, vol. 5, 411–415, 2011.

20. F. H. L. Koppens, D. E. Chang, F. J. Garcia de Abajo, Graphene plasmonics: A platform for strong light–matter interactions, *Nano Lett.* vol. 11, 3370–3377, 2011.

21. D. Pile, Plasmonics: Graphene shrinks light, *Nat. Photonics*, vol. 7, 511, 2013.

22. I. Llatser, C. Kremers, A. Cabellos-Aparicio, J. M. Jornet, E. Alarcon, and D. N. Chigrin, Graphene-based nano-patch antenna for terahertz radiation, *Photon. Nanostruct: Fundam. Appl.* published on-line, 2012.

23. S. Thongrattanasiri, F. H. L. Koppens, and F. J. García de Abajo, Complete optical absorption in periodically patterned graphene, *Phys. Rev. Lett.* vol. 108, 047401, 2012.

24. J. Chen, M. Badioli, P. Alonso-González, S.Thongrattanasiri, F. Huth, J. Osmond, M. Spasenović et al. Optical nano-imaging of gate-tunable graphene plasmons, *Nature.* vol. 487, 77–81, 2012.

25. C. S. R. Kaipa, A. B. Yakovlev, G. W. Hanson, Y. R. Padooru, F. Medina, and F. Mesa, Enhanced transmission with a graphene–dielectric microstructure at low-terahertz frequencies, *Phys. Rev. B* vol. 85, 245407, 2012.

26. A. N. Grigorenko, M. Polini, and K. S. Novoselov, Graphene plasmonics, *Nat. Photonics.* vol. 6, 749–758, 2012.

27. S. A. Maier, *Plasmonics: Fundamentals and Applications*, Springer, New York, 2007.

28. S. A. Maier, M. L. Brongersma, P. G. Kik, S. Meltzer, A. A. G. Requicha, and H. A. Atwater, Plasmonics—A route to nanoscale optical devices, *Adv. Mater.*, vol. 13, 1501–1505, 2001.

29. A. V. Zayats, I. I. Smolyaninov, and A. A. Maradudin, Nano-optics of surface plasmon polaritons, *Phy. Rep.* vol. 408, 131–314, 2005.

30. E. Ozbay, Ekmel, Plasmonics: Merging photonics and electronics at nanoscale dimensions, *Science.* vol. 311, 189–193, 2006.

31. A. Vakil and N. Engheta, Fourier optics on graphene, *Phys. Rev. B* vol. 85, 075434, 2012.

32. Q. Bao and K. P. Loh, Graphene photonics, plasmonics, and broadband optoelectronic devices, *ACS Nano.* vol. 6, 3677–3694, 2012.

33. S. H. Lee, M. Choi, T. T. Kim, S. Lee, M. Liu, X. Yin, H. K. Choi et al. Switching terahertz waves with gate-controlled active graphene metamaterials, *Nat. Materials.* vol. 11, 936–941, 2012.

34. L. Ju, B. Geng, J. Horng, C. Girit, M. Martin, Z. Hao, H. A. Bechtel et al. Graphene plasmonics for tunable terahertz metamaterials, *Nat. Nanotech.* vol. 6, 630–634, 2011.

35. H. Yan, X. Li, B. Chandra, G. Tulevski, Y. Wu, M. Freitag, W. Zhu, P. Avouris, and F. Xia, Tunable infrared plasmonic devices using graphene/insulator stacks, *Nat. Nanotech.* vol. 7, 330–334, 2012.

36. P. Y. Chen and A. Alù, Terahertz metamaterial devices based on graphene nanostructures, *IEEE Trans. Terahertz Sci. Technol.* vol. 3, 745–756, 2013.

37. M. Farhat, C. Rockstuhl, and H. Bağcı, A 3D tunable and multi-frequency graphene plasmonic cloak, *Opt. Express.* vol. 21, 12592–12603, 2013.

38. P. Y. Chen, J. Soric, Y. R. Padooru, H. M. bernety, A. B. Yakovlev, and A. Alù, Nanostructured graphene metasurface for tunable terahertz cloaking, *New J. Phys.* vol. 15, 123029, 2013.

39. P. Y. Chen, C. Argyropoulos, and A. Alù, Terahertz antenna phase shifters using integrally-gated graphene transmission-lines, *IEEE Trans. Antennas Propagat.* vol. 61, 1528–1537, 2013.

40. P. Y. Chen and A. Alù, A terahertz photomixer based on plasmonic nanoantennas coupled to a graphene emitter, *Nanotech.* vol. 24, 455202, 2013.

41. E. Carrasco, J. Perruisseau-Carrier, Reflectarray antenna at terahertz using graphene, vol. 12, 253–256, 2013.

42. R. Filter, M. Farhat, M. Steglich, R. Alaee, C. Rockstuhl, and F. Lederer, Tunable graphene antennas for selective enhancement of THz-emission, *Opt. Express*, vol. 21, 3737–3745, 2013.

43. M. Tamagnone, J. S. Gómez-Díaz, J. R. Mosig, and J. Perruisseau-Carrier, Analysis and design of terahertz antennas based on plasmonic resonant graphene sheets, *J. Appl. Phys.* vol. 112, 114915, 2012.

44. F. T. Chuang, P. Y. Chen, T. C. Cheng, C. H. Chien, and B. J. Li, Improved field emission properties of thiolated multi-wall carbon nanotubes on a flexible carbon cloth substrate, *Nanotech.* vol. 18, 395702, 2007.

45. J. M. Dawlaty, S. Shivaraman, J. Strait, P. George, M. Chandrashekhar, F. Rana, M. G. Spencer, D. Veksler, and Y. Chen, Measurement of the optical absorption spectra of epitaxial graphene from terahertz to visible, *Appl. Phys. Lett.* vol. 93, 131905, 2008.

46. O. Luukkonen, C. Simovski, G. Granet, G. Goussetis, D. Lioubtchenko, A. V. Raisanen, and S. A. Treyakov, Simple and accurate analytical model of planar grids and high-impedance surfaces comprising metal strips or patches, *IEEE Trans. Antenna Propagat.* vol. 56, 1624–1632, 2008.

47. Y. R. Padooru, A. B. Yakovlev, C. S. R. Kaipa, F. Medina, and F. Mesa, Circuit modeling of multiband high-impedance surface absorbers in the microwave regime, *Phys. Rev. B* vol. 84, 035108, 2011.

48. L. B. Whitbourn and R. C. Compton, Equivalent-circuit formulas for metal grid reflectors at a dielectric boundary, *Appl. Opt.* vol. 24, 217–220, 1985.

49. D. M. Pozar, *Chapter 8, Microwave Engineering*, 2nd ed. John Wiley and Sons, New York, 1998.

50. A. Alù and N. Engheta, Achieving transparency with plasmonic and metamaterial coatings, *Phys. Rev. E* vol. 72, 016623, 2005.

51. A. Alù and N. Engheta, Cloaking a sensor, *Phys. Rev. Lett.* vol. 102, 233901, 2009.

52. M. Farhat, P. Y. Chen, S. Guenneau, S. Enoch, R. McPhedran, C. Rockstuhl, and F. Lederer, "Understanding the functionality of an array of invisibility cloaks," *Phys. Rev. B* vol. 84, 235105, 2011.

53. P. Y. Chen, J. Soric, and A. Alù, Invisibility and cloaking based on scattering cancellation, *Adv. Mater.* vol. 24, 281–304, 2012.

54. J. C. Soric, P. Y. Chen, A. Kerkhoff, D. Rainwater, K. Melin, and A. Alù, Demonstration of an ultralow profile cloak for scattering suppression of a finite-length rod in free space, *New J. Phys.* vol. 15, 033037, 2013.

55. J. Q. Hu, Y. Bando, F. F. Xu, Y. B. Li, J. H. Zhan, J. Y. Xu, and D. Golberg, Growth and field-emission properties of crystalline, thin-walled carbon microtubes, *Adv. Mater.* vol. 16, 153–156, 2004.

56. H. S. Skulason, H. V. Nguyen, A. Guermoune, V. Sridharan, M. Siaj, C. Caloz, and T. Szkopek, 110 GHz measurement of large-area graphene integrated in low-loss microwave structures, *Appl. Phys. Lett.* vol. 99, 153504, 2011.

57. S. Abadal, E. Alarcón, M. C. Lemme, M. Nemirovsky, A. Cabellos-Aparicio, Graphene-enabled wireless communication for massive multicore architectures, *IEEE Communications Magazine*, 137–143, 2012.

58. P. Y. Chen and A. Alù, All-graphene terahertz analog nanodevices and nanocircuits, *Proceedings of 7th European Conference on. Antennas and Propagation (EuCAP)*, 697–698, Gothenburg, Sweden, April 8–11, 2013.

59. J. M. Jornet and I. F. Akyildiz, "Graphene-based nano-antennas for electromagnetic nanocommunications in the terahertz band," *Proceedings of 4th European Conference on Antennas and Propagation (EuCAP)*, 1–5, Barcelona, Spain, April 12–16, 2010.

60. J. B. Pendry, A. J. Holden, D. J. Robbins, and W. J. Stewart, Magnetism from conductors and enhanced nonlinear phenomena, *IEEE Trans. Microwave Theory Tech.* vol. 47, 2075–2084, 1999.

61. R. Shelby, D. R. Smith, and S. Schultz, Experimental verification of a negative index of refraction, *Science* vol. 292, 77–79, 2001.

62. H. J. Lezec, J. A. Dionne, and H. A. Atwater, Negative refraction at visible frequencies. *Science* vol. 316, 430–432, 2007.

63. W. Cai, U. K. Chettiar, A. V. Kildishev, and V. M. Shalaev, Optical cloaking with metamaterials, *Nat. Photonics*, vol. 1, 224–227, 2007.

64. D. J. Kern, D. H. Werner, and M. Lisovich, Metaferrites: Using electromagnetic bandgap structures to synthesize metamaterial ferrites, *IEEE Trans. Antenna Propagat.* vol. 53, 1382–1389, 2005.

65. H. Mosallaei and K. Sarabandi, Antenna miniaturization and bandwidth enhancement using a reactive impedance substrate, *IEEE Trans. Antenna Propagat.* vol. 52, 2403–2414, 2004.

66. N. Engheta, Thin absorbers using space-filling-curve high-impedance surfaces, *IEEE AP-S International Symposium*, San Antonio, Texas. Jun. 16–21, 2002.

67. R. Alaee, M. Farhat, C. Rockstuhl, and F. Lederer, A perfect absorber made of a graphene micro-ribbon metamaterial, *Opt. Express* vol. 20, 28017–28024, 2012.

68. P. Y. Chen and A. Alù, Integrated infrared nanodevices based on graphene monolayers, *Proceedings of 2012 IEEE Antennas and Propagation Society International Symposium*, Chicago, Illinois, July 8–14, 2012.

69. X. Liu, T. Tyler, T. Starr, A. F. Starr, N. M. Jokerst, and W. J. Padila, Taming the blackbody with infrared metamaterials as selective thermal emitters, *Phys. Rev. Lett.* vol. 107, 045901, 2011.

70. N. Zheludev, The road ahead for metamaterials, *Science* vol. 328, 582–583, 2010.

71. S. OBrien, D. McPeake, S.A. Ramakrishna, and J. B. Pendry, Near-infrared photonic band gaps and nonlinear effects in negative magnetic metamaterials, *Phys. Rev. B* vol. 69, 241101(R), 2004.

72. I. V. Shadrivov, S. K. Morrison, Y. S. Kivshar, Tunable split-ring resonators for nonlinear negative-index metamaterials, *Opt. Exp.* vol. 14, 9344, 2006.

73. P. Y. Chen and Alù, Optical nanoantenna arrays loaded with nonlinear materials, *Phys. Rev. B* vol. 82, 235405, 2010.

74. P. Y. Chen, M. Farhat, and A. Alù, Bistable and self-tunable negative-index metamaterial at optical frequencies, *Phys. Rev. Lett.* vol. 106, 105503, 2011.

75. M. Amin, M. Farhat, and H. Bagci, A dynamically reconfigurable Fano metamaterial through graphene tuning for switching and sensing applications, *Sci. Rep.* vol. 3, 2105, 2013.

76. A. A. Oliner and T. Tamir, Guided complex wave, part I: Field at an interface, *Proc. IEE* vol. 110, 310–324, 1963.

77. A. B. Yakovlev, G. W. Hanson, and A. Mafi, High-impedance surfaces with graphene patches as absorbing structures at microwaves, *Proceedings of 3rd International Congress on Advanced Electromagnetic Materials in Microwaves and Optics*, 782–784, London, September 1–4, 2009.

78. P. Y. Chen, H. Huang, D. Akinwande, A. Alù, Graphene-based plasmonic platform for reconfigurable terahertz nanodevices, *ACS Photonics* vol. 1, 647–654, 2014.

79. P. Y. Chen, M. Farhat, A. N. Askarpour, M. Tymchenko, and A. Alù, Infrared beam-steering using acoustically modulated surface plasmons over a graphene monolayer, *J Opt.* vol. 16, 094008, 2014.

80. P. Y. Chen, M. Farhat, and H. Bağcı, Graphene metascreen for designing compact infrared absorbers with enhanced bandwidth, *Nanotechnol.* vol. 26, 164002, 2015.

5 Graphene/Polymer Nanocomposites
Crystal Structure, Mechanical and Thermal Properties

Fabiola Navarro-Pardo, Ana Laura Martínez-Hernández, and Carlos Velasco-Santos

CONTENTS

Abstract 77
5.1 Introduction 77
5.2 Polymer Nanocomposites 78
    5.2.1 Modification of Graphene Fillers 78
    5.2.2 Methods of Nanocomposite Preparation 79
5.3 Crystallization Behavior 79
    5.3.1 Isothermal Crystallization 81
    5.3.2 Non-Isothermal Crystallization 82
    5.3.3 Effects on the Crystallinity Degree 84
    5.3.4 Crystal Modifications 85
    5.3.5 Hybrid Crystalline Structures 86
5.4 Mechanical Properties 86
    5.4.1 Effects of Filler Loading 86
        5.4.1.1 Models 89
    5.4.2 Influence of Crystallinity Properties 89
    5.4.3 Other Factors Influencing Mechanical Response 90
5.5 Thermal Properties 91
    5.5.1 Glass Transition Temperature 91
    5.5.2 Thermal Stability 92
    5.5.3 Thermal Conductivity 93
5.6 Concluding Remarks 93
References 94

ABSTRACT

A wide diversity of polymeric matrices have been studied with graphene as reinforcement; in this chapter we paid particular attention to semicrystalline polymers. The presence of a second phase in this type of polymers has the ability to induce crystallization. Knowledge of the crystallization behavior is essential to understand the structure–property relationships in nanocomposites. Furthermore, the crystallinity properties influence the mechanical response of graphene/polymer nanocomposites. The improvements in mechanical properties are also ascribed to the extremely large surface area of graphene. The mechanical performance is significantly dependent on the dispersion and the interface between the graphene sheets and the matrix. Additionally, several investigations have indicated the influence of these parameters on the thermal properties of nanocomposites. Extensive research on different graphene/polymer nanocomposite systems has been conducted including a wide range of properties; however, thermal and mechanical properties related to semicrystalline polymer nanocomposites and their crystal structure are the main issues presented in this chapter.

5.1 INTRODUCTION

The one-atom-thick and two-dimensional (2D) layers of sp^2-bonded carbon atoms of graphene confer this nanomaterial unique properties. Measurements of a single layer of graphene have shown a high intrinsic mechanical strength of 130 GPa, Young's modulus of 1 TPa, and a breaking strength of 42 N/m [1]. A 30-in. monolayer graphene film offers sheet resistances as low as ~125 Ω/sq with 97.4% optical transmittance [2,3]. Regarding thermal properties, the room temperature values of the thermal conductivity reach up to 5.30×10^3 W/m K [4]. These properties make graphene an excellent candidate as advanced reinforcing filler for high-strength, lightweight, and functional polymer nanocomposites [5–8].

There is a wide variety of techniques for synthesizing graphene; Table 5.1 shows the most used approaches by

TABLE 5.1

Commonly Used Methods for Obtaining Graphene Sheets

Method	Characteristics	References
Micromechanical cleavage of graphite and chemical vapor deposition	Offer the ability to obtain high-quality and defect-free graphene nanosheets (GNs)	[8]
Chemical reduction of GO	This is the most promising route to obtain chemically reduced GO (CRGO) due to its simplicity, suitability for high-volume production, and relatively low material cost	[9,10]
Thermal reduction of GO	The high density of defects in thermally reduced GO (TRGO) confers this material greater specific surface area and higher overall heterogeneous electron transfer rates as compared with other forms of reduced GO	[11]

different research groups. Among them, a strategy extensively employed is the chemical modification of graphite through oxidative routes for its subsequent exfoliation to produce graphene oxide (GO) followed by the reduction method [12]. Graphite can also be chemically modified with organic solvents involving a sonication step to produce unoxidized and defect-free suspended graphene sheets (SGs) [9,10,12]. However, this method is less popular because it can only be employed in matrices which can be dissolved in a suspension [13]. A number of works and reviews describe other methods for obtaining graphene sheets [6,9,11,12]. However, the various types of materials described above are the most commonly used in polymer nanocomposites.

Addition of low loadings of these fillers into polymers has shown significant enhancements in mechanical properties, accompanied by the modification of functional properties such as electrical conductivity, thermal conductivity, and barrier behavior [6,14,15]. The crystallinity in the host polymer matrix is also influenced by the addition of these nanomaterials [16–18]. The polymer nanocomposite properties are deeply related to the features of the constituent phases, dispersion, and to the interfacial interaction between the graphene basal plane and the matrix [5,19]. The latter two depend on physicochemical properties of the polymer and graphene, the nanofiller content, the nanocomposite production technique, and the processing conditions during fabrication [20,21].

A few research groups have shown comparative studies between carbon nanotubes (CNTs) and graphene-based materials [18,22–27]. The advantages found in graphene when compared with CNTs are the very high surface area providing higher adhesion with the polymer host [28], higher surface-to-volume ratio due to the inaccessibility of the inner nanotube surface to polymer chains [29], and the fact that graphene can be produced from graphite which is more available and less expensive [30].

Graphene/polymer nanocomposites have potential applications mainly in automotive and aerospace fields [31–35]. Diverse studies have revealed they can also be used in sporting goods, packaging, sensors, actuators, and optical components for electronic devices [2,3,21,30,36,37]. Biomedical applications of graphene have also attracted intensive attention [38].

This chapter contains a brief explanation about the different modifications of graphene fillers used for reinforcement of semicrystalline matrices and the methods of nanocomposite preparation. Special attention to the crystallization behavior

provided by these nanofillers was made. The mechanical and thermal properties found in semicrystalline polymers considering the diverse parameters involved in their response are also discussed in detail.

5.2 POLYMER NANOCOMPOSITES

Polymeric materials with improved properties have always been a topic of interest for several research groups [5–8]. In addition, the reinforcement of polymers with materials of nanometric dimensions has brought extraordinary benefits in the final properties of nanocomposites [39,40]. There is extensive research regarding the study of the effects provided by nanomaterials in a wide variety of polymer matrices [28–52]. In amorphous polymer-based nanocomposites, the polymer molecules can either be in the amorphous bulk matrix or confined within the nanomaterials depending on the interactions between the matrix and them [41]. Semicrystalline polymers are composed of crystalline lamellae embedded into an amorphous phase. Lamellar crystals usually have thicknesses of the order of tens of nanometers, whereas the lateral dimensions are much larger [42]. The incorporation of a second nanometric phase as graphene allows this filler to act as a nucleus and it can also induce the polymer lamellae growth on its surface [43]. The crystallinity features also have influence on other properties such as mechanical response, thermal stability, and thermal conductivity [44,45,53].

5.2.1 MODIFICATION OF GRAPHENE FILLERS

Pristine graphene is characterized for having low compatibility with most polymers as well as for having strong van der Waals interactions among graphene sheets; these features make the filler dispersion not individual and homogeneous [46]. GO offers many advantages over pristine graphene due to the facility to exfoliate and disperse in polar polymer matrices; Table 5.2 lists some of the polymers commonly used in graphene nanocomposites. Among the disadvantages of GO are the significant number of defects in the platelets and their inferior mechanical properties when compared with GNs [10]. In addition, GO requires surface modification to disperse in nonpolar polymers. However, the variety and use of these polymers is limited, as it can be seen from Table 5.3. To achieve a better affinity between the nanocomposite components, several methods have been developed to covalently

TABLE 5.2

Examples of Polar Polymers Used in Graphene Nanocomposites

Polymer	Abbreviation	References
Ethylene-vinyl acetate copolymer	EVA	[24]
Chitosan	–	[50,54–56]
Nafion	–	[57]
Polyamide	PA	[18,22,47,58–66]
Polyaniline	PANI	[21,27,37,44,67,68]
Polybutylene terephthalate	PBT	[69]
Polybutylene succinate	PBS	[70]
Polycaprolactone	PCL	[23,38,58,71,72]
Polyethylene glycol	PEG	[73]
Polyethylene terephthalate	PET	[74,75]
Poly-3-hydroxybutyrate	PHB	[76]
Poly-3-hydroxybutyrate-co-4-hydroxybutyrate	PHBV	[77]
Polyimides	PI	[34,78,79,80]
Polylactic acid	PLA	[81,82]
Poly-L-lactic acid	PLLA	[11,17,25,73,83–89]
Poly(D, L-lactic-co-glycolic acid)	PLGA	[90]
Polyvinyl alcohol	PVA	[15,46,49,91–99]
Polyvinylidene fluoride	PVDF	[47,48,100–103]
Polyurethane	PU	[104–110]
Polyvinyl chloride	PVC	[111]
Thermoplastic polyester elastomer	TPEE	[112]

or non-covalently functionalize the graphene sheets [53,137]. Among the species employed are organic molecules [138,139], polymers [37,54,116,124,125,140–142], and inorganic particles [139,141]. Polymer modification or the use of polymer blends has also been found to be efficient to promote and compatibilize adhesion between fillers and polymers [73,74,116,117,126].

For the sake of homogenizing the diversity of the nomenclature employed for graphene fillers in the literature explored in this chapter, the terms were simplified in the following manner: GNs for the pristine single layers or few layer

TABLE 5.3

Examples of Nonpolar Polymers Used in Graphene Nanocomposites

Polymer	Abbreviation	References
Natural rubber	NR	[113]
Poly-3-butylthiophene	P3BT	[114]
Poly-3-hexylthiophene	P3HT	[45]
Poly-p-phenylene benzobisoxazole	PBO	[115]
Polyethylene	PE	[51,116–123]
Polypropylene	PP	[14,16,38,52,74,89,124–133]
Polyphenylene oxide	PPO	[125]
Polystyrene	PS	[134]
PVDF[a]		[135,136]

[a] Nonpolar in the α-form but polar in the β- and γ-form [136].

graphene. GO, CRGO, TRGO, and SGs for the forms of graphene previously described (Table 5.1). FGSs (functionalized graphene sheets) will be used for any type of functionalized graphene sheets. Some works have further attached polymers to graphene; therefore, the polymers employed are indicated in these cases, i.e. PMMA–FGs.

5.2.2 METHODS OF NANOCOMPOSITE PREPARATION

Polymer nanocomposites can be basically obtained from three methods: solution mixing, in situ polymerization, and melt blending [5,40]. Table 5.4 shows these techniques and a variety of works that have used these approaches to obtain nanocomposites and Table 5.5 shows a variety of nanocomposites obtained by a combination of different methods. As indicated in this table, the majority of research in graphene-based polymer nanocomposites is prepared by solution mixing; this method is an ideal strategy for achieving uniform dispersion between both phases [140]. Good dispersion has also been obtained through in situ polymerization which furthermore enables control over the polymer architecture and the structure of the composites [75]. On the other hand, it is highly desirable to eliminate the need for solvents. Consequently, alternative methods have been developed [22,104]. Melt compounding is a more economic technique, compatible with industrial processes and free of harmful or dangerous solvents [81,127]. In some cases, this method helps breaking the graphene agglomerates through the high shear forces created during processing [22]. There are several studies and reviews involving different approaches for preparing graphene-based nanocomposites indicating the pros and cons of them and hence are not discussed further [7,8,58].

5.3 CRYSTALLIZATION BEHAVIOR

Polymer crystallization is affected by the presence of graphene due to its heterogeneous nucleation effect for crystal growth. Crystallization induced by carbon nanofillers play a crucial

TABLE 5.4

Preparation Methods of Graphene/Polymer Nanocomposites

Method		References
In situ polymerization		[14,21,60,61,67,69,84,104–107,110,143]
Solution mixing	Casting	[15,27,34,46–49,54–56,78,83,85–87,91–94,96–99,108,109,111,119,121,134,136,144,145]
	Coagulation	[11,16,25,52,58,70,76,79,85,88,118,129,130]
	Electrospinning	[18,38,66,90,137]
Melt blending	Compression molding	[116,124,132]
	Extrusion	[112,117,127]
	Melt spinning	[22,59]

TABLE 5.5
Graphene Nanocomposites Obtained by the Combination of Different Methods

Polymer	Nanofiller	Method of Preparation	References
PA6	FGs	In situ polymerization/melt spinning	[62]
PA6	GO	In situ polymerization/melt spinning	[64]
PA6	FGs	In situ polymerization/melt spinning	[59]
PA6	FGs	In situ polymerization /extrusion	[63]
PANI	GNs	In situ polymerization/compression molding	[44]
PBO	TRGO	In situ polymerization/dry–wet spinning	[115]
PET	FGs	In situ-melt polymerization	[75]
PI	FGs	In situ polymerization/casting	[80]
PLLA	FGs	In situ polymerization/casting	[17]
PVDF	FGs	In situ polymerization/casting	[100]
PVDF	FGs	In situ polymerization/casting	[101]
PVDF	FGs	In situ polymerization/casting/annealing	[103]
NR	FGs	Coagulation/extrusion/compression molding	[113]
Liquid crystalline polymer (LCP)	TRGO	Casting/compression molding	[20]
PCL	GO	Casting/compression molding	[71]
PE	FGs	Casting/extrusion	[123]
PE	FGs	Casting/compression molding	[51]
PE	GO	Casting/extrusion	[117]
PLLA	GO	Casting/extrusion	[146]
PP	TRGO	Solution blending/extrusion	[133]
PP/PPO	FGs	Coagulation/extrusion/injection molding	[125]
PVA	FGs	Casting/compression molding	[33]
PA12	FGs	Extrusion/compression molding	[65]
PE	TRGO	Extrusion/compression molding	[121]
PET/PVA	GNs	Extrusion/injection molding	[74]
PLA	TRGO	Extrusion/compression molding	[81]
PP	FGs	Extrusion/compression molding	[126,140]

role in optimizing the macroscopic performance of polymers, such as improving the interfacial adhesion and stress transfer [147]. The dimensional effects of CNTs and graphene on the crystallization of alkane melts have been studied by molecular dynamics (MD) simulations [26]. It was found that both nanostructures provide the nucleation sites for the alkane molecules to become an ordered structure. Furthermore, the effect of the graphene layers on the chain dynamics as a function of chain molecular weight is also very important for the structure–property relationships. Rissanou et al. have simulated the effect of graphene on the mobility of polymer chains [148]. The mechanism of polymer crystallization has been explained as an adsorption step and an orientation step [149,150]. In the first step, the polymer chain begins to attach to the substrate surface. This is followed by the adsorption of more and more CH_x groups onto graphene substrate. As the adsorption progresses, the molecular order in the adsorbed polymer chain gradually grows. The second step involves chain adjustments and the formation of the crystalline structure [150]. Yang et al. [149] found that the PE (polyethylene) chains adsorbed onto the graphene surface adopted a preferred parallel orientation to the graphene plane in the layer-by-layer crystallization of polymer. Additionally, crystallization is found to be strongly dependent on temperature. Figure 5.1 shows the behavior of PE

chains upon different temperature. A PE multilayered structure develops at lower temperatures and a single-layer structure develops at higher temperatures [150]. Another factor that has a profound effect on polymer crystallization is the confinement provided by nanomaterials. Eslami et al. have used a simulation model that describes well the confinement of polymers by graphene [150]. This research group also found that a minimum interphase thickness in this system is associated with local structural properties such as layering of individual superatoms and the hydrogen bonding between amide groups [151]. A study with PA6,6 oligomers showed that the degree of hydrogen-bond formation is governed by layering and geometrical restrictions [152]. The hydrogen bonds weaken in highly confined systems but in well-formed structures at some intermediate distances the hydrogen bonding can even strengthen in comparison with the bulk polymer [152].

Both nucleation and confinement mechanisms provide diverse effects in polymer crystal growth, including acceleration/retardation of crystallization, changes in morphology, and crystal structure modification in a few cases [39,41]. Xu et al. [147] have done a review of the effects that graphene nanofillers have on polymer non-isothermal and isothermal crystallization. The research regarding the crystallization of graphene nanocomposites using techniques such as

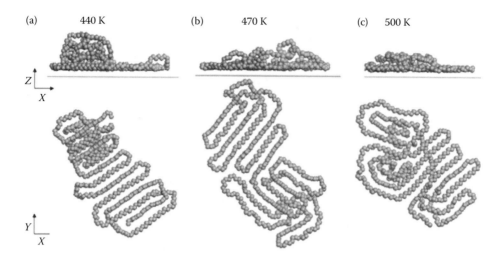

FIGURE 5.1 Final configuration of graphene-induced polymer chain isothermal crystallization at various temperatures for 2000 ps: (a) for $T = 440$ K, (b) for $T = 470$ K, and (c) $T = 500$ K. (Reprinted from *Comput Theor Chem*, 1002, Wang, L.-Z. and L.-L. Duan, Isothermal crystallization of a single polyethylene chain induced by graphene: A molecular dynamics simulation, 59–65, Copyright 2012, with permission from Elsevier.)

differential scanning calorimetry (DSC), wide-angle x-ray diffraction (WAXD), Fourier transform infrared spectroscopy (FTIR), and polarized light optical microscopy (PLOM) is summarized in the following subthemes.

5.3.1 Isothermal Crystallization

Polymer crystallization kinetics has been often studied by isothermal crystallization [83,118,146]. The isothermal data obtained from DSC have been commonly analyzed using the Avrami equation

$$(1 - X_t) = \exp(-Kt^n)$$

where X_t is the fractional crystallization at time t, K the rate constant, and n is the Avrami exponent [39]. The later depends on the nature of the nucleation and growth geometry of the crystals and K involves both nucleation and growth rate parameters [76]. The crystallization of the composites might not result in a simple heterogeneous three–dimensional (3D) crystal growth but usually follows a complicated nucleation mechanism [84]. Wu et al. [82] found spherulitic growth from nuclei initiated at time zero, plate-like growth from nuclei continued over time, or a mixture of these two mechanisms. In other words, the high density of spherulites growing near graphene sheets impinge with each other forming a quasi-2D layer of spherulites [118]. Avrami plots of PLLA/GO nanocomposites exhibited a change in the slopes at the late stage, associated with the formation of new nucleus and the impingement of neighboring spherulites [146]. The decreased n values for the GO/PA4,6 nanocomposites indicated that the nucleation also changed to a 2D crystal growth mechanism [47]. Huang et al. [85] found that at GO contents below 4 wt% PLLA followed a 3D spherulite growth in a heterogeneous nucleating mode; when the nanofillers loading was 4 wt% the Avrami exponent decreased to 2.2,

suggesting a quasi-2D crystal growth behavior. In contrast, similar studies in PLLA and PLLA/GO nanocomposites suggested that the incorporation of GO did not change the crystallization mechanism [83]. Figure 5.2 illustrates the Avrami plots of neat PHB and its nanocomposites. The average values of n obtained from these plots are around 2.4 for neat PHB and 2.6 for PHB/TRGO nanocomposites, indicating that the crystallization mechanism of PHB does not change despite the crystallization temperature (T_c) and the TRGO loading [76]. A comparison between the CNT and CRGO effects on the dimensionality of iPP crystals showed that n was decreased for the CNT/iPP composites compared with the pure polymer. On the other hand, the large lateral areas of 2D CRGO sheets determined that the 2D nanofiller network was looser than the 1D (one-dimensional) nanofiller network at the same content; therefore, the composites containing a relatively low CRGO content (e.g., 20 wt%), resulted in only a slightly decreased n compared with the pure polymer. A rigorous spatial confinement imposed by CRGO to iPP crystallization was found at higher loadings (50–70 wt%) [128].

Several studies have shown the acceleration of the polymer crystallization kinetics due to the incorporation of graphene nanofillers [16,76,83]. Qiu et al. found that the overall isothermal cold crystallization rate of PLLA is increased with increasing GO content, indicating the nucleating effect of GO [83,86]. The overall isothermal melt crystallization rates of PHB were faster in TRGO nanocomposites than in neat PHB and increased at higher TRGO loading [76]. On the other hand, PLLA crystallization was retarded when a high loading of GO was incorporated as compared with the low loading graphene nanocomposites [25]. PLA nanocomposites have also exhibited retarded crystallization due to the incorporation of graphene fillers [17,82]. Figure 5.3 shows the isothermal curves of pure PLA and PLA nanocomposites; the crystallization half times obtained from these thermograms are higher in the nanocomposites than in the neat PLA. These

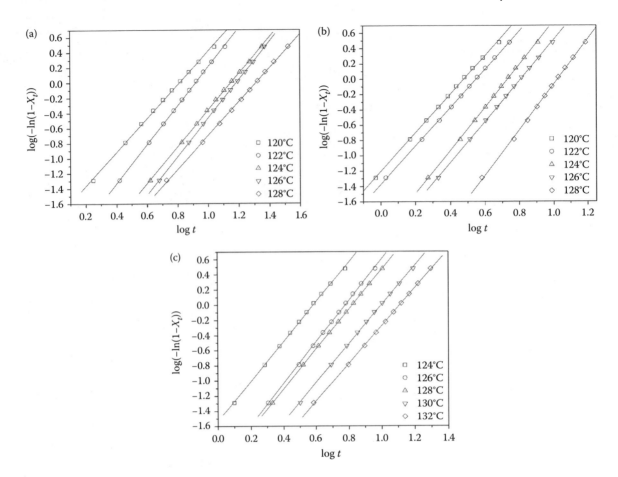

FIGURE 5.2 Avrami plots of (a) neat PHB, (b) 0.5 wt% TRGO/PHB, and (c) 1 wt% TRGO/PHB during isothermal melt crystallization. (Reprinted with permission from Jing, X. and Z. Qiu, Effect of low thermally reduced graphene loadings on the crystallization kinetics and morphology of biodegradable poly(3-hydroxybutyrate). *Ind Eng Chem Res* 51(42):13686–91. Copyright 2012 American Chemical Society.)

results suggest complex effects of graphene on polymer crystallization: a small amount of nanofiller is sufficient to provide nucleation sites for polymer crystallization, whereas the nano-confinement becomes the overwhelming factor when a higher concentration of graphene is used [118]. This has been also attributed to the decreased mobility of the polymer chains due to the increased viscosity resulting from the incorporation of graphene [82].

The isothermal crystallization of graphene nanocomposites has also been studied by FTIR and WAXD; these techniques are highly sensitive to conformational changes and the packing density of molecular chains [16,25]. Xu et al. [16] found that GO promoted the formation of long, ordered structures of PP especially at the early stage of crystallization. WAXD studies suggested that there is a synergistic effect of GO and shear flow on the isothermally crystallized PP nanocomposites [129]. A comparative study between CNTs and GO showed the latter provided the lowest crystallization rate in ethylene-vinyl acetate copolymer composites, indicating that graphene suppressed the polymer crystal growth more critically than CNT networks [24]. Li et al. [25] explained that PLLA chains adsorbed on the surface of graphene sheets need more time to adjust their conformations than in CNTs, making the induction process more complex. A scheme of the

conformational ordering was also proposed (see Figure 5.4). A similar behavior was found by Li and coworkers [87]. The crystallization of alkane melts on CNTs and on the surface of GNs investigated using molecular MD simulations supported the experimental observations by Li et al. [26]. Chen and coworkers studied the behavior of high loadings (50–70 wt%) of CNTs and CRGO in iPP. The nucleation ability was weakened at higher nanomaterial content due to the fact that the dense carbon nanofiller networks caused more serious confinement effects to iPP crystallization [128].

5.3.2 NON-ISOTHERMAL CRYSTALLIZATION

When the crystallization process takes place, polymer chains need enough kinetic energy to overcome the energy barrier and adjust their configurations [150]. The nucleation ability of graphene in a polymer melt has been evaluated using Dobreva's method [127,130,153]. The results obtained have indicated that graphene acts reducing the barrier to nucleation caused by the surface interfacial free energy [127]. Furthermore, the addition of graphene fillers usually shifts the T_c to higher values, suggesting their heterogeneous nucleating effect [38,105,116,125,140] and increasing the graphene filler loading has also gradually increased the T_c [58,140]. The

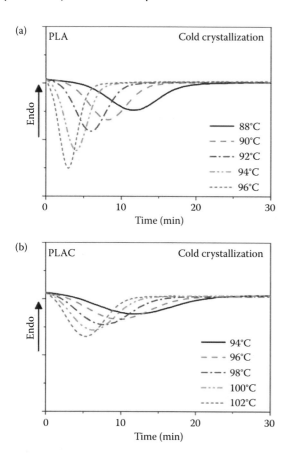

FIGURE 5.3 DSC thermograms of isothermal cold crystallization for (a) neat PLA and (b) PLA nanocomposites at various temperatures. (Reprinted with permission from Wu, D. et al., Crystallization behavior of polylactide/graphene composites. *Ind Eng Chem Res* 52(20):6731–9. Copyright 2013 American Chemical Society.)

functionalization type also has an influence on the T_c; a study demonstrated that increasing the length of grafted chains on FGs restrained the movement of polymer chains resulting in higher T_c [59].

The incorporation of graphene in polymers can broaden their crystallization peaks suggesting slightly retarded crystal growth when compared with the pure polymer [33,47,100]. This feature is shown in Figure 5.5. This behavior was explained by the confinement induced in the polymer by graphene sheets, along with a higher extent of interaction between the two phases [47,58]. The broadening effect has also been attributed to the different nucleation ability of various polymer chains [125].

The most frequently used analysis of non-isothermal crystallization data is the Ozawa equation, which is an extension of the Avrami equation [39]. However, this model is based on the assumption that crystallization occurs at a constant cooling rate and the crystallization originates from a distribution of nuclei that grow as spherulites with constant radial growth rate at a given temperature [153]. Owing to the complexity of the crystalline structures formed in the presence of nanofillers this model has not provided an appropriate description of the non-isothermal crystallization in polymer nanocomposites [127]. Other works have shown the application of different models for describing the crystallization kinetics of graphene nanocomposites [60,127,153]. Chen and coworkers have described the assumptions needed for applying the Avrami and Ozawa equations for non-isothermal crystallization of polymers. In addition, this research group showed that both equations failed to describe the noncrystallization behavior due to the omission of the effects of cooling rate and secondary crystallization. However, using the method developed by Mo and coworkers has been successful to describe

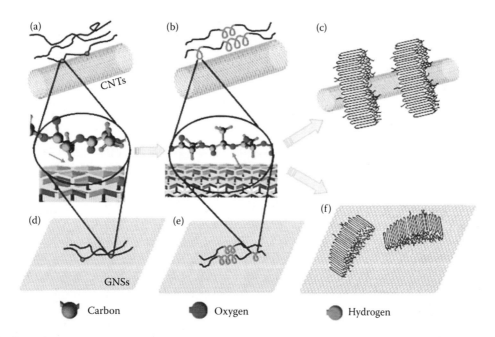

FIGURE 5.4 Conformational ordering and crystallization of PLLA in the presence of CNTs (a–c) and GO (d–f). (Reprinted with permission from Xu, J.-Z., T. Chen, C.-L. Yang et al., Isothermal crystallization of poly(L-lactide) induced by graphene nanosheets and carbon nanotubes: A comparative study. *Macromolecules* 43(11):5000–8. Copyright 2010 American Chemical Society.)

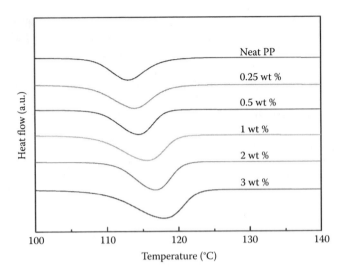

FIGURE 5.5 Typical non-isothermal curves of neat PP and FGs/PP nanocomposites at various loadings. (Reprinted from *Chem Phys Lett*, 570, Yang, X. et al., Functionalization of graphene using trimethoxysilanes and its reinforcement on polypropylene nanocomposites, 125–31, Copyright 2013, with permission from Elsevier.)

the non-isothermal crystallization behavior of other polymers and their nanocomposites [88,119].

Zhang and coworkers found that at low cooling rates (5, 10, and 20 K/min), CRGO/PA6 nanocomposites have lower crystallization rate than pure polymer. The opposite performance was found at 40 K/min cooling rate [60]. The lower crystallization rates were explained by the weakening of the hydrogen bonds of PA6 due to the inclusion of graphene; meanwhile this nanofiller acts as nuclei and accelerate the crystallization. This behavior has been also observed in GO/PLLA nanocomposites [146]. The non-isothermal crystallization behavior of iPP nanocomposites containing 50 wt% CNTs or CRGO caused the strongest nucleation ability of the polymer. However, CNTs acted more effectively as heterogeneous nucleating agents than CRGO, attributed to the 1D topological structure of the former filler, which caused a simpler induction approach to iPP crystallization [128].

5.3.3 EFFECTS ON THE CRYSTALLINITY DEGREE

The addition of graphene fillers can result in a change of the degree of crystallinity of polymers; a variety of works have shown modest increase in the crystallinity of graphene-based nanocomposites [18,47,77,101,118,135,140]. The addition of GO in a PA6 matrix increased the degree of crystallinity in comparison with that of neat PA6 but decreased with increasing the filler content [143]. Das et al. [91] obtained a slight increase in crystallinity, from 42% to 47.5%, when adding 0.6 wt% FGs in PVA. A study showed the presence of FGs increased the crystallinity of PVDF, weakening the destructive effect of polymethyl methacrylate (PMMA) chains on the crystallization of PVDF/PMMA nanocomposites [101]. The same effect was provided when PMMA-GN fibers reinforced a PCL matrix [23]. Some other authors have shown

that crystallinity remains unchanged after the addition of graphene fillers [92,93]. PEG and GO showed to have a synergistic effect on the crystallization behavior of PLLA; GO functioned as a 2D-layer nucleation agent to improve the nucleation density and PEG functioned as a plasticizer to enhance the mobility of the chain segments of PLLA. This resulted in an increased crystallization rate and higher crystallinity [73].

Nanocomposites have also exhibited decreases in the polymers degree of crystallinity [14,23,38,49,51,55,67,70,94,106,107,154]. Partially crystalline PVA became totally amorphous in the presence of FGs [155]. The same behavior was obtained in PVA after the incorporation of 3.5 wt% of GO, suggesting some interaction between the polymer and graphene causes the detrimental effect of interactions among polymer chains [94]. 3D-TRGO inhibited the crystallization of PANI according to WAXD patterns [67]. Deng et al. found that the existence of hydrogen bonding between CRGO and PVA damaged the hydrogen bonding among PVA chains, which caused a decrease in PVA crystallinity [95]. Reduced degree of crystallinity ascribed to interfacial interactions between graphene sheets and polymer has been explained by the reduction of chain flexibility and the retardation of crystallization process even though this nanomaterial serves as heterogeneous nuclei [70]. Diwan et al. [154] showed a monotonous decrease in the crystallinity of polymer electrolyte at GNs loadings of lower than 0.04 wt%; however, at 0.07 wt% an increase of crystallinity was attributed to rolling up of graphene sheets. The addition of FGs in PE was found to decrease the crystallinity up to ~17% for an 8 wt% loading; this behavior was attributed to the formation of a random interface [51]. The in situ polymerization of cyclic butylene terephthalate oligomers in the presence of GNs showed a decrease in the crystallinity when compared with pure polymer, suggesting poor dispersion of graphene into the polymer matrix which obstructed crystalline ordering of the polymer segments [69]. Jeong et al. found that the crystallization of the soft segment of PU was inhibited by FGs, and this was more evident at high nanofiller contents [107]. Similar behavior was found by Coleman and coworkers [108]. On the other hand, a decrease in the endothermic peak of the hard segment of waterborne PU (WPU) was observed as the FGs content increased [106,109]. DSC thermograms showed the synergistical effect of the compatibilizer and GO on the crystallization of PE, depending on the chlorine content of the former. At the same amount of GO, addition of amorphous PE containing 35% chlorine (CPE35) resulted in a further decrease in degree of crystallinity of polymer; this was related to the amorphous nature of CPE35 which hindered the crystalline packing of PE chains. In the case of the compatibilizer containing 25% chlorine (CPE25), the extent of crystallinity in the nanocomposites was always higher than the pure polymer irrespective of the compatibilizer content [117].

PANI nanocomposites containing GNs showed an increase in the crystalline regions of polymer, unlike CNT-based nanocomposites which had a WAXD pattern very similar to that of pure polymer [27]. Electrospun nanofibers reinforced with FGs exhibited superior crystallinity than the ones containing functionalized CNTs (FCNTs) as the filler loading was increased [18]. This was attributed to better dispersion of graphene sheets as compared with CNTs, which induced more regular orders and hence an increase in the crystallinity [18,27].

5.3.4 CRYSTAL MODIFICATIONS

The presence of nanofillers in polymers can result in different crystalline phases or crystal size changes at the nanometric or microscopic scale [18,102,135]. TRGO and GO effectively promoted growth of the γ-form in PA6 [61]. The addition of GNs in PVDF changed the polymorphism from α-phase to γ-phase [135]. Another work showed GO induced the formation of γ-phase when crystallizing from solution, but only α-phase was produced from melt crystallization in PVDF nanocomposites [102]. PMMA grafted GO (PMMA-g-GO) nucleated PVDF crystals and a gradual decrease of α-phase occurred with a simultaneous rise of piezoelectric β-phase [100]. Mohamadi and coworkers found that PMMA can promote the β-phase in PVDF but cannot stabilize it at elevated temperatures causing the formation of α-phase. On the other hand, FGs could stabilize this phase at 90°C and with increasing the annealing temperature to 120°C some of the β-form converts to the γ-form [101]. Figure 5.6 shows the stabilization of the γ-phase and β-phase at high annealing temperatures

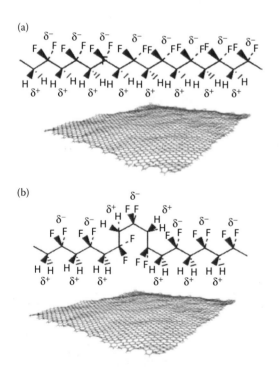

FIGURE 5.6 Stabilization of (a) β-phase with all-trans (TT) and (b) γ-phase with T3G + T3G-conformation. (With kind permission from Springer Science+Business Media: *J Polym Res*, Evaluation of graphene nanosheets influence on the physical properties of PVDF/PMMA blend, 20, 2013, 46, Mohamadi, S., N. Sharifi-Sanjani and A. Foyouhi, Figure 5.6, Copyright 2013.)

by limitation of chain rotation [103]. In another study, the addition of only 0.1 wt% GO demonstrated to be enough to nucleate all PVDF chains into the β-phase [48]. DSC thermograms suggested the stimulation of the γ-form crystal of PA6 by FGs and this was further confirmed by WAXD [62]. The β and γ crystalline forms of PVDF have also been induced with non-covalent FGs [136]. Nguyen et al. found that the addition of FGs to PA6 induced the α-phase in the nanocomposites; the polymer originally crystallized in the γ-phase [63]. The formation of β-crystals in PP can be produced by several nucleating agents or shear-induced crystallization [131]. Xu et al. [129] observed that GO presented a shear amplification effect for nucleating β-crystals in 0.05 wt% GO/PP and 0.1 wt% GO/PP nanocomposites. Yuan et al. [132] also obtained the β-form in PP by incorporating 1 wt% FGs. A number of studies of graphene-based composites have also shown that polymorphism is not affected by the addition of these fillers [18,68,76,83,93,130].

Furthermore, the crystalline structure of polymers in contact with nanomaterials depends on the intermolecular interactions and on the geometry of their surface; it is also very sensitive to the chemical structure of the confined polymer and the confining nanomaterial [150,151]. A decrease in crystal size of PA6,6 nanocomposites was obtained from the incorporation of 0.1 and 0.5 wt% FGs and FCNTs when compared with pure polymer [18]. At the highest loading (1 wt%) FCNTs showed a decrease in crystal size in the direction perpendicular to the (100) plane of PA6,6 and an increase in crystal size in the direction perpendicular to the (010/110) plane. The 1 wt% FGs/PA66 nanocomposites showed a decrease in both directions, indicating that the FGs had a better dispersion than FCNTs attributed to the geometry and content of functional groups of nanomaterials.

PLOM is very useful for observing the morphological changes and density of the crystalline structures in polymers. The addition of nanomaterials leads to an increased number of nucleation sites that act reducing the crystalline structures due to impingement of the spherulites with one another [58,88,127,130]. PLOM showed that TRGO on glass fiber (GF) surface largely improves the nucleation ability of GF for PP crystallization [89]. Wang et al. studied the morphology of PLLA spherulites at different GO loadings; the spherulite nucleation density was enhanced with an increase of filler content indicating good dispersion. However, at high content, GO agglomerated and reduced the nucleation density of PLLA spherulites [86]. The density of PHB spherulites was also improved by TRGO [76]. PLOM micrographs of neat PA4,6 showed spherulites of ~40 μm in diameter and after the incorporation of GO these became barely observable and their size decreased dramatically [47].

Transmission electron microscopy (TEM) and scanning electron microscopy (SEM) have also been used to observe the different morphologies and crystalline structures of nanocomposites [47,114]. The crystalline morphology of GO/PA4,6 nanocomposites is displayed in Figure 5.7, which shows the crystalline lamellae originating from the GO sheets (circled ones) [47].

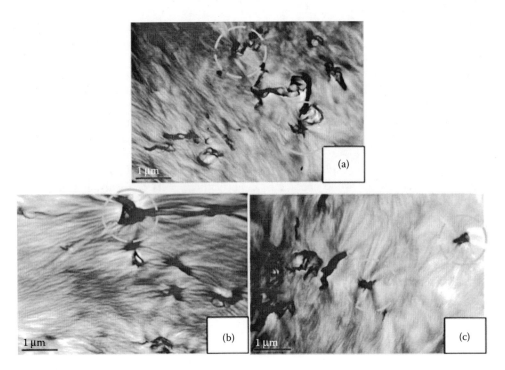

FIGURE 5.7 TEM images of microtomed composite films: (a) 0.5 wt% GO/PA4,6, (b) 1 wt% GO/PA4,6, and (c) 2 wt% GO/PA4,6. (Reprinted from *Polym Test*, 31(7), Chiu, F.-C. and I.-N. Huang, Phase morphology and enhanced thermal/mechanical properties of polyamide 46/graphene oxide nanocomposites, 953–62, Copyright 2012, with permission from Elsevier.)

5.3.5 HYBRID CRYSTALLINE STRUCTURES

Another approach to improve stress transfer from the filler to the polymer is the interfacial crystallization; in this route the filler acts as a nucleus and induces the growth of polymer lamellae on its surface [43]. The mechanism of polymer crystallization on graphene sheets follows epitaxial growth [114,118,120]. The epitaxial crystallization could be due to a unit-cell dimensional match or crystal structure similarity between the crystalline filler and the polymer matrix [43]. Xu and coworkers found PE lamellar crystals formed nanohybrid "shish-kebab" structures on CNTs whereas GO sheets were only decorated with petal-like PE crystals. The high surface curvature and the perfect ordered crystal structures of CNTs favored the periodical growth of PE crystals on CNTs; while on GO the PE chains showed multiple orientations of lamellae due to the lattice matching and complex interactions between both materials [120].

Yang et al. crystallized P3BT on GO through three different approaches, obtaining different hybrid crystalline structures [114]. Zhai and Chunder crystallized P3HT nanowires on CRGO for obtaining 2D supramolecular structures [45]. Li et al. studied the growth of PE crystals on CRGO nanosheets over a broad range of temperatures. As shown in Figure 5.8 the PE lamellae with an average dimension of a few hundreds of nanometers are randomly distributed on the basal plane of CRGO [118]. GF/TRGO induced the formation of transcrystalline structure in PP and PLLA, indicating a new way to control interfacial crystallization [89]. Fan and coworkers reported the synthesis of PANI–GO–CNT hybrid materials by oxidative polymerization of aniline to form a 3D fish scale-like microstructure assembled on the GO sheets and CNTs [68].

5.4 MECHANICAL PROPERTIES

The high levels of stiffness and strength of graphene sheets give the possibility of obtaining polymer nanocomposites with outstanding mechanical properties [1,6]. Stress transfer from high modulus graphene sheets to low modulus polymers is the fundamental issue in reinforcement; as explained earlier, this feature is influenced by the interfacial adhesion and dispersion between both phases. There are also other parameters involved in the mechanical response such as graphene content, crystallinity, physical and chemical characteristics of nanofiller, and the processing techniques [6,44,156]. These parameters are considered in the next subsections and the results obtained by different research groups are displayed in Table 5.6 through 5.8.

5.4.1 EFFECTS OF FILLER LOADING

Given the large surface area of graphene sheets and their rigidity, significant improvements in mechanical response are often seen at low contents [38,59,64,111,125,156]. Polyester nanocomposites containing only 0.05 wt% GO resulted in an increase of 72% in tensile strength accompanied by a 55% enhancement in elongation at break [75]. PA6 fibers achieved an increase of 65% in tensile strength and 290% in modulus by the addition of 0.1 wt% FGs [62]. Feng et al. reported 0.2 wt% CRGO/PLA nanocomposites lead to a 26%

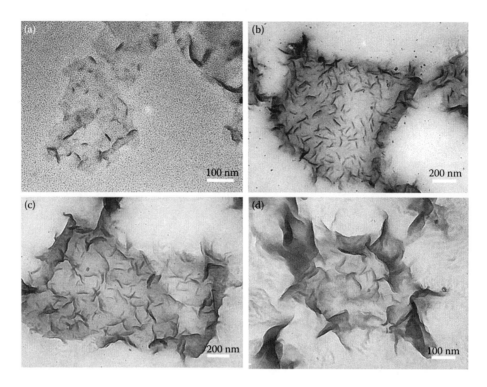

FIGURE 5.8 Bright-field TEM images of CRGO nanosheets induced PE crystallization at different conditions. (a) PE was crystallized at 90°C for 1 h with a PE/CRGO concentration ratio of 1:2. (b–d) PE was crystallized at 100°C overnight with PE/CRGO ratio 1:1, 2:1, and 5:1, respectively. (Reprinted with permission from Cheng, S. et al., Reduced graphene oxide-induced polyethylene crystallization in solution and nanocomposites. *Macromolecules* 45(2):993–1000. Copyright 2012 American Chemical Society.)

TABLE 5.6
Improvement of the Thermo-Mechanical Properties of Graphene-Based Nanocomposites

Nanofiller	Nanofiller Content (wt%)	Polymer	Modulus		T_g (°C)	T_d (°C)		References
GNs	0.5	PA12	457%	Tensile	–	–	–	[22]
GNs	3	PANI	–	–	–	245	$T_{10\%}$	[27]
GNs	1.0	PBT	–	–	–	27	T_{max}	[69]
GNs	1	PCL	65%	Tensile	–	–	–	[23]
GNs	2	PVC	58%	Tensile	25	–	–	[111]
GNs	55	PU	15%	Tensile	–	–	–	[108]
CRGO	2	PBS	27%	DMA (−55°C)	–	16	$T_{5\%}$	[70]
CRGO	0.5	PCL	18%	DMA (25°C)	–	11	$T_{5\%}$	[11]
CRGO	0.5	PU	212%	–	7	–	–	[97]
CRGO	2	PU	202%	DMA (−75°C)	–	40	$T_{5\%}$	[110]
CRGO	0.6	PVA	107%	Tensile	–	–	–	[95]
CRGO	1.8	PVA	940%	Tensile	–	–	–	[96]
CRGO	3.5	PVA	16%	Tensile	14	38	T_{max}	[94]
TRGO	5	LCP	55%	DMA (25°C)	–	–	–	[20]
TRGO	0.2	PBO	178%	Tensile	–	21	T_{max}	[115]
TRGO	2	PCL	292%	DMA (−75°C)	–	–	–	[72]
TRGO	3	PE	147%	Tensile	–	–	–	[121]
TRGO	17.4	PP	50%	Tensile	14	–	–	[14]
TRGO	3	PLA	–	–	–	14	$T_{5\%}$	[81]
TRGO	2	PLLA	–	–	5	38	T_{max}	[84]
TRGO	1	PU	95%	Tensile	–	–	–	[104]
TRGO	3	PU	43%	Tensile	–	–	–	[107]
TRGO	4	PU	334%	Tensile	–	–	–	[109]
TRGO	5	PP	34%	Tensile	–	35	$T_{5\%}$	[133]
TRGO	6	PU	10%	Tensile	–	30	$T_{10\%}$	[106]

TABLE 5.7
Improvement of the Thermo-Mechanical Properties of Functionalized Graphene Oxide-Based Nanocomposites

Nanofiller Content (wt%)	Polymer	Modulus		T_g (°C)	T_d (°C)		References
3	Chitosan	108%	DMA (25°C)	–	62	$T_{10\%}$	[55]
0.5	Chitosan/starch (Ch/S)	929%	DMA (35°C)	–	–	–	[54]
1	PA6	67%	Tensile	–	–	–	[143]
3	PA4,6	71%	DMA (40°C)	–	–	–	[47]
1	PA6,6	139%	DMA (30°C)	6	–	–	[137]
6	PHBV	26%	Tensile	–	–	–	[77]
1	PI	25%	Tensile	–	–	–	[78]
2	PLGA	5212%	DMA (37°C)	35	–	–	[90]
0.6	PVA	35%	Tensile	–	–	–	[91]
0.7	PVA	62%	Tensile	–	6	T_{max}	[98]
3	PVA	29%	Tensile	–	36	T_{max}	[92]
2	PVDF	137%	Tensile	–	20	T_{max}	[48]
4	PU	200%	Tensile	–	50	T_{max}	[145]

raise in tensile strength and an 18% raise in Young's modulus [11]. The presence of 0.3 wt% GO enhanced the tensile strength, Young's modulus, and energy at break of PCL by 95%, 66%, and 416%, respectively; further increase in GO content improved the mechanical properties substantially [38]. The incorporation of 0.6 wt% FGs caused a significant improvement in ultimate tensile strength (35%), elongation to break (200%), impact energy (175%), and toughness (72%) of PA12; although no significant improvement in Young's modulus occurred [65]. The addition of 2 wt% GO resulted in an increase of 137% in Young's modulus and 92% in tensile strength of PVDF nanocomposites [48]. The mechanical response of GO/PA6 nanocomposites also improved as the filler loading was increased [143].

TABLE 5.8
Improvement of the Mechanical Properties of Functionalized Graphene-Based Nanocomposites

Functionalizing Agent	Nanofiller Content	Polymer	Modulus		References
3-Aminopropyltriethoxy silane	0.5 wt%	PLLA	200%	DMA (25°C)	[87]
4,4′-Diphenylmethane diisocyanate	0.1 wt%	TPEE	7.5%	DMA (−70°C)	[112]
4,4′-Oxydianiline	2 wt%	PI	9880%	DMA (225°C)	[80]
4-Aminobenzoic acid	2.5 wt%	PA6	100%	Tensile	[64]
4-Aminophenethyl alcohol	3 wt%	PU	842%	Tensile	[105]
4-Substituted benzoic acid	0.1 wt%	PA6	290%	Tensile	[62]
Dodecyl amine	8 wt%	PE	118%	DMA (50°C)	[51]
Dodecyltrimethoxysilane	17 wt%	PP	17%	Tensile	[33]
Ethylendiamine	0.1 wt%	PA6,6	116%	DMA (30°C)	[137]
Ethylendiamine	1 wt%	PE	131%	Tensile	[122]
Keratin	0.5 wt%	Ch/S	312%	DMA (35°C)	[54]
Maleic anhydride-g-PP	0.5 wt%	PP	15%	DMA (−30°C)	[132]
MPP	1 wt%	PVA	30%	Tensile	[99]
NR latex	5 phr	NR	270%	Tensile	[113]
Octadecylamine	1 wt%	PP	47%	Tensile	[126]
PA6	0.015 wt%	PA6	139%	Tensile	[63]
PE	0.3 wt%	PE	20%	Tensile	[116]
PMMA	3 wt%	PVDF	91%	DMA (−50°C)	[100]
PMMA	30 wt%	PVDF	39%	Tensile	[103]
Polystyrene-b-ethylene-*co*-butylene-b-styrene	2 wt%	PS	73%	Tensile	[134]
PP latex	1 wt%	PP	36%	DMA (25°C)	[140]
PVA	2 wt%	PVA	231%	Tensile	[46]
Tetra-*n*-octylammonium bromide	0.1 wt%	PP	70%	Tensile	[52]
Triethylamine	1 wt%	PI	49%	Tensile	[79]
Triethylamine	17 wt%	PCL	1193%	Tensile	[58]
Triphenylene	2 wt%	PVA	116%	Tensile	[15]

Zhao and coworkers [96] considered that there is a "mechanical percolation" which means that lower than this content, graphene can be fully exfoliated in the polymer matrix and effective reinforcement can be achieved, while further loading than this point will cause the stacking of graphene sheets due to the strong van der Waals forces among them. Wang et al. reported the tensile strength of PVA increased with increasing GO loading. However, the elongation at break of the GO/PVA nanocomposites reached a maximum at 0.5 wt% GO content, as shown in Figure 5.9a. CRGO/PVA nanocomposites also achieved the maximum mechanical reinforcement at the same filler loading (see Figure 5.9b) [97]. Good dispersion at high loadings of graphene in nanocomposites is very difficult to accomplish. However, some authors have shown studies involving loadings higher than 2 wt% and great enhancement in mechanical properties [51,74,100,105]. Nandi and coworkers showed the storage modulus increased a 124%, stress at break a

157%, and Young's modulus a 321% which is a high percent enhancement for an amount of 5 wt% PMMA-FGs in PVDF nanocomposites [100]. The addition of 8 wt% FGs provided a 118% enhancement in the storage modulus of PE [51].

GO and FGs can provide strong interaction with polymeric matrices resulting from the wrinkled surface of graphene sheets that are capable of mechanically interlocking with polymer chains and also due to the hydrogen bonding formed between the functionalities of the graphene sheets and a polar matrix [138,141,157]. The reinforcing effect of FGs was found to be superior to those of FCNTs in terms of Young's modulus of PP nanocomposites [52]. Our investigation group also studied both carbon structures and the results showed a strong dependence in the amount of functional groups on the carbon surface, which provided superior reinforcement in FG nanocomposites and a decreasing behavior in the mechanical response of the FCNT nanocomposites as the nanofiller loading was increased [18]. GO also showed to be a better reinforcement agent in a PA6,6 matrix, when compared with OCNTs [137].

5.4.1.1 Models

Graphene/polymer nanocomposite moduli have been predicted by a variety of models [6,31,34,113,121,144,158]. The Halpin–Tsai model is the most common method used in the literature [46,70,78,85,95,96,140]. This model is based on the assumptions of the perfect stress transfer between the fillers and matrix as well as the perfect 2D plane structure of graphene sheets [81]. Most experimental results have shown agreement with the random distribution of graphene in this model [15,46,64,70,78,96]. Nevertheless, there are few reports showing agreement with the aligned distribution of graphene sheets in the polymer matrix. Song et al. found that at low loadings of PP latex-g-CRGO the experimental modulus agrees well with the theoretical simulation of aligned graphene sheets parallel to the surface of the nanocomposite. However, with further filler loading, Young's modulus gradually approaches the theoretical simulation for the random distribution of graphene sheets [140]. A similar behavior was found by Ha and coworkers in GO/PI nanocomposites [78]. The experimental data of GO/PVA composites was in good agreement with the theoretical results of parallel alignment of graphene sheets to the surface of the nanocomposite film [98]. Li et al. also found a good consistency between the experimental results and the theoretical simulation for the 2D-aligned parallel distribution of FGs/PI nanocomposites, as seen in Figure 5.10 [79].

Computational simulations have also been made for studying the mechanical behavior between CNTs or graphene with polymers [158,159]. MD simulations performed on the graphene–polymer interface system indicated that mechanical interactions between graphene and polymer chains are stronger than those among the polymer chains [159].

5.4.2 Influence of Crystallinity Properties

The values of the critical strain are affected by the polymer chain microstructure, whereas the corresponding values of

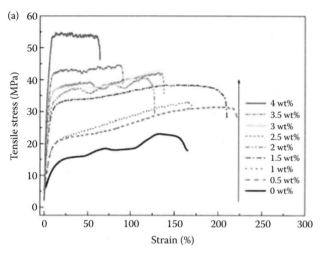

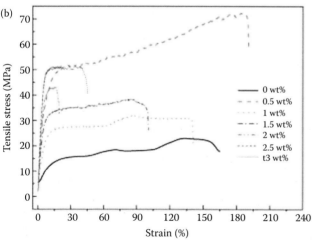

FIGURE 5.9 Typical stress–strain curves of (a) GO/PVA nanocomposites with varying GO loadings and (b) CRGO/PVA nanocomposites with varying graphene loadings. (Reprinted with permission from Wang, J. et al., Preparation of graphene/poly(vinyl alcohol) nanocomposites with enhanced mechanical properties and water resistance. *Polym Int* 60(5):816–22. Copyright © 2011, Society of Chemical Industry.)

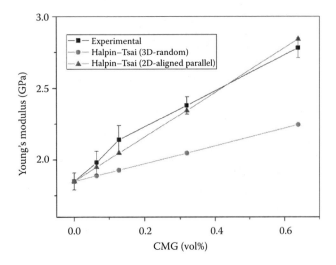

FIGURE 5.10 Young's modulus of the experimental and Halpin–Tsai theoretical model. Two hypotheses were proposed as follows: 3D-random and 2D-aligned parallel distributions of chemically modified graphene (CMG) in the PI matrix, respectively. (Reprinted with permission from Huang, T., R. Lu, C. Su et al., Chemically modified graphene/polyimide composite films based on utilization of covalent bonding and oriented distribution. *ACS Appl Mater Interfaces* 4(5):2699–708. Copyright 2012 American Chemical Society.)

the stress depend on the degree of crystallinity, the amount of structural disorder present in the crystals, and on the relative stability of the crystalline forms [42]. In addition, a polymer–crystalline interface could lead to a stronger response to the strain of nanofillers and better stress transfer than the polymer in the amorphous state [71]. The increase in toughness of PA12 was attributed to higher levels of γ-phase caused by the incorporation of FGs [65]. The addition of 0.6 wt% FGs in PVA exhibited an increase in the elastic modulus attributed not only to the surface area and surface roughness of graphene but also to the slight increase in crystallinity of nanocomposites [91]. PA6,6 nanofibers showed a tendency of increased crystallinity and storage modulus as the content of FGs increased [18,137]. Mohamadi et al. [103] found that PMMA-FGs enhanced the crystallite density of polymer which also acted as reinforcing units resulting in an enhancement of the elastic modulus in PVDF/PMMA nanocomposites. Cai and Song [71] obtained a polymer crystalline layer that improved the stress transfer in SGs/PCL nanocomposites. Park and coworkers found that the substantial increase in modulus was consistent with the increase in crystallinity of PHBV nanocomposites reinforced with CRGO [77].

On the other hand, a decrease in the crystallinity of polymers has also affected the mechanical response. FGs inhibited the hard segment crystallinity of WPU resulting in a reduced tensile modulus of the polymer at room temperature [106]. Comparable results were found in FGs/WPU obtained by in situ polymerization [109]. Jeong and coworkers found a modulus decrease in PU nanocomposites caused by the reduced soft segment crystallinity which overshadows the reinforcing

effect of FGs [107]. Fabbri et al. [69] obtained a decrease in the mechanical properties of the PBT nanocomposite containing 1 wt% GNs when compared with neat polymer; this was attributed to the lower molecular weight and lower degree of crystallinity present in the nanocomposite [69]. The mechanical properties in GO/CPE35/PE nanocomposites were lower than GO/CPE25/PE composites due to the amorphous nature of CPE35 compatibilizer [117]. Annealing treatment produced an unfavorable reorganized crystalline structure of Nafion, resulting in a decrease of the modulus; addition of GO improved effectively the mechanical properties of the nanocomposites [57].

5.4.3 OTHER FACTORS INFLUENCING MECHANICAL RESPONSE

NR reinforced by NR latex-TRGO showed a larger reinforcement effect than that of TRGO due to a more uniform dispersion of the graphene sheets coupled with a larger accessible interfacial surface area [113]. Another study in an LCP also found that the mechanical properties are dependent on the TRGO particle size, with the larger particles causing higher storage modulus than the small-sized ones [20]. Young's modulus of PU nanocomposites decreased considerably as the size of SGs was reduced while the ultimate tensile strength and strain at break were slightly enhanced at the same conditions [108]. The tensile strength, tensile modulus, and ultimate strain of PVA nanocomposites suggested these properties are only slightly affected by the thickness of GO [92].

Huang et al. reported that the addition of melamine polyphosphate (MPP) to PVA resulted in the deterioration of mechanical response. However, after adding PVA-CRGO into MPP/PVA systems, the mechanical properties were improved to some extent [99]. The mechanical properties of GO/PI nanocomposites were enhanced with the use of Mg^{2+} as a cross-linker [78]. PP latex-g-CRGO was incorporated into a PP matrix in order to overcome the stacking of graphene sheets during melt blending and increase the mechanical properties; nevertheless, higher levels of nanofillers content lead to a slight reduction in the storage and Young's moduli due to the plasticization effect of low molecular weight PP latex overwhelming the reinforcement effect of graphene sheets [140]. Young's modulus, yielding strength, and tensile strength of PE were improved by the incorporation of PE-g-GO at very low loadings [116]. The chain length of alkylamine-g-GO also had a dependence on Young's modulus of PP nanocomposites, and longer alkylamine chains had a more pronounced increase in this property [126]. Chitosan/starch (Ch/S) nanocomposites showed the different effects provided by the addition of GO and keratin-g-GO [54]. Figure 5.11 shows a dramatic enhancement of the GO nanocomposites (Ch/S/GO) compared with the keratin-g-GO (Ch/S/GKGO) and neat polymer (Ch/S). The reinforcement provided by TRGO in PU was found to depend strongly on the polymer hard segment content which was attributed to the skeleton-like superstructures observed

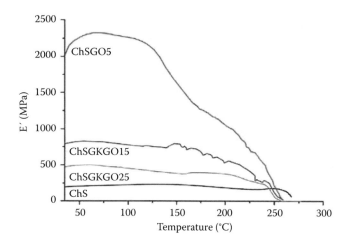

FIGURE 5.11 Temperature versus dynamic storage modulus (E') of neat chitosan/starch (Ch/S) film and the nanocomposite Ch/S/GO, Ch/S/GKGO films at 0.5 wt% nanofiller content. (Reprinted with permission from Rodríguez-González, C. et al., Polysaccharide nanocomposites reinforced with graphene oxide and keratin-grafted graphene oxide. *Ind Eng Chem Res* 51:3619–29. Copyright 2012 American Chemical Society.)

only at high content of hard segments [104]. Kim et al. [122] obtained a promising hybrid via the simultaneous reduction, functionalization, and stitching of GO and –COCl activated MWCNTs using ethylenediamine as a cross-linking agent which effectively reinforced PE.

CRGO/PCL nanocomposites prepared by the solution mixing method exhibited improved tensile strength while covalently linked nanocomposites showed more significant enhancements in the tensile strength and elongation at break [58]. Xu et al. [92] showed that PVA/GO nanocomposites obtained by vacuum filtration displayed better tensile yield strength and Young's modulus values than those obtained by casting. Solvent blending was found to be a better method for dispersing TRGO in PE than melt compounding and the tensile modulus was further increased when PE was functionalized with amine, nitrile, and isocyanate groups [121]. Cryomilled GNs/chitosan nanocomposites exhibited significant improvements in tensile properties compared with raw GNs/chitosan nanocomposites [56]. Yan and coworkers found that chitosan/GO films in the wet state exhibited a higher increase in tensile strength at break than the composites tested in the dry state [55].

5.5 THERMAL PROPERTIES

The thermal properties achieved in polymers by graphene incorporation dominate the macroscopic properties of nanocomposites. Therefore, their study is important in order to control the desired properties depending on their potential application. The most common thermal properties studied in graphene nanocomposites are the glass transition temperature, thermal stability, and thermal conductivity; their behaviors are explained in the following subsections. Tables 5.6, 5.7, and 5.9 also contain the obtained results for thermal properties of graphene-based nanocomposites.

TABLE 5.9

Improvement of the Thermal Properties of Functionalized Graphene-Based Nanocomposites

Content (wt%)	Polymer	T_g (°C)	T_d (°C)		References
0.1	PA6	–	5	T_{max}	[62]
1	PE	–	28	$T_{5\%}$	[122]
8	PE	–	47	$T_{50\%}$	[51]
0.5	PLLA	12	–	–	[87]
0.5	PP	–	75	T_{max}	[132]
1.5	PP/PPO	–	25	$T_{20\%}$	[125]
2	PS	–	26	$T_{10\%}$	[134]
3	PU	–	39	T_{max}	[105]
1	PVA	7	8	T_{max}	[99]
3	PVDF	20	20	T_{max}	[100]
0.1	TPEE	–	17	$T_{10\%}$	[112]

5.5.1 GLASS TRANSITION TEMPERATURE

The glass transition temperature (T_g) of nanocomposites is dependent on the free volume and mobility of the chain segments of polymers. The most common techniques for determining T_g are DSC and dynamical mechanical analysis (DMA). In DMA, T_g can be evaluated using the loss modulus data or the tan δ plot. The former is related to the dissipation of energy as heat and the latter is related to the reduction of vibration of the material, that is, damping [100]. When graphene acts as a barrier for heat flow there is no significant dissipation of heat which causes a very small decrease of T_g; on the other hand, the enhanced stiffness of graphene composites causes a very large damping variation and hence T_g increases significantly when this temperature is obtained from the tan δ plot [100,157].

Some studies have reported that the addition of graphene fillers resulted in a remarkable enhancement of the polymer T_g [49,79,90,94,155]. This property increased from 63.6°C for pure PVA to 72.2°C in a 0.72 vol% GO/PVA nanocomposite [49]. Li et al. obtained an enhancement of 14°C in 1 wt% FGs/PI nanocomposites [79]. PLGA nanofiber containing 2 wt% GO showed a significant enhancement on the T_g (66.7°C), as compared with neat PLGA (32°C) [90]. The incorporation of CRGO in PVA increased gradually the T_g from 76°C for pure polymer to 90°C at a nanofiller content of 3.5 wt% [94]. Li and coworkers found this property was shifted toward higher temperatures with increasing volume fraction of CRGO and this was accompanied by a higher degree of crystallization in PVA nanocomposites [160]. The T_g of PVA-g-GO nanocomposites was further increased as compared with the GO nanocomposites, as shown in Figure 5.12. The good miscibility and the strong interfacial adhesion between the PVA-g-GO and the PVA matrix allowed the GO sheets to be dispersed individually and homogeneously, which could reduce the mobility of polymer chains in a higher degree [46]. Many other studies have shown enhancements in this property as graphene filler loading increased [46,49,90,94,98,100,160].

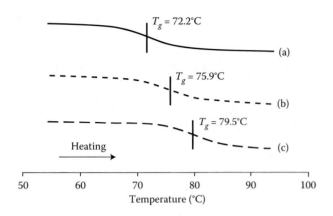

FIGURE 5.12 DSC thermograms showing the T_g at a heating rate of 10°C/min for (a) PVA, (b) 1 wt% GO/PVA, and (c) 1.4 wt% PVA-g-GO/PVA nanocomposites. (Reprinted with permission from Cheng, H.K.F. et al., Poly(vinyl alcohol) nanocomposites filled with poly(vinyl alcohol)-grafted graphene oxide. *ACS Appl Mater Interfaces* 4(5):2387–94. Copyright 2012 American Chemical Society.)

Other works have indicated that there are no significant changes in this property due to the inclusion of graphene nanofillers [23,51,56,83,93,144]. The T_g of PCL loaded with PMMA electrospun fibers containing CNTs or GNs was not affected [23]. The reinforcement effect caused by PVA-CRGO was weakened by the addition of MPP in PVA nanocomposites as shown in T_g results [99]. Appel and coworkers [104] studied the effect of TRGO, CNTs, and carbon black (CB) in PU. The results have shown that there is no influence on the T_g upon incorporation of the fillers; this was attributed to the weak interaction between carbon nanofillers and the PU soft segment. The decrease in the T_g of PU nanocomposites as the FGs content was increased was attributed to the reduced crystallinity of PU soft segment [107]. Reductions in this property were also attributed to aggregation and heterogeneous dispersion of graphene fillers [78,107].

5.5.2 THERMAL STABILITY

The low thermal stability of polymers limits their uses in high-temperature applications. Enhancements in thermal stability can be achieved by the homogeneous dispersion of graphene fillers in the polymer matrix which disturb the diffusion of small molecules by creating a "tortuous path" that retards the progress of the gas molecules through the matrix [19,161]. Moreover, graphene sheets can act as barriers in polymer matrices delay the permeation of oxygen and the escape of volatile degradation products [134]. Thermogravimetric analysis (TGA) allows the evaluation of nanocomposite thermal stability and three parameters have been used to evaluate the degradation behavior: (1) the onset temperature, considered as the temperature at which the system starts to degrade, (2) the degradation temperature (T_d), measured as the temperature at which the maximum degradation rate occurs, and (3) the degradation rate, evaluated in the derivative weight loss as a function of temperature curve [8].

Low loadings of graphene fillers have provided modest enhancements in thermal stability of nanocomposites when tested under nitrogen [11,98]. Zhang et al. [61] reported that the onset and maximum decomposition temperatures of TRGO and GO nanocomposites at loadings lower than 2 wt% are similar to those of neat PA6. The onset temperature of degradation of 0.7 wt% GO nanocomposite was about 3°C higher than that of neat PVA [98]. Thermal stability of cross-linked PE (XLPE) containing GO or FGs was increased slightly at loadings in the range of 0.5–3 wt%; the FGs provided a higher stability effect on the XLPE when compared with the GO [123]. The thermal degradation temperature at 5% weight loss of 0.2 wt% CRGO/PLA nanocomposites increased by ~10°C [11]. The thermal degradation temperature at 10% weight loss of PBO composite fibers containing 2 wt% TRGO was increased by 20°C when compared with pure polymer [115].

PMMA-FGs/PVDF nanocomposites showed increased degradation temperatures as the loading of graphene was increased [100]. The thermal stability of PS nanocomposites also exhibited a gradually enhancing tendency with an increase in content of FGs [134]. PU nanocomposites containing 3 wt% FGs achieved an enhancement of 30°C at a 50% weight loss when compared with pure polymer [105]. A lower amount of CRGO (2 wt%) provided a 40°C increase in thermal stabilization of PU at 5% weight loss [110]. Another work showed dramatic results in this property with an increase of 65°C in the thermal stability of PVDF at 2 wt% GO loading [48]. TRGO retarded thermal degradation of PLA nanocomposites, increasing in 14 K the temperature at 5% weight loss when adding 3 wt% of this nanofiller [81]. The degradation temperature of 3 wt% GO/PVA nanocomposites was 36°C higher than that of pure PVA. Furthermore, the decomposition rate of the nanocomposite was significantly slower than that of the neat polymer [92].

The onset temperature of PLLA was increased from 173°C to 211°C by the incorporation of 2 wt% TRGO; however, the well-dispersed TRGO in the PLLA matrix improved the thermal conductivity which in turn increased the thermal decomposition rate of the polymer [84]. A similar behavior was found in WPU nanocomposites, where the initial weight loss was accelerated in the presence of FGs [109]. PBS nanocomposites containing 2 wt% CRGO were found to have faster thermal degradation rate; this was also attributed to the good heat conductivity of CRGO [70].

The combination of PVA-CRGO and MPP improved the decomposition of PVA composites. In contrast, the effect provided by only PVA-CRGO was more favorable than when adding MPP, indicating this flame retardant decreased the thermal stability of the composites [99]. CNT/PANI and GNs/PANI films showed higher decomposition temperatures than neat polymer; graphene-based nanocomposites displayed the best thermal stability, as shown in Figure 5.13 [27]. Jin and coworkers showed the differences provided by FCNTs and FGs in the thermal degradation behavior of PP nanocomposites. Nanocomposites containing 0.5 wt% FCNTs showed a 16°C increase in this property compared with that of pure polymer. On the other hand, the 0.5 wt% FGs/PP

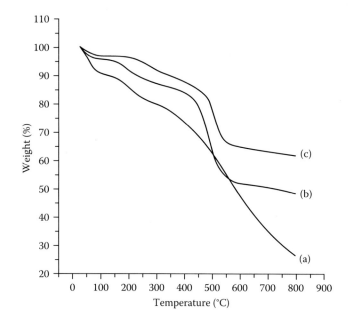

FIGURE 5.13 TGA of: (a) PANI, (b) 3 wt% CNTs/PANI, and (c) 3 wt% GNs/PANI. (Reprinted from *Composites Part B-Eng*, 47, Ansari, M.O. et al., Thermal stability in terms of DC electrical conductivity retention and the efficacy of mixing technique in the preparation of nanocomposites of graphene/polyaniline over the carbon nanotubes/polyaniline, 155–61, Copyright 2013, with permission from Elsevier.)

nanocomposites exhibited an increase of 24°C in the thermal degradation temperature; this was attributed to stronger interfacial adhesion by the larger number of functional groups in FGs [52]. Yuan and coworkers [132] obtained nanocomposites with significantly improved thermal stability, enhancing by 133°C the initial decomposition temperature and by 94°C the temperature at maximum rate of weight loss (T_{max}) in the nanocomposites containing 1 wt% FGs.

The values previously summarized were obtained under nitrogen atmosphere. A review has shown a variety of behaviors in the thermal stability of nanocomposites [162]. The presence or absence of oxygen in the atmosphere in which degradation is carried out has been an important factor for determining whether the fillers have an advantageous influence on the polymer [161,162]. Only a few works have shown the thermal degradation of graphene nanocomposites in an air atmosphere [51,112,140,145]. The incorporation of graphene improved significantly the thermal oxidative stability of PP showing an enhancement of 26°C in the initial degradation temperature when adding 1 wt% of PP latex-CRGO [140]. PE nanocomposites containing 8 wt% FGs displayed an increase of 47°C and 21°C at a 50% and 80% weight loss, respectively [51].

5.5.3 THERMAL CONDUCTIVITY

The excellent thermal conduction property of graphene shows potential for thermal management in polymer nanocomposites [4]. In materials with higher thermal conductivity, heat is distributed through the material and temperature increase at the surface is decelerated [133]. Nevertheless, thermal

conductivity studies of graphene/polymer nanocomposites are limited; this could be because the improvements in this property are not as outstanding as the exponential increase in electrical conductivity [8,156]. In addition, most research has been mainly focused on amorphous matrices as summarized by several reviews [138,141,156,163].

The thermal conductivity of GNs/PANI nanocomposites showed relatively low values even with high GNs content (30 wt%) [44]. According to research conducted by Ruoff et al., CB provided larger improvement in thermal conductivity than TRGO; the greater modulus and aspect ratio of TRGO platelets compared with CB provided higher reinforcement at the cost of a relatively smaller increase of thermal conductivity in the NR nanocomposites [113]. The poor response in this property after the incorporation of these nanofillers can be ascribed to defects and residual functional groups in the structure of graphene which is a key problem limiting its intrinsic thermal conductivity and interfacial thermal resistance [156]. In contrast, PP latex-g-CRGO significantly enhanced the thermal conductivity of PP, and this property monotonously increased with increasing the nanofiller loading; a content of 5 wt% PP latex-g-CRGO resulted in a thermal conductivity of 0.396 W/mK, about two times than that of neat PP (0.201 W/mK) [140]. Furthermore, based on the theoretical and experimental results, graphene nanofillers can improve this property more effectively than CNTs [156].

5.6 CONCLUDING REMARKS

Graphene/polymer nanocomposites are one of the most promising applications of this 2D nanomaterial. Owing to the vast research concerning these nanocomposites, this chapter has been particularly focused on the semicrystalline polymer matrices. Furthermore, most industrial polymers exhibit crystallization and the inclusion of nanofillers affect the structural features of these matrices. Owing to the nanometric nature of graphene, the nanocomposite crystallization behavior is mainly driven by two factors: (1) a small content of these nanofillers provides a high density of nucleation sites for polymer crystallization but (2) the nanoconfinement provided by the homogenously dispersed graphene sheets can hinder the ordering of polymer chains especially at high nanofiller loading. In this context, both competing mechanisms explain the reason that most studies have shown the crystallinity degree in graphene nanocomposites remains very similar to that of pure polymer. On the other hand, several investigations have found modifications in the crystalline phases and crystal dimensions of polymers due to the incorporation of graphene. These changes have been sometimes related to an improved mechanical response of the nanocomposites. The reduced size of the microscopic crystalline domains obtained from the well-dispersed graphene nanosheets in the matrix also has an influence on the nanocomposite properties. Another approach to improve the polymer properties has been made by means of the epitaxial crystallization of polymers on graphene sheets. The hybrid crystalline structures obtained can be assembled in nanocomposites and can provide a better interfacial interaction between these structures and the matrix. There is a variety of research concerning the effects of graphene on the crystallization behavior of polymers; however, there are still many questions regarding the relationship between the crystal structure and the mechanical and thermal properties. Most of the research on these properties has been addressed on the effect provided by the physical and chemical modifications of graphene and the nanofiller content. The reason of the interest in studying these parameters is because these properties are strongly dependent on the dispersion and interaction of graphene sheets within the polymer matrix. Therefore, there are still many challenges for explaining the structure–property relationships of graphene/polymer nanocomposites in order to achieve the highest performance provided by both phases.

REFERENCES

1. Lee, C., X. Wei, J.W. Kysar, and J. Hone. 2008. Measurement of the elastic properties and intrinsic strength of monolayer graphene. *Science* 321(5887):385–8.
2. Bae, S., H. Kim, Y. Lee et al. 2010. Roll-to-roll production of 30-inch graphene films for transparent electrodes. *Nat Nanotech* 5(8):574–8.
3. Avouris, P. 2010. Graphene: Electronic and photonic properties and devices. *Nano Lett* 10(11):4285–94.
4. Balandin, A.A., S. Ghosh, W. Bao et al. 2008. Superior thermal conductivity of single-layer graphene. *Nano Lett* 8(3):902–7.
5. Kuilla, T., S. Bhadra, D. Yao, N.H. Kim, S. Bose, and J.H. Lee. 2010. Recent advances in graphene based polymer composites. *Prog Polym Sci* 35(11):1350–75.
6. Young, R.J., I.A. Kinloch, L. Gong, and K.S. Novoselov. 2012. The mechanics of graphene nanocomposites: A review. *Compos Sci Technol* 72(12):1459–76.
7. Potts, J.R., D.R. Dreyer, C.W. Bielawski, and R.S. Ruoff. 2011. Graphene-based polymer nanocomposites. *Polymer* 52(1):5–25.
8. Verdejo, R., M.M. Bernal, L.J. Romasanta, and M.A. Lopez-Manchado. 2011. Graphene filled polymer nanocomposites. *J Mater Chem* 21(10):3301–10.
9. Compton, O.C. and S.T. Nguyen. 2010. Graphene oxide, highly reduced graphene oxide, and graphene: Versatile building blocks for carbon-based materials. *Small* 6(6):711–23.
10. Park, S. and R.S. Ruoff. 2009. Chemical methods for the production of graphenes. *Nat Nanotechnol* 4(4):217–24.
11. Wong, C.H., A. Ambrosi, and M. Pumera. 2012. Thermally reduced graphenes exhibiting a close relationship to amorphous carbon. *Nanoscale* 4(16), 4972–7.
12. Soldano, C., A. Mahmood, and E. Dujardin. 2010. Production, properties and potential of graphene. *Carbon* 48(8):2127–50.
13. Cao, Y., J. Feng, and P. Wu. 2010. Preparation of organically dispersible graphene nanosheet powders through a lyophilization method and their poly(lactic acid) composites. *Carbon* 48(13):3834–9.
14. Milani, M.A., D. González, R. Quijada et al. 2013. Polypropylene/graphene nanosheet nanocomposites by *in situ* polymerization: Synthesis, characterization and fundamental properties. *Compos Sci Technol* 84:1–7.
15. Das, S., F. Irin, H.S.T. Ahmed et al. 2012. Non-covalent functionalization of pristine few-layer graphene using triphenylene derivatives for conductive poly (vinyl alcohol) composites. *Polymer* 53(12):2485–94.
16. Xu, J.-Z., Y.-Y. Liang, G.-Z. Zhong et al. 2012. Graphene oxide nanosheet induced intrachain conformational ordering in a semicrystalline polymer. *J Phys Chem Lett* 3(4):530–5.

17. Sun, Y. and C. He. 2012. Synthesis and stereocomplex crystallization of poly(lactide) – graphene oxide nanocomposites. *ACS Macro Lett* 1:709–13.
18. Navarro-Pardo, F., G. Martínez-Barrera, A.L. Martínez-Hernández et al. 2012. Nylon 6,6 electrospun fibres reinforced by amino functionalised 1D and 2D carbon. *IOP Conf Ser: Mater Sci Eng* 40:012023.
19. Terrones, M., O. Martín, M. González et al. 2011. Interphases in graphene polymer-based nanocomposites: Achievements and challenges. *Adv Mater* 23(44):5302–10.
20. Biswas, S., H. Fukushima, and L.T. Drzal. 2011. Mechanical and electrical property enhancement in exfoliated graphene nanoplatelet/liquid crystalline polymer nanocomposites. *Composites Part A-Appl Sci Manuf* 42(4):371–5.
21. Wu, Z., X. Chen, S. Zhu et al. 2013. Room temperature methane sensor based on graphene nanosheets/polyaniline nanocomposite thin film. *IEEE Sens J* 13(2):777–82.
22. Chatterjee, S., F.A. Nüesch, and B.T.T. Chu. 2013. Crystalline and tensile properties of carbon nanotube and graphene reinforced polyamide 12 fibers. *Chem Phys Lett* 557:92–6.
23. Lamastra, F.R., D. Puglia, M. Monti et al. 2012. Poly(ε-caprolactone) reinforced with fibres of poly(methyl methacrylate) loaded with multiwall carbon nanotubes or graphene nanoplatelets. *Chem Eng J* 195–196:140–8.
24. Pang, H., G.-J. Zhong, Y. Wang et al. 2012. In-situ synchrotron x-ray scattering study on isothermal crystallization of ethylene-vinyl acetate copolymers containing a high weight fraction of carbon nanotubes and graphene nanosheets. *J Polym Res* 19:9837.
25. Xu, J.-Z., T. Chen, C.-L. Yang et al. 2010. Isothermal crystallization of poly(l-lactide) induced by graphene nanosheets and carbon nanotubes: A comparative study. *Macromolecules* 43(11):5000–8.
26. Yang, J.-S., C.-L. Yang, M.-S. Wang, B.-D. Chen, and X.-G. Ma. 2011. Crystallization of alkane melts induced by carbon nanotubes and graphene nanosheets: A molecular dynamics simulation study. *Phys Chem Chem Phys* 13(34):15476–82.
27. Ansari, M.O., S.K. Yadav, J.W. Cho, and F. Mohammad. 2013. Thermal stability in terms of DC electrical conductivity retention and the efficacy of mixing technique in the preparation of nanocomposites of graphene/polyaniline over the carbon nanotubes/polyaniline. *Composites Part B-Eng* 47:155–61.
28. Bao, Q., H. Zhang, J.-X. Yang et al. 2010. Graphene-polymer nanofiber membrane for ultrafast photonics. *Adv Funct Mater* 20(5):782–91.
29. Stankovich, S., D.A. Dikin, G.H.B. Dommett et al. 2006. Graphene-based composite materials. *Nature* 442(7100): 282–6.
30. Heldt, C.L., A.K. Sieloff, J.P. Merillat et al. 2013. Stacked graphene nanoplatelet paper sensor for protein detection. *Sens Actuators B-Chem* 181:92–8.
31. Mortazavi, B., F. Hassouna, A. Laachachi et al. 2013. Experimental and multiscale modeling of thermal conductivity and elastic properties of PLA/expanded graphite polymer nanocomposites. *Thermochim Acta* 552:106–13.
32. Roy, N., R. Sengupta, and A.K. Bhowmick. 2012. Modifications of carbon for polymer composites and nanocomposites. *Prog Polym Sci* 37(6):781–819.
33. Yang, X., X. Wang, J. Yang, J. Li, and L. Wan. 2013. Functionalization of graphene using trimethoxysilanes and its reinforcement on polypropylene nanocomposites. *Chem Phys Lett* 570:125–31.
34. Wang, Y., Z. Shi, and J. Yin. 2011. Kevlar oligomer functionalized graphene for polymer composites. *Polymer* 52(16):3661–70.

35. Wu, H. and L.T. Drzal. 2012. Graphene nanoplatelet paper as a light-weight composite with excellent electrical and thermal conductivity and good gas barrier properties. *Carbon* 50(3):1135–45.

36. Lammer, H. 2010. Sporting goods with graphene material. US Patent 20100125013, filed November, 2009 and issued May, 2010.

37. He, W., W. Zhang, Y. Li, and X. Jing. 2012. A high concentration graphene dispersion stabilized by polyaniline nanofibers. *Synth Met* 162(13–14):1107–13.

38. Wan, C. and B. Chen. 2011. Poly(ε-caprolactone)/graphene oxide biocomposites: Mechanical properties and bioactivity. *Biomed Mater* 6:055010.

39. Jog, J.P. 2006. Crystallisation in polymer nanocomposites. *Mater Sci Technol* 22(7):797–806.

40. Martínez-Hernández, A.L., C. Velasco-Santos, and V.M. Castaño. 2010. Carbon nanotubes composites: Processing, grafting and mechanical and thermal properties. *Curr Nanosci* 6(1):12–39.

41. Harrats, C. and G. Groeninckx. 2008. Features, questions and future challenges in layered silicates clay nanocomposites with semicrystalline polymer matrices. *Macromol Rapid Commun* 29(1):14–26.

42. De Rosa, C., F. Auriemma, and O.R. De Ballesteros. 2006. A microscopic insight into the deformation behavior of semicrystalline polymers: The role of phase transitions. *Phys Rev Lett* 96:167801.

43. Ning, N., S. Fu, W. Zhang et al. 2012. Realizing the enhancement of interfacial interaction in semicrystalline polymer/filler composites via interfacial crystallization. *Prog Polym Sci* 37(10):1425–55.

44. Lu, Y., Y. Song, and F. Wang. 2013. Thermoelectric properties of graphene nanosheets-modified polyaniline hybrid nanocomposites by an *in situ* chemical polymerization. *Mater Chem Phys* 138(1):238–44.

45. Chunder, A., J. Liu, and L. Zhai. 2010. Reduced graphene oxide/poly(3-hexylthiophene) supramolecular composites. *Macromol Rapid Commun* 31(4):380–84.

46. Cheng, H.K.F., N.G. Sahoo, Y.P. Tan et al. 2012. Poly(vinyl alcohol) nanocomposites filled with poly(vinyl alcohol)-grafted graphene oxide. *ACS Appl Mater Interfaces* 4(5):2387–94.

47. Chiu, F.-C. and I.-N. Huang. 2012. Phase morphology and enhanced thermal/mechanical properties of polyamide 46/graphene oxide nanocomposites. *Polym Test* 31(7):953–62.

48. El Achaby, M., F.Z. Arrakhiz, S. Vaudreuil, E.M. Essassi, and A. Qaiss. 2012. Piezoelectric β-polymorph formation and properties enhancement in graphene oxide—PVDF nanocomposite films. *Appl Surf Sci* 258(19):7668–77.

49. Huang, H.-D., P.-G. Ren, J. Chen, W.-Q. Zhang, X. Ji, and Z.-M. Li. 2012. High barrier graphene oxide nanosheet/poly(vinyl alcohol) nanocomposite films. *J Membr Sci* 409–410:156–63.

50. Rodríguez-González, C., O.V. Kharissova, A.L. Martínez-Hernández et al. 2012. Graphene oxide sheets covalently grafted with keratin obtained from chicken feathers. *Dig J Nanomater Biostruct* 8(1):127–38.

51. Kuila, T., S. Bose, A.K. Mishra, P. Khanra, N.H. Kim, and J.H. Lee. 2012. Effect of functionalized graphene on the physical properties of linear low density polyethylene nanocomposites. *Polym Test* 31(1):31–8.

52. Yun, Y.S., Y.H. Bae, D.H. Kim et al. 2011. Reinforcing effects of adding alkylated graphene oxide to polypropylene. *Carbon* 49(11):3553–9.

53. Dreyer, D.R., S. Park, C.W. Bielawski, and R.S. Ruoff. 2010. The chemistry of graphene oxide. *Chem Soc Rev* 39(1):228–40.

54. Rodríguez-González, C., A.L. Martínez-Hernández, V.M. Castaño et al. 2012. Polysaccharide nanocomposites reinforced with graphene oxide and keratin-grafted graphene oxide. *Ind Eng Chem Res* 51:3619–29.

55. Han, D., L. Yan, W. Chen, and W. Li. 2011. Preparation of chitosan/graphene oxide composite film with enhanced mechanical strength in the wet state. *Carbohydr Polym* 83(2):653–8.

56. Lee, J.H., J. Marroquin, K.Y. Rhee, S.J. Park, and D. Hui. 2013. Cryomilling application of graphene to improve material properties of graphene/chitosan nanocomposites. *Composites B-Eng* 45(1):682–7.

57. Lee, S., B.G. Choi, D. Choi, and H.S. Park. 2014. Nanoindentation of annealed Nafion/sulfonated graphene oxide nanocomposite membranes for the measurement of mechanical properties. *J Membr Sci* 451:40–5.

58. Sayyar, S., E. Murray, B.C. Thompson, S. Gambhir, D.L. Officer, and G.G. Wallace. 2013. Covalently linked biocompatible graphene/polycaprolactone composites for tissue engineering. *Carbon* 52:296–304.

59. Hou, W., B. Tang, L. Lu et al. 2014. Preparation and physicomechanical properties of amine-functionalized graphene/polyamide 6 nanocomposite fiber as a high performance material. *R Soc Chem Adv* 4(10):4848–55.

60. Zhang, F., X. Peng, W. Yan, Z. Peng, and Y. Shen. 2011. Nonisothermal crystallization kinetics of *in situ* nylon 6/graphene composites by differential scanning calorimetry. *J Polym Sci Polym Phys* 49(19):1381–8.

61. Zheng, D., G. Tang, H.-B. Zhang et al. 2012. *In situ* thermal reduction of graphene oxide for high electrical conductivity and low percolation threshold in polyamide 6 nanocomposites. *Compos Sci Technol* 72(2):284–9.

62. Liu, H.-H., W.-W. Peng, L.-C. Hou et al. 2013. The production of a melt-spun functionalized graphene/poly(ε-caprolactam) nanocomposite fiber. *Compos Sci Technol* 81:61–8.

63. Nguyên, L., S.-M. Choi, D.-H. Kim et al. 2014. Preparation and characterization of nylon 6 compounds using the nylon 6-grafted GO. *Macromol Res* 22(3):257–63.

64. Liu, H., L. Hou, W. Peng, Q. Zhang, and X. Zhang. 2012. Fabrication and characterization of polyamide 6-functionalized graphene nanocomposite fiber. *J Mater Sci* 47(23):8052–60.

65. Rafiq, R., D. Cai, J. Jin, and M. Song. 2010. Increasing the toughness of nylon 12 by the incorporation of functionalized graphene. *Carbon* 48(15):4309–14.

66. Pant, H.R., C.H. Park, L.D. Tijing et al. 2012. Bimodal fiber diameter distributed graphene oxide/nylon-6 composite nanofibrous mats via electrospinning. *Colloids Surf A*, 407:121–5.

67. Liu, H., Y. Wang, X. Gou et al. 2013. Three-dimensional graphene/polyaniline composite material for high-performance supercapacitor applications. *Mater Sci Eng B-Solid* 178(5):293–8.

68. Ning, G., T. Li, J. Yan et al. 2013. Three-dimensional hybrid materials of fish scale-like polyaniline nanosheet arrays on graphene oxide and carbon nanotube for high-performance ultracapacitors. *Carbon* 54:241–8.

69. Fabbri, P., E. Bassoli, S.B. Bon, and L. Valentini. 2012. Preparation and characterization of poly (butylene terephthalate)/grapheme composites by in-situ polymerization of cyclic butylene terephthalate. *Polymer* 53(4):897–902.

70. Wang, X., H. Yang, L. Song et al. 2011. Morphology, mechanical and thermal properties of graphene-reinforced poly(butylene succinate) nanocomposites. *Compos Sci Technol* 72(1):1–6.

71. Cai, D. and M. Song. 2009. A simple route to enhance the interface between graphite oxide nanoplatelets and a

semi-crystalline polymer for stress transfer. *Nanotechnology* 20(31):315708.

72. Zhang, J. and Z. Qiu. 2011. Morphology, crystallization behavior, and dynamic mechanical properties of biodegradable poly(ε-caprolactone)/thermally reduced graphene nanocomposites. *Ind Eng Chem Res* 50:13885–91.

73. Yang, J.-H., Y. Shen, W.-D. He et al. 2013. Synergistic effect of poly(ethylene glycol) and graphene oxides on the crystallization behavior of poly(l-lactide). *J Appl Polym Sci* 130(5):3498–508.

74. Inuwa, I.M., A. Hassan, S.A. Samsudin et al. 2014. Characterization and mechanical properties of exfoliated graphite nanoplatelets reinforced polyethylene terephthalate/polypropylene composites. *J Appl Polym Sci* 130:3498–508.

75. Liu, K., L. Chen, Y. Chen et al. 2011. Preparation of polyester/reduced graphene oxide composites via *in situ* melt polycondensation and simultaneous thermo-reduction of graphene oxide. *J Mater Chem* 21(24):8612–7.

76. Jing, X. and Z. Qiu. 2012. Effect of low thermally reduced graphene loadings on the crystallization kinetics and morphology of biodegradable poly(3-hydroxybutyrate). *Ind Eng Chem Res* 51(42):13686–91.

77. Sridhar, V., I. Lee, H.H. Chun, and H. Park. 2013. Graphene reinforced biodegradable poly(3-hydroxybutyrate-co-4-hydroxybutyrate) nano-composites. *Express Polym Lett* 7(4):320–8.

78. Kong, J.-Y., M.-C. Choi, G.Y. Kim et al. 2012. Preparation and properties of polyimide/graphene oxide nanocomposite films with Mg ion crosslinker. *Eur Polym J* 48(8):1394–405.

79. Huang, T., R. Lu, C. Su et al. 2012. Chemically modified graphene/polyimide composite films based on utilization of covalent bonding and oriented distribution. *ACS Appl Mater Interfaces* 4(5):2699–708.

80. Huang, T., Y. Xin, T. Li. 2013. Modified graphene/polyimide nanocomposites: Reinforcing and tribological effects. *Appl Mater Interfaces* 5:4878–91.

81. Kim, I.H. and Y.G. Jeong. 2010. Polylactide/exfoliated graphite nanocomposites with enhanced thermal stability, mechanical modulus, and electrical conductivity. *J Polym Sci Polym Phys* 48(8):850–8.

82. Wu, D., Y. Cheng, S. Feng, Z. Yao, and M. Zhang. 2013. Crystallization behavior of polylactide/graphene composites. *Ind Eng Chem Res* 52(20):6731–9.

83. Wang, H. and Z. Qiu. 2011. Crystallization behaviors of biodegradable poly(l-lactic acid)/graphene oxide nanocomposites from the amorphous state. *Thermochim Acta* 526(1–2):229–36.

84. Yang, J.-H., S.-H. Lin, and Y.-D. Lee. 2012. Preparation and characterization of poly(L-lactide)-graphene composites using the *in situ* ring-opening polymerization of PLLA with graphene as the initiator. *J Mater Chem* 22(21):10805–15.

85. Huang, H.-D., J.-Z. Xu, Y. Fan, L. Xu, and Z.-M. Li. 2013. Poly(l-lactic acid) crystallization in a confined space containing graphene oxide nanosheets. *J Phys Chem B* 117(36):10641–51.

86. Wang, H. and Z. Qiu. 2012. Crystallization kinetics and morphology of biodegradable poly(l-lactic acid)/graphene oxide nanocomposites: Influences of graphene oxide loading and crystallization temperature. *Thermochim Acta* 527:40–6.

87. Li, W., C. Shi, M. Shan et al. 2013. Influence of silanized low-dimensional carbon nanofillers on mechanical, thermomechanical, and crystallization behaviors of poly(L-lactic acid) composites—A comparative study. *J Appl Polym Sci* 130(2):1194–202.

88. Chen, Y., X. Yao, Q. Gu, and Z Pan. 2013. Non-isothermal crystallization kinetics of poly (lactic acid)/graphene nanocomposites. *J Polym Eng* 33(2):163–71.

89. Ning, N., W. Zhang, J. Yan et al. 2013. Largely enhanced crystallization of semi-crystalline polymer on the surface of glass fiber by using graphene oxide as a modifier. *Polymer* 54(1):303–9.

90. Yoon, O.J., C.Y. Jung, I.Y. Sohn et al. 2011. Nanocomposite nanofibers of poly(d, l-lactic-co-glycolic acid) and graphene oxide nanosheets. *Composites Part A-Appl Sci Manuf* 42(12):1978–84.

91. Das, B., K. Eswar Prasad, U. Ramamurty, and C.N.R. Rao. 2009. Nano-indentation studies on polymer matrix composites reinforced by few-layer graphene. *Nanotechnology* 20:125705.

92. Xu, Y., W. Hong, H. Bai, C. Li, and G. Shi. 2009. Strong and ductile poly(vinyl alcohol)/graphene oxide composite films with a layered structure. *Carbon* 47(15):3538–43.

93. Kim, H.M., J.K. Lee, and H.S. Lee. 2011. Transparent and high gas barrier films based on poly(vinyl alcohol)/graphene oxide composites. *Thin Solid Films* 519(22):7766–71.

94. Yang, X., L. Li, S. Shang, and X.-M. Tao. 2010. Synthesis and characterization of layer-aligned poly(vinyl alcohol)/graphene nanocomposites. *Polymer* 51(15):3431–5.

95. Zhou, T., F. Chen, C. Tang et al. 2011. The preparation of high performance and conductive poly (vinyl alcohol)/graphene nanocomposite via reducing graphite oxide with sodium hydrosulfite. *Compos Sci Technol* 71(9):1266–70.

96. Zhao, X., Q. Zhang, D. Chen, and P. Lu. 2010. Enhanced mechanical properties of graphene-based polyvinyl alcohol composites. *Macromolecules* 43(5):2357–63.

97. Wang, J., X. Wang, C. Xu, M. Zhang, and X. Shang. 2011. Preparation of graphene/poly(vinyl alcohol) nanocomposites with enhanced mechanical properties and water resistance. *Polym Int* 60(5):816–22.

98. Liang, J., Y. Huang, L. Zhang et al. 2009. Molecular-level dispersion of graphene into poly(vinyl alcohol) and effective reinforcement of their nanocomposites. *Adv Funct Mater* 19(14):2297–302.

99. Huang, G., H. Liang, Y. Wang, X. Wang, J. Gao, and Z. Fei. 2012. Combination effect of melamine polyphosphate and graphene on flame retardant properties of poly(vinyl alcohol). *Mater Chem Phys* 132(2–3):520–8.

100. Layek, R.K., S. Samanta, D.P. Chatterjee, A.K. Nandi. 2010. Physical and mechanical properties of poly(methyl methacrylate) –functionalized graphene/poly(vinylidine fluoride) nanocomposites: Piezoelectric β-polymorph formation. *Polymer* 51:5846–56.

101. Mohamadi, S. and N. Sharifi-Sanjani. 2011. Investigation of the crystalline structure of PVDF in PVDF/PMMA/graphene polymer blend nanocomposites. *Polym Compos* 32(9):1451–60.

102. Li, Y., J.-Z. Xu, L. Zhu, G.-J. Zhong, and Z.-M. Li. 2012. Role of ion-dipole interactions in nucleation of gamma poly(vinylidene fluoride) in the presence of graphene oxide during melt crystallization. *J Phys Chem B* 116(51):14951–60.

103. Mohamadi, S., N. Sharifi-Sanjani, and A. Foyouhi. 2013. Evaluation of graphene nanosheets influence on the physical properties of PVDF/PMMA blend. *J Polym Res* 20:46.

104. Appel, A.-K., R. Thomann, and R. Mülhaupt. 2012. Polyurethane nanocomposites prepared from solvent-free stable dispersions of functionalized graphene nanosheets in polyols. *Polymer* 53(22):4931–9.

105. Yadav, S.K. and J.W. Cho. 2013. Functionalized graphene nanoplatelets for enhanced mechanical and thermal properties of polyurethane nanocomposites. *Appl Surf Sci* 266:360–7.

106. Raghu, A.V., Y.R. Lee, H.M. Jeong, and C.M. Shin. 2008. Preparation and physical properties of waterborne polyurethane/functionalized graphene sheet nanocomposites. *Macromol Chem Phys* 209(24):2487–93.

107. Nguyen, D.A., Y.R. Lee, A.V. Raghu et al. 2009. Morphological and physical properties of a thermoplastic polyurethane reinforced with functionalized graphene sheet. *Polym Int* 58(4):412–7.

108. Khan, U., P. May, A. O'Neill, and J.N. Coleman. 2010. Development of stiff, strong, yet tough composites by the addition of solvent exfoliated graphene to polyurethane. *Carbon* 48(14):4035–41.

109. Lee, Y.R., A.V. Raghu, H.M. Jeong, and B.K. Kim. 2009. Properties of waterborne polyurethane/functionalized graphene sheet nanocomposites prepared by an *in situ* method. *Macromol Chem Phys* 210(15):1247–54.

110. Wang, X., Y. Hu, L. Song et al. 2011. *In situ* polymerization of graphene nanosheets and polyurethane with enhanced mechanical and thermal properties. *J Mater Chem* 21(12):4222–7.

111. Vadukumpully, S., J. Paul, N. Mahanta, and S. Valiyaveettil. 2011. Flexible conductive graphene/poly(vinyl chloride) composite thin films with high mechanical strength and thermal stability. *Carbon* 49(1):198–205.

112. Ma, L., B. Yu, X. Qian et al. 2014. Functionalized graphene/thermoplastic polyester elastomer nanocomposites by reactive extrusion-based masterbatch: Preparation and properties reinforcement. *Polym Adv Technol* 25(6):605–612.

113. Potts, J.R., O. Shankar, S. Murali, L. Du, and R.S. Ruoff. 2013. Latex and two-roll mill processing of thermally-exfoliated graphite oxide/natural rubber nanocomposites. *Compos Sci Technol* 74:166–72.

114. Zhou, X., Z. Chen, Y. Qu, Q. Su, and X. Yang. 2013. Fabricating graphene oxide/poly(3-butylthiophene) hybrid materials with different morphologies and crystal structures. *RSC Adv* 3(13):4254–60.

115. Jeong, Y.G., D.H. Baik, J.W. Jang, B.G. Min, and K.H. Yoon. 2014. Preparation, structure and properties of poly(p-phenylene benzobisoxazole) composite fibers reinforced with graphene. *Macromol Res* 22(3):279–86.

116. Lin, Y., J. Jin, and M. Song. 2011. Preparation and characterisation of covalent polymer functionalized graphene oxide. *J Mater Chem* 21(10):3455–61.

117. Chaudhry, A.U. and V. Mittal. 2013. High-density polyethylene nanocomposites using masterbatches of chlorinated polyethylene/graphene oxide. *Polym Eng Sci* 53(1):78–88.

118. Cheng, S., X. Chen, Y.G. Hsuan, and C.Y. Li. 2012. Reduced graphene oxide-induced polyethylene crystallization in solution and nanocomposites. *Macromolecules* 45(2):993–1000.

119. Wang, H., P.-G. Ren, J.-Z Xu et al. 2014. Non-isothermal crystallization kinetics of alkyl-functionalized graphene oxide/high-density polyethylene nanocomposites. *Compos Interface* 21(3):203–15.

120. He, L., X. Zheng, Q. Xu, Z. Chen, and J. Fu. 2012. Comparison study of PE epitaxy on carbon nanotubes and graphene oxide and PE/graphene oxide as amphiphilic molecular structure for solvent separation. *Appl Surf Sci* 258(10):4614–23.

121. Kim, H., S. Kobayashi, M.A. Abdurrahim et al. 2011. Graphene/polyethylene nanocomposites: Effect of polyethylene functionalization and blending methods. *Polymer* 52(8):1837–46.

122. Kim, N.H., T. Kuila, and J.H. Lee. 2014. Enhanced mechanical properties of a multiwall carbon nanotube attached prestitched graphene oxide filled linear low density polyethylene composite. *J Mater Chem A* 2(8):2681–9.

123. Hu, W., J. Zhan, X. Wang et al. 2014. Effect of functionalized graphene oxide with hyper-branched flame retardant on flammability and thermal stability of cross-linked polyethylene. *Ind Eng Chem Res* 53(8):3073–83.

124. Wang, D., X. Zhang, J.-W. Zha et al. 2013. Dielectric properties of reduced graphene oxide/polypropylene composites with ultralow percolation threshold. *Polymer* 54(7):1916–22.

125. Cao, Y., J. Feng, and P. Wu. 2012. Polypropylene-grafted graphene oxide sheets as multifunctional compatibilizers for polyolefin-based polymer blends. *J Mater Chem* 22(30):14997–5005.

126. Ryu, S.H. and A.M. Shanmugharaj. 2014. Influence of long-chain alkylamine-modified graphene oxide on the crystallization, mechanical and electrical properties of isotactic polypropylene nanocomposites. *Chem Eng J* 244:552–60.

127. Ferreira, C.I., C. Dal Castel, M.A.S. Oviedo, and R.S. Mauler. 2013. Isothermal and non-isothermal crystallization kinetics of polypropylene/exfoliated graphite nanocomposites. *Thermochim Acta* 553:40–8.

128. Chen, J.-B., J.-Z. Xu, H. Pang et al. 2014. Crystallization of isotactic polypropylene inside dense networks of carbon nanofillers. *J Appl Polym Sci* 131(6):163–71.

129. Xu, J.-Z., C. Chen, Y. Wang et al. 2011. Graphene nanosheets and shear flow induced crystallization in isotactic polypropylene nanocomposites. *Macromolecules* 44(8):2808–18.

130. Xu, J.-Z., Y.-Y. Liang, H.-D. Huang et al. 2012. Isothermal and nonisothermal crystallization of isotactic polypropylene/graphene oxide nanosheet nanocomposites. *J Polym Res* 19:9975.

131. Navarro-Pardo, F., J. Laria, T. Lozano et al. 2013. Shear effect in beta-phase induction of polypropylene in a single screw extruder. *J Appl Polym Sci* 130: 2932–37.

132. Yuan, B., C. Bao, L. Song et al. 2014. Preparation of functionalized graphene oxide/polypropylene nanocomposite with significantly improved thermal stability and studies on the crystallization behavior and mechanical properties. *Chem Eng J* 237:411–20.

133. Dittrich, B., K.-A. Wartig, D. Hofmann, R. Mülhaupt, and B. Schartel. 2013. Flame retardancy through carbon nanomaterials: Carbon black, multiwall nanotubes, expanded graphite, multi-layer graphene and graphene in polypropylene. *Polym Degrad Stab* 98:1495–505.

134. Cao, Y., Z. Lai, J. Feng, and P. Wu. 2011. Graphene oxide sheets covalently functionalized with block copolymers via click chemistry as reinforcing fillers. *J Mater Chem* 21(25): 9271–8.

135. Zha, D.-A., S. Mei, Z. Wang et al. 2011. Superhydrophobic polyvinylidene fluoride/graphene porous materials. *Carbon* 49(15):5166–72.

136. Wang, J., J. Wu, W. Xu, Q., Zhang, and Q. Fu. 2014. Preparation of poly(vinylidene fluoride) films with excellent electric property, improved dielectric property and dominant polar crystalline forms by adding a quaternary phosphorus salt functionalized graphene. *Compos Sci Technol* 91:1–7.

137. Navarro-Pardo, F., G. Martínez-Barrera, A.L. Martínez-Hernández et al. 2013. Effects on the thermo-mechanical and crystallinity properties of nylon 6,6 electrospun fibres reinforced with one dimensional (1D) and two dimensional (2D) carbon. *Materials* 6:3494–513.

138. Cai, D. and M. Song. 2010. Recent advance in functionalized graphene/polymer nanocomposites. *J Mater Chem* 20(37):7906–15.

139. Georgakilas, V., M. Otyepka, A.B. Bourlinos et al. 2012. Functionalization of graphene: Covalent and non-covalent approaches, derivatives and applications. *Chem Rev* 112(11):6156–214.

140. Song, P., Z. Cao, Y. Cai et al. 2011. Fabrication of exfoliated graphene-based polypropylene nanocomposites with enhanced mechanical and thermal properties. *Polymer* 52(18):4001–10.

141. Song, M. and F.D. Cai. 2012. Graphene functionalization: A review in polymer–graphene nanocomposites. In *Polymer–Graphene Nanocomposites* ed. V. Mittal, 1–51. Cambridge: The Royal Society of Chemistry.

142. Bustos-Ramírez, K., A.L. Martínez-Hernández, G. Martínez-Barrera et al. 2013. Covalently bonded chitosan on graphene oxide via redox reaction. *Materials* 6(3):911–26.

143. Zhang, X., X. Fan, H. Li, and C. Yan. 2012. Facile preparation route for graphene oxide reinforced polyamide 6 composites via *in situ* anionic ring-opening polymerization. *J Mater Chem* 22(45):24081–91.

144. Liao, K.-H., Y. Qian, and C.W. MacOsko. 2012. Ultralow percolation graphene/polyurethane acrylate nanocomposites. *Polymer* 53(17):3756–61.

145. Cai D., J. Jin, K. Yusoh, R. Rafiq, M. Song. 2012. High performance polyurethane/functionalized graphene nanocomposites with improved mechanical and thermal properties. *Compos Sci Technol* 72:702–7.

146. Chen, H., W. Zhang, X. Du et al. 2013. Crystallization kinetics and melting behaviors of poly(L-lactide)/graphene oxides composites. *Thermochim Acta* 566:57–70.

147. Xu, J.-Z., G.-J. Zhong, B.S. Hsiao, Q. Fu, and Z.-M. Li. 2014. Low-dimensional carbonaceous nanofiller induced polymer crystallization. *Prog Polym Sci* 39(3):555–93.

148. Rissanou, A.N. and V. Harmandaris. 2014. Dynamics of various polymer-graphene interfacial systems through atomistic molecular dynamics simulations. *Soft Matter* 10(16):2876–88.

149. Yang, H., X.J. Zhao, and M. Sun. 2011. Induced crystallization of single-chain polyethylene on a graphite surface: Molecular dynamics simulation. *Phys Rev E—Stat, Nonlinear Soft Matter Phys* 84:011803.

150. Wang, L.-Z. and L.-L. Duan. 2012. Isothermal crystallization of a single polyethylene chain induced by graphene: A molecular dynamics simulation. *Comput Theor Chem* 1002:59–65.

151. Eslami, H. and F. Müller-Plathe. 2013. How thick is the interphase in an ultrathin polymer film? Coarse-grained molecular dynamics simulations of polyamide-6,6 on graphene. *J Phys Chem C* 117(10):5249–57.

152. Eslami, H. and F. Müller-Plathet. 2009. Structure and mobility of nanoconfined polyamide-6,6 oligomers: Application of a molecular dynamics technique with constant temperature, surface area and parallel pressure. *J Phys Chem B* 113(16):5568–81.

153. Papageorgiou, D.G., G.Z. Papageorgiou, D.N. Bikiaris, and K. Chrissafis. 2013. Crystallization and melting of propylene-ethylene random copolymers. Homogeneous nucleation and β-nucleating agents. *Eur Polym J* 49(6):1577–90.

154. Diwan, P., S. Harms, K. Raetzke, and A. Chandra. 2012. Polymer electrolyte-graphene composites: Conductivity peaks and reasons thereof. *Solid State Ionics* 217:13–8.

155. Salavagione, H.J., M.A. Gómez, and G. Martínez. 2009. Polymeric modification of graphene through esterification of graphite oxide and poly(vinyl alcohol). *Macromolecules* 42(17):6331–4.

156. Du, J. and H.-M. Cheng. 2012. The fabrication, properties, and uses of graphene/polymer composites. *Macromol Chem Phys* 213(10–11):1060–77.

157. Kuila, T., S. Bose, P. Khanra et al. 2011. Characterization and properties of *in situ* emulsion polymerized poly(methyl methacrylate)/graphene nanocomposites. *Composites Part A-Appl Sci Manuf* 42(11):1856–61.

158. Ebrahimi, S., K. Ghafoori-Tabrizi, and H. Rafii-Tabar. 2012. Multi-scale computational modelling of the mechanical behaviour of the chitosan biological polymer embedded with graphene and carbon nanotube. *Comput Mater Sci* 53(1):347–53.

159. Awasthi, A.P., D.C. Lagoudas, and D.C. Hammerand. 2009. Modeling of graphene–polymer interfacial mechanical behavior using molecular dynamics. *Model Simul Mater Sci* 17:015002.

160. Feng, H., Y. Li, and J. Li. 2012. Strong reduced graphene oxide-polymer composites: Hydrogels and wires. *RSC Adv* 2(17):6988–93.

161. Pielichowski, K., A. Leszczy, and J. Njuguna. 2010. Mechanisms of thermal stability enhancement in polymer nanocomposites. In *Optimization of Polymer Nanocomposite Properties*, ed. V. Mittal, 195–210. Weinheim: Wiley-VCH Verlag GmbH & Co.

162. Chrissafis, K. and D. Bikiaris. 2011. Can nanoparticles really enhance thermal stability of polymers? Part I: An overview on thermal decomposition of addition polymers. *Thermochim Acta* 523:1–24.

163. Song, W.-L., W. Wang, L.M. Veca et al. 2012. Polymer/carbon nanocomposites for enhanced thermal transport properties—carbon nanotubes versus graphene sheets as nanoscale fillers. *J Mater Chem* 22:17133.

6 Graphene-Like Structures as Cages for Doxorubicin

Iva Blazkova, Pavel Kopel, Marketa Vaculovicova, Vojtech Adam, and Rene Kizek

CONTENTS

Abstract 99
6.1   Introduction 99
6.2   Graphene and Its Oxide for Tumor Disease Diagnostics 100
     6.2.1   Biosensing of Tumor Cells 100
     6.2.2   Imaging of Tumor Cells 102
6.3   Graphene and Its Oxide for Drug Delivery Systems 102
     6.3.1   Cytostatics 102
     6.3.2   Gene Delivery 103
6.4   Methods for Studying Graphene-Based Biomaterials 104
6.5   Conclusions 105
Acknowledgment 106
References 106

ABSTRACT

Carbon nanomaterials including graphene belong to the most intensively explored in materials science. Extraordinary physicochemical and structural properties, and biocompatibility of graphene and graphene oxide predestine them for many potential applications including photocatalysis, electrochemistry, electronics, and optoelectronics. Moreover, the two-dimensional (2D) layer of sp²-bonded carbon atoms and the high specific surface area are very promising platforms for biomedical applications. In this chapter, we describe recent results on employing graphene and its oxide in the diagnosis and treatment of tumor diseases. Various drug delivery systems are discussed. Moreover, we give a brief discussion on the challenges and perspectives of these materials for future progress in the field of biomedical applications.

6.1   INTRODUCTION

Nanotechnology belongs to the most developing branch of science in the first decade of the twenty-first century. Nanomaterials are interesting because they have specific physicochemical properties and can be applied in many fields of science, electronics, optics, and medicine (Wang et al. 2009; Zhang, Nayak et al. 2012). The most used materials in medicine are quantum dots (Cai et al. 2007; Pollinger et al. 2014), paramagnetic nanoparticles (Thorek et al. 2006; Gong et al. 2009), liposomes (Burger et al. 2002; Lim et al. 2014), microspheres (Haase et al. 2013), polymeric shells (Maeda et al. 2001; Hirsch et al. 2006; Sciallero et al. 2013), and carbon nanotubes (Lacerda et al. 2006; Pumera 2009; Villegas et al. 2014). Some of the nanomaterials are applied as

transporters where cargo can be loaded by different kinds of mechanisms, such as encapsulation, surface absorption, and hydrogen bonding. Loading capacity can be improved by very strong π-stacking interactions as it was found in the binding of aromatic drug molecules to carbon nanotubes (Liu et al. 2007). A variety of carbon-based nanomaterials with potential of transporting properties are summarized in Figure 6.1.

Graphene (GFN), as well as its derivatives graphene oxide (GO), and reduced graphene oxide (rGO) are formed by 2D structure of sp²-hybridized carbon atoms arranged in six-membered rings with a high specific surface area (2630 m²/g). Physicochemical and structural properties of GFN, namely conductivity, high elasticity, mechanical strength, large surface area, and rapid heterogeneous electron transfer, make the material very interesting for many applications (Novoselov et al. 2004; Wei and Qu 2012; Kong and Huang 2014; Li et al. 2014). There are many papers and studies describing the use of GFN for widespread biomedical applications, ranging from biosensors (Shao et al. 2010; Huang et al. 2011; Song et al. 2011; Pandey et al. 2014), drug and gene delivery (Liu et al. 2008; Feng and Liu 2011; Liu et al. 2011; Misra et al. 2012; Zhou and Liang 2014), cell imaging, biological sensing and imaging (Yang et al. 2010; Zhang, Lu et al. 2012; Zhang, Nayak et al. 2012; Mao and Li 2013; Mosaiab et al. 2013; Liu, Gao et al. 2014; Zhang et al. 2014) to biocompatible scaffolds for cell culture. The intensive research on the bioapplications of GFN and its derivatives is due to many fascinating properties, such as high specific surface area, electronic and thermal conductivities, biocompatibility, facile biological/chemical functionalization, and low cost and scalable production (Guo and Dong 2011; Jiang 2011). It was a study of Dai et al. (Liu et al. 2008) in 2008 that started an interest in GO to be used as

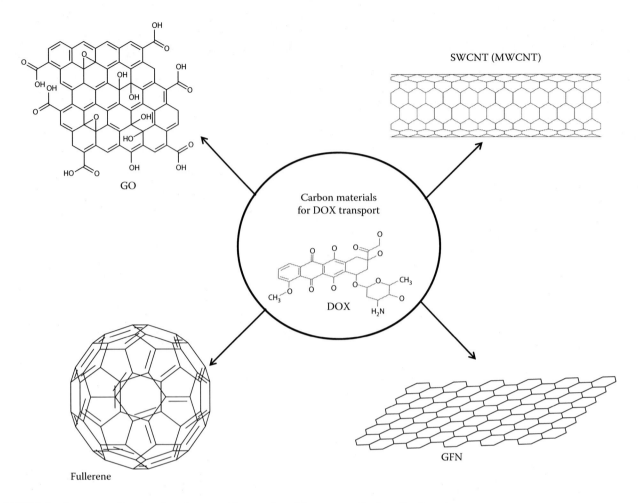

FIGURE 6.1 Summary of carbon-based nanocarriers employed for nanomedical applications.

an efficient nanotransporter for drug delivery. GO, prepared by oxidation of graphite by Hummers' method (Hummers and Offeman 1958), is an ideal nanotransporter for efficient drug and gene delivery. GO used for drug delivery is usually composed of 1–3 layers (1–2 nm thick), with size ranging from a few nanometers to several hundred nanometers (Kovtyukhova et al. 1999; Sun et al. 2008; Loh et al. 2010; Shen et al. 2012). Moreover, the presence of reactive COOH and OH groups enables the formation of composite materials with polymers (Shan et al. 2009; Cha et al. 2014), biomolecules as DNA (Lei et al. 2011; Liu et al. 2013), protein (Zhang et al. 2010; Lee et al. 2011; Tan et al. 2013), quantum dots (Dong et al. 2010; Markad et al. 2013), and iron oxide nanoparticles (Li et al. 2011; Mendes et al. 2012).

6.2 GRAPHENE AND ITS OXIDE FOR TUMOR DISEASE DIAGNOSTICS

GFN and GO are increasingly important nanomaterials, which exhibit great promise in the area of bionanotechnology and nanobiomedicine, such as biological imaging, molecular imaging, drug and gene delivery, and cancer therapy (Kim et al. 2011; Ku and Park 2013; Li and Yang 2013; Zhang, Peng et al. 2013) (Figure 6.2). However, exploration of GFN with

intracellular monitoring and *in situ* molecular probing is still at an early stage. Thus, graphene appears to be an auspicious nanoparticle in medicine and may bring novel opportunities for future disease diagnosis and treatment (Wang et al. 2010; Feng and Liu 2011; Yue et al. 2013). GFN and GO can inhibit the migration and invasion of cancer cells (Zhou et al. 2014). The different inorganic nanoparticles can be attached to the surface of GFN, obtaining functional graphene-based nanocomposites with good optical and magnetic properties useful for multimodal imaging and imaging-guided cancer therapy. GFN materials are used as carriers for antibody, DNA, protein, and drug (Du et al. 2012).

6.2.1 BIOSENSING OF TUMOR CELLS

Molecular biosensing systems can be based on naturally available and artificially designed enzymes, binding proteins, and antibodies (Cissell et al. 2008). GFN is a great material for biosensing and biological imaging (Du et al. 2012) and is a promising material for ultrasensitive nanomaterial-based biosensors. Manipulable multifunctionalized surface chemistry allows realizing sensitive and selective detection of biomolecules (Li and Yang 2013). As fast cancer diagnosis and effective measurement of cancer cells is a main challenge in

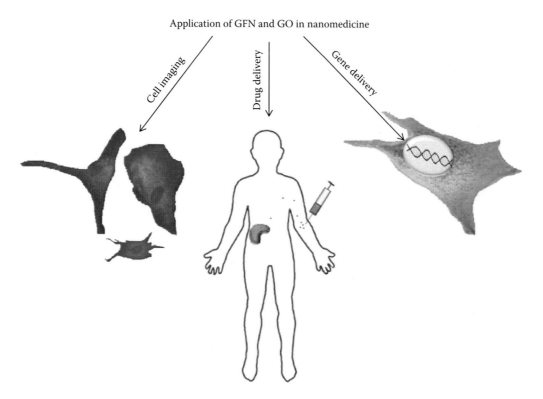

Application of GFN and GO in nanomedicine

Cell imaging

Drug delivery

Gene delivery

FIGURE 6.2 Schematic illustration of application of GFN and GO in nanomedicine.

early cancer diagnosis, miniature multiplex chip was created for *in situ* detection of cancer cells by implementing a novel GO-based Förster resonance energy transfer (FRET) biosensor strategy (Cao et al. 2012).

GFN can be also used in single-molecule label-free biosensing technologies (Kravets et al. 2013). Highly sensitive and label-free detection of the biomarker carbohydrate antigen 15-3 (CA 15-3) is prospect for breast cancer diagnosis and is an electrochemical immunosensor with a highly conductive GFN (i.e., N-doped graphene sheets)-modified electrode, exhibiting considerably increased electron transfer and high sensitivity toward CA 15-3. This strategy is promising for clinical research and diagnostic applications (Li et al. 2013). An electrochemical immunosensor for sensitive detection of cancer biomarker alpha-fetoprotein (AFP) was also seen. The sensor was based on a graphene sheet platform and functionalized carbon nanospheres (CNSs) labeled with horseradish peroxidase-secondary antibodies (HRP-Ab2). The developed immunosensor showed a 7-fold increase in detection signal in comparison to the immunosensor without GFN modification and CNSs labeling (Du et al. 2010).

The GO might be a promising material for targeted drug delivery to the lungs. Compared with other carbon nanomaterials, GO had long blood circulation time (halftime 5.3 ± 1.2 h), and GO showed good biocompatibility with red blood cells and was predominantly deposited in the lungs. Low uptake of GO was observed in the reticuloendothelial system (Zhang et al. 2011) and it was found that nanoparticle vectors can penetrate endothelial barriers to reach tumor sites (Portney and Ozkan 2006).

The novel highly sensitive multiplex electrochemiluminescence (ECL) immunoassay for the simultaneous detection of AFP (marker for hepatocellular and germ cell carcinoma) and CEA (carcinoembryonic antigen) was developed using QDs (quantum dots) as trace tag and GFN as a conducting bridge. This immunosensor is great for simultaneous detection of AFP and CEA with wide linear ranges, low detection limits, good specificity, and acceptable accuracy (Guo et al. 2013). Moreover, novel graphene oxide sheets/polyaniline/CdSe (GO/PANI/CdSe) quantum dot nanocomposites were prepared and used for sensitive ECL biosensing (Hu et al. 2013). The graphene-based cathodic electrogenerated chemiluminescence immunosensor was demonstrated to determine PSA (prostate-specific antigen) in human serum samples (Xu et al. 2011). As an alternative, the modification of the immunosensor surface led to acceleration of electron transfer and high specificity and sensitivity was demonstrated (Wu et al. 2013). The described immunosensor was prepared by covalent immobilization of antibodies on a chitosan/electrochemically rGO film-modified glassy carbon electrode. Cells were captured with a sandwich-type immunoreaction and various QD-coated silica nanoparticle tracers were captured on the surface of the cells.

It is necessary to pay the attention to the toxicity of the GO as soon as it is used for biomedical applications (Zhang et al. 2011). The cytotoxicity characteristics, cellular-uptake mechanism, and intracellular metabolic pathway of GFN and its derivatives are still not well understood (Du et al. 2012; Pumera 2012; Horvath et al. 2013). It was observed that after the exposure of the mice to GO (10 mg/kg body weight) for

14 days, changes including inflammation, cell infiltration, pulmonary edema, and granuloma formation were found. No pathological changes were observed in the examined organs when mice were exposed to 1 mg/kg body weight of GO for 14 days (Zhang et al. 2011). When 80 mg/kg was injected intravenously into mice, there was 100% fatality in the GO-treated group, but there was 100% survival among mice treated with pGO (polyethylene glycol-grafted graphene oxide nanosheets). The pGO nanosheets have superior *in vivo* safety relative to GO (Miao et al. 2013). Oxidative stress and cell wall membrane damage were determined as the main mechanisms involved in the cytotoxicity of rGO sheets (Akhavan et al. 2012). GO is found to be more toxic than rGO of the same size. GO and rGO induce significant increases in both intercellular reactive oxygen species (ROS) levels and messenger RNA (mRNA) levels of heme oxygenase 1 (HO1) and thioredoxin reductase (TrxR) (Das et al. 2013). Better water solubility and stability of GO caused its grafting with phosphorylcholine oligomer. Good biocompatibility and incorporation into cells by endocytosis was then observed (Liu, Zhang et al. 2014). The aspiration of GFN caused only mild oxidation and did not cause inflammation in mice, so it probably could be used in tissue engineering (Schinwald et al. 2014). The ability to stimulate myogenic differentiation by GO shows a potential for skeletal tissue engineering applications (Ku and Park 2013). Biocompatibly coated nano-graphene with ultrasmall sizes can be cleared out from the organism after systemic administration, without noticeable toxicity to the treated mice (Li and Yang 2013). GO has cytoprotective effect; it enables control over gene transfection through region-selective gene delivery only into GO-untreated cells and not into the GO-treated cells. GO can protect cells from internalization of toxic hydrophobic molecules, nanoparticles, and nucleic acids such as siRNA and plasmid DNA by interacting with cell surface lipid bilayers without noticeably reducing cell viability (Na et al. 2013).

6.2.2 Imaging of Tumor Cells

Tumor imaging has become an indispensable tool in the study of cancer biology and in clinical prognosis and treatment (Condeelis and Weissleder 2010). The backbone of biomolecules and biological structures is carbon based, thus, it is no doubt to integrate biological systems with nanocarbons and to use graphene-based scaffold for cell culture (Yang et al. 2013). Owing to fine biocompatibility and low toxicity strongly fluorescent graphene quantum dots (GQDs) are demonstrated to be excellent bioimaging agents in cell imaging. The synthesized GQDs showed high solubility, excellent biocompatibility, and excellent optical properties and could be used directly for intracellular imaging without any surface modification (Zhu et al. 2011; Sun et al. 2013). Graphene-based FRET for cellular imaging was also reported (Wang et al. 2010). A multicomponent nanosystem based on GFN and containing individual cyclodextrins (hosts for functional units) can be used as imaging agents, for anticancer drug delivery, and as tumor-specific ligands. The cyclodextrin-functionalized graphene nanosheet (GNS/-CD) facilitates host–guest chemistry between the nanohybrid and functional payloads (Dong et al. 2013). Moreover, biocompatible nitrogen-doped graphene quantum dots (N-GQDs) may be used as efficient two-photon fluorescent probes for cellular and deep-tissue imaging. N-GQD can achieve a large imaging depth of 1800 μm, significantly exceeding the fundamental two-photon imaging depth limit. The N-GQD is nontoxic to living cells and exhibits great photostability under repeated laser irradiation (Zhang, Lu et al. 2012). Moreover, graphene oxide nanoparticles (GONs) have been shown to be good optical probes because of their strong two-photon luminescence (Zhang, Lu et al. 2012). After the cell labeling by GONs, highly localized and low power/energy therapy can be achieved. They can be functionalized with other targeting molecules, which enable more specific targeting into the different malignant tissues.

6.3 GRAPHENE AND ITS OXIDE FOR DRUG DELIVERY SYSTEMS

6.3.1 Cytostatics

GFN has been widely explored as novel nanocarrier for drug delivery (Li and Yang 2013). Ultrasmall GO nanosheets (less than 50 nm) could be used as the ideal nanocarriers for drug delivery due to the great biocompatibility, lower cytotoxicity, and higher cellular uptake amount compared to the random large GO nanosheets (Zhang, Peng et al. 2013). A triple functionalized drug delivery system was developed by encapsulation of superparamagnetic GO and doxorubicin (DOX) (Kizek et al. 2012; Stiborova et al. 2012) with folic acid (FA)-conjugated chitosan (CHI) (Wang, Zhou et al. 2013). The release of DOX was pH sensitive; the lower pH values lead to weaker hydrogen bonds and degradation of CHI, and thus result in a higher release rate of DOX. This system could be used as a dual-targeted drug nanocarrier by combination of biological (active) and magnetic (passive) targeting capabilities. The micrograph of human fibroblast cells exposed to GFN conjugated to DOX is shown in Figure 6.3. Besides, a multifunctional superparamagnetic graphene oxide–iron oxide hybrid nanocomposite (GO–IONP) can be synthetized and functionalized by a biocompatible polyethylene glycol (PEG) polymer to achieve high stability in physiological solutions. DOX can be loaded onto GO–IONP–PEG, forming a GO–IONP–PEG–DOX complex, which enables magnetically targeted drug delivery (Ma et al. 2012). The targeting peptide (PI)-modified mesoporous silica-coated graphene nanosheet (GSPI) can serve in the field of drug delivery. Conjugation of DOX with chlorotoxin—GO complex increased the toxicity of DOX to cancer cells (Wang et al. 2014). Hyaluronic acid–GO conjugate has very low toxicity to cells and after its connection with DOX selectively delivered the drug and inhibited the growth tumors in mice (Wu et al. 2014). A DOX-loaded GSPI-based system (GSPID) showed heat-stimulative, pH-responsive, and sustained release properties. Combined therapy, chemotherapy, and photothermal therapy

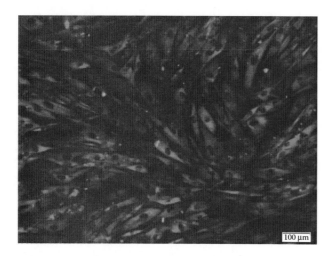

FIGURE 6.3 Fluorescent micrograph of human foreskin fibroblasts exposed to DOX-GFN (30 min incubation, excitation 520–550 nm, emission: 580 nm, exposure time: 104.5 ms, ISO 200, zoom: 100×, temperature: 37°C, and atmosphere: 5% CO_2).

were more toxic to glioma cells compared with single chemotherapy or photothermal therapy (Wang, Wang et al. 2013). Kavitha et al. (2013) presented GO functionalized covalently with pH-sensitive poly[2-(diethylamino) ethyl methacrylate] (PDEA). Common water-insoluble cancer drug camptothecin (CPT) was loaded on GO-PDEA by simple physisorption by π–π stacking and by hydrophobic interaction. Loaded CPT can be released only at the lower (acidic) pH, which is normally found in a tumor environment but not in basic and neutral pH. GO-PDEA is not toxic to N2a cancer cells but the GO-PDEA-CPT complex exhibited high potency in killing N2a cancer cells *in vitro*. The GO-PDEA nanocargo carrier could be a good material for site-specific anticancer drug delivery and controlled release. The interaction of GO, a medicinal drug 10-hydroxy camptothecin (HCPT), and bovine serum albumin (BSA) were studied by Ni et al. (2013). The delivery of HCPT to BSA was improved in the presence of GO. GO enhances the fluorescence response of HCPT to BSA. The low-cost fluorescence biosensing platform was created for fluorescence-enhanced detection of BSA based on GO. Alternatively, daunorubicin (DNR)-loaded graphene–gold nanocomposites (GGN) inducing apoptosis in drug-resistant leukemia cells (K562/A02; KA) were investigated by Zhang, Chang et al. (2013). The monoclonal P-glycoprotein (P-gp) antibodies and DNR anticancer drug linked to GGN were proved to be a good drug delivery vector that induces apoptosis of KA cells and inhibits tumor growth in KA nude mice. Moreover, enhanced cell death was detected by combination treatment of SWNT/GO and Tx (paclitaxel), indicating a synergistic effect and demonstrated the potential of SWNT/GO as co-therapeutic agents with Tx for the treatment of lung cancer (Arya et al. 2013). With the aim to improve its water solubility and biocompatibility, 6-armed PEG was grafted onto GO via an amidation process. Oridonin, a cancer chemotherapy drug, was connected on GO-PEG. The drug loading ratio (105%) was

higher than in other ordinary drug carriers. The GO-PEG/oridonin nanocarrier showed higher cytotoxicity in A549 and MCF-7 cells in comparison to oridonin (Xu et al. 2013). Yang et al. demonstrated that the ^{125}I-labelled PEGylated GO was less adsorbed in the intestine after oral uptake (Shi et al. 2013). On the other side, after intraperitoneal injection high accumulation of PEGylated GO derivatives was observed in the reticuloendothelial system including the liver and spleen. Although, GO and PEGylated GO derivatives would retain in the mouse body over a long period of time, their toxicity to the treated animals is insignificant.

The rGO can be used for *in vivo* tumor vasculature targeting. The targeting of rGO in a breast cancer model was detected, with ^{64}Cu as the positron emission tomography PET label and TRC105 (TRC105 is proangiogenic and play a role in remodeling the vasculature of malignant tumors) as the targeting ligand. CD105 (antigen, transmembrane glycoprotein), the target of TRC105, is specifically heavily expressed on proliferating tumor endothelial cells of many solid tumor types; it makes it suitable for nanomaterial-based tumor targeting. The rGO conjugates exhibited great stability and high specificity for CD105. ^{64}Cu–NOTA–rGO–TRC105 (NOTA—1,4,7-triazacyclononane-1,4,7-triacetic acid) exhibited little extravasation in the 4T1 cell line, showing that tumor vasculature (instead of tumor cell) targeting is a valid and preferred approach for nanomaterials. The rGO conjugate could serve as a promising theranostic agent (integrates imaging and therapeutic components) (Shi et al. 2013).

6.3.2 GENE DELIVERY

As a result of their good solubility and biocompatibility, GNF and GO are promising carriers for gene delivery in nonviral-based gene therapy (Feng et al. 2011; Du et al. 2012). Functionalization of GO by branched polyethylenimine (PEI-GO) proved to be of significantly lower cytotoxicity than PEI 25 kDa. The PEI–GO could effectively deliver plasmid DNA into cells and could be localized in the nucleus (Chen, Liu, Zhang et al. 2011). The GO/PEI/DNA complex was effective for intracellular gene delivery into the widely used HeLa cell cultures (Feng et al. 2011). Single-stranded DNA is immediately adsorbed onto functionalized GFN forming strong molecular interactions that prevent DNase I from approaching the constrained DNA. Constraining a single-stranded DNA probe on GFN improved the specificity of its response to a target sequence. The features and properties of DNA–GFN interactions can be used to construct DNA–GFN nanobiosensors with facile design, excellent sensitivity, selectivity, and biostability. With the low cost of producing GFNs, the use of graphene in both fundamental research and practical applications is promising (Tang, Wu et al. 2010).

The positively charged GO–PEI complexes are able to bind with plasmid DNA (pDNA) for intracellular transfection (Feng et al. 2011). The study on the intracellular uptake of Cy3-labelled pDNA indicated that the supplementation of one of the primary nuclear localized signal peptides called PV7 could effectively assist the GO–PEI to deliver plasmid DNA

TABLE 6.1
Summary of Review Articles Focused on Graphene

Review Main Topic	Title	First Author	Number of Pages	Number of References Cited	Year	Reference
Diagnostics and delivery	Graphene-based nanomaterials: diagnostic applications	Pandey	26	344	2014	Pandey et al. (2014)
	Graphene-based nanomaterials for drug delivery and tissue engineering	Goenka	14	165	2014	Goenka et al. (2014)
Biosensing applications	Fabrication, optimization, and use of graphene field effect sensors	Stine	13	152	2013	Stine et al. (2013)
	Biological and chemical sensors based on graphene materials	Liu	25	312	2012	Y.X. Liu et al. (2012)
	Graphene-based electronic sensors	He	9	106	2012	He et al. (2012)
	Chemical preparation of graphene-based nanomaterials and their applications in chemical and biological sensors	Jiang	15	125	2011	Jiang (2011)
	Graphene and related materials in electrochemical sensing	Ratinac	24	143	2011	Ratinac et al. (2011)
	Electrochemical sensors based on graphene materials	Gan	19	178	2011	Gan and Hu (2011)
	Graphene for electrochemical sensing and biosensing	Pumera	12	56	2010	Pumera et al. (2010)
Material physical properties	Novel graphene-based nanostructures: physicochemical properties and applications	Chernozatonskii	29	285	2014	Chernozatonskii et al. (2014)
	Synthesis and electronic properties of chemically functionalized graphene on metal surfaces	Gruneis	14	100	2013	Burger et al. (2002)
	Photoluminescence properties of graphene versus other carbon nanomaterials	Cao	10	42	2013	Cao et al. (2013)
	Electronic and optical properties of semiconductors and GQDs	Sheng	25	160	2012	Sheng et al. (2012)
	Tuneable electronic properties in graphene	Craciun	19	121	2011	Craciun et al. (2011)
	Electronic properties of graphene nanostructures	Molitor	5	74	2011	Molitor et al. (2011)
	Probing mechanical properties of graphene using Raman spectroscopy	Ferralis	20	103	2010	Ferralis (2010)

directly into the nucleus without common aggregations. The cytotoxicity of GO–PEI was much lower than PEI 10 kDa and PEI 25 kDa against both HeLa cells and 293 T cells. This complex can serve as an alternative strategy for a nuclear-targeted gene delivery (Ren et al. 2012). Moreover, PEI-grafted ultrasmall graphene oxide (PEI-g-USGO) has good transfection efficiencies and very low cytotoxicity. The transfection of plasmid DNA into mammalian cell lines was with up to 95% efficiency and 90% viability (Zhou et al. 2012). Low-molecular mass branched polyethylenimine (BPEI) to GO improved the effective molecular weight of BPEI and also improved DNA binding and condensation and transfection efficiency (Kim et al. 2011). An efficient gene delivery system based on GO chemically functionalized with a nontoxic linear PEI (LP-GO) was reported. Linear PEI-grafted GO conjugates efficiently condensed pDNA and delivered it to the inside of the cells. LP-GO is able to deliver siRNA efficiently

into the cells (Tripathi et al. 2013). The *in vivo* results indicate significant regression in tumor growth and tumor weight after plasmid-based Stat3 siRNA delivered by GO-PEI-PEG treatment and no side effect from GO-PEI-PEG treatment was detected (Yin et al. 2013).

6.4 METHODS FOR STUDYING GRAPHENE-BASED BIOMATERIALS

Currently GFN, GO, and related materials are used as a detection platform for sensitive determination of numerous analytes. The applicability of GFN and GO as a tool for biosensing has been reviewed many times (Pumera 2010; Pumera et al. 2010; Shao et al. 2010; Gan and Hu 2011; Jiang 2011; Ratinac et al. 2011; He et al. 2012; Y.X. Liu et al. 2012; Stine et al. 2013). Mainly due to the excellent electronic properties of these materials numerous studies developing and utilizing

GFN and GO as electrochemical sensors have been presented (Shao et al. 2010; Gan and Hu 2011). A summary of review articles focused on the application of graphene is given in Table 6.1. On the other hand, the overviews of the methods investigating the GFN and mainly its bioconjugates as analytes are not so common. Generally, the reviews are mostly focused on characterization of GFN and GO from the fabrication point of view determining the mechanical (Ferralis 2010), electronic (Craciun et al. 2011; Molitor et al. 2011; Sheng et al. 2012; Gruneis 2013), and optical (Z.B. Liu et al. 2012; Cao et al. 2013) properties, and/or looking for defects (Dresselhaus et al. 2010; Banhart et al. 2011).

A group of basic methods for GFN analysis includes spectroscopic methods such as Raman spectroscopy (Ferrari 2007; Tang, Hu et al. 2010; Saito et al. 2011), impedance spectroscopy (Bonanni and Pumera 2013; Loo et al. 2013), X-ray spectroscopy (Ilkiv et al. 2012a,b; Lee et al. 2012), infrared spectroscopy (Li et al. 2008), and/or UV–vis spectroscopy (Lai et al. 2012; Mak et al. 2012). Another set of methods comprises microscopic techniques such as transmission electron microscopy (Liu, Zan et al. 2011, 2012; Wang et al. 2012), scanning tunneling microscopy (Paredes et al. 2009; Sutter et al. 2009; Andrei et al. 2012), and/or atomic force microscopy (Paredes et al. 2009; Ahmad et al. 2011; Ding et al. 2011) (Table 6.2).

The methods characterizing GFN in terms of its bioconjugation are depending on the nature of the bioconjugate. In case of conjugation with fluorescent partner such as quantum dots or fluorescently labeled biomolecules the appropriate method is based on fluorescence emission with GFN for fluorescence

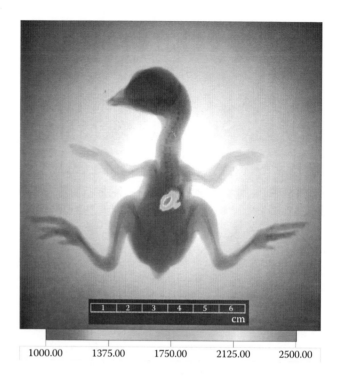

FIGURE 6.4 Fluorescence *in vivo* imaging detection of DOX in the chicken embryo. X-ray image of the embryo and overlaid by fluorescence image of the embryo after the application of fullerenes with DOX into the chicken breast (100 μL fullerene with doxo [500 μg/mL doxo]); parameters: excitation—480 nm, emission—600 nm, exposure time—2 s, binning: 2 × 2 pixels—2 × 2, field of view—11.5 × 11.5 cm, and stop—1.1.

imaging of live cells (Chen, Liu, Hu et al. 2011). Owing to fluorescent properties, DOX interacted with fullerenes can be detected as shown in Figure 6.4.

6.5 CONCLUSIONS

Owing to their unique properties, carbon nanomaterials including fullerenes, nanotubes, graphene, and graphene oxide have already proven their potential, applicability, and benefits in numerous areas of research. Biochemical analyses as well as biomedical applications have improved significantly by utilization of these carbon-based nanoparticles. However, despite their promise, both graphene and carbon nanotubes still face considerable challenges such as heterogeneity and in case of GFN separating the layers and keeping them separated, control the number of layers, and minimizing folding and bending during processing. One challenge connected to biomedical applications of GFN is thorough and profound understanding of interactions between GFN and cells. Especially the cellular uptake mechanism has to be fully understood. However, given its structural features and exceptional physicochemical properties, the design of GFN-based delivery/therapeutic multimodal and multifunctional platform is a new direction to follow. Finally, it has to highlight that these aims can only be reached by the combined effort of researchers from chemistry, biomedicine, materials sciences, and nanotechnology.

TABLE 6.2
Summary of Analytical Methods Used for Graphene Characterization

	Analytical Method	Reference
Spectroscopy	Raman spectroscopy	Tang, Hu et al. (2010)
		Saito et al. (2011)
		Ferrari (2007)
	Impedance spectroscopy	Bonanni and Pumera (2013)
		Loo et al. (2013)
	X-ray spectroscopy	Ilkiv et al. (2012b)
		Ilkiv and Zaulychnyy (2012a)
	IR spectroscopy	Li et al. (2008)
	UV/vis spectroscopy	Lai et al. (2012)
Microscopic techniques	TEM	Liu, Wang et al. (2012)
		Zan et al. (2012)
	Scanning tunneling microscopy	Paredes et al. (2009)
		Sutter et al. (2009)
		Andrei et al. (2012)
	AFM	Paredes et al. (2009)
		Ahmad et al. (2011)
		Ding et al. (2011)

ACKNOWLEDGMENT

The financial support from CYTORES GA CR P301/10/0356 is acknowledged

REFERENCES

Ahmad, M., S. A. Han, D. H. Tien, J. Jung, and Y. Seo. 2011. Local conductance measurement of graphene layer using conductive atomic force microscopy. *J. Appl. Phys.* 110 (5):1–6.

Akhavan, O., E. Ghaderi, and A. Akhavan. 2012. Size-dependent genotoxicity of graphene nanoplatelets in human stem cells. *Biomaterials* 33 (32):8017–8025.

Andrei, E. Y., G. H. Li, and X. Du. 2012. Electronic properties of graphene: A perspective from scanning tunneling microscopy and magnetotransport. *Rep. Prog. Phys.* 75 (5):1–79.

Arya, N., A. Arora, K. S. Vasu, A. K. Sood, and D. S. Katti. 2013. Combination of single walled carbon nanotubes/graphene oxide with paclitaxel: A reactive oxygen species mediated synergism for treatment of lung cancer. *Nanoscale* 5 (7):2818–2829.

Banhart, F., J. Kotakoski, and A. V. Krasheninnikov. 2011. Structural defects in graphene. *ACS Nano* 5 (1):26–41.

Bonanni, A. and M. Pumera. 2013. High-resolution impedance spectroscopy for graphene characterization. *Electrochem. Commun.* 26:52–54.

Burger, K. N. J., R. W. H. M Staffhorst, H. C. de Vijlder, M. J. Velinova, P. H. Bomans, P. M. Frederik, and B. de Kruijff. 2002. Nanocapsules: Lipid-coated aggregates of cisplatin with high cytotoxicity. *Nat. Med.* 8 (1):81–84.

Cai, W. B., A. R. Hsu, Z. B. Li, and X. Y. Chen. 2007. Are quantum dots ready for *in vivo* imaging in human subjects? *Nanoscale Res. Lett.* 2 (6):265–281.

Cao, L., M. J. Meziani, S. Sahu, and Y. P. Sun. 2013. Photoluminescence properties of graphene versus other carbon nanomaterials. *Acc. Chem. Res.* 46 (1):171–180.

Cao, L. L., L. W. Cheng, Z. Y. Zhang, Y. Wang, X. X. Zhang, H. Chen, B. H. Liu, S. Zhang, and J. L. Kong. 2012. Visual and high-throughput detection of cancer cells using a graphene oxide-based FRET aptasensing microfluidic chip. *Lab. Chip.* 12 (22):4864–4869.

Cha, I., Y. Yagi, T. Kawahara, K. Hashimoto, K. Fujiki, S. Tamesue, T. Yamauchi, and N. Tsubokawa. 2014. Grafting of polymers onto graphene oxide by trapping of polymer radicals and ligand-exchange reaction of polymers bearing ferrocene moieties. *Colloids Surf. A-Physicochem. Eng. Aspects* 441:474–480.

Chen, B. A., M. Liu, L. M. Zhang, J. Huang, J. L. Yao, and Z. J. Zhang. 2011. Polyethylenimine-functionalized graphene oxide as an efficient gene delivery vector. *J. Mater. Chem.* 21 (21):7736–7741.

Chen, M. L., J. W. Liu, B. Hu, and J. H. Wang. 2011. Conjugation of quantum dots with graphene for fluorescence imaging of live cells. *Analyst* 136 (20):4277–4283.

Chernozatonskii, L. A., P. B. Sorokin, and A. A. Artukh. 2014. Novel graphene-based nanostructures: Physicochemical properties and applications. *Russ. Chem. Rev.* 83 (3):251–279.

Cissell, K. A., S. Shrestha, J. Purdie, D. Kroodsma, and S. K. Deo. 2008. Molecular biosensing system based on intrinsically disordered proteins. *Anal. Bioanal. Chem.* 391 (5):1721–1729.

Condeelis, J. and R. Weissleder. 2010. *In vivo* imaging in cancer. *Cold Spring Harbor Perspect. Biol.* 2 (12):1–22.

Craciun, M. F., S. Russo, M. Yamamoto, and S. Tarucha. 2011. Tuneable electronic properties in graphene. *Nano Today* 6 (1):42–60.

Das, S., S. Singh, V. Singh, D. Joung, J. M. Dowding, D. Reid, J. Anderson et al. 2013. Oxygenated functional group density on graphene oxide: Its effect on cell toxicity. *Part. Part. Syst. Charact.* 30 (2):148–157.

Ding, Y. H., P. Zhang, H. M. Ren, Q. Zhuo, Z. M. Yang, X. Jiang, and Y. Jiang. 2011. Surface adhesion properties of graphene and graphene oxide studied by colloid-probe atomic force microscopy. *Appl. Surf. Sci.* 258 (3):1077–1081.

Dong, H. F., W. C. Gao, F. Yan, H. X. Ji, and H. X. Ju. 2010. Fluorescence resonance energy transfer between quantum dots and graphene oxide for sensing biomolecules. *Anal. Chem.* 82 (13):5511–5517.

Dong, H. Q., Y. Y. Li, J. H. Yu, Y. Y. Song, X. J. Cai, J. Q. Liu, J. M. Zhang, R. C. Ewing, and D. L. Shi. 2013. A versatile multi-component assembly via β-cyclodextrin host–guest chemistry on graphene for biomedical applications. *Small* 9 (3):446–456.

Dresselhaus, M. S., A. Jorio, A. G. Souza, and R. Saito. 2010. Defect characterization in graphene and carbon nanotubes using Raman spectroscopy. *Philos. Trans. R. Soc. A-Math. Phys. Eng. Sci.* 368 (1932):5355–5377.

Du, D., Z. X. Zou, Y. S. Shin, J. Wang, H. Wu, M. H. Engelhard, J. Liu, I. A. Aksay, and Y. H. Lin. 2010. Sensitive immunosensor for cancer biomarker based on dual signal amplification strategy of graphene sheets and multienzyme functionalized carbon nanospheres. *Anal. Chem.* 82 (7):2989–2995.

Du, D., Y. Q. Yang, and Y. H. Lin. 2012. Graphene-based materials for biosensing and bioimaging. *MRS Bull.* 37 (12):1290–1296.

Feng, L. Z. and Z. A. Liu. 2011. Graphene in biomedicine: Opportunities and challenges. *Nanomedicine* 6 (2):317–324.

Feng, L. Z., S. A. Zhang, and Z. A. Liu. 2011. Graphene based gene transfection. *Nanoscale* 3 (3):1252–1257.

Ferralis, N. 2010. Probing mechanical properties of graphene with Raman spectroscopy. *J. Mater. Sci.* 45 (19):5135–5149.

Ferrari, A. C. 2007. Raman spectroscopy of graphene and graphite: Disorder, electron-phonon coupling, doping and nonadiabatic effects. *Solid State Commun.* 143 (1–2):47–57.

Gan, T. and S. S. Hu. 2011. Electrochemical sensors based on graphene materials. *Microchim. Acta* 175 (1–2):1–19.

Goenka, S., V. Sant, and S. Sant. 2014. Graphene-based nanomaterials for drug delivery and tissue engineering. *J. Controlled Release* 173:75–88.

Gong, Y. Y., N. Pei, Z. Y. Huang, J. B. Ge, W. L. Ma, and W. L. Zheng. 2009. Deep capture of paramagnetic particle for targeting therapeutics. Edited by R. Shi, W. J. Fu, Y. Q. Wang and H. B. Wang, *Proceedings of the 2009 2nd International Conference on Biomedical Engineering and Informatics*, Vols 1–4. New York: Ieee.

Gruneis, A. 2013. Synthesis and electronic properties of chemically functionalized graphene on metal surfaces. *J. Phys.-Condes. Matter* 25 (4):1–7.

Guo, S. J. and S. J. Dong. 2011. Graphene nanosheet: Synthesis, molecular engineering, thin film, hybrids, and energy and analytical applications. *Chem. Soc. Rev.* 40 (5):2644–2672.

Guo, Z. Y., T. T. Hao, S. P. Du, B. B. Chen, Z. B. Wang, X. Li, and S. Wang. 2013. Multiplex electrochemiluminescence immunoassay of two tumor markers using multicolor quantum dots as labels and graphene as conducting bridge. *Biosens. Bioelectron.* 44:101–107.

Haase, M. G., K. Liepe, D. Faulhaber, G. Wunderlich, M. Andreeff, R. Jung, G. B. Baretton, G. Fitze, and J. Kotzerke. 2013. Dose-dependent histological alterations in the rat lung following intravenous application of Re-188-labeled microspheres. *Int. J. Radiat. Biol.* 89 (10):863–869.

He, Q. Y., S. X. Wu, Z. Y. Yin, and H. Zhang. 2012. Graphene-based electronic sensors. *Chem. Sci.* 3 (6):1764–1772.

Hirsch, L. R., A. M. Gobin, A. R. Lowery, F. Tam, R. A. Drezek, N. J. Halas, and J. L. West. 2006. Metal nanoshells. *Ann. Biomed. Eng.* 34 (1):15–22.

Horvath, L., A. Magrez, M. Burghard, K. Kern, L. Forro, and B. Schwaller. 2013. Evaluation of the toxicity of graphene derivatives on cells of the lung luminal surface. *Carbon* 64:45–60.

Hu, X. W., C. J. Mao, J. M. Song, H. L. Niu, S. Y. Zhang, and H. P. Huang. 2013. Fabrication of GO/PANi/CdSe nanocomposites for sensitive electrochemiluminescence biosensor. *Biosens. Bioelectron.* 41:372–378.

Huang, X., Z. Y. Yin, S. X. Wu, X. Y. Qi, Q. Y. He, Q. C. Zhang, Q. Y. Yan, F. Boey, and H. Zhang. 2011. Graphene-based materials: Synthesis, characterization, properties, and applications. *Small* 7 (14):1876–1902.

Hummers, W. S. and R. E. Offeman. 1958. Preparation of graphitic oxide. *J. Am. Chem. Soc.* 80 (6):1339–1339.

Ilkiv, B., S. Petrovska, R. Sergiienko, and Y. Zaulychnyy. 2012a. X-ray spectral investigation of graphene nanosheets deposited on silicon substrate. *Metallofiz. Noveishie Tekhnol.-Met. Phys. Adv. Technol.* 34 (11):1487–1493.

Ilkiv, B., S. Petrovska, R. Sergiienko, T. Tomai, E. Shibata, T. Nakamura, I. Honma, and Y. Zaulychnyy. 2012b. X-ray emission spectra of graphene nanosheets. *J. Nanosci. Nanotechnol.* 12 (12):8913–8919.

Jiang, H. J. 2011. Chemical preparation of graphene-based nanomaterials and their applications in chemical and biological sensors. *Small* 7 (17):2413–2427.

Kavitha, T., S. I. H. Abdi, and S. Y. Park. 2013. pH-Sensitive nanocargo based on smart polymer functionalized graphene oxide for site-specific drug delivery. *Phys. Chem. Chem. Phys.* 15 (14):5176–5185.

Kim, H., R. Namgung, K. Singha, I. K. Oh, and W. J. Kim. 2011. Graphene oxide-polyethylenimine nanoconstruct as a gene delivery vector and bioimaging tool. *Bioconjugate Chem.* 22 (12):2558–2567.

Kizek, R., V. Adam, J. Hrabeta, T. Eckschlager, S. Smutny, J. V. Burda, E. Frei, and M. Stiborova. 2012. Anthracyclines and ellipticines as DNA-damaging anticancer drugs: Recent advances. *Pharmacol. Ther.* 133 (1):26–39.

Kong, X. L. and Y. Huang. 2014. Applications of graphene in mass spectrometry. *J. Nanosci. Nanotechnol.* 14 (7):4719–4732.

Kovtyukhova, N. I., P. J. Ollivier, B. R. Martin, T. E. Mallouk, S. A. Chizhik, E. V. Buzaneva, and A. D. Gorchinskiy. 1999. Layer-by-layer assembly of ultrathin composite films from micron-sized graphite oxide sheets and polycations. *Chem. Mater.* 11 (3):771–778.

Kravets, V. G., F. Schedin, R. Jalil, L. Britnell, R. V. Gorbachev, D. Ansell, B. Thackray et al. 2013. Singular phase nano-optics in plasmonic metamaterials for label-free single-molecule detection. *Nat. Mater.* 12 (4):304–309.

Ku, S. H. and C. B. Park. 2013. Myoblast differentiation on graphene oxide. *Biomaterials* 34 (8):2017–2023.

Lacerda, L., A. Bianco, M. Prato, and K. Kostarelos. 2006. Carbon nanotubes as nanomedicines: From toxicology to pharmacology. *Adv. Drug Deliv. Rev.* 58 (14):1460–1470.

Lai, Q., S. F. Zhu, X. P. Luo, M. Zou, and S. H. Huang. 2012. Ultraviolet-visible spectroscopy of graphene oxides. *AIP Adv.* 2 (3):1–5.

Lee, D. Y., Z. Khatun, J. H. Lee, Y. K. Lee, and I. In. 2011. Blood compatible graphene/heparin conjugate through noncovalent chemistry. *Biomacromolecules* 12 (2):336–341.

Lee, V., R. V. Dennis, B. J. Schultz, C. Jaye, D. A. Fischer, and S. Banerjee. 2012. Soft X-ray absorption spectroscopy studies of the electronic structure recovery of graphene oxide upon chemical defunctionalization. *J. Phys. Chem. C* 116 (38):20591–20599.

Lei, H. Z., L. J. Mi, X. J. Zhou, J. J. Chen, J. Hu, S. W. Guo, and Y. Zhang. 2011. Adsorption of double-stranded DNA to graphene oxide preventing enzymatic digestion. *Nanoscale* 3 (9):3888–3892.

Li, H., J. He, S. J. Li, and A. P. F. Turner. 2013. Electrochemical immunosensor with N-doped graphene-modified electrode for label-free detection of the breast cancer biomarker CA 15-3. *Biosens. Bioelectron.* 43:25–29.

Li, J. and X. Y. Yang. 2013. Applications of novel carbon nanomaterials-graphene and its derivatives in biosensing. *Prog. Chem.* 25 (2–3):380–396.

Li, Q. F., D. A. P. Tang, J. Tang, B. L. Su, G. N. Chen, and M. D. Wei. 2011. Magneto-controlled electrochemical immunosensor for direct detection of squamous cell carcinoma antigen by using serum as supporting electrolyte. *Biosens. Bioelectron.* 27 (1):153–159.

Li, Z., M. Y. He, D. D. Xu, and Z. H. Liu. 2014. Graphene materials-based energy acceptor systems and sensors. *J. Photochem. Photobiol. C-Photochem. Rev.* 18:1–17.

Li, Z. Q., E. A. Henriksen, Z. Jiang, Z. Hao, M. C. Martin, P. Kim, H. L. Stormer, and D. N. Basov. 2008. Dirac charge dynamics in graphene by infrared spectroscopy. *Nat. Phys.* 4 (7):532–535.

Lim, S. K., D. H. Shin, M. H. Choi, and J. S. Kim. 2014. Enhanced antitumor efficacy of gemcitabine-loaded temperature-sensitive liposome by hyperthermia in tumor-bearing mice. *Drug Dev. Ind. Pharm.* 40 (4):470–476.

Liu, B. W., Z. Y. Sun, X. Zhang, and J. W. Liu. 2013. Mechanisms of DNA sensing on graphene oxide. *Anal. Chem.* 85 (16):7987–7993.

Liu, F. Y., Y. L. Gao, H. J. Li, and S. G. Sun. 2014. Interaction of propidium iodide with graphene oxide and its application for live cell staining. *Carbon* 71:190–195.

Liu, X. H., J. W. Wang, Y. Liu, H. Zheng, A. Kushima, S. Huang, T. Zhu et al. 2012. *In situ* transmission electron microscopy of electrochemical lithiation, delithiation and deformation of individual graphene nanoribbons. *Carbon* 50 (10):3836–3844.

Liu, Y., Y. Zhang, T. Zhang, Y. J. Jiang, and X. F. Liu. 2014. Synthesis, characterization and cytotoxicity of phosphoryl-choline oligomer grafted graphene oxide. *Carbon* 71:166–175.

Liu, Y. X., X. C. Dong, and P. Chen. 2012. Biological and chemical sensors based on graphene materials. *Chem. Soc. Rev.* 41 (6):2283–2307.

Liu, Z., X. M. Sun, N. Nakayama-Ratchford, and H. J. Dai. 2007. Supramolecular chemistry on water-soluble carbon nanotubes for drug loading and delivery. *ACS Nano* 1 (1):50–56.

Liu, Z., J. T. Robinson, X. M. Sun, and H. J. Dai. 2008. PEGylated nanographene oxide for delivery of water-insoluble cancer drugs. *J. Am. Chem. Soc.* 130 (33):10876–10877.

Liu, Z., J. T. Robinson, S. M. Tabakman, K. Yang, and H. J. Dai. 2011. Carbon materials for drug delivery and cancer therapy. *Mater. Today* 14 (7–8):316–323.

Liu, Z. B., X. L. Zhang, X. Q. Yan, Y. S. Chen, and J. G. Tian. 2012. Nonlinear optical properties of graphene-based materials. *Chin. Sci. Bull.* 57 (23):2971–2982.

Loh, K. P., Q. L. Bao, G. Eda, and M. Chhowalla. 2010. Graphene oxide as a chemically tunable platform for optical applications. *Nat. Chem.* 2 (12):1015–1024.

Loo, A. H., A. Bonanni, and M. Pumera. 2013. Soldering DNA to graphene via 0, 1 and 2-point contacts: Electrochemical impedance spectroscopic investigation. *Electrochem. Commun.* 28:83–86.

Ma, X. X., H. Q. Tao, K. Yang, L. Z. Feng, L. Cheng, X. Z. Shi, Y. G. Li, L. Guo, and Z. Liu. 2012. A functionalized graphene oxide-iron oxide nanocomposite for magnetically targeted drug delivery, photothermal therapy, and magnetic resonance imaging. *Nano Res.* 5 (3):199–212.

Maeda, H., T. Sawa, and T. Konno. 2001. Mechanism of tumor-targeted delivery of macromolecular drugs, including the EPR effect in solid tumor and clinical overview of the prototype polymeric drug SMANCS. *J. Controlled Release* 74 (1–3):47–61.

Mak, K. F., L. Ju, F. Wang, and T. F. Heinz. 2012. Optical spectroscopy of graphene: From the far infrared to the ultraviolet. *Solid State Commun.* 152 (15):1341–1349.

Mao, X. W. and H. B. Li. 2013. Chiral imaging in living cells with functionalized graphene oxide. *J. Mater. Chem. B* 1 (34):4267–4272.

Markad, G. B., S. Battu, S. Kapoor, and S. K. Haram. 2013. Interaction between quantum dots of CdTe and reduced graphene oxide: Investigation through cyclic voltammetry and spectroscopy. *J. Phys. Chem. C* 117 (40):20944–20950.

Mendes, R. G., A. Bachmatiuk, A. A. El-Gendy, S. Melkhanova, R. Klingeler, B. Buchner, and M. H. Rummeli. 2012. A facile route to coat iron oxide nanoparticles with few-layer graphene. *J. Phys. Chem. C* 116 (44):23749–23756.

Miao, W., G. Shim, S. Lee, Y. S. Choe, and Y. K. Oh. 2013. Safety and tumor tissue accumulation of pegylated graphene oxide nanosheets for co-delivery of anticancer drug and photosensitizer. *Biomaterials* 34 (13):3402–3410.

Misra, S. K., P. Kondaiah, S. Bhattacharya, and C. N. R. Rao. 2012. Graphene as a nanocarrier for tamoxifen induces apoptosis in transformed cancer cell lines of different origins. *Small* 8 (1):131–143.

Molitor, F., J. Guttinger, C. Stampfer, S. Droscher, A. Jacobsen, T. Ihn, and K. Ensslin. 2011. Electronic properties of graphene nanostructures. *J. Phys.-Condes. Matter* 23 (24):1–5.

Mosaiab, T., I. In, and S. Y. Park. 2013. Temperature and pH-tunable fluorescence nanoplatform with graphene oxide and BODIPY-conjugated polymer for cell imaging and therapy. *Macromol. Rapid Commun.* 34 (17):1408–1415.

Na, H. K., M. H. Kim, J. Lee, Y. K. Kim, H. Jang, K. E. Lee, H. Park et al. 2013. Cytoprotective effects of graphene oxide for mammalian cells against internalization of exogenous materials. *Nanoscale* 5 (4):1669–1677.

Ni, Y. N., F. Y. Zhang, and S. Kokot. 2013. Graphene oxide as a nanocarrier for loading and delivery of medicinal drugs and as a biosensor for detection of serum albumin. *Anal. Chim. Acta* 769:40–48.

Novoselov, K. S., A. K. Geim, S. V. Morozov, D. Jiang, Y. Zhang, S. V. Dubonos, I. V. Grigorieva, and A. A. Firsov. 2004. Electric field effect in atomically thin carbon films. *Science* 306 (5696):666–669.

Pandey, A. P., K. P. Karande, M. P. More, S. G. Gattani, and P. K. Deshmukh. 2014. Graphene based nanomaterials: Diagnostic applications. *J. Biomed. Nanotechnol.* 10 (2):179–204.

Paredes, J. I., S. Villar-Rodil, P. Solis-Fernandez, A. Martinez-Alonso, and J. M. D. Tascon. 2009. Atomic force and scanning tunneling microscopy imaging of graphene nanosheets derived from graphite oxide. *Langmuir* 25 (10):5957–5968.

Pollinger, K., R. Hennig, S. Bauer, M. Breunig, J. Tessmar, A. Buschauer, R. Witzgall, and A. Goepferich. 2014. Biodistribution of quantum dots in the kidney after intravenous injection. *J. Nanosci. Nanotechnol.* 14 (5):3313–3319.

Portney, N. G. and M. Ozkan. 2006. Nano-oncology: Drug delivery, imaging, and sensing. *Anal. Bioanal. Chem.* 384 (3):620–630.

Pumera, M. 2009. The electrochemistry of carbon nanotubes: Fundamentals and applications. *Chem.-Eur. J.* 15 (20):4970–4978.

Pumera, M. 2010. Graphene-based nanomaterials and their electrochemistry. *Chem. Soc. Rev.* 39 (11):4146–4157.

Pumera, M. 2012. Graphene, carbon nanotubes and nanoparticles in cell metabolism. *Curr. Drug Metab.* 13 (3):251–256.

Pumera, M., A. Ambrosi, A. Bonanni, E. L. K. Chng, and H. L. Poh. 2010. Graphene for electrochemical sensing and biosensing. *TRAC-Trends Anal. Chem.* 29 (9):954–965.

Ratinac, K. R., W. R. Yang, J. J. Gooding, P. Thordarson, and F. Braet. 2011. Graphene and related materials in electrochemical sensing. *Electroanalysis* 23 (4):803–826.

Ren, T. B., L. Li, X. J. Cai, H. Q. Dong, S. M. Liu, and Y. Y. Li. 2012. Engineered polyethylenimine/graphene oxide nanocomposite for nuclear localized gene delivery. *Polym. Chem.* 3 (9):2561–2569.

Saito, R., M. Hofmann, G. Dresselhaus, A. Jorio, and M. S. Dresselhaus. 2011. Raman spectroscopy of graphene and carbon nanotubes. *Adv. Phys.* 60 (3):413–550.

Schinwald, A., F. Murphy, A. Askounis, V. Koutsos, K. Sefiane, K. Donaldson, and C. J. Campbell. 2014. Minimal oxidation and inflammogenicity of pristine graphene with residence in the lung. *Nanotoxicology* 8 (8):824–832.

Sciallero, C., D. Grishenkov, Svvn Kothapalli, L. Oddo, and A. Trucco. 2013. Acoustic characterization and contrast imaging of microbubbles encapsulated by polymeric shells coated or filled with magnetic nanoparticles. *J. Acoust. Soc. Am.* 134 (5):3918–3930.

Shan, C. S., H. F. Yang, D. X. Han, Q. X. Zhang, A. Ivaska, and L. Niu. 2009. Water-soluble graphene covalently functionalized by biocompatible poly-L-lysine. *Langmuir* 25 (20):12030–12033.

Shao, Y. Y., J. Wang, H. Wu, J. Liu, I. A. Aksay, and Y. H. Lin. 2010. Graphene based electrochemical sensors and biosensors: A review. *Electroanalysis* 22 (10):1027–1036.

Shen, H., L. M. Zhang, M. Liu, and Z. J. Zhang. 2012. Biomedical applications of graphene. *Theranostics* 2 (3):283–294.

Sheng, W. D., M. Korkusinski, A. D. Guclu, M. Zielinski, P. Potasz, E. S. Kadantsev, O. Voznyy, and P. Hawrylak. 2012. Electronic and optical properties of semiconductor and graphene quantum dots. *Front. Phys.* 7 (3):328–352.

Shi, S. X., K. Yang, H. Hong, H. F. Valdovinos, T. R. Nayak, Y. Zhang, C. P. Theuer, T. E. Barnhart, Z. Liu, and W. B. Cai. 2013. Tumor vasculature targeting and imaging in living mice with reduced graphene oxide. *Biomaterials* 34 (12):3002–3009.

Song, Y. J., W. L. Wei, and X. G. Qu. 2011. Colorimetric biosensing using smart materials. *Adv. Mater.* 23 (37):4215–4236.

Stiborova, M., T. Eckschlager, J. Poljakova, J. Hrabeta, V. Adam, R. Kizek, and E. Frei. 2012. The synergistic effects of DNA-targeted chemotherapeutics and histone deacetylase inhibitors as therapeutic strategies for cancer treatment *Curr. Med. Chem.* 19 (25):4218–4238.

Stine, R., S. P. Mulvaney, J. T. Robinson, C. R. Tamanaha, and P. E. Sheehan. 2013. Fabrication, optimization, and use of graphene field effect sensors. *Anal. Chem.* 85 (2):509–521.

Sun, H. J., L. Wu, N. Gao, J. S. Ren, and X. G. Qu. 2013. Improvement of photoluminescence of graphene quantum dots with a biocompatible photochemical reduction pathway and its bioimaging application. *ACS Appl. Mater. Interfaces* 5 (3):1174–1179.

Sun, X. M., Z. Liu, K. Welsher, J. T. Robinson, A. Goodwin, S. Zaric, and H. J. Dai. 2008. Nano-graphene oxide for cellular imaging and drug delivery. *Nano Res.* 1 (3):203–212.

Sutter, E., D. P. Acharya, J. T. Sadowski, and P. Sutter. 2009. Scanning tunneling microscopy on epitaxial bilayer graphene on ruthenium (0001). *Appl. Phys. Lett.* 94 (13):1–3.

Tan, X. F., L. Z. Feng, J. Zhang, K. Yang, S. Zhang, Z. Liu, and R. Peng. 2013. Functionalization of graphene oxide generates a unique interface for selective serum protein interactions. *ACS Appl. Mater. Interfaces* 5 (4):1370–1377.

Tang, B., G. X. Hu, and H. Y. Gao. 2010. Raman spectroscopic characterization of graphene. *Appl. Spectrosc. Rev.* 45 (5):369–407.

Tang, Z. W., H. Wu, J. R. Cort, G. W. Buchko, Y. Y. Zhang, Y. Y. Shao, I. A. Aksay, J. Liu, and Y. H. Lin. 2010. Constraint of DNA on functionalized graphene improves its biostability and specificity. *Small* 6 (11):1205–1209.

Thorek, D. L. J., A. Chen, J. Czupryna, and A. Tsourkas. 2006. Superparamagnetic iron oxide nanoparticle probes for molecular imaging. *Ann. Biomed. Eng.* 34 (1):23–38.

Tripathi, S. K., R. Goyal, K. C. Gupta, and P. Kumar. 2013. Functionalized graphene oxide mediated nucleic acid delivery. *Carbon* 51:224–235.

Villegas, J. C., L. Alvarez-Montes, L. Rodriguez-Fernandez, J. Gonzalez, R. Valiente, and M. L. Fanarraga. 2014. Multiwalled carbon nanotubes hinder microglia function interfering with cell migration and phagocytosis. *Adv. Healthcare Mater.* 3 (3):424–432.

Wang, H., W. Gu, N. Xiao, L. Ye, and Q. Y. Xu. 2014. Chlorotoxin-conjugated graphene oxide for targeted delivery of an anticancer drug. *Int. J. Nanomed.* 9:1433–1442.

Wang, X., L. H. Liu, O. Ramstrom, and M. D. Yan. 2009. Engineering nanomaterial surfaces for biomedical applications. *Exp. Biol. Med.* 234 (10):1128–1139.

Wang, Y., Z. H. Li, D. H. Hu, C. T. Lin, J. H. Li, and Y. H. Lin. 2010. Aptamer/graphene oxide nanocomplex for *in situ* molecular probing in living cells. *J. Am. Chem. Soc.* 132 (27):9274–9276.

Wang, Y., K. Y. Wang, J. F. Zhao, X. G. Liu, J. Bu, X. Y. Yan, and R. Q. Huang. 2013. Multifunctional mesoporous silica-coated graphene nanosheet used for chemo-photothermal synergistic targeted therapy of glioma. *J. Am. Chem. Soc.* 135 (12):4799–4804.

Wang, Z. H., C. F. Zhou, J. F. Xia, B. Via, Y. Z. Xia, F. F. Zhang, Y. H. Li, and L. H. Xia. 2013. Fabrication and characterization of a triple functionalization of graphene oxide with Fe3O4, folic acid and doxorubicin as dual-targeted drug nanocarrier. *Colloid Surf. B-Biointerfaces* 106:60–65.

Wei, W. L. and X. G. Qu. 2012. Extraordinary physical properties of functionalized graphene. *Small* 8 (14):2138–2151.

Wu, H. X., H. L. Shi, Y. P. Wang, X. Q. Jia, C. Z. Tang, J. M. Zhang, and S. P. Yang. 2014. Hyaluronic acid conjugated graphene oxide for targeted drug delivery. *Carbon* 69:379–389.

Wu, Y. F., P. Xue, Y. J. Kang, and K. M. Hui. 2013. Highly specific and ultrasensitive graphene-enhanced electrochemical detection of low-abundance tumor cells using silica nanoparticles coated with antibody-conjugated quantum dots. *Anal. Chem.* 85 (6):3166–3173.

Xu, S. J., Y. Liu, T. H. Wang, and J. H. Li. 2011. Positive potential operation of a cathodic electrogenerated chemiluminescence immunosensor based on luminol and graphene for cancer biomarker detection. *Anal. Chem.* 83 (10):3817–3823.

Xu, Z. Y., Y. J. Li, P. Shi, B. C. Wang, and X. Y. Huang. 2013. Functionalized graphene oxide as a nanocarrier for loading and delivering of oridonin. *Chin. J. Org. Chem.* 33 (3):573–580.

Yang, K., S. A. Zhang, G. X. Zhang, X. M. Sun, S. T. Lee, and Z. A. Liu. 2010. Graphene in mice: Ultrahigh *in vivo* tumor uptake and efficient photothermal therapy. *Nano Lett.* 10 (9):3318–3323.

Yang, M., J. Yao, and Y. X. Duan. 2013. Graphene and its derivatives for cell biotechnology. *Analyst* 138 (1):72–86.

Yin, D., Y. Li, H. Lin, B. F. Guo, Y. W. Du, X. Li, H. J. Jia, X. J. Zhao, J. Tang, and L. Zhang. 2013. Functional graphene oxide as a plasmid-based Stat3 siRNA carrier inhibits mouse malignant melanoma growth in vivo. *Nanotechnology* 24 (10):1–9.

Yue, Z. G., P. P. Lv, H. Yue, Y. J. Gao, D. Ma, W. Wei, and G. H. Ma. 2013. Inducible graphene oxide probe for high-specific tumor diagnosis. *Chem. Commun.* 49 (37):3902–3904.

Zan, R., U. Bangert, Q. Ramasse, and K. S. Novoselov. 2011. Metal-graphene interaction studied via atomic resolution scanning transmission electron microscopy. *Nano Lett.* 11 (3):1087–1092.

Zan, R., U. Bangert, Q. Ramasse, and K. S. Novoselov. 2012. Interaction of metals with suspended graphene observed by transmission electron microscopy. *J. Phys. Chem. Lett.* 3 (7):953–958.

Zhang, G., H. C. Chang, C. Amatore, Y. Chen, H. Jiang, and X. M. Wang. 2013. Apoptosis induction and inhibition of drug resistant tumor growth *in vivo* involving daunorubicin-loaded graphene-gold composites. *J. Mater. Chem. B* 1 (4):493–499.

Zhang, H., C. Peng, J. Z. Yang, M. Lv, R. Liu, D. N. He, C. H. Fan, and Q. Huang. 2013. Uniform ultrasmall graphene oxide nanosheets with low cytotoxicity and high cellular uptake. *ACS Appl. Mater. Interfaces* 5 (5):1761–1767.

Zhang, J. L., F. Zhang, H. J. Yang, X. L. Huang, H. Liu, J. Y. Zhang, and S. W. Guo. 2010. Graphene oxide as a matrix for enzyme immobilization. *Langmuir* 26 (9):6083–6085.

Zhang, L. N., H. H. Deng, F. L. Lin, X. W. Xu, S. H. Weng, A. L. Liu, X. H. Lin, X. H. Xia, and W. Chen. 2014. *In situ* growth of porous platinum nanoparticles on graphene oxide for colorimetric detection of cancer cells. *Analytical Chemistry* 86 (5):2711–2718.

Zhang, T. Y., G. W. Lu, J. Liu, H. M. Shen, P. Perriat, M. Martini, O. Tillement, and Q. H. Gong. 2012. Strong two-photon fluorescence enhanced jointly by dipolar and quadrupolar modes of a single plasmonic nanostructure. *Appl. Phys. Lett.* 101 (5):1–5.

Zhang, X. Y., J. L. Yin, C. Peng, W. Q. Hu, Z. Y. Zhu, W. X. Li, C. H. Fan, and Q. Huang. 2011. Distribution and biocompatibility studies of graphene oxide in mice after intravenous administration. *Carbon* 49 (3):986–995.

Zhang, Y., T. R. Nayak, H. Hong, and W. B. Cai. 2012. Graphene: A versatile nanoplatform for biomedical applications. *Nanoscale* 4 (13):3833–3842.

Zhou, H. J., B. Zhang, J. J. Zheng, M. F. Yu, T. Zhou, K. Zhao, Y. X. Jia, X. F. Gao, C. Y. Chen, and T. T. Wei. 2014. The inhibition of migration and invasion of cancer cells by graphene via the impairment of mitochondrial respiration. *Biomaterials* 35 (5):1597–1607.

Zhou, X., F. Laroche, G. E. M. Lamers, V. Torraca, P. Voskamp, T. Lu, F. Q. Chu, H. P. Spaink, J. P. Abrahams, and Z. F. Liu. 2012. Ultra-small graphene oxide functionalized with polyethylenimine (PEI) for very efficient gene delivery in cell and zebrafish embryos. *Nano Res.* 5 (10):703–709.

Zhou, X. F. and F. Liang. 2014. Application of graphene/graphene oxide in biomedicine and biotechnology. *Curr. Med. Chem.* 21 (7):855–869.

Zhu, S. J., J. H. Zhang, C. Y. Qiao, S. J. Tang, Y. F. Li, W. J. Yuan, B. Li et al. 2011. Strongly green-photoluminescent graphene quantum dots for bioimaging applications. *Chem. Commun.* 47 (24):6858–6860.

7 Mathematical Modeling for Hydrogen Storage Inside Graphene-Based Materials

Yue Chan

CONTENTS

Abstract 111
7.1 Introduction 111
7.2 Theoretical Background 114
7.3 Numerical Results 116
    7.3.1 Numerical Results for Double-Layered Graphene Sheets 116
    7.3.2 Numerical Results for GOFs 117
7.4 Conclusion 120
Acknowledgment 121
References 121

ABSTRACT

In this chapter, applied mathematical modeling in conjunction with statistical mechanics is used to investigate the storage of hydrogen between graphene and inside graphene-oxide frameworks (GOFs), which comprise double-layered graphene sheets and are uniformly separated by the molecular ligands with certain densities. Hydrogen uptake is calculated for graphene and GOFs using the equations of state in both bulk gas and adsorption phases, where the molecular interactions between a hydrogen molecule and the host structure are determined by continuous approximation. First, we verify our numerical results by comparing the hydrogen storage between the graphene sheets using the present mathematical approach with that using computational simulations and experimental results. Then, we determine the optimal hydrogen storage structures for graphene oxide frameworks across wide ranges of temperatures and external pressures. Such variations in different optimal structures could be partially explained by the idea of the geometric effect and the extra energy induced by the ligands. Theoretical methodologies that address the gas storage inside nanostructures range from several fundamental equations such as Langmuir single-layer model and the Brunauer, Emmett, and Teller (BET) multilayer model to more sophisticated computational simulations such as molecular dynamics simulations, Monte Carlo simulations, and ab initio quantum mechanical-based principles. While the fundamental equations provide almost instantaneous estimations for the surface area and the heat of adsorption using the experimental adsorption isotherms, the computationally intensive simulations provide detailed predictions of almost all aspects of adsorption but require substantial prior experiences in the field. The missing block between two extreme methodologies could be filled by the present hybrid mathematical and statistical approach, and other continuum-based models proposed in the current literature (Shi, F., Bhalla, A., Chen, C., Gunaratne, G.H., Jiang, J., Meletis, E.I. 2011. Atomistically-informed continuum model for hydrogen storage on graphene. *Interface* 57–61). The present methodology provides more insights than that of fundamental equations but demands less specific knowledge than that of computational simulations. Most importantly it facilitates rapid numerical results and generates deductive results. The theoretical results obtained from this chapter are ready to be employed for other types of gases, for example, argon, methane, etc. and nanomaterials, for example, nanotubes, zeolites, etc. without conceptual difficulties.

7.1 INTRODUCTION

The 2010 Nobel Prize for Physics was jointly awarded to Andre Geim and Konstantin Novoselov [1] for their groundbreaking discovery of two-dimensional graphene sheet, which is experimentally fabricated by chemical vapor deposition. They also determined the electronic properties of such unique molecular structure. Ever since, graphene has attracted wide attention from the research and industrial communities, and has become one of the hottest topics in natural sciences. Graphene is generally viewed as a two-dimensional sheet comprising carbon atoms, which are arranged in a lattice structure. Electronic states of graphene are formed by the p_z orbital, which is perpendicular to the graphene sheet, and three occupied valence states of carbon atoms, that is, σ bonds through sp^2 hybridization. These unique electronic states contribute to the robustness of the graphene structure [2], for example, graphene possesses superior and novel electronic, mechanical (remarkably high Young's modulus of 1 TPa), thermal, and optical properties. A new class of the graphene-based structure, referred as GOFs (see Figure 7.1), has also been proposed by Burress et al. and could be experimentally

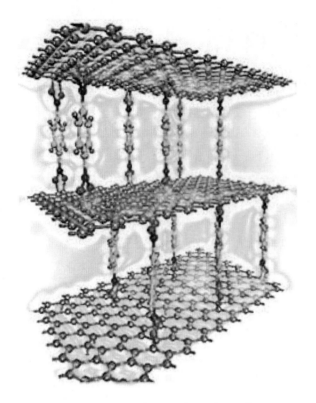

FIGURE 7.1 GOFs where layers of graphene sheets are supported and separated by benzenediboronic acid pillars. Each ligand comprises eight hydrogen, four oxygen, two boron and six carbon atoms. (Burress, J.W. et al.: Graphene oxide framework materials: Theoretical predictions and experimental results. *Angew. Chem. Int. Edn* 2010. 49: 8902–4. Copyright Wiley-VCH Verlag GmbH & Co. KGaA. Reproduced with permission.)

fabricated through a chemical reaction between boronic acids and the hydroxyl group [3]. The benzenediboronic acid pillars, herein ligands, between the graphene sheets separate the double-layered graphene sheets and provide a mechanical support to the molecular structure. With a proper tuning of the ligand densities, the optimal hydrogen uptakes could be determined from the geometric effect such as the free volume or the extra energy induced by the ligands, depending on the combination of temperatures and external pressures. Such optimal ligand densities could provide a guideline for future experimental condensation. In particular for hydrogen storage, graphene-based materials are used because of their abundance in nature, toughness against fatigue and crack, lightweight, low manufacturing cost, and chemically inert and environmentally friendly characteristics leading to numerous commercial applications that are ready for mass production in the marketplace. Hydrogen molecules are covalently bonded by two hydrogen atoms. They are abundant in the universe, and can be extracted and stored in fuel cells to generate energy for mobile vehicles. However, hydrogen gas is highly explosive so that it must be stored under safe conditions leading to stringent requirements for the host materials. Graphene has also recently been proposed as a molecular sieve, especially for the purpose of seawater desalination, which has merits of high operational efficiency and low operational cost.

A challenge of the twenty-first century is to generate, store, and release energy in an affordable, clean, and renewable manner. Therefore, materials' reversible ability in capturing, storing, and releasing gases plays an indispensable role in partially addressing this challenge. Some examples for gas storage and energy generation include carbon capture and storage for coal burning, petroleum and natural gas combustion to combatting climate change; methane capture and storage for climate control; and methane and hydrogen storage for fuel cells generating low and zero carbon emission. This gas separation and storage technology has formed a key strategy in countries, where costs are allocated for adverse human impacts resulting from massive emission of toxic and greenhouse gases. As a result, large power plants, factory plants, and airlines burning coal or fossil fuels become incentive in cutting down toxic and greenhouse gas emissions to minimize their costs. One of the numerous applications is to remove carbon dioxide from gas streams through the membranes comprising nanomaterials and store it underground.

Porous materials such as metal-organic frameworks (MOFs), zeolitic imidazolate frameworks (ZIFs), covalent organic frameworks (COFs), and graphite-based materials are foreseen to be promising candidates to achieve prominent gas storage. It is worth to note that MOFs, unlike other rigid nanomaterials, could alter their molecular topologies to accommodate more gas. Factors such as capacity, operating temperature, thermal and hydrothermal stability, surface kinetic and chemistry; and cyclability are crucial for using such nanomaterials as successful gas absorbents. In addition, porous materials ought to be selected to target-specific applications depending on their intrinsic structures and absorption properties, for example, silicas and mesoporous metal oxides possess pores of the right size but their pore distribution is uneven so as to limit their gas selectivity abilities; zeolites possess periodic porosity but their pores are so small that the infiltration of organic molecules is forbidden, and MOFs possess periodic pores in various size ranges which enable them to absorb different molecules resulting in a wide range of applications such as gas separation, gas storage, ion exchange, catalysis, and magnetism. Experimentally, techniques such as metalation, interpenetration, optimization of pore size, and pore infiltration could be adopted to improve their adsorption enthalpies.

The usage of clean fuels such as hydrogen and methane will cultivate low carbon emissions. Vehicles are responsible for around 20% of the total greenhouse gas emission, and the motor vehicles and other mechanical systems powered by hydrogen gas have the major merit of producing water after combustion. Insufficient storage capacity, slow kinetics, poor reversibility and high dehydrogenation temperatures are the main challenges for an acceptable media to be a candidate for hydrogen storage. Porous materials described above might partially solve the problems, and numerous experimental and theoretical investigations on hydrogen storage using porous materials such as carbon nanotubes, zeolites, and metal-organic frameworks have been performed due to their high adsorption surface areas and intrinsic chemical properties. Notable nanocontainers include 2 wt% for graphene, 3.5 wt% for carbon nanotube, 4.5 wt% for

MOF-5, 9 wt% for pillared graphene [4], and all are measured at 77 K and 1 bar. In particular, graphene offers a considerably higher surface area to make contact with polymer composites for multi-functional purposes.

While experimental discovery has led to dramatic improvements in gas storage using nanomaterials, further data condensation and optimization could certainly speed up this process. Fundamental equations such as Langmuir single-layer model, and the Brunauer, Emmett, and Teller (BET) multilayer model provide the simplest way to calculate adsorption properties, for example, the heat of adsorption, which are accessible for most nonspecialists. Related computational simulation techniques include ab initio principles, molecular dynamics simulations, and Monte Carlo simulations which offer more comprehensive picture for gas absorption. Here, we provide some basic knowledge about these computational simulations techniques: ab initio principles start directly at the level of established laws of quantum physics and do not make any assumptions of empirical models and parameters. An example includes the calculation of materials' electronic structures using Schrodinger's equation within a set of approximations resulting in huge computational times for solving the Schrodinger's equation; molecular dynamics simulations is a computer simulation of physical movement for atoms and molecules in the classical regime. The atoms and molecules are allowed to interact for a period of time, and the trajectories of them are numerically determined by solving the Newtonian equations of motion, where forces between the particles are determined by empirical force fields; Monte Carlo simulations rely on repeated random sampling, which could be sampled by some probability distributions, for example, the Boltzmann distribution to obtain numerical results. The methodology is useful for simulating systems with many coupled degrees of freedom such as fluids, disordered materials, strongly coupled solids, and cellular structures. While simulation techniques such as ab initio principles, molecular dynamics simulations, and Monte Carlo methods provide comprehensive pictures for almost all aspects of adsorption within the specific atomic structures, they generally require substantial prior knowledge to perform the computational simulations in the highest accuracy. Besides, the computational simulations are very limited in both spatial and temporal scales that they can access. Although several methodologies such as mean-field density functional theories and the lattice Boltzmann/gas method could be employed to speed up certain simulations, still the lengthy computational times could never be severely compressed. These obstacles limit their deductive capabilities.

In this chapter, we investigate the hydrogen storage between the graphene sheets and inside the GOFs. Some related experiments and theoretical investigations are summarized in Table 7.1. However, the present mathematical frameworks for graphene and GOFs could be inferred to other gases and well-defined porous nanomaterials without conceptual difficulties. While the carbon atoms are assumed to be smeared across the surface of graphene sheets, hydrogen, oxygen, boron, and carbon atoms are smeared across the envisaged columns

TABLE 7.1

Some References for Related Topics

Materials	References
Graphene	• Hydrogen on graphene: Electronic structure, total energy, structural distortions and magnetism from first-principle calculations [13] • Graphene nanostructures as tunable storage media for molecular hydrogen [14] • Ab initio investigation of molecular hydrogen physisorption on graphene and carbon nanotubes [15] • High-capacity hydrogen storage by metallized graphene [16] • Understanding adsorption of hydrogen atoms on graphene [17] • Enhancement of hydrogen physisorption on graphene and carbon nanotubes by Li doping [18] • Effective pathway for hydrogen atom adsorption on graphene [19] • Synthesis of graphene-like nanosheets and their hydrogen adsorption capacity [20] • Hydrogen adsorption behavior of graphene above critical temperature [21] • Chemical storage of hydrogen in few-layer graphene [22] • Density functional study of adsorption of molecular hydrogen on graphene layers [23] • Ab initio investigation of physisorption of molecular hydrogen on planar and curved graphenes [24] • Uptake of H_2 and CO_2 by graphene [25] • Computational study of hydrogen storage characteristics of covalent-bonded graphenes [26]
Graphene-related materials	• Pillared graphene: A new 3D network nanostructure for enhanced hydrogen storage [4] • Graphene oxide as an ideal substrate for hydrogen storage [27] • GFO materials: Theoretical predictions and experimental results [3] • Hydrogen storage in graphite nanofibers [28]

(see Figure 7.2), so that the continuous approach could be used to approximate the nonbonded van der Waals interactions between a hydrogen molecule and the bulk structure in terms of surface integrals. The continuous approximation has successfully been applied in numerous occasions such as the encapsulation of drug molecules inside single-walled nanotubes as "magic bullets" [5], the diffusion and storage of lithium ions between graphene sheets [6,7], the transportation of single-filed water molecules inside ultra-small radii nanotubes and ultrafiltration using functionalized carbon nanotubes [8–11], and the shuttle memory nanodevice using metallofullerenes and carbon nanotubes [12]. Important parameters such as the heat of absorption, and the absorption and bulk volumes could be determined by the total potential energy computed using continuous approximation and the Boltzmann distribution. The hydrogen uptake can then be obtained from the equations of state for both adsorption and bulk gas phases.

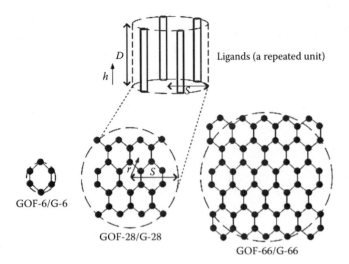

GOF-6/G-6

GOF-28/G-28

GOF-66/G-66

FIGURE 7.2 Repeated unit of GOFs (upper); Bases for GOF-6, GOF-28, and GOF-66, where numbers denote the number of carbon atoms on each base, and G-6, G-28, and G-66 denote the corresponding double-layered graphene sheets, that is, GOFs without ligands (lower).

The structure of the chapter is organized as follows: In Section 7.2, the theoretical background is presented, followed by the numerical results and some discussions in Section 7.3. In the final section, a general conclusion is provided.

7.2 THEORETICAL BACKGROUND

In this section, we present the theoretical background for this chapter. We assume that the interactions between hydrogen molecules and the host material are dominated by van der Waals forces. Therefore, the atomic interactions between a hydrogen molecule and the graphene or GFOs, herein GOFs, are modeled using the nonbonded 6–12 Lennard–Jones potential, where the atom–atom interaction is given by

$$V(\rho) = 4\varepsilon \left[-\left(\frac{\sigma}{\rho}\right)^6 + \left(\frac{\sigma}{\rho}\right)^{12} \right] = -\frac{A}{\rho^6} + \frac{B}{\rho^{12}} \quad (7.1)$$

where ρ, ε, and σ denote the distance, the potential well depth and the Lennard–Jones distance between two atoms, respectively. In addition, A and B denote the attractive and repulsive Hamaker constants, respectively. We comment that following the accounts by Jones, many theoretical efforts have been attempted to improve the Jones' empirical rules taking into account the dielectric properties of the molecular surface. We also comment that strictly speaking, quantum mechanical effects such as the dispersion interactions raising from the fact that the graphene sheet is a Dirac solid might needed to be incorporated into the present model. However, the central spirit of the applied mathematical modeling is to simplify and solve problems in a tractable manner without losing the primary effects. We therefore ignore the above quantum effects due to the fact that these confinements would only

produce a small increment in accuracy but would substantially increase the complexity and computational times of the present mathematical approach. In addition, 6–12 Lennard–Jones potential is chosen due to its simplest form resulting in analytical expressions, which could be easily computed using mathematical software such as MATLAB, Maple, and Mathematica.

First, only double-layered graphene sheets are considered and the carbon atoms are assumed to be smeared across the surface of a circular graphene sheet (see Figure 7.2), so that continuous approximation [29,30] could be used to approximate the total pairwise potential energies between the single hydrogen molecule and the graphene sheets, and the total potential energy could be analytically determined as

$$E_1(r,h) = \eta \int_{S_1} V(\rho) dS_1$$

$$= \eta\pi \left\{ \begin{array}{l} \left[-\dfrac{A}{2}\left(\dfrac{1}{h^4} + \dfrac{1}{(D-h)^4}\right) + \dfrac{B}{5}\left(\dfrac{1}{h^{10}} + \dfrac{1}{(D-h)^{10}}\right) \right] \\[2mm] + \left[\dfrac{A}{2}\left(\dfrac{1}{(h^2+r^2)^2} + \dfrac{1}{((D-h)^2+r^2)^2}\right) \right] \\[2mm] - \dfrac{B}{5}\left(\dfrac{1}{(h^2+r^2)^5} + \dfrac{1}{((D-h)^2+r^2)^5}\right) \right] \end{array} \right\}$$

(7.2)

where r, h, D, η, and $dS_1 = 2\pi r dr$ denote the radius of the circular graphene sheets, the perpendicular distance of the hydrogen molecule measured from the lower graphene sheet, the separation between the two graphene sheets, the atomic density of the graphene sheets, and the surface element of the graphene sheets, respectively. Equation 7.2 is obtained by directly submitting Equation 7.1 into the integrand in Equation 7.2. We comment that more complicated van der Waals models could be used but the total potential energy could only be sought numerically. It is worth to note that the analytical expression for the total potential energy resembles the Steele 10-4-3 potential, which is a variation of the Lennard–Jones potential model for the fluid–wall systems. The potential expressions in analytical forms for other simple geometries such as the spherical and cylindrical configurations could be found in Thornton et al. [31]. We comment that no curve-fitting exercise is adopted here in comparison to Thornton et al. [31]. Here, the potential well depth and the Lennard–Jones distance have to be modified by taking into account the interactions between two different atoms, that is, the hydrogen molecule and the carbon atoms on the graphene sheets, which could be achieved by the simple Lorentz–Berthelot mixing rules, that is, $\varepsilon = (\varepsilon_i \varepsilon_j)^{1/2}$ and $\sigma = (\sigma_i + \sigma_j)/2$, respectively, where i and j denote two distinct atoms.

Now, a repeated unit of GOFs comprising four molecular ligands and two parallel graphene sheets is considered. Furthermore, the hydrogen, oxygen, boron, and carbon atoms are assumed to be smeared across an envisaged one-dimensional column. We consider four types of atoms individually so that the total potential energy between the hydrogen molecule and the four ligands could be approximated by the linear sum of four surface integrals, that is,

$$E_2(r) = \sum_{i=1}^{4} \eta_i \int_{S_2} V(\rho) dS_2$$

$$= \sum_{i=1}^{4} \frac{3\pi\eta_i S}{8} \left[-\frac{2A_i\pi}{(S-r)^5} F\left(\frac{5}{2},\frac{1}{2};1;-\frac{2rS}{(S-r)^2}\right) \right.$$

$$\left. + \frac{21B_i\pi}{16(S-r)^{11}} F\left(\frac{11}{2},\frac{1}{2};1;-\frac{2rS}{(S-r)^2}\right) \right]$$

$$(7.3)$$

where η_i, A_i, B_i, S_2, S, and F denote the atomic number density of hydrogen ($i = 1$), oxygen ($i = 2$), boron ($i = 3$), and carbon ($i = 4$), the attractive constant for the hydrogen molecule and the ith atom at the ligands, the repulsive constant for the hydrogen molecule and the ith atom at the ligands, the surface area of the ligand, the approximate radius of the graphene sheets, and the usual hypergeometric function, respectively. We note that the hypergeometric function can be expanded in terms of the series expression, which is given by

$$F(a,b,;c;z) = \sum_{n=0}^{\infty} \frac{(a)_n (b)_n}{(c)_n} \frac{z^n}{n!}$$

where $(a)_n = (a + n - 1)!/(a - 1)!$ and ! denotes the usual factorial. For $z \ll 0$, the value of the first few terms in the series has already approached the value of the function. The total potential energy of the hydrogen molecule inside GOFs, $E(r, h)$ is therefore given by

$$E(r,h) = E_1(r,h) + E_2(r,h) + \sum_i E_3(|r-r_i|,|h-h_i|) \quad (7.4)$$

where $E_3(|r - r_i|, |h - h_i|)$ denotes the molecular interactions between the ith hydrogen molecule and its nearby hydrogen molecules inside the host structure.

On determining the total potential energy using continuous approximation, statistical mechanics could be employed to investigate the storage properties of the hydrogen gas inside porous materials. Hydrogen gas inside the porous structures can be divided into the adsorption and bulk gas phases. The probability that a gas molecule is absorbed inside the porous material could be expressed as $P_{ad}(r, h) = 1 - \exp(-|E(r, h)|/k_B T)$ using the Boltzmann distribution, where $E(r, h)$ is obtained from Equation 7.4, and k_B and T denote the Boltzmann constant and the temperature, respectively. The adsorption volume, V_{ad} and the bulk volume, V_{bulk} could be subsequently obtained from the integrals over the total free area V_f:

$$V_{ad} = \int_{V_f} P_{ad}(r,h)dV, \quad V_{bulk} = \int_{V_f} \{1 - P_{ad}(r,h)\}dV \quad (7.5)$$

where V_f and $dV = rdr\, d\theta\, dh$ denote the total free area from which $E(r, h) \leq 0$ and the volume element of the cavity, respectively. We comment that the integrals could be numerically determined using Simpson's rule. The heat of adsorption is defined by $Q = |E_{avg}| + (k_B T)/2$, where the average energy is $E_{avg} = (\int_{V_f} E(r, h)dV)/V_f$. The modified equations of state are eventually given by

$$P\left(\frac{V_{ad}}{n_{ad}} - \upsilon\right) = RT \exp\left(-\frac{Q}{RT}\right), \quad P\left(\frac{V_{bulk}}{n_{bulk}} - \upsilon\right) = RT \quad (7.6)$$

where P, R, n_{ad}, n_{bulk}, and υ denote the external pressure, the molar gas constant, the number of moles in the adsorption phase, the number of moles in the bulk gas phase, and the densely occupied molar volume, respectively. We comment that the densely occupied molar volume could be determined from the total potential energy profile between hydrogen molecules and the host structure. In addition, the extra exponential term in Equation 7.6 accounts for the energy well exerted by the graphene sheets and GOFs. The total number of moles of molecules inside the host structure is $n = n_{ad} + n_{bulk}$, where n_{ad} and n_{bulk} are solved using Equation 7.6, and the gravimetric uptake is defined as $\{w_H/(w_H + W)\}*100\%$, where w_H and W denote the weight of hydrogen molecules and the weight of the porous material, respectively.

However, from Equation 7.6, singularity occurs for n when T approaches zero or P approaches infinity, which is not physically feasible. To tackle this problem, further microscopic constraints are imposed so that our calculations terminate when

1. The work done on a hydrogen molecule exerted by the external pressure P together with its kinetic energy is higher than its average binding energy, that is, $PV_{vdW} + (k_B T/2) > |E_{avg}|$, where V_{vdW} denotes the van der Waals volume for the hydrogen molecule.
2. The volume occupied by the hydrogen gas (hard sphere) is larger than the physically available volume of the porous cavity, that is, $nN_A(1.33\pi r_H^3) > V$, where r_H and N_A denote the van der Waals radius of the hydrogen molecule and Avogadro's constant, respectively.

After some simple algebra, the iterations terminate whenever

$$n > \frac{3S^2 D}{4r_H^3 N_A} \quad \text{or} \quad P > \frac{|E_{avg}| - k_B T/2}{(4/3\pi r_H^3)} \quad (7.7)$$

are satisfied. We comment that the first and second constraints determine the maximum hydrogen uptake for the low- and high-temperature limits, respectively. For example, as T

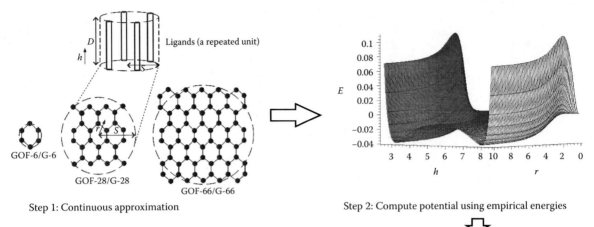

Step 1: Continuous approximation

Step 2: Compute potential using empirical energies

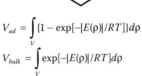

$$P\left(\frac{V_{ad}}{n_{ad}} - \gamma\right) = RT\exp\left(-\frac{Q}{RT}\right)$$

$$P\left(\frac{V_{bulk}}{n_{bulk}} - \gamma\right) = RT$$

$$n = n_{ad} + n_{bulk}$$

$$V_{ad} = \int_V \{1 - \exp[-|E(\rho)|/RT]\}d\rho$$

$$V_{bulk} = \int_V \exp[-|E(\rho)|/RT]d\rho$$

Step 4: Use equations of state to find total number of gas molecules in the system

Step 3: Compute volume for which gas will be absorbed or not

FIGURE 7.3 Schematic diagram for the theoretical background.

increases, the second constraint will be satisfied quicker than the first one for certain Ps. Now, we are equipped with sufficient theoretical tools for performing the following numerical results. The schematic figure for the theoretical background is summarized in Figure 7.3.

7.3 NUMERICAL RESULTS

The numerical results, arising from the theoretical background derived in Section 7.2, are provided in this section. Two types of GFOs, namely double-layered graphene sheets and GFO,s are considered. In particular, four different types of GOFs are investigated, namely GOF-120, GOF-66, GOF-28, and GOF-6, where the numbers denote the number of carbon atoms per repeated unit per graphene sheet, which are indicated in Figure 7.2. The corresponding double-layered graphene sheets, namely G-120, G-66, G-28, and G-6 are also investigated, and the numerical results are compared with the recent theoretical and experimental results for verification of the present mathematical approach.

7.3.1 NUMERICAL RESULTS FOR DOUBLE-LAYERED GRAPHENE SHEETS

First, the hydrogen uptake between two graphene sheets is investigated. An interlayer separation distance $D = 11$ Å is chosen, from which the largest hydrogen storage is realized for GOFs using computational simulation. The values of all parameters used in this chapter are provided in Table 7.2. For zero-order approximation, that is, we ignore the hydrogen–hydrogen interactions so that the hydrogen–host structure interactions define the total potential energy. Equation 7.4

reduces to E_1, and the total potential energy of the hydrogen molecule inside G-120 is obtained and plotted in Figure 7.4. From Figure 7.4, the hydrogen molecule is more probably to be situated in two potential wells, that is, -0.04 eV at $h = 3.2$ and 7.8 Å, and the total potential energy decreases asymptotically in the r-direction reflecting the diminishing effect of van der Waals forces in distance. The total potential energies of the hydrogen molecule between G-66, G-28, and G-6 can be determined in a similar manner, and the minimum potential energies are given by -0.036, -0.032, and -0.006 eV, respectively. In addition, the average potential energies for G-120,

TABLE 7.2
Table of Parameters Adopted in This Chapter

Description	Parameter	Value
Attractive constant H_2-graphene	A	9.610 eVÅ^6
Repulsive constant H_2-graphene	B	8695.500 eVÅ^{12}
Attractive constant H_2-H	A_1	3.592 eVÅ^6
Repulsive constant H_2-H	B_1	1362.523 eVÅ^{12}
Attractive constant H_2-O	A_2	7.512 eVÅ^6
Repulsive constant H_2-O	B_2	5093.263 eVÅ^{12}
Attractive constant H_2-B	A_3	13.144 eVÅ^6
Repulsive constant H_2-B	B_3	12394.747 eVÅ^{12}
Attractive constant H_2-C	A_4	13.531 eVÅ^6
Repulsive constant H_2-C	B_4	12493.006 eVÅ^{12}
Number density of graphene	η	0.381 Å
Radius for GOF/G-120	S	9.840 Å
Radius for GOF/G-66	S	7.387 Å
Radius for GOF/G-28	S	4.920 Å
Radius for GOF/G-6	S	1.230 Å
vdW radius for H_2	r_H	1.516 Å

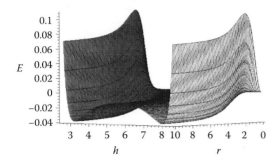

FIGURE 7.4 Total potential energy of hydrogen molecule situated inside G-120 with $D = 11$ Å, where E is in the unit of eV, and h and r are in the unit of Å. From which important parameters such as average energy E_{avg}, the heat of absorption Q, the free volume V_f, the adsorption volume V_{ad}, and the bulk volume V_{bulk} can be numerically determined.

G-66, G-28, and G-6 are given by −0.0170, −0.0151, −0.0112, and −0.00130 eV, respectively. It is intuitive that larger the size of the graphene sheets, the lower the total potential energy is. Owing to the sufficiently large separation distance D, the total potential energies become zero when $h = 3.2$ and 7.8 Å for all proposed graphene sheets, from which we can determine the free volume V_f and hence the heat of absorption Q, which is crucial to depict the adsorption behavior of hydrogen molecules between the graphene sheets and inside GOFs. Given that, we can calculate the adsorption volume V_{ad} and the bulk volume V_{bulk} using Equation 7.5, and then Equation 7.6 can be used to deduce the hydrogen uptake of the double-layered graphene sheets as a function of temperature T and pressure P.

Now, we fix the temperature T at 77 K and the gravimetric uptake, which is defined as the percentage of the weight of the absorbed hydrogen gas over the total weight of the system including the weight of the absorbed hydrogen gas, is plotted against pressure in Figure 7.5. The gravimetric hydrogen

uptakes between two graphene sheets increase from 0 wt% for G-6 to 1.85 wt% for G-120 as the size of the graphene sheets increases. The zero gravimetric uptake for G-6 reveals an insufficient binding energy generated by G-6 to attract hydrogen molecules, whereas the gravimetric uptake for G-120 approaches that of an infinite graphene sheet owing to the fact that the van der Waals forces diminish rapidly and asymptotically in the r-direction. In addition, the gravimetric uptakes for the proposed graphene sheets also increase linearly with external pressure. The results of the zero-order approximation are in good agreement with the experimental data, that is, Srinivas et al. [20] obtain 1.2 wt% and Ghosh et al. [25] obtain 1.7 wt%, and the theoretical investigation, that is, Dimitrakakis et al. [4] obtain 2 wt%, all are measured at 77 K and 1 bar. These agreements reveal that the interactions between the hydrogen molecule and the host structure dominate the total potential energy. Now, the temperature is fixed at 77 K and the pressure is increased up to 80 bars to investigate all isotherms, which are given in Figure 7.6. We observe that for certain pressures and beyond, the gravimetric uptakes level off and reach the plateaus resulting from the microscopic enforcements of Equation 7.7. We could also fix the pressure and vary the temperature, and the methodology of obtaining the hydrogen uptake could be deduced in a similar manner, which indicates how rapid the present mathematical approach facilitates numerical results.

7.3.2 NUMERICAL RESULTS FOR GOFs

We could either accept the zero-order approximation or determine the hydrogen–hydrogen interactions, that is, E_3 in Equation 7.4 by other means, for example, through interpolating experimental or simulations results, or using Monte-Carlo simulations. For simplicity, the former methodology is adopted and E_2 in Equation 7.4 is used by taking into account

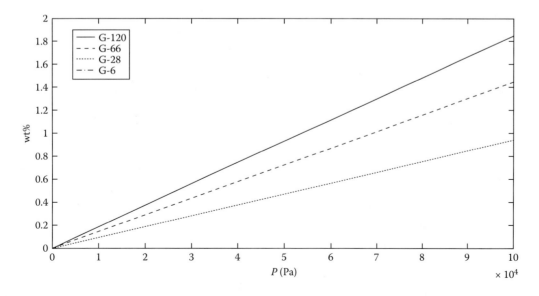

FIGURE 7.5 Total gravimetric uptakes for G-6, G-28, G-66, and G-120 at 77 K and for pressure up to 1 bar. There exists a linear relationship between gravimetric uptakes and external pressure.

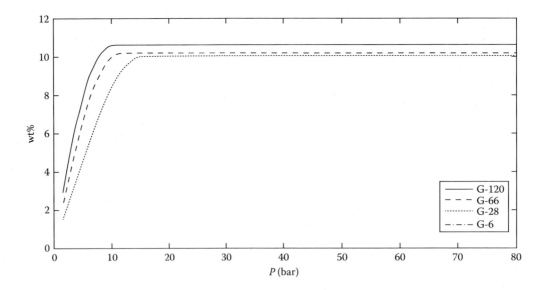

FIGURE 7.6 Total gravimetric uptakes for G-6, G-28, G-66, and G-120 at 77 K and for pressure up to 80 bars. There exists a linear relationship between gravimetric uptakes and pressure for sufficiently low pressures, nonlinear relationship for intermediate pressures, and plateau is formed when the pressure is higher than 20 bars for all proposed double-layered graphene sheets.

the additional atomic interactions arising from the ligands. The total potential energy for GOF-6 is calculated and is given in Figure 7.7. Owing to the strong repulsive energy generated by the ligands, no hydrogen uptake occurs for GOF-6. Furthermore, the total potential energy for GOF-120 is calculated and is given in Figure 7.8. The total potential energy for GOF-120 features spires about the four ligands and features deeper potential wells in comparison to that of G-120, that is, Figure 7.4. We also comment that the total potential energies for GOF-28 and GOF-66 possess similar potential landscapes, and the average energies E_{avg} are −0.0225, −0.0266, and −0.0487 eV for GOF-120, GOF-66, and GOF-28, respectively. The corresponding minimum energies are −0.05, −0.06, and −0.065 eV, respectively. We comment that GOF-28 possesses the lowest minimum and average energies among the proposed GOFs.

For convenience, we partition the external pressure into the low-pressure limit, that is, from 0 to 1 bar, the intermediate-pressure regime, that is, from 1 bar to 10 bars, and the high-pressure limit, that is, larger than 10 bars. The gravimetric uptakes for GOFs are determined in a similar manner as in the

case of the double-layered graphene sheets, and the numerical result of the hydrogen uptakes for the proposed GOFs, at 77 K and the pressure up to 10 bars are given in Figure 7.9. Owing to the extremely strong repulsive energy generated by the dense ligands, there is no hydrogen uptake for GOF-6. In the low-pressure limit, GOF-28 possesses the optimal gravimetric uptake of 6.33 wt%, and the corresponding gravimetric uptakes of GOF-66 and GOF-120 are given by 2 and 1.68 wt%, respectively, all are measured at 77 K and 1 bar. On comparing Figures 7.5 and 7.9, due to the low ligand densities, the gravimetric uptakes of GOF-120 and GOF-66 approach that of G-120 and G-66, respectively, where we could virtually assume that the ligand densities are zero. This reflects that the molecular effect arising from the ligands diminishes when the ligand density is sufficiently low at low temperatures and pressures.

The optimal hydrogen uptake for GOF-28 at low temperatures and pressures could be partially explained by scrutinizing the average/minimum potential energies among GOFs. From the preceding paragraph, the average potential energy for GOF-28 is −0.0487 eV, which is much lower than −0.0266

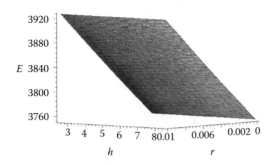

FIGURE 7.7 Total potential energy for GOF-6. Strong repulsive force expels hydrogen molecules away from the host structure.

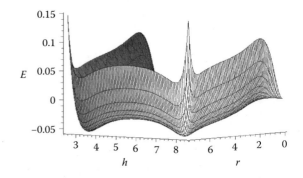

FIGURE 7.8 Total potential energy for GOF-120 where spires are formed near ligands.

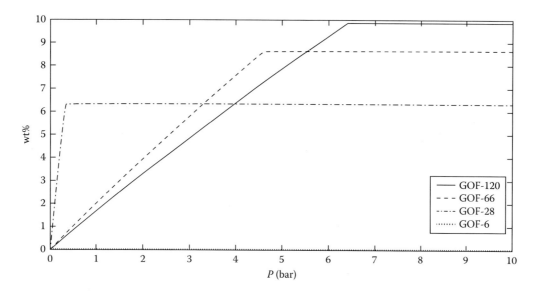

FIGURE 7.9 Total gravimetric uptakes for GOF-6, GOF-28, GOF-66, and GOF-120 at 77 K and for pressure up to 10 bars.

and −0.0225 eV for GOF-66 and GOF-120, respectively. There is always a competition between the kinetic energy of the hydrogen gas and the potential energy generated by the host molecular structure to determine whether the hydrogen molecules would stay or escape from the host structure, and the lowest average potential energy for GOF-28 provides a deeper binding energy to keep the hydrogen molecules inside the host structure. For GOFs with low ligand densities, such as GOF-66 and GOF-120, the role of the ligands acts mainly as a mechanical support for the host structures. However, there is a peculiar effect for the case of GOF-28, where the density of ligands is just right that the ligands neither reject hydrogen molecules nor provide merely mechanical support for the case of the extremely high or low ligand densities, respectively. The extra energy induced by the ligands inside GOF-28 boosts up the heat of absorption significantly so as to enhance its hydrogen storage, although the free volume V_f of GOF-28 is smaller than that of G-28.

The theoretical gravimetric uptake for GOF-28 reaches the highest value of 6.33 wt% at 77 K and 1 bar, which is larger than the reported hydrogen uptakes for most porous materials, for example, 4.5 wt% for MOF-5, 2 wt% for graphene, 3.5 wt% for carbon nanotubes, but falls below that of the pillared graphene, that is, 9 wt%. However, it is noteworthy that the pillared graphene is geometrically intricate and extremely hard to be fabricated resulting in low commercial value.

However, as the pressure increases, the effect arising from the extra induced energy diminishes as more hydrogen molecules could be manipulatively squeezed into the host structure by the strong pressure and the geometric effect takes place, that is, the higher the free volume, the larger the hydrogen uptake. A mixed performance of GOF-28, GOF-66, and GOF-120 occurs in the intermediate-pressure regime (see Figure 7.9). However, in the high-pressure limit, GOF-120 turns out to be the optimal structure because it possesses the largest

free volume. The saturated hydrogen uptakes of GOF-120 and GOF-66 approaches that of G-120 and G-66 but are less than that of G-120 and G-66 simply due to the fact that the free volumes of GOF-120 and GOF-66 are smaller than that of G-120 and G-66 (see Figures 7.6 and 7.9).

In contrast to computational simulations, though they provide more flexibilities and information than the present mathematical approach, they take extremely long times to compute a single result by varying even a single parameter, for example, temperature or pressure. In addition, they become intractable when the system sizes approach microscale and when the evolution time exceeds microseconds. In the present mathematical approach, since all information and parameters are absorbed into the two equations of the state, that is, Equation 7.6, the computational times are almost similar while the underlying parameters are varied. Again to demonstrate this, we consider two more realistic scenarios where the usual and high bulk temperatures of 300 and 500 K are adopted, and the external pressure is extended from 10 to 800 bars to observe all isotherms. The numerical results are given in Figure 7.10.

For the proposed daily temperature, that is, $T = 300$ K, GOF-28 still reveals its advantage in the low-pressure limit as the extra energy induced by the ligands continues to enhance the hydrogen uptake for GOF-28 at low pressures. It is inclusive to tell which GOFs are optimal in the intermediate pressure regime. However in the high-pressure limit, again the geometric effect takes place and GOF-120 becomes the best container. The maximum gravimetric uptakes for all proposed GOFs in room temperature asymptotically approach that of GOFs at 77 K but could only be achieved at higher pressures. Therefore, the hydrogen adsorption characteristics for GOFs are qualitatively the same for all proposed GOFs at 77 and 300 K.

However, at sufficiently high temperatures, for example, at 500 K, the adsorption characteristics differ dramatically. At that temperature, the kinetic energies of the hydrogen

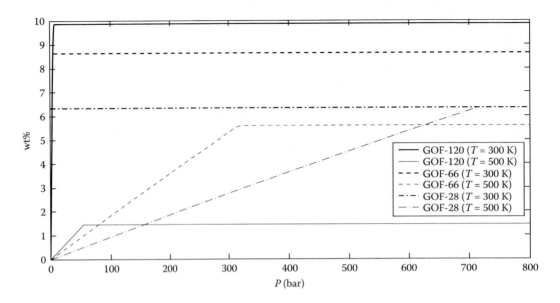

FIGURE 7.10 Total gravimetric uptakes for GOF-6, GOF-28, GOF-66, and GOF-120 at 300 and 500 K, and for pressure up to 800 bars.

molecules are so high that the extra induced energy produces marginal forces to attract more hydrogen molecules even in the low-pressure limit, and the geometric effect takes place. It turns out that GOF-120 possesses the largest gravimetric uptake in both low and intermediate pressure regimes. In the high-pressure limit, the maximum gravimetric uptakes of both GOF-66 and GOF-120 at 500 K are smaller than that of GOF-66 and GOF-120 at 77 and 300 K, and it is surprising to observe that GOF-28 still reaches the largest gravimetric uptake of 6.33 wt% at sufficiently high pressure. This could be explained from the fact that the very strong pressure pushes more hydrogen molecules toward the boundary of GOF-28 and they become more likely to interact with the ligands. The adsorption characteristics of the proposed GOFs at different temperatures and pressures are summarized in Table 7.3. However, it still poses a huge challenge for experimentalists to fabricate GOFs in a perfect form due to the formation of metal clusters inside GOFs although several experimental techniques have been developed to tackle the cluster formation problem [32].

7.4   CONCLUSION

In this chapter, continuous approximation and equations of state for both the bulk gas and adsorption phases deduced from statistical mechanics are used to calculate the total potential energies and the hydrogen uptakes for the double-layered graphene sheets and GOFs, respectively. The zero-order approximation results for the hydrogen storage between two parallel graphene sheets are shown to be consistent with several experimental and theoretical results which increase our confidence in using the present mathematical approach to investigate the hydrogen uptakes for GOFs. Unlike graphene sheets, where the geometric sizes of the graphene sheets determine their hydrogen adoption properties, unexpected effects occur in GOFs, where the extra induced energies and the geometric effects alternatively affect their adsorption properties depending on the combination of temperatures and pressures: For low and intermediate temperatures up to 400 K and in the low-pressure limit, the extra induced energy dominates and the ligands do not merely act as mechanical support to the host structure but also enhance the hydrogen storage inside GOF-28. In particular, we determine 6.33 wt% for GOF-28 at 77 K and 1 bar. However, in the high-pressure limit, the geometric effect overrides the extra induced energy as hydrogen gas could be manipulatively squeezed into GOFs and GOF-120 turns out to have the largest hydrogen uptake. For temperatures higher than 400 K and in the low-pressure limit, the geometric effect takes place and GOF-120 becomes the ideal candidate for hydrogen storage. However, in the high-pressure limit, the extra induced energy overrides the geometric effect, and GOF-28 possesses the optimal storage structure. The present approach adopted in this chapter is ready to be employed in other gases and nanostructures. In spite of the recent triumph for computational simulations in this area, the computational times increase dramatically in both temporal and spatial scales which limit their deductive

TABLE 7.3
Table of Adsorption Characteristics for Different Temperatures and Pressures

	Low Temperatures	Intermediate Temperatures	High Temperatures ($T > 400$ K)
Low-pressure limit (0–1 bar)	GOF-28	GOF-28	GOF-120
Intermediate pressure regime (1 bar–10 bars)	Mix	Mix	GOF-120
High-pressure limit (>10 bars)	GOF-120	GOF-120	GOF-28

capacities. Applied mathematical approach presented in this chapter might provide supplements to the current computational simulations and provide better insights than those fundamental equations.

ACKNOWLEDGMENT

The author acknowledges Dr. Barry Cox from the University of Adelaide, Australia, and Dr. Aaron Thornton from CSIRO, Australia, for their original contributions on the present hybrid mathematical and statistical model. He also acknowledges all his ex-colleagues from the nanomechanics group from both the University of Adelaide and University of Wollongong, Australia.

REFERENCES

1. Novoselov, K.S., Geim, A.K., Morozov, S.V., Jiang, D., Zhang, Y., Dubonos, S.V., Grigorieva, I.V., Firsov, A.A. 2004. Electric field effect in atomically thin carbon films. *Science* 306: 666–9.
2. Zhang, T., Xue, Q., Zhang, S., Dong, M. 2012. Theoretical approaches to graphene and graphene-based materials. *Carbon* 7: 180–200.
3. Burress, J.W., Gadipelli, S., Ford, J., Simmons, J.M., Zhou, W., Yildirim, T. 2010. Graphene oxide framework materials: Theoretical predictions and experimental results. *Angew. Chem. Int. Edn* 49: 8902–4.
4. Dimitrakakis, G.K., Tylianakis, E., Froudakis, G.E. 2008. Pillared graphene: A new 3D network nanostructure for enhanced hydrogen storage. *Nano Lett.* 8: 3166–3170.
5. Hilder, T.A., Hill, J.M. 2007. Carbon nanotubes as drug nanocapsules. *Curr. Appl. Phys.* 8: 258–61.
6. Chan, Y., Hill, J.M. 2010. Modelling interaction of atoms and ions with graphene. *Micro Nano Lett.* 5: 247–50.
7. Chan, Y., Hill, J.M. 2011. Lithium ion storage between graphenes. *Nanoscale Res. Lett.* 6: 203.
8. Chan, Y., Hill, J.M. 2011. A mechanical model for single-file transport of water through carbon nanotube membrane. *J. Membr. Sci.* 372: 57–65.
9. Chan, Y., Hill, J.M. 2012. Modeling on ion rejection using membrane comprising ultra-small radii carbon nanotubes. *Eur. Phys. J. B* 85: 56.
10. Chan, Y., Hill, J.M. 2013. Ion selectivity using membranes comprising functionalized carbon nanotubes. *J. Math. Chem.* 51: 1258–73.
11. Chan, Y. 2013. Mathematical modeling on ultra-filtration using functionalized carbon nanotube. *Appl. Mech. Mater.* 328: 664–8.
12. Chan, Y., Lee, R.K.F., Hill, J.M. 2010. Metallofullerenes in composite carbon nanotubes as a nanocomputing memory device. *IEEE Trans. Nanotechnol.* 10: 947–52.
13. Boukhvalov, D.W., Katsnelson, M.I., Lichtenstein, A.I. 2008. Hydrogen on graphene: Electronic structure, total energy, structural distortions and magnetism from first-principles calculations. *Phys. Rev. B* 77: 035427.
14. Patchkovskii, S., Tse, J.S., Yurchenko, S.N., Zhechkov, L., Heine, T., Seifert, G. 2005. Graphene nanostructures as tunable storage media for molecular hydrogen. *Proc. Natl. Acad. Sci.* 102: 10439–10444.
15. Henwood, D., Carey, J.D. 2007. Ab initio investigation of molecular hydrogen physisorption on graphene and carbon nanotubes. *Phys. Rev. B* 75: 245413.
16. Ataca, C., Akturk, E., Ciraci, S., Ustunel, H. 2008. High-capacity hydrogen storage by metallized graphene. *Appl. Phys. Lett.* 93: 043123.
17. Casolo, S., Lovvik, O.M., Martinazzo, R., Tantardini, G.F. 2009. Understanding adsorption of hydrogen atoms on graphene. *J. Chem. Phys.* 130: 054704.
18. Cabria, I., Lopez, M.J., Alonso, J.A. 2005. Enhancement of hydrogen physisorption on graphene and carbon nanotubes by Li doping. *J. Chem. Phys.* 123: 204721.
19. Miura, Y., Kasai, H., Dino, W.A., Nakanishi, H., Sugimoto, T. 2003. Effective pathway for hydrogen atom adsorption on graphene. *J. Phys. Soc. Jpn.* 72: 995–997.
20. Srinivas, G., Zhu, Y., Piner, R., Skipper, N., Ellerby, M., Ruoff, R. 2010. Synthesis of graphene-like nanosheets and their hydrogen adsorption capacity. *Carbon* 48: 630–635.
21. Ma, L.P., Wu, Z.S., Li, J., Wu, E.D., Ren, W.C., Cheng, H.M. 2008. Hydrogen adsorption of graphene above critical temperature. *Int. J. Hydrogen Energy* 34: 2329–2332.
22. Subrahmanyam, K.S., Kumar, P., Maitra, U., Govindaraj, A., Hembram, K.P.S.S., Waghmare, U.V., Rao, C.N.R. 2011. Chemical storage of hydrogen in few-layer graphene. *Proc. Natl. Acad. Sci.* 108: 2674–2677.
23. Arellano, J.S., Molina, L.M., Rubio, A., Alonso, J.A. 2000. Density functional study of adsorption of molecular hydrogen on graphene layers. arXiv: physics/0002015.
24. Okamoto, Y., Miyamoto, Y. 2001. Ab initio investigation of physisorption of molecular hydrogen on planar and curved graphenes. *J. Phys. Chem. B* 105: 3470–3474.
25. Ghosh, A., Subrahmanyam, K.S., Krishna, K.S., Datta, S., Govindaraj, A., Pati, S.K., Rao, C.N.R. 2008. Uptake of H$_2$ and CO$_2$ by graphene. *J. Phys. Chem. C* 112: 15704–15707.
26. Park, N., Hong, S., Kim, G., Jhi, S.H. 2007. Computational study of hydrogen storage characteristics of covalent-bonded graphenes. *JACS.* 129: 8999–9003.
27. Wang, L., Lee, K., Sun, Y.Y., Lucking, M., Chen, Z., Zhao, J.J., Zhang, S.B. 2009. Graphene oxide as an ideal substrate for hydrogen storage. *ACS Nano.* 3: 2995–3000.
28. Chambers, A., Park, C., Baker, R.T.K., Rodriguez, N.M. 1998. Hydrogen storage in graphite nanofibers. *J. Phys. Chem. B* 102: 4253–4256.
29. Cox, B.J., Thamwattana, N., Hill, J.M. 2007. Mechanics of atoms and fullerenes in single-walled carbon nanotubes. I. Acceptance and suction energies. *Proc. R. Soc. A* 463: 461.
30. Cox, B.J., Thamwattana, N., Hill, J.M. 2007. Mechanics of atoms and fullerenes in single-walled carbon nanotubes. II. Oscillatory behavior. *Proc. R. Soc. A* 463: 477.
31. Thornton, A.W., Furman, S.A., Nairn, K.M., Hill, A.J., Hill, J.M., Hill, M.R. 2013. Analytical representation of micropores for predicting gas adsorption in porous materials. *Micropor. Mesopor. Mat.* 167: 188–197.
32. Wang, L., Lee, K., Sun, Y.Y., Lucking, M., Chen, Z.F., Zhao, J.J., Zhang, S.B.B. 2009. Graphene oxide as an ideal substrate for hydrogen storage. *ACS Nano* 3: 2995.

8 Morphology of Cylindrical Carbon Nanostructures Grown by Catalytic Chemical Vapor Deposition Method

S. Ray, M. Jana, and A. Sil

CONTENTS

Abstract 123
8.1 Introduction 123
8.2 Structure of Cylindrical CNS 128
8.3 Theoretical Studies on the Formation of CNS 130
8.4 Morphology of CNS by CCVD Method 134
    8.4.1 Growth of CNS Using Metal-Based Catalysts 136
8.5 Conclusions: The Emerging Picture 140
References 142

ABSTRACT

Rolled graphene sheets result in cylindrical carbon nanostructures (CNSs) of different morphologies like single- and multi-walled nanotubes, nanofibers, or tapes. In catalytic chemical vapor deposition (CCVD) method one may obtain all these morphologies of cylindrical nanostructures by decomposition of carbon bearing precursor gas on the surface of the catalyst nanoparticles, followed by the formation of nanostructures. Depending on the size and composition of the catalyst particles as well as the prevailing growth conditions in the deposition chamber including the flow rate of the carbon bearing gas, the growth of a specific morphology of cylindrical nanostructure is favored. There are a large number of investigations using catalysts of nanoparticles of transition metals such as cobalt, nickel, iron, and their oxides. A number of other metals have also been used. Recent in situ investigations inside environmental transmission electron microscope (ETEM) and the molecular dynamic (MD) simulations have revealed the possible steps leading to the formation of CNSs. But still there are unresolved issues pertaining to the necessity of surface melting of catalyst nanoparticles claimed to be responsible for substantial change in shape of the catalyst nanoparticles inside growing nanotubes and to the necessity of carbide formation as an intermediate step as well as the possibility of survival of oxide nanoparticles in the reducing environment during growth of nanostructures. There are diverse claims with regard to these issues based on experimental findings by different research workers. The studies carried out so far indicate that one may produce cylindrical CNSs of specific morphology by CCVD method if one tailors the catalyst nanoparticles and controls the growth condition appropriately. However, more research is necessary to attain the capability of controlled production of a specific morphology, which is extremely important from technological standpoint.

8.1 INTRODUCTION

Cylindrical carbon nanostructures (CNSs) of different morphologies have been grown by catalytic chemical vapor deposition (CCVD) method using catalyst nanoparticles of metals, alloys, and oxides on a wide variety of substrates inside a reaction chamber maintained at an elevated temperature. Carbon-bearing precursor gas/vapor alone or mixed with other gases flow into the chamber at a given pressure where the precursor gas decomposes on the catalyst nanoparticles leading to the growth of nanostructures on it. Figure 8.1 [1] shows a schematic diagram of the experimental set-up used in CCVD.

Several experiments have been carried out by different research workers who have obtained nanostructures of different morphologies. Table 8.1 lists the outcome of some of these studies. The carbon nanostructures grown include cylindrical morphology such as single-walled carbon nanotube (SWCNT), multiwalled carbon nanotube (MWCNT), and carbon nanofiber (CNF)/nanotape.

CNSs, in general, include all the carbon entities at the nanoscale. There are different approaches used to classify CNSs. Based on dimensionality, the classification of CNSs is given in Table 8.2. The classification could be based on any criteria of interest in the context of specific applications. For the development of super capacitor, the arrangement and density of dangling bonds are important for the capacitance since the space charge at the edges of graphene basal sheets and the dangling bonds dominate the total capacitance of

TABLE 8.1
Selected Results Summarizing Morphology of Carbon Nanostructures Grown Using Different Catalysts and Carbon Bearing Gas Mixture in CCVD

Carbon Bearing Gas Mixture	Type of Catalyst		Substrate	Size (nm)	Growth Condition in CCVD		Type of CNS/Mode of Growth	Reference
	Chemical	Morphology			Temperature (°C)	Pressure of Gas		
$C_2H_2 + NH_3$	Ni	p	SiO_2 coated Si	<6	650–700	20 sccm	SWCNT/B	[2]
$C_2H_2 + NH_3 + H_2$	Fe, Co, Ni	p	SiO_2	<5	700	0.2 Pa	SWCNT/B	[3]
$CH_4 + Ar + H_2$	Al/Fe	f	Mo	2/1, 4/1, 6/1	750–900	2000 cm³ min⁻¹	SWCNT/B	[4]
$CO + H_2 + He$	Co, Mo, Co–Mo	p	SiO_2	<6	700	100 cm³ min⁻¹	SWCNT/-	[5]
CH_4	Fe–Mo	p	Al_2O_3 coated on Si	3–16	900	–	SWCNT/-	[6]
$C_2H_2 + H_2$	Ni_xFe_{1-x}	p	No	1–6	600	0.5 sccm	SWCNT/-	[7]
$C_2H_4 + C_2H_2 + H_2$	Au	p	Si/SiO_2	–	800/900	20 sccm (C_2H_4), 30 sccm (CH_4)	SWCNT/-	[8]
CH_4	Mn	p	Type1: SiO_2 (1 μm) coated Si, Type 2:10 nm SiO_2 or Al_2O_3 on type 1 substrate	3.1	900	500 sccm	SWCNT/-	[9]
Ethanol	Au, Ag, Pt, Pd, Fe, Co, Ni, Cu	f	SiO_2, Al-hydroxide, sapphire	3	850	–	SWCNT/-	[10]
$CH_4 + H_2$	Cu	p	Si	10	825–850	800 sccm	SWCNT/-	[11]
$C_2H_2 + H_2$	Fe/Ti/Fe	f	SiO_2 coated Si	0–2/10/0.5–2	700	40 sccm	SWCNT-MWCNT/-	[12]
Ethylene	Fe, Ni	f	Si	2–3	500–1400	5 sccm	SWCNT/-	[13]
Ethanol + H_2	Diamond	p	Graphite	4–5	850	–	SWCNT/-	[14]
CH_4/ethanol	$RuCl_3$	N/A	SiO_2 coated Si	–	900	200 sccm	SWCNT/-	[15]
CH_4	SiO_2, Al_2O_3, TiO_2, rare earth oxides	p	SiO_2 coated Si	<5	900	–	SWCNT/-	[16]
$CH_4 + H_2$	SiO_2	f	Si or SiO_2	30	900	500 sccm	SWCNT/-	[17]
$CH_4 + H_2$	Iron oxide	p	SiO_2	1–2, 3–5	900	200 sccm	SWCNT/B	[18]
$CH_4 + Ar$	Fe/Mo oxide	p	silica-alumina	~1	900	6000 cm³ min⁻¹	SWCNT/B	[19]
$C_2H_2 + NH_3$	Fe and Ni oxide	p	SiO_x	5	615	2 × 10⁻⁷ mbar	SWCNT/B	[20]
$CH_4 + H_2$	$Mg_{0.8} M_y M'_z Al_2O_4$ (M, M' = Fe, Co, Ni, y + z = 0.2)	p	–	–	1070	250 sccm	SWCNT/-	[21]
Ethanol + H_2	Oxides of Zn	f	Si	–	900	–	SWCNT/-	[22]
Ethanol	SiC, Ge, Si	p	Si for SiC and SiC for Ge, Si	5	850	–	SWCNT/ DWCNT/-	[23]
$CH_4 + C_2H_2 + H_2$	$FeSi_2$	p	Si	–	900	1500 (CH_4), 30 sccm (C_2H_2)	DWCNT/-	[24]
C_6H_{12}	Fe	p	Al_2O_3	5	800	–	MWCNT/B/T	[25]
C_6H_{12}	Fe–Pt	p	SiO_2/Al_2O_3	1.6, 3.6	590–890	50 mbar	MWCNT/B	[26]
$CH_4 + Ar + H_2$	Al/Fe	f	Au	2/1, 4/1, 6/1	700–800	2000 cm³ min⁻¹	MWCNT/T	[4]
$C_2H_2 + NH_3 + H_2$	Fe, Co, Ni	p	SiO_2	15	700	0.2 Pa	MWCNT/T	[3]
$C_2H_2 + H_2$	Ni	p	SiO_2	10–50	550–650	5 sccm	MWCNT/T	[27]

(Continued)

TABLE 8.1 (Continued)
Selected Results Summarizing Morphology of Carbon Nanostructures Grown Using Different Catalysts and Carbon Bearing Gas Mixture in CCVD

Carbon Bearing Gas Mixture	Type of Catalyst			Growth Condition in CCVD			Type of CNS/Mode of Growth	Reference
	Chemical	Morphology	Substrate	Size (nm)	Temperature (°C)	Pressure of Gas		
C_2H_2	Ni	f	Ti / TiO_x	30/50 / 10/50	540	1000 Pa	MWCNT/B (for Ti substrate) / MWCNT/T Base tip (for TiO_x substrate)	[28]
C_2H_2	Ni, Ni–Au	f	Perforated SiO_2	2	520	0.4 Pa	MWCNT/B	[29]
C_2H_2, C_2H_4, methanol, or benzene + N_2	Au	p	SiO_2–Al_2O_3	3–10	450–800	30 sccm	MWCNT/T & B	[30]
$C_2H_2/CH_4 + N_2 + H_2$	Ni, Co, Fe, W, Pt, Al, Ir, Ti, Pd In, Mg, K, Cs, Na, Mn, Mo	f / p	SiO_2 over Si	1,2,5	600–1000	100–1000 mL min^{-1}	MWCNT/T	[31]
$C_2H_4 + Ar + H_2 + H_2O$	Fe/Gd	f	$Al_2O_3/SiO_2/Si$	1.5	730–820	75 sccm	MWCNT/B	[32]
Ethylene + $H_2 + N_2$	Co	p	Al_2O_3	–	550–700	300–500 sccm	MWCNT/-	[33]
$C_2H_2 + NH_3$	Fe, Ni	f	Ti coated Si	20	750	7 Torr	MWCNT/CNF/T	[34]
Benzene or methanol	Fe, Co, Ni, Cr, Mn, Zn, Cd, Ti, Zr, La, Cu, V, Gd-based metallocenes or chlorides	N/A	SiO_2	N/A	800–900	–	MWCNT/-	[35]
Ethanol or ethanol/ triethyl borate	SiO_2	p	Quartz	–	900	–	MWCNT/CNF/T	[36]
$CH_4 + H_2$	NiO	p	$MgAl_2O_4$	5–20	500–540	< 2.0 mbar	MWCNT/CNF/T	[37]
Methane + H_2	Fe_2O_3	p	Al_2O_3, MgO	18–26	750	1 L_N min^{-1}	MWCNT/-	[38]
$C_2H_2 + NH_3$	Fe and Ni oxide	p	SiO_x	5	480–700	2×10^{-7} mbar	CNF/T	[20]
$C_2H_4 + H_2$	Iron oxide	p	Fe_2O_3	–	500–800	–	CNF/-	[39]
$C_2H_2 + NH_3$	Oxides of cobalt doped with copper	p	Nanoporous alumina	225–580	640	20 sccm	CNF/ MWCNT/ Carbon nanotape/-	[40]
$C_2H_2 + NH_3$	Ni	p	SiO_2 coated Si	7–30	650–700	20 sccm	Bamboo-like MWCNT/T	[2]
$C_2H_2 + NH_3$	Fe	f	SiO_x	100	550–950	40–80 sccm	Bamboo-like CNT/T	[41]
$C_2H_2 + NH_3 + Ar$	Co	f	SiO_2	200	950	20–80 sccm	Bamboo-like CNT/B	[42]
$H_2 + N_2$	Bamboo	–	–	–	600–900	–	bamboo-like CNT, herringbone CNT/B	[43]
$C_2H_2 + Ar + H_2$	Sn/Fe	p	$BaSrTiO_3$	–	750	20–40 sccm	Carbon nanocoil/T	[44]
$C_2H_2 + H_2$	Iron oxide	p	Silica	–	715	–	Double helices, Nanobraids/-	[45]
$C_2H_2 + H_2$	Ni/Cu	p	Bi-metallic layer of Ni/Cu	2/8	511	6 sccm	Carbon octopi/T	[46]

Note: Abbreviations: p—particles, f—film, B—base growth, and T—tip growth.

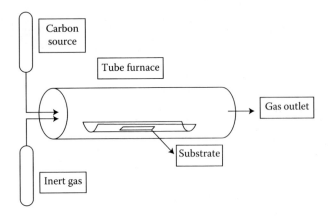

FIGURE 8.1 Schematic diagram of a typical CCVD set-up. (Reprinted with permission from Oncel, C. and Yurum, Y. 2006. *Fullerenes, Nanotubes, and Carbon Nanostructures* 14:17–37.)

these materials. Stoner and Glass [47] have put forward a classification of graphitic carbon-based materials including nanostructures, on the basis of linear edge density per unit area, ρ_L, as shown in Table 8.3. The table starts with highly oriented pyrolytic graphite having basal planes on the surface ($HOPG_B$) and one-dimensional (1D) edges at grain boundaries and steps, have density lower than 10 cm^{-1} but the density increases to the range between 10^6 and 10^7 cm^{-1} when the surface is textured with the edges ($HOPG_E$). The spatial arrangement of the edges is still planar. Textured nanosheets have a high density of edges but the edges are still not arranged spatially in three dimensions as in activated carbon (AC), which is a mixture of powders having exposed surface of edge and basal planes. Edge density in carbon nanotubes (CNTs) is similar to fibers and nanorods but on texturing the fibers, edge density could be increased. If the CNTs are aligned (a-CNT) and the edges are arranged in 3D (three dimension), edge density could be increased to as high as 10^5 cm^{-1} but it is still lower than that in AC. The edge density in aligned CNTs could be increased further if there are periodic rings around, as it happens in bamboo CNTs (b-CNT). The edge density could be increased to as high as 10^{10} cm^{-1} in aligned graphenated CNTs (g-CNT), since the edges are arranged in 3D. The spatial arrangement of the edges is extremely important in determining its density, and various graphitic carbon-based materials

are represented on an edge triangle as shown in Figure 8.2 [48], where the corners show maximum density for a given dimension of the spatial arrangement of edges. The edge density varies from 10^0 at the 1D corner to 10^{10} as indicated by dotted curved lines marked with the value of exponent. The edge density of 1D arrangement of edges is lower than that in 2D but 3D arrangement has the maximum density, as is evident from Table 8.3.

It is possible to attain still higher edge density in g-CNT$_S$ if the CNT columns could be joined across by edges and it will have the claim to occupy the 3D corner position, marked with question, till its development.

The present review is limited to cylindrical morphologies of CNS grown by CCVD method. One would like to understand the precise conditions, under which the growth of a specific cylindrical morphology is favored over others even within the ambit of a given deposition method like CCVD. However, the primary obstacle to consolidating these results to a precise understanding is the inadequate reporting of the conditions under which the experiments have been carried out as is evident from Table 8.1.

The source of carbon for the growth of nanostructures by CCVD is either a carbon-bearing gas or liquid vapor. Ethylene, acetylene, methane, carbon monoxide, and ethanol are the most commonly used sources of carbon [49]. Methane, carbon monoxide, and ethanol vapor are often used for growing SWCNT. The decomposition of these carbon-bearing compounds in the reaction chamber of CCVD also produces hydrogen and oxygen, which influence the morphology of nanostructures growing from carbon. Bachmann et al. [48] have tried to correlate the morphology of carbon obtained in various major CVD processes with the specific interest to identify the relative proportion of constituent carbon, hydrogen, and oxygen in the flowing gas mixture, which favors the growth of diamond films. The results on morphology have been plotted in a ternary C–H–O map as shown in Figure 8.3a. The three corners of the diagram represent carbon, hydrogen, and oxygen flowing alone and the lines joining any two corners represent the ratio of standard volume in terms of the corner elemental species (denoted by C, H, and O) flowing per unit time to the standard volume flow rate of both the species as $X_{H/\Sigma} = H/(C + H)$, varying along the line joining carbon and hydrogen corners and so on, as shown in Figure 8.3a. There is CO line representing a mixture of CO with hydrogen

TABLE 8.2
CNSs and Their Dimensionality

Dimensionality	Zero Dimensional	One Dimensional	Two Dimensional	Three Dimensional
CNSs	Fullerenes: C_{60}, C_{70}; Higher fullerenes: C_{76}, C_{78}, C_{82}, C_{94}, C_{96}; concentric shelled onion	Nanotubes: single-walled (SWCNT); double-walled (DWCNT); multiwalled (MWCNT); bamboo structures; Nano fibers (CNF); Nano tapes; Double helices; Nanobraids	Graphene; diamond films	Nano-crystalline diamond; fullerite

TABLE 8.3

Microstructural Classification of CNS Based on Linear Edge Density of Dangling Bonds

Structure Schematic	Examples	Description	Edge Density (ρ_L in cm^{-1})
	HOPG$_B$ (basal plane textured)	Low charge density. Primarily graphene basal plane exposure. Edges are 1D structures at grain boundaries or steps	<10
	Graphite, SWNT, MWNT	Mostly basal plane exposure with moderate density of 1D edges exposed via steps or grain boundaries	10–10^3
	Textured fibers	Mixture of edge and basal plane exposure. The edge exposure is organized along the length of the tube or fiber	10–10^5
	Nanosheets (NS)	Textured graphene nanosheets with both edge and basal plane exposure	10^4–10^6
	HOPG$_E$ (edge-textured)	Mostly edge exposure of graphene or nanosheets packed into a 2D array	10^6–10^7
	Activated carbon (AC)	Mesoporous structures organized into a 3D network of particles with mixture of edge and basal planes exposed	10^6–10^9
	Aligned CNTs (a-CNT)	Mostly basal, plane exposure, organized in 3D via vertically aligned CNTs	10–10^5
	Bamboo CNTs (b-CNT)	Similar to a-CNTS, but with rings or edge defects around the circumference due to periodic bamboo-like sections	10–10^9
	Graphenated CNTs (g-CNTs)	Graphene foliates with both edge and basal plane exposure, depending on density of foliates, organized in 3D structure	10^7–10^{10}

Source: Stoner, B. R. and Glass, J. T. 2012. *Diamond and Related Materials* 23:130–4.

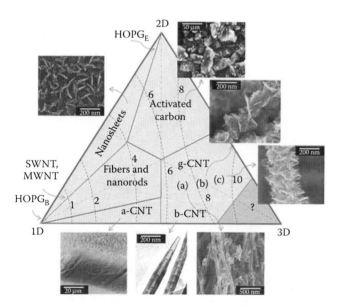

FIGURE 8.2 Classification of CNSs in a ternary map based on dimensional organization of dangling bonds and their micrographs as observed under SEM/TEM—(i) nanosheets (NS), (ii) AC, (iii) aligned CNT, (iv) bamboo structured CNT, (v) graphenated aligned CNT with increasing foliate density from (a) to (c) in the map, and (vi) fibers and nanorods. Dotted curved line gives logarithm of linear iso-density contour of dangling bonds indicated by number with density in units of cm^{-1}. (Reprinted with permission from Stoner, B. R. and Glass, J. T. 2012. *Diamond and Related Materials* 23:130–4.)

and their relative amounts in the flowing gas vary along the line. Similarly, there are lines representing mixtures of acetylene, ethylene, and the like with oxygen. In this diagram, the boundary of the domain for diamond growth has been identified separating it from non-diamond carbon growth on one side and no growth region on the other.

The authors, apart from revealing the influence of constituents of gas mixture on morphology, have also demonstrated an interesting rational approach for consolidating the results into a map but ignoring other important parameters including the temperature and pressure, as many investigators failed to report these parameters. Some of the results on the morphology of CNS reported in Table 8.1 are from experiments using gas mixture with elemental constituents of carbon, hydrogen, and nitrogen. It is thus possible to consolidate the results in a C–H–N ternary diagram following the approach of Bachmann et al. [48] in order to separate the regions favoring the growth of specific cylindrical morphology. In Table 8.1 there are blank columns indicating missing information in a given study. Elemental constituents of the gas mixture may not be adequate in determining the morphology and, the temperature, pressure, and the chemical nature of the catalyst may have important roles in influencing the morphology of CNS. The results of McCaldin et al. [39] clearly bring out the influence of temperature in Figure 8.3b. However, apart from temperature and the chemical nature of the catalyst, there may be other variables that are important

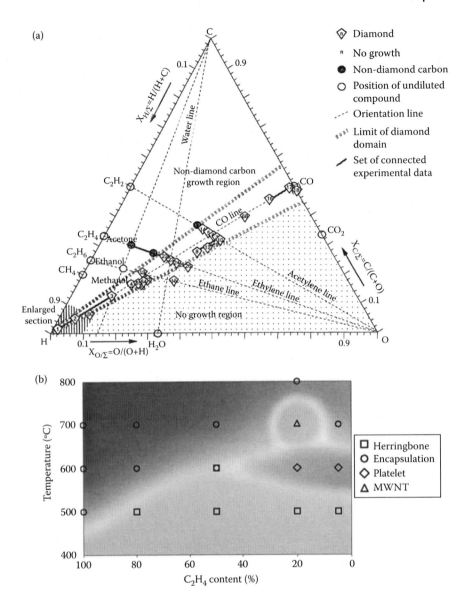

FIGURE 8.3 (a) Composition of flowing gas influencing the morphology of carbon in CVD (Reprinted with permission from Bachmann, P. K., Leers, D., and Lydtin, H. 1991. *Diamond and Related Materials* 1:1–12.) and (b) temperature influencing the morphology of cylindrical CNS. (Reprinted with permission from McCaldin, S. et al. 2006. *Carbon* 44:2273–80.)

in deciding the morphology. The present review is an attempt to identify those variables and present a holistic picture identifying the gaps.

8.2 STRUCTURE OF CYLINDRICAL CNS

The structure of the tube could be characterized in terms of the arrangement of the hexagons in the tube as shown in Figure 8.4 [50].

A more general description of the structure may be given by an imaginary construction shown in Figure 8.5 [51]. If one unfolds the cylindrical tube after sectioning along the dotted line at the end of the circumferential vector AA′ as shown in the figure of the tube and maps it by matching the hexagons on a hexagonal mesh, the mapped circumferential vector AA′ in

the unfolded sheet is limited by the extended boundary lines AB and A′B′. The vector $C_h = AA'$, is called chiral vector, and the chiral angle θ is the angle that strip direction (parallel to the tube axis) makes with the closest C–C bond direction. Both the chiral vector and the chiral angle characterize the structure of the tube. The magnitude of C_h gives the width of the strip and θ gives the direction. If A and A′ and B and B′ are equivalent sites then the tube is seamless and the tube contains only hexagons exclusively. In terms of the two vectors $\vec{a}_1$ and $\vec{a}_2$ of the hexagon one may write, $\vec{C}_h = n\vec{a}_1 + m\vec{a}_2$. The set of numbers (n, m) characterizes the atomic arrangement of the CNT and determines its metallic, semiconducting, or semimetallic properties. When chiral vector is such that ∥m–n∥ is 3k, when k is an integer, the SWCNT is metallic and it is semiconducting when ∥m–n∥ = 3k ± 1. The values of (n, m) for

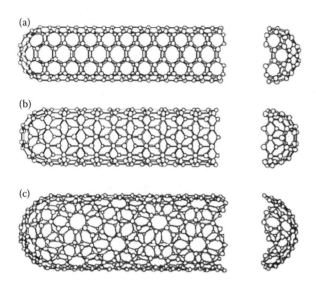

FIGURE 8.4 Models for different types of single-shell nanotubes—(a) armchair tube, (b) zigzag tube, and (c) chiral tube. (Reprinted with permission from Dresselhaus, M. S., Dresselhaus, G., and Saito, R. 1995. *Carbon* 33:883–91.)

a tube may be determined from the peaks of optical absorption spectra due to electronic transitions between pairs of van Hove singularities of semiconducting and metallic SWCNTs [52]. There is a connection between the set (n, m) and the diameter d of the CNT: $d = a_{c-c}[3(m^2 + mn + n^2)]^{1/2}/\pi = |\vec{C}_h|/\pi$ where a_{c-c} is the length of carbon–carbon bond.

If the unfolded graphene strip has tube axis parallel to a set of C–C bonds ($\theta = 0°$), the tube has "zigzag" structure as shown in Figure 8.4b and the CNT is characterized by $(n, 0)$ for any positive integer n. If the unfolded strip has tube axis perpendicular to a set of C–C bonds (i.e., $\theta = \pm30°$), the tube has "armchair" structure as shown in Figure 8.4a and the CNT is characterized by (n, n). Both types of tubes are called "achiral" tubes so as to distinguish them from "chiral" tubes, which have an arbitrary chiral angle, θ, between $0°$ and $\pm30°$. In the achiral tubes, the carbon atoms are densely

arranged either along a circle perpendicular to the tube axis or along the lines parallel to the tube axis while in chiral tubes the carbon atoms are densely arranged along helices, justifying the name chiral. The structural description in terms of chiral vector and chiral angle is adequate for SWCNT. But for MWCNT, which is an assembly of coaxial tubes, all the tubes in the assembly may have the same chiral angle to make them "isochiral" or the tubes may have different chiral angle making them "polychiral." There are three types of models proposed for multiwalled tubes—(i) coaxial cylinders with circular cross section, (ii) coaxial cylinders with polygonal cross section, and (iii) single graphene sheet folded in scroll. The model of cylindrical coaxial tubes of graphene sheets is consistent with most of the observations. However, it has been argued that serpentine carbon fibers have scroll-type structure and this question could not be settled on the basis of diffraction results alone.

The multiwalled tubes of coaxial cylinders with circular cross section when viewed normal to the tube axis will appear as assembly of parallel folded graphene sheets with spacing $c/2$ and the circumference of successive tubes will differ by πc, corresponding to eight to nine additional rows of hexagons parallel to tube axis in each successive outward layer compared to an adjacent inner layer, depending on the chiral angle. Since each successive sheet nucleates over an inner layer in random locations the local structure in a volume element will be disordered to an extent. The diffraction response of electron beam along the normal will result in sharp spots $00.l$ ($l = 2, 4, \ldots$) and these will not be affected by turbostatic or translational disorders. The $hk.0$ spots may be streaked along the nearest [00.l] directions as a result of 1D disorder while the streaking of these spots tangent to the circles is attributed to diffuse coronae in planes perpendicular to tube axis as shown in Figure 8.6a [53]. For achiral tubes, the incident beam normal to the tube axis will see parallel hexagons at the top and the bottom across the tube and hexagonal pattern of $hk.0$ spots will be formed. However, for chiral tubes, the spots from the top and the bottom will not coincide. Since the hexagon at the top and the bottom will now differ by twice the chiral angle, 2θ, the pattern of spots will be superposition of those coming from the top and the bottom as shown in Figure 8.6a for $\theta = \eta$. The streaking of $hk.0$ spots outward may also be due to decreasing of apparent interplanar spacing with increasing inclination of the incident beam away from the tube axis, which may be indicated by shifting of the spots with increasing inclination of the incident beam from the normal to tube axis obtained by tilting the specimen as shown in Figure 8.6b. The spots marked by A and B move closer with increasing tilt and become one in Figure 8.6b(iv) when the tilt is $10°$. Similarly, the spots C and D approach each other and become one at a tilt of $30°$ in Figure 8.6b(vi).

High-resolution transmission electron microscopy (HRTEM) has been employed to examine the structure of cylindrical multiwalled nanotubes. Most high-resolution images of multi-walled tubes exhibit (0002) fringes

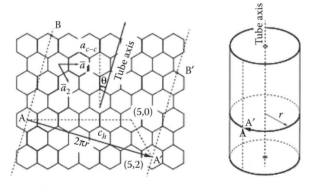

FIGURE 8.5 Mapping of the unfolded CNT to characterize its structure to define chiral vector C_h and chiral angle θ. (Reprinted with permission from Amelinckx, S., Lucas, A., and Lambin, P. 1999. *Reports on Progress in Physics* 62:1471–524.)

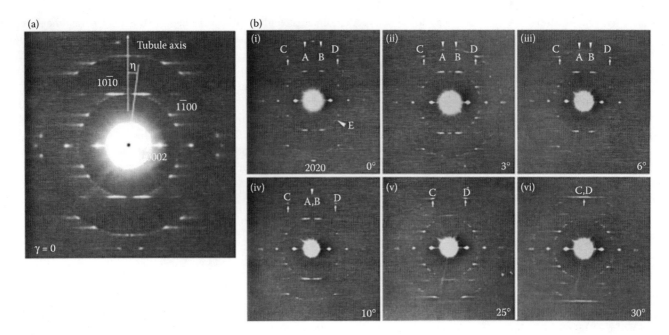

FIGURE 8.6 (a) Typical electron diffraction pattern of the multiwall polychiral CNT and (b) evolution of an electron diffraction pattern on tilting the specimen about an axis perpendicular to the tube axis. (Reprinted with permission from Amelinckx, S., Lucas, A., and Lambin, P. 1999. *Reports on Progress in Physics* 62:1471–524; Zhang, X. B. et al. 1994. *Ultramicroscopy* 54:237–49.)

clearly spaced at $d_{0002} = 0.34$ nm as shown in Figure 8.7a. Sometimes, in predominantly achiral tubes one may observe 1010 fringes with spacing of 0.21 nm as shown in Figure 8.7b.

The defects in multiwalled tubes such as anomalous spacing, dislocations, bends, and terminating caps may be observed by HRTEM. For a detailed understanding of the electron diffraction pattern and HRTEM images of the cylindrical CNT, the reader is referred to the review by Amelinckx et al. [51].

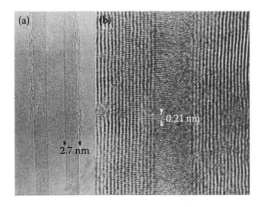

FIGURE 8.7 (a) Basal plane lattice fringes of multiwall tube (Reprinted with permission from Amelinckx, S., Lucas, A., and Lambin, P. 1999. *Reports on Progress in Physics* 62:1471–524.) and (b) high-resolution image of curved fringes of 0.21 mm spacing inside the tube, resolved next to the basal fringes of 0.34 mm spacing. (Reprinted with permission from Amelinckx, S., Lucas, A., and Lambin, P. 1999. *Reports on Progress in Physics* 62:1471–524; Zhang, X. B. et al. 1994. *Ultramicroscopy* 54:237–49.)

8.3 THEORETICAL STUDIES ON THE FORMATION OF CNS

There are a number of molecular dynamics (MD) simulation studies that reveal the atomic picture of the initial stages of growth and a summary may be obtained in the review article by Yu et al. [54]. Gomez-Gualdron et al. [55] have carried out MD simulation using a classical reactive force field and followed the evolution of graphitic carbon rings at the interface of the nanoparticles of nickel catalyst providing a hexagonal template of stable void sites. In the conclusion of this study, it has been stated that the growth of cylindrical nanotube takes place in the following steps: (1) carbon dissolution into the metal catalyst nanoparticle, (2) carbon segregation on the nanoparticle surface, (3) formation of chains, isolated rings (usually branched), and concatenated rings (usually branched) on the metal surface, (4) "merging" of carbon rings to form a cap, (5) lifting-off of the cap from the substrate, and (6) incorporation of carbon to the rim of the cap at the nanoparticle surface leading to a growing tube. Thus, the process conceived is dissolution of the carbon in the catalyst nanoparticle and their precipitation in the form of nanostructures of chains, isolated ring, and concatenated rings. Generally, chains form earlier to rings, which forms before concatenated ring and it becomes more apparent in catalyst particles of larger surface area. The adhesion between the substrate and the nanoparticle of the catalyst influences the number of carbon chains/rings formed. For the catalyst of a nickel cluster of 32 atoms initially annealed for 0.5 ns, 15 carbon atoms deposited on it, result in two chains (C_{11} and C_4) when the energy of adhesion between the substrate and the catalyst, E_{adh}, is −1.39 eV while three chains form (C_6, C_5, and C_4) when $E_{adh} = -0.16$ eV. Higher mobility of metal atoms at lower adhesion creates more favorable sites for the formation of clusters of carbon. These rings and chains merge to result in a fullerene cap. If one cuts fullerene C_{60}, through an equatorial plane normal to the fivefold symmetry, there are six pentagons in each of the hemisphere and C_{70} forms by adding a belt of hexagons at the equator. A single-walled nanotube is formed by having the top hemisphere followed by continuing addition of hexagons in cylindrical belt [51].

Simulation studies by Gomez-Gualdron et al. [55] also reveal the influence of pregrowth annealing time of the catalyst and the adhesion between substrate and the catalyst nanoparticle on the chiral angle. For relatively weak metal–substrate adhesion (-0.16 eV $< E_{adh} < -0.43$ eV) there is a tendency to favor near-armchair nanotubes particularly for larger pregrowth annealing times. However, at shorter pregrowth annealing times within this range of adhesion, there is a tendency to form near-zigzag nanotubes. At higher adhesion between the catalyst particle and the substrate (-0.70 eV $< E_{adh} < -1.39$ eV), simulations using Ni_{32} resulted mostly in near-zigzag nanotubes but the quality is relatively poor. The chirality trends are better expressed in catalyst particles of intermediate size (Ni_{80}) than at smaller sizes (Ni_{32}).

Gomez-Gualdron et al. [55] have investigated the lifting of the cap by the addition of carbon atoms at the rim of the cap in contact with the metal nanoparticle and the tube grows by root growth mechanism. Their study indicates that (i) there is correlation between nanotube diameter and the size of catalyst nanoparticle, (ii) cap lift-off takes place faster for stronger catalyst particle–substrate adhesion, and (iii) there is change in the shape of catalyst metallic nanoparticle during the growth of nanotube, which is more for weaker particle–substrate adhesion as revealed by change in height of the mass center of the catalyst particle. A strong adhesion

limits the mobility of the atoms in the catalyst particle. There is also the presence of capillary forces, which may overcome adhesion allowing the particle move inside the tube. Now, from these studies, it appears that the carbon atoms, adsorbed on the surface are involved in rearranging themselves at the initial stage leading to the formation of the cap. Yet, Gomez-Gualdron et al. [55] talk about dissolution as the first step of the sequence of steps leading to the formation of nanotube. If dissolution is considered, the question about the state of atoms on the surface of the catalyst nanoparticle of metal is relevant as it decides the extent of carbon dissolution. Are they in solid state or liquid state? Also, solid state diffusion will be relatively slow compared to that in the liquid state. If the first step is adsorption on the solid surface followed by interaction between the adsorbed carbon atoms and the metal atoms of the template, that is, the surface of the metallic nanoparticle, then the scenario of arrival of large number of carbon atoms at a faster rate due to decomposition of carbon-bearing gas in the gas mixture may easily lead to the deposition of carbon over the catalyst nanoparticles, commonly called carbon nanobeads and the growth of cylindrical nanostructure will be prevented. However, the assumed process of dissolution will introduce a mechanism of transport away from the surface to inside the nanoparticle of metallic catalyst and the formation of nanobeads may take place at a relatively higher rate of decomposition/deposition. The rate of transport of carbon atoms from the surface will be higher for liquid state as the diffusion coefficient will be relatively higher than that in solid state. Yu et al. [54] have carried out MD simulation using a jellium model for the catalyst of metallic nanoparticle, where the solid state atomic arrangement is obliterated and it is observed that there is formation of cap and nanotube as shown in Figure 8.8. Thus, the necessity

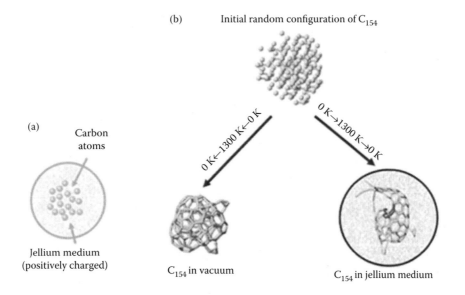

FIGURE 8.8 (a) A schematic presentation of the jellium model. The background in the sphere represents the jellium medium and the circles represent C atoms. (b) C_{154} cluster with initial random configuration (top), annealed at 1300 K and slowly cooled down to 0 K in vacuum (bottom left) and in the jellium medium (bottom right), respectively. The open-end of the tubular-like structure is marked by dark color. (Reprinted with permission from Yu, M., Wu, S. Y., Jayanthi, C. S. 2009. *Physica E* 42:1–16.)

of an atomic template on the surface of catalyst may not be an essential requirement for the formation of hemispherical cap of carbon.

An initial random configuration of 154 carbon atoms annealed at 1300 K are allowed to cool down to 0 K while evolving in vacuum in the presence of jellium particle of metallic catalyst as explained schematically in Figure 8.8. There is an optimized charge transfer of 0.1e from jellium to carbon and it is substantially more near defects. A pentagon is more stable in jellium and the cluster evolved to an open tubular structure where there is significantly more charge transfer between the jellium and the carbon atoms at the open end. However, in vacuum the random cluster evolves into a more compact cluster. Shibuta and Maruyama [56–58] have carried out individually and together a number of MD simulation studies, and in their study of carbon and nickel atoms clustering together, tubular growth has been observed during cooling because metal atoms present at the open end of the carbon cluster prevent its closure to become a compact fullerene kind of cluster. Thus, it appears that the tube nucleation starts with the formation of a fullerene type of hemispherical cap and the required pentagons for this purpose, are more stable on the catalyst nanoparticle of metal. However, a more crucial role of the catalyst nanoparticle involves substantial charge transfer of carbon to metal atoms at the open end paving the way for tubular growth by further addition of hexagons of carbon at this end. It may be remembered that even after a layer of hexagon is added to the hemisphere, the closure could take place leading to C_{70}. The presence of metallic catalyst does not allow this closure so that tubular growth may continue and it happens by metal atoms forming a local metal–carbon binary structure to cause increased charge transfer from carbon to metal at the end but these metal atoms could move forward and remain situated at the end when new carbon atoms join the hexagonal mesh for axial growth. This role of the catalyst demands some mobility of metal atoms to fulfill its catalytic role. The simulation studies outlined above also indicates that formation of hexagonal sp^2 bonded carbon cluster may not require any structural template as it forms in the absence of catalyst in fullerene type of compact closed structure. Shibuta and Elliot [59] observed that graphene bonds broken due to impact of the crystalline metal surface reformed readily on Ni $(111)_{fcc}$ and Co $(0001)_{hcp}$ surfaces but not so readily on Fe $(100)_{bcc}$ surface and claimed that there is some epitaxial effect of close packed 2D (two-dimensional) surface. The study of interaction between graphene and metal indicates relatively strong interaction between carbon and iron so as to create a surface with energy twice that of those forming with either cobalt or nickel, out of which nickel forms the lowest energy surface, providing better epitaxial effect to form graphene sheet. Monte Carlo simulation [60] shows that this interaction has consequence in the growth of nanotube and strong carbon–metal interaction leads to double-walled tube while weak interaction results in single-walled nanotube. However, if the tubular growth is taking place at

the rim of the lifted cap in contact with metal cluster by addition of carbon, the epitaxial effect of the metal surface may not be relevant at that stage and the consequence of the epitaxial effect will be on the structure of the cap, to which tubular growth has to conform. It is interesting to observe that the jellium model for the metal cluster smearing its atomic arrangement, could still lead to the formation of hexagonal mesh and tubular growth as it could still allow charge transfer.

The simulation studies reported above examine either carbon from vapor phase clustering on the surface of catalyst nanoparticles of metal during cooling or both carbon and metal atoms in the vapor phase cooling to binary clusters. These pictures do not correspond to the scenario of VLS (vapor–liquid–solid) theory requiring the presence of three phases. The metal cluster, which forms on cooling binary mixture of metal–carbon vapor, may have an intermediate state of liquid marked by short-range order in the absence of long-range order. However, no attention has been given to determine the state of matter in the metal cluster as the vapor is cooled from high temperature.

Nanoparticles of metal particles have interesting melting behavior marked by size-dependent melting point depression and also by surface melting over a solid core caused by surface curvature and high surface area to volume ratio. The melting behavior is thus distinctly different from bulk metals. It is well established that melting temperature of nanoparticle decreases with decreasing size, which is determined by the number of atoms in the particle. Theoretical investigations on thermodynamic model of size-dependent melting of nanoparticles have been reviewed by Nanda [61] and three different models are proposed so far—(a) homogeneous melting (HMM), (b) liquid nucleation at the surface and its growth to consume the solid core (LNG), and (c) liquid skin melting (LSM) and the results of estimated melting temperature with size for each of these models are shown in Figure 8.9.

The prediction of LSM model matches better for the experimental results on tin nanoparticles [62,63]. Thermodynamic considerations suggest that melting may start at the surface as a skin, when γ_{SV} exceeds sum of γ_{LV} and γ_{SL} where γ is the interface energies between the phases indicated by S, L, and V in the subscript, respectively, for solid, liquid, and vapor phases. In other words, the change in interface energy, $\Delta\gamma = \gamma_{SV} - (\gamma_{LV} + \gamma_{SL}) > 0$. For gold, tin, and lead, $\Delta\gamma$ is positive and during heating, the melting starts at the surface at temperatures lower than the bulk melting temperature. The thickness of the molten skin increases with increasing temperature till the entire particle melts. MD simulation of melting behavior has been carried out for a number of metals and alloys including some transition metals. Before the study of melting of nanoparticle, there is need to address the following issues: (i) how can we recognize the signature of melting and solidification in a cluster of atoms, and (ii) what is the structure of the solid corresponding to the global minimum of energy? Melting/solidification is a first-order phase

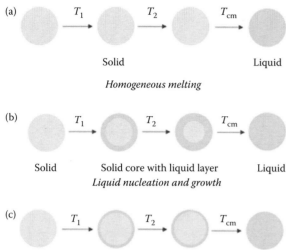

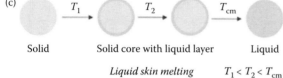

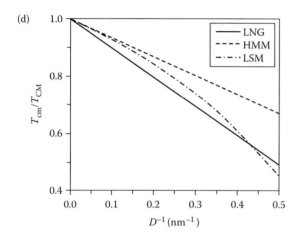

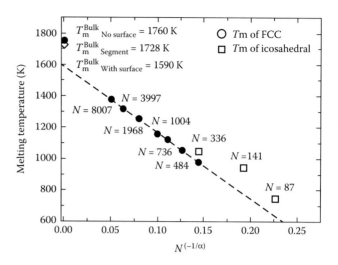

FIGURE 8.10   Dependence of melting temperature of nickel cluster on cluster size. The dashed line shows the best fit to a linear function of $N^{(-1/3)}$. This leads to a predicted value of 1590 K for large N, well below the value of 1760 K calculated for the bulk system (no surface). The melting temperatures for icosahedral nickel clusters are all higher than that predicted by the linear fit. (Reprinted with permission from Qi, Y. et al. 2001. *Journal of Chemical Physics* 115:385–94.)

where r_{ij} is the bond length between atoms marked i and j. Qi et al. [64] have carried out MD simulation for the melting of nickel clusters and observed that the melting temperature decreases more or less linearly with $N^{(-1/3)}$ as shown in Figure 8.10, when N is the number of atoms in the fcc cluster.

The interface energies of bulk nickel indicate that $\Delta\gamma = -2$ mJ m^{-2} and so there should not be surface melting. But Qi et al. [64] have observed that interface energies change with the size of particle and $\Delta\gamma$ becomes positive leading to a liquid-like skin at the surface indicating surface melting. These authors have also observed that for clusters larger than 500 atoms, the latent heat of melting also decreases with size as $N^{(-1/3)}$. Interestingly, there is a size of cluster for which the latent heat becomes zero and it leads to a phase transition spreading over a finite temperature range because of size-dependent liquid skin and thermodynamic fluctuations.

The size-dependent melting of nanoparticles leads one to expect that the diffusion rate of carbon in the cluster of atoms in the catalyst may also depend on their size. Thus, it appears that the size of catalyst particle has important bearing on the mode of assimilation of carbon atoms decomposed from carbon-bearing precursor gas. Depending on the size, if the cluster has solid bulk-like compact structure, there may be surface adsorption primarily, whereas in a cluster with liquid-like structure there is likely to be dissolution of carbon inside. If the rate of arrival of carbon to the growing nanotube is more than that required to sustain growth rate, there is deposit of amorphous carbon on the surface of the growing tube. If the rate of decomposition is so high that the rate of deposition of

FIGURE 8.9   (a–c) Different hypothesis of melting of nanoparticles, and (d) melting temperature of nanoparticles of diameter D following the three hypotheses. (Reprinted with permission from Nanda, K. K. 2009. *Pramana-Journal Physics* 72:617–28.)

transition accompanied by a discontinuous change in enthalpy and there should be a corresponding sharp peak in heat capacity. MD simulations evaluate heat capacity by the following expression:

$$C_V = \frac{<E^2> - <E>^2}{nk_BT^2}$$

where $< >$ indicates average at temperature T, n is the number of atoms in the cluster, and k_B is the Boltzmann constant. Further, one also looks for an abrupt change in Lindemann index, a measure of rms (root mean square) bond fluctuation, evaluated by

$$\delta = \frac{2}{n(n-1)} \sum_{i<j} \frac{<r_{ij}^2> - <r_{ij}>^2}{r_{ij}}$$

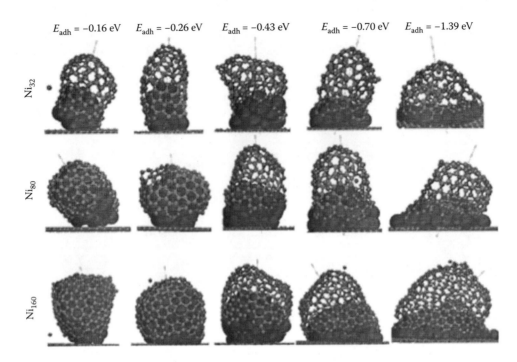

$E_{adh} = -0.16$ eV $E_{adh} = -0.26$ eV $E_{adh} = -0.43$ eV $E_{adh} = -0.70$ eV $E_{adh} = -1.39$ eV

Ni_{32}

Ni_{80}

Ni_{160}

FIGURE 8.11   The simulated growth showing the CNS formed (gray atoms) after 0.5 ns on metallic nanoparticle (dark atoms) annealed for 0.10 ns, adhering to a substrate for increasing adhesion energy from left to right and for increasing size of the metallic nanoparticle from top to bottom. (Reprinted with permission from Gomez-Gualdron, D. A. et al. 2012. *ACS Nano* 6:720–35.)

carbon on the catalyst is more than the rate of its assimilation inside, there will be loose deposit of carbon on the catalyst surface leading to the formation of nanobeads as observed commonly. If there is any role of catalyst particle in decomposition, this role is impaired with the formation of carbon cover as the accessibility of precursor gas to the particle decreases. Surface melting may be helpful in faster assimilation of carbon by dissolution depending on solubility of carbon in the catalyst and the thickness of the molten layer, both of which may possibly depend on size and temperature. Thus, the size-dependent structural state within the catalyst particle at the prevailing growth temperature may be immensely important for the growth of cylindrical nanostructure.

The growth of nanotube is essentially the organization of carbon atoms assimilated within the metal cluster but molecular simulation studies have confined it to the surface of metal cluster of nanometer size leading to cap formation and its lift-off by addition of hexagonal mesh of carbon atoms in the rim. It does not allow carbon atoms inside the cluster to capture the difference in scenario for the evolution of different morphologies such as SWCNT, MWCNT, or CNF. However, the adhesion between the catalyst particle and the substrate delays lift-off of the cap as shown in Figure 8.11 but in addition, it has been claimed that in the limiting cases of strong and weak adhesion one observes the two reported modes of growth—root growth and tip growth as shown in Figure 8.12a. For the tip growth, there is no supporting simulation showing energetic feasibility of bending graphene sheets without a cap. The cap easily bends hexagonal mesh to a cylinder as observed for C_{70} molecules as well. MD simulation

results [65–68] demonstrate a cap formation to be a necessary step for nanotube growth, which is contradicting the tip growth mechanism proposed. If one could conceive of cap formation separating the substrate and the catalyst particle by transport of carbon through the liquid phase of the catalyst at its surface, one could explain bending of graphene to form wall of a tube as well as movement of catalyst by capillary action during growth. The feasibility of this scenario needs to be examined. Figure 8.12b shows an interesting schematic of growth incorporating periodic formation of cap so as to result in bamboo structured CNT [69].

8.4   MORPHOLOGY OF CNS BY CCVD METHOD

The three methods used extensively for the synthesis of CNS are arc discharge [70,71], laser ablation [72], and CCVD [73]. While arc discharge and laser ablation grow CNS by evaporating graphite, CCVD method described briefly at the beginning, involves thermal decomposition of the precursor, carbon bearing vapors or gases, leading to nucleation and growth of CNS on a catalyst nanoparticle and it is the most popular method for producing CNS because of its low cost and scalability for mass production [19]. There are several variations of CCVD method. The pyrolysis of a suitable compound vapor at a suitable temperature liberates metal nanoparticle in situ, acting as the catalyst for CCVD and such a process is known as floating catalyst method. Alternatively, catalyst particles may be formed by heating thin film of catalyst metal on a substrate so that the film balls up into nanoparticles, which may be used as catalyst for CNS growth.

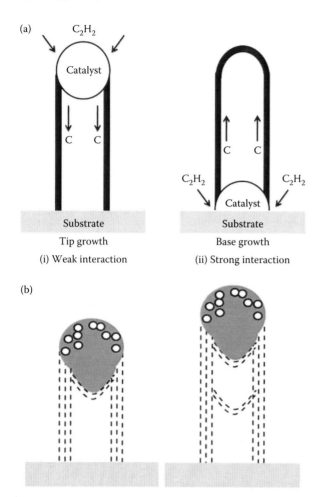

FIGURE 8.12 Catalytic growth model for CNS for (a) weak interaction (tip-growth (i)), strong interaction (base-growth (ii)), and (b) bamboo structured growth. (Reprinted with permission from Wang, T. and Wang, B. 2006. *Applied Surface Science* 253:1606–10.)

The first step toward the formation of a CNS is the release of carbon from a source or precursor material. Except methane, the decomposition of commonly used precursors is exothermic and these are less stable than graphite at all temperatures. The decomposition of ethylene, acetylene, and ethanol is thermodynamically feasible at all temperatures under atmospheric pressure and takes place when there is adequate kinetic activation. It is not clear whether catalyst nanoparticles have any contributing role toward this activation. It has been claimed that methane adsorbs on the surface of iron nanoparticles, which also catalyzes decomposition of this gas during growth of SWCNT [52]. Methane is more stable than graphite till 700°C and thermodynamic feasibility of its decomposition is only above this temperature. Carbon monoxide, often used for growing SWCNT over Co, Ni, or Fe [52], is more stable than graphite above 700°C. However, the equilibrium decomposition temperature may change with changing partial pressure of the gases involved in the reaction [49].

The metallic nanoparticles, which act as catalyst, could be categorized as those with no tendency to form stable carbide and those which are carbide formers. The solubility of carbon in the bulk metals, which do not form stable carbides, is not very large at typical growth temperatures. Nanoparticles of transition metals such as Co, Ni, Fe are the most commonly used catalyst in CCVD for the growth of CNT. Using larger nanoparticles (size > 3 nm), MWCNT has been grown on catalyst nanoparticles of Na, K, Cs, Mg, Al, Mn, In, W, Ti, Pd, Pt, and Au [31,38]. SWCNT was grown using smaller nanoparticles of late transition metals—Pd, Pt, and Ru [10,15]; noble metals—Cu, Ag, and Au [8,10,11]; early transition metals—Mn, Cr, Mo [9,74,75]; elements of the carbon family—diamond, Si, Ge, Sn, and Pb [14,23,75]; lanthanides—Gd and Eu [76]; and compounds—$FeSi_2$, TiC, SiC, SiO_2, Al_2O_3, TiO_2, Er_2O_3, and ZnO [16,17,22,23,35,36]. At 700°C, the limit of solubility of carbon is below detection limit in Zn, 0.01 at% in Cu, 0.3–0.5 at% in Ni, 1.2 at% in fcc Co, and 0.1 at% in bcc iron. In fcc iron at higher temperature, the solubility of carbon increases substantially (4 at% at 800°C) [49]. For the growth of nanostructures, a small solubility of carbon in the catalyst may be adequate but a high rate of diffusion of the dissolved carbon is relevant for reasonable rate of growth. Iron, nickel, and cobalt form carbides that decompose at lower temperatures but Ti and Mo are strong carbide formers. But the behavior of metals toward carbon, in terms of solubility and the rate of diffusion may change substantially from that of bulk, when size decreases to nanometer range and the arrangement of atoms starts deviating from that of bulk, often leading to a change in structure of the nanoparticle as well.

Mixed catalysts and alloys [77] have also been used in CCVD. Liquid state offers a higher rate of diffusion. However, the solubility in nanoparticles and melting temperatures are significantly different from that of bulk. The catalyst and its melting could have a role in deciding morphology of the nanostructure growing over it.

In Section 8.2 it has already been explained that the structure of the SWCNT is characterized by chirality, which determines the electronic properties of the tube. Chirality is related to the diameter of the tube, which is believed to be related to the size of the catalyst nanoparticle over which it is growing. Zhu et al. [78] has used HRTEM images to find a correlation between the chirality of the nanotube and the structure of the catalyst nanoparticle. There are attempts to grow tubes of controlled chirality by controlling the size of the catalyst but it is yet to succeed. Liu et al. [79] have grown tubes of controlled chirality by separating the seed tubes of the desired chirality by DNA-based method, for their further growth to long SWCNTs of known chirality. The chirality (7, 6) was grown on quartz substrate using methane while (6, 5) was grown on the same substrate using either methane or ethanol vapor. The armchair SWCNT (7, 7), which is metallic, was grown on both quartz and Si/SiO_2 substrates [79].

The catalytic growth of CNSs takes place by dissolution of carbon and their precipitation on a moving catalyst particle surface [80], yet there is no unanimity on CNS growth mechanism. However, interesting new experiments provide considerable insight toward a better understanding of the growth of CNS.

8.4.1 GROWTH OF CNS USING METAL-BASED CATALYSTS

During growth of CNSs, one generally gets a mix of amorphous carbon deposits, SWCNT, MWCNT, and other morphologies. The selectivity of one of these morphologies is increased by choosing appropriate catalysts and growth conditions, which are also chosen to ensure good yield CNS of desired morphology. As mentioned earlier, smaller catalyst nanoparticles help in growth of SWCNT while larger nanoparticles result in other nanostructures such as MWCNT or CNF and so on as inferred from Table 8.1. Apart from morphology and yield, the other important issues related to growth of SWCNT are purity and defect. SWCNT is widely investigated for their potential applications based on their electronic properties. Purity means freedom from adsorbed carbonaceous impurities on the surface of the SWCNT degrading its electronic properties. Similarly, structural defects like pentagonal or heptagonal rings in place of hexagonal ones, also damage the electronic properties. After growth it is possible to improve purity by purification processes but often defects are introduced in this step. Therefore, there is stress on direct growth of high purity SWCNT. Lim et al. [81] have studied a number of monometallic catalysts such as iron, cobalt, nickel, molybdenum, copper, and platinum and observed that iron has considerably higher yield, which has been attributed to high efficiency of decomposition of ethanol vapor (source of carbon in their experiments) and high solubility of carbon in iron. But the decomposition also leads to formation of amorphous carbon degrading the purity of SWCNT. Thereafter, the same investigators carried out experiments with iron-based bi-metallic catalyst like FeCo, FeNi, FeMo, FeCu, FePt, and it was observed that addition of inactive elements such as copper to iron lowers the yield but enhances the purity more than any of the other elements. It is believed that lowering the solubility of carbon in FeCu has resulted in lower amorphous carbon on the surface of SWCNT. Takagi et al. [10] used the same monometallic catalysts as Lim et al. [81] but the catalyst nanoparticles were made by balling up thin metal film on a substrate of aluminum hydroxide in the chamber of CCVD in air at 850–950°C. Then, the chamber was evacuated and ethanol vapor was allowed to flow resulting in growth of CNT on catalyst nanoparticles. This route resulted in as high a yield for Cu catalyst as it was for iron. These investigators believe that heat treatment in air is responsible for enhanced yield for copper catalyst. Yuan et al. [75] have grown horizontally aligned SWCNT along X-direction on ST-cut quartz wafer using a number of metal catalysts including Mo, Cu, Mn, Cr, Sn, Mg, Al apart from those commonly used. They observed that yield as well as degree of alignment depends also on the flow rate of precursor gas.

Controlled defects in SWCNT could be used for tuning the electronic properties as required. The defects are often preferred in applications, which require functionalization of the tube as it is easier to attach molecules to these defect sites. The tubes may collapse to develop a cone as a consequence of these defects and growth may eventually terminate by closing the end. The defects may also create elbow connection to change the chirality of the tube. Picher et al. [82] have observed that long tubes with lower density of defects are predominant in samples grown at high temperatures and lower pressure of precursor gas while growth of short tubes with defects dominate at lower temperatures and relatively higher pressure.

There has been effort to grow SWCNT with controlled chirality using monometallic catalyst. Abdullahi et al. [52,83] have prepared the catalyst by chemical route from solution of iron- and cobalt-bearing compounds by drying, calcinations, and reduction. Adding nanoactive MgO powders in solution of metal compounds results in a slurry, which is subjected to the same processing route to prepare Fe–MgO and Co–MgO catalysts. It has been claimed that dispersion of metals in MgO lattice could result in better selectivity for SWCNT over other morphologies when used for growing nanostructure in CCVD flowing methane or methane/helium mixture of gas at temperatures between 800°C and 900°C. Further, these catalyst nanoparticles also resulted in narrow range of diameter distribution and chirality.

Some research workers claim that nanoparticle of metal catalyst first forms carbide when carbon-bearing gas/vapor comes in contact with "hot" metal nanoparticles in CCVD chamber and the growth of CNS takes place by decomposition of carbide to release carbon for the formation of CNS. Interestingly, Ni et al. [84], while acknowledging high decomposition temperature of methane and catalytic role of Co, Ni, or Fe (on MgO substrate) in the decomposition of precursor gas, claims that growth temperatures should also be higher than the decomposition temperature of the respective metal carbides. They have reported a direct relationship between CNT formation and metal carbide decomposition for Ni, Co, and Fe catalysts. It has been claimed that metal carbide formation, which in their view is an essential intermediate step, precedes the CNT or CNF nucleation. Baker and many others proposed that pure metal is the active catalyst [85,86], while Oberlin and many others detected the encapsulated particles to be iron carbide [72,87]. Metal carbides have sometimes been detected in experiments involving iron catalyst nanoparticles. During in situ growth of CNT on iron nanoparticle inside ETEM (environmental TEM), Hoffman et al. [20] observed extensive formation of carbon–iron bonds by x-ray photoelectron spectroscopy (XPS) initially as indicated by XPS peak at ~282.6 eV but this peak is replaced soon by C–C peak (~284.5 eV) corresponding to sp^2 bonded graphitic carbon as shown in Figure 8.13. There is only a transitory presence of carbidic carbon (~283.4 eV) before graphitic carbon forms. If carbide formation and decomposition would be an essential intermediate step, one would have observed the presence of carbide phase throughout the growth of CNT. C–Fe bonds could form initially when carbon presence is low to form sp^2 bonded carbon on the surface.

Similar formation of iron carbide (Fe_3C) during CNT synthesis using Fe catalyst nanoparticles has also been reported by others [88]. In 2008, Yoshida et al. [89] performed atomic-scale in situ observation of acetylene decomposition on Fe catalyst at 600°C. Electron diffraction analysis of the metal

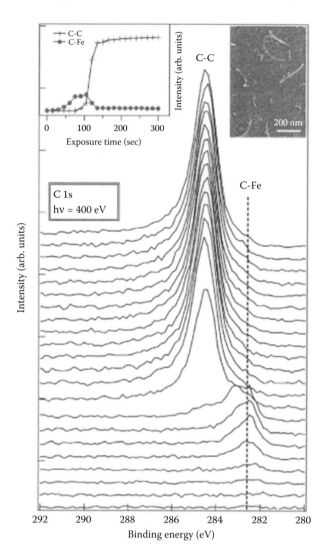

FIGURE 8.13 Time-resolved evolution of C 1s core level during Fe exposure to C_2H_2 at 580°C. Insets show SEM image of part of the probed area and the time evolution of the chemisorbed (dots) and graphitic (crosses) carbon peaks. (Reprinted with permission from Hofmann, S. et al. 2007. *Nano Letters* 7:602–8.)

clusters in each frame was reported to match with that of iron carbide corresponding to cementite (Fe_3C) as shown in Figure 8.14. Accordingly, the authors concluded that the active catalyst was in "fluctuating solid state" of "iron carbide."

Sharma et al. [29] believe that metastable phase of carbide may form under a dynamic equilibrium for the following reaction $3Ni + C \leftrightarrow Ni_3C$ where the continuous flux of carbon, generated by decomposition of C_2H_2, favors the formation of Ni_3C while at high temperature (>520°C) decomposition is favored. It is not surprising that metal carbides form prior to CNT growth as there is high carbon affinity of 3d transition metals reflected by their low enthalpy of carbide formation [90]. Confusion persists because lattice constants of pure metal and their carbide are sometimes very close. For example, the planes fcc Ni (111) or Ni_3C (113) have the same "d" spacing and diffraction peaks may appear at the same angle. Moreover, for nanoparticles, some deviation in the lattice

constants from that of the bulk crystal may be expected due to limited number of atoms in nanoparticles. Wirth et al. [91], based on their in situ electron microscopy and x-ray photoluminescence spectroscopic analyses, argued that the catalyst exists in pure metallic form: right from the CNT nucleation to the growth termination as shown in Figure 8.15.

Takagi et al. [10] have used non-carbide forming catalysts such as gold, silver, and copper to grow SWCNT by CCVD. Thus, carbide formation may not be an essential step in the growth of CNS.

It has been shown in Table 8.1 that CNS may grow even when nanoparticles of oxides of transition metals such as Co, Ni, or Fe, and so on are used as catalyst in CCVD. Since the environment during growth of CNS is highly reducing due to decomposition of carbon-bearing gas, it may be questioned whether oxide as such may survive in such an environment or are being reduced to metal nanoparticles, which act as catalyst. Baker et al. [85] claimed that FeO appears to be a much better catalyst than metallic iron for the formation of filamentary carbon and inferred that there was no reduction of oxide before growth of CNS. The oxide catalysts have often been subjected to pretreatment in reducing environment containing hydrogen and this step may actually reduce the oxides to metals. Jana et al. have used doped cobalt [40] and nickel oxide [92] catalyst in the CCVD set-up at a temperature lower than that existing during growth of CNS and observed that the oxide catalyst reduces to metal. However, the reduced catalyst may not have the crystal structure corresponding to that of bulk metallic phase at room temperature as the energetics for stability in nanoparticles could be different from that in bulk structure, particularly for closely competing structures for stability like fcc and hcp in Co and Ni. There could be evolution to nonequilibrium structures during reduction of oxide catalysts. This matter needs thorough investigation and eventually may provide the explanation for different results for filamentary carbon obtained with iron and iron oxide catalysts. Doping may also contribute to the stability of one or the other structure in the reduced nanoparticles of catalyst. Hernadi et al. [93] also considered that prior reduction pretreatment of iron oxide is not necessary since the hydrocarbon atmosphere is able to reduce the catalyst to the required extent under prevailing conditions during growth of CNS.

Jana et al. [40] in their pioneering investigation of the doped cobalt oxide catalyst nanoparticles by DTA (differential thermal analysis) observed a broad endothermic peak ending with a sharp peak, which is claimed to correspond to bulk melting of Co_3O_4 as shown in Figure 8.16. The designations shown in the curve correspond to different levels of doping and the average size of particles is indicated in bracket. For example, designation CoCu01 represents cobalt oxide doped with copper oxide in order to replace cobalt atoms by 01 wt.% of copper. The oxide nanoparticles have wide size distribution and thus, the start of melting could begin at the start of the broad endothermic peak. It has been further observed that there is no growth of CNS below the temperature at the start of melting. There is only carbon black surrounding the catalyst nanoparticles resulting in carbon nanobeads [40]. If the

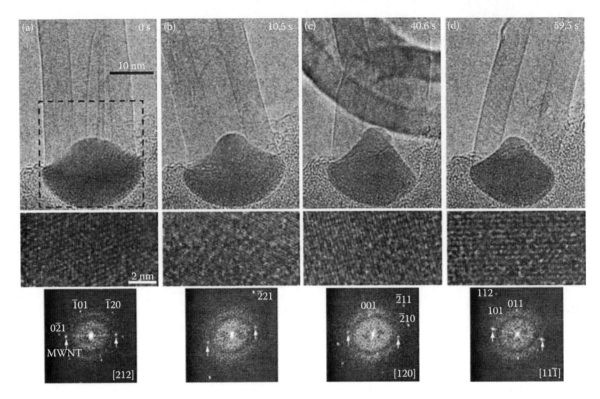

FIGURE 8.14 (a–d) Growth of MWCNT from iron nanoparticles at time shown in each image and enlarged images of the dotted square region in (a) shown below and Fourier transforms. SAD pattern shows presence of carbide along with spots corresponding to MWCNT marked by the arrows. (Reprinted with permission from Yoshida, H. et al. 2008. *Nano Letters* 8:2082–6.)

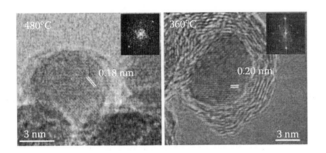

FIGURE 8.15 TEM image of Ni catalyst nanoparticle at 480°C and 360°C showing fringes of lattice planes. (Reprinted with permission from Wirth, C. T., Hofmann, S., and Robertson, J. 2009. *Diamond and Related Materials* 18:940–5.)

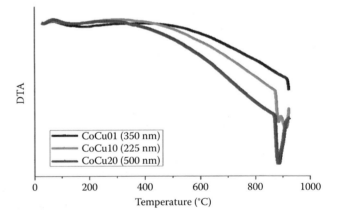

FIGURE 8.16 Typical DTA curves for cobalt oxide-based nanoparticles with different levels of doping and size. (Reprinted with permission from Jana, M., Sil, A., and Ray, S. 2011. *Carbon* 49:5142–9.)

catalyst gets fully covered by decomposed carbon, there is no formation of cylindrical nanostructures.

It is possible that melting of oxide helps to transport carbon from the surface to the inside and facilitate faster reduction while preventing formation of carbon black at the surface, isolating the catalyst nanoparticle from the environment. The melting allows relatively faster diffusion of carbon inside and faster reduction of the oxide in order to facilitate formation of CNS. Since the oxides are chemically stable during heating in air some people have put forward the argument that the broad peak may not be due to melting but because of solid state sintering and agglomeration. When the particles were examined under the microscope after DTA studies and compared

with the as synthesized particles, one observes rounding of surface contour merging locally in fused mass as shown in Figure 8.17 indicating surface melting and liquid phase sintering of particles.

Yosida et al. [89], while observing in situ growth of CNT inside ETEM, noted significant change in the shape of catalyst nanoparticles during growth of both SWCNT and MWCNT growing, respectively, on relatively smaller and larger catalyst

FIGURE 8.17 FESEM micrographs of (a) as synthesized oxide nanoparticles of NiCu10 and (b) the same after DTA.

dissolved in the liquid phase of the catalyst and diffuses to the growing CNS. Tibbetts [95] has claimed that the bending of grapheme sheet to tubular growth is energetically favorable due to large anisotropy in surface energy, that is, high surface energy of other planes of graphite compared to the basal plane. But the formation of a hemispherical cap, as described earlier, provides a mechanism for bending of graphene layer forming during growth of cylindrical tube. However, the energy of bending is relatively more as the diameter is smaller and beyond some smaller inner radius, formation of another layer over the smallest one may not be energetically favorable due to "weak maximum" of the difference of chemical

particles. Hoffman et al. [20] have also observed "dynamic reshaping" of the nickel catalyst nanoparticles during growth of CNS for which there is no plausible explanation. It appears that there is surface melting of metallic catalyst nanoparticles and the surface liquid moves under the action of capillary forces during growth of CNS forming a sharp conical head as shown in Figure 8.18a through d. Wang and Wang have also observed similar conical head formation on nickel catalyst, which is claimed to be in liquid or quasi-liquid state as shown in Figure 8.18e.

Baker et al. [94] in their in situ experiments with nickel catalyst and acetylene at 600°C inside TEM observed the change in shape of the catalyst during the growth of CNF and attributed it to VLS mechanism of growth. Their estimate of activation energy for growth is similar to the activation energy of carbon diffusion in liquid nickel, and fiber growth is observed to be diffusion controlled. In the VLS growth, carbon resulting from decomposition of carbon-bearing gas or vapor, gets

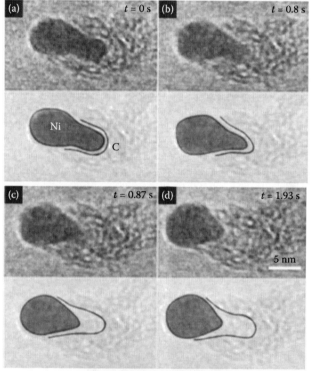

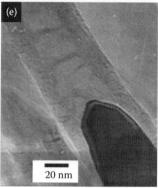

FIGURE 8.18 (a–d) ETEM image sequence showing a growing CNF at different reaction times and the lower drawings indicate schematically the deformation of Ni catalyst and resulting C–Ni interface (Reprinted with permission from Hofmann, S. et al. 2007. *Nano Letters* 7:602–8.) and (e) TEM images of bamboo structured CNTs grown in the presence of nitrogenous gases. (Reprinted with permission from Wang, T. and Wang, B. 2006. *Applied Surface Science* 253:1606–10.)

potential. This may explain the growth of SWCNT on smaller catalyst particles while MWCNT grows on relatively larger ones. This analysis has presumed diffusion of carbon along the surface of the catalyst as shown in Figure 8.19 and the basis for this assumption is its activation energy being similar to that of bulk diffusion of carbon in iron [96]. But the surface and the bulk diffusion are expected to have distinctly different activation energies and so, this conclusion of surface diffusion may not be correct.

Jana et al. [92] have studied the growth of cylindrical CNS with cobalt and nickel oxide doped with different amounts of copper oxide and observed that the size and shape of the catalyst and the extent of doping are influential in determining the growth morphology giving rise to MWCNT, CNF, and Tape. The results are summarized in Figure 8.20a and explained in terms of the extent of surface melting of the catalyst, which increases with decreasing size and increasing doping. The start of melting of the smallest oxide nanoparticles, T_s, has been measured by DTA and the growth temperature in CCVD, T_g, used in this study is 640°C. A morphology map of $\Delta = |(T_s - T_g)|/T_s$ and the minimum size of the oxide nanoparticle in a catalyst sample has been plotted from these results as shown in Figure 8.20b.

For the smallest size of oxide particles, when Δ increases, one may observe a morphology change from MWCNT to CNF with mixed morphology around the transition region. The observation of the growth of SWCNT at high Δ at smaller

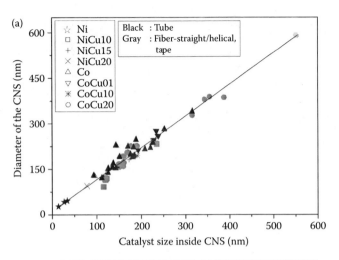

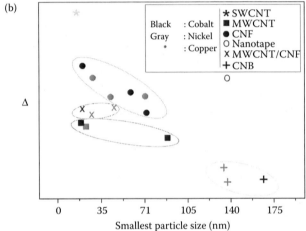

FIGURE 8.20 (a) Variation of the diameter of cylindrical CNS with the size of cobalt and nickel oxide-based catalyst generating it and (b) morphology map of CNSs grown by CCVD using cobalt, nickel, and copper-based oxide nanoparticles.

particle size is possibly due to lack of space for graphitic growth. On increasing minimum particle size up to around 100 nm, the transition from MWCNT to CNF is still observed with increasing Δ. When the minimum size of nanoparticles increases further to around 150 nm, the oxide growth is anisotropic leading to the formation of tape at similar values of Δ where smaller particle sizes lead to CNF. Higher particle size and lower Δ, lead to the formation of nanobeads.

8.5 CONCLUSIONS: THE EMERGING PICTURE

The emerging picture for the growth of cylindrical CNSs by CCVD method is summarized below, in order to crystallize the outstanding questions.

The composition of the gas/vapor source of carbon appears to influence the morphology of carbon deposits as shown in Figure 8.3a and it has been claimed [75] that the structure of the tube is influenced by the flow rate of precursor carbon-bearing gas/vapor. These aspects have not been explored adequately in the context of growth of cylindrical CNS. Although there are a number of investigations using different carbon-bearing

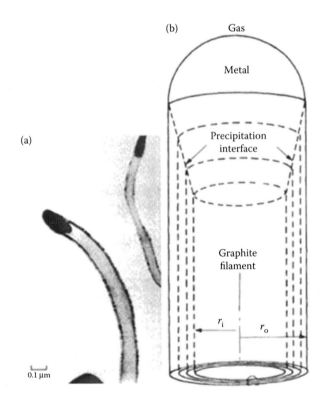

FIGURE 8.19 Schematic illustration of (a) typical carbon filament grown from natural gas on an iron catalyst particle and (b) simplified growth model of this fiber showing inner, outer diameters, and the precipitation interface. (Reprinted with permission from Tibbetts, G. G. 1984. *Journal of Crystal Growth* 66:632–8.)

gas along with other gases as shown in Table 8.1, these studies often use different temperatures in the reaction chamber of CCVD during growth but there is no systematic investigation to ascertain even the role of temperature in deciding the morphology of carbon deposits as demonstrated in Figure 8.3b. A variety of monometallic, bimetallic, and compound catalyst nanoparticles, with size distributed over a range, have been used in CCVD to get a mix of morphologies of CNSs. The selectivity of the desired morphology has been increased by varying growth conditions. The yield, purity, and defects in the nanostructures are other important considerations for arriving at the right conditions for growth. Purity could be controlled by decreasing the activity of a catalyst like iron toward carbon by addition of inactive metal such as copper. There should be more investigations to evolve approaches related to growth for the control of purity and defects as well as for selectivity and yield.

The structure of cylindrical CNS in terms of chirality, chiral vector and chiral angle, is important in deciding the properties including electronic properties of single-walled and multiwalled tubes. The diameter of the tube is related to the magnitude of the chiral vector. TEM, selected area diffraction (SAD), and peaks of optical absorption are important tools to extract information about chirality. HRTEM is routinely used to examine the fringes from the walls of multiwalled tubes. MD simulation studies has revealed the role of catalyst, its pre-growth annealing time and adhesion with the substrate, on which the catalyst is situated, in deciding chirality, particularly for the intermediate size of particles. Experimental studies indicate a correlation between the chirality of the nanotube and the structure of the catalyst nanoparticle. There is a need to undertake further experimental studies to understand the extent of relation between chirality and the catalyst, its chemical nature, size, heat treatment, and adhesion with the substrate, as revealed by MD simulation studies and the experiments carried out so far.

The good performance of transition metal catalysts of Co and Ni gave us the impression that the closely packed crystalline plane with hexagonal atomic arrangement may be providing a template for the formation of hexagonal rings of carbon on the walls of cylindrical CNS. The catalytic ability of other metals not possessing hexagonal arrangement of atoms and MD simulation with the jellium model indicates that the template effect may not be so crucial. The critical role of the catalyst is to keep an end open preventing closure of hemispherical cap during growth as it happens in C_{70} molecules. The metal atoms at the end allow carbon atoms to join the carbon side during growth of cylindrical structure by shifting continually to the open end of the growing tube. The growth of the hemispherical cap over the catalyst may also be very important as it makes the curvature of the graphene sheet to a cylindrical surface feasible both from thermodynamic and kinetic standpoints. The size of the catalyst particle and its adhesion with the substrate also have an important role in lifting off the hemispherical cap from the catalyst as revealed by MD simulation. The lifting-off takes place faster for relatively smaller particles and also, for relatively stronger adhesion

between the catalyst–substrate surfaces. There is a need to verify these results through suitably designed in situ experiments.

The catalyst may help decomposition of carbon-bearing gas on its surface and these atoms may get adsorbed on the surface of the solid catalyst. It could either be physisorbed or chemisorbed on the solid surface of the catalyst. Under optimum condition, the deposition rate of carbon may release as much carbon as could be accommodated on the surface of the catalyst and there will be no loose deposit of carbon. Under steady state, deposition rate should release as much of carbon as could be adsorbed and released to the end of cylindrical nanotube for growth and the maximum growth will be limited by the capacity of adsorption on the surface. This scenario may not allow a large growth rate as observed experimentally. Also, this scenario does not explain large change in shape of the catalyst and its smooth surface as observed in Figures 8.14 and 8.18 during growth of CNS. The observation of broad peak in DTA of cobalt oxide and doped cobalt oxide nanoparticles followed by sharp peak corresponding to a temperature near bulk melting point of cobalt oxide as shown in Figure 8.16 clearly indicates the possibility of surface melting of nanoparticles as confirmed by their fusing together by liquid-phase sintering (Figure 8.17b). However, these results are for nanoparticles of oxide catalysts, which are observed to reduce to metal in the reaction chamber of CCVD method either during prior treatment with gas mixture containing hydrogen or during the growth of nanoparticles in the absence of prior treatment.

The melting of metallic nanoparticles has been extensively investigated by modeling as summarized in Figure 8.9 and also by MD simulation of melting of specific metals and alloys. The results for nickel are shown in Figure 8.10. Therefore, it may not be surprising to observe surface melting at least for metallic catalyst particles at the temperatures (Table 8.1) used for the growth of cylindrical nanostructures explaining the change in shape and the smooth contour of the surface of the catalyst particles observed during the growth of nanostructures. This is in conformity with the observation that the activation energy for growth is similar to that for diffusion of carbon in liquid nickel and diffusion of carbon in liquid is the rate limiting step for the growth of CNSs, grown by CCVD with nickel catalyst. In spite of this observation, there is some confusion about the transport of carbon either through the surface or inside the liquid phase, created by surface melting of the catalyst or melting of the entire particle. If the template effect is not important as indicated by the jellium model, the cluster of rings and chains leading to hemispherical cap could form even without catalyst as it happens in fullerene. The catalytic role of metal may be confined to preventing closure of the end of the cap to allow cylindrical growth and metal atoms in the liquid phase of metal catalyst could play this role more effectively due to their higher mobility compared to the atoms in solid. Further studies are needed to clarify the nature of catalytic role of metals in the formation of cylindrical CNS.

Some of the metal catalysts used for growth of cylindrical nanostructure are strong carbide formers and so, carbon released on the surface of the catalyst by decomposition of carbon-bearing vapor/gas may form carbide and carbon released from carbide in a dynamic situation may supply the carbon for the formation and growth of nanotube. However, this may happen for specific catalysts but it is not an essential step for the formation of CNS. Even for iron catalyst, some results show transient C–Fe XPS peaks as shown in Figure 8.13 while others show the presence of carbide (cementite) in the SAD during growth as shown in Figure 8.14. C–Fe bonds possibly forms when sufficient carbon atoms are not available to form carbon clusters. Now it is known that nanoparticles of non-carbide forming elements such as gold, silver, and copper may catalyze the formation of cylindrical nanostructure and so, carbide formation cannot be an essential step.

In CCVD method the carbon-bearing gas/vapor in the feed gas mixture appears to decompose on the catalyst releasing carbon. If the catalyst is an oxide it will be reduced by hydrogen during pretreatment or by hydrogen, carbon, and so on during growth. The difference in filamentary carbon grown with iron and iron oxide catalyst particles may not be due to the absence of reduction but could be the result of arriving at different metal structures, which affect the chirality of CNSs and hence, their properties. For cylindrical growth, melting to create a layer of liquid on the surface of the catalyst may help faster transport of carbon to the growing cylindrical nanostructure by dissolution. Dissolution may remove carbon released on the surface at a relatively faster rate compared to that for solid, without getting the catalyst covered by loose carbon deposits as it happens leading to the formation of carbon nanobeads. But the question remains as to why a particular cylindrical morphology like SWCNT, MWCNT, CNF, or tape, forms under a given condition.

In oxide catalyst of mixed nanoparticle sizes, lower particle size and surface melting leads to cylindrical nanostructures. For very low sizes (<10 nm), the cylindrical structure appears to go through transition from MWCNT to CNF and from CNF to SWCNT with the increasing extent of melting. But higher particle size and absence of melting drives to the growth of nanobeads. There is hardly any study for metal catalysts addressing the selection of morphology. More studies are necessary to confirm these trends observed for oxide catalysts and to understand the reasons more clearly.

REFERENCES

1. Oncel, C. and Yurum, Y. 2006. Carbon nanotube synthesis via the catalytic CVD method: A review on the effect of reaction parameters. *Fullerenes, Nanotubes, and Carbon Nanostructures* 14:17–37.
2. Lin, M., Tan, J. P. Y., Boothroyd, C., Loh, K. P., Tok, E. S., and Foo, Y-L. 2007. Dynamical observation of bamboo-like carbon nanotube growth. *Nano Letters* 7:2234–8.
3. Gohier, A., Ewels, C. P., Minea, T. M., and Djouadi, M. A. 2008. Carbon nanotube growth mechanism switches from tip- to base-growth with decreasing catalyst particle size. *Carbon* 46:1331–8.
4. Matthews, K. D., Lemaitre, M. G., Kim, T., Chen, H., Shim, M., and Zuo, J-M. 2006. Growth modes of carbon nanotubes on metal substrates. *Journal of Applied Physics* 100:044309. http://dx.doi.org/10.1063/1.2219000
5. Alvarez, W., Kitiyanan, B., Borgna, A., and Resasco, D. 2001. Synergism of Co and Mo in the catalytic production of single-wall carbon nanotubes by decomposition of CO. *Carbon* 39:547–58.
6. Li, Y., Liu, J., Wang, Y., and Wang, Z.L. 2001. Preparation of monodispersed Fe-Mo nanoparticles as the catalyst for CVD synthesis of carbon nanotubes. *Chemistry of Materials* 13:1008–14.
7. Chiang, W. H. and Sankaran, R. M. 2009. Linking catalyst composition to chirality distributions of as-grown single-walled carbon nanotubes by tuning Ni_xFe_{1-x} nanoparticles. *Nature Materials* 8:882–6.
8. Bhaviripudi, S., Mile, E., Steiner III, S. A., Zare, A. T., Dresselhaus, M. S., Belcher, A. M., and Kong, J. 2007. CVD synthesis of single-walled carbon nanotubes from gold nanoparticle catalysts. *Journal of the American Chemical Society* 129:1516–7.
9. Liu, B., Ren, W., Gao, L., Li, S., Liu, Q., Jiang, C., and Cheng, H.-M. 2008. Manganese catalyzed surface growth of single-walled carbon nanotubes with high efficiency. *The Journal of Physical Chemistry C* 112:19231–5.
10. Takagi, D., Homma, Y., Hibino, H., Suzuki, S., and Kobayashi, Y. 2006. Single walled carbon nanotube growth from highly activated metal nanoparticles. *Nano Letters* 6:2642–5.
11. Zhou, W. H. Z., Wang, J. Z. Y., Jin, Z. S. X., Zhang, Y. Y. C, and Li, Y. 2006. Copper catalyzing growth of single-walled carbon nanotubes on substrates. *Nano Letters* 6:2987–90.
12. Zhong, G., Xie, R., Yang, J., and Robertson, J. 2014. Single-step CVD growth of high-density carbon nanotube forests on metallic Ti coatings through catalyst engineering. *Carbon* 67:680–7.
13. Rao, R., Pierce, N., Liptak, D., Hooper, D., Sargent, G., Semiatin, S. L., Curtarolo, S., Harutyunyan, A. R., and Maruyama, B. 2014. Revealing the impact of catalyst phase transition on carbon nanotube growth by *in situ* Raman spectroscopy. *ACS Nano* 7:1100–7.
14. Takagi, D., Kobayashi, Y., and Homma, Y. 2009. Carbon nanotube growth from diamond. *Journal of the American Chemical Society* 131:6922–3.
15. Qian, Y., Wang, C., Ren, G., and Huang, B. 2010. Surface growth of singlewalled carbon nanotubes from ruthenium nanoparticles. *Applied Surface Science* 256:4038–41.
16. Huang, S., Cai, Q., Chen, J., Qian, Y., and Zhang, L. 2009. Metal-catalyst-free growth of single-walled carbon nanotubes on substrates. *Journal of the American Chemical Society* 131:2094–5.
17. Liu B., Ren W., Gao L., Li S., Pei S., Liu C., Jiang, C., and Cheng, H. M. 2009. Metal-catalyst free growth of single-walled carbon nanotubes. *Journal of the American Chemical Society* 131:2082–3.
18. Li, Y., Kim, W., Zhang, Y., Rolandi, M., Wang, D., and Dai, H. 2001. Growth of single-walled carbon nanotubes from discrete catalytic nanoparticles of various sizes. *Journal of Physical Chemistry B* 105:11424–31.
19. Cassell, A. M., Raymakers, J. A., Kong, J., and Dai, H. 1999. Large scale CVD synthesis of single-walled carbon nanotubes. *Journal of Physical Chemistry B* 103:6484–92.
20. Hofmann, S., Sharma, R., Ducati, C., Du, G., Mattevi, C., Cepek, C., Cantoro, M. et al. 2007. *In situ* observations of catalyst dynamics during surface-bound carbon nanotube nucleation. *Nano Letters* 7:602–8.

21. Flahaut, E., Govindaraj, A., Peigney, A., Laurent, C., Rousset, A., and Rao, C. 1999. Synthesis of single-walled carbon nanotubes using binary (Fe, Co, Ni) alloy nanoparticles prepared *in situ* by the reduction of oxide solid solutions. *Chemical Physics Letters* 300:236–42.

22. Gao, F., Zhang, L., and Huang, S. 2010. Zinc oxide catalyzed growth of single-walled carbon nanotubes. *Applied Surface Science* 256:2323–6.

23. Takagi, D., Hibino, H., Suzuki, S., Kobayashi, Y., and Homma, Y. 2007. Carbon nanotube growth from semiconductor nanoparticles. *Nano Letters* 7:2272–5.

24. Qi, H., Qian, C., and Liu, J. 2007. Synthesis of uniform double-walled carbon nanotubes using iron disilicide as catalyst. *Nano Letters* 7:2417–21.

25. Rummeli, M. H., Schaffel, F., Kramberger, C., Gemming, T., Bachmatiuk, A., Kalenczuk, R. J., Rellinghaus, B., Buchner, B., and Pichler, T. 2007. Oxide-driven carbon nanotube growth in supported catalyst CVD. *Journal of American Chemical Society* 129:15772–3.

26. Schunemann, C., Schaffel, F., Bachmatiuk, A., Queitsch, U., Sparing, M., Rellinghaus, B., Lafdi, K., Schultz, L., Buchner, B., and Rummeli, M. H. 2011. Catalyst poisoning by amorphous carbon during carbon nanotube growth: Fact or fiction? *ACS Nano* 5:8928–34.

27. Abdi, Y., Koohsorkhi, J., Derakhshandeh J., Mohajerzadeh, S., Hoseinzadegan, H., Robertson, M. D., Bennett, J. C., Wu, X., and Radamson, H. 2006. PECVD-grown carbon nanotubes on silicon substrates with a nickel-seeded tip-growth structure. *Material Science and Engineering C* 26:1219–23.

28. Horibe, M., Nihei, M., Kondo, D., Kawabata, A., and Awano, Y. 2004. Influence of growth mode of carbon nanotubes on physical properties for multiwalled carbon nanotube films grown by catalystic chemical vapor deposition. *Japanese Journal Applied Physics* 43:7337–41.

29. Sharma, R., Chee, S-W., Herzing, A., Miranda, R., and Rez, P. 2011. Evaluation of the role of Au in improving catalytic activity of Ni nanoparticles for the formation of one-dimensional carbon nanostructures. *Nano Letters* 11:2464–71.

30. Yamada, M., Kawana, M., and Miyake, M. 2006. Synthesis and diameter control of multi-walled carbon nanotubes over gold nanoparticle catalysts. *Applied Catalysis A* 302:201–7.

31. Esconjauregui, S., Whelan, C. M., and Maex, K. 2009. The reasons why metals catalyze the nucleation and growth of carbon nanotubes and other carbon nanomorphologies. *Carbon* 47:659–69.

32. Cho, W., Schulz, M., and Shanov, V. 2014. Growth termination mechanism of vertically aligned centimeter long carbon nanotube arrays. *Carbon* 69:609–20.

33. Pooperasupong, S., Caussat, B., Serp, P., and Damronglerd, S. 2014. Synthesis of Multi-walled carbon nanotubes by fluidized-bed chemical vapor deposition over Co/ Al$_2$O$_3$. *Journal of Chemical Engineering of Japan* 47:28–39.

34. Gao, J., Zhong, J., Bai, L., Liu, J., Zhao, G., and Sun, X. 2014. Revealing the role of catalysts in carbon nanotubes and nanofibers by scanning transmission x-ray microscopy. *Scientific Reports* 4:1–6.

35. Deck, C. P. and Vecchio, K. 2006. Prediction of carbon nanotube growth success by the analysis of carbon-catalyst binary phase diagrams. *Carbon* 44:267–75.

36. Bachmatiuk, A., Borrnert, F., Grobosch, M., Schaffel, F., Wolff, U., Scott, A., Zaka, M. et al. 2009. Investigating the graphitization mechanism of SiO$_2$ nanoparticles in chemical vapor deposition. *ACS Nano* 3:4098–104.

37. Helveg, S., Lopez-Cartes, C., Sehested, J., Hansen, P. L., Clausen, B. S., Rostrup-Nielsen, J. R., Abild-Pedersen, F., and Norskov, J. K. 2004. Atomic-scale imaging of carbon nanofibre growth. *Nature* 427:426–9.

38. Torres, D., Pinilla, J. L., Lazaro, M. J., Moliner, R., and Suelves, I. 2014. Hydrogen and multiwall carbon nanotubes production by catalytic decomposition of methane: Thermogravimetric analysis and scaling-up of Fe-Mo catalysts. *International Journal of Hydrogen Energy* 39:3698–709.

39. McCaldin, S., Bououdina, M., Grant, D. M., and Walker, G. S. 2006. The effect of processing conditions on carbon nanostructures formed on an iron-based catalyst. *Carbon* 44:2273–80.

40. Jana, M., Sil, A., and Ray, S. 2011. Tailoring of surface melting of oxide based catalyst particles by doping to influence the growth of multi-walled carbon nano-structures. *Carbon* 49:5142–9.

41. Lee, C. J. and Park, J. 2001. Growth model for bamboo like structured carbon nanotubes synthesized using thermal chemical vapor deposition. *Journal of Physical Chemistry* B105:2365–8.

42. Lee, C. J. and Park, J. 2001. Growth and structure of carbon nanotubes produced by thermal chemical vapor deposition. *Carbon* 39:1891–6.

43. Ye, X., Yang, Q., Zheng, Y., Mo, W., Hu, J., and Huang, W. 2014. Biotemplate synthesis of carbon nanostructures using bamboo as both the template and the carbon source. *Materials Research Bulletin* 51: 366–71.

44. Sun, J., Koos, A. A., Dillon, F., Jurkschat, K., Castell, M. R., and Grobert, N. 2013. Synthesis of carbon nanocoil forests on BaSrTiO$_3$ substrates with the aid of a Sn catalyst. *Carbon* 60:5–15.

45. Liu, J., Zhang, X., Zhang, Y., Chen, X., and Zhu, J. 2003. Nano-sized double helices and braids: Interesting carbon nanostructures. *Materials Research Bulletin* 38:261–7.

46. Saavedra, M. S., Sims, G. D., McCartney, L. N., Stolojan, V., Anguita, J. V., Tan, Y. Y., Ogin, S. L., Smith, P. A., and Silva, S. R. P. 2012. Catalysing the production of multiple arm carbon octopi nanostructures. *Carbon* 50:2141–6.

47. Stoner, B. R. and Glass, J. T. 2012. Carbon nanostructures: A morphological classification for charge density optimization. *Diamond and Related Materials* 23:130–4.

48. Bachmann, P. K., Leers, D., and Lydtin, H. 1991. Towards a general concept of diamond chemical vapour deposition. *Diamond and Related Materials* 1:1–12.

49. Jourdain, V. and Bichara, C. 2013. Current understanding of the growth of carbon nanotubes in catalytic chemical vapour deposition. *Carbon* 58:2–39.

50. Dresselhaus, M. S., Dresselhaus, G., and Saito, R. 1995. Physics of carbon nanotubes. *Carbon* 33:883–91.

51. Amelinckx, S., Lucas, A., and Lambin, P. 1999. Electron diffraction and microscopy of nanotubes. *Reports on Progress in Physics* 62:1471–524.

52. Abdullahi, I., Sakulchaicharoen, N., and Herrera, J. E. 2014. Selective synthesis of single-walled carbon nanotubes on Fe–MgO catalyst by chemical vapor deposition of methane. *Diamond and Related Materials* 41:84–93.

53. Zhang, X. B., Zhang, X. F., Amelinckx, S., Van Tendeloo, G., and Van Landuyt, J. 1994. The reciprocal space of carbon tubes: A detailed interpretation of the electron diffraction effects. *Ultramicroscopy* 54:237–49.

54. Yu, M., Wu, S. Y., and Jayanthi, C. S. 2009. A self-consistent and environment-dependent Hamiltonian for large-scale simulations of complex nanostructures. *Physica E* 42:1–16.

55. Gomez-Gualdron, D. A., McKenzie, G. D., Alvarado, J. F. J., and Balbuena, P. B. 2012. Dynamic evolution of supported metal nanocatalyst/carbon structure during single-walled carbon nanotube growth. *ACS Nano* 6:720–35.

56. Shibuta, Y. and Maruyama, S. 2002. Molecular dynamics simulation of generation process of SWNTs. *Physica B: Condensed Matter* 323:187–9.

57. Shibuta, Y. and Maruyama, S. 2003. Molecular dynamics simulation of formation process of single-walled carbon nanotubes by CCVD method. *Chemical Physics Letters* 382:381–6.

58. Shibuta, Y. and Maruyama, S. 2007. A molecular dynamics study of the effect of a substrate on catalytic metal clusters in nucleation process of single-walled carbon nanotubes. *Chemical Physics Letters* 437:218–23.

59. Shibuta, Y. and Elliott, J. A. 2009. A molecular dynamics study of the graphitization ability of transition metals for catalysis of carbon nanotube growth via chemical vapor deposition. *Chemical Physics Letters* 472:200–6.

60. Elliott, J. A., Hamm, M., and Shibuta, Y. 2009. A multiscale approach for modeling the early stage growth of single and multiwall carbon nanotubes produced by a metal-catalyzed synthesis process. *Journal of Chemical Physics* 130:034704–12.

61. Nanda, K. K. 2009. Size-dependent melting of nanoparticles: Hundred years of thermodynamic model. *Pramana-Journal Physics* 72:617–28.

62. Wronski, C. R. M. 1967. The size dependence of the melting point of small particles of tin. *British Journal of Applied Physics* 18:1731–7.

63. Lai, S. L., Guo, J. Y., Petrova, V., Ramanath, G., and Allen, L. H. 1996. Size-dependent melting properties of small tin particles: Nanocalorimetric measurements. *Physical Review Letters* 77:99–102.

64. Qi, Y., Cagin, T., Johnson, W. L., and Goddard, W. A. 2001. Melting and crystallization in Ni nanoclusters: The mesoscale regime. *Journal of Chemical Physics* 115:385–94.

65. Ding, F. and Bolton, K. 2006. The importance of supersaturated carbon concentration and its distribution in catalytic particles for single-walled carbon nanotube nucleation. *Nanotechnology* 17:543–8.

66. Page, A. J., Ohta, Y., Irle, S., and Morokuma, K. 2010. Mechanisms of single-walled carbon nanotube nucleation, growth, and healing determined using QM/MD methods. *Accounts of Chemical Research* 43:1375–85.

67. Ribas, M. A., Ding, F., Balbuena, P. B., and Yakobson, B. I. 2009. Nanotube nucleation versus carbon-catalyst adhesion-probed by molecular dynamics simulations. *The Journal of Chemical Physics* 131:224501-1-7.

68. Charlier, J-C., Amara, H., and Lambin, Ph. 2007. Catalytically assisted tip growth mechanism for single-wall carbon nanotubes. *ACS Nano* 1:202–7.

69. Wang, T. and Wang, B. 2006. Study on structure change of carbon nanotubes depending on different reaction gases. *Applied Surface Science* 253:1606–10.

70. Iijima, S. 1991. Helical microtubules of graphitic carbon. *Nature* 354:56–8.

71. Ebbesen, T. W. and Ajayan, P. M. 1992. Large-scale synthesis of carbon nanotubes. *Nature* 358:220–2.

72. Thess, A., Lee, R., Nikolaev, P., Dai, H. J., Dai, H., Petit, P., Robert, J. et al. 1996. Crystalline ropes of metallic carbon nanotubes. *Science* 273:483–7.

73. Cheng, H. M., Li, F., Su, G., Pan H. Y., He, L. L., Sun, X., and Dresselhaus, M. S. 1998. Large-scale and low-cost synthesis of single-walled carbon nanotubes by the catalytic pyrolysis of hydrocarbons. *Applied Physics Letters* 72:3282–4.

74. Li, Y., Cui, R., Ding, L., Liu, Y., Zhou, W., Zhang, Y., Jin, Z., Peng, F., and Liu, J. 2010. How catalysts affect the growth of single walled carbon nanotubes on substrates. *Advanced Materials* 22:1508–15.

75. Yuan, D., Ding, L., Chu, H., Feng, Y., McNicholas, T. P., and Liu, J. 2008. Horizontally aligned single-walled carbon nanotube on quartz from a large variety of metal catalysts. *Nano Letters* 8:2576–9.

76. Swierczewska, M., Rusakova, I., and Sitharaman, B. 2009. Gadolinium and europium catalyzed growth of single-walled carbon nanotubes. *Carbon* 47:3139–42.

77. Dupuis, A. C. 2005. The catalyst in the CCVD of carbon nanotubes—A review. *Progress in Materials Science* 50:929–61.

78. Zhu, H., Suenaga, K., Wei, J., Wang, K., and Wua, D. 2008. A strategy to control the chirality of single-walled carbon nanotubes. *Journal of Crystal Growth* 310:5473–6.

79. Liu, J., Wang, C., Tu, X., Liu, B., Chen, L., Zheng, M., and Zhou, C. 2012. Chirality-controlled synthesis of single-wall carbon nanotubes using vapour-phase epitaxy. *Nature Communications* 3:1–7.

80. Oberlin, A., Endo, M., and Koyama, T. 1976. Filamentous growth of carbon through benzene decomposition. *Journal of Crystal Growth* 32:335–49.

81. Lim, H. E., Miyata, Y., Nakayama, T., Chen, S., Kitaura, R., and Shinohara, H. 2011. Purity-enhanced bulk synthesis of thin single-wall carbon nanotubes using iron–copper catalysts. *Nanotechnology* 22:395602–7.

82. Picher, M., Navas, H., Arenal, R., Quesnel, E., Anglaret, E., and Jourdain, V. 2012. Influence of the growth conditions on the defect density of single-walled carbon nanotubes. *Carbon* 50:2407–16.

83. Abdullahi, I., Sakulchaicharoen, N., and Herrera, J. E., 2013. Selective growth of single-walled carbon nanotubes over Co–MgO catalyst by chemical vapor deposition of methane. *Diamond and Related Materials* 38:1–8.

84. Ni, L., Kuroda, K., Zhou, L-P., Ohta, K., Matsuishi, K., and Nakamura, J. 2009. Decomposition of metal carbides as an elementary step in carbon nanotube synthesis. *Carbon* 47:3054–62.

85. Baker, R. T. K., Alonzo, J. R., Dumesic, J. A., and Yates, D. J. C. 1982. Effect of the surface state of iron on filamentous carbon formation. *Journal of Catalysis* 77:74–84.

86. Yang, K. L. and Yang, R. T. 1986. The accelerating and retarding effects of hydrogen on carbon deposition on metal surfaces. *Carbon* 24:687–93.

87. Ducati, C., Alexandrou, I., Chhowalla, M., Robertson, J., and Amaratunga, G. A. J. 2004. The role of the catalytic particle in the growth of carbon nanotubes by plasma enhanced chemical vapor deposition. *Journal of Applied Physics* 95:6387–91.

88. Sharma, R., Moore, E. S., Rez, P., and Treacy, M. M. J. 2009. Site-specific fabrication of Fe particles for carbon nanotube growth. *Nano Letters* 9:689–94.

89. Yoshida, H., Takeda, S., Uchiyama, T., Kohno, H., and Homma, Y. 2008. Atomic-scale in-situ observation of carbon nanotube growth from solid state iron carbide nanoparticles. *Nano Letters* 8:2082–6.

90. Meschel, S. V. and Kleppa, O. J. 1997. Standard enthalpies of formation of some 3d transition metal carbides by high temperature calorimetry. *Journal of Alloys and Compounds* 257:227–33.

91. Wirth, C. T., Hofmann, S., and Robertson, J. 2009. State of the catalyst during carbon nanotube growth. *Diamond and Related Materials* 18:940–5.

92. Jana, M., Sil, A., and Ray, S. 2013. Carbon Nanostructures-development and application in the anode of Li-ion battery. PhD dissertation, Indian Institute of Technology, Roorkee.

93. Hernadi, K., Fonseca, A., Nagy, J. B., Bernaerts, D., and Lucas, A.A. 1996. Fe-catalyzed carbon nanotube formation. *Carbon* 34:1249–57.

94. Baker, R. T. K., Barber, M. A., Harris, P. S., Feates, F. S., and Waite, R. J. 1972. Nucleation and growth of carbon deposits from the nickel catalyzed decomposition of acetylene. *Journal of Catalysis* 26:51–62.

95. Tibbetts, G. G. 1984. Why are carbon filaments tubular? *Journal of Crystal Growth* 66:632–8.

96. Baker, R. T. K., Harris, P. S., Thomas, R. B., and Waite, R. J. 1973. Formation of filamentous carbon from iron, cobalt and chromium catalyzed decomposition of acetylene. *Journal of Catalysis* 30:86–95.

9 sp² to sp³ Phase Transformation in Graphene-Like Nanofilms

Long Yuan, Zhenyu Li, and Jinlong Yang

CONTENTS

Abstract 147
9.1 Chemically Induced sp² to sp³ Phase Transformation in Graphene Systems 147
    9.1.1 Functionalized Single Graphene Sheet...147
    9.1.2 Functionalized Bilayer Graphene 150
    9.1.3 Functionalized Graphene/h-BN Heterostructures 153
9.2 Chemically Induced sp² to sp³ Phase Transformation in SiC Films...153
    9.2.1 Chemical Functionalized 2D SiC Sheets..153
    9.2.2 Hydrogenated Wurtzite SiC Nanofilms for Bipolar Magnetic Semiconductors Application.......155
9.3 Conclusions and Perspectives 157
References..157

ABSTRACT

Due to its unique electronic structure, graphene has become a widely studied new material. It is desirable to enrich the physics and chemistry of graphene-based materials by electronic structure engineering. An important property-tuning means is introducing an sp² to sp³ phase transformation. In this chapter, first-principles studies of functionalization induced sp² to sp³ phase transformations of graphene-like nanofilm are reviewed. Diamondized graphene bilayer and graphene–BN hybrid bilayer have a tunable electronic structure, from magnetic semiconductor to nonmagnetic metal. More interestingly, hydrogenated wurtzite SiC nanofilm is a two-dimensional bipolar magnetic semiconductor material, where its spin polarization can be conveniently controlled by a gate voltage. The tunable electronic and magnetic properties demonstrated here pave new avenues to construct graphene-based electronics and spintronics devices.

Graphene, a one-atom thickness honeycomb sp²-hybridized carbon nanofilm, has attracted considerable attention due to its many unique physical properties such as half-integer quantum Hall effect, massless Dirac fermion behavior, ambipolar electric field effect, and high carrier mobility.[1–5] Since graphene is a semimetal with a zero bandgap, exploring effective ways to open a bandgap is indispensable for its applications in electronic devices. Hence, physical or chemical-based graphene electronic structure engineering has attracted lots of research interests. For example, graphene's electronic properties could be tuned by using appropriate substrate,[6–10] substitutional atom doping,[11–17] molecular doping,[18–24] or by patterning graphene into nanoribbons.[25–30] At the same time, some other two-dimensional (2D) structures similar to graphene have also been synthesized. Despite their structure similarity, properties of these 2D systems can be very different from graphene. For example, sp²-hybridized SiC sheet is a semiconductor with a large direct energy gap.

To broaden their applications, it is very desirable to have the capability of electronic structure engineering for graphene-like 2D structures. Chemical functionalization is one of the most important means to tune properties of materials. Many graphene-related nanomaterials with novel electronic and magnetic properties are predicted and/or synthesized on the basis of chemical functionalization. Especially, interesting physics and chemistry has been shown when chemical functionalization induces sp² to sp³ phase transformation, which is the main topic to be reviewed in this chapter.

9.1 CHEMICALLY INDUCED sp² TO sp³ PHASE TRANSFORMATION IN GRAPHENE SYSTEMS

9.1.1 FUNCTIONALIZED SINGLE GRAPHENE SHEET

In 2008, Sofo et al.[31] first predicted the existence of graphane, an extended 2D hexagonal hydrocarbon network (Figure 9.1a) by using the first-principles calculations. Graphane could be regarded as a fully hydrogenated single-layer graphene with formula CH, where all carbon atoms are sp³-hybridized and hydrogen atoms are bonded to carbon atoms on both sides of the plane in an alternating manner. Graphane is predicted to be stable when compared with other hydrocarbons such as benzene, cyclohexane, and polyethylene. It has two favorable configurations: the chair-like configuration with the hydrogen atoms alternating on both sides of the plane and the boat-like configuration with the hydrogen atoms alternating in pairs. Binding energy calculations show that the chair-like configuration is more stable. Because hydrogenation destroys the origin π-band network of the pristine graphene, a large bandgap is opened in graphane. The chair-like configuration has a

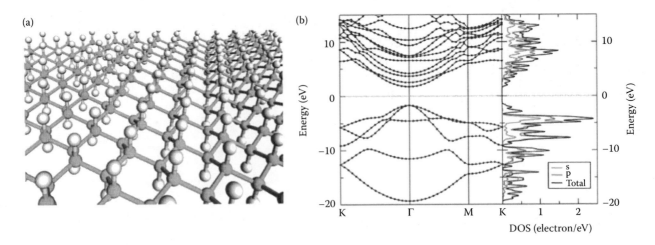

FIGURE 9.1 (a) Structure of graphane in the chair conformation. The carbon atoms are shown in gray and the hydrogen atoms in white. (b) Band structure and density of states for the chair-typed graphane. (Reprinted with permission from Sofo, J. O., A. S. Chaudhari, G. D. Barber, *Phys. Rev. B*, 75, 153401. Copyright 2007 by the American Physical Society.)

bandgap of 3.5 eV at the GGA (generalized gradient approximations) level (Figure 9.1b), and it is corrected to be 5.4 eV with GW calculations. Sofo's predictions are confirmed soon by an independent experiment. In 2009, Elias et al.[32] synthesized graphane by reacting graphene with hydrogen plasma, which could transform this zero-gap semiconductor into an insulator.

Zhou et al.[33,34] showed that when half of the hydrogen atoms in the graphane sheet were removed, as shown in Figure 9.2, the resulting semihydrogenated graphene sheet (referred as graphone) is a ferromagnetic semiconductor with a small indirect gap of 0.46 eV. They found that when half of the carbon atoms are hydrogenated, strong σ-bonds are formed between C and H atoms and the π-bonding network is broken, the hydrogenated C atoms become sp³-hybridized. The unhydrogenated C atoms are still sp²-hybridized leaving the electrons in these C atoms localized and unpaired and producing ferromagnetism.

Besides hydrogenated graphene, fluorinated graphene is another important graphene derivative.[35,36] Fluorographene has a similar geometric structure with graphane.[37–38] All the C atoms are sp³-hybridized, with each C atom covalently bonded to one fluorine atom. The chair configuration is

more stable than the boat configuration. Its electronic property is also similar to graphane. The chair configuration has a direct bandgap of 3.20 eV predicted at the GGA level, and 7.42 eV at the more accurate GW level. Jeon et al.[39] treated graphene with XeF_2 and produced a partially fluorinated graphene with covalent C–F bonding and local sp³-carbon hybridization. After spectroscopy characterization, the synthesized fluorinated graphene is an insulator with a bandgap of 3.8 eV.

Liu et al.[40] studied the electronic and magnetic properties of the semifluorinated graphene via first-principles calculations. Different from semihydrogenated graphene, semifluorinated graphene exhibits magnetic and metallic behavior. The magnetism mainly ascribed to the exchange splitting of dangling bond of the unfluorinated C atoms with a coupling with impurity state induced by F adatom.

By using DFT and nonequilibrium Green's function (NEGF) method, Li et al.[41] explore the possibility of functionalized graphene as a high-performance 2D spintronics device. They considered two types of functionalization. One is functionalization on one side of graphene (the number of functional groups equals half the number of C atoms), and the other is different functionalizations on the two sides of

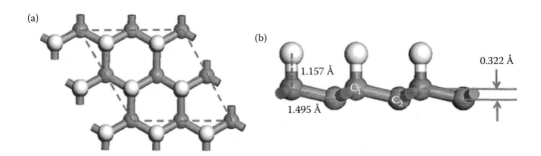

FIGURE 9.2 Optimized geometric structure of graphone. (a) Top view with the rhombus marked in gray shows the supercell, (b) side view shows C–C and C–H distances. (Reprinted with permission from Zhou, J. et al. *Nano Lett.*, 9, 3867. Copyright 2009 American Chemical Society.)

graphene (the number of each kind of functional groups equals half the number of C atoms). Both boat and chair conformations have been considered. When graphene is functionalized with O on one side and H on the other in the chair conformation, a ferromagnetic metal with a spin-filter efficiency up to 54% at finite bias is obtained. The semifluorinated graphene in the chair conformation is an antiferromagnetic semiconductor. They designed a spin device based on it by introducing a magnetic field to stabilize its metallic ferromagnetic state. The resulting room-temperature magnetoresistance is up to 2200%, which is 1 order of magnitude larger than available experimental values.

When graphane is cut into one-dimensional (1D) hydrocarbon, it will yield graphane nanoribbons, in which all the C atoms are still sp^3-hybridized. Graphane nanoribbons possess totally different properties compared to the sp^2-hybridized graphene nanoribbons (GNRs). In GNRs, edge states play an important role in determining their electronic and magnetic properties. However, in graphane nanoribbons,[42] it is the inner C and H atoms that dominate the electronic properties of graphane nanoribbons, as shown in Figure 9.3. The zigzag

and armchair graphane nanoribbons are both nonmagnetic semiconductors. Their bandgaps are highly width dependent, which decrease monotonically as a function of ribbon width due to quantum confinement effect. The formation energy of both kinds of graphane nanoribbons increases with increasing ribbon width.

Zhang et al.[43] investigated the bandgap modulation of graphane nanoribbons under uniaxial elastic strain. They found that the bandgap of graphane nanoribbons has a linear relationship with uniaxial strain, and the bandgap is more sensitive to compressive than tensile strain. Moreover, the calculated Young's modulus of graphane nanoribbons is much smaller than that of the graphene nanoribbon (GNR). As for 1D zigzag and armchair fluorographene nanoribbons,[44] similar to graphane nanoribbons, they are all nonmagnetic semiconductors with width-dependent bandgaps. Energy gaps of fluorographene nanoribbons are smaller than those of the corresponding graphane nanoribbons with the same width.

Results of studies introduced in this section are summarized in Table 9.1.

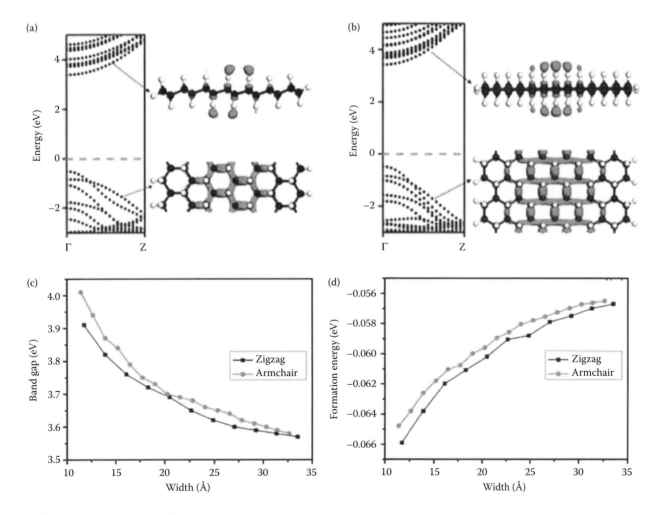

FIGURE 9.3 Band structures (left) and partial charge densities for (a) 7 zigzag and (b) 13 armchair graphane nanoribbons; variation of the bandgap (c) and the formation energy (d) of zigzag and armchair graphane nanoribbons as a function of ribbon width. (Reprinted with permission from Li, Y. et al., *J. Phys. Chem.* C, 113, 15043. Copyright 2009 American Chemical Society.)

TABLE 9.1
Summary of Previous Research on Functionalized Graphene Sheet

Functionalized single-layer graphene	Hydrogenated single-layer graphene[31–34]
	Fluorinated single-layer graphene[35–41]
Functionalized GNRs	Graphane nanoribbons[42–44]
Functionalized bilayer graphene	Hydrogenated bilayer graphene[49,52–53]
	Diamondization of graphene bilayer[50–56]
Functionalized graphene/h-BN heterostructures	Hybrid diamondization of graphene/ BN heterostructures[51,60,64]

9.1.2 FUNCTIONALIZED BILAYER GRAPHENE

Compared with single-layer graphene, bilayer graphene displays some unique properties. For example, bilayer graphene could open an energy gap by inducing an external electric field.[45–48] But the induced bandgap by this means is usually very small. It is interesting to know how electronic structure of bilayer graphene can be tuned by functionalization with chemical groups.

Ab initio density-functional theory calculations revealed that when both sides of the bilayer are hydrogenated, the weak van der Waals bonding is replaced by strong chemical bonds between the two layers, as shown in Figure 9.4. When the hydrogenation rate reaches its limit of 50%, all C atoms become sp³-hybridized, with similar geometry structures as graphane.[49]

A recent experiment has shown evidence for the room-temperature phase transformation from hydroxylated sp²-hybridized bilayer graphene to hydroxylated sp³-hybridized diamondized nanofilms via increasing pressure,[50] as shown in Figure 9.5. Ab initio calculations showed that the resulting 2D material, which is called diamondol, is a ferromagnetic insulator, with a bandgap of 0.6 eV and a magnetic moment

of 1 Bohr per unit cell. For the majority spin, the bandgap is indirect and much larger than that of the minority spin.

Motivated by this interesting experiment and previous theoretical investigations, based on first-principles calculations, we systematically studied functionalization-induced diamondization of graphene bilayer.[51] When we decorated bilayer graphene on single side by using different chemical groups such as H, OH, and F, they all could transform into semifunctionalized diamondized nanofilms. In all these diamondized structures, the C atoms in the first layer become fully sp³-hybridized. In the second layer, only half the carbon atoms are sp³-hybridized, with the other C atoms remaining sp²-hybridized. An interesting magnetic property is obtained for these diamondized structures. Because half the carbon atoms in the second layer remain sp²-hybridized, where a localized electron is expected, which gives about 1 μ_B magnetic moment per unit cell. In all three diamondized nanofilms, the ferromagnetic coupling is energetically more favorable.

By using the mean-field theory, we could estimate the Curie temperature. The Curie temperatures of the single-side hydrogenated, hydroxylated, and fluorinated diamondized nanofilms were estimated to be 435, 378, and 438 K, respectively, when treated as 2D systems. Band structure calculations show that the single-side hydrogenated and fluorinated diamondized nanofilms are indirect bandgap semiconductors, with bandgap values of 0.72 and 0.80 eV, respectively. The single-side hydroxylated diamondized nanofilm has a direct bandgap of 0.86 eV. Hence, the electronic properties of bilayer graphene can be modified with different chemical groups, with tunable electronic structures.

Minimum energy path (MEP) calculations provide more details about the transformation process of the functionalized graphene bilayer. Take single-side fluorinated graphene bilayer as an example, as shown in Figure 9.6a, in the initial state, when half of the carbon atoms in the top layer are fluorinated, a strong σ-bond between the C atoms and F atoms is formed and the original extensive π-bonding network of graphene is broken, leaving the electrons in the unfluorinated

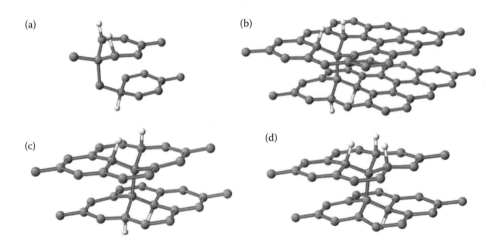

FIGURE 9.4 Formations of a chemical bond between the two layers in bilayer graphene in (a) a 2 × 2, (b) a 4 × 4, and [(c) and (d)] a 3 × 3 supercell. The light colored atoms are H and the others C. (Reprinted with permission from Leenaerts, O., B. Partoens, F. M. Peeters, *Phys. Rev.* B, 80, 245422. Copyright 2009 by the American Physical Society.)

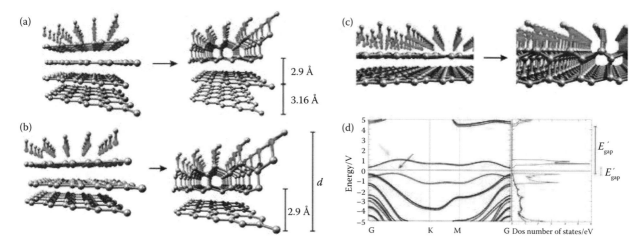

FIGURE 9.5 The initial and optimized geometries for (a) four, (b) three, and (c) two graphene layers. The distances from the diamondol structure and subsequent graphene layers are also shown in (a) and (b), where the height d of the unit cell is also specified. (d) The spin-dependent electronic dispersion and the density of states (DOS) for the unit cell of the final structure in (c). (Barboza, A. P. M. et al.: *Adv. Mater.* 2011. 23. 3014. Copyright Wiley-VCH Verlag GmbH & Co. KGaA. Reproduced with permission.)

C atoms localized and unpaired. So the fluorinated top graphene layer exhibits a buckled structure with a thickness of 0.31 Å. The computed equilibrium distance between the two layers is 3.10 Å. In the transition state, the thickness of the top and bottom layers are 0.42 and 0.21 Å, respectively. The distance between the two layers is 2.01 Å. In the final state, the thickness of the top and bottom layers are 0.48 and 0.36 Å, respectively, reaching their maximum value. The calculated transformation barrier is 0.22 eV per unit cell for fluorinated diamondized nanofilms. For hydrogenated and hydroxylated diamondized nanofilms, they show similar behaviors, with a barrier about 0.23 eV per unit cell. This means that different chemical groups have little effect on the transformation barrier.

Fully functionalized bilayer graphene is also studied.[51] For surface decoration with the same chemical groups such as H, OH, and F, after geometry optimization, homogeneous diamondized nanofilms could be formed. In all these diamondized structures, C atoms are fully sp³-hybridized. There is no explicit magnetism in fully functionalized films. Band structure calculations show that all three films are direct semiconductors with bandgaps of 3.06, 3.25, and 4.03 eV for fully hydrogenated, hydroxylated, and fluorinated nanofilms, respectively, much larger than the semifunctionalized

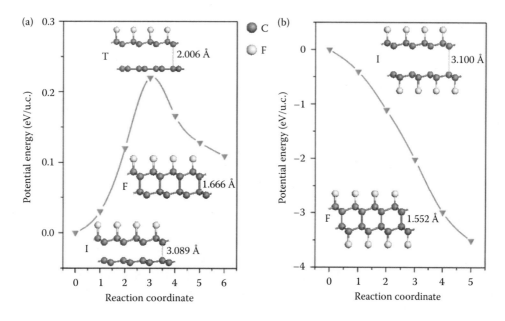

FIGURE 9.6 Minimum energy path for transition processes (a) from single-side functionalized bilayer to diamondized film, (b) from double-side functionalized bilayer to diamondized film. (Reprinted with permission from Yuan, L. et al. *Phys. Chem.* Chem. Phys., 2012, 14, 8179.)

diamondized nanofilms. The transformation is proved to be a barrierless process, as shown in Figure 9.6b. Therefore, external pressure may not be required to synthesize these double-side functionalized films.

Wang and coworkers[52] further investigated the conformation and electronic properties of fully hydrogenated bilayer graphene. They studied its two conformations: the chair conformation and the boat conformation (Figure 9.7). They found that the fully hydrogenated chair conformation is the lowest energy conformation, with a direct bandgap of 3.24 eV at the LDA (local density approximation) level. The removing of hydrogen from one side of bilayer graphane results in a ferromagnetic semiconductor. They also found that the bandgap is tunable with external electric fields. Applying a perpendicular electric bias between two fully hydrogenated graphene bilayer leads to transition from semiconductor to metal.

Zhang et al.[53] utilized first principles calculations to investigate the biaxial strain-dependent electronic properties of the fully hydrogenated bilayer graphene. They studied two types of the fully hydrogenated bilayer graphene: the Bernal stacking (termed B-type), where in which one half of the carbon atoms in one layer lie directly above the carbon atoms of the adjacent layer, while the other half that are hydrogenated lie over the centers of hexagonal rings of the adjacent layer; the hexagonal stacking (termed H-type), where all carbon atoms of one layer lie above the carbon atoms of the other layer. The B-type and H-type have almost the same binding energy. Further, the energy barriers from B- to H-type and from H- to B-type are estimated to be 502 and 487 meV per carbon atom, respectively. Therefore, they concluded that both B- and H-type configurations can be regarded as stable states. They also found that the bandgap of both type of the fully hydrogenated bilayer graphene could be tuned continuously by the biaxial strain. Moreover, compressive strain can induce the semiconductor-to-metal transition of this hydrogenated system. Such a tunable bandgap and semiconductor-to-metal

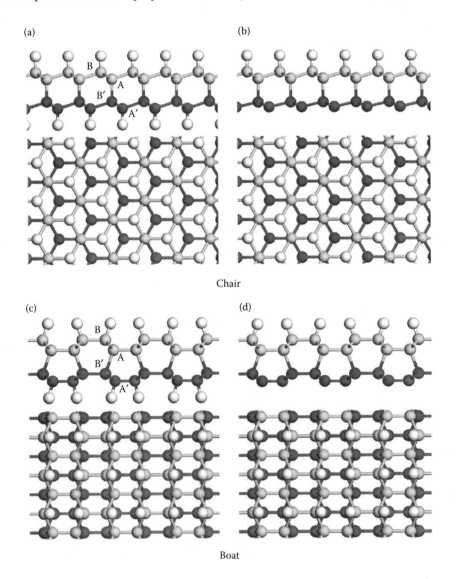

(a)   (b)

Chair

(c)   (d)

Boat

FIGURE 9.7 Side views of (a) fully hydrogenated and (b) semihydrogenated chair conformations of bilayer graphene. Top and side views of (c) fully hydrogenated and (d) semihydrogenated boat structures of bilayer graphene. Hydrogen atoms are colored with white. (Reprinted with permission from Samarakoon, D. K., X. Wang, *ACS Nano*, 4, 4126. Copyright 2010 American Chemical Society.)

transition are mainly contributed by the bonding state between px and py orbitals of carbon atoms and the antibonding state between pz orbital of hydrogenated carbon atoms and s orbital of hydrogen atoms.

Sivek et al.[54] further examined the stability and electronic properties of fully fluorinated bilayer graphene. Compared with previously investigated fully hydrogenated bilayer graphene, bilayer fluorographene is found to be a much more stable material, which is consistent with our result. The electronic band structure of bilayer fluorographene is similar to that of monolayer fluorographene, but its electronic bandgap is significantly larger (about 1 eV). They also calculated the effective masses around the Γ point for fluorographene and bilayer fluorographene and found that they are isotropic. The computed 2D Young's modulus of bilayer fluorographene is approximately 300 N m⁻¹, almost as strong as graphene, much stronger than monolayer fluorographene.

Hu et al.[55] studied electronic structures of two types of stoichiometrically fully fluorinated bilayer graphene as well as those under biaxial compressive and tensile strains. Under biaxial compressive strains, the bandgaps of both types of fully fluorinated bilayer graphene can be reduced and could transform them from semiconductor to metals. However, under biaxial tensile strains, both types of fully fluorinated bilayer graphene are found to be able to undergo a direct-to-indirect bandgap transition.

As for multilayer 2D diamond nanofilms, Li et al.[56] studied effects of semihydrogenation (SH) and full-hydrogenation (FH) on the structural evolution of the 2D diamond nanofilms. The calculated formation energies of different layers are all negative suggesting that hydrogenation plays an important role in stabilizing the 2D diamond nanofilms. Moreover, the absolute data of formation energy are decreased monotonously with the increase of layer numbers. For the FH systems, the direct bandgaps are predicted in the region of 2.54–3.55 eV at the GGA level and decreased with increasing layer numbers. For the SH cases, a ferromagnetism characteristic is presented determining by the unpaired electrons on the un-hydrogenated side, and the spin-related bandgaps are in an infrared region of 0.74–1.17 eV at the GGA level, which are also strongly dependent on layer numbers.

9.1.3 Functionalized Graphene/h-BN Heterostructures

Recently, direct growth of graphene on hexagonal boron nitride (h-BN) has been realized experimentally.[57–59] We studied the electronic and magnetic properties of graphene–BN bilayers modified by different functional groups such as H, OH, and F.[51] After geometry optimization, new hybrid diamondized films are obtained. In these films, all C and B atoms become sp³-hybridized and N atoms remain sp²-hybridized. Band structures calculations show that they are all non-magnetic metals. Examination of the states near the Fermi level shows that the N 2p and C 2p states contribute to the band crossing the Fermi level. Electron transfer between the bottom BN layer and the top graphene layer is the origin of the metallicity.

Chen et al.[60] studied fully hydrogenated graphene/BN bilayer and graphene/BN multilayers to exploit the effects of the heterogeneous interface and external bias voltage on controlling their electronic properties. Two stable structures are obtained for hydrogenated C/BN bilayer. Both of them are direct bandgap semiconductors, with the valence band maximum (VBM) and conduction band minimum (CBM) at the Γ point. The energy gaps are 1.74 and 1.47 eV at the LDA level, respectively. The VBM is mainly distributed on the graphene layer showing an σ-bond characteristic. However, the CBM is mainly delocalized on the surface of hybrid films, which is known as the nearly free electron (NFE) states.[61–63] They also investigated the bandgap modulation with external electric field. Under a positive electric field, the bandgap decreases and disappears at a critical value. Under a negative field, the bandgap increases gradually. The bandgap could also be tuned by adjusting the layer numbers.

Zhang et al.[64] demonstrated that a few layers of graphene sandwiched between h-BN layers can undergo spontaneous transformation into hybrid cubic BN-diamond nanofilms upon fluorination, in which all atoms are sp³-hybridized. This transformation is spontaneous for a sufficiently thin multilayer and will suffer an energy barrier with increasing the total number of multilayers, as shown in Figure 9.8. Increasing the ratio of the h-BN layer in the initial multilayer can greatly reduce the energy barrier to promote the transformation of the hybrid diamond film. The electronic properties of the nanofilms can be tuned by controlling the ratio of the BN component and film thickness, which can yield narrow-gap semiconductors for novel electronic applications. In addition, the energy gaps in the nanofilms can be modulated linearly by applying external electric fields.

9.2 CHEMICALLY INDUCED sp² TO sp³ PHASE TRANSFORMATION IN SiC FILMS

Different form graphene, theoretical calculations predicted SiC as a semiconductor with a large direct energy gap of 2.55 eV at the GGA level and 3.90 eV with a more accurate GW approximation.[65,66] It also has a desirable large exciton binding energy.[67] This gives it high internal quantum efficiency, which cannot be easily reached using pristine graphene. A recent experiment synthesized ultrathin graphitic SiC nanosheets through sanitation of wurtzite SiC nanofilms and exhibited outstanding light-emitting ability.[68]

Just like graphene, the electronic properties and magnetic properties of 2D SiC are affected by vacancy defects, adatoms, substitutional, impurities, or by patterning SiC into nanoribbons.[65,69–73] Beyond this, chemical functionalization plays an important role to tune the physical properties of SiC sheets.

9.2.1 Chemical Functionalized 2D SiC Sheets

Xu et al.[74] performed first-principles calculations to investigate the formation energies, electronic, and magnetic properties of

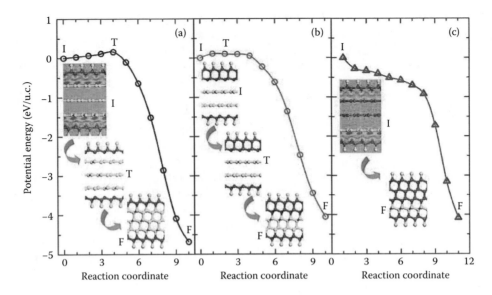

FIGURE 9.8 Calculated MEP for (a) $C_3(BN)_2$, (b) $C_2(BN)_3$, and (c) $C_1(BN)_4$ to the corresponding hybrid diamond nanofilms. Insets illustrate the geometry structures of the initial state (I), transition state (T), and final state (F). (Reprinted with permission from Zhang, Z., X. Zeng, W. Guo, *J. Phys. Chem.* C, 115, 21678. Copyright 2011 American Chemical Society.)

semihydrogenated SiC sheet, as shown in Figure 9.9. Because the C sites and Si sites in the 2D SiC sheets are not equivalent, they studied the SH on either C sites or Si sites, respectively. Both semihydrogenated SiC structures have favorable formation energies, indicating a great possibility of realizing them in future experiments. However, they exhibit totally different geometric, electronic, and magnetic properties. When Si atoms are hydrogenated, all the Si atoms become sp³-hybridized and C atoms remain sp²-hybridized. The formed semihydrogenated SiC sheet is a ferromagnetic semiconductor with an indirect bandgap of 0.85 eV at the GGA level; when C atoms are hydrogenated, all the C atoms become sp³-hybridized and Si atoms remain sp²-hybridized. The formed semihydrogenated SiC sheet is an antiferromagnetic semiconductor with an indirect bandgap 1.19 eV. The semihydrogenated SiC

sheet with the C atoms hydrogenated is found to be more stable than the sheet with the Si atoms hydrogenated.

Ma et al.[75] investigated the magnetic and electronic properties of the semifluorinated SiC sheet. To determine the most stable structure of the semifluorinated structures, they considered all possible adsorbed structures, and found that the configuration in which F atoms absorbed on the Si sites is most stable. The semifluorinated SiC sheet is an antiferromagnetic semiconductor with a bandgap of about 1 eV at the GGA level.

Garcia et al.[76] performed a first-principles investigation on the structural and electronic properties of fully hydrogenated or fluorinated SiC sheet. All the atoms become sp³-hybridized due to chemical functionalization. The analysis on the formation energies show that the fully hydrogenated and fluorinated SiC sheet are very stable and should be easily synthesized in

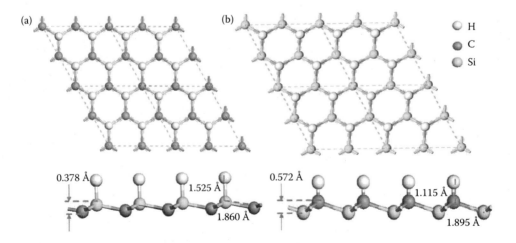

FIGURE 9.9 The optimized structures of the semihydrogenated SiC sheets: (a) top and side views of the H–SiC structure; (b) top and side views of the H–CSi structure. The dashed rhombus outlines the supercell used in calculations. (Reprinted with permission from Xu, B. et al. *Appl. Phys. Lett.*, 96, 143111. Copyright 2010, American Institute of Physics.)

TABLE 9.2

Summary of Electronic and Magnetic Properties of Chemical Functionalized SiC Sheets[a]

	Bandgap (eV)	Magnetism
Semihydrogenated SiC (C cites)[74]	0.85	Ferromagnetic
Semihydrogenated SiC (Si cites)[74]	1.19	Antiferromagnetic
Semifluorinated SiC (Si cites)[75]	1.0	Antiferromagnetic
Fully hydrogenated SiC[77]	4.04	Nonmagnetic
Fully fluorinated SiC[77]	1.94	Nonmagnetic
Hydrogenated bilayer SiC[78]	0.834	Ferromagnetic

[a] All DFT calculations are performed at the GGA level.

the laboratory. Both are nonmagnetic semiconductors with a bandgap of 4.04 and 1.94 eV at the GGA level for fully hydrogenated and fluorinated SiC sheet, respectively.

Wang and coworkers[77] further investigated the structural properties of fully hydrogenated SiC sheet. Based on structure optimization and phonon dispersion analysis, they found that both chair- and boat-like configurations are dynamically stable, and the chair-like conformer is energetically more favorable than the boat-like one. The chair- and boat-like configurations are revealed to be nonmagnetic semiconductors with direct bandgaps of 3.84 and 4.29 eV, respectively. Charge distributions show that bonding in chair- and boat-like conformers are all characterized with covalency.

Results of studies introduced in this section are summarized in Table 9.2.

9.2.2 Hydrogenated Wurtzite SiC Nanofilms for Bipolar Magnetic Semiconductors Application

Recently, a new kind of spintronics material, bipolar magnetic semiconductors (BMS), has been proposed.[78] As shown in Figure 9.10, in a BMS material, the valence and conduction bands possess opposite spin polarization when approaching the Fermi level. Therefore, completely spin-polarized currents with reversible spin polarization can be created simply by applying a gate voltage. By using first-principles electronic and transport calculations, Li et al. predict that the semihydrogenated single-wall carbon nanotube (SWCNT) is an ideal BMS candidate. For applications in integrated circuits, a 2D BMS material is more desirable. Considering the larger spin–orbit interaction and thus larger possible magnetic anisotropy energy of Si than C, SiC-based 2D materials are very attractive for spintronics applications. In a recent study, we predict that hydrogenated wurtzite SiC nanofilm is a 2D BMS material.[79]

The formation energy of semihydrogenated SiC sheet with C atoms hydrogenated is smaller than semihydrogenated SiC sheet with Si atoms hydrogenated, indicating that the H atom prefers to absorb on C sites, which is consistent with previous theoretical work.[74] Three types of stacking arrangements in the graphitic-like SiC multilayered structures are checked. C–Si ordering is the most energetically favorable one, as also suggested previously.[80]

An MD (molecular dynamics) simulation at 200 K lasted for 2 ps is used to study hydrogenation at the C sites on the top layer of the SiC bilayer, as shown in Figure 9.11a. After the simulation, we find that the graphitic-like SiC bilayer could be easily transformed back into the wurtzite structure via surface hydrogenation. To investigate the thermal stability, an MD simulation at 450 K lasted for 10 ps is performed. The formed hydrogenated wurtzite nanofilms are very stable during the simulation. The optimized structure at 0 K is shown in Figure 9.11b. In the second layer, only the C atoms are sp³-hybridized, while the Si atoms remain sp²-hybridized. MEP of the hydrogenation is shown in Figure 9.11d. The system only needs to overcome a small barrier of 0.03 eV per atom to complete the transformation and the formed wurtzite nanofilm is more energetically favorable than the initial structure. Therefore, it should be feasible to synthesize the hydrogenated wurtzite nanofilm in experiments under mild conditions.

Si atoms in the second layer remain sp²-hybridized, leaving the electrons in the unhydrogenated Si atoms localized and unpaired, which gives about 1 μ_B magnetic moment per unit cell, as shown in Figure 9.11c, magnetic coupling calculations show that ferromagnetic state is energetically more

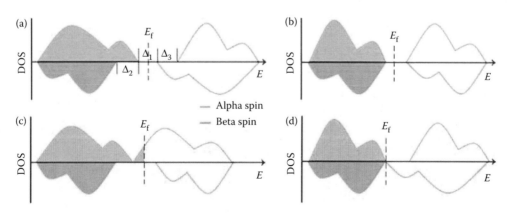

FIGURE 9.10 Schematic density of states of (a) bipolar magnetic semiconductor, (b) general magnetic semiconductor, (c) half metal, and (d) spin gapless semiconductor. (Reprinted with permission from Li, X. et al. *Nanoscale*, 2012, 4, 5680.)

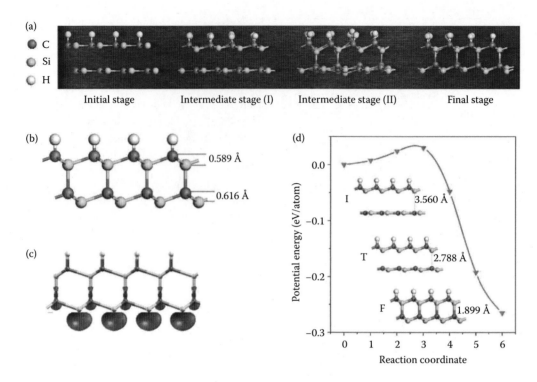

FIGURE 9.11 Formation of the hydrogenated wurtzite nanofilm from the semihydrogenated SiC bilayer. (a) Snapshots of the initial, intermediate, and final stage of the transformation. (b) Optimized geometry of H-(CSi)$_2$ at 0 K. (c) Spin density distribution of H-(CSi)$_2$ (isosurface value: 0.05 e/Å^3). (d) Calculated MEP of the transformation, insets show the atomic structures of initial (I), transition (T), and final (F) states along the energy path. (Reprinted with permission from Yuan, L., Z. Li, and J. L. Yang, *Phys. Chem. Chem. Phys.*, 2013, 15, 497.)

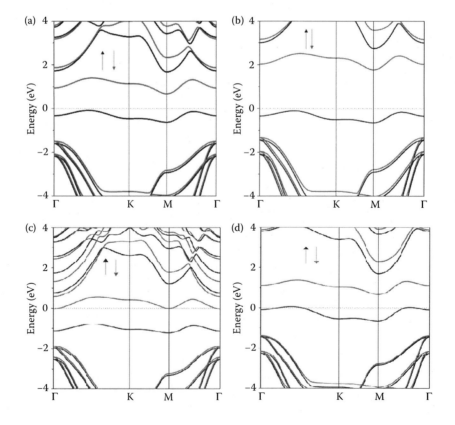

FIGURE 9.12 Electronic band structures of H-(CSi)$_2$ based on (a) PBE functional and (b) HSE functional. PBE band structures with doping concentrations of (c) 0.02 electron per atom and (d) 0.03 hole per atom. The Fermi level is set at zero. (Reprinted with permission from Yuan, L., Z. Li, and J. L. Yang, *Phys. Chem. Chem. Phys.*, 2013, 15, 497.)

favorable than the antiferromagnetic and nonmagnetic states. The formed hydrogenated wurtzite nanofilm is a typical BMS material with an indirect bandgap of 0.834 eV at the GGA level, as shown in Figure 9.12a. A test calculation with the HSE (Heyd–Scuseria–Ernzerhof) functional also gives a similar band structure with an enlarged bandgap of 1.874 eV, as shown in Figure 9.12b. Half-metallicity with opposite spin polarization can be obtained via electron or hole doping, as shown in Figure 9.12c and d.

Under an electric field as strong as 0.4 V Å^{-1}, the bandgap is only slightly affected by the field. Therefore, it is a robust BMS material even under a strong electric field. On the other hand, the hydrogenated wurtzite nanofilm should be on a suitable substrate in applications. A substrate can apply a stress onto the material. So, it is interesting to see whether the BMS electronic structure is robust under an external strain. We find that the BMS characteristics are well preserved under high-level external strains. A real system of a semihydrogenated SiC bilayer on a substrate, the hydrogenated (0001) SiC surface, is also tested. The BMS character of the electronic structure is kept intact.

9.3 CONCLUSIONS AND PERSPECTIVES

During the past several years, people are devoting more time and effort on developing novel graphene-related functional materials. In this brief review, we focus on the chemical functionalization induced sp² to sp³ phase transformation in graphene and SiC systems. Functional nanomaterials with fantastic properties can be obtained after phase transformation, with promising technological potential in the fields of electronics and spintronics. However, in spite of so much theoretical investigations, experimental research on the chemical functionalization induced sp² to sp³ phase transformation in graphene and SiC systems are still relatively rare. Therefore, it is urgent that the synthesis issues should be addressed so that the formed graphene and SiC derivatives through phase transformation can finally accomplish their highly promising technical potential.

REFERENCES

1. Novoselov, K. S., A. K. Geim, S. V. Morozov, D. Jiang, Y. Zhang, S. V. Dubonos, I. V. Grigorieva, A. A. Firsov, *Nature*, 2005, 438, 197.
2. Zhang, Y., Z. Jiang, J. P. Small, M. S. Purewal, Y.-W. Tan, M. Fazlollahi, J. D. Chudow, J. A. Jaszczak, H. L. Stormer, P. Kim, *Phys. Rev. Lett.*, 2006, 96, 136806.
3. Novoselov, K. S., A. K. Geim, S. V. Morozov, D. Jiang, Y. Zhang, S. V. Dubonos, I. V. Grigorieva, A. A. Firsov, *Science*, 2004, 306, 666.
4. Berger, C., Z. Song, X. Li, X. Wu, N. Brown, C. Naud, D. Mayou, T. Li, J. Hass, A. N. Marchenkov, E. H. Conrad, P. N. First, W. A. de Heer, *Science*, 2006, 312, 1191.
5. Neto, A. H. C., F. Guinea, N. M. R. Peres, K. S. Novoselov, A. K. Geim, *Rev. Mod. Phys.*, 2009, 81, 109.
6. Zhou, S. Y., G.-H. Gweon, A. V. Fedorov, P. N. First, W. A. de Heer, D.-H. Lee, F. Guinea, A. H. Castro Neto, A. Lanzara, *Nat. Mater.*, 2007, 6, 770.
7. Giovannetti, G., P. A. Khomyakov, G. Brocks, P. J. Kelly, J. vanden Brink, *Phys. Rev. B.*, 2007, 76, 073103.
8. Varchon, F., R. Feng, J. Hass, X. Li, B. Ngoc Nguyen, C. Naud, P. Mallet, J.-Y. Veuillen, C. Berger, E. H. Conrad, L. Magaud, *Phys. Rev. Lett.*, 2007, 99, 126805.
9. Romero, H. E., N. Shen, P. Joshi, H. R. Gutierrez, S. A. Tadigadapa, J. O. Sofo, P. C. Eklund, *ACS Nano*, 2008, 2, 2037.
10. Kang, Y.-J., J. Kang, K. J. Chang, *Phys. Rev. B*, 2008, 78, 115404.
11. Yu, S., W. Zheng, C. Wang, Q. Jiang, *ACS Nano*, 2010, 4, 7619.
12. Lherbier, A., X. Blase, Y.-M. Niquet, F. Triozon, S. Roche, *Phys. Rev. Lett.*, 2008, 101, 036808.
13. Wei, D., Y. Liu, Y. Wang, H. Zhang, L. Huang, G. Yu, *Nano Lett.*, 2009, 9, 1752
14. Ci, L. et al. *Nat. Mater.*, 2010, 9, 430.
15. Gan, Y., L. Sun, F. Banhart, *Small*, 2008, 4, 587.
16. Wang, H. et al. *Nano Lett.*, 2012, 12, 141.
17. Zheng, B., P. Hermet, L. Henrard, *ACS Nano*, 2010, 4, 4165.
18. Das, B., R. Voggu, C. S. Rout, C. N. R. Rao, *Chem. Commun.*, 2008, 5155.
19. Kozlov, S. M., F. Vines, A. Gorling, *Adv. Mater.*, 2011, 23, 2638.
20. Leenaerts, O., B. Partoens, F. M. Peeters, *Phys. Rev. B*, 2008, 77, 125416.
21. Prado, M. C., R. Nascimento, L. G. Moura, M. J. S. Matos, M. S. C. Mazzoni, L. G. Cancado, H. Chacham, B. R. A. Neves, *ACS Nano*, 2011, 5, 394.
22. Zhang, W. et al. *ACS Nano*, 2011, 5, 7517.
23. Park, J., S. B. Jo, Y.-J. Yu, Y. Kim, J. W. Yang, W. H. Lee, H. H. Kim, B. H. Hong, P. Kim, K. Cho, K. S. Kim, *Adv. Mater.*, 2012, 24, 407.
24. Wehling, T. O., K. S. Novoselov, S. V. Morozov, E. E. Vdovin, M. I. Katsnelson, A. K. Geim, A. I. Lichtenstein, *Nano Lett.*, 2008, 8, 173.
25. Kobayashi, Y., K. Fukui, T. Enoki, K. Kusakabe, Y. Kaburagi, *Phys. Rev. B*, 2005, 71, 193406.
26. Niimi, Y., T. Matsui, H. Kambara, K. Tagami, M. Tsukada, H. Fukuyama, *Phys. Rev. B*, 2006, 73, 085421.
27. Son, Y., M. L. Cohen, S. G. Louie, *Phys. Rev. Lett.*, 216803, 97, 2006.
28. Liu, G. et al. *ACS Nano*, 2012, 6, 6786.
29. Campos-Delgado, J. et al. *Nano Lett.*, 2008, 8, 2773.
30. Jia, X. et al. *Science*, 2009, 323, 1701.
31. Sofo, J. O., A. S. Chaudhari, G. D. Barber, *Phys. Rev. B*, 2007, 75, 153401.
32. Elias, D. C. et al. *Science*, 2009, 323, 610.
33. Zhou, J., Q. Wang, Q. Sun, X. S. Chen, Y. Kawazoe, P. Jena, *Nano Lett.*, 2009, 9, 3867.
34. Zhou, J., M. M. Wu, X. Zhou, Q. Sun, *Appl. Phys. Lett.*, 2009, 95, 103108.
35. Robinson, J. T. et al. *Nano Lett.*, 2010, 10, 3001.
36. Nair, R. R., W. C. Ren, R. Jalil, I. Riaz, V. G. Kravets, L. Britnell, P. Blake et al., *Small*, 2010, 6, 2877.
37. Leenaerts, O., H. Peelaers, A. D. Hernández-Nieves, B. Partoens, F. M. Peeters, *Phys. Rev. B*, 2010, 82, 195436.
38. Samarakoon, D. K., Z. Chen, C. Nicolas, X.-Q. Wang, *Small*, 2011, 7, 965.
39. Jeon, K., Z. Lee, E. Pollak, L. Moreschini, A. Bostwick, C. Park, R. Mendelsberg et al., *ACS Nano*, 2011, 5, 1042.
40. Liu, H. Y., Z. F. Hou, C. H. Hu, Y. Yang, Z. Z. Zhu, *J. Phys. Chem. C*, 2012, 116, 18193.
41. Li, L., R. Qin, H. Li, L. Yu, Q. Liu, G. Luo, Z. Gao, J. Lu, *ACS Nano*, 2011, 5, 2601.

42. Li, Y., Z. Zhou, P. Shen, Z. Chen, *J. Phys. Chem. C*, 2009, 113, 15043.

43. Zhang, Y., X. Wu, Q. Li, J. Yang, *J. Phys. Chem. C*, 2012, 116, 9356.

44. Tang, S., S. Zhang, *J. Phys. Chem. C*, 2011, 115, 16644.

45. Min, H., B. Sahu, S. K. Banerjee, A. H. MacDonald, *Phys. Rev. B*, 2007, 75, 155115.

46. Castro, E. V., K. S. Novoselov, S.V. Morozov, N. M. R. Peres, J. M. B. Lopes dos Santos, J. Nilsson, F. Guinea, A. K. Geim, A. H. Castro Neto, *Phys. Rev. Lett.*, 2007, 99, 216802.

47. Oostinga, J. B., H. B. Heersche, X. Liu, A. F. Morpurgo, L. M. K. Vandersypen, *Nat. Mater.*, 2007, 7, 151.

48. Mak, K. F., C. Lui, J. Shan, T. F. Heinz, *Phys. Rev. Lett.*, 2009, 102, 256405.

49. Leenaerts, O., B. Partoens, F. M. Peeters, *Phys. Rev. B*, 2009, 80, 245422.

50. Barboza, A. P. M., M. H. D. Guimaraes, D. V. P. Massote, L. C. Campos, *Adv. Mater.*, 2011, 23, 3014.

51. Yuan, L., Z. Li, J. Yang, J. Hou, *Phys. Chem. Chem. Phys.*, 2012, 14, 8179.

52. Samarakoon, D. K., X. Wang, *ACS Nano*, 2010, 4, 4126

53. Zhang, Y., C. Hu, Y. Wen, S. Wu, Z. Zhu, *New J. Phys.*, 2011, 13, 063047.

54. Sivek, J., O. Leenaerts, B. Partoens, F. M. Peeters, *J. Phys. Chem. C*, 2012, 116, 19240.

55. Hu, C. H., Y. Zhang, H. Y. Liu, S. Q. Wua, Y. Yang, Z. Z. Zhu, *Comput. Mater. Sci.*, 2012, 65, 165.

56. Li, J., H. Li, Z. Wang, G. Zou, *Appl. Phys. Lett.*, 2013, 102, 073114.

57. Xue, J., J. Sanchez-Yamagishi, D. Bulmash, P. Jacquod, A. Deshpande, K. Watanabe, T. Taniguchi, P. Jarillo-Herrero, B. J. LeRoy, *Nat. Mater.*, 2011, 10, 282.

58. Liu, Z., L. Song, S. Zhao, J. Huang, L. Ma, J. Zhang, J. Lou, M. Ajayan, *Nano Lett.*, 2011, 11, 2032.

59. Decker, R., Y. Wang, V. W. Brar, W. Regan, H. Z. Tsai, Q. Wu, W. Gannett, A. Zettl, M. F. Crommie, *Nano Lett.*, 2011, 11, 2291.

60. Chen, X. F., J. S. Lian, Q. Jiang, *Phys. Rev. B*, 2012, 86, 125437.

61. Margine E. R. and V. H. Crespi, *Phys. Rev. Lett.*, 2006, 96, 196803.

62. Hu, S. L., J. Zhao, Y. D. Jin, J. L. Yang, H. Petek, J. G. Hou, *Nano Lett.*, 2010, 10, 4830.

63. Lu, N., Z. Y. Li, J. L. Yang, *J. Phys. Chem. C*, 2009, 113, 16741.

64. Zhang, Z., X. Zeng, W. Guo, *J. Phys. Chem. C*, 2011, 115, 21678.

65. Bekaroglu, E., M. Topsakal, S. Cahangirov, S. Ciraci1, *Phys. Rev. B*, 2010, 81, 075433.

66. Lü, T., X. Liao, H. Wang, J. Zheng, *J. Mater. Chem.*, 2012, 22, 10062.

67. Lin, X., S. Lin, Y. Xu, A. A. Hakro, T. Hasan, B. Zhang, B. Yu, J. Luo, E. Li, H. Chen, *J. Mater. Chem. C*, 2013, 1, 2131.

68. Lin, S. S., *J. Phys. Chem. C*, 2012, 116, 3951.

69. Lou, P., *Phys. Status Solidi B*, 2012, 249, 91.

70. Costa, C. D., J. M. Morbec, *J. Phys.: Condens. Matter*, 2011, 23, 205504.

71. Sun, L., Y. Li, Z. Y. Li, Q. Li, Z. Zhou, Z. Chen, J. L. Yang, J. G. Hou, *J. Chem. Phys.*, 2008, 129, 174114.

72. Ding Y., Y. Wang, *Appl. Phys. Lett.*, 2012, 101, 013102.

73. Lou, P., J. Y. Lee, *J. Phys. Chem. C*, 2010, 114, 10947.

74. Xu, B., J. Yin, Y. D. Xia, X. G. Wan, Z. G. Liu, *Appl. Phys. Lett.*, 2010, 96, 143111.

75. Ma, Y., Y. Dai, M. Guo, C. Niu, L. Yu, B. Huang, *Appl. Surf. Sci.*, 2011, 257, 7845.

76. Garcia, J. C., D. B. de Lima, L. V. C. Assali, J. F. Justo, *J. Phys. Chem. C*, 2011, 115, 13242.

77. Wang, X., J. Wang, *Physics Letters A*, 2011, 375, 2676.

78. Li, X., X. Wu, Z. Li, J. Yang, J. G. Hou, *Nanoscale*, 2012, 4, 5680.

79. Yuan, L., Z. Li, J. L. Yang, *Phys. Chem. Chem. Phys.*, 2013, 15, 497.

80. Yu, M., C. S. Jayanthi, S. Y. Wu, *Phys. Rev. B*, 2010, 82, 075407.

10 Symmetry and Topology of Graphenes

A. R. Ashrafi, F. Koorepazan-Moftakhar, and O. Ori

CONTENTS

Abstract 159
10.1 Introduction 159
10.2 Symmetry 160
10.3 Topology 162
10.4 Conclusions 164
References 164

ABSTRACT

Graphene is an allotrope of carbon. These are two-dimensional materials consisting of a single layer of carbon atoms arranged in a regular hexagonal pattern similar to graphite. The mathematics of these objects constitutes a significant part of mathematical physics. Suppose X is a physical object that can be described by a finite set $\{x_1, x_2, \ldots, x_n\}$ of real numbers. A permutation f: X $\longrightarrow$ X is a symmetry of X if $d(f(x_i), f(x_j)) = d(x_i, x_j)$, where $d(-,-)$ is a given distance function. In this chapter, we first study the symmetry structure of some classes of graphene lattice and then apply this information to compute distance-based topological indices of some classes of graphenes.

10.1 INTRODUCTION

Graphene is a form of carbon in which carbon atoms are arranged in a regular hexagonal pattern [1]. The discovery of graphene is a result of low-dimensional quantum field theory [2]. The aim of this chapter is to investigate the mathematical properties of this two-dimensional material [3,4]. To do this, we first describe some mathematical notions containing symmetry and topology of objects that will be kept throughout. Then some useful computational methods are presented, and finally we describe how we can apply these methods to discover the geometry and topology of graphene from a computational point of view.

A *graph* is a mathematical structure containing two types of elements: *vertices* and *edges*. Suppose G is a graph. The set of all vertices and edges of G are denoted by V(G) and E(G), respectively. G is called connected if for any 2-element subset $\{x,y\}$ of vertices, we can connect x and y by a sequence $x_0 = x$, $e_1, x_1, \ldots, e_n, y = x_n$ of vertices and edges in such a way that x_{i-1} and x_i are end vertices of the edge e_i. A *component* for G is a maximum connected subset of G. A vertex v in G is said to be *cut vertex*, if removing v increases the number of components of G. If G does not have cut vertex then we say G is *2-connected*. A *molecular graph* is a graph in which atoms are vertices and chemical bonds are edges of the graph. The degree of vertices in a molecular graph is at most four. The

graph G is said to be planar if there exists a two-dimensional representation for G such that edges without any of the edges crossing over.

A nonempty set H together with a binary operation $\blacklozenge$: $H \times H \longrightarrow H$ is a *group* if the following three conditions are satisfied:

1. For all elements a, b, c $\in$ H, we have $a \blacklozenge (b \blacklozenge c) = (a \blacklozenge b) \blacklozenge c$.
2. There exists an element e $\in$ H such that for each element x $\in$ H, $x \blacklozenge e = e \blacklozenge x = x$.
3. Every element x $\in$ H has an inverse x^{-1} such that $x \blacklozenge x^{-1} = x^{-1} \blacklozenge x = e$.

Suppose G is a graph. A one-to-one function β from V(G) onto V(G) is called an *automorphism* if for each two vertex x and y in V(G), xy $\in$ E(G) if and only if $\beta(x)\beta(y) \in$ E(G). The set of all automorphisms of G is denoted by Aut(G). It is not difficult to prove that Aut(G) constitutes a group under composition of functions. A topological index for G is a number that is invariant under Aut(G). Clearly, the set of vertices and edges of G are topological indices of G. Suppose G is connected and x, y are vertices of G. The *topological distance* between x and y is the length of a minimal path connecting them. This number is denoted by d(x,y). The maximum value of d(x,y) over all vertices of G is diameter and the minimum is radius of G. These numbers are denoted by D(G) and R(G), respectively. A topological index defined as a function in terms of the topological distance function $d(-,-)$ is called a distance-based topological index.

Suppose G is a hexagonal system, e = xy, f = uv $\in$ E(G), and w $\in$ V(G). Define $d(w,e) = \text{Min}\{d(w,x), d(w,y)\}$. We say that e is parallel to f if $d(x,f) = d(y,f)$ and we write e $\parallel$ f.

Theorem 10.1: $\parallel$ Is a Reflexive and Symmetric but Not Transitive.

Proof. Reflexivity is clear. We first assume that e = xy is parallel to f = uv. Then, $d(x,f) = d(y,f)$. If $d(x,u) = d(x,v)$ then we obtain a cycle of odd length containing the edge f, contradict

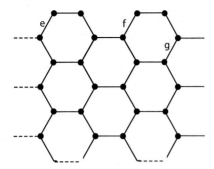

FIGURE 10.1 Polyhex nanotorus.

by bipartivity of hexagonal systems. Hence, $d(x,u) \neq d(x,v)$ and similarly $d(y,u) \neq d(y,v)$. Without loss of generality we can assume that $d(x,u) < d(x,v)$. Then, by assumption $d(y,v) < d(y,u)$ and we can see that $d(x,u) = d(y,v)$, $d(x,v) = d(y,u)$. On the other hand, $d(x,u) < d(x,v)$ and $d(y,v) < d(y,u)$ imply that $d(x,v) = d(x,u) + 1$ and $d(y,u) = d(y,v) + 1$. This shows that $d(u,e) = Min\{d(u,x),d(u,y)\} = Min\{d(x,v) - 1,d(y,v) + 1\} = Min\{d(y,u) - 1, d(x,u) + 1\} = Min\{d(y,v), d(x,v)\} = d(v,e)$, which shows that ‖ is symmetric. To prove that ‖ is not transitive, we consider the graph of a polyhex nanotorus depicted in Figure 10.1. In this graph, $e \parallel f$ and $f \parallel g$ but e is not parallel to g.

Question 10.1: Under What Condition(s) Is Parallelism an Equivalence Relation?

Suppose G is a hexagonal system and $e \in E(G)$. We define $P(e)$ to be the set of all edges parallel to e and $N(e) = |P(e)|$.

In this chapter, we study some well-known classes of bounded graphene lattice. Throughout this chapter our notations are standard and taken mainly from [5,6].

10.2 SYMMETRY

Suppose A is a k-dimensional figure. A continuous one-to-one and onto function α from the k-dimensional Euclidean space R^k into R^k is called a *symmetry* if and only if α^{-1} is continuous and $\alpha(A) = A$. The set of all symmetry elements of A is denoted by Sym(A). It is easy to see that Sym(A) is closed under composition of functions and identity function is an element of Sym(A).

A hexagonal system is a 2-connected plane graph such that its each inner face is a regular hexagon of side length 1. Suppose G is a symmetry group containing two symmetry elements a and b such that $a^2 = b^2 = (ab)^n = $ identity. Then, we say that G has dihedral symmetry group. The order of this group is obviously $2n$. For example, the symmetry group of a regular n-gon has dihedral type. A group generated by one element of order n is denoted by Z_n. An abelian group generated by two elements of order two is denoted by K_4, the Klein four-group.

Let H_n be a hexagonal system consisting of one central hexagon which is surrounded by $n - 1$ layers of hexagonal cells when $n \geq 1$ (Figure 10.5). H_n is a molecular graph, corresponding to benzene ($n = 1$), coronene ($n = 2$), circumcoronene ($n = 3$), circum-circumcoronene ($n = 4$), and so on. We are now ready to state our second result as follows:

Theorem 10.2 The Hexagonal Graph H_n Has Dihedral Group Symmetry of Order 12.

In Figure 10.3, a two-dimensional shape of the hexagonal-parallelogram P(n, k) is depicted. It can be verified that $|V(P(n,k))| = 2(n + 1)(k + 1) - 2$, $|E(P(n,k))| = 3nk + 2(n + k) - 1$, and $D(P(n,k)) = 2(n + k) - 1$. From this figure one can see that the symmetry group in the case that $n = k$ or $n \neq k$ are different, see [7,8] for details.

Theorem 10.3: The Hexagonal-Parallelogram P(n, k) Has a Klein Symmetry Group, If n = k, and a Cyclic Symmetry Group of Order Two, If n ≠ k.

Following Shiu et al. [9], a hexagonal rectangle is called hexagonal jagged-rectangle, or simply HJR, if the number of hexagonal cells in each row is alternative between n and $n - 1$. Obviously, there are three types of HJR. If the top and bottom rows are longer we shall call it HJR of type I and denote by $I^{n,m}$. If the top and bottom rows are shorter we shall call it HJR of type K and denote by $K^{n,m}$. The last one is called HJR of type J and denoted by $J^{n,m}$, Figures 10.4 through 10.6.

Theorem 10.4: The HJR of Type I and K Have Klein Symmetry Group and the HJR of Type J Has Cyclic Symmetry Group of Order 2.

Consider the hexagonal triangle graph T(n) of Figure 10.7, containing j hexagons in the jth row, $1 \leq j \leq n$. This graph is related to the atomic structure of bipod-shaped nanocrystals. Since the graph G has an equilateral figure, $|E(G)| = 3(2 + 3 + 4 + \cdots + (n + 1)) = 3/2(n^2 + 3n)$. By a simple argument we have

Theorem 10.5: The Molecular Graph of T[n] Has a Non-Abelian Symmetry Group of Order 6.

In the end of this section, generator sets for the symmetry group of some small shape of these hexagonal systems, Figures 10.2 through 10.7 are presented. We start by T[5]. Define

$$X_1 = (1,21)(2,20)(3,19)(4,16)(5,17)(6,22)(7,15)$$
$$(8,12)(9,13)(10,18)(23,28)(24,31)(25,30)(27,29)$$
$$(32,35)(33,38)(34,37)(40,43)(41,42)(44,46)$$

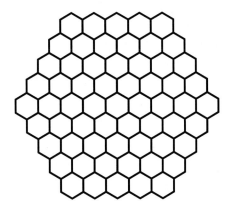

FIGURE 10.2   Hexagonal graph H_5.

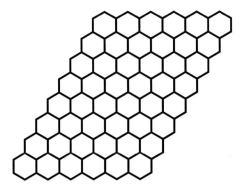

FIGURE 10.3   Hexagonal-parallelogram graph.

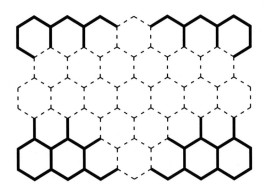

FIGURE 10.4   Graph $I^{n,m}$.

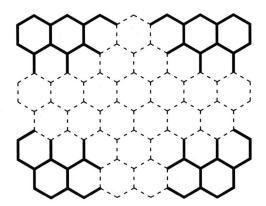

FIGURE 10.5   Graph $J^{n,m}$.

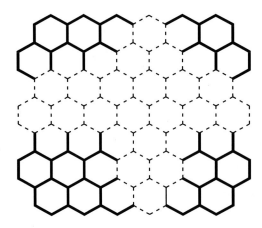

FIGURE 10.6   Graph $K^{n,m}$.

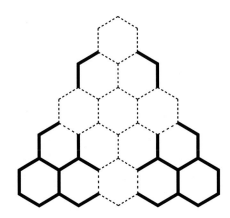

FIGURE 10.7   Graph of $T[n]$.

$$X_2 = (1,3)(4,6)(7,25)(8,24)(9,23)(11,34)(12,33)(13,32)$$
$$(14,27)(15,41)(16,40)(17,39)(18,36)(19,46)(20,45)$$
$$(21,44)(22,43)(28,35)(30,42)(31,38)$$

Then, a simple code by GAP (groups, algorithms, and programming) [10] shows that the $Sym(T[5])$ is generated by $\{X_1, X_2\}$. We now consider the hexagonal system $I^{7,5}$. Consider the following two permutations:

$$Y_1 = (1,29)(2,28)(3,27)(4,24)(5,25)(6,30)(7,23)(8,20)$$
$$(9,21)(10,26)(11,19)(12,16)(13,17)(14,22)(31,43)$$
$$(32,42)(33,41)(34,40)(35,39)(36,38)(44,60)(45,59)$$
$$(46,58)(47,57)(48,56)(49,55)(50,54)(51,53)(61,73)$$
$$(62,72)(63,71)(64,70)(65,69)(66,68)(74,90)(75,89)$$
$$(76,88)(77,87)(78,86)(79,85)(80,84)(81,\ 83)$$
$$Y_2 = (1,75)(2,74)(3,61)(4,62)(5,77)(6,76)(7,63)(8,64)$$
$$(9,79)(10,78)(11,65)(12,66)(13,81)(14,80)(15,67)$$
$$(16,68)(17,83)(18,82)(19,69)(20,70)(21,85)(22,84)$$
$$(23,71)(24,72)(25,87)(26,86)(27,73)(28,90)(29,89)$$
$$(30,88)(31,46)(32,47)(33,48)(34,49)(35,50)(36,51)$$
$$(37,52)(38,53)(39,54)(40,55)(41,56)(42,57)(43,58)$$
$$(44,45)(59,60)$$

Then we can apply again computer algebra system GAP to prove that Sym($I^{7,5}$) can be generated by Y_1 and Y_2. Consider the molecular graph of H_5. To calculate a generating set for the symmetry group of this hexagonal system, we assume that

$Z_1 = (1,2,3,4,5,6)(7,10,13,16,19,22)(8,11,14,17,20,23)$
$(9,12,15,18,21,24)(25,30,35,40,45,50)(26,31,36,41,$
$46,51)(27,32,37,42,47,52)(28,33,38,43,48,53)(29,34,$
$39,44,49,54)(55,91,84,77,70,63)(56,92,85,78,71,64)$
$(57,88,81,74,67,60)(58,93,86,79,72,65)(59,94,87,80,$
$73,66)(61,95,89,82,75,68)(62,96,90,83,76,69)(97,143,$
$134,125,116,107)(98,144,135,126,117,108)(99,140,131,$
$122,113,104)(100,145,136,127,118,109)(101,146,137,$
$128,119,110)(102,147,138,129,120,111)(103,148,139,$
$130,121,112)(105,149,141,132,123,114)(106,150,142,$
$133,124,115)$

$Z_2 = (1,5)(2,4)(7,19)(8,18)(9,17)(10,16)(11,15)(12,14)$
$(20,24)(21,23)(25,27)(28,54)(29,53)(30,52)(31,51)$
$(32,50)(33,49)(34,48)(35,47)(36,46)(37,45)(38,44)$
$(39,43)(40,42)(56,59)(57,58)(60,93)(61,96)(62,95)$
$(63,91)(64,94)(65,88)(66,92)(67,86)(68,90)(69,89)$
$(70,84)(71,87)(72,81)(73,85)(74,79)(75,83)(76,82)$
$(78,80)(97,100)(98,103)(99,102)(104,147)(105,150)$
$(106,149)(107,145)(108,148)(109,143)(110,146)$
$(111,140)(112,144)(113,138)(114,142)(115,141)$
$(116,136)(117,139)(118,134)(119,137)(120,131)$
$(121,135)(122,129)(123,133)(124,132)(125,127)$
$(126,130)$

Then we can see that the symmetry group of H_5 is generated by above permutations. As a final example we consider the molecular graph of $P(5, 5)$. Then the following two permutations generate the symmetry group of this hexagonal system:

$T_1 = (1,6)(2,5)(3,4)(7,18)(8,17)(9,16)(10,15)(11,26)(12,25)$
$(13,24)(14,23)(19,20)(21,28)(22,27)(29,30)(31,50)$
$(32,49)(33,48)(34,47)(35,52)(36,51)(37,54)(38,53)$
$(39,70)(40,69)(41,68)(42,67)(43,66)(44,65)(45,64)$
$(46,63)(55,56)(57,62)(58,61)(59,60)$
$T_2 = (1,60)(2,61)(3,56)(4,55)(5,58)(6,59)(7,38)(8,37)$
$(9,46)(10,57)(11,36)(12,35)(13,44)(14,45)(15,62)$
$(16,63)(17,54)(18,53)(19,30)(20,29)(21,28)(22,27)$
$(23,64)(24,65)(25,52)(26,51)(31,66)(32,67)(33,48)$
$(34,47)(39,68)(40,69)(41,70)(42,49)(43,50)$

10.3 TOPOLOGY

Let G be a connected graph on n vertices and $e = uv \in E(G)$. Define $n_u(e)$ to be the number of vertices closer to u than v and $m_u(e)$ to be the number of edges closer to u than v. The quantities $n_v(e)$ and $m_v(e)$ are defined analogously. Then the Wiener index W(G) [11], the edge Wiener index $W_e(G)$ [12,13], the PI index [14], and the Szeged index Sz(G) [15], are respectively defined as follows:

I. $W(G) = \sum_{\{u,v\} \subseteq V(G)} d(u,v)$,
II. $W_e(G) = \sum_{\{f,g\} \subseteq E(G)} d_e(f,g)$,
III. $PI(G) = |E|^2 - \sum_{e \in E(G)} (m_u(e) + m_v(e))$,
IV. $Sz(G) = \sum_{uv \in E(G)} n_u(e) n_v(e)$.

Define $N(e) = |E| - (m_u(e) + m_v(e))$, where e is an edge of the graph G. Then one can easily see that $PI(G) = |E|^2 - \sum_{e \in E(G)} N(e)$. We start our topology discussion with the hexagonal graph T[n]. Since an arbitrary edge e of the jth row of T(n) has exactly $j + 1$ parallel edges, $\sum_{e \in G} N(e) = 3[2^2 + 3^2 + \ldots + (n + 1)^2] = 1/2[2n^3 + 18n^2 + 13n] = n^3 + 9n^2 + 13/2n$. Therefore,

Theorem 10.6: [16] $PI(T[n]) = |E|^2 - \sum_{e \in E(G)} N(e) = 1/4[9n^4 + 50n^3 + 63n^2 - 26n]$.

To compute the Szeged index of this graph, we first notice that $|V(G)| = 3 + 5 + \ldots + (2n + 1) = n^2 + 4n + 1$. We have

Theorem 10.7: [16] $Sz(T[n]) = 1/4[n^6 + 12n^5 + 49n^4 + 84n^3 + 58n^2 + 12n]$.

Alizadeh and Klavžar [8] introduced an interpolation method to obtain closed formulas for some distance-based topological indices of families of hexagonal graphs. They discuss the method on 2-parametric families of graphs. We mention here one of the main results of the mentioned paper regarding Wiener, Szeged, and edge Wiener index of hexagonal graph H_n. It is advantageous here to mention that the Wiener and Szeged indices of this graph for the first time was reported in [9,16], respectively.

Theorem 10.8: [8] $W(H_n) = 164/5n^5 - 6n^3 + n/5$ and $W_e(H_n) = 369/5n^5 - 123/2n^4 + 15n^3 - 3/2n^2 + 6/5n$.

Since the jth row of the graph H_n has exactly $n + j$ vertical edges, $|V(G)| = 6n^2$ and $|E(H_n)| = 3\{2[(n + 1) + (n + 2) + \ldots + (2n - 1)] + 2n\} = 9n^2 - 3n$. If e is an arbitrary edge of the jth row of this graph, $1 \leq j \leq n - 1$, then e has exactly $n + j$ parallel edges and we have

Theorem 10.9: [14] $PI(H_n) = 81n^4 - 68n^3 + 12n^2 - n$ and $Sz(H_n) = 54n^6 - 3/2n^4 + 3/2n^2$.

We are now ready to consider the topology of hexagonal-parallelogram P(n, k). Alizadeh and Klavžar [8] proved that

Theorem 10.10: The Edge Wiener and Szeged Indices of P(n,k) Can Be Computed by the Following Formulas:

$$W_e(P(n,k)) = \left(3k^2 + 4k + \frac{4}{3}\right)n^3 + \left(\frac{2}{3}k^3 + 9k^2 + 3k - 2\right)n^2$$
$$+ \left(\frac{3}{4}k^4 + 4k^3 + \frac{13}{4}k^2 - \frac{3}{2}k + \frac{2}{3}\right)n$$
$$- \left(\frac{3}{20}k^5 - \frac{5}{4}k^3 + 2k^2 - \frac{9}{10}k\right),$$

$$Sz(P(n,k)) = \left(2k^3 + 6k^2 + \frac{16}{3}k + \frac{4}{3}\right)n^3 + (6k^3 + 12k^2 + 6k)n^2$$
$$+ \left(\frac{17}{3}k^3 + 7k^2 + 2k - \frac{1}{3}\right)n + \left(\frac{1}{6}k^4 - k^3 - \frac{1}{6}k\right).$$

We now calculate the PI index of this graph in the cases that $m = n$. It is clear that $|E(P(n,n))| = 3n^2 + 4n - 1$ and $|V(P(n,n))| = 2n^2 + 4n$. Let A, B, and C denote the set of all vertical, left oblique, and right oblique edges of P(n,n). Then, $\sum_{e \in E(Q_{n,n})} N(e) = \sum_{e \in A} N(e) + \sum_{e \in B} N(e) + \sum_{e \in C} N(e)$ and we have

Theorem 10.11: [16] The PI Index of P(n,k) Is Computed as Follows:

$$PI(P(n,k)) = \begin{cases} 9k^2n^2 + 12k^2n + 11kn^2 - 2kn - 5/3k^3 \\ \quad + 4k^2 - 16/3k + 4n^2 - 5n + 3 \qquad n < k \\ 9k^2n^2 + 10k^2n + 11kn^2 - 4kn + 1/3k^3 \\ \quad + 4k^2 - 19/3k + 4n^2 - 6n + 2 \qquad n \geq k \end{cases}.$$

The Wiener index of this molecular graph is reported by Klavžar et al. in [17].

Theorem 10.12: [17] The Wiener Index of P(n,k) Is Computed by the Following Formula:

$$W(P(n,k)) = \frac{4}{3}(k^2 + 2k + 1)n^3 + \frac{2}{3}(k^3 + 9k^2 + 8k)n^2$$
$$+ \frac{1}{3}(k^4 + 8k^3 + 16k^2 + 2k - 1)n$$
$$- \frac{1}{15}(k^5 - 20k^3 + 4k).$$

In the end of this chapter, the Wiener and PI indices of HJR hexagonal systems are investigated.

Theorem 10.13: [9] The Wiener Index of HJR Can Be Computed as Follows:

$$W(I^{n,m}) = \begin{cases} \frac{1}{15}m(80mn^3 + 80m^2n^2 + 120mn^2 - 20n^2 \\ \quad + 40m^3n + 80m^2n + 30mn - 20n - 8m^4 \\ \quad + 20m^3 + 30m^2 - 20m - 7) \qquad 1 \leq m \leq n \\ \frac{1}{15}(-8n^5 + 40mn^4 - 20n^4 + 80mn^3 \\ \quad - 10n^3 + 16m^3n^2 + 10mn^2 + 20n^2 + 160m^3n \\ \quad - 60mn + 18n + 40m^3 - 25m) \qquad n \leq m \end{cases}$$

$$W(I^{n,m}) = \begin{cases} \frac{2}{15}(40m^2n^3 + 40mn^3 + 10n^3 + 40m^3n^2 \\ \quad + 120m^2n^2 + 20mn^2 - 15n^2 + 20m^4n \\ \quad + 80m^3n + 45m^2n - 15mn + 5n - 4m^5 \\ \quad + 5m^3 - 45m^2 - 31m) \qquad 1 \leq m \leq n \\ \frac{2}{15}(-4n^5 + 20mn^4 + 40mn^3 - 5n^3 + 80m^3n^2 \\ \quad + 120m^2n^2 + 65mn^2 + 8m^3n - 45mn + 9n \\ \quad + 20m^3 - 30m^2 - 35m) \qquad n-1 \leq m \end{cases}$$

$$W(I^{n,m}) = \begin{cases} \frac{1}{15}(80m^2n^3 + 160mn^3 + 80n^3 + 80m^3n^2 \\ \quad + 360m^2n^2 + 220mn^2 - 60n^2 + 40m^4n \\ \quad + 240m^3n + 270m^2n - 40mn + 10n - 8m^5 \\ \quad - 20m^4 - 50m^3 - 280m^2 - 197m - 15) \qquad 1 \leq m \leq n \\ \frac{1}{15}(-8n^5 + 40mn^4 + 20n^4 + 80mn^3 - 10n^3 \\ \quad + 160m^3n^2 + 480m^2n^2 + 490mn^2 + 100n^2 \\ \quad + 160m^3n - 360mn - 132n + 40m^3 - 120m^2 \\ \quad - 55m + 45) \qquad n-1 \leq m \end{cases}$$

Theorem 10.14: [16] The PI Index of HJR Is as Follows:

$$PI(I^{n,m}) = \begin{cases} 36m^2n^2 + 4m^2n - 14mn^2 - 4mn \\ \quad + 8/3m^3 + m^2 - 11/3m + 2n^2 \qquad n \geq m \\ 36m^2n^2 + 12m^2n - 14mn^2 - 4mn \\ \quad - 16/3m^3 + 9m^2 - 11/3m + 2n^2 \qquad n < m \end{cases}$$

$$PI(J^{n,m}) = \begin{cases} 36m^2n^2 + 4m^2n + 22mn^2 - 30mn \\ \quad + 8/3m^3 + 5m^2 - 11/3m + 4n^2 - 10n + 6 \qquad n > m \\ 36m^2n^2 + 12m^2n + 22mn^2 - 22mn \\ \quad - 16/3m^3 - 3m^2 - 17/3m + 4n^2 - 8n + 6 \qquad n \leq m \end{cases}$$

$$\mathrm{PI}(K^{n,m}) = \begin{cases} 36\,m^2 n^2 + 84\,m^2 n - 14\,mn^2 - 92\,mn \\ \quad + 16/3\,m^3 + 49\,m^2 - 271/3\,m + 2\,n^2 & n < m \\ \quad + 16\,n + 36 \\ 36\,m^2 n^2 + 88\,m^2 n - 14\,mn^2 - 92\,mn \\ \quad + 4/3\,m^3 + 53\,m^2 - 271/3\,m + 2\,n^2 & n \geq m \\ \quad + 16\,n + 36 \end{cases}$$

10.4 CONCLUSIONS

In this chapter, the symmetry and topology of some graphene-like structures are described. We focused on symmetry and distance-based topology on some important structures in nanoscience. Only distance-based invariants with close relationship with experiment are taken into account. However, it is possible to describe a molecular structure by a polynomial or matrix as numbers. In [18–20], Ori and his coworkers considered some different problems related to the topology of graphene-like structures by polynomial approaches. Then they computed the limit of related counting polynomial when the number of carbon atoms tends to infinity. By this method, it is possible to discover symmetries for these nanostructures in large systems.

REFERENCES

1. A. K. Geim and K. S. Novoselov, The rise of graphene, *Nat. Mater.* **6**, 2007, 183–191.
2. I. V. Fialkovsky and D. V. Vassilevich, Quantum field theory in graphene, *Int. J. Mod. Phys.* A **27**, 2012, 1260007, 12 pp.
3. H. Yousefi-Azari, A. R. Ashrafi and N. Sedigh, On the Szeged index of some benzenoid graphs applicable in nanostructures, *Ars Combin.* **90**, 2009, 55–64.
4. H. Yousefi-Azari, B. Manoochehrian and A. R. Ashrafi, PI and Szeged indices of some benzenoid graphs related to nanostructures, *Ars Combin.* **84**, 2007, 255–267.
5. F. Harary, *Graph Theory*, Addison-Wesley Publishing Co., Reading, MA, 1969.
6. H. E. Rose, *A Course on Finite Groups*, Universitext, Springer-Verlag, London, 2009.
7. J. R. Dias, Benzenoid series having a constant number of isomers, *J. Chem. Inf. Comput. Sci.* **30**, 1990, 61–64.
8. Y. Alizadeh and S. Klavžar, Interpolation method and topological indices: 2-parametric families of graphs, *MATCH Commun. Math. Comput. Chem.* **69**, 2013, 523–534.
9. W. C. Shiu, C. S. Tong and P. C. B. Lam, Wiener number of hexagonal jagged-rectangles, *Discrete Appl. Math.* **122**, 2002, 251–261.
10. The GAP Team, *GAP, Groups, Algorithms and Programming*, Lehrstuhl De für Mathematik, RWTH, Aachen, 1995.
11. H. Wiener, Structural determination of the paraffin boiling points, *J. Am. Chem. Soc.* **69**, 1947, 17–20.
12. M. H. Khalifeh, H. Yousefi-Azari, A. R. Ashrafi and S. G. Wagner, Some new results on distance-based graph invariants, *Eur. J. Combin.* **30**, 2009, 1149–1163.
13. H. Yousefi-Azari, M. H. Khalifeh and A. R. Ashrafi, Calculating the edge Wiener and Szeged indices of graphs, *J. Comput. Appl. Math.* **235**, 2011, 4866–4870.
14. P. V. Khadikar, On a novel structural descriptor PI, *Natl. Acad. Sci. Lett.* **23**, 2000, 113–118.
15. I. Gutman, A formula for the Wiener number of trees and its extension to graphs containing cycles, *Graph Theory Notes N. Y.* **27**, 1994, 9–15.
16. B. Manoochehrian, H. Yousefi-Azari and A. R. Ashrafi, PI polynomial of some benzenoid graphs, *MATCH Commun. Math. Comput. Chem.* **57**, 2007, 653–664.
17. S. Klavžar, I. Gutman and A. Rajapakse, Wiener numbers of pericondensed benzenoid hydrocarbons, *Croat. Chem. Acta* **70**, 1997, 979–999.
18. J. Sedlar, D. Vukičević, F. Cataldo, O. Ori and A. Graovac, Compression ratio of Wiener index in 2-d rectangular and polygonal lattices, *Ars Math. Contemp.* **7**, 2014, 1–12.
19. M. V. Putz and O. Ori, Bondonic characterization of extended nanosystems: application to graphene's nanoribbons, *Chem. Phys. Lett.* **548**, 2012, 95–100.
20. F. Cataldo, O. Ori and S. Iglesias-Groth, Topological lattice descriptors of graphene sheets with fullerene–like nanostructures, *Mol. Simul.* **36**, 2010, 341–353.

Section II

Modified Graphene

11 N-Doped Graphene for Supercapacitors

Dingsheng Yuan and Worong Lin

CONTENTS

Abstract 167
11.1 Introduction 167
11.2 Nitrogen-Doped Species and Contribution of Active Sites in Graphene 167
11.3 Doping Methods of Nitrogen into Graphene 170
    11.3.1 Nitrogen Doped into Graphene by CVD 170
    11.3.2 Arc Discharge 170
    11.3.3 Hydrothermal and Solvothermal Reaction 172
    11.3.4 Other Deduced Methods 173
11.4 N-Doped Graphene for Energy Storage in Ultracapacitors 174
11.5 Summary and Perspectives 177
Acknowledgments 177
References 177

ABSTRACT

Nitrogen-doped carbon nanomaterials have attracted increasing research interest in energy storage and conversion due to of their electrocatalysis and pseudocapacitance characteristics, which are the potential excellent candidates as the electrode materials in lithium–O_2 and zinc–O_2 batteries, ultracapacitors, and fuel cells. Among these, N-doped graphene is the most considerable carbon material. In this chapter, the nitrogen contents, species, synthesis methods, and application on nitrogen-doped graphene are summarized. The practical applications of NG were mainly focused on the ultracapacitors for energy storage. Nitrogen atoms have been doped in pristine graphene, resulting in electron structure change and forming redox active sites. Therefore, electrochemical performance of NG has been visibly improved for potential applications as high-energy density, ultrahigh-power density, and long cycling stability capacitors in the micro-device, vehicle, lift, and the other devices at high rates.

11.1 INTRODUCTION

Carbon is a fascinating element in the periodic table, which is composed of a few allotropes. To date, graphene is the "youngest" allotrope in the carbon family. Simultaneously, graphene is also the basic building block of other important carbon allotropes. Since Geim group isolated "free" and "perfect" graphene sheets and demonstrated the unprecedented electronic properties of graphene in 2004, considerable attention has been given to the study on graphene owing to its high mobility, Hall Effect, transparency, mechanical strength, and thermal conductivity [1,2].

In recent years, nitrogen atoms introduced into the honeycomb lattice of graphene have gained increasing research interest. A nitrogen atom has one extra electron in comparison

with a carbon atom. For the N-doped graphene (NG), the extra electron energy level lies around the nitrogen atom and gives a rise to a donor state near the Fermi level [3]. N-configuring graphene is attributed to n-type doping, in which the electron structure was changed, resulting in N-graphene different from the pristine graphene. The content and doping species (graphitic-N or quarternary-N (N-Q), pyrrolic-N (N-5), pyridinic-N (N-6), cyanide (-CN), and pyridinic-N-oxide) of nitrogen and the synthesis methods of NG are summarized and listed in Table 11.1. Due to the special characteristics of NG, many researchers and scientists are focusing their attention on N-doped graphene involving catalyst carriers, metal-free electrocatalysts for oxygen reduction reaction (ORR), ultracapacitors, biosensors, and electronic devices. In this chapter, the NG was focused on the ultracapacitors (UCs).

11.2 NITROGEN-DOPED SPECIES AND CONTRIBUTION OF ACTIVE SITES IN GRAPHENE

The species of nitrogen in NGs contain graphitic-N or quarternary-N, pyrrolic-N, pyridinic-N, cyanide (-CN), and pyridinic-N-oxide, as shown in Figure 11.1. However, a specific NG includes two or three, even four kinds of N-doping species, depending on N-containing carbon precursors and synthesis methods of nitrogen doped [10,24]. Nitrogen as rich electron element has been doped in the honeycomb lattice of graphene to take the place of the carbon element that corresponds to nitrogen atoms bonded to three carbon atoms, named as N-Q, changing the electron state in network and improving the conductivity. Therefore, the graphene with N atom exhibits metallic properties helpful with the electron transport [25]. Generally, N-5 and N-6 lie at the edges of graphene, where nitrogen atom is bonded to two carbon

TABLE 11.1

N Content of Nitrogen-Doped Graphene by Different Methods under Different Dopants

No.	Synthesis Method	Precursors	Types of N (N content, at.%)	Reference
1	CVD	CH_4, H_2, and NH_3	N-G, N-6, and -CN (0.44%)	[4]
2	CVD	Ar/H_2, hexane, and acetonitrile	N-6, N-5, and N-G (9%)	[5]
3	CVD	Ar/H_2, pyridine vapor, and methane	N-6 and N-G (2.4%)	[6]
4	CVD	H_2, Ar, and 1,3,5-triazine	N-6, N-5, and N-G (5.6%)	[7]
5	CVD	CH_4 and NH_3	N-6 and N-G (0.25%)	[8]
6	CVD	CH_4 and NH_3	N-6, N-5, and N-G (8.9%)	[9]
7	PECVD	GO and nitrogen plasma	N-6, N-5, and N-G (1.68%–2.51%)	[10]
8	Arc-discharge	Graphite electrodes in the presence of He and pyridine	N-6 and N-G (1.4%)	[11]
9	Arc-discharge	Graphite rod in He and NH_3	N-6 and N-G (1%)	[12]
10	Arc-discharge	Graphite evaporation in H_2 and pyridine	(0.6–1.0 wt%)	[13]
11	Thermal annealing	Graphite oxide with melamine	N-6, N-5, and N-G (10.1%)	[14]
12	Thermal annealing	Graphene oxide, H_2, and N_2H_4	N-6, N-5, N-G, and –CN (3 wt%)	[15]
13	Thermal annealing	GO and ammonium	N-6 and N-5	[16]
14	Thermal annealing	Graphite oxide with NH_3	N-6, N-5, and N-G (3%–5%)	[17]
15	Hydrothermal	Graphene oxide, NH_3, and hydrazine hydrate	N-6, N-5, and pyridinic-N oxide (7%)	[18]
16	Hydrothermal	GO solution, N_2H_4, and ammonia	N-6, N-5, N-G, -CN, and pyridine-N oxides (7.4%)	[19]
17	Solvothermal	Cyanuric chloride, lithium nitride, and tetrachloromethane	N-6, N-5, and N-G (16.4%)	[20]
18	Nitrogen plasma treatment	Graphene and nitrogen atmosphere	N-6, N-5, and N-G (1.35%)	[21]
19	Pyrolysis of graphene nanoribbons	Pyrolyzing GNR/PANI	N-6, N-5, N-G, and pyridine-N oxides (8.3 wt%)	[22]
20	Microwave Irradiation	GO with $NaNH_2$	N-6, N-5, N-G, and –CN (4%–8%)	[23]

atoms, creating cavities and defects. The nitrogen atom of N-6 contributes one p-electron to the aromatic π-system and has a lone electron pair in the plane of the ring, whereas as for pyrrolic-N, the nitrogen atom contributes two p-electrons to the π-system, and a hydrogen atom is bound in the plane of the ring [26,27]. Due to high electron density, N-5 and N-6 act as the acceptor and can accept the donors, thereby causing the redox reactions for contributing to pseudocapacitance [10,24].

The species of nitrogen in graphene are usually detected by x-ray photoelectronic spectroscopy (XPS). The binding energy of N-6, N-5, N-Q, and pyridinic-N-oxide are centered at 398, 400, 401, 402–405 eV, respectively. However, Jeong et al. reported on a plasma-enhanced chemical vapor deposition (PECVD) nitrogen-doping process. Nitrogen atoms were expected to successfully replace carbon atoms in the original graphene sheets and form three species of N-6, N-5, and N-Q verified by scanning photoemission microscopy (SPEM) [10].

Figure 11.2 shows their N-configuration mappings within single sheets of NG using SPEM. They demonstrated that this technique was able to detect N-configurations at various spots within a single sheet of NG so that N-configurations at basal planes and edges were clearly distinguishable. From their distribution data in Figure 11.2, N-configurations between basal planes and edges can be addressed. N-doping took place even at the basal plane as shown in Figure 11.2g. This observation suggests that defects at basal planes generated by the plasma process can initiate N-doping at basal planes. Their observation was different from the previous predictions of Dai's groups, related to thermal doping processes that N-doping would take place only at edges due to their higher reactivities [17,28]. As the plasma process progresses further, at basal planes the N-5 portion increased but the N-Q portion decreased: this trend might be due to increased basal plane defects that did not allow the N-Q formation but the N-5 formation. For the same reason, associated with the necessity of neighboring defects for the N-5 formation, the N-5 portion appeared higher at edges than at basal planes (see Figure 11.2g through i). At all the plasma durations, there were some spot-to-spot deviations in N-configuration distributions. Simultaneously, for comparison, the authors also provided N-configuration data, which were obtained for bulk-scale

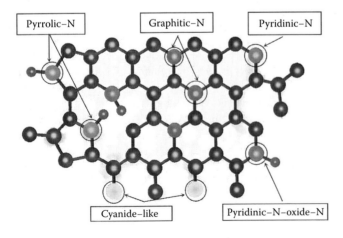

FIGURE 11.1 Nitrogen types in the NG.

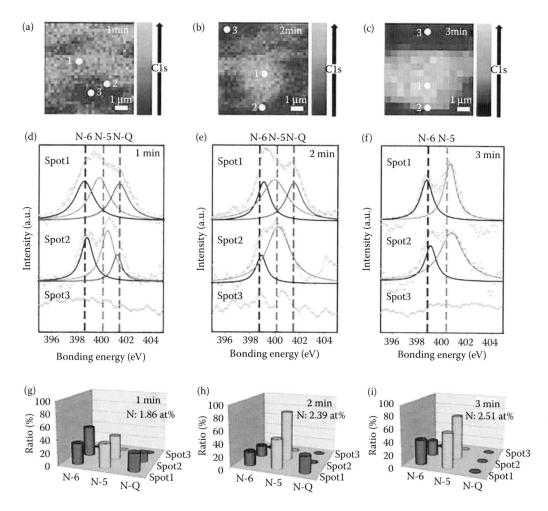

FIGURE 11.2 Nitrogen configuration mappings within single sheets of NG using SPEM. (a–c) C 1s mappings for the 1, 2, and 3 min treated NG obtained by SPEM. The spots where N 1s data were taken are denoted. The size of the white dot at each spot was reflective of the actual SPEM beam size. (d–f) The N 1s data taken from the denoted spots in panels a–c. Each N 1s peak was deconvoluted into three subpeaks corresponding to the N-Q (black), N-5 (light gray), and N-6 (gray) configurations. The amplitudes of the signals taken at all spot 3 were within a noise level. (g–i) The distributions of the three N-configurations measured at the spots denoted in panels a–c. (Reprinted with permission from Jeong, H. M. et al. Nitrogen-doped graphene for high-performance ultracapacitors and the importance of nitrogen-doped sites at basal planes. *Nano Letters*. 11:2472–7. Copyright 2011 American Chemical Society.)

samples using typical XPS measurements. These bulk-scale data turned out to be consistent with the local mapping data in that throughout the plasma process the N-5 portion continuously increased, but the N-Q portion continuously decreased, and the N-6 portion remained almost steady [10].

Raman spectrum is used to analyze the structure change of N-doped graphene, which is an effective method to detect the defects of NG [11,29,30]. It is important to know how it correlates with the concentration of carriers or dopants. The shift in the G-band frequency measured by Raman spectroscopy has many physical contributions. Rao groups [11] considered calculations combined with experiments to uncover their relative magnitudes. They demonstrated that the shifts in vibrational frequencies of graphene with N-doping indeed had opposite signs if one took into account only the changes in bond length obtained at a fixed lattice constant. However, the relaxation or the change of lattice constant is highly asymmetric: the lattice constant decreased slightly with 2% N-substitution

in single layer graphene, resulting in a slight increase in frequency, respectively. The calculated results were contrary to their measured trends. When they adopted the dynamic corrections followed as References 31 and 32, the frequency shifts became positive, in good agreement with their experimental observations reported. Similar changes in the G-bands of bilayer graphene were estimated from their calculations with N-substitution, assuming the same magnitude of dynamic corrections as in the monolayer. Rao et al. concluded that N-graphene can be synthesized to exhibit n-type semiconducting electronic properties that can be systematically tuned with the concentration of N, and that they can be characterized with Raman spectroscopy. Interestingly, elemental doping and electrochemical doping produce similar shifts in the Raman G-band, but molecular charge-transfer gives rise to different effects. Realization of such n-type conducting graphene bilayers should be usable in a variety of devices similar to those based on semiconductors [11]. In addition, graphene-doped nitrogen will cause the defects to increase and the intensity of D band is increased comparable with graphene.

11.3 DOPING METHODS OF NITROGEN INTO GRAPHENE

The different contents and doping species of nitrogen were doped into honeycomb lattice of graphene via chemical vapor deposition (CVD) [4–10,33], arc discharge [11–13], nitrogen plasma process [21], electrothermal reaction [28] or thermal annealing graphite oxide (GO) [14–17], hydrothermal and solvothermal reaction [18–20,34] and denotation process [35], and so on. In general, graphene directly grown or graphite oxide (GO) was employed as the host material and some small molecules such as N_2, NH_3, N_2H_4, acetonitrile, urea, melamine ($C_3H_6N_6$) [33], and dicyandiamide (N_4H_4) [36], were used as the doping reagents. The different methods present the different advantages and disadvantages in the preparation of NG. The main methods are summarized as follows.

11.3.1 Nitrogen Doped into Graphene by CVD

The chemical vapor deposition is popularly employed to prepare graphene and N-doped graphene. Substitutional doping with heteroatoms via CVD provides an effective route for simple and stable tuning of doping levels in graphene. CVD are reliable and straightforward methods to fabricate single- and few-layers NG. However, these processes suffered from rigorous conditions and complicated equipment, accompanied by low yields and high cost. In addition, the nitrogen precursors of pyridine and the NH_3 are also adverse to the environment and subsequent treatments [34]. CVD is usually limited to the use of gaseous raw materials, making it difficult to apply the technology to a wider variety of potential feedstocks. N-doped graphene is obtained by being introduced to a doping gas (NH_3) into the CVD systems during graphene growth [9] or through the treatment of synthesized graphene oxide (GO) with N_2 through plasma-enhanced CVD [10].

To realize graphene-based electronics, Liu and co-workers prepared NG via a CVD process by using a 25 nm thick of Cu film on a Si substrate as the catalyst, where CH_4 and NH_3 were introduced into the flow as the C source and N source at 800°C, respectively. This was the first experimental example of the substitutionally doped graphene, which was hard to be produced by other methods. The CVD method was a nondestructive route to produce graphene and realize the substitutional doping, as the doping accompanies with the recombination of the carbon atoms into the graphene in the CVD process. The electric measurement presented NGs like an n-type semiconductor, indicating that the doping can modulate the electrical properties of graphene. This research provided a new type of graphene experimentally, which was required for the applications of graphene [9].

Recently, large-area mosaic N-doped graphene [37] was successfully grown on annealed copper foil loaded inside a homemade low-pressure CVD system using methane and acetonitrile, as shown in Figure 11.3. Yan et al. employed a well-controlled CVD process for direct growth of mosaic NG, wherein acetonitrile vapor was introduced as a nitrogen-containing carbon precursor. Mosaic graphene was produced in large-area monolayers with spatially modulated, stable, and uniform doping, and showed considerably high room temperature carrier mobility of 2500 cm²/(V s) in nitrogen-doped portion. The unchanged crystalline registry during modulation doping indicated the single-crystalline nature of p–n junctions. Efficient hot carrier-assisted photocurrent was generated by laser excitation at the junction under ambient conditions. They indicated that this was a facile avenue for large-scale synthesis of single-crystalline graphene p–n junctions, allowing for batch fabrication and integration of high-efficiency optoelectronic and electronic devices within the atomically thin film.

Interestedly, Sun et al. developed a solid carbon sources (poly methyl methacrylate, PMMA) and solid doping reagents (melamine, $C_3H_6N_6$) to prepare N-doped graphene via one step without any changes to the CVD system [33]. $C_3H_6N_6$ was mixed with PMMA and spin-coated onto the Cu surface. In order to keep the nitrogen-atom concentration in the systems, they had used conditions similar to those employed for the growth of PMMA-derived graphene at 800°C for 10 min, except that the growth was done at atmospheric pressure. N-doped graphene had an N content of 2%–3.5%. The D peak of this material was always presented in the Raman spectra, because the heteroatoms break the graphene symmetry and thereby introduce the defects that were detected by Raman analysis. XPS analysis indicated that only one type of N was presented at 399.8 eV, corresponding to N-Q in graphene.

Kang's group presented that the dried GO was reduced by a plasma-enhanced CVD process [10]. In this process, the GO was first reduced by a hydrogen plasma process. Subsequently, after the reduction step, nitrogen plasma was introduced onto the graphene surfaces to obtain NG, as shown in Figure 11.4. Three species of N-6, N-5, and N-Q had been successfully doped into the original graphene sheets confirmed by SPEM and XPS. The N contents of 1.86%, 2.39%, and 2.51% were doped into graphene, corresponding to the different plasma treatment time of 1, 2, and 3 min, respectively.

11.3.2 Arc Discharge

Since Krastchmer and Hoffman had first used arc-discharge method to prepare C60 in 1990, it was widely used to prepare carbon nanomaterials such as fullerene, carbon nanotubes, and graphene. The electric arc oven for synthesis of graphene mainly comprises two electrodes and a steel chamber cooled by water. The cathode and anode are both pure graphite rods. The current in the discharge process is maintained at 100–150 A. Up to now, the atmospheres for arc evaporation of graphite rods are H_2, NH_3, and He, air. As the rods are brought close together, discharge occurs resulting in the formation of plasma. As the anode is consumed, the rods are kept at a constant distant from each other of about 1–2 mm by rotating the cathode. When the discharge ends, the soot generated is

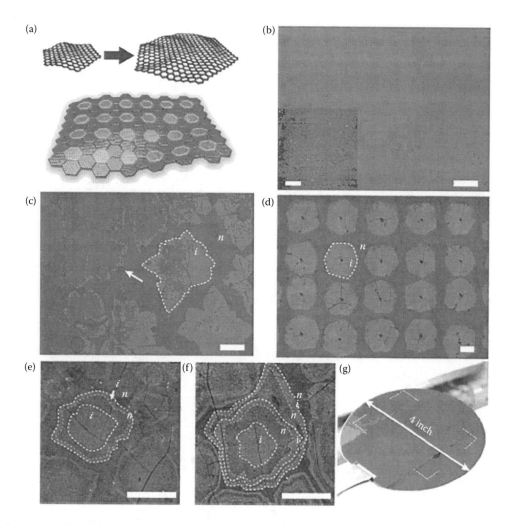

FIGURE 11.3 Synthesis and morphology of modulation-doped graphene. (a) Top: schematic drawing of modulation doping growth. The nucleation and expanding stages were grown with alternate forming gases. Bottom: schematic diagram of mosaic graphene, where pixels with different colors represent graphene with different doping levels. (b) OM image of mosaic graphene transferred onto 300 nm SiO_2/Si substrate. Scale bar, 10 mm. Inset: AFM image of mosaic graphene. Scale bar, 5 mm. (c, d) Typical SEM images of random and ordered mosaic graphene, respectively. Intrinsic and N-doped portions can be identified by the contrast. Scale bars, 2 mm. (e, f) Typical SEM images with two and three modulation cycles, respectively. These high-order structures behave as irregular homocentric disks with radial doping modulations. Scale bars, 2 mm. (g) Large-area mosaic graphene transferred onto a 4-in. silicon wafer. (Reprinted by permission from Macmillan Publishers Ltd. *Nat Commun.*, Yan, K. et al. Modulation-doped growth of mosaic graphene with single-crystalline p–n junctions for efficient photocurrent generation. 3:1280, copyright 2012.)

collected under ambient conditions. Only the soot deposited on the inner wall of the chamber is collected, avoiding the substance at the bottom of the chamber, for the latter tends to contain other graphitic particles [38]. The kinds of method for the carbon nanomaterial also suffer energy consumption and high cost.

Nitrogen-doped graphene had been prepared by carrying out arc discharge in the mixing atmosphere of He and NH_3 [12], H_2 and pyridine, H_2 and ammonia, or transformation of nanodiamond in the presence of pyridine [11,13,39]. Li et al. synthesized the graphene sheets with mainly 2–6 layers and the sizes of 100–200 nm. The multilayered graphene sheets can be purified by a simple heat treatment process. The content of N atoms on the multilayered graphene sheets can be tuned by simply changing the proportion of NH_3 in the atmosphere. The high-resolution XPS spectra of

N1s at 399.59 eV is pyrrolic-like N-doping, rather than pyridine-type described by Reference 12. The element analysis showed that the content of N was about ~1% for the sample produced in the atmosphere of pure NH_3. Rao's groups systematically reported on a series of synthesis and characterization of N-doped graphene by arc-discharge method [11,13,39]. They concluded that NG can be synthesized to exhibit n-type semiconducting electronic properties that can be systematically tuned with the concentration of N, and that they can be characterized with Raman spectroscopy. The elemental doping and electrochemical doping produce similar shifts in the Raman G-band, but molecular charge-transfer gives rise to different effects (see Raman analysis in Section 11.2). Realization of such n-type conducting graphene bilayers should be usable in a variety of devices similar to those based on semiconductors.

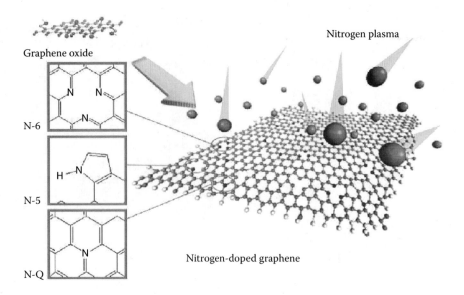

FIGURE 11.4 Nitrogen-doped graphene by the plasma treatment. A schematic illustration of the plasma doping process. By the plasma process with physical momentum, nitrogen atoms replace the existing carbon atoms. (Inset) Possible nitrogen configurations by the doping treatment. (Reprinted with permission from Jeong, H. M. et al. Nitrogen-doped graphene for high-performance ultracapacitors and the importance of nitrogen-doped sites at basal planes. *Nano Letters.* 11:2472–7. Copyright 2011 American Chemical Society.)

11.3.3 HYDROTHERMAL AND SOLVOTHERMAL REACTION

Solvothermal processes are known for their simple operation, mild synthesis conditions, and capability to deliver relatively large quantities [20]. Deng et al. applied a solvothermal process for synthesis of N-doped graphene based on the reaction of tetrachloromethane with lithium nitride below 350°C for the first time. The schematic illustration mechanism for solvothermal synthesis of NG is presented in Figure 11.5. They believed that this facile method will enable production of N-doped graphene at a larger scale because the yield is only limited by the capacity of the autoclave. N-5, N-6, and N-Q had been successfully doped. Most importantly, the high N/C ratio (16.4%) was obtained with the additional cyanuric chloride in the above-mentioned experimental process.

The hydrothermal process as an environmental-friendly and low-cost method had been considered. Fu's groups prepared nitrogen-doped graphene nanosheets with nitrogen level as high as 10.13 at.% via a hydrothermal reaction of graphene oxide (GO) and urea (see Figure 11.6) [34]. The nitrogen level and species could be conveniently controlled by easily tuning the experimental parameters, including the mass ratio between urea and GO and the hydrothermal temperature. In the fabrication, the nitrogen-enriched urea played a pivotal role in forming NG with a high nitrogen level. N-doping and reduction of GO were achieved simultaneously under hydrothermal reaction. The N-dopant of urea could release NH_3 in a sustained manner, accompanied by the released NH_3 reacting with the oxygen functional groups of the GO and then the nitrogen atoms doped into graphene skeleton, leading to the formation of NG. Three N-type of N-5, N-6, and N-Q were doped into carbon honeycomb lattice.

In addition, NG had been achieved to reduce GO by using N_2H_4 and ammonia as reducing reagents via a hydrothermal

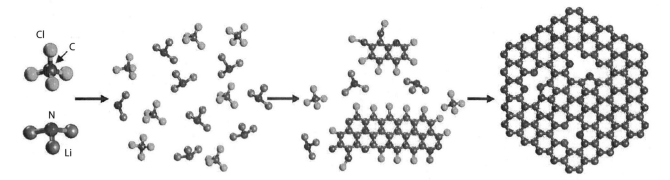

FIGURE 11.5 Scheme of a proposed mechanism for solvothermal synthesis of N-doped graphene via the reaction of CCl_4 and Li_3N, where black balls represent C atoms, dark for N, light gray for Cl, and dark gray for Li. (Reprinted with permission from Deng, D. H. et al. Toward N-doped graphene via solvothermal synthesis. *Chemistry of Materials.* 23:1188–93. Copyright 2011 American Chemical Society.)

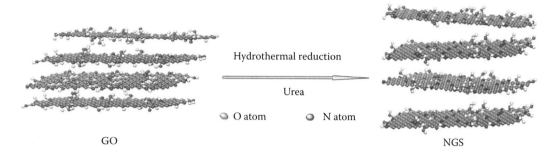

FIGURE 11.6 Schematic procedures for preparation of NG. (Sun, L. et al. 2012. Nitrogen-doped graphene with high nitrogen level via a one-step hydrothermal reaction of graphene oxide with urea for superior capacitive energy storage. *Rsc Advances*. 2:4498–506. Reproduced by permission of The Royal Society of Chemistry.)

approach [19]. Yoon and co-workers discussed that the structure and surface chemistry of the resulting graphene sheets were strongly dependent on the hydrothermal temperature. Nitrogen content was up to 5 wt% in the NG. Four N-doped species of N-5, N-6, N-Q, and pyridine-N oxide were detected by XPS spectra. The pyridine structure, located at the 398.8 eV, showed an increasing tendency with the increase of hydrothermal temperature, while the N contents show inconspicuous changes for the other N-doped species. However, some researchers considered that N_2H_4 cannot be stored due to its toxicity and risk of explosion. To the best of our knowledge, the autoclave was used in the hydrothermal and solvothermal reaction so that the autoclave could be exploded because of high pressure. We consider that the large-scale carbon nanomaterials obtained by these methods are fancied if the larger autoclaves were used for scalable synthesis. Additionally, if the organic reagents were employed as the precursors to synthesize the carbon nanomaterials, the as-synthesized carbon materials with various morphologies are co-existed, resulting in purification and separation difficultly.

11.3.4 OTHER DEDUCED METHODS

Some similar or deduced methods from CVD and PECVD and hydrothermal or solvothermal reactions have also been reported, for example, electrothermal reaction [28] or thermal annealing graphite oxide (GO) [17], nitrogen plasma process [21], denotation process [35], and supercritical reaction [40].

Dai's groups developed electrothermal reaction for preparing NG. Individual graphene nanoribbons (GNRs) were covalently functionalized by nitrogen species through high-power electrical joule heating in ammonia gas, leading to n-type electronic doping consistent with theory. The formation of the carbon–nitrogen bond should occur mostly at the edges of graphene where chemical reactivity was high. NG had been successfully applied to fabricate an n-type graphene field-effect transistor that operates at room temperature [28]. They also used a chemical method to obtain bulk quantities of N-doped reduced GO sheets through thermal annealing of GO in ammonia [17]. The conclusions covered annealing temperatures that affect N-doping level. Nitrogen doped in GO occurred at a temperature as low as 300°C, while the highest

doping level of ~5% N was achieved at 500°C. N-doping was accompanied by the reduction of GO with decreases in oxygen levels from ~28% in as-made GO down to ~2% in 1100°C NH_3 reacted GO. N-6 and N-Q in N-graphene were detected by XPS, where the N binding configurations of doped GO found pyridinic N in the doped samples and N-Q increased in GO annealed with the increase of temperatures. Oxygen groups in GO were found responsible for reactions with NH_3 and C–N bond formation. Prereduced GO with fewer oxygen groups by thermal annealing in H_2 exhibited greatly reduced reactivity with NH_3 and a lower N-doping level. Electrical measurements of individual GO sheet devices demonstrated that GO annealed in NH_3 exhibited higher conductivity than those annealed in H_2, suggesting more effective reduction of GO by annealing in NH_3 than in H_2. The N-doped reduced GO showed clearly n-type electron doping behavior with the Dirac point (DP) at negative gate voltages in three terminal devices [17].

Wang et al. used nitrogen plasma treatment of graphene to synthesize successfully N-doped graphene [21]. To prepare N-doped graphene, graphene was fixed on the glassy carbon electrode, followed by being placed in the plasma chamber, which was back filled with a nitrogen atmosphere. By controlling the exposure time, the N percentage in host graphene can be regulated, ranging from 0.11% to 1.35%. Interestingly, N-doped graphene electrodes were stored at room temperature and used directly for the electrochemical glucose biosensing.

A denotation process was explored to synthesize NG at low temperature using a denotation process with cyanuric chloride and trinitrophenol as reagents [35]. A gram-scale NG was easily obtained in the laboratory. The authors sound that this method allowed scalable synthesis, standing as a complementary route for large-scale growth of NG. The existence of nitrogen substitution with an N/C atomic ratio of 12.5% in the graphene sheets, which was greater than those reported in the literature, was confirmed by XPS analysis where N-6, N-5, and N-Q were verified into carbon lattice, as shown in Figure 11.7. NG was demonstrated to use as a metal-free electrode with excellent electrocatalytic activity and long-term operation stability for ORR.

Additionally, a supercritical reaction was introduced to obtain nitrogen-doped few-layer graphene sheets in

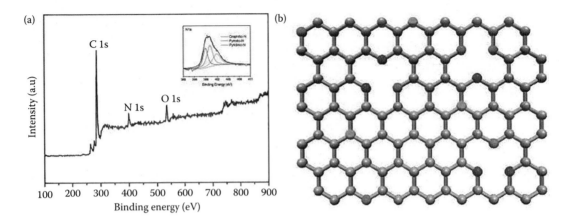

FIGURE 11.7 XPS survey of the synthesized NG. Inset shows the high-resolution N 1s spectrum (a). Schematic representation of the N-doped graphene. The light gray, gray, black, and dark gray spheres represent the C, "graphitic" N, "pyridinic" N, and "pyrrolic" N atoms in the NG, respectively (b). (Reprinted with permission from Feng, L. Y., Chen, Y. G., Chen, L. Easy-to-operate and low-temperature synthesis of gram-scale nitrogen-doped graphene and its application as cathode catalyst in microbial fuel cells. *Acs Nano.* 5:9611–8. Copyright 2011 American Chemical Society.)

acetonitrile at low temperature of 310°C. Tuning reaction time from 2 to 24 h, the N-doping level of NG increased from 1.57% to 4.56 at.%. NG contains N-6, N-5, and graphitic-N species [40].

11.4 N-DOPED GRAPHENE FOR ENERGY STORAGE IN ULTRACAPACITORS

Ultracapacitors (UCs) or supercapacitors store and release electrical energy based on the electrostatic interactions between ions in the electrolyte and electrodes, and this interaction occurs in the so-called electrical double layers (EDLs) near the electrode surfaces [10]. Among carbon-based materials, graphene, porous carbon materials, and carbon nanotubes have mainly attracted considerable interests as electrode materials for electrochemical energy storage because of their high surface area, electrical conductivity, chemical stability, and cycling stability. In order to further improve the power and energy densities of the capacitors, carbon-based composites combining electrical double layer capacitors (EDLC)-capacitance and pseudo-capacitance have been developed [41]. Based on the previous reports [42,43], nitrogen-enriched carbon materials used as supercapacitor electrodes is mainly attributed to nitrogen groups that can enhance the capacitance by additional Faradaic redox reactions and improve the wettability of carbon surface. The mechanism of redox reaction in alkaline electrolyte [44] was proposed as the following Equation 11.1,

$$-C=NH + H_2O + e^- \rightarrow C - NH_2 + OH^- \qquad (11.1)$$

NG is one of the most attractive carbon materials for UCs. Nitrogen introduced into the honeycomb-like carbon network changes the graphene structure and electronic state. Therefore, the N-content and N-doping species and active sites will influence the electrochemical energy storage performance of NG. In more detail, the pyridinic-N and pyrrolic-N play important

roles for improving pseudo-capacitance by the redox reaction, where this is due to the fact that the nitrogen atoms provide a pair of electrons and change the electron donor/acceptor characteristics of carbon materials.

Jeong and co-authors reported a wonderful research on NG for UCs. They studied on the N-content and N-doping species and their distribution in detail and also on how to affect the electrochemical performance [10]. NG can be designed in the form of a binder-free ink that can be applied onto various flexible substrates such as paper and textiles, as shown in Figure 11.8 [10]. The devices were built on conductive paper exhibit with capabilities of maintaining capacitances at large current densities to those of UCs based on nickel substrates (Figure 11.8d). After 10,000 cycles, UCs based on nickel and paper substrates showed 95.3% and 99.8% of initial capacitances, respectively (Figure 11.8e). The capacitance retention still reached 70% after 230,000 cycles measured over a period of ~3 months. The excellent cycle life confirmed again that the charging and discharging process based on the electrostatic interactions were very robust over a large number of cycles and NG UCs should be able to function as long-term energy storage devices. When operated in 1 mol/L TEABF₄, a power density up to ~8 × 10⁵ W/kg and an energy density up to ~48 W h/kg were achieved. Especially, NG UCs exhibited superior power densities, being ascribed to the significant N-Q portion that had been known for the conductivity enhancement.

NG ultracapacitors' capacitances (~280 F/g electrode) were about four times larger than those of pristine graphene. The authors considered that the improved capacitances of the NG UCs were attributed to the N-configuration distributions (see SPEM analysis in Figure 11.2) and their binding energies with ions in the electrolyte. The binding energy enhancement (in negative scale) would contribute to the capacitance increase because larger binding energy results in a larger number of ions to be accommodated on the electrode surface even for the given surface area of electrode. The theoretic calculation also illustrated that N-6 at basal planes played a major

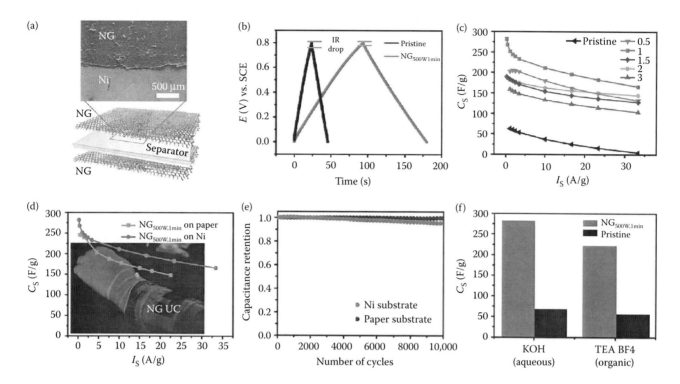

FIGURE 11.8 Ultracapacitors based on NG and their electrochemical testing. (a) A schematic illustration of the assembled UC structure alongside a scanning electron microscopy image showing a top view of the device. (b) Charging and discharging curves measured by galvanostatic characterization. Black and gray lines correspond to the pristine graphene and the $NG_{500W, 1min}$, respectively. The IR drops at the top cut off potentials are also denoted. (c) Gravimetric capacitances of UCs based on various NG and pristine graphene measured at a series of current densities. The numbers in the legend indicate the plasma durations in minutes. (d) Gravimetric capacitances of UCs built on nickel and paper substrates measured at a series of current densities. (Inset) A photograph showing that a wearable UC wrapped around a human arm can store the electrical energy to light up a LED. (e) The cycling tests for the UCs based on Ni and paper substrates up to 10,000 cycles. (f) The specific capacitances measured in aqueous and organic electrolytes. For both electrolytes, the specific capacitances of NG and pristine graphene are compared. The data in (b–e) were measured with the 6 M KOH electrolyte. All of the data presented in this figure are based on the NG UCs with mass loadings of ~1 mg/cm². (Reprinted with permission from Jeong, H. M. et al. Nitrogen-doped graphene for high-performance ultracapacitors and the importance of nitrogendoped sites at basal planes. *Nano Letters.* 11:2472–7. Copyright 2011 American Chemical Society.)

role in the improved capacitances for all the electrolyte cases as N-6 at basal planes exhibited the largest binding energy. Simultaneously, N-5 could also contribute further to the capacitance increase due to the large binding energies at both basal planes and edges [10]. Thus, they affirmed that the improved capacitance was primarily based on the N-doping effect clearly different from those reported works on the improved capacitance of carbon materials based on the contributions of the surface areas and functional groups via using the plasma treatment [45,46].

Sun et al. prepared NG of high nitrogen (10.13 at.%) and large surface area (593 m²/g) via hydrothermal reaction [34]. The specific capacitance of as-synthesized NG reached 326 F/g at 0.2 A/g and superior cycling stability were obtained after 5000 cycles, which even maintained initial capacity in 6 mol/L KOH electrolyte. The energy density of 25.02 Wh/kg could be achieved at power density of 7980 W/kg by a two-electrode symmetric capacitor test. A series of experiment results demonstrated that not only the N-content but also the N-type were very significant for the capacitive behaviors, which the N-6 and N-5 played major roles for improving pseudo-capacitance by the redox reaction, while N-Q could

enhance the conductivity of the materials which was favorable to the transport of electrons during the charge/discharge process. The hydrothermal reaction was a simple and low-cost approach to obtain high N-content NG materials, which could be used as advanced electrodes in high-performance supercapacitors.

The 3-dimensional (3D) N-doped graphene framework (GF) was recently reported on excellent electrochemical performances for ultrafast UCs [47,48]. Zhao et al. prepared ultralight 3D N-doped graphene frameworks that exhibited versatile functions such as adsorption and ultracapacitor and ORR [47]. In the UCs, GF electrodes performed by three-electrode system had presented high-performance capacitive behaviors, as shown in Figure 11.9. Cyclic voltammogram (CV) curves of GF electrode exhibited an approximately rectangular shape at different scan rates, characteristic of the ideal double-layer capacitor (Figure 11.9a). The galvanostatic charge–discharge curves gave a symmetric triangle feature without obvious potential drop (IR drop) at large current densities in Figure 11.9b. At a current density of 1 A/g, GF presents a specific capacitance of 484 F/g at the operating potential of 0.8 V (see Figure 11.9c). Accordingly, the theoretical EDL capacitance

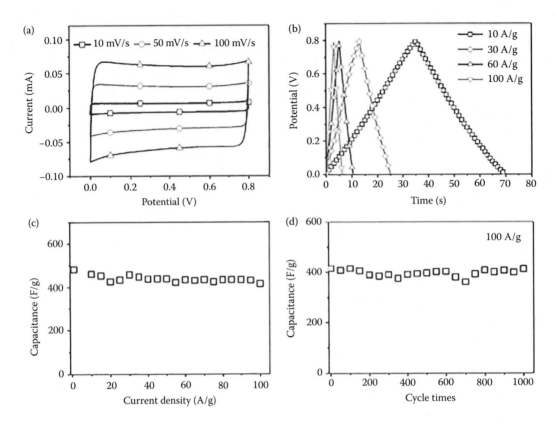

FIGURE 11.9 (a) CVs of a GF electrode in 1 mol/L LiClO$_4$ aqueous solution. (b) The galvanostatic charge–discharge curves at a current density of 10, 30, 60, and 100 A/g. (c) The specific capacitances calculated from the discharge curves under different current density. (d) Cyclic stability of the GF-based capacitor with a current density of 100 A/g. (Zhao, Y. et al.: A versatile, ultralight, nitrogen-doped graphene framework. *Angew Chem Int Edit.* 2012. 51:11371–5. Copyright Wiley-VCH Verlag GmbH & Co. KGaA. Reproduced with permission.)

of 550 F/g for pure graphene was reported by Reference 49. To date, the authors considered this as the highest value achieved for graphene electrodes. Even at the high current density of 100 A/g, the capacitance still reaches 415 F/g, which is about 150% that of N-doped graphene film (~280 F/g) [10], and far above that of B and N co-doped 3D graphene (240 F/g) [36]. GFs exhibited some unique advantages as a novel type of electrode material for high performance supercapacitors. For the practical application, the 2-electrode performance for GFs should be carried out.

In 2013, Yu's groups also reported a unique and convenient hydrothermal process for controlled synthesis and structural adjustment of 3D nitrogen-doped graphene hydrogels (GN-GHs), which can be readily scaled-up for mass production of GN-GHs by using organic amine and graphene oxide as precursors [48]. In this work, the organic amine was illustrated not only as nitrogen sources to obtain the nitrogen-doped graphene but also as an important modification to control the assembly of graphene sheets in the 3D structures. A two-electrode system was employed to test the electrochemical performances of the supercapacitor test cells of the GN-GHs in aqueous electrolyte at very fast scan rates (even 5 V/s) and large current densities (240 A/g). This material presented the electrochemical performance for which the high power density of 205.0 kW/kg can be obtained at the ultrafast charge/discharge rate of 185.0 A/g, and 95.2% of its

capacitance was retained for 4000 cycles at a current density of 100.0 A/g. Nevertheless, some drawbacks are that the visible IR drops were still found and specific energy density should be improved though very high specific power densities were obtained. The as-synthesized GN-GHs have potential applications as ultrahigh-power density capacitors in the vehicle, lift, and the other devices at high rates. However, the hydrothermal reaction relies on the autoclave so that the safety and large-scale production are still a challenge.

Wen et al. successfully fabricated highly crumpled nitrogen-doped graphene nanosheets (C-NGNSs) with a high pore volume of 3.42 cm^3/g from the GO and cyanamide (NH$_2$CN) at 900°C. The C-NGNSs exhibited significant improvement in terms of various performance parameters of supercapacitors (e.g., capacity, rate, and cycling) due to the abundant wrinkled structures, high pore volume, nitrogen doping, and improved electrical conductivity [50]. C-NGNSs were carried out on performing the electrochemical energy storage by 3-electrode system in 6 mol/L KOH aqueous electrolyte. The electrochemical measurement indicated that C-NGNSs-900 was an optimal material. Therefore, the energy storage of C-NGNSs-900 was studied in detail by a symmetrical two-electrode system in 1.0 mol/L [Bu$_4$N]BF$_4$ acetonitrile solution. In organic electrolyte, the operating potential window of 3.0 V was presented. The C-NGNSs-900 possessed a specific capacitance approaching 248.4 F/g at 5 mV/s. Even this showed only a slight decrease to 183.3 F/g at the high scan rate of 1000 mV/s. A similar specific capacitance of 245.9 F/g was obtained by the galvanostatic charge/discharge at 1.0 A/g. In addition, the voltage drop at the initiation of the discharge is extremely small (0.089 V), even at a high current density of 10 A/g, indicating a very low equivalent series resistance (ESR) in the symmetric supercapacitor and the potential of C-NGNSs-900 for high-power operations. In the cycling stability measurement, the C-NGNSs-900 supercapacitors maintained around 96.1% of the corresponding initial specific capacity. Additionally, the aqueous supercapacitor for C-NGNSs-900 was also investigated in 6.0 mol/L KOH aqueous electrolytes, which exhibited 302 F/g at a scan rate of 5 mV/s and a broad operating potential of 2.0 V in aqueous electrolytes. In the aqueous electrolyte, 2.0 V operating potential is very unusual.

11.5 SUMMARY AND PERSPECTIVES

In this chapter, the nitrogen-doped contents, species, distribution maps, and methods in graphene were summarized. The practical applications of NG were mainly focused on the ultracapacitors for energy storage. Nitrogen atoms have been introduced in the pristine graphene, resulting in the electron structure change and forming redox active sites. Therefore, electrochemical performances of NG have been visibly improved for potential applications as high energy density, ultrahigh-power density, and long cycling stability capacitors in the micro-device, vehicle, lift, and the other devices at high rates. However, low-cost and scalable synthesis methods should be developed as the specific energy density of NG is still low for practical application. In future, nitrogen-doped graphene should be anchored by battery materials such as metal oxides and metal hydroxides and conducting polymers with very high pseudo-capacitance, improving the specific energy density and sacrificing a part of cycle life of UCs. However, NG combined with metal oxides, metal hydroxides, and conducting polymers are fabricated in the asymmetric ultracapacitors of aqueous and organic solid-state electrolyte, enhancing the operating potentials. Safe hybrid ultracapacitors with high capacity should be the focus of development.

ACKNOWLEDGMENTS

This work was supported by National Natural Science Foundation of China (21031001 and 21376105) and the Project Sponsored by the Scientific Research Foundation for the Returned Overseas Chinese Scholars, State Education Ministry.

REFERENCES

1. Novoselov, K. S., Geim, A. K., Morozov, S. V., Jiang, D., Zhang, Y., Dubonos, S. V., Grigorieva, I. V., Firsov, A. A. 2004. Electric field effect in atomically thin carbon films. *Science*. 306:666–9.
2. Wan, X. J., Huang, Y., Chen, Y. S. 2012. Focusing on energy and optoelectronic applications: A journey for graphene and graphene oxide at large scale. *Accounts Chem Res*. 45:598–607.
3. Kvashnin, D. G., Sorokin, P. B., Bruning, J. W., Chernozatonskii, L. A. 2013. The impact of edges and dopants on the work function of graphene nanostructures: The way to high electronic emission from pure carbon medium. *Appl Phys Lett*. 102:183112–1
4. Schiros, T., Nordlund, D., Palova, L., Prezzi, D., Zhao, L. Y., Kim, K. S., Wurstbauer, U. et al. 2012. Connecting dopant bond type with electronic structure in N-doped graphene. *Nano Letters*. 12:4025–31.
5. Reddy, A. L. M., Srivastava, A., Gowda, S. R., Gullapalli, H., Dubey, M., Ajayan, P. M. 2010. Synthesis of nitrogen-doped graphene films for lithium battery application. *Acs Nano*. 4:6337–42.
6. Jin, Z., Yao, J., Kittrell, C., Tour, J. M. 2011. Large-scale growth and characterizations of nitrogen-doped monolayer graphene sheets. *Acs Nano*. 5:4112–7.
7. Lu, Y. F., Lo, S. T., Lin, J. C., Zhang, W. J., Lu, J. Y., Liu, F. H., Tseng, C. M., Lee, Y. H., Liang, C. T., Li, L. J. 2013. Nitrogen-doped graphene sheets grown by chemical vapor deposition: Synthesis and influence of nitrogen impurities on carrier transport. *Acs Nano*. 7:6522–32.
8. Lv, R., Li, Q., Botello-Mendez, A. R., Hayashi, T., Wang, B., Berkdemir, A., Hao, Q. Z. et al. 2012. Nitrogen-doped graphene: Beyond single substitution and enhanced molecular sensing. *Sci Rep-Uk*. 22:586.
9. Wei, D. C., Liu, Y. Q., Wang, Y., Zhang, H. L., Huang, L. P., Yu, G. 2009. Synthesis of N-doped graphene by chemical vapor deposition and its electrical properties. *Nano Letters*. 9:1752–8.
10. Jeong, H. M., Lee, J. W., Shin, W. H., Choi, Y. J., Shin, H. J., Kang, J. K., Choi, J. W. 2011. Nitrogen-doped graphene for high-performance ultracapacitors and the importance of nitrogen-doped sites at basal planes. *Nano Letters*. 11:2472–7.
11. Panchokarla, L. S., Subrahmanyam, K. S., Saha, S. K., Govindaraj, A., Krishnamurthy, H. R., Waghmare, U. V., Rao, C. N. R. 2009. Synthesis, structure, and properties of boron- and nitrogen-doped graphene. *Advanced Materials*. 21:4726–30.
12. Li, N., Wang, Z. Y., Zhao, K. K., Shi, Z. J., Gu, Z. N., Xu, S. K. 2010. Large scale synthesis of N-doped multi-layered graphene sheets by simple arc-discharge method. *Carbon*. 48:255–9.
13. Subrahmanyam, K. S., Panchakarla, L. S., Govindaraj, A., Rao, C. N. R. 2009. Simple method of preparing graphene flakes by an arc-discharge method. *J Phys Chem C*. 113:4257–9.
14. Sheng, Z. H., Shao, L., Chen, J. J., Bao, W. J., Wang, F. B., Xia, X. H. 2011. Catalyst-free synthesis of nitrogen-doped graphene via thermal annealing graphite oxide with melamine and its excellent electrocatalysis. *Acs Nano*. 5:4350–8.
15. Seo, S., Yoon, Y., Lee, J., Park, Y., Lee, H. 2013. Nitrogen-doped partially reduced graphene oxide rewritable nonvolatile memory. *Acs Nano*. 7:3607–15.
16. Nethravathi, C., Rajamathi, C. R., Rajamathi, M., Gautam, U. K., Wang, X., Golberg, D., Bando, Y. 2013. N-doped graphene–VO2(B) nanosheet-built 3D flower hybrid for lithium ion battery. *Acs Appl Mater Inter*. 5:2708–14.
17. Li, X. L., Wang, H. L., Robinson, J. T., Sanchez, H., Diankov, G., Dai, H. J. 2009. Simultaneous nitrogen doping and reduction of graphene oxide. *Journal of the American Chemical Society*. 131:15939–44.
18. Bag, S., Roy, K., Gopinath, C. S., Raj, C. R. 2014. Facile single-step synthesis of nitrogen-doped reduced graphene oxide–Mn3O4 hybrid functional material for the electrocatalytic reduction of oxygen. *Acs Appl Mater Inter*. 6:2691–8.

19. Long, D. H., Li, W., Ling, L. C., Miyawaki, J., Mochida, I., Yoon, S. H. 2010. Preparation of nitrogen-doped graphene sheets by a combined chemical and hydrothermal reduction of graphene oxide. *Langmuir: The ACS Journal of Surfaces and Colloids.* 26:16096–102.

20. Deng, D. H., Pan, X. L., Yu, L. A., Cui, Y., Jiang, Y. P., Qi, J., Li, W. X. et al. 2011. Toward N-doped graphene via solvothermal synthesis. *Chemistry of Materials.* 23:1188–93.

21. Wang, Y., Shao, Y. Y., Matson, D. W., Li, J. H., Lin, Y. H. 2010. Nitrogen-doped graphene and its application in electrochemical biosensing. *Acs Nano.* 4:1790–8.

22. Liu, M. K., Song, Y. F., He, S. X., Tjiu, W. W., Pan, J. S., Xia, Y. Y., Liu, T. X. 2014. Nitrogen-doped graphene nanoribbons as efficient metal-free electrocatalysts for oxygen reduction. *Acs Appl Mater Inter.* 6:4214–22.

23. Lee, K. H., Oh, J., Son, J. G., Kim, H., Lee, S.-S. 2014. Nitrogen-doped graphene nanosheets from bulk graphite using microwave irradiation. *Acs Appl Mater Inter.* 6:6361–8.

24. Ouyang, W., Zeng, D., Yu, X., Xie, F., Zhang, W., Chen, J., Yan, J. et al. 2014. Exploring the active sites of nitrogen-doped graphene as catalysts for the oxygen reduction reaction. *International Journal of Hydrogen Energy.* 39:15996–6005.

25. Kapteijn, F., Moulijn, J., Matzner, S., Boehm, H.-P. 1999. The development of nitrogen functionality in model chars during gasification in CO_2 and O_2. *Carbon.* 37:1143–50.

26. Pels, J., Kapteijn, F., Moulijn, J., Zhu, Q., Thomas, K. 1995. Evolution of nitrogen functionalities in carbonaceous materials during pyrolysis. *Carbon.* 33:1641–53.

27. Casanovas, J., Ricart, J. M., Rubio, J., Illas, F., Jiménez-Mateos, J. M. 1996. Origin of the large N 1s binding energy in X-ray photoelectron spectra of calcined carbonaceous materials. *Journal of the American Chemical Society.* 118:8071–6.

28. Wang, X. R., Li, X. L., Zhang, L., Yoon, Y., Weber, P. K., Wang, H. L., Guo, J., Dai, H. J. 2009. N-doping of graphene through electrothermal reactions with ammonia. *Science.* 324:768–71.

29. Das, A., Pisana, S., Chakraborty, B., Piscanec, S., Saha, S. K., Waghmare, U. V., Novoselov, K. S. et al. 2008. Monitoring dopants by Raman scattering in an electrochemically topgated graphene transistor. *Nat Nanotechnol.* 3:210–5.

30. Chen, W., Chen, S., Qi, D. C., Gao, X. Y., Wee, A. T. S. 2007. Surface transfer p-type doping of epitaxial graphene. *Journal of the American Chemical Society.* 129:10418–22.

31. Lazzeri, M., Mauri, F. 2006. Nonadiabatic Kohn anomaly in a doped graphene monolayer. *Phys Rev Lett.* 97:266407–4.

32. Ando, T. 2006. Anomaly of optical phonon in monolayer graphene. *J Phys Soc Jpn.* 75:124701–4.

33. Sun, Z. Z., Yan, Z., Yao, J., Beitler, E., Zhu, Y., Tour, J. M. 2010. Growth of graphene from solid carbon sources. *Nature.* 468:549–52.

34. Sun, L., Wang, L., Tian, C. G., Tan, T. X., Xie, Y., Shi, K. Y., Li, M. T., Fu, H. G. 2012. Nitrogen-doped graphene with high nitrogen level via a one-step hydrothermal reaction of graphene oxide with urea for superior capacitive energy storage. *Rsc Advances.* 2:4498–506.

35. Feng, L. Y., Chen, Y. G., Chen, L. 2011. Easy-to-operate and low-temperature synthesis of gram-scale nitrogen-doped graphene and its application as cathode catalyst in microbial fuel cells. *Acs Nano.* 5:9611–8.

36. Wu, Z. S., Winter, A., Chen, L., Sun, Y., Turchanin, A., Feng, X. L., Mullen, K. 2012. Three-dimensional nitrogen and boron co-doped graphene for high-performance all-solid-state supercapacitors. *Adv. Mater.* 24:5130–5.

37. Yan, K., Wu, D., Peng, H. L., Jin, L., Fu, Q., Bao, X. H., Liu, Z. F. 2012. Modulation-doped growth of mosaic graphene with single-crystalline p–n junctions for efficient photocurrent generation. *Nat Commun.* 3:1280.

38. Li, N., Wang, Z., Shi, Z. 2011. Synthesis of graphenes with arc-discharge method. *Physics and Applications of Graphene—Experiments*, InTech, Chapter 2, pp. 23–36.

39. Rao, C. N. R., Sood, A. K., Subrahmanyam, K. S., Govindaraj, A. 2009. Graphene: The new two-dDimensional nanomaterial. *Angew Chem Int Edit.* 48:7752–77.

40. Qian, W., Cui, X., Hao, R., Hou, Y. L., Zhang, Z. Y. 2011. Facile preparation of nitrogen-doped few-layer graphene via supercritical reaction. *Acs Appl Mater Inter.* 3:2259–64.

41. Wang, H. L., Dai, H. J. 2013. Strongly coupled inorganic-nanocarbon hybrid materials for energy storage. *Chem. Soc. Rev.* 42:3088–113.

42. Yuan, D. S., Zhou, T. X., Zhou, S. L., Zou, W. J., Mo, S. S., Xia, N. N. 2011. Nitrogen-enriched carbon nanowires from the direct carbonization of polyaniline nanowires and its electrochemical properties. *Electrochem. Commun.* 13:242–6.

43. Min, B. G., Sreekumar, T. V., Uchida, T., Kumar, S. 2005. Oxidative stabilization of PAN/SWNT composite fiber. *Carbon.* 43:599–604.

44. Yuan, D. S., Zeng, F. L., Yan, J., Yuan, X. L., Huang, X. J., Zou, W. J. 2013. A novel route for preparing graphitic ordered mesoporous carbon as electrochemical energy storage material. *Rsc Adv.* 3:5570–6.

45. Roy, S., Bajpai, R., Soin, N., Bajpai, P., Hazra, K. S., Kulshrestha, N., Roy, S. S., McLaughlin, J. A., Misra, D. S. 2011. Enhanced field emission and improved supercapacitor obtained from plasma-modified bucky paper. *Small.* 7:688–93.

46. Yoon, B. J., Jeong, S. H., Lee, K. H., Kim, H. S., Park, C. G., Han, J. H. 2004. Electrical properties of electrical double layer capacitors with integrated carbon nanotube electrodes. *Chem Phys Lett.* 388:170–4.

47. Zhao, Y., Hu, C. G., Hu, Y., Cheng, H. H., Shi, G. Q., Qu, L. T. 2012. A versatile, ultralight, nitrogen-doped graphene framework. *Angew Chem Int Edit.* 51:11371–5.

48. Chen, P., Yang, J. J., Li, S. S., Wang, Z., Xiao, T. Y., Qian, Y. H., Yu, S. H. 2013. Hydrothermal synthesis of macroscopic nitrogen-doped graphene hydrogels for ultrafast supercapacitor. *Nano Energy.* 2:249–56.

49. Xia, J. L., Chen, F., Li, J. H., Tao, N. J. 2009. Measurement of the quantum capacitance of graphene. *Nat Nanotechnol.* 4:505–9.

50. Wen, Z. H., Wang, X. C., Mao, S., Bo, Z., Kim, H., Cui, S. M., Lu, G. H., Feng, X. L., Chen, J. H. 2012. Crumpled nitrogen-doped graphene nanosheets with ultrahigh pore volume for high-performance supercapacitor. *Adv. Mater.* 24:5610–6.

12 Electrical and Optical Properties and Applications of Doped Graphene Sheets

Ki Chang Kwon and Soo Young Kim

CONTENTS

Abstract 179
12.1 Introduction 179
12.2 Methods of Doping Graphene 183
    12.2.1 Substitution Doping 183
    12.2.2 Chemical Doping 186
    12.2.3 Surface Charge Transfer 189
12.3 Device Applications of Doped Graphene 191
    12.3.1 Field-Effect Transistors 191
    12.3.2 Organic Photovoltaic Cells 193
    12.3.3 Light-Emitting Diodes 195
12.4 De-Doping Phenomena 197
12.5 Conclusion 201
Acknowledgments 202
References 202

ABSTRACT

Graphene, a single layer of sp^2-bonded carbon atoms, has attracted much attention due to its unique physical properties. The electronic properties of graphene sheets have recently attracted strong experimental and theoretical interest. Recently, large-area graphene sheets were successfully synthesized by chemical vapor deposition. One of the most attractive applications of large-scale graphene is as flexible transparent conducting films for electronic devices. For application to a large-area, flexible transparent conducting electrode, many researchers have focused on the synthesis of graphene with a high transmittance (>90%) and low sheet resistance (<200 Ω/□). Transmittance and sheet resistance have become acceptable for replacing commercial transparent conducting electrodes such as indium tin oxide (ITO). Even though graphene was reported to be used as a transparent electrode in optoelectronic devices, graphene-based devices have usually underperformed relative to ITO-based ones. One of these problems was induced by the energy level difference between the graphene electrode and the other active layers in electronic devices. Therefore, modulation of the work function in the graphene layer is crucial for improving device performances. This chapter reviews the electrical and optical properties of graphene from the viewpoint of transparent conducting electrodes.

12.1 INTRODUCTION

Graphene, a single layer of honeycomb-structured sp^2-bonded carbon atoms, has potential applications in a variety of electronic devices because of its superior electrical and mechanical properties, high mobility of charge carriers (200,000 cm^2/V s), anomalous quantum Hall Effect, and ballistic transport properties (Table 12.1) [1–6]. These excellent properties enable graphene to be applied to biosensors, pH sensors, gas sensors, thin film transistors, and transparent electrodes [7–10]. Furthermore, there have been many attempts to develop graphene sheets for application as interlayers or composite materials in electronic devices [11–13].

It is necessary to establish methods of synthesizing high-quality graphene in order to use graphene in a broader range of industrial applications. A simple method of mechanically exfoliating graphene was developed when graphene was first discovered. The "scotch tape method" is the original top-down method of mechanically exfoliating graphite, which forms graphene layers stacked together by van der Waals forces (Figure 12.1). Graphite can be mechanically cleaved by applying an external force of more than 300 μN/μm^2. Although this method produces the highest quality graphene, it also produces very low yield and is not readily scalable. Therefore, a number of alternative methods of producing graphene, such as growing epitaxial graphene layers on silicon carbide (SiC) surfaces, thermal reduction of graphene oxide (GO) to graphene, and chemical vapor deposition (CVD) have been investigated to synthesize large uniform graphene layers (The graphene synthesis techniques are summarized in Figure 12.2.) [14].

Although thermally growing epitaxial graphene layers on single crystalline SiC surfaces yields high-quality, high-purity graphene films, it is difficult to transfer the synthesized

TABLE 12.1
Properties of Graphene

Properties	Values	Reference
Young's modulus	~1100 Gpa	[1]
Elasticity modulus	~1000 Gpa	[1]
Tensile strength	125 Gpa	[1]
Carrier density	10^9 A/cm^2	[2]
Carrier mobility	~200,000 cm^2/V s	[3]
Specific surface area	2630 m^2/g	[2]
Thermal conductivity	~5000 W/mK	[1]
Transparency	~97.3% for monolayer (each layer = 2.7%)	[1]

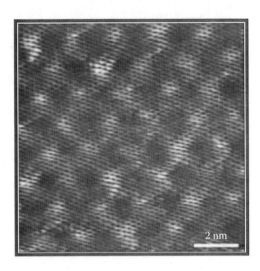

FIGURE 12.3 Atomically resolved scanning tunneling microscope image of a CVD-epitaxial graphene layer grown on a 4H-SiC(0001) substrate, taken over an area of 10×10 nm^2. (Reprinted with permission from Strupinsli, W. et al., Graphene epitaxy by chemical vapor deposition on SiC. *Nano Letters* 11:1786–1791, Copyright 2011 American Chemical Society.)

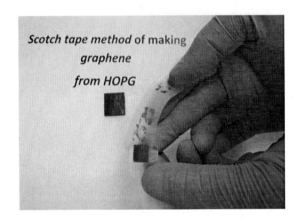

FIGURE 12.1 Mechanical exfoliation of graphene using scotch tape from HOPG. (Reprinted from *Progress in Materials Science*, 56, Singh, V. et al., Graphene based materials: Past, present and future, 1178–1271, Copyright 2011, with permission from Elsevier.)

graphene films onto any other substrates (Figure 12.3). Therefore, this method of synthesizing graphene films for application in electronic and optoelectronic devices has very limited usage [15].

In 1957, Hummers developed a method of fabricating graphene oxide (GO) by mixing graphite with sodium nitrate, sulfuric acid, and potassium permanganate, which is the well-known Hummers' method, whereby strong acids and oxidants are used to oxidize graphite. Figure 12.4 shows the progress of fabrication of GO and reduced graphene oxide (rGO) thin films. The chemical synthesis of GO offers several advantages: it is a low-temperature solution-based process and it can produce high yields of GO. Further, the degree of oxidation can be optimized by varying the reaction conditions (i.e., temperature, pressure, etc.) and the amount of oxidants used. A method of perfectly reducing chemically synthesized GO should be developed to obtain graphene flakes. Thermal annealing and chemical reductants have reportedly been used to reduce GO [16,17]. Although thermally annealing GO at high temperatures is an efficient method of producing uniformly reduced GO layers, the plastic substrates required for flexible devices could not bear high-temperature annealing.

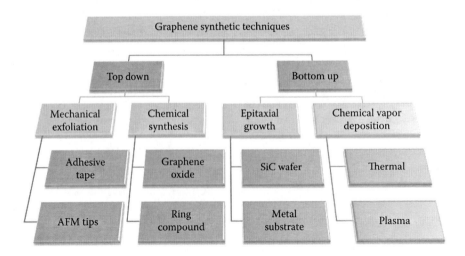

FIGURE 12.2 Graphene synthesis technique.

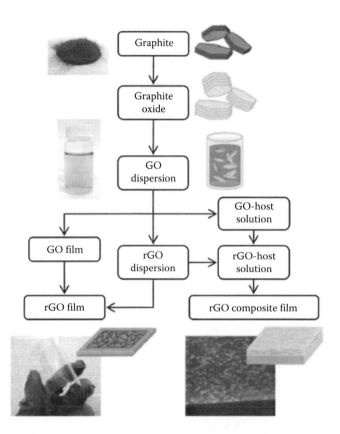

FIGURE 12.4 Process scheme for fabricating rGO-based thin films. The schematic illustrations show the structure of the material at each stage of the process. (Eda, G., Chhowalla, M.: Chemically derived GO: Towards large-area thin-film electronics and optoelectronics. *Advanced Materials*. 2010. 22. 2392–2415. Copyright Wiley-VCH Verlag GmbH & Co. KGaA. Reproduced with permission.)

Chemically reducing GO involves too many tedious steps and explosive chemicals such as hydrazine. Further, the chemical reduction of GO is often incomplete, and therefore, the properties of chemically reduced GO are often more degraded than those of thermally reduced GO. For example, the concentration of charge carriers and carrier mobility of chemically reduced GO are usually less than those for thermally reduced GO; moreover, the number of defects on the surface of chemically reduced GO is usually higher than that on the surface of thermally reduced GO [18].

CVD is the most efficient method of preparing high-quality, large, uniform crystalline graphene carbon networks. The photographic image of thermal CVD system is displayed in Figure 12.5. Graphene has previously been synthesized using transition metals (Ru, Ir, Ni, etc.) as catalysts [19–21]. Among the transition metals, copper shows the lowest solubility with carbon in the binary phase diagram at ~1000°C, indicating that copper thin films could be the best catalyst for synthesizing monolayer graphene sheets. Therefore, copper foils are usually used to synthesize monolayer graphene sheets [22]. The mechanism for synthesizing graphene with CVD involves three steps as shown in Figure 12.6: in the first step, the substrate should be pre-annealed at high temperature to obtain uniform graphene sheets. Hydrogen gas is used

FIGURE 12.5 Photographs of the roll-based production of graphene films. Copper foil wrapping around a 7.5-in. quartz tube to be inserted into an 8-in. quartz reactor. The lower image shows the stage in which the copper foil reacts with CH_4 and H_2 gases at high temperatures. (Reprinted by permission from Macmillan Publishers Ltd. *Nature Nanotechnology*, Bae, S. et al., Roll-to-roll production of 30-inch graphene films for transparent electrodes, copyright 2010.)

during preheating to produce a large, smooth copper surface. In the second step, a mixture of hydrocarbon, hydrogen, and argon is flowed into a CVD quartz tube to synthesize the graphene sheet after the preheating step. The carbon atoms are detached from the hydrocarbon atoms in this step because of the catalytic reaction between the copper and carbon atoms. Finally, the quartz tube is cooled to room temperature under vacuum to precipitate the carbon atoms from copper–carbon solid solutions [23,24].

To produce large, high-quality graphene sheets, not only the methods of synthesizing graphene but also the methods of transferring graphene onto other substrates are important. Therefore, researchers who use CVD to synthesize graphene have investigated not only methods of synthesizing graphene but also methods of transferring it onto other substrates (Figure 12.7) [25–28]. Transferring graphene onto arbitrary substrates to obtain graphene sheets with few defects and wrinkles attracted significant attention when the CVD-grown graphene sheets were first synthesized. The detailed procedure for transferring graphene onto other substrates is

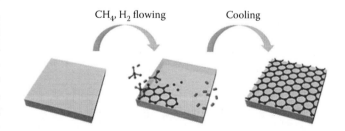

FIGURE 12.6 The graphene synthesis process by the CVD method. (Reprinted from *Solar Energy Materials and Solar Cells*, 109, Kwon, K. C. et al., Extension of stability in organic photovoltaic cells using UV/ozone-treated graphene sheets, 148–154, Copyright 2013, with permission from Elsevier.)

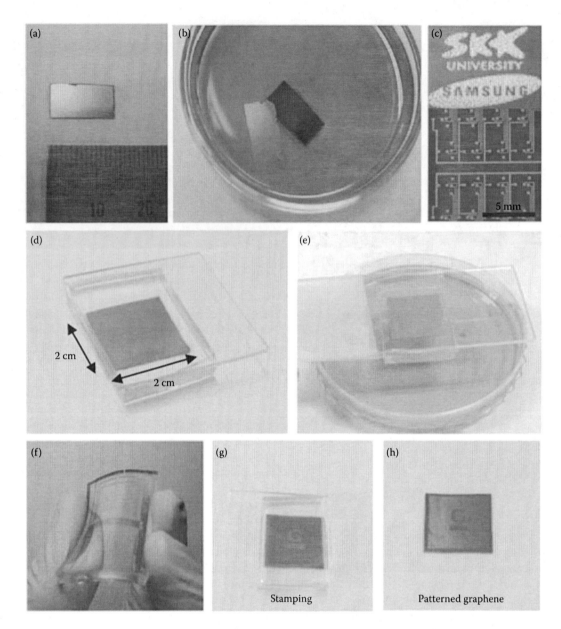

FIGURE 12.7 Transfer processes for large-scale graphene films. (a) A centimeter-scale graphene film grown on a Ni(300 nm)/SiO$_2$(300 nm)/ Si substrate. (b) A floating graphene film after etching the nickel layers in 1 M FeCl$_3$ aqueous solution. After the removal of the nickel layers, the floating graphene film can be transferred by direct contact with substrates. (c) Various shapes of graphene films can be synthesized on top of patterned nickel layers. (d) and (e) The dry-transfer method based on a PDMS stamp is useful in transferring the patterned graphene films. After attaching the PDMS substrate to graphene (d), the underlying nickel layer is etched and removed using FeCl$_3$ solution (e). (f) Graphene films on the PDMS substrates are transparent and flexible. (g) and (h) The PDMS stamp makes conformal contact with a silicon dioxide substrate. Peeling back the stamp (g) leaves the film on the SiO$_2$ substrate (h). (Reprinted by permission from Macmillan Publishers Ltd. *Nature*, Kim, K. S. et al., Large-scale pattern growth of graphene films for stretchable transparent electrodes, 457:706–710, copyright 2009.)

as follows: poly[methyl methacrylate] (PMMA, 46 mg/mL dissolved in chlorobenzene) is first spin coated onto a CVD-grown graphene-coated copper foil, which is then baked at 180°C for 1 min. O$_2$ plasma is then used to etch graphene on the other side of the copper foil. The sample is then immersed in a bath consisting of either ferric chloride (1 M FeCl$_3$) or ammonium persulfate (1 M (NH$_4$)$_2$S$_2$O$_8$) at room temperature for at least 12 h to etch the copper foil. The PMMA-coated graphene remaining on the copper foil is then carefully dipped into a bath of deionized water for more than seven times to remove any residual etchant. The PMMA-coated graphene sheets are then transferred onto an arbitrary substrate. The PMMA is then removed by immersing the substrate into an acetone bath at 50°C for 30 min after the PMMA/graphene layer has completely adhered to the target substrate [25–28]. Large, high-quality graphene sheets can be transferred onto arbitrary substrates through this method. Therefore, CVD has become the most popular method of producing very large, uniform graphene sheets with low sheet resistance (<200 Ω/□) and high transmittance (>90%) [29–34].

Replacing commercial indium tin oxide (ITO) electrodes with large graphene-film-based ones is one of the goals of producing large, high-quality graphene sheets. However, there are two major problems that must be solved before graphene sheets can be used to replace ITO films: first, the moderate work function (~4.2 eV) of graphene induces differences between energy levels of the active layers in devices and the graphene electrodes. Such differences between the energy levels mean graphene-electrode-based devices are considerably less efficient than ITO-based-electrode ones. Second, the sheet resistance of graphene is higher than that of ITO, which lowers the efficiency of graphene-based devices despite the high transparency of graphene [35–44].

Many researchers have developed methods of doping graphene to solve these problems. As a result, three major methods of intrinsically doping graphene film have been reported. The first method involves introducing foreign atoms into the carbon networks in graphene films to form substitutional impurities while graphene is being synthesized. B and N atoms are natural candidates for doping graphene because their sizes are similar to that of C atoms and because their hole-acceptor and electron-donor characteristics are optimal for B- and N-doping. The second method involves using the dipole moments introduced by self-assembled monolayers (SAMs) physisorbed or chemisorbed onto the graphene surface. SAMs with electron-donating or electron-withdrawing functional groups could localize electrons on the graphene surface because of the difference between the electron affinities of the SAMs and the graphene networks. Such dipole moments near the graphene surface could adjust the Fermi level of graphene, resulting in n- or p-doped graphene. The third method is known as the surface charge transfer method in which the difference between the electroreduction potentials of the graphene sheets and the other materials on graphene surface induces the charge on the surface of graphene to be transferred to the other materials or vice versa. Charge transfer is determined by the relative positions of the Fermi level of graphene and the density of states (DOS) of the highest occupied molecular orbital (HOMO) and lowest unoccupied molecular orbital (LUMO) of the dopants.

There are many reports on graphene doping, which comprise an extensive body of experimental and theoretical research. To date, the degree of graphene doping has been extensively characterized using several methods such as x-ray photoemission spectroscopy (XPS), Raman spectroscopy, ultraviolet photoemission spectroscopy (UPS), scanning electron microscopy (SEM), and transmission electron microscopy (TEM). The type of dopant in graphene can be determined from the specific element peaks in the XPS spectra. The degree of doping can be calculated from the ratio of the areas under the peaks corresponding to the dopant element and carbon atoms. Raman spectroscopy is widely used to measure the D band (1350 cm⁻¹), G band (1580 cm⁻¹), and the 2D band (2700 cm⁻¹) of graphene, which are related to the various disorders, doping state, and the number of layers of graphene [45,46]. From the UPS spectra, we can determine the change in the work function of doped graphene modified

using chemisorbed or physisorbed materials. The work function is determined from the secondary electron threshold, $\Phi = h\nu - E_{th}$, where $h\nu$ and E_{th} represent the photon energy of the excitation light source (He I discharge lamp, 21.2 eV) and the secondary electron threshold energy, respectively. SEM and TEM play important roles in measuring the surface and crystalline structure of the graphene sheets synthesized for application in electronic devices. Furthermore, the elements on the graphene surface can be analyzed using the energy dispersive spectra obtained with SEM equipment. High-resolution TEM (HR-TEM) images show the hexagonal graphene lattice and the atomic-scale carbon networks in graphene sheets. Graphene defects and the substitution doping of graphene can be studied from the HR-TEM images. Additional evidence supporting the hexagonal lattice of graphene can be obtained using supplemental equipment and selected-area electron diffraction (SAED).

This chapter reviews the experimental and theoretical studies on graphene doping. P- and n-type graphene doping will be discussed from the perspective of the three major methods of doping previously described: (i) substitution, (ii) chemical, and (iii) surface charge transfer doping. In addition, the effect of thermal annealing on doped graphene also will be discussed based on previous reports.

12.2 METHODS OF DOPING GRAPHENE

Methods of graphene doping are summarized in Figure 12.8.

12.2.1 SUBSTITUTION DOPING

Doping is the most feasible method of modifying the electrical and electronic properties of materials used in semiconductor technology. Electron donors and acceptors must be added to materials in order to modify their electronic properties, which is a well-known method in the semiconductor field. Since the discovery of graphene, there have been many attempts to substitute the carbon atoms in the graphene with foreign atoms in order to modulate the electronic properties of graphene.

Substitution doping is reportedly a useful method of opening the bandgap of graphene sheets. Among the potential dopants, boron (B) and nitrogen (N) are excellent candidates for chemically doping carbon-based materials because their atomic sizes are comparable to that of carbon and because they have three and five valence electrons, respectively, available to form strong valence bonds with carbon atoms, which would impart n- and p-doping to graphene. B- and N-doping of graphene have been theoretically and practically investigated. It has previously been shown that dopant atoms can modify the electronic band structure of graphene, opening an energy bandgap between the valence and conduction bands. Therefore, graphene could be changed into p-type or n-type semiconductors by substituting the C atoms with B or N ones [47,48].

Two methods of producing semiconducting graphene by substituting C atoms with B and N ones have been reported to date: direct synthesis and post-treatment. In the direct

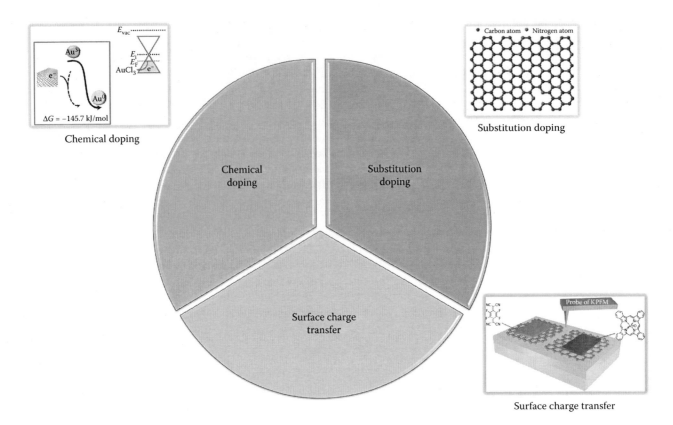

FIGURE 12.8 Methods of graphene doping.

synthesis, graphene is doped with B and N atoms as it is being synthesized. CVD has been extensively used with the precursors of carbon and gases such as boric acid (HBO_3, B_2O_3, and H_2B_6) [49–53] as shown in Figure 12.9 and ammonia (NH_3) [52–57] to form networks of doped carbon atoms as shown in Figure 12.10. Panchakarla et al. arc-discharged graphite

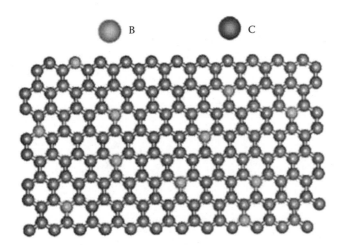

FIGURE 12.9 Schematic structure of a B-doped graphene with B content of ~6.8 at.%, as detected in the synthesized graphene. (Reprinted with permission from Tang, Y.-B. et al., Tunable bandgaps and p-type transport properties of boron-doped graphenes by controllable ion doping using reactive microwave plasma. *ACS Nano* 6, 1970–1978. Copyright 2012 American Chemical Society.)

electrodes in a mixture of H_2 and B_2H_6 and arc-discharged boron-stuffed graphite electrodes to synthesize B-doped graphene. They also arc-discharged the electrodes in mixtures of either H_2 and pyridine or H_2 and ammonia and transformed the resulting nanodiamonds in pyridine to synthesize N-doped graphene. They showed that the electronic properties of the B- and N-doped graphene could be systematically tuned by altering the concentrations of B and N [52,53]. Wu et al. and Wei et al. used CVD with H_2, CH_4, and heteroatoms containing B_2O_3, BH_3, NH_3, and pyridine sources to synthesize B- and N-doped graphene on 25-μm-thick copper foil [49–51,54–57]. They concluded that the degree of doping in the synthesized graphene sheet was related to the number of heteroatom sources and the reaction (growth) times used.

Alternatively, graphene can be thermally annealed at high temperature or treated with plasma containing heteroatoms in the vapor or liquid phases to prepare graphene in advance. Thermal annealing was initially introduced to remove the residual supporting polymer layers such as PMMA and to reduce the GO. From the perspective of doping, thermal annealing can activate the doped atoms in graphene. Sheng et al. [52] used high-temperature annealing to dope the boron atoms derived from B_2O_3 vapor into the graphene framework. Further, NH_3 annealing and N^+-ion irradiation were used both before and after graphene had been synthesized at room temperature under vacuum to controllably n-type dope the graphene. This method can be used to optimize the degree of doping by varying the number of nitrogen atoms

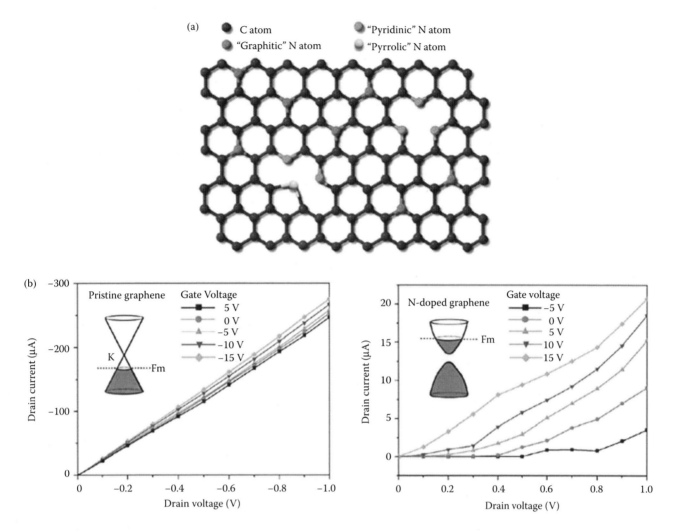

FIGURE 12.10 (a) Schematic representation of the N-doped graphene. The spheres represent the C, "graphitic" N, "pyridinic" N, and "pyrrolic" N atoms in the N-doped graphene, respectively. (b) I_{ds}/V_{ds} (I_{ds}: drain current, V_{ds}: drain voltage) characteristics at various V_g (gate voltage) for the pristine graphene and the N-doped graphene FET (field effect transistor) device, respectively. The insets are the presumed band structures. (Reprinted with permission from Wei, D. et al., Synthesis of N-doped graphene by chemical vapor deposition and its electrical properties. *Nano Letters* 9, 1752–1758. Copyright 2009 American Chemical Society.)

[58]. The results of the XPS characterization and electrical measurements supported the formation of N-doped graphene through the reduction of GO annealed in NH_3 gas. They showed that the heteroatoms contained in the gas can act as dopants in the network of carbon atoms when graphene is thermally annealed at high temperature (above 300°C for n-type doping in N_2) [59].

Although plasma processing is another efficient method of regulating the semiconducting properties of carbon materials during post-treatment, it could induce high-density defects, which would deteriorate the electrical properties of graphene sheets. Wang et al. [60], Soin et al. [61], and Shao et al. [62] reported the effect of N_2 plasma treatment on graphene at room temperature. The functionalized graphene was exposed to the N_2 plasma, which then showed properties consistent with those of nitrogen-doped graphene. The content of the nitrogen dopant in graphene was easily optimized by varying the plasma strength or exposure time. However, treating graphene with high-energy plasma could introduce high-density defects, which would deteriorate the electrical performances of graphene. Ion irradiation with heteroatoms in gas has previously been used to dope graphite and carbon nanotubes (CNTs) [50,51]. Although the structure of graphene is similar to those of graphite and CNTs, the relations between the number of defects, surface charge, and deterioration of the electronic structure have only been investigated for the ion-irradiated graphene.

It is important to obtain evidence that supports the existence of dopants in the hexagonal lattice of graphene and the synthesis of doped graphene. XPS is a powerful method of determining which elements are present in a sample. The presence of dopants in graphene could be supported by specific peaks in the survey spectra. Further, concentration of dopants in graphene could be calculated by comparing areas under each peak in the XPS spectra. The chemical and electrical states of each element in graphene are usually analyzed by separating the peaks in each high-resolution spectrum. The G band in Raman spectra is sensitive to chemical doping, and there is a useful empirical rule for determining the type of graphene doping. The G band in the Raman spectra downshifts and stiffens for graphene containing molecules with electron-donating groups or for n-type doped graphene. In contrast, the G band in the Raman spectra upshifts and softens for graphene containing molecules with electron-withdrawing groups or for p-type doped graphene. Furthermore, Raman spectroscopy is a powerful method of determining the doped state of graphene sheets. The G band in Raman spectra is sensitive to other dopants. It is well known that the G band stiffens and upshifts for samples containing both electron- and hole-doping. The full width at half maximum (FWHM) of the G bands decreased because of both types of doping. The G band in the Raman spectra for n-type doped graphene sheets (i.e., graphene doped with electron-donating groups) downshifts and stiffens. In contrast, the G band in the Raman spectra for p-type doped graphene sheets (i.e., graphene doped with electron-withdrawing groups) upshifts and softens. However, the G band in the Raman spectra would be different for graphene doped using substitution doping, according to previous reports [53]. Nitrogen- and boron-doped graphene showed upshifted G bands in their corresponding Raman spectra [63].

Although substitution doping foreign atoms into graphene is an effective method of intrinsically modifying the properties of graphene, the environmental stability of chemically doped graphene is reportedly poor, and the dopant elements could be easily released from the surface of graphene sheets. Therefore, more effort is needed to solve these problems.

Devices are usually fabricated with doped graphene to determine whether graphene was well doped. Field-effect transistors (FETs) have been fabricated to investigate the effects of B and N atoms on graphene. The type of semiconductor can be identified from the Dirac points in the electrical properties of thin-film transistors (TFTs). The Dirac point of pristine graphene was displayed in the positive gate voltage area, indicating that pristine graphene was a p-type semiconductor. The Dirac point of the TFTs composed of N-doped graphene was located in negative gate voltage area, showing that N-doped graphene was an n-type semiconductor. These data show that nitrogen atoms act as n-type dopants in the hexagonal lattice of carbon atoms in graphene. The details will be further discussed in the following chapters.

12.2.2 CHEMICAL DOPING

Aligning the bands between electrodes and semiconductors is crucial to optimizing device performance. However, it is very difficult to find a conductive material that also has the desired work function. Therefore, it is preferable to dope the material or treat the surface of the material in order to adjust the work function. Various methods of chemically doping materials have attracted significant attention since the beginning of the era of silicon semiconductors. Many scientists and engineers are trying to determine the most effective method of doping

materials because doping can be used to efficiently modify the electrical properties of target materials. Further, chemical doping is the easiest method of optimizing the electronic properties of target materials without damaging the materials or generating basal plane reactions. For these reasons, chemical doping has been used to modify the work functions of carbon nanomaterials including graphene and CNTs.

Using chemical dopants to modulate the Fermi level of graphene sheets has been experimentally and theoretically demonstrated, indicating that the work function of graphene could be engineered as desired. Gold chloride ($AuCl_3$), which is commonly used to dope organic conductors, is reportedly a p-type dopant for graphene [64]. The gold cations (Au^{3+}) on the surface of graphene sheets spontaneously reduce to gold nanoparticles (Au^0) by taking electrons from the π-orbitals of the graphene lattice when the graphene is either immersed in or spincast with solutions containing $AuCl_3$ or $HAuCl_4 \cdot 3H_2O$ [65]. The redox reaction strongly depends on the reduction potentials of the carbon atoms in graphene and Au^{3+} ions. According to the Nernst equation, Au^{3+} ions have a positive reduction potential (or a negative Gibbs free energy) as shown in Figure 12.11 [66]. The larger the reduction potential of an ion, greater is the tendency that the ion will accept electrons from the π-orbitals in carbon atoms. Therefore, the electrons near the Dirac point of pristine graphene are depleted, increasing the work function of graphene. Doped graphene the desired work function of which is higher than that of pristine graphene can be obtained by varying the concentration of $AuCl_3$. The detailed chemical reaction is shown in the following equations:

$$Graphene + 3AuCl_3 \rightarrow Graphene + AuCl_2^- + Au(I) + AuCl_4^- \tag{12.1}$$

$$3AuCl_2^- \rightarrow Au^0 + 2AuCl_4^- + 2Cl^- \tag{12.2}$$

$$AuCl_4^- + Graphene \rightarrow Graphene^+ + Au^0 + 4Cl^- \tag{12.3}$$

Kim et al. [67] used $AuCl_3$ to develop a method of improving the conductivity of graphene film. They obtained the lowest sheet resistance of 150 Ω/□ and transmittance of 87% at 550 nm when the concentration of $AuCl_3$ was optimized. Other groups have reported that four layers of graphene doped with $AuCl_3$ showed excellent stability during bending, a sheet resistance of 54 Ω/□, and a transmittance of 85% at 550 nm [68]. Benayad et al. [69] investigated the role of Au^{3+} ions in the reduced GO doped with $AuCl_3$. They demonstrated that the work function and sheet resistance of reduced GO could be modulated by varying the concentration of the Au^{3+} ion. Kwon et al. [70] reported the effect of metal chloride on the work function of sheets consisting of a few layers of graphene. They used $AuCl_3$, $IrCl_3$, $MoCl_3$, $OsCl_3$, $PdCl_2$, and $RhCl_3$ as p-type dopants. The work functions for the graphene sheets doped with the metal in metal chloride solutions were higher than 4.6 eV. On the other hand, the sheet resistance of the

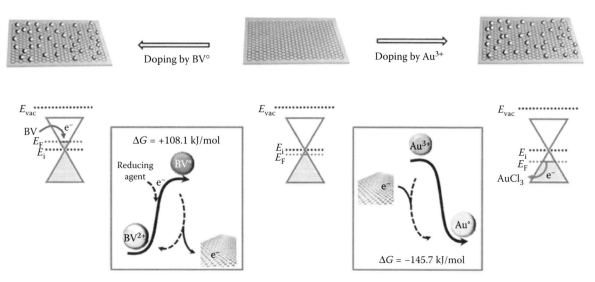

FIGURE 12.11 Modulation of electronic structure by differences of reduction potentials. (Reprinted with permission from Shin, H.-J. et al., Control of electronic structure of graphene by various dopants and their effects on nanogenerator. *Journal of American Chemical Society* 132, 15603–15609. Copyright 2010 American Chemical Society.)

same doped graphene sheets was ~30%–55% lower than the undoped graphene sheet. Furthermore, the work function increased from 4.2 for the undoped graphene sheet to 4.7–5.1 eV for the doped graphene sheets. $AuBr_3$, Au_2S, $Au(OH)_3$, and $AuCl_3$ were used as the p-type dopants to determine the effects of the anions in the Au complexes on the work function and sheet resistance of doped graphene sheets [71]. Each dopant was dissolved in an appropriate solvent (nitromethane, methanol, ethanol) and solutions containing the dopants were spincoated onto the surface of graphene that had previously been transferred onto a substrate. The work function of the doped graphene sheets varied with the order of the electronegativity of anions, that is, the electronegativity of the dopant anions and the negative Gibbs free energy of the Au atoms play important roles in graphene doping.

Similarly, viologen (benzyl viologen, 1,1′dibenzyl-4,4′bipyridinium dichloride) is a well-known n-type dopant used for organic and graphene molecules [66,72,73]. For alkyl viologens (AVs), the reduction of AV^{2+} to AV^0 is not spontaneous because of the positive Gibbs free energy of the redox reaction. Therefore, a reduction agent such as sodium borohydride ($NaBH_4$) was used instead of AV^{2+} to directly prepare AV^0 [66]. After the graphene was doped with AV^0, the AV^0 spontaneously donated electrons to the graphene in order to stabilize it and transform into AV^{2+}. The work function of the doped graphene was consequently lower than that of pristine graphene. The viologen molecules are an efficient n-type dopant and do not generate any basal plane reactions through this mechanism. Alkali metal carbonates are another efficient n-type dopant for graphene. Kwon et al. [74] used alkali carbonate molecules in deionized water to dope graphene. They showed that the sheet resistance of n-type doped graphene increased and that the work function of n-type doped graphene decreased because electrons from the alkali carbonate molecules were donated to graphene.

However, other researchers have tried to dope graphene only by treating the surface of graphene with acid vapor. Pristine graphene has a low carrier density and a high sheet resistance. The $FeCl_3$ or PMMA residues remaining on the surface of graphene acted as unintentional dopants when the graphene was transferred onto another substrate. Therefore, the Fermi level of graphene mostly did not reside at the Dirac point of graphene when the transferred graphene was exposed to air. Therefore, the surface of graphene was treated to try to remove the residues on the graphene sheets and to stimulate the transfer of charge from graphene to the other materials. HNO_3 is a p-type dopant for graphitic materials. An electron is transferred from graphene to nitric acid as a charge-transfer complex is formed according to the following chemical reaction on the surface of the graphene sheet (Figure 12.12) [75]:

$$6HNO_3 + 25C \rightarrow C_{25}^+NO_3^- \cdot 4HNO_3 + NO_2 + H_2O \quad (12.4)$$

This reaction results in a shift in the Fermi level, which increases the carrier concentration and conductivity of graphene layers. This effect was first observed in single-walled carbon nanotubes (SWCNTs). It has been previously demonstrated that treating the surface of SWCNTs with nitric acid increased the conductivity of SWCNT films. Zheng et al. [76] reported that HNO_3 is a surface treatment material that can be used to efficiently modify the Fermi level of graphene sheets. The sheet resistance and transmittance of the HNO_3-treated graphene sheets were 90 $\Omega/\square$ and 80% at 550 nm. They also reported the effects of various acids and acid/halogen atom combinations on the properties of graphene. Considerable research has been conducted to enhance the electrical conductivities of carbon-nanomaterial-based thin films. Carbon-based materials the surfaces of which were first pretreated with HNO_3 were subsequently dipped into a bath containing

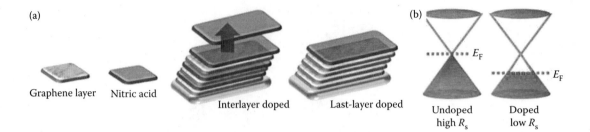

FIGURE 12.12 (a) Schematic illustrating the two different doping methods pursued here. In the interlayer-doped case, the sample is exposed to nitric acid after each layer is stacked, whereas in the last-layer-doped case, the film is exposed to nitric acid after the final layer is stacked. (b) Illustration of the graphene band structure, showing the change in the Fermi level due to chemical p-type doping. (Reprinted with permission from Kasry, A. et al., Chemical doping of large-area stacked graphene films for use as transparent, conducting electrodes. *ACS Nano* 4, 3839–3844. Copyright 2010 American Chemical Society.)

either $SOCl_2$ or $SOBr_2$ under a gentle flow of nitrogen. The electrical conductivity of graphene thin films improved from 2031.3 to 1598.0 $\Omega/\square$, and the transparency of graphene thin films improved through a series of chemical treatments after the GO was thermally reduced. (The graphite oxide was 38.7 nm thick.) HNO_3, $SOCl_2$, and $SOBr_2$ surface treatments were intended to form acyl chloride or bromide functional groups on the surface of graphene and CNTs. The acid treatment removed the impurities on the surface of the graphene film and on the substrate surface, while the strongly electronegative –Br or –SOBr functional groups acted as electron acceptors, increasing the density of holes in graphene [77,78].

Another method involving the use of organic molecules to modify the surface of graphene was also reported. Organic molecules that act as electron donors or acceptors may modify the electronic structure of graphene sheets, giving rise to significant changes in their electrical properties. Dong et al. reported the effects of organic molecules on the properties of mechanically exfoliated graphene sheets. They used 1,5-naphtalene diamine (Na-NH₂), 9,10-dimethylanthracene (An-CH₃), 9,10-dibromoanthracene (An-Br), and tetrasodium 1,3,6,8-pyrenetetrasulfonic acid (TPA) as organic dopants shown in Figure 12.13. These molecules can stably bind to single-layer graphene films through strong π–π interactions between the aromatic rings of the organic molecules and graphene. Aromatic molecules with electron-donating groups cause the n-type doping (i.e., they increase electron density) of graphene, while those with electron-withdrawing groups impose p-type doping (i.e., they decrease the electron density) on single-layer graphene [79]. Other research groups have investigated the effects of the dopants bis(trifluoromethanesulfonyl)amide[((CF₃SO₂)₂NH)] (TFSA), poly(ethylene imine) (PEI), and diazonium salts on the surface of graphene. The TFSA dopant is environmentally stable because it is hydrophobic; therefore, the interaction between TFSA and graphene surface stabilized the latter. The TFSA molecules on graphene sheets can accept electrons from the graphene sheets, imparting p-type properties to graphene [80]. PEI and diazonium salts are used as complementary molecular dopants on graphene to suppress the formation of either electrons or holes in order to optimize the conductivity of graphene. Furthermore, simulations and experiments involving these dopants demonstrate that both PEI and diazonium salts behave as long-range scattering factors on graphene [81].

Raman spectroscopy is usually used to investigate the effects of chemical dopants on the properties of graphene

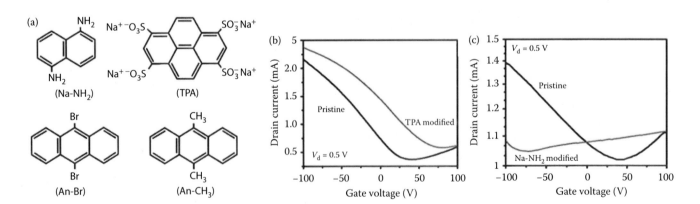

FIGURE 12.13 (a) Chemical structures of the aromatic molecules used as dopants. Transfer curves (I_d [drain current] versus V_g [gate voltage]) for single-layer graphene FETs (b) before and after TPA doping (drain voltage, $V_d = 0.5$ V) and (c) before and after Na–NH₂ doping ($V_d = 0.5$ V). (Dong, X. et al.: Doping single-layer graphene with aromatic molecules. *Small*. 2009. 5. 1422–1426. Copyright Wiley-VCH Verlag GmbH & Co. KGaA. Reproduced with permission.)

because it can be used to rapidly and nondestructively examine the intrinsic physical properties of various carbon nanostructures including flat and one-atom-thick carbon crystalline layers. The bands in Raman spectra can reveal the number of stacked graphene layers [82–84] and changes in the concentration of charge carriers induced by a static electrical field [85,86]. The G bands in Raman spectra are sensitive to the chemical doping of graphene. There is a useful empirical rule for determining the type of graphene doping: the G band in Raman spectra downshifts and stiffens for graphene-containing molecules with electron-donating groups, or n-type doped graphene, and upshifts and softens for graphene-containing molecules with electron-withdrawing groups, or p-type doped graphene [85,86]. These properties will help us to identify graphene and to determine the degree by which it is doped.

12.2.3 SURFACE CHARGE TRANSFER

The last major method of doping graphene discussed in this chapter is surface charge transfer, which has been investigated to modify the bandgap of graphene sheets. Many researchers have used chemical materials whose electronegativities were higher or lower than that of carbon to try to dope graphene sheets in order to improve the electrical properties of graphene sheets at room temperature. The electrons in the graphene sheets were depleted through the transfer of charge between the dopant and carbon atoms in the graphene lattice when the electronegativities of the dopants were higher than that of the carbon atoms. In contrast, the dopants donated their electrons to the graphene sheet, thereby inducing n-type doping of graphene when the electronegativities of the dopants were lower than that of the carbon atoms. Surface charge transfer doping occurs when the charge is transferred from the adsorbed dopant (or graphene) to the graphene (or dopant). Charge transfer is determined by the relative positions of the Fermi level of graphene and the DOS of the HOMO and LUMO levels of the dopant. If the HOMO level of a dopant is located above the Fermi level of graphene, the charge transfers from the dopant to the graphene layer, and the dopant acts as a donor. If the LUMO is located below the Fermi level of graphene, the charge transfers from the graphene layer to the dopant, and the dopant acts as an acceptor.

Colleti et al. [87] have studied the properties of graphene modified by tetrafluorotetracyanoquinodimethane (F4-TCNQ). F4-TCNQ is an effective p-type dopant, which acts as a strong electron acceptor. It has high electron affinity (E_{ea} = 5.24 eV) and has been used as a state-of-the-art p-type dopant in electronic devices. When it was used in organic light-emitting diodes (OLEDs), the hole-injection barrier was reduced by forming a narrow space-charge region near the metal contact, thereby improving the device performance. It has recently been theoretically and experimentally suggested that F4-TCNQ shows a p-type doping effect on graphene (Figure 12.14). Modifying the graphene surface with F4-TCNQ is expected to favor the transfer of electrons from graphene to the F4-TCNQ. It has previously been shown that a charge-transfer complex is formed between the graphene

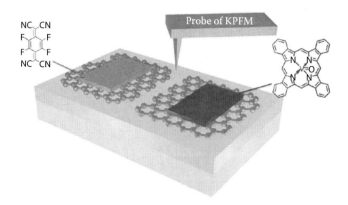

FIGURE 12.14 Both n- and p-type doped exfoliated graphene sheets are presented by virtue of adsorbing organic molecules. Flat organic layers are uniformly grown on graphene sheets by the technique. (Reprinted with permission from Wang, X. et al., Quantitative analysis of graphene doping by organic molecular charge transfer. *Journal of Physical Chemistry C* 115, 7596–7602. Copyright 2011 American Chemical Society.)

film and the F4-TCNQ molecular layers. The electrons are removed from the graphene layer through the cyano groups of the F4-TCNQ molecule. Therefore, a depletion layer is formed in graphene, thereby achieving p-type doping of graphene in which the concentration of holes in graphene can be optimized by varying the thickness of the F4-TCNQ layer and by covering the entire surface of graphene with F4-TCNQ. As F4-TCNQ remains stable under ambient conditions and at elevated temperatures and because it can be readily applied using wet chemistry, surface charge transfer is an attractive method of doping graphene with F4-TCNQ, and it appears feasible that this method can be incorporated into existing technological processes [88,89]. Ishikawa et al. [90] reported that TCNQ is a powerful electron acceptor as shown in Figure 12.15. They showed that electrons are transferred from graphene into the TCNQ molecules, leading to p-type doping of graphene. This is a new and inexpensive method of fabricating charge-transfer-modified graphene sheets.

The halogenation of graphene provides another method of tuning the chemical functionality of graphene sheets in order to engineer the bandgap of graphene sheets. The halogenation of graphene has significant advantages over the hydrogenation or oxidation of graphene. The high electronegativity of halogen atoms enables the graphene to be efficiently doped and the bandgap of graphene to be efficiently opened. Graphene can be fluorinated, chlorinated, or brominated to improve its performance. Graphene has been either treated with fluorine or chlorine plasma [91], or exposed to F_2 at high temperatures [92] to attach F or Cl atoms to its basal plane. However, as previously mentioned, plasma can damage the surface of graphene through ion bombardment, and the plasma reaction induces high temperatures; therefore, more caution is needed when plasma processing is used to dope graphene. Lee et al. used a solid source of fluorine and laser irradiation to fluorinate graphene under controlled conditions, which would not damage the graphene surface [93]. The fluoropolymer CYTOP was

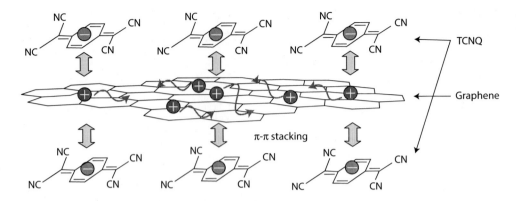

FIGURE 12.15 Schematic image of doping graphene by adsorbed TCNQ molecules. (Reprinted from Ishikawa, R. et al., 2011. Doping graphene films via chemically mediated charge transfer. *Nanoscale Research Letters* 6:111, permission is not needed.)

used as the solid source of fluorene, and a Raman laser beam operating at 488 nm was used to decompose the CYTOP into the fluorine precursor. Irradiating the CYTOP layer with the laser produced active fluorine radicals that reacted with the sp²-hybridized carbon atoms in graphene to form C–F bonds. Therefore, electrons were transferred from the carbon atoms in the graphene sheets to the fluorine atoms because fluorine atoms are much more electronegative than carbon atoms. This finding indicates that the CYTOP layer is an efficient p-type polymer supporting layer. Furthermore, a CYTOP layer could be used instead of a PMMA layer as a polymer supporting layer during the transfer of surface charge from graphene to the dopant layer. The authors reported that the charge transferred from graphene to the dopant and graphene

was simultaneously doped when the CYTOP layer was used as shown in Figure 12.16 [94]. The charge cannot spontaneously transfer from graphene to the CYTOP because CYTOP is an insulating polymer and because the electron affinity of CYTOP is not higher than the work function of graphene. The authors claimed that doping graphene with the CYTOP layer can be induced by the electrostatic potential created by the dipole moment between graphene and the CYTOP layer.

SAMs have recently received considerable attention because an ultrathin layer can be uniformly constructed on an oxide surface through the methods of self-assembly. Park et al. [95] used SAMs to optimize the work functions of graphene electrodes in order to improve the performance of organic field-effect transistors (OFETs). The work function of

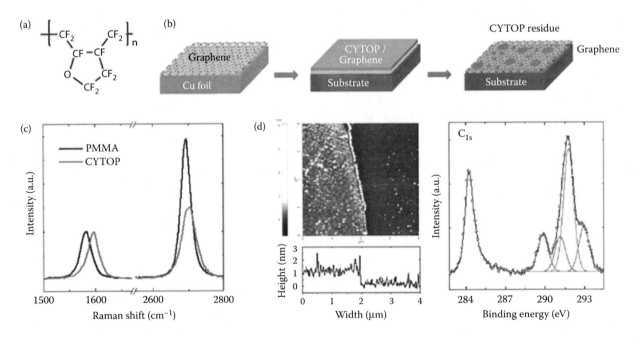

FIGURE 12.16 (a) Chemical structure of CYTOP. (b) Schematic of the graphene transfer process with CYTOP supporting layer. (c) Raman spectra of the graphene films transferred onto SiO₂/Si substrates with PMMA or CYTOP as the supporting layer. (d) AFM image (4 × 4 μm) (left) and XPS C1s spectrum (right) of the graphene film having an ultrathin CYTOP residue layer on the top surface. Bottom inset of the AFM image shows the cross-sectional profile. (Reprinted with permission from Lee, W. H. et al., Simultaneous transfer and doping of CVD-grown graphene by fluoropolymer for transparent conductive films on plastic. *ACS Nano* 6, 1284–1290. Copyright 2012 American Chemical Society.)

graphene increased by ~0.25 eV when graphene was located on the CH_3-SAM-modified SiO_2 instead of on SiO_2. This result implies that extra holes are incorporated into graphene, lowering the Fermi level of graphene from the Dirac point. The work function for the graphene fabricated on the NH_2-SAM-modified SiO_2 substrate decreased by about 0.35 eV. It seems that the lone pair of electrons in NH_2-SAM induces extra electrons in graphene, raising the Fermi level of graphene above the Dirac point. These results suggest that the types of graphene doping could be optimized using SAMs with various functional groups.

The chemisorption or physisorption of organic compounds can also induce graphene doping through the transfer of surface charge. Zhang et al. [96] reported that the adsorptions of 2,3-dichloro-5,6-dicyano-1,4-benzoquinone (DDQ) and tetrathiafulvalene (TTF) had induced hybridization between the molecular levels and the valence band of graphene, transforming the zero-gap semiconducting graphene into a metallic graphene. The DDQ and TTF molecules on the surface of graphene play the roles of organic electron acceptors and donors, respectively. This result showed that functionalizing the surface of graphene with organic noncovalent acceptor and donor molecules could significantly change the electronic structure of graphene. Jung et al. [97] investigated the effects of the physical adsorption of halogen atoms on graphene sheets through a similar concept. Br_2 and I_2 vapors were exposed at room temperature to a few layers of graphene sheets. Br and I atoms are more electronegative than the carbon atoms in graphene sheets. Therefore, the surface charge should transfer from the graphene to the halogen atoms, producing p-type doped graphene. Adsorption-induced charge-transfer doping (including intercalation) shows great potential for developing adjustable doping patterns at high densities in laterally large graphene samples without disrupting the π-electrons. Song et al. [98] reported a unique and facile method of homogeneously and stably p-type doping graphene. Graphene was hybridized with ZnO thin films fabricated using MeV electron beam irradiation (MEBI) under ambient conditions. The MEBI method enables uniformly thick, ultraflat-surfaced ZnO thin films to be facilely fabricated on graphene. The effect of the uniform ZnO layers on the graphene sheets was investigated using Raman spectroscopy. The ratio of the intensities of the two-dimensional (2D) peaks to the G peaks in the Raman spectra significantly decreased when ZnO had been formed on graphene. This behavior can be explained by the difference between the work functions of the ZnO thin film (5.1–5.3 eV) and graphene (4.5–4.8 eV), whereby graphene was p-type doped when the electron charge was transferred from graphene to ZnO.

The properties of doped graphene are summarized in Table 12.2.

12.3 DEVICE APPLICATIONS OF DOPED GRAPHENE

Device applications of doped graphene are summarized in Figure 12.17.

12.3.1 FIELD-EFFECT TRANSISTORS

OFETs have generated much research interest because they are inexpensive to fabricate, can be fabricated using a variety of materials, are flexible, and can be used in a wide span of potential applications [99–101]. The development of graphene-based FETs immediately followed the first observation of field effects in graphene in 2004. Graphene has recently received much attention as a potential electrode material for organic electronic devices [1–6]. The zero bandgap of graphene induces a high off-current and a low on/off ratio, on the order of 10^2 or lower, which is much lower than those (10^3 to 10^6) required for application in computer logic [102,103]. Although the charge carrier density of undoped graphene layers is higher than that of silicon, the zero bandgap semiconductor nature of graphene severely limits the on/off ratio that can be achieved for graphene-based devices. Therefore, graphene technology cannot yet compete with mainstream silicon technologies [104]. In other words, devices fabricated with channels produced from large-area graphene cannot be switched off and are therefore unsuitable for logical applications because the bandgap of graphene is zero.

Therefore, much research has been conducted toward opening the bandgap of graphene. The width of graphene molecules can be decreased to use the nanosize effect in order to induce an energy gap in graphene [105]. E-beam lithography and oxygen plasma etching can be used to fabricate graphene nanoribbons (GNRs) down to a width of 10 nm [106]. Although GNRs have been used to investigate the properties of graphene-based FETs, this method is still inadequate for fabricating wafer-scale FETs and is very expensive because it requires ultrahigh vacuum conditions.

There are some reports about using heteroatom substitution doping to solve these problems. Substitution doping is one of the most feasible methods of optimizing the semiconducting properties of graphene and is intentionally used to tailor the electrical properties of intrinsic semiconductors. Heteroatoms can be used as dopants to modify the electronic band structure of graphene and open an energy gap between the valence and conduction bands [101].

Guo et al. [58] used N^+-ion irradiation at room temperature on mechanical exfoliated single-layer graphene to synthesize nitrogen-doped graphene as shown in Figure 12.18. The graphene sheet was irradiated with 30-keV N^+ ions at five different positions. Graphene-based back-gate FETs were fabricated on a 300-nm SiO_2/p^{2+}Si substrate with Cr/Au (5 nm/70 nm) source/drain electrodes in order to identify the electronic properties of graphene. The transfer curve for the pristine-graphene-based FET was measured under atmospheric and vacuum conditions. The x- and y-axis of the transfer curve represent the gate voltage (V_g) and conductance, respectively. Graphene exhibited bipolar transistor property. The minimal conductance shown in the transfer curve corresponds to the Dirac point (V_{Dirac}) for graphene. The device operated as a p-channel FET for $V_g < V_{Dirac}$ and as an n-channel FET for $V_g > V_{Dirac}$ [54]. The pristine graphene FET operating in air showed p-type behavior because of the physisorbed oxygen

TABLE 12.2
Properties of Doped Graphene

Doping Method	Graphene Synthetic Method	Dopants	Properties	Reference
Substitution doping	CVD	HBO_3, $CO(NH_2)_2$	• Atomic ratio Boron: 0% → 4.3% Nitrogen: 0% → 4.8% • Gate voltage B-doped G: +5 V → ~+ 30 V N-doped G: +5 V → ~− 23 V	[49]
	CVD	C_3H_9B(TMB)	• Atomic ratio Boron: 0% → ~12.5% • Gate voltage B-doped G: +3 V → ~+ 8 V	[50]
	GO powder	B_2O_3	• Atomic ratio Nitrogen: 0% → ~3.2% • CV curve enhanced.	[52]
	CVD	NH_3	• Atomic ratio Nitrogen: 0% → ~12.5% • Gate voltage N-doped G: +3 V → ~ +8 V	[55]
Chemical doping	CVD	$AuCl_3$ DDQ BV	• Sheet resistance $AuCl_3$: 201 Ω/□ → 149 Ω/□ DDQ: 201 Ω/□ → 190 Ω/□ BV: 201 Ω/□ → 260 Ω/□ • Work function $AuCl_3$: 4.5 eV → 4.8 eV DDQ: 4.5 eV → 4.7 eV BV: 4.5 eV → 4.0 eV	[66]
	CVD	$AuCl_3$ $IrCl_3$ $MoCl_3$ $OsCl_3$ $PdCl_2$ $RhCl_3$	• Sheet resistance $AuCl_3$: 1100 Ω/□ → 500 Ω/□ $IrCl_3$: 1100 Ω/□ → 600 Ω/□ $MoCl_3$: 1100 Ω/□ → 720 Ω/□ $OsCl_3$: 1100 Ω/□ → 700 Ω/□ $PdCl_2$: 1100 Ω/□ → 520 Ω/□ $RhCl_3$: 1100 Ω/□ → 620 Ω/□ • Work function $AuCl_3$: 4.2 eV → 4.9 eV $IrCl_3$: 4.2 eV → 4.8 eV $MoCl_3$: 4.2 eV → 4.7 eV $OsCl_3$: 4.2 eV → 4.6 eV $PdCl_2$: 4.2 eV → 4.85 eV $RhCl_3$: 4.2 eV → 5.14 eV	[70]
	CVD	HNO_3 $SoCl_2$ (acid)	• Sheet resistance HNO_3: 725 Ω/□ → 657 Ω/□ $SoCl_2$: 425 Ω/□ → 103 Ω/□ • Raman shift HNO_3: 1591 cm⁻¹ → 1595 cm⁻¹ $SoCl_2$: 1592 cm⁻¹ → 1600 cm⁻¹	[68] (HNO_3) [77] ($SoCl_2$)
Surface charge transfer	SiC epitaxy	F4-TCNQ C_{60}	• Work function F4-TCNQ: 4.0 eV → 5.25 eV C_{60}: 4.0 eV → 4.15 eV • C 1s spectra F4-TCNQ: 284.4 eV → 283.9 eV C_{60}: 284.4 eV → 284.9 eV	[88]
	GO powder	TCNQ	• Sheet resistance TCNQ: 8×10^6 Ω/□ → 5×10^4 Ω/□ • Transmittance at 550 nm TCNQ: 87% → 86.5%	[90]
	Mechanical exfoliation	Cl_2 H_2 CF_4 (plasma)	• Raman shift Slightly upshifted for all samples. • Gate voltage After plasma, gate voltage was upshifted to higher voltage area.	[92]

(Continued)

TABLE 12.2 (*Continued*)
Properties of Doped Graphene

Doping Method	Graphene Synthetic Method	Dopants	Properties	Reference
	CVD	CYTOP	• Raman shift CYTOP: 1582 cm^{-1} → 1597 cm^{-1} • Gate voltage After CYTOP removal, gate voltage was upshifted to higher voltage area.	[94]

molecules [57]. The Dirac point of the n-doped graphene produced by both N^{+}-ion treatment and NH$_3$ annealing was shown at the negative V_g, which is consistent with the N-doped graphene. The hole and electron mobilities for their N-doped graphene FET were calculated as approximately 6000 cm^2/V s [54,57,58].

Park et al. [95] used graphene as an active channel layer or electrode materials to study the effects of surface charge transfer doping on the electrical properties of graphene as shown in Figure 12.19. A SAM produced with various functionalized molecules was used to modify the electronic structure of graphene. Au was used as the source/drain electrodes, and graphene was used as the channel material. The Dirac point drastically changed according to the surface characteristics. The Dirac point voltage of the graphene transferred onto the 300-nm SiO$_2$ wafer was at the +43-V gate voltage, indicating that the pristine graphene on the SiO$_2$ wafer was a p-type semiconductor. The Dirac voltage of the FET produced using –NH$_2$-surface-treated graphene sheets was at –110 V, indicating that the –NH$_2$ functionalization had changed the graphene from a p-type semiconductor into an n-type one. The Dirac point of the –CH$_3$-terminated graphene shifted to a slightly positive region (+10 V), which means that the –CH$_3$ functional group had changed the graphene into a p-type semiconductor.

Graphene is potentially well suited to radio frequency applications because of its promising carrier transport properties and its purely 2D structure. Furthermore, graphene doped with either B or N atoms could open the bandgaps of semiconducting graphene sheets, indicating that graphene-based FETs could play a key role in fabricating switching devices such as

complementary metal oxide semiconductor (CMOS) applications. Therefore, various methods of doping are essential for the application of graphene in the FET industry.

12.3.2 ORGANIC PHOTOVOLTAIC CELLS

Organic photovoltaic (OPVs) cells have attracted much interest in both scientific and experimental research because they are lightweight, flexible, inexpensive, and solution processable, all of which give OPVs tremendous advantages over conventional photovoltaic cells [107–109]. The OPV cells offer a significant advantage over silicon technology from the perspective of mechanical flexibility. The theoretical and experimental research conducted to enhance the efficiency of OPV cells include the synthesis of new materials in the active layer, optimization of the device structures, and modification of fabrication techniques. As a result, the performance of OPV cells has recently increased dramatically and power conversion efficiencies of more than 7% have been attained using low-bandgap polymers in bulk heterojunction structures [110–112]. ITO has mainly been used as the transparent conducting electrode in OPV structures. However, the use of ITO as an anode material may have limited applications in the OPV cells because indium is a very scarce natural resource and because of diffusion of ions into the active layers of the cells. Furthermore, ITO electrodes show low stability in acidic solutions, and the low mechanical strength of ITO may prevent its application to flexible devices.

In these respects, CVD-synthesized large-area graphene films are a promising material for application as flexible transparent conducting electrodes, and many efforts have been devoted to replacing ITO with graphene layers. Enhancing the conductivity of graphene to a level comparable with that of ITO remains one of the major challenges for using graphene electrodes in organic optoelectronic devices. Various methods of doping graphene have been investigated to decrease the sheet resistance of graphene sheets, as mentioned in the preceding chapters.

Hsu et al. [113] used layer-by-layer molecular doping to construct graphene/TCNQ/graphene films for use as anodes in OPV cells as shown in Figure 12.20. The TCNQ molecules acted as a p-type dopant. They used Au-assisted graphene transfer without an organic solvent to prevent damage to or dissolution of the TCNQ molecules during the transfer. Poly(3-hexylthiophene)/phenyl-C61-butyric acid methyl ether (P3HT/PCBM) bulk heterojunction OPV cells were fabricated based on the multilayered graphene/TCNQ anodes and

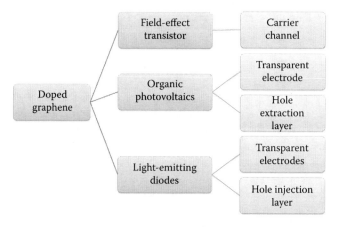

FIGURE 12.17 Device applications of doped graphene.

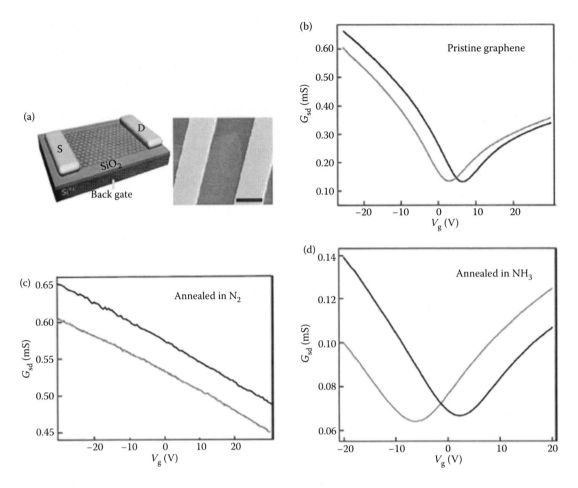

FIGURE 12.18 Comparison of the transport property of FET using different graphene samples (left and right curves were measured in air and in vacuum, respectively). (a) The scheme of the graphene-based FET device (left) and a typical SEM image of the device (right). The scale bar is 1 μm. (b) The G_{sd}–V_g curves of pristine graphene. (c) The G_{sd}–V_g curves of graphene annealed in N_2 after irradiation. (d) The G_{sd}–V_g curves of graphene annealed in NH_3 after irradiation. G_{sd}–V_g curves in (b–d) are measured at $V_{sd} = 0.03$ V. (Reprinted with permission from Guo, B. et al., 2010. Controllable N-doping of graphene. *Nano Letters* 10 4975–4980. Copyright 2010 American Chemical Society.)

were characterized. The OPV fabricated using the graphene/ TCNQ/graphene/TCNQ/graphene anode showed the optimum power conversion efficiency of ~2.58%, where the stable, low sheet resistance of the graphene stacks was achieved without acid doping. Their proposed anode may be suitable for application in next-generation flexible devices requiring high transparency and reliable electrical conductivity. Lee et al. [114] treated the surface of graphene with HNO_3 and $SOCl_2$ to fabricate flexible OPV cells with doped graphene electrodes. The CVD-grown multilayer graphene films were chemically doped and used as conductive and flexible electrodes. The optimal sheet resistance of the multilayer-doped graphene films was twofold less than that of a pristine multilayer graphene film. The power conversion efficiency of the OPV cells based on the doped multilayer graphene electrodes reached approximately 2.6%, which is high among the power conversion efficiencies previously reported for OPV cells produced with graphene electrodes. The performance of the devices fabricated with the doped multilayer graphene electrodes remained nearly constant, even under various bending conditions, indicating that the doped graphene electrode could be an appropriate candidate for application in flexible

OPV cells. Wang et al. [115] demonstrated that the device efficiency of OPV cells fabricated with a graphene anode and an MoO_3 (2 nm)/poly(3,4-ethylenedioxythiophene):poly(styrenesulfonate) (PEDOT:PSS) hole extraction layer was 2.5%, which is 83.3% of the power conversion efficiency achieved for the ITO-based OPV cell. They used four layers of graphene treated with hydrochloric and nitric acids as the anode in their OPV cells as shown in Figure 12.21. The authors mentioned that the sheet resistance of graphene should be reduced to further improve the power conversion efficiency in graphene-based OPV cells. Modifying the work function and the surface free energy are also important aspects of optimizing device performance. Therefore, engineering the interface between the graphene electrode and the interlayer in the OPV structure is the key to improving the power conversion efficiency in graphene-based OPV cells.

Many researchers have recently used graphene sheets as a hole-extraction layer in OPV cells. GO and rGO nanosheets have gained much interest because they are inexpensive and solution processable. Liu et al. [37] functionalized GO to produce both hole- and electron-extraction materials for application in bulk heterojunction OPV cells. They simply neutralized the

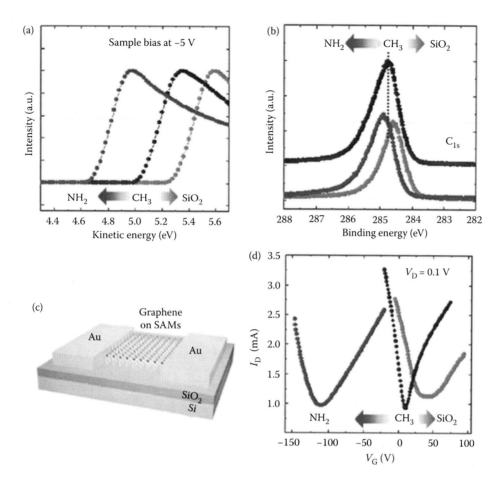

FIGURE 12.19 (a) UV photoemission spectra (at the secondary electron emission region) of graphene films on different SAM modified SiO$_2$/Si substrates. (b) XPS carbon 1s peaks (C1s) of graphene films on different SAM-modified SiO$_2$/Si substrates. (c) Schematic of the graphene FETs having SAMs as an insertion layer. (d) Current–voltage transfer characteristics of graphene FETs on different SAM-modified SiO$_2$/Si substrates. CH$_3$: octadecyltrichlorosilane; NH$_2$: aminotriethoxysilane; SiO$_2$: untreated; V_D: drain voltage; V_G: gate voltage; I_D: drain current. (Reprinted with permission from Park, J. et al., Work-function engineering of graphene electrodes by self-assembled monolayers for high-performance organic filed-effect transistors. *Journal of Physical Chemistry Letters* 2, 841–845. Copyright 2011 American Chemical Society.)

charge of the –COO$^-$ groups in GO with Cs$_2$CO$_3$ (GOCs) to tune the electronic structure of GO, thereby making the GO derivatives useful as both hole- and electron-extraction layers in bulk heterojunction OPV cells. They used GO and GOCs instead of PEDOT:PSS and LiF as the hole- and electron-extraction layers, respectively. Although the power conversion efficiency of the reference cell fabricated with the ITO/PEDOT:PSS/P3HT:PCBM/LiF/Al structure was 3.15%, that of the target cell fabricated with the ITO/GO/P3HT:PCBM/GOCs/Al structure was 3.67%. They demonstrated for the first time that simply neutralizing the charge of the –COOH groups in GO with Cs$_2$CO$_3$ could reverse the charge extraction in bulk heterojunction OPV cells. GO acted as an excellent hole-extraction layer while its Cs-based derivative, GOCs, was an excellent electron-extraction material. However, the performance of the GO-based OPV cell was highly dependent on the film thickness because GO is an insulator.

In this respect, Kwon et al. [116] reported the effect of ultraviolet (UV)-ozone-treated CVD-derived graphene as a hole-extraction layer on the performance of OPV cells as shown in Figure 12.22. They used CVD-derived graphene sheets treated with UV light and ozone for various lengths of time as

hole-extraction layers. According to their experimental data, the work function of graphene increased from 4.3 to 4.85 eV for the graphene treated with UV light and ozone for 9 min. The device fabricated with graphene-treated UV light and ozone for 5 min showed the best efficiency of 3.0% because the longer UV–ozone treatment damaged the graphene sheet. The power conversion efficiency of the PEDOT:PSS-based device rapidly decreased to 0% when it was exposed to moist, humid conditions for 14 h while the UV–ozone-treated device continued to operate for 26 h under the same conditions, indicating that more stable and efficient passivation can be achieved by replacing conventional PEDTO:PSS with UV–ozone-treated graphene.

12.3.3 Light-Emitting Diodes

Developing high-power gallium-nitride (GaN)-based light-emitting diodes (LEDs) is important for future solid-state lighting applications. The amount of contact resistance between the GaN and current-spreading layers must be low to achieve high-power LEDs [117,118]. Some reports have

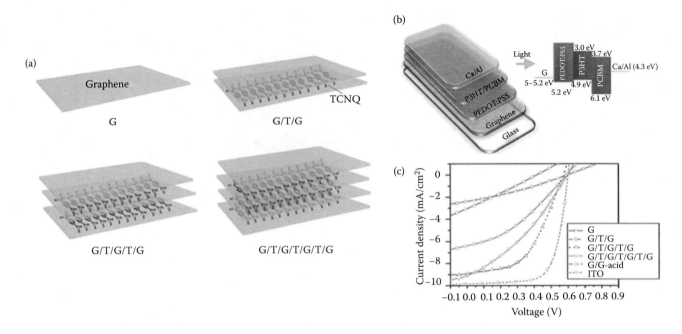

FIGURE 12.20 (a) Four anode structures. G and T represent a graphene layer and a thermally evaporated TCNQ layer, respectively. (b) Scheme for the solar cell device with G as an anode, where the graphene anode is sitting on a glass substrate and PEDOT:PSS is spun on top of the graphene anode for enhancing hole transport. Ca/Al is used as the reflective cathode. A band diagram for the solar cell is also shown. (c) Measured current density (J)-applied bias (V) curves of the fabricated devices. (Reprinted with permission from Hsu, C.-K. et al., Layer-by-layer graphene/TCNQ stacked films as conducting anodes for organic solar cells. *ACS Nano* 6, 5031–5039, Copyright 2012 American Chemical Society.)

previously described the improvement in the lateral current-spreading layers in GaN-based LEDs. The oxidized Ni/Au and ITO systems in particular offer comparatively low contact resistance and high optical transparency [119,120]. However, certain obstacles limit their use at high output powers. When Ni/Au contacts oxidize, NiO detaches from the p-type GaN layer and forms a phase of insulating amorphous Ni–Ga–O at high transverse current densities, affecting device consistency [119]. Furthermore, directly depositing ITO onto p-type GaN produces poor ohmic contact [120]. Additional obstacles to using ITO as a lateral current-spreading layer in GaN-based LEDs are the high cost of In and low chemical resistance of ITO. Researchers have replaced ITO and Ni/Au with graphene sheets to try to solve these problems.

Usage of pristine graphene, whose work function is approximately 4.5 eV, as a current-spreading electrode in the p-type GaN poses intrinsic limitations on current injection because the work function of pristine graphene is lower than that of p-type GaN [121]. Several methods of chemically doping graphene have been introduced to overcome this problem.

Chandramohan et al. [121] used work-function-tuned multilayer graphene to investigate the performance of InGaN/GaN-multiquantum-well-based blue LEDs used as a current-spreading layer (Figure 12.23). The electrode was formed by inserting a thin layer of gold metal between graphene and p-type GaN, and the electrode was then rapidly thermally annealed. The graphene electrode was doped with AuCl$_3$ through chemical charge transfer to tune the work function of graphene and decrease its sheet resistance and current barrier injection height. The forward voltage of the graphene-based LED prepared using the AuCl$_3$-doped multilayer graphene

film and a thin Au interlayer was slightly lower than that of the ITO-based LED.

Jo et al. [39] applied CVD-synthesized multilayer graphene films as current-spreading layers in GaN-based LEDs. They used multilayer graphene whose sheet resistance was ~620 Ω/□ and ~85% of whose transmittance was in the range 400–800 nm. The results showed that the multilayer graphene electrodes in GaN LED operated as a lateral current-spreading layer. The forward voltages of the LEDs prepared with the multilayer graphene and ITO electrodes were ~5.6 and 3.8 V, respectively, at 20 mA input. They concluded that the graphene electrode may degrade the performance of GaN LEDs because the sheet resistance of the graphene electrode was higher than that of the ITO one. Increasing the conductivity and tuning the work function of the graphene film could improve the performance of graphene-based GaN LEDs, and the graphene must be doped to achieve these objectives. These findings provide encouraging evidence that graphene can be applied to GaN-based LEDs as an alternative transparent electrode to ITO. Using doped graphene sheets and interface engineering to fabricate LEDs may lead LEDs with higher output power.

Lee et al. [122] fabricated GaN LEDs with very thin metal/graphene electrodes as transparent and current injection layers in p-type GaN. The contact resistance of the LEDs was reduced from 5.5 to 0.6 Ω/cm^2 by inserting a Ni/Au layer between the monolayer graphene and the p-type GaN, while the optical transmittance of the LEDs exceeded ~78% for visible light. The GaN LEDs prepared with the metal/graphene electrodes demonstrated uniform blue light emission over a large area because of the improved current-spreading and injection characteristics of the electrodes. This result suggests

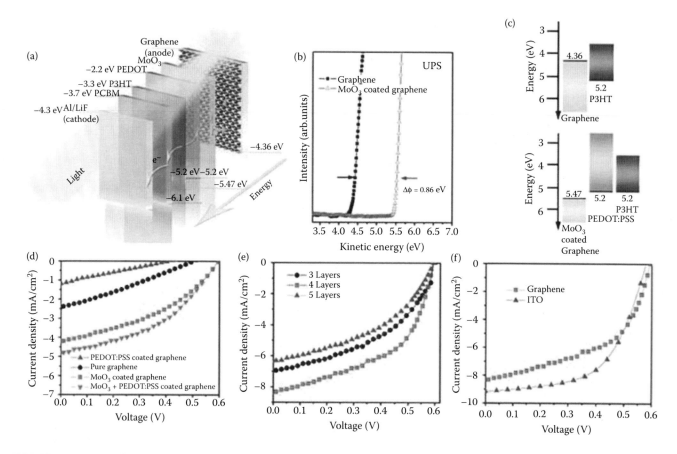

FIGURE 12.21 (a) Schematic diagram of a photovoltaic device structure. (b) UPS spectrum showing the secondary cutoff of four layers of graphene before and after modification with MoO_3. (c) Energy level diagram at the anode side of the photovoltaic cells without (top) and with (bottom) the MoO_3 + PEDOT:PSS layer. (d) Current density–voltage (J–V) characteristics of devices under light illumination. Anode/P3HT:PCBM/LiF/Al (anode = 4 layers of as-grown graphene and the same coated by PEDOT:PSS only, MoO_3 only, or MoO_3 + PEDOT:PSS). (e) 3–5 layers of acid-doped graphene/MoO_3 + PEDOT:PSS/P3HT:PCBM/LiF/Al. (f) Anode/PEDOT:PSS/P3HT:PCBM/LiF/Al (anode is ITO or MoO_3-coated graphene). (Wang, Y. et al.: Interface engineering of layer-by-layer stacked graphene anodes for high-performance organic solar cells. *Advanced Materials*. 2011. 23. 1514–1518. Copyright Wiley-VCH Verlag GmbH & Co. KGaA. Reproduced with permission.)

that the contact resistance between graphene and p-type GaN plays a key role in improving the performance of GaN-based LEDs, and graphene must be doped to achieve this objective.

12.4 DE-DOPING PHENOMENA

Since researchers have started investigated the methods of doping graphene, the practical application of doped graphene has been limited to laboratory-scale electronic devices such as LEDs, OPVs, and OLEDs. Although doping has improved the electrical characteristics and increased the work functions of graphene sheets, the effects of ambient conditions on doped graphene sheets have not yet been thoroughly investigated. Some requirements such as (i) controllability in the doping area, (ii) stability under ambient conditions, and (iii) comparability with Si-based device technologies must be taken into account to optimize the electrical properties of graphene before the doped graphene sheets could be applied to industrial electronic devices. Although there have been several reports on the degradation of doped CNTs exposed to ambient conditions for various lengths of time, only few such reports have been conducted on the degradation of doped graphene sheets.

Polyethyleneimine (PEI) [123], poly(acrylic acid) (PAA) [124], b-nicotinamide adenine dinucleotide (NADH) [125], and viologen molecules [126] have previously been used as dopant materials in CNTs. However, such CNTs degraded under ambient conditions. Furthermore, organic dopants are vulnerable to degradation at temperatures as low as 150°C, which is fairly low for device applications. Some researchers have studied the effects of thermal and environmental stress on doping and de-doping of CNTs to investigate the mechanisms by which CNTs degrade under ambient conditions.

Yoon et al. [127] reported the effect of thermal annealing on $AuCl_3$-doped SWCNTs. The chlorination and desorption of chlorine atoms were shown for SWCNTs thermally annealed at various temperatures. The CNTs thermally annealed at 160°C were chlorinated into $AuCl_3$-doped CNT surfaces. However, the chlorine atoms desorbed from the CNTs when the CNTs were annealed at more than 289°C. The results of the UV–vis-near infrared (NIR) spectroscopy, Raman scattering, x-ray photoemission spectroscopy (XPS), and electrical measurements showed that the electronic and electrical properties of the $AuCl_3$-doped SWCNTs could be modulated with annealing temperature. The Au^{3+} ions imparted strong p-type

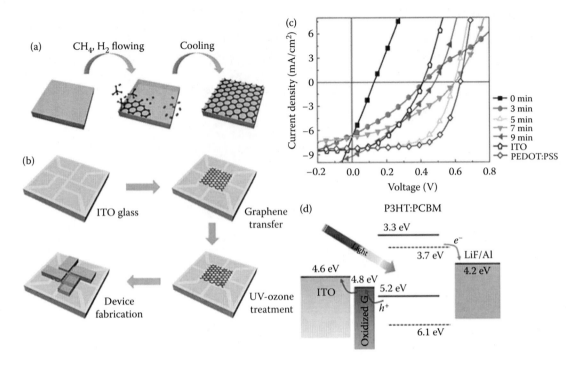

FIGURE 12.22 The schematic process to investigate the effect of UV/ozone-treated graphene sheet as an hole extraction layer (HEL) in OPV cells: (a) the graphene synthesis procedure by CVD method, (b) the fabrication process of the OPV cells with the UV/ozone-treated graphene sheet as the HEL, (c) current density–voltage characteristics of the OPV cells without HEL and with pristine, 3, 5, 7, and 9 min-treated graphene, and PEDOT:PSS HELs, and (d) the schematic band diagram of the OPV cells. (Reprinted from *Solar Energy Materials and Solar Cells*, 109, Kwon, K. C. et al., Extension of stability in organic photovoltaic cells using UV/ozone-treated graphene sheets, 148–154, Copyright 2013, with permission from Elsevier.)

doping to the CNTs at the low-temperature region because of the large difference between the reduction potentials of the Au³⁺ ions and CNTs. However, the p-type doping was suppressed, and the n-type doping was properly thermally induced from the CNT-Cl annealed at higher temperatures.

Kim et al. [128] have also reported the role of anions in thermally annealing AuCl₃-doped CNTs as shown in

Figure 12.24. AuCl₃ was used as a p-type dopant. Gold ions are usually obtained by dissolving AuCl₃ powder in nitromethane or deionized water. The Au³⁺ ions in the solution are believed to play a key role in extracting electrons from CNTs because of the very large reduction potential and negative Gibbs free energy of the Au³⁺ ions, which theoretically act as a precursor for further Cl adsorption, and the chemisorbed

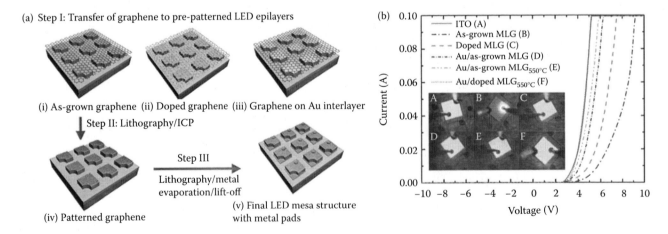

FIGURE 12.23 (a) Schematic illustration of the key steps involved in the fabrication of GaN LEDs with multilayer graphene electrode. (b) Current–voltage characteristics of GaN LEDs with different electrode schemes studied. The inset shows the optical image of the corresponding LEDs during light emission at an injection current of 0.5 mA. (Reprinted with permission from Chandramohan, S. et al., Work-function-tuned multilayer graphene as currentspreading electrode in blue light-emitting diodes. *Applied Physics Letters* 100:023502. Copyright 2012, American Institute of Physics.)

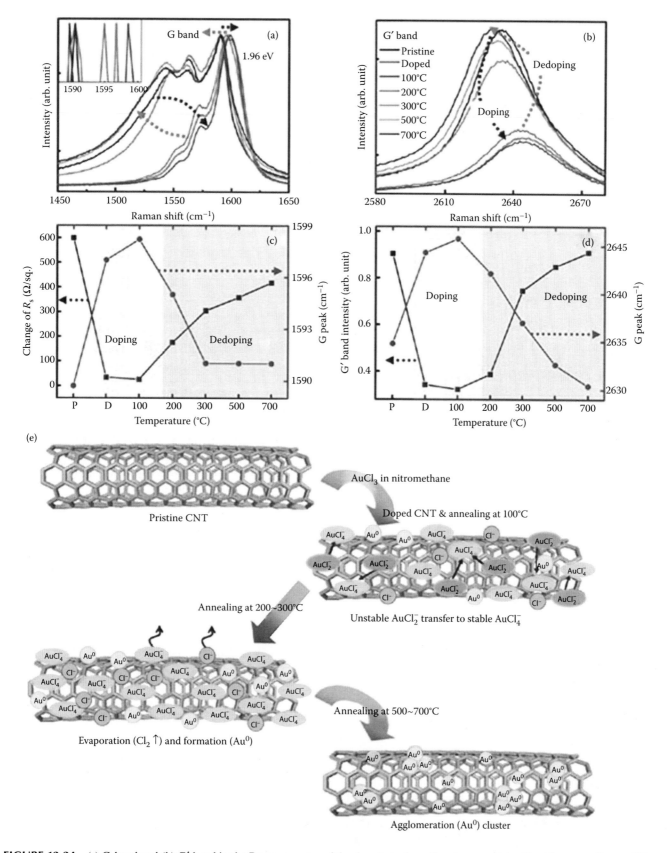

FIGURE 12.24 (a) G-band and (b) G′-band in the Raman spectra of the doped single wall carbon nanotube film after annealing at different temperatures. The inset shows the peak position shift. (c) The relationship between sheet resistance and G-band peak position and (d) the relationship between G′-band intensity and G′ band peak position at an excitation energy of 1.96 eV. (e) Cl-associated doping/dedoping mechanism. (Reprinted with permission from Kim, S. M. et al., Role of anions in the AuCl₃-doping of carbon nanotubes. *ACS Nano* 2, 1236–1242. Copyright 2011 American Chemical Society.)

Cl ions then directly extract electrons from the CNTs. The effect of thermal annealing on the AuCl$_3$-doped CNTs was investigated at various temperatures and under atmospheric conditions in Ar. XPS was used to identify Cl ions on the thermally annealed doped CNTs. The Cl atoms were completely desorbed from the CNTs thermally annealed at 500°C, and the sheet resistance of the film prepared with those CNTs was approximately 40% higher than that of the film prepared with unannealed doped CNTs. The TEM images showed that thermal annealing had aggregated the Au nanoparticles and that the size of the nanoparticles increased as a function of the annealing temperature. On the basis of these results, they assumed that the doping and de-doping mechanisms occurred through the following chemical reactions:

i. Ionization of CNTs and reduction of Au^{3+} to Au^0

$$2CNTs + 2AuCl_3 \rightarrow 2CNTs^+ + AuCl_2^- (Au\ I) + AuCl_4^- (Au\ III)$$
$$(12.5)$$

$$3AuCl_2^- \rightarrow 2Au^0 \downarrow + AuCl_4^- (Au\ III) + 2Cl^- \quad (12.6)$$

$$AuCl_4^- + 3CNTs \rightarrow 3CNTs^+ + Au^0 + 4Cl^- \quad (12.7)$$

ii. Neutralization of CNTs by Cl$^-$ anions

$$CNT^+ + Cl^- \rightarrow CNT - Cl \quad (12.8)$$

$$CNT^+ + AuCl_4^- \rightarrow CNT - AuCl_4 \quad (12.9)$$

iii. Desorption of Cl atoms and de-doping

$$2CNTs - Cl \rightarrow 2CNT + Cl_2 \uparrow \quad (12.10)$$

$$CNT - AuCl_4 \rightarrow CNT - Au + 2Cl_2 \uparrow \quad (12.11)$$

They concluded that the doping level of the CNT films was strongly correlated with the number of adsorbed Cl$^-$ anions, which was further supported by the change in sheet resistance and the shifts in the peaks of XPS and Raman spectra. The results showed that Cl adsorption plays a more important role in doping CNTs, while the reduction of Au^{3+} to Au^0 plays an important role as an intermediate precursor to accommodate subsequent Cl adsorption.

The mechanisms by which AuCl$_3$ doped and de-doped graphene are believed to be similar to those by which CNTs are doped and de-doped. There are a few similar reports on the mechanisms of doping and de-doping graphene thermally annealed at various temperatures [129]. Kwon et al. [130] reported the mechanism by which metal-chloride-doped graphene sheets degraded by plotting the degradation as a function of annealing temperature as shown in Figure 12.25. The sheet resistance of the thermally annealed doped graphene sheets increased from 500–700 Ω/□ to 10 kΩ/□, and the transmittance of the thermally annealed doped graphene sheets

decreased from 95% to 87%–91% at 550 nm. Furthermore, the work function of the doped graphene decreased from 4.7–5.1 to 4.2–4.5 eV after annealing. XPS was used to identify chloride anions and atoms in the thermally annealed doped graphene. The results of the XPS analysis showed that the chloride anions and atoms had completely disappeared from the doped graphene thermally annealed at 400°C. The SEM images demonstrated that annealing gathered the unstable metal cations and aggregated the metal particles. The degree by which the graphene sheets were doped was strongly related to not only the metal cations but also to the chloride anions. They concluded that the aggregation of metal particles and desorption of chloride ions from the carbon atoms in graphene degraded the properties of graphene by plotting the aggregation and desorption as functions of annealing temperature.

Furthermore, Kwon et al. [71] used a similar concept to investigate the effects of anions in Au complexes on the doping and degradation of graphene sheets. They used Au(OH)$_3$, Au$_2$S, AuBr$_3$, and AuCl$_3$ powders as dopants, each of which forms different anions in Au complexes. The sheet resistance decreased from 950 Ω/□ for the undoped graphene sheets to 820, 600, 530, and 300 Ω/□ for the graphene sheets doped using Au(OH)$_3$, Au$_2$S, AuBr$_3$, and AuCl$_3$, respectively. The work function increased from 4.3 eV for the undoped graphene to 4.6, 4.8, 5.0, and 4.9 eV for the graphene doped using Au(OH)$_3$, Au$_2$S, AuBr$_3$, and AuCl$_3$, respectively. However, thermal annealing dramatically increased the sheet resistances and decreased the work functions of the doped graphene sheets. Each Au complex had dissolved in the solvent to form Au^{3+}, its corresponding anions (X$^-$), and the highly reactive AuX$^-$ species, which can easily form AuX$_2^-$, Au^0, and X$^-$ on the graphene surface. However, not all of the Au complexes had completely dissolved into Au^{3+} and X$^-$ in the solvent because of the solubility and coordination of Au^{3+} and X$^-$ with dopant materials. The proposed mechanism by which the Au complexes doped graphene is as follows. The positively charged reactive carbon centers are first combined with anions. Au^{3+} cations on the graphene sheets then withdraw electrons from the graphene sheets to stabilize the sheets, leading to the formation of strong chemical bonds between the anions and the reactive carbon centers in the graphene sheets. Therefore, the electrons are likely extracted to the Au^{3+} cations and to the Br$^-$, S^{2-}, OH$^-$, and Cl$^-$ anions, whose electronegativities are higher than that of carbon atoms. However, thermally annealing the doped graphene at 400°C broke the graphene-anion covalent bonds and aggregated the Au nanoparticles. The sublimation or evaporation of the anions through thermal annealing seems to affect the density of electrons on the surface of the doped graphene, enabling it to recover to the surface state of pristine graphene. As a result, the work function of the thermally annealed doped graphene sheets returned to that of pristine graphene. Further, the sheet resistance of the thermally annealed doped graphene sheets dramatically increased as a function of temperature. The degree of change in sheet resistance was strongly related to the strength of the covalent bonds between the carbon atoms in graphene and the dopant anions. Among all the Au complex dopants, AuCl$_3$

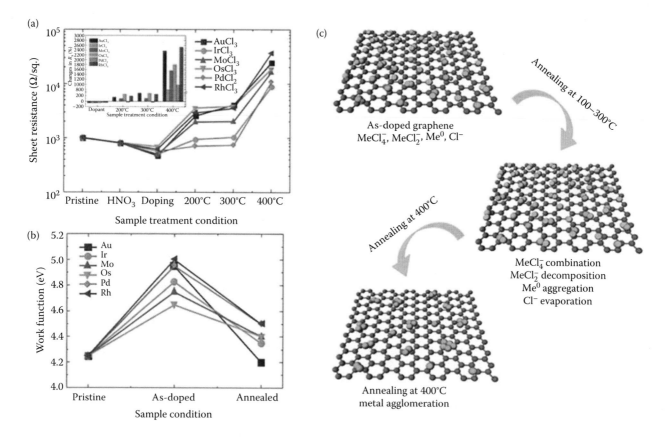

FIGURE 12.25 (a) The variation of sheet resistance as a function of doping and annealing temperature. The inset graph shows the changes in sheet resistance in percentage terms, with the sheet resistance of pristine graphene set at zero percent. (b) The work function variation graph was investigated as a function of the sample condition. All of the annealed samples show lower work function compared with their as-doped sample values. (c) The proposed mechanism of tannealing-induced degradation of doped graphene. The spheres indicate chlorine anions and metal cations, respectively. (Kwon, K. C. et al., 2013. Role of ionic chlorine in the thermal degradation of metal chloride-doped graphene sheets. *Journal of Materials Chemistry C*, 109:253–259. Reproduced by permission of The Royal Society of Chemistry.)

formed the weakest bond; therefore, the degree of change in sheet resistance and work function of the $AuCl_3$-doped graphene were the largest. The results suggested that the degree by which graphene was doped was related to the electronegativity of anions in the Au complex and that the degradation of graphene sheets was strongly related to not only the annealing temperature but also the strength of the bond between the Au^{3+} cation and the corresponding anions.

12.5 CONCLUSION

In this chapter, we discussed various methods of doping graphene, the application of doped graphene to electronic devices, and the de-doping phenomenon. Graphene must be doped to enhance the electrical properties of graphene sheets.

Substitution doping is used to introduce foreign atoms such as boron (p-type doping) or nitrogen (n-type doping) into graphene films while they are being synthesized, thus forming interstitial and substitutional impurities in the networks of carbon atoms in the films. Although substitution doping is an efficient method of modifying and determining the electronic structures of graphene sheets, some methods such as ion implantation and high-intensity plasma require the use of high power and thus produce many defects in graphene networks.

These defects induce the degradation of the electrical properties of p- or n-type doped graphene.

Chemical and surface charge transfer doping are the best methods of enhancing and modifying the electrical properties of graphene sheets. The Fermi level of pristine graphene can be shifted by transferring the charge from graphene sheets to the adsorbed dopant species or vice versa. Such adsorption-induced chemical doping can effectively optimize the Fermi level of graphene sheets without producing basal plane reactions, thereby preventing any damage to the carbon network in the graphene sheets. The method of transferring the surface charge has great potential in modifying the electrical and electronic properties of graphene. Graphene sheets can be modified to n- and p-type by doping graphene with electron-donating and electron-withdrawing functional groups, respectively, which can efficiently and easily optimize the electronic structure of graphene through the dipole moments formed between the adsorbed species and the carbon atoms of graphene.

Many researchers have tried to apply doped graphene sheets as electrodes or active channel materials in electronic devices. FET devices are usually fabricated with graphene as the channel material to understand the electronic structure of both pristine and doped graphene sheets. The performances of doped-graphene-based OPV devices are lower than those of ITO-based ones because the sheet resistance of graphene is higher than that of ITO and the surface of graphene is highly hydrophobic. Therefore, developing methods of doping and engineering interfaces are the most important aspects of improving the performance of graphene-based OPV cells. Doped graphene sheets have already been used as current-spreading electrodes, replacing the Ni/Au and ITO layers in LEDs. Combining layer-by-layer doped graphene with intercalated thin metal layers shows promise for replacing Ni/Au and ITO electrodes with doped graphene ones.

The two major obstacles to the practical application of graphene-based electronic devices are (i) the instability of doped graphene under ambient conditions and thermal stresses and (ii) the incompatibility of doped graphene with Si- or ITO-based device technologies. Despite the excellent performance of graphene, doped graphene degrades under atmospheric conditions and during thermal annealing. Therefore, graphene degradation should be overcome in order to practically apply doped graphene sheets to transparent conducting electrodes.

ACKNOWLEDGMENTS

This work was supported in part by the Basic Science Research Program (2011-0008994) and the Mid-Career Research Program (2011-0028752) through the National Research Foundation (NRF) of Korea funded by the Ministry of Education Science and Technology and in part by the Center for Green Airport Pavement Technology (CGAPT) of Chung-Ang University, Republic of Korea.

REFERENCES

1. Geim, A. K. 2009. Graphene: Status and prospects. *Science* 324:1530–1534.
2. Geim, A. K., Novoselov, K. S. 2007. The rise of graphene. *Nature Materials* 6:183–191.
3. Du, X., Skachko, I., Barker, A., Andrei, E. Y. 2008. Approaching ballistic transport in suspended graphene. *Nature Nanotechnology* 3:491–495.
4. Novoselov, K. S., Jiang, Z., Zhang, Y., Morozov, S. V., Stormer, H. L., Zeitler, U., Maan, J. C., Boebinger, G. S., Kim, P., Geim, A. K. 2007. Room-temperature quantum Hall effect in graphene. *Science* 315:1379.
5. Zhang, Y., Tan, Y.-W., Stormer, H. L., Kim, P. 2005. Experimental observation of the quantum Hall effect and Berry's phase in graphene. *Nature* 438:201–204.
6. Gunlycke, D., Lawler, H. M., White, C. T. 2007. Room-temperature ballistic transport in narrow graphene strips. *Physics Review B* 75:085418–085422.
7. Robinson, J. T., Perkins, F. K., Snow, E. S., Wei, Z., Sheehan, P. E. 2008. Reduced graphene oxide molecular sensors. *Nano Letters* 8:3137–3140.
8. Szafranek, B. N., Schall, D., Otto, M., Neumaier, D., Kurz, H. 2011. High on/off ratios in bilayer graphene field effect transistors realized by surface dopants. *Nano Letters* 11:2640–2643.
9. Meric, I., Dean, C. R., Young, A. F., Baklitskays, N., Tremblay, N. J., Nuckolls, C., Kim, P., Shepard, K. L. 2011. Channel length scaling in graphene field-effect transistors studied with pulsed current-voltage measurements. *Nano Letters* 11:1093–1097.
10. Pang, S., Hernandez, Y., Feng, X., Mullen, K. 2011. Graphene as transparent electrode material for organic electronics. *Advanced Materials* 23:2779–2795.
11. Yun, M.-J., Yeo, J.-S., Kim, J. Jeong, H.-G., Kim, D.-Y., Noh, Y.-J., Kim, S.-S., Ku, B.-C., Na, S.-I. 2011. Solution-processable reduced graphene oxide as a novel alternative to PEDOT:PSS hole transport layers for highly efficient and stable polymer solar cells. *Advanced Materials* 23:4923–4928.
12. Li, S.-S., Tu, K.-H., Lin, C.-C., Chen, C.-W., Chhowalla, M. 2010. Solution-processable graphene oxide as an efficient hole transport layer in polymer solar cells. *ACS Nano* 4:3169–3174.
13. Han, T.-H., Lee, Y., Choi, M.-R., Woo, S.-H., Bae, S.-H., Hong, B. H., Ahn, J.-H., Lee, T.-W. 2012. Extremely efficient flexible organic light-emitting diodes with modified graphene anode. *Nature Photonics* 6:105–110.
14. Zhang, Y., Small, J. P., Pontis, W. V., Kim, P. 2005. Fabrication and electric-field-dependent transport measurements of mesoscopic graphite devices. *Applied Physics Letters* 86:073104–073106.
15. First, P. N., de Heer, W. A., Seyller, T., Berger, C., Stroscio, J. A., Moon, J.-S. 2010. Epitaxial graphenes on silicon carbides. *MRS Bulletin* 35:296–305.
16. Becerril, H. A., Mao, J., Liu, Z., Stoltenberg, R. M., Bao, Z., Chen, Y. 2008. Evolution of solution-processed reduced graphene oxide films as transparent conductors. *ACS Nano* 2:463–470.
17. Pei, S., Zhao, J., Du, J., Ren, W., Cheng, H.-M. 2010. Direct reduction of graphene oxide films into highly conductive and flexible graphene films by hydrohalic acids. *Carbon* 48:4466–4474.
18. Zhu, Y., Murali, S., Cai, W., Li, X., Suk, J. W., Potts, J. R., Ruoff, R. S. 2010. Graphene and graphene oxide: Synthesis, properties, and applications. *Advanced Materials* 22:3906–3924.
19. Sutter, P. W., Flege, J.-I., Sutter, E. A. 2008. Epitaxial graphene on ruthenium. *Nature Materials* 7:406–411.
20. Coraux, J., N'Diaye, A. T., Busse, C., Michely, T. 2008. Structural coherency of graphene on Ir(111). *Nano Letters* 8:565–570.
21. Obraztsov, A. N., Obraztsova, E. A., Tyurnina, A. V., Zolotukhin, A. A. 2007. Chemical vapour deposition of thin graphite films of nanometer thickness. *Carbon* 45:2017–2021.
22. Mattevi, C., Kim, H., Chhowalla, M. 2011. A review of chemical vapour deposition of graphene on copper. *Journal of Materials Chemistry* 21:3324–3334.
23. Yu, Q., Lian, J., Siriponglert, S., Li, H., Chen. Y. P., Pei, S.-S. 2008. Graphene segregated on Ni surfaces and transferred to insulators. *Applied Physics Letters* 93:113103–113105.
24. Verma, V. P., Das, S., Lahiri, I., Choi, W. 2010. Large-area graphene on polymer film for flexible and transparent anode in field emission device. *Applied Physics Letters* 96:203108–203110.
25. Li, X., Zhu, Y., Cai, W., Borysiak, M., Han. B., Chen, D., Piner, R. D., Colombo, L., Ruoff, R. S. 2009. Transfer of large-area graphene films for high-performance transparent conductive electrodes. *Nano Letters* 9:4359–4363.
26. Lee, Y., Bae, S., Jang, H., Jang, S., Zhu, S.-E., Sim, S. H., Song, Y. I., Hong, B. H., Ahn, J.-H. 2010. Wafer-scale synthesis and transfer of graphene films. *Nano Letters* 10:490–493.
27. Bae, S., Kim, H., Lee, Y., Xu, X., Park, J.-S., Zheng, Y., Balakrishnan, J. et al. 2010. Roll-to-roll production of 30-inch graphene films for transparent electrodes. *Nature Nanotechnology* 5:574–578.

28. Caldwell, J. D., Anderson, T. J., Culbertson, J. C., Jernigan, G. G., Hobart, K. D., Kub, F. J., Tadjer, M. J. et al. 2010. Technique for the dry transfer of epitaxial graphene onto arbitrary substrates. *ACS Nano* 4:1108–1114.

29. Blake, P., Brimicombe, P. D., Nair, R. R., Booth, T. J., Jiang, D., Schedin, F., Ponomarenko, L. A. et al. 2008. Graphene-based liquid crystal device. *Nano Letters* 8:1704–1708.

30. Hernandez, Y., Nocolosi, V., Lotya, M., Blighe, F. M., Sun, Z., De, S., McGovern, I. T. et al. 2008. High-yield production of graphene by liquid-phase exfoliation of graphite. *Nature Nanotechnology* 3:563–568.

31. Park, J., Ruoff, R. S. 2010. Chemical methods for the production of graphenes. *Nature Nanotechnology* 4:217–224.

32. Kim, K. S., Zhao, Y., Jang, H., Lee, S. Y., Kim, J. M., Kim. K. S., Ahn, J.-H., Kim, P., Choi, J.-Y., Hong, B. H. 2009. Large-scale pattern growth of graphene films for stretchable transparent electrodes. *Nature* 457:706–710.

33. Reina, A., Jia, X., Ho, J., Nezich, D., Son, H., Bulovic, V., Dresselhaus, M. S., Kong, J. 2009. Large area, few-layer graphene films on arbitrary substrates by chemical vapour deposition. *Nano Letters* 9:30–35.

34. Li, X., Cai, W., An, J., Kim, S., Nah, J., Yang, D., Piner, R. et al. 2009. Large-area synthesis of high-quality and uniform graphene films on copper foils. *Science* 324:1312–1314.

35. Wu, J., Becerril, H. A., Bao, Z., Liu, Z., Chen, Y., Peumans, P. 2008. Organic solar cells with solution-processed graphene transparent electrodes. *Applied Physics Letters* 92:263302–263305.

36. Gao, Y., Yip, H.-L., Chen, K.-S., O'Malley, K. M., Acton, O., Sun, Y., Ting, G., Chen, H., Jen, A. K.-Y. 2011. Surface doping of conjugated polymers by graphene oxide and its application for organic electronics devices. *Advanced Materials* 23:1903–1908.

37. Liu, J., Xue, Y., Gao, Y., Yu, D., Durstock, M., Dai, L. 2012. Hole and electron extraction layers based on graphene oxide derivatives for high-performance bulk heterojunction solar cells. *Advanced Materials* 24:2228–2233.

38. Park, Y., Choi, K. S., Kim, S. Y. 2012. Graphene oxide/PEDOT:PSS and reduced graphene oxide/PEDOT:PSS hole extraction layers in organic photovoltaic cells. *Physica Status Solidi A* 209:1363–1368.

39. Jo, G., Choe, M., Cho, C.-Y., Kim, J. H., Park, W., Lee, S., Hong. W.-K. et al. 2010. Large-scale patterned multi-layer graphene films as transparent conducting electrodes for GaN light-emitting diodes. *Nanotechnology* 21:175201–175206.

40. Park, H., Howden, R. M., Barr, M. C., Bulovic, V., Gleason, K., Kong, J. 2012. Organic solar cells with graphene electrodes and vapour printed poly(3,4-ethylenedioxythiophene) as the hole transport layers. *ACS Nano* 6:6370–6377.

41. Choe, M., Lee, B. H., Jo, G., Park, J., Park, W., Lee, S., Hong, W.-K. et al. 2010. Efficient bulk-heterojunction photovoltaic cells with transparent multi-layer graphene electrodes. *Organic Electronics* 11:1864–1869.

42. Wan, X., Long, G., Huang, L., Chen, Y. 2011. Graphene—A promising material for organic photovoltaic cells. *Advanced Materials* 23:5342–5358.

43. Seo, T. H., Lee, K. J., Oh, T. S., Lee, Y. S., Jeong, H., Park, A. H., Kim, H. et al. 2011. Graphene network on indium tin oxide nanodot nodes for transparent and current spreading electrode in InGaN/GaN light emitting diode. *Applied Physics Letters* 98:251114–251116.

44. Cox, M., Gorodetsky, A., Kim, B., Kim, K. S., Jia, Z., Kim, P., Nuckolls, C., Kymissis, I. 2011. Single-layer graphene cathodes for organic photovoltaics. *Applied Physics Letters* 98:123303–123305.

45. Sun, T., Wang, Z. L., Shi, Z. J., Ran, G. Z., Xu, W. J., Wang, Z. Y., Li, Y. Z., Dia, L., Qin, G. G. 2010. Multilayered graphene used as anode of organic light-emitting devices. *Applied Physics Letters* 96:133301–133303.

46. Reich, S., Thomsen, C. 2004. Raman spectroscopy of graphite. *Philosophical Transactions of the Royal Society of London A* 362:2271–2288.

47. Podila, R., Rao, R., Tsuchikawa, R., Ishigami, M., Rao, A. M., 2012. Raman spectroscopy of folded and scrolled graphene. *ACS Nano* 6:5784–5790.

48. Ci, L., Song, L., Jin, C., Jariwala, D., Wu, D., Li, Y., Srivastava, A. et al. 2010. Atomic layers of hybridized boron nitride and graphene domains. *Nature Materials* 9:430–435.

49. Wu, T., Shen, H., Sun, L., Cheng, B., Liu, B., Shen, J. 2012. Nitrogen and boron doped monolayer graphene by chemical vapor deposition using polystyrene, urea and boric acid. *New Journal of Chemistry* 36:1385–1391.

50. Tang, Y.-B., Yin, L.-C., Yang, Y., Bo, X.-H., Cao, Y.-L., Wang, H.-E., Zhang, W.-J. et al. 2012. Tunable band gaps and p-type transport properties of boron-doped graphene by controllable ion doping using reactive microwave plasma. *ACS Nano* 6:1970–1978.

51. Faccio, R., Fernandez-Werner, L., Pardo, H., Goyenola, C., Ventura, O. N., Mombru, A. W. 2010. Electronic and structural distortions in graphene induced by carbon vacancies and boron doping. *Journal of Physical Chemistry C* 114:18961–18971.

52. Sheng, Z.-H., Gao, H.-L., Bao, W.-J., Wang, F.-B., Xia, X.-H. 2012. Synthesis of boron doped graphene for oxygen reduction reaction in fuel cells. *Journal of Materials Chemistry* 22:390–395.

53. Panchakarla, L. S., Subrahmanyam, K. S., Saha, S. K., Govindaraj, A., Krishnamurthy, H. R., Waghmare, U. V., Rao, C. N. R. 2009. Synthesis, structure, and properties of boron- and nitrogen-doped graphene. *Advanced Materials* 21:4726–4730.

54. Wei, D., Liu, Y., Wang, Y., Zhang, H., Huang, L., Yu, G. 2009. Synthesis of n-doped graphene by chemical vapor deposition and its electrical properties. *Nano Letters* 9:1752–1758.

55. Jin, J., Yao, J., Kittrell, C., Tour, J. M. 2011. Large-scale growth and characterizations of nitrogen-doped monolayer graphene sheets. *ACS Nano* 5:4112–4117.

56. Hwang, J. O., Park, J. S., Choi, D. S., Kim, J. Y., Lee, S. H., Lee, K. E., Kim, Y.-H., Song, M. H., Yoo, S., Kim, S. O. 2012. Work function-tunable, n-doped reduced graphene transparent electrodes for high-performance polymer light-emitting diodes. *ACS Nano* 6:159–167.

57. Wang, X., Li, X., Zhang, L., Yoon, Y., Weber, P. K., Wang, H., Guo, J., Dai, H. 2009. N-doping of graphene through electrothermal reactions with ammonia. *Science* 324:768–771.

58. Guo, B., Liu, Q., Chen, E., Zhu, H., Fang, L., Gong, J. R. 2010. Controllable n-doping of graphene. *Nano Letters* 10:4975–4980.

59. Li, X., Wang, H., Robinson, J. T., Sanchez, H., Diankov, G., Dai, H. 2009. Simultaneous nitrogen doping and reduction of graphene oxide. *Journal of the American Chemistry Society* 131:15939–15944.

60. Wang, Y., Shao, Y., Matson, D. W., Li, J., Lin, Y. 2010. Nitrogen-doped graphene and its application in electrochemical biosensing. *ACS Nano* 4:1790–1798.

61. Soin, N., Roy, S. S., Roy, S., Hazra, K. S., Misra, D. S., Lim, T. H., Hetherington, C. J., McLaughlin, J. A. 2011. Enhanced and stable field emission form *in situ* nitrogen-doped few-layered graphene nanoflakes. *Journal of Physical Chemistry C* 115:5366–5372.

62. Shao, Y., Zhang, S., Engelhard, M. H., Li, G., Shao, G., Wang, Y., Aksay I. A., Lin, Y. 2010. Nitrogen-doped graphene and its electrochemical applications. *Journal of Materials Chemistry* 20:7491–7696.

63. Liu, H., Zhu, D. 2011. Chemical doping of graphene. *Journal of Materials Chemistry* 21:3335–3345.

64. Shi, Y., Kim, K. K., Reina, A., Hofmann, M., Li, L.-J., Kong, J. 2010. Work-function engineering of graphene electrode via chemical doping. *ACS Nano* 4:2689–2694.

65. Kong, B.-S., Geng, J., Jung, H.-T. 2009. Layer-by-layer assembly of graphene and gold nanoparticles by vacuum filtration and spontaneous reduction of gold ions. *Chemistry Communications* 2174–2176.

66. Shin, H.-J., Choi, W. M., Choi, D., Han, G. H., Yoon, S.-M., Park, H.-K., Kim, S.-W. et al. 2010. Control of electronic structure of graphene by various dopants and their effects on nanogenerator. *Journal of the American Chemistry Society* 132:15603–15609.

67. Kim, K. K., Reina, A., Shi, Y., Park, H., Li, L.-J., Kong, J. 2010. Enhancing the conductivity of transparent graphene films via doping. *Nanotechnology* 21:285205–285210.

68. Gunes, F., Shin, H.-J., Biswas, C., Han, G. H., Kim, E. S., Chae, S. J., Choi, J.-Y., Lee, Y. H. 2010. Layer-by-layer doping of few-layer graphene film. *ACS Nano* 4:4595–4600.

69. Benayad, A., Shin, H.-J., Park, H. K., Yoon, S.-M., Kim, K. K., Jin, M. H., Jeong, H.-K., Lee, J. C., Choi, J.-Y., Lee, Y. H. 2009. Controlling work function of reduced graphite oxide with Au-ion concentration. *Chemical Physics Letters* 475:91–95.

70. Kwon, K. C., Choi, K. S., Kim, S. Y. 2012. Increased work function in few-layer graphene sheets via metal chloride doping. *Advanced Functional Materials* 22:4724–4731.

71. Kwon, K. C., Kim, B. J., Lee, J.-L., Kim, S. Y. 2013. Effect of anions in Au complexes on doping and degradation of graphene. *Journal of Materials Chemistry C* 1:2463–2469.

72. Krishnamurthy, S., Lightcap, I. V., Kamat, P. V. 2011. Electron transfer between methyl viologen radicals and graphite oxide: Reduction, electron storage and discharge. *Journal of Photochemistry and Photobiology A: Chemistry* 221:214–219.

73. Jeong, H. K., Kim, K., Kim, S. M., Lee, Y. H. 2010. Modification of the electronic structures of graphene by viologen. *Chemical Physics Letters* 498:168–171.

74. Kwon, K. C., Choi, K. S., Kim, B. J., Lee, J.-L., Kim, S. Y. 2012. Work-function decrease of graphene sheet using alkali metal carbonates. *Journal of Physical Chemistry C* 116:26586–26591.

75. Kasry, A., Kuroda, M. A., Martyna, G. J., Tulevski, G. S., Bol, A. A. 2010. Chemical doping of large-area stacked graphene films for use as transparent, conducting electrodes. *ACS Nano* 4:3839–3844.

76. Zheng, Q. B., Gudarzi, M. M., Wang, S. J., Geng. Y., Li, Z., Kim, J.-K. 2011. Improved electrical and optical characteristics of transparent graphene thin films produced by acid and doping treatments. *Carbon* 49:2905–2916.

77. Wassei, J. K., Cha, K. C., Tung, V. C., Yang, Y., Kaner, R. B. 2011. The effects of thionyl chloride on the properties of graphene and graphene-carbon nanotube composites. *Journal of Materials Chemistry* 21:3391–3396.

78. Parekh, B. B., Fanchini, G., Eda, G., Chhowalla, M. 2007. Improved conductivity of transparent single-wall carbon nanotube thin films via stable postdeposition functionalization. *Applied Physics Letters* 90:121913–121915.

79. Dong, X., Fu, D., Fang, W., Shi, Y., Chen, P., Li, J.-L. 2009. Doping single-layer graphene with aromatic molecules. *Small* 12:1422–1426.

80. Miao, X., Tongay, S., Petterson, M. K., Berke, K., Rinzler, A. G., Appleton, B. R., Heberd, A. F. 2012. High efficiency graphene solar cells by chemical doping. *Nano Letters* 12:2745–2750.

81. Farmer, D. B., Golizadeh-Mojarad, R., Perebeions, V., Lin, Y.-M., Tulevski, G. S., Tsang, J. C., Avouris, P. 2009. Chemical doping and electron-hole conduction asymmetry in graphene devices. *Nano Letters* 9:388–392.

82. Gupta. A., Chen, G., Joshi, P., Tadigadapa, S., Eklund, P. C. 2006. Raman scattering from high-frequency phonons in supported n-graphene layer films. *Nano Letters* 6:2667–2673.

83. Ferrari, A. C., Meyer, J. C., Scardaci, V., Casiraghi, C., Lazzeri, M., Mauri, F., Piscanec, S. et al. 2006. Raman spectrum of graphene and graphene layers. *Physical Review Letters* 97:187401–187404.

84. Graf, D., Moliter, F., Ensslin, K., Stampfer, C., Jungen, A., Hierold, C., Wirtz, L. 2007. Spatially resolved Raman spectroscopy of single- and few-layer graphene. *Nano Letters* 7:238–242.

85. Das, A., Pisana, S., Chakranorty, B., Piscanec, S., Saha, S. K., Waghmare, U. V., Novoselov, K. S. et al. 2008. Monitoring dopants by Raman scattering in an electrochemically top-gated graphene transistor. *Nature Nanotechnology* 3:210–215.

86. Yan, J., Zhang, Y., Kim, P., Pinczuk, A. 2007. Electric field effect tuning of electron–phonon coupling in graphene. *Physical Review Letters* 98:166802–166805.

87. Coletti, C., Riedl, C., Lee, D. S., Krauss, B., Patthey, L., von Klitzing, K., Smet, J. H., Starke, U. 2010. Charge neutrality and band-gap tuning of epitaxial graphene on SiC by molecular doping. *Physical Review B* 81:235401–235408.

88. Chen, W., Chen, S., Qi, D. C., Gao, X. Y., Wee, A. T. S. 2007. Surface transfer p-type doping of epitaxial graphene. *Journal of American Chemistry Society* 129:10418–10422.

89. Wang, X., Xu, J.-B., Xie, W., Du, J. 2011. Quantitative analysis of graphene doping by organic molecular charge transfer. *Journal of Physical Chemistry C* 115:7596–7602.

90. Ishikawa, R., Bando, M., Morimoto, Y., Sandhu, A. 2011. Doping graphene films via chemically mediated charge transfer. *Nanoscale Research Letters* 6:111–115.

91. Li, B., Zhou, L., Wu, D., Peng, H., Yan, K., Zhou, Y., Liu, Z. 2011. Photochemical chlorination of graphene. *ACS Nano* 5:5957–5961.

92. Wu, J., Xie, L., Li, Y., Wang, H., Ouyang, Y., Guo, J., Dai, H. 2011. Controlled chlorine plasma reaction for noninvasive graphene doping. *Journal of the American Chemistry Society* 133:19668–19671.

93. Lee, W. H., Suk, J. W., Chou, H., Lee, J., Hao, Y., Wu, Y., Piner, R., Akinwande, D., Kim, K. S., Ruoff, R. S. 2012. Selective-area fluorination of graphene with fluoropolymer and laser irradiation. *Nano Letters* 12:2374–2378.

94. Lee, W. H., Suk, J. W., Lee, J., Hao, Y., Park, J., Yang, J. W., Ha, H.-W. et al. 2012. Simultaneous transfer and doping of CVD-grown graphene by fluoropolymer for transparent conductive films on plastic. *ACS Nano* 6:1284–1290.

95. Park, J., Lee, W. H., Huh, S., Sim, S. H, Kim, S. B., Cho, K., Hong, B. H., Kim, K. S. 2011. Work-function engineering of graphene electrodes by self-assembled monolayers for high-performance organic field-effect transistors. *Journal of Physical Chemistry Letters* 2:841–845.

96. Zhang, Y.-H., Zhou, K.-G., Xie, K.-F., Zeng, J., Zhang, H.-L., Peng, Y. 2010. Tuning the electronic structure and transport properties of graphene by noncovalent functionalization: Effects of organic donor, acceptor, and metal atoms. *Nanotechnology* 21:065201–065207.

97. Jung, N., Kim, N., Jockusch, S., Turro, N. J., Kim, P., Brus, L. 2009. Charge transfer chemical doping of few layer graphenes: Charge distribution and band gap formation. *Nano Letters* 9:4133–4137.

98. Song, W., Kim, Y., Kim, S. H., Kim, S. Y., Cha, M.-J., Song, I., Jung, D. S. et al. 2013. Homogeneous and stable p-type doping of graphene be MeV electron beam-stimulated hybridization with ZnO thin films. *Applied Physics Letters* 102:053103–053105.

99. Lee, T.-W., Byun, Y., Koo, B.-W., Kang, I.-N., Lyu, Y.-Y., Lee, C. H., Pu, L., Lee, S. Y. 2005. All-solution-processed n-type organic transistors using a spinning metal process. *Advanced Materials* 17:2180–2184.

100. Forrest, S. R. 2004. The path to ubiquitous and low-cost organic electronic appliances on plastic. *Nature* 428:911–918.

101. Guo, B., Fang, L., Zhang, B., Gong, J. R. 2011. Graphene doping: A review. *Inscience Journal: Nanotechnology* 1:80–89.

102. Schwierz, F. 2010. Graphene transistors. *Nature Nanotechnology* 5:487–496.

103. Novoselov, K. S., Geim, A. K., Morozov, S. V., Jiang, D., Zhang, Y., Dubonos, S. V., Grigorieva, I. V., Firsov, A. A. 2004. Electric field effect in atomically thin carbon films. *Science* 306:666–669.

104. Kim, K., Choi, J.-Y., Kim, T., Cho, S.-H., Chung, H.-J. 2011. A role for graphene in silicon-based semiconductor devices. *Nature* 479:338–344.

105. Han, M. Y., Ozylmaz, B., Zhang, Y., Kim, P. 2007. Energy band-gap engineering of graphene nanoribbons. *Physical Review Letters* 98:206805–206808.

106. Han, M. Y., Brant, J. C., Kim, P. 2010. Electron transport in disordered graphene nanoribbons. *Physical Review Letters* 104:056801–056804.

107. Brabec, C. J., Gowrisanker, S., Halls, J. J. M., Laird, D., Jia, S., Williams, S. P. 2010. Polymer-fullerene bulk-heterojunction solar cells. *Advanced Materials* 22:3839–3856.

108. Brabec, C. J., Sariciftci, N. S., Hummelen, J. C. 2001. Plastic solar cells. *Advanced Functional Materials* 11:15–26.

109. Blom, P. W. M., Mihailetchi, V. D., Koster, L. J. A., Markov, D. E. 2007. Device physics of polymer: Fullerene bulk heterojunction solar cells. *Advanced Materials* 19:1551–1566.

110. Kim, J. Y., Kim, S. H., Lee, H.-H., Lee, K., Ma, W., Gong, X., Heeger, A. J. 2006. New architecture for high-efficiency polymer photovoltaic cells using solution-based titanium oxide as an optical spacer. *Advanced Materials* 18:572–576.

111. Ma, W., Yang, C., Gong, X., Lee, K., Heeger, A. J. 2005. Thermally stable, efficient polymer solar cells with nanoscale control of the interpenetrating network morphology. *Advanced Functional Materials* 15:1617–1622.

112. Shrotriya, V., Li, G., Yao, Y., Chu, C.-W., Yang, Y. 2006. Transition metal oxides as the buffer layer for polymer photovoltaic cells. *Applied Physics Letters* 88:073508–073510.

113. Hsu, C.-L., Lin, C.-T., Huang, J.-H., Chu, C.-W., Wei, K.-H., Li, L.-J. 2012. Layer-by-layer graphene/TCNQ stacked films as conducting anodes for organic solar cells. *ACS Nano* 6:5031–5039.

114. Lee, S., Yeo, J.-S., Ji, Y., Cho, C., Kim, D.-Y., Na, S.-I., Lee, B. H., Lee, T. 2012. Flexible organic solar cells composed by P3HT:PCBM using chemically doped graphene electrodes. *Nanotechnology* 23:344013–344018.

115. Wang, Y., Tong, S. W., Xu, X. F., Ozyilmaz, B., Loh, K. P. 2011. Interface engineering of layer-by-layer stacked graphene anodes for high-performance organic solar cells. *Advanced Materials* 23:1514–1518.

116. Kwon, K. C., Dong, W. J., Jung, G. H., Ham, J., Lee, J.-L., Kim, S. Y. 2013. Extension of stability in organic photovoltaic cells using UV/ozone-treated graphene sheets. *Solar Energy Materials and Solar Cells* 109:148–154.

117. Song, J. O., Ha, J.-S., Seong, T.-Y. 2010. Ohmic-contact technology for GaN-based light-emitting diodes: Role of p-type contact. *IEEE Transactions on Electron Devices* 57:42–59.

118. Smalc-Koziorowska, J., Grzanka, S., Litwin-Staszewska, E., Piotrzkowki, R., Nowak, G., Leszczynski, M., Perlin, P., Talik, E., Kozubowski, J., Krukowski, S. 2010. Ni–Au contacts to p-type GaN—Structure and properties. *Solid-State Electronics* 54:701–709.

119. Jung, S.-P., Ullery, D., Lin, C.-H., Lee, H. P., Lim, J.-H., Hwang, D.-K., Kim, J.-Y., Yang, E.-J., Park, S.-J. 2005. High-performance GaN-based light-emitting diode using high-transparency Ni/Au/Al-doped ZnO composite contacts. *Applied Physics Letters* 87:181107–181109.

120. Chang, K.-M., Chu, J.-Y., Cheng, C.-C. 2005. Investigation of indium-tin-oxide ohmic contact to p-GaN and its application to high-brightness GaN-based light-emitting diodes. *Solid-State Electronics* 49:1381–1386.

121. Chandramohan, S., Kang, J. H., Katharria, Y. S., Han, N., Beak, Y. S., Ko, K. B., Park, J. B., Kim, H. K., Suh, E.-K., Hong, C.-H. 2012. Work-function-tuned multilayer graphene as current spreading electrode in blue light-emitting diodes. *Applied Physics Letters* 100:023502–023504.

122. Lee, J. M., Jeong, H. Y., Choi, K. J., Park, W. I. 2011. Metal/graphene sheets as p-type transparent conducting electrodes in GaN light emitting diodes. *Applied Physics Letters* 99:041115–041117.

123. Shim, M., Javey, A., Kam, N. W. S., Dai, H. 2001. Polymer functionalization for air-stable n-type carbon nanotube field-effect transistors. *Journal of the American Chemistry Society* 123:11512–11513.

124. Siddons, G. P., Merchin, D., Back, J. H., Jeong, J. K., Shim, M. 2004. Highly efficient gating and doping of carbon nanotubes with polymer electrolytes. *Nano Letters* 4:927–931.

125. Kang, B. R., Yu, W. J., Kim, K. K., Park, H. K., Kim, S. M., Park, Y., Kim, G. et al. 2009. Restorable type conversion of carbon nanotube transistor using pyrolytically controlled antioxidizing photosynthesis coenzyme. *Advanced Functional Materials* 19:2553–2559.

126. Kim, S. M., Jang, J. H., Kim, K. K., Park, H. K., Bae, J. J., Yu, W. J., Lee, I. H. et al. 2009. Reduction-controlled viologen in bisolvent as an environmentally stable n-type dopant for carbon nanotubes. *Journal of the American Chemistry Society* 131:327–331.

127. Yoon, S.-M., Kim, U. J., Beneyad, A., Lee, I. H., Son, H., Shin, H.-J., Choi, W. M. et al. 2011. Thermal conversion of electronic and electrical properties of $AuCl_3$-doped single-walled carbon nanotubes. *ACS Nano* 5:1353–1359.

128. Kim, S. M., Kim, K. K., Jo, Y. W., Park, M. H., Chae, S. J., Duong, D. L., Yang, C. W., Kong, J., Lee., Y. H. 2011. Role of anions in the $AuCl_3$-doping of carbon nanotubes. *ACS Nano* 5:1236–1242.

129. Duong, D. L., Lee, I. H., Kim, K. K., Kong, J., Lee, S. M., Lee, Y. H. 2010. Carbon nanotube doping mechanism in a salt solution and hygroscopic effect: Density functional theory. *ACS Nano* 4:5430–5436.

130. Kwon, K. C., Kim, B. J., Lee, J.-L., Kim, S. Y. 2013. Role of ionic chlorine in the thermal degradation of metal chloride-doped graphene sheets. *Journal of Materials Chemistry C* 1:253–259.

13 Chemical Modifications of Graphene via Covalent Bonding

Liang Cui, Dongjiang Yang, and Jingquan Liu

CONTENTS

Abstract 207
13.1 Introduction 207
13.2 Covalent Modifications of Graphene 208
    13.2.1 Modifications of Graphene via Free Radical Addition Reactions 208
        13.2.1.1 Modifications of Graphene via Diazonium Salt Reactions 208
        13.2.1.2 Other Free Radical Reactions 211
    13.2.2 Functionalization of Graphene via Cyclization Reactions 211
        13.2.2.1 Functionalization with Arynes 212
        13.2.2.2 Functionalization with Nitrenes 212
        13.2.2.3 Functionalization with Carbenes 214
        13.2.2.4 Nucleophilic Addition 214
    13.2.3 Modifications of Graphene via Substitutional Reaction and Chemical Doping 215
    13.2.4 Modifications of Graphene via Residual Groups 215
13.3 Covalent Modification of GO 215
13.4 Applications of Covalent Functionalized Graphene 217
    13.4.1 Electronic and Optic Devices 217
    13.4.2 Polymer Nanocomposites 217
    13.4.3 Biomaterials 218
    13.4.4 Other Applications 218
13.5 Conclusions and Perspectives 218
Acknowledgments 218
References 218

ABSTRACT

Graphene, which consist of a single atom sheet conjugated sp^2 carbon atoms, has triggered intensive interest in various areas due to its excellent properties and inexpensive sources (graphite) since its discovery in 2004. However, its hydrophobicity and easy aggregation, which can be changed by chemical modifications, make it hard to process. Covalent binding of molecules to graphene represents an interesting alternative for the development of novel graphene derivatives with a compendium of interfacial interactions. Covalent linking, the most convenient method in terms of solution processing, but certainly not the best method where electronic application is concerned, is stronger than traditional non-covalent linking, such as van der Walls force, hydrogen bonding, and so on. This feature article provides an overview of the strategies currently employed to covalently modify graphene. We focus on the modifications of pure graphene other than graphene oxide as the latter has plenty of functional groups and can easily be modified. The modification methods of graphene and graphene oxide, as well as the applications of functionalized graphene and graphene oxide, are summarized.

13.1 INTRODUCTION

Graphene's two-dimensional, highly conjugated, and single-atom layered structure endow it with many unique properties, for example, high Young's modulus, surface area and fracture strength, and fast mobility of charge carriers [1,2]. These remarkable properties, along with the ballistic electronic transport and quantum Hall effect [3,4] cause graphene the world's thinnest, strongest, and stiffest material, as well as being a significant potential material in electronic and optic devices, polymer composite materials, and biomedical materials [5–7]. Graphene has proved to be a more promising carbon allotrope for some applications than other nanocarbon family members such as 1-dimensional nanotubes and 0-dimensional fullerenes.

A number of methods have been employed to prepare graphene nanosheets, among which the mostly applied methods include the exfoliation of graphite [8], reduction of graphene oxide (GO) [9], and chemical vapor deposition (CVD) [10,11]. The exfoliation and CVD methods can be used to obtain less-defected, large-sized pristine graphene that is preferred for the fabrication of electronic devices. However, these two methods are limited by the low production and high cost at

TABLE 13.1
Covalent Modification and Property of Graphene

Modified Agent	Typical Parameters of Experiments	Properties	Reference
Diazonium salts	Stirring at room temperature	Solubility	[17,35–37]
		Conductivity	[15,16,38,40–42]
		Surface potential	[43]
Benzoyl peroxide (BPO)	Intense laser irradiation	Conductivity Hole doping	[52,53]
Trifluoromethylphenylene (CF3Ph)	Electrochemistry method	Water contact angles	[54]
Styrene	High-intensity ultrasound	Dispersibility	[55]
Carbene	Stirring at room temperature	Dispersibility	[76–80]
Nitrene	Refluxing in organic solvent	Dispersibility	[18,56,67–70]
Aryne	Diels–Alder reaction	Dispersibility	[59–63]
Cyclopropane adducts	Microwave irradiation	Dispersibility	[82]
Diborane, urea	Hydrothermal	Electronics	[85–89]
Functional groups	GO	Dispersibility	[91,92,95,99,101]

current stage. The most economical and mass production method to prepare graphene is reduction of GO that is usually prepared by the oxidation of graphite [12]. However, the reduced graphene oxide (RGO) has some drawbacks such as more defects, small size, and some residual groups which are concerned for electronic application. Other methods, such as arc discharge, unzipping carbon nanotubes, epitaxial growth on silicon carbide are also reported [13].

However, there are two major problems that need to be tackled for both pristine graphene and RGO before their applications. The first is the low dispersibility of graphene in common organic and inorganic solvents. In some cases the good dispersion of graphene in common solvents is a crucial move toward the formation of homogeneous nanocomposites. The second is the zero bandgap of graphene, which is the obstacle for the application in some electronic devices such as transistors. Chemical modifications are convenient protocols to address both of these issues through the introduction of variable chemical decoration [14]. Chemical functionalization not only prevents the irreversible agglomeration graphene sheets via π–π stacking interactions, the bandgap could also be opened up as required for the fabrication of specific electronics.

Generally speaking, graphene can be modified with covalent and noncovalent methods, both of which have their advantages and disadvantages [15]. Noncovalent methods include π–π stacking interactions [16,17], electrostatic interaction, hydrogen bonding, coordination bonds, and van der Waals forces [18]. Noncovalent modification can maximally preserve graphene's natural structure; however, the interactions between functionalities and graphene surface are relatively weak. Therefore, it is not suitable for some applications where strong interactions are required.

In this chapter, we focus on the chemical modification of graphene using covalent bonding. Chemical covalent functionalization can improve the properties of graphene including opening its bandgap, tuning conductivity, and improving solubility and stability. The covalent functionalization of graphene was also discussed from both experimental and theoretical aspects. Since graphene is usually prepared by the reduction of

GO, incomplete reduction should leave some oxygen-containing functional groups, through which covalent modifications can be conducted. Therefore, covalent modification of GO and the resulting applications are also included (Table 13.1).

13.2 COVALENT MODIFICATIONS OF GRAPHENE

Since graphene consists of sp^2 hybridized carbon atoms covalent modification of graphene can be achieved via opening the unsaturated double bond, resulting in sp^3 hybridized carbon network. This can be reached by two general routes: The first is to form one covalent bond between the modifier and graphene surface via free radical reactions, substitution, or rearrangement and the second is to form two covalent bonds via cyclization reaction between the modifier and graphene surface via cycloaddition or nucleophilic addition. The cycloaddition can be obtained using different functionalities: carbene, nitrene, or aryne. The common covalent modifications involve radical reaction [19–21], cyclization reaction [22,23], residual group reaction of RGO and functional groups reactions of GO [24] (Figure 13.1). Other covalent methods are also used to modify graphene such as defects reaction [25,26]. Tessonnier and coworkers developed a new approach based on Hansen's solubility theory to chemically modify RGO through the attachment of decyl groups on a variety of defect sites already present on RGO surfaces depending on the constraints of target application. They grafted alkyl chains onto RGO sheets and found that the attached alkyl chains also act as spacers, which improve the exfoliation of the sheets and prevent them from reagglomeration [26].

13.2.1 MODIFICATIONS OF GRAPHENE VIA FREE RADICAL ADDITION REACTIONS

13.2.1.1 Modifications of Graphene via Diazonium Salt Reactions

Diazonium salts react with graphene through a free radical mechanism which has been previously achieved on fullerene, graphite, and carbon nanotubes. Pinson and coworkers were

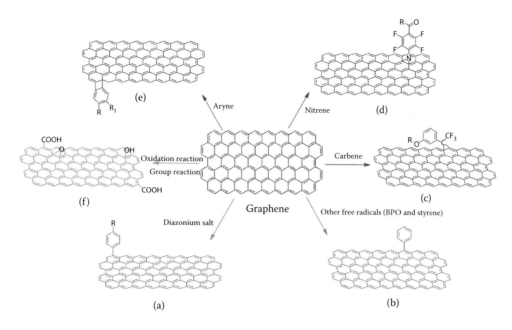

FIGURE 13.1 Modification illustration of graphene with radicals (a and b), cyclization reaction (c, d, and e), and group reaction (f). (Reprinted with permission from Park, J. Covalent functionalization of graphene with reactive intermediates. *Acc Chem Res* 46:181–89. Copyright 2013 American Chemical Society.)

the first to describe the reaction mechanism between diazonium salt and carbon–carbon double bond [27] and recently other groups also have found that aryl groups can be covalently attached to graphene surfaces through free radical reactions [17,19,28–30]. This methodology has been successfully applied to open the bandgap of graphene and to improve its solubility. The intensive interest in using aryl diazonium salts as modifiers lies in their ease of preparation, rapid (electro) reduction, large choice of reactive functional groups, and strong ary-surface covalent bonding [31,32].

Modification of graphene via diazonium salt reactions could be used to improve the solubility of graphene, prevent aggregation in different solvents, and improve the processability as well as enhance the interactions with organic polymers [33–38]. Lomeda and coworkers [21] were the first to functionalize RGO with high amounts of varying aryl diazonium salts, and found that the functionalized RGO can disperse readily in polar aprotic solvents, allowing alternative avenues for simple incorporation into different polymer matrices. Other groups also reported stable graphene dispersions through diazonium reactions [39,40]. Specifically, Zhu and coworkers [41] developed two routes for the functionalization of graphene nanoribbons, the diazonium salt reaction (route I) and the *in situ* functionalization of GNRs using organic nitrite and aromatic amine (route II), allowing the nanoribbons to become soluble in organic solvents (Figure 13.2a and b).

Since diazonium salt reactions will partially destroy the π-conjugated structure of graphene, they can be employed to

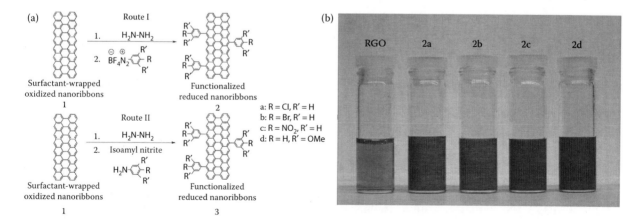

FIGURE 13.2 (a) Syntheses routes I and II for the production of FG nanoribbons; (b) FG dispersion with different concentrations in DMF. From left to right, the samples are as follows: unfunctionalized control (reduced only), 2a, 2b, 2c, 2d. (Reprinted with permission from Zhu, Y. Covalent functionalization of surfactant-wrapped graphene nanoribbons. *Chem Mater* 21:5284–91. Copyright 2009 American Chemical Society.)

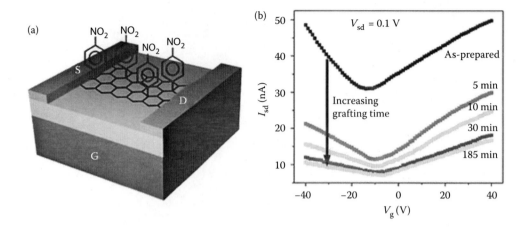

FIGURE 13.3 (a) Electronic devices consisted of monolayer GNRs contacted with Pt source (S) and drain (D) electrodes. (b) I_{sd}–V_g curves recorded at $V_{sd} = 0.1$ V for the same electronic device in panel after several consecutive grafting experiments. (Reprinted with permission from Sinitskii, A. Kinetics of diazonium functionalization of chemically converted graphene nanoribbons. *ACS Nano* 4:1949–54. Copyright 2010 American Chemical Society.)

manipulate the electrical and electronic properties of graphene [20,42–44]. Covalent functionalization of the surface or the edge of epitaxial graphene could also provide a novel route for introducing patterning that can modulate the energy bandgap, affect electron scattering, and direct current flow by producing dielectric regions in a graphene wafer. A mixed two-step functionalization method was designed to modify graphene prepared using the CVD method by Sun and coworkers [43]. The first step is the controlled hydrogenation of graphene and the second is the activation of hydrogenated graphene with 4-bromophenyldiazonium tetrafluoroborate. It was found that the density of the sp^3 C functional groups on graphene's basal plane can be controlled from 0.4% to 3.5% with this two-step covalent functionalization process. This methodology permits modulation of the electronic properties of graphene's basal planes and could hold promise for specifically patterned optoelectronic and sensor devices based on this exciting new material.

Conductivity of graphene can also be manipulated with diazonium reactions. Liu and coworkers [19] functionalized RGO with monoaryl and bi-diazonium salts and compared their electrical conductivity. They found that the electrical

conductivity of these functionalized graphene (FG) decreased with the increase in the amount of diazonium salts which is due to the destruction of the conjugated structure and the increasing distance among graphene layers. The kinetics of graphene functionalization with diazonium salts by probing the electrical properties of graphene nanoribbons (GNRs), either in vacuum after the grafting, or *in situ* in the solution was studied by Sinitskii and coworkers [45]. A simple device was fabricated on a 200-nm-thick thermal SiO$_2$ over heavily doped p-type Si that was used as a back gate (G) (Figure 13.3a). After the diazonium treatment, the conductivity of GNR devices gradually decreased with grafting time over the entire V_g-range, which can be explained by the covalent attachment of 4-nitrophenyl groups to the GNRs, resulting in the transition of graphene carbon atoms from sp^2 to sp^3 hybridization. What's more, the reaction of GNRs with 4-nitrobenzene diazonium tetrafluoroborate is reasonably fast, such that more than 60% of the maximum change in electrical properties is observed within 5 min of grafting at room temperature (Figure 13.3b).

However, different conclusions were reported by Huang and coworkers when graphene was functionalized via nitrophenyl

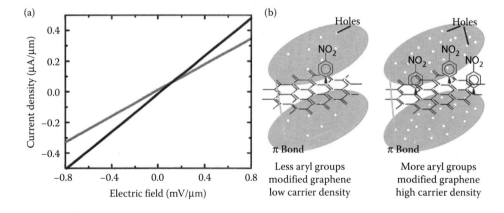

FIGURE 13.4 (a) *I*–*V* curves of pristine and modified graphene. (b) Illustration of charge transfer effect introduced by nitrophenyl groups covalently bonding to graphene basal plane. (Reprinted with permission from Huang, P. Graphene covalently binding aryl groups: Conductivity increases rather than decreases. *ACS Nano* 5:7945–49. Copyright 2011 American Chemical Society.)

groups through the reaction of diazonium salts. Electric transport measurements demonstrated that modified graphene is more conductive than intrinsic graphene [46] (Figure 13.4a). The enhancement of graphene's conductivity was contributed to the charge transfer effect as shown in Figure 13.4b. In the low reaction extent, fewer electrons transfer from the graphene conjugated π-bond to the covalent bonds between graphene and nitrophenyl groups, leaving fewer holes as charge carriers. In this situation, the scattering effect of nitrophenyl groups is dominant. In high reaction extent, much more holes are formed in the conjugated π-bond and this effect overruns the scattering effect; hence the conductivity is enhanced.

Diazonium functionalization was also used to control the surface potential of graphene. Stark and coworkers [47] used photoresist to create a patterned surface on highly oriented pyrolytic graphite (HOPG), and then selectively functionalized the top layer by exposing the diazonium reagent to the unmasked areas. Functionalization with *p*-nitrophenyl group induced a change in the surface potential, measured by Kelvin probe force microscopy, of $\Delta \Psi = -74 \pm 16$ mV relative to HOPG. They also studied the effect of the substituent on the phenyl group by functionalizing HOPG top layer using various aryl diazonium salts. Further modifications of diazonium FG were also carried out simply via the addition of different polymers. Atom transfer radical polymerization (ATRP) and radical addition fragmentation transfer (RAFT) polymerization methods are often used to graft the polymer to diazonium functionalized RGO. Yang and coworkers [48] modified the diazonium functionalized RGO via click chemistry and RAFT polymerizations. They grafted poly(*N*-isopropylacrylamide) (PNIPAM) to RGO and found that the RGO/PNIPAM nanocomposites presented a lower critical solution temperature at 33.2°C. Polystyrene chains were grafted onto the surface of small reduced graphene sheets [49] and resulted in a grafting content as high as 82 wt% and a 15°C enhancement in T_g.

In addition to the radical mechanism, the edges of graphene nanoribbons and defects in graphene basal plane are easier to react with diazonium salts [50,51] as the chemical reactivity of graphene is influenced by its physical structure, including the number of layers, the type and structure of edges, and the degree of strain and grain boundaries [52]. Lim and coworkers [53] investigated the site-dependent and spontaneous functionalization of mechanically exfoliated graphene (MEG) using 4-bromobenzene diazonium tetrafluoroborate (4-BBDT) and its doping effect. It was found that 4-BBDT molecules were spontaneously functionalized noncovalently on basal plane of MEG, while they were covalently bonded to the edge of MEG, which was successfully explicated by Raman spectroscopy. Sharma and coworkers [28] found that single graphene sheets are almost 10 times more reactive than bilayers or multilayers of graphene before and after chemical reaction and that the reactivity of edges is at least two times higher than that of the bulk single graphene sheet. A similar conclusion was obtained by Sun and coworkers [54]; they functionalized thermally the expanded graphite with 4-bromophenyl addends using the *in situ* diazonium formation procedure and sonication to obtain chemically assisted exfoliated graphene (CEG) sheets which had higher solubility than pristine graphene without any stabilizer additive and a majority of the Br signals came from the edges of the CEG indicating that the basal planes were not highly functionalized. Further study by Wu and coworkers have shown that transferring graphene onto an Si wafer substrate decorated with SiO₂ nanoparticles induced local regions of mechanical strain and increased the chemical reactivity of at least some of the carbon atoms at these sites. In particular, *in situ* generated aryl radicals were found to couple to the graphene with a higher degree of local curvature, as confirmed by micro-Raman mapping spectroscopy [55].

13.2.1.2 Other Free Radical Reactions

The functionalization of graphene via free radical addition approach has also been achieved by both photochemical and thermal treatments. In addition to diazonium salts, benzoyl peroxide, fluorinated aryl iodonium salts and styrene can also react with graphene by phenyl radicals. Barron and coworkers [56,57] described a photochemical and thermal reaction between graphene and benzoyl peroxide (BPO). The intense laser irradiation was used to produce free phenyl radicals and significant defects in the basal plane of graphene. The attachment of the phenyl groups was directly indicated by the appearance of this D band at 1343 cm⁻¹, which is due to the formation of sp³ carbon atoms in the basal plane of graphene. This reaction was also carried out on a graphene sheet placed in a field effect transistor (FET) device and found that the conductivity significantly decreased due to the increase of sp³ carbon atoms after the covalent addition of phenyl groups. What's more, an increase in the level of hole doping was also observed which is attributed to the physisorption of benzoyl peroxide. Regarding the mechanism of radical generation, they suggest that a hot electron initiates an electron transfer from photoexcited graphene to the physisorbed benzoyl peroxide. The short-lived benzoyl peroxide radical anion is then decomposed to produce the phenyl radicals, which react with the sp² graphene carbon atoms. Phenyl radicals can also be obtained by electrochemistry method. Chan and coworkers [58] demonstrated the electrochemically driven covalent bonding of trifluoromethylphenylene (CF3Ph) onto epitaxial graphene and found increased water contact angles and work functions.

Another interesting radical chemistry of graphene by irradiating graphite powder in styrene with high-intensity ultrasound was also reported [59]. The ultrasound allowed graphite to exfoliate as well as styrene to polymerize onto the graphene flakes. The reaction was proposed to involve radicals generated during sonication of styrene, which reacted with the graphene to form polystyrene-FG. The covalent bond formation was supported by the appearance of the Raman D band that is indicative of the conversion of sp² into sp³ carbon, and the high stability of the polystyrene-FG that could be dispersed in organic solvents without precipitation.

13.2.2 Functionalization of Graphene via Cyclization Reactions

Apart from free radical addition reactions, graphene can also be functionalized through cyclization reactions. The aryl diazonium salt-based reactions exhibit a higher reactivity at the edges than the interior of the sheets. Instead, the cyclization reactions seem to take place not only at the edges but also at the internal C=C bonds [60–62]. The characteristic of cyclization reaction is to form a cycle between graphene and functionalities and two common reactions can be adopted: cycloaddition and nucleophilic addition reaction. A number of molecules can attach to the graphene surface or edges through cyclization mechanism; nitrenes, carbenes, and arynes are some of them.

13.2.2.1 Functionalization with Arynes

The formation of a four-electron cycloaddition on graphene sp² carbon network goes through an aryne or benzyne intermediate via an elimination–addition mechanism. Arynes are an uncharged reactive intermediate generated by the abstraction of two ortho-substituents on arenas. A common reaction of arynes is the Diels–Alder reaction with dienes. Unlike the radical addition chemistry discussed above, the Diels–Alder reactions are expected to give rise to [4 + 2] cycloaddition products. The products of the Diels–Alder reaction with graphene readily undergo a thermal retro-Diels–Alder reaction; the versatility of this covalent carbon–carbon bond formation chemistry and the dual behavior of graphene as either diene or dienophile in this chemistry (Figure 13.5) allows covalent grafting of a wide variety of functional groups as dienes or dienophiles and consequently provides a convenient platform for the application of this chemistry in various post-grafting modifications of graphene for sensing and advanced material applications. For graphene as diene, tetracyanoethylene and maleic anhydride have been introduced as dienophile and the reactivity was probed by Raman analyses. The analyses also indicated the effects of graphene type and temperature on the success and reversibility of the cycloaddition, which prompted for an optimized reaction temperature for each graphene sample. Subsequent application of graphene as dienophile was demonstrated by cycloaddition with 9-methylanthracene and 2, 3-dimethoxy-1, 3-butadiene. The hybrid graphene materials were also highlighted to show nonmetallic behavior over a temperature range of 100–300 K [63–66]. Another arynes, pyridynes, can also be used to covalently modify graphene sheet through cycloaddition reaction [67].

However, the formation of arynes typically requires harsh conditions (strong base, metals), a fluoride-induced decomposition methodology on an o-trimethylsilyl-phenyl triflate provides a mild and neutral condition instead. The presence of a fluoride ion induced a desilylation step (driven by the formation of a strong F–Si bond) to provide a carbanion with a filled sp² orbital in the plane of the ring. This proceeded by an elimination of the triflate group to give benzyne, which is electrophilic. Subsequent nucleophilic attack by the graphene carbon network resulted in a [2 + 2] cycloaddition. Zhong and coworkers [68] also developed a novel and convenient approach to functionalize graphene by using aryne cycloaddition with easily accessible 2-triflatophenyl silane benzyne precursors. 1,2-(trimethylsilyl)phenyl triate was also used to modify graphene sheets through aryne mechanism with the addition of CsF [4] (Figure 13.1e). The resulting highly functionalized and thermally stable aryne-modified graphene sheets can be well dispersed in various solvents. Moreover, the electronic properties of aryne-modified graphene may be changed due to the functionalization with aryne groups. The degree of functionalization was determined as 1 per 17 carbons from the 36.5% weight loss measured by TGA, or 1 per 16 carbons from the integration of F1s and C1s peaks by XPS.

13.2.2.2 Functionalization with Nitrenes

The nitrene chemistry has become a powerful method for chemical modification of carbon nanotubes and fullerenes and subsequently applied to graphene [69]. Nitrene can be generated from a thermal or photodecomposition of an azide group (Figure 13.1d) and a variety of azides have been tested to functionalize graphene such as azomethine ylides [22,60,70], azidotrimethylsilane (ATS) [71], azidophenylalanine [72], and perfluorophenyl azides (PFPAs) [73]. Azomethine ylides can be generated from N-methyl glycine and reacted onto graphene [74]. Quintana and coworkers [22] functionalized graphene using the 1,3-dipolar cycloaddition of azomethine ylides at 125°C. Their results showed that reaction has taken place not just at the edges but also at the internal C=C bonds of graphene. A different reaction condition was carried out by Georgakilas and coworkers; they found that graphene can also be functionalized with azomethine ylides by refluxing in DMF at 145–150°C for 96 h [60].

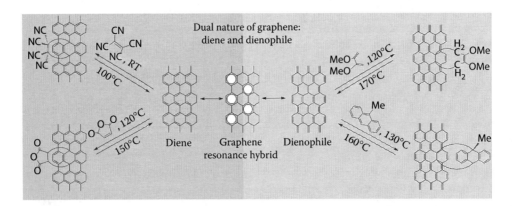

FIGURE 13.5 Graphene as diene–dienophile and their respective reactivities. (Reprinted with permission from Sarkar, S. 2012. Chemistry at the Dirac point: Diels–Alder reactivity of graphene. *Acc Chem Res* 45:673–82. Copyright 2012 American Chemical Society.)

Azidotrimethylsilane (ATS) can react with graphene in vacuum, and a bandgap of 0.66 eV was acquired for the FG [71]. The degree of functionalization was relatively small, as shown by the intensity ratio of N/C ≈ 1/53 from the N1s and C1s core-level XPS spectra. The ATS-FG was stable toward thermal treatment at 250°C for 5 min. Further annealing at 850°C for 5 min resulted in the bandgap closure and restoration of the metallic behavior of graphene. Storm and coworkers demonstrated a high yield method of functionalizing graphene nanosheets through nitrene addition of azidophenylalanine exfoliated microcrystalline graphite [72]. They refluxed solvent-exfoliated graphene flakes together with Boc-protected azidophenylalanine in o-dichlorobenzene for 4 days and the final product showed a high density of phenylalanine as determined by TGA (69 wt % mass loss). However, when the azidophenylalanine was protected at carboxyl and amino ends (Boc-Phe-OMe), the experiment showed a higher extent of functionalization (78% mass loss) that may be because of multilayers, rather than a single layer of phenyl azide deposited on graphene.

Perfluorophenyl azides (PFPAs) is another common functionality to modify graphene, which overcome the lack of biomolecular chemistry of alkyl and phenyl nitrenes and have become a class of popular and highly efficient photoaffinity labeling agents [73]. Liu and coworkers [75] treated graphene with PFPAs and introduced well-defined functional groups to pristine graphene by thermal or photochemical activation. This was accomplished in one step by heating the solution of o-dichlorobenzene-exfoliated graphene flakes and PFPAs at 90°C for 3 days (Figure 13.6a). The obtained FG has fine-tuned solubility and surface properties which greatly enhance the processability of graphene-based materials. The electronic

characteristics of PFPA-FG were also studied by Suggs and coworkers [76]. Their results showed that that the [2 + 1] cyclo-addition preserves the sp² hybridization network of the carbons on graphene. However, the π conjugation of graphene near the Fermi level is greatly disturbed by functionalization, which leads to the opening of a bandgap dependent on the concentration of the addend. This kind of reaction also offers an efficient method for the formation of stable graphene on technologically significant substrates such as silicon wafers. Silicon wafers were first functionalized with PFPAs by silanization and graphene was then pressed onto the PFPA surface and was heated to covalently attach the bottom layer of the HOPG on the wafer. Additionally, because the PFPAs reaction can be triggered by photons and electrons, lithography techniques can be applied where the location and spatial features of the attached graphene can be controlled [77]. On the basis of a similar methodology, Barron and coworkers demonstrated the addition of an amino acid side chain, phenylalanine, via a [2 + 1] cycloaddition to modify the solubility of exfoliated graphene materials. More importantly, the phenylalanine chains (consist of –COOH and –NH₂ arms) provided additional anchors for further functionalization. It was found that the functionalization occurred not only at the edge planes, but mostly on the basal planes as supported by TGA analyses whereby TGA showed 69% by mass functionalization while the cumulative edge planes would have only accounted for 2%–3% by mass.

Cycloaddition is different from condensation reaction. Quintana and coworkers performed two well-established organic reactions (1,3-dipolar cycloaddition and amide-condensation reactions) on graphene sheets [61,78] (Figure 13.6b). Their experimental results indicated that 1,3-dipolar cycloaddition

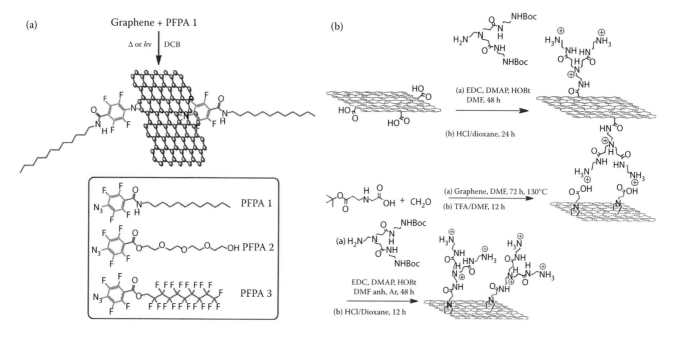

FIGURE 13.6 (a) Solution-phase functionalization of pristine graphene with PFPAs. (Reprinted with permission from Liu, L.-H. Derivitization of pristine graphene with well-defined chemical functionalities. *Nano Lett* 10:3754–56. Copyright 2010 American Chemical Society.) (b) Functionalization of graphene via amide-condensation reactions (upper) and 1, 3-dipolar cycloaddition (lower). (Reprinted with permission from Quintana, M. Organic functionalization of graphene in dispersions. *Acc Chem Res* 46:138–48. Copyright 2013 American Chemical Society.)

occurs probably next to defects, inducing a higher dispersibility in DMF, while in the amide-condensation reaction the carboxylic groups are mainly present at the edges of pristine graphene exfoliated in DMF. A comparative study of the addition of azomethine and carbonyl ylide to graphene models was reported by Cao and coworkers [79]. Reaction energetics has been obtained which shows that edge areas of graphene are much more favorable reaction sites than the center sites. Azomethine ylide cannot directly react at the center area, while carbonyl ylides are promising reagents for functionalization of graphene.

13.2.2.3 Functionalization with Carbenes

The highly reactive carbene is capable of undergoing insertion reaction to C–H bonds and [1 + 2] cycloaddition reaction to C=C bonds in high yield, if there is no competing reaction paths (Figure 13.1c). Chloroform [80] and diazirines [81] are two carbene precursors that have been successfully used to functionalize RGO. Chloroform, upon treating with a base, forms dichlorocarbene that can be transferred to an organic layer by a phase transfer catalyst. Carbene in the form of dichlorocarbene has been introduced on the graphene sp² carbon network. Carbene formation was achieved by the conventional α-elimination method from a mixture of chloroform in strong base (NaOH). Carbene generated by this method usually exists as singlet carbene since the starting central carbon atom of the chloroform precursor inherited two paired electrons from the broken C–H σ-bond. Moreover, the electron-rich substituents (chlorides) readily stabilize the singlet spin state. Since carbene is highly reactive, it would readily react in its initial spin state, which means that it is unlikely that it will switch to a triplet state. In this case, the singlet carbene which behaves like an electrophile reacts spontaneously with the graphene sp² carbon atoms in a concerted manner.

Diazirines are three-membered heterocyclic rings that have an sp³ carbon atom bonded to an azo group. Analogous to azides, diazirines decompose on heating or irradiation to release molecular nitrogen and give the electron-deficient carbene species. Diazirines have found applications in surface modification and in the synthesis of functional materials: CNTs [82], diamond, fullerene, and RGO [80]. Similar to azides, diazirines are stable in the absence of light and can rapidly form carbine upon light activation [83]. Compared with phenyl azides, however, diazirines can be time consuming to synthesize and less efficient as some carbenes require a second photon to generate [84]. 3-Aryl-3(trifluoromethyl) diazirine is one of the most frequently used diazirine compounds due to the lack of intramolecular rearrangement of the corresponding carbene [73,83]. Diazirines are also used to prepare graphene–polymer and graphene–nanoparticle hybrids. Alkyne-modified graphene was prepared and then reacted with azido-terminated polystyrene to graft polymer chains to graphene surface [85]. The obtained PS-grafted graphene sheets exhibited good dispersibility and full exfoliation in common organic solvents. Workentin and coworkers immobilized Au nanoparticles on RGO using 3-aryl-3(trifluoromethyl)-diazirine. Au nanoparticles were functionalized with mixed thiols of 1-decanethiol and a thiol-derivatized diazirine, and were UV irradiated in the presence of RGO in THF to give Au nanoparticles–graphene conjugate [81] (Figure 13.7a).

13.2.2.4 Nucleophilic Addition

The Bingel reaction originated from the cyclopropanation chemistry of fullerene. It utilizes a halide derivative of the diethyl malonate moiety in the presence of a base such as 1,8-diazabicyclo[5.4.0]undec-7-ene or sodium hydride. It has

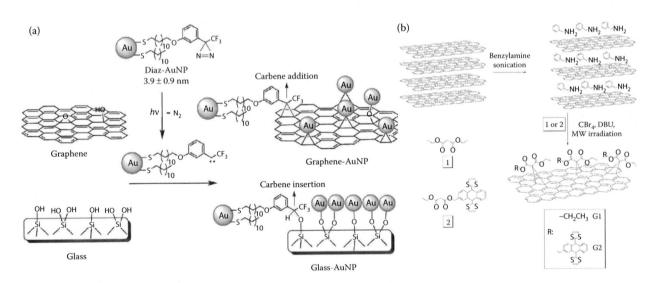

FIGURE 13.7 (a) Cartoon illustration of the carbene insertion/addition approach utilized for the covalent attachment of AuNPs onto graphene and glass. (Reprinted with permission from Ismaili, H. Light-activated covalent formation of gold nanoparticle–graphene and gold nanoparticle–glass composites. *Langmuir* 27:13261–68. Copyright 2011 American Chemical Society.) (b) Functionalization of graphene with diethyl malonate (1) and extended tetrathiafulvalene (2) moieties based on the Bingel reaction. (Reprinted with permission from Economopoulos, S.P. Exfoliation and chemical modification using microwave irradiation affording highly functionalized graphene. *ACS Nano* 4:7499–507. Copyright 2010 American Chemical Society.)

ever since found its usefulness in graphene chemistry given the ease of its reaction conditions. The halide–malonate moiety is typically generated *in situ* in a mixture of tetrahalomethane and a base. The base abstracts a proton from the halide–malonate to provide an enolate which subsequently nucleophilically attacks a C=C bond on the graphene carbon framework. The resulting carbanion undergoes a subsequent nucleophilic substitution which displaces the halide atom to provide a cyclopropane adduct via intramolecular ring closure.

On the basis of the Bingel cyclopropanation reaction, Tagmatarchis and coworkers successfully introduced cyclopropane adducts (diethyl malonate and extended tetrathiafulvalene (exTTF) moieties) onto the exfoliated graphene sp^2 carbon network with microwave irradiation (Figure 13.7b) [86]. In this work, the group also introduced the usage of benzylamine solvent to exfoliate graphene layers from graphite under ultrasonication treatment. The hybrid graphene materials were reported to be functionalized up to 23% and displayed good dispersibility in organic solvents such as *m*-dichloromethane, *o*-dichlorobenzene, dimethylformamide (DMF), and toluene.

13.2.3 Modifications of Graphene via Substitutional Reaction and Chemical Doping

The functionalizations of graphene carbon networks based on electrophilic substitution reactions have been reported as well owing to the electron-rich nature of graphene. This has resulted in the high reactivity of graphene toward strong electrophiles. Substitution reactions have been successfully achieved via the Friedel–Crafts acylation and hydrogen–lithium exchange methods. Friedel–Crafts acylation remains the only method to introduce aryl ketone groups onto the graphene platform which has been successfully performed by Pumera and coworkers [87]. Yan and coworkers [88] highlighted the functionalization of a graphene carbon network based on initial hydrogen–lithium exchange. In the effort toward producing a solid basic catalyst with a triethylamine moiety, an electrophilic substitution method was utilized.

Chemical doping is a kind of substitutional doping. Substitutional doping refers to the substitution of carbon atoms in the honeycomb lattice of graphene by atoms with different numbers of valence electrons such as nitrogen and boron, which play the role of n-type and p-type doping in carbon-based materials, respectively [89]. Li and coworkers developed a simple chemical method to obtain bulk quantities of N-doped RGO sheets through thermal annealing of GO in ammonia [90]. The motivation of doping is to control the type and concentration of charge carriers of graphene and further to modulate the electronic properties of graphene [91]. When nitrogen atoms are incorporated into the basal plane of graphene, they donate electrons into graphene leading to n-type doping of graphene while graphene doped with boron would exhibit p-type behavior. Panchokarla and coworkers [92] synthesized boron-doped graphene using arc discharge of graphite electrodes in the presence of H$_2$, He, and diborane (B$_2$H$_6$) and found that the B-doped graphene exhibits

higher conductivity and the Fermi level shifts 0.65 eV below the Dirac point by calculations, indicating the p-type doping of graphene. A CVD technique was also designed to produce the N-doped graphene, and the electrical properties of the N-doped graphene were measured to be an n-type semiconductor (Figure 13.8) [93].

13.2.4 Modifications of Graphene via Residual Groups

Graphene can be prepared through the reduction of GO and thus functional groups such as carboxyl, hydroxyl, and epoxy group might be left since the reduction is usually difficult to proceed completely. Recently, Boukhvalov et al. (35) and Gao et al. (36) used density functional theory to investigate the structure of graphene after deoxygenation. Interestingly, both groups confirmed that it was impossible to remove completely the oxygen-containing functional groups from the graphene surface using chemical or thermal reduction, or even a combination of the two. Experimentally, to the best of our knowledge, the results of characterization of graphene fabricated from GO show the existence of the residual oxygen-containing functional groups on graphene. However, these unremovable residual functional groups would be very useful in some cases. Some advantages can be expected with the covalent modification of graphene via its residual oxygen-containing functional groups. For example, the graphene conjugated structure will not be destructed during functionalization. Although using the abundant oxygen-containing functional groups of GO to react with chemical molecules, followed by subsequent chemical reduction, can be used to prepare FG, the degree of reduction may be somewhat limited. This is because the oxygen-containing functional groups are not just the reactive sites, but also, more importantly, they are the restorable sites [94]. Ma and coworkers successfully grafted maleic anhydride (MA) treated poly (oxyalkylene)amines (POA) onto graphene surface through two methods: free radical grafting and residual groups on RGO without damaging its structure (Figure 13.9). The approach described herein could improve the reinforcement effect of graphene in other applications, such as in the preparation of functional graphene/polymer composites. Poly (vinyl alcohol) (PVA) has also been used to covalently functionalize multilayer graphene nanosheets (MLGNs) by the reaction between the residual carboxyl groups on graphene and the hydroxyl groups of PVA chains. The modified graphene demonstrated enhanced dispersion capacity in DMSO and hot water [95].

13.3 COVALENT MODIFICATION OF GO

As the most important graphene derivatives, GO has been used widely as a starting material for the synthesis and modifications of processable graphene. There are different methods for the production of GO from natural graphite and the modified Hummers' method is mostly applied in preparing GO [96,97]. The surfaces of GO sheets are highly oxygenated, bearing hydroxyl, epoxide, diol, ketone, and carboxyl

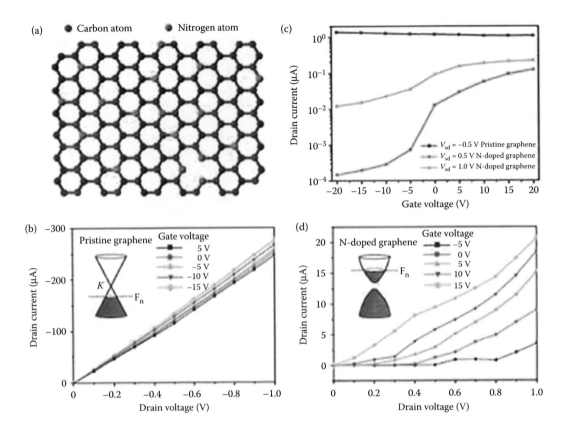

FIGURE 13.8 (a) Schematic representation of the N-doped graphene. (b) Transfer characteristics of pristine graphene and N-doped graphene. (c) and (d) I_{ds}/V_{ds} characteristics at various V_g for the pristine graphene and the N-doped graphene field effect transistor device, respectively. The insets are the presumed band structures. (Reprinted with permission from Wei, D. Synthesis of N-doped graphene by chemical vapor deposition and its electrical properties. *Nano Lett* 9:1752–58. Copyright 2009 American Chemical Society.)

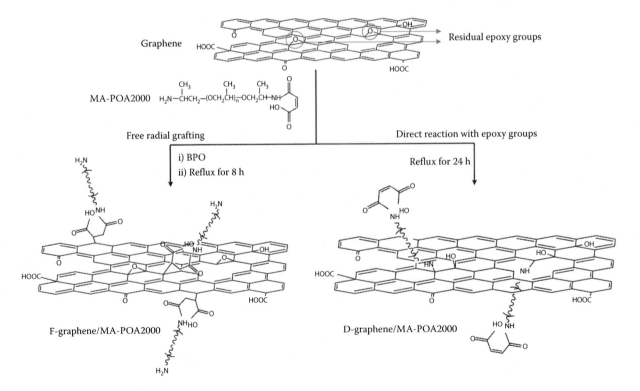

FIGURE 13.9 Preparation of F-graphene/MA-POA2000 and D-graphene/MA-POA2000. (Reprinted with permission from Hsiao, M.-C. Preparation of covalently functionalized graphene using residual oxygen-containing functional groups. *ACS Applied Materials and Interfaces* 2:3092–99. Copyright 2010 American Chemical Society.)

functional groups that can alter the van der Waals interactions significantly and lead to a range of solubility in water and organic solvents. It was revealed that carboxylic acid exists mostly at their edges (according to the widely accepted Lerf–Klinowski model), and epoxy and hydroxyl groups on the basal planes, which renders it a good candidate for use in various applications through chemical functionalization. Chemical modification of GO has been a promising route to achieve mass production of chemically modified graphene platelets.

A wide range of modifications utilizing carboxylic acids and hydroxyl has been applied to GO through esterification reaction of GO by DCC chemistry or the acylation reaction between the carboxyl acid groups of GO and ROH alkylamine [98–102]. Reactions through the epoxy group of GO are also used to covalently modify GO. Nucleophilic ring-opening reaction can occur between epoxy groups of GO and the amine groups of an amine-terminated organic molecule [103]. The surface of GO can be modified *in situ* with octadecylamine (ODA) in dichlorobenzene using the same reaction to give ODA–GO nanosheets, which could well disperse in organic solvents [104]. Another extension of this concept involves the stabilization of GO nanosheets by amine-terminated ionic liquids (ILs). The resultant IL-FG composite nanosheets could be dispersed well in water, N,N-dimethylformamide (DMF), and dimethyl sulfoxide (DMSO) due to the enhanced solubility and electrostatic inter-sheet repulsion provided by the ionic liquid units [105].

13.4 APPLICATIONS OF COVALENT FUNCTIONALIZED GRAPHENE

13.4.1 ELECTRONIC AND OPTIC DEVICES

There has been much interest in FG because tunable electrical properties of graphene are attainable by various functionalization methods. Chemically modified graphene has been widely used in electronic and optic device, such as in solar cells, capacitors, and field effect transistors [106,107]. Liu and coworkers prepared a functionalized organic-soluble graphene, which is then blended with poly (3-hexylthiophene) (P3HT) [106]. The as-prepared composite was used as the active layer in bulk heterojunction (BHJ) polymer photovoltaic cells. The FG can be well dispersed in P3HT to form a donor/acceptor structure with 10 wt% doping. Adding FG into P3HT induces great quenching of the photoluminescence of the P3HT, indicating strong electron/energy transfer from the P3HT to the graphene. The interaction between the graphene and the P3HT makes this composite work well as the active layer in the BHJ photovoltaic devices.

Graphene was also explored as a supercapacitor electrode material by Wang and coworkers [108]. They fabricated supercapacitors using RGO materials and found that the RGO prepared using the gas-based hydrazine reduction at room temperature gave remarkable results with specific capacitance of 205 F/g, energy density of 28.5 Wh/kg, and power density of 10 kW/kg. However, graphene-based field effect transistors

cannot be turned off effectively due to the absence of a bandgap in pristine graphene. Therefore, the problems associated with the bandgap of pristine graphene should be solved before using the material as a transistor. FG nanocomposites can be used in supercapacitors to solve this problem [109,110]. Dong and coworkers fabricated graphene field effect transistors with mechanically exfoliated single-layer graphene (SLG) and bilayer graphene (BLG) sheets and then modified with bromophenyl groups through diazonium reaction. The tested results indicated that chemical modification is a simple approach to tailor the electrical properties of graphene sheets with different numbers of layers [111].

13.4.2 POLYMER NANOCOMPOSITES

Graphene is one of the strongest materials ever measured and the breaking strength and Young's modulus of graphene reaches 42 N m^{-1} and 1 TPa, respectively. However, the strength of graphene/polymer composites is mainly dependent on the interaction between graphene and polymer chains rather than that of graphene itself. Therefore, it is believed that the FG can do a better job than pristine graphene to mechanically reinforce composites. The strong interaction between the FG and polymeric matrices could result from (1) the wrinkled surface of extremely thin graphene sheets that is capable of mechanically interlocking with polymer chains and (2) the hydrogen bonding formed between the oxygen functionalities of the FG and polymeric matrices. The formation of the chemical bonding with the FG is a more effective approach to engineer a strong interface for some polymers with particular functionalities. Cai and coworkers used this strategy to achieve highly stiff PU nanocomposites. The chemical links between the FG and a PU matrix occurred due to the reaction of hydroxyl and carboxyl groups with the isocyanate group on the end of PU chains. Young's modulus of the PU has been improved by nearly nine times with the addition of 4.4 wt% FG, which indicated the high efficiency of stress transfer across the chemical interface [112].

Hydrogen bonding plays an important role in the polymer/graphene composite materials. Layek and coworkers studied the physical properties of sulfonated graphene/poly(vinyl alcohol) (PVA/SG) composites, and found that PVA/SG composite exhibits different morphology at different compositions of the hybrids (SG1, SG3 SG5) due to supramolecular organization. The highest increase of stress (177%), strain at break (45%), and toughness (235%) over PVA is observed in the dendritic SG3. Young's modulus increases progressively with increasing SG concentration in the composites showing the highest increase (180%) over PVA for the SG5 system. The storage modulus of SG3 shows the highest (1005%) increase over PVA at 30°C. Moreover an increase of 10 orders of magnitude of dc conductivity over PVA and 10-fold increase in dc conductivity in the dendritic SG3 than any other morphology are achieved [33]. Diazonium salt-FG can also be used to enhance the mechanical property of PVA, and an 181% improvement in tensile strength has been achieved with 2.0 wt.% of FG by Yu and coworkers [113].

13.4.3 Biomaterials

FG can be used as biomaterials, such as biosensors [114,115], tissue engineering [116], and drug- and gene-delivery system [117]. The advancement of the FG provides a fascinating opportunity for development of biotechnology because of their structures, components, and properties. In comparison with its allotropes, CNT, graphene exhibits some merits such as low cost, two external surfaces, facile fabrication and modification, and absence of toxic metal particles. FG can be used in drug- and gene-delivery systems. Wei and coworkers [118] designed a drug-delivery system using functionalized RGO with *p*-aminobenzoic acid through diazonium salt reaction followed by covalent binding of water-soluble PEI for excellent water solubility and folic acid (FA) molecules to specifically target CBRH7919 cancer cells. The drug-delivery system had higher loading, and the loading ratio of elsinochrome A and doxorubicin reached 45.56% and 28.62%, respectively.

13.4.4 Other Applications

FG can also be used in gas sensors and memory devices. Yuan and coworkers designed a simple chemo-resistor-type NO_2 sensor based on chemically modified graphene nanosheets (CMGs) with high sensitivity, selectivity, and reversibility. These sensors exhibit 16.4 and 4.3 times stronger responses compared with conventional RGO-based sensors and can spontaneously recover their initial states by flowing N_2 gas, without UV/IR light illumination or thermal assistance [119]. Cui and coworkers prepared a nonvolatile memory device using gold nanoparticles (AuNPs) and RGO. AuNPs were chemically bound to the RGO channel through a π-conjugated molecular linker, 4-mercapto-benzenediazonium tetrafluoroborate (MBDT) salt, which was newly synthesized and used as a molecular bridge to connect the AuNPs and RGO [120].

13.5 CONCLUSIONS AND PERSPECTIVES

In this chapter, we summarize the chemical covalent functionalization methods of graphene. Covalent modifications disrupt the extended conjugation of π-electrons in graphene, resulting in bandgap opening and change in conductivity. Functional groups can also be introduced onto graphene to manipulate its chemical properties or allow further conjugation of additional molecules and materials. Covalent modification further improves the solubility and processability of graphene materials. However, there are still several challenges for the covalent modifications of graphene. The reactivity and regioselectivity of the basal plane versus the under-coordinated edges, the zigzag versus armchair configuration are fundamental questions that warrant comprehensive studies. The high reactivity of the reactive intermediates makes it difficult to control the reaction kinetics. Side reactions often occur, making it challenging to control the composition of the surface coating and grafting density. For instance, radicals can combine or react with species other than the substrate. Nitrenes and carbenes often react with

solvents or among themselves. Thus, it is equally important to improve current methods to increase the ease of functionalization as well as discover further applications of FG.

ACKNOWLEDGMENTS

JL acknowledges the NSF of China (51173087), NSF of Shandong (ZR2011EMM001), NSF of Qingdao (12-1-4-2-(2)-jch), and Taishan Scholars Program for the financial support.

REFERENCES

1. Novoselov, K.S. 2004. Electric field effect in atomically thin carbon films. *Science* 306:666–69.
2. MEYER, J. 2009. Carbon sheets an atom thick give rise to graphene dreams. *Science* 324:875–77.
3. Novoselov, K. 2005. Two-dimensional gas of massless Dirac fermions in graphene. *Nature* 438:197–200.
4. Zhang, Y. 2005. Experimental observation of the quantum Hall effect and Berry's phase in graphene. *Nature* 438:201–04.
5. Goenka, S. 2014. Graphene-based nanomaterials for drug delivery and tissue engineering. *J Controlled Release* 173:75–88.
6. Li, H.-M. 2014. Photonic properties of graphene device. In Ting Yu (ed.). *Two-Dimensional Carbon: Fundamental Properties, Synthesis, Characterization, and Applications.* Pan Stanford Publishing, pp. 279–295.
7. Sun, Y. 2013. Graphene/polymer composites for energy applications. *J Polym Sci, Part B: Polym Phys* 51:231–53.
8. Liu, Z. 2013. Preparation of graphene/polymer composites by direct exfoliation of graphite in functionalised block copolymer matrix. *Carbon* 51:148–55.
9. Wei, Z. 2010. Nanoscale tunable reduction of graphene oxide for graphene electronics. *Science* 328:1373–76.
10. Obraztsov, A.N. 2009. Chemical vapour deposition: Making graphene on a large scale. *Nat Nanotechnol* 4:212–13.
11. Yan, Z. 2014. Chemical vapor deposition of graphene single crystals. *Acc Chem Res* 47:1327–37.
12. Zhu, Y.W. 2010. Graphene and graphene oxide: Synthesis, properties, and applications. *Adv Mater (Weinheim, Ger)* 22:3906–24.
13. Edwards, R.S. 2013. Graphene synthesis: Relationship to applications. *Nanoscale* 5:38–51.
14. Sreeprasad, T. 2013. How do the electrical properties of graphene change with its functionalization? *Small* 9:341–50.
15. Genorio, B. 2014. Functionalization of graphene nanoribbons. *J Phys D: Appl Phys* 47:094012.
16. Liu, J. 2010. Synthesis, characterization, and multilayer assembly of pH sensitive graphene-polymer nanocomposites. *Langmuir* 26:10068–75.
17. Koehler, F.M. 2010. Selective chemical modification of graphene surfaces: Distinction between single- and bilayer graphene. *Small* 6:1125–30.
18. Liu, J. 2012. Strategies for chemical modification of graphene and applications of chemically modified graphene. *J Mater Chem* 22:12435–52.
19. Liu, J. 2012. Using molecular level modification to tune the conductivity of graphene papers. *J Phys Chem C* 116:17939–46.
20. Huang, P. 2013. Diazonium functionalized graphene: Microstructure, electric, and magnetic properties. *Acc Chem Res* 46:43–52.
21. Lomeda, J.R. 2008. Diazonium functionalization of surfactant-wrapped chemically converted graphene sheets. *J Am Chem Soc* 130:16201–06.

22. Quintana, M. 2010. Functionalization of graphene via 1, 3-dipolar cycloaddition. *ACS Nano* 4:3527–33.
23. Zhao, J.-X. 2012. Chemical functionalization of graphene via aryne cycloaddition: A theoretical study. *J Mol Model* 18:2861–68.
24. Dreyer, D.R. 2010. The chemistry of graphene oxide. *Chem Soc Rev* 39:228–40.
25. Boukhvalov, D. 2008. Chemical functionalization of graphene with defects. *Nano Lett* 8:4373–79.
26. Tessonnier, J.-P. 2012. Dispersion of alkyl-chain-functionalized reduced graphene oxide sheets in nonpolar solvents. *Langmuir* 28:6691–97.
27. Delamar, M. 1992. Covalent modification of carbon surfaces by grafting of functionalized aryl radicals produced from electrochemical reduction of diazonium salts. *J Am Chem Soc* 114:5883–84.
28. Sharma, R. 2010. Anomalously large reactivity of single graphene layers and edges toward electron transfer chemistries. *Nano Lett* 10:398–405.
29. Bekyarova, E. 2009. Chemical modification of epitaxial graphene: Spontaneous grafting of aryl groups. *J Am Chem Soc* 131:1336–37.
30. Jahan, M. 2010. Structure-directing role of graphene in the synthesis of metal–organic framework nanowire. *J Am Chem Soc* 132:14487–95.
31. Mahouche-Chergui, S. 2011. Aryl diazonium salts: A new class of coupling agents for bonding polymers, biomacromolecules and nanoparticles to surfaces. *Chem Soc Rev* 40:4143–66.
32. Shih, C.-J. 2013. Disorder imposed limits of mono- and bilayer graphene electronic modification using covalent chemistry. *Nano Lett* 13:809–17.
33. Layek, R.K. 2012. The physical properties of sulfonated graphene/poly (vinyl alcohol) composites. *Carbon* 50:815–27.
34. Wang, D. 2011. Graphene functionalized with Azo polymer brushes: Surface-initiated polymerization and photoresponsive properties. *Adv Mater (Weinheim, Ger)* 23:1122–25.
35. Jin, Z. 2009. Mechanically assisted exfoliation and functionalization of thermally converted graphene sheets. *Chem Mater* 21:3045–47.
36. Kuilla, T. 2010. Recent advances in graphene based polymer composites. *Prog Polym Sci* 35:1350–75.
37. Huang, W. 2012. High-performance nanopapers based on benzenesulfonic functionalized graphenes. *ACS Nano* 6: 10178–85.
38. Liu, M. 2014. Diazonium functionalization of graphene nanosheets and impact response of aniline modified graphene/bismaleimide nanocomposites. *Mater Des* 53:466–74.
39. Si, Y. 2008. Synthesis of water soluble graphene. *Nano Lett* 8:1679–82.
40. Jia, B. 2012. Graphene nanosheets reduced by a multi-step process as high-performance electrode material for capacitive deionisation. *Carbon* 50:2315–21.
41. Zhu, Y. 2009. Covalent functionalization of surfactant-wrapped graphene nanoribbons. *Chem Mater* 21:5284–91.
42. Niyogi, S. 2011. Covalent chemistry for graphene electronics. *J Phys Chem Lett* 2:2487–98.
43. Sun, Z. 2011. Towards hybrid superlattices in graphene. *Nat Commun* 2:559.
44. Hossain, M.Z. 2010. Scanning tunneling microscopy, spectroscopy, and nanolithography of epitaxial graphene chemically modified with aryl moieties. *J Am Chem Soc* 132: 15399–403.
45. Sinitskii, A. 2010. Kinetics of diazonium functionalization of chemically converted graphene nanoribbons. *ACS Nano* 4:1949–54.
46. Huang, P. 2011. Graphene covalently binding aryl groups: Conductivity increases rather than decreases. *ACS Nano* 5:7945–49.
47. Koehler, F.M. 2009. Permanent pattern-resolved adjustment of the surface potential of graphene-like carbon through chemical functionalization. *Angew Chem, Int Ed* 48:224–27.
48. Yang, Y. 2012. Synthesis of PNIPAM polymer brushes on reduced graphene oxide based on click chemistry and RAFT polymerization. *Journal of Polymer Science Part A: Polymer Chemistry* 50:329–37.
49. Fang, M. 2009. Covalent polymer functionalization of graphene nanosheets and mechanical properties of composites. *J Mater Chem* 19:7098–98.
50. Jiang, D.-E. 2006. How do aryl groups attach to a graphene sheet? *J Phys Chem B* 110:23628–32.
51. Zhu, H. 2012. Microstructure evolution of diazonium functionalized graphene: A potential approach to change graphene electronic structure. *J Mater Chem* 22:2063–68.
52. Paulus, G.L. 2013. Covalent electron transfer chemistry of graphene with diazonium salts. *Acc Chem Res* 46:160–70.
53. Lim, H. 2010. Spatially resolved spontaneous reactivity of diazonium salt on edge and basal plane of graphene without surfactant and its doping effect. *Langmuir* 26:12278–84.
54. Sun, Z. 2010. Soluble graphene through edge-selective functionalization. *Nano Res* 3:117–25.
55. Wu, Q. 2012. Selective surface functionalization at regions of high local curvature in graphene. *Chem Commun (Cambridge, UK)* 49:677–79.
56. Liu, H. 2009. Photochemical reactivity of graphene. *J Am Chem Soc* 131:17099–101.
57. Hamilton, C.E. 2010. Radical addition of perfluorinated alkyl iodides to multi-layered graphene and single-walled carbon nanotubes. *Nano Research* 3:138–45.
58. Chan, C.K. 2013. Electrochemically driven covalent functionalization of graphene from fluorinated aryl iodonium salts. *J Phys Chem C* 117:12038–44.
59. Xu, H. 2011. Sonochemical preparation of functionalized graphenes. *J Am Chem Soc* 133:9148–51.
60. Georgakilas, V. 2010. Organic functionalisation of graphenes. *Chem Commun (Cambridge, UK)* 46:1766–68.
61. Quintana, M. 2011. Selective organic functionalization of graphene bulk or graphene edges. *Chem Commun (Cambridge, UK)* 47:9330–32.
62. Zhang, X. 2012. Preparation of dispersible graphene through organic functionalization of graphene using a zwitterion intermediate cycloaddition approach. *RSC Adv* 2:12173–76.
63. Bekyarova, E. 2013. Effect of covalent chemistry on the electronic structure and properties of carbon nanotubes and graphene. *Acc Chem Res* 46:65–76.
64. Yuan, J. 2012. One-step functionalization of graphene with cyclopentadienyl-capped macromolecules via Diels–Alder "click" chemistry. *J Mater Chem* 22:7929–36.
65. Sarkar, S. 2012. Chemistry at the Dirac point: Diels–Alder reactivity of graphene. *Acc Chem Res* 45:673–82.
66. Sarkar, S. 2011. Diels–Alder chemistry of graphite and graphene: Graphene as diene and dienophile. *J Am Chem Soc* 133:3324–27.
67. Zhong, X. 2014. Pyridyne cycloaddition of graphene: "External" active sites for oxygen reduction reaction. *J Mater Chem A* 2:897–901.
68. Zhong, X. 2010. Aryne cycloaddition: Highly efficient chemical modification of graphene. *Chem Commun (Cambridge, UK)* 46:7340–42.
69. Servant, A. 2014. Graphene for multi-functional synthetic biology: The last 'zeitgeist'in nanomedicine. *Bioorg Med Chem Lett* 24:1638–49.

70. Zhang, B. 2011. Conjugated polymer-grafted reduced graphene oxide for nonvolatile rewritable memory. *Chemistry—Eur J* 17:13646–52.

71. Choi, J. 2009. Covalent functionalization of epitaxial graphene by azidotrimethylsilane. *J Phys Chem C* 113:9433–35.

72. Strom, T.A. 2010. Nitrene addition to exfoliated graphene: A one-step route to highly functionalized graphene. *Chem Commun (Cambridge, UK)* 46:4097–99.

73. Park, J. 2013. Covalent functionalization of graphene with reactive intermediates. *Acc Chem Res* 46:181–89.

74. Chua, C.K. 2013. Covalent chemistry on graphene. *Chem Soc Rev* 42:3222–33.

75. Liu, L.-H. 2010. Derivitization of pristine graphene with well-defined chemical functionalities. *Nano Lett* 10:3754–56.

76. Suggs, K. 2011. Electronic properties of cycloaddition-functionalized graphene. *J Phys Chem C* 115:3313–17.

77. Liu, L.-H. 2010. A simple and scalable route to wafer-size patterned graphene. *J Mater Chem* 20:5041–46.

78. Quintana, M. 2013. Organic functionalization of graphene in dispersions. *Acc Chem Res* 46:138–48.

79. Cao, Y. 2010. Computational assessment of 1, 3-dipolar cycloadditions to graphene. *J Mater Chem* 21:1503–08.

80. KiangáChua, C. 2012. Introducing dichlorocarbene in graphene. *Chem Commun (Cambridge, UK)* 48:5376–78.

81. Ismaili, H. 2011. Light-activated covalent formation of gold nanoparticle–graphene and gold nanoparticle–glass composites. *Langmuir* 27:13261–68.

82. Lawrence, E.J. 2011. 3-Aryl-3-(trifluoromethyl) diazirines as versatile photoactivated "linker" molecules for the improved covalent modification of graphitic and carbon nanotube surfaces. *Chem Mater* 23:3740–51.

83. Brunner, J. 1980. 3-Trifluoromethyl-3-phenyldiazirine. A new carbene generating group for photolabeling reagents. *J Biol Chem* 255:3313–18.

84. Ismaili, H. 2010. Diazirine-modified gold nanoparticle: Template for efficient photoinduced interfacial carbene insertion reactions. *Langmuir* 26:14958–64.

85. Sun, S. 2010. Click chemistry as a route for the immobilization of well-defined polystyrene onto graphene sheets. *J Mater Chem* 20:5605–07.

86. Economopoulos, S.P. 2010. Exfoliation and chemical modification using microwave irradiation affording highly functionalized graphene. *ACS Nano* 4:7499–507.

87. Chua, C.K. 2012. Friedel–Crafts acylation on graphene. *Chemistry, an Asian journal* 7:1009.

88. Yuan, C. 2012. Amino-grafted graphene as a stable and metal-free solid basic catalyst. *J Mater Chem* 22:7456–60.

89. Maiti, U.N. 2014. 25th Anniversary article: Chemically modified/doped carbon nanotubes and graphene for optimized nanostructures and nanodevices. *Adv Mater (Weinheim, Ger)* 26:40–67.

90. Li, X. 2009. Simultaneous nitrogen doping and reduction of graphene oxide. *J Am Chem Soc* 131:15939–44.

91. Deng, D. 2011. Toward N-doped graphene via solvothermal synthesis. *Chem Mater* 23:1188–93.

92. Panchakarla, L. 2009. Synthesis, structure, and properties of boron- and nitrogen-doped graphene. *Adv Mater (Weinheim, Ger)* 21:4726–30.

93. Wei, D. 2009. Synthesis of N-doped graphene by chemical vapor deposition and its electrical properties. *Nano Lett* 9:1752–58.

94. Hsiao, M.-C. 2010. Preparation of covalently functionalized graphene using residual oxygen-containing functional groups. *ACS Appl Mater Interfaces* 2:3092–99.

95. MonicaáVeca, L. 2009. Polymer functionalization and solubilization of carbon nanosheets. *Chem Commun (Cambridge, UK)* 2565–67.

96. Park, S. 2009. Chemical methods for the production of graphenes. *Nat Nanotechnol* 4:217–24.

97. Hummers, W.S. 1958. Preparation of graphitic oxide. *J Am Chem Soc* 80:1339–39.

98. Karousis, N. 2010. Porphyrin counter anion in imidazolium-modified graphene-oxide. *Carbon* 48:854–60.

99. Salavagione, H.J. 2009. Polymeric modification of graphene through esterification of graphite oxide and poly (vinyl alcohol). *Macromolecules* 42:6331–34.

100. Stankovich, S. 2006. Synthesis and exfoliation of isocyanate-treated graphene oxide nanoplatelets. *Carbon* 44:3342–47.

101. Tang, J. 2014. Effect of photocurrent enhancement in porphyrin–graphene covalent hybrids. *Mater Sci Eng C* 34:186–92.

102. Wan, Y.-J. 2014. Grafting of epoxy chains onto graphene oxide for epoxy composites with improved mechanical and thermal properties. *Carbon* 69:467–80.

103. Li, Z.-F. 2014. Covalently-grafted polyaniline on graphene oxide sheets for high performance electrochemical supercapacitors. *Carbon* 71:257–67.

104. Wang, S. 2008. Band-like transport in surface-functionalized highly solution-processable graphene nanosheets. *Adv Mater (Weinheim, Ger)* 20:3440–46.

105. Yang, H. 2009. Covalent functionalization of polydisperse chemically-converted graphene sheets with amine-terminated ionic liquid. *Chem Commun (Cambridge, UK)* (26): 3880–82.

106. Bonaccorso, F. 2009. Graphene photonics and optoelectronics. *Nat Photon* 4:611–22.

107. Wang, H.-X. 2011. Photoactive graphene sheets prepared by "click" chemistry. *Chem Commun (Cambridge, UK)* 47:5747–49.

108. Wang, Y. 2009. Supercapacitor devices based on graphene materials. *J Phys Chem C* 113:13103–07.

109. Xia, F. 2010. Graphene field-effect transistors with high on/off current ratio and large transport band gap at room temperature. *Nano Lett* 10:715–18.

110. Mishra, A.K. 2011. Functionalized graphene-based nanocomposites for supercapacitor application. *J Phys Chem C* 115:14006–13.

111. Dong, X. 2012. The electrical properties of graphene modified by bromophenyl groups derived from a diazonium compound. *Carbon* 50:1517–22.

112. Cai, D. 2010. Recent advance in functionalized graphene/polymer nanocomposites. *J Mater Chem* 20:7906–15.

113. Yu, D.S. 2014. Enhanced properties of aryl diazonium salt-functionalized graphene/poly (vinyl alcohol) composites. *Chem Eng J* 245:311–22.

114. Hosseini, H. 2014. A novel bioelectrochemical sensing platform based on covalently attachment of cobalt phthalocyanine to graphene oxide. *Biosensors and Bioelectronics* 52:136–42.

115. Jiang, L. 2014. Amplified impedimetric aptasensor based on gold nanoparticles covalently bound graphene sheet for the picomolar detection of ochratoxin. *A Anal Chim Acta* 806:128–35.

116. Sayyar, S. 2013. Covalently linked biocompatible graphene/polycaprolactone composites for tissue engineering. *Carbon* 52:296–304.

117. Wei, G. 2014. Functional materials from the covalent modification of reduced graphene oxide and β-cyclodextrin as a drug delivery carrier. *New J Chem* 38:140–45.

118. Wei, G. 2012. Covalent modification of reduced graphene oxide by means of diazonium chemistry and use as a drug-delivery system. *Chemistry—Eur J* 18:14708–16.

119. Yuan, W. 2013. High-performance NO₂ sensors based on chemically modified graphene. *Adv Mater (Weinheim, Ger)* 25:766–71.

120. Cui, P. 2011. Nonvolatile memory device using gold nanoparticles covalently bound to reduced graphene oxide. *ACS Nano* 5:6826–33.

14 Functionalization and Vacancy Effects on Hydrogen Binding in Graphene

A. Tapia, C. Cab, and G. Canto

CONTENTS

Abstract 221
14.1 Introduction 221
    14.1.1 Structural Properties 221
    14.1.2 Electronic Properties 222
    14.1.3 Synthesis 222
    14.1.4 Modified Graphene 222
14.2 Hydrogen Adsorption on Graphene 222
    14.2.1 Hydrogen Chemisorptions on Graphene 223
    14.2.2 Hydrogen Physisorption on Graphene 224
14.3 Defects in Graphene 225
    14.3.1 Mono-Vacancy Defects in Graphene 225
        14.3.1.1 Properties of Graphene with MVs 226
14.4 Hydrogen Adsorption on Graphene with MVs 227
14.5 Hydrogen Adsorption on Graphene with Alkali Metals 228
14.6 Conclusions 229
Acknowledgments 229
References 229

ABSTRACT

Graphene is a form of carbon with an hexagonal structure and one atom thickness. Geometrically, if graphene is bent, it could be considered as a base structure for constructing other important carbon nanomaterials, such as nanotubes and fullerenes. Following the isolation and extraction of graphene in 2004, the study of the electronic properties of graphene has intensified and researchers have proposed modifications and new experimental arrangements. The nature of the modifications can be structural or chemical, such as the presence of vacancies and doping with atoms and molecules. This chapter discusses how the defects and alkaline functionalization on graphene can modulate its physical and chemical properties.

14.1 INTRODUCTION

Many studies have focused on scientific and technological development of carbon nanomaterials owing to its peculiarities. Carbon can form nanotubes and fullerenes, which are allotropes in one and zero dimensions, respectively. Furthermore for some years, two-dimensional (2D) carbon nanomaterial as graphene (a monolayer of graphite) has received wide attention, due to its remarkable electrical and structural properties [1]. The technological trend is toward the use of materials with nanoscale dimensions. Materials with this scale are projected to the development of new technology. Studies show that structural and electronic characteristics for this material would have revolutionary technological applications, since it is the thinnest material known and the strongest so far measured, and their charge carriers exhibit giant intrinsic mobility as they have a mass effective zero, allowing several micrometers travel without dispersion at room temperature [1]. Following isolation or extraction in 2004 [2], study of chemical modifications or defects for new electronic properties has intensified in recent years.

14.1.1 STRUCTURAL PROPERTIES

Graphene is a 2D form of carbon, presenting an hexagonal structure with one atom of thickness (see Figure 14.1), geometrically, if graphene is bended it could be considered as a base structure for constructing other important carbon nanomaterials like nanotubes and fullerenes.

The graphene structure is considered to be formed by the superposition of two triangular networks named A and B (see Figure 14.1). The hexagonal shape of the lattice is directly related to the sp^2 hybridization of carbon atoms and the unit cell contains two carbon atoms. Two lattice vectors ($\mathbf{a}_1$, $\mathbf{a}_2$) generate the hexagonal structure and the bond length between adjacent carbon atoms is close to 1.42 Å [2]. Such honeycomb

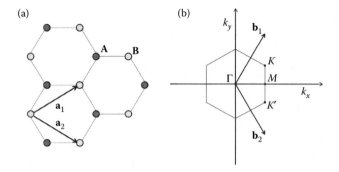

FIGURE 14.1 (a) The hexagonal structure of graphene formed by two overlapping triangular networks (**A**, **B**) and network vectors $\mathbf{a}_1$ and $\mathbf{a}_2$. (b) The lattice in reciprocal space with high symmetry points (K or K') and reciprocal vectors $\mathbf{b}_1$ and $\mathbf{b}_2$.

structure provides graphene its known remarkable mechanical properties like the Young's modulus of 1 TPa and the intrinsic strength of graphene sheet (free of defects) of 42 N/m [3].

14.1.2 ELECTRONIC PROPERTIES

Carbon is a nonmetallic element and a poor conductor of electricity, but graphene being an allotrope of carbon presents properties as a variable-gap semiconductor or semimetal with very small overlap [2]. Besides, the concentration of the charge conductors, electrons and holes, can be modulated by an external electric field to values of 10^{13} cm^{-2}, and mobilities reach values of 15,000 cm^2/V s even under environmental conditions [2]. In the Brillouin zone for graphene, the Fermi level lies at the crossing point between the cone-like dispersions [4]. Due to functional energy form, the electrons can move at extremely high speeds as relativistic particles behave practically like they have no mass and the average electron mobility at room temperature may become higher than in silicon transistors [2].

14.1.3 SYNTHESIS

Micromechanic exfoliation was one of the first techniques utilized to obtain mono and multilayered graphene [2], however this technique is expensive and is reserved for high-quality samples for research, and not for producing large amounts of this nanomaterial. With techniques like the SiC thermal decomposition and chemical vapor deposition (CVD) it is possible to get a very small amount of graphene and, therefore, are not feasible for large-scale applications. Actually, there are probably a dozen methods being used and developed to prepare graphene of various dimensions, shapes, and quality [5].

Applications of graphene as flexible electronics, high-frequency transistors, and photodetectors, demand specific properties and these methods will evolve in the future to fulfill the requirements. Some scalable methods for mass production are liquid-phase exfoliation of graphite and exfoliation of graphite oxide [5]. Solvent techniques use a liquid phase to obtain the monolayer and multilayer of a few micrometers of graphene flakes. In general, graphene is obtained as a mixture

of multilayer flakes for applications such as sensors, transparent electrodes, and conductive composites [6].

14.1.4 MODIFIED GRAPHENE

A few years ago, graphene was considered only as a very interesting theoretical system. Studies showed that because of their structural and electronic characteristics, this material would have revolutionary technological applications [1]. From isolation of this material, the research around their outstanding properties has intensified in recent years. But current researches are aimed at obtaining new behaviors and understanding graphene properties, to achieve control and enforcement of the same. For this reason, graphene researchers propose modifications and new experimental arrangements. The modifications can have a structural nature, limiting the size of the graphene sheets by cutting with various geometries. An example of such arrangements is called graphene ribbons or flakes (flakes, ribbons). Other common modifications are the presence of one or more vacancies in the material by loss of carbon atoms and doping with carbon species and related chemicals by atoms and molecules as adsorbates (adatoms). Experimental studies of the adsorption of gases such as CO, NH$_3$, and NO$_2$ at very low concentrations were performed, which reported that only a few molecules are adsorbed, but generate a measurable modification in the electrical resistivity [7]. In general, the modification of the graphene properties by adsorbates and incorporation of defects, generate a vast sea of possibilities that can lead to the discovery of new manufacturing processes and devices applications.

14.2 HYDROGEN ADSORPTION ON GRAPHENE

There are two strong motivations for studying the adsorption of hydrogen on graphene: The first is environmental conditions related to the energy consumption of humanity. The study of hydrogen production processes and materials capable of storing this volatile element, have become topics of great importance from the point of view of basic science and technology, since hydrogen as fuel has very attractive properties like clean combustion and renewability despite that it is not found free in nature. The second motivation is focused on the electronic area, where the manufacturing process is geared toward smaller devices that consume less energy and could work at higher frequencies.

The atomic hydrogen adsorption on graphene was predicted in 2007 [8] and 2 years later, it was observed experimentally that hydrogen adsorption is favorable and stable under normal physical conditions [9], so it may even be considered graphene as an hydrogen storage system.

Studies have also been performed on the storage of hydrogen in carbon-based materials such as nanotubes, nanofibers, and intercalated graphite compounds [10–13]. In these systems, the binding energy is much weaker and the molecules are not dissociated.

Hydrogen adsorption on graphene can be classified into two types. The first is the chemical adsorption or chemisorption that consists in bonding forces depending on the system-specific chemical that may require an activation energy (atomic hydrogen adsorption). The second type is the physical adsorption or physisorption and depends on the van der Waals (vdW) forces between the adsorbent and the adsorbate (molecular hydrogen adsorption).

Hydrogenation of graphene usually requires high-energy conditions, which are achieved at high temperature [14,15]. As a result of the adsorption of hydrogen, the electronic properties of graphene may have a forbidden energy gap whose value varies depending on the hydrogen concentrations [16]. When both sides of graphene are hydrogenated a composite called graphane is obtained [9] and if the hydrogenation comprises a single face, the graphone is achieved [17]. The results of recent research in hydrogenated graphene, induce a new era in which electronics would be based on graphene, replacing silicon [18,19].

14.2.1 Hydrogen Chemisorptions on Graphene

The early experimental works of atomic hydrogen adsorption were on a graphite surface [20,21], here the adsorption and formation mechanisms of the hydrogen molecules and clusters were studied. Also, graphite surfaces were considered in the first theoretical studies of atomic hydrogen chemisorption, where the barrier adsorption and adsorption energy of 0.2 and 0.57 eV were predicted, respectively [22,23].

By means of density functional theory (DFT) and using the generalized gradient approximation (GGA) for the exchange and correlation functional, Ivanovskaya and coworkers [24] calculated the chemisorption reaction path for an hydrogen atom on graphene; in its work, they demonstrated the sensitivity of the formation energy and structural characteristics upon the hydrogenation coverage. In the limit of a single hydrogen atom interacting with graphene, a chemisorption barrier of 0.2 eV and desorption barrier of 1.17 eV, were predicted [24].

Also, other theoretical studies that have been conducted on the subject using ab initio calculations reported that the physical properties of graphene are modified considerably when atomic hydrogen atoms are adsorbed over carbon atoms [8,16,20–23]. The adsorption of atomic hydrogen modifies the electronic and structural geometry of graphene around the carbons in which the hydrogen atoms bond, and hence an hybridization transition from the sp^2 to the sp^3, is obtained, which leads to a strong reconstruction in the graphene sheet [16,20,25]. Figure 14.2 shows the structural distortion of the carbon atoms in the graphene plane as a results of the adsorption of atomic hydrogen.

The adsorption of atomic hydrogen introduces a bandgap in the electronic states of graphene and an increase in the amount of hydrogen adsorbed at random positions increases the width of the bandgap in graphene [27]. Photoemission intensity experiments have confirmed the transition to a state of insulation in the graphene due to the atomic hydrogen adsorption [28].

The chemisorption of one hydrogen atom in a sublattice of graphene induces a localized magnetic moment. Due to the hybridization transformation in the adsorbent atom from sp^2 to sp^3, it leaves an unpaired electron belonging to π bond, which remains in the neighboring carbon atoms (other sublattice of graphene) and induces a magnetic moment (1 μ_B) [16,29].

Hydrogen chemisorption on graphene produces an electronic imbalance that makes one of the p orbitals to be no longer available for taking part in the π bands system. Also, there is a gap due to the absence of the π band, so that the graphene is transformed into a semiconductor system as shown in Figure 14.3. There is one midgap state for each spin species and the degeneracy is lifted [29].

Hydrogen could not be found freely under environmental condition, however the chemisorption of hydrogen dimers on graphene has been studied theoretically from the molecular physisorption process and there are similar studies

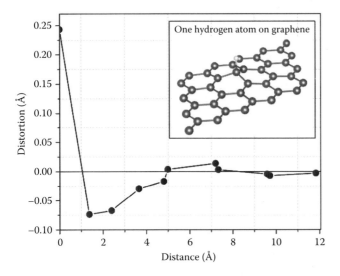

FIGURE 14.2 Deviation of the carbon atoms of the graphene plane (distortion) as a function of distance of adsorption site. (Data taken from S. Patchkovskiim, J. S. Tse, S. N. Yurchenko et al., *Proc. Natl. Acad. Sci. U. S. A.*, 102, 10439, 2005.) The picture inserted represents the structural distortion on graphene by hydrogen chemisorption.

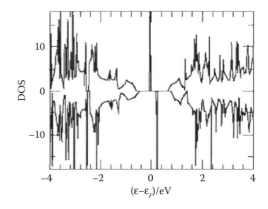

FIGURE 14.3 DOS for spin up and down of graphene system with a hydrogen atom chemisorbed. (Reprinted with permission from S. Casolo, O. M. Løvvik, R. Martinazzo, *J. Chem. Phys.* 130, 054704. Copyright 2009, American Institute of Physics.)

considering graphite surfaces [18,21,30–35]. The stability of small groups of adsorbed hydrogen on graphene is critical to clearly understand the hydrogen saturation process; and so, dimers, trimers, and other clusters of atomic hydrogen have also been studied [35]. Experimentally, it is found that the formation of clusters of hydrogen atoms on graphite surface is favored [21].

The positions of hydrogen dimers over graphene were named *ortho*, *meta*, and *para* according to the chemisorption configuration of each pair of hydrogen atoms and such configurations are shown in Figure 14.4.

Theoretically, once the first hydrogen atom is adsorbed, the adsorption of second atom is more favorable, showing an energetic sensitivity on the position of second hydrogen atom (*ortho* > *para* > *meta*) [33–35]. Table 14.1 summarizes some results of the adsorption energy of hydrogen dimers studied with DFT methodology for different carbon arrangements.

Total magnetic moment of dimers in *ortho*, *meta*, and *para* configuration depends on the adsorption sites that could belong to one or other of the subnets, for example, in *ortho* and *para* configurations both hydrogen atoms are adsorbed in sites of different subnets (see Figure 14.1a) and the total magnetic moment is 0 μ_B, while in *meta* configuration the magnetic moment is 2 μ_B, since hydrogen atoms are adsorbed in the same subnet [16,33].

Theoretical studies of hydrogen-coated graphene reported that at different hydrogen concentrations some domains adsorptions are formed; which can create and modulate the electronic gap. However, for higher concentrations of hydrogen the composite material presents a semiconductor character [28,36]. Also, the full hydrogenation from both sides of graphene (graphane) was studied theoretically [8]. According to

the positions of adsorption of hydrogen atoms, two configurations of coating or adsorption ("chair" and "boat") proved to be energetically favorable, showing forbidden energy gaps of 3.5 and 3.7 eV, respectively [8]. The first experimental evidence for the existence of graphane were obtained by Elias et al. [9].

14.2.2 HYDROGEN PHYSISORPTION ON GRAPHENE

Physisorption happens with hydrogen in molecular form (H_2) is dependent on vdW interactions; the binding energy is small, so the hydrogen molecule can be easily desorbed. The H_2 binding energy was theoretically evaluated in the range of 0.01–0.07 eV for pristine graphene [26,37,38]. The binding energy of H_2 on graphene is very weak and requires low temperatures and high pressures to ensure reasonable storage stability, however, empirical arguments suggest that under these physical conditions a uniform monolayer of H_2 can be formed on the nanotube surface, corresponding to a gravimetric density of 3.3 wt% (weight percentage of hydrogen stored to the total weight of the system) [39,40]. Experimentally, a value of 3 wt% was obtained for multilayer graphene samples obtained by exfoliation of graphite oxide [41].

For the physisorption phenomena, the binding energy and equilibrium distance are among the most important parameters in order to predict the more favored configuration for hydrogen adsorption. Mirnezhad and colleagues [42] studied the physisorption of H_2 on graphene by means of DFT with both, GGA and local density approximation (LDA) functionals. In their work, they considered the positions that are shown in Figure 14.5, where the points A, B, and C, correspond to the molecule with the axis perpendicular to the surface of graphene, while the molecular positions denoted as D, E, F, and G have the axis parallel to the plane of graphene. From their results with LDA functional, the positions D and E presented lower adsorption energies with 85.2 and 85 meV, respectively, both with an equilibrium distance of 2.7 Å, while for GGA functional, the adsorption energies of 14.3 and 14.4 meV have been predicted with a distance of 3.5 Å for the same adsorption positions (D and E) on graphene [42]. Similar results were obtained in other theoretical studies [43,44]. Although

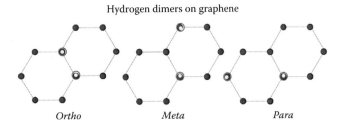

Hydrogen dimers on graphene

Ortho   *Meta*   *Para*

FIGURE 14.4 Configurations of the chemisorption of hydrogen dimers on graphene.

TABLE 14.1
Adsorption Energies for Atomic Hydrogen Dimers on Carbon Nanomaterials

Reference	Carbon System	*Ortho* (eV)	*Para* (eV)	*Meta* (eV)
[35]	Graphene	2.73	2.68	1.58
[34]	Graphite surface	2.75	2.72	1.65
[33]	Circum pyrene ($C_{42}H_{16}$)	2.11	2.1	1.07

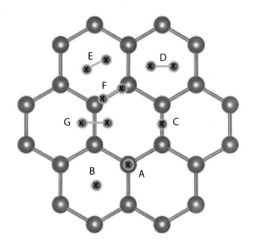

FIGURE 14.5 Adsorption sites for a hydrogen molecule on graphene.

there have not been experimental reports about the adsorption energies of H_2 on graphene yet, some studies have been determined that the adsorption energy of H_2 on graphite is 41.6 meV [45] and 35.55 meV [46], with separation distances of 2.87 and 3.27 Å, respectively.

Due to the low interactions between hydrogen molecules and graphene, the structural properties of graphene remain unchanged under the effect of the physical adsorption [42]. Similarly, the electronic properties remain without substantial change.

14.3 DEFECTS IN GRAPHENE

Defects in bulk crystals have been studied extensively for many decades, however crystals in two dimensions have been considered only recently, when single layers of graphene were isolated for the first time by mechanical exfoliation [2]. The ground state of pure carbon systems is formed by the strong sp^2-bonding network at the conjugated lattice of the graphene, however it is not perfect and it could have many possible topological defects in the layer. The graphene properties in a particular grade depend greatly on the material quality, type of defects, substrate, and so forth, which are strongly affected by the production method [5]. There are several types of defects in graphene and can be generated by the production method or be deliberately introduced [47]. It is not necessary a high energy to form them and can exist in equilibrium inside of graphene [48].

The physical and chemical properties of graphene samples with high perfection level in the atomic lattice are outstanding, but structural defects, which may appear during growth or processing, deteriorate the performance of graphene-based devices. Therefore, to achieve nanodevices made with this material, the control of its defects is very important to develop future technology, for example, one carbon atom itself does not own magnetic moment; however, the defects in the grapheme sheet could induce a magnetism that could be applied in spintronics [49–52]. Therefore, these defects can have a strong influence on the electronic, optical, thermal, and mechanical properties, which could generate a vast variety in those physical and chemical properties, defining this nanomaterial as a very promising and important material for the technology development [47].

14.3.1 MONO-VACANCY DEFECTS IN GRAPHENE

The mono-vacancy (MV) in graphene is considered as a typical defect and is formed by a missing lattice atom. This loss of periodical symmetry induces in the graphene a Jahn–Teller distortion that leads to the saturation of two of the three dangling bonds toward the missing atom and promotes the out-of-plane displacement of the dangling bond atom by elastic effect mediated by the lattice, as a consequence of the electronic repulsions between the paired electrons in the new bond and the atom opposite with its dangling bond. When the MV is generated, and the structural reconstructions are allowed, the graphene lose its D_{3h} symmetry and gets a new C_s symmetry [53]. Theoretical researches with the use of first-principles calculations predict that the energy of the C_s configuration is ~0.2 eV, which is lower than the nonreconstructed one [53,54].

The atomic structure and diffusion of vacancy defects in graphene have been theoretically investigated by means of different methodologies and approaches [54–58], revealing that a pentagonal and an enneahedral ring of carbon atoms are formed at a single vacancy defect [54,56–58]. This structural arrangement has been reported in some experimental studies, which provide evidence of MV in graphene by means of images with atomic resolution, obtained with a transmission electron microscopy (TEM) [48,59–63] and scanning tunneling microscopy (STM) [52,64], see Figure 14.6.

(a)   (b)

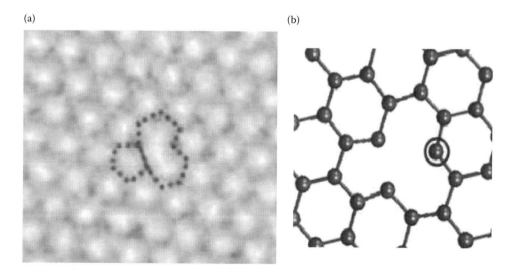

FIGURE 14.6   (a) Elementary MV in graphene observed under electron irradiation by TEM. Atomic bonds are superimposed on the defective areas in the bottom row. (Reprinted with permission from J. Kotakoski A. V. Krasheninnikov U. Kaiser et al., *Phys. Rev. Lett.*, 106, 105505, Copyright 2011 by the American Physical Society.) (b) Structure of a MV in graphene representing inside of a circle the dangling bond atom.

14.3.1.1 Properties of Graphene with MVs

In order to understand the defect role and their respective concentrations in the physical and chemical properties of graphene, several theoretical studies with different methodologies have been performed, proposing answers about the questions. Some of them are focused on the mechanical properties, reporting that Young's modulus is reduced with a linear dependence to the MV concentrations in graphene [65–67]. In one of these studies, also was discussed that the thermal conductivity in graphene is much more sensitive and relies dramatically on the MV concentrations [65].

The strain induced in the graphene around of MV defect was studied by means of the DFT, with interesting results that show that strain fields reach far into the unperturbed hexagonal network. Due to the presence of MV in the graphene, the bonds in the same direction of the new bond that formed the pentagon, get length increments, while the remaining bonds surrounding the vacancy exhibit a length decrement [68]. These structural changes produce two separate strain fields around a single vacancy [69]. The first field corresponds to the region where the bonds are shortened due to a new bond at the pentagon. The second field, that is oriented perpendicular to the first one, corresponds to the region where the bonds are stretched. As a result of the interactions between strain fields of MVs in graphene, a partial compensation of the strain (and thus gain in energy when strain fields with opposite signs overlap) results in attraction between the vacancies, while the overlap of the fields of the same sign gives rise to repulsion [69].

Concentrations of vacancies change the structural properties of graphene, and the electronic and magnetic properties are not exempt from the influence of these defects [70–72]. Structural defects, in general, give rise to localized electronic states, generate a net magnetic moment, produce flat bands associated with defects and an increase in the density of states (DOS) at the Fermi level, and eventually a development of magnetic ordering [73].

A calculation of the DOS with the use of DFT under relaxed graphene with defects in different configurations, reveals that the MV defect in graphene induces a transition from semi-metal to metal in the electronic properties, regardless of considerations of the structural reconstruction [53,70–72,74]. However, the final symmetry can influence on the DOS at the Fermi level increasing or decreasing its value [53]. The increments of vacancy concentrations in graphene do not vanish the electronic states at the Fermi level and increase its magnitude at the Fermi level and near it for concentrations below 10% [70,75]. De Laissardière and Mayou [75] presents the DOS as function of MV concentration in graphene, calculated by means of tight-binding (TB) scheme that can reproduce the ab initio calculations, the authors with this TB methodology predict that for large MV concentrations (>30%), the DOS at the Fermi level start to diminish, and could reach a zero value when the MV concentrations are larger than 50% [75].

In another theoretical study with DFT methodology and considering vdW interaction, a metallic character in graphene was found, induced by the MV presence, with a peak in the DOS near to Fermi level [72], similar to that predicted by De Laissardière and Mayou [75], which according to its predictions the π states are the most contributing.

The MV induced an appearing of states in the vicinity of the Fermi level, whose distribution will determine the material's magnetism with a strong dependence on the local bonding environment [71,76]. Vacancies are arguably considered the primary source of magnetic moments and can induce an extended π magnetic moments in graphene, which present long-range interactions and are capable of magnetic ordering [72], however, the synthesis of graphene with magnetic properties creates the possibility of being applied in the development of spintronic devices, due to its characteristics as small spin–orbit coupling, long-spin scattering length, and ballistic transport [77,78].

Yu and collaborators [79] proposed a generic geometric rule for graphene, which predicts that any two zigzag edges will be ferromagnetic-coupled if they are at an angle of 0° or 120° and antiferromagnetic-coupled for angles of 60° or 180°. According to this unified rule, a magnetic state in the graphene with MV is expected, however the concentrations and arrangements of MVs in graphene can decrease or even suppress this magnetic state [51,78–82].

Using DFT with the GGA, Yazyev and Helm [51] have calculated magnetic moments with values around 1.15–1.53 μ_B per vacancy, with a dependence of MV concentrations, while Palacios and Ynduráin [72], using the same methodology and considering vdW interactions describe the total magnetic moment behavior as a function of supercell size (see Figure 14.7). From Figure 14.7 we can see a tendency of the magnetic moment to move toward 1 μ_B (lower limit imposed by the unpaired electron of the σ dangling bond) when the supercell size increases, resulting in a magnetic moment for an isolated vacancy.

Some theoretical studies also analyzed the effects of sublattice configuration of MV in graphene and predicted that

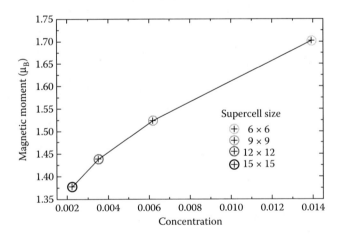

FIGURE 14.7 Magnetic moment (circle) as a function of the MV concentration (inverse of the supercell size) in a graphene monolayer. (Data taken from J. J. Palacios, F. Ynduráin, *Phys. Rev. B*, 85, 245443, 2012.)

only when these defects are located in the same hexagonal sublattice on graphene, can ferromagnetism be produced originating from a coupling between the magnetic moments [51,80]. For many MV defects, the sublattice configuration has an important role, when the MV number on one sublattice is different from that on the other sublattice, local magnetic moments are induced, even when they are equal, it is still possible that local moments could be induced [81].

From a DFT study using spin-polarized to analyze the magnetism in graphene induced by MV presence, Singh and Kroll [83] found that the MV defects break the symmetry in the π-electron system of carbon atoms in graphene and gives rise to the magnetic quasilocalized states in the graphene. When a C atom is missing, one dangling bond persists and induces magnetism, due to the appearance of unpaired π electrons of C, which play a dominant role in the magnetization of the graphene. Here the peripheral carbon atoms far from MV defects do not have obvious spin polarization, reaffirming that only atoms containing unpaired electrons have significant magnetization [76].

Table 14.2 presents the values calculated for the energy of vacancy formation (E_f), magnetic moment (B), new bond length (d_{nb}) in the MV, and the out-of-plane displacement of the dangling bond atom (dz_{db}). In the current literature, only the experimental value for the energy of vacancy formation was measured and reported, however here the theoretical predictions are presented and compared considering differences in the atoms number and approach level in the methodology.

14.4 HYDROGEN ADSORPTION ON GRAPHENE WITH MVs

The existence of hydrogen may take an important role in the properties of graphene with MV and its effect cannot be completely ruled out due to the various experimental issues [82]. Hydrogen atoms can easily adsorb on vacancy dangling bonds when the defects appear in graphite samples during He and H ions irradiation [73]. The MV presence can enhance the binding of atomic and molecular hydrogen in graphene due to its reactivity changes, which may be achieved by disturbing the charge distribution or changing the local electronic structure [38], so makes possible to attach metal atoms much more firmly than to pristine graphene and supplies a tool for tailoring the electronic structure of graphene [68].

When an hydrogen atom encounters an empty vacancy in graphene, then it saturates the dangling bond and is pinned at a height of 1.25 Å above the plane, providing a moment of 2.3 μ_B, with an adsorption energy of 5.66 eV. If a second hydrogen is present on the MV, then this atom will bond to other side of vacancy with an adsorption energy of 3.2 eV and a distance of 0.76 Å below the plane, causing in the first H atom a displacement of 0.89 Å above the plane, decreasing the magnetic moment of the system to 1.2 μ_B, located on the dangling bond that remains. Addition of a third H atom completes the decoration of the vacancy edges, saturates the remaining dangling bond, and depending on the vacancy arrangements can destroy the magnetism of the vacancy [73] or only the metallic character keeping the magnetism [72].

TABLE 14.2

Values Calculated for the Energy of Vacancy Formation (E_f), Magnetic Moment (B), New Bond Length (d_{nb}), and the Out-of-Plane Displacement of the Dangling Bond Atom (dz_{db}) for Different Number of Carbon Atoms (N) and Approach Level in the Methodology (V_{IC})

References	Arrangement (N)	V_{IC}	E_f (eV/atom)	B (μ_B)	d_{nb} (Å)	dz_{db} (Å)
[54]	Sheet $C_{120}H_{27}$	LDA	7.4		2.1	0.47
[71]	Supercell (32)	LDA	7.83	1.08		
[71]	Supercell (32)	vdW	7.36	1.02	1.84	
[71]	Supercell (72)	vdW	7.44			
[57]	Supercell (200)	LDA	7.65		2.15	0.43
[53]	Supercell (50)	LDA			2.02	0.43
[72]	Supercell (72)	vdW		1.71	2.05	
[72]	Supercell (450)	vdW		1.39	1.93	
[83]	Supercell (48)	GGA	7.77	0.89	2.40	
[83]	Supercell (60)	GGA	7.80	0.45	2.42	
[83]	Supercell (80)	GGA	7.80	0.93	2.40	
[83]	Supercell (100)	GGA	7.80	1.14	2.40	
[50]	Supercell (128)	GGA	7.7	1.04	2.02	0.18
[78]	Supercell (288)	GGA	7.5	1.00		
[76]	Supercell (72)	GGA	7.655	1.33	1.95	0.184
[84]	Supercell (32)	LDA	8.00			
[85]	Graphite experimental		7.0 ± 0.5			

According to theoretical studies, if two dangling bonds of the MV are saturated by hydrogen atoms, an energy gap is opened, making the system become semiconducting [82,86]. As suggested, in their case the magnetic order may be due to instability in the p-electron system with respect to a long-range polarization characterized by alternation in the spin direction between adjacent carbon atoms [86].

Lei and collaborators [84] described by mean of DFT with LDA a different picture in relation to energetic interactions between molecular and atomic hydrogen and the graphene with MVs. They report that the barrier energy for the dissociation of a hydrogen molecule on a perfect graphene is 2.38 eV, however when a vacancy is present, the height of the barrier is greatly decreased to 0.63 eV. When a single H atom is adsorbed on the carbon atoms around the vacancy, a strong C–H bond is formed in very stable way with a binding energy of 2.1 eV, doing very hard to recombine and form the molecule again (barrier energy is around 4 eV) [73], despite the fact that the binding energy of 0.17 eV for molecular physisorption of hydrogen on MV was reported, each MV in graphene can interact in a more favorable way up to two molecules (four hydrogen atoms), resulting in the hydrogen chemisorption on the carbon atoms neighboring in the vacancy [73].

14.5 HYDROGEN ADSORPTION ON GRAPHENE WITH ALKALI METALS

A useful system for hydrogen storage must be between ~0.20 and ~0.40 eV in the binding energy. Unfortunately, graphene provides a weaker interaction, pushing to the search for complexes that show a stronger interaction than clean graphene. In order to tackle this problem, defects such as vacancies or the use of adatom has been studied. However, in such systems it must be warned for the clustering of adatoms and its undesirable effect of a reduction in the storage capacity due to the segregation of adatoms.

Krasnov et al. [87] have shown that although indeed light transition metal decorated SWNT presents potential material for the hydrogen storage, care should be taken to avoid the metal clustering on support material, to achieve and maintain higher hydrogen capacity. In another hand, some theoretical studies have tried to gain knowledge about the intrinsic properties of the interactions between graphene and several chemical species. Two interesting studies considering both alkali metals and transition metals was conducted by Chan et al. [88] and Liu et al. [89]. They calculate structural, energetic, and electronic properties and conclude similar results, the interaction of alkali metals with graphene can provide an ionic bonding with very small distortion on graphene structure and with a tendency to form 2D layers on the surface of graphene. The result is in agreement with experiments observing 2D alkali layer formation in graphite [90]. While for transition metals, they found a strong covalent bonding with distortion in graphene.

On the basis of the behavior of alkali metals on graphene, the next step has been to study the effect on the adsorption properties of hydrogen. Tapia et al. studied the influence of

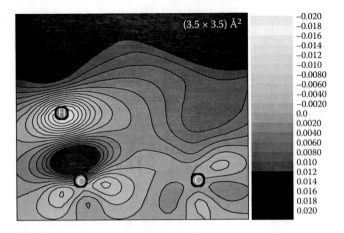

FIGURE 14.8 Result of the charge density due to influence of the K presence for *para* dimer. The changes between carbon and hydrogen regions are shown in a grayscale with dark and light representing charge density increments and decrements, respectively. (Generated with data from A. Tapia, C. Acosta, R.A. Medina-Esquivel et al., *Comput. Mater. Sci.*, 50, 2427, 2011.)

potassium on the hydrogen adsorption on graphene by means of DFT in order to gain knowledge about structural parameters, bonding and magnetic properties for H atoms interacting with potassium on graphene [91]. They found a charge transfer from K atom toward graphene even when the H atom pairs are adsorbed (see Figure 14.8).

The binding energy per H atom is increased in the most stable arrangement. Tapia et al. suggest that the hydrogen atom binding energy on graphene layer increase up to 82% due to the preadsorption of potassium [91].

Graphane has been previously predicted by Sofo et al. [8] and synthesized by Elias et al. [9], attracting attention to change its thermodynamic properties in order to get the right kinetic to absorb/desorb hydrogen. Two branches followed are (i) to get systems similar to graphane with enhanced properties by means of alkali metals on graphene's surface and (ii) to add alkali metals to graphane directly. In the former, Medeiros et al. [92] simulate a system like graphane but with all the hydrogen substituted by alkali metals in order to know the changes in electronic and structural properties between H-graphene (graphane), and Li, K, and Na on graphene. They do not support the sp^2 to a sp^3 transition of carbon orbitals claimed by Yang [93]. Medeiros suggest that carbon atoms relaxed stay outside of the alkali tetrahedron. They claimed that this difference is due to the π-acceptor character of lithium that was not taken into account previously and shown a lower adsorption energy than in graphene. In the latter, Antipina et al. [94] explore graphane with chemically bonded alkali metals (Li, Na, and K) for a variety of concentrations and the effect on the binding energy, for the alkali metals and the dependence of binding energy of the nth hydrogen molecule on the alkali – metal + graphene complex. They found that 20% metal concentration is energetically favorable for Li, Na, and K. The binding energy for hydrogen decay with the number of molecules up to 0.20 eV, with four molecules per alkali metal (see Figure 14.9).

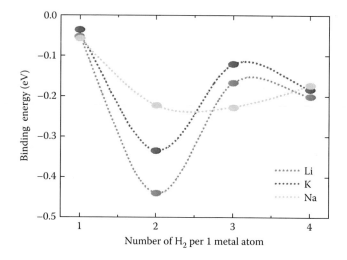

FIGURE 14.9 Dependence of binding energy of the nth hydrogen molecule on the metal–graphene complex upon the number of adsorbed H_2 molecules per the metal atom. (Data taken from L. Y. Antipina, P. V. Avramov, S. Sakai et al., *Phys. Rev. B*, 86, 085435, 2012.)

This complexes provide gravimetric capacities as high as 12.2, 10.3, and 8.56 wt% for lithium, sodium, and potassium, respectively, with binding energies around −0.20 eV. All of them exceed the Department of Energy (DOE) target of 5.5 wt% even in the case of three hydrogen atom per alkali metal atom. Also, there are efforts to simulate bigger complexes, an example is the study of Neek-Amal et al. [95]. They simulate complexes arrangements on graphene at 50 K and found that the transition atoms considered aggregate and make various size nanoclusters distributed randomly on graphene surface. In contrast, alkali atoms make one atomic layer on graphene sheets. The potassium atoms almost deposit regularly on the surface at low temperature providing an enhanced electrical conductivity [95]. In conclusion, graphene doped with potassium atoms are a promising material in the hydrogen storage problem due to the gravimetric capacity, molecular binding energy, and a well-distributed centers for molecular adsorption, that is, alkali atoms.

14.6 CONCLUSIONS

Graphene are among the most important materials of modern nanoscience and nanotechnology, including molecular electronics and chemistry, playing a remarkable role in the development of future technology. The energy storage and electronics developments are two strong motivations for studying the adsorption of hydrogen on graphene, however, its properties in a particular grade depend strongly on the material quality. The defects and alkaline functionalization on graphene can modulate its physical and chemical properties, the MV presence can enhance the binding of atomic and molecular hydrogen on graphene, due to the reactivity of dangling bonds, while preadsorbed alkali metals induce located zones that favored the hydrogen binding.

ACKNOWLEDGMENTS

The author is grateful to R. A. Medina-Esquivel PhD for their helpful assistance.

REFERENCES

1. A. K. Geim, *Science*, 324, 1530, 2009.
2. K. S. Novoselov, A. K. Geim, S. V. Morozo et al., *Science*, 306, 666, 2004.
3. C. Lee, X. Wei, J. W. Kysar et al., *Science*, 321, 385, 2008.
4. T. Ando, *NPG Asia Mater.*, 1, 17, 2009.
5. K. S. Novoselov et al., *Nature*, 490, 192, 2012.
6. Y. Hernandez, V. Nicolosi, M. Lotya et al., *Nat. Nanomater.*, 3, 563, 2008.
7. F. Schedin, A. K. Geim, S. V. Morozov et al., *Nature*, 6, 652, 2007.
8. J. O. Sofo, A. S. Chaudhari, G. D. Barber et al., *Phys. Rev. B*, 75, 153401, 2007.
9. D. C. Elias, R. R. Nair, T. M. G. Mohiuddin et al., *Science*, 323, 610, 2009.
10. A. C. Dillon, K. M. Jones, T. A. Bekkedahl et al., *Nature*, 386, 377, 1997.
11. G. Canto, P. Ordejón, C. Hansong et al., *New Journal of Physics*, 5, 1241, 2003.
12. Q. Wang, J. K. Johnson, *Int. J. Thermophys.*, 19, 835, 1998.
13. H. Cheng, G. Pez, G. Kresse, J. Hafner, *J. Phys. Chem. B*, 105, 736, 2001.
14. J. J. Palacios, J. Fenández-Rossier, *Phys. Rev. B*, 77, 195248, 2008.
15. A. Savchenko, *Science*, 323, 5914, 2009.
16. D. W. Boukhvalov, M. I. Katsnelson, A. I. Lichtenstein, *Phys. Rev. B*, 77, 035427, 2008.
17. J. Zhou, Q, Wang, Q. Sun et al., *Nano Lett.*, 9, 3867, 2009.
18. C. Berger, Z. Song, T. Li et al., *J. Phys. Chem. B*, 108, 19912, 2004.
19. P. Avouris, Z. Chen, V. Perebeinos, *Nat. Nanotechnol.*, 2, 605, 2007.
20. L. Hornekær, Ž. Šljivančanin, W. Xu et al., *Phys. Rev. Lett.*, 96, 156104, 2006.
21. L. Hornekær, E. Rauls, W. Xu et al., *Phys. Rev. Lett.*, 97, 186102, 2006.
22. L. Jeloica, V. Sidis, *Chem. Phys. Lett.*, 300, 157, 1999.
23. X. Sha, B. Jackson, *Surf. Sci.*, 496, 318, 2002.
24. V. V. Ivanovskaya, A. Zobelli, D. Teillet-Billy et al., *Eur. Phys. J. B*, 76, 481, 2010.
25. D. W. Boukhvalov, M. I. Katsnelson, *J. Phys.: Condens. Matter*, 31, 34405, 2009.
26. S. Patchkovskiim, J. S. Tse, S. N. Yurchenko et al., *Proc. Natl. Acad. Sci. U. S. A.*, 102, 10439, 2005.
27. K. S. Choi, C. H. Park, *J. Korean Phys. Soc.*, 54, 939, 2009.
28. M. Z. S. Flores, P. A. S. Autreto, S. B. Legoas et al., *Nanotechnology*, 20, 465704, 2009.
29. S. Casolo, O. M. Løvvik, R. Martinazzo, *J. Chem. Phys.* 130, 054704, 2009.
30. Y. Miura, H. Kasai, W. Diño et al., *J. Appl. Phys.*, 93, 3395, 2003.
31. F. Costanzo, P. L. Silvestrelli, F. Ancilotto, *J. Chem. Theory Comput.*, 8, 1288, 2012.
32. Y. Ferro, F. Marinelli, A. Allouche, *J. Chem. Phys.*, 116, 8124, 2002.
33. Y. Ferro, D. Teillet-Billy, N. Rougeau et al., *Phys. Rev. B*, 78, 085417, 2008.
34. Ž. Šljivančanin, E. Rauls, L. Hornakær et al., *J. Chem. Phys.*, 131, 084706, 2009.

35. Ž. Šljivančanin, M. Andersen, L. Hornekaer et al., *Phys. Rev B*, 83, 205426, 2011.
36. P. Chandrachud, B. S. Pujari, S. Haldar et al., *J. Phys.: Condens. Matter*, 22, 465502, 2010.
37. V. Tozzini, V. Pellegrini, *J. Phys. Chem. C*, 115, 25523, 2011.
38. Y. Kwon, *J. Korean Phys. Soc.*, 57, 778, 2010.
39. A. Züttel, P. Sudan, Ph. Mauron et al., *Int. J. Hydrogen Energy*, 27, 203, 2002.
40. V. Tozzini, V. Pellegrini, *Phys. Chem. Phys.*, 15, 80, 2013.
41. A. Ghosh, K. S. Subrahmanyam, K. S. Krishna et al., *J. Phys. Chem. C*, 112, 15704, 2008.
42. M. Mirnezhad, R. Ansari, M. Seifi, H. Rouhi, M. Faghihnasiri, *Solid State Commun.*, 152, 842, 2012.
43. D. Henwood, J. D. Carey, *Phys. Rev. B*, 75, 245413, 2007.
44. J. S. Arellano, L. M. Molina, A. Rubio et al., *J. Chem. Phys.*, 112, 8114, 2000.
45. G. Vidali, G. Ihm, H. Kim, *Surf. Sci. Rep.*, 12, 133, 1991.
46. E. L. Pace, A. R. Siebert, *J. Phys. Chem.*, 63, 1398, 1959.
47. F. Banhart, J. Kotakoski et al. ACS Nano, 5, 26, 2011.
48. A. Hashimoto, K. Suenaga, A. Gloteret et al., *Nature*, 430, 870, 2004.
49. E. Santos, D. Sánchez-Portal, A. Ayuela, *Phys. Rev. B*, 81, 125433, 2010.
50. Y. Ma, P. O. Lehtinen, A.S. Foster et al., *New J. Phys.*, 6, 68, 2004.
51. O. V. Yazyev, L. Helm, *Phys. Rev. B*, 75, 125408, 2007.
52. M. Topsakal, E. Aktürk, H. Sevinçli, *Phys. Rev. B*, 78, 235435, 2008.
53. H. Amara, S. Latil, V. Meunier et al., *Phys. Rev. B*, 76, 115423, 2007.
54. El-Barbary, R. H. Telling, C. P. Ewels et al., Phys. Rev. B, 68, 144107, 2003.
55. E. Kaxiras, K. C. Pandey, *Phys. Rev. Lett.*, 61, 2693, 1988.
56. G. D. Lee, C. Z. Wang, E. Yoon et al., *Phys. Rev. Lett.*, 95, 205501, 2005.
57. K. Yamashita, M. Saito, T. Oda, *Jpn J. Appl. Phys.*, 45, 6534, 2006.
58. M. Saito, K. Yamashita, T. Oda, *Jpn J. Appl. Phys.*, 46, L1185, 2007.
59. M. H. Gass, U. Bangert et al., *Nat. Nanotechnol.*, 3, 676, 2008.
60. J. C. Meyer, C. Kisielowski et al., *Nano Lett.*, 8, 3582, 2008.
61. J. H. Warner, M. H. Rümmeli, L. Ge et al., *Nat. Nanotech.*, 4, 500, 2009.
62. Ç. Ö. Girit, J. C. Meyer, R. Erni et al., *Science*, 323, 5922, 2009.
63. J. Kotakoski, A. V. Krasheninnikov, U. Kaiser et al., *Phys. Rev. Lett.*, 106, 105505, 2011.
64. M. M. Ugeda, I. Brihuega, F. Guinea et al., *Phys. Rev. Lett..*, 104, 096804, 2010.
65. F. Hao, D. Fang, Z. P. Xu, *Appl. Phys. Lett.*, 99, 041901, 2011.
66. A. Tapia, R. Peón-Escalante, C. Villanueva et al., *Comput. Mater. Sci.*, 55, 255, 2012.
67. N. Jing, Q. Xue, C. Ling, *RSC Adv.*, 2, 9124, 2012.
68. A. V. Krasheninnikov, R. M. Nieminen, *Theor. Chem. Acc.* 129, 625, 2011.
69. J. Kotakoski, A. V. Krasheninnikov, K. Nordlund, *Phys. Rev. B*, 74, 245420, 2006.
70. C. Carpenter, A. Ramasubramaniam, D. Maroudas, *Appl. Phys. Lett.*, 100, 203105, 2012.
71. P. A. Denis, R. Faccio, F. Iribarne, *Comput. Theor. Chem.*, 995, 1, 2012.
72. J. J. Palacios, F. Ynduráin, *Phys. Rev. B*, 85, 245443, 2012.
73. P. O. Lehtinen, A. S. Foster et al., *Phys. Rev. Lett.*, 93, 187202, 2004.
74. V. Ferro, A. Allouche, Phys. Rev. B, 75, 155438, 2007.
75. G. T. De Laissardière, D. Mayou, *Mod. Phys. Lett. B*, 25, 1019, 2011.
76. X. Q. Dai, J. H. Zhao, M. H. Xie et al., *Eur. Phys. J. B*, 80, 343, 2011.
77. O. Yazyev, *Rep. Prog. Phys.*, 73, 056501, 2010.
78. X. Y. Cui, R. K. Zheng, Z. W. Liu et al., *Phys. Rev. B*, 84, 125410, 2011.
79. D. Yu, E. M. Lupton, H. J. Gao et al., *Nano Res.*, 1, 497, 2008.
80. Y. Zhang, S. Talapatra, S. Kar et al., Phys. Rev. Lett., 99, 107201, 2007.
81. H. Kumazaki, D. S. Hirashima, *Low Temp. Phys.*, 34, 805, 2008.
82. X. Yang, H. Xia, X. Qin et al., *Carbon*, 47, 1399, 2009.
83. R. Singh, P. Kroll, *J. Phys.: Condens. Matter*, 21, 196002, 2009.
84. Y. Lei, S. A. Shevlin et al., *Phys. Rev. B*, 77, 134114, 2008.
85. P. A. Thrower, R. M. Mayer, *Phys. Status Solidi A*, 47, 11, 1978.
86. L. Pisani, B. Montanari, N. M. Harrison, *New J. Phys.*, 10, 033002, 2008.
87. P. O. Krasnov, F. Ding, A. K. Singh et al., *J. Phys. Chem. C*, 111, 17977, 2007.
88. K. T. Chan, J. B. Neaton, M. L. Cohen, *Phys. Rev. B*, 77, 235430, 2008.
89. X. Liu, C. Z. Wang, Y. X. Yao et al., *Phys. Rev. B*, 83, 235411, 2011.
90. M. Caragiu, S. Finberg, *J. Phys.: Condens. Matter.* 17, R995, 2005.
91. A. Tapia, C. Acosta, R.A. Medina-Esquivel et al., *Comput. Mater. Sci.*, 50, 2427, 2011.
92. P. V. C. Medeiros, F. de Brito Mota et al., *Nanotechnology*, 21, 115701, 2010.
93. Y. Chih-Kai, *Appl. Phys. Lett.*, 94, 163115, 2009.
94. L. Y. Antipina, P. V. Avramov, S. Sakai et al., *Phys. Rev. B*, 86, 085435, 2012.
95. M. Neek-Amal, R. Asgari, M. R. Rahimi Tabar, *Nanotechnology*, 20, 135602, 2009.

15 Modifications of Electronic Properties of Graphene by Boron (B) and Nitrogen (N) Substitution

Debnarayan Jana, Palash Nath, and Dirtha Sanyal

CONTENTS

Abstract 231
15.1 Introduction 231
15.2 Basic Methodology 234
    15.2.1 DFT and Many Body Physics 234
    15.2.2 DFT and Its Various Approximations 235
        15.2.2.1 Local Density Approximations 235
        15.2.2.2 Generalized Gradient Approximations 236
15.3 Electronic Properties Modification of Graphene 236
    15.3.1 Strained Graphene: Band Structure Modification 237
    15.3.2 Doped Graphene: Electronic Properties Modifications 238
15.4 Boron and Nitrogen Doping: Electronic Properties Modification 239
15.5 Future Directions 243
15.6 Conclusions 244
References 244

ABSTRACT

Graphene is a zero bandgap purely two–dimensional honeycomb-like carbon (C)-based nanomaterial. Graphene could be used as a next-generation nanoelectronic device once a finite bandgap can be introduced by proper substitution of other elements in graphene. A large number of theoretical as well as experimental studies have been carried out in the recent past to produce bandgap in graphene-based system, especially by substitutional doping in the hexagonal carbon network of graphene. Boron (B) and nitrogen (N) are two very good candidates for substitutional doping in the graphene network because the atomic radii of these two elements are very close to the atomic radius of carbon. Graphene with B doping behaves like a p-type material whereas with N doping it becomes n-type material. One would expect to fabricate p–n junction nanodevices with co-doping of both B and N elements in the graphene system at suitable position with proper concentration. In the present chapter, it has been attempted to review the bandgap opening at the Dirac point (linear band crossing point) of graphene by substitution of B and N by using density functional theory. We also discuss the bandgap opening by application of strain. Since the optical properties are closely related to the band structure we hope these studies will shed light on the designing optoelectronic device based on graphene.

15.1 INTRODUCTION

Graphene is a purely two-dimensional (2D) honeycomb-like crystal of carbon atoms (Novoselov et al., 2004; Geim and Novoselov, 2007; Castro Neto et al., 2009; Loh et al., 2010; Choi et al., 2010; Kuila et al., 2012 ; Novoselov et al., 2012). Two adjacent carbon atoms of the graphene planar structure build up the unit cell of the crystal. Out of four valence electrons of the carbon atoms, three are used for the planar C–C sp² hybridized bonding. Thus, each carbon atom contains single p_z electron. Due to the presence of this single loosely bound electron, graphene exhibits fascinating electronic and electrochemical properties. Before the discovery of graphene, it was believed that a 2D crystal structure is thermodynamically unstable (Mermin, 1968; Landau and Lifshitz, 1980). In 2004, Geim discovered the first 2D crystal structure, graphene, experimentally (Geim, 2009). Graphene can be considered to be the mother of other carbon allotropes, namely graphite (three dimensional [3D]) which is nothing but stacks of graphene layers. Carbon nanotubes (CNT) is the rolled up graphene nanosheet (one-dimensional system [1D]) and fullerenes (zero-dimensional system) is the spherical-shaped graphene sheet with certain number of pentagons in the hexagonal network. Studies on different properties of graphene nanostructures are fundamentally important to get the idea of other graphene-based carbon allotropes.

After the discovery of layered graphene nanostructures, there have been growing intensive interest due to its fascinating electronic and physical properties (Katsnelson, 2007; Kuila et al., 2012). The unbound p_z electrons of the graphene system behave effectively like massless Dirac fermions (Katsnelson, 2007; Ando, 2009, Casolo et al. 2011, Das-Sarma et al., 2011). The energy momentum dispersion relation exhibit linear relationship and band crossing is present at the Fermi level. Dirac cone is observed at the Fermi level of pristine planar graphene systems (Novoselov et al., 2005). Graphene exhibits ballistic transport of electrons and holes even at room temperature. Also, fractional quantum Hall effect (FQHE) has been observed in graphene systems (Novoselov et al., 2007). Apart from these, graphene can be promising material for nanoelectronics in the near future. It is believed that it can even replace silicon-based electronic devices. There are several advantages in using graphene-based system. First of all, it is a 2D material and therefore it can be used as surface grown nanoelectronic device. Second, it is cost-effective to fabricate a few layers of graphene, graphene nanosheet, and so on. The length scale of graphene nanostructures can be made very small thereby reducing the size of the device significantly, which is one of the major demand as well as challenge in device industries. The electron mobility of graphene is very high compared with silicon and germanium. The electron mobility of graphene is ~20,000 cm²/V s (Katsnelson, 2007), which is one order of magnitude greater than the modern Si-based transistor. Graphene can be used as spin valve devices as it possesses negligible spin–orbit coupling. There is also opportunity to use it as gas sensor, due to large surface-to-volume ratio (because of 2D extensions) of graphene; it can also interact with foreign body easily. However, for the purpose of gas sensor, doped graphene systems are more efficient than pristine graphene.

The charge carriers in graphene behave like ultra-relativistic particles with a speed of $c/300$ (where c is the velocity of light in free space) controlled by the Fermi velocity suitable for 2D electron systems. This particular characteristic feature about graphene is the key factor for various physical properties associated with graphene. Below we present some of the novel interesting properties in comparison with conventional 2D semiconductors.

- Conventional semiconductors possess some finite nonzero bandgap while graphene has a nominal gap of zero.

- In a conventional semiconductor, the study of electron and hole is done with differently doped materials. However, in graphene the nature of charge carriers changes drastically from electron to hole and vice versa at the Dirac points.
- In conventional semiconductor, the Fermi level lies in the bandgap whether doped or undoped. But in graphene, the position of the Fermi level is either in the valence or conduction band.
- Graphene has a linear dispersion relation in contrast to quadratic one for normal semiconductors ($E = \hbar^2 k^2 / 2m$). Because of the linear dispersion of graphene around a Dirac point, it is not possible to define the effective mass such as done in conventional semiconducting material.
- The dispersion branches in graphene interact very weakly with one another giving rise to chirality. This chiral effect is quantified by a pseudo-spin quantum number associated with the charge carriers. This chirality can be understood from Dirac-like Hamiltonian in terms of the Pauli matrix. This chirality gives rise to a distinctive behavior such as Klein tunneling akin to conventional semiconductor.
- The traditional 2D electron gas (2DEG) in a quantum well or heterostructure possess as an effective thickness between 5 and 50 nm. However, in graphene being a single layer of thickness of carbon atoms has a thickness of ~0.3 nm only. Naturally, conduction in z direction for graphene is more compared to conventional 2DEG.
- It has been observed that graphene possesses a finite minimum conductivity even in case of vanishing charge carriers. This is important from the point of view of field effect transistors (FET).

In Table 15.1, we show a comparison of various physical parameters of graphene with other conventional semiconductors.

Apart from these positive sides, the major drawback of graphene system is that, it exhibits zero bandgap at the Fermi level. Bandgap engineering in graphene can be done in several ways namely (i) by applying electric field perpendiculars to bilayers of graphene, (ii) hydrogenation, (iii) graphene–substrate interaction (hexagonal boron nitride [h-BN]), (iv) doping or alloying, (v) applying strains, and (v) introduction of Stone–Wales (SW) defect. It would be nice if by alloying or doping

TABLE 15.1

Comparisons of Graphene Electronic Properties with Respect to Other 2D Semiconductors

Electronic Property	Si	Ge	GaAs	AlGaN/ GAN 2DEG	Graphene
Bandgap (at 300 K) (eV)	1.1	0.67	1.43	3.3	0
Ratio of effective mass to electron's mass	1.08	0.55	0.067	0.19	0
Electron's mobility (at 300 K) (cm⁻² V⁻¹ S⁻¹)	1350	3900	4600	1500–2000	$(2–3)10^5$

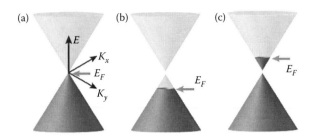

FIGURE 15.1 Schematic picture of band diagram for (a) pristine graphene, (b) p-type graphene, and (c) n-type graphene. There is no bandgap for (a) where the Fermi energy is at the crossing point. However, there is small but finite nonzero bandgap for (b) and (c).

some proper materials one can introduce ~1 eV bandgap at the Fermi level in the graphene-based system and the bandgap can be tunable properly. Suitable doping in the pristine graphene structure turns it into p-type as well as n-type material such that the Fermi level shifts downward to the valence band and upward to the conduction band as shown in Figure 15.1. To overcome these difficulties as well as to study the electronic and electrochemical properties of graphene-based systems different kinds of attempts have been carried out extensively in the last decade. Substitutional as well as interstitial position doping of different atoms in the graphene structure yields different kinds of modifications of the electronic properties, mainly the band structure and density of states (DOS). The Fermi energy (E_F) can vary with doping concentration. First principles calculations on the bandgap of graphene layer of h-BN indicate (Kaplan et al., 2014) that the bandgap is quite sensitive to the external perpendicular electric field as well as the separation between the graphene and the h-BN layers. Based on density functional theory (DFT) calculations, it has been found (Quhe et al., 2012) that the bandgap of single layer graphene (SLG) sandwiched between two h-BN single layers can be opened to 0.34 eV in the presence of a strong electric field while 0.16 eV without the application of external electric field. This tunable and sizeable bandgap along with structural integrity might render this structure a promising candidate for FET. First-principles calculations on the graphene supported on functionalized h-BN with hydrogen and fluorine atoms have indicated (Tang et al., 2013) that a finite bandgap of 79 meV can be opened due to significant charge transfer from graphene to substrate. This bandgap can, however, be tuned by changing the interlayer spacing, increasing the number of functionalized BN layers, and applying external electric field.

DFT calculations have predicted a bandgap up to 1.21 eV in fully chlorinated graphene (CCl) (Sachin and Ciraci, 2012) and a high mobility of 1535 cm²/V s was achieved (Zhang et al., 2013). DFT study on the electronic properties of BN-doped graphene monolayer, bilayer, trilayer, and multilayer systems have been attempted and theoretical computations have indicated (Kaloni et al., 2014) the high mobility of charge carriers from their effective masses. Moreover, the bandgaps varied from 0.02 to 2.43 eV depending on the doping concentrations. DFT-based tight binding (TB) framework has been used to compute the adsorption energy of various

nucleobases in BN substituted graphene nanoribbons for modeling biosensor device (Bhattacharya et al., 2013).

A first principles calculation on the bandgap of graphene/graphitic carbon nitride (g-C_3N_4) bilayers has indicated (Li et al., 2014) that the gap can be up to 108.5 meV quite large enough at room temperature ($k_B T = 26$ mev). This may be quite helpful for using graphene as suitable candidate for novel spintronic device. Recently, directly synthesized BCN graphene by solvo-thermal process has shown (Jung et al., 2014) a high potential for FET application with an optical bandgap of 3.3 eV. Besides, high-quality atomic layer film with boron, nitrogen, and carbon containing atomic layer having full range of composition have been synthesized (Gong et al., 2014) at the nanometre scale to produce fabrication layered structure in 2D integrated circuits.

Electronic properties of graphene are also significantly modified in the presence of intrinsic as well as extrinsic disorder in the hexagonal network. Surface ripples and topological defects belong to intrinsic disorder whereas extrinsic disorders can be of different form; mainly, adatoms, vacancies charges on the top of graphene sheet, etc. (Castro Neto et al., 2009). One of the major topological defects is SW defect, that is, paired combination of pentagon and heptagon in hexagonal lattice. Such defects produce long-range deformation in the structure and give rise to electronic properties modification (Castro Neto et al., 2009). Edge states and ribbon width of graphene nanoribbons (GNRs) also play a crucial role to determine its electronic properties. The first-principles analysis on the electronic and vibrational spectrum of graphene by SW defect shows (Shirodkar and Waghmare, 2012) the presence of defect bands in electronic DOS spectra and softening of G band while hardening of D band. Recently, a real space recursion method appropriately suitable for disordered SW defects on graphene revealed (Chowdhury et al., 2014) the existence of local defect state above the Dirac point and widening of the defect structure with increasing disorder strength. Thus, the presence of defect bands or widening of defect structure in DOS can be appropriately used in experiment to characterize SW defects in graphene.

Moreover, the application of in-plane strain in planar graphene structure along different directions breaks the sublattice symmetry and shows a bandgap opening at the Dirac point; here the bandgap magnitude can be controlled with the strength of the strain.

In this chapter, an attempt has been taken to discuss mainly the effect of nitrogen (N) and boron (B) doping in the graphene nanostructures. The choice of N/B as dopant has several advantages, for example, these two atoms are the nearest neighbor of the C atom in the periodic table; boron is trivalent while nitrogen is pentavalent. The atomic radii of these two atoms are comparable to that of the carbon atom, therefore substitutional doping of B/N in the graphene does not distort the structure significantly. Doping of nitrogen (N) atom in the system can be considered as n-type doping whereas boron (B) atom doping can be considered as p-type doping. Both B and N atom doping with same concentration in the graphene system keeps the system isoelectronic but can be considered as electron–hole co-doping in the system. p–n junction devices can be fabricated with proper co-doping of B and N atom in the graphene.

15.2 BASIC METHODOLOGY

The many particle problems in quantum physics appear when one is interested to examine atoms, molecules, or solids as charged particle systems. Because of the presence of large number of atoms and the long-time scales, an effective theoretical calculation needs naturally high-speed computational scheme. One of the main computational schemes used in material science is the DFT, which obviously uses density function rather than the wave function. It is regarded as one of the most successful theories of many body physics applied to electronic structure of matter and other related aspects of matter. In chemistry and physics, it is nowadays used routinely to compute the binding energy, bond length, and bandgap of stable materials. In conventional electronic structure calculation, such as Hartree–Fock (HF) or any other theory, one uses complicated many particle wave function. There are $3N$ dimensional Schrödinger equation for wave function, where N being the number of particles in the system. However, DFT can be regarded as an alternative way (Parr and Yang, 1989; Dreizler and Gross, 1995; Kohn, 1999) of computing the electronic structure of N particle system in terms of 3D density $n(\vec{r})$. It is interesting to note that density $n(\vec{r})$ (real and positive) remains functions of only three variables, no matter what complex system does exist in nature. In this sense, DFT uses density as the basic variable for calculation of physical properties in many particle physics. Besides, density is very simple to deal with both conceptually as well as in computational framework. There are quite a few interesting review articles available (Capelle, 2006; Ao and Jiang, 2009) dealing with various aspects of DFT.

Traditional wave functional methods (either variational or perturbative) can be applied to smaller system with high accuracy. In fact, for smaller number of atoms, the wave functional approach is quite successful (Kohn, 1999). However, for an interacting system having large number of degrees of freedom, one faces an exponential wall (Kohn, 1999) in the traditional wave method. More importantly, even if one has solved the problem of the wave function $\psi(\vec{r})$ somehow, still, for recording that wave function in the computer, one requires quite a large number of bits. However, this problem can be overcome for the density, as the number of bits to be recorded is very small. Nevertheless, one has to pay in approximating certain quantities as discussed later. One can illustrate these points through a simple example (Kohn, 1999; Jana, 2008). One can consider a real space representation of ψ on a mesh. In this mesh, we assume each coordinate is discretized by 30 mesh points. Hence, for N electrons, ignoring spin ψ is a function of $3N$ variables and a total of 30^{3N} values are required to describe it on the above mesh. In contrast, density $n(\vec{r})$, being a function of three variables, requires 30^3 values on the same mesh. Therefore, for $N = 10$, the conventional wave function approach requires $30^{30}/30^3 = 30^{27}$ times more storage space than density. Therefore, DFT represents an efficient approach

to describe the quantum effects in the electronic structure of a many body system.

15.2.1 DFT AND MANY BODY PHYSICS

The properties of a material are uniquely determined by behaviors of electrons that comprise bond strengths (Kohn, 1999; Ao et al., 2009) of the relevant system. The central theorem (Hohenberg and Kohn, 1964; Hohenberg et al., 1990) of DFT is due to Hohenberg–Kohn (HK). The theorem states that the ground state density $n_0(\vec{r})$ of a bound interacting electronic system in some external potential $v(\vec{r})$ determines this potential uniquely. This statement is mathematically rigorous. The ground state energy functional thus can be written as

$$E_0 = \left\langle \psi_0[n_0] \left| T + v(\vec{r}) + V_{ee} \right| \psi_0[n_0] \right\rangle$$

$$= \int n_0(\vec{r})v(\vec{r})d^3r + T[n_0] + V_{ee}[n_0] \qquad (15.1)$$

$$= \int n_0(\vec{r})v(\vec{r})d^3r + F[n_0]$$

Note that although $T[n_0]$ and $V_{ee}[n_0]$ are universal functions for all the systems, the external potentials differ from system to system. The functional $F[n_0]$ is independent of external potentials. If one is able to minimize the above energy functional to get the ground state density $n_0(\vec{r})$, then all the physical observable quantities can be computed.

In DFT, the ground state of an interacting electron gas is mapped onto the ground state of a (fictitious) noninteracting electron gas, which experiences an effective potential. This potential is in contrast to other techniques such as HF that solves a full set of wave functions in the Schrödinger equation. This mapping in principle gives exact ground-state properties (such as the cohesive energy, lattice parameters, and phonon spectra). In this regard, Kohn–Sham (KS) (Kohn and Sham, 1965) first put forward the prescription for practical DFT to calculate E_0 from $n_0(\vec{r})$ without finding the wave function. If one defines the change in kinetic energy $\Delta T = T[n] - T_s[n]$ and the change in interaction energy

$$\Delta V_{ee} = V_{ee}[n] - \frac{1}{2}\iint \frac{n(\vec{r}_1)n(\vec{r}_2)}{r_{12}}d^3r_1d^3r_2$$

with respect to noninteracting system, then the total energy functional of the real physical interacting system can be rewritten as

$$E[n] = T_s[n] + \int v(\vec{r})n(\vec{r})d^3r + \frac{1}{2}\iint \frac{n(\vec{r}_1)n(\vec{r}_2)}{r_{12}}d^3r_1d^3r_2$$

$$+ \Delta T[n] + \Delta V_{ee}[n] = T_s[n] + \int v(\vec{r})n(\vec{r})d^3r + \frac{1}{2}$$

$$\iint \frac{n(\vec{r}_1)n(\vec{r}_2)}{r_{12}}d^3r_1d^3r_2 + E_{xc}[n] \qquad (15.2)$$

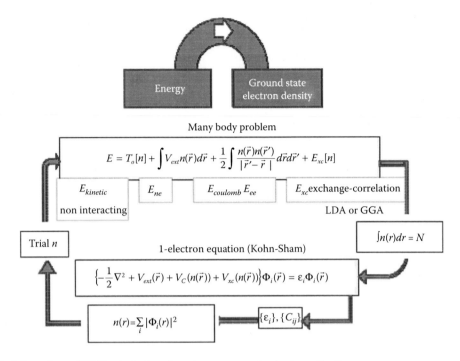

FIGURE 15.2 Typical flow chart of DFT calculations have been summarized schematically.

The first term ($T_s[n]$) is from the noninteracting kinetic energy of the particles while the second term is due to external potential. However, the third term is simply the classical electrostatic interelectronic repulsion energy. The most important is the exchange correlation energy E_{xc} coming from the interaction energy between the electrons and this part is the most troublesome one in DFT and needs the appropriate approximation for each particular problem in quantum mechanical system. This E_{xc} essentially depends on the density $n(\vec{r})$. In most of the situation, the contribution of this term to the ground-state energy compared to others is very small; however, it is not easy to calculate. Better and better approximation to this quantity will bring the better estimation of molecular properties. Besides, there exists a particular reason for rewriting in the above fashion; the total energy functional is expressed in terms of the modest number of physical quantities, which are directly evaluated from the density. The philosophy of this method is to rewrite the energy functional in terms of fictitious noninteracting system plus some unknown exchange correlation energy functional. In fact, this prescription allows one to write simple KS equations (Kohn et al., 1965) of the fictitious noninteracting system with ε_i as the ith energy eigen value

$$\left[-\frac{\hbar^2}{2m}\nabla^2 + v_s(\vec{r})\right]\Phi_i(\vec{r}) = \varepsilon_i\Phi_i \qquad (15.3)$$

These equations will yield a typical density $n(\vec{r}) = \sum_i |\Phi_i(\vec{r})|^2$. These equations are to be solved self-consistent or iterative way by guessing a suitable initial density. With the initial guess of the density, one calculates the corresponding effective potential and then solves again the orbital wave function to yield a new density. This method is frequently repeatedly until the convergence is ensured. Hence, the minimization condition of DFT in terms of density $n_0(\vec{r})$ reduces to the minimization of the KS orbital Φ_i. KS equations (Kohn et al., 1965) are exact and yield the exact density—no approximations are made here. These are single particle equations—much easier to solve than coupled Schrödinger equations for large systems. But there is an unknown exchange interaction to be approximated. The typical flow chart of DFT calculations has been summarized in Figure 15.2.

15.2.2 DFT AND ITS VARIOUS APPROXIMATIONS

To solve numerically, the ground state density by variational and self-consistent ways, we have to invoke to approximation for exchange interaction term. These approximations, however, were originated outside the regime of DFT and basically reflect the physics of electronic structure. The most common approximations used in the literature are discussed below.

15.2.2.1 Local Density Approximations

This is the most common approximation used in DFT for the exchange interaction term. The functional is a simple integral over a function of the electron density at each point over the real 3D space. This approximation is exact for a special case of uniform electronic system. The exchange term in this local density approximations (LDA) (Parr et al., 1989; Dreizler et al., 1995; Kohn, 1999; Capelle, 2006; Ao et al., 2009) can be written as

$$E_{ex}^{LDA} = \int n(\vec{r})\varepsilon_{ex}[n]d^3r \qquad (15.4)$$

where $\varepsilon_{ex}[n]$ is the exchange correlation energy per particle of a uniform electron gas having density $n(\vec{r})$. The net exchange correlation energy in this approximation is underestimated by 7%. In LDA, the overall shape and position of highest occupied molecular orbital and lowest unoccupied molecular orbital is good, but the gap is generally underestimated by a factor of 2. Nevertheless, the estimation of bond length is fine. However, it completely fails for highly correlated systems such as solid nickel oxide (NiO). Although LDA has been successful in dealing with many systems, however, there are newer and sophisticated approximations beyond this approximation.

15.2.2.2 Generalized Gradient Approximations

This approximation is based on the fact that the real systems are far away from uniform ones. So, naturally slow variation or the gradient of density has to be taken into the consideration of the exchange term. In this approximation (Parr et al., 1989; Dreizler et al., 1995; Kohn, 1999; Capelle, 2006; Ao et al., 2009), the exchange and correlation energies are separately placed to yield the exchange energy term as

$$E_X^{GGA} = \int n(\vec{r}) \varepsilon_x^{uni}[n] F_X[s(r)] d^3r \qquad (15.5)$$

where $F_X[s(r)]$ is the factor of exchange enhancement caused by the generalized gradient approximations (GGA) and $s = |\vec{\nabla} n|/2k_F n$ is a dimensionless density gradient parameter being dependent on the Fermi wave vector k_F of the material. This dimensionless parameter s basically measures the gradient of the density on the density itself while $\varepsilon_x^{uni}[n]$ is the exchange energy of the uniform electron gas. The full form of the exchange correlation according to GGA involving both the density and gradient of density can be written as

$$E_{XC}^{GGA} = \int n(\vec{r}) \varepsilon_x^{uni}[n] F_{XC}[r_s, s] d^3r \qquad (15.6)$$

Here, the exchange energy has been renormalized to the scale of r_s, the radius of a sphere containing one electron and given by $(4\pi/3)r_s^3 = n^{-1}$. Different varieties and flavor arise in GGA due to the form of $F_X[s(r)]$ required for the truncation of the series of the energy as a function of density and gradient of the density. The typical Perdew, Burke, and Enzerhof (PBE) (Perdew et al., 1966) GGA scheme for $[s(r)]$ is given by

$$F_X(s) = 1.804 - \frac{0.804}{1 + 0.24302 s^2} \qquad (15.7)$$

Note that the values of $F_X(0)$ indicate the exchange and correlation energy of uniform electron gas as a function of r_s. As an illustrative example, LDA predicts the ground state of iron (Fe) crystal as face-centered cube (FCC) while GGA gives the correct body-centered cube (BCC) phase. The modern DFT schemes involve $DFT + U$ or $DFT + GW$ approximation for better estimation of physical parameters for comparison with experiments. Here, U is the Hubbard type of interaction, G and

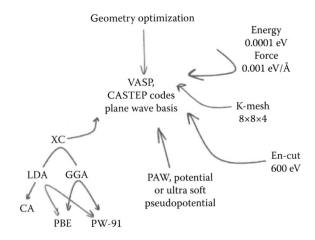

FIGURE 15.3 An involved practical computation of DFT including the various approximation schemes has been illustrated in this figure.

W denote Green's function and screened Coulomb potential, respectively. The GW approximation is based on an expansion of exchange-correlated self-energy in terms of Green's function (G) and dynamically screened electron–electron interaction (W). It is important to remember that GW approximation is very much dependent on the LDA results because LDA results serve as the starting point for the perturbation theory involved in the approximation. It has been noticed that the GW predictions are quite correct and provide naturally an ingenuous tool for analyzing semiconducting nanostructures. The interested reader should look into the other related better approximations in the literatures (Kawazoe, 2003). To summarize, we have noticed that although DFT theory is exact, it however, contains an exchange correlation term, which needs to be approximated in real computational calculations. Moreover, the failure of DFT lies in the wrong choice of local potential. With the advent of increasingly accurate sophisticated exchange correlation potential, the use of DFT in various calculations related to material science is increasing day by day due to development of codes related to linear scaling with the number of atoms. The details of the band structure computation via DFT have been recently reviewed by Jana et al. (2013). An involved practical computation of DFT including the various approximation schemes has been illustrated in Figure 15.3.

15.3 ELECTRONIC PROPERTIES MODIFICATION OF GRAPHENE

One of the main motivations behind the research work with graphene systems is to open up a finite bandgap and to control the bandgap, otherwise it is not possible to use a graphene-based system as a nanoelectronic device. There are several attempts to incorporate bandgap of graphene nanostructures by doping of different kinds of elements and functional group, altering the topology of the planar structures such as SW kind of defects as well as straining the planar structure of the graphene nanosheet. In pure graphene, the electrons at the Fermi

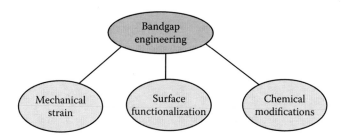

FIGURE 15.4 Schematic representation of different ways to tailor the electronic and electrochemical properties of graphene.

level are highly mobile and cannot be controlled easily. The doping is essential to control these electrons for applications in nanoelectronics. In fact, a bandgap of the order of 0.3–0.7 eV is quite ideal for device application in the sense that one can have faster yet controlled transmission of electrons in the transistor. We illustrate in Figure 15.4, the various methods for modifying the bandgap in graphene. Substitutional doping can readily affect the electronic properties of monolayer graphene. The individual dopants in the monolayer can be characterized by scanning tunneling microscopy (STM), Raman Spectroscopy, and x-ray spectroscopy along with first principle calculations.

15.3.1 STRAINED GRAPHENE: BAND STRUCTURE MODIFICATION

In this section, we will discuss the effect of planar strain on graphene nanosheet to alter its electronic properties (Lu and Guo, 2010; Naumov and Bratkovsky, 2011), mainly the electronic band structure and DOS. The hexagonal symmetry of the graphene structure yields a zero bandgap at the Fermi level (E_F). Therefore, breaking the hexagonal symmetry of the planar graphene structure one can introduce bandgap in the system. Mainly three kinds of planar strains are important for breaking the symmetry of the structures: (i) symmetrical strain (Figure 15.5a), (ii) asymmetric uniaxial strain along the zigzag direction (Figure 15.5b), and (iii) asymmetric uniaxial strain along the armchair direction (Figure 15.5c). Symmetrical strain means the strains along both zigzag and armchair directions are equal. Such a strain, however, does not break the hexagonal symmetry of the structure while the other two asymmetrical strains break the hexagonal symmetry of the graphene structure and give rise to opening of bandgap at the Fermi level. As far the symmetrical strain does not break the hexagonal symmetry however, DFT-based calculations show that uniform strain can alter the position of EF. It has been observed that a downward shift of E_F with respect to such strain and the variation is almost linear up to 5% of strain (ε). In Figure 15.6, the change of E_F, that is, $E_F(\varepsilon) - E_F(0)$ from pristine one has been depicted as a function of uniform strain. As the hexagonal symmetry is broken for uniaxial strains, the hopping integral in the TB approximation among the nearest sublattice sites are different. While for symmetrical strain the hopping integral remains same for all nearest sublattice sites but its amplitude decreases with increase of lattice constant.

(a)

(b)

(c)

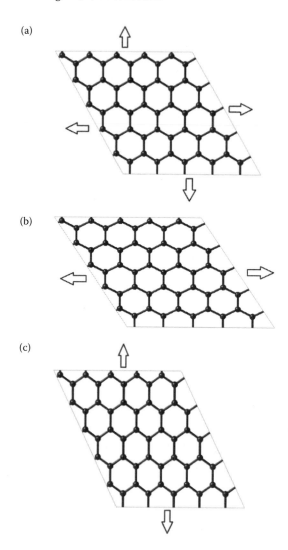

FIGURE 15.5 (a) Schematic representation of uniform strain along both armchair and zigzag direction. (b) Schematic representation of uniaxial strain along zigzag direction. (c) Schematic representation of uniaxial strain along armchair direction.

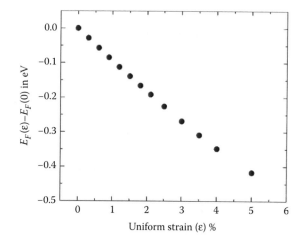

FIGURE 15.6 The variation of Fermi energy ($E_F(\varepsilon) - E_F(0)$) has been plotted against the amount of uniform in-plane strain (ε) to the pristine graphene nanosheet. It falls off linearly with the strength of strain.

For uniaxial strain two different values of hopping integral are required resulting a finite bandgap opening at the Fermi level.

Recent experimental work by Ni et al. (2008) reveals that planar uniaxial strain of graphene nanosheet yields an opening of bandgap about 300 meV in the elastic regime of strain (~1.0%) (see Figure 15.7). This experimental result is also consistent with their *ab initio* DFT calculation. The Poisson ratio in the elastic limit of strain obtained to be 0.186. A detailed theoretical work was carried out by Gui et al. (2008) to study the effect of different kinds of planar strains on the electronic properties of graphene nanosheet. As already discussed, they do not get any bandgap opening for symmetrical strain but in this case the pseudo gap (determined at M point) decreases linearly with the applied strain. Maximum of 0.170 eV bandgap opening is observed for zigzag strain of 4.91% while maximum of 0.486 eV amount of bandgap opening has been observed for 13.1% of armchair strain at the Dirac point. Beyond these two limiting strain value the bandgap however decreases. For smaller strain (<2.0%) the Poisson ratio remains constant at 0.1732 for both zigzag and armchair strain. It has been observed (Gui et al., 2008) that the Poisson ratio falls off with the strain beyond the elastic limit differently for different types of uniaxial strain.

Recently, Topsakal et al. (2010) studied the bandgap tuning by applying uniform strain on the recently discovered graphene derivative called graphane. Graphane is basically hydrogenated graphene structure; all the dangling bonds of carbon atoms of graphene are hydrogenated. Due to the hydrogenation it is buckle shaped and unlike graphene it is a wide bandgap direct semiconductor. The strain energy initially increases with the uniform strain up to $\varepsilon \sim 0.3$ (where ε is the strain) in the elastic regime. However, in the plastic

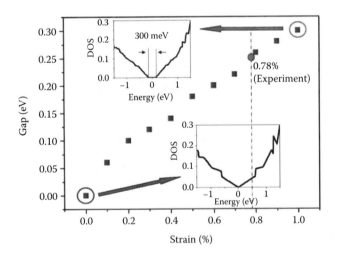

FIGURE 15.7 The variation of bandgap of strained graphene with the uniaxial tensile strain on graphene. The insets show the calculated DOS of unstrained (right-bottom corner) and 1% tensile strained graphene (top-left corner). The dotted line indicates the calculated bandgap of graphene under the highest strain (0.78%) in the experiment. (Reprinted with permission from Ni, Z. H. et al. 2008. *ACS Nano* 2:2301–2305.)

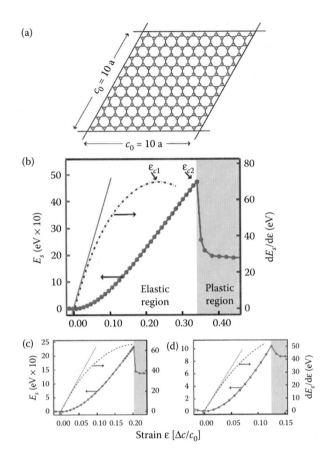

FIGURE 15.8 2D graphane under uniform expansion. (a) Initial atomic arrangement of a (10 × 10) graphane supercell. (b) Shows the variation in strain energy (E_S) and its derivative. The gray shaded region indicates the plastic range. Strains corresponding to two critical points in the elastic range are labeled as ε_{c1} and ε_{c2}. (c) Similar to (b) for a single hydrogen vacancy. (d) For C–H divacancy. (Reprinted with permission from Topsakal, M., S. Cahangirov, and S. Ciraci. 2010. *Appl. Phys. Lett.* 96:091912;1–3.)

deformation regime ($\varepsilon > 0.3$) strain energy falls off suddenly. The variation of strain energy and its derivative with respect to strain has been depicted in Figure 15.8. The bandgap of this wide bandgap material is also tunable by application of in-plane strain. DFT-GGA calculation reveals that it has a bandgap of ~3.54 eV but with GW correction Topsakal et al. (2010) obtained the bandgap to be ~5.66 eV without any strain. Initially, the bandgap increases with strain monotonically up to $\varepsilon \sim 0.15$ and then decreases monotonically with the strain in the elastic regime, as depicted in Figure 15.9.

15.3.2 DOPED GRAPHENE: ELECTRONIC PROPERTIES MODIFICATIONS

The major parts of the research interest toward the graphene systems are to study the modifications of the electronic and electrochemical properties of graphene due to the presence of impurity such as atoms and functional groups in the pristine systems. With the impurity type and concentration, the electronic properties of graphene can be modified. Recent study

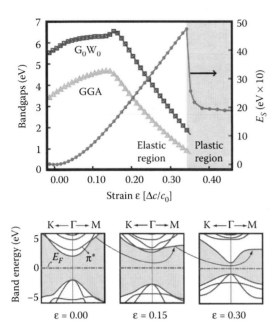

FIGURE 15.9 The variation in energy bandgaps of graphane with uniform expansion. The bandgaps obtained both from GGA (gray points) and *GW* (black points) calculations increase with increasing strain up to $\varepsilon = 0.15$, then decreases up to the yielding point. The black points indicate the variation of strain energy with strain (right side scale). (Reprinted with permission from Topsakal, M., S. Cahangirov, and S. Ciraci. 2010. *Appl. Phys. Lett.* 96:091912;1–3.)

shows that the position-dependent adatoms also alter the electronic DOS and band structure of the system.

With B and N doping, it is expected that the local structure and the associated charge density will be modified in an essential way. The doping of hole forces the 2D band of graphene to shift toward blue end while with electron, the redshifts (Rao and Voggu , 2010) are noticed. Recently, Raman spectroscopy has been employed to record the G-band shifts in tetrathiafulvalene (TTF) and tetracyanoethylene (TCNE) functionalized in graphene (Rao et al., 2010). It has been observed that G band shifts toward the higher frequencies for B concentration while for N concentration in the lower frequencies (Panchakarla et al., 2009). It would be quite interesting to employ experimental and theoretical optical studies on effects of TTF and TCNE on B- and N-doped graphene.

The N atom is used widely for chemical doping of carbon-related materials because of two important properties. One is the comparable atomic size and another is that it has five valence electrons for bonding with carbon atoms. The typical methods for doping N in graphitic structure involve plasma arc-discharge and chemical vapor deposition (CVD). These processes are, however, of high cost with uncontrolled doping structure. Nitrogen plasma treatment of graphene can yield various concentration of N in 2D graphene by controlling the exposure time (Wang et al., 2010). In fact, N-doped graphene (Usachov et al., 2011, 2013) can show efficient electro-catalytic activity for the reduction of H_2O_2. Besides, N-doped graphene can perform glucose biosensing in a low concentration of

0.01 N. Ultraviolet (UV) photoemission spectroscopy clearly revealed the raising of density of π states near the Fermi level in pyridinic N-doped graphene (Luo et al., 2011) and this type of doping is highly responsible for oxygen reduction reaction (ORR) activity. The CVD method (Wei et al., 2009) can be used to prepare N-doped graphene. These large area syntheses of N-graphene films can be easily transferred to different substrates for various characterization and measurements. These doped films exhibit remarkable performance for ORR associated with alkaline fuel cell (Qu et al., 2010). These methods can be repeated for new catalytic materials for future applications to fuel cells or even beyond the fuel cells.

15.4 BORON AND NITROGEN DOPING: ELECTRONIC PROPERTIES MODIFICATION

Defect-mediated/controlled graphene involving B and N need some careful investigation to look for some specific applications. This is relevant for emerging trends in device application as well as spectroscopic developments. The theoretical and experimental approach should also pay attention to suitable control of B and N in the low doping region. In this context, it is to mention that N doping along with native point defects modify the electronic properties of graphene (Huo et al., 2013). Until date, major emphasis have been in the electronic and electronic energy loss spectroscopy (EELS) of these doped systems; however, the mechanical and electrochemical aspects have not been explored thoroughly with the variation of dopant concentrations. On the other hand, theoretical efforts have been attempted to correlate the atomic structure and the electronic transport taking into account the electron–electron and electron–phonon interaction. However, one needs further investigation and involved calculations to match the experimental results with the theoretical ones. In recent years, topological defects are seen to play a key role in the formation of structures. Thus, it is necessary to explore these ideas in doped systems for suitable application in electronic switches or in transport processes. In Raman spectroscopy, it is highly desirable to know the variation Raman breathing mode (RBM) with different concentration of N for various flavors of doped graphene. It is also possible to dope graphene with B and N and it is desirable to know the variation of electronic and optical properties as a function of the concentration. Most of the theoretical studies have been focused on graphite-like substitution of boron and nitrogen atoms into the carbons in the honeycomb lattice of graphene. Further works considering other configurations such as pyridine like or pyrrole like in the graphene network must be developed to compare theoretical results with experimental ones. Although some interesting breakthrough works have been done so far, however, it is expected that some new electronic, mechanical, optical, and vibration effects may emerge after doping and functionalization with other unexplored heteroatoms. Below we discuss the two possible directions where B or N will definitely play a significant role in controlling the various physical properties in carbon-related systems.

It is well known from theoretical study that N-doped CNTs are indeed good support for noble metal nanoparticles because of the enhancement of the metal-CNT binding energy. The graphene layers inside the N-doped CNT act as an additional anchoring point for metal nanoparticles such as Pt or Ru. N-doped CNT-graphene hybrid nanostructures show (Lv et al., 2011) highly conductive nature and turns out to be an ideal catalyst support material. Recently, a thorough study by Zhao et al. (2011) on the electronic structure of N-doped graphene film using CVD on copper substrate indicates that the electron–hole asymmetry in local density of states (LDOS) is significantly stronger on the N atom in accord with DFT calculations. Moreover, STM results (Zhao et al., 2011) reveal that each graphitic N dopant actually contributes on an average 0.42 ± 0.07 mobile charge carriers to the graphene lattice.

Nanocomposites-based nitrogen doped graphene sheets can show potential applications in wastewater treatment, fuel cells, and nanodevices. In particular, recently graphene zinc selenide (GN-ZnSe) nanocomposites have been synthesized (Chen et al., 2012) in the hydrothermal process to show enhanced electro chemical (EC) and photo catalytic activity. The choice of ZnSe material is basically for two reasons; firstly, it is a direct bandgap material (2.7 eV at room temperature), secondly, it is a good candidate for short wavelength lasers and blue laser diodes.

N-doped reduced graphene film was prepared (Hwang et al., 2012) from the spin casting of graphene oxide dispersion. This method is supposed to be an efficient method to exploit a low-cost solution processing without substrate transfer. The work function of the prepared film can be precisely tailored by the doping level of N. This can be used as a transparent electrode in polymer light-emitting diodes (PLED) compared to fluorine-doped tin oxide (FTO). The sheet resistance of such N-doped reduced graphene film was shown to attain a lowest value at 80% transmittance. Recently, a powerful method for N-doped graphene sheets has been developed by Yang et al. (2012), which involves functionalization by self-assembled molecules. The highly porous, high specific surface area along with high percentage yield of pyridinic graphitic nitrogen structures are seen to be promising candidate in electrocatalytic activity for ORR. N-doped graphene quantum dots with oxygen-rich functional groups can emit blue luminescence and possess superior electrocatalytic activity (Li et al., 2012). Moreover, they can also be used as metal-free ORR catalysts in fuel cell. These properties and applications of such doped systems have opened up the possibility of potential applications in biomedical imaging and other interesting optoelectronic devices.

Compared to standard SiO_2 substrate, it has been pointed out that BN substrate for monolayer graphene is a better option for nano-optoelectonic devices. The essential reason behind this is that BN significantly improves the mobility of the charge carriers. Moreover, the charge homogeneity of such a system is reduced quite largely (Dekker et al., 2011) in comparison with SiO_2 substrate. The gate-dependent differential conductance (dI/dV) STM spectra of graphene/BN have showed (Dekker et al., 2011) sharp spectroscopic feature than

graphene/SiO_2. This study can help one to explore the relation between Dirac point energy and the gate voltage enabling one to calculate the capacitance per unit area. Most recently, DFT in the framework of nonequilibrium Green's function has been invoked (Deng et al., 2012) to study the transport properties of BN-doped trigonal graphene. The calculations have indicated (Deng et al., 2012) that the $I–V$ characteristics are indeed sensitive to geometrical factors and the observed rectification can be understood from the formation of a potential barrier in BN region just like p–n junction.

Ab initio DFT calculations on N/B substitutional doped graphene systems exhibit the n-type (N doped) and p-type (B doped) behavior. An upward shift of E_F as well as downward shift of E_F has been observed for N- and B-doped graphene systems, respectively, and it depends on doping concentration (n). N and B doping concentrations have been varied within 0%–18% and Figure 15.10 clearly demonstrates the change of Fermi level ($\Delta E_F = E_F(n) - E_F(0)$) with respect to pristine one with the doping concentration. In low doping concentration, a sudden change has been observed whereas there is no significant change in E_F for high doping concentration. However, substitutional doping of N atom is energetically favorable rather than B atom substitutional doping; it is possibly due to the fact that the deformation of structure is less for N atom. Formation energy of such doped systems can be evaluated using the formula, $E_f = (E_d + nE_C - E_p - nE_a)$, where E_d, E_C, E_p, and E_a denotes the total energies of doped system, isolated C atom, pristine system, and isolated adatom, respectively. In Figure 15.11 the variation of formation energy calculated from DFT calculations of N/B doped systems are depicted with respect to doping concentrations, which clearly indicates that N doping is favorable than B doping. Pristine graphene is zero bandgap semimetal; but *ab initio* DFT-GGA calculation with 32 C atom super-cell reveals that the lattice symmetry can be broken by isoelectronic B–N co-doping in the graphene nanosheet, which yields a finite bandgap

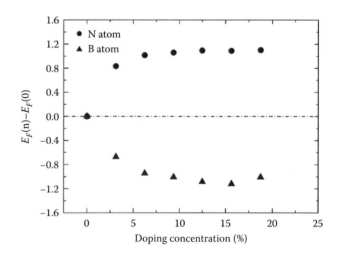

FIGURE 15.10 Fermi energy deviation, that is, $E_F(n) - E_F(0)$ in N/B doped systems as a function of doping concentration (n). N atoms yield electron doping effect, whereas B atom substitution is considered as hole doping.

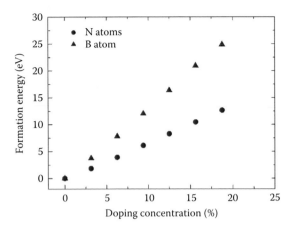

FIGURE 15.11 Formation energy increases almost linearly with doping concentration of N/B substitutional doping in graphene nanosheet. N atoms doping are energetically favorable than B atoms doping.

opening (Nath et al., 2014a) at the Fermi level ~0.3 eV (see Figure 15.12). First-principle DFT calculations have clearly demonstrated (Fan et al., 2012) that B–N domains in graphene plane can eventually open up a bandgap around $K(K')$ points. Even the random doping with B or N also induces (Fan et al., 2012) a small gap at the Dirac points. The modulation of the gap is although insensitive to the size of BN domain however critically dependent on the concentration of BN. The calculation of DOS confirms the nonexistence of defect states near the Fermi level. The absence of localized defect states in the band structure is due to the strong interaction between B(N) with C atoms and small difference of electronegativity. The numerical study also reveals that at low concentration of B or N, the shift of the Fermi level (ΔE_F) is related with the concentration (n) through a simple equation of the form $\Delta E_F \propto \sqrt{n}$. This fact, along with spin-dependent DFT calculation indicates that the introduction of

BN domain in graphene does not induce any ferromagnetism in the system. This study will be helpful to microelectronic community for device fabrication. Recent *ab initio* theoretical work based on DFT shows that boron and nitrogen double doping, that is, B–B, N–N, and N–B doping in the graphene interacts differently in the system (Al-Aqtash et al., 2012). B–B and N–N interaction in the system is repulsive whereas N–B interaction is attractive. The interaction energy varies inversely proportional to the distance between the dopants. For separation distance greater than or equal to 4 Å, the interaction strength is almost zero. B–N pair yields the most energetically favorable configuration than the others. This observation eventually supports the existence of hexagonal boron nitride patch in graphene system. Recently, *ab initio* calculations (Nath et al., 2014b) on the optical properties of B, N, and B–N co-pair doped graphene have indicated that the static real dielectric constant for N-doped system is increased to 45 times compared to pristine one and no significant shift in Fermi energy is noticed for N–B co-doping.

Brito et al. (2012) have employed *ab initio* calculations to study B and N doping in graphene into the grain boundary (GB) region. The study indicates that B and N doping into the defective region is an energetically favorable process. Further, the electronic properties of B (N) doping in graphene having GB structures show p(n)-type behaviors. However, B–N co-doping induces small modifications in their electronic structures. B–N co-doping in the GB region, however, does not produce any significant changes in the electronic band structure compared to the undoped one. The *ab initio* calculations with graphene doped with BN nanoribbons and patches show (Shinde and Kumar, 2011) an opening of bandgap and the bandgap turns out to be directly suitable for nanoelectronic applications. Moreover, the bandgap interestingly depends on the narrowest region of the graphene independent of the concentration and shape of the patches. The alternation of band structures for B–N nanoribbons and patches in the graphene has been depicted in Figure 15.13 for different doping concentration and configurations. With the same doping concentration bandgap is different for different configurations, that is, either B–N nanoribbons or B–N patches. A detailed comparison of bandgap is provided in Table 15.2. The band structure calculations of BN-doped graphene by Tochikawa et al. (2011) have suggested a large red shift of the bandgap in contrast to normal graphene. It is noteworthy to mention at this juncture that the infinite graphene has a zero bandgap while BN-doped graphene (C_6BN) has a bandgap of 1.25 eV (Liu and Shen, 2009; Gorjizadeh and Kawazoe, 2010). The spin-polarized DFT methods were used (Ma et al., 2011) to study the electronic and magnetic properties of quasi 1D zigzag triwing graphene (ZZ-TWG) ribbons with boron and nitrogen. The asymmetric BN doping in these systems give rise to half-metallicity suppressing the spin polarization of the doped wings. Heavily doping, however, turns a metallic ZZ-TWG into a semiconductor. This study is indicative of choosing a good half-metallic nanostructure for future spintronic devices. It will be interesting to explore the optical properties of these systems as

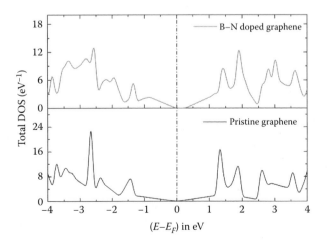

FIGURE 15.12 Bandgap opening is observed for B–N co-doped graphene nanosheet, whereas pristine graphene exhibits exactly zero bandgap at the Fermi level showing a linear variation of DOS with energy.

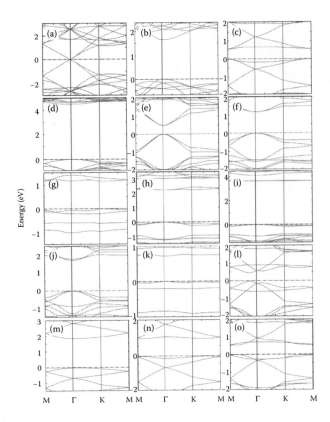

FIGURE 15.13 Energy bands along symmetry directions of the Brillouin zone (BZ) of the B–N doped graphene supercells corresponding to different doping configurations (i.e., for boron nitride nanoribbon and patches) and the dashed line shows the top of the valence band. (Reprinted with permission from Shinde, P. P. and V. Kumar. 2011. *Phys. Rev. B* 84:125401;1–6.)

TABLE 15.2

Bandgap Opening in Graphene Nanosheet with Different B–N Doping Concentrations and Configuration

System	BN (%)	Bandgap (eV)
a	0.00	0.00
b	50.00	1.65
c	50.00	0.58
d	100.00	4.53
e	8.33	0.49
f	33.67	1.18
g	44.45	1.16
h	66.67	2.27
i	91.67	3.66
j	44.45	1.74
k	55.56	0.72
l	16.67	0.47
m	83.33	1.86
n	66.67	1.00
o	33.33	0.54

Source: Shinde, P. P. and V. Kumar. 2011. *Phys. Rev. B* 84:125401;1–6.

a function of concentration in different polarizations of the applied electromagnetic field.

Using first principle calculations Fan et al. (2011) reported that the bandgap and electron effective mass (EEM) can be tuned systematically by tuning the interlayer gap and stacking arrangement of graphene-boron nitride heterobilayers (C-BN HBLs). The graphene-boron nitride heterobilayers have smaller electron effective mass than that of graphene bilayers, which means higher carrier mobility. The binding energy is highly sensitive to the stacking arrangement of graphene-BN heterobilayers, the variation of binding energy is depicted in Figure 15.14. The most stable configuration is pattern II (see Figure 15.14), where N atoms are top on the center of graphene hexagon and B atoms are on the top of graphene C atoms. This is because of cation (B atom)–π attractive interaction and anion (N atom)–π repulsive interaction. The N atoms

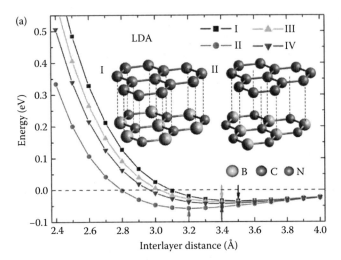

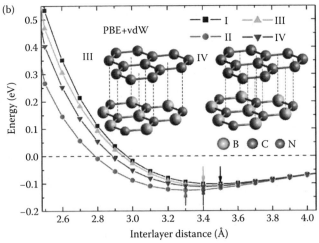

FIGURE 15.14 Binding energy per unit cell of graphene-boron nitride heterobilayers with different patterns as a function of interlayer spacing using (a) LDA and (b) PBE combined with Tkatchenko–Scheffler–van der Waals (TS–vdW) correction. The inset shows the four inequivalent stacking patterns (I–IV) of the heterobilayers. (Reprinted with permission from Fan, Y. et al. 2011. *Appl. Phys. Lett.* 98:083103;1–3.)

prefer to be located at the top of the hexagonal center of graphene network where π-electron density is very low, whereas B atoms prefer to be located on the top of C atom because of its electropositive nature. In all cases, the binding energies are minimum when the interlayer separation is ~3.2 to 3.5 Å depending on the stacking pattern. With the increase of separation from ~2.3 to 4.2 Å energy bandgap decreases monotonically but differently for different pattern, but for greater than 3.2 Å interlayer gap all exhibits almost zero bandgap as shown in Figure 15.15. Thus, hetero bilayer can modify electronic properties of the nanoscale graphene-based systems. Furthermore, AA-stacked bilayer graphene system exhibits strong modifications of its optical properties with the variation of its interlayer spacing (Nath et al., 2015). Besides, the electronic band structure of bilayer graphene system is highly sensitive to applied external transverse electric field when it is asymmetrically doped (Zhang et al., 2009). Recent DFT study reveals that interlayer spacing between sandwiched graphene layer and BN nanosheet appreciably alters the opto-electronic properties of such systems (Das et al., 2015).

Instead of common normal bonds in graphene, graphyne is made from double and triple bonded units of two carbon atoms. Recently, DFT has been employed to explore the electronic band structure of bilayer graphene and graphyne doped with B and N in the presence of external electric field (Majidi and Karami, 2013). The calculation reveals the opening of bandgap and the magnitude can be enhanced by the application of electric field. This can be used in carbon-based electronic device since the bandgap can be tailored by the concentration of impurity and the strength of the applied electric field. An experimental up-to-date review has been compiled by Cooper et al. (2012). Although for bulk materials, there are quite a few techniques to measure the electronic band structure of a material, however, graphene being a few layers of carbon atoms, angle resolved photo emission spectroscopy (ARPES) turns out to be the most efficient method for measuring small bandgap in the system due to doping.

The presence of any impurity or defect generally modifies the electronic band structure and the role of the defects or impurities can be visualized through STM. Extended defects such as GB can suitably control the functionality of graphene sheet (Banhart et al., 2011). Spin-polarized first principle calculations have been carried out by Xu et al. (2010) on the BN-doped super lattice to visualize the effect of geometric shape and size of embedded B–N Nanon. The involved calculations clearly indicate that the bandgap of the graphene super lattice increases with the B–N Nanon regardless of its shape. Interestingly, with the exchange of B and N atoms in the super lattice, the valence and conduction bands are inverted with respect to the Fermi level due to inherent electron–hole symmetry. Besides, the mid-gap states do appear in the super lattice for triangular B–N nanodot. A low pressure CVD method on Cu foil has been employed (Chang et al., 2013) for synthesis of large area few layer BN co-doped graphene. In this work, it has been demonstrated clearly that a small amount of h-BN domain in graphene is enough to modify the electronic structure with a significant bandgap. In particular, for low BN samples, with the increase of BN concentration, a monotonically increasing bandgap up to 600 meV has been noticed. However, a further enhancement of BN concentration after 6% decreases the bandgap. This is due to the quantum confinement effect in graphene domains.

15.5 FUTURE DIRECTIONS

A recent surge of interest in graphene is due to two main reasons. First, high-quality graphene can be produced in the laboratory quite easily. However, there needs to be some correlation among various methods for large-scale production, quality control, and repeatability of doping concentration. Even for this miracle material, equivalence of various preparation methods has not been explored in detail with exfoliation technique. The properties of particular sample of graphene depend strongly on the defects and substrates in various production schemes. Second, graphene can be chemically functionalized, which can alter the electronic band structure significantly and hence the optical properties. In practice, a small amount of B–N domain can effectively alter the semimetallic character of the pristine graphene. Tuning of bandgap essentially depends on doping concentration as well as ad-atoms positions in the hexagonal network of graphene. Therefore, in order to modify the band structure of graphene, controlled doping is essential to fabricate a graphene-based device. However, except some remarkable Raman spectroscopic studies in B–N doped graphene, linear nonlinear optical properties involving dielectric

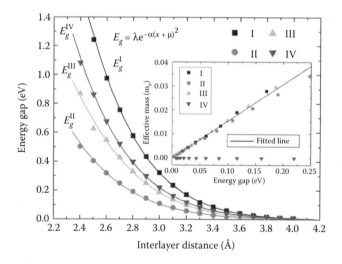

FIGURE 15.15 Variation in the energy gap of graphene-boron nitride heterobilayers as a function of interlayer spacing. The bandgap varies with spacing (x) as $E_g = \lambda \exp\{-\alpha(x+\mu)^2\}$; the constants λ, α, and μ are 16.49, 0.59, and 0.41 for pattern I, 10.00, 0.48, and 0.07 for pattern II, 3.84, 0.73, and 0.96 for pattern III, 25.95, 0.47, and 0.19 for pattern IV, respectively obtained from fitting. In the inset the variation of the effective masses versus bandgap is depicted. (Reprinted with permission from Fan, Y. et al. 2011. *Appl. Phys. Lett.* 98:083103;1–3.)

constant, optical conductivity, optical absorption, EELS have not been explored in detail at all.

CVD turns out to be a promising method for producing large area uniform polycrystalline graphene film containing B–N domains. In spite of the presence of defects, GB, many thick layers in these films, they can be used as transparent conductive coating. For these applications, these films need to be characterized. In such a case, DFT calculations can be helpful for optical characterizations involving polarizations and frequencies. In fact, it is highly desirable to grow large-scale graphene in a cost-effective way on arbitrary surfaces at low enough temperature with minimal number of defects. In that sense, graphene will be competitive with conventional semiconductor materials such as Si or GaAs.

In photonic device applications, graphene can be used in a wide spectral window ranging from UV to infrared compared to very limited detected spectral width of conventional photo-detectors. Besides owing to its high operating bandwidths, graphene can be used in high-speed data communications. The sensitivity of the photo-detectors can be improved by incorporating plasomonic nanostructures, which enhance the typical local optical electric field. Doping in graphene can significantly alter the position of the Fermi level. This fact can suitably be used as an important property of transparent electrode in quantum dot and solar cell. Besides, chemical functionalization may improve the selectivity of graphene sensors suitable for biosensing. The use of graphene sensors in bioapplications relies on the fact of multifunctionality, large surface area, easy chemical functionalization, and chemical purity.

Besides, the techniques used in graphene can also be suitably used to create many new 2D materials such as BN, NbSe$_2$, TaS$_2$, and so on. Using these 2D materials, one can create heterostructures made up of two, three, or four layers. The results obtained from theoretical study will enable one to compare with experimental observations.

15.6 CONCLUSIONS

Unlike other 2D atomic crystals, graphene is a zero bandgap purely 2D honeycomb-like carbon (C)-based nanomaterial. However, due to the absence of a finite nonzero bandgap, graphene cannot be used for device applications related to graphene-based electronics. In this situation, chemical doping such as B or N seems to be promising as there is a possibility of opening of bandgap at Dirac points. B and N are two very good candidates for substitutional doping in the graphene network because the atomic radii of these two elements are very close to the atomic radius of carbon. Graphene with B doping behaves like a p-type material, whereas with N doping it becomes n-type material. One would expect to fabricate p–n junction nanodevices by co-doping of both B and N elements in the graphene system by suitable position and concentration. In this chapter, it has been attempted to review the bandgap opening and controlling at the Dirac point (linear band crossing point) of graphene by substitution of B and N by using density functional theory. Applications of strain

could open up gap at Dirac points. In fact, these modifications of the electronic band structure will eventually alter the optical properties of doped systems. Besides, combining defects with these doped systems can help to build some novel 2D materials particularly suitable for device applications such as electronics, optoelectronics, photonics, and spintronics. Works in these directions are welcome in the next few years.

REFERENCES

Al-Aqtash, N., K. M. Al-Tarawneh, T. Tawalbeh, and I. Vasiliev. 2012. *Ab initio* study of the interactions between boron and nitrogen dopants in graphene. *J. Appl. Phys.* 112:034304;1–7.

Ando, T. 2009. The electronic properties of graphene and carbon nanotubes. *NPG Asia Mater.* 1:17–21.

Ao, Z. M. and Q. Jiang. 2009. Density functional theory calculations on interface structures and adsorption properties of graphenes: A review. *Open Nanosci. J.* 3:34–55.

Banhart, F., J. Kotakoski, and A. V. Krasheninnikov. 2011. Structural defects in graphene. *ACS Nano* 5:26–41.

Bhattacharya, B., N. B. Singh, and U. Sarkar. 2013. Interaction of boron–nitrogen substituted graphene nanoribbon with nucleobases; the idea of biosensor. *Soft Nanosc. Lett.* 3:43–45.

Brito, W. H., R. Kagimura, and R. H. Miwa. 2012. B and N doping in graphene rules by grain boundary defects. *Phys. Rev. B* 85:035404;1–6.

Capelle, K. 2006. A bird's eye view of density functional theory. *arXiv:cond-mat/0211443v5.*

Casolo, S., R. Martinazzo, and G. F. Tantardini. 2011. Band engineering in graphene with superlattices of substitutional defects. *J. Phys. Chem. C* 115:3250–3256.

Castro Neto, A. H., F. Guinea, N. M. R. Peres, K. S. Novoselov, and A. K. Geim. 2009. The electronic properties of graphene. *Rev. Mod. Phys.* 81:109–162.

Chang, C.-K., S. Kataria, C.-C. Kuo et al. 2013. Band gap engineering of chemical vapor deposited graphene by *in situ* BN doping. *ACS Nano* 7:1333–1341.

Chen, P., T.-Y. Xiao, H.-H. Li et al. 2012. Nitrogen-doped graphene/ZnSe nanocomposites: Hydrothermal synthesis and their enhanced electrochemical and photocatalytic activities. *ACS Nano* 6:712–719.

Choi, W., I. Lahiri, R. Seelboyina, and Y. S. Kang. 2010. Synthesis of graphene and its applications: A review. *Crit. Rev. Solid State Mater. Sci.* 35:52–71.

Chowdhury, S., S. Baidya, D. Nafday, T. Saha-Dasgupta, A. Mookerjee, D. Jana, M. Kabir, S. Halder, and B. Sanyal. 2014. Real space technique for the study of random extended defects in solids: Signature of disordered Stone Wales defects on the electronic structure of graphene. *Physica E* 61:191.

Cooper, D. R., B. D'Anjou, N. Ghattamaneni et al. 2012. Experimental review of graphene. *ISRN Condensed Matter Physics* Article ID 501689:1–56.

Das, R., S. Chowdhury, and D. Jana. 2015. A first principles approach to magnetic and optical properties in single layer graphene sandwiched between boron nitride nanolayers. *Mater. Res. Express.* 2: 075601: 1–12.

Das-Sarma, S., S. Adam, E. H. Hwang, and E. Rossi. 2011. Electronic transport in two dimensional graphene. *Rev. Mod. Phys.* 83:407–470.

Dekker, R., Y. Wang, V. W. Brar et al. 2011. Local electronic properties of graphene on a BN substrate via scanning tunneling microscopy. *Nano Lett.* 11:2291–2295.

Deng, X. Q., Z. H. Zhang, G. P. Tang, Z. Q. Fan, M. Qiu, and C. Guo. 2012. Rectifying behaviors induced by BN-doping in trigonal graphene with zigzag edges. *Appl. Phys. Lett.* 100:063107;1–3.

Dreizler, R. and E. Gross. 1995. *Density Functional Theory.* New York: Plenum Press.

Fan, X., Z. Shen, A. Q. Liu, and J. L. Kuo. 2012. Band gap opening of graphene by doping small boron nitride domains. *Nanoscale* 4:2157–2165.

Fan, Y., M. Zhao, Z. Wang, X. Zhang, and H. Zhang. 2011. Tunable electronic structures of graphene/boron nitride heterobilayers. *Appl. Phys. Lett.* 98:083103;1–3.

Geim, A. K. 2009. Graphene: Status and prospects. *Science* 324:1530–1534.

Geim, A. K. and K. S. Novoselov. 2007. The rise of graphene. *Nat. Mater.* 6:183.

Gong, Y., G. Shi, Z. Zhanga et al. 2014. Direct chemical conversion of graphene to boron and nitrogen and carbon-containing atomic layers. *Nat. Commun.* 5:1–8.

Gorjizadeh, N. and Y. Kawazoe. 2010. Chemical functionalization of graphene nanoribbons. *J. Nanomater.* 2010:513501;1–7.

Gui, G., J. Li, and J. Zhong. 2008. Band structure engineering of graphene by strain: First-principles calculations. *Phys. Rev. B* 78:075435;1–6.

Hohenberg, P. C. and W. Kohn. 1964. Inhomogeneous electron gas. *Phys. Rev.* 136:B864–B871.

Hohenberg, P. C., W. Kohn, and L. J. Sham. 1990. The beginnings and some thoughts on the future. *Adv. Quantum Chem.* 21:7–26.

Hou, Z., X. Wang, T. Ikeda, K. Terakura, M. Oshima, and M. Kakimoto. 2013. Electronic structure of N-doped graphene with native point defects. *Phys. Rev. B* 87:165401.

Hwang, J. O., J. S. Park, D. S. Choi et al. 2012. Work function tunable, N-doped reduced graphene transparent electrodes for high-performance polymer light-emitting diodes. *ACS Nano* 6:159–167.

Jana, D. 2008. On some basic aspects of density functional theory. *Phys. Teach.* 50:21–28.

Jana, D., C.-L. Sun, L.-C. Chen, and K.-H. Chen. 2013. Effect of chemical doping of boron and nitrogen on the electronic, optical and electro-chemical properties of carbon nanotubes. *Prog. Mater. Sci.* 58:565–635.

Jung, S.-M., E. K. Lee, M. Choi et al. 2014. Direct solvo-thermal synthesis of B/N doped graphene. *Angew. Chem. Int. Ed.* 53:2398–2401.

Kaloni, T. P., R. P. Joshi, N. P. Adhikari, and U. Schwingenschlogl. 2014. Band gap tuning in BN-doped graphene systems with high mobility. ArXiv:1402.0122v2.

Kaplan, D., G. Recine, and V. Swaminathan, 2014. Electrically dependent bandgaps in graphene on hexagonal boron nitride. *Appl. Phys. Lett.* 104:133108–5.

Katsnelson, M. I. 2007. Graphene: Carbon in two dimensions. *Mater. Today* 10:20–27.

Kawazoe, Y. 2003. Realization of prediction of materials properties by *ab initio* computer simulation. *Bull. Mater. Sci.* 26:13–17.

Kohn, W. 1999. Electronic structure of matter–wave functions and density functionals. *Rev. Mod. Phys.* 71:1253–1266.

Kohn, W. and L. J. Sham. 1965. Self-consistent equations including exchange and correlation effects. *Phys. Rev.* 140:A1133–A1138.

Kuila, T., S. Bose, A. K. Mishra, P. Khanra, N. H. Kim, and J. H. Lee. 2012. Chemical functionalization of graphene and its applications. *Prog. Mater. Sci.* 57:1061–1105.

Landau, L. D. and E. M. Lifshitz. 1980. *Statistical Physics. Part I.* Oxford: Pergamon Press.

Li, X., Y. Dai, Y. Ma, S. Han, and B. Huang. 2014. Graphene/g-C_3N_4 bilayer: Considerable band gap opening and effective band structure engineering. *Phys. Chem. Chem. Phys.* 16:4230–4235.

Li, Y., Y. Zhao, H. Cheng, Y. Hu, G. Shi, L. Dai, and L. Qu. 2012. Nitrogen-doped graphene quantum dots with oxygen-rich functional groups. *J. Am. Chem. Soc.* 134:15–18.

Liu, L. and Z. Shen. 2009. Band gap engineering of graphene: A density functional theory study. *Appl. Phys. Lett.* 95:252104;1–3.

Loh, K. P., Q. Bao, P. K. Ang, and J. Yang. 2010. The chemistry of graphene. *J. Mater. Chem.* 20:2277–2289.

Lu, Y. and J. Guo. 2010. Band gap of strained graphene nanoribbons. *Nano Res.* 3:189–199.

Luo, Z., S. Lim, Z. Tian et al. 2011. Pyridinic N-doped graphene: Synthesis, electronic structure and electro-catalytic property. *J. Mater. Chem.* 21:8038–8044.

Lv, R., T. Cui, M.-S. Jun et al. 2011. Open-ended, N-doped carbon-nanotube-graphene hybrid nanostructures as high-performance catalyst support. *Adv. Funct. Mater.* 21:999–1006.

Ma, L., H. Hu, L. Zhu, and J. Wang. 2011. Boron and nitrogen doping induced half-metallicity in zigzag triwing graphene nanoribbons. *J. Phys. Chem. C* 175:6195–6199.

Majidi, R. and A. R. Karami. 2013. Electronic properties of BN-doped bilayer graphene and graphyne in the presence of electric field. *Molecular Phys.* 111: 3194–3199.

Mermin, N. D. 1968. Crystalline order in two dimensions. *Phys. Rev.* 176:250–254.

Nath, P., D. Sanyal, and D. Jana. 2014a. Semi-metallic to semi-conducting transition in graphene nanosheet with site specific co-doping of boron and nitrogen. *Physica E* 56:64–68.

Nath, P., S. Chowdhury, D. Sanyal, and D. Jana. 2014b. Ab-initio calculation of electronic and optical properties of nitrogen and boron doped graphene nanosheet. *Carbon* 73:275–282.

Nath, P., D. Sanyal, and D. Jana. 2015. Ab-initio calculation of optical properties of AA-sacked bilayer graphene with tunable interlayer separation. *Current Appl. Phys.* 15: 691–697.

Naumov, I. I. and A. M. Bratkovsky. 2011. Gap opening in graphene by simple periodic inhomogeneous strain. *Phys. Rev. B* 84:245444–245446.

Ni, Z. H., T. Yu, Y. H. Lu, Y. Y. Wang, Y. P. Feng, and Z. X. Shen. 2008. Uniaxial strain on graphene: Raman spectroscopy study and band-gap opening. *ACS Nano* 2:2301–2305.

Novoselov, K. S., A. K. Geim, S. V. Morozov et al. 2004. Electric field effect in atomically thin carbon films. *Science* 306: 666–669.

Novoselov, K. S., A. K. Geim, S. V. Morozov et al. 2005. Two-dimensional gas of massless Dirac fermions in graphene. *Nature* 438:197–200.

Novoselov, K. S., Z. Jiang, Y. Zhang et al. 2007. Room temperature quantum Hall effect in graphene. *Science* 315:1379.

Novoselov, K. S., V. I. Falko, L. Colombo, P. R. Gellert, M. G. Scawab, and K. Kim. 2012. A roadmap for graphene. *Nature* 490:192–200.

Panchakarla, L. S., K. S. Subrahmanyam, S. K. Saha, A. Govindraj, H. R. Krishnamurthy, U. V. Waghmare, and C. N. R. Rao. 2009. Synthesis, structure and properties of boron and nitrogen doped graphene. *Adv. Mater.* 21:4726–4730.

Parr, R. G. and W. Yang. 1989. *Density Functional Theory of Atoms and Molecules.* Oxford: Oxford University.

Perdew, J. P., K. Burke, and M. Enzerhof. 1966. Generalized gradient approximation made simple. *Phys. Rev. Lett.* 77:3865–3868.

Qu, L., Y. Lin, J.-B. Baek, and L. Dai. 2010. Nitrogen-doped graphene as efficient metal-free electrocatalyst for oxygen reduction in fuel cells. *ACS Nano* 4:1321–1326.

Quhe, R., J. Zheng, G. Luo et al. 2012. Tunable and sizable band gap of single-layer graphene sandwiched between hexagonal boron nitride. *NPG Asia Mater.* 4(e6):1–10.

Rao, C. N. R. and R. Voggu. 2010. Charge-transfer with graphene and nanotubes. *Mater. Today* 13:34–40.

Sachin, H. and S. Ciraci. 2012. Chlorine adsorption on graphene: Chlorographene. *J. Phys. Chem. C* 116:24075–24083.

Shinde, P. P. and V. Kumar. 2011. Direct band gap opening in graphene by BN doping: *Ab initio* calculations. *Phys. Rev. B* 84:125401;1–6.

Shirodkar, S. N. and U. V. Waghmare. 2012. Electronic and vibrational signatures of Stone-Wales defects in graphene: First-principles analysis. *Phys. Rev. B* 86:165401–165410.

Tachikawa, H., T. Iyama, and K. Azumi. 2011. Density functional theory study of boron- and nitrogen-atom doped graphene chips. *J. Appl. Phys.* 50:01BJ03;1–4.

Tang, S., J. Yu, and L. Liu. 2013. Tunable doping and band gap of graphene on functionalized hexagonal boron nitride with hydrogen and fluorine. *Phys. Chem. Chem. Phys.* 15:5067–5077.

Topsakal, M., S. Cahangirov, and S. Ciraci. 2010. The response of mechanical and electronic properties of graphane to the elastic strain. *Appl. Phys. Lett.* 96:091912;1–3.

Usachov, D., O. Vilkov, A. Gruneis et al. 2011. Nitrogen-doped graphene: Efficient growth, structure, and electronic properties. *Nano Lett.* 11:5401–5407.

Usachov, D., A.V. Fedorov, O. Vilkov et al. 2013. Synthesis and electronic structure of nitrogen doped graphene. *Phys. Solid State* 55:1325–1332.

Wang, Y., Y. Shao, D. W. Matson, J. Li, and Y. Lin. 2010. Nitrogen-doped graphene and its application in electrochemical biosensing. *ACS Nano* 4:1790–1798.

Wei, D., Y. Liu, Y. Wang, H. Zhang, L. Huang, and G. Yu. 2009. Synthesis of N-doped graphene by chemical vapor deposition and its electrical properties. *Nano Lett.* 9:1752–1758.

Xu, B., Y. H. Lu, Y. P. Feng, and J. Y. Lin. 2010. Density functional theory study of BN-doped graphene superlattice: Role of geometrical shape and size. *J. Appl. Phys.* 108:073711;1–7.

Yang, S.-Y., K.-H. Chang, Y.-L. Huang et al. 2012. A powerful approach to fabricate nitrogen doped graphene sheets with high specific area. *Electrochem. Commun.* 14:39–42.

Zhang, X., A. Hsu, H. Wang, Y. Song, J. Kong, M. S. Dresselhaus, and T. Palacios. 2013. Impact of chlorine functionalization on high-mobility chemical vapor deposition grown graphene. *ACS Nano* 7:7262–7270.

Zhang, Y., T.-T. Tang, C. Girit et al. 2009. Direct observation of a widely tunable bandgap in bilayer graphene. *Nature* 459:820–823.

Zhao, L., R. He, K. T. Rim et al. 2011. Visualizing individual nitrogen dopants in monolayer graphene. *Science* 333:999–1003.

Section III

Characterization

16 Electronic Structure and Topological Disorder in sp^2 Phases of Carbon

Y. Li and D. A. Drabold

CONTENTS

Abstract...249
16.1 Introduction 249
16.2 Computational Methods 250
    16.2.1 Empirical Potentials..250
    16.2.2 Tight-Binding Approximation 250
    16.2.3 *Ab Initio* Methods 250
16.3 Crystalline Graphene 251
16.4 Fullerenes 251
16.5 Carbon Nanotubes 252
16.6 Schwartzite 253
16.7 Amorphous Graphene..254
    16.7.1 Experimental Results...254
    16.7.2 Amorphous Graphene Models...255
    16.7.3 Symmetry Breaking...255
        16.7.3.1 Random Normal Distortion 256
        16.7.3.2 Molecular Dynamics Simulations...256
    16.7.4 Conformational Fluctuation...257
    16.7.5 Phonon Calculation...259
        16.7.5.1 Vibrational Density of States 259
        16.7.5.2 Vibrational Modes 259
16.8 Conclusion 260
Acknowledgments...261
References...261

ABSTRACT

Carbon has shown itself to be the most flexible of atoms, crystallizing in divergent phases such as diamond and graphite, and being the constituent of the entire zoo of (locally) graphitic balls, tubes, capsules, and possibly negative curvature analogs of fullerenes, the schwartzites. We compare the various sp^2-bonded forms of carbon, and describe conditions that open or close the optical gap. We also explore topological disorder in three-coordinated networks including odd-membered rings in amorphous graphene, as seen in some experimental studies. We start with the Wooten–Weaire–Winer (WWW) models due to Kumar and Thorpe, and then carry out *ab initio* studies of the topological disorder. The structural, electronic, and vibrational characteristics are explored. We show that topological disorder qualitatively changes the electronic structure near the Fermi level. The existence of pentagonal rings also leads to substantial puckering in an accurate density functional simulation. The vibrational modes and spectra have proven to be interesting, and we present evidence that one might detect the presence of amorphous graphene from a vibrational signature.

16.1 INTRODUCTION

Carbon-based semiconductors are one of the hottest topics in condensed matter science. Although silicon-based electronics have achieved tremendous success, scientists and engineers are always seeking alternative materials. One of the main reasons is that the size of silicon-based transistors, which are the building blocks of electronics, is reaching basic limits. One challenge of these short-length scales is the requirement of rapid heat dissipation. Nowadays, remarkable improvements in growth techniques allow scientists to build carbon structure with reduced dimensionality in high precision. The advances in computational tools and theoretical models make it possible to investigate and make plausible predictions about the electronic, vibration, or optical properties of carbon materials.

Single-layer graphene was first isolated by Novoselov et al. [1] using mechanical exfoliation. Graphene's two-dimensional structure, which consists only of hexagons, gives rise to its unique and interesting electronic properties and promising potential for applications [2,3]. However, different categories of defects have to be taken into account for applications. It has been shown that these defects may lead to various graphitic arrangements, associated with a menagerie of local minima on the sp^2 carbon energy landscape. Among these analogs of graphite, we will briefly consider the properties of crystalline graphene, fullerene, carbon nanotubes, and schwarzite, and focus mainly on amorphous graphene.

16.2 COMPUTATIONAL METHODS

To model complex sp^2 carbon systems, the interatomic potential is the fundamental tool to accurately represent the structural electronic and vibrational properties. There are three commonly used interatomic potential in materials theory: empirical potentials, potentials from the tight-binding approximation, and *ab initio* methods [4].

16.2.1 EMPIRICAL POTENTIALS

Empirical potentials are based upon classical chemical concepts. The potential energy function can be expressed as the summation of individual contributions due to various kinds of atomic interactions, which can be categorized into three types [5]: bonded terms, nonbonded terms, and correction terms. By carefully considering different contributions and fitting the experimental data, known properties of the reference materials can be reproduced. Such potentials usually face limits on their ability to represent structures very different from what they were fitted to.

16.2.2 TIGHT-BINDING APPROXIMATION

Another commonly used method is the tight-binding scheme. In this approximation, the electrons are considered as tightly bound to the nuclei, and have limited influence on nearby atoms. The single orbital Hamiltonian can be simplified as (following [6])

$$H = \sum_{\vec{R}} U_{\vec{R}} \mid \vec{R} \rangle\langle \vec{R} \mid + \sum_{\vec{R}\vec{R}'} t_{\vec{R}\vec{R}'} \mid \vec{R} \rangle\langle \vec{R}' \mid + t_{\vec{R}'\vec{R}} \mid \vec{R}' \rangle\langle \vec{R} \mid \quad (16.1)$$

Here $\vec{R}'$ represents the nearest neighbors of $\vec{R}$. The first term in Equation 16.1 gives the potential of electron at a lattice site and the second term describes the interaction energy due to hopping between nearest neighbors. For a Bravais lattice, due to symmetry the hopping matrix elements t are same for all lattice sites. For disordered system, they depend on atomic coordinates, which means their values vary from site to site. The free parameters in tight-binding Hamiltonian are obtained by fitting to density functional or experimental results.

The method can make useful predictions about electronic structure, and with appropriate care in the fitting, total energy and forces. Tight binding suffers from transferability issues as empirical potentials do, though usually less severely, since the tight-binding approximation does grapple at least approximately with the electronic structure.

16.2.3 AB INITIO METHODS

The third well-developed set of methods is characterized as *ab initio*. Since full quantum mechanical calculations are computationally intensive, certain levels of approximations are required for a practical implementation. The dominant approach is density functional theory (DFT). The basic idea used in the density functional setting is to replace the original many-particle wave functions with the ground state charge density as the fundamental variable. This well-accepted theory was introduced by Kohn and Sham (1965). This approximation introduces a set of N single-electron orbitals $\mid \psi_l(\vec{r}) \rangle$, so that the Schrödinger equation (Kohn–Sham equation) can be written as

$$-\frac{\hbar^2}{2m}\nabla^2\psi_l(\vec{r}) + \left[U(\vec{r}) + \int d\vec{r}\, \frac{e^2\rho(\vec{r}')}{\mid \vec{r} - \vec{r}' \mid} + \frac{\partial \mathcal{E}_{xc}[\rho]}{\partial_n} \right]$$
$$\times \psi_l(\vec{r}) = \mathcal{E}_l \psi_l(\vec{r}) \quad (16.2)$$

In Equation 16.2 the exchange correlation $\epsilon_{xc}[\rho(\vec{r})]$ due to a superposition of the potential associated with each of the nuclei is still unknown.

Two common approximations have met with widespread use. The simpler approximation is the local density approximation (LDA). This approximation LDA can be expressed as

$$\mathcal{E}_{xc}[\rho] = \int \epsilon_{xc}[\rho(\vec{r})]\rho(d\vec{r}) \quad (16.3)$$

Here, $\epsilon_{xc}[\rho(\vec{r})]$ is the known exchange-correlation energy per particle of the homogeneous electron gas. The exchange energy is given by a simple analytic form $\epsilon_x[\rho] = -(3/4)(3\rho/\pi)^{1/2}$ [6] and the correlation energy has been calculated to great accuracy for the homogeneous electron gas with Monte Carlo methods [7].

The other approximation is the generalized gradient approximation (GGA), which define the functional as [8]

$$\mathcal{E}_{xc}[\rho] = \int \epsilon_{xc}\left[\rho(\vec{r}), \mid \nabla\rho(\vec{r}) \mid \right]\rho d\vec{r}$$
$$= \int \epsilon_{xc}[\rho(\vec{r})], F_{xc}\left[\rho(\vec{r}), \mid \nabla\rho(\vec{r}) \mid \right]\rho d\vec{r} \quad (16.4)$$

Here, $\epsilon_{xc}[\rho(\vec{r})]$ is the exchange-correlation functional of the homogeneous electron gas and F_{xc} is dimensionless, and based upon three widely used forms of Becke (B88) [9], Perdew and Wang (PW91) [10], and Perdew, Burke, and Enzerhof (PBE) [11].

Both empirical potential and tight-binding approximation are computationally cheap relative to *ab initio*, but suffer from a lack of transferability and reliability in arbitrary bonding environments. *Ab initio* methods are applicable to many systems, but at a significant computational price. Nowadays, SIESTA and Vienna *ab initio* simulation package (VASP) are two widely used *ab initio* programs to calculate band structure, electronic density of states (EDOS), total energies, forces, and other quantities. SIESTA uses pseudopotentials and both LDA and GGA functionals [12]. While VASP is based on pseudopotentials, it employs a plane-wave basis and offers various density functionals [13].

16.3 CRYSTALLINE GRAPHENE

Crystalline graphene is just one layer of graphite, in which carbon atoms are arranged on a perfect honeycomb lattice. After experimental isolation in 2004, graphene's electronic properties have been predicted theoretically [1,14–16]. Since there have been extensive studies on crystalline graphene, here we will briefly discuss the electronic properties. To calculate the band structure of crystalline graphene, we employed a single-ζ (SZ) basis set with or without Harris functional [17], a double-ζ polarized (DZP) basis set with SIESTA, and also VASP to compute the eight lowest-energy bands. For both SZ and DZP calculations by SIESTA, 20 k-points along each special symmetry lines were taken, and for VASP 50 k-points along each line were sampled. The result from SIESTA using SZ basis and Harris functional is essentially identical with the one based on DZP basis for the four occupied bands. These results of SIESTA with SZ basis and Harris functional and of VASP show excellent agreement with published results for each code, respectively [18,19], as shown in Figure 16.1. For energies above the Fermi level, agreement of results for the four unoccupied bands are rather poor, which can be amended by carefully choosing the energy cutoff to minimize the total energy as shown by Machon et al. [21]. While this is presumably irrelevant for ground-state studies, these artifacts would be significant for transport or optics.

The comparison of density of states of crystalline graphene between DFT using SIESTA and tight-binding methods is shown in Figure 16.2. The tight-binding result is calculated based on Equation 14 in Reference 2, where $t' = 0$, and

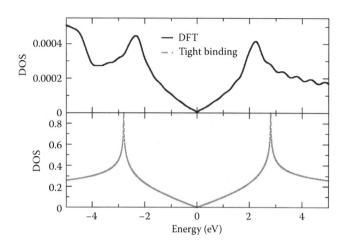

FIGURE 16.2 Density of states of 800-atom crystalline graphene using both DFT and tight-binding methods. The Fermi energy is 0 eV. Solid line represents the result of SIESTA. The density of states due to tight binding is shown by the dashed line.

$t = 2.8$ eV. Both of these two results around Fermi level can be approximated as $\rho(E) \sim |E|$. The broadening of the DFT DOS is due to incomplete Brillouin Zone sampling.

16.4 FULLERENES

In 1985, Kroto et al. [22] found that there exist cage-like molecules containing purely threefold carbon atoms (*sp²* hybridization), which are named fullerenes. This discovery stimulated extensive investigations into this molecular graphite allotrope. Generally speaking, fullerenes refer to a family of closed carbon cages formed by 12 pentagons and various numbers of hexagons, which can be prepared by the vaporization of graphite in an electric arc at low pressure [23]. In this section, the electronic properties of C_{60} and C_{240} (see Figure 16.3) will be discussed. Their structures were optimized by SIESTA with SZ basis and Harris functional without any symmetry constraints.

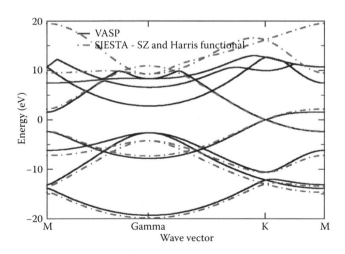

FIGURE 16.1 Density functional band structure of crystalline graphene. The result of VASP is given by solid line. The results by SIESTA using SZ basis and Harris functional are represented by the dash-dotted curve. (Y. Li et al.: *Phys. Status Solidi B.* 2011. **248**. 2082. Copyright Wiley-VCH Verlag GmbH & Co. KGaA. Reproduced with permission.)

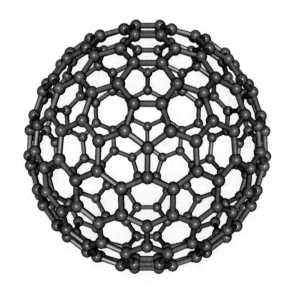

FIGURE 16.3 Optimized structure of C_{240}.

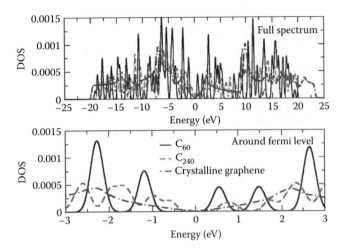

FIGURE 16.4 DOS of C_{60}, C_{240}, and crystalline graphene. The C_{240} upper panel shows the whole spectrum, and DOS around Fermi level is given in the lower panel.

TABLE 16.1
The HOMO–LUMO Gap and Total Energy Relative to Crystalline Graphene of Various Fullerene and Schwarzite Models

Allotropes	Models	>Gap (eV)	E_{tot}/N_{atom} (eV)
Fullerene	C_{60}	1.724	0.402
	C_{240}	1.231	0.132
	G-384 schw	0.183	0.188
Schwarzite	P-536 schw	0.151	0.112
	P-792 schw	0.086	0.090
	P-984 schw	0.394	0.077

The comparison between DOS of these two fullerenes and crystalline graphene are shown in Figure 16.4. It appears that the curved topology of fullerene opens a gap around the Fermi level. According to Figure 16.4 and Table 16.1, the HOMO (highest occupied molecular orbital)–LUMO gaps of C_{60} and C_{240} decrease with increasing numbers of atoms, which is consistent with other calculations [24]. The evolution of the electronic and phonon density of icosahedral fullerenes as a function of size, and their ultimate convergence to graphene is published elsewhere [25].

16.5   CARBON NANOTUBES

Carbon nanotubes can be visualized as a graphene sheet rolled into a cylinder. There are three different types of carbon nanotubes due to different ways of rolling the graphene sheet. Distinct geometry of these three types gives rise to varied electronic behavior. Their structures can be characterized by a chiral vector $\vec{C}_h$ as shown in Figure 16.5. Since carbon nanotubes are derived from crystalline graphene, the geometric properties of a carbon nanotube are commonly described by the ones of graphene.

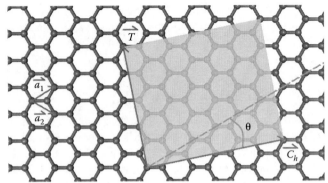

FIGURE 16.5 The chiral vector $\vec{C}_h$ shown in honeycomb lattice. $\vec{T}$ is the translation vector, representing the axial direction of the carbon nanotube. Shaded region represents the unit cell of carbon nanotube and Θ is the chiral angle. $\vec{a}_1$ and $\vec{a}_2$ are the lattice vectors of original honeycomb lattice. (From R. Saito, G. Dresselhaus, and M. S. Drsselhaus, *Physical Properties of Carbon Nanotubes*, 1st ed. (Imperical College Press, London, 1998).)

Recall that the two unit vectors in the honeycomb lattice are defined as $\vec{a}_1 = \left(\left(\sqrt{3a}/2 \right), (a/2) \right)$ and $\vec{a}_2 = \left(\left(\sqrt{3a}/2 \right), -(a/2) \right)$ where $a \approx 1.42$ Å. To represent the geometry of carbon nanotube according to the original honeycomb lattice, chiral vector $\vec{C}_h$ defines the diameter of carbon nanotube, and translation vector $\vec{T}$ defines the axial direction along the nanotube. Both of them are expressed as

$$\vec{C}_h = n\vec{a}_1 + m\vec{a}_2$$

$$\vec{T} = \frac{2m+n}{d_r}\vec{a}_1 + \frac{-2n+m}{d_r}\vec{a}_2 \qquad (16.5)$$

Here both m and n are integers and $n > m$. d_r is the highest common divisor of $(2n + m, 2m + n)$. (n, m) values are crucial to the properties of nanotubes. Nanotubes with the same number of unit vector indices (n, n) are called armchair nanotubes. Chiral vector indices $(n, 0)$ with $m = 0$ represents the zigzag nanotubes. Besides these two cases, if the chiral vector indices (n, m) are $n \neq m \neq 0$, the nanotube is called chiral, with a screw symmetry along the axis of the tube [27]. A few examples of these three types of nanotubes are shown in Figure 16.6.

To evaluate the electronic properties of carbon nanotubes, we use three carbon nanotubes with different chiral vector indices: (4,4) tube with $n = m = 4$, (6,0) tube with $n = 6, m = 0$, and (4,3) tube $n = 4, m = 3$. Carbon nanotubes exhibit either metallic behavior or semiconductor depending on their chiral indices. Theoretical derivations show that if the indices (n, m) of nanotube satisfy the greatest common divisor of $(n - m, 3)$ is 3, the given carbon nanotube behaves like metal, otherwise, it will be a semiconductor [26]. The Γ point DOS calculation results using SIESTA with SZ basis are shown in Figure 16.7. Consistent with the theory, the (6,0) tube has more states around the Fermi level, and obviously is metallic, and (4,3) tube exhibits a gap and is a semiconductor. However, the (4,4) tube that should be metallic exhibits a gap.

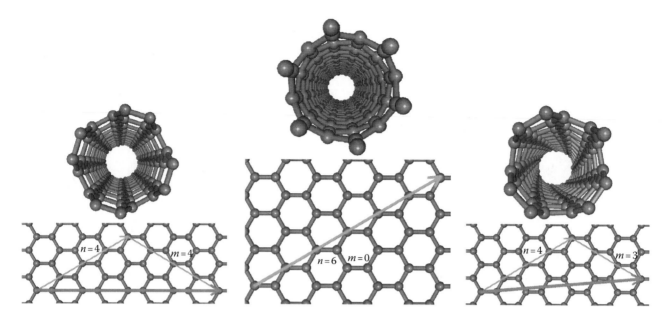

FIGURE 16.6   Three examples of carbon nanotubes with chiral vector indices (4,4), (6, 0), and (4,3), respectively.

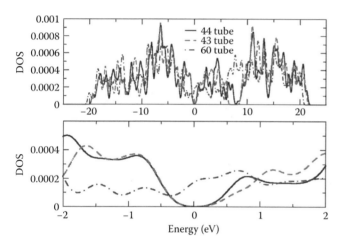

FIGURE 16.7   Normalized DOS of (4,4) tube, (4,3) tube, and (6,0) tube. Fermi level is 0 eV. The full spectrums are shown in the higher panel and lower panel shows in the structures around the Fermi level.

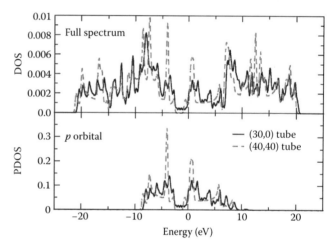

FIGURE 16.8   Comparison between DOS and PDOS of (30,0) tube and (40,40) tube.

In our case, we found that SIESTA calculations with SZ basis always tend to overestimate the gap, due to incomplete basis set and Brillouin Zone sampling. By carefully choosing the basis set and fully integrating over the Brillouin Zone, the gap would be reduced. Details of this calculation will be discussed elsewhere.

We also compute the projected density of states (PDOS) using a similar approach for carbon nanotubes. Here we use one zigzag (30,0) tube and one armchair (40,40) tube. According to the law of greatest common divisor of $(n - m, 3)$, (30,0) tube should be metallic, which is consistent with our results as shown in Figure 16.8. Also, by comparing the contributions from all three sp^2 orbitals and the p orbital, PDOS on p orbital have significant weight around Fermi energy. And in Figure 16.8, the PDOS curves of p orbitals in (30,0) tube and (40,40) tube have identical shape with the DOS around

the Fermi level. Thus, the electronic properties around the Fermi level are determined by the interaction between p orbitals. This result is in fine agreement with bandstructure calculations, in which the two π bands, which are due to the interaction between p orbitals, determine the metallic behavior of carbon nanotubes [26].

16.6 SCHWARTZITE

Unlike fullerenes and carbon nanotubes, which have positive Gaussian curvature due to the presence of 5-membered rings, schwarzites have negative curvature, which are induced by 7- and 8-membered rings as shown in Figure 16.9. As in crystalline graphene, all the atoms of schwarzite are three fold [29]. To study the electronic properties of schwarzite, four models are used: gycoid 384-atom (G-384 schw), primitive 536-atom

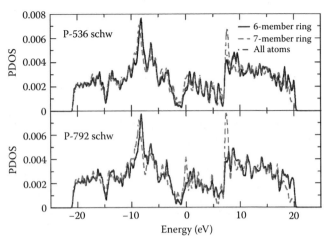

FIGURE 16.11 PDOS of P-536 and P-792 schw models on 7- and 6-member rings, and the DOS represented by the dot-dashed lines.

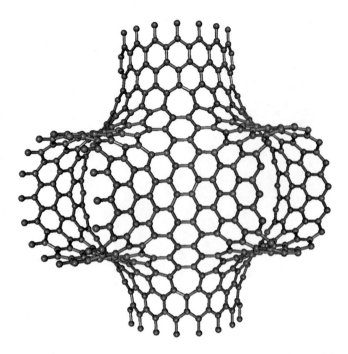

FIGURE 16.9 Structure of primitive 792-atom schwarzite model. Only half of this model is shown here. (From H. Terrones and M. Terrones, *New Journal of Physics* **5**, 126.1–126.37, 2003.)

(P-536 schw), primitive 792-atom (P-792 schw), and primitive 984-atom (P-984 schw) schwarzite models. DOS of these four models are calculated by SIESTA using SZ basis and Harris functional with at least $2 \times 2 \times 2$ Monkhorst-Pack grid [30]. The results are given in Figure 16.10.

As shown in Figure 16.10, with increasing schwarzite unit cell, DOS curve near Fermi level approaches to the shape obtained for crystalline graphene as shown in Figure 16.2. According to Table 16.1, increasing the size of the schwarzite cell also leads to decline of HOMO–LUMO (lowest unoccupied molecular orbital) gap.

The total energy per atom also decreases with increasing unit cell size. And the difference in total energy between

crystalline graphene and large schwarzite cell approaches 0.077 eV, which suggests that there is a real hope to prepare real schwarzite samples.

The calculations of Γ point PDOS of both P-536 and P-792 schw models have been performed. Since the primitive schwarzites contain only 6- and 7-member rings, and the negative curvature is introduced by 7-member rings, here we compare the PDOS on the atoms within 7-member rings and the ones with 6-member rings, as shown in Figure 16.11. It appears that for both of these two models, 7-member rings are responsible for the structure of the DOS around the Fermi level.

16.7 AMORPHOUS GRAPHENE

As mentioned in the introduction, crystalline graphene and associated materials have extraordinarily interesting electronic properties. The electronic, thermal, and vibrational properties of graphene depend sensitively on the perfection of the honeycomb lattice. Thus, it is worthwhile investigating defects in graphene. Although extensive efforts have been devoted in curved graphene derivatives such as carbon nanotubes and fullerenes, little attention has been given to nonhexagonal defects and their electronic and vibrational properties in a planar graphene. In this section, details about the progress in producing real amorphous graphene samples in experiment, techniques on preparing computational models, and calculation results about electronic and vibrational properties of amorphous graphene will be discussed.

16.7.1 EXPERIMENTAL RESULTS

From the 1980s, the progress in the growth engineering and characterization techniques made it possible to grow low-dimensional materials under tight control. Recent electron bombardment experiments have been able to create amorphous graphene pieces [31,32]. Clear images of regions of amorphous graphene have been taken by Meyer et al. [33], following the method described in Reference 34.

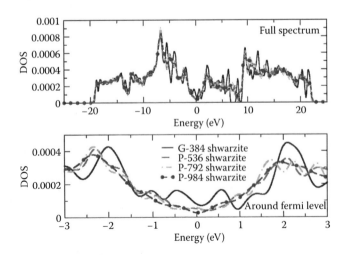

FIGURE 16.10 Normalized DOS of four schwarzite models. Fermi energy is 0 eV.

TABLE 16.2

Ring Statistics of 800 a-g, 836 a-g1, and 836 a-g2 Models (%)

Ring Size	800 a-g	836 a-g1	836 a-g2
5	33.5	25	24
6	38	53	52
7	24	19	25
8	4.5	3	0

Source: Y. Li et al.: *Phys. Status Solidi B.* 2011. **248**. 2082. Copyright Wiley-VCH Verlag GmbH & Co. KGaA. Reproduced with permission.

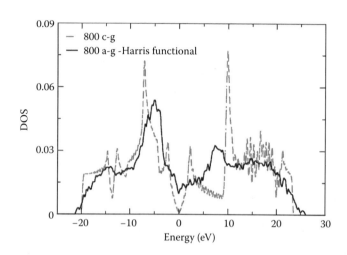

FIGURE 16.12 Top view of 800-atom crystalline and 836-atom amorphous graphene. (Y. Li et al.: *Phys. Status Solidi B.* 2011. **248**. 2082. Copyright Wiley-VCH Verlag GmbH & Co. KGaA. Reproduced with permission.)

Recently, Kawasumi et al. successfully embedded non-hexagonal rings into a crystalline graphene subunit in experiment, synthesized by stepwise chemical methods, isolated, purified, and fully characterized the material spectroscopically [35]. They reported the multiple odd-membered-ring defects in this subunit lead to nonplanar distortion, as shown in Figure 2 in [35], which is consistent with our published results, as described in the following sections.

16.7.2 Amorphous Graphene Models

To investigate the electronic and vibrational properties of amorphous graphene, three models are employed: 800 atom model (800 a-g), two 836 atom models (836 a-g1 and 836 a-g2). All these models are prepared by introducing Stone–Wales (SW) defects [36] into perfect honeycomb lattice and a WWW annealing scheme [37] with varying concentration of 5-, 6-, and 7-member rings [20]. Their ring statistics are shown in Table 16.2. All the atoms in these models are three fold, forming a practical realization of the continuous random network (CRN) model, proposed by Zachariasen [38]. A comparison between crystalline graphene and 836 a-g1 model is shown in Figure 16.12.

The electronic DOS of the planar 800 a-g model is compared to a DOS of the crystalline 800 c-g model in Figure 16.13 due to SIESTA. The electronic structure of the 800 a-g model is vastly different from the crystalline graphene near the Fermi level due to the presence of ring defects, as first reported by Kapko et al. [39].

16.7.3 Symmetry Breaking

The planar symmetry of the original amorphous graphene models is extremely sensitive to external distortions. In the

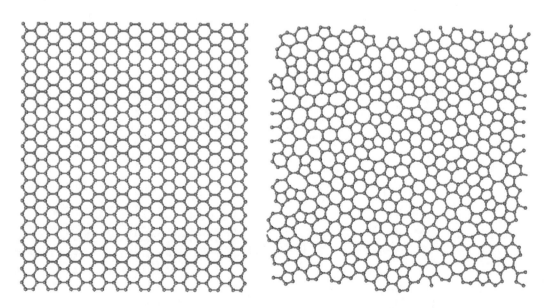

FIGURE 16.13 DOS of 800-atom amorphous and crystalline graphene, the Fermi energy is 0 eV. (Y. Li et al.: *Phys. Status Solidi B.* 2011. **248**. 2082. Copyright Wiley-VCH Verlag GmbH & Co. KGaA. Reproduced with permission.)

following sections, two different approaches inducing curvature into the original flat graphene models will be discussed.

16.7.3.1 Random Normal Distortion

Small random distortions have been applied to all the atoms in these three models along the normal direction of the graphene plane. Disordered configurations are relaxed by SIESTA, using SZ basis and Harris functional. The planar symmetry breaks with curvature above or below the initial plane starting from a flat sheet [20]. The final structures depend on the initial conditions, and share a consistent pattern of relaxation as shown in Table 16.3.

As shown in Table 16.3, initially all the atoms have been randomly moved along normal direction by a small displacement $[-\delta r, +\delta r]$, as the first column of this table. The relaxation results are shown in the second and third columns, where the mean displacement of the system away from the original flat plane is given in the second column; and the third column shows the total energy per atom in the relaxed models. Taking 800 a-g model as an example, the EDOS around the Fermi level of all the puckered and original flat models are shown in Figure 16.14. The influence of puckering on DOS around Fermi level is modest. The puckered system, after relaxation with $\delta r = 0.05$ Å is shown in Figure 16.21. Whereas, the mean bond length remains around 1.42 Å, the changes in ring statistics after relaxation are not significant either, as shown in Figure 16.15. This implies that it would be difficult to detect puckering using diffraction experiments or the associated radial distribution function.

To further test the relation between puckering and the initial distortion, various initial conditions have been employed, and puckering structures arise and remain at the same regions [20]. It suggests that rings with certain members may give rise to the puckering. With this motivation, we searched for regions with highest or lowest variations along the normal direction (puckered or smooth regions), as shown in Figure 16.16.

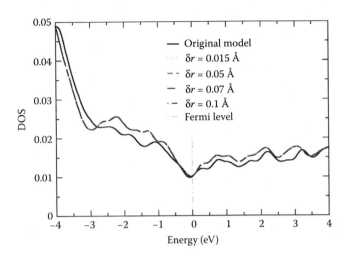

FIGURE 16.14 The EDOS of both flat and relaxed puckered amorphous graphene models. The Fermi level is corrected to 0 eV. (Y. Li et al.: *Phys. Status Solidi B.* 2011. **248**. 2082. Copyright Wiley-VCH Verlag GmbH & Co. KGaA. Reproduced with permission.)

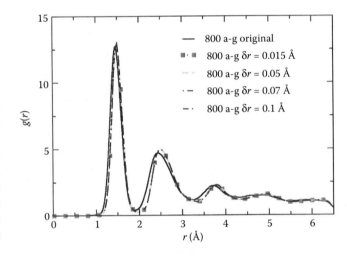

FIGURE 16.15 Comparison of RDF of flat and puckered 800 a-g models. (Y. Li et al.: *Phys. Status Solidi B.* 2011. **248**. 2082. Copyright Wiley-VCH Verlag GmbH & Co. KGaA. Reproduced with permission.)

As illustrated in Figure 16.16, puckered regions tend to have a higher ratio of pentagons to heptagons. The most distorted bonds do not belong to these pentagons, instead they are within the hexagons or heptagons connecting two pentagons. Thus, the puckering regions are strongly associated with pentagons.

16.7.3.2 Molecular Dynamics Simulations

Symmetry breaking can also be achieved by molecular dynamics (MD) simulations. Calculations are performed with SIESTA with 800 a-g and 836 a-g1 using pseudopotentials and LDA with a SZ basis and Harris-functional at constant volumes. Details about these calculations are discussed in Reference 40. It is observed that in every case, even at $T = 20$ K, the system puckers breaking the planar symmetry, reflecting the extremely shallow minimum for the flat structure. In other words, weak thermal disorder is sufficient to

TABLE 16.3
The Influence of δr on 800 a-g, 836 a-g1, and 836 a-g2 Systems Relative to the Initial Flat Model

Models	δr (Å)	$\bar{\delta r'}$ Å	E_{tot}/N_{atom} (eV)
	0.01	0.510	−0.086
800 a-g	0.05	0.526	−0.100
	0.07	0.527	−0.100
	0.01	2.53E-3	0.0
836 a-g1	0.05	1.402	−0.102
	0.07	1.401	−0.102
	0.01	2.72E-3	0.0
836 a-g2	0.05	1.183	−0.090
	0.07	1.180	−0.090

Source: Y. Li et al.: *Phys. Status Solidi B.* 2011. **248**. 2082. Copyright Wiley-VCH Verlag GmbH & Co. KGaA. Reproduced with permission.

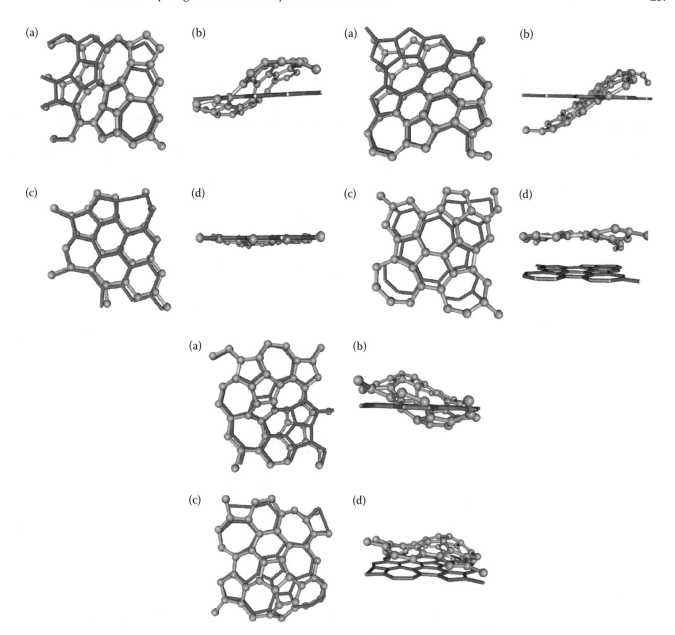

FIGURE 16.16 Enlarged plot of puckered and smooth regions of three amorphous models. (a) The top view of puckered region. (b) The side view of puckered region. (c) The top view of smooth region. (d) The side view of smooth region. (Y. Li et al.: *Phys. Status Solidi B*. 2011. **248**. 2082. Copyright Wiley-VCH Verlag GmbH & Co. KGaA. Reproduced with permission.)

induce puckering. The relation between total energy of the system and the maximum separation of atoms along the normal direction (magnitude of puckering) is shown in Figure 16.17. It indicates that loosing planar symmetry will lower the total energy of the system, which is consistent with our results due to random distortions as shown in Table 16.3.

16.7.4 Conformational Fluctuation

As mentioned in the previous section, several constant temperature MD simulations have been performed. The main purpose of this section is to investigate the nature of puckering minima of amorphous graphene, which can be induced by different initial conditions as observed before [20] and is also

a universal phenomenon in amorphous materials [41]. Here we employ the method proposed by Fedders and Drabold [41], which is akin to the conformational space annealing approach mentioned in Reference 42. Starting with a perfectly relaxed model (in our case 800 a-g and 836 a-g1), during each MD simulation we let the network evolve for 8.0 ps at four different mean temperatures of 20, 500, 600, or 900 K using Berendsen thermostat (velocity rescaling) to achieve the target temperatures. Then snapshots have been taken every 0.15 ps and quenched to find the metastable minima (or unique basins on potential energy landscape (PEL) [42]).

To investigate the conformational changes between these quenched minima, two auto correlation functions are used as defined in [41]

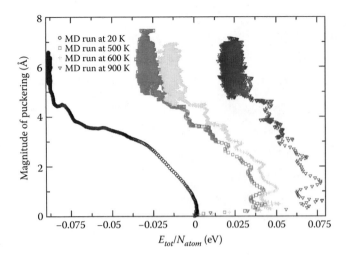

FIGURE 16.17 Correlation between the total energy per atom and magnitude of puckering for constant temperature MD simulations. Zero total energy refers to the total energy of original flat 800 a-g model. (Y. Li and D. A. Drabold: *PSSB*. 2013. **250**, 1012–1019. Copyright Wiley-VCH Verlag GmbH & Co. KGaA. Reproduced with permission.)

$$\Delta\theta(t_1, t_2) = \sum_i \left(\frac{(\theta_i(t_1) - \theta_i(t_2))^2}{N} \right)^{1/2} \tag{16.6}$$

$$\Delta r(t_1, t_2) = \sum_i \left(\frac{(r_i(t_1) - r_i(t_2))^2}{N} \right)^{1/2} \tag{16.7}$$

where θ_i is the ith bond angle, and r_i represents the distance between ith nearest-neighbor pair. Time t_1 and t_2 refers to two distinct quenched snapshots. These autocorrelation functions provide an insight into how thermal MD simulations lead to transition between various energy basins. Taking 800 a-g model as example, its autocorrelation functions are shown in Figure 16.18. It appears $\Delta\theta(t_1, t_2)$ and $\Delta r(t_1, t_2)$ from MD runs at three different temperatures are qualitatively similar. They all increase linearly initially and fluctuate about a constant eventually. The continuity of these curves suggests that there exists a continuum of states, which differ structurally in moderate ways. The results using 836 a-g1 have great consistency with 800 a-g.

The total energy distributions of all these quenched snapshots from MD runs at different temperature are shown in Figure 16.19. Excepting quenched configurations from 20 K MD, the total energy distributions exhibit several peaks. The annealing process at the beginning of each MD simulation is responsible for the minor peaks on the right side of Figure 16.19. The major peaks (labeled as 3, 2, 1 in Figure 16.19) are contributed by the distinct puckering supercells sampled while MD simulations achieving equilibrium at constant T. Correspondingly in Figure 16.18, autocorrelation functions reach an asymptotic state after around 7.0 ps. Thus, each of these three peaks in Figure 16.19 demonstrates a basin on the PEL of amorphous graphene. A series of nearly degenerate quenched minima are trapped in distinct basins on the

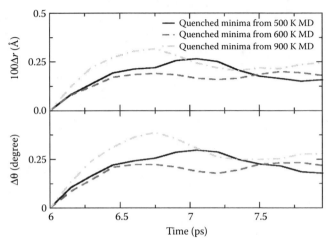

FIGURE 16.18 Time variation of two autocorrelation functions. This figure shows autocorrelation functions of $\Delta\theta(t_1, t_2)$ and $100\Delta r(t_1, t_2)$ with $t_1 = 6.0$ ps and t_2 varying from 6.0 to 7.95 ps. (Y. Li and D. A. Drabold: *PSSB*. 2013. **250**, 1012–1019. Copyright Wiley-VCH Verlag GmbH & Co. KGaA. Reproduced with permission.)

PEL, forming a continuum metastable state around inherent structures. Details of the topological variations are discussed in Reference 40. The average total energy variation between these basins corresponding to the three major peaks in Figure 16.19 is around 1.405×10^{-3} eV, and these quenched configurations belonging to distinct basins share identical local bonding. The only difference is that they pucker in unique ways. Parallel calculations of 836 a-g1 reveal close agreement with the results of 800 a-g as stated above.

Analogous calculations using a 64-atom model of a-Si reveal that PEL of 3D (three-dimensional) system (a-Si) is smooth consisting of one general basin [40]. By investigating the correlation between topology and energy level of these quenched minima, it appears that lower total energy (more

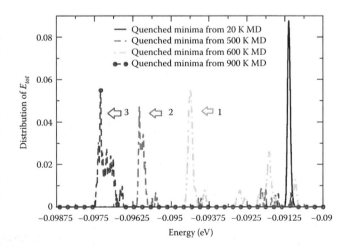

FIGURE 16.19 Total energy distribution functions of quenched supercells from MD runs under 20, 500, 600, and 900 K. The total energy of original flat 800 a-g is considered as 0 eV. (Y. Li and D. A. Drabold: *PSSB*. 2013. **250**, 1012–1019. Copyright Wiley-VCH Verlag GmbH & Co. KGaA. Reproduced with permission.)

stable state) corresponds with more puckered (higher surface roughness) configurations with minute variation in bond lengths and angles from the original flat 800 a-g model.

16.7.5 PHONON CALCULATION

To study vibrational properties of amorphous graphene, calculations of the dynamical matrix, eigenvalues and eigenvectors are performed for the original flat 800 a-g and 836 a-g1 models, and their puckered derivatives. The dynamical matrix was constructed from finite difference calculations (using six orthogonal displacements of 0.04 Bohr for each atom). Phonon calculation has also been performed for an 800-atom crystalline graphene model (800 c-g). Although there exist certain topological difference between 800 a-g and 836 a-g1, their vibrational properties are consistent with each other, as well as their puckered derivatives. In what follows, we will emphasize on the results of 800 a-g model.

For the puckered configurations of 800 a-g model, we focus on two quenched configurations with certain regions puckering in opposite directions, designated "pucker-up" and "pucker-down" models, as shown in Figure 16.20.

16.7.5.1 Vibrational Density of States

The vibrational density of states (VDOS) of 800 c-g model, pucker-up and pucker-down 800 a-g models are shown in

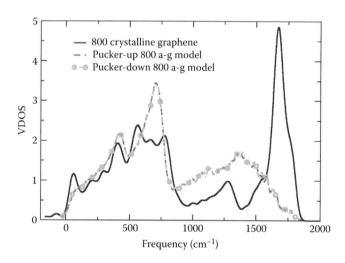

FIGURE 16.20 VDOS of 800 crystalline graphene, pucker-up and -down a-g models. (Y. Li and D. A. Drabold: *PSSB*. 2013. **250**, 1012–1019. Copyright Wiley-VCH Verlag GmbH & Co. KGaA. Reproduced with permission.)

Figure 16.21. The VDOS of crystalline graphene shows great agreement with a published result [43]. In Figure 16.21, the spectrum of crystalline graphene reaches a minimum at a frequency near 1375 cm⁻¹, whereas the spectrum of two puckered amorphous configurations achieve a local maximum. Thus, Raman scattering experiments could provide a way to distinguish crystalline and amorphous phases of graphene. While dealing with a mixed sample containing 3D amorphous carbon, this alternation from ordered graphene near 1375 cm⁻¹ might not be very helpful, since in various phases of a-C (with varying sp^2/sp^3 ratio) many modes have been observed near the relevant energy [44–46]. Between pucker-up and -down 800 a-g models there is no significant difference in the spectrum.

16.7.5.2 Vibrational Modes

In both the original flat 800 a-g and 836 a-g1 models, there exist eigenvectors with imaginary eigenvalues, whose components along the normal direction of the graphene plane is much larger than the other ones (at least four orders of magnitude higher than longitudinal ones). Two examples of these imaginary-frequency modes are given in Figure 16.22. It appears these modes are localized on pentagons in the configurations. According to previous discussion in Section 16.7.3.1, pentagons are the origin of puckering and symmetry breaking. Thus, these imaginary-frequency modes give us clean evidence about the pentagon-induced instability of the flat amorphous graphene models.

In the puckered models, modes with very low frequency, around 14–20 cm⁻¹, have been observed. These modes remind us of "floppy modes" proposed by Phillips and Thorpe [47,48]. These low-frequency modes are rather extended, and have weight on pentagonal puckering regions and large rings, as shown in Figure 16.23. This is consistent with the results obtained by Fedders and Drabold in a-Si:H [41]. The lowest-frequency acoustic phonon modes in an identical sized crystalline graphene model are almost twice larger than the energy scale of these low-frequency modes in amorphous graphene, which is around a few meV. As pointed out by the theory of "two-level systems," there exists a distribution of low-energy excitations, caused by tunneling of atoms between nearly degenerate equilibrium states [49,50]. As mentioned in Section 16.7.4, the energy variations between minima within one basin is in the order of 10⁻⁴ eV, and the energy difference between basins is in the order of 10⁻³ eV [40]. Then the energy of these low-frequency modes (meV) is sufficient to cause conformational variations around one minima (within one basin), but not high enough to overcome the barrier between different basins on PEL.

FIGURE 16.21 Side view of pucker-up and -down 800 a-g models. Gray balls and sticks illustrate pucker-up model, and pucker-down supercell is represented by the mesh. (Y. Li and D. A. Drabold: *PSSB*. 2013. **250**, 1012–1019. Copyright Wiley-VCH Verlag GmbH & Co. KGaA. Reproduced with permission.)

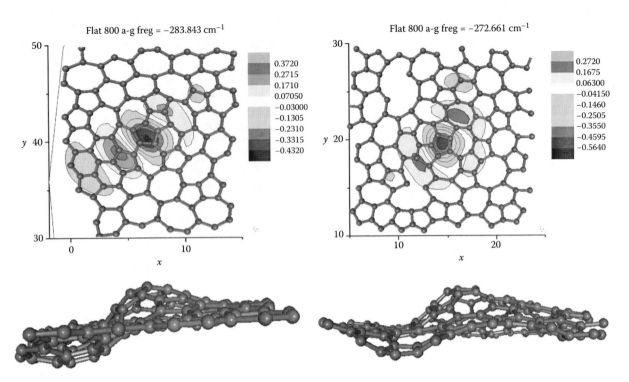

FIGURE 16.22 Two examples of imaginary-frequency modes in flat 800 α-g model. The contour plot represents the component of eigenvector along the direction transverse to the plane. (Y. Li and D. A. Drabold: *PSSB*. 2013. **250**, 1012–1019. Copyright Wiley-VCH Verlag GmbH & Co. KGaA. Reproduced with permission.)

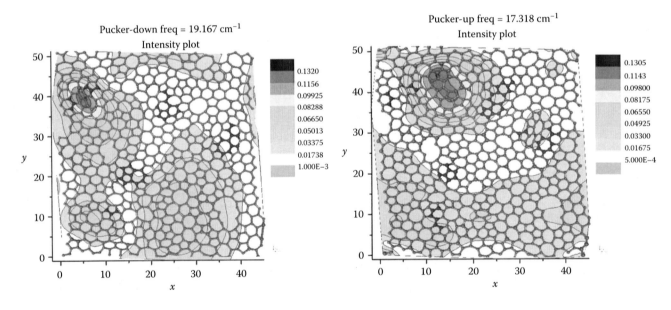

FIGURE 16.23 Examples of low-frequency modes in pucker-down and -up 800 α-g models. The contour plots represent the intensity of eigenvectors on each atom. (Y. Li and D. A. Drabold: *PSSB*. 2013. **250**, 1012–1019. Copyright Wiley-VCH Verlag GmbH & Co. KGaA. Reproduced with permission.)

In the high-frequency range, we observe some highly localized high-frequency modes, which is due to pentagonal defects [40]. This is consistent with the observation of Biswas et al. [51] and Fedders et al. [52], who have concluded that strain and topological defects are active at high frequencies.

16.8 CONCLUSION

In summary, positive curvature opens up a HOMO–LUMO gap in fullerenes. For carbon nanotubes, their electronic properties strongly depend on the chiral indices. However, the influence of negative curvature on DOS is reduced by increasing schwarzite unit cell size, and the difference of DOS between large schwarzite and crystalline graphene diminishes. For both positively and negatively curved carbon allotropes, the bigger the closed cage is, the lower the total energy per atom will be.

For amorphous graphene, the presence of pentagons and heptagons induces extensive states around Fermi level, and pentagons increase the sensitivity of whole system to external distortions leading to puckered states. A series of MD and quenching simulations suggest these puckered states correspond to distinct local minima on PEL of amorphous graphene, whose boundaries can be overcome by heating up the system. Each basin on PEL is associated with a continuum of bond length, bond angles, and energy scale, which is consistent with the studies of a-Si:H in 1996 [42].

Vibrational calculations reveal the existence of localized imaginary-frequency modes in flat 800 a-g model, which are localized on pentagons and play the key role in breaking planar symmetry. Delocalized low-frequency phonon modes, akin to floppy mode, have substantial weight on pentagonal puckering regions and are triggered by the transition between adjacent basins on the PEL.

ACKNOWLEDGMENTS

We wish to acknowledge Dr. M. F. Thorpe at Arizona State University and his group members, who have prepared all the amorphous graphene models, Dr. Mingliang Zhang and Dr. Binay Prasai for their suggestions, and Dr. Humberto Terrones at Penn State University for his kindly help. The calculations of radial distribution function are done by ISAACS [53]. This work is supported by the Army Research Office.

REFERENCES

1. K. S. Novoselov, A. K. Geim, S. V. Morozov, D. Jiang, Y. Zhang, S. V. Dubonos, I. V. Gregorieva, and A. A. Firsov, *Science* **306**, 666, 2004.
2. A. H. Castro Neto, F. Guinea, N. M. R. Peres, K. S. Novoselov, and A. K. Geim, *Rev. Mod. Phys.* **81**, 109, 2009.
3. N. Bilton, Bend it, charge it, dunk it: Graphene, the material of tomorrow, *The New York Times*, 2014. Retrieved from http://bits.blogs.nytimes.com/2014/04/13/bend-it-charge-it-dunk-it-graphene-the-material-of-tomorrow/?_php=true&_type=blogs&_r=0.
4. D. A. Drabold, *Euro. Phys. J. B* **68**, 1–21, 2009.
5. M. L. Cohen and V. Heine, *Solid State Physics*, Vol. 24 (Academic Press, New York, 1970), p. 37.
6. N. W. Ashcroft and N. D. Mermin, *Solid State Physics* (Thomson Learning, Inc., New York, 1976), p. 177.
7. D. M. Ceperley and B. J. Alder, *Phys. Rev. Lett.* **45**, 566, 1980.
8. R. M. Martin, *Electronic Structure: Basic Theory and Practical Methods* (Cambridge University Press, Cambridge, England, 2004).
9. A. D. Becke, *Phys. Rev. A* **38**, 3098, 1988.
10. J. P. Perdew and Y. Wang, *Phys. Rev. B* **45**, 13244, 1992.
11. J. P. Perdew, K. Burke, and M. Ernzerhof, *Phys. Rev. Lett.* **77**, 3865, 1996.
12. E. Artacho, E. Anglada, O. Dieguez, J. D. Gale, A. Garcia, J. Junquera, R. M. Martin et al., *J. Phys.: Condens. Matter* **20**, 064208, 2008.
13. G. Kresse and J. Furthmuller, *Phys. Rev. B* **54**, 11169, 1996.
14. K. S. Novoselov, D. Jiang, F. Schedin, T. J. Booth, V. V. Khotkevich, S. V. Morozov, and A. K. Geim, *Proc. Natl. Acad. Sci. U.S.A.* **102**, 10451, 2005.
15. Y. Zhang, J. W. Tan, H. L. Stormer, and P. Kim, *Nature (London)* **438**, 201, 2005.
16. K. S. Novoselov, A. K. Geim, S. V. Morozov, D. Jiang, M. I. Katsnelson, L. V. Grigorieva, S. V. Dubonos, and A. A. Firsov, *Nature (London)* **438**, 197, 2005.
17. J. Harris, *Phys. Rev. B* **31**, 1770, 1985.
18. K. V. Christ and H. R. Sadeghpour, *Phys. Rev. B* **75**, 195418, 2007.
19. G. Gui, J. Li, and J. Zhong, *Phys. Rev. B* **78**, 075435, 2008.
20. Y. Li, F. Inam, A. Kumar, M. Thorpe, and D. A. Drabold, *Phys. Status Solidi B* **248**, 2082, 2011.
21. M. Machon, S. Reich, C. Thomsen, D. Sanchez-Portal, and P. Ordejon, *Phys. Rev. B* **66**, 155410, 2002.
22. H. W. Kroto, J. R. Heath, S. C. Obrien, R. F. Curl, and R. E. Smalley, *Nature* **318**, 162, 1985.
23. W. Kratschmer, L. D. Lamb, K. Fostiropoulos, and D. R. Huffman, *Nature* **347**, 354, 1990.
24. M. Yu, I. Chaudhuri, C. Leahy, S. Y. Wu, and C. S. Jayanthi, *J. Chem. Phys.* **130**, 184708, 2009.
25. D. A. Drabold, P. Ordejon, J. Dong, and R. M. Martin, *Solid State Commun.* **96** 833 1995.
26. R. Saito, G. Dresselhaus, and M. S. Drsselhaus, *Physical Properties of Carbon Nanotubes*, 1st ed. (Imperical College Press, London, 1998).
27. G. S. Diniz, Electronic and transport properties of carbon nanotubes: Spin-orbit effects and external, Doctoral dissertation, Ohio University, Athens, Ohio, USA, 2012.
28. H. Terrones and M. Terrones, *New J. Phys.* **5**, 126.1–126.37, 2003.
29. R. Phillips, D. A. Drabold, T. Lenosky, G. B. Adams, and O. F. Sankey, *Phys. Rev. B* **46**, 1941, 1992.
30. H. J. Monkhorst and J. D. Pack, *Phys. Rev. B* **13**, 5188, 1976.
31. J. Kotakoski, J. C. Meyer, S. Kurasch, D. Santos-Cottin, U. Kauser, and A. V. Krasheninnikov, *Phys. Rev. B* **83**, 245420, 2011.
32. J. Kotakoski, A. V. Krasheinnikov, U. Kaiser, and J. C. Meyer, *Phys. Rev. Lett.* **106**, 105505, 2011.
33. J. C. Meyer, C. Kisielowski, R. Erni, M. D. Rossell, M. F. Crommie, and A. Zettl, *Nano Lett.* **8**, 3582, 2008.
34. J. C. Meyer, C. O. Girit, M. F. Crommie, A. Zettl, *Appl. Phys. Lett.* **92**, 123110, 2008.
35. K. Kawasumi, Q. Zhang, Y. Segawa, L. T. Scott, and K. Itami, *Nat. Chem* **5**, 739–744, 2013.
36. A. J. Stone and D. J. Wales, *Chem. Phys. Lett.* **128**, 501, 1986.
37. F. Wooten, K. Winer, and D. Weaire, *Phys. Rev. Lett.* **54**, 1392, 1985.
38. W. H. Zachariasen, *J. Am. Chem. Soc.* **54**, 3841, 1932.
39. V. Kapko, D. A. Drabold, and M. F. Thorpe, *Phys. Status Solidi B* **247**, 1197–1200, 2010.
40. Y. Li and D. A. Drabold, *PSSB*, **250**, 1012–1019, 2013.
41. P. A. Fedders and D. A. Drabold, *Phys. Rev. B* **53**, 3841, 1996.
42. D. J. Wales, *Energy Landscapes with Applications to Clusters, Biomolecules and Glasses* (Cambridge University Press, New York, 2003).
43. F. Liu, P. Ming, and J. Li, *Phys. Rev. B* **76**, 064120, 2007.
44. M. A. Tamor and W. C. Vassell, *J. Appl. Phys.* **76**, 3823, 1994.
45. A. C. Ferrari and J. Robertson, *Phys. Rev. B* **61**, 14095, 2000.

46. D. A. Drabold, P. A. Fedders, and P. Stumm, *Phys. Rev. B* **49**, 16415, 1994; D. A. Drabold, P. A. Fedders, and M. P. Grumbach, *Phys. Rev. B* **54**, 5480, 1996.
47. J. C. Phillips, *J. Non-Cryst. Solids* **43**, 37, 1981.
48. M. F. Thorpe, *J. Non-Cryst. Solids* **57**, 355, 1983.
49. P. W. Anderson, B. I. Halperin, and C. M. Varma, *Philos. Mag.* **25**, 1, 1972.
50. W. A. Phillips, *J. Low Temp. Phys.* **7**, 351, 1972.
51. R. Biswas, M. Bouchard, W. A. Kamitakahara, G. S. Grest, and C. M. Soukoulis, *Phys. Rev. Lett.* **60**, 2280, 1988.
52. P. A. Fedders, D. A. Drabold, and S. Klemm, *Phys. Rev. B* **45**, 4048, 1992.
53. S. L. Roux and V. Petkov, *J. Appl. Crystallogr.* **43**, 181–185, 2010.

17 3D Macroscopic Graphene Assemblies

Marcus A. Worsley, Juergen Biener, Michael Stadermann, and Theodore F. Baumann

CONTENTS

Abstract 263
17.1 Introduction 263
17.2 Top-Down Synthesis: Polymer-Derived Nanographene 264
17.3 Bottom-Up Synthesis: GO-Derived Graphene Assemblies 266
17.4 Applications 268
    17.4.1 Hydrogen Storage 268
    17.4.2 Batteries 270
    17.4.3 Supercapacitors 271
    17.4.4 Capacitive Deionization 272
    17.4.5 Charging Induced Macroscopic Effects: Actuation and Electrode Resistivity 272
    17.4.6 Catalysis 273
References 273

ABSTRACT

Integration of 2D graphene sheets into 3D macroscale architectures is critical in realizing the full potential of graphene sheets in macroscale devices and applications. Several fabrication strategies for building these architectures have emerged, and they fall into two categories: (1) top-down and (2) bottom-up approaches. Top-down approaches involve synthesis of the graphene and network simultaneously from polymer or gaseous precursors. Techniques such as sol–gel chemistry and chemical vapor deposition growth fall into this category. Bottom-up techniques begin with a graphene precursor such as graphene oxide (GO) sheets and then uses various strategies to assemble them into a network. Chemical and hydrothermal gelation/reduction of GO suspensions falls into this category. This chapter summarizes the various strategies for fabricating these novel structures and shows how they produce graphene architectures with unique properties such as high electrical conductivities, large surface areas, exceptional mechanical properties, a wide range of densities, and more. Owing to these unique properties, incorporation of these novel assemblies in macroscale systems results in enhanced and/or novel performance in a wide range of applications, including energy storage, catalysis, actuators, electronics, and desalination.

17.1 INTRODUCTION

Individual graphene sheets possess a number of remarkable properties, including extremely low electrical and thermal resistivity [1], large carrier mobility [2], high surface area [3], and exceptional mechanical elasticity [4]. As such, graphene and graphene-based materials hold technological promise in the areas of energy storage [5,6], electronics [7,8], composites [9], actuators [10], and sensors [11,12]. Realizing the full potential of graphene in these applications, however, requires the design of bulk multifunctional architectures that retain the exceptional properties of graphene. Unfortunately, fabricating 3D macroscopic graphene assemblies that retain the properties of 2D graphene sheets has not been trivial. Early synthetic schemes relied on physical interactions (i.e., van der Waals forces) between graphene sheets to stabilize the network structure. This approach produced low-density (<100 mg/cm³) 3D graphene assemblies with fairly low Young's moduli of ~10^2 kPa and electrical conductivities of ~5×10^{-1} S/m [13–17]. Though the individual graphene sheets may still exhibit more exceptional values on the microscale, it is the link between the sheets that dominates the properties on the macroscale. The first covalently bonded 3D graphene assemblies showed comparatively higher electrical conductivities [18,19] and Young's moduli [18], but the surface areas of these materials were well below 1000 m^2/g—less than half the theoretical value for a single graphene sheet. Thus, strengthening the interactions between the sheets translated into an improvement in bulk properties, but the sheets were not sufficiently exfoliated in the 3D assembly to take advantage of the huge surface area present in the 2D sheet. To address these issues, a number of different approaches have been tested for the fabrication of 3D macroscopic graphene assemblies [13–16,18–21] (Table 17.1). We recently reported the synthesis of 3D graphene assemblies with high electrical conductivity [22–25], modulus [23,25,26], and surface area [23,24,26,27]. These graphene assemblies were produced using two different strategies: (1) top-down and (2) bottom-up. Top-down approaches involve synthesis of the graphene and network simultaneously from polymer or gaseous precursors. Techniques such as

TABLE 17.1
Summary of Graphene Macroassemblies

Scheme	Gel Type	Precursor	Reference
Top-down	Aerogel	RF	[26,27]
Top-down	CVD foam	CH_4 + Ni foam	[19]
Bottom-up	Aerogel	GO	[18,23,24]
Bottom-up	Xerogel	GO	[25]
Bottom-up	Aerogel	GO + RF	[22]
Bottom-up	Aerogel	GO + metal ion	[15]
Bottom-up	Hydrogel	GO	[13,14]
Bottom -up	Cryogel	GO	[17]
Bottom-up	Cryogel	GO + CNT	[16]

sol–gel chemistry and chemical vapor deposition growth fall into this category. Bottom-up techniques, on the other hand, begin with a graphene precursor such as graphene oxide (GO) sheets and then uses various strategies to assemble them into a network. Chemical and hydrothermal gelation/reduction of GO suspensions falls into this category. This chapter summarizes the various strategies for fabricating these novel structures and discusses how to produce graphene architecture with unique properties such as high electrical conductivities, large surface areas, exceptional mechanical properties, a wide range of densities, and more. Furthermore, the incorporation of these novel assemblies in macroscale systems is demonstrated, giving enhanced and/or novel performance in a wide range of applications, including energy storage, desalination, actuators, electronics, and catalysis.

17.2 TOP-DOWN SYNTHESIS: POLYMER-DERIVED NANOGRAPHENE

One approach to the design of 3-D graphene materials utilizes highly cross-linked organic polymer gels as scaffold structures for the construction of monolithic graphene assemblies [26–28]. With this "top down" approach, highly cross-linked organic polymer gels are converted into 3-D graphene assemblies through a series of high-temperature carbonization and activation process (Figure 17.1). One advantage to this approach is that the bulk properties of the graphene assembly (i.e., surface area, porosity, mechanical strength) can be controlled by the temperature and atmosphere used during carbonization as well as the composition of the polymer.

In addition, this process is relatively inexpensive, scalable, and yields mechanically robust, centimeter-sized monolithic samples that are comprised almost entirely of interconnected networks of single-layer graphene nanoplatelets, and were therefore dubbed "nanographenes" (nG).

The starting point for fabrication of these materials is a macroporous polymer structure prepared by the catalyzed sol–gel polymerization of organic precursors. While a variety of multifunctional monomers have been utilized for sol–gel polymerization chemistry [29–33], the polymer gels used for the nanographene are derived from resorcinol (1,3-dihydroxybenzene) and formaldehyde. In this particular system, the resorcinol serves as the multifunctional monomer that contains three reactive sites at the 2, 4, and 6 positions on the benzene ring [34]. The resorcinol reacts with formaldehyde in aqueous solution, initiated by a catalyst to form mixtures of addition and condensation products. The formation of the resorcinol–formaldehyde (RF) reaction products involves two main steps: (1) addition of the formaldehyde to resorcinol to form hydroxymethyl derivatives and (2) subsequent condensation of these hydroxymethyl derivatives to form methylene and methylene ether bridges between resorcinol molecules. These reactions then lead to the formation of colloidal polymer species that crosslink to form the extended 3-D gel network. The amount and type of catalyst used in the polymerization reaction dictates the size, shape, and connectivity of the primary network particles and, therefore, can be used to influence the structural properties of the resultant polymer. It was observed that acid catalysts, such as acetic acid, yield a macroporous structure that exhibits enhanced mechanical integrity relative to porous polymer structures prepared with basic catalysts, such as sodium carbonate.

The monolithic polymer gel is then dried and converted into graphitic carbon foam through pyrolysis under an inert atmosphere. This carbonization step yields a porous sp^2-bonded carbon network (density ~550 kg/m^3) comprising both amorphous regions and multilayer graphene nano-platelets (Figure 17.1) [26]. The monolithic nanographene is obtained by preferentially etching away the more reactive amorphous carbon components and partial etching of multilayer graphite components. The challenge to this process step is to preserve the macroporous architecture of the foam while uniformly etching throughout the bulk of the material, yielding large and uniform monoliths. The required homogeneous etching is achieved by thermal activation, a process that involves

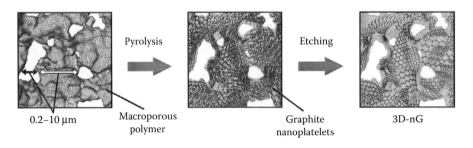

0.2–10 µm   Macroporous polymer   Pyrolysis   Graphite nanoplatelets   Etching   3D-nG

FIGURE 17.1 Schematic of the polymer-based top-down synthesis approach.

controlled carbon burn-off in an oxidizing atmosphere at elevated temperatures. For etching of the polymer-derived carbons, carbon dioxide using the Boudouard equilibrium $(C + CO_2 \rightleftharpoons 2CO)$ proved to be most effective as the low reactivity of CO_2 provides uniform preferential removal of more reactive components throughout monolithic samples. A typical activation is carried out under flowing CO_2 at 950°C for several hours (Figure 17.1). The yield of nanographene from the CO_2 activation process is typically ~25%–30%, based on the weight of the starting carbon foam. Not surprisingly, the surface area of the nanographene increases with the degree of activation, yielding materials with BET surface areas in excess of 3000 m²/g [27]. Despite the high surface area, the polymer-derived nanographene has a relatively high density (~200 kg/m³, ~9% of the full density of bulk graphite) that makes the material mechanically very robust as demonstrated by nanoindentation tests that revealed modulus E and Meyer hardness values of 300–800 MPa and 20–80 MPa, respectively [26]. These values are 30–100 times higher than the highest values reported for graphene aerogels prepared by other methods (see next section).

The conversion of the graphitic carbon foam to nanographene can be followed by x-ray diffraction (XRD). The absence of the stacking-related (002) diffraction peak in the XRD pattern for activated nanographene material indicates the transition from a structure containing graphite nano-platelets to one consisting of single layer graphene (Figure 17.2). The in-plane crystallite size (L_a) of the graphene sheets is 2–5 nm as obtained by a Debye–Scherrer analysis of the (100) diffraction peak. These results are further confirmed by high-resolution transmission electron microscopy (HRTEM) and Raman spectroscopy. Examination of the material by HRTEM reveals short length, linear features which, due to their extremely small lateral extension and spacing, can be identified as individual graphene sheets viewed edge-on

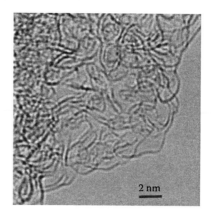

FIGURE 17.3 HRTEM images of nanographene.

(Figure 17.3). Consequently, the skeletal network of the material consists almost entirely of curved and intertwined monolayer graphene sheets; only in a few regions stacks of multiple (mostly just twin) graphene layers are observed. The Raman features for the graphene assembly are similar to those observed for commercial graphene samples grown by chemical vapor deposition (Figure 17.4). Analysis of the D/G band ratio (~3) suggests L_a values around ~5 nm (based on the Tuinstra–Koenig relationship). The broadness of the Raman D band indicates considerable bond disorder such as the presence of 5- and 7-membered rings within individual sheets. These defects explain the curved appearance of the graphene sheets in HRTEM images.

Filled and empty electronic states can be probed by soft-x-ray emission (SXE) and x-ray absorption spectroscopy (XAS), respectively. The SXE/XAS spectra for polymer-derived nanographene are very similar to those recorded from highly oriented pyrolytic graphite (HOPG) (Figure 17.5). This confirms that this material is composed primarily of sp²-bonded carbon, and that the states near the Fermi level have π/π* character. Significantly, the spectra collected from HOPG and the nanographene exhibit a comparable degree of overlap between filled (SXE) and empty (XAS) states at the Fermi level, suggesting that the nanographene is a semi-metal

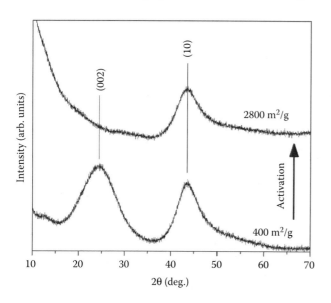

FIGURE 17.2 XRD data confirming the transformation of the initial multilayer graphene component to one that is dominated by single-layer graphene.

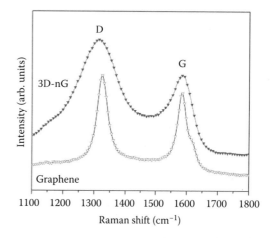

FIGURE 17.4 Comparison of the Raman spectra in the D/G band region from 3D-nG and multiple-layer graphene.

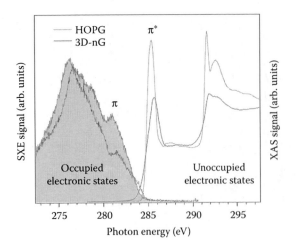

FIGURE 17.5 Valence and conductance band structure of 3D-nG ($2800\ m^2\ g^{-1}$) and a freshly cleaved highly oriented pyrolytic graphite standard probed by SXE (left) and XAS (right), respectively. The SXE excitation energy was set at 300.5 eV.

although one must exercise caution in inferring ground state properties from the XAS and SXE spectra due to the core–hole effects.

17.3 BOTTOM-UP SYNTHESIS: GO-DERIVED GRAPHENE ASSEMBLIES

Another approach to graphene assembly utilizes organic sol–gel chemistry [31] to covalently cross-link GO sheets. Instead of using sol–gel chemistry to glue the GO units together, these 3D graphene architectures are prepared by using the chemical functionality of GO to directly cross-link the network structure. The various functional groups (e.g., epoxide, hydroxide) abundant in GO sheets serve as chemical cross-linking sites for the 3D macroassembly network. Upon thermal reduction,

these cross-links are transformed into conductive carbon bridges that provide structural support for the assembly, while also limiting aggregation of the individual graphene sheets. A key feature of this approach is a 3D graphene assembly that is highly crystalline. The crystallite size depends on the size of the GO units, which is usually in the order of microns [24]. As such, these materials exhibited electrical conductivities and mechanical properties that were significantly higher than those of physically cross-linked structures of comparable density. In addition, the bulk properties of the graphene assemblies (porosity, surface area, conductivity) can be controlled through the cross-linking chemistry [24].

These 3D graphene assemblies are prepared by gelation of an aqueous GO suspension (1–2 wt%) under basic conditions. In a glass vial, 3 mL of the GO suspension are mixed with 500 µL concentrated NH_4OH. The vial is then sealed and placed in an oven at 85°C overnight. The resulting wet gel is washed in deionized water to purge NH_4OH followed by an exchange of water with acetone inside the pores. Supercritical CO_2 is used to dry the gels that are then converted into the final 3D graphene macroassembly by pyrolysis at 1050°C under nitrogen. Typical densities of the black monoliths are in the range of 60–100 mg/cm³.

Solid-state nuclear magnetic resonance (NMR) characterization has been used to gain insight into the types of functional groups in GO involved in the cross-linking process as well as to follow the reduction of GO to graphene (Figure 17.6a). The GO powder contains significant epoxide and hydroxyl functionality as evidenced by numerous peaks between 50 and 75 ppm, as well as carbonyl groups (168 ppm) and sp² carbon (123 ppm) in its ¹³C NMR spectrum. These peaks and assignments are consistent with the existing literature [35]. After gelation, the epoxide, hydroxyl, and carbonyl peaks are virtually eliminated and an aliphatic carbon peak (26 ppm) appears. The disappearance of the large peaks

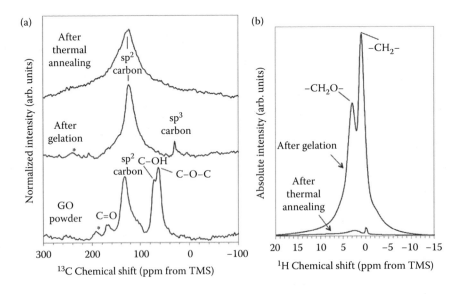

FIGURE 17.6 ¹³C and ¹H NMR spectra for GO powder, GO after initial gelation, and 3D graphene macroassembly. (Worsley, M.A. et al., Mechanically robust 3D graphene macroassembly with high surface area. *Chemical Communications*, 2012. Reproduced by permission of The Royal Society of Chemistry.)

between 50 and 75 ppm in the gel suggests that epoxide and hydroxyl groups are involved in the cross-linking mechanism. Conversely, the emergence of the aliphatic carbon (sp³) peak suggests that –CH₂- and/or –CH₂O- are formed during gelation. The –CH₂- and –CH₂O- moieties likely function as the cross-links that support the initial 3D GO network similar to the cross-links formed in RF sol–gel chemistry to form organic gels [34,36]. ¹H NMR spectra (Figure 17.6b) for the sample after gelation also support the presence of –CH₂- and –CH₂O- moieties with peaks at 0.9 and 3.1 ppm. The presence of –CH₂O- moieties is further supported by energy-dispersive x-ray (EDX) analysis, which measured 11 at% oxygen remaining in the initial aerogel. After pyrolysis, only the sp² carbon peak remains, suggesting that the sp³ carbon cross-links were thermally converted into conductive sp² carbon junctions, again analogous to the carbonization process that occurs during the pyrolysis of RF-based gels. The ¹H NMR spectrum (Figure 17.6b) supports the conversion of the –CH₂- and –CH₂O- moieties by virtual elimination of those peaks in the thermally treated sample. Finally, the reduction of carbon is confirmed by an oxygen content of less than 2 at%, as determined by EDX, in the final graphene assembly. The low

heteroatom concentration and predominance of sp² carbon is also supported by XAS results [24].

Field emission scanning electron microscopy (FE-SEM) has been used to determine the microstructure of the 3D graphene. These FE-SEM images (Figure 17.7) show that the 3D graphene monolith has a sheet-like microstructure similar to that reported in other graphene assemblies. In particular, the morphology resembles that of an RF-free graphene assembly reported to have a surface area in excess of 1000 m²/g, presumably due to minimal thickness of graphene sheets [24]. Nitrogen porosimetry results for the material are consistent with the morphology revealed by FE-SEM. The nitrogen adsorption/desorption isotherm shown in Figure 17.8a is Type IV, indicative of a mesoporous material. The observation of a Type 3 hysteresis loop (IUPAC classification) at high relative pressure is consistent with other 3D graphene materials, but the increased magnitude of the loop is indicative of a much larger pore volume than those reported for other graphene assemblies. The BET surface area for this type of graphene macroassembly is ~1300 m²/g or roughly half of the theoretical value expected for a single graphene sheet. The surface area is thus much higher than those of graphene assemblies made using sol–gel chemistry [22] suggesting that layering/overlapping of sheets can be significantly reduced by direct cross-linking. The reduction in layering of sheets is also consistent with an XRD pattern that lacks a strong (002) peak at ~28° (graphite interlayer spacing). The pore size distribution (Figure 17.8b) shows that much of the pore volume (4.0 cm³/g) lies between 3 and 10 nm, with a peak pore diameter at 6 nm. Finally, these textural properties can be modified via synthetic parameters (e.g., concentration of GO, RF, catalyst, etc.) [24].

The mechanical behavior of the 3D graphene macroassemblies has been determined by flat-punch nanoindentation. Figure 17.9 shows the stress vs. strain plot, revealing a mechanical behavior qualitatively similar to that of a carbon nanotube (CNT)-based assembly reported by Shin et al. [37].

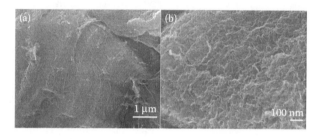

FIGURE 17.7 FE-SEM images of the 3D graphene macroassembly at (a) low and (b) high magnification. (Worsley, M.A. et al., Mechanically robust 3D graphene macroassembly with high surface area. *Chemical Communications*, 2012. Reproduced by permission of The Royal Society of Chemistry.)

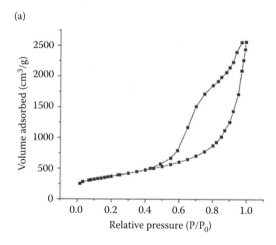
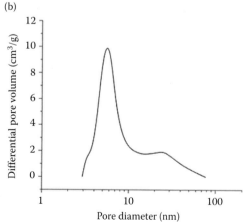

FIGURE 17.8 (a) Nitrogen adsorption/desorption isotherm and (b) pore size distribution for the 3D graphene macroassembly. (Worsley, M.A. et al., Mechanically robust 3D graphene macroassembly with high surface area. *Chemical Communications*, 2012. Reproduced by permission of The Royal Society of Chemistry.)

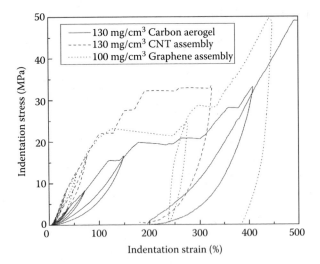

FIGURE 17.9 Representative load-displacement (stress-strain) curve of the 3D graphene assembly (with a density of 100 mg/cm³) indented with a flat punch tip with a diameter of 62 μm. Indentation was performed as a series of loading cycles with increasing maximum loads and complete unloading at the end of each load cycle. (Worsley, M.A. et al., Mechanically robust 3D graphene macroassembly with high surface area. *Chemical Communications*, 2012. Reproduced by permission of The Royal Society of Chemistry.)

Loading is characterized by an initial linear–elastic region, followed by a pronounced nonlinear–elastic region. Both shape and volume of the monolith are completely restored after the load is removed. Failure is indicated by a sudden jump of the strain at a constant stress (a "pop-in" event). Stress and strain at the initial stage of the first pop-in event can be assigned to the failure of stress and failure strain, respectively [37]. The graphene assembly has a Young's modulus of 51 ± 12 MPa, which is orders of magnitude higher than those reported for graphene assemblies made by other methods [13–15,18]. In addition to being extraordinarily stiff, the 3D graphene monoliths exhibit super-compressive behavior with failure strains of 57 ± 21% and a complete recovery for lower strains. The failure stress is 10.4 ± 3.9 MPa. These values of failure stress and strain are comparable to those of CNT-based aerogels of the same density (100 mg/cm³). These remarkable mechanical properties seem to be a consequence of the robustness and preponderance of sp² carbon cross-links between graphene sheets in combination with the excellent mechanical properties of the graphene sheets themselves.

Bulk electrical conductivity of the 3D graphene macroassembly, evaluated by the four-probe method, was measured at 100–200 S/m. This is consistent with carbon junctions cross-linking graphene sheets [22] and is orders of magnitude higher than for graphene assemblies made via physical cross-links. Cyclic voltammetry (CV) was used to characterize the energy storage capabilities of the graphene assemblies in aqueous electrolyte (5 M KOH). At low scan rates, the CVs exhibit the typical rectangular shape expected for pure double-layer capacitors like conventional carbon aerogels, as well as CNT and graphene assemblies. Analysis of the CVs

measured at low scan rates reveals a maximum capacitance of 165 F/g. Remarkably, the 500 μm thick 3D graphene electrode was able to maintain greater than 50% of its maximum capacitance (89 F/g) up to 100 mV/s, indicating an exceptionally fast charge/discharge capability. The 3D graphene has a maximum energy density of 27 Wh/kg and a maximum power density approaching 10 kW/kg. Further optimization of the electrodes, such as using thinner electrodes (e.g., 100 μm vs. 500 μm thickness), and electrolyte (e.g., inorganic vs. aqueous) could push the power and energy densities to ~10² kW/kg and ~10² Wh/kg, respectively. These observations illustrate the potential of 3D graphene for energy storage applications.

In summary, we have developed 3D graphene macroassembly materials using two general methods: (1) top-down and (2) bottom-up. The top-down methods used a sol–gel chemistry to simultaneously synthesize a network of nanographene. The bottom-up method began with GO sheets that were directly cross-linked into a macroassembly and reduced to graphene. Both methods produced novel structures with unique properties such as high electrical conductivities, large surface areas, and exceptional mechanical properties. In the next section, the incorporation of these novel assemblies into macroscale systems is demonstrated, giving enhanced and/or novel performance in a wide range of applications such as hydrogen storage, batteries, supercapacitors, desalination, actuators, electronics, and catalysis.

17.4 APPLICATIONS

17.4.1 HYDROGEN STORAGE

One area of carbon research that has received significant attention is the use of porous carbon materials as sorbents for hydrogen [38–42]. Safe and efficient storage of hydrogen is considered one of the main challenges associated with utilization of this fuel source in the transportation sector [43]. Two important criteria required for effective hydrogen physisorption are (1) a high surface area that exposes a large number of sorption sites to ad-atom or ad-molecule interaction [44] and (2) sufficiently deep potential wells so that the storage material can be utilized at reasonable operating temperatures. Porous carbons are promising candidates for hydrogen physisorption due to their lightweight frameworks and high accessible surface areas. The low hydrogen binding energies, however, that are typical of carbonaceous sorbents (~6 kJ/mol H_2), require that cryogenic temperatures (77 K) be utilized for storage of hydrogen in these materials. In general, the amount of surface excess hydrogen adsorbed on porous carbons at 77 K and ~3.5 MPa varies linearly with BET surface area, and the gravimetric uptake is ~1 wt% H_2 per 500 m²/g of surface area [42,45].

The ultra-high surface area, polymer-derived nanographene assemblies were originally developed for use as hydrogen physisorbents [27,46]. These materials, as described above, were prepared via the top-down approach, simultaneously building the network and graphene from polymer precursors. Activation with CO_2 then yields nanographene sorbents with BET surface areas in excess of 3000 m²/g. Presumably, edge termination sites constitute a substantial fraction of the surface area in these activated nanographenes, as is the case for traditional high surface area activated carbons. Hydrogen uptake at 77 K in the activated nanographenes scaled linearly with the BET surface area up to 2500 m²/g, yielding gravimetric densities up to 5 wt% H_2 (Figure 17.10), comparable to the highest values measured in porous carbons [42]. Above 2500 m²/g, the differential increase in the hydrogen storage capacity is smaller than expected, likely due to increase in the pore size in this ultra-high surface area material. Previous studies have shown that size and shape of the pores in hydrogen physisorbents play a critical role in hydrogen uptake, with the optimal structure having slit-shaped pores with diameters between 0.7 and 1 nm [47,48]. As shown in Figure 17.11, the activation process not only increases the surface area of the sorbent, but also shifts the pore size distribution to larger values. In addition to gravimetric capacity, volumetric capacity is an equally important consideration in the design of functional hydrogen sorbents. Depending on the density of the nanographene, the volumetric capacity of these materials can range from 10 to 29 g H_2/L [46]. While these values are on par with those of other porous carbon materials, further optimization of nanographene pore structure is required for increased hydrogen energy density.

The hydrogen binding enthalpies measured for the activated nanographenes were ~6 kJ/mol, as would be expected

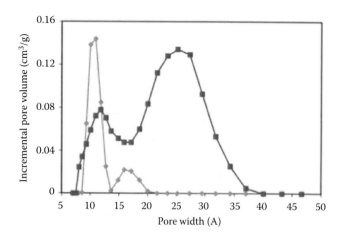

FIGURE 17.11 Pore-size distribution for activated nanographenes with BET surface areas of 1500 (light gray) and 3100 (dark gray) m²/g, as determined from the N_2 adsorption isotherm using the density functional theory (DFT) method. (Biener, J. et al., Advanced carbon aerogels for energy applications. *Energy and Environmental Science*, 2011. **4**(3): p. 656–667. Reproduced by permission of The Royal Society of Chemistry.)

for a carbon-based sorbent. As mentioned above, the low-binding energies associated with porous carbons are an obstacle to meeting capacity requirements at reasonable operating temperatures (>273 K). Previous work has shown that hydrogen adsorption energies near 15 kJ/mol, over the full range of surface coverage, are necessary to meet this requirement [49]. Numerous approaches have thus been employed to improve the thermodynamics of hydrogen binding in porous carbons while retaining large surface areas for sorption. The hydrogen spillover effect, for example, has been suggested as a mechanism to increase the reversible hydrogen storage capacity at room temperature in metal-loaded carbon nanostructures [50–53]. The spillover process involves the dissociative chemisorption of molecular hydrogen on a supported metal catalyst surface (e.g., platinum or nickel), followed by the diffusion of atomic hydrogen onto the surface of the carbon support. Alternatively, substitutional doping of carbon with boron or other light elements has also been presented as a promising route toward increasing hydrogen-binding energy in these sorbent materials [54,55]. The flexibility associated with nanographene synthesis allows for the incorporation of such modifiers into the carbon framework. For example, gas and solution phase deposition techniques have been developed that allow for the uniform incorporation of metal nanoparticles into the carbon framework. The performance of these modified nanographenes as next-generation hydrogen storage materials is currently being evaluated.

Beyond their use as hydrogen sorbents, nanographenes have also been used to enhance the performance of other solid-state hydrogen storage materials, specifically complex hydrides, such as alanates (AlH_4^-), amides (i.e., NH_2^-), and borohydrides (BH_4^-). Complex hydrides offer a number of advantages for the storage of hydrogen, including high gravimetric and volumetric hydrogen capacities [56,57]. The thermodynamics and kinetics associated with reversible hydrogen storage in these

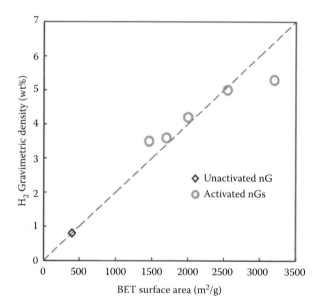

FIGURE 17.10 Excess gravimetric density (wt% H_2) saturation value at 77 K as a function of BET surface area for the activated nanographene (nG). The dotted line shows the correlation of 1 wt% H_2 per 500 m²/g. It is important to note that surface excess hydrogen values are a measure of H_2 adsorbed *on* the surface of the nanographene only and do not account for free hydrogen gas in the pores of the nanographene. Therefore, total gravimetric hydrogen capacities in these nanographene materials are higher than the surface excess values. (Biener, J. et al., Advanced carbon aerogels for energy applications. *Energy and Environmental Science*, 2011. **4**(3): p. 656–667. Reproduced by permission of The Royal Society of Chemistry.)

systems, however, present significant obstacles to their utilization as storage materials at reasonable operating temperatures and pressures. Incorporation of complex hydrides into the free pore volume of a nanoporous solid or scaffold has been shown to improve the rates of both hydrogenation and dehydrogenation for these materials (Figure 17.12). The enhanced kinetics can be attributed to the shorter diffusion distances for hydrogen as well as the other elements (i.e., Li, B, etc.) in the nanostructured hydride. Practical application of the scaffold approach requires the design of porous solids with small pore sizes to physically confine the nanostructured hydride as well as large accessible pore volumes to minimize the gravimetric and volumetric capacity penalties associated with the use of the scaffold in a storage system. In addition, these scaffold materials should be chemically inert, mechanically robust, and capable of managing thermal changes associated with the cycling of the incorporated hydride. While a variety of porous matrices have been investigated as scaffolds [58–62], nanographenes have emerged as one of the most promising candidates due to their large pore volume and tunable porosities, as well as for the ability to modify the surface characteristics of carbon framework. As described above, the pore structure of nanographenes can be controlled through the sol–gel polymerization conditions (i.e., R/C ratio, concentration of reactants), allowing for the fabrication of monolithic scaffolds with large internal pore volumes distributed over nanometer-sized pores. Recent work by several groups has demonstrated the promise of the nanographene scaffolding approach with a number of different hydrogen storage materials. For example, incorporation of $LiBH_4$ into nanographene scaffolds was shown to significantly enhance the rate of hydrogen exchange in the material relative to that of bulk $LiBH_4$, and that the effect is inversely correlated with the average pore size of the scaffold (i.e., smaller pores yielding faster kinetics) [63]. In fact, the rate of dehydrogenation for $LiBH_4$ was shown to increase by as much as 50 times at 300°C in a nanographene with 13 nm pores. Similar improvements in dehydrogenation kinetics were also reported for $NaAlH_4$ incorporated into a small pore carbon scaffold [64,65]. In addition, the $NaAlH_4$ within the scaffold, unlike bulk $NaAlH_4$, could be readily rehydrogenated to full capacity at ~160°C under 100 bar H_2. The desorption kinetics of other solid-state hydrogen storage materials, such as MgH_2 and NH_3BH_3, were also affected considerably when embedded within the pores of a carbon scaffold [66–70]. Despite these promising results, additional efforts in this area are necessary to better understand the influence of the porous carbon textural properties and surface chemistry on the performance of the incorporated hydride. Further optimization of the nanographene architecture is also required for the design of robust scaffolds that can accommodate larger weight fractions of the complex hydride. These structural refinements present a challenging trade-off in terms of porosity and mechanical properties. Increasing the pore volume in these scaffolds, while maintaining small pore sizes, requires that the walls defining the pore structure be very thin. The thickness of the wall structure, in turn, determines the mechanical integrity of the material. This aspect of scaffold design is an important consideration, as these materials need to have sufficient mechanical strength to withstand the stresses associated with infiltration and cycling of the hydride. The recent development of bottom-up graphene assemblies with both large pore volumes, small pore sizes, and mechanical robustness represent a promising option for improving the performance of metal and chemical hydride systems that is currently under evaluation.

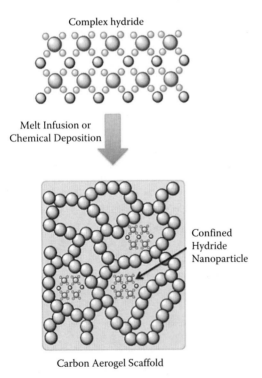

Complex hydride

Melt Infusion or
Chemical Deposition

Confined
Hydride
Nanoparticle

Carbon Aerogel Scaffold

FIGURE 17.12 Scaffolding approach involves the confinement of nanostructured hydride material within the free pore volume of a nanoporous scaffold material, such as a carbon aerogel. (Biener, J. et al., Advanced carbon aerogels for energy applications. *Energy and Environmental Science*, 2011. **4**(3): p. 656–667. Reproduced by permission of The Royal Society of Chemistry.)

17.4.2 Batteries

3D graphene assemblies show promise in several battery applications. Recently, graphene and metal-oxide functionalized graphene has been heavily investigated for use as anode in Li-ion batteries. These materials boast very high capacities of up to 1000 mAh/g, but in some cases show lower voltages and lower charge efficiencies. Even on metal-oxide functionalized anodes, the carbon appears to be involved in the charge storage, but the storage mechanism yet remains to be explained.

Graphene assemblies may also find future use in batteries as current collectors with tunable porosities [26,71,72] or scaffolds for 3D intercalated batteries [73]. In the first function, they provide advantages for the power density of the device: the tunable porosities can be used to minimize diffusion resistance while maintaining a constant surface area, thus reducing internal resistance as well as polarization potentials and increasing the power density of the device. At the same time, the aerogel can function as a low-density current collector that is directly functionalized and eliminates the need for binder and filler, which increases conductance of electrode as well as the energy density. Graphene assemblies also provide a means to utilize lithium intercalation materials usually limited by large volumetric changes, such as Sn or Si, which would be coated onto the current collector in very thin layers or as nanoparticles or nanowires, and thus subjected to lower stresses [74,75]. The drawback to using macroscopic assemblies is that they are not compatible with current battery fabrication processes, and thus are not likely to be implemented unless substantial performance improvements can be demonstrated.

17.4.3 Supercapacitors

The most common application for 3D graphene in energy storage is in the electrical double-layer capacitor (EDLC) [76]. In these devices, charge is stored in the form of ions accumulated on the surface of the material (Figure 17.13), thus creating an intermediate between batteries and electrostatic capacitors [77]. The energy density is lower than that of a battery, but higher than that of an electrostatic capacitor. The inverse is true for power density. While the high-power density and cyclability make supercapacitors very attractive, their low-energy density and high cost has limited their application so far.

Owing to their high surface areas and good conductivity, 3D graphene macroassemblies are ideal candidates for supercapacitors. However, surface area does not drive capacity alone:

pore size plays an important role as well, and, if the assembly is used as a monolith, its density is important, too. As the pore size decreases, the area-specific capacitance decreases as well [78] as the access of solvated ions gets increasingly sterically hindered and part of the surface can no longer be utilized. Once the pores become smaller than the hydrated ion radius, the specific capacitance increases again due to the distortion and reduction in size of the ion solvation sheath of ions in sub-nm pores [79]. Since the resistance of the device is dominated by the electrolyte, pore size, tortuosity, length, and shape of the pores all affect the resistance of the capacitor and thus its power density. Recent experiments have demonstrated that, while meso- and macro-scale pores are generally important for mass transport and power density, the macropore size can be reduced to 100 nm and less before there is a measurable impact on the resistance of the electrolyte.

In general, 3D graphene assemblies are ideally polarizable high capacitance electrodes. The specific capacitance depends on the synthesis route of the material. Materials that are derived from RF aerogels, whether they include fillers such as CNTs and graphene or not, appear to all fall on a similar curve (Figure 17.14). Materials derived from cross-linked GO do not fall on this curve, and instead maintain a specific capacitance of 12 $\mu F/cm^2$. This difference may result from different pore-size distributions: the GO-derived material has significant populations of both meso- (10–100 nm) and micropores (<2 nm), while the RF-derived material has primarily sub-nanometer and macropores (>100 nm), and increases in

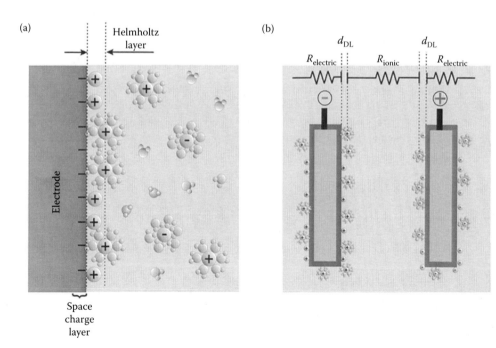

FIGURE 17.13 (a) Electric double layer (EDL) model showing the accumulation of cations at a negatively polarized electrode surface in contact with an electrolyte. In the case of an ideally polarizable electrode, no charge transfer takes place across the electrode/electrolyte interface, leading to a capacitor-like separation of positive and negative charges. The atomic-scale dimensions of the EDL results in much higher capacities compared to that of a conventional parallel-plate capacitor. (b) An EDLC consists of two EDL capacities in series (one at each electrode). The electrode needs to be chemically inert, exhibit a high surface area, and allow for fast diffusional mass and electronic transport. (Biener, J. et al., Advanced carbon aerogels for energy applications. *Energy and Environmental Science*, 2011. **4**(3): p. 656–667. Reproduced by permission of The Royal Society of Chemistry.)

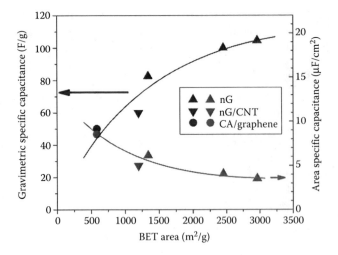

FIGURE 17.14 Gravimetric and area-specific capacitance vs. surface area for activated nanographene (nG) (▲) as well as activated nG/CNT (▼) and carbon aerogel (CA)/graphene (●) composite materials. The gravimetric specific capacitance increases and the area specific capacitance decreases with increasing surface area. (Biener, J. et al., Advanced carbon aerogels for energy applications. *Energy and Environmental Science*, 2011. **4**(3): p. 656–667. Reproduced by permission of The Royal Society of Chemistry.)

surface area are attained by creating more small pores while gradually widening the larger pores.

One major drawback of 3D graphene materials for the EDLC application is their low density (<500 mg/cm³) that limits the achievable volumetric energy density. At these low densities, the energy density of the device is dominated by the mass of electrolyte that fills the pores, rather than the mass weight of the active material. A strategy to increase the material density that was recently shown to be effective with GO-derived material is compression of the material. The macroassemblies could be compressed from 0.07 to 0.7 g/cm³ without any decrease in specific capacitance or power density.

17.4.4 Capacitive Deionization

3D graphene macroassemblies also have considerable potential to significantly improve desalination efficiency through a technique known as capacitive deionization (CDI). Like the supercapacitor, CDI relies on the formation of an EDL to store charge (Figure 17.13). In CDI, however, the goal of the charge storage is not energy storage, but charge removal from the electrolyte: the electrolyte is sea water or brackish water *flowing* between the electrode pairs, and the charging of the electrodes and formation of the double layer significantly *depletes* ions concentration in the electrolyte. CDI has potentially significant advantages over other, more conventional, water desalination techniques such as reverse osmosis (RO): CDI does not require high pressures or large amount of infrastructure for efficient desalination, the materials are more robust than RO membranes, and the energy cost scales with the salinity of the feed solution.

Electrodes for CDI have different requirements than for supercapacitors. Capacitance is still important, as it determines the maximum salt concentration that can be desalinated with a single charge. However, electrode resistance is a lot more important here, since the energy consumption of the method is directly proportional to device resistance. Finally, for the recently developed flow-through electrode CDI, in which the feed passes directly through the electrodes rather than between them, the macropore size is also of critical importance, since it determines the pressure required for the process. 3D graphene macroassemblies fulfill all of these requirements and are ideally suited as CDI electrodes.

17.4.5 Charging Induced Macroscopic Effects: Actuation and Electrode Resistivity

In the supercapacitor/desalination applications discussed above it was assumed that graphene-based electrodes are ideally polarizable with no charge transfer across the electrode/electrolyte interface, and that the only effect of the electrode/

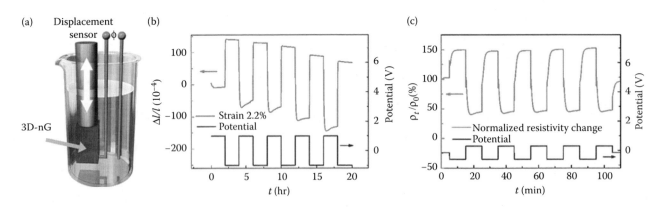

FIGURE 17.15 Interfacial charging induced macroscopic electrode response. (a) Schematic presentation of in situ strain measurements performed in a dilatometer equipped with an electrochemical cell; (b) potential (black, right ordinate) induced macroscopic strain response Δl/l (gray, left ordinate) vs. time *t* in a 0.7 M NaF electrolyte solution, potential jumps between −1 V and +1 V; (c) potential (black, right ordinate) induced electrode resistivity (gray, left ordinate) changes vs. time in 0.1 M NaClO₄ aqueous electrolyte solution, potential range from −0.6 V to 0.3 V vs. Ag/AgCl. (Biener, J. et al.: Macroscopic 3D nanographene with dynamically tunable bulk properties. *Advanced Materials*. 2012. **24**. 5083–5087. Copyright Wiley-VCH Verlag GmbH & Co. KGaA. Reproduced with permission.)

electrolyte interface polarization is an electron/cation accumulation at negative applied potentials and an electron depletion/anion accumulation at positive applied potentials. Potential-induced effects on the electronic structure of the electrode beyond this simple electron accumulation/depletion are usually not considered despite the presence of very strong interfacial electric fields at the charged electrode/electrolyte interface. But even if the electronic structure does not change, electron accumulation/depletion can change the physical and chemical properties of the interfacial electrode surface layer as electrons are added to or removed from electronic states with bonding or antibonding character which, for example, can lead to a macroscopic strain response. This effect that effectively converts electric energy into mechanical work and thus could be used for actuation [26,72] has first been observed for nanoporous metals [80]. More recently, a charge-induced macroscopic strain response has also been observed for graphene-based electrodes. Utilization of this charge-induced macroscopic strain effect for actuation requires a material that is mechanically robust and stiff to achieve a high mechanical work density, and has a high surface area for large strain amplitudes. This requirement makes 3D graphene assemblies very promising materials for this application as they combine mechanical robustness with an ultra-high surface area: The exceptional mechanical properties of the structural building block graphene makes them mechanically robust and stiff, and the 2D structure of graphene provides the required ultra-high surface area. In the case of monolithic, polymer-derived nanographene materials reversible strain amplitudes as large as 2.2% (corresponding to a volume strain of 6.6%) have been achieved while cycling the potential between -1.0 V and $+1.0$ V (Figure 17.15) [10]. The 3D nanographene electrode expands during negative charging and contracts during positive charging. Both sign and the maximum value of the strain amplitude are in agreement with graphite intercalation data, where it was found that the C–C in-plane bond length increases if electrons are added into the graphite π bands and vice versa [81]. The strain amplitude is considerably larger than those reported for CNT arrays [82], nanoporous metals [80], and piezoelectric materials [83]. Potential induced electron accumulation/depletion does also affect the conductivity of graphene-derived electrodes, and reversible charging induced conductivity changes of up to 300% have been reported [26]. This effect can be understood in terms of charge accumulation/depletion induced shifts of the Fermi level toward regions with higher density of states.

17.4.6 CATALYSIS

3D graphene assemblies are also very promising nanomaterials for catalytic and electrocatalytic applications where they function as both advanced support and active catalyst itself. Most studies have focused on electrocatalysis, with special emphasis on the oxygen reduction reaction (ORR) that is of critical importance for fuel cells and metal-air batteries. But 3D graphene assemblies have also been used as highly efficient and transparent counter electrodes for dye-sensitized solar cells [84] as well as catalyst materials for various oxidation

reactions including the oxidation of nitric oxide and the ozonation of organic compounds. As advanced support materials, 3D graphene provides excellent electrical conductivity, high surface area, good chemical stability and, if required, optical transparency, all combined with a tunable open pore morphology that can be optimized for fast mass transport. For example, nitrogen-doped 3D graphene aerogels decorated with Fe_3O_4 nanoparticles have shown excellent electrocatalytic performance for the oxygen-reduction reaction [85]. 3D graphene assemblies have also been used as metal-free catalysts mostly via modification by doping with heteroatoms such as B, N, and P [86–88]. Graphitic and pyridinic (edge sites) nitrogen centers seem to be responsible for the reported enhancement of the ORR activity of N-doped graphene-based catalysts [86,87]. Additional doping with B and P seems to further increase the ORR activity of N-doped graphene-based catalysts, and it was proposed that this is caused by reducing the band gap between the highest occupied (HOMO) and the lowest unoccupied molecular orbitals (LUMO) [86]. However, it needs to be emphasized that even undoped GO-derived graphene materials show catalytic activity due to the presence of a large number of oxygen-containing functional groups and structural defects [89]. The advantage of 3D graphene over other more traditional carbon supports such as commercial carbon blacks is that their surface area, pore size, and pore volume can be tailored independently from each other [23,24]. Compared to polymer-derived carbon aerogels, 3D graphene also offers superior chemical stability due to its larger domain size, which also reduces electric losses in the electrode. If loading with another active catalyst material is desired, 3D graphene assemblies can be uniformly decorated with metal or metal oxide films/nanoparticles using atomic layer deposition [90] or wet chemistry deposition techniques [91–95].

REFERENCES

1. Li, X.L. et al., Highly conducting graphene sheets and Langmuir–Blodgett films. *Nature Nanotechnology*, 2008. **3**(9): p. 538–542.
2. Geim, A.K. and K.S. Novoselov. The rise of graphene. *Nature Materials*, 2007. **6**(3): p. 183–191.
3. Peigney, A. et al., Specific surface area of carbon nanotubes and bundles of carbon nanotubes. *Carbon*, 2001. **39**(4): p. 507.
4. Lee, C. et al., Measurement of the elastic properties and intrinsic strength of monolayer graphene. *Science*, 2008. **321**(5887): p. 385–388.
5. Pumera, M., Graphene-based nanomaterials for energy storage. *Energy and Environmental Science*, 2011. **4**(3): p. 668–674.
6. Zhu, Y.W. et al., Carbon-based supercapacitors produced by activation of graphene. *Science*, 2011. **332**(6037): p. 1537–1541.
7. Xia, F.N. et al., Graphene field-effect transistors with high on/off current ratio and large transport band gap at room temperature. *Nano Letters*, 2010. **10**(2): p. 715–718.
8. Schwierz, F., Graphene transistors. *Nature Nanotechnology*, 2010. **5**(7): p. 487–496.
9. Zhang, L.S. et al., Mono dispersed SnO(2) nanoparticles on both sides of single layer graphene sheets as anode materials in Li-ion batteries. *Journal of Materials Chemistry*, 2010. **20**(26): p. 5462–5467.

10. Shao, L.-H. et al., Electrically tunable nanoporous carbon hybrid actuators. *Advanced Functional Materials*, 2012. **22**(14), p. 3029–3034.

11. Schedin, F. et al., Detection of individual gas molecules adsorbed on graphene. *Nature Materials*, 2007. **6**(9): p. 652–655.

12. Bai, H. et al., A pH-sensitive graphene oxide composite hydrogel. *Chemical Communications*, 2010. **46**(14): p. 2376–2378.

13. Sheng, K.X. et al., High-performance self-assembled graphene hydrogels prepared by chemical reduction of graphene oxide. *New Carbon Materials*, 2011. **26**(1): p. 9–15.

14. Xu, Y. et al., Self-assembled graphene hydrogel via a one-step hydrothermal process. *ACS Nano*, 2010. **4**(7): p. 4324.

15. Tang, Z.H. et al., Noble-metal-promoted three-dimensional macroassembly of single-layered graphene oxide. *Angewandte Chemie-International Edition*, 2010. **49**(27): p. 4603–4607.

16. Sun, H.Y., Z. Xu, and C. Gao, Multifunctional, ultra-flyweight, synergistically assembled carbon aerogels. *Advanced Materials*, 2013. **25**(18): p. 2554–2560.

17. Wang, J. and M. Ellsworth, Graphene aerogels. *ECS Transactions*, 2009. **19**(5): p. 241.

18. Zhang, X. et al., Mechanically strong and highly conductive graphene aerogel and its use as electrodes for electrochemical power sources. *Journal of Materials Chemistry*, 2011. **21**(18): p. 6494–6497.

19. Chen, Z.P. et al., Three-dimensional flexible and conductive interconnected graphene networks grown by chemical vapour deposition. *Nature Materials*, 2011. **10**(6): p. 424–428.

20. Liu, F. and T.S. Seo, A controllable self-assembly method for large-scale synthesis of graphene sponges and free-standing graphene films. *Advanced Functional Materials*, 2010. **20**(12): p. 1930–1936.

21. Qiu, L. et al., Biomimetic superelastic graphene-based cellular monoliths. *Nature Communications*, 2012. **3**. Article number: 1241.

22. Worsley, M.A. et al., Synthesis of graphene aerogel with high electrical conductivity. *Journal of the American Chemical Society*, 2010. **132**(40): p. 14067–14069.

23. Worsley, M.A. et al., Mechanically robust 3D graphene macroassembly with high surface area. *Chemical Communications*, 2012. **48**, 8428–8430.

24. Worsley, M.A. et al., High surface area, sp²-cross-linked three-dimensional graphene monoliths. *The Journal of Physical Chemistry Letters*, 2011. **2**(8): p. 921–925.

25. Worsley, M.A. et al., Toward macroscale, isotropic carbons with graphene-sheet-like electrical and mechanical properties. *Advanced Functional Materials*, 2014. **24**(27): p. 4259–4264.

26. Biener, J. et al., Macroscopic 3D nanographene with dynamically tunable bulk properties. *Advanced Materials*, 2012. **24**(37): p. 5083–5087.

27. Baumann, T.F. et al., High surface area carbon aerogel monoliths with hierarchical porosity. *Journal of Non-Crystalline Solids*, 2008. **354**(29): p. 3513–3515.

28. Biener, J. et al., Advanced carbon aerogels for energy applications. *Energy and Environmental Science*, 2011. **4**(3): p. 656–667.

29. Fricke, J. and T. Tillotson, Aerogels: Production, characterization, and applications. *Thin Solid Films*, 1997. **297**(1–2): p. 212–223.

30. Li, W.C., A.H. Lu, and S.C. Guo, Characterization of the microstructures of organic and carbon aerogels based upon mixed cresol-formaldehyde. *Carbon*, 2001. **39**(13): p. 1989–1994.

31. Pekala, R.W. et al., Aerogels derived from multifunctional organic monomers. *Journal of Non-Crystalline Solids*, 1992. **145**(1–3): p. 90–98.

32. Pekala, R.W. et al., New organic aerogels based upon a phenolic-furfural reaction. *Journal of Non-Crystalline Solids*, 1995. **188**(1–2): p. 34–40.

33. Quignard, F., R. Valentin, and F. Di Renzo, Aerogel materials from marine polysaccharides. *New Journal of Chemistry*, 2008. **32**(8): p. 1300–1310.

34. Al-Muhtaseb, S.A. and J.A. Ritter, Preparation and properties of resorcinol–formaldehyde organic and carbon gels. *Advanced Materials*, 2003. **15**(2): p. 101–114.

35. Gao, W. et al., New insights into the structure and reduction of graphite oxide. *Nature Chemistry*, 2009. **1**(5): p. 403–408.

36. Pekala, R.W. and F.M. Kong, Resorcinol-formaldehyde aerogels and their carbonized derivatives. *Abstracts of Papers of the American Chemical Society*, 1989. **197**: p. 113-POLY.

37. Shin, S.J. et al., Mechanical deformation of carbon nanotube-based aerogels. *Carbon*, 2012.

38. McNicholas, T.P. et al., H-2 storage in microporous carbons from PEEK precursors. *Journal of Physical Chemistry C*, 2010. **114**(32): p. 13902–13908.

39. Guan, C. et al., Characterization of a zeolite-templated carbon for H-2 storage application. *Microporous and Mesoporous Materials*, 2009. **118**(1–3): p. 503–507.

40. Xia, K.S. et al., Activation, characterization and hydrogen storage properties of the mesoporous carbon CMK-3. *Carbon*, 2007. **45**(10): p. 1989–1996.

41. Yang, Z.X., Y.D. Xia, and R. Mokaya, Enhanced hydrogen storage capacity of high surface area zeolite-like carbon materials. *Journal of the American Chemical Society*, 2007. **129**(6): p. 1673–1679.

42. Panella, B., M. Hirscher, and S. Roth, Hydrogen adsorption in different carbon nanostructures. *Carbon*, 2005. **43**(10): p. 2209–2214.

43. Schlapbach, L. and A. Züttel, Hydrogen-storage materials for mobile applications. *Nature*, 2001. **414**(6861): p. 353–358.

44. Denbigh, K., *The Principles of Chemical Equilibrium*, 1971, Cambridge University Press, New York.

45. Chahine, R. and P. Benard, in *Advances in Cryogenic Engineering*, P. Kittel, Editor. 1998, Plenum Press, New York. p. 1257.

46. Kabbour, H. et al., Toward new candidates for hydrogen storage: High-surface-area carbon aerogels. *Chemistry of Materials*, 2006. **18**(26): p. 6085–6087.

47. Patchkovskii, S. et al., Graphene nanostructures as tunable storage media for molecular hydrogen. *Proceedings of the National Academy of Sciences of the United States of America*, 2005. **102**(30): p. 10439–10444.

48. Murata, K. et al., Adsorption mechanism of supercritical hydrogen in internal and interstitial nanospaces of single-wall carbon nanohorn assembly. *Journal of Physical Chemistry B*, 2002. **106**(43): p. 11132–11138.

49. Bhatia, S.K. and A.L. Myers, Optimum conditions for adsorptive storage. *Langmuir*, 2006. **22**(4): p. 1688–1700.

50. Stadie, N.P. et al., Measurements of hydrogen spillover in platinum doped superactivated carbon. *Langmuir*, 2010. **26**(19): p. 15481–15485.

51. Li, Y.W. and R.T. Yang, Hydrogen storage on platinum nanoparticles doped on superactivated carbon. *Journal of Physical Chemistry C*, 2007. **111**(29): p. 11086–11094.

52. Yoo, E. et al., Atomic hydrogen storage in carbon nanotubes promoted by metal catalysts. *Journal of Physical Chemistry B*, 2004. **108**(49): p. 18903–18907.

53. Robell, A.J., E.V. Ballou, and M. Boudart, Surface diffusion of hydrogen on carbon. *Journal of Physical Chemistry*, 1964. **68**(10): p. 2748–2753.

54. Jin, Z. et al., Solution-phase synthesis of heteroatom-substituted carbon scaffolds for hydrogen storage. *Journal of the American Chemical Society*, 2010. **132**: p. 15246–15251.

55. Kim, Y.H. et al., Nondissociative adsorption of H_2 molecules in light-element-doped fullerenes. *Physical Review Letters*, 2006. **96**(1): p. 016102.

56. Orimo, S.I. et al., Complex hydrides for hydrogen storage. *Chemical Reviews*, 2007. **107**(10): p. 4111–4132.

57. Züttel, A. et al., LiBH4 a new hydrogen storage material. *Journal of Power Sources*, 2003. **118**(1–2): p. 1–7.

58. Zheng, S.Y. et al., Hydrogen storage properties of space-confined NaAlH4 nanoparticles in ordered mesoporous silica. *Chemistry of Materials*, 2008. **20**(12): p. 3954–3958.

59. de Jongh, P.E. et al., The preparation of carbon-supported magnesium nanoparticles using melt infiltration. *Chemistry of Materials*, 2007. **19**(24): p. 6052–6057.

60. Balde, C.P. et al., Facilitated hydrogen storage in $NaAlH_4$ supported on carbon nanoribers. *Angewandte Chemie-International Edition*, 2006. **45**(21): p. 3501–3503.

61. Gutowska, A. et al., Nanoscaffold mediates hydrogen release and the reactivity of ammonia borane. *Angewandte Chemie-International Edition*, 2005. **44**(23): p. 3578–3582.

62. Heung, L.K. and C.G. Wicks, Silica embedded metal hydrides. *Journal of Alloys and Compounds*, 1999. **293**: p. 446–451.

63. Gross, A.F. et al., Enhanced hydrogen storage kinetics of LiBH4 in nanoporous carbon scaffolds. *Journal of Physical Chemistry C*, 2008. **112**(14): p. 5651–5657.

64. Stephens, R.D. et al., The kinetic enhancement of hydrogen cycling in NaAlH4 by melt infusion into nanoporous carbon aerogel. *Nanotechnology*, 2009. **20**(20): p. 204018.

65. Schüth, F., B. Bogdanovic, and A. Taguchi. 2003. *Patent Application WO2005014469.*

66. Nielsen, T.K. et al., Confinement of MgH2 nanoclusters within nanoporous aerogel scaffold materials. *ACS Nano*, 2009. **3**(11): p. 3521–3528.

67. Zhang, S. et al., The synthesis and hydrogen storage properties of a MgH2 incorporated carbon aerogel scaffold. *Nanotechnology*, 2009. **20**(20): p. 204027.

68. Bogerd, R. et al., The structural characterization and H-2 sorption properties of carbon-supported Mg1-xNix nanocrystallites. *Nanotechnology*, 2009. **20**(20): p. 204019.

69. Gross, A.F. et al., Fabrication and hydrogen sorption behaviour of nanoparticulate MgH_2 incorporated in a porous carbon host. *Nanotechnology*, 2009. **20**(20): p. 204005.

70. Sepehri, S. et al., Spectroscopic studies of dehydrogenation of ammonia borane in carbon cryogel. *Journal of Physical Chemistry B*, 2007. **111**(51): p. 14285–14289.

71. Sakamoto, J.S. and B. Dunn, Hierarchical battery electrodes based on inverted opal structures. *Journal of Materials Chemistry*, 2002. **12**: p. 2859–2861.

72. Shao, L.H. et al., Electrocapillary maximum and potential of zero charge of carbon aerogel. *Physical Chemistry Chemical Physics*, 2010. **12**(27): p. 7580–7587.

73. Long, J.W. et al., Three-dimensional battery architectures. *Chemical Reviews*, 2004. **104**(10): p. 4463–4492.

74. Bazin, L. et al., High rate capability pure Sn-based nano-architectured electrode assembly for rechargeable lithium batteries. *Journal of Power Sources*, 2009. **188**(2): p. 578–582.

75. Cheah, S.K. et al., Self-supported three-dimensional nano-electrodes for microbattery applications. *Nano Letters*, 2009. **9**(9): p. 3230–3233.

76. Pekala, R.W. et al., Carbon aerogels for electrochemical applications. *Journal of Non-Crystalline Solids*, 1998. **225**(1): p. 74–80.

77. Conway, B.E., Transition from "Supercapacitor" to "Battery" behavior in electrochemical energy storage. *Journal of the Electrochemical Society*, 1991. **138**(6): p. 1539–1548.

78. Gamby, J. et al., Studies and characterisations of various activated carbons used for carbon/carbon supercapacitors. *Journal of Power Sources*, 2001. **101**(1): p. 109–116.

79. Chmiola, J. et al., Anomalous increase in carbon capacitance at pore sizes less than 1 nanometer. *Science*, 2006. **313**(5794): p. 1760–1763.

80. Weissmüller, J. et al., Charge-induced reversible strain in a metal. *Science*, 2003. **300**(5617): p. 312–315.

81. Sun, G.Y. et al., Dimensional change as a function of charge injection in graphite intercalation compounds: A density functional theory study. *Physical Review B*, 2003. **68**: p. 125411.

82. Baughman, R.H. et al., Carbon nanotube actuators. *Science*, 1999. **284**(5418): p. 1340–1344.

83. Zhang, Q.M., V. Bharti, and X. Zhao, Giant electrostriction and relaxor ferroelectric behavior in electron-irradiated poly(vinylidene fluoride-trifluoroethylene) copolymer. *Science*, 1998. **280**(5372): p. 2101–2104.

84. Cheng, W.-Y., C.-C. Wang, and S.-Y. Lu, Graphene aerogels as a highly efficient counter electrode material for dye-sensitized solar cells. *Carbon*, 2013. **54**(0): p. 291–299.

85. Wu, Z.-S. et al., 3D Nitrogen-doped graphene aerogel-supported Fe3O4 nanoparticles as efficient electrocatalysts for the oxygen reduction reaction. *Journal of the American Chemical Society*, 2012. **134**(22): p. 9082–9085.

86. Choi, C.H. et al., B, N- and P, N-doped graphene as highly active catalysts for oxygen reduction reactions in acidic media. *Journal of Materials Chemistry A*, 2013. **1**(11): p. 3694–3699.

87. Lai, L. et al., Exploration of the active center structure of nitrogen-doped graphene-based catalysts for oxygen reduction reaction. *Energy and Environmental Science*, 2012. **5**(7): p. 7936–7942.

88. Lin, Z. et al., 3D Nitrogen-doped graphene prepared by pyrolysis of graphene oxide with polypyrrole for electrocatalysis of oxygen reduction reaction. *Nano Energy*, 2013. **2**(2): p. 241–248.

89. Chen, D., H. Feng, and J. Li, Graphene oxide: Preparation, functionalization, and electrochemical applications. *Chemical Reviews*, 2012. **112**(11): p. 6027–6053.

90. King, J.S. et al., Ultralow loading Pt nanocatalysts prepared by atomic layer deposition on carbon aerogels. *Nano Letters*, 2008. **8**(8): p. 2405–2409.

91. Han, T.Y.J. et al., Synthesis of ZnO coated activated carbon aerogel by simple sol-gel route. *Journal of Materials Chemistry*, 2011. **21**(2): p. 330–333.

92. Worsley, M.A. et al., Carbon scaffolds for stiff and highly conductive monolithic oxide-carbon nanotube composites. *Chemistry of Materials*, 2011. **23**(12): p. 3054–3061.

93. Worsley, M.A. et al., Route to high surface area TiO2/C and TiCN/C composites. *Journal of Materials Chemistry*, 2009. **19**(38): p. 7146–7150.

94. Worsley, M.A. et al., High surface area nanocarbon-supported metal oxide aerogels. *Abstracts of Papers of the American Chemical Society*, 2010. **239**. Meeting Abstract: 19-CATL.

95. Worsley, M.A. et al., Synthesis and characterization of monolithic, high surface area SiO2/C and SiC/C composites. *Journal of Materials Chemistry*, 2010. **20**(23): p. 4840–4844.

18 3D-AFM-Hyperfine Imaging of Graphene Monolayers Deposit on YBCO-Superconducting Surface

Khaled M. Elsabawy

CONTENTS

Abstract 277
18.1 Introduction 277
18.2 Experiments and Discussion 279
    18.2.1 Synthesis and Characterization of 123-YBCO Superconducting Substrate 279
    18.2.2 Phase Identification 280
    18.2.3 Superconducting Measurements 281
    18.2.4 Raman Spectroscopy Measurements 281
    18.2.5 Graphene Deposition on 123-YBCO Substrate 281
    18.2.6 Raman and Nano-Structural Features of Graphene Monolayers 282
    18.2.7 New Trend of AFM-Nano-Imaging View for Accurate Surface Topology 282
    18.2.8 DC-Electrical Conductivity Measurements 285
18.3 Conclusions 286
References 286

ABSTRACT

The current investigations show a new approach in applying AFM microscopy to visualize real 3D-high-resolution imaging of graphene layers deposited via chemical vapor deposition (CVD) technique (ICVD) on 123-YBCO superconducting surface. The accurate analysis of graphene surface topology indicated that the graphene thickness varies throughout the surface sample, which confirms that the rate of graphene deposition via CVD process is not unified. According to this information, we expect that conduction quality could be changed from point to point in the same surface of graphene deposit. Many of the structural and electrical parameters will be discussed and interpreted.

18.1 INTRODUCTION

Graphene has several interesting physical, electrical, and structural features owing to its exotic chemical properties and proposed applications in field-effect transistors, high-speed analog electronics, ultra-sensitive chemical detectors, interconnects, and spintronic devices [1–9]. One of the major hurdles in graphene research is the difficulty of accurately counting the number of atomic layers in samples obtained by either mechanical exfoliation from bulk graphite or grown by some other means. The ability to see graphene on Si/SiO_2 substrates with a certain thickness of oxide layer in an optical microscope was instrumental in the initial boom in graphene

research. At the same time, optical inspection has proven to be a rather difficult—if not impossible—technique to definitively identify the number of layers of graphene. Typically, single and bilayer graphene flakes are outnumbered by much thicker graphene flakes, which make the search for graphene a formidable task. Atomic force microscopy (AFM) alone may not clearly identify the number of graphene layers. Other alternatives include low-temperature transport studies or cross-sectional transmission electron microscopy (TEM). The major disadvantages of these methods include lengthy and involved experimental or sample preparation procedures.

Recently, micro-Raman spectroscopy has become a conventional technique for the identification and characterization of graphene layers [10–13]. It is a fast, nondestructive, high-throughput, and unambiguous approach. The Raman spectrum of graphene is very sensitive to the number of atomic layers and the presence of disorder or defects, which allows for accurate graphene characterization. Most Raman spectroscopic studies of graphene have been carried out for graphene on standard Si/SiO_2 substrates with 300 nm thickness of the oxide layer. These substrates ensured graphene visibility under optical microscopes.

Two techniques have shown a great potential for the accomplishment of such a task: the thermal decomposition of SiC surfaces [3] and [4], and CVD on catalytic metal surfaces [5] and [6]. However, they both have inherent drawbacks. The former allows the synthesis of epitaxial graphene on both (0 0 0 1) and (0 0 0 $\bar{1}$) polar faces of hexagonal SiC, but requires high temperatures (above 1300°C) and can obviously not be extended to another substrate. In addition, the controlled growth of a desired number of graphene layers with complete thickness homogeneity is still a challenge. While the synthesis of homogeneous mono- or bilayer graphene on the (0 0 0 $\bar{1}$) C-face is difficult to achieve due to a fast sublimation of Si atoms, the formation of uniform and continuous few-layer graphene on the (0 0 0 1) Si-face has not been demonstrated so far. In the case of CVD synthesis on metals, the required transfer onto nonmetallic substrates often leads to contamination of the graphene surface and the generation of structural defects, which degrade the electronic properties of the material. In recent works it was demonstrated, that CVD can be used to grow graphene directly on nonmetallic substrates [7–9].

It is well known that graphene is a single atomic layer of graphite, which can be epitaxially grown on substrates or exfoliated, and is now very well known to have outstanding properties [14–23]. Because silicon carbide (SiC), a IV–IV compound semiconductor, has a wide bandgap (ranging from 2.4 to 3.3 eV depending on polytype) [24,25], epitaxial graphene grown on a SiC substrate is promising for future electronics applications [26–29] and has indeed been added to the Roadmap of Semiconductor Technology. Such epitaxial growth on hexagonal 6H/4H-SiC(0001)/(000$\bar{1}$) Si- and C-faces, and cubic 3C-SiC(100) has been shown to achieve graphene layers or nanoribbons [17–24,30,31]. Their atomic/electronic structures and transport properties have been determined using advanced experimental techniques, and state-of-the-art theoretical calculations [30–32]. On the C-face, graphene multilayers can be grown with each layer decoupled from one another [29,30], leading to unprecedented high carrier mobility of up to 250,000 cm²/Vs at room temperature [29–31] and 200,000 cm²/Vs elevated temperatures.

In contrast, a single graphene layer epitaxially grown on a C face SiC substrate has significantly lower mobility of 20,000 cm²/Vs [29], possibly due to the poor quality of the graphene/SiC interface as a result of harsh graphene growth conditions. Strain is the driving force in SiC surface ordering, leading to complex surface reconstructions for both Si and C faces [32–39] leading to self-organized Si [40–44] and C [45] nanostructures but also to result in nano-crack defects [46]. Indeed, so far little is known about graphene/SiC interfaces where defects could affect the carrier transport properties.

The absence of an effective technology of fabrication of graphene is a serious obstacle holding down progress in experimental studies of graphene and graphene-based structures. Each of the methods employed presently in producing graphene films consisting of one or several graphene layers is plagued by severe limitations.

Graphene films prepared on the surface of silicon carbide by an appropriate thermal treatment in vacuum or a neutral atmosphere cannot be separated from the substrate that mediates their electronic structure through interlayer coupling. Apart from this, the size of the uniform crystalline parts of such graphene films depends directly on the quality of the silicon carbide substrate surface and the accuracy with which it is oriented with respect to the basal crystal planes. Graphene

film preparation on the surface of metals through segregation of the carbon dissolved in them or decomposition of vapors of hydrocarbons, a method enjoying presently considerable attention (it is by this approach that monolayer carbon films were originally obtained [3]), is also not free of fundamental limitations. In the first case, this method does not allow monitoring of the number of the deposited atomic layers, while in the second, it restricts the size of the graphene islands with a regular structure to that of crystallites on the metal surface.

In general, the isolation of monolayer graphene as a free-standing material by different synthesizing techniques has enabled the investigation of the special physical properties arising from its linear band dispersion at the Dirac point. In addition, properties such as high charge carrier mobility, good optical transparency, and mechanical toughness makes it a promising material for microelectronic devices, transparent electrodes in optoelectronics, and a huge array of other potential applications that exploit the mechanical strength of this thinnest possible material. Some of these applications, for example, the use of graphene in TEM grids, are already commercial products. To develop the full potential of graphene, the materials science of graphene has to be developed and the progress over the last few years has been breathtaking. Now scientists have a very good basic understanding of how to synthesize large-scale wafers of graphene? what imperfections to expect in graphene? and what are the properties of these imperfections? Perhaps, most importantly for device applications, we have also started to understand the processing methods of graphene and are developing methods to modify graphene's properties. Graphene is the ultimate "surface material" and therefore surface characterization methods will remain central in future studies of graphene. To date, most true surface science studies have been carried out on graphene supported on their growth substrate. This is because it allows the formation of a well-defined large area single crystal necessary for many surface science investigations. Furthermore, although the transfer of graphene from metal supports to other materials has become common practice for device fabrication, there is still an issue of contamination of the graphene with, for example, polymer residues. Furthermore, a wafer-sized, single-grain graphene sample transfer to insulating/dielectric materials remains challenging. Although graphene on weak interacting metal supports is a good model system for investigating properties of graphene by surface science studies, future surface studies should be more focused on graphene on relevant substrates for applications, that is, mainly dielectric substrate materials. Truly free-standing graphene may be ideal for transport measurements at low temperatures. However, it has been shown that for room-temperature devices, suspended graphene is less ideal because of flexural phonons increasing charge carrier scattering. Therefore, it is likely that microelectronic devices will be fabricated on supported graphene, which also seems somewhat easier to achieve. Currently, the best-insulating support material for graphene is *h*-BN. Unlike SiO₂ or other initially tried insulating substrates, *h*-BN is very uniform. Because *h*-BN is a layered 2D material, similar to graphene, it does not have any "dangling" bonds or charge inhomogeneities. This is crucial in order to maintain the high charge carrier mobility of graphene. Therefore, in the near term we may expect more studies of graphene/*h*-BN hybrid materials. For surface science studies and also for applications the successful large area synthesis of such heterostructures will be critical. If such samples with sufficient size can be prepared, many of the graphene modification investigations we have summarized for metal or SiC-supported graphene may be studied on *h*-BN as the support material [20–24].

Nandi et al. [25] fabricated polypyrrole-based graphene nanostructure by using cetyl trimethylammonium bromide (CTAB) surfactant with variation of graphene loading and characterized them by XRD, FT-IR, SEM, TEM, AFM, Raman, and so on. They noticed that the variable range hopping (VRH) model applied for explication of temperature-dependent DC conductivity scrutinized from 20 to 300 K. They reported a 3D–2D effective dimensional crossover observed at ≈100 K with variation of hopping length, which can be controlled by temperatures. Below ≈50 K another cross-over between Mott and ES (Efrost and Shklovskii) type VRH mechanism had also been explored as resistivity increases with lowering of temperature. Here, charge transport occurred via VRH between intact graphene island and polypyrrole chain.

Liu et al. [26] synthesized monolayer tunable periodic graphene nanostructures with the dimensions ranging from ~20 nm to several hundreds of nanometers by e-beam lithography by controlling in exposure dose and etching time.

Bhaskar et al. [27] improved nanoscale conduction capability of a MoO_2/graphene composite for high-performance anodes in lithium ion batteries. A MoO_2/graphene composite as a high-performance anode for Li ion batteries was synthesized by a one pot *in situ* low-temperature solution phase reduction method. Electron microscopy and Raman spectroscopy results confirmed that 2D graphene layers entrap MoO_2 nanoparticles homogeneously in the composite. Conductive AFM reveals an extraordinarily high nanoscale electronic conductivity for MoO_2/graphene, greater by 8 orders of magnitude in comparison to bulk MoO_2.

Scientists such as Mortazavi et al. [28] have developed a multiscale scheme using molecular dynamics (MD) and finite element (FE) methods for evaluating the effective thermal conductivity of graphene epoxy nanocomposites. They proposed that thermal conductivity of single-layer graphene decline by around 30% in epoxy matrix for two different hardener chemicals.

Another challenge in designing the properties of graphene is to control modifications. We have, for example, shown that defects, like carbon vacancies and grain boundaries in graphene can have interesting properties. However, in order to utilize these defects we have to be able to define their structure accurately (e.g., multiple defect structures have been reported for di-vacancies in graphene), density and location in the graphene. Similarly, for graphene nanoribbons better control over ribbon-widths and edge structure is needed. Modifications of graphene by adsorbates should also be localized on micro

to tens of nanometer scale; for example, for defining p- or n-type-doped regions (e.g., molecular adsorbates), or electron confinement regions (e.g., confining pristine graphene within hydrogen- or fluorine-modified graphene). In addition, in order to be able to locally modify graphene, the question of stability of the formed patterns needs to be addressed. On a one-to-few nanometers length scale, Moiré-patterns formed by graphene on metals have been successfully used as templates for preferential adsorption of metals, hydrogen, and some organic molecules. These self-organization mechanisms are fascinating from a pure chemical point of view, but it is less likely that these patterns remain stable if the graphene is transferred from the metal support to another substrate and therefore these self-organization patterns may not easily be exploitable for electronic applications. On the contrary, these graphene-based templates for self-organization of metal clusters may be useful systems for fundamental studies in cluster science. Thus, although graphene for electronic applications is taking the spot light, there are many other interesting aspects in fundamental research for which (supported) graphene can provide a rich resource for researchers for the various aspects of the materials science of nanostructures. The fast moving pace of graphene research will ensure that this remains an exciting field for years to come and new developments will soon add to the materials reviewed in these investigations.

18.2 EXPERIMENTS AND DISCUSSION

18.2.1 SYNTHESIS AND CHARACTERIZATION OF 123-YBCO SUPERCONDUCTING SUBSTRATE

The pure YBCO ($YBa_2Cu_3O_7$) with general formula, $YBa_2Cu_3O_z$, was prepared by solution route and sintering procedures by using two different solution precursors.

The pure 123-YBCO was selected as substrate owing to its highest thermal and structural stability at high temperature in addition to its superior conduction properties.

A. The first precursor was formed from the appropriate amounts (Y_2O_3), which dissolved in few drops of concentrated nitric acid forming yttrium nitrate that was finally diluted to 100 mL by adding distilled water and the pH adjusted to neutral by adding ammonia solution.

B. The second precursor was for the appropriate amounts of $BaCO_3$ and CuO, each of highly pure chemical grade purity, which dissolved also in few drops of concentrated nitric acid and the net solution was 100 mL using distilled water and the pH adjusted to neutral by adding ammonia solution.

The mixtures (A + B) were poured into 1 L round-bottomed flask while 0.4 M urea/NH_3 solution was added drop wise, carefully, with continuous stirring forming heavy gelatinous precipitate from metal-hydroxide plus hydroxylated oxides. The precipitates were filtered and dried and then calcined at 800°C under a compressed O_2 atmosphere for 30 h and later reground and pressed into pellets (thickness 0.2 cm and diameter 1.2 cm) under 8 ton/cm². Sintering was carried out under oxygen stream at 925°C for 100 h. The samples were slowly cooled down (10°C/h) till 500°C and annealed there for 20 h under oxygen stream. The furnace was shut off and cooled slowly down to room temperature. Finally, the materials were kept in vacuum desiccator over silica gel dryer. A levitation preliminary superconductivity test was thoroughly applied for the achievement of superconductive phase and hence superconductivity.

18.2.2 PHASE IDENTIFICATION

The x-ray diffraction (XRD) measurements were carried out at room temperature on fine ground samples using $Cu-K_\alpha$ radiation source, Ni-filter, and a computerized STOE diffractometer/Germany with two-theta-step scan technique.

Figure 18.1 displays XRD recorded for pure synthesized 123-YBCO superconducting sample. The analysis of two theta values, interspacing distances dÅ indicated that the sample mainly belongs to orthorhombic superconducting phase with P_6/mmm space group (see Figure 18.1b). The unit cell dimensions were calculated using the most intense x-ray reflection

FIGURE 18.2 SE-micrograph recorded for pure 123-YBCO substrate.

peaks [0 1 1] and found to belong to orthorhombic crystal form with lattice parameters $a = 3.8234$ Å, $b = 3.8662$ Å, and $c = 11.7941$ Å for the pure 123-YBCO phase, which is fully in agreement with those mentioned in the literature [47–49].

Scanning electron microscopy (SEM) measurements were carried out using small pieces of the prepared samples by using a computerized SEM camera with elemental analyzer unit (PHILIPS-XL 30 ESEM/USA).

Figure 18.2 shows the SEM micrographs recorded for pure YBCO applied on the ground powders that was prepared by solution route precursor. The average particle size was calculated and found to be in between 0.26 and 0.87 μm. The EDX examinations for random spots in the same sample confirmed and are consistent with our XRD analysis for pure YBCO, such that the differences in the molar ratios EDX estimated for the same sample is emphasized and an evidence for the existence of 123-YBCO superconductive phase with good approximate molar ratios.

From Figure 18.3 it is difficult to observe inhomogeneity within the micrograph because the powders used are very fine and the particle size estimated too small.

The grain size for 123-YBCO phase was calculated according to

Scherrer's formula [50],

$$B = 0.87\lambda/D\cos\theta \qquad (18.1)$$

where D is the crystalline grain size in nanometers, θ, half the diffraction angle in degree, λ the wavelength of x-ray source ($Cu-K_\infty$) in nanometers, and B, degree of widening of diffraction peak, which is equal to the difference of full-width at half-maximum (FWHM) of the peak at the same diffraction angle between the measured sample and standard one. From SEM mapping, the estimated average grain size was found to be 1.29–1.73 μm, which is relatively large in comparison with that calculated applying Scherrer's formula for pure 123-phase (D ~ 0.44 μm). This indicates that, the actual grain size in the material bulk is smaller than that detected on the surface morphology.

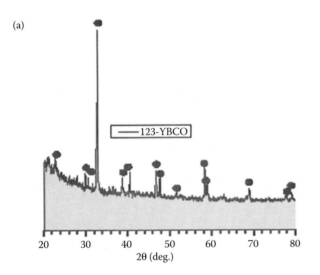

(a)

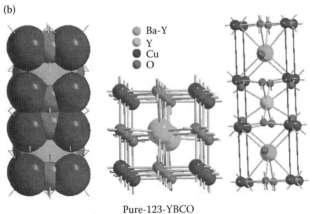

(b)

Ba-Y
Y
Cu
O

Pure-123-YBCO

FIGURE 18.1 (a) XRD-profile recorded for pure 123-YBCO substrate. (b) Orthorhombic unit cell of pure 123-YBCO substrate.

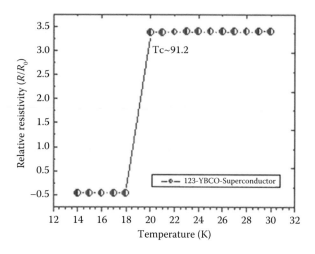

FIGURE 18.3 Superconducting *resistivity-temperature* curve of 123-pure-YBCO.

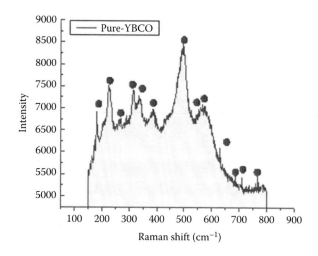

FIGURE 18.4 Raman spectrogram recorded for pure 123-YBCO substrate.

18.2.3 SUPERCONDUCTING MEASUREMENTS

The resistivity of the synthesized sample was undertaken as a function of cryogenic temperature down to 30 K using liquid helium refrigerator using four terminal probe—circuit. Powdered samples were used for measurements in order to reveal the content of superconductivity inside the bulk of sample.

From Figure 18.3 it is evident that 123-pure-YBCO sample exhibits HTc ~ 91.2 K corresponding to 123-phase, which is annealed in oxygen and is noticeable clearly in our XRD as a major phase. This confirmed magnetically the existence of 123-YBCO in highly pure phase.

18.2.4 RAMAN SPECTROSCOPY MEASUREMENTS

The measurements of Raman spectra were carried out on the finally ground powders with laser wavelength = 632.8 nm (He–Ne laser with power = 1 mW) and laser power applied to the site of the sample = 0.4 mW with microscope objective = ×20, accumulation time = 1000–4000 s, up to more than an hour.

Figure 18.4 shows the Raman spectrogram recorded for pure-YBCO superconductors. From the modes frequencies that are listed and compared with some references [51–53] (see Table 18.1) one can indicate that 123-YBCO phase is the dominating phase present in polycrystalline YBCO superconductors. From Raman spectrograms it is obvious that the fundamental vibrational modes were identical with those reported in references [51,52] with small exception where a strong peak lies at 588–595 cm⁻¹, which is attributable to the impurity phase $BaCuO_2$ [52]. The vibrational Raman active-phonon mode lies at ~480 ± 10 cm⁻¹ originally due to the apical oxygen O_4 (A_{1g}) [52]. This vibrating mode became flat and very broad so as to be noticeable in the pure YBCO. This flattening can be attributed to disturbances and changes that occurred in the interatomic distances of apical oxygen O_4 to the two copper layers (Cu_2). As is known, the vibrational mode lying at 330 ± 5 cm⁻¹ is the out-of-phase B_{1g} of the

TABLE 18.1

Mode Frequencies of Raman Spectra Recorded for Pure-123 YBCO in the Present Work in Contrast with Some References

References		
Reference 17	References 19 and 20	Pure-123-YBCO
229	229	224.6s
336	336	314m
500	575	337.8m
575	592	386.28b
440	633	438w
		495.37s
		571.2b

Note: s = strong, m = moderate, b = broad, and w = weak.

couple O_2–O_3, which confirms the existence of 123-YBCO in pure phase [52,53] (Table 18.1).

18.2.5 GRAPHENE DEPOSITION ON 123-YBCO SUBSTRATE

The well-characterized polished 123-YBCO pellet with dimensions 1 cm × 1 cm was applied as substrate to deposit few layers of graphene by injection-CVD process (ICVD). The high crystalline and the large area growth of multilayer graphene are desirable as a resistive barrier for formulated YBCO-graphene. For CVD of the graphene layers, a freshly prepared 1 cm × 1 cm YBCO-polished surface substrate was heated and stabilized at the desired synthesis temperature first. Then, propylene gas (C_3H_6), which served as the carbon source, was introduced into the chamber. By doing so, the gas pressure was adjusted to 4 × 10⁻⁶ mbar using a leak valve. In the incipient reaction of propylene with the 1 cm × 1 cm YBCO-polished surface, a graphene monolayer is formed as evidenced by the appearance of a single π band in angle

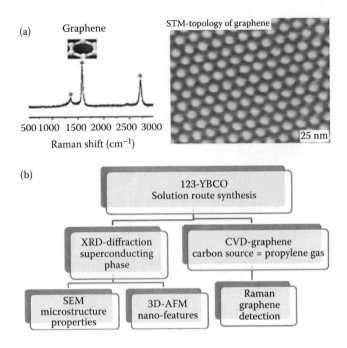

FIGURE 18.5 (a) Characterization of graphene by both Raman spectroscopy and STM imaging (atomic resolution mode) with honeycomb structure, respectively. (b) Schematic diagram shows different steps of synthesizing graphene-loaded-123-superconducting YBCO surface.

resolved photoemission spectroscopy (ARPES) as reported in Reference [54] and as we confirmed by both Raman and STM imaging (Atomic resolution mode Reference [55]) see Figure 18.5a. During the whole synthesis procedure both the C_3H_6 gas pressure and the substrate temperature is monitored. This complete set of information allowed us to monitor the full growth process dependent on time and the details of CVD synthesis are described elsewhere [56–59].

For this synthesis temperature (635°C) the sample can be cooled down any time after the graphene layers are complete (i.e., t > 1600 s) without inducing a change in the graphene layer. This time of completion of one layer graphene is nearly double that recorded in Reference [54] due to our processing temperature being lower (635°C) than that reported in Reference [54], which was 900°C (Figure 18.5b).

18.2.6 RAMAN AND NANO-STRUCTURAL FEATURES OF GRAPHENE MONOLAYERS

AFM: High-resolution AFM is used for testing morphological features and topological map (Veeco-di Innova Model-2009-AFM-USA). The applied mode was tapping noncontacting mode. For accurate mapping of the surface topology AFM-raw data were forwarded to the Origin-Lab version 9-USA program to visualize more accurate three- and two-dimension surface of the sample under investigation. This process is a new trend to obtain high-resolution 3D-mapped surface for a very small area of 0.01 μm².

Figure 18.6a represents high-resolution AFM image captured by using Veeco-di Innova Model-2009-AFM-USA

applying tapping noncontacting mode, for testing morphological features and topological map of the graphene-deposit layers.

Figure 18.6a displays 3D-AFM-tapping micrograph for scanned area = 0.2x0.2y0.2z μm, it was noticed that the surface of graphene is not unified in heights (*z-axis*) and has a lot of surface gradient with different heights. These surface heights gradient may be attributed to two different factors: the first one is a mechanical error of smoothing the YBCO-substrate surface; the second is more probable to take place that graphene deposition rate is not the same throughout the process of CVD. This confirms that the graphene deposition process is experimental-condition-dependent and time-dependent.

Figure 18.6a shows deflection centers, which dispersed regularly throughout the whole scanned area as hemi spiral shape. These deflection dot centers could be benefitted to interpret the pinning centers promotion (which is responsible for current carrier mobility and conduction mechanism). Furthermore, deflection centers could also supply a qualitative information for grain formation rate, which is considered a new trend for AFM investigations to obtain conclusive information not only surface topology features but also good qualitative deductions for materials bulk constituents.

Figure 18.6a displays regular 2D-tapping mode image of deposit graphene, which routinely gives some information about roughness of the surface and the grain size existed on the surface and near the surface layers (Figure 18.9).

Figure 18.7 displays Raman spectrogram recorded for graphene deposit layers showing sharp inflection peaks at ~1500, 2722 cm⁻¹, which corresponds to graphene as nano-sheets (few layers of graphene) these results are in partial agreement with some references in literature such as References [60–62]. Ferrari [62] reported that focus on the origin of the D and G peaks and the second order of the D peak on the Raman spectrograms and confirmed that the G and 2D Raman peaks change in shape, position, and relative intensity with number of graphene layers. This reflects the evolution of the electronic structure and electron–phonon interactions. Furthermore, Raman spectroscopy can be efficiently used as an analytical tool to monitor number of layers, quality of layers, doping level and confinement of synthesized carbon compounds as graphite and graphene.

18.2.7 NEW TREND OF AFM-NANO-IMAGING VIEW FOR ACCURATE SURFACE TOPOLOGY

For more accuracy of surface topology the raw data of AFM-instrument were forwarded to the Origin-Lab version 9-USA program to visualize more accurately three- and two-dimension surface of the sample under investigation. This process is a new trend to get a high-resolution 3D-mapped surface for very small area of 0.01 μm².

Figure 18.8a displays 2D-AFM-grayscale photo-imaging for graphene layers, which enable us to map the topology with accurate numerical values as in Figure 18.8a. The image describes the surface topology of graphene monolayer with

(a)

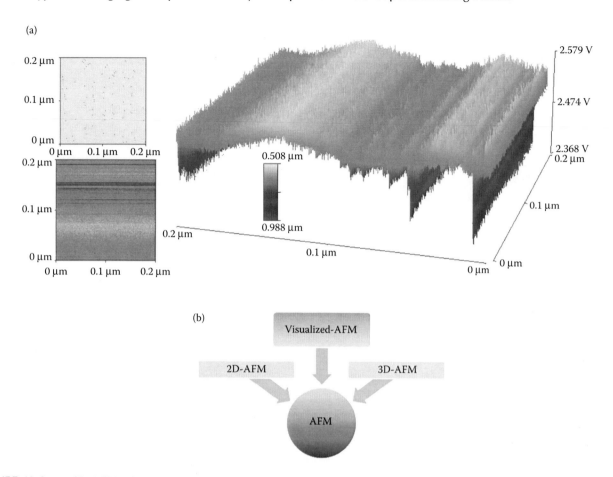

(b)

FIGURE 18.6 (a) 3D-AFM micrograph recorded for graphene deposit layer on YBCO-substrate. Deflection points centers located on the scanned graphene (Area = 0.2x0.2y0.2z μm). 2D-imaging of graphene surface (0.2 × 0.2 μm). (b) Schematic diagram shows different sequences of AFM-investigations.

dimension 0.25 × 0.25 μm (scanned area =0.225 μm²). This method draws real visualized image of graphene surface as shown in Figure 18.8a.

The two diagonal axes (x and y-axes) in the real image can be shifted up and down and in each shift one can see the corresponding horizontal and vertical view as in Figure 18.8a

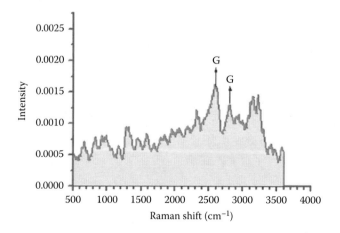

FIGURE 18.7 Raman spectrogram recorded for graphene deposit layers.

and by accurate scanning to the whole surface we get a three-dimensional image as in Figure 18.8b.

The 3D imaging could be converted into a 2D-AFM image as shown in Figures 18.8c and 18.8d in which surface features can be interpreted accurately.

Graphene surface in Figure 18.8b through 18.8d can be divided into three zones: light gray color represents ~23% of the whole scanned area, which is ~0.225 μm², the surface heights in this zone ranged between 4.375 and 4.64 μm as in the key image. The maximum height in the whole image is a shiny gray, which represents 4% = 0.01 μm² that has the highest height on the scanned area with height$_{max}$. = 4.64 μm, the second partition is gray in color with a ratio ~12% of the whole area = 0.0275 μm² and the third part has area percent ~0.0225 μm², which represents (9%) with heights ranged in between 4.375 and 4.450 μm. The second zone represents ~ (dark gray), which represents ~23% (0.0575 μm²) from the whole scanned area with heights gradient ranged in between 4.225 and 4.30 μm. The third zone occupies ~53% = 0.1325 μm² from the whole scanned area with heights gradient lying between 4.0 and 4.22 μm. This gradient is represented by light black color. In contrast with regular AFM-image for graphene surface loaded on YBCO-substrate, from Figure 18.9 one can notify how many differences are there

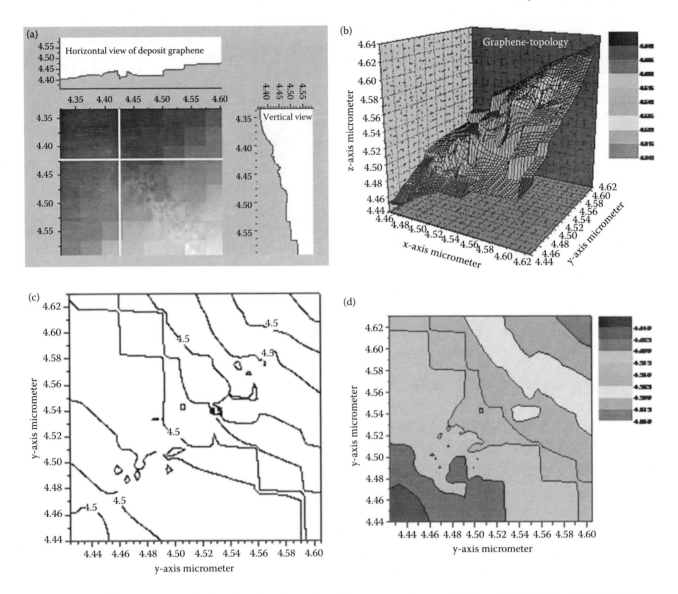

FIGURE 18.8 (a) 2D-AFM-grayscale photo-imaging for graphene layers, scanned area = 0.225 μm². (b) 3D-Visualized-AFM-Contuor-imaging for graphene, scanned area = 0.225 μm². (c) Mapped image of graphene. (d) 2D-AFM-contour image of graphene layers.

between our new trend of AFM-imaging analysis and the regular image.

The new trend of transformation of AFM-raw data into 3D-origin lab figure has many advantages than regular image supplied from AFM instrument. The most important one is accurate numerical values that describe the topology

of the scanned area. As an example, see Figure 18.10, which explains *xz-axes* sector in the graphene surface with very accurate Cartesian numerical values for very tiny scanned area (0.2 × 0.2 μm), which is so difficult in normal AFM. From Figure 18.16 one can deduce maximum heights that equals 389 nm and minimum height ~27 nm. These accurate

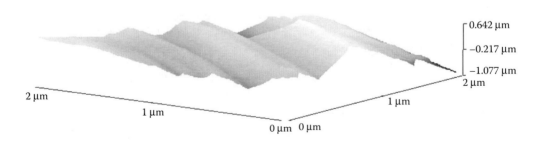

FIGURE 18.9 Normal AFM-tapping mode imaging of graphene layers.

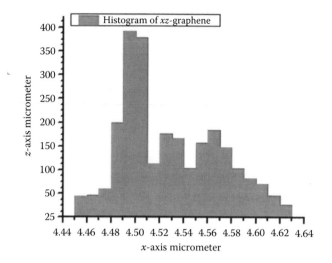

FIGURE 18.10   Histogram of *xz*-graphene surface.

numerical values are considered the main difference between both methods of calculations: normal and our new trend of calculations.

18.2.8   DC-ELECTRICAL CONDUCTIVITY MEASUREMENTS

Figure 18.11a through c shows the relation between DC-electrical conductivity (log σ) against reciprocal of absolute temperature (1000/T) K^{-1} as a function of spot location on the graphene-deposit layer.

The DC-electrical conductivity measurements were performed for the same surface of the sample at three random different spots namely (A, B, and C), respectively. As in Figure 18.11a through c the three measurements display the same behavior (conductor and semiconductor) but the resistive barrier varies from spot to another spot, which confirm that the deposit graphene thickness is not unified throughout the surface due to CVD controlling parameters. These observations were also confirmed via AFM investigations. These fruitful observations are in partial agreement with Bernardo et al. [63], who investigated electrical conductivity values of different carbon compounds, namely, graphene, multi-wall carbon nanotubes, carbon black, and graphite powder. They reported [63] that the bulk conductivity depends not only on the intrinsic material properties but is also strongly affected by the number of particle contacts and the packing density, which is a function in thickness. Conductivities at high pressure (5 MPa) for the graphene, nanotube, and carbon black show lower values (~10^2 S/m) as compared to graphite (~10^3 S/m). For nanotube, graphene and graphite particles, the conductive behavior during compaction was governed by mechanical particle arrangement/deformation mechanisms while for carbon black this behavior was mainly governed by the increasing particle contact area.

Others such as [64] indicated that the thermal conductivity was controlled by many types of defective graphene including single vacancy, double vacancy, and Stone–Wales defects, which can greatly reduce the thermal conductivity of

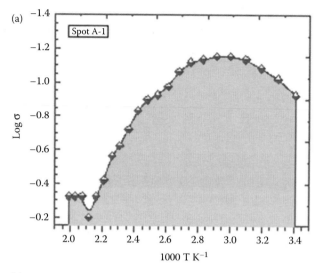

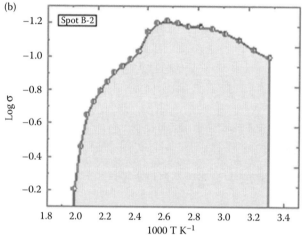

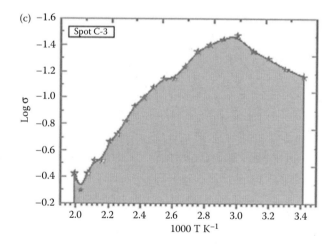

FIGURE 18.11   Shows the relation between DC-electrical conductivity as a function of spot location on the graphene-deposit layer.

graphene. The amount of reduction depends strongly on the density and type of defects at small density level. However, at higher defect density level, the thermal conductivity of defective graphene decreases slowly with increasing defect density and shows marginal dependence on the defect type. The thermal conductivity was found to become less sensitive to temperature with increasing defect density.

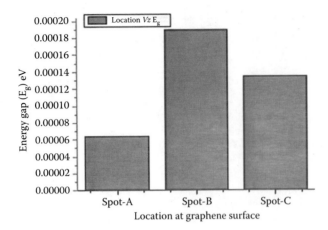

FIGURE 18.12 Values of energy gaps (E_g) versus spot location.

Also we found that the energy gap (E_g) and the number of (e^-) in conduction band (N_{cb}) of graphene were different in values on the same surface of graphene but at different location, which confirm that resistive layers of graphene (thickness) are different from point to point confirming that number of graphene layers vary according to CVD rate and experimental conditions specially CVD-controlling parameters Figures 18.12 and 18.13, respectively. These results are in full agreement with Hai et al. [65], who formulated a correlation between thermal conductivities values and number of layers of graphene (size of graphene layers, i.e., the thermal conductivity can be controlled by changing number and size of graphene layers).

$$\rho = \rho_o\, e^{-\,\Delta Eg/KT} \qquad (18.2)$$

$$N_{cb} = AT^{3/2}e^{-\,Eg/2KT} \qquad (18.3)$$

Where in Equation 18.2 ρ and ρ_o are conductance and specific conductance, respectively. K is Boltzmann constant and T is absolute temperature in kelvins.

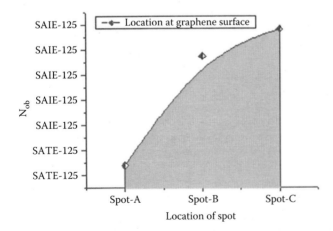

FIGURE 18.13 Number of electrons in conduction band (N_{cb}) versus spot location.

Furthermore, Kim et al. [66] reported that there is a strong correlation between energy gap and number of layers deposited through CVD process on the basis of interspacing distance between substrate (YBCO in our studies) and graphene, which will vary as the number of layers change. Many investigators such as Hu and Poulikakos [67] have studied the interfacial thermal resistance, which plays a pivotal role in determining the thermal, transport properties, and performance of nanostructured materials; the near interface region can significantly contribute to the overall contact resistance in the case of strongly coupled thermal interface materials. They reported that the overall interfacial thermal transport is dominated by the resistance at the adjoining near-interface region and not the interface itself. Accordingly, the application of YBCO-superconducting substrate could be beneficial in this point of view depending on resistance barrier of interfacial layer between graphene and the YBCO-substrate will be minimum.

18.3 CONCLUSIONS

This chapter can be summarized in the following points:

1. AFM alone is not enough to identify the number of graphene layers.
2. The success of visualization of raw data is a new trend of AFM-nano-imaging view to give accurate 3D-mapped surface topology with precise numerical values.
3. Graphene deposition rate is not the same through the process of CVD and 3D-AFM investigations confirmed that graphene layer epitaxially grown on YBCO-substrate have height gradient controlled via applied experimental parameters.
4. Raman spectroscopy together with AFM can be efficiently used as analytical tool to monitor number of layers, quality of layers, and confinement of synthesized graphene.
5. The thermal conductivity can be varied by changing the number and size of graphene layers.
6. Deflections dots centers can be beneficial to map and contour the surface of a sample which enables us to understand a lot of conduction mechanisms and pinning centers locations.

REFERENCES

1. Geim, A.K., Novoselov, K.S. 2007. *Nature* 6:183–87.
2. Castro, E.V., Novoselov, K.S., Morozov, S.V., Peres, N.M.R., Lopes dos Santos, J.M.B., Nilsson, J., Guinea, F., Geim, A.K., Castro Neto, A.H. 2007. *Phys. Rev. Lett.* 99:216802.
3. Novoselov, K.S., Geim, A.K., Morozov, S.V., Jiang, D., Zhang, Y., Dubonos, S.V., Grigorieva, I.V., Firsov, A.A. 2004. *Science* 306:666–71.
4. Miao, F., Wijeratne, S., Zhang, Y., Coskun, U.C., Bao, W., Lau, C.N. 2007. *Science* 317:1530–37.
5. Schedin, F., Geim, Morozov, A.K., Hill, S.V., Blake, P., Katsnelson, M.I., Novoselov, K.S. 2007. *Nat. Mater.* 6:652.
6. Yazyev, O., Katsnelson, M.I. 2008. *Phys. Rev. Lett.* 100:047209.

7. Wang, X., Ouyang, Y., Wang, H., Guo, J., Dai, H. 2008. *Phys. Rev. Lett.* 100:206803.
8. Chen, J.H., Jang, C., Xiao, S., Ishigami, M., Fuhrer, M.S. 2008. *Nature Nano.* 3:206.
9. Vovoselov, K.S., McCann, E., Morozov, S.V., Fal'ko, V.I., Katsnelson, M.I., Zeitler, U., Jiang, D., Schedin, F., Geim, A.K. 2006. *Nat. Phys.* 2:177.
10. Ferrari, A.C., Meyer, J.C., Scardaci, V., Casiraghi, C., Lazzeri, M., Mauri, F., Piscanec, S., Jiang, D., Novoselov, K.S., Roth, S., Geim, A.K. 2006. *Phys. Rev. Lett.* 97:187401.
11. Gupta, A., Chen, G., Joshi, P., Takigadapa, P., Eklund, C. 2006. *Nano Lett.* 6:2667; 2007. *Appl. Phys. Lett.* 91:201904.
12. Calizo, I., Teweldebrhan, D., Bao, W., Miao, F., Lau, C.N., Balandin, A.A. 2008. *J. Phys. C* 109:012008.
13. Calizo, I., Bao, W., Miao, F., Lau, C.N., Balandin, A.A. 2007. *Appl. Phys. Lett.* 91:071913.
14. Forbeaux, I., Themlin, J., Debever, J. 1998. *Phys. Rev. B* 58:16396; Van Bommel, A.J., Crombeen, J.E., Van Tooren, A. 1975. *Surf. Sci.* 48:463.
15. Novoselov, K.S. 2004. *Science* 306:666–67.
16. Berger, C. 2004. *J. Phys. Chem. B* 108:19912; 2006. *Science* 312:1191.
17. Emtsev, K.V. 2009. *Nat. Mater.* 8:203.
18. Sprinkle, M., Soukiassian, P., de Heer, W.A., Berger, C., Conrad, E.H. 2009. *Phys. Status Solidi* RRL 3:A91.
19. Camara, N. 2009. *Phys. Rev. B* 80:125410.
20. Boeckl, J., Mitchel, W.C., Clarke, E., Weijie, L. 2010. *Mater. Sci. Forum* 645:573.
21. Hibino, H., Kageshima, H., Nagase, M. 2010. *J. Phys. D Appl. Phys.* 43:374005; Tanaka, S., Morita, K., Hibino, H. 2010. *Phys. Rev. B* 81:041406(R).
22. Sprinkle, M. 2010. *J. Phys. D Appl. Phys.* 43:374006.
23. Veuillen, J. 2010. *J. Phys. D Appl. Phys.* 43:374008.
24. Soukiassian, P., Enriquez, H. 2004. *J. Phys. Condens. Matter* 16:16118 and references therein.
25. Nandi, D., Nandi, S., Prasun K. Pal, Arup K. Ghosh, Amitabha De., Uday C.G. 2014. *Applied Surface Science* 293:90–96.
26. Liu, L.Z., Tian, S.B., Long, Y.Z., Li, W.X., Yang, H.F., Li, J.J., Gu, C.Z. 2014. *Vacuum*, 105:21–25.
27. Bhaskar, A., Deepa, M., Rao, T.N., Varadaraju, U.V. 2012. *J. Power Sources* 216:169–78.
28. Mortazavi, B., Benzerara, O., Meyer, H., Bardon, J., Ahzi, S. 2013. *Carbon* 60:356–65.
29. Heinz, K., Bernhardt, J., Schardt, J., Starke, U. 2004. *J. Phys. Condens. Matter* 16 S:17058.
30. Virojanadara, C. 2008. *Phys. Rev. B* 78:245403; Riedl, C., Coletti, C., Starke, U. 2010. *Phys. D: Appl. Phys.* 43:374009.
31. Aristov, V.Y. 2010. *Nano Lett.* 10:992.
32. Suemitsu, M., Fukidome, H. 2010. *J. Phys. D Appl. Phys.* 43:374012.
33. Orlita, M. 2008. *Phys. Rev. Lett.* 101:267601.
34. Miller, D.L. 2009. *Science* 324:924–25.
35. De Heer, W.A. 2010. *J. Phys. D Appl. Phys.* 43:374007.
36. Malle, P. 2007. *Phys. Rev. B* 76:41403R.
37. Hiebel, F., Mallet, P., Varchon, F., Magaud, L., Veuillen, J.Y. 2008. *Phys. Rev. B* 78:153412.
38. Hass, J. 2008. *Phys. Rev. Lett.* 100:125504.
39. Sprinkle, M. 2009. *Phys. Rev. Lett.* 103:226803.
40. Xiaosong, W. 2009. *Appl. Phys. Lett.* 95:223108.
41. Semond, F., D'angelo, M. 2004. *Phys. Rev. Lett.* 77:045317.
42. Soukiassian, P., Tejeda, A. 1997. *Phys. Rev. Lett.* 78:907.
43. Derycke, V., Soukiassian, P., Mayne, G., Dujardin, T. 2000. *Surf. Sci. Lett.* 446:101–03.
44. Soukiassian, P., Semond, A., Mayne, F., Dujardin, G. 1997. *Phys. Rev. Lett.* 79:2498; Douillard, L., Aristov, V. Yu., Semond, F., Soukiassian, P. 1998. *Surf. Sci. Lett.* 401:L395; Aristov, V. Yu., Douillard, L., Soukiassian, P. 1999. *Surf. Sci. Lett.* 440:L825.
45. Derycke, V., Soukiassian, P., Dujardin, G., Gautier, J. 1998. *Phys. Rev. Lett.* 81:5868.
46. Amy, F., Soukiassian, P., Brylinski, C. 2004. *Appl. Phys. Lett.* 85:926; Soukiassian, P., Amy, F., Brylinski, C., Mentes, T.O., Locatelli, A. 2007. *Mater. Sci. Forum*, 556:481.
47. Starowicz, P., Sakolowski, J., Balandaan, M., Szytula, A. 2001. *Physica C* 363:80.
48. Awana, V.P.S., Ansari, M.A., Nigam, R., Gupta, A., Samanta, S.B., Saxena, R.B., Kishan, H. et al. 2005. *Supercond. Sci. Technol.* 18:716.
49. Manthiram, A., Lee, S.J., Goodenough, J.B. 1988. *J. Solid State Chem.* 73:278.
50. Duong, C.H., Vu, L.D., Hong, L.V. 2001. *J. Raman Spectrosc.* 32:827–31.
51. Song, X., Daniels, G., Feldmann, D., Gurevich, A., Larbalestier, D. 2005. *Nat. Mater.* 4:470.
52. Mannhart, J., Muller, D.A. 2005. *Nat. Mater.* 4:431–35.
53. Giri, R., Awana, V.P.S., Singh, H.K., Tiwari, R.S., Srivastava, O.N., Gupta, A., Kumarswami, B.V., Kishan, H. 2005. *Physica C* 419:101–08.
54. Grüneis, A., Kummer, K., Vyalikh, D.V. 2009. *New J. Phys.* 6:11.
55. Elsabawy, K.M. 2011. *RSC Adv.* 1:964–67.
56. Grüneis, A., Vyalikh, D. 2008. *Phys. Rev. B* 5:232.
57. Nagashima, A., Tejima, N., Oshima, C. 1994. *Phys. Rev. B.* 50:17487.
58. Gamo, Y., Nagashima, A., Wakabayashi, M., Oshima, C. 1997. *Surf. Sci.* 4:345–46.
59. Preobrajenski, A.B., Vinogradov, A.S., Martensson, N. 2008. *Phys. Rev. B* 34:313.
60. Qianweni, L., Wang, W., Quin, Q. 2011. *Mater. Lett.* 65(15–16):2410–412.
61. El-Shazly, M., Duraia, M., Tokmoldin, S. 2011. *Vacuum* 86(2):232–34.
62. Ferrari, A.C. 2007. *Solid State Commun.* 143(2):47–57.
63. Bernardo, M., Ghislandi, M., Tkalya, E., Koning, C.E., Gijsbertus, D. 2012. *Powder Technol.* 221:351–58.
64. Zhang, Y.Y., Cheng, Y., Pei, Q., Xiang, Y. 2012. *Phys. Lett. A*, 376(8):3668–672.
65. Hai, C., Zhi, G., Xiang, H. 2012. *Chem. Mater. J.* 376, 4:525–28.
66. Kim, D., Hashmi, A., Hwang, C., Hong, L., 2013. *J. Surf. Sci.* 610:27–32.
67. Hu, M., Poulikakos, D. 2013. *J. Surf. Sci.* 62:205–13.

19 Phonon Spectrum and Vibrational Thermodynamic Characteristics of Graphene Nanofilms

Alexander Feher, Sergey Feodosyev, Igor Gospodarev, Eugen Syrkin, and Vladimir Grishaev

CONTENTS

Abstract 289
19.1 Introduction 289
19.2 Graphite Lattice Dynamics.. 290
    19.2.1 Graphite Crystal Structure and Character of Force Constants between Its Atoms.............. 290
    19.2.2 Force Constants and Flexural Rigidity of the Layers... 291
19.3 Phonon Spectrum and Vibrational Characteristics of Graphene Nanofilms.................................. 294
    19.3.1 Reconstruction and Relaxation at Nanofilms Formation 294
    19.3.2 Spectral Densities and Mean-Square Amplitudes of Atomic Displacements 295
    19.3.3 Phonon Heat Capacity of Graphite and Graphene Nanofilms: Its "Non-Debye" Behavior 298
19.4 Distinctive Properties of Impurities Effect on the Phonon Spectrum of the Graphene Layers 299
    19.4.1 Localization of the Vibrations on Point Defects 299
    19.4.2 Phonon Spectrum of Graphite Intercalated by Metals 300
19.5 Conclusion 302
Acknowledgment 303
References.. 303

ABSTRACT

This chapter describes phonon spectra and vibrational thermodynamic characteristics of graphene nanofilms. The relaxation of interlayer distances and force constants in the process of nanofilm formation is investigated. Moreover, the influence of the nanofilm structure (ABAB, ABCABC) on the thermodynamic properties of these systems is studied. Characteristics of the Van Hove singularities for phonon modes having different polarization are investigated.

The harmonic character of lattice vibrations and the stability of nanofilms persisting up to temperatures exceeding room temperature are confirmed by the calculations of the mean square amplitude of atomic displacements along the different crystallographic directions. It is shown that the features of the specific heat temperature dependence are caused by the nonsound phonon dispersion. Such phonons determine the main contribution to the low-temperature heat capacity for the structures considered. We carry out an analysis of the influence of local defects (vacancy type) and extended defects (step type between bigraphene and trigraphene) on the vibrational characteristics of carbon nanofilms.

19.1 INTRODUCTION

Investigation of the phonon spectrum features of graphite and carbon nanofilms is useful at least for two reasons: in addition to an obvious fundamental interest also from a practical viewpoint. One of them is the prospect of using the wide acoustic phonon spectrum zone of such materials in modern nanotechnology. Strong anisotropy of acoustic properties and "non-sound" behavior of the dispersion of elastic waves normally polarized to the layers are characteristic for these materials. This fact leads to some unique features of the phonon density behavior.

Investigation of the phonon spectrum features of graphene nanofilms is very important in understanding the electron–phonon interaction in such systems (Endlich et al. 2013, Dahal and Batzill 2014, Park et al. 2014) as well as for considerations on probable mechanisms of superconductivity in intercalated graphene (Fedorov et al. 2014, Yang et al. 2014). The unique nature of the two-dimensional phonon transport translates into unusual heat conduction in graphene and related materials (Nika and Balandin 2012, Pop et al. 2012, Serov et al. 2013, Wang et al. 2014). Optical phonons are used for counting the number of atomic planes and stacking order in Raman experiments with few-layer graphene (Ferrari and Basko 2013, Herziger and Maultzsch 2013, Shang et al. 2013, Lui and Heinz 2013).

Another important reason is the analysis of dynamic stability of carbon nanostructures, which is necessary for the "fine tuning" of the process of their synthesis and for determining the working conditions of devices constructed on their basis.

The flat graphene monolayer cannot exist in a free condition. Atoms of carbon are connected with a substrate by the same van der Waals interaction as graphene monolayers in a graphite crystal. This interaction involves only the lowest-frequency region of the graphene phonon spectrum (~2%). Its change practically does not influence the interatomic interactions in the layer plane. Therefore, it is necessary to begin the research of phonon spectra and vibrational characteristics of graphene and carbon nanofilms (both in free condition and as absorbed on a substrate) by studying the phonon spectrum of graphite. We may note that the vibrational characteristics of graphite are already well studied, experimentally.

In this chapter, the vibrational characteristics of carbon nanofilms for bigraphene and trigraphene with different structural types (ABA and ABC) are studied. The formation of nanofilms determines the relaxation of the interlayer distances and force constants. The analysis of the influence of local and extended defects on the vibrational characteristics of some carbon nanostructures is carried out. In particular, defects of the "step" type between bigraphene and trigraphene, and the influence of a substrate on the phonon spectrum of the graphene monolayer adsorbed are being analyzed.

This chapter consists of three parts:
In the first part, an adequate and at the same time simple enough model of the graphite lattice is constructed. Force constants characterizing interatomic interaction are defined without preliminary assumptions on the character of chemical bonds. Widely accepted experimental acoustic and optical characteristics as well as the data on the nonelastic dispersion of neutrons and soft x-ray radiation on the crystal lattice of graphite are used. The reasons causing flexural rigidity of graphene monolayers are analyzed and it is shown that this flexural rigidity is high, but it weakly depends on the interlayer interactions. We point to the determining role of the flexural rigidity of graphene monolayers for maintaining the dynamic stability of graphite and for the formation of basic distinctive features of its phonon spectrum and vibrational characteristics.

In the second part, vibrational characteristics of graphene and carbon nanofilms of bigraphene and trigraphene are considered, in general. The relaxation of the interlayer distances and force constants are determined in the process of nanofilm formation. The calculated root-mean-square amplitudes of atomic displacements along various crystallographic directions convincingly proved the stability of the considered nanostructures to temperatures exceeding room temperatures and negligible anharmonic corrections in the phonon spectrum, confirming the harmonic character of the lattice vibrations. The stability of the structure with the "step on surface" type defects, which might be important for the "management" of the electronic spectrum of carbon nanosystems is shown. The character of the Van Hove features in phonon modes for different polarizations is analyzed. A "non-Debye" character of the temperature dependences of heat capacity, caused by a nonsound dispersion of the phonons giving the basic contribution to the low-temperature heat capacity of studied structures is shown.

The third part of this chapter is devoted to the analysis of the influence of some defects on the vibrational characteristics of some carbon nanostructures. Localized and quasilocalized levels of graphite, originating from both impurities and point defects, are investigated. Phonon spectra of metal intercalated graphite are studied. Local spectral densities of carbon and metal atoms are calculated and analyzed.

19.2 GRAPHITE LATTICE DYNAMICS

19.2.1 Graphite Crystal Structure and Character of Force Constants between Its Atoms

A strongly anisotropic layered graphite crystal consists of the so-called graphene monolayers whose atoms form regular hexagons. Such a two-dimensional lattice is a complex lattice comprising two close-packed 2D triangular sublattices ($\bigcirc$ and $\bullet$), the atoms of one sublattice occupying the centers of mass of the triangles of the other lattice. In stacks with more than one graphene layer, two consecutive layers are normally oriented in such a way that the atoms in one of the two sublattices of the honeycomb structure of one layer are directly above one-half of the atoms in the neighboring layer. The second set of atoms in one layer sits on the top of the (empty) center of a hexagon in the other layer. This is the most common arrangement of nearest-neighbor layers observed in nature (ABA or Bernal stacking, see Figure 19.1). The Bravais vectors, lying in the basal plane, can be chosen as $r_1 = (a_0\sqrt{3}/2;\ a_0/2;\ 0)$ and $r_2 = (a_0\sqrt{3}/2;\ -a_0/2;\ 0)$, where the parameter $a_0 \approx 2.45$ Å. The period of the lattice along the c axis is equal to twice the interlayer distance $r_3 = (0;\ 0;\ c_0)$, where the parameter $c_0 \approx 6.7$ Å. Thus, the unit cell of graphite contains four atoms. We note that the atoms belonging to different sublattices in the 2D lattice of graphene, containing two atoms per unit cell, are physically equivalent.

The local Green's functions characterizing the contributions of each atom to the phonon density of states (DOS) and the vibrational characteristics are the same for the atoms in different sublattices ($G^{\circ}(\omega) = G^{\bullet}(\omega)$). This equivalence breaks down in the graphite lattice because the interlayer interactions are different for each of the sublattices. Indeed, as is clearly seen in Figure 19.1, the atoms of different sublattices from the basal plane are arranged differently with respect to the atoms of neighboring planes and therefore interact differently with them. Each atom of the sublattice $\bullet$ has in its nearest layers two neighboring atoms from the same sublattice, located at a distance of $\Delta_4 = c_0/2 \approx 3.35$ Å, and six neighboring atoms from the sublattice $\bigcirc$, located at $\Delta_5 = \sqrt{\Delta_4^2 + \Delta_1^2} \approx 3.64$ Å; each atom of the sublattice $\bigcirc$ has 12 neighboring atoms which are located at a distance Δ_5 (six from the sublattice $\bullet$ and six from the sublattice $\bigcirc$). That is, the atoms belonging to different sublattices of one graphene monolayer present in the graphite lattice are actually nonequivalent: the local Green's functions corresponding to these atoms and the vibrational

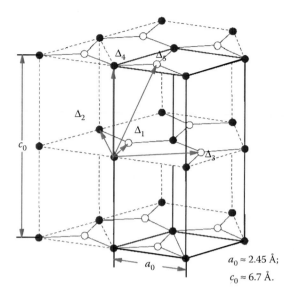

$a_0 \approx 2.45$ Å;

$c_0 \approx 6.7$ Å.

FIGURE 19.1 Crystal lattice of graphite (ABA-stacking).

characteristics determined by them, for example, the root-mean-square displacements of these atoms along different crystallographic directions, will thus be different.

Strong anisotropy of the interatomic interaction and other properties that are characteristic for graphite are due to the facts that the difference of the distances between the nearest neighbors in a layer and in neighboring planes is quite considerable and that the bonding forces in different crystallographic directions are of different types. For example, the interaction between the nearest neighbors in the basal plane, which lie at a distance of $\Delta_1 = a_0/\sqrt{3} \approx 1.415$ Å, is of the covalent type, whereas it is of the van der Waals type between the atoms located at the distances $\Delta_2 = a_0$ and $\Delta_3 = 2a_0/\sqrt{3} \approx 2.83$ Å (second and third neighbors in the basal plane) as well as between the atoms lying in neighboring layers at distances Δ_4 and Δ_5 from each other. In addition, graphite possesses metallic conductivity, which somewhat alters the interatomic interaction, primarily between the nearest neighbors.

Since the coordinate z and the coordinates x and y of the basal plane in the crystal lattice of graphite are transformed according to different irreducible representations of the point symmetry group D_{6h} of the crystal, the force-constants matrix can be represented in the form

$$\Phi_{ik}(\mathbf{r},\mathbf{r'}) = \Phi_{ik}(\mathbf{r}-\mathbf{r'}) \equiv \Phi_{ik}(\Delta) = -(1-\delta_{iz}\delta_{kz}) \cdot$$

$$\left[\alpha(\Delta) \cdot \frac{\Delta_i \Delta_k}{\Delta^2} - \beta_x(\Delta) \cdot \delta_{ik}\right] - \beta_z(\Delta) \cdot \delta_{iz}\delta_{kz}. \quad (19.1)$$

To describe the weak interlayer interaction it is natural to take into account the interaction of atoms located in nearest-neighbor layers. For the nearest neighbors in the basal plane ($\Delta = \Delta_1$), whose mutual interaction is determined by the superposition of the covalent and metallic bonds, the matrices (19.1) will be characterized by three parameters. The interatomic interaction between the more distant neighbors

($\Delta = \Delta_2$; $\Delta = \Delta_3$; $\Delta = \Delta_4$ and $\Delta = \Delta_5$) can be assumed to be a van der Waals interaction and can be described by an isotropic pair potential $\varphi(\Delta) = \varphi(\Delta)$. The corresponding force-constants matrices can be represented in the form (19.1), with $\beta_x(\Delta) = \beta_z(\Delta) = \beta(\Delta) = \varphi'(\Delta)/\Delta$ and $\alpha(\Delta) \equiv \varphi''(\Delta) - \beta(\Delta)$.

For describing the weak interlayer interaction it is impossible to be restricted only to interactions between atoms which are apart by Δ_4, as in this case the counteraction against the shift of a whole layer will be ensured exclusively by noncentral forces whose origin remains unknown so far. Taking into account the interaction between the atoms, which are apart by Δ_5 (exceeding Δ_4 by approximately 8%), allows us to explain the origin of noncentral forces, to describe the relaxation of the interlayer distances and interlayer interactions at the formation of carbon nanofilms and also to include central forces into the description of interlayer interactions. Thus, in describing the bonds between the atoms of one layer it is natural to consider the interaction between the first and third neighbors, that is the interaction between all atoms that are closer to each, other than Δ_4. Further, we will show that only by taking into account the interactions between more distant neighbors will we be able to describe the flexural stiffness of graphene layers. This stiffness has a key role in providing the stability of the graphite crystal structure and determines the peculiarities of its vibrational characteristics.

Thus, in the proposed model (Gospodarev et al. 2009) the dynamics of the graphite lattice is described by means of 11 force constants: five describing central interaction and six describing noncentral interactions between atoms.

19.2.2 FORCE CONSTANTS AND FLEXURAL RIGIDITY OF THE LAYERS

We can obtain five equations for force constants using known experimental data on the elasticity module connected with the structure of crystal and its matrices of force constants (see, e.g., Kossevich 1984)

$$\begin{cases} c_{iklm} = \dfrac{1}{V_0}(b_{imkl} + b_{kmil} - b_{lmki}); \\ b_{iklm} = -\dfrac{1}{2}\sum_\Delta \Phi_{ik}(\Delta) \cdot \Delta_l \cdot \Delta_m. \end{cases} \quad (19.2)$$

The condition for the symmetry of the elastic moduli tensor with respect to the permutation of pairs of indices for the symmetry of the Voigt matrix $C_{ik} = C_{ki}$ is a condition for the transition in the long-wavelength limit into the equations of the theory of elasticity. Since the coordinates in the basal plane ab and along the c axis are transformed according to different irreducible representations of the point symmetry group D_{6h} of graphite, the relation $C_{13} = C_{31}$ is not completely fulfilled and provides an additional equation for the force constants:

$$\beta_{1z} + 6\beta_2 + 4\beta_3 = \frac{2}{3\Delta_1^2}\left[\Delta_4^2\beta_4 - 9(\Delta_1^2 - \Delta_4^2)\beta_5\right]. \quad (19.3)$$

Four of the five elastic moduli (C_{11}, C_{22}, C_{33}, and C_{44}) are reliably determined by the sound velocity experiments along the high-symmetry crystallographic directions from both acoustic (e.g., Blakslee et al. 1970) and neutron (Nicklow et al. 1972) experiments. The elastic modulus C_{13} is determined by more complicated methods (e.g., from measurements of Young's modulus and Poisson's ratio), which, considering the smallness of C_{13}, leads to such discrepancies between its values, which attain the order of the values themselves. Consequently, the expression (19.2) for C_{13} cannot be used to determine the force constants. The experimentally determined values of the elastic module of graphite are presented in Table 19.1.

Using the relation (19.3), the following equations are obtained for the elastic moduli of graphite from Equation 19.2:

$$C_{11} = \frac{\sqrt{3}}{12\Delta_4}\left[3(\alpha_1 + 6\alpha_2 + 4\alpha_3) + 4(\beta_{1x} + 6\beta_2 + 4\beta_3)\right.$$

$$\left. + 9\frac{\Delta_1^2}{\Delta_5^2}\alpha_5 + 12\beta_5\right]; \tag{19.4}$$

$$C_{66} = \frac{\sqrt{3}}{12\Delta_4}\left[\alpha_1 + 6\alpha_2 + 4\alpha_3 + 4(\beta_{1x} + 6\beta_2 + 4\beta_3)\right.$$

$$\left. + 3\frac{\Delta_1^2}{\Delta_5^2}\alpha_5 + 12\beta_5\right]; \tag{19.5}$$

$$C_{33} = \frac{2\Delta_4\sqrt{3}}{9\Delta_1^2}\left(\alpha_4 + 9\frac{\Delta_4^2}{\Delta_5^2}\alpha_5 + \beta_4 + 9\beta_5\right); \tag{19.6}$$

$$C_{44} = \Delta_4\sqrt{3}\left[\frac{\alpha_5}{\Delta_5^2} + \frac{2(\beta_4 + 9\beta_5)}{9\Delta_1^2}\right]. \tag{19.7}$$

Remaining equations can be obtained from the neutron diffraction data (Nicklow et al. 1972), Raman scattering data (Dresselhaus and Dresselhaus 1981), and inelastic x-ray scattering data (Maultzsch et al. 2004). Thus, the following expressions are valid for the frequencies $\omega_{TO}(\Gamma)$ and $\omega_{LO}(\Gamma)$ (see Figure 19.2) in the model proposed:

$$m\omega_{TO}^2(\Gamma) = Q + 2\beta_4 + 2T - \sqrt{(Q - 2\beta_4)^2 + (T - 2\beta_4)^2}; \tag{19.8}$$

TABLE 19.1
Elastic Moduli of Graphite

	C_{11}	C_{66}	C_{33}	C_{44}	C_{13}
References			10^{10} N/m^2		
Nicklow et al. (1972)	106 ± 2	44 ± 2	3.65 ± 0.1	0.4 ± 0.04	–
Blakslee et al. (1970)	106	44	3.7	0.37 ± 0.02	1.5

$$m\omega_{LO}^2(\Gamma) = G + 2F + R - \sqrt{(G - F)^2 + (F - R)^2}, \tag{19.9}$$

where m is the mass of carbon atom. Following notation was introduced in Equations 19.8 and 19.9:

$$F \equiv 3\left(\frac{\Delta_4^2}{\Delta_5}\alpha_5 + 2\beta_5\right); G \equiv 2(\alpha_4 + \beta_4);$$

$$Q \equiv 3\left(\frac{\alpha_1 + \alpha_3}{2} + \beta_{1x} + \beta_3\right); T \equiv 3\left(\frac{\Delta_1^2}{\Delta_5}\alpha_5 + 2\beta_5\right);$$

$$R \equiv -6\beta_2 - 12\beta_3 + \frac{2\Delta_4^2}{\Delta_1^2}\beta_4 + 9 \cdot \frac{2\Delta_4^2 - \Delta_1^2}{\Delta_1^2}\beta_5.$$

The following data obtained in Nicklow et al. 1972 are $\omega_{TO}(\Gamma)/2\pi \approx 1{,}44$ THz and $\omega_{TO}(\Gamma)/2\pi \approx 3{,}76$ THz (corresponding points are denoted in the bottom part of Figure 19.2 as I and II), corresponding to the data obtained in Reich and Thomsen (2004).

The relations for the Raman frequency $\omega_{E_{2g_2}}/2\pi \approx 47.64$ THz (point III in the top part of Figure 19.2) and the frequencies $\omega_{E_{1u}}(\Gamma) \approx 22.6$ THz and $\omega_{A_{2u}}/2\pi \approx 26.04$ THz (point IV) (Dresselhaus and Dresselhaus 1981, Maultzsch et al. 2004), manifested in the infrared light scattering, have the form:

$$m\omega_{E_{2g_2}}^2(\Gamma) = Q + 2\beta_4 + 2T + \sqrt{(Q - 2\beta_4)^2 + (T - 2\beta_4)^2}; \tag{19.10}$$

$$m\omega_{E_{1u}}^2(\Gamma) = Q + T; \tag{19.11}$$

$$m\omega_{A_{2u}}^2(\Gamma) = 2(R + G). \tag{19.12}$$

Force constants α_1, α_2, and α_3 enter into Equations 19.4 and 19.5 only as a linear sum $\alpha_1 + 6\alpha_2 + 4\alpha_3$ and into Equations 19.8 and 19.11 only as $\alpha_1 + \alpha_3$. This indeterminacy can be eliminated by using, for example, the data of Maultzsch et al. (2004) and Nicklow et al. (1972) for the frequencies of the in-plane polarized acoustic vibrations at the points K and M of the Brillouin zone of graphite. So, the frequencies are $\omega_{TA_\parallel}(M)/2\pi \approx (22.35 \pm 5\%)$ THz and $\omega_{LA_\parallel}(M)/2\pi \approx 39.7 \pm 5\%)$ THz. In the frame of the considered model we can write:

$$m\omega_{TA_\parallel}(M) = 4(\beta_{1z} + 2\beta_2) + o(\alpha_4, \beta_4); \tag{19.13}$$

$$m\omega_{LA_\parallel}(M) = 2\alpha_1 + 6\alpha_2 + 3\alpha_3 + 2\beta_{1x} + 8\beta_2 + 6\beta_3 + o(\alpha_4, \beta_4) \tag{19.14}$$

Thus, the values of all force constants that characterize the interaction between atoms of graphite can be unambiguously obtained from Equations 19.3 and 19.14. These values are given in Table 19.2. Further, we will use designations $\alpha_i \equiv \alpha(\Delta_i)$ and $\beta_i \equiv \beta(\Delta_i)$ ($i = 1, 2, 3, 4, 5$).

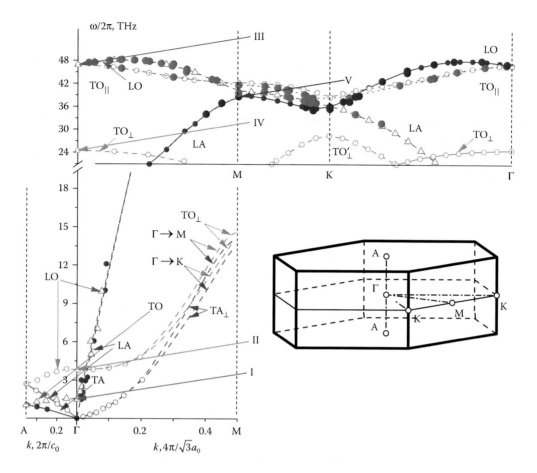

FIGURE 19.2 Dispersion curves of graphite along highly symmetrical directions in reciprocal space obtained experimentally, that is, by inelastic neutron scattering (Nicklow et al. 1972) (bottom part) and by inelastic scattering of soft x-ray radiation (Maultzsch et al. 2004) (top part). The first Brillouin zone of graphite with designations of the corresponding highly symmetrical directions is also shown.

TABLE 19.2
Force Constants of Graphite

	Δ				
	Δ_1	Δ_2	Δ_3	Δ_4	Δ_5
α, N/m	337.88	50.48	19.65	2.58	0.37
β, N/m	$\beta_x = 170.86$	–	–	–	–
	$\beta_z = 96.38$	10.15	–8.66	-65.37×10^{-3}	35.26×10^{-3}

Let us note that the noncentral interaction of the nearest neighbors in a layer plane exceeds, by two orders of magnitude, the central interlayer interaction. It means that the role of the noncentral interlayer interactions in the formation of the restoring force acting on an atom, which is displaced from the equilibrium position in the c axis direction, is dominating. Thus, β_{1x} and β_{1z} are positive, corresponding to the attraction of the nearest neighbors, while β_2 and β_3 are negative, corresponding to the repulsion between more remote atoms.

This, at first sight paradoxical fact, can be explained by different types of interactions—that between the nearest neighbors (very short-range covalent bonds) and that between more remote atoms (van der Waals bonds). The equilibrium value of the potential describing the van der Waals interaction is $r_0 \in (\Delta_4, \Delta_5)$, that is, it considerably exceeds both Δ_2 and Δ_3. Therefore, in graphene layers the interaction of the nearest neighbors is a covalent attraction, while the interaction with the second and third neighbors is van der Waals repulsion. This fact plays a key role in the formation of flexural rigidity of graphene monolayers, which in turn determines the stability of crystal structures of graphite and carbon nanofilms, and, thanks to the square-law of the dispersion $\omega(k)$ of flexural fluctuations (see, e.g., the top part of Figure 19.2), causes a non-Debye behavior of the low-temperature vibrational thermodynamic characteristics of considered systems.

In Qiang et al. (2009) an analytic formula is derived for the elastic bending modulus of monolayer graphene based on an empirical potential for solid-state carbon atoms. Theoretically, bending modulus of a monolayer graphene has been predicted on the basis of empirical potentials (Huang 2006) and *ab initio* calculations (Ribeiro et al. 2009). The fact that an atomically thin graphene monolayer has a finite bending modulus is in contrast to classical theories for plates and shells (Landau and Lifshitz 1970). For the correct evaluation of the graphene monolayer flexural stiffness it is necessary to consider either the lattice discreteness (Syrkin et al. 2009) or use the nonlocal theory of elasticity (Gelfgat 1980, Kunin 1975).

Elastic moduli C_{33} and C_{44}, corresponding to the displacement along the c axis and determining the velocity of sound, which propagates or is polarized along given direction, are 30–300 times lower than elastic moduli C_{11} and C_{66}, determining the velocity of sound propagating and polarized in basic planes (Blakslee et al. 1970, Nicklow et al. 1972). Therefore, if the vibrations polarized along the c axis would have, at small frequencies, sound and not quasiflexural character, the root-mean-square average displacements of atoms in the given direction would attain—already at low temperatures—the values corresponding to the melting of a crystal. That is, already the fact of the existence of solid graphite at room temperature gives evidence that these vibrations are essentially determined by noncentral interatomic interactions and indicates the presence of elastic stresses in graphene layers forming the crystal lattice of graphite.

In Syrkin et al. (2009) limiting transition from the equations of dynamics of a lattice to the equations of the theory of elasticity was carried out within the above-mentioned model. For the acoustic mode polarized along the c axis and propagating in the plane of layers (i.e., a quasiflexural mode) the following expression was received:

$$\omega_3^2(k_x, k_y, 0) \approx \frac{C_{44}}{\rho} \cdot k^2 + \frac{-6\beta_3}{m} \cdot \frac{\beta_{1z} + 2\beta_3}{\beta_{1z} + \beta_3} \cdot \frac{(ak)^4}{64} \quad (19.15)$$

where the elastic module C_{44} is defined by the expression (19.7), ρ is the graphite density and $k^2 = k_x^2 + k_y^2$. Having substituted into Equation 19.15 the values of parameters β_i from Table 19.2, we obtain for the flexural rigidity of graphene the following expression:

$$\kappa = \frac{a^2}{8} \cdot \sqrt{\frac{-6\beta_3}{m} \cdot \frac{\beta_{1z} + 2\beta_3}{\beta_{1z} + \beta_3}} \approx 1,21 \times 10^{-6} \text{ m}^2/\text{s}. \quad (19.16)$$

Thus, the formula (19.15) corresponds to the quasiflexural phonon mode shown on the top part of Figure 19.2.

In the following section we discuss the analysis of the phonon spectrum and vibrational characteristics of graphite, comparing them to the phonon spectrum and vibrational characteristics of graphene nanofilms.

19.3 PHONON SPECTRUM AND VIBRATIONAL CHARACTERISTICS OF GRAPHENE NANOFILMS

19.3.1 RECONSTRUCTION AND RELAXATION AT NANOFILMS FORMATION

When considering the atomic dynamics of superthin films of graphite we note that the flat form of a free graphene monolayer is not stable since even at $T = 0$ the root-mean-square displacement of atoms in the direction, perpendicular to the

layer, logarithmically diverges. Therefore, in this section the phonon spectrum and the root-mean-square amplitudes of vibrations in films consisting of two and three graphene monolayers are analyzed.

It has been shown in the previous section that in the graphite consisting of weakly bonded graphene monolayers, the interlayer interaction involves both central and noncentral forces. This distinguishes graphite from other layered crystals formed by weakly bonded fragments, containing few monolayers (e.g., from transition metal dichalcogenides; Galetich et al. 2009). Therefore in graphite, the surface formation cannot be described within the limits of the Lifshits–Rozentsveig model (Lifshits and Rozentsveig 1948), but is characterized by the surface reconstruction and surface relaxation. It is also quite natural to assume that the breakage of interlayer bonds, that is weak van der Waals interactions, should not change both the distance between atoms in graphene layers and the force constants characterizing intralayer interaction. The surface reconstruction and relaxation will be reduced only to changes of interlayer distances and force constants α_4 and β_4, and α_5 and β_5, characterizing the interlayer interaction.

The fulfillment of the condition $\sigma_{iz} n_z = 0$ leads to the flat form of layers and to the same relation between force constants and lattice parameters as the condition $C_{13} = C_{31}$. For the case of thin films consisting of N monolayers the mentioned condition can be written as

$$\beta_{1z} + 6\beta_2 + 4\beta_3 = \frac{N}{N-1} \cdot \frac{2}{3\Delta_2^2} \left[\Delta_4^2 \beta_4 - 9(\Delta_1^2 - \Delta_4^2)\beta_5 \right]. \quad (19.17)$$

Distances Δ_4 and Δ_5 in the crystal lattice of graphite differ by less than 8%. Hence, neither one can be an equilibrium distance for a pair potential describing the interlayer van der Waals interaction. The equilibrium distance r_0 for this potential lies between these values: $\Delta_4 < r_0 < \Delta_5$. The small, of the order of the amplitudes of the harmonic atomic vibrations along the c axis, difference of the distances Δ_4 and Δ_5 from r_0 makes it possible to describe the interlayer van der Waals interaction by the Lennard–Jones potential (see, e.g., Kossevich 1984):

$$\varphi_\perp(r) = \varphi_{L-J}(r) = 4\varepsilon \left[\left(\frac{\sigma}{r}\right)^{12} - \left(\frac{\sigma}{r}\right)^6 \right]. \quad (19.18)$$

The parameters of this potential can be determined from the force constants α_4, β_4, α_5, and β_5 obtained in section I: $\sigma \equiv r_0/\sqrt[6]{2} \approx 3.092$ Å; $\varepsilon \approx 152.3$ K.

Starting with Equations 19.17 and 19.18, for considered thin films of graphite it is easy to find both interlayer distance $\tilde{\Delta}_4$, and force constants $\tilde{\alpha}_4$, $\tilde{\beta}_4$, $\tilde{\alpha}_5$, and $\tilde{\beta}_5$, which describe interlayer interaction in such objects. For a two-layer film (bigraphene) $\tilde{\Delta}_4 \approx 3.636$ Å, $\tilde{\alpha}_4 \approx 372.82 \times 10^{-3}$ N/m; $\beta_4 \approx$

35.10×10^{-3} N/m; $\tilde{\Delta}_5 \equiv \sqrt{\tilde{\Delta}_4^2 + \Delta_1^2} \approx 3.902$ Å; $\tilde{\alpha}_5 \approx -87.44 \times 10^{-3}$ N/m; $\tilde{\beta}_5 \approx 41.43 \times 10^{-3}$ N/m. For a three-layer film (trigraphene): $\tilde{\Delta}_4 \approx 3.453$ Å; $\tilde{\alpha}_4 \approx 1585.10 \times 10^{-3}$ N/m; $\tilde{\beta}_4 \approx -15.34 \times 10^{-3}$ N/m; $\tilde{\Delta}_5 \approx 3.713$ Å; $\tilde{\alpha}_5 \approx 162.60 \times 10^{-3}$ N/m; $\tilde{\beta}_5 \approx 40.66 \times 10^{-3}$ N/m.

19.3.2 Spectral Densities and Mean-Square Amplitudes of Atomic Displacements

The phonon DOS $g(\omega)$ and the spectral densities $\rho_i^{(s)}(\omega)$, corresponding to the displacements of the surface atoms of the sublattice s along the crystallographic direction i were calculated, for the model proposed in the preceding section for the crystal lattice of graphite, by means of Jacobi's matrices (Haydock et al. 1972, Peresada 1968, Peresada et al. 1975). As shown in Kosevich et al. (1994), in strongly anisotropic layered crystals the interaction of the vibration modes polarized along the directions of the strong and weak bonds is proportional to the squared ratio of the weak interlayer to the strong intralayer interaction. In graphite this ratio is $(C_{33}/C_{11})^2 \sim 10^{-3}$. Consequently, for frequencies $\omega > \omega_{TO}(\Gamma)$, when the isofrequency surfaces of the vibrational branches polarized in the plane of the layer become open along the c axis, the phonon spectrum acquires a practically two-dimensional character, and the functions $\rho_{ab}(\omega)$ and $\rho_c(\omega)$ are the phonon densities of states of graphene monolayers for independent atomic vibrations in the plane of the layer and in the direction perpendicular to it.

Curve 1 in Figure 19.3a represents the calculated results for the bulk graphite phonon DOS $g(\omega)$. Figure 19.3b and c show partial contributions to $g(\omega)$ from atomic displacements in basal planes,

$$\rho_{ab}(\omega) = \frac{1}{6}\left[\rho_a^{(\circ)}(\omega) + \rho_a^{(\bullet)}(\omega) + \rho_b^{(\circ)}(\omega) + \rho_b^{(\bullet)}(\omega)\right]. \quad (19.19)$$

and Figure 19.3d and e show the contributions from displacements along the c axis,

$$\rho_c(\omega) = \frac{1}{6}\left[\rho_c^{(\circ)}(\omega) + \rho_c^{(\bullet)}(\omega)\right]. \quad (19.20)$$

The spectral densities $\rho_i^{(s)}(\omega)$ are normalized to unity and $g(\omega) = \rho_{ab}(\omega) + \rho_c(\omega)$.

In an ideal crystal lattice the spectral densities $\rho_i^{(s)}(\omega)$ normalized to unity fulfill the relation

$$\rho_i^{(s)}(\omega) = \frac{V_0}{(2\pi)^3}\sum_{\sigma=1}^{3q}\oint_{\omega_\sigma(k)=\omega}\frac{\left|e_i^{(s)}(\mathbf{k},\sigma)\right|^2}{\left|\nabla_\mathbf{k}\omega_\sigma(\mathbf{k})\right|}dS_{\mathbf{k},\sigma}, \quad (19.21)$$

where V_0 and q are the unit-cell volume and the number of atoms per unit cell, index σ enumerates the vibrational modes, $e^{(s)}(\mathbf{k},\sigma)$ are polarization vectors and the integration extends over the isofrequency surfaces in reciprocal space. The phonon DOS $g(\omega)$ is given by

$$g(\omega) = \frac{V_0}{(2\pi)^3}\sum_{\sigma=1}^{3q}\oint_{\omega_\sigma(k)=\omega}\frac{dS_{\mathbf{k},\sigma}}{\left|\nabla_\mathbf{k}\omega_\sigma(\mathbf{k})\right|} = \frac{1}{3q}\sum_{i=1}^{3}\sum_{s=1}^{q}\rho_i^{(s)}(\omega). \quad (19.22)$$

Function $\rho_{ab}(\omega)$ (Figure 19.3b and c) contains a kink at $\omega = \omega_{TO}(\Gamma)$, that is, a singularity, analogous to the three-dimensional van Hove singularity corresponding to the transition from the quadratic dependence of DOS at low frequencies, which is characteristic for crystal lattices, to a linear dependence characteristic for two-dimensional lattices (see Figure 19.3c). Other Van Hove singularities in this function have logarithmic form, which is characteristic for two-dimensional structures (see, e.g., Kossevich 1984). The isofrequency surfaces with $\omega > \omega_{TO}(\Gamma)$ are cylindrical and can be regarded as isofrequency lines in the 2D reciprocal space. The logarithmic van Hove singularities correspond to the rates of change of the topology of these isofrequency lines.

The function $\rho_c(\omega)$ (Figure 19.3d and e) acquires two-dimensional character for $\omega > \omega_{LO}(\Gamma)$. Its form corresponds to the DOS of the two-dimensional scalar model, that is, in terms of the density it is analogous to the electronic DOS of graphene (see, e.g., Novoselov et al. 2005, Skrypnyk and Loktev 2008). So, the function $\rho_c(\omega)$ contains a singularity, which is analogous to the so-called Dirac singularity in the electron DOS of graphene. This singularity likewise corresponds to the K point of the Brillouin zone.

Curves 2 in Figure 19.3 display the phonon densities of states for bigraphene (in Figure 19.3a) as well as the contributions to them from atomic displacements along the layers (in Figure 19.3b and c) and in the direction perpendicular to them (in Figure 19.3d and e).

It is evident that in a wide frequency range $\omega > \omega_{LO}(\Gamma)$ the densities of states of a film and a bulk sample are practically identical. In the frequency interval where the phonon spectrum of graphite exhibits three-dimensional behavior and the interaction between the vibrational modes polarized in the plane of the layers and in a direction perpendicular to the layers is quite strong, an appreciable difference is observed in the behavior of the corresponding spectral densities of bigraphene and the bulk sample.

In bigraphene, instead of the transverse phonon modes TA and TO, propagating along the c axis, there will be two discrete levels corresponding to the in-phase and antiphase displacements of layers in basal plane and along the c axis. The frequencies of these levels are denoted in Figure 19.3c and e as $\omega_a^{(-)}$ and $\omega_a^{(+)}$, respectively. The frequencies $\omega_a^{(+)}$ and $\omega_c^{(+)}$ correspond to the same atomic displacements as the frequencies $\omega_{TO}(\Gamma)$ and $\omega_{LO}(\Gamma)$ in the case of a bulk sample. The decrease of values $\omega_a^{(+)}$ and $\omega_c^{(+)}$ as compared to the frequencies $\omega_{TO}(\Gamma)$ and $\omega_{LO}(\Gamma)$ is due to the surface relaxation. The spectral density $\rho_{ab}(\omega)$ of bigraphene, as seen in Figure 19.3c, exhibits kinks at $\omega = \omega_a^{(-)}$ and $\omega = \omega_a^{(+)}$. For $\omega > \omega_a^{(+)}$, the spectral density acquires a two-dimensional form ($\rho_{ab}(\omega) \sim \omega$). The spectral density $\rho_c(\omega)$ of bigraphene, as seen in Figure 19.3e,

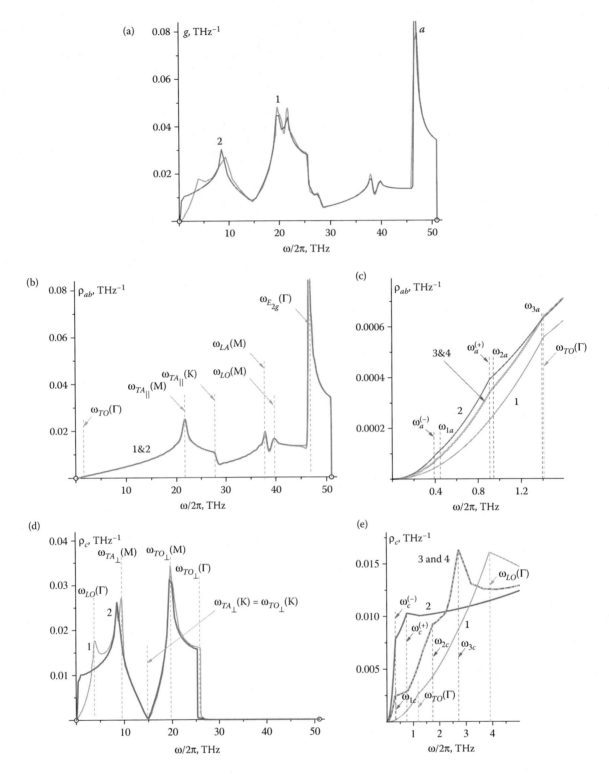

FIGURE 19.3 The phonon DOS of bulk graphite and bilayer graphene (part *a*, curves 1 and 2, respectively), as well as the partial contributions to them from the atomic displacements along the basal plane (parts *b* and *c*) and along axis *c* (parts *d* and *e*). In parts *c* and *e* the corresponding dependence for trigraphene (curves 3 for ABA-stacking, curves 4 for ABC-stacking) are shown.

exhibits kinks at $\omega = \omega_c^{(-)}$ and $\omega = \omega_c^{(+)}$. For $\omega \geq \omega_c^{(+)}$ there is a range around 2 THz on the spectral density of $\rho_c(\omega)$, in which this function weakly depends on the frequency. The spectral density takes the form corresponding to that of the flexural waves propagating in the plane. The flexural wave has

a quadratic dispersion relation in the long-wave length region and the DOS of the flexural mode at $\omega \to 0$ converges to the constant value of ~κ^{-1} (κ is the flexural rigidity (19.16)).

In stacks with more than one graphene layer, the relative position of two neighboring layers allows for two different orientations of the third layer. If we label the positions of the first two atoms as A and B, the third layer can be of type A, leading to the sequence ABA, or it can fill a third position different from A and B, that is C. There are no more nonequivalent positions where a new layer can be placed, so that thicker stacks can be described in terms of these orientations. Regions with the stacking ABC (rhombohedral stacking) have also been observed in different types of graphite. Finally, graphene films with stacking order in which all atoms in one layer occupy positions directly above the atoms in the neighboring layers (AA—hexagonal stacking) have been considered theoretically and have been grown on a (111) diamond (Lee et al. 2008), SiC (0001) (Borysiuk et al. 2011), and Ru (0001) (Papagno et al. 2012).

The ABC-stacking has not four (as ABA) but eight atoms in its unit cell and, accordingly, its phonon spectrum contains twice more branches. This circumstance suggests an appreciable difference of behavior of phonon densities in the low-frequency region. Naturally, this difference can be manifested only at frequencies below the frequency of the first van Hove peculiarity, corresponding to the transition from the closed isofrequency surfaces to open ones along the c axis. This is manifested also in the temperature dependences of vibrational thermodynamic performances at low temperatures. The study of the influence of the structure on phonon spectra and vibrational performances of carbon nanofilms is of particular interest.

The spectral densities corresponding to atomic displacements along different crystallographic directions for a trigraphene of both ABA and ABC stacking are shown in Figure 19.3c and e (curve 3 for ABA, curve 4 for ABC).

It is clearly visible that

1. The frequency of the transition into the quasi two-dimensional behavior of spectral densities (for bulk graphite it is the frequency of the first van Hove peculiarity) decreases with decreasing number of layers for $\rho_c(\omega)$, but it practically does not change for $\rho_{ab}(\omega)$;
2. For spectral densities $\rho_{ab}(\omega)$, the quasi two-dimensional behavior leads to $\rho_{ab}(\omega) \sim \omega$, that is, they show two-dimensional Debye character. In the long-wave limit the spectral density $\rho_c(\omega)$ is comparable to the densities of states in a quasiflexural mode with the quadratic dispersion relation and it approaches to a constant value, which is inversely proportional to flexural rigidity of layers;
3. In spectral densities of nanofilms at frequencies below the frequency of the transition to the quasi two-dimensional behavior additional singularities appear, caused by the transformation of phonon branches with the wave vector along the c axis into discrete levels with frequencies $\omega_{a,c}^{(-)}$ and $\omega_{a,c}^{(+)}$ for bigraphene and $\omega_{1a,c}$, $\omega_{2a,c}$ and $\omega_{3a,c}$ for trigraphene;
4. There is no essential difference between the spectral densities $\rho_{ab}(\omega)$ and $\rho_c(\omega)$ for trigraphenes with ABA-stacking and ABC-stacking (see Figure 19.3c and e).

Let us note that if with the frequency approaching to zero the spectral density does not converge to zero, the mean-square displacements of atoms will diverge even at zero temperature. This circumstance causes the instability of linear chains and flat monolayers. It is apparent from Figure 19.3, that $\rho_c(\omega)$ of graphene nanofilms (especially of bigraphene) weakly depends on ω even at very low frequencies ($\omega \geq \omega_c^{(+)}$), which could lead to large mean-square displacements of atoms in the direction perpendicular to layers. It, at first sight, calls into question the stability of the lattice of a bigraphene and the applicability of the harmonic approach for the description of the lattice vibrations. In Figure 19.4, the temperature dependences of the mean-square amplitudes of atomic displacements of bigraphene, trigraphene, and bulk graphite along layers and in the perpendicular direction to them are presented.

The mean-square amplitudes atomic displacements of atoms s along crystallographic direction i are expressed in terms of the spectral density $\rho_i^{(s)}(\omega)$ as (Kossevich 1984, Peresada 1968)

$$\left\langle \left| u_i^{(s)} \right| \right\rangle \equiv \sqrt{\left\langle \left(u_i^{(s)} \right)^2 \right\rangle_T} ; \left\langle \left(u_i^{(s)} \right)^2 \right\rangle_T = \frac{\hbar}{2m_s} \int_D \frac{d\omega}{\omega} \cdot \mathrm{cth}\left(\frac{\hbar\omega}{2kT} \right) \cdot \rho_i^{(s)}(\omega)$$

(19.23)

We note that in graphite the mean-square amplitudes of atomic vibrations along the weak coupling direction (see Figure 19.4) at room temperature, calculated by means of spectral densities $\rho_{ab}(\omega)$, are approximately 0.12 Å, which is about 3% of the corresponding interatomic distance. Therefore, it is quite justified to calculate the vibrational performances of graphite by using harmonic approach.

It is evident from Figure 19.4, that the amplitudes of the atomic displacements along the graphene layers are

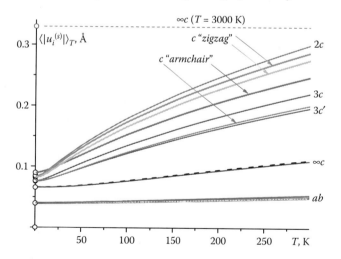

FIGURE 19.4 Temperature dependence of the mean-square amplitudes of atoms along different crystallographic directions (along the ab plane and along the c axis) in the graphene nanofilms consisting of two and three graphene monolayers, as well as "trigraphen–bigraphene" atomic steps of the "zigzag" and "armchair" type.

practically independent of the thickness of the sample. The mean-square amplitudes of atomic displacements along the c axis increase strongly with the decreasing film thickness. Thus, the room-temperature amplitude of the transverse vibrations of an atom in the central layer in trigraphene (curve 3c′) is twice the corresponding value of the bulk sample (curve ∞c). Curves 3c (atom in the surface layer in trigraphene) and 2c (atom in bigraphene) are shifted to even higher values. In Figure 19.4, the horizontal line marks the mean-square amplitudes of atomic vibrations along the c axis in bulk graphite at $T = 3000$ K. This temperature is by approximately 1000 K lower than the melting temperature of graphite, $T_{me} \approx 4000 \pm 50$ K. Therefore, at $T = 3000$ K the crystal lattice of graphite possesses a sufficient margin of stability. Since at room temperatures the value of the mean-square amplitudes of atomic vibrations of bigraphene and trigraphene lie appreciablly below the dashed line, bigraphene and trigraphene possess an adequate margin of stability at room temperature.

In the same figure, the temperature dependences of the mean-square amplitudes of atoms within the defects of the "step on surface" type are shown. Such defects essentially change the electronic spectrum of carbon nanofilms. Their formation can be used as a method for preparation of carbon nanosystems with desired electronic characteristics. The dependences $\langle | u_i^{(s)} | \rangle$ given in Figure 19.4 for two basic configurations of such a defect (*zigzag* and *armchair*) show that such defects are stable up to 500 K.

It is necessary to note that graphene monolayers cannot exist in a free state as flat two-dimensional structures, since in that case there is a full splitting of the bulk phonon modes polarized in the layer plane and flexural modes. Therefore, the spectral density of the flexural mode with the dispersion law $\omega = \kappa k^2$ (formula (19.15) for $C_{44} = 0$) leads to the divergence of mean-square displacements even at $T = 0$. So the graphene monolayers can exist in the flat shape only when adsorbed on some substrate. For studying the electronic spectra of graphene monolayers dielectric substrates are usually used, in which bonds between carbon atoms and substrate atoms have van der Waals character. The contribution of the substrate to the phonon spectrum of graphene is manifested in intertwinning of longitudinal modes with quasiflexural ones, which then will take their usual form (19.15). So, the phonon spectrum of the given heterostructure practically corresponds to the phonon spectrum of a graphene monolayer and will not differ essentially from $\omega_{TO}(\Gamma)$ for longitudinal mode and from $\omega_{LO}(\Gamma)$ for quasiflexural mode. The influence of substrate on the phonon spectrum of a graphene monolayer practically disappears at frequencies exceeding ω_{TO} and ω_{LO}.

19.3.3 PHONON HEAT CAPACITY OF GRAPHITE AND GRAPHENE NANOFILMS: ITS "NON-DEBYE" BEHAVIOR

Experimental study of the low-temperature heat capacity is an important and reliable source of information on the

quasiparticle excitations and, in particular, on the phonon spectra. Calorimeter experiments are, as a rule, very accurate and, unlike the majority of optical and ultrasonic experiments, do not require good quality single crystals.

It is evident from both the experimental data (Nicklow et al. 1972, Maultzsch et al. 2004), and the calculations given above, that the phonon spectrum band for graphite and carbon nanofilms is very wide. Calculated Debye temperature for graphite is about 2500 K and therefore the phonon heat capacity increases with temperature much more slowly than the majority of solids. At room temperatures the phonon heat capacity is still far from saturation and the electronic contribution to the total heat capacity cannot be considered as negligibly small (when compared to the phonon contribution). Therefore, the analysis of the temperature dependences of the graphite and carbon nanofilm phonon heat capacity, based on microscopic calculations is needed for correct interpretation of calorimeter measurements.

The temperature dependence of the molar isochoric phonon heat capacity $C_V(T)$ of a system whose all atoms have three degrees of freedom is described through its phonon density $g(\omega)$ as follows (see, e.g., Kossevich 1984):

$$C_V(T) = 3R \cdot \int_D \left(\frac{\hbar\omega}{2k_BT} \right)^2 \cdot \sinh^{-2}\left(\frac{\hbar\omega}{2k_BT} \right) g(\omega)d\omega, \qquad (19.24)$$

where k_B is the Boltzmann constant and $R \equiv N_A k_B$ is the gas constant (N_A is the Avogadro number). Results of calculations for graphite, bigraphene, and trigraphene (ABA, ABC) are shown in Figure 19.5. This figure shows the temperature dependences of the phonon heat capacity of graphite and graphene nanofilms, and also the contributions to these quantities from atomic vibrations along the basal plane and along the c axis. We use the same designation (numbering of curves, colors) as in Figure 19.3.

Usually, when explaining the temperature dependence of the phonon heat capacity at low temperatures, it is assumed

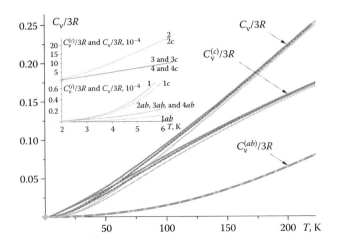

FIGURE 19.5 Phonon heat capacities of graphite and graphene nanofilms. The colors and numbering correspond to the curves in Figure 19.3.

that the main contribution to the phonon heat capacity comes from the long-wave low-frequency phonons with the sound dispersion relation $\omega(\mathbf{k}) = sk$, where s is the sound velocity depending on the direction of the sound propagation and polarization. For such phonons, the DOS has a so-called Debye form:

$$g^{(q)}(\omega) = \frac{q\omega^{q-1}}{[\omega_D^{(q)}]^q}, \qquad (19.25)$$

where q is the dimensionality of the lattice (number of degrees of freedom), ω_D is the Debye frequency averaged through all directions and polarizations. Then, the phonon heat capacity is given by the expression

$$C_D(T) = qR\left\{D_q\left(\frac{\Theta_D^{(q)}}{T}\right) - \frac{\Theta_D^{(q)}}{T}D_q'\left(\frac{\Theta_D^{(q)}}{T}\right)\right\} \qquad (19.26)$$

where $\Theta_D^{(q)} \equiv \hbar\omega_D^{(q)}/k_B$, $D_q(\Theta_D^{(q)}/T)$ is the so-called Debye function and

$$D_q(x) \equiv \frac{q}{x^3}\int_0^x \frac{z^q\,dz}{e^z - 1}. \qquad (19.27)$$

In the literature, we usually find these expressions for $q = 3$, see, for example Landau and Lifshits (1969).

At low temperatures $T \ll \Theta_D^{(q)}$, the heat capacity is proportional to T^q. In Figure 19.5, the contribution to the heat capacity from displacements of atoms along the basal plane at low temperatures is really close to the quadratic one, which is in good agreement with the quasi two-dimensional type of the spectral densities $\rho_{ab}(\omega)$ (Figure 19.3). We note that the presence of the symmetry axis of the sixth order in graphene layers makes them elastic isotropic, which consequently leads to the agreement of $C_V^{(ab)}(T)$ with the Debye model. However, the main contribution to the low-temperature heat capacity of graphite and, especially, of graphene nanofilms comes, as is apparent from Figure 19.5, from atomic displacements along the c axis. These displacements propagate in the plane of layers as quasiflexural waves with nonacoustic dispersion relation (19.15) and their contribution to the phonon heat capacity cannot be described by the Debye approximation for any q value. The possibility of the decisive contribution of the quasiflexural vibrations to the heat capacity of chain and layered structures was mentioned for the first time in Lifshits (1952).

It is also evident from Figure 19.5 that with the decreasing number of layers the phonon heat capacity increases, the main part of which is caused by vibrations of atoms in the perpendicular direction to the layer and no appreciable difference in the behavior of the phonon heat capacity of trigraphenes with different structures is observed.

To summarize, we note that the anomalies in thermal expansion of layered crystals, in particular in graphite, the

so-called "membrane effect," that is negative values of the thermal expansion in the layer plane, are not the result of the presence of quasiflexural modes with dispersion relation (19.15) in these crystals. The origin of this abnormality is a much faster increase with temperature of the mean-square displacements of atoms along the c axis than in the direction along layer (Feodosyev et al. 1998). Such a "membrane effect" is observed in a number of compounds in which the quasiflexural mode in phonon spectra is manifested very weakly (e.g., in niobium diselenide; Gospodarev et al. 2013) or is even absent at all (HTSC of 123 type; Eremenko et al. 2006).

19.4 DISTINCTIVE PROPERTIES OF IMPURITIES EFFECT ON THE PHONON SPECTRUM OF THE GRAPHENE LAYERS

19.4.1 LOCALIZATION OF THE VIBRATIONS ON POINT DEFECTS

As was noted in the previous section, function $\rho_c(\omega)$ at $\omega > \omega_{LO}(\Gamma)$ gets two-dimensional character, that is it actually represents the DOS of the quasiflexural branch of a graphene monolayer. Thus, the shape of this function corresponds to the phonon DOS of a two-dimensional scalar model and so it is fully analogous to the electronic density of graphene (Novoselov et al. 2005). In particular, on the function $\rho_c(\omega)$ near to frequencies $\omega_{TA\perp}(K) \approx \omega_{TO\perp}(K)$ V-shaped peculiarities appear, similar to the Dirac-like peculiarities on the density of the graphene electronic states. Figure 19.6 presents the spectral density (normalized to unity) of vibrations polarized along the c axis (i.e., $3\rho(\omega)$, curve 1) and the real part of the corresponding local Green's function Re $G_c(\omega)$ (curve 2) is related to this function by the Kramers–Kronig relation. The behavior of Re$G_c(\omega)$ near $\omega_{TA\perp}(K) \approx \omega_{TO\perp}(K)$ and at $\omega > \omega_{TO}(\Gamma)$ gives foundation to suppose the existence of the Lifshits equation solutions (see, e.g., Kossevich 1984, Lifshits

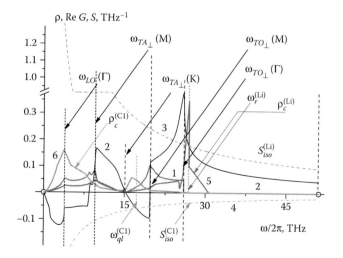

FIGURE 19.6 Formation of localized levels of impurity atoms in graphene.

1945), which determines the levels localized at the impurity atom:

$$\text{Re } G_c(\omega) = S(\omega,\{\Lambda_i\}). \qquad (19.28)$$

In Equation 19.28, the function $S(\omega\{\Lambda_i\})$ describes the defect influence ($\{\Lambda_i\}$ is a set of parameters characterizing defect).

In the frequency range $[\omega_{TO\perp}(\Gamma), \omega_{max}]$ (ω_{max} is the maximum frequency of the band of quasicontinuous spectrum) vibrations polarized along the c axis are practically absent (their contribution is exponentially small) and the function Re $G_c(\omega)$ has nearly hyperbolic shape, characteristic for the Green's function outside the quasicontinuous band (when the given function is real). Therefore, in this frequency range sharp resonant levels caused by weak- or strong-bonded impurities may appear. These levels are similar to local vibrations that arise under the influence of such defects outside the quasicontinuous band (Lifshits 1945).

As an illustration, Figure 19.6 shows the solutions of Equation 19.28 for the case of light (curve 3) and heavy (curve 4) isotopic substitution impurities (in isotopic approach these solutions describe the impurities Li and Cl, respectively), and also the local spectral density of the impurity atoms (curves 5 and 6, respectively). Function $S(\omega,\{\Lambda_i\})$ in case of an isotopic substitution impurity depends, in addition to the frequency, on the mass of defect, $\varepsilon(m'-m)/m$ (m' and m are mass of impurity atom and of the atom of the host lattice, respectively) (see, e.g., Kossevich 1984).

The behavior of curve Re $G_c(\omega)$ gives ground to guess that also other defects (e.g., intercalated metals) may stimulate the formation of the localized vibrations with frequencies near to $\omega_{TA\perp}(K) \approx \omega_{TO\perp}(K)$ and may lead to considerable increase in the number of phonons near this frequency.

19.4.2 PHONON SPECTRUM OF GRAPHITE INTERCALATED BY METALS

Graphite structures, intercalated by metals are interesting because the temperature of superconducting transition T_c for such compounds essentially depends on the type of intercalator. For example, T_c for C_6Yb is 6.5 K while for C_6Ca it is 11.5 K. Since the electronic spectra of these compounds probably do not depend on the type of the intercalated metal, T_c variations in such compounds are dominantly determined by the peculiarities of their phonon spectra. Electronic spectra of C_6Ca and C_6Yb should not differ qualitatively even though their structures are different. In C_6Yb intercalated metal forms a HCP structure and the unit cell of this compound contains 14 atoms. In C_6Ca intercalated metal forms a BCC a lattice and the unit cell consists of seven atoms. However, the densities of states in these closely packed lattices differ only slightly from each other (at identical values of force constants or overlapping integrals a difference appears, starting from the fourth moment). Overlapping metal–metal and metal–carbon integrals differ only slightly (as calcium and ytterbium belong to the same group in Periodic table) and hence the differences in electronic spectra of these compounds are not expected.

The presence of a singularity in the phonon spectrum of the quasiflexural vibrations in graphite, analogous to the Dirac singularity in the electronic spectrum of graphene, can lead to anomalies in the phonon DOS at frequencies near the frequency of this singularity. Quasiflexural vibrations are analogous to the quasilocal vibrations often appearing in the low-frequency region of the quasicontinuous spectrum of various lattices under the influence of strongly or weakly bonded impurities. Models of such lattices are well known (see, e.g., Kossevich 1984).

Data on ultrasonic, optical, and other properties of C_6Ca and C_6Yb, which would allow us to calculate the parameters of interatomic interactions (as it was done in the first section) are currently lacking. It has led us to make some assumptions which will not quantitatively affect the behavior of the studied spectral characteristics.

Metal intercalated graphite consists of weakly bonded (by van der Waals forces) graphene monolayers, between which triangular lattices of metal with a period of $a_0\sqrt{3}$ are placed. Atoms of both sublattices of graphene monolayers are placed one below the other. The distance between metal layers is in both compounds $c' \approx 4,5$ Å, see, for example, Emery et al. (2005) and Weller et al. (2009).

We neglect the interaction between carbon and metal atoms situated in different layers. The interaction of metal atoms within one layer is considered to be a central force, that is, the matrices of force constants have form (19.1) for $\beta_z(\Delta) = \beta_x(\Delta) = 0$, where $\Delta = a_0\sqrt{3}$. The value of $\alpha(a_0\sqrt{3})$ can be obtained from Young's modulus for Ca and Yb polycrystals ($E_{ca} \approx 2.6 \times 10^{10}$ N/m² and $E_{Yb} \approx 1.185 \times 10^{10}$ N/m²). The value of $\alpha(a_0\sqrt{3})$ in monolayers of calcium is ≈ 4.0 N/m, and in monolayers of ytterbium ≈ 2.75 N/m. The distance between the nearest atoms of metal and carbon is $r_{C-Me} \equiv \sqrt{(c'/2)^2 + a_0^2/3} \approx 2.66$, that is, it is larger than between the second-nearest carbon neighbors $\alpha_0 \approx 2.45$ Å, but lower than between the third neighbors $(2a_0/\sqrt{3} \approx 2,83$ Å). Therefore, it is natural to assume that the potential describing this interaction may be considered as pair and isotropic, that is $\beta_z(r_{C-Me}) = \beta_x(r_{C-Me}) = \beta_x(r_{C-Me})$.

As atomic spacings in graphene monolayers do not vary due to metal intercalation, the force constants describing corresponding interatomic interactions do not change, too. Thus, the quantity $\beta(r_{C-Me})$ can be found from the condition $C_{13} = C_{31}$, which in this case becomes

$$\beta_z\left(\frac{a_0}{\sqrt{3}}\right) + 6\beta(a_0) + 4\beta\left(\frac{2a_0}{\sqrt{3}}\right) = 2\left[\left(\frac{c'}{2a_0}\right)^2 - \frac{1}{3}\right]\beta(r_{C-Me})$$

$$(19.29)$$

from which $\beta(r_{C-Me}) \approx 0.31$ N/m, both for the interaction of carbon with calcium and of carbon with ytterbium. Experimental data for the determination of the force constants $\alpha(r_{C-Me})$,

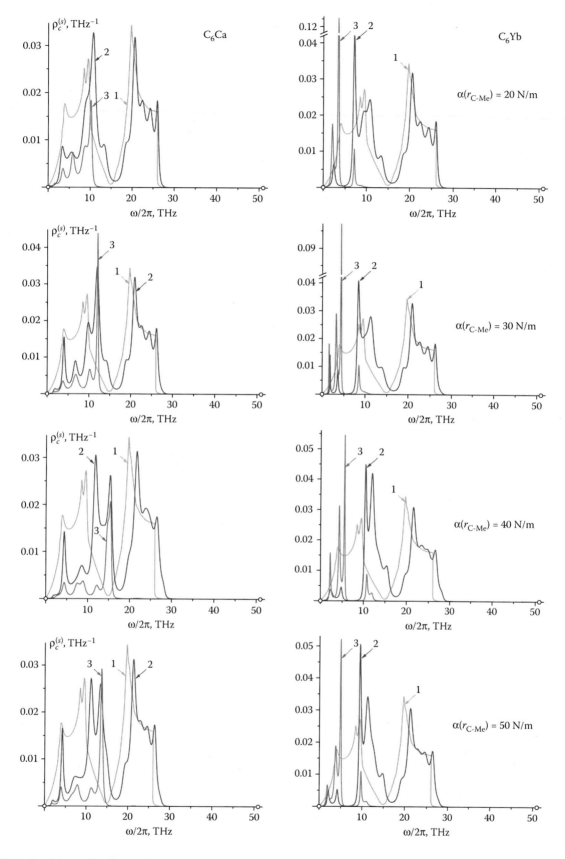

FIGURE 19.7 Partial contributions to the phonon DOS of the intercalated graphite from the displacements (along the *c* axis) of carbon (curves 2) and metal (curves 3) atoms. Curves 1 correspond to pure graphite. Top to bottom $\alpha(r_{C-Me})$ = 20, 30, 40, and 50 N/m.

characterizing central-force interaction between atoms of metal and carbon are not available yet.

Therefore it is supposed, based on the distance between atoms of carbon and metal in graphite, that the value of this quantity will lie in the range from $\alpha(r_{C-Me}) \approx 19.65$ N/m to $\alpha(a) \approx 50.48$ N/m. From this range, we picked up four values of this quantity, $\alpha(r_{C-Me}) = 20.0, 30.0, 40.0,$ and 50.0 N/m.

Figure 19.7 shows the frequency dependences of partial contributions to the density of phonon states from displacements of metal and carbon atoms in the direction perpendicular to layers. The areas below dependences corresponding to intercalating metal are hatched. In Figure 19.7, the left set shows dependences for C_6Ca, the right one for C_6Yb, the force constant $\alpha(r_{C-Me})$ increases from top to bottom. We see that for C_6Ca, sharp resonance peaks appear on partial contributions from both intercalating metal and carbon. These peaks are shifted, with the increase of $\alpha(r_{C-Me})$, toward the center of the frequency range $[\omega_{TA_\perp}(M), \omega_{TO_\perp}(M)]$, leading to the increase of phonon DOS near the Brillouin zone's K–point, through which Fermi level of electrons in graphene passes. For C_6Yb compound (Yb has more than four times larger atomic mass than Ca) the resonance peaks appear at lower frequencies and an apparent increase of the phonon state density at frequencies near the Brillouin zone's K-point is observed only for anomalously large values of $\alpha(r_{C-Me}) \sim 40–50$ N/m. Note that the temperature of superconducting transition for C_6Ca is almost 1.8 times higher than that for C_6Yb.

Figures 19.8a and b present the total phonon DOS for both intercalated compounds C_6Ca and C_6Yb (curves 2) as well as contributions to these functions from displacements of intercalating metal atoms (curves 3) and the carbon atoms (curves 4).

For comparison, in Figure 19.8 the phonon DOS of pure graphite is shown by curve 1. It may be seen that the metal intercalation appreciably changes the phonon spectrum not only at low frequencies, corresponding to the bands of quasicontinuous spectrum of intercalating metals, but also in the high-frequency region $[\omega_{TA_\perp}(M), \omega_{TO_\perp}(M)]$. The presence of

an intercalator essentially affects the vibrational spectrum of carbon, mainly for vibrations polarized along the c axis (perpendicularly to graphene layers). In the case of intercalation by lighter atoms of calcium this effect is manifested more strongly.

Note that the peculiar features of phonon spectra and vibrational characteristics of graphene systems are one of the effects of the strong anisotropy of the interatomic interactions. The strong anisotropy of interatomic interactions is characteristic for many compounds and could be responsible for important properties of their electronic (in particular superconducting), magnetic, and other characteristics. It determines the character of the phase transition and critical phenomena in some structures (Sirenko 2012).

Such unusual behavior of the phonon subsystem in the metal intercalated graphite can essentially influence the electronic properties, in particular the temperature of superconducting transition T_c. Some qualitative considerations about this topic are given in Feher et al. (2009).

It is possible to state unambiguously that for the increase of T_c it is necessary to have high phonon frequencies, large electron–phonon interaction and a considerable density of electron states on the Fermi surface. Such properties are manifested in, for example, compounds of metals with light elements—such compounds are, for example, hydrides, borides, carbides, and nitrides—since the phonon spectra show high-frequency modes corresponding to the vibrations of light atoms (H, B, C, N). According to Maksimov et al. (2004), Nagamatsu et al. (2001), and Pickett (2006) electrons in MgB_2 strongly interact only with two quasiflexural modes.

19.5 CONCLUSION

We can conclude that our model of the graphite lattice reflects all distinctive peculiarities of its phonon spectrum and allows to calculate the vibrational characteristics with high accuracy. All force constants in the proposed model are determined from reliable experimental data obtained by various

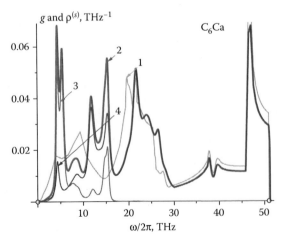
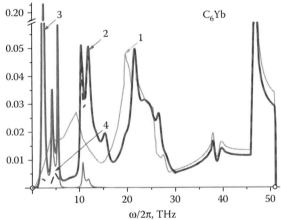

FIGURE 19.8 Phonon DOS of intercalated graphite (curves 2) and the contributions to them from the displacements of metal (curves 3) and carbon atoms (curves 4). The value of parameter $\alpha(r_{C-Me})$ is 50 N/m. Curves 1 are the phonon DOS of pure graphite.

experimental methods, such as acoustic, neutron diffraction, and optical measurements and scattering of the soft x-rays.

Graphene monolayers possess large flexural stiffness, which is clearly manifested in the dispersion relation of the transversal phonons polarized along the c axis. A "non-Debye" behavior of the low-temperature heat capacity of graphene nanofilms is caused by the deviation from the acoustic dispersion law for the vibrations polarized along the c axis. On a large part of the quasicontinuous spectrum of graphite the isofrequency surfaces are open along the c axis, that is at the frequencies exceeding the frequency of the first van Hove singularity dispersion relations and phonon densities of graphite are identical with those for graphene monolayers.

The transition frequency to the quasi two-dimensional behavior (for bulk graphite it is the frequency of the first van Hove singularity) the spectral density $\rho_c(\omega)$ decreases with the decrease in the number of layers, but the spectral density $\rho_{ab}(\omega)$ practically does not change. For spectral densities $\rho_{ab}(\omega)$ the quasi two-dimensional behavior means that $\rho_{ab}(\omega) \sim \omega$, that is, it has a two-dimensional Debye form. In the long-wave limit the spectral density $\rho_c(\omega)$ is comparable with the densities of states in a quasi-flexural mode with quadratic dispersion relation and it approaches to a constant value, which is inversely proportional to the flexural rigidity of layers. In the spectral densities of nanofilms at frequencies below the transition frequency to the quasi two-dimensional behavior additional singularities appear, caused by the transformation of graphite phonon branches of graphite with the wave vector along the c axis into discrete levels. With the decrease in the number of layers the phonon heat capacity increases and this increase is mainly caused by the perpendicular vibrations of atoms.

There is no essential difference between the spectral densities $\rho_{ab}(\omega)$ and $\rho_c(\omega)$ for trigraphene with ABA-stacking and with ABC-stacking. Also, there is no observable difference in the phonon heat capacity of trigraphenes with different stacks.

A specific form of the spectral density of the corresponding real part of the Green function of graphene monolayers leads to the formation, under the influence of defects, of localized states with the frequencies exceeding the maximum frequency in the quasiflexural branch. It also leads to the formation of the quasilocalized vibrations with frequencies near to the quasiflexural mode frequency in the K-point of the Brillouin zone. Therefore, intercalation of graphite by metals substantially enriches the phonon DOS, the frequency corresponding to the K-point of the first Brillouin zone of pure graphite. The amount of such enrichment correlates with the changes of the superconducting transition temperature in C_6Ca and C_6Yb.

ACKNOWLEDGMENT

This work is one of the outputs of the Research and Education at UPJŠ—heading toward Excellent European Universities project, ITMS project code 26110230056, supported by the Operational Program Education funded by the European Social Fund (ESF).

REFERENCES

Blakslee, O.L., D.G. Proctor, and G.B. Spence. 1970. Elastic constants of compression-annealed pyrolitic graphite. *J. Appl. Phys.* 41:3373–3382.

Borysiuk, J., J. Soltys, and J. Piechota. 2011. Stacking sequence dependence of graphene layers on SiC ($000\overline{1}$)—Experimental and theoretical investigation. *J. Appl. Phys.* 109:093523.

Dahal, A. and M. Batzill. 2014. Graphene–nickel interfaces: A review. *Nanoscale* 6:2548–2562.

Dresselhaus, M.S. and G. Dresselhaus. 1981. Intercalation compounds of graphite. *Adv. Phys.* 30:139–326.

Emery N., C. Herold, M. d'Astuto et al. 2005. Superconductivity of bulk CaC_6. *Phys. Rev. Lett.* 95:087003.

Endlich, M., A. Molina-Sanchez, L. Wirtz, and J. Kroger. 2013. Screening of electron-phonon coupling in graphene on Ir(111). *Phys. Rev. B* 88:205403.

Eremenko, V.V., I.A. Gospodarev, V.V. Ibulaev, V.A. Sirenko, S.B. Feodosyev, and M. Yu. Shvedun. 2006. Anisotropy of $Eu_{1+x}(Ba_{1-y}Ry)_{2-x}Cu3O_{7-d}$ lattice parameters variation with temperature in quasi-harmonic limit. *Low Temp. Phys.* 32:1560–1565.

Fedorov, A.V., N.I. Verbitskiy, D. Haberer et al. 2014. Observation of a universal donor-dependent vibrational mode in graphene. *Nat. Comm.* 5:3257.

Feher, A., I.A. Gospodarev, V.I. Grishaev et al. 2009. Effects of defects on quasiparticle spectra of graphite and graphene. *Low Temp. Phys.* 35:679–686.

Feodosyev, S.B., I.A. Gospodarev, and E.S. Syrkin. 1998. Anomalies of linear expansion coefficient in highly anisotropic crystals at low temperature. *Phys. Stat. Sol. (b)* 150:K19–K23.

Ferrari, A.C. and D.M. Basko. 2013. Raman spectroscopy as a versatile tool for studying the properties of graphene. *Nat. Nanotech.* 8:235–246.

Galetich, I.K., I.A. Gospodarev, V.I. Grishaev et al. 2009. Vibrational characteristics of the niobium dichalcogenide. Bulk samples and nano-films. *Superlattices Microstruct.* 45:564–575.

Gelfgat, I.M. 1980. Account of space dispersion at studying the non-Rayleigh surface wave in crystals. *Solid State Phys.* 22: 2815–2816.

Gospodarev, I.A., K.V. Kravchenko, E.S. Syrkin, and S.B. Feodos'ev. 2009. Quasi-two-dimensional features in the phonon spectrum of graphite. *Low Temp. Phys.* 35:589–595.

Gospodarev, I.A., A.V. Eremenko, K.V. Kravchenko et al. 2013. Distinctive features of thermal expansion of niobium diselenide. *Phys. Solid State* 55:898–904.

Haydock, R., V. Heine, and M.J. Kelly. 1972. Electronic structure based jn the local atomic environment for tight-binding bands. *J. Phys. C.* 5:2845–2858.

Herziger, F. and J. Maultzsch. 2013. Influence of the layer number and stacking order on out-of plane phonons in few-layer graphene. *Phys. Status Sol. (b)* 250:2697–2701.

Huang, Y., J. Wu, and K.C. Hwang. 2006. Thickness of graphene and single-wall carbon nanotubes. *Phys. Rev B* 74:033524.

Kosevich, A.M., E.S. Syrkin, and S.B. Feodosyev. 1994. Peculiar features of phonon spectra of low-dimensional crystals. *Phys. Low-Dim. Struct.* 3:47–51.

Kossevich, A.M. 1984. *The Crystal Lattice. Phonons, Solitons, Dislocations.* Berlin: Wiley-VCH Verlag GmbH.

Kunin, I.A. 1975. *Theory of Elastic Media with Microstructure.* Moscow: "Nauka" (in Russian).

Landau, L.D. and E.M. Lifshitz. 1970. *Theory of Elasticity (Vol. 7 of the Course of Theoretical Physics).* Oxford: Pergamon Press.

Landau, L.D. and E.M. Lifshits. 1969. *Stastistical Physics (Vol. 5 of a Course of Theoretical Physics).* Oxford: Pergamon Press.

Lee, J.-K., S.-C. Lee, J.-P. Ahn et al. 2008. The growth of AA graphite on (111) diamond. *J. Chem. Phys.* 129:234709.

Lifshits, I.M. 1945. On theory of regular excitations. *Rep. AS USSR.* 48:83–86 (in Russian).

Lifshits, I.M. 1952. About heat properties of chain and layered structures. *Zh. Exp. Teor. Fiz.* 22: 475–478 (in Russian).

Lifshits, I.M. and L.N. Rozentsveig. 1948. Dynamics of crystal lattice filling the semi-space. *Zh. Exper. Teor. Fiz.* 18: 1012–1022 (in Russian).

Lui, C.H. and T.F. Heinz. 2013. Measurement of layer breathing mode vibrations in few-layer graphene. *Phys. Rev. B* 87:121404(R).

Maksimov, E.G., M.V. Magnitskaya, S.V. Ebert, and S.Yu. Savrasov. 2004. Ab initio calculations of the superconducting transition temperature for NbC at various pressures. *JETP Lett.* 80:548–551.

Maultzsch, J., S. Reich, C. Thomsen, H. Requardt, and P. Ordejyn. 2004. Phonon dispersion in graphite. *Phys. Rev. Lett.* 92:075501.

Nagamatsu, J., N. Nakagawa, T. Muranaka, Y. Zenitani, and J. Akimitsu. 2001. Superconductivity at 39K in magnesium diboride. *Nature* 410:63–64.

Nicklow, R., N. Wakabayashi, and H.G. Smith. 1972. Lattice dynamics of pirolitic graphite. *Phys. Rev. B* 5:4951–4962.

Nika, D.L. and A.A. Balandin. 2012. Two-dimensional phonon transport in graphene. *J. Phys.: Condens. Matter* 24:233203.

Novoselov, K.S., A.K. Geim, S.V. Morozov et al. 2005. Two-dimensional gas of massless Dirac fermions in graphene. *Nature* 438:197–200.

Papagno, M., D. Pacilé, D. Topwal et al. 2012. Two distinct phases of bilayer graphene films on Ru(0001). *ACS Nano* 6:9299–9304.

Park, C.H., N. Bonini, T. Sohier et al. 2014. Electron–phonon interactions and the intrinsic electrical resistivity of graphene. *Nano Lett.* 14:1113–1119.

Peresada, V.I. 1968. New computational method in theory of crystal lattice. *Condensed Matter Phys.* (Kharkov). 2:172–210 (in Russian).

Peresada, V. I., V.N. Afanas'ev, and V.S. Borovikov. 1975. On calculation of density of states of single-magnon perturbations in an ferromagnetics. *Soviet Low Temp. Phys.* 1:227–232.

Pickett, W.E. 2006. Design for a room-temperature superconductor. *J. Superconduct. Nov. Magn.* 19:291–297.

Pop, E., V. Varshney, and A.K. Roy. 2012. Thermal properties of graphene: Fundamentals and applications. *MRS Bull.* 37:1237.

Qiang L., M. Arroy, and R. Huang. 2009. Elastic bending modulus of monolayer graphene. *J. Phys. D: Appl. Phys.* 42:102002.

Reich, S. and Thomsen, C. 2004. Raman spectroscopy of graphite. *Philos. Trans. R. Soc. Lond.* A 362:2271–2288.

Ribeiro, R.M., Vitor M. Pereira, N.M.R. Peres, P.R. Briddon, and A.H. Castro Neto. 2009. Strained graphene: Tight-binding and density functional calculations. *New J. Phys.* 11:115002.

Shang, J., C. Cong, J. Zhang, Q. Xiong, G. G. Gurzadyan, and T. Yu. 2013. Observation of low-wavenumber out-of-plane optical phonon in few-layer grapheme. *J. Raman Spectrosc.* 44:70–74.

Serov, A.Y., Z-Y. Ong, and E. Pop. 2013. Effect of grain boundaries on thermal transport in graphene. *Appl. Phys. Lett.* 102:033104.

Sirenko, V.A. Critical phenomena in uniaxial superconductorsand antiferromagnets. 2012. *Low Temp. Phys.* 38:799–806.

Skrypnyk, Yu.V. and V.M. Loktev. 2008. Spectral function of graphene with short-range impurity centers. *Low Temp. Phys.* 34:818–826.

Syrkin, E.S., S.B. Feodos'ev, K.V. Kravchenko, A.V. Eremenko, B.Ya. Kantor, and Yu. A. Kosevich 2009. Flexural rigidity of layers and its manifestation in the vibrational characteristics of strongly anisotropic layered crystals. Characteristic frequencies and stability conditions in quasi-two-dimensional systems. *Low Temp. Phys.* 35:158–165.

Wang, Y., Z. Song, and Z. Xu. 2014. Characterizing phonon thermal conduction in polycrystalline graphene. *J. Mater. Res.* 29:262–372.

Weller, T.E., M. Ellerby, S.S. Saxena, R.P. Smith and N.T. Skipper. 2009. Superconductivity in the intercalated graphite compounds C_6Yb and C_6Ca. *Nat. Phys.* 1:39–41.

Yang, S.-L., J. A. Sobota, C. A. Howard et al. 2014. Superconducting graphene sheets in CaC6 enabled by phonon-mediated interband interactions. *Nat. Commun.* 5:3493.

20 Tuning Atomic and Electronic Properties of Graphene by Selective Doping

Cecile Malardier-Jugroot, Michael N. Groves, and Manish Jugroot

CONTENTS

Abstract... 305
20.1 Rationale on Graphene and Selective Doping.. 305
20.2 Graphene: Structure and Synthesis.. 306
20.3 Doping of Graphene 307
    20.3.1 N-Doped Graphene... 307
    20.3.2 B-Doped Graphene... 308
    20.3.3 O-Doped Graphene... 308
20.4 Electronic Properties of Doped Graphene: Application for Proton Exchange Membrane Fuel Cells 309
    20.4.1 CNTs as a Catalyst Support.. 309
    20.4.2 Graphene as a Catalyst Support..310
    20.4.3 Doping in Graphene to Increase Platinum–Carbon Interaction Strength 312
    20.4.4 Catalytic Efficiency of Pt on Doped Graphene for Fuel Cell Application 314
    20.4.5 N-Doped Graphene as Metal-Free Catalyst for Fuel Cell Application 315
20.5 Concluding Remarks 315
References..315

ABSTRACT

A comprehensive review on the atomic and electronic properties of graphene and the effect of dopants on those properties as well as potential applications for different graphene structures is presented. Changes are observed experimentally on the properties of pure graphene due to the introduction of dopant. Several methods are used to obtain doped graphene, which in turn leads to changes in electronic properties of the graphene surfaces and can be predicted and characterized by powerful numerical methods such as density functional theory simulations. The changes in electronic properties characterized by natural bond orbital analysis and density of state description of graphene highlights important differences compared to carbon nanotubes. Doping of graphene as carbon support also exhibits a significant effect on the durability and efficiency of catalysts for instance in fuel cells applications. Critical reaction including the complete oxygen reduction reaction can be explored by quantum-based methods such as ab initio molecular dynamics simulations. The predictions extensively compared to experimental characterization of the different surfaces lead to a complete summary of the synthesis and characterization of the systems. Major applications of doped graphene and related changes in electronic properties of the graphene surface to the specific applications are reviewed and discussed.

20.1 RATIONALE ON GRAPHENE AND SELECTIVE DOPING

Summary Table

The recent discovery of graphene has sparked wide interest given the customization of its properties	Novoselov et al. (2012), Wei and Liu (2010), Konwg et al. (2014)

Carbon structures have attracted interest in the last decades starting with the discovery of fullerene structures and carbon nanotubes (CNTs) (Dresselhaus et al. 1996). These new architectures have a very high surface area due to their nanoscale dimensions. It can potentially help develop unique properties compared to bulk materials, such as very high strength, and controllable conductivity. These unique properties would also allow the development of novel and high impact application such as DNA sequencing, sensors, nanoelectronics, and related applications.

The interest in two-dimensional (2D) carbon materials such as graphene is more recent, due to the fact that this material was thought to be unstable in its free form. Indeed, the growth of 2D crystal structures was believed to be unstable due to the thermal fluctuations occurring during the crystal growth process, which would favor a three-dimensional (3D) crystal to the growing 2D structure (Geim 2009). Graphene

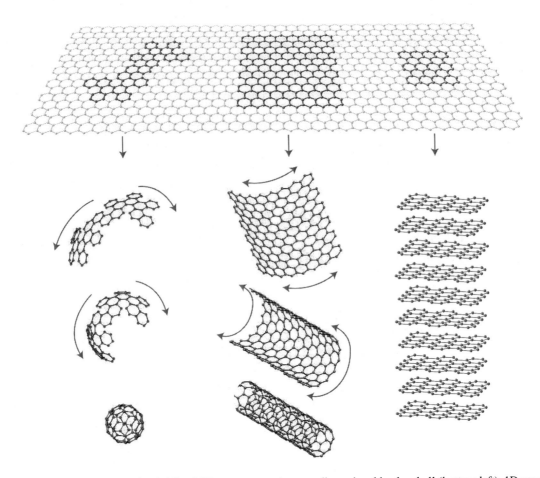

FIGURE 20.1   Graphene (top) is the sp²-hybridized 2D precursor to the zero-dimensional buckey ball (bottom left), 1D nanotube (bottom middle), and 3D graphite (bottom right). Even though each of these structures is composed of the exact same material with similar bonds, they each possess unique characteristics as a catalyst support. (Reprinted with permission from Geim, A. K. and K. S. Novoselov. 2007. *Nature Materials* 6:183–91.)

is a flat surface composed of sp² hybridized carbon atoms, which can be isolated from graphite by layering the graphite surface to a single sheet of graphene with an atomic thickness of 0.33 nm as shown in Figure 20.1. This material is the thinnest material produced, however it is very stable, durable, and possesses unique optical, mechanical, thermal, and electronic properties (Novoselov et al. 2012; Kong et al. 2014). Indeed, graphene was characterized to be 200 times stronger than structural steel and could produce extremely fast transistors. The interest in graphene-related materials has also been proven by the significant increase in scientific papers and patent applications in recent years. By isolating the first free graphene flakes, A. Geim and K. Novoselov earned the 2010 Nobel Prize in Physics.

Interestingly, from a strategic point of view, in 2013 the European Commission chose graphene as one of Europe's first 10-year, 1000 million Euro future energy technology (FET) flagships. This effort involves initially 126 research groups in 17 countries focusing on areas of graphene production and applications such as transport, communications, sensors, and energy technologies with a budget of €54 million for 30 months. In addition, large companies such as Samsung,

IBM, and Nokia have been investing in graphene research to develop novel applications like flexible touchscreens or gigahertz transistors. In parallel, high-level initiatives are also observed in the United States, Japan, and other regions of the world with potential breakthrough high-technology applications being targeted.

20.2   GRAPHENE: STRUCTURE AND SYNTHESIS

Summary Table

The method of graphene synthesis can influence its electrical and structural properties	Wei and Liu (2010), Areshkin and White (2007), Avouris and Dimitrakopoulos (2012), Jiao et al. (2010)

Graphene can be synthesized by several methods ranging from chemical methods to plasma-based techniques. Graphene was first produced mechanically from graphite in 2004 (Novoselov et al. 2004), although prior experiments had been successfully performed but the sheets had irregular properties (Choi et al. 2010).

The different attractive properties of graphene are directly linked to the shape, size, edges, and layer structures (Wei and Liu 2010). Indeed, graphene nanoribbons smaller than 10 nm are expected to be semiconductors due to the quantum confinement effect (Son et al. 2006). Larger nanoribbons could be either metallic or semiconductors depending on the edge configuration of the graphene nanoribbons. These important tunable properties of graphene could lead to an all graphene nanocircuit depending on the semiconductor or metallic character of the different components of the circuit (Areshkin and White 2007). In a quest to control the properties of pure graphene, several synthetic methods have been developed including small-scale as well as bench-scale methods. Graphene can be obtained by mechanically splitting 3D graphite into 2D graphene sheets. This method was the first successful method used to isolate graphene (Geim 2009). However, this method is very time consuming and could not be implemented on a large scale. Liquid exfoliation can also be performed but the defect levels are high and cannot be controlled. Graphene can also be grown epitaxially on metals but can potentially suffer from structural defects (Avouris and Dimitrakopoulos 2012). Plasma-enhanced models are also relevant for the synthesis of graphene as well as doped-graphene, for instance N-doped type (Bo et al. 2013). Interestingly, the main advantage of plasma method is that it can be synthesize on virtually any selected substrate. Moreover, instead of doping graphene during growth, a post-processing method can also be considered by exposure to a plasma (Lin et al. 2010). Plasma processes can also be tailored to favor less harsh environmental conditions (low temperature, short treatment time, and reduced toxicity) for the doping reaction (Wang et al. 2012).

Another method used for the synthesis of graphene is chemical vapor deposition (CVD). This method is very similar to the growth of CNTs, where the catalyst pattern favors the growth of the graphene layer (Wei and Liu 2010). The control of the edge configuration can be obtained by etching of the graphene layer. The precise shapes are produced using a resist, which will protect the area of the graphene layer that will not be etched. Then the etching is performed using an oxidative plasma etching.

Graphene sheets with very precise conformation can also be produced by opening a carbon nanotube (CNT). This method was used to obtain aligned graphene nanoribbons (Jiao et al. 2010). The CNTs (either single or small multiwall) are deposited on a SiO_2/Si substrate, then the CNT are coated with a layer of poly(methyl methacrylate), the upper layer of protective PMMA is removed to expose the CNT to Ar plasma to produce the unzipped graphene nanoribbons. This method produces nanoribbons with very smooth edges and narrow width. As can be revealed by the increased number of research articles in literature on the development of controlled synthesis of graphene, a slight change in the conformation and size of graphene can have a large impact on its properties. This possibility to tune the properties of graphene opens a wide range of applications.

20.3 DOPING OF GRAPHENE

Summary Table

Doping of graphene with nitrogen, boron, and oxygen can occur through chemical vapor deposition, electrothermal reactions with ammonia, microwave plasma reaction, microwave-induced annealing, and the chemical reduction of graphite oxide	Reddy et al. (2010), Qu et al. (2010), Wei et al. (2009), Dato et al. (2008), Xu et al. (2008), Li et al. (2009), Wang et al. (2009), Kim et al. (2014)
Doping of graphene with boron can occur through chemical vapor deposition, and by arc-discharge method in a hydrogen atmosphere	Ayala et al. (2008), Palnitkar et al. (2010)
Doping of graphene with oxygen may occur with exposure to ozone and light by cycloaddition, or by using graphene oxide as a starting material	Ghosh et al. (2010), Suggs et al. (2011), Lota et al. (2011), Zhu et al. (2010), Wang et al. (2009)

The different applications and tremendous opportunities for graphene due to its unique properties could be significantly increased if the properties could be tuned and controlled for specific applications. The properties can be altered by a conformational change but also using functionalization of the material or by doping. The dopants change the electronic structure of the CNT without significantly modifying its morphology. For instance, pure CNTs vary between semiconducting to metallic while both boron and nitrogen doped nanotubes are exclusively metallic (Glerup et al. 2003).

20.3.1 N-Doped Graphene

In addition, N-doped carbon structures such as CNTs have been shown experimentally to improve the durability and uniformity of the interaction between the carbon structure and metal nanoclusters such as platinum (Pt) nanoclusters as shown in Figure 20.2 (Saha et al. 2009). This property will allow smaller metal catalysts to be deposited on carbon nanoarchitectures, while maintaining or improving the durability of the catalyst therefore increasing significantly the surface area of the metal catalyst compared to traditional deposition methods. The increase in surface area is expected to show significant improvement for applications such as sensors, catalytic activity, and electrochemical applications. Indeed, it has been shown experimentally that Pt_3Co alloy dispersed on an N-doped graphene exhibited a power density four times higher than commercial Pt/C cathode (Vinayan et al. 2012). This improved efficiency of the fuel cell was attributed to the uniform dispersion of the metal nanoclusters on N-doped graphene and to the increased catalytic activity of both the PtCo alloy and pure N-doped graphene. It is important to note that pure N-doped graphene also displays interesting catalytic activities. N-doped graphene has been studied experimentally as catalyst for the oxygen reduction reaction (ORR) in fuel cells and compared to Pt/C cathodes (Shao et al. 2010). This study determined that the metal-free

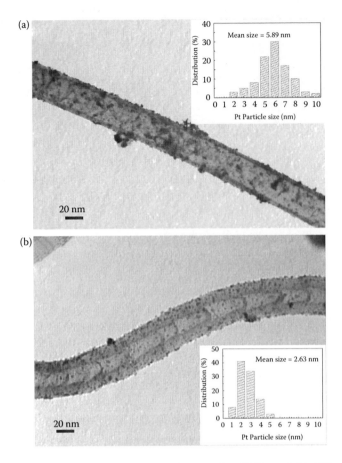

(a)

(b)

Mean size = 5.89 nm

Pt Particle size (nm)

Mean size = 2.63 nm

Pt Particle size (nm)

FIGURE 20.2 TEM images of (a) regular CNTs and (b) CN_x grown on carbon paper after deposition of Pt nanoparticles. Insets: size distribution of Pt particles. (Reprinted from *Electrochemistry Communications*, 11, Saha, M. S. et al., 3-D composite electrodes for high performance PEM fuel cells composed of Pt supported on nitrogen-doped carbon nanotubes grown on carbon paper, 438–41, Copyright 2009, with permission from Elsevier.)

N-doped graphene greatly increases the catalytic activity toward ORR compared to metal-free graphene. This catalyst exhibits a lower catalytic activity compared with Pt/C catalyst freshly deposited, however after accelerated degradation testing, the metal-free N-doped graphene exhibits higher catalytic activity than Pt/C catalyst emphasizing the improved durability of the N-doped graphene. The long-term stability of N-doped graphene and improved tolerance to poison effect when compared to Pt/C was also observed for the ORR in alkaline fuel cell with N-doped graphene synthesize by CVD (Qu et al. 2010) and N-doped CNT (Gong et al. 2009). The N-doped graphene can be produced by various large-scale methods including CVD (Wei et al. 2009; Qu et al. 2010; Reddy et al. 2010) electrothermal reactions with ammonia (Wang et al. 2009), microwave plasma reaction (Dato et al. 2008), microwave-induced annealing (Kim et al. 2014), and the chemical reduction of graphite oxide (Xu et al. 2008; Li et al. 2009). The catalytic properties of N-doped graphene have also been emphasized in the fast direct electron transfer kinetics for glucose oxidase and reduction of hydrogen

peroxide an unwanted side reaction in the fuel cell ORR. Furthermore, the N-doped graphene was shown to exhibit high sensibility and selectivity for glucose biosensing (Wang et al. 2010).

20.3.2 B-DOPED GRAPHENE

In addition to N-doped surface, Boron-doped carbon surface are expected to tune the properties of pure graphene. Boron-doped carbon structures have been synthesized directly through chemical vapor deposition (Ayala et al. 2008) and by arc-discharge method in a hydrogen atmosphere (Palnitkar et al. 2010) and are considered stable. B-doped carbon structures have attracted interest in the field of electronics (Panchakarla et al. 2009). Furthermore, boron-doped graphene is more conductive than pure graphene due to a larger density of states generated near the Fermi level. This property allowed an improved hole-collecting ability for better photovoltaic efficiency in solar cells (Lin et al. 2011).

20.3.3 O-DOPED GRAPHENE

Doping graphene with oxygen is expected to have a significant increase of the carbon–surface/metal interactions for the synthesis of more durable catalysts with optimized metal surface area and a decrease in the amount of metal needed for an optimal catalytic efficiency (Groves et al. 2012). Therefore, O-doped graphene is a very promising system and some efforts are developed to efficiently control their synthesis. For instance, oxygen doping of the sidewall of a CNT can be achieved by exposing them to ozone and light by cycloaddition (Ghosh et al. 2010; Lota et al. 2011; Suggs et al. 2011).

Another structure of graphene, graphene oxide, has shown some promise for applications as field effect transistors, sensors, transparent conductive films, clean energy devices (Zhu et al. 2010), and in polymer composites for supercapacitors (Wang et al. 2009).

These tunable properties of novel materials based on doped graphene are expected to open opportunities for efficient and integrated nanodevices. It is therefore very important to develop new methods to synthesize graphene and doped graphene with precise geometries, controlled at the nanoscale level, which could be transferable to large-scale synthesis. Equally important for the development of new materials is understanding the influence of the doping process on the electronic properties of graphene. Therefore, the next part of the chapter will focus on molecular simulation studies using predictive numerical tools on the electronic structures of pure graphene and four doped graphene surfaces (O, B, N, and Be). The predictions and novel properties of these materials will be compared to pure graphene for a specific application dependent on carbon support for durability and efficiency: proton exchange membrane fuel cells.

20.4 ELECTRONIC PROPERTIES OF DOPED GRAPHENE: APPLICATION FOR PROTON EXCHANGE MEMBRANE FUEL CELLS

Summary Table

Reduced platinum group metal loading necessary for commercialization of PEMFC	Gasteiger et al. (2004), Wang (2005), Shao et al. (2007)
Carbon fullerenes, including graphene, are a promising catalyst support due to their electronic and structural properties	Serp et al. (2003), Lee et al. (2006), Stankovich et al. (2006), Si and Samulski (2008a)
Graphene doping reduces agglomeration of deposited metal through modifications in the local molecular orbitals	Jafri et al. (2010), Brownson et al. (2011), Groves et al. (2009, 2012), Tang et al. (2013)
Doped graphene has also been shown to act as a metal-free catalyst through a reduction of the HOMO–LUMO gap	Zhang and Xia (2011), Shao et al. (2010), Matter et al. (2006), Maldonado and Stevenson (2005), Niwa et al. (2009), Tsetseris et al. (2014)

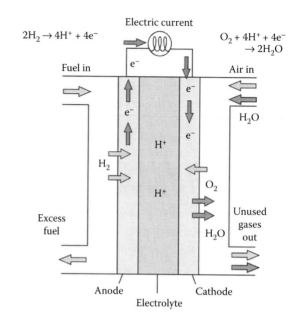

FIGURE 20.3 A diagram of a proton exchange membrane fuel cell.

Economic and environmental arguments motivate the development of other energy-generating systems, which do not negatively alter the global ecosystem. Similar to the mix of fossil fuel-based technologies that propelled humanity through the industrial revolution to the present, there will be no one breakthrough that will form the solution to this issue. Instead, a portfolio of ecologically and commercially viable power sources will be necessary so that a variety of applications can be satisfied. One of the focus applications for tuning graphene properties has been geared toward fuel cells applications and more specifically to improve the catalytic efficiency and catalyst durability for proton exchange membrane fuel cells (PEMFCs).

The PEMFC is one possible technology that would fit these criteria. A PEMFC is a refuelable battery, which intakes hydrogen and oxygen gas to produce a current. Its only exhaust is water. A schematic of a proton exchange membrane fuel cell is shown in Figure 20.3. Hydrogen gas enters on the anode side and is split into two protons and two electrons as indicated in the half-cell reaction. The electrons are collected and passed through the circuit producing electricity. When the protons pass through the proton conduction barrier to the cathode they react with oxygen gas and form water as indicated by the cathode half-cell reaction. The reaction of the overall process is

$$2H_2 + O_2 = 2H_2O \quad E^\circ = 1.23\,V$$

where E° is the standard reduction potential.

These units would be useful in portable power generation in applications ranging from mobile electronics to automobiles. Other applications might include stationary power sources for individual buildings (Knights et al. 2004). One of the principal reasons why it is not widely used is due to the cathode half-cell reaction. The best currently available catalyst is platinum, which is extremely expensive. Much less platinum is required on the anode than on the cathode, which was found to be where the rate limiting step for the reaction resides (Gasteiger et al. 2004; Wang 2005). Furthermore, there are durability issues where the platinum tends to agglomerate under operating conditions causing the overall reaction to slow down (Shao et al. 2007). As a result, a twofold problem occurs where the system costs more than what it is competing to replace up front and also would need to be replaced sooner. This means that an increase in durability and activity of the cathode would help bring this technology to greater use and could be achieved by including high surface area nanomaterials such as graphene.

20.4.1 CNTs as a Catalyst Support

CNTs, being one-dimensional (1D) carbon structures formed by sp^2 hybridized carbon, were tested as a catalyst support prior to graphene, given that it only recently became possible to isolate individual sheets from graphite (Novoselov et al. 2004). The synthesis of graphene, which had previously been thought to be thermodynamically unstable (Geim and Novoselov 2007), was initially focused on nanoelectrical applications and condensed matter physics, but it soon became apparent that graphene could also serve as a catalyst support. The work with CNTs for fuel cell applications only predate using graphene by a couple of years, however, it is still useful to examine their properties given that CNTs and graphene share similar properties.

CNTs are also very interesting catalyst supports, which are mechanically strong, chemically inert, possess a high surface area, and can be electrically conductive (Serp et al. 2003). According to a review (Lee et al. 2006) several methods for platinum deposition have been created, since the synthesis method can greatly affect the catalyst morphology, utilization, and activity. This effort is buoyed by the fact that CNTs have a superior electric conductivity and higher surface area than the typical carbon black supports commonly used. One of their positive attributes, their chemical inertness, is also a shortcoming. Their resistance to the environmental operating conditions in a fuel cell also makes it more difficult to deposit a stable, well-dispersed layer of platinum on its surface. Therefore, the deposition technique causes such variation in the properties of the catalyst layer.

One of the first instances where CNTs were incorporated into a proton exchange membrane fuel cell was by Rajalakshmi et al. (2005). As CNTs are inert, adsorption of platinum to their exterior wall requires the application of strong acids such as H_2SO_4 or HNO_3. This functionalizes the nanotube walls with hydroxyl, carboxyl, and carbonyl groups that are necessary to anchor the metal ions (Yu et al. 1998). A study by Rajalakshmi et al. demonstrates the necessity to create these metal-bonding sites given that they measured an increase in performance (~40 mV) and platinum dispersion between CNTs pretreated with a strong solution of HNO_3 from those without the treatment (Rajalakshmi et al. 2005). This process was enhanced by sonication of the nanotubes (Xing 2004).

Functionalization of the CNTs to promote platinum dispersion and adhesion has proven to be an important step in producing superior electrodes. Another way of achieving this property on CNTs is to substitute another element for carbon, such as nitrogen or boron, in the lattice. Shortly after the CNT's initial discovery (Iijima 1991) there were already reports of substituting boron and nitrogen for carbon in these structures (Stephan et al. 1994; Weng-Sich et al. 1995; Zhang et al. 1997). The dopants change the electronic structure of the CNT without significantly modifying its morphology. For instance, pure CNTs vary between semiconducting to metallic while both boron and nitrogen doped nanotubes are exclusively metallic (Glerup et al. 2003). These dopants can facilitate the same role as functionalization with strong acids without the intermediate step since doping can occur during synthesis.

Simply doping the CNT with nitrogen has been shown to increase its activity for the ORR with and without any catalyst metal (Matter et al. 2006; Sidik et al. 2006). When platinum is used as a catalyst, the nitrogen doping has been credited with increasing the binding energy between the platinum and the support as well as increasing its activity for the ORR (Shao et al. 2007). The nitrogen atom promotes the delocalized π bonds around the carbon lattice given that it is an electron donor, which contributes to reducing the electrical resistance of the support (Maiyalagan and Viswanathan 2005). These π sites contribute to the basicity of the carbon surface, which in turn benefit the platinum–support interaction (Coloma et al. 1994; Antolini 2003).

Due to these benefits, nitrogen doping of CNT supports for platinum catalysts in proton exchange membrane fuel cells

has been increasingly explored. Indeed, a study by Saha et al. (2009) compared nitrogen-doped CNTs, and pure CNTs to conventional E-TEK platinum/carbon electrodes. Their motivation included looking for alternatives to acid treatments of CNTs to functionalize their inert surface, which can weaken their electrical and structural properties. Both the pure and nitrogen-doped CNTs were created with an aerosol-assisted chemical vapor deposition process directly onto carbon paper. This was to ensure that all the grown nanotubes as well as the deposited platinum were in electrical contact with the external circuit. It was observed that the nitrogen-doped CNTs possessed smaller, better dispersed platinum nanoparticles than was observed on pure CNTs. This can be seen in Figure 20.2. The platinum dispersion is credited with the higher catalytic activity measured when used as a cathode in a proton exchange membrane fuel cell when compared to the pure CNT support and the standard platinum/carbon electrode.

The effect of nitrogen atomic percent on particle size and stability was characterized by Chen et al. (2009) who reported that for CNT samples with a nitrogen atomic percent (at%) of 1.5, 5.4, and 8.4 produced deposited platinum particle sizes of 5.0 ± 2.3, 4.8 ± 2.0, and 4.2 ± 1.7 nm, respectively. Their pure CNT case had a larger average platinum nanoparticle size with a greater distribution at 6.2 ± 3.4 nm. This positive trend with particle size also held up for stability. The CNT sample with the largest nitrogen content was able to retain 42.5% of its initial electrochemical surface area compared to 26.6% (5.4 at% N), 20.2% (1.5 at% N), 11.2% (pure CNTs), and 4.6% (standard platinum carbon electrode) after 4000 cycles undergoing voltammetric cycling between 0.6 and 1.2 V in oxygen saturated sulfuric acid. This test was meant to mimic the conditions on the cathode of a proton exchange membrane fuel cell and to measure the agglomeration of the platinum. The final particle sizes followed the same trend as before where in all cases the particle size roughly doubled. The rationale for this effect was again attributed to the nitrogen atom having an extra electron to enhance the delocalized π bond interaction with the platinum nanoparticle.

20.4.2 GRAPHENE AS A CATALYST SUPPORT

When it was initially shown that stable graphene was possible, applications were focused on nanoelectrical applications and condensed matter physics. However, it soon became apparent that graphene could also serve as a catalyst support. The many benefits that this carbon geometry exhibits are similar to CNTs and include thermal conductivity and mechanical stability of graphite in-plane (its fracture strength should match CNTs with an equivalent number of defects), but the most interesting is their high electrical conductivity (Stankovich et al. 2006) where the charge carriers have been described as massless (Geim and Novoselov 2007).

One other major benefit of graphene over CNTs is the ability to fabricate these 2D structures at industrial scales. The many layers of graphite, a readily available material, are not easily dissociated to yield a suspension of graphene plates. Graphene oxide, a heavily oxidized version of graphite bearing hydroxyl and epoxide functional groups on its basal planes and carboxyl and carbonyl groups along its edges, is highly hydrophilic, which swells and separates in water. This results in a stable suspension of graphene sheets and is a good starting place to fabricate composite materials. Graphene oxide is electrically insulating but its conductivity can be restored by chemical reduction (Stankovich et al. 2006).

Unfortunately, chemical reduction also immediately removes the oxygen groups, which were key in dispersing the graphite. Furthermore, this reaction being achievable only in water, limits other additives that are hydrophobic and will not disperse to form nanocomposites. To circumvent this drawback, a three-step reduction process was developed (Si and Samulski 2008b) that replaces the oxygen-based groups with SO_3H groups, which keep the sheets separated. Sodium borohydride was first used to remove the majority of oxygen-based functionality before sulfonating using diazonium salt of sulfanilic acid. A reduction step with hydrazine removes the remainder of the oxygen-based groups. Sodium borohydride cannot be used for this final reduction step since it causes complete precipitation of graphite. This process restores the conductivity of the resulting graphene plates (1250 S m^{-1}) to a quarter of the conductivity of the original graphite (6120 S m^{-1}). To put this into perspective, the conductivity of the graphene flakes before the final hydrazine reduction is only 17 S m^{-1} due to the oxygen functionalization disrupting the π conjugated system. This lightly sulfonated product can be dispersed in water and mixed with other solvents including methanol, acetone, and acetonitrile.

This process was then used to create a stacked platinum nanoparticle/graphene sheet structure used to demonstrate that it can function as a cathode for the proton exchange membrane fuel cell (Si and Samulski 2008a). One large advantage that graphene sheets display over graphite is their incredibly large surface area given that each face of each sheet is exposed. When the aqueous solution is dried the sheets agglomerate, thus greatly reducing the surface area. Even stacks of only ten graphene sheets revert to properties more analogous to graphite. To prevent this, platinum nanoparticles are formed on both sides of the suspended sheets so that the platinum can force a gap between each sheet retaining exposure and the unique properties of the graphene as illustrated in Figure 20.4. This composite material was synthesized by reducing chloroplatinic acid onto the dispersed lightly sulfonated graphene flakes with methanol. The resulting flakes were precipitated out of solution with sulfuric acid, filtered from the solution and then dried. The presence of surfactant 3-(N,N-dimethyldodecylammonio) propanesulfonate during the reduction of the platinum inhibits large-scale agglomeration and creates nanoparticle sizes of 3–4 nm. The material had a measured surface area of 862 m^2 g^{-1} determined using the Brunauer, Emmett, and Teller (BET) method, which uses the adsorption of a gas to determine this quantity (Brunauer et al. 1938). This is about a third the surface area of single graphene sheet (~2600 m^2 g^{-1}). This implies that on average three graphene sheets separate each layer of platinum. When used in a proton exchange membrane fuel cell with hydrogen and oxygen as the reactant gases a current density of 300 mA cm^{-2}

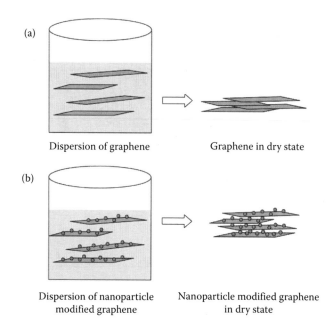

FIGURE 20.4 Schematic of graphene sheets and nanoparticle-modified graphene sheets in its dispersion and dry state. (a) Graphene sheets, isolated graphene sheets exist in its dispersion; in the dry state, graphene sheets aggregate and stack into a layered structure like graphite. (b) Nanoparticle modified graphene sheets; isolated sheets exist in its dispersion; in the dry state the nanoparticles with diameters spanning several nanometers act as nanoscale spacers to increase the interlayer spacing between graphene sheets, thus making both faces of graphene accessible. (Reprinted with permission from Si, Y. and E. T. Samulski. Exfoliated graphene separated by platinum nanoparticles. *Chemistry of Materials* 20:6792–7. Copyright 2008 American Chemical Society.)

was measured at 0.65 V and 65°C. It was not compared to any traditional support since this result was only meant for demonstration purposes. It was recommended that further optimization of their graphene/platinum catalyst needs to be optimized for this application before it could yield improvements over more traditional supports (Si and Samulski 2008a). For instance, it has been suggested that Pt–graphene supports should have greater long-term stability than Pt–CNT systems (Kauffman and Star 2010).

Analogous to CNT support development, nitrogen doping in the context of a graphene support was also explored (Jafri et al. 2010). In this work, graphene platelets were prepared by rapidly heating heavily oxidized graphite to 1050°C in under a minute in an argon atmosphere. This causes the rapid evolution of carbon dioxide between the graphene sheets and its liberation is strong enough to overcome their van der Waals attraction. It also removes the oxygen species and restores conductivity to the flakes (Schniepp et al. 2006). Once separated, they exposed them to a nitrogen plasma, which added about 3 at% of nitrogen to the graphene. Hexachloroplatinic acid was then reduced onto the nitrogen-doped graphene flakes using sodium borohydride to create a 0.5 mg of platinum per cm^2 deposition. Electrochemical measurements were then taken by forming this product into a cathode for a proton exchange membrane fuel cell. It was compared to graphene flakes formed from the same rapid heating process but then functionalized in nitric and sulfuric acids (similar to previous CNT functionalization) before reducing a similar amount of platinum onto them. The result parallels the CNT result in that the power output from the nitrogen doped graphene flakes was 50 mW cm^{-2} superior to the pure graphene flakes (Jafri et al. 2010). It was inferred that the process creates pyrrolic nitrogen defects, which serve as anchoring sites that improve support-catalyst binding or electrical conductivity. This enhances the catalytic properties of the graphene structure (Brownson et al. 2011).

Synthesis of these nitrogen-doped graphene structures demonstrates that they are useful for fuel cell applications. Given this success, it is important to understand how the dopants modify the support and generate these effects to be able to optimize their role. Physical techniques such as Raman spectroscopy have proven to be a useful characterization tool of graphene (Ferrari et al. 2006) and doped graphene (Casiraghi 2009). However, ab initio methods such as density functional theory (Hohenberg and Kohn 1964) are incredibly insightful to determine what new materials are worth pursuing (Shao et al. 2007).

20.4.3 Doping in Graphene to Increase Platinum–Carbon Interaction Strength

It has been reported in the literature (Wang 2005; Shao et al. 2007; Lepro et al. 2008; Wu et al. 2008) that nitrogen-doped carbon supports can increase the durability and activity of a Pt catalyst. The effects of nitrogen doping in graphene flakes were examined by Groves et al. (2009), such as the increase in interaction strength between platinum and graphene when doped with nitrogen. Using ab initio methods (Frisch et al. 2004), the multipurpose hybrid functional B3LYP (Becke 1993), and effective core potential basis set Lanl2DZ (Hay and Wadt 1985a,b; Wadt and Hay 1985), six graphene flakes are examined: a pristine graphene flake composed of 42 carbon atoms and 16 hydrogen atoms that serve to terminate the edges, as well as five nitrogen-doped configurations. These five configurations serve to determine how the number of nitrogen atoms, as well as their proximity to the binding carbon atom changes the platinum–surface interaction. One is a single nitrogen substitution, which is illustrated in Figure 20.5. There are two nitrogen substitutions where the nitrogen–nitrogen distance is 5.15, 3.79, and 2.50 Å. The final one has one-quarter of the carbon atoms exchanged with nitrogen atoms uniformly over the flake. This final one serves both to see what effect high nitrogen content has on the interaction, but also serves to keep these simulations as close as possible to what realistically can be fabricated. The single nitrogen-doped case corresponds to a very low nitrogen content while the two nitrogen doped cases serve to identify how proximity to the binding carbon plays a role.

To determine the root cause of the enhanced platinum–substrate interaction observed experimentally in nitrogen doped graphene means that platinum will also need to be

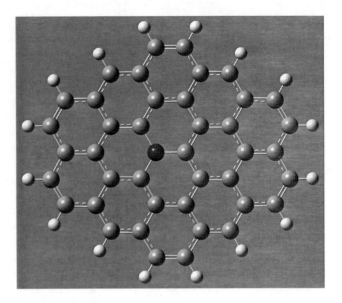

FIGURE 20.5 The single nitrogen-doped graphene flake.

added to the flake. For these calculations, a single platinum atom was used. This was to reduce the computational cost of the simulations but can also still provide an accurate description of physical systems (Sidik and Anderson 2002). Changes due to doping are local to the atoms that are adjacent to dopant (Acharya and Turner 2008). Consequently, large platinum clusters may not even be helpful when examining the effects of doping.

The result of the calculations is that the binding energy increases between the platinum atom and the surface with an increase in the number of nitrogen atoms incorporated into the surface as well as a decrease in the dopant's proximity to the binding carbon atom. The resulting binding energies are reported in Table 20.1. It is interesting to note that there appears to be a limit on the positive effect that nitrogen doping has on the system based on the 2.50 Å two nitrogen-doped system and the quarter-doped case. There is only a 0.011 eV gain by adding the extra nine nitrogen atoms. The proximity

TABLE 20.1

Interaction Strength Calculated between a Nitrogen-Doped Graphene Flake and a Single Platinum Atom

Nitrogen Arrangement	Platinum Binding Energy (eV)
None	−1.271
One	−1.695
Two N (5.16 Å)	−2.038
Two N (3.79 Å)	−2.328
Two N (2.50 Å)	−2.510
Quarter N	−2.521

Source: Groves, M. N. et al. 2009. *Chemical Physics Letters* 481:214–9.

of the nitrogen atoms in the quarter-doped case may not be as close as in the 2.50 Å two nitrogen-doped case, however, the small increase in binding energy may not outweigh the large cost in stability to the graphene flake due to the large number of dopants. The addition of nitrogen atoms increases the number of dislocations in the lattice (Terrones et al. 2002; Trasobares et al. 2002; Maldonado and Stevenson 2005) due to nitrogen's propensity to form pentagonal structures. While this has been shown to increase the hardness and elasticity of the resulting structure (Sjostrom et al. 1995) there can be penalties to its activity at least in CNT systems with direct methanol fuel cells (Maiyalagan 2008). At least in CNT systems, for nitrogen substitutions up to 8.4 at%, a decrease in deposited platinum nanoparticle size and a greater resistance to agglomeration was observed using cyclic voltammetry (Chen et al. 2009).

What the nitrogen doping accomplishes in the graphene flakes is to locally disrupt the delocalized double bonds characteristic of the aromatic rings, which allows for the adjacent carbon atom to form a true bond with the platinum atom. When no dopant is present, the platinum–graphene interaction can be described as a donation from the graphene π orbital into an empty platinum orbital (Cotton and Wilkinson 1988, 72). This is reciprocated by the platinum atom donating into the empty π* orbital. When nitrogen is introduced, the π orbital seems to be disrupted allowing the adjacent carbon atom to use an sp³ hybridized orbital to interact with the platinum atom. This is depicted in the molecular orbitals shown in Figure 20.6.

The nitrogen atoms also serve to increase the positive charge on the binding carbon atom.

When no dopant is present the binding carbon atom is essentially zero. When one nitrogen atom is present it increases to $0.4424e$ and when two nitrogen atoms are adjacent to the binding carbon atom it rises again to $1.079e$. This increase in positive charge is attractive to the platinum atom given its large number of electrons that is available to bond with the surface (Groves et al. 2009).

Moreover, other second row dopants in graphene flakes, including beryllium, boron, and oxygen (Groves et al. 2012), change its structure and properties, which can have positive effects on the ORR. In this work, the Pople basis set 6-31G(d) is used to describe the light elements (Hehre et al. 1972; Frisch et al. 1984) while the Stuttgart—Dresden pseudopotential (SDD) is effective core potential for the platinum atom (Dolg et al. 1987; Andrae et al. 1990). This basis set combination had been shown to describe transition metal complexes better than the Lanl2DZ basis set (Buhl and Kabrede 2006; Buhl et al. 2008). The B3LYP hybrid functional was also used for this work. As was calculated with the nitrogen-doped case, all the dopants modified the surface in a way that increased the interaction energy with the platinum atom. This interaction turns out to be stronger the further the element was positioned away from carbon in both directions along the second row. The substrate that yielded the strongest binding energy was the oxygen doped where a sevenfold increase over the undoped case was calculated (Groves et al. 2012). All the binding energies from this study can be found in Table 20.2.

These three new dopants also interrupted the usual molecular orbital configuration in graphene. Depending on whether the dopant came from a group with a larger or smaller number of valence electrons than carbon determined its interaction with the platinum. Indeed, nitrogen and oxygen possess increasingly full valence shells relative to carbon, which are satisfied by in-surface interactions. Beryllium and boron possess relatively empty valence shells, which can also still satisfy bonding arrangements in the surface. The difference between the two types of dopants, to the right, and to the left of carbon,

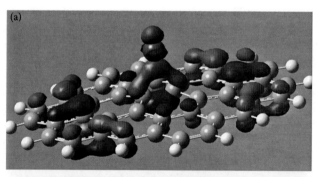

FIGURE 20.6 Molecular orbitals showing the change in the interaction between the platinum atom and the surface. (a) Pure graphene flake. (b) Single nitrogen doped graphene flake. (Reprinted from *Chemical Physics Letters*, 481, Groves, M. N. et al., Improving platinum catalyst binding energy to graphene through nitrogen doping, 214–9, Copyright 2009, with permission from Elsevier.)

TABLE 20.2

Interaction Strength Calculated between a Single Second Row Doped Graphene Flakes and a Single Platinum Atom

Dopant	Platinum Binding Energy (eV)	PBC Pt BE (eV)
None	−0.37	−4.05
Nitrogen	−0.72	−4.56
Boron	−0.99	−4.78
Beryllium	−1.86	−5.92
Oxygen	−2.59	−6.80

Source: Groves, M. N., C. Malardier-Jugroot, and M. Jugroot. 2012. *The Journal of Physical Chemistry C* 116:10548–56.

is that when the valence shell is relatively empty the platinum will interact with the dopant and the adjacent carbon atom. When the valence shell is full, the platinum interacts almost exclusively with the adjacent carbon atom (Groves et al. 2012).

This has two interesting implications. The first is that as the valence shell becomes increasingly full/empty, a defect in the graphene lattice becomes more pronounced, which allows the platinum atom to bind more strongly with the surface. As was explained previously in the nitrogen-doped case, the nitrogen atom only needs three bonds to fill its valence shell. This creates a binding site on the adjacent carbon atom that used to be filled by the delocalized double bond and has characteristics of an sp^3 hybridized orbital. Oxygen only requires two bonds to fill its valence shell. This leave one carbon atom with an open binding site, which it uses to interact very strongly with the platinum atom. This bond has an sp^2-like character since the platinum–carbon bond resembles a typical graphene carbon–carbon bond. This trend is also seen in boron and beryllium: as elements are selected moving to the left of carbon, the binding energy increases due to a modification in how the adjacent carbon atom interact with the dopant (Groves et al. 2012). This presumably has a limit since it is not likely that lithium or fluorine is stable as dopants in graphene.

The second implication centers around how multiple doping can enhance the interaction energy. Nitrogen and oxygen dopants do not directly participate in the platinum–surface interaction. They instead cause the adjacent carbon atom to become a more favorable binding site. As a result, multiple dopants can further enhance the adjacent carbon since the dopants localize the binding site next to them. This was seen in the multiple nitrogen doped case. Boron and beryllium localize the binding site between themselves and the adjacent carbon atom. This means that multiple doping from these elements do not lead to as large of an increase in binding energy. Specifically in boron doping, as the numbers of dopants are increased and their proximity to the binding carbon decreases there is not as large of an increase in the interaction energy as in equivalent nitrogen-doped cases. This is due to the binding site not being located in one place, but instead over many locations (Groves et al. 2009). The reduction of platinum loading predicates a small platinum nanoparticle size. The stronger local sites that can be created with elements to the right of carbon might be a better fit for more durable catalyst layers than the more dispersed sites that the elements to the left of carbon provide.

The second row dopant calculations have also been performed using periodic boundary conditions in order to simulate large graphene structures. Binding energies based on using the self-consistent field energies were reported and a similar trend emerges where the undoped case yields the lowest interaction energy while the oxygen-doped case is calculated to be the most stable structure (Groves et al. 2012). A complete list of results is shown in Table 20.2. The simulation used the 6-31G(d)/SDD basis set combination in Gaussian 09 (Frisch et al. 2009). In addition, the VSXC functional (Voorhis and Scuseria 1998) was used instead of B3LYP since it had been shown to be more efficient at simulating accurate results in CNTs (Wang et al. 2008).

Other, larger dopants have also been examined. In this case, creating a divacancy is recommended in order to better trap the dopant (Tsetseris et al. 2014). These dopant types have been examined for platinum adsorption as well as CO oxidation, such as single-doped silicon, phosphorus, and sulfur in graphene sheet and compared with boron, nitrogen, oxygen dopants (Tang et al. 2013; Zhang et al. 2014). A study by Tang et al. used the density functional theory (DFT) Vienna ab initio simulation package (VASP) (Kresse and Furthmuller 1996a,b) to examine a 32 carbon atom structure and verified that for single-doped systems where a carbon atom is substituted by another second row atom, there is an increase in binding energy between the graphene surface and a single platinum atom. Oxygen doping still provides the strongest interaction energy followed by boron, and then nitrogen.

The third row elements, silicon, phosphorus, and sulfur also show an increase in binding energy over the nondoped case. However, the sulfur-doped structures had a calculated platinum–substrate interaction energy that was between nitrogen and boron. The strongest interaction is provided by phosphorus, followed closely by silicon. Both of these cases had a stronger binding energy than even the oxygen doped case. These larger dopants cannot sit in the lattice, but instead sit above the graphene sheet in the defect. This does not seem to affect the formation energy relative to the oxygen doped case.

Calculating the density of states of the silicon, boron, and oxygen-doped cases also reveals that tuning of where the Dirac point exists is possible in graphene. Given that silicon has similar electronic characteristics as carbon, the density of states around the Fermi level is zero. For boron doping the Fermi level is 0.77 eV below the Dirac point, which indicates that the boron is acting as an electron acceptor making the graphene a p-type conductor. Conversely, the oxygen-doped case causes an upward shift in the Fermi level. The large binding energies between platinum and the doped graphene surface all come from doped graphene surfaces, which act as electron donors allowing the platinum 5d orbitals to form a strong bond with it (Tang et al. 2013).

20.4.4 CATALYTIC EFFICIENCY OF PT ON DOPED GRAPHENE FOR FUEL CELL APPLICATION

Currently there is a limited body of experimental work examining the effect of doped graphene as a catalyst support. In terms of the ORR, nitrogen doping is by far the most popular given its many synthesis pathways. The results indicate that nitrogen doping of a graphene support for a platinum catalyst increases its activity and also resists platinum agglomeration better than pristine graphene (Jafri et al. 2010; Zhang et al. 2010) as indicated by the theoretical studies. In addition to platinum, other nonprecious metal combinations such as FeCN (Tsai et al. 2011) and Co_3O_4 (Liang et al. 2011) have been fabricated with nitrogen-doped graphene also with positive results. Theoretical studies also predicted that the activity of the Pt catalyst toward ORR could be improved significantly by using other dopants. With only few experimental examples of the doped graphene systems, it is expected that novel graphene-doped surfaces may lead to even greater gains in activity and durability for fuel cell applications as predicted by simulation.

20.4.5 N-Doped Graphene as Metal-Free Catalyst for Fuel Cell Application

Platinum's role in the ORR has always been viewed as important, however, graphene surfaces with nitrogen doping without a metal catalyst have also been shown to effectively catalyze this reaction. According to another DFT study, the addition of nitrogen to the graphene flake shrinks the highest occupied molecular orbital (HOMO)–lowest unoccupied molecular orbital (LUMO) gap thereby making a more reactive surface (Zhang and Xia 2011). This allows OOH groups, an important ORR intermediate, to adsorb onto nitrogen-doped graphene surfaces without the presence of any metal catalyst. It was determined that the spin densities and atomic charge densities of the carbon atoms are indicators to determine whether they are catalytically active sites. These active sites were carbon atoms that were adjacent to nitrogen dopants and thus were directly affected by the substitution.

The method used to substitute nitrogen into the lattice also can affect how beneficial it is toward the four electron ORR reaction. There are generally four different substitutions: pyridinic, pyrrolic, quaternary, and nitrogen oxides involving apyridinic configuration. The pyridinic structure is where the nitrogen is at the edge of a graphene plane and is bonded to two carbon atoms. It donates one p-electron to the aromatic π-system. Pyrrolic nitrogen is also bonded to two carbon atoms but donates two p-electrons to the aromatic π-system. The quaternary configuration refers to nitrogen substituted in the center of the lattice for a carbon atom. Finally, the nitrogen oxide structure is where the nitrogen is in a pyridinic arrangement where it is bound to two carbon atoms with the additional constraint that it must also be bound to an oxygen atom (Shao et al. 2010). Identifying which species exist in a given sample is straightforward using x-ray photoelectron spectroscopy (XPS) (Qu et al. 2010; Shao et al. 2010) and scanning tunneling microscopy (Wang et al. 2012) but the role of each of these types in promoting the reaction is still being explored. A study by Matter, Zhang, and Ozkan as well as Maldonado and Stevenson show that pyridinic and pyrrolic nitrogen structures are the main factors for the increase in catalytic activity (Maldonado and Stevenson 2005; Matter et al. 2006) while a separate study (Niwa et al. 2009) showed that quaternary nitrogen surfaces are the important ingredients.

Looking more closely at the pyridinic nitrogen case, Luo et al. were able to synthesize these types of graphene layers varying in nitrogen concentration from 0 to 16 at% (Luo et al. 2011). They were able to confirm this exclusive type of doping using XPS. They found that the nitrogen changed the valence structure of the graphene. This included raising the density of π states near the Fermi level as well as lower the work function. However, they also concluded that the pyridinic nitrogen doping was not suitable for the two electron ORR. They also found that there was a limit to the amount of pyridinic

nitrogen that could be added to the surface. They conclude that at the larger concentrations of nitrogen doping the lone pairs of the nitrogen atoms repelled the electronegative oxygen molecules, which were trying to interact with the adjacent positively charged carbon atoms. As a result, a balance between the attractive carbon–oxygen interaction that promotes the ORR and the repulsive nitrogen–oxygen interaction needs to be considered when designing a metal-free nitrogen doped graphene catalyst layer.

20.5 CONCLUDING REMARKS

The numerous interesting properties of graphene have only just recently begun to be explored. Interestingly, it has already been demonstrated that by modifying the lattice by replacing carbon atoms with other dopants, thereby modifying the electronic structure of the surface, the surface can be tailored to specific applications. For instance, in the context of proton exchange membrane fuel cells, nitrogen doping was shown to contribute to a more durable catalyst layer. This increase in durability can be attributed to the nitrogen atom locally disrupting the delocalized double bond allowing adjacent carbon atoms to form sp^3 hybridized bonds with the platinum atom. Additional nitrogen atoms adjacent to the bonding carbon atom intensify this effect. Other second-row dopants can also provide a stronger binding energy with platinum. Indeed, an oxygen-doped surface, which was calculated to have the strongest binding energy showed that the carbon atom bonded with the platinum atom using an sp^2-hybridized bond as if it were a part of the surface. This is due to the fact that the valence shell of the oxygen is already full by bonding with the other two adjacent carbon atoms leaving the third with a dangling bond. This increased interaction strength creates a more durable catalyst layer where the deposited platinum can remain dispersed and resist agglomeration. The extension of the active lifetime of the cathode will reduce the overall cost of a fuel cell unit and will bring it closer to wide-scale commercialization. In addition, the inclusion of dopants on the graphene surface is expected to be beneficial to the intrinsic catalytic activity of Pt. Understanding the effect of selective doping on the graphene electronic structure is therefore crucial for the development of novel applications, and quantum-based simulation methods offer valuable insight into small-scale interactions and provide a powerful predictive tool that will help the tailoring of graphene surfaces to specific applications.

REFERENCES

Acharya, C. K. and C. H. Turner. 2008. CO oxidation with Pt(111) supported on pure and boron-doped carbon: A DFT investigation. *Surface Science* 602:3595–602.

Andrae, D., U. HauBermann, M. Dolg, H. Stoll, and H. PreuB. 1990. Energy-adjusted ab initio pseudopotentials for the second and third row transition elements. *Theoretical Chemistry Accounts* 77:123–41.

Antolini, E. 2003. Formation, microstructural characteristics and stability of carbon supported platinum catalysts for low temperature fuel cells. *Journal of Material Science* 38:2995–3005.

Areshkin, D. A. and C. T. White. 2007. Building blocks for integrated graphene circuits. *Nano Letters* 7:3253–9.

Avouris, P. and C. Dimitrakopoulos. 2012. Graphene: Synthesis and applications. *Materials Today* 15:86–97.

Ayala, P., W. Plank, A. Gruneis, E. I. Kauppinen, M. H. Rummeli, H. Kuzmanyb, and T. Pichler. 2008. A one step approach to B-doped single-walled carbon nanotubes. *Journal of Material Chemistry* 18:5676–81.

Becke, A. D. 1993. Density-functional thermochemistry. III. The role of exact exchange. *Journal of Chemical Physics* 98:5648–52.

Bo, Z., Y. Yang, J. Chen, K. Yu, J. Yana, and K. Cena. 2013. Plasma-enhanced chemical vapor deposition synthesis of vertically oriented graphene nanosheets. *Nanoscale* 5:5180–204.

Brownson, D., D. Kampouris, and C. Banks. 2011. An overview of graphene in energy production and storage applications. *Journal of Power Sources* 196:4873–85.

Brunauer, S., P. H. Emmett, and E. Teller. 1938. Adsorption of gases in multimolecular layers. *Journal of the American Chemical Society* 60:309–19.

Buhl, M., C. Reimann, D. A. Pantazis, T. Bredow, and F. Neese. 2008. Geometries of third-row transition-metal complexes from density-functional theory. *Journal of Chemical Theory and Computation* 4:1449–59.

Buhl, M. and H. Kabrede. 2006. Geometries of transition-metal complexes from density-Functional theory. *Journal of Chemical Theory and Computation* 2:1282–90.

Casiraghi, C. 2009. Doping dependence of the Raman peaks intensity of graphene close to the Dirac point. *Physical Review B* 80:233407.

Chen, Y., J. Wanga, H. Liu, R. Li, X. Sun, S. Ye, and S. Knights. 2009. Enhanced stability of Pt electrocatalysts by nitrogen doping in CNTs for PEM fuel cells. *Electrochemistry Communications* 11:2071–6.

Choi, W., I. Lahiri, R. Seelaboyinaand, and Y. S. Kang. 2010. Synthesis of graphene and its applications: A review. *Critical Reviews in Solid State and Materials Sciences* 35:52–71.

Coloma, F., A. Sepulvedaescribano, J. L. G. Fierro, and F. Rodriguezreinoso. 1994. Preparation of platinum supported on pregraphitized carbon blacks. *Langmuir* 10:750–5.

Cotton, F. A. and G. Wilkinson. 1988. *Advanced Inorganic Chemistry* (5 ed.). Toronto: John Wiley & Sons.

Dato, A., V. Radmilovic, Z. Lee, J. Phillips, and M. Frenklach. 2008. Substrate-free gas-phase synthesis of graphene sheets. *Nano Letters* 8:2012–6.

Dolg, M., U. Wedig, H. Stoll, and H. Preuss. 1987. Energy-adjusted ab initio pseudopotentials for the first row transition elements. *Journal of Chemical Physics* 86:866–72.

Dresselhaus, M. S., G. Dresselhaus, and P. C. Eklund. 1996. *Science of Fullerenes and Carbon Nanotubes: Their Properties and Applications*. San Diego, CA: Elsevier Academic Press.

Ferrari, A. C., J. C. Meyer, V. Scardaci, C. Casiraghi, M. Lazzeri, F. Mauri, S. Piscanec et al. 2006. Raman spectrum of graphene and graphene layers. *Physical Review Letters* 97:187401.

Frisch, M., J. Pople, and J. Binkley. 1984. Self-consistent molecular orbital methods 25. Supplementary functions for Gaussian basis sets. *Journal of Chemical Physics* 80:3265–9.

Frisch, M. J., G. W. Trucks, H. B. Schlegel, G. E. Scuseria, M. A. Robb, J. R. Cheeseman, G. Scalmani et al. 2009. *Gaussian 09*. Wallingford, CT: Gaussian Inc.

Frisch, M. J., G. W. Trucks, H. B. Schlegel, G. E. Scuseria, M. A. Robb, J. R. Cheeseman, J. A. Montgomery Jr. et al. 2004. *Gaussian 03*. Wallingford, CT: Gaussian, Inc.

Gasteiger, H. A., J. E. Panels, and S. G. Yan. 2004. Dependence of PEM fuel cell performanceon catalyst loading. *Journal of Power Sources* 127:162–71.

Geim, A. K. 2009. Graphene: Status and prospects. *Science* 324:1530–4.

Geim, A. K. and K. S. Novoselov. 2007. The rise of graphene. *Nature Materials* 6:183–91.

Ghosh, S., M. Bashilo, R. A. Simonette, K. M. Beckingham, and R. B. Weisman. 2010. Oxygen doping modifies near-infrared band gaps in fluorescent single-walled carbon nanotubes. *Science* 330:1656–9.

Glerup, M., M. Castignolles, M. Holzinger, G. Hug, A. Loiseaub, and P. Bernier. 2003. Synthesis of highly nitrogen-doped multi-walled carbon nanotubes. *Chemical Communications* 39:2542–3.

Gong, K., F. Du, Z. Xia, M. Dustock, and L. Dai. 2009. Nitrogen-doped carbon nanotube arrays with high electrocatalytic activity for oxygen reduction. *Science* 323:760–4.

Groves, M. N., A. S. W. Chan, C. Malardier-Jugroot, and M. Jugroot. 2009. Improving platinum catalyst binding energy to graphene through nitrogen doping. *Chemical Physics Letters* 481:214–9.

Groves, M. N., C. Malardier-Jugroot, and M. Jugroot. 2012. Improving platinum catalyst durability with a doped graphene support. *The Journal of Physical Chemistry C* 116:10548–56.

Hay, P. J. and W. R. Wadt. 1985a. Ab initio effective core potentials for molecular calculations. Potentials for K to Au including the outermost core orbitals. *Journal of Chemical Physics* 82:299–310.

Hay, P. J. and W. R. Wadt. 1985b. Ab initio effective core potentials for molecular calculations. Potentials for the transition metal atoms Sc to Hg. *Journal of Chemical Physics* 82:270–83.

Hehre, W., R. Ditchfield, and J. Pople. 1972. Self-consistent molecular orbital methods. XII. Further extensions of Gaussian-type basis sets for use in molecular orbital studies of organic molecules. *Journal of Chemical Physics* 56:2257–61.

Hohenberg, P. and W. Kohn. 1964. Inhomogeneous electron gas. *Physical Review* 136:B864–71.

Iijima, S. 1991. Helical microtubules of graphitic carbon. *Nature* 354:56–8.

Jafri, R. I., N. Rajalakshmi, and S. Ramaprabhu. 2010. Nitrogen doped graphene nanoplatelets as catalyst support for oxygen reduction reaction in proton exchange membrane fuel cell. *Journal of Materials Chemistry* 20:7114–7.

Jiao, L., L. Zhang, L. Ding, J. Liu, and H. Dai. 2010. Aligned graphene nanoribbons and crossbars from unzipped carbon nanotubes. *Nano Research* 3:387–94.

Kauffman, D. and A. Star. 2010. Graphene versus carbon nanotubes for chemical sensor and fuel cell applications. *Analyst* 135:2790–7.

Kim, Y., D.-H. Cho, S. Ryu, and C. Lee. 2014. Tuning doping and strain in graphene by microwave-induced annealing. *Carbon* 67:673–9.

Knights, S. D., K. M. Colbow, J. St-Pierre, and D. P. Wilkinson. 2004. Aging mechanisms and lifetime of PEFC and DMFC. *Journal of Power Sources* 127:127–34.

Kong, X.-K., C.-L. Chen, and Q.-W. Chen. 2014. Doped graphene for metal-free catalysis. *Chemical Society Reviews* 43:2841–57.

Kresse, G. and J. Furthmuller. 1996a. Efficiency of ab-initio total energy calculations for metals and semiconductors using a plane-wave basis set. *Computational Materials Science* 6:15–50.

Kresse, G. and J. Furthmuller. 1996b. Efficient iterative schemes for ab initio total-energy calculations using a plane-wave basis set. *Physical Review B* 54:11169–86.

Lee, K., J. Zhang, H. Wang, and D. P. Wilkinson. 2006. Progress in the synthesis of carbon nanotube- and nanofiber-supported Pt electrocatalysts for PEM fuel cell catalysis. *Journal of Applied Electrochemistry* 36:507–22.

Lepro, X., E. Terres, Y. Vega-Cantu, F. J. Rodriguez-Macias, H. Muramatsu, Y. A. Kim, T. Hayahsi, M. Endo, T. R. Miguel, and M. Terrones. 2008. Efficient anchorage of Pt clusters on N-doped carbon nanotubes and their catalytic activity. *Chemical Physics Letters* 463:124–9.

Li, X., H. Wang, J. T. Robinson, H. Sanchez, G. Diankov, and H. Dai. 2009. Simultaneous nitrogen doping and reduction of graphene oxide. *Journal of the American Chemical Society* 131:15939–44.

Liang, Y., Y. Li, H. Wang, J. Zhou, J. Wang, T. Regier, and H. Dai. 2011. Co_3O_4 nanocrystals on graphene as a synergistic catalyst for oxygen reduction reaction. *Nature Materials* 10: 780–6.

Lin, T., F. Huang, J. Liang, and Y. Wang. 2011. A facile preparation route for boron-doped graphene, and its CdTe solar cell application. *Energy and Environmental Science* 4:862–5.

Lin, Y.-C., C.-Y. Lin, and P.-W. Chiu. 2010. Controllable graphene N-doping with ammonia plasma. *Applied Physics Letters* 96:133110.

Lota, G., K. Fic, and E. Frackowiak. 2011. Carbon nanotubes and their composites in electrochemical applications. *Energy and Environmental Science* 4:1592–605.

Luo, Z., S. Lim, Z. Tian, J. Shang, L. Lai, B. MacDonald, C. Fu, Z. Shen, T. Yu, and J. Lin. 2011. Pyridinic N doped graphene: Synthesis, electronic structure, and electrocatalytic property. *Journal of Material Chemistry* 21:8038–44.

Maiyalagan, T. 2008. Synthesis and electro-catalytic activity of methanol oxidation on nitrogen containing carbon nanotubes supported Pt electrodes. *Applied Catalysis B: Environmental* 80:286–95.

Maiyalagan, T. and B. Viswanathan. 2005. Template synthesis and characterization of well-aligned nitrogen containing carbon nanotubes. *Materials Chemistry and Physics* 93:291–5.

Maldonado, S. and K. J. Stevenson. 2005. Influence of nitrogen doping on oxygen reduction electrocatalysis at carbon nanofiber electrodes. *Journal of Physical Chemistry B* 109:4707–16.

Matter, P. H., L. Zhang, and U. S. Ozkan. 2006. The role of nanostructure in nitrogen containing carbon catalysts for the oxygen reduction reaction. *Journal of Catalysis* 239:83–96.

Niwa H., K. Horiba, Y. Harada, M. Oshima, T. Ikeda, K. Terakura, J. Ozaki, and S. Miyata. 2009. X-ray absorption analysis of nitrogen contribution to oxygen reduction reaction in carbon alloy cathode catalysts for polymer electrolyte fuel cells. *Journal of Power Sources* 187:93–7.

Novoselov, K. S., V. I. Fal'ko, L. Colombo, P. R. Gellert, M. G. Schwab, and K. Kim. 2012. A roadmap for graphene. *Nature* 490:192–200.

Novoselov, K. S., A. K. Geim, S. V. Morozov, D. Jiang, Y. Zhang, S. V. Dubonos, I. V. Grigorieva, and A. A. Firsov. 2004. Electric field effect in atomically thin carbon films. *Science* 306:666–9.

Palnitkar, U. A., R. V. Kashid, M. A. More, D. S. Joag, L. S. Panchakarla, and C. N. R. Rao. 2010. Remarkably low turn-on field emission in undoped, nitrogen-doped, and boron-doped graphene. *Applied Physics Letters* 97:063102.

Panchakarla. L. S., K. S. Subrahmanyam, S. K. Saha, A. Govindaraj, H. R. Krishnamurthy, U. V. Waghmare, and C. N. R. Rao. 2009. Synthesis, structure, and properties of boron- and nitrogen-doped graphene. *Advanced Materials* 21:4726–30.

Qu, L., Y. Liu, J.-B. Baek, and L. Dai. 2010. Nitrogen-doped graphene as efficient metal-free electrocatalyst for oxygen reduction in fuel cells. *ACS Nano* 4:1321–6.

Rajalakshmi, N., H. Ryu, M. M. Shaijumon, and S. Ramaprabhu. 2005. Performance of polymer electrolyte membrane fuel cells with carbon nanotubes as oxygen reduction support material. *Journal of Power Sources* 140:250–7.

Reddy, A. L. M., A. Srivastava, S. R. Gowda, H. Gullapalli, M. Dubey, and P. M. Ajayan. 2010. Synthesis of nitrogen-doped graphene films for lithium battery application. *ACS Nano* 4:6337–42.

Saha, M. S., R. Li, X. Sun, and S. Ye. 2009. 3-D composite electrodes for high performance PEM fuel cells composed of Pt supported on nitrogen-doped carbon nanotubes grown on carbon paper. *Electrochemistry Communications* 11:438–41.

Schniepp, H. C., J.-L. Li, M. J. McAllister, H. Sai, M. Herrera-Alonso, D. H. Adamson, R. K. Prud'homme, R. Car, D. A. Saville, and I. A. Aksay. 2006. Functionalized single graphene sheets derived from splitting graphite oxide. *Journal of Physical Chemistry B* 110:8535–39.

Serp, P., M. Corrias, and P. Kalck. 2003. Carbon nanotubes and nanofibers in catalysis. *Applied Catalysis A: General* 253:337–58.

Shao, Y., G. Yin, and Y. Gao. 2007. Understanding and approaches for the durability issues of Pt-based catalysts for PEM fuel cells. *Journal of Power Sources* 171:558–66.

Shao. Y., S. Zhang, M. H. Engelhard, G. Li, G. Shao, Y. Wang, J. Liu, I. A. Aksay, and Y. Lin. 2010. Nitrogen-doped graphene and its electrochemical applications. *Journal of Material Chemistry* 20:7491–6.

Si, Y. and E. T. Samulski. 2008a. Exfoliated graphene separated by platinum nanoparticles. *Chemistry of Materials* 20:6792–7.

Si, Y. and E. T. Samulski. 2008b. Synthesis of water soluble graphene. *Nano Letters* 8:1679–82.

Sidik, R. and A. Anderson. 2002. Density functional theory study of O_2 electroreduction when bonded to a Pt dual site. *Journal of Electroanalytical Chemistry* 528:69–76.

Sidik, R. A., A. B. Anderson, N. P. Subramanian, S. P. Kumaraguru, and B. N. Popov. 2006. O_2 reduction on graphite and nitrogen-doped graphite: Experiment and theory. *Journal of Physical Chemistry B* 110:1787–93.

Sjostrom, H., S. Stafstrom, M. Boman, and J. E. Sundgren. 1995. Superhard and elastic carbon nitride thin films having fullerene like microstructure. *Physical Review Letters* 75:1336–9.

Son, Y. W., M. L. Cohen, and S. G. Louie. 2006. Energy gaps in graphene nanoribbons. *Physical Review Letters* 97:216803.

Stankovich, S., D. A. Dikin, G. H. B. Dommett, K. M. Kohlhaas, E. J. Zimney, E. A. Stach, R. D. Piner, S. T. Nguyen, and R. S. Ruoff. 2006. Graphene-based composite materials. *Nature* 442:282–6.

Stephan, O., P. M. Ajayan, C. Colliex, P. Redlich, J. M. Lambert, P. Bernier, and P. Lefin. 1994. Doping graphitic and carbon nanotube structures with boron and nitrogen. *Science* 266:1683–5.

Suggs, K., V. Person, and X.-Q. Wang. 2011. Band engineering of oxygen doped single-walled carbon nanotubes. *Nanoscale* 3:2465–68.

Tang, Y., Z. Yang, X. Dai, D. Ma, and Z. Fu. 2013. Formation, stabilities, and electronic and catalytic performance of platinum catalyst supported on non-metal-doped graphene. *Journal of Physical Chemistry C* 117:5258–68.

Terrones, M., P. M. Ajayan, F. Banhart, X. Blase, D. L. Carroll, J. C. Charlier, R. Czerw et al. 2002. N-doping and coalescence of carbon nanotubes: Synthesis and electronic properties. *Applied Physics A* 74:355–61.

Trasobares, S., O. Stephan, C. Colliex, W. K. Hsu, H. W. Kroto, and D. R. M. Walton. 2002. Compartmentalized CN_x nanotubes: Chemistry, morphology and growth. *Journal of Chemical Physics* 116:8966–72.

Tsai, C.-W., M.-H. Tu, C.-J. Chen, T.-F. Hung, R.-S. Liu, W.-R. Liu, M.-Y. Lo et al. 2011. Nitrogen-doped graphene nanosheet-supported non-precious iron nitride nanoparticles as an efficient electrocatalyst for oxygen reduction. *RSC Advances* 1:1349–57.

Tsetseris, L., B. Wang, and S. T. Pantelides. 2014. Substitutional doping of graphene: The role of carbon divacancies. *Physical Review B* 89:035441.

Vinayan, B. P., R. Nagar, N. Rajalakshmi, and S. Ramaprabhu. 2012. Novel platinum–cobalt alloy nanoparticles dispersed on nitrogen-doped graphene as a cathode electrocatalyst for PEMFC applications. *Advanced Functional Materials* 22:3519–26.

Voorhis, T. and G. Scuseria. 1998. A novel form for the exchange-correlation energy functional. *Journal of Chemical Physics* 109:400–10.

Wadt, W. R. and P. J. Hay. 1985. Ab initio effective core potentials for molecular calculations. Potentials for main group elements Na to Bi. *Journal of Chemical Physics* 82:284–98.

Wang, B. 2005. Recent development of non-platinum catalyst for oxygen reduction reaction. *Journal of Power Sources* 152:1–15.

Wang, C. D., M. F. Yuen, T. W. Ng, S. K. Jha, Z. Z. Lu, S. Y. Kwok, T. L. Wong et al., 2012. Plasma-assisted growth and nitrogen doping of graphene films. *Applied Physics Letters* 100:253107–11.

Wang, H., Q. Hao, X. Yang, L. Lu, and X. Wang. 2009. Graphene oxide doped polyaniline for supercapacitors. *Electrochemistry Communications* 11:1158–61.

Wang, H., T. Maiyalagan, and X. Wang. 2012. Review on recent progress in nitrogen-doped graphene: Synthesis, characterization, and its potential applications. *ACS Catalysis* 2:781–94.

Wang, H. W., B. C. Wang, W. H. Chen, and M. Hayashi. 2008. Localized Gaussian type orbital—Periodic boundary condition—Density functional theory study of infinite-length single-walled carbon nanotubes with various tubular diameters. *Journal of Physical Chemistry A* 112:1783–90.

Wang, X., X. Li, L. Zhang, Y. Yoon, P. K. Weber, H. Wang, J. Guo, and H. Dai. 2009. N-doping of graphene through electrothermal reactions with ammonia. *Science* 324:768–71.

Wang, Y., Y. Shao, D. W. Matson, J. Li, and Y. Lin. 2010. Nitrogen-doped graphene and its application in electrochemical biosensing. *ACS Nano* 4:1790–8.

Wei, D. and Y. Liu. 2010. Controllable synthesis of graphene and its applications. *Advanced Materials* 22:3225–41.

Wei, D., Y. Liu, Y. Wang, H. Zhang, L. Huang, and G. Yu. 2009. Synthesis of N-doped graphene by chemical vapor deposition and its electrical properties. *Nano Letters* 9:1752–8.

Weng-Sich, Z., K. Cherrey, N. G. Chopra, X. Blase, Y. Miyamoto, A. Rubio, M. L. Cohen, S. G. Louie, A. Zettl, and R. Gronsky. 1995. Synthesis of BxCyNz nanotubules. *Physical Review B* 51:11229–32.

Wu, G., D. Li, C. Dai, D. Wang, and N. Li. 2008. Well-dispersed high-loading Pt nanoparticles supported by shell-core nanostructured carbon for methanol electrooxidation. *Langmuir* 24:3566–75.

Xing, Y. 2004. Synthesis and electrochemical characterization of uniformly-dispersed high loading Pt nanoparticles on sonochemically-treated carbon nanotubes. *Journal of Physical Chemistry B* 108:19255–9.

Xu, Y., H. Bai, G. Lu, C. Li, and G. Shi. 2008. Flexible graphene films via the filtration of water-soluble noncovalent functionalized graphene sheets. *Journal of the American Chemical Society* 130:5856–7.

Yu, R., L. Chen, Q. Liu, J. Lin, K.-L. Tan, S. C. Ng, H. S. O. Chan, G.-Q. Xu, and T. S. A. Hor. 1998. Platinum deposition on carbon nanotubes via chemical modification. *Chemistry of Materials* 10:718–22.

Zhang, L. and Z. Xia. 2011. Mechanisms of oxygen reduction reaction on nitrogen-doped graphene for fuel cells. *Journal of Physical Chemistry C* 115:11170–6.

Zhang, L., J. Niu, M. Li, and Z. Xia. 2014. Catalytic mechanisms of sulfur-doped graphene as efficient oxygen reduction reaction catalysts for fuel cells. *Journal of Physical Chemistry C* 118:3545–53.

Zhang, L.-S., X.-Q. Liang, W.-G. Song, and Z.-Y. Wu. 2010. Identification of the nitrogen species on N-doped graphene layers and Pt/NG composite catalyst for direct methanol fuel cell. *Physical Chemistry Chemical Physics* 12:12055–9.

Zhang, Y., H. Gu, K. Suenaga, and S. Iijima. 1997. Heterogeneous growth of B–C–N nanotubes by laser ablation. *Chemical Physics Letters* 279:264–9.

Zhu, Y., S. Murali, W. Cai, X. Li, J. Suk, J. R. Potts, and R. Ruoff. 2010. Graphene and graphene oxide: Synthesis, properties, and applications. *Advanced Materials* 22:3906–24.

21 Scanning Electron Microscopy of Graphene

Yoshikazu Homma, Katsuhiro Takahashi, Yuta Momiuchi,
Junro Takahashi, and Hiroki Kato

CONTENTS

Abstract 319
21.1 Introduction 319
21.2 Secondary Electron Contrast of Monolayer 320
21.3 Graphene Imaging by Scanning Electron Microscopy 320
    21.3.1 Insulator Surface 320
    21.3.2 Metal Surface 321
    21.3.3 Graphene Thickness Dependence 322
21.4 *In Situ* Observation of Graphene Segregation 322
    21.4.1 Monolayer Graphene 322
    21.4.2 Multilayered Graphene 323
21.5 Summary 324
References 324

ABSTRACT

This chapter focuses on the imaging of graphene by scanning electron microscopy (SEM). Although a monolayer of graphene can be observed by SEM, there are different mechanisms for image formation of the monolayered material. Image formation mechanism is discussed for insulator and metal substrates. Applications to *in situ* SEM observation are shown for graphene segregation on a polycrystalline nickel substrate.

21.1 INTRODUCTION

The outbreak of the graphene research was triggered by the paper that showed monolayer graphene could be obtained by repeating mechanical exfoliation of graphite (Novoselow and Geim 2004). Another important aspect of the paper is that graphene on a SiO₂ film of 300 nm thick can be observed with an optical microscope and the thickness of graphene layers can be identified. It is critically important that monolayer graphene can be seen with an optical microscope, which enables onc to prepare graphene specimen easily. For *in situ* observation of graphene formation process, on the other hand, an optical microscope is not the best choice. An observation method capable at high temperatures or in a gas environment is necessary. Electron microscopy has been utilized for this purpose. Table 21.1 summarizes publications on electron microscopy of graphene on substrate surfaces. Low energy electron microscopy (LEEM) is an excellent method for observing formation of graphene in ultrahigh vacuum. Graphene formation

processes by thermal decomposition of SiC (Hibino et al. 2008, 2010) and segregation from metals (Sutter et al. 2008, 2009; Loginova et al. 2009; Wofford et al. 2010; Odahara et al. 2011) have been observed by LEEM. However, because several tens kilovolts are applied to the specimen, LEEM cannot be used for vapor phase deposition in the pressure range where vacuum discharge occurs. In addition, because of the spherical aberration of the electron optics, the field of view is limited to 100 μm or less, which is a disadvantage of LEEM when a large area needs to be observed.

We have used scanning electron microscopy (SEM) for the observation of monolayer growth and sublimation on Si and GaAs surfaces (Homma 2011). SEM is advantageous in the wide field of view ranging up to millimeter, and capability of gas introduction in the observation chamber. For single-walled carbon nanotubes (SWCNTs), the formation process of suspended SWCNTs between micro-pillars (Homma et al. 2006) and the extension process of SWCNTs on SiO₂ substrates (Takagi et al. 2006; Wako et al. 2007) were observed by repeating chemical vapor deposition (CVD) and SEM observation alternately in the SEM chamber. For graphene, monolayer segregation processes were observed on polycrystalline Ni surfaces by SEM (Takahashi et al. 2012). Recently, *in situ* SEM was applied to graphene growth on polycrystalline Cu by chemical vapor deposition (Kidambi et al. 2013).

Apart from *in situ* observation, SEM is now widely used for the observation of graphene. Monolayer graphene can indeed be observed by SEM (Hiura et al. 2010; Chen et al. 2011; Kochat et al. 2011; Wood et al. 2011; Yang et al. 2012; Zhao et al. 2013). However, the formation mechanism of

TABLE 21.1

Electron Microscopy Imaging of Graphene

Method	Substrate	Reference
LEEM	SiC(0001)	Hibino (2008, 2010)
	Ru(0001)	Sutter (2008)
	Pt(111)	Sutter (2009)
	Ir(111)	Loginova (2009)
	Cu foil	Wofford (2010)
	Ni(111)	Odahara (2011)
SEM	SiO₂	Hiura (2010)
		Kochat (2011)
	Cu foil	Chen (2011)
		Wood (2011)
		Yang (2012)
		Zhao (2013)
In situ SEM	Ni foil	Takahashi (2012)
	Cu foil	Kidambi (2013)
	Cu foil	Wang, H. et al. (2015)
	Cu foil	Wang, Z-J. et al. (2015)

TABLE 21.2

Origin of Secondary Electron Contrast of Monolayers

Origin of Contrast	Example	Reference	Relationship with Graphene Imaging
Topographic effect of edge	Atomic steps	Homma (1991, 1993a, 1996)	Monolayer graphene on metal
Atomic configuration	Si(001)2 × 1 & 1 × 2	Homma (1993b)	
Elemental composition	InAs(001)4 × 2 & 2 × 4	Yamaguchi (1995)	Oxidation of non-graphene covered surface
Work function	Au, Ag submonolayers on Si Si(111)7 × 7 & 1 × 1	Endo (1993a,b) Homma (1993b)	Multilayer graphene
Charging	Carbon nanotube on insulator	Homma (2004)	Graphene on insulator

secondary electron image of the monolayered material is by no means simple, and it is not well understood. In this chapter, the SEM observations of graphene are reviewed, and the image formation mechanism is discussed.

21.2 SECONDARY ELECTRON CONTRAST OF MONOLAYER

Single monatomic layer can be imaged by SEM. There are many factors forming the secondary electron (SE) contrast of monolayer, such as topographic contrast of the edge, elemental composition, work function, and charging, as summarized in Table 21.2. The topographic contrast appears even for monatomic-high steps due to secondary electron yield change at the step edge and anisotropy of detection efficiency of secondary electron detector. The behavior of the edge contrast in terms of the relationship between electron beam incidence and step edge, and that between detector position (or detection efficiency depending on the direction of emitted secondary electrons) and step direction are all the same for monatomic steps and macroscopic steps (Homma et al. 1991, 1993a). The difference is only the SE intensity at the steps. Usually, monatomic step contrast is faint.

An example of the SE contrast of monatomic layer caused by elemental difference is the contrast between the As-stabilized InAs 2 × 4 (001) surface and In-stabilized InAs 4 × 2 (001) surface (Yamaguchi et al. 1995). This contrast depends on which element, As or In, comes to the top most layer in alternate layers of As and In. Au and Ag sub-monolayers on Si also produce SE contrast (Endo and Ino 1993a,b). In these cases, the local work function may also change, but the presence of large atomic number elements on the Si surface may contribute to a large SE yield. Graphene, on the other hand, consists of carbon and the SE yield from the carbon monolayer is low.

As shown later, monolayer graphene does not greatly change the SE emission from the substrate underneath it. When graphene layer increases from monolayer to five layers or more, the surface becomes gradually darker with an increase in the layer number. This contrast change can be explained by the change in the work function, which is 3.9 eV for the monolayer graphene and 4.6 eV for the graphite surface (Oshima and Nagashima 1997).

The SE contrast of graphene on the substrate depends on the electrical property of the substrate. On an insulator surface, SE emission is reduced when the surface is positively charged, whereas it is enhanced when the surface is negatively charged (Joy and Joy 1995). When a monolayer graphene covers the insulator surfaces, the charging can be compensated by conduction through the graphene layer, thus causing a contrast change between the graphene-covered surface and the bare insulator surface.

Atomic layer contrast also depends on the atomic configuration in the atomic layer, such as 1 × 2 and 2 × 1 reconstructed structures on the Si(001) surface (Homma et al. 1993b). However, even if a clean graphene surface is observed using an appropriate SE detection condition, the different domains of three hold symmetry graphene might not be easily observable as SE images.

21.3 GRAPHENE IMAGING BY SCANNING ELECTRON MICROSCOPY

21.3.1 INSULATOR SURFACE

On an insulator such as SiO₂, the charging of the surface strongly influences the SE contrast.

Hiura et al. reported detailed SE contrast change depending on the graphene thickness and the primary electron energy using mechanically exfoliated few-layer graphene on

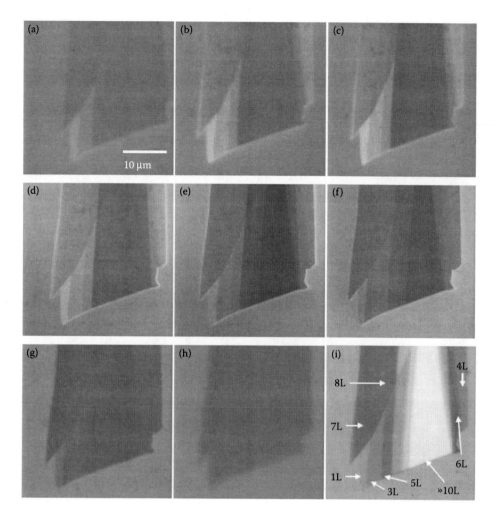

FIGURE 21.1 SE contrast of few-layer graphene on SiO₂ (300 nm) depending on the primary electron energy. Primary electron energy: (a) 0.5, (b) 0.8, (c) 1.0, (d) 1.4, (e) 2.0, (f) 3.0, (g) 5.0, (h) 20.0 keV. An optical microscope image is shown in (i). The numbers with L denote the number of graphene layers. (Reprinted with permission from *The Japan Society of Applied Physics*, Hiura, H., H. Miyazaki, and K. Tsukagoshi. 2010. *Appl. Phys. Exp.* 3: 095101.)

a 300 nm-thick SiO₂ on Si (Hiura et al. 2010). For such a thin insulator film, the surface charging state depends on the primary electron energy, that is, the penetration depth of primary electrons. Because the threshold electron energy for penetrating through the 300 nm-thick SiO₂ film is around 3 keV, the SE contrast of graphene changes largely between lower and higher energies than 3 keV. Below 3 keV (Figure 21.1a through e), since the surface of SiO₂ is positively charged due to SE emission, further SE emission from SiO₂ is suppressed, while the area covered with graphene and its periphery emit SEs owing to the electron supply through the graphene. This is the same mechanism of SWCNT imaging on an insulator surface (Homma et al. 2004). Monolayer graphene itself does not affect greatly the SE emission from underneath SiO₂. Thus, the contrast change reflects the charging state of the monolayer-graphene covered surface and the bare SiO₂ surface. On the other hand, primary electrons of 3 keV or higher can penetrate through 300 nm-thick SiO₂ film. Due to the electron–hole pair generation in the primary electron range, SiO₂ has some conductivity and electrons are supplied to SiO₂

from the Si substrate (Homma et al. 2004). The surface charging is, thus, diminished and SE emission recovers. As a result, monolayer-graphene-covered regions no more appear brighter than the bare SiO₂ surface (Figure 21.1f through h). Even the monolayer-graphene covered region appears slightly darker than the bare SiO₂ surface. This means that monolayer graphene is not completely transparent to the SEs emitted from underneath region. For graphene layers of three or thicker, the SE intensity decreases with an increase in the thickness.

21.3.2 Metal Surface

On metal surfaces, charging contrast is usually absent. On a polycrystalline surface, crystal orientation of each grain causes SE contrast due to electron channeling or work function difference, which makes recognition of few-layer graphene difficult. Figure 21.2a shows an SE image of few-layer graphene on a polycrystalline Ni surface. The graphene layer was segregated during the cooling process in the SEM chamber and observed without exposing to air (Takahashi et al. 2012).

100 µm

FIGURE 21.2 SEM images of few-layer graphene grown on polycrystalline Ni observed (a) without exposure to air and (b) after exposure to air. The number of graphene layers increases from left to right. (Reprinted from *Surf. Sci.* 606, Takahashi, K. et al. *In situ* scanning electron microscopy of graphene growth on polycrystalline Ni substrate, 728–32, Copyright 2012, with permission from Elsevier.)

The thickness of the graphene layer increases from left to right of the image. The SE intensity decreases with an increase in the number of graphene layers. Monolayer graphene is located at the left half, but the grain contrast of the Ni surface makes it difficult to distinguish the monolayer-graphene covered surface from the bare Ni surface. Figure 21.2b shows the SE image of the same surface after exposure to air. Brighter regions appear after air exposure. These are non-graphene covered regions oxidized in air, where the SE yield increases. The area covered with graphene remains unoxidized (Chen et al. 2011), and now the difference between graphene-covered and bare Ni surfaces is clear.

Similar oxidation contrasts could be obtained on other metal surfaces such as Co, Cu, and Pd. SE images of monolayer graphene can be found in some papers (see Table 21.1). Those are owing to the contrast difference between graphene-covered (unoxidized) and bare metal surfaces oxidized in air.

21.3.3 GRAPHENE THICKNESS DEPENDENCE

The SE intensity from few-layer graphene is dependent not only on the layer thickness (or the change in the work function), but also on the range of primary electrons and SEs emitted from the underneath material. In Figures 21.1 and 21.2,

the regions with three graphene layers or more appear dark. The SE yield reaches to that of graphite as the layer thickness increases. To distinguish the thickness of few-layer graphene, the primary electron energy of 1.5–2 keV is appropriate. The images in Figure 21.2 were observed with the 1.45 keV primary electron beam and an in-lens detector. When using primary electron energy of 1 keV or less, it is difficult to resolve the difference in five layers or more (Figure 21.1a through c). On the other hand, with the electron energy of 5 keV or larger, the difference of SE intensity with the layer numbers becomes small (Figure 21.1g and h). For an insulating film, the SE contrast of few-layer graphene is also dependent on the relationship between the thickness of insulator and the primary electron range as shown in Figure 21.1.

21.4 *IN SITU* OBSERVATION OF GRAPHENE SEGREGATION

21.4.1 MONOLAYER GRAPHENE

When carbon doped Ni is slowly cooled from 900°C, carbon atoms precipitate to the surface of Ni and form monolayer graphene (Shelton et al. 1974). The segregation process of graphene can be observed by SEM. Figure 21.3 shows monolayer graphene segregation process from a polycrystalline Ni surface at around 800°C. The Ni specimen 0.5 mm × 30 mm and 0.5 mm thick was heated by passing direct current through it. Above 400°C, the edge contrast of monolayer graphene becomes prominent, thus, monolayer graphene can be distinguished from Ni grains in SE images. Figure 21.3a is the Ni surface before graphene segregation, and only Ni grain contrast is seen. Starting from Figure 21.3b, graphene nucleated and extended to cover the Ni surface. Monolayer graphene appears as if it had steric edges due to the topographic contrast. There are two types of edge contrast: dark contrast for edges facing toward top and right of the SE image; bright contrast for edges facing toward bottom and left of the SE image. The edge contrasts are the same as those for macroscopic steps, and influenced by the relationship between electron beam incidence and step edge, and/or detection efficiency depending on the direction of emitted secondary electrons as schematically shown in Figure 21.4 (Homma et al. 1991, 1993a). In the case of Figure 21.3, because normal incidence of the primary electron beam and an in-lens type SE detector that corrects SEs through the objective lens were used, the bright and dark contrasts are due to anisotropy of detection efficiency depending on the direction of emitted secondary electrons (Figure 21.4b). Note that the SE images in Figure 21.3 are obtained at low magnification, and the full width of the image corresponds to 0.47 mm. Even though such a wide area is imaged, the monolayer edge contrast is extremely clear. Although monatomic steps of GaAs and Si have been imaged by SEM, much higher magnifications are necessary for direct imaging using the edge contrast (Homma et al. 1991, 1996). For monolayer graphene, the strength of edge contrast depends on temperature. For Ni, the edge contrast becomes prominent at around

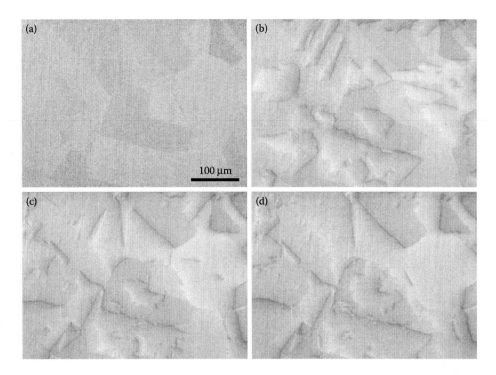

FIGURE 21.3 *In situ* observation of monolayer-graphene segregation process. (a) Ni surface before graphene segregation. (b) Just after graphene nucleation. (c) 3 min after graphene nucleation. (d) 7 min after graphene nucleation. A Ni foil commercially available was used without further polishing.

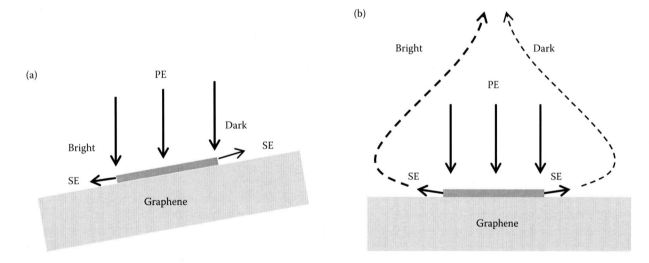

FIGURE 21.4 Schematic illustration of edge contrast. (a) Effect of primary electron (PE) incidence: the edge in the forward scattering direction appears bright, while that in the back scattering direction appears dark. (b) Effect of anisotropy of detection efficiency: the edge located in the direction of higher SE detection efficiency appears bright.

400°C and higher. It is surprising that the edge contrast is clear even on the unpolished polycrystalline surface where the surface roughness is much larger than the monolayer height. Similar edge contrast was observed for Co, Pd, and Pt, and the optimum temperature range for obtaining edge contrast somewhat depends on the metal species. On the other hand, edge contrast was not prominent or missing on the Cu surface (Wang 2015). The origin of the prominent edge contrast and its temperature dependence is still under investigation.

21.4.2 MULTILAYERED GRAPHENE

Figure 21.5 shows the successive graphene layer segregation after the monolayer graphene segregation shown in Figure 21.3 at the same temperature. Due to the temperature gradient in the specimen, the right side of the SE image was at lower temperature. The second layer is formed at the regions indicated by arrows in Figure 21.5a and b. Note that the edge contrast does not appear for the second layer, because the second layer precipitates underneath the first layer. The second layer

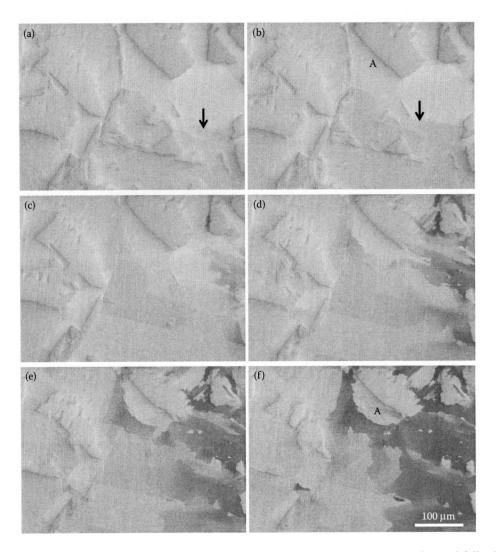

FIGURE 21.5 *In situ* observation of multilayer-graphene segregation process. Those images were observed following Figure 21.3d. Elapsed time after graphene nucleation (Figure 21.3b) is: (a) 8 min, (b) 9 min, (c) 11 min, (d) 13 min, (e) 15 min, and (f) 17 min.

appears slightly darker. The third layer that is further darker appears at the right edge of Figure 21.5c. The darker regions expand and the right end of the SE image becomes much darker indicating the increase in the layer number in Figure 21.5d. The right half of the surface is covered with three layers or thicker graphene layers. Interestingly, after the first layer, no further layers are segregated at the region denoted by "A."

Graphene nucleation preferentially occurs on the (111) and (110) Ni grains in the segregation case (Takahashi et al. 2012). By decreasing the specimen temperature, graphene expands to other grains, but second and later layers also segregate underneath.

21.5 SUMMARY

Monolayer graphene can be imaged by SEM, although it is almost transparent to the secondary electrons generated in the underneath material. There are several image formation mechanisms. On an insulator surface, charging state difference causes the SE contrast; darker on the positively charged insulator surface and brighter on the monolayer-graphene covered

surface. On an air-exposed metal surface, oxidized and non-oxidized surfaces are responsible for the SE contrast; generally brighter on the oxidized metal surface and darker on the monolayer-graphene covered (non-oxidized) surface. At elevated temperatures, graphene edges exhibit prominent topographic contrast: either brighter or darker contrast depending on the relationship between the edge direction and the anisotropy in SE detection efficiency. For thicker layers of graphene, the SE intensity decreases with an increase in the number of layers. Optimum electron energies exist for discriminating the number of layers regarding the layer resolution and the layer depth. With a primary electron energy of 1.5–2 keV, SE images are sensitive to the number of layers in few-layer graphene.

REFERENCES

Chen, S., L. Brown, M. Levendorf, W. Cai, S.-Y. Ju, J. Edgeworth, X. Li et al. 2011. Oxidation resistance of graphene-coated Cu and Cu/Ni alloy. *ACS Nano* 5: 1321.
Endo, A. and S. Ino. 1993a. Observation of the Ag/Si(111) system using a high-resolution ultra-high vacuum scanning electron microscope. *Surf. Sci.* 293: 165–82.

Endo, A. and S. Ino. 1993b. Observations of the Au/Si(111) system with a high-resolution ultrahigh-vacuum scanning electron microscope. *Jpn. J. Appl. Phys.* 32: 4718–25.

Hibino, H., H. Kageshima, F. Maeda, M. Nagase, Y. Kobayashi, and H. Yamaguchi. 2008. Microscopic thickness determination of thin graphite films formed on SiC from quantized oscillation in reflectivity of low-energy electrons. *Phys. Rev. B* 77: 075413.

Hibino, H., H. Kageshima, and M. Nagase. 2010. Epitaxial few-layer graphene: Towards single crystal growth. *J. Phys. D: Appl. Phys.* 43: 374005.

Hiura, H., H. Miyazaki, and K. Tsukagoshi. 2010. Determination of the number of graphene layers: Discrete distribution of the secondary electron intensity stemming from individual graphene layers. *Appl. Phys. Exp.* 3: 095101.

Homma, Y. 2011. Measuring nucleation and growth processes in thin films. In *Thin Film Growth: Physics, Materials Science and Applications*, ed. Z. Cao, 3–21. Cambridge: Woodhead.

Homma, Y., M. Tomita, and T. Hayashi. 1991. Secondary electron imaging of monolayer steps on a clean Si(111) surface. *Surf. Sci.* 258: 147–52.

Homma, Y., M. Tomita, and T. Hayashi. 1993a. Atomic step imaging on silicon surfaces by scanning electron microscopy. *Ultramicrosco.* 52: 187–92.

Homma, Y., M. Suzuki, and M. Tomita. 1993b. Atomic configuration dependent secondary electron emission from reconstructed silicon surfaces. *Appl. Phys. Lett.* 62: 3276–8.

Homma, Y., J. Osaka, and N. Inoue. 1996. Secondary electron imaging of nucleation and growth of GaAs. *Surf. Sci.* 357–358: 441–45.

Homma, Y., S. Suzuki, Y. Kobayashi, M. Nagase, and D. Takagi. 2004. Mechanism of bright selective imaging of single-walled carbon nanotubes on insulators by scanning electron microscopy. *Appl. Phys. Lett.* 84: 1750–2.

Homma, Y., D. Takagi, and Y. Kobayashi. 2006. Suspended architecture formation process of single-walled carbon nanotubes. *Appl. Phys. Lett.* 88: 023115.

Joy, D. C. and C. S. Joy. 1995. Dynamic charging in the low voltage SEM. *J. Microsco. Soc. Am.* 1: 107–12.

Kidambi, P. R., B. C. Bayer, R. Blume, Z.-J. Wang, C. Baehtz, R. S. Weatherup, M.-G. Willinger, R. Schloegl, and S. Hofmann. 2013. Observing graphene grow: Catalyst – graphene interactions during scalable graphene growth on polycrystalline copper. *Nano Lett.* 13: 4769–77.

Kochat, V., A. N. Pal, E. S. Sneha, A. Sampathkumar, A. Gairola, S. A. Shivashankar, S. Raghavan, and A. Ghosh. 2011. High contrast imaging and thickness determination of graphene with in-column secondary electron microscopy. *J. Appl. Phys.* 110: 014315.

Loginova, E., S. Nie, K. Thürmer, N. C. Bartelt, and K. F. McCarty. 2009. Defects of graphene on Ir(111): Rotational domains and ridges. *Phys. Rev. B* 80: 085430.

Novoselov, K. S., A. K. Geim, S. V. Morozov, D. Jiang, Y. Zhang, S. V. Dubonos, I. V. Grigorieva, and A. A. Firsov. 2004. Electric field effect in atomically thin carbon films. *Science* 306: 666–9.

Odahara, G., S. Otani, C. Oshima, M. Suzuki, T. Yasue, and T. Koshikawa. 2011. In-situ observation of graphene growth on Ni(111). *Surf. Sci.* 605: 1095–8.

Oshima, C. and A. Nagashima. 1997. Ultra-thin epitaxial films of graphite and hexagonal boron nitride on solid surfaces. *J. Phys.: Condens. Matter* 9:1–20.

Shelton, J. C., H. R. Patila, and J.M. Blakely. 1974. Equilibrium segregation of carbon to a nickel (111) surface: A surface phase transition. *Surf. Sci.* 43: 493–520.

Sutter, P., J. T. Sadowski, and E. Sutter. 2009. Graphene on Pt(111): Growth and substrate interaction. *Phys. Rev. B* 80: 245411.

Sutter, P. W., J.-I. Flege, and E. A. Sutter. 2008. Epitaxial graphene on ruthenium. *Nat. Mater.* 7: 406–11.

Takagi, D., Y. Homma, S. Suzuki, and Y. Kobayashi. 2006. *In situ* scanning electron microscopy of single-walled carbon nanotube growth, *Surf. Interface Anal.* 38: 1743–6.

Takahashi, K., K. Yamada, H. Kato, H. Hibino, and Y. Homma. 2012. *In situ* scanning electron microscopy of graphene growth on polycrystalline Ni substrate. *Surf. Sci.* 606: 728–32.

Wako, I., T. Chokan, D. Takagi, S. Chiashi, and Y. Homma. 2007. Direct observation of single-walled carbon nanotube growth processes on SiO₂ substrate by *in situ* scanning electron microscopy. *Chem. Phys. Lett.* 449: 309–13.

Wang, H., C. Yamada, and Y. Homma. 2015. Scanning electron microscopy imaging mechanisms of chemical vapor deposition grown graphene on Cu substrate revealed by *in situ* observation. *Jpn. J. Appl. Phys.* 54: 050301.

Wang, Z.-J., G. Weinberg, Q. Zhang, T. Lunkenbein, A. Klein-Hoffmann, M. Kurnatowska, M. Plodinec et al. 2015. Direct observation of graphene growth and associated copper substrate dynamics by in situ scanning electron microscopy. *ACS Nano* 9: 1506–19.

Wofford, J. M., S. Nie, K. F. McCarty, N. C. Bartelt, and O. D. Dubon. 2010. Graphene islands on Cu foils: The Interplay between shape, orientation, and defects. *Nano Lett.* 10: 4890–6.

Wood, J. D., S. W. Schmucker, A. S. Lyons, E. Pop, and J. W. Lyding. 2011. Effects of polycrystalline Cu substrate on graphene growth by chemical vapor deposition. *Nano Lett.* 11: 4547–55.

Yamaguchi, H., Y. Homma, and Y. Horikoshi. 1995. In-situ observation of phase transition and transition-induced step-bunching on InAs (001) surfaces. *Appl. Phys. Lett.* 66: 1626–8.

Yang, F., Y. Liu, W. Wu, W. Chen, L. Gao, and J. Sun. 2012. A facile method to observe graphene growth on copper foil. *Nanotechnology* 23: 475705.

Zhao, P., A. Kumamoto, S. Kim, X. Chen, B. Hou, S. Chiashi, E. Einarsson, Y. Ikuhara, and S. Maruyama. 2013. Self-limiting chemical vapor deposition growth of monolayer graphene from ethanol. *J. Phys. Chem. C* 117: 10755–63.

22 Tunneling Current of the Contact of the Curved Graphene Nanoribbon with Metal and Quantum Dots

Mikhail B. Belonenko, Natalia N. Konobeeva,
Alexander V. Zhukov, and Roland Bouffanais

CONTENTS

Abstract 327
22.1 Introduction 327
22.2 Graphene and Its Hamiltonians 329
22.3 Mathematical Rules 330
22.4 Basic Equations and Spectrum of Electrons..333
22.5 Tunneling Characteristics 335
22.6 Friedmann Model 337
22.7 Conclusion 338
Acknowledgments..338
References..338

ABSTRACT

In this chapter, we present the overview of our recent studies on the electron spectrum and the density of states of long-wave electrons in curved graphene nanoribbons, based on the Dirac equation in a curved space–time. The current–voltage characteristics for the contact of nanoribbon–quantum dots and nanoribbon–metal have been revealed. The dependence of the specimen properties on its geometry was analyzed. Also the regions with negative differential conductivity were found.

22.1 INTRODUCTION

Few decades ago, it has been well realized that the gauge invariance plays a key role in the quantum field theory (QFT) description of fundamental interactions between elementary particles. The recent comprehensive review by Vozmediano et al. [1] presents a detailed picture of the relation between QFT and the condensed matter physics of graphene. In this chapter, we only briefly overview certain key points before going directly to the matter of our study.

The mathematical concept of a non-Abelian gauge field introduced first in QFT for a description of the electroweak interaction, followed by the experimental discovery of the W and Z bosons, is an example of the most impressive achievements of theoretical physics. Before introducing the various gauge fields associated with the physics of graphene and in order to clarify their specific nature, we will make a brief

description of the classical concept of gauge invariance and of the associated gauge fields [1].

The concept of gauge invariance has been naturally introduced in classical electrodynamics. In particular, the electromagnetic field (**E**, **B**) is expressed in terms of the potentials (Φ, **A**) through

$$E = -(\nabla \Phi + \partial_t \mathbf{A}), \quad \mathbf{B} = \nabla \times \mathbf{A}. \tag{22.1}$$

The fields do not change under the transformation

$$\mathbf{A} \to \mathbf{A} + \nabla \chi, \quad \Phi \to \Phi - \partial t \chi, \tag{22.2}$$

where χ is an arbitrary scalar function of coordinate. This invariance was shown to remain applicable to the quantum mechanics of a charged spinless particle in an electromagnetic field provided that the wave function was simultaneously transformed as

$$\Psi \to \Psi \exp(ie\chi). \tag{22.3}$$

The relativistic wave equation for a spinless particle with charge e interacting with electromagnetic fields is derived by first performing the substitution $p_\mu \to p_\mu - eA_\mu$, where $A_\mu = (A^0 = \Phi, A)$ is the 4-vector electromagnetic potential, and then performing the usual substitution $p_\mu \to i\hbar\partial_\mu$. A formal solution for the wave function of a particle interacting with the electromagnetic potential A_μ can be written in terms of the solution without interaction

$$\psi = \exp\left[-ie\int A^\mu dx_\mu\right]\psi_0. \qquad (22.4)$$

Quantum dynamics, that is, the form of the quantum equation, remains unchanged by the transformations (22.2) if the wave function of the particle is multiplied by a local (space–time-dependent) phase.

The first example of a QFT gauge model is four-dimensional quantum electrodynamics (QED). A free spin 1/2 Dirac fermion with charge e and mass m is described by the action integral

$$S_\psi = \int d^4x\overline{\psi}[\gamma^\mu\partial_\mu + m]\psi, \qquad (22.5)$$

which is invariant under the global $U(1)$ group of transformations:

$$\psi(x) \to U\psi(x), \quad \overline{\psi}(x) \to U^*\overline{\psi}(x), \quad U = \exp(ie\chi), \qquad (22.6)$$

where χ is a constant. Gauge invariance requires invariance of the action under the local group of transformations obtained by replacing $\chi \to \chi(x)$. This can be achieved by replacing the derivative in (22.5) by the covariant derivative $D_\mu = \partial_\mu + ieA_\mu$. Under a local $U(1)$ transformation defined by Equation 22.6 with a space–time dependent function $\chi(x)$, $A\mu(x)$ transforms as $A_\mu \to A_\mu - \partial_\mu\chi$, a generalization of Equation 22.2.

The invariance of Maxwell's equations allows a formulation of classical electromagnetism in terms of 4-vectors and tensors. The equations can be written in a covariant way by introducing the electromagnetic tensor $F_{\mu\nu}$ as follows:

$$F_{0i} = E_i, \quad F_{ij} = -\epsilon_{ijk}B_k, \qquad (22.7)$$

and the 4-current $J^\mu = (\rho, \mathbf{J})$ made of charge density and current. In terms of these geometric objects, the four Maxwell equations reduce to

$$\partial_\lambda F_{\mu\nu} + \partial_\mu F_{\nu\lambda} + \partial_\nu F_{\lambda\mu} = 0, \qquad (22.8)$$

$$\partial_\mu F_{\mu\nu} = J_\nu. \qquad (22.9)$$

The conservation of the current $\partial_\nu J^\nu = 0$ follows from the antisymmetry of $F_{\mu\nu}$. The first equation is a Bianchi identity. It can be integrated by introducing a gauge field A_μ, such that $F_{\mu\nu} = \partial_\mu A_\nu - \partial_\nu A_\mu$. It is readily verified that two gauge fields related by the gauge transformation $A_{0\mu} = A_\mu - \partial_\mu\Omega$ give rise to the same electromagnetic tensor field. Maxwell's equations can be derived from the action

$$S(A, J) = \int d^4x[F_{\mu\nu}F^{\mu\nu} + J_\mu A^\mu], \qquad (22.10)$$

which coincides with the full action in quantum electrodynamics.

The concepts of gauge fields and covariant derivatives can be interpreted in the terms of differential geometry. In general, the gauge field has a mathematical interpretation as a Lie connection and is used to construct covariant derivatives acting on fields, whose form depends on the representation of the group under which the field transforms. The field tensor $F_{\mu\nu}$ is a curvature 2-form given by the commutator of two covariant derivatives. It is an element of the Lie algebra associated with the gauge group. The gauge connection generates parallel transport of the geometric objects under gauge transformations. The generalization of $U(1)$ to non-Abelian groups such as SU(N) is straightforward: the main modification arises in the definition of the field strength (22.6) that becomes $F_{\mu\nu} = \partial_\mu A_\nu - \partial_\nu A_\mu + [A_\mu, A_\nu]$.

General relativity can be also interpreted as a gauge theory, where gauge invariance is invariance under diffeomorphisms (local smooth changes of coordinates) in the space–time manifold. The connection, which generates parallel transport, plays the role of the gauge field. Gauge invariance corresponds to the independence of field equations from the choice of the local frame. The spin connection plays the role of the gauge field.

The gauge invariance allows fixing some conditions on the gauge potentials that will not affect the physical properties. In classical electromagnetism, the gauge-fixing problem is simply the problem of choosing a representative in the class of equivalent potentials, convenient for practical calculations or most suited to the physical nature of a particular problem under consideration. In non-relativistic problems, one of the most popular choices is the Coulomb gauge, $\nabla\mathbf{A}(t, \mathbf{x}) = 0$, whose relativistic counterpart is $\partial_\mu A_\mu(t, \mathbf{x}) = 0$ ($\mu = \{0,1,2,3\}$), called Landau or Lorentz's gauge. The freedom of a gauge condition choice is related to the full gauge invariance of the action. When fictitious gauge fields are generated by analogy with the gauge formalism but there is no dynamics associated to them it can happen that the gauge potentials are fixed by the physics involved and no extra conditions can be imposed. Gauge fields were introduced in condensed matter in the early works of References 2 and 3, but now this question is very popular among many researchers [4–7].

The problem of modified graphene properties has attracted a considerable attention (see References 8–12 for instance) because "pure" graphene has no energy gap in the band structure, and, therefore, the creation of different structures (for example, analogs of transistors) is extremely difficult. However, the situation becomes more promising when various modifications of the specimen are introduced. As an example, we consider the modified graphene, for example, graphene nanoribbon, which has quantized electron energy spectrum due to the limited space in one dimension, which in turn can lead to the formation of an energy gap. Furthermore, it is well known that graphene has a naturally wave-like curved surface due to the instability of the planar structure of its sheets [13,14]. All of the above reasons have stimulated the study of different modifications of curved graphene [15,16].

The long-wave approximation—widely used to describe the properties of electrons in graphene—leads to an analog of the Dirac equation, which in turn makes it easy to produce generalization to the case when the graphene surface is curved. Note that, in this case, the degeneracy in the Dirac points disappears and, therefore, it becomes possible to create various structures with different band gaps. Consideration of the Dirac equation for curved graphene [15] also shows a change in the density of states of electrons, and, therefore, it makes it possible to change the whole set of electrical characteristics of a graphene sample. Apparently, the easiest way to experimentally verify those changes in the density of states is to study the tunneling current [17], for example, through the contact with quantum dots. Reducing the size of the particles leads to the manifestation of a very unusual properties of the material from which it is made. The reason for this are quantum-mechanical effects originated from the spatial limitation of charge carriers movements: carriers' energy becomes discrete in this case. The number of energy levels depends on the size of the potential well, the potential barrier height, and the mass of the charge carrier. An increase in the well size leads to an increase in the number of energy levels. Movement of charge carriers can be restricted in one coordinate (forming quantum films), in two coordinates (quantum wires or strands), or in all three areas (quantum dots). Quantum dots are still a rather "young" object of study, but their use in various fields of science and technology is obviously extremely promising (from the design of new lasers and the generation of new displays to building qubits) [18–21].

22.2 GRAPHENE AND ITS HAMILTONIANS

From the chemical point of view, the main element of any graphite compound is a sheet of graphene, which can be regarded as benzene hexagons whose hydrogen atoms are replaced by carbon atoms in the adjacent cells, hexes. The carbon atoms in graphene form a honeycomb-like structure according to the sp2 hybridization. This structure cannot be regarded as a Bravais lattice, since two adjacent cells are not equivalent from the crystallographic point of view.

Let us consider the structure of graphene with two sublattices A and B (Figure 22.1), where a_1 and a_2 are the basis

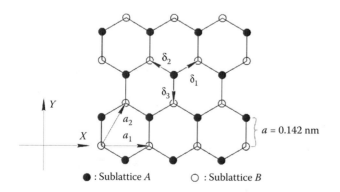

● : Sublattice *A*   ○ : Sublattice *B*

FIGURE 22.1 Crystal lattice of graphene.

vectors; δ_1, δ_2, and δ_3 are the vectors connecting a site of the sublattice A with the nearest neighbor sites of the sublattice B.

This study is required to identify the characteristics of the electronic structure of graphene, the presence of a gap, that will properly take into account the initial conditions of the problem.

Construction of a microscopic model describing the interaction of electrons in graphene has been done within a framework of the Hückel approximation. The Hamiltonian of the electron system was considered in the framework of the Hubbard model for a single-layer graphene. The model takes into account the Coulomb interaction between electrons, which leads to a substantial change in dispersion, and hence in the optical response of the system. Moreover, the account of the electron interaction Hamiltonian leads to a change in the spectrum of elementary excitations of the model:

$$E(\bar{p}) = \frac{\varepsilon(\bar{p})}{2} + \frac{U}{2} \mp \frac{1}{2}\sqrt{\varepsilon^2(\bar{p}) - 2\varepsilon(\bar{p})U(1 - 2n_0) + U^2},$$

(22.11)

where U is the Coulomb repulsion between electrons trapped at a single site; $\varepsilon(\bar{p})$ is the dispersion, which describes the interaction of electrons and phonons in graphene without the Coulomb repulsion, and n_0 is the average number of the on-site electrons.

Account for impurities in the case of doped graphene was carried out in the framework of the Anderson model, where only the hybridization of electronic subsystems is considered. The latter allowed us to avoid the complexities associated with the lack of a gap in the graphene. The resulting Hamiltonian reads

$$H = H_h + H_{im} + H_{hyb}$$

$$H_h = -\sum_{j\Delta\sigma} t_\Delta \left(a_{j\sigma}^+ b_{j+\Delta\sigma} + b_{j+\Delta\varsigma}^+ a_{j\sigma} \right)$$

$$+ U \sum_j \left(a_{j\sigma}^+ a_{j\sigma} a_{j-\sigma}^+ a_{j-\sigma} + b_{j\sigma}^+ b_{j\sigma} b_{j-\sigma}^+ b_{j-\sigma} \right)$$

$$H_{im} = \sum_j \left(\tilde{\varepsilon} d_{j\sigma}^+ d_{j\sigma} + \tilde{\varepsilon} d_{j-\sigma}^+ d_{j-\sigma} + U_1 d_{j\sigma}^+ d_{j\sigma} d_{j-\sigma}^+ d_{j-\sigma} \right)$$

$$H_{hyb} = V \sum_{j\sigma} \left(a_{j\sigma}^+ d_{j\sigma} + d_{j\sigma}^+ a_{j\sigma} \right),$$

(22.12)

where $a_{j\sigma}^+, a_{j\sigma}, b_{j\sigma}^+, b_{j\sigma}$ are the creation and annihilation operators of electrons with spin σ on two mutually dual carbon sublattices, so that the electrons jump only between the sublattices; t_Δ is the hopping integral between neighboring sites in the sublattices; U is the constant of the Coulomb repulsion of electrons trapped at a single site; $d_{j\sigma}^+, d_{j\sigma}$ are the creation and annihilation operators of the impurity electrons with spin σ; $\tilde{\varepsilon}$ is the impurity level energy; U_1 is the constant of the

Coulomb repulsion of the impurity electrons; V is the overlap integral between the wave functions of the impurity electrons and the π-electrons of carbon, forming the bands. Estimations based on the semi-empirical quantum-chemical method MNDO have shown that typical values for these parameters are $t_\Delta \approx 2$ eV, $U \approx 12$ eV, $U \approx 12$ eV, and $V \approx 2$ eV.

Since the properties of the model described by the Anderson Hamiltonian is quite complicated, we assume that $U \to \infty$, and that all the average values are spatially homogeneous. It should be noted that the approximation $U \to \infty$ is consistent with the quantum-mechanical calculations for graphene-like structures. The spectrum of elementary excitations can be represented by

$$E_\sigma(p) = \frac{1}{2}$$

$$\left[\varepsilon(p) + \varepsilon - n^\sigma + \sqrt{(-\varepsilon(p) + \varepsilon - n^\sigma)^2 + 4(1 - n_\sigma^{im})|V|^2} \right],$$

(22.13)

where V is the hybridization parameter, $\varepsilon(p)$ is the electron spectrum for graphene, determined from the diagonalized Hamiltonian H_h, n^σ, and n_σ^{im} are the parameters determined by the problem stability conditions.

Let us consider the calculation of the energy eigenvalues for electrons in the crystal lattice of graphene with adsorbed atomic hydrogen [22], which is regarded as an impurity. Such a choice of impurity is motivated by the fact that, in this case, the Coulomb interaction energy of the electrons in the adsorbed atom is zero, as there is only one electron in atomic hydrogen. The hybridization potential $V_{\kappa a}$ in the Anderson Hamiltonian can be estimated from a quantum-chemical approach, as it is defined by the overlap integral of the wave functions of the s-orbital (the hydrogen atom), and p$_z$-orbitals (carbon atoms in graphene):

$$V = \frac{1}{2}(\beta_H + \beta_C)S_{HC},$$

$$S_{HC} = \int \Psi_{1s}(\mathbf{r})\Psi_{2p_z}(\mathbf{r})d\mathbf{r},$$

(22.14)

$$\Psi_{1s} = \frac{1}{\sqrt{\pi}}\left(\frac{z}{a_0}\right)^{\frac{3}{2}}e^{-\rho}, \quad \rho = \frac{zr}{a_0}, \quad z(H) = 1;$$

$$\Psi_{2p_z} = \frac{1}{4\sqrt{2\pi}}\left(\frac{z}{a_0}\right)^{\frac{3}{2}}\rho e^{-\frac{\rho}{2}}\cos\theta, \quad \rho = \frac{zr}{a_0}, \quad z(C) = 6;$$

where S_{HC} is the overlap integral of the wave functions, β_H and β_C are the parameters derived from the semi-empirical quantum-chemical method MNDO [23], $\beta_H = -6.99$ eV, $\beta_C = -7.93$ eV, a_0 is the Bohr radius, and z is the atomic charge.

An estimate of the hybridization potential gives a value of $V_{\kappa a} = -1.43$ eV. The energy value is negative, therefore a stable

state is formed, which is important for practical applications. To estimate the energy of adsorbed atoms ε_a, the method of images is used, based on the fact that the surface of the conductor is equipotential [24]. As a result, we obtain

$$\tilde{\varepsilon}_a = I + \frac{1}{4\pi\varepsilon_0}\frac{e^2}{4l},$$

where $I = -13.6$ eV is the ionization potential of a hydrogen, e is the elementary charge, ε_0 is the dielectric constant, $l = 1.2$ Å is the distance from the center of the adatom to the plane of its image on the substrate, which is of the order of the atomic radius of the adatom (the length of the adsorption bond). To describe the spectrum of elementary excitations of graphene, we use the classical mathematical technique of Green's functions. The expression of Green's function for the lattice with adsorbed atomic defect can be written as follows:

$$\ll c_{k\sigma} | c_{k\sigma}^+ \gg = \frac{i}{2\pi}\frac{(\omega - \varepsilon_a)}{(\omega - \varepsilon_a)(\omega - \varepsilon_k) - |V_{ka}|^2}, \quad (22.15)$$

where $c_k\sigma$, $c_{k\sigma}^+$ are the creation and annihilation Fermi operators, and ω is the energy variable.

The analytical expression for Green's function of the crystal lattice of graphene (22.15) allows us to determine the eigenvalues of the electron energy in the crystal, caused by the adsorption of atomic hydrogen. The eigenvalues of the electron energy of the crystal lattice with attached atomic defects are given by the poles of Green's function:

$$E(k) = \frac{1}{2}\left[\varepsilon_a + \varepsilon_k \pm \sqrt{(\varepsilon_a - \varepsilon_k)^2 + 4|V_{ka}|^2}\right], \quad (22.16)$$

where ε_k is the band structure of the "pure" graphene.

In the case of double-layer graphene, the system has been considered in the framework of tight-binding model for π-electrons using a nearest-neighbor approximation with intraplane and interplane hopping integrals t_0, while the electrostatic potential U was applied between the two layers of graphene. The band structure of bilayer graphene, obtained from this tight binding approximation, gives us the following dispersion relation:

$$E_p^{\pm\pm}(U) = \pm\sqrt{\varepsilon(p)^2 + \frac{t_0^2}{2} + \frac{U^4}{4} \pm \sqrt{\frac{t_0^4}{4} + (t_0^2 + U^2)\varepsilon(p)^2}}.$$

(22.17)

22.3 MATHEMATICAL RULES

In this work, the transition to curvilinear coordinates has been used. Therefore, it is necessary to do a little mathematical retreat, which will contribute to the understanding of the calculations made in the following paragraphs.

We consider the transition from a coordinate system x^0, x^1, x^2, x^3 to another one $x^{0'}, x^{1'}, x^{2'}, x^{3'}$ by means of the following transformation [25]:

$$x^i = f^i(x^{0'}, x^{1'}, x^{2'}, x^{3'}), \qquad (22.18)$$

where f^i are some smooth functions. When the coordinates are transformed according to Equation 22.18, their differentials transforms read [25]

$$dx^i = \frac{\partial x^i}{\partial x'^k} dx'^k. \qquad (22.19)$$

It should be noted that here and below, a repeated index implies summation over that index. A contravariant 4-vector is any set of four variables A^i, which are defined through their differentials at the curvilinear transition

$$A^i = \frac{\partial x^i}{\partial x'^k} A'^k. \qquad (22.20)$$

Derivatives of some scalar after the coordinate conversion are calculated as follows:

$$\frac{\partial \varphi}{\partial x^i} = \frac{\partial \varphi}{\partial x'^k} \frac{\partial x'^k}{\partial x^i}. \qquad (22.21)$$

A covariant 4-vector is any set of four variables A_i, which are converted as derivatives of a scalar using the coordinate transform:

$$A_i = \frac{\partial x'^k}{\partial x^i} A'_k. \qquad (22.22)$$

Similarly, the 4-tensors of various ranks are defined. Thus, the contravariant 4-tensor of the second rank A^{ik} is the set of 16 variables that transform as the multiplication of two contravariant vectors, that is, according to the following law:

$$A^{ik} = \frac{\partial x^i}{\partial x'^l} \frac{\partial x^k}{\partial x'^m} A'^{lm}. \qquad (22.23)$$

A covariant tensor of the second rank A_{ik} is converted by the law:

$$A_{ik} = \frac{\partial x'^l}{\partial x^i} \frac{\partial x'^m}{\partial x^k} A'_{lm}, \qquad (22.24)$$

and the mixed 4-tensor A_k^i by the formula:

$$A_k^i = \frac{\partial x^i}{\partial x'^l} \frac{\partial x'^m}{\partial x^k} A'^l_m. \qquad (22.25)$$

These definitions are natural extensions of the definitions of 4-vectors and 4-tensors for the Galilean coordinates, according to which the differentials dx^i are also contravariant vectors, and the derivatives $\partial \phi / \partial x^i$ are the covariant 4-vectors.

The construction rules of 4-tensors by the multiplication or its simplification by other 4-tensors in curvilinear coordinates are the same as for the Galilean coordinates. Definition of the unit 4-tensor δ_k^i also does not change: its components are $\delta_k^i = 0$ for $i \neq k$, and $\delta_k^i = 1$ for $i = k$.

The square of the length element in the curvilinear coordinates is a quadratic form of the differentials dx^i:

$$ds^2 = g_{ik} dx^i dx^k, \qquad (22.26)$$

where g_{ik} are the coordinate functions; g_{ik} are symmetric in indices i and k:

$$g_{ik} = g_{ki}, \qquad (22.27)$$

Since the multiplication (simplified) g_{ik} on a contravariant tensor $dx^i dx^k$ is a scalar, then g_{ik} is a covariant tensor, which is called the metric tensor. Two tensors A_{ik} and B^{ik} are said to be the inverse of each other, if and only if $A_{ik} B^{kl} = \delta_i^l$. Obviously, the only variables which can determine the relationship between the one and the other are the components of the metric tensors. Such a relationship is given by the following expression

$$A^i = g^{ik} A_k, \quad A_i = g_{ik} A^k. \qquad (22.28)$$

In a Galilean coordinates system, this tensor has the components:

$$g_{ik}^{(0)} = g^{ik(0)} = \begin{pmatrix} 1 & 0 & 0 & 0 \\ 0 & -1 & 0 & 0 \\ 0 & 0 & -1 & 0 \\ 0 & 0 & 0 & -1 \end{pmatrix}. \qquad (22.29)$$

Now, we consider the covariant differentiation. We define the transformation formula for differentials dA_i. Since a covariant vector is calculated by the following formula:

$$A_i = \frac{\partial x'^k}{\partial x^i} A'_k ,$$

we readily obtain

$$dA_i = \frac{\partial x'^k}{\partial x^i} dA'_k + A'_k d\frac{\partial x'^k}{\partial x^i} = \frac{\partial x'^k}{\partial x^i} dA'_k + A'_k \frac{\partial^2 x'^k}{\partial x^i \partial x^l} dx^l.$$

We now undertake the definition of a tensor which in curvilinear coordinates plays the same role as $\partial A_i / \partial x^k$ in Galilean coordinates. In other words, we must transform $\partial A_i / \partial x^k$ from Galilean to curvilinear coordinates. In curvilinear coordinates, in order to obtain a differential of a vector which behaves like a vector, it is necessary that the two vectors to be subtracted from each other be located at the same point in space. In other words, we must somehow "translate" one of the vectors (which are separated infinitesimally from each other) to the point where the second is located, after which we determine the difference of the two vectors which we now refer to as one and the same point in space. The operation of translation itself must be defined so that in Galilean coordinates the difference shall coincide with the ordinary differential dA_i. The difference in the components of the two vectors after translating one of them to the point where the other is located will not coincide with their difference before the translation (i.e., with the differential dA_i) [25]. Therefore, to compare two infinitesimally separated vectors we must subject one of them to a parallel translation to the point where the second is located. Let us consider an arbitrary contravariant vector; if its value at the point x^i is A^i, then at the neighboring point $x^i + dx^i$ is equal to $A^i + dA^i$. We subject the vector A^i to an infinitesimal parallel displacement to the point $x^i + dx^i$. We denote the change in the vector which results from this by δA^i. Then, the difference DA^i between the two vectors which are now located at the same point is

$$DA^i = dA^i - \delta A^i, \qquad (22.30)$$

$$\delta A^i = -\Gamma^i_{kl}A^k dx^l, \qquad (22.31)$$

where Γ^i_{kl} are some functions of coordinates, whose form depends on the choice of the coordinate system. In a Galilean coordinate system all of Γ^i_{kl} are equal to zero.

From this, it is already clear that the quantities Γ^i_{kl} do not form a tensor, since a tensor which is equal to zero in one coordinate system is equal to zero in every other one. In a curvilinear space, it is of course impossible to make all Γ^i_{kl} vanish over all of space. But we can choose a coordinate system for which Γ^i_{kl} become zero over a given infinitesimal region (see the end of this section). The quantities Γ^i_{kl} are called Christoffel symbols. In addition to the quantities Γ^i_{kl}, we shall later also use quantities $\Gamma_{i,kl}$, defined as follows:

$$\Gamma_{i,kl} = g_{im}\Gamma^m_{kl}. \qquad (22.32)$$

Conversely,

$$\Gamma^i_{kl} = g^{im}\Gamma_{m,kl}. \qquad (22.33)$$

It is also easy to relate the changes in the components of a covariant vector under a parallel displacement to the Christoffel symbols. To do this, we note that under a parallel displacement, a scalar is unchanged. In particular, the scalar product of two vectors does not change under a parallel displacement. Let A_i and B^i are some covariant and contravariant vectors. Then from $\delta(A_i B^i) = 0$, we have

$$B^i \delta A_i = -A_i \delta B^i = \Gamma^i_{kl} B^k A_i dx^l.$$

Hence, in view of the arbitrariness of B^i, we obtain that

$$\delta A_i = \Gamma^k_{il} A_k dx^l, \qquad (22.34)$$

which determines the change of the covariant vector.

Substituting (22.31) and $dA^i = (\partial x'^k / \partial x^i) x^i$ in formula (22.32), we obtain

$$DA^i = \left(\frac{\partial A^i}{\partial x^l} + \Gamma^i_{kl} A^k \right) dx^l. \qquad (22.35)$$

Similarly, we find for the covariant vector

$$DA_i = \left(\frac{\partial A_i}{\partial x^l} - \Gamma^k_{il} A_k \right) dx^l. \qquad (22.36)$$

Tensors defined by the following formulas (22.35) and (22.36) are called covariant derivatives of the vectors A^i and A_i. We will denote them by $A^i_{;k}$ and $A_{i;k}$. Thus,

$$DA^i = A^i_{;l} dx^l, \quad DA_i = A_{i;l} dx^l, \qquad (22.37)$$

while the covariant derivatives themselves are

$$A^i_{;l} = \frac{\partial A^i}{\partial x^l} + \Gamma^i_{kl} A^k, \qquad (22.38)$$

$$A_{i;l} = \frac{\partial A_i}{\partial x^l} - \Gamma^k_{il} A_k. \qquad (22.39)$$

In a Galilean coordinate system, all coefficients $\Gamma^i_{kl} = 0$ and covariant derivatives are reduced to ordinary differentiation.

It is also easy to calculate the covariant derivative of a tensor. To do this, we must determine the change in the tensor under an infinitesimal parallel displacement. For example, let us consider any contravariant tensor, expressible as a product of two contravariant vectors $A^i B^k$. Under parallel displacement,

$$\delta(A^i B^k) = A^i \delta B^k + B^k \delta A^i = -A^i \Gamma^k_{lm} B^l dx^m - B^k \Gamma^i_{lm} A^l dx^m.$$

By virtue of the linearity of this transformation we must also have, for an arbitrary tensor A^{ik},

$$\delta A^{ik} = -(A^{im}\Gamma^k_{ml} + A^{mk}\Gamma^i_{ml}) dx^l. \qquad (22.40)$$

Hence, we find covariant derivative of the tensor in the following form:

$$A^{ik}_{;l} = \frac{\partial A^{ik}}{\partial x^l} + \Gamma^i_{ml} A^{mk} + \Gamma^k_{ml} A^{im}. \qquad (22.41)$$

Quite similarly, we obtain the covariant derivative of the mixed tensor A_k^i and the covariant tensor A_{ik} in the form

$$A_{k;l}^i = \frac{\partial A_k^i}{\partial x^l} - \Gamma_{kl}^m A_m^i + \Gamma_{ml}^i A_k^m$$

$$A_{ik;l} = \frac{\partial A_{ik}}{\partial x^l} - \Gamma_{il}^m A_{mk} - \Gamma_{kl}^m A_{im}.$$

(22.42)

One can similarly determine the covariant derivative of a tensor of arbitrary rank. In doing so, one finds the following rule of covariant differentiation: to obtain the covariant derivative of the tensor $A:::$ with respect to x^l, we add to the ordinary derivative $\partial A:::/\partial x^l$ for each covariant index $i(A:i:)$ a term $-\Gamma_{il}^k$. One can easily verify that the covariant derivative of a product is found by the same rule as for ordinary differentiation of products. In doing so, we must consider the covariant derivative of a scalar as an ordinary derivative, that is, as the covariant vector $\phi_k = \partial\phi/\partial x$, in accordance with the fact that for a scalar $\delta\phi = 0$, and therefore $D\phi = d\phi$. For example, the covariant derivative of the product $A_i B_k$ is given by

$$(A_i B_k)_{;l} = A_{i;l} B_k + A_i B_{k;l}.$$

If in a covariant derivative we raise the index signifying the differentiation, we obtain the so-called contravariant derivatives:

$$A_i^{;k} = g^{kl} A_{i;l}, \quad A^{i;k} = g^{kl} A_{;l}^i.$$

Now, we have the formulas for transforming the Christoffel symbols from one coordinate system to another. These formulas can be obtained by comparing the laws of transformation of the two sides of the equations defining the covariant derivatives, and requiring that these laws be the same for both sides. It is straightforward to get [25]

$$\Gamma_{kl}^i = \Gamma_{np}^{'m} \frac{\partial x^i}{\partial x'^m} \frac{\partial x'^n}{\partial x^k} \frac{\partial x'^p}{\partial x^l} + \frac{\partial^2 x'^m}{\partial x^k \partial x^l} \frac{\partial x^i}{\partial x'^m}. \quad (22.43)$$

It can be seen that values Γ_{kl}^i behave like a tensor only under linear transformations of the coordinates (when the second term disappears in the expression (22.43)). The relationship between the Christoffel symbols and the metric tensor, and its first coordinate derivatives can be written in the following form [25]:

$$\Gamma_{\mu\nu}^\alpha = \frac{1}{2} g^{\alpha\beta} \left(\frac{\partial g_{\mu\beta}}{\partial x^\nu} + \frac{\partial g_{\beta\nu}}{\partial x^\mu} - \frac{\partial g_{\mu\nu}}{\partial x^\beta} \right). \quad (22.44)$$

At this point, we turn back to the main goal of our study. The Dirac equation in a carbon nanosystem (CNS), taking into account the curvature of the surface, can be obtained as follows. We introduce a set of orthogonal vectors e_α on the manifold, described by the metric tensor $g_{\mu\nu}$, transforming on the group SO(2): $g_{\mu\nu} = e_\mu^\alpha e_\nu^\beta \delta_{\alpha\beta}$, where e_μ^α is the dyadic coefficients [26], $\alpha,\beta = 1,2$ are the orthonormal indices, and $\mu,\nu = 1,2$ are the coordinate indices. Dyads can be selected with certain gauge freedom, resulting in the emergence of a SO(2)-field ω_μ, which is a spin connection. It should be subjected to a condition analogous to the metric tensor without torsion:

$$D_\mu e_\nu^\alpha := \partial_\mu e_\nu^\alpha - \Gamma_{\mu\nu}^\lambda e_\lambda^\alpha + (\omega_\mu)_\beta^\alpha e_\nu^\beta = 0$$

(elongated derivative of the expression, which has metric and spin indices can be formally written in the following form: $D_\mu = \partial_\mu + \Gamma_\mu + \omega_\mu$), then the spin connection can be defined as

$$(\omega_\mu)^{\alpha\beta} = e_\nu^\alpha D_\mu e^{\beta\nu}.$$

Thus, the Dirac equation, taking into account the curvature of the surface, takes the form:

$$i\gamma^\alpha e_\alpha^\mu (\nabla_\mu - i a_\mu^k - i W_\mu) \psi^k = E\psi^k,$$

where $a_\mu^k, k = K, K_-$ are the Dirac points, W_μ are the gauge fields (defect fields), γ^α are the SU(2)-matrices (special unitary matrices of the second order) 2×2, which can be selected as $\gamma_i = -i\sigma_i$, and $\nabla_\mu = \partial_\mu + \Omega_\mu$, where $\Omega_\mu = (1/8)\omega_\mu^{\alpha\beta}[\gamma_\alpha, \gamma_\beta]$.

22.4 BASIC EQUATIONS AND SPECTRUM OF ELECTRONS

We consider a graphene nanoribbon, which is curved along the toroidal and the helical surfaces, as represented in Figure 22.2. Properties of electrons in graphene nanoribbons in the long-wave approximation and in the vicinity of the Dirac points will be described on the basis of the Dirac equation generalized for the case of a curved space–time [1]:

$$\gamma^\mu (\partial_\mu - \Omega_\mu)\Psi = 0, \quad (22.45)$$

where ∂_μ is the partial derivative with respect to coordinate μ, Ω_μ is the component of the spin connection, $\Psi = (\phi/\phi)$ is the wave function (column vector) consisting of wave functions describing the electrons from different sublattices near the Dirac point.

As is well known [1,27], if we are given the metric tensor

$$ds^2 = g_{\alpha\beta} dx^\alpha dx^\alpha$$

$$g_{\alpha\beta} g^{\beta\gamma} = \delta_\alpha^\gamma, \quad (22.46)$$

(δ_α^γ—delta is the Kronecker symbol) then we can define the field frames (tetrads):

$$g_{\alpha\beta} = e_\alpha^a e_\beta^b \eta_{ab}$$

$$g^{\alpha\beta} = e_a^\alpha e_b^\beta \eta^{ab} \quad (22.47)$$

$$\eta_{ab} \eta^{bc} = \delta_a^c,$$

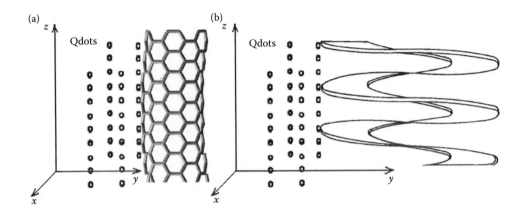

FIGURE 22.2 Geometry of a problem: (a) toroidal nanoribbon, (b) helical nanoribbon. (Adapted from M.B. Belonenko et al. *J. Nanotechnol.* **2011**, ID 161849, 2011.)

where for the two-dimensional curved surfaces, we have $\eta_{ab} = \mathrm{diag}(1, -1, -1)$. Then

$$\Omega_\mu = \frac{1}{4}\gamma_a\gamma_b e^a_\lambda g^{\lambda\sigma}(\partial_\mu e^b_\sigma - \Gamma^\lambda_{\mu\sigma}e^b_\lambda)$$

$$\Gamma^\lambda_{\mu\sigma} = \frac{1}{2}g^{\lambda\nu}(g_{\sigma\nu,\mu} + g_{\nu\mu,\sigma} - g_{\mu\sigma,\nu}) \qquad (22.48)$$

$$\gamma^\mu = e^\mu_a\gamma_a.$$

Using the torus and the helical parameterization

$$x = (R + r\cos x_1)\cos x_2$$

$$y = (R + r\cos x_1)\sin x_2 \qquad (22.49)$$

$$z = r\sin x_1,$$

$$\begin{cases} x = x_1\cos x_2 \\ y = x_1\sin x_2 \,, \\ z = h\cdot x_2 \end{cases} \qquad (22.50)$$

we find that the metrics on the torus surface and the helicoid are given by

$$ds^2 = dx_0^2 - r^2 dx_1^2 - (R + r\cos x_1)^2 dx_2^2, \qquad (22.51)$$

$$ds^2 = dx_0^2 - dx_1^2 - (h^2 + x_1^2)dx_2^2. \qquad (22.52)$$

Note that all the Christoffel symbols are equal to zero, except Γ^2_{12} and Γ^1_{22}. For the torus, we have $\Omega_0 = 0$; $\Omega_1 = 0$; $\Omega_2 = (1/2)\gamma_1\gamma_2 f'/r$ ($f = R + r\cos x_1$; $f' = \partial f/\partial x_1$), while in the case of the helicoid $\Omega_0 = 0$; $\Omega_1 = 0$;

$$\Omega_2 = \frac{1}{2}\gamma_1\gamma_2\frac{x_1}{(h^2 + x_1^2)^{1/2}}.$$

Choosing $\gamma_0 = \sigma_3$; $\gamma_1 = -i\sigma_2$; $\gamma_2 = -i\sigma_1$, where σ are the Pauli matrices, we obtain the following system of equations:

$$\begin{cases} V_F^{-1}\partial_t\varphi = -\frac{1}{r^2}\partial_{x_1}\Psi - \frac{i}{f^2}\partial_{x_2}\Psi + \frac{f'}{2f^2 r}\Psi \\ V_F^{-1}\partial_t\varphi = -\frac{1}{r^2}\partial_{x_1}\varphi + \frac{i}{f^2}\partial_{x_2}\varphi + \frac{f'}{2f^2 r}\varphi \end{cases}, \qquad (22.53)$$

$$\begin{cases} V_F^{-1}\partial_t\varphi + \partial_{x_1}\Psi + \frac{i}{h^2 + x_1^2}\partial_{x_2}\Psi - \frac{x_1}{2(h^2 + x_1^2)^{3/2}}\Psi = 0 \\ -V_F^{-1}\partial_t\Psi - \partial_{x_1}\varphi + \frac{i}{h^2 + x_1^2}\partial_{x_2}\varphi - \frac{x_1}{2(h^2 + x_1^2)^{3/2}}\varphi = 0 \end{cases}. \qquad (22.54)$$

Here, V_F is the Fermi velocity for planar graphene, $\partial_0 = V_F^{-1}\partial_t$. Note that since the metrics (22.51) and (22.52) admit two Killing vectors corresponding to the translations along x_0, x_2, the solutions (22.53) and (22.54) can be found in the form $\begin{pmatrix}\varphi \\ \Psi\end{pmatrix} \to \begin{pmatrix}\varphi(x_1) \\ \Psi(x_1)\end{pmatrix}e^{iEt - ikx_2}$, which finally gives

$$\Psi'' = \left(-\frac{E^2 r^4}{V_f^2} + \frac{k_n^2 r^4}{f^4}\right)\Psi + \frac{rf'}{2f^2}\Psi'$$

$$+ \left(\frac{2k_n r^2 f'}{f^3} + \frac{rf''}{2f^2} - \frac{rf'^2}{f^3} - \frac{r^2 f'^2}{4f^4}\right)\Psi, \qquad (22.55)$$

$$\Psi'' = \left(-\frac{\varepsilon^2}{V_f^2} + \frac{k^2}{(h^2 + x_1^2)^2}\right)\Psi$$

$$+ \left(-\frac{kx_1}{(h^2 + x_1^2)^{5/2}} + \frac{x_1^2}{4(h^2 + x_1^2)^3}\right)\Psi. \qquad (22.56)$$

Note that the wave vector k is found from the boundary conditions at the ends of the nanoribbon. In our particular case, we have chosen the armchair-type ribbon [9], and therefore

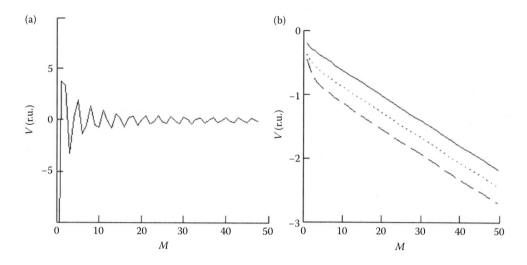

FIGURE 22.3 Dependence of the correction to the energy V caused by the perturbation $\hat{V}$ on atoms number along nanoribbon axis M: (a) for torus ($r/R = 0.1$, $n = 1$); (b) for helicod ($h = 1.5$): (i) $n = 1$—solid line; (ii) $n = 2$—dotted line; (iii) $n = 3$—dashed line. (Adapted from M.B. Belonenko et al. *J. Nanotechnol.* **2011**, ID 161849, 2011.)

$$k_n = \frac{2\pi}{3a_0}\left(\frac{2M+1+n}{2M+1}\right), \quad (22.57)$$

where a_0 is the distance between the atoms in the carbon lattice, M is the number of atoms along the nanoribbon axis, and n is the quantum number. We can consider Equations 22.55 and 22.56 as Schrödinger equations with perturbation

$$\hat{V}_{\text{Torus}} = \left[\left(\frac{2k_n r^2 f'}{f^3} + \frac{rf''}{2f^2} - \frac{rf'^2}{f^3} - \frac{r^2 f'^2}{4f^4}\right) + \frac{rf'}{2f^2}\partial_x\right],$$

$$\hat{V}_{\text{Helicoid}} = \left(-\frac{kx_1}{(h^2+x_1^2)^{5/2}} + \frac{x_1^2}{4(h^2+x_1^2)^3}\right).$$

In this particular case, the spectrum of perturbation reads

$$E = \pm\sqrt{k_n^2 + k_y^2}. \quad (22.58)$$

Expanding the functions in the denominator as a Taylor series up to the second order, we calculate the first perturbation correction term to the spectrum, $\hat{V}_{\text{Torus}}$ and $\hat{V}_{\text{Helicoid}}$, as follows

$$E_1 = \int \Psi^* \hat{V}\Psi dx, \quad \Psi_n = A\cdot\text{Sin}(k_n x_1).$$

The integration is performed from 0 to $L = (3M + 1)a_0$ and the corrections are as follows:

$$E_1 = \frac{2}{L}\left\{-\frac{r^2}{4R^2}\text{Sin}(L) + \frac{r^2}{8R^2}\left(1-\frac{1}{k_n}\right)\right.$$

$$\left.\frac{\text{Sin}(2k_n-1)L}{(2k_n-1)} + \frac{r^2}{8R^2}\left(1+\frac{1}{k_n}\right)\frac{\text{Sin}(2k_n+1)L}{(2k_n+1)}\right\}, \quad (22.59)$$

$$E_1 = \frac{2}{L}\left\{\begin{array}{l}\left(-\frac{kL^2}{4h^5} + \frac{k}{4h^5 k_n^2} + \frac{h^{-2/3}L^3}{24}\right)\\[2mm] + \text{Sin}(2k_n L)\left(\frac{kL}{4h^5 k_n} + h^{-2/3}\left(\frac{1}{32k_n^3} - \frac{L^2}{16k_n}\right)\right)\\[2mm] + \text{Cos}(2k_n L)\left(\frac{-k}{4h^5 k_n^4} + \frac{h^{-2/3}L}{16k_n^2}\right)\end{array}\right\}. \quad (22.60)$$

The dependence of the perturbations on the atom numbers along the nanoribbons M is presented in Figure 22.3.

The dependence shown in Figure 22.3a is rather complex, which is associated with the quantization of the electron spectrum in graphene nanoribbons in relation with Equation 22.57. It should be noted that the dependence of the energy gap in carbon nanotubes of zigzag type is pretty similar, which also arises from the quantization of the electron spectrum in the direction along the circumference of the nanotube. The calculations show (Figure 22.3a) that the value of the helicoids parameterization h influences most strongly the correction to the energy (as well as its sign). The dependence of the energy correction on the ratio r/R is demonstrated in Figure 22.3.

As expected, the dependence shown in Figure 22.4 shows that with increasing curvature of the graphene nanoribbon (i.e., with increasing ratio r/R), the absolute value of the correction to the energy of the electrons increases [28].

22.5 TUNNELING CHARACTERISTICS

The Hamiltonian of the system of electrons can be written in the following form:

$$H = \sum_p E_p^A a_p^+ a_p + \sum_q E_q^B b_q^+ b_q + \sum_{pq} T_{pq}(a_p^+ b_q + b_q^+ a_q), \quad (22.61)$$

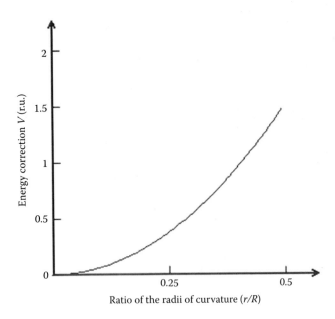

FIGURE 22.4 Dependence of the energy correction V, caused by the perturbation $\bar{V}$, on the ratio of the radii of curvature r/R ($M = 20$, $n = 1$). (Reprinted from *Sol. State Commun.*, **151**, M.B. Belonenko, N.G. et al., 1147, Copyright 2011, with permission from Elsevier.)

where $a_p^+; a_p$ are the electron creation and annihilation operators with momentum p in the carbon nanoribbons; E_p^A is the electron spectrum of the carbon nanoribbons (22.58) while taking into account Equations 22.59 and 22.60; T_{pq} is the matrix element of the tunneling operator between p and q states; $b_q^+; b_q$ are the electron creation and annihilation operators with momentum q in a substance which is in contact with a carbon nanoribbon; E_p^B is the electron spectrum of another substance. It should be noted that p and q are multi-indices in formula (22.61). Hence, for graphene nanoribbon (further, we consider an arm-chair nanoribbon only) $p = (p_y, n)$, $n = 0,1,...,M - 1$. Multi-index q is determined by the substance which is in contact with the carbon nanoribbon, and, for example, for quantum dots $q = (p_x, p_y, p_z)$, whereas for graphene $q = (p_x, p_y)$. A consideration of the external electric field $\vec{E}$ (and choosing $\vec{E} = -(1/c)(\partial \vec{A}/\partial t)$) can be carried out by the canonical transformation $p \rightarrow p - eA/c$.

The tunneling current is considered to be given by

$$J = ie \sum_{pq} (a_p^+ b_q - b_q^+ a_p). \qquad (22.62)$$

With a gauge transformation [29,30], we have

$$a_p \rightarrow S^{-1} a_p S$$

$$S = \exp\left(ieVt \sum_p a_p^+ a_p \right),$$

where V is the applied voltage, and e is the electron charge. Formally, it is possible to reduce a problem of calculation of

the current–voltage characteristics to the calculation of the operator response

$$J_t = ie \sum_{pq} (a_p^+ b_q e^{ieVt} - b_q^+ a_q e^{-ieVt})$$

on the external influence [29,30]

$$H_t = \sum_{pq} T_{pq} \left(a_p^+ b_q e^{ieVt} + b_q^+ a_q e^{-ieVt} \right).$$

The solution was obtained within the framework of the Kubo theory:

$$J = 4\pi e |T|^2 \int_{-\infty}^{\infty} dE\, v_A(E + eV) v_B(E) (n_f(E) - n_f(E + eV),$$

$$v_A(E) = \sum_p \delta(E - E_p^A); v_B(E) = \sum_q \delta(E - E_q^B); \qquad (22.63)$$

where $\delta(x)$ is the Dirac delta function, $v_{A(B)}(E)$ is the tunneling density of states; $n_f(E)$ is the equilibrium number of fermions with energy E. The approximation of a "rough" contact is used thereafter, so that $T_{pq} = T$ (this imposes certain restrictions on the contact geometry, i.e., the case discussed below means that nanoribbon should be perpendicular to the contact material surface). For definiteness, we choose the dispersion law for the graphene nanoribbons given by Equations 22.58 through 22.60, and the dispersion law for the quantum dots as the contact material being

$$E_q^A = E_0 - \Delta \cos(p), \qquad (22.64)$$

where E_0 is the electron energy of a quantum well, Δ is the tunneling integral determined by the overlap of electron wave functions in the adjacent wells, and the momentum p is directed along the axis Z.

Equation 22.63 under study has been solved numerically. The current–voltage characteristic of the contact is presented in Figure 22.5.

Figure 22.5 shows the asymmetric behavior of current versus voltage applied to the contact. This is due to both the peculiarities of the electronic structure (density of states) of the metal and graphene nanoribbons, and the processes of carrier recombination in the transition contact, which dominate over the thermal processes when $V > 0$. The resulting dependence may have important practical applications in the study of nanocontacts and the design of tunnel diodes based on graphene nanoribbons. Also, the region with negative differential resistance was observed for some values of V. The presence of such region allows the use of a tunnel diode as a high-speed switch.

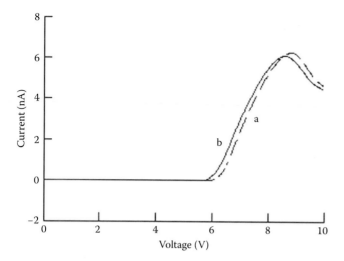

FIGURE 22.5 Current–voltage characteristic of the contact: curved graphene nanoribbon–quantum dots: (a) for torus; (b) for helicoid. (Adapted from M.B. Belonenko et al. *J. Nanotechnol.* **2011**, ID 161849, 2011.)

22.6 FRIEDMANN MODEL

Now, we consider graphene nanoribbons in the Friedmann model of non-stationary universe [25,31]. Properties of electrons in the graphene nanoribbon in the long-wave approximation in the vicinity of the Dirac point will be described on the basis of the generalized Dirac equation for a curved space–time [1] (see Equation 22.45).

For a strained/curved graphene, constantly under the influence of an external variable mechanical force, the effect of this force leads to a periodic change in the distance between the atoms of graphene, which, in turn, leads to a change in the Fermi velocity, vF. Using the analogy with a curved space–time, we can say that this force leads to a periodic change of spatial intervals, which, as is well known, is adequately described in the frame of the Friedmann non-stationary model. The metrics in the Friedmann non-stationary model of the universe has the form:

$$ds^2 = dt^2 - e^{f(t)}(dx^2 + dy^2),\qquad (22.65)$$

where $e^{f(t)} = 1 + a \sin(\omega_0 t)$.

Here, a stands for the relative amplitude of the strain, while ω_0 is the characteristic frequency of oscillatory deformation. There are only four non-zero Christoffel symbols:

$$\Gamma^0_{11} = \frac{1}{2} e^f f';\quad \Gamma^0_{22} = \frac{1}{2} e^f f';\quad \Gamma^1_{01} = \frac{f'}{2};\quad \Gamma^2_{02} = \frac{f'}{2}.$$

Thus

$$\Omega_0 = 0;\quad \Omega_1 = -\frac{\gamma_0\gamma_1 f' e^{f/2}}{4};\quad \Omega_2 = -\frac{\gamma_0\gamma_2 f' e^{f/2}}{4}.$$

Let us choose $\gamma_0 = \sigma_3$; $\gamma_1 = -i\sigma_2$; $\gamma_2 = -i\sigma_1$, where σ are the Pauli matrices. Then, we obtain the following system of equations:

$$V_F^{-1}\partial_t\varphi + e^{-f}\partial_{x_1}\psi + \frac{f' e^{-f/2}}{4}\varphi + i e^{-f}\partial_{x_2}\psi - \frac{i f' e^{-f/2}}{4}\varphi = 0$$

$$V_F^{-1}\partial_t\psi - e^{-f}\partial_{x_1}\varphi - \frac{f' e^{-f/2}}{4}\psi + i e^{-f}\partial_{x_2}\varphi - \frac{i f' e^{-f/2}}{4}\psi = 0$$

$$(22.66)$$

(here, we explicitly introduced the Fermi velocity for the flat graphene via $\partial_0 = V_F^{-1}\partial_t$). It should be noted that the solution of the system (22.66) can be found in the form:

$$\begin{pmatrix}\varphi\\\psi\end{pmatrix} \to \begin{pmatrix}\varphi\\\psi\end{pmatrix} e^{ip_x x + ip_y y},$$

then

$$V_F^{-1}\partial_t\varphi + i p_x e^{-f}\psi + \frac{f' e^{-f/2}}{4}\varphi - p_y e^{-f}\psi - \frac{i f' e^{-f/2}}{4}\varphi = 0$$

$$V_F^{-1}\partial_t\psi - i p_x e^{-f}\varphi - \frac{f' e^{-f/2}}{4}\psi - p_y e^{-f}\partial_{x_2}\varphi - \frac{i f' e^{-f/2}}{4}\psi = 0.$$

$$(22.67)$$

From the Equation 22.67, it is easy to obtain the following equation for the function φ:

$$\varphi_t + g\varphi = 0,$$

$$g = \alpha V_F f' e^{-f/2},\quad \alpha = \frac{1-i}{4}.$$

Then the substitution: $\varphi \to \varphi \cdot e^{-\int g\,dt}$ applied in the set of Equation 22.67 yields the non-linear Shrödinger equation with the excitation term (second one), known as the Mathieu equation:

$$\varphi_{tt} + \varphi_t(f' + g^* - g) + |p|^2 e^{-2f}\varphi = 0.$$

Let us choose the trial unexcited function in the form: $\varphi(t) = \varphi_0 e^{i\omega t}$, and besides $f = 0$. In the non-perturbed case, we obtain the spectrum:

$$\omega^2 = |p|^2 .\qquad (22.68)$$

Let us calculate the first energy correction, $\hat{V}$:

$$E_1 = \int \Psi^* \hat{V}\Psi dx,\quad \Psi_n = A \cdot \mathrm{Sin}(k_n x_1),$$

$$\hat{V} = (f' + g^* - g)\partial_t.$$

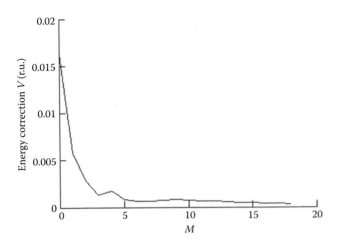

FIGURE 22.6 Dependence of the energy correction V, caused by the perturbation, on the number of atoms M along the axis of the nanoribbon ($n = 1$) in Friedman model.

The integration is done from 0 to $L = (3M + \text{т}1)a_0$ and results in

$$E_1 = \frac{2}{L}\left\{\frac{k_n \sin(2k_n t)}{4} \cdot \frac{a\omega_0 \cos(\omega_0 t)}{1 + a\sin(\omega_0 t)}\left(1 + \frac{1}{\sqrt{1 + a\sin(\omega_0 t)}}\right)\right\}.$$

$$(22.69)$$

The dependence of the energy correction on the atom numbers M is demonstrated in Figure 22.6.

This dependence has a step-like form, which is associated with the quantization of the electron spectrum in graphene nanoribbons according to Equation 22.69. Note that this is similar to the dependence of the energy gap in zigzag-type carbon nanotubes [32], which also arises from the quantization of the electron spectrum in the direction along the circumference of the nanotube.

Furthermore, it is worth characterizing the dependence of V on the parameters ω_0 and n. This dependence is shown in

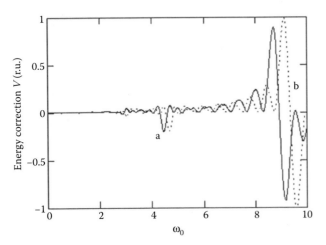

FIGURE 22.7 Dependence of the energy correction V, caused by the perturbation, on the parameter ω_0 for $M = 20$. (a) $n = 1$; (b) $n = 3$. (Adapted from A.V. Zhukov et al. *JETP Lett.* **97**, 400, 2013.)

Figure 22.7, and it demonstrates that with the increase of the characteristic frequency ω_0, we observe a periodic change of the correction to the energy of the electrons. With increasing quantum number n, we observe a shift to the right and a remarkable increase in the amplitude [31].

Also, we constructed the current–voltage characteristics of the contact between a curved nanoribbon and a metal. This dependence shows the asymmetric behavior of current versus voltage applied to the contact, as in the case where we do not take into account the Friedmann model.

22.7 CONCLUSION

In this chapter, we first briefly overviewed the formal relation between the quantum field theory, the general relativity, and the condensed matter of graphene-based nanostructures. Proceeding to the more practical applications, we summarized our studies on the tunnel characteristics between a curved/strained graphene and a metal, considering different geometrical configurations of a graphene sheet. As a result, we have demonstrated that the above contacts behave similarly to classical diodes, which may have important practical applications in the study of nanocontacts and the design of tunnel diodes based on graphene nanoribbons. Moreover, applying the non-stationary Friedmann model leads us to the same qualitative conclusion.

ACKNOWLEDGMENTS

This work was partially supported by the Russian Foundation for Basic Research under Project No. 08-02-00663, No. 12-02-31654 and by the Federal Target Program "Scientific and pedagogical manpower" for 2010–2013. A. V. Zhukov and R. Bouffanais are financially supported by the SUTD-MIT International Design Centre (IDC).

REFERENCES

1. M.A.H. Vozmediano, M.I. Katsnelson, F. Guines. Gauge fields in graphene. *Phys. Rep.* **496**, 109, 2010.
2. I. Dzyaloshinskii, G.E. Volovik. On the concept of local invariance in the theory of spin glasses. *J. de Phys.* **39**, 693, 1978.
3. I. Dzyaloshinskii, G.E. Volovik. Poisson brackets in condensed matter. *Ann. Phys.* **125**, 67, 1980.
4. A.A. Pacheco Sanjuan, Z. Wang, H. Pour Imani, M. Vanević, S. Barraza-Lopez. Graphene's morphology and electronic properties from discrete differential geometry. *Phys. Rev. B.* **89**, 121403(R), 2014.
5. F.M.D. Pellegrino, L. Chirolli, Rosario Fazio, V. Giovannetti, Marco Polini. Theory of integer quantum Hall polaritons in graphene. *Phys. Rev. B.* **89**, 165406, 2014.
6. J.V. Sloan, A.A. Pacheco Sanjuan, Z. Wang, C. Horvath, S. Barraza-Lopez. Strain gauge fields for rippled graphene membranes under central mechanical load: An approach beyond first-order continuum elasticity. *Phys. Rev. B.* **87**, 155436, 2013.
7. D.-W. Zhang, C.-J. Shan, F. Mei, M. Yang, R.-Q. Wang, S.-L. Zhu. Valley-dependent gauge fields for ultracold atoms in square optical superlattices. *Phys. Rev. A.* **89**, 015601, 2014.

8. Y. Zhang, J.W. Tan, H.L. Stormer, P. Kim. Experimental observation of the quantum Hall effect and Berry's phase in graphene. *Nature* **438**, 201, 2005.

9. S. Stankovich, D.A. Dikin, G.H.B. Dommett, K.M. Kohlhaas, E.J. Zimney, E.A. Stach, R. D. Piner, S.T. Nguyen, R.S. Ruoff. Graphene-based composite materials. *Nature* **442**, 282, 2006.

10. A. Zhang, Z. Dai, L. Shi, Y. Ping Feng, C. Zhang. Energy gap opening and quenching in graphene under periodic external potentials. *J. Chem. Phys.* **133**, 224705, 2010.

11. J. Dauber, B. Terrés, C. Volk, S. Trellenkamp, C. Stampfer. Reducing disorder in graphene nanoribbons by chemical edge modification. *Appl. Phys. Lett.* **104**, 083105, 2014.

12. S. Dubey, V. Singh, A.K. Bhat, P. Parikh, S. Grover, R. Sensarma, V. Tripathi, K. Sengupta, M.M. Deshmukh. Tunable superlattice in graphene to control the number of Dirac points. *Nano Lett.* **13**, 3990, 2013.

13. A. Cortijo, M.A.H. Vozmediano. Effects of topological defects and local curvature on the electronic properties of planar graphene. *Nucl. Phys. B.* **763**, 293, 2007.

14. A. Cortijo, M.A.H. Vozmediano. Electronic properties of curved graphene sheets. *EPL.* **77**, 47002, 2007.

15. D.V. Kolesnikov, V.A. Osipov. Electronic structure of negatively curved graphene. *JETP Lett.* **87**, 419, 2008.

16. L. Brey, H.A. Fertig. Electronic states of graphene nanoribbons studied with the Dirac equation. *Phys. Rev. B.* **73**, 235411, 2006.

17. M.B. Belonenko, N.G. Lebedev, N.N. Yanyushkina. Tunneling through the carbon nanotube/graphene interface exposed to a strong oscillating electric field. *J. Nanophotonics.* **4**, 041670, 2010.

18. N. Cobayasi. *Introduction in Nanotechnology.* Moscow: BINOM, 2007.

19. X. Gao, E. Nielsen, R.P. Muller, R.W. Young, A.G. Salinger, N.C. Bishop, M.P. Lilly, M.S. Carroll. QCAD simulation and optimization of semiconductor quantum dots. *J. Appl. Phys.* **114**, 164302, 2013.

20. D. Solenov, S.E. Economou, T.L. Reinecke. Excitation spectrum as a resource for efficient two-qubit entangling gates. *Phys. Rev. B.* **89**, 155404, 2014.

21. A. Ayachi, W. Ben Chouikha, S. Jaziri, R. Bennaceur. Telegraph noise effects on two charge qubits in double quantum dots. *Phys. Rev. A.* **89**, 012330, 2014.

22. M.B. Belonenko, A.S. Popov, N.G. Lebedev, A.V. Pak, A.V. Zhukov. Extremely short optical pulse in a system of nanotubes with adsorbed hydrogen. *Phys. Lett. A.* **375**, 946, 2011.

23. N.F. Stepanov. *Quantum Mechanics and Quantum Chemistry.* Moscow: Mir, 2001.

24. G.A. Mironova. *Condensed Matter. From Structural Units to Living Matter.* Moscow: MGU, 2004.

25. L.D. Landau, E.M. Lifshitz. *The Classical Theory of Fields.* 4th Ed. Oxford: Butterworth-Heinemann, 2000.

26. V.A. Osipov, E.A. Kochetov, M. Pudlak. Electronic structure of carbon nanoparticles. *J. Exp. Theor. Phys.* **96**, 140, 2003.

27. N.D. Birrel, P.C. W. Davies. *Quantum Fields in Curved Space.* Cambridge: Cambridge University Press, 1982.

28. L.S. Levitov, A.V. Shitov. *Green's Functions. Tasks with Answers.* Moscow: Fizmatlit, 2003, 392 p.

29. M.B. Belonenko, N.G. Lebedev, A.V. Zhukov, N.N. Yanyushkina. Electron spectrum and tunneling current of the toroidal and helical graphene nanoribbon-quantum dots contact. *J. Nanotechnol.* **2011**, ID 161849, 2011.

30. M.B. Belonenko, N.G. Lebedev, N.N. Yanyushkina, A.V. Zhukov, M. Paliy. Electronic spectrum and tunneling current in curved graphene nanoribbons. *Sol. State Commun.* **151**, 1147, 2011.

31. A.V. Zhukov, R. Bouffanais, N.N. Konobeeva, M.B. Belonenko. On the electronic spectrum in curved graphene nanoribbons. *JETP Lett.* **97**, 400, 2013.

32. M.S. Dresselhaus, G. Dresselhaus, P. Avouris. *Carbon Nanotubes: Synthesis, Structure, Properties, and Applications.* Berlin: Springer, 2001.

23 Using Few-Layer Graphene Sheets as Ultimate Reference of Quantitative Transmission Electron Microscopy

Wang-Feng Ding, Bo Zhao, and Fengqi Song

CONTENTS

Abstract 341
23.1 Quantitative High-Angle Annular Dark-Field Scanning Transmission Microscopy Imaging Technique 342
23.2 Quantifying Graphene Sheets under Transmission Electron Microscopy/Scanning Transmission Microscopy 344
23.3 Verification of the Graphene Mass Standard 346
23.4 Weighing Carbon Clusters Using Graphene as the Mass Standard 349
23.5 Electron Scattering in Graphene Sheets 351
23.6 Summary 354
References 355

ABSTRACT

Transmission electron microscope (TEM) is an equipment where electrons interact with materials under delicate controls. The states of electrons after interactions could be measured quantitatively with so much precision that atomic resolution can be achieved not only in image taking, but also in element analysis. However, to accomplish the latter requires an intensity reference with well-defined accuracy for signals acquired. In this chapter, we show that the discrete layers of few-layer graphene sheets (GSs) enable exact determination of sheet thickness, making it a promising reference for quantitative analysis. Since large-scale GSs could be readily prepared nowadays, they bear the convenience of being both the supporting films of TEM samples and the reference for the intensity of signals.

With the discovery of graphene, scientists have found numerous applications of graphene attributing to its exceptional geometric and electronic structures (Geim 2009; Neto et al. 2009; Rao et al. 2009). In the field of transmission electron microscopy (TEM), the first thought occurs to researchers is that one- or few-layer GSs serve ideally as the supporting films (Booth et al. 2008) for specimens of interest, as the trivial thickness of GSs provides ultra-low background contrast, and their ballistic electron transport prevents charge accumulation on TEM specimens. Recent developments of chemical vapor deposition (CVD) method allow quick productions of macroscopic-sized GSs, which can be directly deposited on TEM grids (Dai et al. 2011, 2012). Such grids are commercially available nowadays.

Table 23.1 presents an overview of the literature on GSs utilized as the supporting or wrapping films for TEM samples.

The advantages of GSs as the supporting films are magnified when materials composed of light elements, such as biological samples, are investigated. The liquid cell developed in the past few years greatly enhanced the capability of TEM to view dynamic processes at atomic scale. Traditional approaches of preparing such a liquid cell specimen involve complicated processes and specific instruments. With the large-scale graphene, one saves all the troubles by encapsulating liquid drops between two GSs, which are further transferred to TEM grids for investigation.

As we know, TEM has added to the knowledge of materials down to the atomistic level, not only by presenting us the submicron world with direct visions, but also by quantitative analysis of molecules and nanoparticles (Williams and Carter 2009). As one of the modes available on most TEMs, scanning transmission electron microscopy (STEM) complemented with high-angle annular dark-field (HAADF) detectors has been proved reliable in quantitative imaging, by which the in-depth analysis of atomic clusters and nanoparticles have been achieved (Howie 1979; Isaacson et al. 1979; Singhal et al. 1997). There are two ways using HAADF-STEM to obtain accurate mass mapping: absolute and relative measurements. The former, as developed by Howie, requires an accurate knowledge of the electron scattering cross-section (Ω) as a function of the number of atoms, and the number of electrons in the focused electron beam. This involves an extensive calibration of the operational parameters of the microscope, and requires additional hardware to be fit into the microscope. In practice, both the experimental and the theoretical determination of the absolute cross-section are extremely challenging. Hence, this method has been taken up only by a few dedicated groups worldwide until now. Nevertheless, all the problems

341

TABLE 23.1
Summary of Graphene Sheets Complemented TEM Measurements

Roles of GSs	Research Team	Performances
Supporting films	Tanaka et al. (2009)	A single adatom imaged directly on free-standing GSs
	Pantelic et al. (2011)	Oxidative GSs for biological TEM supports with trivial background noise
	Westenfelder et al. (2011)	*In situ* high-resolution TEM electrical investigations
	Buckhout-White et al. (2012)	High-contrast imaging of unstained DNA nanostructures on suspended GSs
Wrapping films	Yuk et al. (2012)	High-resolution imaging of colloidal nanocrystal growth
	Chen et al. (2013)	Viewing the 3D motion of DNA–Au nanoconjugates

are solved if a known mass standard can be established. In the TEM observation of biological samples, molecules and organisms with known weight, such as ferritin and tobacco mosaic virus (TMV), have been used as mass standard for quantitative measurements (Sen et al. 2007). The problem for these mass standards is not only their poor accuracy but also mass loss due to radiation damage or mass gain due to the filling of the central cavity. Recently, size-selected Au clusters have been used as a mass standard for nanoparticles (Young et al. 2008; Wang et al. 2010). This has opened another avenue for weighing nanoparticles on supports and gaining insight into their fine structure and shapes.

In principle, this relative measurement method could be extended to all elements. However, it is difficult for light elements, as the contrast of STEM images reduces substantially with the atomic number of elements. As a result, the carbon film background intensity contributes a significant source of random noise, leading to a poor signal–noise ratio. In this context, GS has been proposed as a mass standard, as well as a supporting film for TEM specimens. On one hand, the thickness of the GS can be measured with considerable accuracy by counting its layer number. This makes it an accurate mass standard for light elements. On the other hand, GSs are facile to prepare or purchase for most labs and can be readily used as the substrates for TEM specimens. In a relative mass measurement, all the operational parameters of STEM should be kept strictly constant for the reference standard and samples. Hence, it is convenient to have the GS as the mass standard for carbon clusters and organic molecules which rest on the GS surface.

In this chapter, we first introduce a quantitative HAADF-STEM technique for mass and 3D morphology mapping of complex nanoparticles. This technique was initially applied to atomic clusters with a few heavy atoms, and later extended its application to nanoparticles consisting of thousands of atoms. Then, we show that the principle also holds for light

elements such as carbon. By applying a relationship between the HAADF intensity and sample thickness obtained for GSs, we successfully mapped the mass distribution of carbon nanoparticles of complex structures. Finally, an investigation on the elastic electron scattering in GSs is performed, and the mean free paths of incident electrons with different energies are obtained. The collection angle of HAADF detector is discussed in terms of simulations based on a multi-slice algorithm.

23.1 QUANTITATIVE HIGH-ANGLE ANNULAR DARK-FIELD SCANNING TRANSMISSION MICROSCOPY IMAGING TECHNIQUE

A TEM is a device utilizing electron interactions with specimens to obtain the inner structure and chemical information of the samples. Electrons traveling through the specimen would be scattered into all directions by the electrons or nucleus of the target, or transmitted uninterrupted. In the scanning transmission microscopy (STEM) mode, a convergent beam of electrons is focused upon specimens and electrons are collected on the other side of the specimens by a circular/annular plate. An annular dark-field detector avoids the direct electron beam in the center and collects scattered electrons in a certain solid angle, while a circular bright-field detector is placed right on the axis to receive the direct beam.

In traditional dark-field imaging, namely dark-field transmission electron microscopy (DFTEM), an objective aperture is placed in the diffraction plane so as to only collect electrons scattered through that aperture, excluding the main beam. An annular dark-field detector, on the other hand, collects electrons from an annulus around the beam, sampling far more scattered electrons than can pass through an objective aperture. This gives an advantage in terms of signal collection efficiency and allows main beam to pass to an electron energy loss spectrometry (EELS) detector, allowing both types of the measurement to be performed simultaneously. The distance between the specimen and the detector is called camera length (CL), which is tuned by the focal strength of the objective lens rather than moving them physically. By varying the CL, you change the collection angle of your detector. A HAADF image formed only by very high angle, incoherently scattered electrons—as opposed to Bragg scattered electrons—is highly sensitive to variations in the atomic number of atoms in the sample (Z-contrast image). Such an imaging technique is known as HAADF-STEM (Williams and Carter 2009).

Figure 23.1 shows schematically the configuration of convergent electron beam scattering in HAADF-STEM. With the electron beam scanning over a selected area of the specimen, a HAADF image is obtained in which the intensity of each pixel corresponds to the number of electrons collected at the very spot during a constant period. So the spatial resolution is limited by the size of the focused incident probe. In a thin TEM specimen, we can always approximate all scattering within the sample to a single scattering event. Typical values of the mean free path of electrons at TEM voltages (80–300 kV) are of the order of tens of nanometers, which

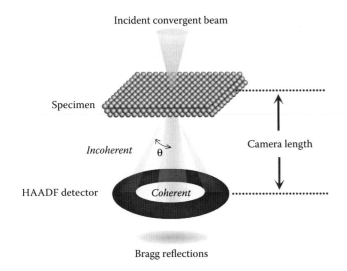

Incident convergent beam

Specimen

Incoherent

θ

Camera length

HAADF detector

Coherent

Bragg reflections

FIGURE 23.1 Schematic configuration of the electron collection by a tunable high-angle annular dark-field (HAADF) detector in the scanning transmission electron microscope (STEM).

is also the order of the thickness of most TEM specimens. Hence single scattering is a safe assumption, which makes it easier to interpret our images and spectroscopic data. As the signal from HAADF detector is mainly due to elastic scattering, and phase-contrast effects are negligible, the intensity of the signal is then quantitatively proportional to both the mass of the specimen that is contained within a defined volume illuminated by the probe, and the atomic number of the atoms within the specimen. Moreover, as the mass is computed from an image, nanoparticle mass can be directly correlated with its shape (Sousa and Leapman 2012).

After Crewe and coworkers in the early 1970s first used STEM "Z-contrast" to image single atoms with high atomic number (Crewe et al. 1970), researchers demonstrated that this technique is reliable for the mass measurement of atomic clusters with a few atoms. They showed that the integrated HAADF intensity grows linearly with the increased number of atoms in the clusters, which establishes the basis of STEM-based mass spectrometry (Howie 1979; Isaacson et al. 1979). Later, Young and coworkers have extended the applicable mass range to the size as far as ~6500 atoms of Au clusters (Young et al. 2008). In the following part of this section, we describe Young et al.'s work to demonstrate the application of the quantitative HAADF-STEM imaging performed on the same instrument.

In Figure 23.2a through c, three representative HAADF-STEM images of clusters with size of 147, 2057, and 6525 Au atoms are shown, respectively. The clusters were prepared in a radio frequency magnetron sputtering and gas aggregation cluster beam source, and were mass-selected with a resolution of $M/\Delta M \approx 25$ before soft-landing on graphite. The STEM images were obtained using an FEI Tecnai F20 TEM at 200 kV. A Fischione 3000 HAADF detector was used to record images at an inner collection angle of 50 mrad and outer collection angle of 150 mrad. The incident probe size was around 4 Å. Efforts had been made to keep the imaging

conditions constant, especially the emission current. Au_{309} was used as a check between different runs, and the change in integrated intensity was found to be smaller than the statistical errors.

Figure 23.2a through c display intuitively the brightest points in the center of clusters, corresponding to the thickest volume through which electrons are transmitted. To establish the relationship between HAADF intensities and the number of atoms in the clusters, one has to integrate the HAADF intensity over clusters of each size, and make background subtraction to eliminate the influence of the graphite substrate. Figure 23.2d gives the integrated HAADF intensity as a function of the number of atoms within the clusters. It can be seen that initially the intensity increases linearly with the number of atoms, while deviations are observed for large clusters. A log–log fit (inset) displays a relationship, $I = 68852N^{0.92315}$, with a fitting residual of $R = 0.99733$. The large error bars of bigger clusters in Figure 23.2d can be attributed not only to microscope stability, substrate uniformity, thickness, and local structure, but also to strong dynamical scattering in large nanoparticles. The deviation from linearity and the large error bars of the mass–HAADF intensity relationship together would lead to bad accuracy in the mass spectrometry. Hence, quantitative intensity analysis has only been performed for N up to 6500.

Once the calibration curve in Figure 23.2d has been obtained, it can be applied to measure Au nanoparticles prepared via different approaches. Figure 23.3a and b are STEM images of Au nanoparticles prepared by wet chemistry method and by thermally evaporating Au onto TEM grid under high vacuum condition, respectively. In the first case, colloidal Au nanoparticles purchased from Agar Scientific were drop cast onto an amorphous-carbon coated TEM grid. In the second case, Au nanoparticles were deposited on the same TEM grid by evaporating 99.99% Au in a vacuum of 10^{-7} mbar. It can be seen that most of particles are circular in 2D projection, except for a few closely spaced clusters that seem to be touching. The histograms of size distributions for both types of nanoparticles are shown in Figure 23.3c and d.

Besides the above size information we have obtained in 2D projection, the depth information along the electron beam direction is also included in the intensity of HAADF-STEM images. By analyzing the intensity distribution within an Au nanoparticle, its 3D morphology can be mapped. By applying the calibration curve in Figure 23.2d, Young et al. found that results from the thermally evaporated Au nanoparticles are consistent with the hemispherical model, while the results from the colloidal Au nanoparticles agree with the spherical model, as shown in Figure 23.3e and f.

The above case has demonstrated that the size-selected Au clusters provide an excellent mass standard, which allows us to obtain a true calibration curve for the particle mass–HAADF intensity relationship over a large mass range. However, practically this is difficult for light elements. Because STEM intensity is proportional to Z^{α}, where Z is the atomic number of the elements, and α is in the range of 1.5–1.9 depending on the collection angle of the detector, specimen thickness

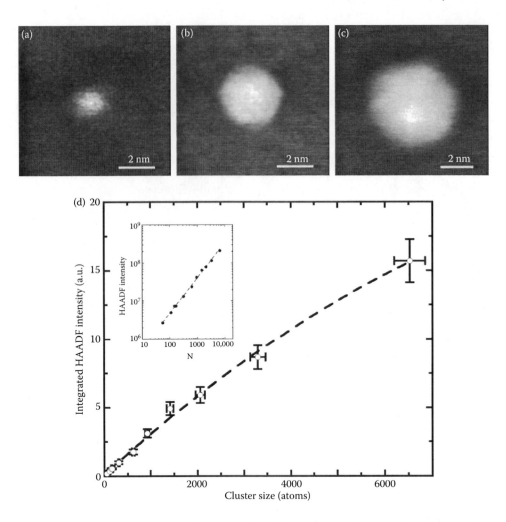

FIGURE 23.2 HAADF-STEM images and their intensity analysis. (a)–(c) Representative images of Au clusters with selected mass of 147, 2057, and 6525 atoms on graphite supports. The images were obtained at 900k× magnification at a resolution of 1024 × 1024 pixels and they were low-pass filtered. (d) The HAADF intensity integrated over clusters as a function of number (N) of Au atoms in the clusters. The error bars were taken from the standard deviation of distribution measured by STEM. The inset shows the logarithmic fit of the intensity versus N, $I = 68862N^{0.92315}$. The residual of the fit $R = 0.99733$. (Reprinted with permission from Young, N. P. et al. Weighing supported nanoparticles: Size-selected clusters as mass standard in nanometrology. *Physical Review Letters*. 101:246103. Copyright 2008 by the American Physical Society.)

and Debye–Waller factor of the atomic species (Krivanek et al. 2010). As a result, the light elements would have a much weaker STEM image contrast than heavy elements. This makes the subtraction of substrate signals difficult, especially when atomic-level accuracy is required.

In the following sections, we propose free-standing GSs as the mass standard for the quantitative measurements of light elements in HAADF-STEM. The GS calibration bears a number of merits, for example, it does not require complex high vacuum equipment for size selection, GSs can replace amorphous carbon films on the TEM grids as supporting films for straight contrast comparison in STEM images, the discrete number of GSs enables an accurate calibration of STEM intensity to be performed. As an example of applications, the calibrated STEM intensity from GSs is applied to gain insight into carbon nanoparticles of complex structures.

23.2 QUANTIFYING GRAPHENE SHEETS UNDER TRANSMISSION ELECTRON MICROSCOPY/ SCANNING TRANSMISSION MICROSCOPY

Graphene is a single atomic plane of graphite. It is the thinnest known material in the universe and the strongest ever measured. Ideally, graphene is a single-layer of carbon atoms, but graphene sheets with two or more layers are being investigated with equal interest. We usually define three different types as: single-layer graphene (SG), bilayer graphene (BG), and few-layer graphene (FG, number of layers ≤ 10). Although graphene is expected to be perfectly flat, ripples are induced by thermal fluctuations. For GSs up to micro-sizes, they can be easily folded like a broad piece of pleated cloth. Figure 23.4 shows an overview of large pieces of GSs supported by holey Formvar film covered Cu TEM grids. In the

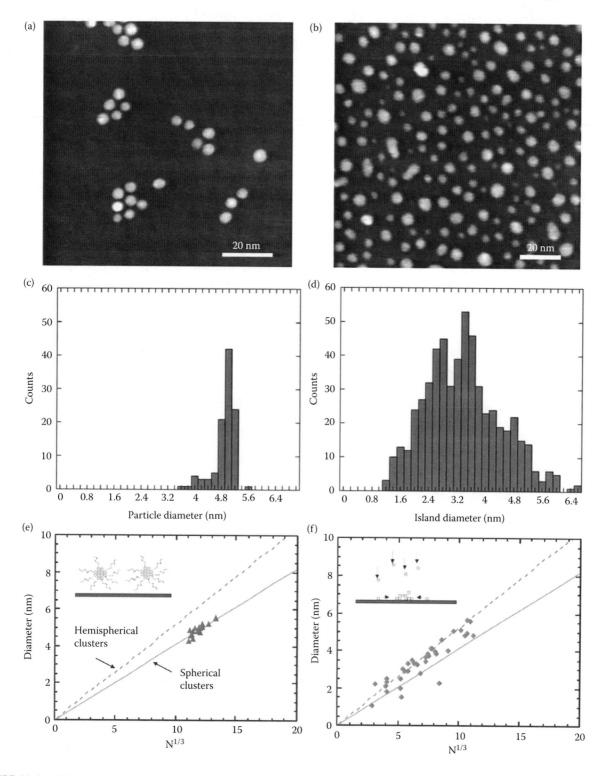

FIGURE 23.3   HAADF-STEM images taken from Au nanoparticles deposited on a Cu TEM grid coated with amorphous carbon thin film. (a) The nanoparticles were prepared by wet chemistry method. (b) The nanoparticles were prepared by thermally evaporating Au onto the TEM grid held at room temperature under high vacuum condition. The corresponding size distribution is shown in (c) and (d), respectively. Relationship between particle diameter and $N^{1/3}$ for Au nanoparticles prepared by two different methods: (e) colloidal nanoparticles, and (f) nanoparticles by thermal evaporation. Spherical cluster approximation (the solid line) and hemispherical cluster approximation (the dashed line) are shown to allow comparison of overall particle geometry. Schematics of growth mechanism of Au nanoparticles are also shown in the insets. (Reprinted with permission from Young, N. P. et al. Weighing supported nanoparticles: Size-selected clusters as mass standard in nanometrology. *Physical Review Letters*. 101:246103. Copyright 2008 by the American Physical Society.)

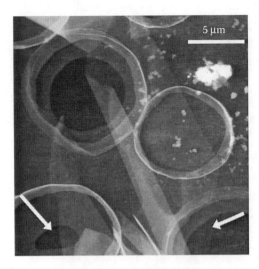

FIGURE 23.4 HAADF-STEM image showing an overview of graphene sheets (GSs) supported by holey Formvar film coated TEM grid. The bright patches on the right side of the images are contamination left on the samples during preparation and handling of the specimen. The arrows indicate areas where graphene are freely suspended on the holey film. (Reprinted from *Ultramicroscopy*, 110, Song et al., Freestanding graphene by scanning transmission electron microscopy, 1460–1464, copyright 2010, with permission from Elsevier.)

HAADF-STEM image, the step-like changes in the contrast of GSs give a direct sight of the GS folding.

To establish direct correlation between the HAADF-STEM intensity and the number of layers, N, in the GSs, we apply an independent layer counting method. In TEM images, the image contrast of the folded edge of each graphene layer within GS is enhanced, leaving parallel dark lines with constant distance in the folding area. Such dark lines are utilized to count the layers of GSs with considerable accuracy. The principle is the same as that for counting the number of walls of multi-wall carbon nanotubes in TEM images. The validity of this method has been cross-checked using Raman spectrum (Ferrari et al. 2006), nanobeam electron diffraction (Meyer et al. 2007) and electron energy loss spectroscopy (Eberlein et al. 2008). To avoid the effect of substrates, only GSs freely suspending over the holes were chosen for the quantitative study. Figure 23.5a and b are two typical HRTEM images of such folded edges, with 3 and 28 layers, respectively. Their HAADF-STEM images taken from the corresponding area are shown in Figure 23.5c and d. The intensity of the HAADF signals along the dashed lines in the STEM images are plotted in Figure 23.5e and f, respectively, which clearly show the step-wise intensity variation between vacuum and the GSs.

In Figure 23.6a and b, the HAADF intensities are plotted as a function of the GS layer numbers. The monotonic relationship is apparent. Here, the HAADF intensity values of GSs with certain number of layers were obtained by examining relatively clean areas (with uniform contrast) over at least 100×100 pixels. A histogram of the pixel intensity distribution for each particular layer thickness was also plotted

to determine the average value. The mean value of the histogram and the standard deviation for each individual layer thickness are displayed in figures. The error bars here include contributions from both contamination and the background dark counts.

It is shown in Figure 23.6a, a linear relationship between the HAADF-STEM intensity and the GS thickness up to nine layers through the origin. This result suggests that the plural scattering is negligible in this ultra-thin film region. This simplifies the subsequent data analysis procedures since simple kinematical scattering models are applicable. Nevertheless, to see how far the single scattering model can be extended to, much thick GSs were also searched purposely, as shown in Figure 23.6b. The linearity persists up to a large number of layers at around 50. Similar scale of error bars is observed in HAADF intensity as for the thin GSs, while larger errors come from counting layer numbers for thick GSs. As counting from the screen is not practical for layer numbers larger than ~10, all counting was done from the after-processed micrograph plates. The errors originate mainly from the Fresnel fringes due to slightly different focal conditions.

23.3 VERIFICATION OF THE GRAPHENE MASS STANDARD

Calibration is the crucial step in implementing the atomic-level precision. This process primarily consists of seeking mass standards with known masses and careful improving of the signal–noise ratio for the specimens. In the above section, we have established a relationship between the HAADF-STEM intensity and the thickness of GSs, which has the potential of studying carbon nanoparticles with different morphologies. Before the application of this calibration, its validity is checked reversely by amorphous carbon nanoparticles here.

The carbon nanoparticles with determined masses were collected from a carbon cluster beam generated in a radiofrequency sputtering chamber of a gas-aggregation cluster source (Han et al. 2007). The liquid nitrogen cooling, which normally used along the drift zone, was not applied. Amorphous nature of the carbon nanoparticle is shown by high-resolution TEM, electron diffraction, and EELS, which eliminates the influence of Bragg's scattering in electron collection. For clusters with diameters of 3.5–8 nm, quantum mechanical effects dependent on the size are of little relevance, and spherical shapes are energetically favored, particularly for the room temperature beams emerging from the gas aggregation chamber. This means that the carbon nanoparticles are spherical when they are flying within the beam before deposition, and the value of masses can be calculated from their spherical shapes.

The stability of the spherical shape when depositing the nanoparticles was confirmed by molecular dynamics simulations. Amorphous carbon nanoparticles comprise multiple hybridization states which may potentially change during impact. Accordingly, we employ the environment-dependent interaction potential for carbon (Marks 2000), which has been

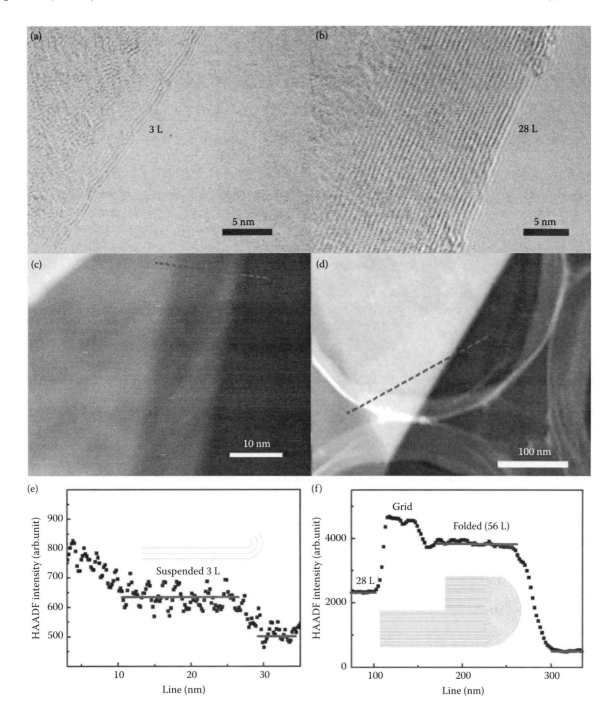

FIGURE 23.5 Graphene Sheet (GS) thickness analysis. High-resolution TEM images of the folded GS edges, where the dark lines indicate (a) 3 and (b) 28 layers of graphene. (c, d) HAADF-STEM images of the corresponding areas. (e, f) Intensity line profiles averaged over a few pixels, taken from the positions indicated in (c) and (d). (Reprinted from *Ultramicroscopy*, 110, Song et al., Freestanding graphene by scanning transmission electron microscopy, 1460–1464, copyright 2010, with permission from Elsevier.)

extensively applied to amorphous carbon in simulations of quenching and deposition. Figure 23.7a shows an amorphous C_{1441} nanoparticle being deposited onto a diamond substrate containing 23,120 atoms at a kinetic energy of 0.25 eV/atom. The spherical shape is maintained as shown in Figure 23.7b. In the experiment, free deposition conditions without any accelerating potentials were employed to deposit the reference

clusters for calibration. The free deposition is believed to guarantee the soft landing of the clusters with little transformation as confirmed by the simulation. Under the assumption that all of the reference clusters keep their spherical shapes after deposition, their lateral diameter is equal to their vertical height. So their masses can be calculated based on the cluster geometry and density as determined by EELS.

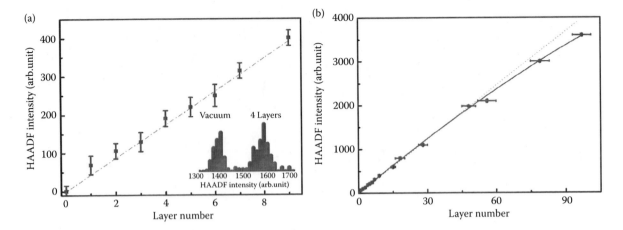

FIGURE 23.6 Calibration curves. (a) HAADF intensities are plotted against the layer number up to 9. The data and the error bars are taken from the mean value and the corresponding standard deviation from the selected layer numbers of graphene. The linear fitting is a guide for the eye. The inset shows an intensity histogram for a suspended 4-layer graphene. (b) An extended range of intensity–thickness relation is fitted by the single scattering approximation (solid line): $I = I_0 (1 - e^{-\tau/\lambda})$. The linear fitting in (a) is shown here by the dashed line as a comparison. (Reprinted from *Ultramicroscopy*, 110, Song et al., Freestanding graphene by scanning transmission electron microscopy, 1460–1464, copyright 2010, with permission from Elsevier.)

To extract the contribution of reference clusters to the HAADF intensity, the intensity of the support film is set as the zero level, and the portion above the zero level is sampled as the signal from the carbon clusters. As shown in Figure 23.8a, a small region in the center of the particle is selected and the intensity of each pixel is averaged to give the HAADF maximum, while the intensity of all the pixels from a cluster is summed as the HAADF integral.

The HAADF maximum and the HAADF integral are both plotted against the cluster diameter in Figure 23.8b, and the HAADF integral is plotted against atom numbers in (c). All curves are shown to increase monotonically even if the atomic columns are unresolved by the TEM. The linearity up to 8 nm, or $3.4 \pm 0.6 \times 10^4$ atoms in equivalence, is observed for both the HAADF maximum curve in Figure 23.8b and the calibration curve in Figure 23.8c. Since the cluster diameters are much larger than the size of the electron probe (<8 Å), the HAADF integral sums over all the probed points of a selected

cluster. As a result, the HAADF integral curve in Figure 23.8b depends on the projected area of the cluster and increases in a polynomial fashion in contrast to the linear growth in the HAADF maximum curve.

The calibration curve demonstrates that this metrology technique can achieve near atomic-level precision. The slope of the curve in Figure 23.8c is the cross-section constant σ_c, and the linear portion extends to atom counts of very few atoms. In addition, the linear fitting is essentially the average of σ_c over the whole linear region, further reducing measurement errors arising from the coarse determination of N and the HAADF intensity. This results in a value of σ_c with an error of 19%, which is also valid for small N.

By applying the calibration curves obtained above, the thickness is measured for GSs suspended on holey carbon film as shown in Figure 23.9a. The cleaved GS is composed of several regions of 3, 10, 28, and 35 layers (L). The lateral extent of the 3 L block is less than 100 nm, which is invisible to normal

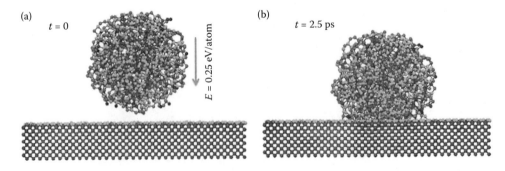

FIGURE 23.7 Molecular dynamic simulation of free deposition of amorphous carbon clusters. (a) and (b) are snapshots of a cluster of C_{1441} before incidence and after collision, respectively. The simulation time is 0 and 2.5 ps. The spherical shape was well maintained after the deposition at 0.25 eV/atom. The gray depth from dark to light of the atoms represent atoms with sp, sp^2, and sp^3 hybridization, respectively. (Reprinted with permission from Song et al., Calibrating the atomic balance by carbon nanoclusters. *Appl. Phys. Lett.* 96:033103. Copyright 2010, American Institute of Physics.)

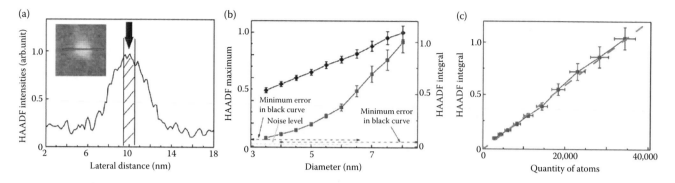

FIGURE 23.8 The calibration curves. (a) Extraction of a reference cluster of 6 nm. (b) The HAADF intensities against the particles' diameters. The upper curve (black) in (b) shows the HAADF maximum of the sampling clusters and the other curve shows the integral HAADF intensities over all the sampling points of a selected cluster. The EELS measurement gives a sp^3 ratio of 52% with the error of 20%. Therefore, the density of 2.77 g/cm³ is used for the nanoclusters. The atom number N is then calculated as shown in (c), which gives a linear calibration curve of the integral HAADF intensities plotted against N. (Reprinted with permission from Song et al., Calibrating the atomic balance by carbon nanoclusters. *Appl. Phys. Lett.* 96:033103. Copyright 2010, American Institute of Physics.)

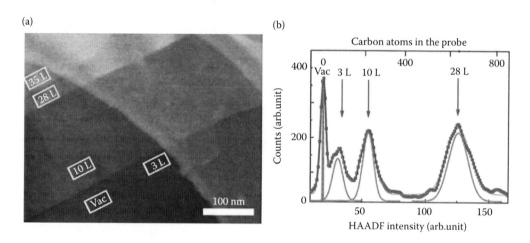

FIGURE 23.9 Measuring suspended graphene sheets (GSs). (a) GSs suspended on a holey carbon film. The darker part is a hole of vacuum region. The layer numbers are marked with 3, 10, 28, and 35 L. The HAADF signals over the region of vacuum, 3, 10, and 28 L are extracted for statistics over the HAADF intensities, yielding a histogram of the dotted curve in (b). The smooth line as a result of Gaussian peak fitting almost coincides with the dotted curve. Based on the ideal Gaussian fitting, the full width at half maximum of the vacuum peak is subtracted from that of the GSs so as to show the possible improvement by detector optimization. This gives the peaks inserted below. (Reprinted with permission from Song et al., Calibrating the atomic balance by carbon nanoclusters. *Appl. Phys. Lett.* 96:033103. Copyright 2010, American Institute of Physics.)

optical techniques. In order to evaluate the layer resolution, all the HAADF signals over the regions of vacuum, 3, 10, and 28 L were extracted for statistics, yielding a histogram curve in Figure 23.9b. The full width at half maximum decreases with decreasing layer number, and results of 3 ± 1.3 L and 10 ± 1.5 L are obtained. The error ratio is further reduced in measuring sheets with larger L, for example, 28 ± 2.6 L. The number of atoms probed by the STEM image can be calculated by considering a cylindrical volume with the diameter of the electron probe and the height of the specimen. In the 3 L GS measurement with the determined σ_c, the probe with the diameter of 8 Å counts the minimum volume of 54 ± 23 carbon atoms. A monolayer of graphene was also measured with a more focused probe (5 Å in diameter), corresponding

to a precision of 5.3 carbon atoms. These results are consistent with the known structure of the GS.

In conclusion, HAADF intensities from carbon atoms have been calibrated and their linearity has been demonstrated to be valid up to 34,000 atoms. With this calibration, the thicknesses of GSs are measured down to one layer. It confirms that GSs and carbon clusters can be used as mass standards for each other interchangeably.

23.4 WEIGHING CARBON CLUSTERS USING GRAPHENE AS THE MASS STANDARD

The HAADF intensity versus GS thickness relationship displayed in Figure 23.6a and b readily provides a mass calibration for the HAADF-STEM imaging. To achieve this, carbon nanoparticles to be studied should be STEM-imaged strictly under the same conditions as one did with the GSs. Then the HAADF intensity of the sample can be compared with that of the GS, and the mass of the sample can be calculated accordingly.

In the experiment, carbon nanoparticles were prepared with a magnetron plasma aggregation cluster source. The nanoparticles were soft-deposited in a high vacuum chamber. It is shown that these particles have an overall spherical shape while the core comprises of several small seeds (see Figure 23.10). Energy-dispersive x-ray (EDX) spectroscopy analysis

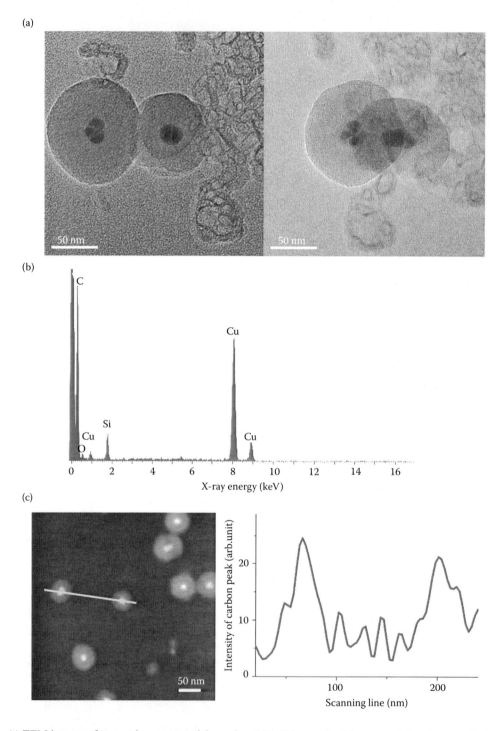

FIGURE 23.10 (a) TEM images of two carbon nanoparticles at 0 and 35 tilting angle. It is apparent that the overall shape of the particles is spherical while the core comprises of several small seeds. (b) Energy-dispersive x-ray spectrum of carbon nanoparticles. (c) A STEM image of carbon nanoparticles (left) and variations of the carbon intensity peak in EDX spectrums obtained by scanning along the solid line across two particles (right). (Reprinted from *Appl. Phys. Lett.*, 96, Song et al., Calibrating the atomic balance by carbon nanoclusters. 033103, copyright 2010, with permission from Elsevier.)

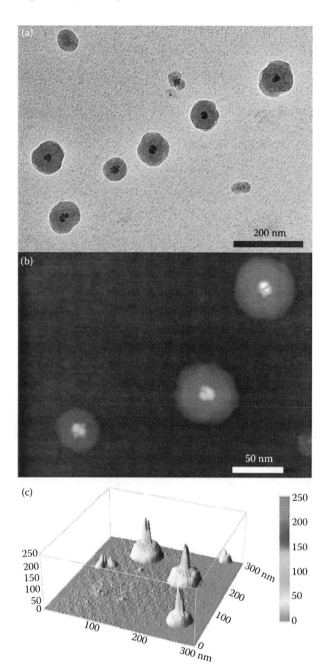

FIGURE 23.11 Complex inhomogeneous structures of carbon nanoparticles. (a) A TEM image of carbon nanoparticles prepared by plasma sputtering. (b) A HAADF-STEM image of the carbon nanoparticles (not the same area as in (a) but from the same specimen). (c) A quasi-3D view of the mass distribution of the carbon nanoparticles. (Reprinted from *Appl. Phys. Lett.*, 96, Song et al., Calibrating the atomic balance by carbon nanoclusters. 033103, copyright 2010, with permission from Elsevier.)

has been applied for assessing the compositions of carbon nanoparticles. Figure 23.10b shows an EDX spectrum taken from the specimen in which the carbon nanoparticles were supported by the Formvar film coated Cu TEM grid. Apart from carbon, peaks for Cu, Si, and O signals can be identified. While Cu and Si signals are likely from the support and the detector, respectively, the small oxygen signal may come from the specimen contamination. Further EDX analysis combined with STEM was performed as shown in Figure 23.10c. The EDX spectrums for discrete sites along the solid line were recorded by scanning across two nanoparticles. It is found that only the carbon signal is enhanced at the location of the particles, which means that the high contrast in the TEM/STEM images within the particles are not due to any heavy elements. An EELS analysis of the carbon atom state shows that both the core and the shell have similar graphite-like chemical bonding. Nanobeam electron diffraction indicates that the nanoparticles are amorphous.

Figure 23.11a and b show a typical bright field TEM image and a STEM image of the carbon nanoparticles. By applying the calibration curves in Figure 23.6, a quasi-3D mass density mapping of the carbon nanoparticles can be performed. The result is shown in Figure 23.11c with false colors. It is apparent that all of these particles have complicated core–shell structures. Their cores are formed by dense nuclear seeds, surrounded by porous shells. The densities of the cores and shells can be approximated as 2.3 ± 0.3 and 0.45 ± 0.06 g/cm^3 on the rough assumption that they are both spherical. The former is close to the value of graphite density, 2.09–2.23 g/cm^3. The uncertainty of the calculated density is estimated from that of the volume containing the atoms interacting with the incident electron beam. The revelation of the core–shell structures of carbon nanoparticles shows the potential of mapping biological samples quantitatively with the HAADF-STEM technique.

23.5 ELECTRON SCATTERING IN GRAPHENE SHEETS

The mean free paths (λ), or the cross-sections (Ω), of electron scattering in various materials are the fundamental data of the radiation physics, particularly in the current material science during the quantitative analysis using electron microscopes. They are characteristic measures of scattering. By definition, they are related as $1/\lambda = N\Omega$, where N is the number of atoms per unit volume. The value of λ is known to depend on several parameters, such as electron energy, geometry of the illumination, and collection optics. Comparing to early Rutherford scattering experiments (Rutherford 1911), modern TEMs are complemented with delicate controls of the electron beam and digital recording of detected signals, making the TEM an ideal platform for the quantitative analysis of electron scattering.

Numerous studies have so far been dedicated to the measuring of λ (or Ω), and a modern TEM allows several approaches to achieve this. For example, Angert et al. measured the elastic and inelastic electron scattering cross-sections of amorphous layers of carbon and vitrified ice using electron spectroscopic diffraction (Angert et al. 1996). Due to their method of determining layer thickness, an experimental error of 20%–25% was found. Chou et al. used an off-axis electron holography approach to obtain the inelastic λ of silicon and poly(styrene) (Chou and Libera 2003). Iakoubovskii and coworkers measured inelastic electron scattering λ with a 200 kV TEM for the majority of stable elemental solids and their oxides (Iakoubovskii et al. 2008). Their results give an insight into the electronic structures of atoms and solids. In this section, we measure the λ of electrons scattered in the GSs using the HAADF-STEM technique. The accuracy of the measurement and its dependence on the collection angle of HAADF detector and electron energies are discussed in terms of multi-slice simulations.

The electron scattering in solids is a complicated process. However, if only elastic, single scattering is concerned, the intensity of incoherent electrons received by a HAADF detector can be simplified as a relation (Yen et al. 2004):

$$I = I_0(1 - e^{-\tau/\lambda}), \tag{23.1}$$

where λ is the mean free path of electrons with certain energy transmitting through specimens, τ is the thickness of the sample, and I_0 is the intensity of the incident beam. In a thin specimen where τ is much smaller than the mean free path λ, the relation can be approximated as

$$I = I_0 \frac{\tau}{\lambda}, \tag{23.2}$$

which is the very linear relationship between GS thickness and HAADF-STEM intensity we found in the above section. In most cases, however, the samples we concern are not uniformly distributed, such as atomic clusters and nanoparticles. Nevertheless, as long as the linear relation holds, we can always integrate the intensity over the interest area in a HAADF-STEM image for the mass mapping. The accuracy of such a weighing method depends critically on the mass standard we applied.

As one extends the thickness of GSs studied to over 50 layers, the relationship eventually deviates from linearity and Equation 23.1 is applicable. Hence, if the HAADF intensities for GSs of various thicknesses are available, the data can be fitted into Equation 23.1 and the values of λ can be obtained.

In order to produce free-standing GSs with different number of layers from a few to over a hundred, a high-yield method of splitting expandable graphite was employed (Liao et al. 2011). The suspension of GSs was finally drop-cast onto a TEM grid covered with holey Formvar films. Once the solvent was evaporated, GSs would be held on the grid. This procedure was repeated several times to increase the density of the captured GSs.

The electron scattering was carried out in a Tecnai F20 TEM/STEM with a field emission gun operated at 80–200 kV. The inner angle of the HAADF detector was tuned by the CL. Here, an inner collection angle of 37 mrad was chosen. By controlling the focal strength of the second condenser lens (C2) and its aperture, the convergent angle of the incident electron beam was approximately set to a constant value of 15 mrad. For each beam energy, regular TEM/STEM alignments need to be done before image-taking. Hence, we can only assure that the operational parameters of STEM are constant for

electrons with the same energy. Comparing the HAADF intensities generated by electrons of different energies makes no sense, even for the same area of GSs. Nevertheless, the ultimate values of λ obtained by fitting the data to Equation 23.1 are independent of the STEM configurations.

The HAADF-intensity versus GS thickness relationship was obtained by following the standard procedures as in the above sections. With the GS thickness extending to 162 layers, we have introduced larger experimental uncertainties. As shown in Figure 23.12, the exact counting of dark lines in the TEM images becomes difficult due to the blurred edges of thick GSs. Hence, the relationship is only extended up to 162 layers in the present study.

Figure 23.13 displays the experimental HAADF intensities plotted as a function of the GS thickness for scattered electrons with energy 80, 120, 160, and 200 keV. For an easy comparison, the intensities in the figure are normalized by the fitting parameter I_0. The curvature of the fitting curves clearly indicates the ability of electrons to penetrate the samples. For each data point in the figure, the x-error bar comes from the counting of dark lines in the TEM image, while y-error bar comes from the standard deviation from averaging the intensity of the HAADF intensities in the STEM image. By fitting the thickness–intensity relationship to the high-angle elastic scattering approximation (the solid line in Figure 23.13), we have obtained the values λ for 80, 120, 160, and 200 keV electrons. The fitting results and experimental errors are listed in Table 23.2. The relative error finds its smallest value of 6.2% at 80 keV, and its largest value of 15.0% at 120 keV. The average error is 9.5% in the present study.

The electrons scattered can be grouped in different ways. Elastic scattering with no loss of energy usually occurs at the range of 1–10°. Elastic electrons are mostly coherent at relatively low angles, but become incoherent at higher angles. Inelastic electrons, on the other hand, are almost always incoherent and at relatively low angles (<1°). Therefore, the types of electrons that contribute to the HAADF intensity depend crucially on the collection angle of the HAADF detector, θ.

In order to rule out the inelastic scattering and coherent electrons, a higher collection angle would be preferred. However, a larger θ would also weaken the HAADF signals, leading to trivial signal–noise ratio. Thus, the collection angle θ should be carefully chosen so that coherent electrons are filtered out, yet the counts of HAADF intensity are reliable for quantitative analysis.

In Figure 23.14a, there are four STEM images of the same free-standing GS captured using different collection angles θ (in mrad). It is apparent that a large θ results in weak HAADF intensity for the ultra-thin GS, whereas a very small θ would cause the contrast of STEM images to reverse. With the increase of CL, the outer angle, as well as the inner angle of the annular detector shrinks, which results in receiving much more electrons near the axial beam spot and less at high angles. This leads to the reversed contrast of the STEM image as for $\theta = 8$ mrad.

To get a better knowledge of the relationship between HAADF intensity and collection angle, a series of STEM

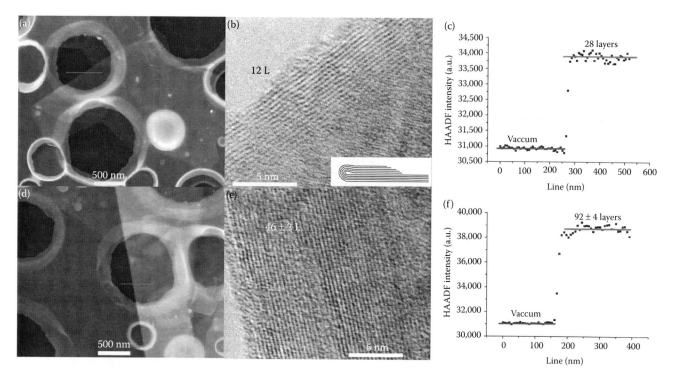

FIGURE 23.12 (a, d) HAADF images of free-standing graphene sheets (GSs) suspended across holes of the substrate. Their high-resolution TEM images show the dark lines, indicating the number of layers of (b) 12 and (e) 46 ± 2, respectively. (c, f) Intensity line profiles averaged over a few pixels, taken from the solid lines marked in (a) and (d). (Reprinted with permission from American Scientific Publisher, Ding et al. 2012. *J. Nanosci. Nanotechnol.* 12:6494–6498.)

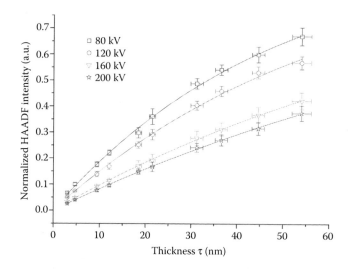

FIGURE 23.13 Experimental HAADF intensities are plotted against the thicknesses of the graphene sheets (GSs) using the electron beam with the energies of 80, 120, 160, and 200 keV, respectively. The curves are fitted according to the formula $I = I_0 (1 - e^{-\tau/\lambda})$ (the solid lines). The inset diagram shows a relation between the electron energy and its mean free path. (Reprinted with permission from American Scientific Publisher, Ding et al. 2012. *J. Nanosci. Nanotechnol.* 12:6494–6498.)

TABLE 23.2

Experimental Scattering Mean Free Paths (λ) and Cross-Sections (Ω) for Electron Scattering in Graphene Sheets at 80, 120, 160, and 200 keV

	80 keV	120 keV	160 keV	200 keV
λ (nm)	48.2 ± 3.0	61.4 ± 9.2	97.9 ± 7.6	115.6 ± 10.2
Ω (10^{-4} nm^2)	1.84 ± 0.11	1.44 ± 0.18	0.90 ± 0.08	0.77 ± 0.07

Source: Ding et al. 2012. *J. Nanosci. Nanotechnol.* 12: 6494–6498.

Note: λ is related to Ω as $1/\lambda = N\Omega$, where N is the atomic density. A graphite density of 2.267 g/cm^3 is used for the GSs.

image simulations were performed suing a multi-slice algorithm (Kirkland 1998; Koch 2002). All the simulation parameters were set in accordance with experimental configurations. Figure 23.14b displays a simulated angular distribution of 200 keV electrons scattered by a 10-layer GS. It is seen that the intensity of electrons oscillates dramatically at low angles (<20 mrad), corresponding to a convergent-electron beam diffraction pattern of the GS crystallization. At high angles (>35 mrad) the curve decreases monotonically, approaching to zero. It is apparent that, a collection angle larger than 35 mrad can be seen as a high angle in the HAADF-STEM imaging for GSs at 200 kV.

Further calculations were made on 200 keV electrons scattered by GSs of different thickness. In each single calculation, detectors with inner collection angles from 35 to 40 mrad were used simultaneously, which, of course, cannot be realized in real experiments. The results presented in Figure 23.14c show that although the electron intensity drops substantially with the increase of θ, by fitting the data to the relationship $I = I_0$

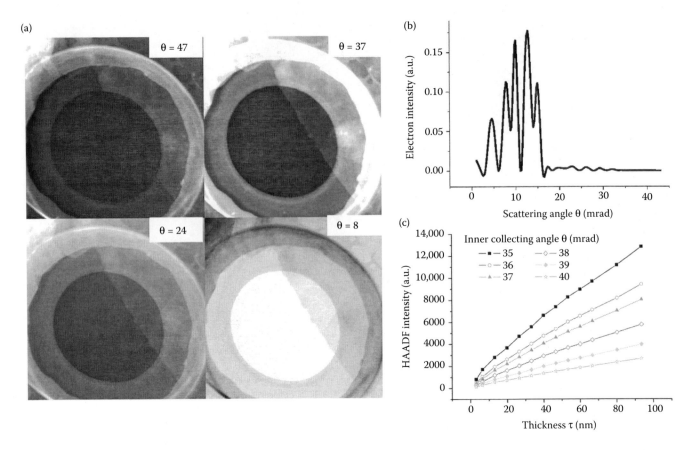

FIGURE 23.14 (a) Typical HAADF-STEM images of the same GS taken at different inner angles of HAADF collection θ (mrad). A series of simulations based on the multi-slice algorithm were performed; (b) the angular distribution of 200 keV electrons scattered by a 10-layer GS; (c) 200 keV electrons scattered by GSs of various layers were collected using detectors with different inner angles (θ = 35–40 mrad). (Reprinted with permission from American Scientific Publisher, Ding et al. 2012. *J. Nanosci. Nanotechnol.* 12:6494–6498.)

$(1 - e^{-\tau/\lambda})$, the resultant values of λ have a relatively small difference within 5.6 nm. The multi-slice simulations prove that θ = 37 mrad employed here is a reasonable value.

23.6 SUMMARY

The basic principle of the HAADF-STEM mass measurement technique relies on the linear relationship between the HAADF intensity and the projected specimen mass at each image pixel. The constant of linearity between image intensity and mass can be calculated from first principles taking into account such factors as the probe current, the collection angle subtended by the HAADF detector, the electron energy, and the elastic and inelastic scattering cross-sections for atoms in the specimen. However, a much simpler approach is to use a reference standard with known mass under the same imaging conditions.

In this chapter, we demonstrate that GSs are ideal mass standard for the mass and 3D morphology mapping of carbon nanoparticles using a HAADF-STEM technique. The HAADF intensity in a STEM image is linearly related to the mass of a nanoparticle under certain sizes. So the mass of the nanoparticle can be weighed if a reference with known mass is available. This method has worked out for a number of heavy elements with atomic level precision.

Compared to our previous attempt using size-selected carbon nanoclusters as mass standard, GS has many advantages as an alternate choice. GS has a nearly perfect 2D crystalline structure, its thickness can be well controlled, and it is very robust. Potentially, it can also replace the conventional amorphous carbon films as TEM supports. As the thickness of GSs can be measured precisely by counting the number of layers, a strict relationship between the HAADF intensity and the thickness of GSs is established. By applying the linear part of this relationship, carbon nanoparticles with complicated core–shell inhomogeneous structures are weighed and their morphologies are mapped. Furthermore, the intensity–thickness relationship up to over 100 layers complies with a simple formula, $I = I_0 (1 - e^{-\tau/\lambda})$. By fitting experimental data into this formula, mean free paths of incident electrons of different energies are obtained with well-controlled experimental errors.

The advance shown in the present chapter is feasible partly due to our recent work in exploring the potential in quantification STEM analysis, and partly due to the recent surge of interests in graphene in the scientific community, making this material routinely available through various physical and chemical preparation routes. It is envisaged that the current proposal of using GSs for mass standard would be interesting not only for the material scientists who work on light elements, but also potentially for biologists as an alternative to the commonly used tobacco virus as mass standard, particularly for their structural investigations of small biological molecules.

REFERENCES

Angert, I., C. Burmester, C. Dinges, H. Rose, and R. R. Schroder. 1996. Elastic and inelastic scattering cross-sections of amorphous layers of carbon and vitrified ice. *Ultramicroscopy* 63:181.

Booth, T. J., P. Blake, R. R. Nair, D. Jiang, E. W. Hill, U. Bangert, A. Bleloch, M. Gass, K. S. Novoselov, M. I. Katsnelson, and A. K. Geim. 2008. Macroscopic graphene membranes and their extraordinary stiffness. *Nano Letters* 8:2442–2446.

Buckhout-White, S., J. T. Robinson, N. D. Bassim, E. R. Goldman, I. L. Medintz, and M. G. Ancona. 2012. TEM imaging of unstained DNA nanostructures using suspended graphene. *Soft Matter* 9:1414–1417.

Chen, Q., J. M. Smith, J. Park, K. Kim, D. Ho, H. I. Rasool, A. Zettl, and A. P. Alivisatos. 2013. 3D motion of DNA-Au nanoconjugates in graphene liquid cell electron microscopy. *Nano Letters* 13:4556–4561.

Chou, T. M. and M. Libera. 2003. Mean free paths for inelastic electron scattering in silicon and poly(styrene) nanospheres. *Ultramicroscopy* 94:31.

Crewe, A. V., J. Wall, and J. Langmore. 1970. Visibility of single atoms. *Science* 168:1338–1340.

Dai, G.-P., P. H. Cooke, and S. Deng. 2012. Direct growth of graphene films on TEM nickel grids using benzene as precursor. *Chemical Physics Letters* 531:193–196.

Dai, G.-P., J.-M. Zhang, and S. Deng. 2011. Synthesis and characterization of nitrogen-doped monolayer and multilayer graphene on TEM copper grids. *Chemical Physics Letters* 516:212–215.

Ding et al. 2012. Scaling the dynamic electron scattering in imaging the graphene sheets by the high-angle annular dark-field microscopy. *Journal of Nanoscience and Nanotechnology.* 12:6494–6498.

Eberlein, T., U. Bangert, R. R. Nair, R. Jones, M. Gass, A. L. Bleloch, K. S. Novoselov, A. Geim, and P. R. Briddon. 2008. Plasmon spectroscopy of free-standing graphene films. *Physical Review B* 77:233406.

Ferrari, A. C., J. C. Meyer, V. Scardaci, C. Casiraghi, M. Lazzeri, F. Mauri, S. Piscanec, D. Jiang, K. S. Novoselov, S. Roth, and A. K. Geim. 2006. Raman spectrum of graphene and graphene layers. *Physical Review Letters* 97:187401.

Geim, A. K. 2009. Graphene: Status and prospects. *Science* 324:1530.

Han, M., C. Xu, D. Zhu, L. Yang, J. Zhang, Y. Chen, K. Ding, F. Song, and G. Wang. 2007. Controllable synthesis of two-dimensional metal nanoparticle arrays with oriented size and number density gradients. *Advanced Materials* 19:2979–2983.

Howie, A. 1979. Image contrast and localized signal selection techniques. *Journal of Microscopy* 117:11–23.

Iakoubovskii, K., K. Mitsuishi, Y. Nakayama, and K. Furuya. 2008. Mean free path of inelastic electron scattering in elemental solids and oxides using transmission electron microscopy: Atomic number dependent oscillatory behavior. *Physcial Review B* 77:104102.

Isaacson, M., D. Kopf, M. Ohtsuki, and M. Utlaut. 1979. Atomic imaging using the dark-field annular detector in the stem. *Ultramicroscopy* 4:101.

Kirkland, E. J. 1998. *Advanced Computing in Electron Microscopy.* New York: Plenum Press.

Koch, C. T. 2002. *Determination of Core Structure Periodicity and Point Defect Density along Dislocations.* Arizona State University, Phoenix.

Krivanek, O. L., M. F. Chisholm, V. Nicolosi, T. J. Pennycook, G. J. Corbin, N. Dellby, M. F. Murfitt. et al. 2010. Atom-by atom structural and chemical analysis by annular dark-field electron microscopy. *Nautre* 464:571.

Liao, K., W. Ding, B. Zhao, Z. Li, F. Song, Y. Qin, T. Chen, J. Wan, M. Han, G. Wang, and J. Zhou. 2011. High-power splitting of expanded graphite to produce few-layer graphene sheets. *Carbon* 49:2862–2868.

Marks, N. A. 2000. Generalizing the evironment-dependent interaction potential for carbon. *Physcial Review B* 63:035401.

Meyer, J. C., A. K. Geim, M. I. Katsnelson, K. S. Novoselov, T. J. Booth, and S. Roth. 2007. The structure of suspended graphene sheets. *Nature* 446:60–63.

Neto, A., H. Castro, F. Guinea, N. M. R. Peres, K. S. Novoselov, and A. K. Geim. 2009. The electronic properties of graphene. *Reviews of Modern Physics* 81:109–162.

Pantelic, S. S., J. W. Suk, Y. Hao, R. S. Ruoff, and H. Stahlberg. 2011. Oxidative doping renders graphene hydrophilic, facilitating its use as a support in biological TEM. *Nano Letters* 11:4319–4323.

Rao, C. N. R., A. K. Sood, K. S. Subrahmanyam, and A. Govindaraj. 2009. Graphene: The new two-dimensional nanomaterial. *Angewandte Chemie International Edition* 48:7752–7777.

Rutherford, E. 1911. The scattering of alpha and beta particles by matter and the structure of the atom. *Philosophical Magazine* 21:669–688.

Sen, S., U. Baxa, M. N. Simon, J. S. Wall, R. Sabate, S. J. Saupe, and A. C. Steven. 2007. Mass analysis by scanning transmission electron microscopy and electron diffraction validate predictions of stacked β-solenoid model of HET-s Prion Fibrils. *Journal of Biological Chemistry* 282:5545–5550.

Singhal, A., J. C. Yang, and J. M. Gibson. 1997. STEM-based mass spectroscopy of supported Re clusters. *Ultramicroscopy* 67:191–206.

Sousa, A. A. and R. D. Leapman. 2012. Development and application of STEM for the biological sciences. *Ultramicroscopy* 123:38–49.

Tanaka, T., Y. Abe, H. Sawada, E. Okunishi, Y. Kondo, Y. Tanishiro, and K. Takayanagi. 2009. Adatom on graphene, directly imaged by aberration corrected TEM at 300 kV. *Microscopy and Microanalysis* 15:1476–1477.

Wang, Z. W., O. Toikkanen, F. Yin, Z. Y. Li, B. M. Quinn, and R. E. Palmer. 2010. Counting the atoms in supported, monolayer-protected gold clusters. *Journal of the American Chemical Society* 132:2854–2855.

Westenfelder, B., J. C. Meyer, J. Biskupek, G. Algara-Siller, L. G. Lechner, J. Kusterer, U. Kaiser, C. E. Krill III, E. Kohn, and F. Scholz. 2011. Graphene-based sample supports for *in situ* high-resolution TEM electrical investigation. *Journal of Physics D: Applied Physics* 44:055502.

Williams, D. B. and C. Barry Carter. 2009. *Transmission Electron Microscopy: A Textbook for Materials Science.* 2nd ed. Springer, New York.

Yen, B. K., B. E. Schwickert, and M. F. Toney. 2004. Origin of low-friction behavior in graphite investigated by surface x-ray diffraction. *Applied Physics Letters* 84:4702.

Young, N. P., Z. Y. Li, Y. Chen, S. Palomba, M. Di Vece, and R. E. Palmer. 2008. Weighing supported nanoparticles: Size-selected clusters as mass standard in nanometrology. *Physical Review Letters* 101:246103.

Yuk, J. M., J. Park, P. Ercius, K. Kim, D. J. Hellebusch, M. F. Crommie, J. Y. Lee, A. Zettl, and A. Paul Alivisatos. 2012. High-resolution EM of colloidal nanocrystal growth using graphene liquid cells. *Science* 336:61–64.

Section IV

Recent Advances

24 Computational Modeling of Graphene and Carbon Nanotube Structures in the Terahertz, Near-Infrared, and Optical Regimes

M. F. Pantoja, D. Mateos Romero, H. Lin, S. G. Garcia, and D. H. Werner

CONTENTS

Abstract 359
24.1 Introduction 359
24.2 Theoretical Derivation of the Conductivity 360
    24.2.1 Graphene 360
    24.2.2 Carbon Nanotubes 365
24.3 Computational Models for THz and Optical Regimes 366
    24.3.1 Differential-Equation-Based Formulation 366
    24.3.2 Integral Equation-Based Formulation 368
24.4 Conclusions 371
References 371

ABSTRACT

This chapter introduces some computational electromagnetic techniques that can simulate the interaction of electromagnetic waves with graphene and carbon nanotubes from a macroscopic perspective. The link with the electromagnetic fields is achieved by employing appropriate Maxwell's equations in conjunction with material parameter models derived from quantum mechanics description. The modeling tools will enable a new generation of electromagnetic devices to be computationally explored, while providing a realistic perspective on future possibilities for graphene-based technology.

24.1 INTRODUCTION

Interest in graphene as innovative material to design electromagnetic devices has skyrocketed in recent years (Novoselov et al. 2004). Whenever technology has made it possible, graphene-based devices have been manufactured and characterized, but the unavailability of manufacturing processes (e.g., nanofabrication techniques) or the monetary cost of the prototypes have limited this exploration. To overcome these issues, it is highly desirable to develop computational electromagnetics techniques that can simulate graphene and carbon nanotubes from a macroscopic perspective. Also, the compatibility of this procedure with the actual numerical techniques employed in other disciplines will better enable the study of the nanodimensional properties of matter and their influences on practical devices.

This chapter introduces numerical methods for modeling the interaction of electromagnetic waves with graphene and carbon nanotubes. It is important to note that most of the designs for electromagnetic devices that are based on graphene have been proposed for operation in the terahertz, infrared, and/or optical regimes. Consequently, the numerical description of graphene and carbon nanotubes presented in this chapter is based on constitutive parameter models that are valid in those regimes of the electromagnetic spectrum. The link with the electromagnetic fields is achieved by employing the appropriate Maxwell's equations in conjunction with the material parameter models developed in Section 24.2. Taking into account the maturity of the numerical simulators in the field of computational electromagnetics (Sadiku 2010), it is not feasible to explore all possible approaches in a single chapter. However, the algorithms presented here, which start from the frequency- or time-domains and the differential- or integral-equation formulations, will serve to demonstrate how the exotic material properties of graphene and carbon nanotubes can be incorporated into standard computational electromagnetic modeling codes. These modeling tools will enable a new generation of electromagnetic devices to be computationally explored, while providing a realistic perspective on future possibilities for graphene-based technology.

A short survey of innovative devices based on graphene is presented in Table 24.1. An intense research activity to fill the so-called THz gap has been performed (Hartman et al. 2014). Devices able to operate in this regime include filters

TABLE 24.1
Contributions Including Innovative Electromagnetic Devices Based on Graphene

Regime	Reference
THz	Gomez-Díaz and Perruisseau-Carrier (2012)
	Abadal et al. (2013)
	Andryieuski and Lavrinenko (2013)
	Chen and Alù (2013)
	Filter et al. (2013)
	Gomez-Diaz et al. (2013)
	Othman et al. (2013)
	Xu et al. (2013)
	Amanatiadis et al. (2014)
	Correas-Serrano et al. (2014)
	Li et al. (2014a,b)
	Wang et al. (2014)
	Zhu et al. (2014)
Mid-IR&Optical	Li et al. (2013)
	Li et al. (2014a,b)
	Luo et al. (2013)
	Tamagnone et al. (2014)
	Zhu et al. (2014)

(Li et al. 2014b), emitters, and detectors (Abadal et al. 2013; Chen and Alù 2013; Filter et al. 2013; Gomez-Díaz et al. 2013; Zhou et al. 2014), absorbers (Andryieuski and Lavrinenko 2013; Xu et al. 2013), propagating media (Gomez-Díaz and Perruisseau-Carrier 2012; Othman et al. 2013; Amanatiadis et al. 2014; Correas-Serrano et al. 2014), and lenses (Wang et al. 2014). Mid-infrared and optical devices including graphene are waveguides (Farhat et al. 2013; Li et al. 2014a), switches and directional couplers (Li et al. 2013; Wang et al. 2012), filters (Li et al. 2014a,b), nonreciprocal devices (Tamagnone et al. 2014), antennas (Zhu et al. 2014), and other plasmonic devices (Luo et al. 2013).

24.2 THEORETICAL DERIVATION OF THE CONDUCTIVITY

As it has been pointed out in the introduction that the main objective of this chapter is to provide a formulation that incorporates the nanodimensional electronic transport properties of graphene into a macroscopic material model suitable for use with Maxwell's equations. To this end, effective models are developed for the constitutive parameters $\{\varepsilon, \mu, \sigma\}$ of graphene and carbon nanotubes (Maksimenko et al. 1999), which are valid in the terahertz, near-infrared, and optical regimes. Taking into account the two-dimensional character of graphene and carbon nanotubes, which make the electrical permittivity and magnetic permeability approximately equal to those of free space, the basis of the proposed technique is to develop a theoretical formulation for the associated conductivities.

In this sense, several methods have been presented in the literature for modeling graphene (Charlier et al. 2007; Castro Neto et al. 2009) from a bottom-up perspective, most of them based on the pioneering work of Wallace (1947). In short, they

employ the second quantization of quantum mechanics, supported by the electronic band description, which uses basis functions that account for the number of particles occupying each energy state in the complete set of single-particle states (Maggiore 2005). However, this approach is extremely complex and difficult to implement in practice. For this reason, the formalism presented here is based on the first quantization, which is more intuitive because it depends exclusively on single-particle wave functions. No matter what intermediate formulation is employed, the key step for incorporating the microscopic electronic transport of carriers into the macroscopic conductivity is through Kubo's equation (Kubo 1956). Approximations of these equations to simpler forms can be made by assuming different regimes or under certain conditions with regard to the physical parameters (e.g., temperature or chemical potential). Moreover, the formulation presented here allows carbon nanotubes to be represented as a particular case of graphene sheets, in which the geometrical periodicity results in a quantization of the transversal momentum.

24.2.1 GRAPHENE

Graphene is an allotrope of carbon arranged in a honeycomb structure made out of hexagons whose vertices are occupied by carbon atoms sharing covalent bonds. Electronic properties of graphene can be derived from the band theory of solids. To this end, a brief description is presented on the geometrical characterization and energy bands in graphene. Then, a first quantization procedure is introduced to achieve the conductivity at any frequency, and finally further approximations for the infrared and optical regimes are considered, which allow simpler mathematical expressions to be derived for the conductivity.

Geometrically, a hexagonal lattice is a particular case of a rhombic lattice with rectangles that are $\sqrt{3}$ times as high as they are wide (Kittel 2004). In this case, a unit cell contains two nonequivalent atoms (noted as A and B), each one forming a sublattice of identical primitive vectors $\vec{a}_i$:

$$\vec{a}_1 = a\left(\frac{\sqrt{3}}{2}, \frac{1}{2}\right)$$
$$\vec{a}_2 = a\left(\frac{\sqrt{3}}{2}, -\frac{1}{2}\right) \tag{24.1}$$

where a is the modulus of the lattice vector, which is related to the carbon–carbon distance a_{CC} as $a = \sqrt{3}a_{CC}$. The first Brillouin zone is also hexagonal (Wallace 1947), with reciprocal-lattice vectors $\vec{b}_i$ (see Figure 24.1):

$$\vec{b}_1 = \frac{2\pi}{a}\left(\frac{1}{\sqrt{3}}, 1\right)$$
$$\vec{b}_2 = \frac{2\pi}{a}\left(\frac{1}{\sqrt{3}}, -1\right) \tag{24.2}$$

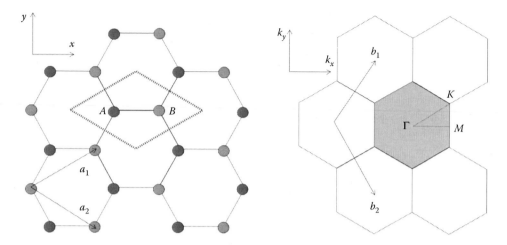

FIGURE 24.1   Arrangement of carbon atoms in graphene: Unit cell (left) and corresponding first Brillouin zone (right).

In this way, three of the four electrons located in the valence band of any carbon atom are in sp^2 hybridization, that is, the $2s$ orbital is mixed with $2p_x$ and $2p_y$ orbitals to form a total of three σ covalent bonds with neighboring carbon atoms. The fourth electron, whose $2p_z$ orbital remains independent of the σ bonds, forms a π covalent bond. The conductivity is mainly related to this latter electron, because σ bands (depicted as dotted lines in Figure 24.2, ranging from -1.1 to -0.5 Hartrees) are below the Fermi level (located in Figure 24.2 around -0.3 Hartrees, where the π and π^* bands collide), and are thus unlikely to transition from the valence to the conduction band. For this reason, a carbon atom in graphene has only one electron in the valence band.

The energy of the π band can be calculated through the tight-binding approximation (Wallace 1947), in which it is assumed that only the interactions between electrons of neighboring atoms are significant. To this end, the wave function of the orbital $2p_z$ in an isolated atom is denoted by $X(\vec{r})$. Using lattice symmetry, the wave function of any equivalent A-atom can be written as $X(\vec{r} - \vec{r}_A)$, where $\vec{r}_A$ is the position vector, and similarly for B-atoms as $X(\vec{r} - \vec{r}_B)$. Using the Bloch theorem [1], the periodic Bloch waves for the sublattices A and B are

$$\varphi_1(\vec{r}) = \frac{1}{\sqrt{N}} \sum_A e^{i\vec{k}\cdot\vec{r}_A} X(\vec{r} - \vec{r}_A)$$

$$\varphi_2(\vec{r}) = \frac{1}{\sqrt{N}} \sum_B e^{i\vec{k}\cdot\vec{r}_B} X(\vec{r} - \vec{r}_B)$$

(24.3)

where N is the number of unit cells of the lattice. Then, the total wave function has the form

$$\varphi(\vec{r}) = c_1\varphi_1(\vec{r}) + c_2\varphi_2(\vec{r})$$

(24.4)

with constants c_1 and c_2 associated with the normalized wave function φ.

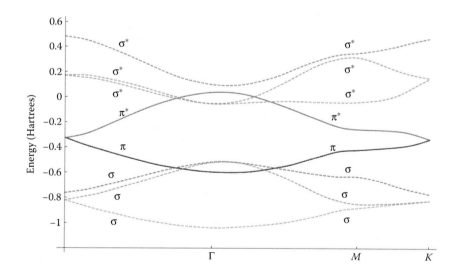

FIGURE 24.2   Energy band structure in graphene.

Let H be the Hamiltonian of the lattice that allows, among other things, the calculation of the energy of an electron in a particular quantum state (Bransden and Joachain 2000). Taking $\{\varphi_1, \varphi_2\}$ as the basis of the space formed by the lattice wave functions, which satisfies $H\varphi_i = E\varphi_i$ where E is the energy of the electron in the φ_i state, then H can be expressed as

$$H = \begin{pmatrix} H_{11} & H_{12} \\ H_{21} & H_{22} \end{pmatrix} \qquad (24.5)$$

where $H_{ij} = \int \varphi_i^* H\varphi_j d\vec{r} = H_{ji}^*$.

Also, it is usual in the tight-binding approach to neglect the overlapping between orbitals $2p_z$ of different atoms, that is, $\int X^*(\vec{r} - \vec{r}_A)X(\vec{r} - \vec{r}_B)d\vec{r} = 0$ for any $A \neq B$. Then, the diagonal terms of H are

$$H_{11} = H_{22} = \frac{1}{N}\sum_{A,A'} e^{-i\vec{k}\cdot(\vec{r}_A - \vec{r}_{A'})} \int X^*(\vec{r} - \vec{r}_A)HX(\vec{r} - \vec{r}_{A'})d\vec{r} =$$

$$\int X^*(\vec{r} - \vec{r}_A)HX(\vec{r} - \vec{r}_A)d\vec{r} = \varepsilon_0 \qquad (24.6)$$

where geometrical symmetry has been employed to allow $H_{11} = H_{22}$, and ε_0 corresponds to the energy of an electron on the $2p_z$ orbital in carbon. Regarding the overlap coming from different sublattices, only nearest neighbors are considered, which leads to the condition that $\int X^*(\vec{r} - \vec{r}_A)HX(\vec{r} - \vec{r}_B)d\vec{r} = 0$ if A and B are not nearest atoms, and the off-diagonal terms of the Hamiltonian are

$$H_{12} = H_{21} = \frac{1}{N}\sum_{A,B} e^{-i\vec{k}\cdot(\vec{r}_A - \vec{r}_B)} \int X^*(\vec{r} - \vec{r}_A)HX(\vec{r} - \vec{r}_B)d\vec{r} =$$

$$-t(1 + e^{i\vec{k}\cdot\vec{a}_1} + e^{i\vec{k}\cdot\vec{a}_2}) \qquad (24.7)$$

where $t = -e^{i\vec{k}\cdot(\vec{r}_A - \vec{r}_B)}\int X^*(\vec{r} - \vec{r}_A)HX(\vec{r} - \vec{r}_B)d\vec{r}$ and $\int X^*(\vec{r} - \vec{r}_{A,B})HX(\vec{r} - \vec{r}_{A',B'})d\vec{r} = 0$, for any $A \neq A'$ and $B \neq B'$, is applied.

Taking into account that only energy gaps are needed to characterize the electrical conduction, the value of ε_0 can be assumed to be a reference energy. In this way, $|t|$ is experimentally measured as approximately equal to 2.8 eV, and the Hamiltonian is rewritten as

$$H = \begin{pmatrix} 0 & w(\vec{k}) \\ w^*(\vec{k}) & 0 \end{pmatrix} \qquad (24.8)$$

where $w(\vec{k}) = -t(1 + e^{i\vec{k}\cdot\vec{a}_1} + e^{i\vec{k}\cdot\vec{a}_2})$. The associated eigenvalues and eigenvectors of the matrix are

$$\varepsilon_{\vec{k},s} = s|t|\sqrt{1 + 4\cos\left(\frac{\sqrt{3}}{2}k_x a\right)\cos\left(\frac{1}{2}k_y a\right) + 4\cos^2\left(\frac{1}{2}k_y a\right)}$$

$$(24.9)$$

and

$$\vec{\xi}_{\vec{k},s} = \frac{1}{\sqrt{2}}\begin{pmatrix} -s\dfrac{t}{|t|}e^{i\varphi_{\vec{k}}} \\ 1 \end{pmatrix} \qquad (24.10)$$

with $s = \pm 1$, and

$$\tan\varphi_{\vec{k}} = \frac{\operatorname{Im}(1 + e^{i\vec{k}\cdot\vec{a}_1} + e^{i\vec{k}\cdot\vec{a}_2})}{\operatorname{Re}(1 + e^{i\vec{k}\cdot\vec{a}_1} + e^{i\vec{k}\cdot\vec{a}_2})}$$

Figure 24.3 shows a three-dimensional picture of the dispersion relation arising from Equation 24.9. The $s = -1$ solution corresponds to the bonding orbitals π, which are filled

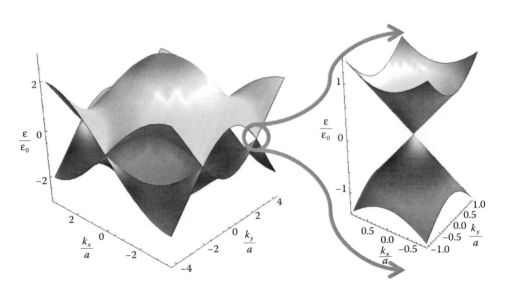

FIGURE 24.3 (Left) Energy of electrons in π and π^* bands. (Right) Zoom near Fermi points.

in the fundamental state, and the $s = 1$ solution forms the antibonding orbitals π^* for excited states of the graphene. A similar graph restricted to the $\Gamma - K$ and $\Gamma - M$ directions of the first Brillouin zone is presented in Figure 24.2, which also includes the σ and σ^* orbitals as reference. At low temperatures, only the σ and π bands are occupied, and the Fermi energy is reached at K vertices of the Brillouin zone, which also form the Fermi surface for graphene. Furthermore, the choice of $\varepsilon_0 = 0$ as the reference for energies implies that the Fermi energy can be considered as zero in the remainder of the chapter.

Therefore, the dispersion analysis can be carried out by considering a small perturbation $\delta\vec{k}$ near the Fermi points of graphene, in the form $\vec{k} = \vec{K} + \delta\vec{k}$. In this way, a series expansion of Equation 24.9 leads to

$$\varepsilon_{\vec{k},s} = s v_F \hbar \, |\, \delta\vec{k}\, | \tag{24.11}$$

where the Fermi velocity for graphene is $v_F = (\sqrt{3}a/2\hbar)|t| \approx 10^6$ m/s. It is worth remarking that Equation 24.11, which predicts both the ballistic transport and isotropic properties for graphene and illustrated as a zoom in the right part of Figure 24.3, has a linear dependence on $|\delta\vec{k}|$. Of course, those properties cannot be assumed in general, for example, in the ultraviolet regime, but they will hold for the situations considered in this chapter.

Regarding the distribution of electrons in the energy bands, the Fermi–Dirac distribution is employed under the assumption that electrons behave as fermions (Sutton 1993):

$$f_{\vec{k},s} = (1 + e^{(\varepsilon_{\vec{k},s} - \mu_c)/k_B T})^{-1} \tag{24.12}$$

where k_B is the Boltzmann constant, T is the temperature of the graphene, and μ_c is the chemical potential that can be tuned through the application of external fields (Figure 24.4 shows the Fermi distribution as a function of $\vec{k}$ for $\mu_c = 0$).

To derive the conductivity of the graphene, electromagnetic interactions are considered by the total linear momentum $\hbar k + (e/c)\vec{A}$, where $\vec{A}$ is the electromagnetic potential vector related to a harmonic electric field $\vec{E}$ as $\vec{A} = (c/i\omega)\vec{E}$. Assuming a relatively small electromagnetic momentum, the perturbation method can be applied such that mechanical and electromagnetic terms can be separated in the Hamiltonian as (Zhang et al. 2008)

$$H = H_0 + H' \tag{24.13}$$

where H_0 corresponds to H in Equation 24.8, and H' is

$$H' = \frac{e}{\hbar c} \begin{pmatrix} 0 & -ite^{i\vec{k}\cdot\vec{a}_1} \\ it^* e^{-i\vec{k}\cdot\vec{a}_1} & 0 \end{pmatrix} \vec{A}\cdot\vec{a}_1$$

$$+ \frac{e}{\hbar c} \begin{pmatrix} 0 & -it^* e^{i\vec{k}\cdot\vec{a}_2} \\ ite^{-i\vec{k}\cdot\vec{a}_2} & 0 \end{pmatrix} \vec{A}\cdot\vec{a}_2 \tag{24.14}$$

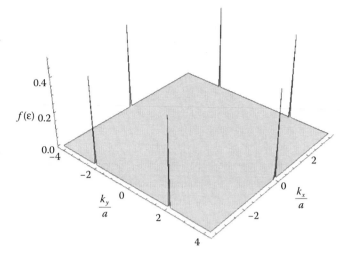

FIGURE 24.4 Fermi distribution of occupied states for electrons in the case where $\mu_c = 0$.

which can be rewritten as

$$H' = \frac{1}{c}\vec{A}\cdot\vec{J} \tag{24.15}$$

The components of the current $\vec{J}$ in Equation 24.15 are

$$J_x = \frac{2ev_F}{\sqrt{\alpha(\vec{k})}} \begin{pmatrix} -J_{x1} & iJ_{x2} \\ -iJ_{x2} & J_{x1} \end{pmatrix} \tag{24.16}$$

with $\alpha(\vec{k}) = 1 + 4\cos\left(\dfrac{\sqrt{3}}{2}k_x a\right)\cos\left(\dfrac{1}{2}k_y a\right) + 4\cos^2\left(\dfrac{1}{2}k_y a\right)$,

$$J_{x1} = \sin\left(\frac{\sqrt{3}}{2}k_x a\right)\cos\left(\frac{1}{2}k_y a\right),$$

and $J_{x2} = 1 + \cos\left(\dfrac{\sqrt{3}}{2}k_x a\right)\cos\left(\dfrac{1}{2}k_y a\right) + \cos\left(k_y a\right)$. Also:

$$J_y = \frac{2ev_F}{\sqrt{3\alpha(\vec{k})}} \begin{pmatrix} -J_{y1} & -iJ_{y2} \\ iJ_{y2} & J_{y1} \end{pmatrix} \tag{24.17}$$

with $J_{y1} = \cos\left(\dfrac{\sqrt{3}}{2}k_x a\right)\sin\left(\dfrac{1}{2}k_y a\right) + \sin\left(k_y a\right)$,

and $J_{y2} = \sin\left(\dfrac{\sqrt{3}}{2}k_x a\right)\sin\left(\dfrac{1}{2}k_y a\right)$.

Next, Kubo's formulation (Kubo 1956) may be applied to obtain the components of the dynamic conductivity tensor as

$$\sigma_{\mu\nu} = \frac{1}{V}\sum_{\vec{k}}\frac{1}{\hbar\omega}\int dt\, e^{i\omega t}\langle J_\mu(t)J_\nu(0) - J_\nu(0)J_\mu(t)\rangle \tag{24.18}$$

where V is the volume of the system and $\langle \rangle$ indicates the trace of the matrix resulting from the product of the corresponding time-varying current densities $\vec{J}_x$ and $\vec{J}_y$ and the auxiliary matrix of Fermi–Dirac distribution ρ can be defined as

$$\rho = \begin{pmatrix} f_{\vec{k},1} & 0 \\ 0 & f_{\vec{k},-1} \end{pmatrix} \quad (24.19)$$

In our case, Equation 24.18 yields

$$\sigma_{xx} = 4i\frac{e^2 v_F^2}{\pi^2 \omega}\int d\vec{k}\,\frac{(J_{x2})^2}{\alpha(\vec{k})}g(\vec{k})\left(f_{\vec{k},1} - f_{\vec{k},-1}\right)$$

$$\sigma_{yy} = 4i\frac{e^2 v_F^2}{3\pi^2 \omega}\int d\vec{k}\,\frac{(J_{y2})^2}{\alpha(\vec{k})}g(\vec{k})\left(f_{\vec{k},1} - f_{\vec{k},-1}\right) \quad (24.20)$$

$$\sigma_{xy} = \sigma_{yx} = 0$$

where $g(\vec{k}) = [1/(\hbar\omega + 2\varepsilon_{\vec{k},1}) - 1/(\hbar\omega - 2\varepsilon_{\vec{k},-1})]$ and the approximation $(1/V)\Sigma_{\vec{k}} \approx 4\int d\vec{k}/(2\pi)^2$ has been invoked.

For low energies, which is the case in the terahertz, near infrared, and optical regimes, the occupied states will be located near the Fermi points K. For this reason, the series expansion $\vec{k} = \vec{K} + \delta\vec{k}$ can be repeated leading to the approximations $(1 + \cos((\sqrt{3}/2)k_x a)\cos((1/2)k_y a) + \cos(k_y a))^2 \approx (3/16)k_y^2 a^2$ and $\sin^2((\sqrt{3}/2)k_x a)\sin^2((1/2)k_y a) \approx (9/16)k_x^2 a^2$, which implies that

$$\sigma = \sigma_{xx} = \sigma_{yy} = i\frac{e^2}{2\pi^2 \hbar^2 \omega}$$

$$\int d\vec{k}\left[\frac{1}{\hbar\omega + 2\varepsilon_{\vec{k},1}} - \frac{1}{\hbar\omega - 2\varepsilon_{\vec{k},1}}\right](f_{\vec{k},1} - f_{\vec{k},-1}) \quad (24.21)$$

The integral in Equation 24.21 can be carried out through a change of variables such that $\varepsilon = \varepsilon_{\vec{k},1} = v_F \hbar |\vec{k}|$ and $\theta = \tan^{-1}(k_y/k_x)$

$$\sigma = i\frac{e^2 v_F^2}{2\pi^2 \hbar^2 \omega}\int_0^{2\pi} d\theta \int_0^\infty d\varepsilon\left[\frac{1}{\hbar\omega + 2\varepsilon} - \frac{1}{\hbar\omega - 2\varepsilon}\right]\varepsilon[f(\varepsilon) - f(-\varepsilon)]$$

$$(24.22)$$

which can be rewritten through integration by parts and applying the property that $\varepsilon((\partial f(\varepsilon)/\partial\varepsilon) - (\partial f(-\varepsilon)/\partial\varepsilon))$ is an even function. Following this procedure results in

$$\sigma = -i\frac{e^2\omega}{\pi}\left[\frac{1}{(\hbar\omega)^2}\int_0^\infty d\varepsilon\left(\frac{\partial f(\varepsilon)}{\partial\varepsilon} - \frac{\partial f(-\varepsilon)}{\partial\varepsilon}\right)\varepsilon - \int_0^\infty d\varepsilon\,\frac{f(-\varepsilon) - f(\varepsilon)}{(\hbar\omega)^2 - (2\varepsilon)^2}\right] \quad (24.23)$$

Scattering of electrons can be taken into account by introducing a complex frequency $\omega \to \omega + 2i\Gamma$ (Hanson 2008), where Γ is related to the relaxation time τ for the scattering of electrons in the form $\Gamma = (2\tau)^{-1}$. Using this fact, Equation 24.23 may be rewritten as

$$\sigma = -i\frac{e^2(\omega + 2i\Gamma)}{\pi\hbar^2}\left[\frac{1}{(\omega + 2i\Gamma)^2}\int_0^\infty d\varepsilon\left(\frac{\partial f(\varepsilon)}{\partial\varepsilon} - \frac{\partial f(-\varepsilon)}{\partial\varepsilon}\right)\varepsilon - \int_0^\infty d\varepsilon\,\frac{f(-\varepsilon) - f(\varepsilon)}{(\omega + 2i\Gamma)^2 - (2\varepsilon/\hbar)^2}\right] \quad (24.24)$$

which is identical to the lower frequency approximation (including the optical and terahertz regime) presented in Gusynin et al. (2007).

The first term of Equation 24.24 can be identified as a first intraband contribution, which has the following analytical solution:

$$\sigma_{\text{intra}} = -i\frac{e^2}{\pi\hbar^2(\omega + 2i\Gamma)}\int_0^\infty d\varepsilon\left(\frac{\partial f(\varepsilon)}{\partial\varepsilon} - \frac{\partial f(-\varepsilon)}{\partial\varepsilon}\right)\varepsilon =$$

$$i\frac{e^2 k_B T}{\pi\hbar^2(\omega + 2i\Gamma)}\left[\frac{\mu_c}{k_B T} + 2\ln(e^{-(\mu_c/k_B T)} + 1)\right] \quad (24.25)$$

The second term of Equation 24.24 represents an interband contribution, which can be approximated using ($k_B T \ll |\mu_c|$, $k_B T \ll \hbar\omega$), as

$$\sigma_{\text{inter}} = -i\frac{e^2(\omega + 2i\Gamma)}{\pi\hbar^2}\int_0^\infty d\varepsilon\,\frac{f(-\varepsilon) - f(\varepsilon)}{(\omega + 2i\Gamma)^2 - (2\varepsilon/\hbar)^2} \approx$$

$$i\frac{e^2}{4\pi\hbar}\ln\left(\frac{2|\mu_c| - (\omega + 2i\Gamma)\hbar}{2|\mu_c| + (\omega + 2i\Gamma)\hbar}\right) \quad (24.26)$$

Equation 24.25 accounts only for the intraband response because $(\partial f(-\varepsilon)/\partial\varepsilon) = -(\partial f(\varepsilon)/\partial\varepsilon)$, which implies that it can be rewritten as $\sigma = -i2e^2[\pi\hbar^2(\omega + 2i\Gamma)]^{-1}\int_0^\infty d\varepsilon(\partial f(\varepsilon)/\partial\varepsilon)\varepsilon$. Hence, only the cases where $s = 1$ or -1 are employed in the calculation, which is not true for Equation 24.26. Also, it is important to remark that for the terahertz regime, it is only necessary to consider Equation 24.25, whereas both

Equations 24.25 and 24.26 should be taken into account at optical frequencies.

24.2.2 CARBON NANOTUBES

The dispersion relation (24.9) remains valid for carbon nanotubes, which can be thought of as the enrollment form of graphene. However, the enrollment enforces a geometrical periodicity and the transversal momentum is quantized, which simplifies the derivation of the conductivity.

Figure 24.5 presents an extended graphene sheet. Any enrolled form may be characterized by a vector $\vec{C}_H = m\vec{a}_1 + n\vec{a}_2$ joining identical carbon atoms, which will be located at the same point in the carbon nanotube. Then, $\vec{C}_H$ represents the circumference of the carbon nanotubes, $|\vec{C}_H| = \sqrt{3}a_{CC}\sqrt{m^2 + mn + n^2}$, and can also be taken as a basis vector in the unit cell of the carbon nanotubes. Furthermore, the translation vector $\vec{T}$ is defined to join the reference point O with the nearest carbon atom, perpendicular to $\vec{C}_H$. Hence, $\vec{T}$ can be expressed as

$$\vec{T} = t_1\vec{a}_1 + t_2\vec{a}_2 \qquad (24.27)$$

where $t_1 = (2m + n)/d_R$ and $t_2 = (m + 2n)/d_R$, with $d_R = \gcd(2m + n, m + 2n)$. In this way, the axis of the carbon nanotube is defined along the $\vec{T}$ direction, referenced as z, and the corresponding perpendicular component along $\vec{C}_H$, denoted as ϕ.

The wave functions in carbon nanotubes are periodic according to $\vec{C}_H$. Therefore, by applying Bloch's theorem

$$\varphi(\vec{r} + \vec{C}_H) = e^{i\vec{k}\cdot\vec{C}_H}\varphi(\vec{r}) = \varphi(\vec{r}) \qquad (24.28)$$

it follows that

$$\sqrt{3}\frac{n+m}{2}k_xa + \frac{n-m}{2}k_ya = 2\pi s \qquad (24.29)$$

For armchair carbon nanotubes ($m = n$), Equation 24.29 reduces to

$$k_x = k_\phi = \frac{s}{m}\frac{2\pi}{\sqrt{3}a}$$
$$k_y = k_z \qquad (24.30)$$

and for zigzag carbon nanotubes ($m = 0$ or equivalently $m = -n$), we obtain

$$k_y = k_\phi = \frac{s}{m}\frac{2\pi}{a}$$
$$k_x = k_z \qquad (24.31)$$

Consequently, Equation 24.24 still holds for carbon nanotubes, but the transversal momentum quantization has to be taken into account for the integration. To this end we apply the identity $2\pi\int_0^\infty d\varepsilon = \hbar^2\int d\vec{k}v_F^2$, which leads to an expression for the intraband conductivity

$$\sigma_{zz}(\omega) = -i\frac{4e^2}{(2\pi)^2(\omega + 2i\Gamma)}\int d\vec{k}\,\frac{\partial f(\varepsilon)}{\partial\varepsilon}v_F^2 \qquad (24.32)$$

where the identity $(\partial f(\varepsilon)/\partial\varepsilon) - (\partial f(-\varepsilon)/\partial\varepsilon) = 2(\partial f(\varepsilon)/\partial\varepsilon)$ has been applied.

For zigzag carbon nanotubes (Figure 24.6), Equation 24.31 can be considered in Equation 24.32, leading to

$$\sigma_{zz}(\omega) = -i\frac{8\pi e^2}{(2\pi)^2(\omega + 2i\Gamma)ma}\sum_{s=1}^{m}\int dk_z\,\frac{\partial f(\varepsilon)}{\partial\varepsilon}v_F^2 \qquad (24.33)$$

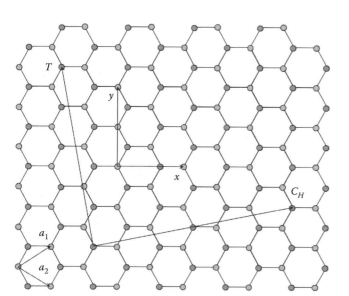

FIGURE 24.5 Geometrical structure of carbon nanotubes.

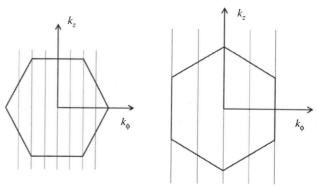

FIGURE 24.6 Quantization of transversal momentum for (left) zigzag and (right) armchair carbon nanotubes.

Only Fermi points ($s = m/3$ and $2m/3$) have to be considered for the case of metallic zigzag ($m = 3n$), and the summation can be reduced to

$$\sigma_{zz}(\omega) \approx -i\frac{8\pi e^2}{(2\pi)^2(\omega+2i\Gamma)ma}\left(4\int_0^{2\pi/a}dk_z\frac{\partial f(\varepsilon)}{\partial\varepsilon}v_F^2\right)$$

$$= i\frac{2\sqrt{3}e^2t}{m\pi\hbar^2(\omega+2i\Gamma)} \qquad (24.34)$$

Following the same procedure, the conductivity of armchair carbon nanotubes (Figure 24.6) can be derived. The resulting expression is found to be

$$\sigma_{zz}(\omega) = i\frac{2e^2t}{m\pi\hbar^2(\omega+2i\Gamma)} \qquad (24.35)$$

24.3 COMPUTATIONAL MODELS FOR THz AND OPTICAL REGIMES

Once the electronic transport properties of graphene and carbon nanotubes have been established, the question remains as to how to implement these characteristics in computational models based on Maxwell's equations. For this purpose, two alternative approaches may be taken into consideration; a differential- or an integral-based formulation. Both of these methodologies may be formulated in terms of frequency or time-domain techniques. Then, the resulting equations may be solved by employing several computational procedures that have been developed in the last few decades, with the two most popular being the moment-method solution for integral equations and the finite-difference discretization for differential equations. Other popular techniques, such as finite-element or Monte-Carlo algorithms, can also be successfully applied and present some computational advantages for particular cases. Therefore, surface and volumetric models have been applied to model graphene in FDTD (finite-difference time-domain) considering intraband (Lin et al. 2012a,b; Li et al. 2013; Nayyeri et al. 2013; Wang et al. 2013; Bouzianas et al. 2014), and both intraband and interband responses (Nayyeri et al. 2013), in frequency-domain integral equations (Shapoval et al. 2013), by using spectral domain techniques (Ardakani et al. 2013), by the transverse resonance method (Gomez-Diaz et al. 2013) and circuital models (Correas-Serrano et al. 2013), and by Green's functions (Tamagnone et al. 2014). However, this chapter focuses only on the moment-method and finite-difference algorithms because they tend to be the most computationally efficient while at the same time producing stable and accurate results.

24.3.1 DIFFERENTIAL-EQUATION-BASED FORMULATION

Any transient electromagnetic wave ($\vec{E}(\vec{r},t), \vec{H}(\vec{r},t)$) propagating along a graphene or a carbon nanotube structure obeys the Faraday- and Ampère–Maxwell equations (Taflove 2005), which can be expressed in the form

$$\frac{\partial\vec{H}(\vec{r},t)}{\partial t} = -\frac{1}{\mu}\nabla\times\vec{E}(\vec{r},t) - \frac{\sigma^*}{\mu}\vec{H}(\vec{r},t) \qquad (24.36)$$

$$\frac{\partial\vec{E}(\vec{r},t)}{\partial t} = -\frac{\sigma}{\varepsilon}\vec{E}(\vec{r},t) + \frac{1}{\varepsilon}\nabla\times\vec{H}(\vec{r},t) \qquad (24.37)$$

where the four constitutive parameters $\{\varepsilon, \mu, \sigma, \sigma^*\}$ account for the specific properties of the material. In our case, the two-dimensional character of graphene can be considered by an electric permittivity and magnetic permeability equal to free space, no magnetic losses, and an electric conductivity derived based on the procedures previously discussed (i.e., $\varepsilon_0, \mu_0, \sigma, 0$).

However, numerical simulations based on an FDTD procedure for solving Equations 24.36 and 24.37 cannot be carried out by a direct substitution of conductivities (24.25) and (24.26) for graphene, or their quantized form of (24.34) and (24.35) for carbon nanotubes, because of the stability issues inherent in the numerical algorithm (Taflove 2005). To avoid these undesired instabilities, an equivalent volume conductivity ε_{eq} is defined, assuming a very small thickness of graphene Δ, in the form $\varepsilon_{eq} = \varepsilon_0 + (\sigma/j\omega\Delta)$. Of course, the complex relative permittivity values vary as a function of frequency under different conditions of $\{\mu_c, \Gamma, T\}$ (Figure 24.7), but a time-domain counterpart $\varepsilon_{eq}(t)$ of the frequency-dependent permittivity $\varepsilon_{eq}(\omega)$ is required as a first step in the derivation of a numerical procedure for solving Equations 24.36 and 24.37. A proposed way to achieve this is by formulating a sum of P partial fractions in terms of complex conjugate pole–residue pairs as follows:

$$\varepsilon_{eq} = \varepsilon_0\varepsilon_\infty + \varepsilon_0\sum_{p=1}^{P}\left(\frac{c_p}{j\omega-a_p} - \frac{c_p^*}{j\omega-a_p^*}\right) \qquad (24.38)$$

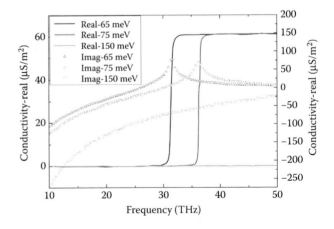

FIGURE 24.7 Surface conductivity of graphene for different values of μ_c. (Lin, H. et al., FDTD modeling of graphene devices using complex conjugate dispersion material model, *IEEE Microwave and Wireless Component Letters*, © 2012 IEEE.)

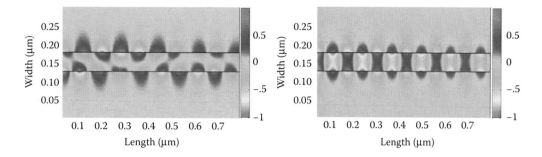

FIGURE 24.8 Electric field of two parallel graphene sheets for symmetric (left) and antisymmetric (right) modes. (Lin, H. et al., FDTD modeling of graphene devices using complex conjugate dispersion material model, *IEEE Microwave and Wireless Component Letters*, © 2012 IEEE.)

where the relative permittivity at infinite frequency ε_∞, the pth pole a_p, and residue c_p, are found by employing heuristic techniques (Haupt 2007). Equation 24.38 presents some desirable numerical properties: (a) it complies with the Kramers–Kronig relationships, which provides it with a physical meaning; (b) it is unconditionally stable because the poles are located in the left complex semi-plane; and (c) it is versatile for modeling intraband or interband responses, because the poles and residuals can describe a Drude or a Lorentz–Drude type of response.

Furthermore, it can be implemented in the time domain by a convolutional or an auxiliary differential equation (ADE) formulation (Han 2006). To this end, the FDTD updating equation for the electric field (the magnetic field-update equation remains unchanged) is

$$\vec{E}^{n+1} = \vec{E}^n + \frac{\Delta t}{\varepsilon_0 \varepsilon_\infty + \sum_{p=1}^{P} \mathrm{Re}\{\beta_p\}}$$

$$\left[\nabla \times \vec{H}^{n+1/2} - \sum_{p=1}^{P} \mathrm{Re}\{(1+k_p)\vec{J}_p^n\} \right] \quad (24.39)$$

where $\vec{J}_p^n$ are auxiliary currents introduced by the complex conjugate pole–residue pairs. These auxiliary currents are updated by employing

$$\vec{J}_p^{n+1} = k_p \vec{J}_p^n + \frac{\beta_p}{\Delta t}(\vec{E}^{n+1} - \vec{E}^n) \quad (24.40)$$

where the updating coefficients k_p and β_p depend on the poles and residues as

$$k_p = \frac{1 + a_p \Delta t/2}{1 - a_p \Delta t/2}$$

$$\beta_p = \frac{\varepsilon_0 c_p \Delta t}{1 - a_p \Delta t/2} \quad (24.41)$$

As an example of results achieved by utilizing this method, simulations for a terahertz parallel plate waveguide formed

by two graphene sheets separated by a distance of 50 nm are shown in Figure 24.8. This figure depicts two examples of propagating modes in the waveguide when a steady-state condition is reached. Thus, the use of numerical codes enables the calculation of propagation parameters, such as the wavelength or the propagation constant, for symmetric (right part) and antisymmetric (left part) modes.

When carbon nanotubes are considered, the former procedure is no longer adequate because the small diameter of the enrollment (roughly on the order of tenths of nanometers or less) will necessitate excessively fine meshes, which would require large supercomputers to carry out practical simulations. To avoid this, it is possible to apply a thin-wire formulation modified to include carbon nanotube structures as illustrated in Figure 24.9. The usual procedure for including thin-wires in FDTD consists of introducing an additional incell inductance $\langle L_{incell} \rangle$ and capacitance $\langle C_{incell} \rangle$ per unit length, both related by the transmission-line equation $\langle C_{incell} \rangle = \varepsilon\mu\langle L_{incell} \rangle$. Considering a carbon nanotube (eventually treated as a thin-wire of radius a and constitutive parameters

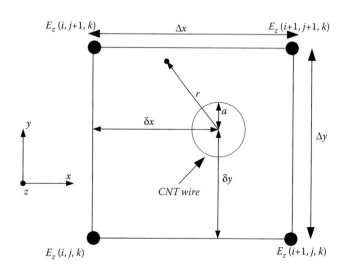

FIGURE 24.9 FDTD model of a cell including a carbon nanotube. (Lin, H. et al., A FDTD thin-wire model for modeling carbon nanotube dipoles at THz regime, *IEEE Antennas and Wireless Propagation Letters*, © 2012 IEEE.)

ε_0, μ_0) placed in a mesh of size $\{\Delta x, \Delta y\}$, the incell inductance may be expressed as (Berenger 2000)

$$\langle L_{incell,A}\rangle = \frac{\mu_0}{4\pi}\left[\ln\frac{\Delta x^2+\Delta y^2}{a^2}+\frac{\Delta y}{\Delta x}\arctan\frac{\Delta x}{\Delta y}+\frac{\Delta x}{\Delta y}\arctan\frac{\Delta y}{\Delta x}+\frac{\pi a^2}{\Delta x\Delta y}-3\right] \quad (24.42)$$

or (Boutayeb 2006)

$$\langle L_{incell,B}\rangle = \begin{cases} \langle L_{incell,A}\rangle - 0.57\frac{\mu_0}{4\pi} & \frac{a}{\min(\Delta x,\Delta y)}<0.3 \\ \langle L_{incell,A}\rangle\frac{\Delta x\Delta y}{\Delta x\Delta y-\pi a^2} & \frac{a}{\min(\Delta x,\Delta y)}\geq 0.3 \end{cases} \quad (24.43)$$

To account for the conductivity of the carbon nanotube, an equivalent circuit model of the conductivity can be employed, which has the form

$$\rho_{CNT}=\frac{1}{\sigma_{CNT}}=R+sL_K+\frac{1}{sC_q} \quad (24.44)$$

where R, L_K, and C_q are, respectively, the resistance, kinetic inductance, and quantum capacitance per unit length. In this way, the thin-wire model for CNTs reduces to

$$(\langle L_{incell}\rangle+L_k)\frac{\partial I_z}{\partial t}+RI_z+\frac{1}{\langle C_{incell}\rangle+C_q}\frac{\partial Q_z}{\partial z}=\langle E_z\rangle \quad (24.45)$$

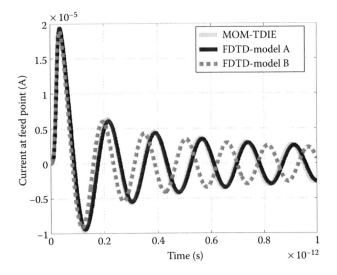

FIGURE 24.10 Current at center of a carbon nanotube dipole of length 20 μm. (Lin, H. et al., A FDTD thin-wire model for modeling carbon nanotube dipoles at THz regime, *IEEE Antennas and Wireless Propagation Letters*, © 2012 IEEE.)

This expression may be easily implemented into the FDTD update scheme by discretizing the current I_z and charge Q_z per unit length following the classical procedure described in Holland (1981).

Results of this method are presented for a dipole of length 20 m and radius 2.712 nm, fed with a Gaussian voltage source at its center at frequencies up to 1 THz. Figure 24.10 plots the time-domain current at the center of the antenna. The propagation of a surface plasmon resonance on the carbon nanotube can be observed, with a low-frequency resonance mostly related to the kinetic inductance, and a small amplitude of the current due to the large resistance of the ballistic transport of carriers. Also, it becomes apparent that the inductance Equation 24.42 provides results consistent with a formulation based on the method-of-moments.

24.3.2 INTEGRAL EQUATION-BASED FORMULATION

Maxwell's equations can be cast in the form of integral equations, such as the electric field integral equation (EFIE), the magnetic field integral equation (MFIE), or the combined field integral equation (CFIE). Each of them can be classified in different ways: based on the dimensionality of the integrals (volumetric, surface, or linear integral equations), in terms of the domain (frequency or time), or depending on the specific type of unknowns in the equation (current or current–charge integral equations) (Volakis 2012). The usual criteria employed to determine which type of integral equation is most appropriate are related to the nature of the problem. For graphene sheets a surface-based integral equation could be employed, while for carbon nanotubes a linear version could be thought of as the most appropriate form (and preferably taking into account that required computational resources are related to the dimensionality of the problem). For the domain of solution, frequency-domain techniques are computationally advantageous when single-frequency electromagnetic sources, or at least narrowband sources, are considered. However, more information for the analysis of physical phenomena is achieved when time-domain methods are utilized, which are typically more useful for emerging electromagnetic/optical technologies, which employ dispersive materials. Regarding the unknowns, the usual choice is the electric current or the current density, which is primarily due to better stability and accuracy of the simulations. Finally, the specific choice of the integral equation depends on the geometry of the problem, that is, closed or open geometries. Taking into account that the actual manufacturing processes are fairly mature for producing relatively simple graphene sheets or carbon nanotubes, which can be thought of as open surfaces, the application of the EFIE is justified.

Therefore, a surface EFIE that takes into account the finite conductivity of graphene can be formulated by applying appropriate surface impedance boundary conditions (Yuferev 2010). Following the equivalent model for graphene as a volumetric conductivity ε_{eq}, a graphene layer acts like a thin metal film when $\text{Re}\{\varepsilon_{eq}\}<0$ and electromagnetic waves at any point $\vec{r}$ on the surface of graphene can be represented as

$$\hat{n} \times \vec{E}^i(\vec{r}) = \hat{n} \times \left(j\omega\mu_0 \int_S \vec{J}_s(\vec{r}')G(\vec{r},\vec{r}')dS' - \right.$$

$$\left. \frac{1}{j\omega\varepsilon_0}\nabla\int_S \nabla'_s \cdot \vec{J}_s(\vec{r}')G(\vec{r},\vec{r}')dS' + \frac{\vec{J}_s(\vec{r})}{\sigma(\vec{r})} \right) \quad (24.46)$$

where $\hat{n}$ denotes the normal vector to the surface of the graphene layer, $G(\vec{r},\vec{r}') = e^{-ik|\vec{r}-\vec{r}'|}/|\vec{r}-\vec{r}'|$ is the Green's function connecting the source point $\vec{r}'$ with the field point $\vec{r}$, $\vec{E}^i(\vec{r})$ is the incident field that can be external for scattering problems or internal for radiation problems, $\vec{J}_s(\vec{r}')$ is the unknown current density ($\nabla'_s \cdot$ signifies the surface divergence with respect to the primed coordinates), and $\sigma(\vec{r})$ is the conductivity that can be tuned locally by changing the chemical potential. The ability to tune the conductivity of graphene layers opens up the doors to a wide range of possible electromagnetic devices such as surface waveguides (Vakil and Engheta 2011), optoelectronic devices (Bao 2012), frequency multipliers (Wang 2009), or high-frequency transistors (Lin 2009). Solutions of Equation 24.46 can be achieved by a method-of-moments Galerkin procedure using Rao–Wilton–Glisson basis functions (as long as the graphene remains in planar layers then higher order basis functions will not increase the solution accuracy) (Volakis 2012).

The time-domain counterpart of Equation 24.46 can be derived using the inverse Fourier transform

$$\hat{n} \times \vec{E}^i(\vec{r},t) = \hat{n} \times \left(\frac{\mu_0}{4\pi}\int_S \left[\frac{1}{R}\frac{\partial}{\partial t}\vec{J}_s\left(\vec{r}',t-\frac{R}{c}\right) \right]dS' - \right.$$

$$\frac{1}{4\pi\varepsilon_0}\int_S \left[\frac{1}{R}\int_0^{t-(R/c)} (\nabla'_s \cdot \vec{J}_s(\vec{r}',\tau))d\tau \right]dS' \right)$$

$$\left. + \hat{n} \times \left(L^{-1}\left\{ \frac{\vec{J}_s(\vec{r})}{\sigma(\vec{r})} \right\} \right) \quad (24.47) \right.$$

where $R = |\vec{r}-\vec{r}'|$ is the distance between source and field points, c is the velocity of light, L^{-1} represents the inverse Laplace transform, and $t-(R/c)$ is the retarded time that provides causality to the electromagnetic wave propagation. Similar to the case with differential formulations in the frequency-domain, stability issues also appear when time-domain solutions are invoked. For this reason, the term including conductivity in Equation 24.47 is modeled as an expansion of Lorentz–Drude series to avoid instabilities. Again, the dispersive permittivity can be represented as intraband and interband equivalent conductivities (σ^{ib} and σ^{ib}, respectively), from the generic formula

$$\sigma = \sigma^{ib} + \sigma^{eb} = \frac{f_0\omega_p^2}{s+\Gamma_0} + \sum_{j=1}^{K} \frac{s\varepsilon_0 f_j\omega_p^2}{s^2+s\Gamma_j+\omega_j^2} \quad (24.48)$$

where the intraband (Drude) term contains ω_p corresponding to the plasma frequency associated with the graphene layer. Also, intraband transitions are characterized by an oscillator strength f_0 and a damping constant Γ_0. The term corresponding to the interband contribution obeys a simple semiquantum model expressed in a Lorentz form, where K represents the number of oscillators needed to achieve a reasonable fit to the analytical conductivity. Each one of these oscillators is described by three parameters corresponding to their frequency ω_j, strength f_j, and damping constant Γ_j. Figure 24.11 presents a comparison of the contributions between typical interband and intraband responses, which shows how the intraband response predominates as the frequency gets closer to the terahertz regime while the optical response is governed only by the interband conductivity.

To achieve a numerically efficient procedure, Equation 24.48 can be rewritten in terms of a circuit equivalent model by defining a set of inductances L_j, capacitances C_j, and resistances R_j, such that

$$L_j = \frac{1}{\varepsilon_0 f_j\omega_p^2}$$

$$C_j = \frac{1}{\omega_j^2 L_j} \quad (24.49)$$

$$R_j = \Gamma_j L_j$$

which leads to

$$\sigma = \frac{1}{R_0 + sL_0} + \sum_{j=1}^{K} \frac{1}{R_j + sL_j + (1/sC_j)} \quad (24.50)$$

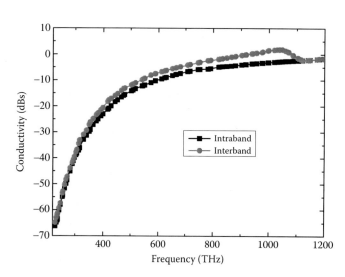

FIGURE 24.11 Comparison of interband and intraband conductivities in the optical regime.

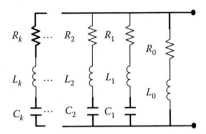

FIGURE 24.12 An equivalent circuit representation of the Lorentz–Drude model.

Figure 24.12 shows a schematic representation of the circuit model, which corresponds to the equivalent conductivity σ of the graphene sheet. Numerical models require one *RL* circuit to represent the Drude term and at least eight *RLC* circuits to represent the Lorentzian responses. It should be noted that the equivalent circuit model is remarkably efficient and does not add a significant computational burden to the solution, as well as provides some physical insight through an inspection of the values corresponding to the specific circuit elements. What is more, it can be implemented to avoid the employment of any numerical inverse Laplace transform in the code, which may suffer either from inaccuracies, as a result of a numerical truncation of the time-domain response, or from extremely poor computational performance if the complete response is considered, because of the large number of calculations required by the convolution operator. By employing the *RLC* equivalent circuit model (Pantoja et al. 2012), the contribution of each term can be carried out numerically by a trapezoidal integration or a finite difference approximation of the equivalent circuit response.

Electromagnetic scattering or radiation problems involving carbon nanotubes can also be solved based on an EFIE. Carbon nanotubes usually have a reduced radius/length ratio and, at those frequencies in which the axial current is much larger than its azimuthal counterpart, they can be effectively modeled using a thin-wire approximation. This thin-wire approach can also be simplified for achieving high computational efficiency by considering a particular case of the exact Green's function usually referred to as the approximate or reduced kernel. This simplification takes advantage of the cylindrical symmetry of the sources and avoids the singularities that arise in the general case, by treating the total current as a filament on the wire axis and enforcing the boundary condition on the wire surface. The use of this approach leads to a modified frequency-domain form of Pocklington's EFIE in a vacuum (Harrington 1993)

$$\hat{n} \times \vec{E}^i(\vec{r}) = \hat{n} \times \left[\frac{1}{\sigma(\vec{r})} \vec{I}(\vec{r}) + \frac{i}{4\pi\varepsilon_0\omega} \int_{C'} \frac{\omega^2}{c^2} \frac{e^{-jkR}}{R} \vec{I}(\vec{r}')ds' - \right.$$

$$\left. \frac{i}{4\pi\varepsilon_0\omega} \int_{C'} I(\vec{r}') \nabla \frac{\partial}{\partial s'} \left(\frac{e^{-jkR}}{R} \right) ds' \right]$$

(24.51)

while the corresponding time-domain EFIE is given by (Miller 1980)

$$\hat{n} \times \vec{E}^i(\vec{r},t) = \hat{n} \times \left[L^{-1} \left\{ \frac{1}{\sigma(\vec{r})} \vec{I}(\vec{r}) \right\} + \frac{1}{4\pi\varepsilon_0} \int_{C'} \frac{1}{c^2 R} \frac{\partial}{\partial t} \vec{I}(\vec{r}',t')ds' \right.$$

$$+ \frac{1}{4\pi\varepsilon_0} \int_{C'} \frac{\vec{R}}{R^3} \left(\int_0^{t'} \frac{\partial}{\partial r'} I(\vec{r}',\tau)d\tau \right) ds' \right]$$

$$\left. + \hat{n} \times \left[\frac{1}{4\pi\varepsilon_0} \int_{C'} \frac{\vec{R}}{cR^2} \frac{\partial}{\partial r'} I(\vec{r}',t')ds' \right] \right.$$

(24.52)

where $\vec{I}$ represents the unknown current along the arclength C' of the thin-wire embedded in a vacuum, and $t' = t - (R/c)$ accounts for the retarded time between the source and field point. It bears remarking that Equations 24.51 and 24.52 are not valid in the upper part of the visible spectrum, where the axial current no longer dominates because the skin effect begins to become appreciable (Hanson 2006). For such cases, simulations of carbon nanotubes can be made by employing Equations 24.46 and 24.47.

Results are presented for a carbon nanotube-based dipole, modeled in the terahertz regime with an *RL* equivalent circuit corresponding to the Drude model of the intraband conductivity. The total length L of the dipole is 20 μm, with a wire radius of 2.712 nm. Figure 24.13 displays the results for the input impedance over a range of frequencies up to 1 THz. These results demonstrate good agreement between the frequency-domain and time-domain EFIE solutions, and they are consistent with other published results (Hanson 2005). Taking into account the length of the dipole, these results confirm the existence of resonances at lower frequencies in carbon nanotube dipoles compared to their perfect electric

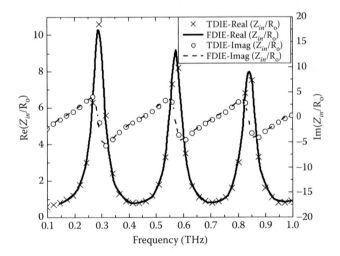

FIGURE 24.13 Input impedance of a carbon nanotube dipole of length 20 μm. (Pantoja, M. F. et al., TDIE modeling of carbon nanotubes dipoles at microwave and terahertz bands, *IEEE Antennas and Wireless Propagation Letters*, © 2010 IEEE.)

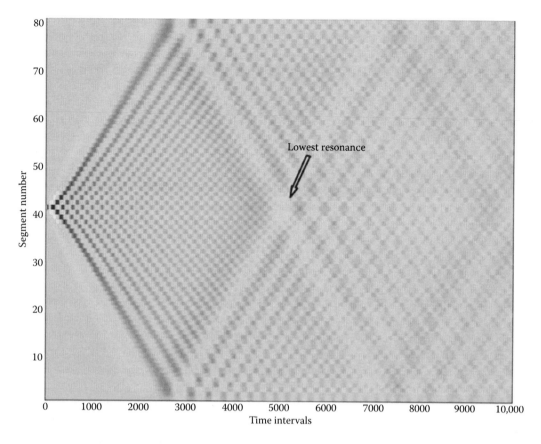

FIGURE 24.14 Space–time diagram of currents in a carbon nanotube dipole of length 20 μm. (Pantoja, M. F. et al., TDIE modeling of carbon nanotubes dipoles at microwave and terahertz bands. *IEEE Antennas and Wireless Propagation Letters,* © 2010 IEEE.)

conductor (PEC) counterparts of the same length, assuming that they could exist in this range of frequencies. The reason for this fact is found in the slower velocity of the propagation of waves for the carbon nanotubes compared to the PEC wire. This effect can be analyzed by depicting a space–time representation of the currents along the antenna. Figure 24.14 shows the currents along the carbon nanotube dipole as a function of temporal (X axis) and length (Y axis) intervals, where the formation of lower frequency resonances can be observed, mainly due to the inductive propagation of carriers at these frequencies. The electric current reaches the ends of the antenna, forming a traveling wave, which returns to the feeding point (at ~5000 time intervals), corresponding to the lowest resonant frequency of 0.24 THz shown in Figure 24.14.

24.4 CONCLUSIONS

This chapter describes a highly efficient and accurate numerical technique for simulating electromagnetic wave interactions with graphene and carbon nanotubes in the terahertz, near-infrared, and optical regimes. The methods presented are intended for implementation in research and commercial software, and for this reason efficient formulations were developed based on equivalent constitutive parameters. The chapter begins by describing the electronic properties of

graphene with a semiclassical procedure of quantum mechanics, known as first quantization, in which carbon atoms are considered by using quantum wave functions and the electromagnetic field is treated classically. As a result, isotropic conductivity models that include interband and intraband effects were presented, which cover both the terahertz and optical regimes. The second part of the chapter introduces a methodology to account for graphene materials in two of the most widely employed algorithms in computational electromagnetics, the finite-difference time-domain, and the method-of-moments formulations. Special attention has been paid to include the complementary time-domain and frequency-domain formulations, because both of them can be useful for simulating complex environments or for modeling the unusual physical properties exhibited by graphene and carbon nanotubes. Results have been included, which illustrate both time-domain and frequency-domain responses in the terahertz, near-infrared, and optical regimes, paving the way for the exploration of a new generation of micro- and nanoscale electromagnetic devices.

REFERENCES

Abadal, S., Alarcón, E., Cabellos-Aparicio, A., Lemme, M., Nemirovsky, M. 2013. Graphene-enabled wireless communication for massive multicore architectures, *IEEE Communications Magazine* 51:137–143.

Amanatiadis, S.A., Karamanos, T.D., Kantartzis, N.V. 2014. Surface plasmon polariton waves onto graphenes surface over an anisotropic metamaterial substrate, *Proceedings of SPIE—The International Society for Optical Engineering* 9125:91250H.

Andryieuski, A., Lavrinenko, A.V. 2013. Graphene metamaterials based tunable terahertz absorber: Effective surface conductivity approach, *Optics Express* 21:9144–9155.

Ardakani, H.H., Kashani, Z.G., Amirkalaee, M.K., Rashed-Mohassel, J. 2013. Fourier transform analysis of graphene-based multilayer structures, *IET Microwaves, Antennas and Propagation* 7:1084–1091.

Bao, Q., Loh, K. 2012. Graphene photonics, plasmonics, and broadband optoelectronic devices, *ACS Nano* 6:3677–3694.

Berenger, J.P. 2000. A multiwire formalism for the FDTD method, *IEEE Transactions on Electromagnetic Compatibility* 42:257–264.

Boutayeb, H. 2006. *Numerical Methods in Electromagnetism: FDTD. Part I.* Available online: http://www.creer.polymtl.ca/Halim_Boutayeb/chapter2FDTD.ppt.

Bouzianas, G.D., Kantartzis, N.V., Yioultsis, T.V., Tsiboukis, T.D. 2014. Consistent study of graphene structures through the direct incorporation of surface conductivity, *IEEE Transactions on Magnetics* 50:6749110.

Bransden, B.H., Joachain, C.J. 2000. *Quantum Mechanics*, 2nd Ed., United Kingdom: Addison-Wesley.

Castro Neto, A., Guinea, F., Peres, N.M.R., Novoselov, K.S., Geim, A.K. 2009. The electronic properties of graphene, *Review of Modern Physics* 81:109–162.

Charlier, J.C., Blasé, X., Roche, S. 2007. Electronic and transport properties of nanotubes, *Review of Modern Physics* 79:677–732.

Chen, P.-Y., Alù, A. 2013. A terahertz photomixer based on plasmonic nanoantennas coupled to a graphene emitter, *Nanotechnology* 24:455202.

Correas-Serrano, D., Gomez-Diaz, J.S., Alvarez-Melcon, A. 2014. On the influence of spatial dispersion on the performance of graphene-based plasmonic devices, *IEEE Antennas and Wireless Propagation Letters* 13:345–348.

Correas-Serrano, D., Gomez-Diaz, J.S., Perruisseau-Carrier, J., Alvarez-Melcon, A. 2013. Spatially dispersive graphene single and parallel plate waveguides: Analysis and circuit model, *IEEE Transactions on Microwave Theory and Techniques* 61:4333–4344.

Farhat, M., Guenneau, S., Bagci, H. 2013. Exciting graphene surface plasmon polaritons through light and sound interplay, *Physical Review Letters* 111:237404.

Filter, R., Farhat, M., Steglich, M., Alaee, R., Rockstuhl, C., Lederer, F. 2013. Tunable graphene antennas for selective enhancement of THz-emission, *Optics Express* 21:3737–3745.

Gomez-Díaz, J.S., Esquius-Morote, M., Perruisseau-Carrier, J. 2013. Plane wave excitation-detection of non-resonant plasmons along finite-width graphene strips, *Optics Express* 21:24856–24872.

Gomez-Diaz, J.S., Mosig, J.R., Perruisseau-Carrier, J. 2013. Effect of spatial dispersion on surface waves propagating along graphene sheets, *IEEE Transactions on Antennas and Propagation* 61:3589–3596.

Gomez-Díaz, J.S., Perruisseau-Carrier, J. 2012. Propagation of hybrid transverse magnetic-transverse electric plasmons on magnetically biased graphene sheets, *Journal of Applied Physics* 112:124906.

Gusynin, V.P., Sharapov, S.G., Carbotte, J.P. 2007. Magneto-optical conductivity in graphene, *Journal of Physics: Condensed Matter* 19:026222–026225.

Han, M., Dutton, R.W., Fan, S. 2006. Model dispersive media in finite-difference time-domain method with complex-conjugate pole-residue pairs, *IEEE Microwave and Wireless Components Letters* 16:119–121.

Hanson, G. 2005. Fundamental transmitting properties of carbon nanotube antennas, *IEEE Transactions on Antennas and Propagation* 53:3426–3435.

Hanson, G. 2006. On the applicability of the surface impedance integral equation for optical and near infrared copper dipole antennas, *IEEE Transactions on Antennas and Propagation* 54:3677–3685.

Hanson, G. 2008. Dyadic Green's functions and guided surface waves for a surface conductivity model of graphene, *Journal of Applied Physics* 103:064302–064308.

Harrington, R.F. 1993. *Field Computation by Moment Methods*, New Jersey, USA: IEEE Press.

Hartmann, R.R., Kono, J., Portnoi, M.E. 2014. Terahertz science and technology of carbon nanomaterials, *Nanotechnology* 25:322001.

Haupt, R., Werner, D.H. 2007. *Genetic Algorithms in Electromagnetics*, New Jersey, USA: Wiley-Interscience.

Holland, R., Simpson, L. 1981. Finite-difference analysis of EMP coupling to thin struts and wires, *IEEE Transactions on Electromagnetic Compatibility* 23:88–97.

Kittel, C. 2004. *Introduction to Solid State Physics*, 8th Ed., New Jersey, USA: Wiley.

Kubo, R. 1956. A general expression for the conductivity tensor, *Canadian Journal of Physics* 34:1274–1277.

Li, H., Wang, L., Huang, Z., Sun, B., Zhai, X., Li, X. 2013. Mid-infrared, plasmonic switches and directional couplers induced by graphene sheets coupling system, *Europhysics Letters* 104:37001.

Li, H.-J., Wang, L.-L., Huang, Z.-R., Sun, B., Zhai, X., Li, X.-F. 2014a. Simulations of multi-functional optical devices based on a sharp 90° bending graphene parallel pair, *Journal of Optics (United Kingdom)* 16:015004.

Li, H.-J., Wang, L.-L., Zhang, H., Huang, Z.-R., Sun, B., Zhai, X., Wen, S.-C. 2014b. Graphene-based mid-infrared, tunable, electrically controlled plasmonic filter, *Applied Physics Express* 7:024301.

Lin, H., Pantoja, M.F., Angulo, L. Alvarez, J., Martin, R.G., Garcia, S.G. 2012a. FDTD modeling of graphene devices using complex conjugate dispersion material model, *IEEE Microwave and Wireless Component Letters* 22:612–614.

Lin, H., Pantoja, M.F., Garcia, S.G., Bretones, A.R., Martin, R.G. 2012b. A FDTD thin-wire model for modeling carbon nanotube dipoles at THz regime, *IEEE Antennas and Wireless Propagation Letters* 11:708–711.

Lin, Y.M., Jenkins, K.A., Valdes-Garcia, A., Small, J.P., Farmer, D.B., Avouris, P. 2009. Operation of graphene transistors at gigahertz frequencies, *Nano Letters* 9:422–426.

Luo, X., Qiu, T., Lu, W., Ni, Z. 2013. Plasmons in graphene: Recent progress and applications, *Materials Science and Engineering R: Reports* 74:351–376.

Maggiore, M. 2005. *A Modern Introduction to Quantum Field Theory*, Oxford: Oxford University Press.

Maksimenko, S.A., Slepyan, G.Y., Lakhtakia, A., Yevtushenko, O., Gusakov, A.V. 1999. Electrodynamics of carbon nanotubes: Dynamic conductivity, impedance boundary conditions, and surface wave propagation, *Physical Review B* 60:17136–17149.

Miller, E.K. 1980. Direct time-domain techniques for transient radiation and scattering from wires, *Proceedings of the IEEE* 68:1396–1423.

Nayyeri, V., Soleimani, M., Ramahi, O.M. 2013. Modeling graphene in the finite-difference time-domain method using a surface boundary condition, *IEEE Transactions on Antennas and Propagation* 61:4176–4182.

Nayyeri, V., Soleimani, M., Ramahi, O.M. 2013. Wideband modeling of graphene using the finite-difference time-domain method, *IEEE Transactions on Antennas and Propagation* 61:6107–6114.

Novoselov, K.S., Geim, A.K., Morozov, S.V., Jiang, D., Zhang, Y., Dubonos, S.V., Grigorieva, I.V., Firsov, A.A. 2004. Electric field effect in atomically thin carbon films, *Science* 306:666–669.

Othman, M.A.K., Guclu, C., Capolino, F. 2013. Graphene-dielectric composite metamaterials: Evolution from elliptic to hyperbolic wave vector dispersion and the transverse epsilon-near-zero condition, *Journal of Nanophotonics* 7:13008.

Pantoja, M.F., Bray, M.G., Werner, D.H., Werner, P.L., Bretones, A.R. 2012. A computationally efficient method for simulating metal-nanowire dipole antennas at infrared and longer visible wavelengths, *IEEE Transactions on Nanotechnology* 11:239–246.

Pantoja, M.F., Werner, D.H., Werner, P.L., Bretones, A.R. 2010. TDIE modeling of carbon nanotubes dipoles at microwave and terahertz bands, *IEEE Antennas and Wireless Propagation Letters* 9:32–35.

Sadiku, M. 2010. *Numerical Techniques in Electromagnetics*, 2nd Ed., New Jersey, USA: CRC Press.

Saito, R., Dresselhaus, G., Dresselhaus, M.S. 2003. *Physical Properties of Carbon Nanotubes*, London, U.K.: Imperial College Press.

Shapoval, O.V., Gomez-Diaz, J.S., Perruisseau-Carrier, J., Mosig, J.R., Nosich, A.I. 2013. Integral equation analysis of plane wave scattering by coplanar graphene-strip gratings in the THz range, *IEEE Transactions on Terahertz Science and Technology* 3:666–674.

Sutton, A.P. 1993. *Electronic Structure of Materials*, Oxford, U.K.: Clarendon.

Taflove, A., Hagness, S.C. 2005. *Computational Electrodynamics: The Finite-Difference Time-Domain Method*, 3rd Ed., Massachussets, USA: Artech House Publishers.

Tamagnone, M., Fallahi, A., Mosig, J.R., Perruisseau-Carrier, J. 2014. Fundamental limits and near-optimal design of graphene modulators and non-reciprocal devices, *Nature Photonics* 8:556–563.

Vakil, A., Engheta, N. 2011. Transformation optics using graphene, *Science* 332(6035):1291–1294.

Volakis, J., Serkel, K. 2012. *Integral Equation Methods for Electromagnetics*, North Caroline, USA: SciTech Publishing.

Wallace, P.R. 1947. The band theory of graphite, *Physical Review* 71:622–634.

Wang, B., Zhang, X., Yuan, X., Teng, J. 2012. Optical coupling of surface plasmons between graphene sheets, *Applied Physics Letters* 100(13):131111–131114.

Wang, G., Liu, X., Lu, H., Zeng, C. 2014. Graphene plasmonic lens for manipulating energy flow, *Scientific Reports* 4:4073.

Wang, H., Nezich, D., Kong, J., Palacios, T. 2009. Graphene frequency multipliers, *IEEE Electron Device Letters* 30: 547–549.

Wang, X.-H., Yin, W.-Y., Chen, Z. 2013. Matrix exponential FDTD modeling of magnetized graphene sheet, *IEEE Antennas and Wireless Propagation Letters* 12:1129–1132.

Wang, X.-H., Yin, W.-Y., Chen, Z.Z. 2013. One-step leapfrog ADI-FDTD method for simulating electromagnetic wave propagation in general dispersive media, *Optics Express* 21:20565–20576.

Xu, B.-Z., Gu, C.-Q., Li, Z., Niu, Z.-Y. 2013. A novel structure for tunable terahertz absorber based on graphene, *Optics Express* 21:23803–23811.

Yuferev, S., Ida, N. 2010. *Surface Impedance Boundary Conditions: A Comprehensive Approach*, Florida, USA: CRC Press.

Zhang, C., Chen, L., Ma, Z. 2008. Orientation dependence of the optical spectra in graphene at high frequencies, *Physical Review B* 77:241402–241404.

Zhou, T., Cheng, Z., Zhang, H., Le Berre, M., Militaru, L., Calmon, F. 2014. Miniaturized tunable terahertz antenna based on graphene, *Microwave and Optical Technology Letters* 56:1792–1794.

Zhu, Z., Joshi, S., Moddel, G. 2014. High performance room temperature rectenna IR detectors using graphene geometric diodes, *IEEE Journal on Selected Topics in Quantum Electronics* 20:6799997.

25 Design and Properties of Graphene-Based Three Dimensional Architectures

Chunfang Feng, Ludovic F. Dumée, Li He, Zhifeng Yi,
Zheng Peng, and Lingxue Kong

CONTENTS

Abstract..375
25.1 Introduction 375
25.2 Carbon-Based Single-Phase 3D Architectures..376
    25.2.1 Graphene Foams...376
        25.2.1.1 Growing Graphene Foams with CVD 379
        25.2.1.2 Formation of Graphene Foams from GO...380
    25.2.2 Carbon Nanotube Webs, Bucky-Papers, and Yarns 382
25.3 Graphene-Based 3D Composites..384
    25.3.1 Graphene and CNTs 384
        25.3.1.1 *In Situ* Growth of Graphene and CNTs 384
        25.3.1.2 Solution Mixing of CNTs and Graphene...388
        25.3.1.3 Layer-by-Layer Self-Assembly of Graphene and CNTs 388
    25.3.2 Graphene and Metals/Metal Oxide Nanoparticles 389
        25.3.2.1 *In Situ* Synthesis of Metal/Metal Oxide Nanoparticles on Graphene 389
        25.3.2.2 Electrostatic Self-Assembly of Graphene and Metal/Metal Oxide Nanoparticles...........................389
    25.3.3 Graphene and Polymers..390
        25.3.3.1 Graphene-Based Polymer Nanowires 392
        25.3.3.2 Graphene-Based Polymer Hydrogels 392
        25.3.3.3 Graphene-Based Polymer Template Remove..393
25.4 Carbon Nanotube-Based 3D Composites 393
25.5 Conclusions..396
References 396

ABSTRACT

This chapter presents a comprehensive overview on the most recent advances in the formation of graphene-based three-dimensional (3D) architectures. Composite 3D graphene materials were combined with other nanomaterials such as carbon nanotubes or metal/metal oxide nanoparticles and incorporated across organic or inorganic matrices to produce high-performance materials of highly diverse structures and morphologies applicable to an extended range of potential applications. The fabrication routes, structures, and main properties of the 3D architectures, including electrochemical properties, thermal stability, and mechanical properties will be presented and critically analyzed. The potential of graphene-based 3D architectures in strategic areas such as electrodes, sensors, energy storage, catalysts, and biomedical applications will also be discussed.

25.1 INTRODUCTION

Graphene-based materials, as novel two-dimensional (2D) nanomaterial networks [1], have attracted tremendous interest since the early 1990s for their outstanding electric and thermal conductivity, high mechanical properties, and nearly frictionless surface properties [2–4], providing new insights into corrosion control, separation, mechanical engineering, and electronics miniaturization [1–9]. The superior electrochemical properties of graphene originate from the hexagonal crystalline sp^2 carbon structure and offer outstanding electrons and phonons transport capacities up to $15,000\ cm^2\ V^{-1}\ s^{-1}$ [1] at room temperature while being theoretically 100 times mechanically stronger than steel [10].

The fabrication of purely graphene-based electrochemical devices solely from graphene sheets is a difficult task due to issues arising from the agglomeration of graphene platelets during processing of graphene suspensions [11] and typically leads to lower-than-predicted properties to the lack of covalent bonds between individual sheets [12]. Therefore, in order to expand the applications of graphene into other technological frontiers, continuous 3D graphene structures have to be designed [13–17]. To date, strategies to fabricate 3D graphene architectures have been successfully demonstrated (Table 25.1). Among these architectures, 3D graphene structures consisting of a single continuous graphene phase have been successfully synthesized through chemical vapor deposition (CVD) or reduction of graphene oxide (GO). These routes lead to the formation of pure graphene materials, such as graphene foams (GF) [13,18–24]. Carbon nanotubes (CNTs), on the other hand, which are made of rolled up graphene planes, share most of the outstanding properties of pure graphene sheets. The preparation of 3D self-supporting CNT macro-structures, such as CNT yarns, webs, or bucky-papers (BPs) has also been the focus of intense research to fully take advantage of the natural properties of these carbon-based materials [12,25–30]. Carbon-based CNT yarns, webs, and BPs 3D structures have, for instance, been used as actuators [31,32], electrical wires and cables [33], membranes or reinforcements in conductive textile composites [34].

The development of 3D graphene architectures also offers an outstanding opportunity to improve current interfaces between electrically conductive and mechanically reinforcing phases within composite structures [2,4,35,36]. An alternative route to generate 3D graphene-based composites is to assemble graphene or GO sheets with CNTs to form 3D macro-structures by *in situ* growth, self-assembly or filtration solution. This process can be easily scaled-up to form lightweight functional materials, which are mechanically robust and highly electrically conductive. The porosity, thickness, and density of these materials may also be easily controlled by varying the morphology of the graphene nanomaterial building blocks [9,12,37–42]. 3D graphene and CNT hybrid electrodes were used, for instance, as sensors to detect dopamine with a high sensitivity of 471 mA/Mcm^2 [24], among the highest sensitivity thresholds ever produced. It was also demonstrated that graphene and CNT materials were promising 3D structures for solar cells [43,44], fuel cells [39,45], and flexible conductors [14,46].

Graphene-based 3D composite nanostructures were also prepared through integration of minute amounts of metal or metal oxide nanoparticles (NPs) at the interface between the graphene nano-building blocks. These hybrid graphene composites were synthesized through doping, *in situ* growth or electrostatic adsorption, to effectively enhance the specific surface area, conductivity, and mechanical stability of the incorporated graphene sheets [21,47]. The adsorption of metal or metal oxide NPs on the surface of graphene sheets can prevent their aggregation and form conductive and cohesive interfaces between graphene sheets [48]. The potential of the 3D graphene hybrid architectures were also evaluated for sensing and as electrodes integrated across super-capacitors and exhibited [17,22,41]. The electrochemical performance of the hybrid composites were shown to be higher

than individual graphene and the inorganic matrix due to the higher specific surface area of graphene-based hybrids. The theoretical capacitances of the pure graphite and graphene are 372 and 600 mAh/g, respectively [36]. Graphene sheets, when combined with metal oxide NPs, offer enhanced electrical capacitances shown to reach as high as 1000 mAh/g opening new routes in sustainable energy storage [49]. The improvement of the electrical capacitance is attributed to the special 3D structures of the graphene, which offer higher specific surface area and better interfaces between the sheets leading to higher electrical conductivity. The large specific surface area of graphene (up to 2630 m^2/g) and the excellent conductivity are also very promising for electrochemical sensing applications [36].

Graphene architectures have also been incorporated into polymer matrices in order to improve the electrochemical properties, flexibility, and thermal stability of the bare polymers [50,51]. Connectivity between graphene nanomaterials or between graphene and their reinforcing additives or surrounding matrix are challenging issues that must be solved in order to fully benefit from the natural pure graphene properties [52]. Building organized 3D architectures by incorporating graphene into functional polymers has also been a frontier as 3D graphene-based polymer skeleton provides an ideal platform toward strong mechanical strengths, superior thermal stability and flexibility, and fast electron transport kinetics [3,7,35,51–54]. High graphene loadings within these composite materials have led to very promising structures in electronics devices, energy storage, sensors, catalysts, biomedical applications, water purification, and composite reinforcement [3,7,35,51–54].

In this chapter, the different fabrication routes to prepare 3D graphene-based architectures will be presented. The properties of these novel meta-materials either made of a unique graphene phase or composed of graphene nano- or macro-particles in a composite will be described and the prospects of their potential in day-to-day applications be critically discussed in light of recent scientific developments.

25.2 CARBON-BASED SINGLE-PHASE 3D ARCHITECTURES

In this section, the most recent progress in graphene-based 3D structures including GF, graphene hydrogels, porous graphene composites and CNT webs, bucky-papers, and yarns have been generally summarized in Figure 25.1. The methods to fabricate 3D graphene architectures such as *in situ* growth, freeze drying, spinning, vacuum filtration, solution mixing, self-assembly, and template removal has been extensively studied and properties related to the structures are briefly introduced.

25.2.1 GRAPHENE FOAMS

So far, single-phase graphene 3D architectures have been obtained either by direct *in situ* growth of graphene [18] or

TABLE 25.1

Fabrication Routes and Applications of Graphene-Based Architectures

Dimensionality	Ingredients and Structures		Fabrication Routes	Matrix/Substrate Incorporation	Applications		References
					Products	Properties	
2D	Graphene sheets		CVD	PDMS stamp	Stretchable electrodes	Tensile strain: 30%; electron mobility: 3,700 cm^2 V^{-1} s^{-1}	[55]
				Dry transfer-printing on a PET substrate	Transparent electrodes	Sheet resistance: 30 Ω/square; 90% optical transmittance	[56]
				Transferred to Si/SiO$_2$ coated with a Pd film	Hydrogen sensor	When graphene sensors respond to 0.05% (500 ppm) hydrogen, the response time is 213 s.	[57]
			Physical exfoliation	0.07 wt% graphene melt mixed with PET	Reinforced composites	A 40% increase in strength and a 13% increase in modulus.	[58]
			Chemical exfoliation and reduction	Spin-coating GO thin-films on quartz	Transparent electrodes	Sheet resistance: 100–1000 Ω/square; 80% transmittance for 550 nm light	[59]
				Micropatterns of GO solution on PET and reduction	Sensor	Detecting the presence and dynamic cellular secretion of biomolecules	[60]
Single phase 3D structures	Graphene foams		CVD	Fixed onto a glass slide, silver painting as an electrical lead and insulated with rubber	Electrochemical Sensing	Sensitivity: 619.6 μA mM^{-1} cm^{-2}	[18]
			Vacuum filtration	Spread out onto a PET substrate	Supercapacitor	The calculated specific capacitance is 110 F g^{-1} for a two-electrode cell.	[22]
			Freeze drying	N/A	Electrodes	The specific capacitance is 128 F g^{-1}.	[61]
	CNT	Webs	Spin CNT forests and align	Placed directly onto the device substrate	Solar cells	A fill factor of 0.47 and power conversion efficiency of 1.66%	[62]
		BPs	Vacuum filtration	PTS as dispersant solution	Membranes	The electrical conductivities: 30 ~ 220 S cm^{-1}; Membrane flux: 2400 ± 1300 L m^{-2} h^{-1} bar^{-1}	[63]
		Yarns	Spin CNT suspended solution or CNT web	Incorporate with nanoparticles or polymers	Multifunctional textiles	Increased strength and electronic connectivity.	[64–66]

(Continued)

TABLE 25.1 (*Continued*)
Fabrication Routes and Applications of Graphene-Based Architectures

Dimensionality	Ingredients and Structures	Fabrication Routes	Matrix/Substrate Incorporation	Applications			References
				Products	Properties		
3D composites	Graphene and	CNT forests/ single bunch	*In situ* growth	N/A	Electron field emitters	Turn-on and threshold electric fields 2.9 and 3.3 V μm^{-1}	[67]
		Random CNTs	Solution mixing	Dropped onto graphite substrate	Supercapacitor	Energy density 21.74 Wh kg^{-1} and power density 78.29 kW kg^{-1}	[17]
		CNT layers	Layer-by-layer/ electrostatic self-assembly	Dropped onto silicon or quartz substrate	Electrode	Sheet resistance of 8 kΩ/sq with a transmittance of 81%	[12]
	Graphene and metal/metal oxide NPs		*In situ* synthesis by chemical reduction	N/A	Colorimetric detection of cancer cells	Distinguished cancer cells by naked-eye observation	[68]
	Graphene and	polymer nanowires	*In situ* polymerization of polymer monomer	Mixed with PTFE and pressed onto gold grid	Energy storage	92% of initial capacitance kept by PANI–GO nanocomposite	[69]
		polymer hydrogels	Violent shaking	N/A	Drug delivery system	Controllably released drugs in physiological pH	[70]
		polymer colloidal particles	polymer template remove	PS and EVA	Conductive polymer composites	High electrical conductivity of 1083.3 S/m	[71]

Note: CVD = chemical vapor deposition, PDMS = polydimethylsiloxane, GO = graphene oxide, CNTs = carbon nanotubes, PET = polyethylene terephthalate, PTS = pthalocyanine tetrasulfonic acid, PTFE = polytetrafluoroethylene, PS = polystyrene, and EVA = ethylene vinyl acetate.

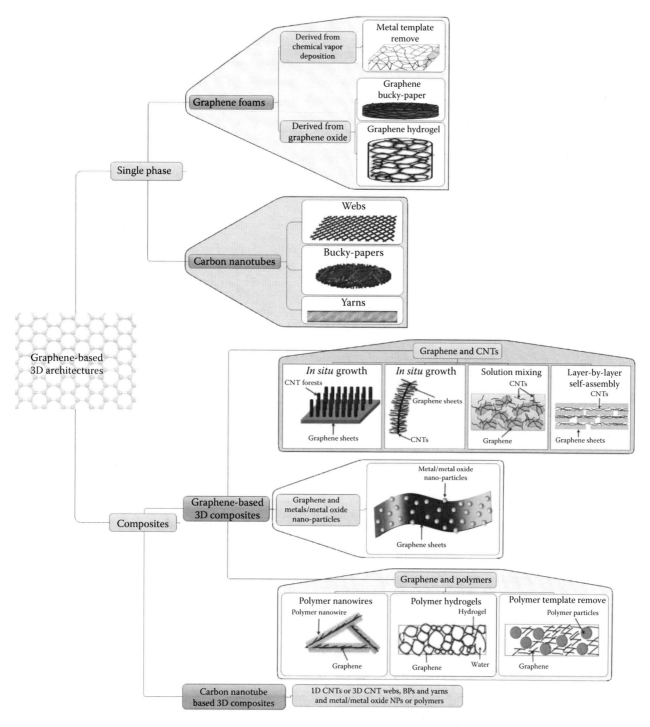

FIGURE 25.1 Summary of graphene-based 3D architectures.

derived from GO sheets [22]. Although *in situ* growth processes, such as CVD, provide graphene of high purity and crystallinity, the yield of these techniques is typically low [6]. However, low-grade graphene could be mass produced from the reduction of GO [72]. Freestanding 3D GFs are the most studied structure to date. GF are alveolus-like graphene semiporous frameworks made of interconnected nano to macro pores [19]. The structure of GFs particularly offers superior prospects within the field of electrochemical processes and

separation [18,24] for its high specific surface area, strong mechanical properties, and porous nature. The detailed description and comparison of the methodologies to fabricate 3D GFs will be presented in this section.

25.2.1.1 Growing Graphene Foams with CVD

The processing of highly crystalline 2D defect-free graphene sheets has been extensively performed by CVD on flat metallic foils, such as copper [73,74]. An alternative to these flat

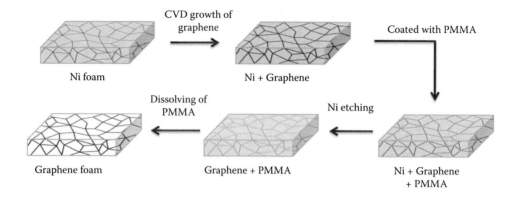

FIGURE 25.2 Flow-chart of synthesis process of freestanding GF by CVD.

substrates were recently investigated in the form of porous metal foams [13], allowing for the growth of a continuous uniform 3D graphene network on the surface of the metallic foam. GFs were produced on preformed foam templates, such as nickel (Figure 25.2) [13]. Preformed nickel foam sheets were inserted into a quartz tube and heated to a high temperature under Ar and H_2 environment to remove oxides formed on the surface of the nickel foam. Carbon sources, such as ethanol or methane, can then be introduced into the quartz tube at atmospheric pressure, to provide necessary carbon feed-stock precursor to the CVD growth [75]. After several minutes of reaction, the samples were quickly cooled to ambient temperature and were ready for the next step—etching nickel foam with $FeCl_3$ or HCl. To prevent the structural collapse of GF during etching process, a small amount of poly (methyl methacrylate) (PMMA) solution was needed to coat the nickel foam grown with graphene (Ni-GN) and dried as a protective film. Then, the Ni-GN sample coated with PMMA (Ni-GN-PMMA) was put into a hot HCl or $FeCl_3$/HCl solution until the nickel layer was completely etched. PMMA was finally dissolved with acetone to obtain freestanding 3D porous GFs [13].

Pore size, porosity, and morphology of the GFs are directly related to these of the initial porous nickel foam template (Figure 25.3a). In addition to its naturally high porosity and pore interconnectivity, nickel foam is a commonly used template to assist in the construction of 3D GF [19]. Since CVD-grown GFs are high crystalline, their conductivity is similar to that of otherwise flat graphene sheets and in the order of 10 S/cm [20]. These GFs exhibit superior electro-chemical properties and can be used as advanced sensing and capacitance platforms [18]. Dong et al. [18] grew graphene on a nickel foam template via CVD to fabricate 3D GFs (Figure 25.3b) and compared the electrochemical performance of glassy carbon (GC) with 3D GF. Cyclic voltammograms obtained on these samples showed that the resistance of the GF (~20.7 Ω) is up to 3 times lower than that of traditional GC electrodes (~69.7 Ω), primarily due to a greater specific surface area of the exposed graphene, opening the route to the production of higher current density electrodes with applications for high sensitivity molecular sensing.

25.2.1.2 Formation of Graphene Foams from GO

Single-phase 3D graphene architectures can also be synthesized through a thermo-chemical process without the assistance of any template by reducing the preformed GO BP. During a leavening process [22], GO BPs were thermo-chemically reduced with hydrazine monohydrate in an autoclave at 90°C (Figure 25.4a). Hydrazine monohydrate is a strong reducing agent, which can reduce functional groups on the surface of GO and simultaneously condense adjoining carbon atoms from sp^2 to sp^3 crystalline structures. H_2O gas and CO_2 are released during the reduction process in which the GO layers are exfoliated and diffused into the compact GO BP. The combined exfoliation and gas diffusion lead to the formation of macro-pores across the structure. It can be seen that although GO films are made of tightly stacked layers and have a smooth surface (Figure 25.4b), leavened GFs present a very rough surface as well as loose macro-structure (Figure 25.4c). This leavening phenomenon leads to large expansion between graphene sheets and to the formation of macro-porosity for pore size ranging of sub-micrometer to several micrometers.

Alternatively, single-phase 3D graphene architectures were processed by thermal or chemical reduction of GO in solution followed by freeze-drying [61,76–78]. This technique produces a stable, rough, and highly porous and sponge-like structure [79]. The sponge-like graphene is obtained by direct sublimation of the solvent, typically water, from the frozen GO suspension. The size and shape of obtained 3D graphene architectures can be controlled by changing the volume and shape of the initial containers. For example, a cylindrical and spherical 3D graphene sponge could be formed by using containers of different shapes (Figure 25.5). The single-phase 3D graphene architectures fabricated by this technique [61] have strong mechanical properties and can support more than 14,000 times their own weight under compression and a high specific capacitance of 128 F/g.

GFs have also been prepared through nucleate boiling of rGO [23]. The resulted GFs can be synthesized on any substrate without etching process, making it very competitive when compared with CVD growth. Partially reduced rGO sheets were stacked onto a substrate such as silicon wafer prior

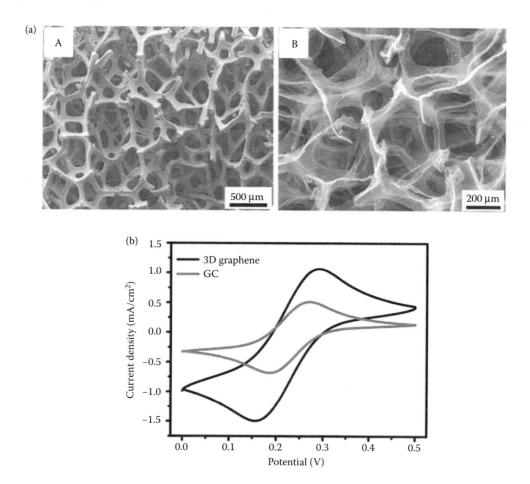

FIGURE 25.3 (a) SEM images of 3D GF with Ni foam (A) and without Ni foam (B). (Reproduced with permission from Cao, X. et al., Preparation of novel 3D graphene networks for supercapacitor applications. *Small*, 2011. **7**(22): p. 3163–3168.) (b) Cyclic voltammograms of GC and 3D GF. (Dong, X. et al.: 3D Graphene foam as a monolithic and macroporous carbon electrode for electrochemical sensing. *CS Applied Materials and Interfaces.* 2012. **4**. p. 3129–3133. Copyright Wiley-VCH Verlag GmbH & Co. KGaA. Reproduced with permission.)

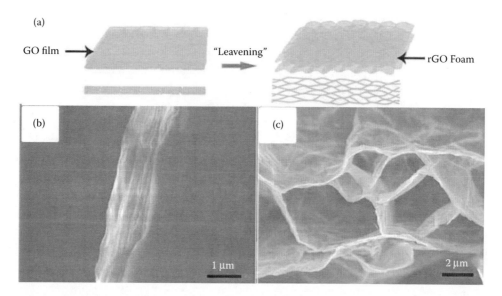

FIGURE 25.4 Schematic of the leavening process to form GFs from GO BP (a) and SEM images of cross section of GO (b) film and GF (c). (Niu, Z. et al.: A leavening strategy to prepare reduced graphene oxide foams. *Advanced Materials.* 2012. **24**. p. 4144–4150. Copyright Wiley-VCH Verlag GmbH & Co. KGaA. Reproduced with permission.)

FIGURE 25.5 Photos of different shaped single-phase graphene 3D architectures by reduction of GO follows with a freeze-drying process. (Tang, Z. et al.: Noblemetal-promoted three-dimensional macroassembly of single-layered graphene oxide. *Angewandte Chemie*. 2010. **122**. p. 4707–4711. Copyright Wiley-VCH Verlag GmbH & Co. KGaA. Reproduced with permission.)

to being placed onto a plate-heater (Figure 25.6a). During the heating process, boiled water bubbles were formed at the interface between the layers of stacked rGO. With the help of condensation between adjacent carboxylic groups remaining on the partially rGO, porous architectures can therefore be obtained by varying the evaporation rate, exposed surface area, and time of exposure to the boiling water (Figure 25.6b). The properties of pure graphene can be largely recovered by further reducing or annealing the rGO once the desired

structure is formed (Figure 25.6c). The advantage of this method is low cost, low energy consumption, and universal adaption on various substrates (glass, copper foil, and PDMS). More importantly, the final products possess high conductivity (11.8 S/cm) and low resistance (91.2 Ω m) compared to that of graphene (10 S/cm) obtained by CVD [23].

GFs as a novel graphene-based 3D architecture have been fabricated through two main techniques and offer excellent platforms to produce complex graphene textures. CVD-grown GFs offer an electrical conductivity very similar to graphene sheets and have versatile morphology by simply changing the original foam template. The drawbacks of the CVD method, however, lie in its high capital and operational costs, and low production throughput. However, GFs derived from thermochemical reduction of GO is highly cost-effective and a certain control of the degree of reduction is possible, opening the route of prefunctionalized GF structures. One drawback of this reducing method is that the structure of reduced GFs is more likely random, leading to high defect contents, which limits the final electrical conductivity of the structure.

25.2.2 Carbon Nanotube Webs, Bucky-Papers, and Yarns

Promising self-supporting 3D architectures made of solely CNTs were fabricated in the form of CNT webs, yarns, and BPs. These structures have been considered for a range

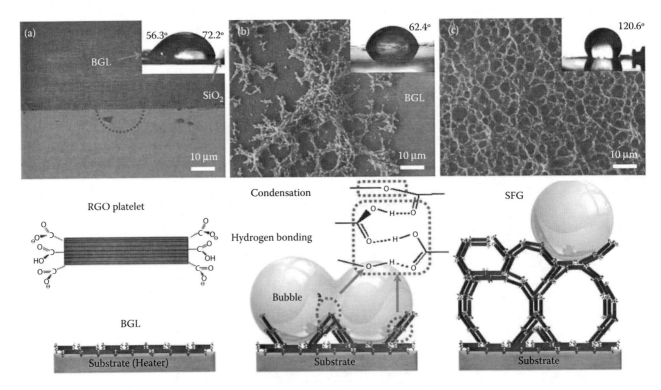

FIGURE 25.6 Formation process of GF by a simple nucleate boiling method: (a) base graphene layer (BGL) stacked onto the SiO₂ surface; (b) bubbles were formed under a boiling condition and GF structure started to assemble; and (c) the morphology of finally formed GF. (Reprinted by permission from Macmillan Publishers Ltd. *Sci. Rep.*, Ahn, H.S. et al., Self-assembled foam-like graphene networks formed through nucleate boiling, **3**, copyright 2013.)

of applications, including sensors [27,80,81], membranes [63,82,83], actuators [84], composites [28,85], advanced textiles [65], and biological scaffolds [30].

CNT webs can be typically drawn from spinnable CNT forests to form thin, highly aligned CNT materials [86]. On the other hand, CNT yarns can be formed either via wet-spinning of a CNT suspended solution or by directly dry-spinning webs of CNTs. The first route involves dispersing or dissolving the CNTs into acidic solutions or surfactants followed by wet-spinning into fibers [26,87,88]. The second approach is a solid state process where the CNTs are directly spun from either a spinnable forest of CNTs [25,66,64,89,90,91] or at the outlet of a CVD reactor [92–94]. The CNTs across these yarns hold together through van der Walls forces and local twist/entanglement, providing outstanding mechanical stiffness with high breaking stresses of up to 2 GPa [90,95–99]. Their mechanical stability enables the fabrication, by knitting or weaving, of highly organized micron-sized architectures made solely of CNTs [25,90].

CNT BPs generally refer to a mat of randomly entangled CNTs generally prepared by vacuum filtration [17,18]. CNTs within the webs and BPs are held together through strong van der Waals interactions, leading to a cohesive and highly porous structure [82]. Although BP properties are highly dependent on the type and morphology of CNTs used and

on the processing conditions, they have been shown to (i) be mechanically robust and flexible with bulk modulus of up to ~1 GPa [83,100], (ii) have high thermal and chemical stability, (iii) exhibit outstanding porosity between 70% and 91% [27,29,82], and high specific surface area of up to 2200 m²/g [83,101]. They are also simpler and cheaper to fabricate and offer an ideal platform for the incorporation of other additives and polymers. These properties make BPs, made of solely graphene material, a promising material for sensors [27,81], electrodes [102], separation [63,82,83], and composites [85,103]. The properties of these self-supporting CNT architectures are highly dependent on their internal structure [28,29,82,83].

Although naturally dense, CNT webs can be further densified by solvent evaporation enhancing CNT–CNT interactions, which further increases the webs electrical conductivity by up to 30% [62]. Self-supporting web structures as thin as ~100 nm were successfully prepared, corresponding to less than 10 CNTs in cross section (Figure 25.7).

The structure, CNT interspacing, and porosity of CNT yarns were revealed by focus ion beam milling (FIB) and subsequent SEM imaging (Figure 25.8). A clear core-sheath arrangement is observed for this yarn with a receding gradient of CNT packing density from the center to the surface of the yarns. Close to the surface, the CNT bundles are easily resolved with porosity close to 70% while at the yarn center,

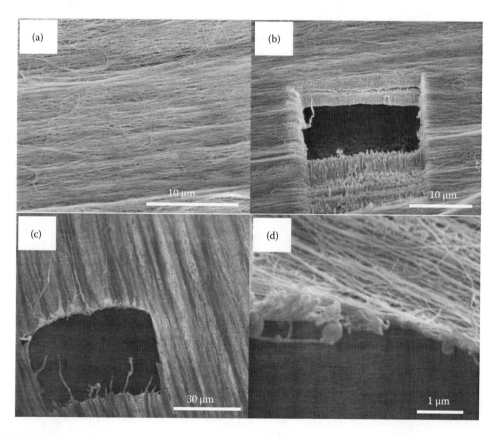

FIGURE 25.7 CNT alignment within CNT bucky-ribbons formed from (a, b) 50 layers of stacked CNT web densified with acetone, and (c, d) 5 layers of CNT web densified with acetone. The thickness of the web in (d) is <100 nm thick corresponding to 8–10 CNTs in cross section. (Reprinted from *Progress in Natural Science: Materials International*, **22**(6), Dumée, L. et al., *In situ* small angle X-ray scattering investigation of the thermal expansion and related structural information of carbon nanotube composites, p. 673–683, Copyright 2012, with permission from Elsevier.)

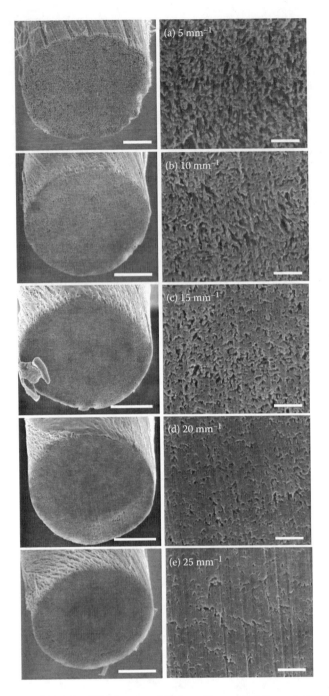

FIGURE 25.8 SEM images of FIB sections milled through CNT yarns with a series of twist densities. The left image corresponds to the whole cross section (scale bar represents 5 µm) while higher magnification images were taken at the yarn center on the right images (scale bar represents 500 nm). (Reproduced with permission from Sears, K. et al., *Carbon*, 2010. **48**: p. 4450–4456.)

bundles merge into near-single block matrix with very low porosity, close to 10%. The versatility of this process offers great perspective to prepare macro-structures made of nano-textured architectures with precise porosity, specific surface area and controlled symmetry [104].

Asymmetric CNT yarn structures were also prepared by rub densification of CNT webs [105]. Rub-densified yarns are formed by drawing the CNT web through two padded rollers,

which rotate and move laterally (Figure 25.9). Different from the direct spinning technique, highly asymmetric structures with high packing density sheath and a low density core were formed. The modulus of these twistless yarns was found to be of 41 GPa [105] against the 14 GPa [104] obtained for the standard spinning process. This great improvement was attributed to the greater packing density and lower twist.

CNT BPs exhibit extremely high porosities, up to 90%, and offer a cheaper alternative to prepare webs and yarns as they can be processed from any type of CNTs and do not require specific spinnable CNTs. As shown in Figure 25.10, the porosity and CNT arrangement across BPs prepared from different CNTs can be readily accessed by Ga ion-FIB milling of sections. The BPs shown in Figure 25.10, were processed by vacuum filtration from suspensions consisting of thin (~10 nm diameter) CNTs, thick (~50 nm diameter) CNTs, or a mixture of the two (1:1 ratio) [106]. Clearly, the type of CNTs used affects the BP porosity and structure. Interestingly, as seen in Figure 25.10c and d, a layered structure was formed for the mixture, and may indicate inhomogeneous mixing or different settling process during the filtration process.

25.3 GRAPHENE-BASED 3D COMPOSITES

3D graphene-based architectures can be directly processed from graphene or by assembling and linking reduced graphene oxide (rGO) sheets through covalent or non-covalent interconnection with other materials such as NPs, nanotubes, and polymers. GO is a simple and cheap precursor to synthesize graphene, which essentially consists of a process of dehybridizing exfoliated graphene sheets. A number of surface functional groups including carbonyl, epoxy, or hydroxyl can be present on their surface, on the sites of defects across the crystalline graphitic planes [107]. Although these functional groups provide potential reactive sites for further surface functionalization and further particulate interactions, their presence may sharply reduce the electronic and thermal conductivity when compared to pristine graphene [108,109]. For this reason, the overview of synthesis routes of graphene-based 3D structures along with the properties related to these structures will be presented in the following contexts.

25.3.1 GRAPHENE AND CNTs

The fabrication of graphene/CNT composites has been reported through a number of strategies and shown to be promising for the design of supercapacitors [110], electrochemical sensors [111], and fuel cells [39].

25.3.1.1 *In Situ* Growth of Graphene and CNTs

The *in situ* CVD growth of either graphene layers on the surface of CNTs [112,113] or CNTs on the surface of graphene sheets [37,67,114] has been reported as a viable route to combine these two forms of carbon materials. Complex 3D networks were fabricated by either growing graphene sheets on the top of vertically aligned CNTs [113] while leaf-stem

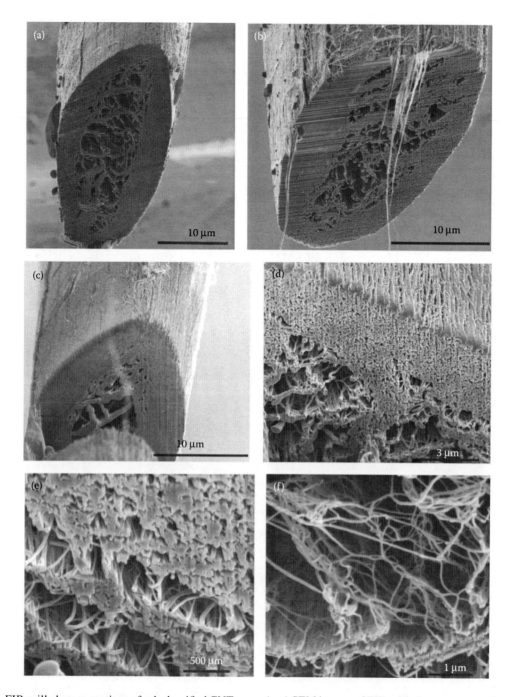

FIGURE 25.9   FIB-milled cross-sections of rub-densified CNT yarns (a–c) SEM images of FIB-milled yarn cross sections. (d and e) SEM images of high CNT density areas. (f) SEM images of large voids in the yarn core. (Reprinted from *Carbon*, **50**, Miao, M. et al., Production, structure and properties of twistless carbon nanotube yarns with a high density sheath., p. 4973–4983, Copyright 2012, with permission from Elsevier.)

structures, where graphene sheets acted as nano-leaves onto CNTs (Figure 25.11a and b) [112].

The *in situ* growth of CNTs onto graphene sheets was also found to exfoliate graphene sheets and prevent the restacking of graphene into graphite [37]. The deposition of the catalyst, onto the surface of GO sheets, used for the CNT growth was assisted by microwave to achieve an even and homogeneous coating and the GO sheets were thermally reduced into rGO during the CNT growth. The length of CNTs used [37] was from 0.2 to 3 µm. When applying this composite as anode in

Li-ion battery, the results showed that graphene/CNT composite with the shorted CNT length (0.2–0.5 µm) possessed higher reversible capacities of 573 mAh/g and 520 mAh/g at low and high constant current, respectively, than those with other lengths (below 500 mAh/g and 400 mAh/g). Another study [114] employed CVD to process 3D pillar-shaped graphene/CNT composite as electrode materials, which indicates similar relationship between the length of CNT and specific capacitance. When finely controlling the length of CNTs grown on the surface of graphene sheets, the length of CNT

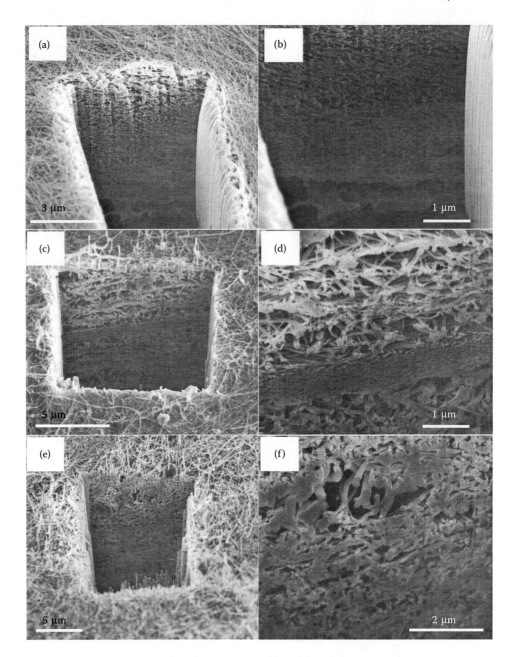

FIGURE 25.10 Revealing porosity of bucky-papers processed from (a, b) thin (~10 nm diameter) CNTs (c, d) a mixture of thin (~10 nm diameter) and thick (~50 nm diameter) CNTs, and (e, f) thick (~50 nm diameter) CNTs. The degree of dispersion is clearly visible from the SEM cross section as a layered structure is formed within the mixture suggesting inhomogeneous CNT mixing. (Reprinted from *Journal of Membrane Science*, **351**(1–2), Dumee, L.F. et al., Characterization and evaluation of carbon nanotube bucky-Paper membranes for direct contact membrane distillation, p. 36–43, Dumée, L. et al. *ICOM08*. In *ICOM08. Preparation of Carbon Nanotube Bucky-Paper Memrbanes for Desalination by Membrane Distillation*, Honolulu, Hawaii, USA, copyright 2010, with permission from Elsevier.)

was directly proportional to the pyrolysis time (Figure 25.12a, b, and c). The length of CNTs grown on the graphene was controlled by adjusting the growth time and concentration of feedstock gas (Figure 25.12d). The length of CNT is from 5 up to 23 μm. The capacitance of the composite graphene/CNT was also found to be related to CNT length [114], with the CNT length of 10 μm offering a specific capacitance of 1400 F/g that is much higher than that of 20 μm CNT length (1050 F/g). These two observations illustrate a broad range of

CNTs grown on graphene sheets from 0.2 to 23 μm, which really provides other researchers with a good reference to achieve desired conductive properties by controlling the length of the CNTs.

Apart from the above studies, the arrangement of the CNT arrays on graphene sheets can also be regulated by polymeric template. Block copolymer can form uniform and well-ordered nanoporous structure, which can assist the growth of CNTs in a controllable manner. Lee et al. [42] reported to

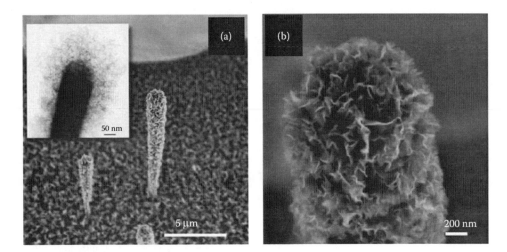

FIGURE 25.11 (a) SEM image of graphene sheets grown on vertically aligned CNTs; the inset is a TEM image of the tip of a graphene/CNT, and (b) a close view of the graphene/CNT tip in panel (c). (Reproduced with permission from Yu, K. et al., *The Journal of Physical Chemistry Letters*, 2011. **2**(13): p. 1556–1562.)

grow CNTs onto stacked GO platelets under a mixed gas of $C_2H_2/H_2/NH_3$ by PE-CVD in a nanoporous polymer template. The specific process and the SEM images of the diameter of CNTs and well-structured CNT arrays are presented in Figure 25.13. First, a GO solution was coated on a SiO_2 substrate followed by patterning catalysts on dried GO surface with the assist of nanoporous template, and then vertically aligned CNTs were grown at the catalyst points at a high temperature of 600°C. This process will benefit the circuit manufacturer due to the tunable structure and pattern of CNT arrays on graphene sheets, which is an obvious advantage to the randomly grown CNTs. The underlying GO layer was finally thermal

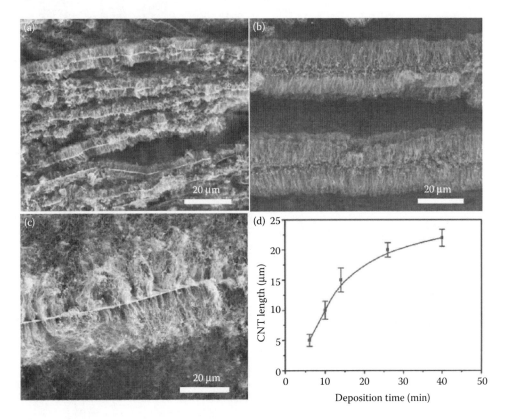

FIGURE 25.12 (a–c) SEM images of the thermally expanded graphene layers intercalated with vertically aligned CNTs for different times of 5, 10, and 30 min, respectively, and (d) the length of CNT as a deposition time. (Reprinted with permission from Du, F. et al., Preparation of tunable 3D pillared carbon nanotube-graphene networks for high-performance capacitance. *Chemistry of Materials*, p. 4810–4816. Copyright 2011 American Chemical Society.)

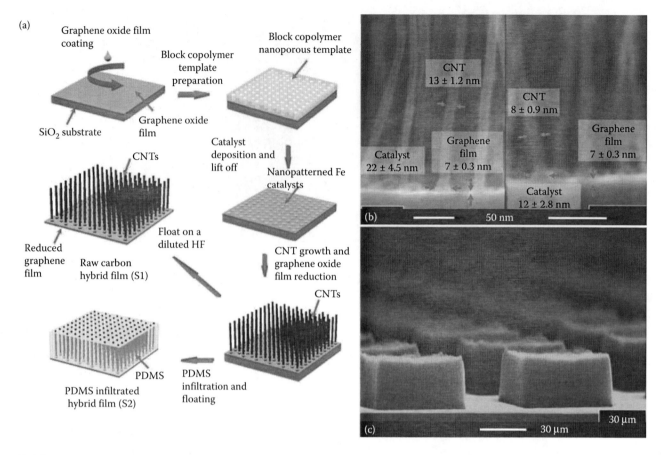

FIGURE 25.13 (a) Schematic illustration of the synthesis process for graphene/CNTs composite films. (b) SEM image of growth of CNTs at the catalyst domains by PE-CVD. (c) SEM image of pattern CNT on graphene film. (Lee, D.H. et al.: Versatile carbon hybrid films composed of vertical carbon nanotubes grown on mechanically compliant graphene films. *Advanced Materials*. 2010. **22**. p. 1247–1252. Copyright Wiley-VCH Verlag GmbH & Co. KGaA. Reproduced with permission.)

reduced to graphene at this high temperature and the electrical conductivity of these carbon hybrids was restored to around 18 S/cm. Such a flexible and conductive film can be directly transferred to any substrate and is readily integrated into a field-emitting device.

25.3.1.2 Solution Mixing of CNTs and Graphene

Although intrinsically hydrophilic due to the presence of carboxylic or alcohol groups on GO, CNTs can conjugate with graphene onto hydrophobic areas by π–π interactions [38]. GO, which is easily dispersible in water, can therefore, acting as a surfactant, play a role in suspending CNTs in aqueous solutions. A proportional relationship between the GO concentration in solution and the CNT solubility was demonstrated [38], with lower GO concentration leading to worse CNT dispersion (Figure 25.14a and b). 3D graphene and CNT conductive networks were successfully synthesized by using this simple solution mixing method [17]. The results reveal that graphene sheet (GS)/CNT weight ratio of 9/1 can prevent restacking of RGO and form a porous network, while further addition of CNTs only leads to severe CNT clustering and aggregation. This is likely due to the fact that large amount of CNTs (7/3, 5/5, and 3/7) tends to form CNT bundles without

efficient space to disperse. The impact of the CNT aggregation on the electrical properties can, in turn, be seen in the measurement of specific current (Figure 25.14c). The GS/CNT ratio of 9/1and 8/2 possess even higher specific current than other composite with the ratio of 7/3, 5/5, and 3/7 and even higher than reduced GO. This could benefit from the synergistic effects between graphene and CNT [9].

25.3.1.3 Layer-by-Layer Self-Assembly of Graphene and CNTs

LBL has also been used to prepare graphene/CNT composites. GO particles are negatively charged in water due to the presence of carbonyl and hydroxyl groups on its surface. Therefore, negatively charged GO can absorb positively charged functionalized CNTs, such as amine-functionalized CNT (Figure 25.15a), by LBL self-assembly [12]. Alternatively, 3D graphene/CNT films were also prepared by grafting positively charged poly(ethylene imine) (PEI) onto graphene and negatively charged CNTs (Figure 25.15b) [115].

1D (CNTs) and 2D (graphene) nanomaterials have been integrated to form 3D graphene architectures. The solution mixing methods are simple routes to fabricate cheap and mechanically strong structures by simply altering the ratio

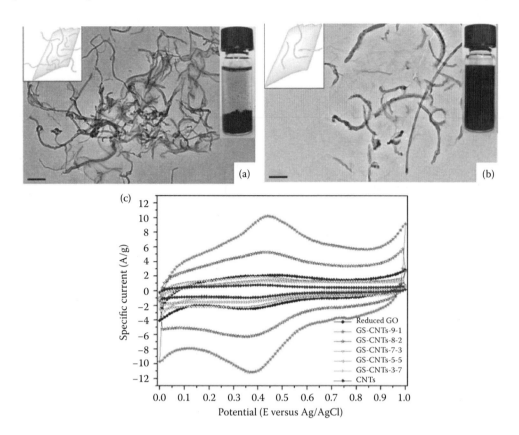

FIGURE 25.14 TEM images [(a) 1/2 and (b) 2/1] and cyclo-voltammetric profiles (c) of GO/ multiwalled CNTs solution with different proportion of GO to multiwalled CNTs. (Reprinted with permission from Zhang, C. et al., Graphene oxide-assisted dispersion of pristine multiwalled carbon nanotubes in aqueous media. *The Journal of Physical Chemistry C*, **114**(26): p. 11435–11440. Copyright 2010 American Chemical Society.)

between graphene sheets and CNTs, while the *in situ* growth of CNTs on graphene allows for the preparation of well-organized and structured architectures with preferential electrical and mechanical anisotropy.

25.3.2 Graphene and Metals/Metal Oxide Nanoparticles

Owing to their sp² crystalline structure, graphene sheets can strongly coordinate with metals [40], opening new routes to their decoration with pure metal or metal oxide NPs [79,116–119]. Metal NPs on the other hand can prevent the aggregation of graphene sheets and enhance the structural stability. The combination of graphene and metals offers perspectives to mechanically reinforce graphene BPs and improve the electrochemical properties of the graphene architectures [48,119–121]. In fact, some of the natural unique properties offered by these metal NPs, such as their catalytic activity, can be enhanced, owing to the high electron mobility once grown or deposited on graphene [121].

25.3.2.1 *In Situ* Synthesis of Metal/Metal Oxide Nanoparticles on Graphene

Gold NPs were, for instance, grown within clusters of graphene sheets dispersed in octadecylamine by chemical reduction of AuCl$_4^-$ with NaBH$_4$ [117]. The electronic properties of

the graphene sheets were shown to stabilize the gold NPs and prevent coalescence, leading to a narrow (10–20 nm) and well-dispersed suspension of gold NPs (Figure 25.16a, b, and c). Silver NPs were also grown at 75°C under N$_2$ protection onto GO deposited onto silica wafers without the need of reducing agents [48]. The synergetic reaction was shown to be spontaneous and led to the simultaneous reduction of silver ions and GO into silver NPs and graphene, respectively. While being reduced, GO provided both the electrons to induce reduction of the silver ions and a support for the growth of Ag NPs [48]. This synthesis route was clearly demonstrated as silver NPs only grew homogeneously (6.0 ± 3.6 nm) on the rGO and no nanoparticles were found on the bare silica wafer surface (Figure 25.16d and e), illustrating that graphene reduction can be used to target other chemical reactions and synthesis in one-pot reactions [122].

25.3.2.2 Electrostatic Self-Assembly of Graphene and Metal/Metal Oxide Nanoparticles

Electrostatic self-assembly, which relies on the interactions between oppositely charged surfaces, appears to be a promising technique to control the structure and properties of self-assembled materials [47,71,118,123–126]. With the help of electrostatic force, graphene can be assembled onto a variety of substrates under a facile condition, which provides more combination to the area related to graphene-based materials.

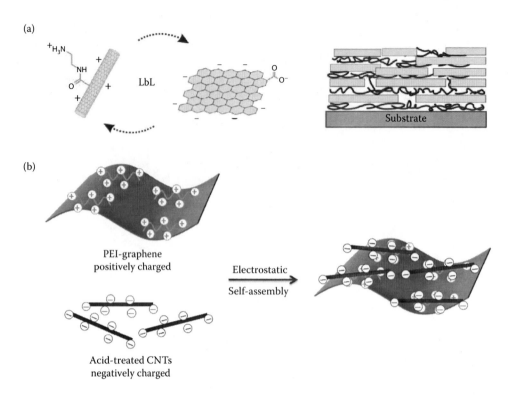

FIGURE 25.15 Schematic illustration of hybrid layer-by-layer films of GO/ amine doped CNT (a). (Reproduced with permission from Hong, T.-K. et al., Transparent, flexible conducting hybrid multilayer thin films of multiwalled carbon nanotubes with graphene nanosheets. *ACS Nano*, **4**(7): p. 3861–3868. Copyright 2010 American Chemical Society.) and PEI-graphene/acid-treated CNT (b). (Reprinted with permission from Yu, D. et al., Self assembled graphene/carbon nanotube hybrid films for supercapacitors. *The Journal of Physical Chemistry Letters*, **1**(2): p. 467–470. Copyright 2009 American Chemical Society.)

3D graphene and metal/metal oxide structures have been fabricated by electrostatic self-assembly in solution [40,123–125] and electrostatic layer-by-layer (LBL) self-assembly [116,118,126]. Gold NPs were deposited by coupling with cationic polyelectrolyte poly(diallyldimethyl ammonium chloride) (PDDA) onto graphene [125]. The hybrid surface was modified by adsorbing citrate ions, leading to the assembly of a positively charged graphene with negatively charged gold Au nanoparticles, which were shown to be very homogeneous in size and well dispersed onto the surface of graphene sheets (Figure 25.17a). In order to demonstrate the electrochemical changes induced by the surface charge shift, glassy carbon electrodes were impregnated with either the graphene/gold hybrid or the gold NPs and used as a sensor to detect hydrogen peroxide in aqueous solution. The results showed that the graphene/gold hybrid electrode possesses a higher sensitivity than the one modified by Au nanoparticles only (Figure 25.17b), due to synergistic effects of graphene and nanoparticles and the conductive surface between nanoparticles constructed by graphene [5].

Electrostatic LBL self-assembly is another way to fabricate 3D graphene/metal oxide composites with highly controllable thicknesses. Multilayer graphene and manganese dioxide (MnO_2) composite were successfully prepared. Poly(sodium 4-styrenesulfonate) (PSS) was first grafted onto graphene (PSS-GN) in order to help stabilize the dispersion and suspension of naturally highly hydrophobic graphene

particles in water [126]. In addition, in order to improve the electric condition between the LBL deposited films, PDDA was used as a positively charged linker between the PSS-GN and MnO_2 individual layers. The specific capacitance of a ten PSS-GN/PDDA/MnO_2 layered-structure showed that 90% of specific capacity was maintained after 1000 cycles of charge/discharge, making the graphene hybrid structures promising materials for supercapacitors. Although based on weak interactions and generally stable over a narrow range of pH, LBL is also very promising for decorating or prefunctionalizing graphene 3D structures.

Metal NPs and graphene nanosheets can be thoroughly combined to produce discrete but yet uniform graphene metal composites. The major advantage of electrostatic self-assembly technique is that the structure and properties of graphene and metal/metal oxide composites can be altering the processing conditions while electrostatic self-assembly provides a simpler and more efficient technique to prepare low cost, but low mechanical stability composites. The versatility of this mechanism could lead to the formation of highly conductive and mechanically reinforced graphene-based structures that are able to smartly incorporate both graphene and metal properties.

25.3.3 GRAPHENE AND POLYMERS

Potential applications of 3D graphene-based polymer composites are in such fields as electronics devices, energy storages,

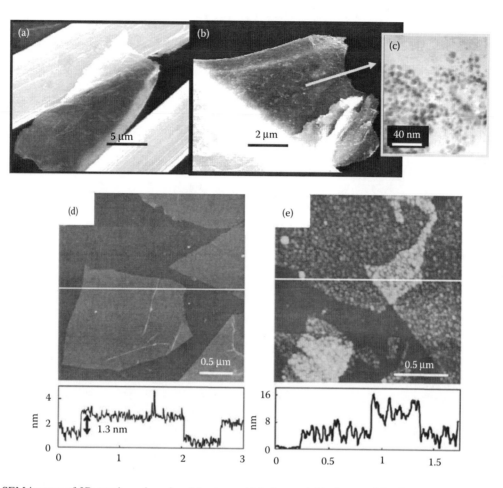

FIGURE 25.16 SEM images of 3D graphene-based architectures: (a) before and (b) after modification with gold nanoparticles, (c) high-resolution TEM image of the sample in B. (Reproduced with permission from Muszynski, R. et al., Decorating graphene sheets with gold nanoparticles. *The Journal of Physical Chemistry C*, **112**(14): p. 5263–5266. Copyright 2008 American Chemical Society.) AFM topographic image and height profile of single-layer GO (d) and Ag nanoparticles grown on single-layer graphene surface (e). (Reprinted with permission from Zhou, X. et al., *In situ* synthesis of metal nanoparticles on single-layer graphene oxide and reduced graphene oxide surfaces. *The Journal of Physical Chemistry C*, **113**(25): p. 10842–10846. Copyright 2009 American Chemical Society.)

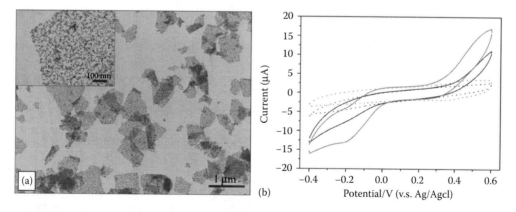

FIGURE 25.17 (a) TEM image of graphene/Au composites and the inset image of enlarged area to show the morphology of Au nanoparticles on graphene; (b) current/voltage curves of graphene/Au (light gray) and Au (black) modified electrode with the absence (solid) and presence (dots) of hydrogen peroxide in aqueous solution. (Reprinted with permission from Fang, Y. et al., Self-assembly of cationic polyelectrolytefunctionalized graphene nanosheets and gold nanoparticles: A two-dimensional heterostructure for hydrogen peroxide sensing. *Langmuir*, **26**(13): p. 11277–11282. Copyright 2010 American Chemical Society.)

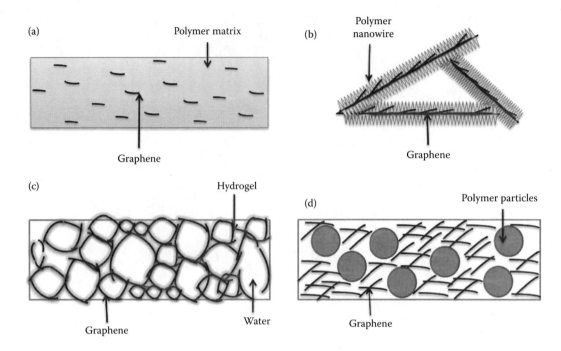

FIGURE 25.18 Schematic of the main strategies for the construction of graphene-based polymer architectures: (a) mixed matrix, (b) nanowire, (c) hydrogel, (d) polymer template remove.

sensors, catalysts, biomedical applications, water purification, and composite reinforcement [3,7,35,50–54,127–129].

The addition of graphene into polymer matrices was initially targeted at the processing of more mechanically robust materials [130]. The graphene particles were dispersed and suspended into the polymer phase to form a mixed matrix material graphene (Figure 25.18a). Smooth, uniform, and flexible GO/Poly(vinyl alcohol) (PVA) composite films were processed and exhibited strong mechanical strength (tensile modulus ~ 6.1 ± 0.5 GPa as opposed to only ~ 2.1 ± 0.2 GPa for PVA alone) [131]. This film was synthesized by mixing GO with PVA by vacuum filtration of the aqueous solutions. However, a critical issue encountered in this process was the irreversible aggregation or restacking of graphene sheets in polymer matrix. Thus, the inherent properties of graphene sheets cannot be fully transferred to the graphene-based polymer macroscopic materials. A number of 3D graphene polymer composite macro-structures have been recently processed. These structures include porous films, scaffolds, and networks, and will pave a way toward high-performance graphene-based materials for day-to-day applications. In this section, the main strategies for the construction of 3D graphene/polymer architectures are briefly presented: nanowire (Figure 25.18b), hydrogel (Figure 25.18c), and polymer template removal (Figure 25.18d).

25.3.3.1 Graphene-Based Polymer Nanowires

Hierarchical graphene nanocomposites were previously fabricated by combining 2D GO nanosheets with 1D poly(aniline) (PANI) nanowires into a 3D structures (Figure 25.19) [69]. The PANI nanowire arrays were grown vertically on GO sheets by dilute polymerization. This graphene-based material has

shown twice higher electrochemical capacitance (555 F/g at a discharge current of 0.2 A/g) and higher stability (2000 consecutive cycles) than each individual component (227 F/g at a discharge current of 0.2 A/g) because of an optimized ionic transport pathway, which led to an improved capacitance performance. Hydrogen peroxide was used to improve the growth of PANI on GO surface in order not to introduce impurities into the composite [132]. This novel composite exhibited high specific capacitance and energy density and had good long-term stability with capacitance retention of 118% of the initial value after the 500 cycles, which is much higher than only PANI (70%). Poly(pyrrole) (PPy) nanowires have also been successfully grown on the surface of GO sheets by *in situ* polymerization of pyrrole monomer in the presence of GO suspension [133]. The GO/PPy composites exhibited a predominant specific capacitance of 633 F/g at current density of 1 A/g. Mao and coauthors [134] also prepared PPy nanowires in the presence of cetyl-trimethyl-ammonium bromide stabilized graphene for high-performance supercapacitor electrode (GCR). This GCR/PPy had a higher specific capacitance of 492 F/g than pure PPy (227 F/g).

25.3.3.2 Graphene-Based Polymer Hydrogels

Hydrogels are networks of polymer chains that are hydrophilic, which can incorporate with a significant amount of water retained in the network structure [135]. Porous 3D graphene hydrogel networks were fabricated by assembling graphene and a tri-block copolymer PEO-b-PPO-b-PEO (F127) with the assistance of α-cyclodextrin (α-CD) [136]. The GO-based hydrogels were formed through supramolecular interactions such as hydrogen bonding, coordination, electrostatic interaction and π–π stacking [8,137]. Different

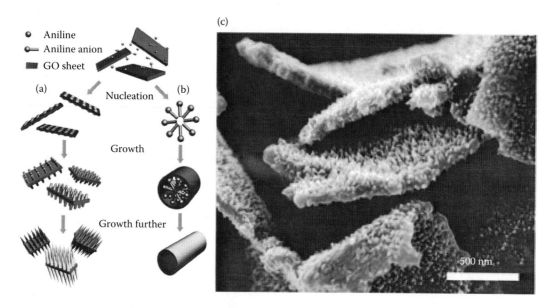

FIGURE 25.19 Schematic illustration of nucleation and growth mechanism of PANI nanowires: (a) heterogeneous nucleation on GO nanosheets; (b) homogeneous nucleation in bulk solution; (c) SEM images of PANI–GO nano-composites. (Reprinted with permission from Xu, P. et al., *In situ* growth of noble metal nanoparticles on graphene oxide sheets and direct construction of functionalized porous-layered structure on gravimetric microsensors for chemical detection. *Chemical Communications*, **48**(87): p. 10784–10786. Copyright 2012 American Chemical Society.)

graphene hydrogel systems have been described in the literature, including poly(vinyl alcohol) (PVA), poly(sodium acrylate) (PSA), poly(vinyl pyrrolidone) (PVP), poly(ethylene oxide) (PEO), polydimethyldiallylammonium chloride (PDDA), polyethylenimine (PEI), cetyltrimethyl ammonium bromide (CTAB), tetramethylammonium chloride (TMAC), and block copolymer surfactants such as F127 [70,136–138]. These graphene hydrogel systems are attractive due to the high surface area-to-volume ratio of the particles [139]. Bai and coauthors [70] demonstrated the ability to self-assemble GO and PVA into a hydrogel, which can be used for pH-controlled selective drug release. In this system, the formation of the hydrogels relies on the assembly of GO sheets and the cross-linking of PVA chains. This GO-based PVA hydrogel is, however, pH sensitive and can form gels in acidic conditions but remain liquid under alkaline conditions. When the concentration of GO is high enough ($C_{GO} > 3$ mg L^{-1}), a loose network instead of an amorphous precipitation can be formed [137]. The size of GO sheets also has a large influence on GO gelation properties. Micron-sized sheets can easily form a stable hydrogel, and no gelation occurs if GO sheets are smaller than 1 μm [137].

25.3.3.3 Graphene-Based Polymer Template Remove

Polystyrene (PS) colloidal particles were also used to guide the formation of hollow micrometer-sized graphene porous structures [140]. The porous structures could be easily constructed by dissolving or pyrolyzing the PS colloidal particles. This template-directed remove method can produce a high-order 3D architecture, which has the potential to be used in drug release, catalysis, electronics, and biotechnology [71,140]. Amine groups grafted PS colloidal particles

with surface positively charged can easily assemble with negatively charged GO by electrostatic interactions [71]. This PS template removed 3D graphene structure shows a high electrical conductivity of 1083.3 S/m, which is superior to graphene-PS composite prepared by solvent blending method by 250 times.

25.4 CARBON NANOTUBE-BASED 3D COMPOSITES

The combination of 1D CNTs, or 3D CNT yarns, webs, and BPs to form 3D carbon-based architectures creates a versatile material that can be fully integrated into complex porous, hybrid, or composite devices [2,8,51]. Graphene and/or CNTs mixed architectures, or composite structures, by combining polymer [50,51] or metal/metal oxide NPs [21,47,48] bridges to graphene sheets or CNT fibers.

CNT sheets and yarns were also demonstrated to be excellent hosting materials for loading and blending high ratios of guest NPs via biscrolling [65]. Using this biscrolling technique various functionalities such as superconducting yarn loaded with magnesium diboride, flexible yarn for Li-ion batteries loaded with GO nanoribbons, and photo-catalytic yarn loaded with TiO$_2$ particles can be fabricated as seen in Figure 25.19 [65]. All of the yarns were formed by twist-spinning a rectangular strip of CNT sheet in either air or liquid. The internal structures appear highly complex, and different morphologies and porosities can be clearly observed depending on the additive NPs and processing conditions used. For example, as shown in Figure 25.20a and b, a layer of Ti is deposited onto the initial CNT sheet by electron beam evaporation and a highly porous structure and sheet folding is

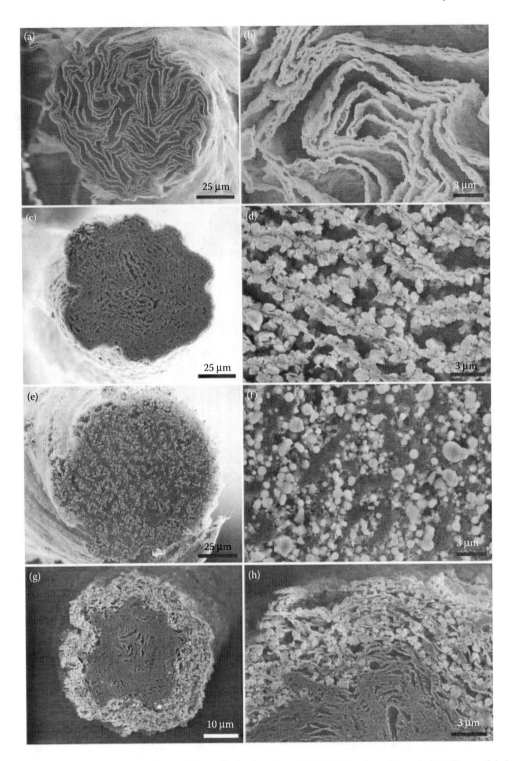

FIGURE 25.20 SEM images of the cross sections of bi-scrolled yarns. (a and b) 70% Titanium decorated CNT yarn fabricated by electron beam evaporation (25 nm Ti) and symmetrical twist insertion in liquid. (c and d) 93% TiO₂ decorated CNT yarn made by filtration-based deposition and symmetrical twist in liquid. (e and f) TiO₂ CNT yarn made by aerosol-based deposition and twist insertion in air. (g and h) TiO₂ CNT yarn made by patterned filtration-based deposition on preformed CNT webs and asymmetrical twist insertion in liquid. (From Lima, M.D. et al., Biscrolling nanotube sheets and functional guests into yarns. *Science*, 2011. **331**(6013): p. 51–55. Reprinted with permission of AAAS.)

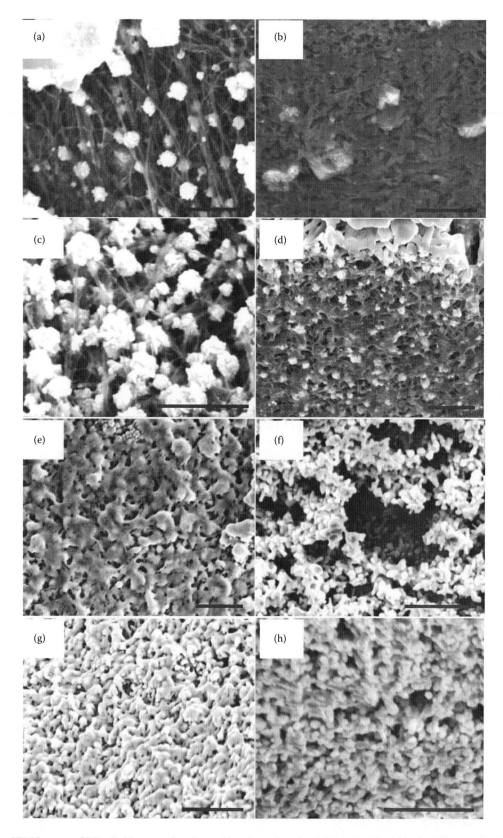

FIGURE 25.21 SEM images of BPs-Gold composites formed by electroless deposition of gold within the BP structure. The plating times are 1 (a, b), 5 (c, d), 20 (e, f) and 26 (g, h). The SEM images are of the surface (left column) and of cross sections (right column) milled with Ga ions in an FIB. All scale bars correspond to 500 nm. (Dumee, L. et al., Activation of gold decorated carbon nanotube hybrids for targeted gas adsorption and enhanced catalytic oxidation. *Journal of Materials Chemistry*, 2012. **22**(18): p. 9374–9378. Reproduced by permission of The Royal Society of Chemistry.)

formed. In contrast, a yarn formed with TiO_2 powder (Figure 25.20c and d) shows a less-porous structure and evidence of a scroll formation can be discerned [65]. Core-sheath structures were also achieved through biscrolling as seen in Figure 25.20d [65].

Composite BPs have also been investigated for various applications, such as sensing and separation. FIB milling, performed to reveal the internal structure of these composite BPs, is particularly important to gain a better understanding of how additives alter the final structure and hence, their properties. For example, metal NPs were grown by electroless deposition within BPs for targeted gas adsorption and catalytic activity [27]. FIB cross sections of these materials showed the formation of discrete nanoparticles at short plating times (1 h). With increasing plating time clustering of Au NPs is observed (5 h) until at long plating times (20–26 h) the CNTs are uniformly coated with a continuous network of Au (Figure 25.21). This progression of Au plating was also used to confirm the porosity and pore size experimentally obtained for the Au-plated BPs by He-porosimetry and perm-porometry, respectively.

25.5 CONCLUSIONS

The design and formation of graphene-based 3D architectures offer highly promising insights in electrochemical, biomedical, separation, and advanced composite areas. These 3D architectures, composed of either single phase graphene such as GFs and graphene hydrogels or of graphene-based composites, offer highly versatile morphologies and tunable properties due to their potentially high specific surface areas and synergistic effect with another phase. The key potential applications of the graphene-based 3D architectures include flexible electronics, chemical adsorption and separation and low weight composite materials with advanced thermoelectrical properties. Although 3D graphene-based architectures offer outstanding opportunities in materials and engineering science, the control of the interface between graphene sheets and/or host ligand is a critical issue requiring further study and development in order to offer sustainable alternatives to current materials.

REFERENCES

1. Geim, A.K. et al., The rise of graphene. *Nature Materials*, 2007. **6**(3): p. 183–191.
2. Chen, D. et al., Graphene-based materials in electrochemistry. *Chemical Society Reviews*, 2010. **39**(8): p. 3157–3180.
3. Sham, A.Y.W. et al., A review of fundamental properties and applications of polymer-graphene hybrid materials. *Soft Matter*, 2013. **9**: p. 6645–6653.
4. Liu, J. et al., Graphene-based materials for energy applications. *MRS Bulletin*, 2012. **37**(12): p. 1265–1272.
5. Wu, Z.-S. et al., Graphene/metal oxide composite electrode materials for energy storage. *Nano Energy*, 2012. **1**(1): p. 107–131.
6. Choi, W. et al., Synthesis of graphene and its applications: A review. *Critical Reviews in Solid State and Materials Sciences*, 2010. **35**(1): p. 52–71.
7. Kim, H. et al., Graphene/polymer nanocomposites. *Macromolecules*, 2010. **43**(16): p. 6515–6530.
8. Bai, H. et al., Graphene oxide/conducting polymer composite hydrogels. *Journal of Materials Chemistry*, 2011. **21**(46): p. 18653–18658.
9. Biswas, C. et al., Graphene versus carbon nanotubes in electronic devices. *Advanced Functional Materials*, 2011. **21**(20): p. 3806–3826.
10. Lee, C. et al., Measurement of the elastic properties and intrinsic strength of monolayer graphene. *Science*, 2008. **321**(5887): p. 385–388.
11. Liang, L.-J. et al., Dispersion of graphene sheets in aqueous solution by oligodeoxynucleotides. *Chem. Phys. Chem*, 2013. **14**(8): p. 1626–1632.
12. Hong, T.-K. et al., Transparent, flexible conducting hybrid multilayer thin films of multiwalled carbon nanotubes with graphene nanosheets. *ACS Nano*, 2010. **4**(7): p. 3861–3868.
13. Cao, X. et al., Preparation of novel 3D graphene networks for supercapacitor applications. *Small*, 2011. **7**(22): p. 3163–3168.
14. Chen, Z. et al., Three-dimensional flexible and conductive interconnected graphene networks grown by chemical vapour deposition. *Nature Materials*, 2011. **10**(6): p. 424–428.
15. Liang, Q. et al., A three-dimensional vertically aligned functionalized multilayer graphene architecture: An approach for graphene-based thermal interfacial materials. *ACS Nano*, 2011. **5**(3): p. 2392–2401.
16. Wang, J.-Z. et al., Graphene-encapsulated Fe3O4 nanoparticles with 3D laminated structure as superior anode in lithium ion batteries. *Chemistry—A European Journal*, 2011. **17**(2): p. 661–667.
17. Yang, S.-Y. et al., Design and tailoring of a hierarchical graphene-carbon nanotube architecture for supercapacitors. *Journal of Materials Chemistry*, 2011. **21**(7): p. 2374–2380.
18. Dong, X. et al., 3D Graphene foam as a monolithic and macroporous carbon electrode for electrochemical sensing. *ACS Applied Materials and Interfaces*, 2012. **4**(6): p. 3129–3133.
19. Brownson, D.A.C. et al., Freestanding three-dimensional graphene foam gives rise to beneficial electrochemical signatures within non-aqueous media. *Journal of Materials Chemistry A*, 2013. **1**(19): p. 5962–5972.
20. Yavari, F. et al., High sensitivity gas detection using a macroscopic three-dimensional graphene foam network. *Sci. Rep.*, 2011. **1**: Article number 166.
21. Dong, X. et al., Hybrid structure of zinc oxide nanorods and three dimensional graphene foam for supercapacitor and electrochemical sensor applications. *RSC Advances*, 2012. **2**(10): p. 4364–4369.
22. Niu, Z. et al., A leavening strategy to prepare reduced graphene oxide foams. *Advanced Materials*, 2012. **24**(30): p. 4144–4150.
23. Ahn, H.S. et al., Self-assembled foam-like graphene networks formed through nucleate boiling. *Sci. Rep.*, 2013. **3**: Article number 1396.
24. Dong, X. et al., Synthesis of graphene-carbon nanotube hybrid foam and its use as a novel three-dimensional electrode for electrochemical sensing. *Journal of Materials Chemistry*, 2012. **22**(33): p. 17044–17048.
25. Atkinson, K.R. et al., Properties and applications of dry-spun carbon nanotube yarns. *Advances in Science and Technology*, 2008. **60**: p. 11–20.
26. Behabtu, N. et al., Strong, light, multifunctional fibers of carbon nanotubes with ultrahigh conductivity. *Science*, 2013. **339**: p. 182–186.
27. Dumee, L. et al., Activation of gold decorated carbon nanotube hybrids for targeted gas adsorption and enhanced catalytic

oxidation. *Journal of Materials Chemistry*, 2012. **22**(18): p. 9374–9378.

28. Dumee, L. et al., Carbon nanotube based composite membranes for water desalination by membrane distillation. *Desalination and Water Treatment*, 2010. **17**(1–3): p. 72–79.

29. Dumee, L. et al., A preliminary study on the effect of macro cavities formation on properties of carbon nanotube bucky-paper composites. *Materials*, 2011. **4**(3): p. 553–561.

30. Edwards, S.L. et al., Tubular micro-scale multiwalled carbon nanotube-based scaffolds for tissue engineering. *Biomaterials*, 2009. **30**: p. 1725–1731.

31. Li, C. et al., Sensors and actuators based on carbon nanotubes and their composites: A review. *Composites Science and Technology*, 2008. **68**(6): p. 1227–1249.

32. Baughman, R.H. et al., Carbon nanotube actuators. *Science*, 1999. **284**(5418): p. 1340–1344.

33. Jarosz, P. et al., Carbon nanotube wires and cables: Near-term applications and future perspectives. *Nanoscale*, 2011. **3**(11): p. 4542–4553.

34. Xie, X. et al., Three-dimensional carbon nanotube-textile anode for high-performance microbial fuel cells. *Nano Letters*, 2011. **11**(1): p. 291– 296.

35. Das, T.K. et al., Graphene-based polymer composites and their applications. *Polymer-Plastics Technology and Engineering*, 2013. **52**(4): p. 319–331.

36. Lim, H.N. et al. 2013. Chapter 12: Inorganic nanostructures decorated graphene. In Sukarno Olavo Ferreira (Ed.), *Advanced Topics on Crystal Growth*. Rijeka, Croatia: InTech.

37. Chen, S. et al., Carbon nanotubes grown *in situ* on graphene nanosheets as superior anodes for Li-ion batteries. *Nanoscale*, 2011. **3**(10): p. 4323–4329.

38. Zhang, C. et al., Graphene oxide-assisted dispersion of pristine multiwalled carbon nanotubes in aqueous media. *The Journal of Physical Chemistry C*, 2010. **114**(26): p. 11435–11440.

39. Yun, Y.S. et al., Porous graphene/carbon nanotube composite cathode for proton exchange membrane fuel cell. *Synthetic Metals*, 2011. **161**(21–22): p. 2460–2465.

40. Hong, W.J. et al., Preparation of gold nanoparticle/graphene composites with controlled weight contents and their application in biosensors. *Journal of Physical Chemistry C*, 2010. **114**(4): p. 1822–1826.

41. Fan, Z. et al., A three-dimensional carbon nanotube/graphene sandwich and its application as electrode in supercapacitors. *Advanced Materials*, 2010. **22**(33): p. 3723–3728.

42. Lee, D.H. et al., Versatile carbon hybrid films composed of vertical carbon nanotubes grown on mechanically compliant graphene films. *Advanced Materials*, 2010. **22**(11): p. 1247–1252.

43. Wang, G. et al., Graphene/polyaniline nanocomposite as counter electrode of dye-sensitized solar cells. *Materials Letters*, 2012. **69**: p. 27–29.

44. Hong, W. et al., Transparent graphene/PEDOT–PSS composite films as counter electrodes of dye-sensitized solar cells. *Electrochemistry Communications*, 2008. **10**(10): p. 1555–1558.

45. Maiyalagan, T. et al., Electrodeposited Pt on three-dimensional interconnected graphene as a free-standing electrode for fuel cell application. *Journal of Materials Chemistry*, 2012. **22**(12): p. 5286–5290.

46. Qiu, L. et al., Biomimetic superelastic graphene-based cellular monoliths. *Nat Commun*, 2012. **3**: p. 1241.

47. Cong, H.-P. et al., Macroscopic multifunctional graphene-based hydrogels and aerogels by a metal ion induced self-assembly process. *ACS Nano*, 2012. **6**(3): p. 2693–2703.

48. Zhou, X. et al., *In situ* synthesis of metal nanoparticles on single-layer graphene oxide and reduced graphene oxide surfaces. *The Journal of Physical Chemistry C*, 2009. **113**(25): p. 10842–10846.

49. Wang, G. et al., Sn/graphene nanocomposite with 3D architecture for enhanced reversible lithium storage in lithium ion batteries. *Journal of Materials Chemistry*, 2009. **19**(44): p. 8378–8384.

50. Huang, X. et al., Graphene-based composites. *Chemical Society Reviews*, 2012. **41**(2): p. 666–686.

51. Sun, Y. et al., Graphene/polymer composites for energy applications. *Journal of Polymer Science Part B: Polymer Physics*, 2013. **51**(4): p. 231–253.

52. Du, J. et al., The fabrication, properties, and uses of graphene/polymer composites. *Macromolecular Chemistry and Physics*, 2012. **213**(10–11): p. 1060–1077.

53. Potts, J.R. et al., Graphene-based polymer nanocomposites. *Polymer*, 2011. **52**(1): p. 5–25.

54. Kuilla, T. et al., Recent advances in graphene based polymer composites. *Progress in Polymer Science*, 2010. **35**(11): p. 1350–1375.

55. Kim, K.S. et al., Large-scale pattern growth of graphene films for stretchable transparent electrodes. *Nature*, 2009. **457**(7230): p. 706–710.

56. Bae, S. et al., Roll-to-roll production of 30-inch graphene films for transparent electrodes. *Nature Nanotechnology*, 2010. **5**(8): p. 574–578.

57. Wu, W. et al., Wafer-scale synthesis of graphene by chemical vapor deposition and its application in hydrogen sensing. *Sensors and Actuators B: Chemical*, 2010. **150**(1): p. 296–300.

58. Paton, K.R. et al., Scalable production of large quantities of defect-free few-layer graphene by shear exfoliation in liquids. *Nature Materials*, 2014. **13**: p. 624–630.

59. Becerril, H.A. et al., Evaluation of solution-processed reduced graphene oxide films as transparent conductors. *ACS Nano*, 2008. **2**(3): p. 463–470.

60. He, Q. et al., Centimeter-long and large-scale micropatterns of reduced graphene oxide films: Fabrication and sensing applications. *ACS Nano*, 2010. **4**(6): p. 3201–3208.

61. Zhang, X. et al., Mechanically strong and highly conductive graphene aerogel and its use as electrodes for electrochemical power sources. *Journal of Materials Chemistry*, 2011. **21**(18): p. 6494–6497.

62. Sears, K. et al., Aligned carbon nanotube webs as a replacement for indium tin oxide in organic solar cells. *Thin Solid Films*, 2013. **531**(0): p. 525–529.

63. Sweetman, L.J. et al., Synthesis, properties and water permeability of SWNT buckypapers. *Journal of Materials Chemistry*, 2012. **22**(27): p. 13800–13810.

64. Jiang, K. et al., Nanotechnology: Spinning continuous carbon nanotube yarns. *Nature*, 2002. **419**(6909): p. 801.

65. Lima, M.D. et al., Biscrolling nanotube sheets and functional guests into yarns. *Science*, 2011. **331**(6013): p. 51–55.

66. Kuznetsov, A.A. et al., Structural model for dry-drawing of sheets and yarns from carbon nanotube forests. *ACS Nano*, 2011. **5**(2): p. 985–993.

67. Nguyen, D.D. et al., Controlled growth of carbon nanotube-graphene hybrid materials for flexible and transparent conductors and electron field emitters. *Nanoscale*, 2012. **4**(2): p. 632–638.

68. Zhang, L.-N. et al., *In situ* growth of porous platinum nanoparticles on graphene oxide for colorimetric detection of cancer cells. *Analytical Chemistry*, 2014. **86**(5): p. 2711–2718.

69. Xu, J. et al., Hierarchical nanocomposites of polyaniline nanowire arrays on graphene oxide sheets with synergistic effect for energy storage. *ACS Nano*, 2010. **4**(9): p. 5019–5026.

70. Bai, H. et al., A pH-sensitive graphene oxide composite hydrogel. *Chemical Communications*, 2010. **46**(14): p. 2376–2378.

71. Wu, C. et al., Highly conductive nanocomposites with three-dimensional, compactly interconnected graphene networks via a self-assembly process. *Advanced Functional Materials*, 2013. **23**(4): p. 506–513.

72. Singh, V. et al., Graphene based materials: Past, present and future. *Progress in Materials Science*, 2011. **56**(8): p. 1178–1271.

73. Ismach, A. et al., Direct chemical vapor deposition of graphene on dielectric surfaces. *Nano Letters*, 2010. **10**(5): p. 1542–1548.

74. Li, X. et al., Large-area graphene single crystals grown by low-pressure chemical vapor deposition of methane on copper. *Journal of the American Chemical Society*, 2011. **133**(9): p. 2816–2819.

75. Ibrahim, I. et al., CVD-grown horizontally aligned single-walled carbon nanotubes: Synthesis routes and growth mechanisms. *Small*, 2012: **8**(13): p. 1973–1992.

76. Yang, X. et al., Bioinspired effective prevention of restacking in multilayered graphene films: Towards the next generation of high-performance supercapacitors. *Advanced Materials*, 2011. **23**(25): p. 2833–2838.

77. Wang, Z.-l. et al., Facile, mild and fast thermal-decomposition reduction of graphene oxide in air and its application in high-performance lithium batteries. *Chemical Communications*, 2012. **48**(7): p. 976–978.

78. Xufeng, Z. et al., Graphene foam as an anode for high-rate Li-ion batteries. *IOP Conference Series: Materials Science and Engineering*, 2011. **18**(6): p. 062006.

79. Tang, Z. et al., Noble-metal-promoted three-dimensional macroassembly of single-layered graphene oxide. *Angewandte Chemie*, 2010. **122**(27): p. 4707–4711.

80. Zhao, H. et al., Carbon nanotube yarn strain sensors. *Nanotechnology*, 2010. **21**: p. 305502.

81. Rein, M.D. et al., Sensors and sensitivity: Carbon nanotube buckypaper films as strain sensing devices. *Composites Science and Technology*, 2011. **71**(3): p. 373–381.

82. Dumee, L.F. et al., Characterization and evaluation of carbon nanotube Bucky-Paper membranes for direct contact membrane distillation. *Journal of Membrane Science*, 2010. **351**(1–2): p. 36–43.

83. Sears, K. et al., Recent developments in carbon nanotube membranes for water purification and gas separation. *Materials*, 2010. **3**(1): p. 127–149.

84. Foroughi, J. et al., Torsional carbon nanotube artificial muscles. *Science*, 2011. **334**(6055): p. 494–497.

85. Gou, J. et al., Single-walled nanotube bucky paper and nanocomposite. *Polymer International*, 2006. **55**(11): p. 1283–1288.

86. Dumée, L. et al., *In situ* small angle X-ray scattering investigation of the thermal expansion and related structural information of carbon nanotube composites. *Progress in Natural Science: Materials International*, 2012. **22**(6): p. 673–683.

87. Ericson, L.M. et al., Macroscopic, neat, single-walled carbon nanotube fibers. *Science*, 2004. **305**: p. 1447–1450.

88. Vigolo, B. et al., Macroscopic fibers and ribbons of oriented carbon nanotubes. *Science*, 2000. **290**: p. 1331–1334.

89. Huynh, C.P. et al., Understanding the synthesis of directly spinnable carbon nanotube forests. *Carbon*, 2010. **48**: p. 1105–1115.

90. Zhang, M. et al., Multifunctional carbon nanotube yarns by downsizing an ancient technology. *Science*, 2004. **306**(5700): p. 1358–1361.

91. Zhang, X. et al., Ultrastrong, stiff, and lightweight carbon-nanotube fibers. *Advanced Materials*, 2007. **19**(23): p. 4198–4201.

92. Koziol, K. et al., High-performance carbon nanotube fiber. *Science*, 2009. **318**: p. 1892–1895.

93. Li, Y.-L. et al., Direct spinning of carbon nanotube fibers from chemical vapor deposition synthesis. *Science*, 2004. **304**: p. 276–278.

94. Motta, M. et al., Mechanical properties of continuously spun fibers of carbon nanotubes. *Nano Letters*, 2005. **5**(8): p. 1529–1533.

95. Liu, K. et al., Carbon nanotube yarns with high tensile strength made by a twisting and shrinking method. *Nanotechnology*, 2010. **21**: p. 045708 (1–7).

96. Miao, M. et al., Poisson's ratio and porosity of carbon nanotube dry-spun yarns. *Carbon*, 2010. **48**(2802–2811): p. 2802.

97. Min, J. et al., High performance carbon nanotube spun yarns from a crosslinked network. *Carbon*, 2013. **52**: p. 520–527.

98. Tran, C.D. et al., Improving the tensile strength of carbon nanotube spun yarns using a modified spinning process. *Carbon*, 2009. **47**(11): p. 2662–2670.

99. Tran, C.D. et al., Manufacturing polymer/carbon nanotube composite using a novel direct process. *Nanotechnology*, 2011. **22**: p. 145302(1–9).

100. Zaeri, M.M. et al., Mechanical modelling of carbon nanomaterials from nanotubes to buckypaper. *Carbon*, 2010. **48**(13): p. 3916–3930.

101. AIST. Development of Materials with Large Specific Surface Areas by Using Single-walled Carbon Nanotubes. 2010; Available from: http://www.aist.go.jp/aist_e/latest_research/2010/20100212/20100212.html.

102. Roy, S. et al., Plasma modified flexible bucky paper as an efficient counter electrode in dye sensitized solar cells. *Energy and Environmental Science*, 2012. **5**(5): p. 7001–7006.

103. Dumee, L.F. et al., A high volume and low damage route to hydroxyl functionalization of carbon nanotubes using hard X-ray lithography. *Carbon*, 2013. **51**: p. 430–434.

104. Sears, K. et al., Focused ion beam milling of carbon nanotube yarns to study the relationship between structure and strength. *Carbon*, 2010. **48**: p. 4450–4456.

105. Miao, M. et al., Production, structure and properties of twistless carbon nanotube yarns with a high density sheath. *Carbon*, 2012. **50**: p. 4973–4983.

106. Dumée, L. et al. ICOM08. in ICOM08. *Preparation of Carbon Nanotube Bucky-Paper Membranes for Desalination by Membrane Distillation*. Honolulu, Hawaii.

107. Hummers, W.S. et al., Preparation of graphitic oxide. *Journal of the American Chemical Society*, 1958. **80**(6): p. 1339.

108. Gómez-Navarro, C. et al., Electronic transport properties of individual chemically reduced graphene oxide sheets. *Nano Letters*, 2007. **7**(11): p. 3499–3503.

109. Karim, M.R. et al., Graphene oxide nanosheet with high proton conductivity. *Journal of the American Chemical Society*, 2013. **135**(22): p. 8097–8100.

110. Cheng, Q. et al., Graphene and carbon nanotube composite electrodes for supercapacitors with ultra-high energy density. *Physical Chemistry Chemical Physics*, 2011. **13**(39): p. 17615–17624.

111. Woo, S. et al., Synthesis of a graphene–carbon nanotube composite and its electrochemical sensing of hydrogen peroxide. *Electrochimica Acta*, 2012. **59**(0): p. 509–514.

112. Yu, K. et al., Carbon nanotube with chemically bonded graphene leaves for electronic and optoelectronic applications. *The Journal of Physical Chemistry Letters*, 2011. **2**(13): p. 1556–1562.

113. Rout, C.S. et al., Synthesis of chemically bonded CNT-graphene heterostructure arrays. *RSC Advances*, 2012. **2**(22): p. 8250–8253.

114. Du, F. et al., Preparation of tunable 3D pillared carbon nanotube-graphene networks for high-performance capacitance. *Chemistry of Materials*, 2011. **23**(21): p. 4810–4816.

115. Yu, D. et al., Self-assembled graphene/carbon nanotube hybrid films for supercapacitors. *The Journal of Physical Chemistry Letters*, 2009. **1**(2): p. 467–470.

116. Kong, B.-S. et al., Layer-by-layer assembly of graphene and gold nanoparticles by vacuum filtration and spontaneous reduction of gold ions. *Chemical Communications*, 2009(16): p. 2174–2176.

117. Muszynski, R. et al., Decorating graphene sheets with gold nanoparticles. *The Journal of Physical Chemistry C*, 2008. **112**(14): p. 5263–5266.

118. Wang, D. et al., Ternary self-assembly of ordered metal oxide – graphene nanocomposites for Electrochemical Energy Storage. *ACS Nano*, 2010. **4**(3): p. 1587–1595.

119. Zhang, H. et al., *In situ* controllable growth of noble metal nanodot on graphene sheet. *Journal of Materials Chemistry*, 2011. **21**(34): p. 12986–12990.

120. Lin, J. et al., Graphene nanoribbon and nanostructured SnO$_2$ composite anodes for lithium ion batteries. *ACS Nano*, 2013. **7**(7): p. 6001–6006.

121. Xu, P. et al., *In situ* growth of noble metal nanoparticles on graphene oxide sheets and direct construction of functionalized porous-layered structure on gravimetric microsensors for chemical detection. *Chemical Communications*, 2012. **48**(87): p. 10784–10786.

122. Dey, R.S. et al., A rapid room temperature chemical route for the synthesis of graphene: Metal-mediated reduction of graphene oxide. *Chemical Communications*, 2012. **48**(12): p. 1787–1789.

123. Wang, D. et al., Self-assembled TiO2–graphene hybrid nanostructures for enhanced Li-Ion insertion. *ACS Nano*, 2009. **3**(4): p. 907–914.

124. Chen, W. et al., Self-assembly and embedding of nanoparticles by *in situ* reduced graphene for preparation of a 3D graphene/nanoparticle aerogel. *Advanced Materials*, 2011. **23**(47): p. 5679–5683.

125. Fang, Y. et al., Self-assembly of cationic polyelectrolyte-functionalized graphene nanosheets and gold nanoparticles: A two-dimensional heterostructure for hydrogen peroxide sensing. *Langmuir*, 2010. **26**(13): p. 11277–11282.

126. Li, Z. et al., Electrostatic layer-by-layer self-assembly multilayer films based on graphene and manganese dioxide sheets as novel electrode materials for supercapacitors. *Journal of Materials Chemistry*, 2011. **21**(10): p. 3397–3403.

127. Georgakilas, V. et al., Functionalization of graphene: Covalent and non-covalent approaches, derivatives and applications. *Chemical Reviews*, 2012. **112**(11): p. 6156–6214.

128. Salavagione, H.J. et al., Recent advances in the covalent modification of graphene with polymers. *Macromolecular Rapid Communications*, 2011. **32**(22): p. 1771–1789.

129. Bai, H. et al., Functional composite materials based on chemically converted graphene. *Advanced Materials*, 2011. **23**(9): p. 1089–1115.

130. Xu, Z. et al., *In situ* polymerization approach to graphene-reinforced Nylon-6 composites. *Macromolecules*, 2010. **43**(16): p. 6716–6723.

131. Bai, H. et al., Non-covalent functionalization of graphene sheets by sulfonated polyaniline. *Chemical Communications*, 2009(13): p. 1667–1669.

132. Luo, Z. et al., Polyaniline uniformly coated on graphene oxide sheets as supercapacitor material with improved capacitive properties. *Materials Chemistry and Physics*, 2013. **139**(2–3): p. 572–579.

133. Li, J. et al., Synthesis of graphene oxide/polypyrrole nanowire composites for supercapacitors. *Materials Letters*, 2012. **78**(0): p. 106–109.

134. Mao, L. et al., Cetyltrimethylammonium bromide intercalated graphene/polypyrrole nanowire composites for high performance supercapacitor electrode. *RSC Advances*, 2012. **2**(28): p. 10610–10617.

135. Schexnailder, P. et al., Nanocomposite polymer hydrogels. *Colloid and Polymer Science*, 2009. **287**(1): p. 1–11.

136. Zu, S.-Z. et al., Aqueous dispersion of graphene sheets stabilized by pluronic copolymers: Formation of supramolecular hydrogel. *The Journal of Physical Chemistry C*, 2009. **113**(31): p. 13651–13657.

137. Bai, H. et al., On the gelation of graphene oxide. *The Journal of Physical Chemistry C*, 2011. **115**(13): p. 5545–5551.

138. Zeng, Y. et al., Significantly enhanced water flux in forward osmosis desalination with polymer-graphene composite hydrogels as a draw agent. *RSC Advances*, 2013. **3**(3): p. 887–894.

139. Li, C. et al., Three-dimensional graphene architectures. *Nanoscale*, 2012. **4**(18): p. 5549–5563.

140. Vickery, J.L. et al., Fabrication of graphene–polymer nanocomposites with higher-order three-dimensional architectures. *Advanced Materials*, 2009. **21**(21): p. 2180–2184.

26 Electronic Structure of Graphene-Based Materials and Their Carrier Transport Properties

Wen Huang, Argo Nurbawono, Minggang Zeng, Gaurav Gupta, and Gengchiau Liang

CONTENTS

Abstract..401
26.1 Introduction 401
26.2 Graphene..402
    26.2.1 Graphene Fundamentals...403
    26.2.2 Density of States and Carrier Conductivity of Graphene..405
    26.2.3 Graphene p–n Junction...406
        26.2.3.1 Negative Refraction 406
        26.2.3.2 Klein Tunneling 407
    26.2.4 QHEs in Graphene...408
    26.2.5 Phonon and Heat Conduction for Graphene 409
    26.2.6 Graphene with Grain Boundary 410
26.3 Graphene Nanoribbons..410
    26.3.1 GNR Fundamentals...410
    26.3.2 Magnetism in ZGNRs and Magnetic-Field-Effect in AGNRs..412
    26.3.3 Thermal and Thermoelectric Properties for GNRs...416
26.4 Summary 417
References..418

ABSTRACT

Graphene, with the amazing physical and electronic properties, has numerous unique properties compared to the existing materials. It is a two-dimensional monolayer of carbon atoms arranged in a hexagonal lattice. Three in-plane sp^2-hybridized σ-bonds and one out-of-plane dangling π-bond at each carbon site empowers graphene with structural flexibility and unique electronic band structure with linear dispersion relations at the Dirac points. This chapter reviews the fundamentals of graphene-like electronic structure, density of states and conductivity, and the peculiar transport properties such as negative refraction, Klein tunneling, and quantum Hall effects (QHE). The heat conduction and grain boundary effects have also been discussed. For nanoelectronic applications, one-dimensional strips of graphene called graphene nanoribbons (GNRs) are of great interests. This chapter also reviews the electronic properties of GNRs, which greatly depend on the boundary effects and quantum confinement. They can be either semiconducting or metallic according to the chirality. By applying the external electrical and magnetic field, their properties can be tuned to exploit the spintronic and magnetoresistive effects. The thermal and thermoelectric properties of various GNR structures are also reviewed in this chapter.

26.1 INTRODUCTION

Decades ago, before the isolation of monolayer graphene, the variety of allotropes of carbon, such as diamond, graphite, buckminsterfullerene, and carbon nanotube (CNT), had also been studied extensively in different scientific areas [1]. Diamond and graphite are two well-known bulk forms of carbon. Diamond is the hardest known material in nature because of its stable and strong covalent bonds. Graphite is thermodynamically the most stable form of carbon. It is a layered stack of graphene and can conduct electricity because of the electron delocalization within the layer. In 1985, zero-dimensional buckminsterfullerenes, also called buckyballs with 60 carbon atoms assembling a football, were discovered by Kroto, Curl, and Smalley [2], who were awarded Nobel Prize in Chemistry in 1996. Besides the spherical buckyball clusters, the fullerene family also contains other types of carbon structures such as nanotubes. CNT [3] is a one-dimensional (1D) fullerene with cylindrical nanostructure. It can be considered as a graphene sheet rolled up along a specific angle [1]. CNTs have extraordinary electrical and optical properties that make them valuable for nanotechnology. In 2004, monolayer graphene was finally discovered by Geim and Novoselov [4], and their groundbreaking experimental work was awarded with Nobel Prize in Physics in 2010. Due to similarity in properties of graphene and CNT, the researchers could immediately benefit from their experience on the latter to expedite the advances in graphene, which has since then drawn intensive attention in both pure sciences and engineering.

Each carbon atom in graphene is at a distance of about $a = 1.42$ Å (bond length) from its three nearest neighbors, conjoined via sp^2-hybridized σ covalent bonds, and has a dangling π-bond perpendicular to the graphene plane. The σ-bonds provide graphene a structural feasibility, while π-bonds are responsible for most of the peculiar electronic properties of graphene. Graphene is a semimetal or zero bandgap semiconductor, in which conduction and valence band meet at the Dirac points (K and K') on the edge of the first Brillouin zone. Near the Dirac points, low-energy electrons are characterized as massless fermionic particles that have a linear dispersion described by the Dirac equation, with the Fermi velocity $v_F \sim 10^6$ m s^{-1} [5]. These massless Dirac fermions can evince extraordinarily high electron mobility of ~200,000 cm^2 V^{-1} s^{-1}, which is an intrinsic limit at room temperature because of the various scattering mechanisms [6]. The large cyclotron energies of relativistic Dirac fermions, moreover, result in new physical phenomena like anomalous QHE [7,8], pseudospin [9], and electro-optical behaviors [10,11]. Besides electronic properties, graphene also has high thermal conductivity because of the out-of-plane vibrational modes in phonon dispersion, which can enable fast heat dissipation in integrated circuits and new functional devices. The experimentally measured thermal conductivity for suspended single-layer graphene is in the range of ~4400–5800 W m^{-1} K^{-1} at room temperature [12]. Furthermore, because of the structural flexibility of graphene, the effects of substrates, defects, and strains [13,14] can be exploited for new possibilities.

In addition, for nanoscale electronic applications, a strip of graphene with finite termination in lateral direction, namely, graphene nanoribbon (GNR), can open the bandgap in the semimetal providing the potential to replace silicon. GNRs can also be treated as unwrapped CNTs. The advancements in fabrication process for both top-down and bottom-up approaches have been introduced to easily narrow-down the graphene to fabricate GNRs with atomic precision for different widths and complex shapes [15,16], which has motivated the studies for different GNR structures. The boundary effects of different edge types (AGNRs and ZGNRs) and quantum confinement have great impact on electrical properties of GNRs. AGNRs are semiconducting with oscillatory decreasing bandgaps [17] for all three families ($3p$, $3p + 1$, and $3p + 2$) as the width increases, while all ZGNRs have very small bandgaps [18]. The magnetic ordering that exists at the edge of ZGNRs serves as a basis for novel spintronics device applications [19], whereas controlling the bandgap through magnetic

field gives rise to the magnetoresistive effect in AGNRs. Hence, the exploitation of magnetic field and electrical field to influence the properties of GNRs should result in interesting applications. However, the thermal conductivity of GNRs is reduced because of the dimensional confinement, making it not as good as graphene for heat sinks. However, low thermal conductivity together with high electrical conductivity and Seebeck coefficient are necessary for efficient thermoelectric conversion between electricity and heat. In addition, thermal conductivity of GNRs can be further suppressed by introducing defects and scatterings, which make it more promising for thermoelectric applications [20].

In this chapter, we summarize the basic physics and fundamentals of graphene and GNRs, along with the theoretical and experimental work on their transport properties. For graphene, density of states, carrier conductivity, negative refraction, Klein tunneling, QHEs, heat conduction, and grain boundary effects are reviewed, while magnetic field, thermal, and thermoelectric properties are reviewed for GNRs. For details of possible applications of graphene and GNRs, readers are encouraged to peruse further discussions in this book series, Volume 6, Chapter 19.

26.2 GRAPHENE

The structure of graphene with three in-plane sp^2 hybridized σ-bonds and one out-of-plane dangling π-bond provides unique properties to it that are promising for novel device applications. In this section, the fundamentals of graphene are introduced followed by a review on graphene transport properties, such as density of states, carrier conductivity, negative refraction, Klein tunneling, QHEs, heat conduction, and grain boundary effects, and some key findings are listed in Table 26.1.

TABLE 26.1
Key Findings of Graphene Transport Properties

Properties	Key Findings	References
Conductivity	Linearly proportional to the carrier density. The ionic impurity scattering dominates the carrier transport characteristics.	[21]
p–n junction	Negative refraction and Klein tunneling effects	[22–24]
AQHE	Due to its relativistic-like energy dispersion behaviors. Graphene is the only material whose QHE is observable at room temperature.	[7,25,26]
Thermal conductivity	Very high thermal conductivity of ~2000–5000 W m^{-1} K^{-1}	[12,27–30]
Structural defects of grain boundary	The electronic properties are different for different types from metallic to semiconducting based on the theory of transverse momentum conservation. All types have excellent thermal conductivity.	[31–35]

26.2.1 GRAPHENE FUNDAMENTALS

A graphene monolayer is basically made of carbon atoms in 2D hexagonal structures [5]. For theoretical analysis, monolayer graphene structure can be described by two sublattices, A and B atoms, which are arranged in triangular lattice as shown in Figure 26.1a. The unit lattice vectors are given by

$$\vec{a}_1 = \frac{a}{2}\left(3,\sqrt{3}\right) \text{ and } \vec{a}_2 = \frac{a}{2}\left(3,-\sqrt{3}\right) \quad (26.1)$$

where the constant a is around 1.42 Å for graphene, which is the carbon–carbon bond length. The nearest-neighbor hopping vectors in Figure 26.1a from lattice B to lattice A are given by

$$\vec{\delta}_1 = \frac{a}{2}\left(1,\sqrt{3}\right), \quad \vec{\delta}_2 = \frac{a}{2}\left(1,-\sqrt{3}\right), \quad \text{and} \quad \vec{\delta}_3 = -a(1,0) \quad (26.2)$$

The corresponding reciprocal space of graphene appears to be rotated by 30° compared to its real space lattice as shown in Figure 26.1b. The reciprocal lattice vectors are given by

$$\vec{b}_1 - \frac{2\pi}{3a}\left(1,\sqrt{3}\right) \quad \text{and} \quad \vec{b}_2 - \frac{2\pi}{3a}\left(1,-\sqrt{3}\right) \quad (26.3)$$

The most important high symmetry points in the first Brillouin zone corners are called K and K'. They play many important roles in the low-energy physics of graphene as we will discuss later. The coordinates of the symmetry points are given by

$$K = \frac{2\pi}{a}\left(\frac{1}{3},\frac{1}{3\sqrt{3}}\right) \quad \text{and} \quad K' = \frac{2\pi}{a}\left(\frac{1}{3},-\frac{1}{3\sqrt{3}}\right) \quad (26.4)$$

Every carbon atom in graphene has four hybridized sp^2 orbitals, which form three σ covalent bonds with its nearest neighbors, and a dangling π orbital oriented perpendicular to the graphene plane. The σ-bonds are responsible for the

graphene's in-plane structural strength and stiffness, resulting in graphene's lack of out-of-plane stiffness creating ripples, which exist across graphene layers. The σ bands are fully occupied bands, and they form deep valence bands in graphene. Most of the important physics of graphene's electronic and transport properties can be well represented by a simple nearest-neighbor tight-binding model of the π orbitals of carbon atoms. The π orbitals are half-filled orbitals, and they have strong Coulomb energy properties, which exhibit strong tight-binding character. A nearest-neighbor tight-binding Hamiltonian for graphene π orbitals is given by

$$H = -t\sum_{<ij>\sigma}(a_{i\sigma}^{\dagger}b_{j\sigma} + H.c.) \quad (26.5)$$

where $a_{i\sigma}^{(\dagger)}$ operator annihilates (or creates) electron i of spin σ at sublattice A, and similarly with operators $b_{j\sigma}^{(\dagger)}$ on sublattice B. The hopping energy t is around 2.5–3 eV in general [1], and a small correction for second-nearest-neighbor hopping (hopping between atoms of the same sublattice) will add to small corrections in asymmetry between the valence and conduction band dispersions. For an infinite graphene layer, this Hamiltonian can be easily solved exactly by expanding the basis vector in terms of Bloch states, where the corresponding states of electron one lattice vector away differ by a Bloch phase factor $e^{i\vec{k}\cdot\vec{a}}$. After summing up all the phases and diagonalizing the Hamiltonian, the valence and conduction π band energy dispersion is given by

$$E(k) = \pm t\sqrt{3 + 2\cos(\sqrt{3}k_y a) + 4\cos\left(\frac{\sqrt{3}}{2}k_y a\right)\cos\left(\frac{3}{2}k_x a\right)} \quad (26.6)$$

For half-filled bands or Fermi energy level at zero energy, the positive sign in Equation 26.6 applies to the upper (conduction) band or sometimes called the π* band, and the negative sign applies to the lower (valence) band or π band. The dispersion relation shows six corners in the Brillouin zone that can be labeled as alternating K and K' points as shown in Figure 26.2a. These K and K' points in the Brillouin zone are called the Dirac points. The corresponding density of states is also shown in Figure 26.2b, which can be calculated from

$$DOS(E) = \int_{BZ} \delta(E(k) - E)dk \quad (26.7)$$

At low-energy excitations around the Dirac points, the physics of graphene quasiparticles can be described as relativistic Dirac fermions. This is because low-energy excitations around K and K' points in the Brillouin zone are approximately linear with the momentum wave vector, and they are isotropic in all directions along the graphene layer as depicted in Figure 26.3a, which form six Dirac cones. To obtain the dispersion around these Dirac points, one can perform a

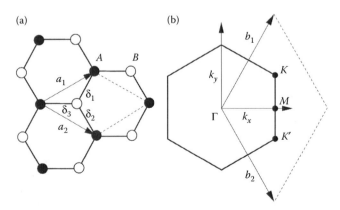

FIGURE 26.1 (a) Real space lattice structure of monolayer graphene with sublattice A and B. (b) Reciprocal space of graphene appears tilted by 30° relative to real space. The Brillouin zone of graphene is shown with its high symmetry points.

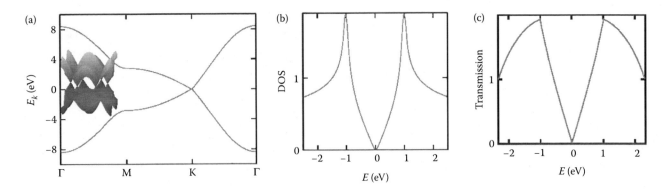

FIGURE 26.2 (a) Band structure of graphene based on π orbital of the tight-binding model. Inset is the band structure visualization in 2D Brillouin zone. (b) Density of states (DOS) of graphene. (c) Transmission of graphene. (The results are benchmarked with G. Liang et al. *Journal of Applied Physics*, 102;2007:054307.)

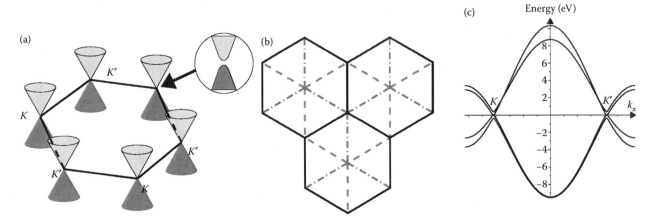

FIGURE 26.3 (a) Dirac cones of graphene at low-energy excitations around K and K' points. Inset shows the details of band gap opening at K point. (b) Common stacking order of multilayer graphene (graphite). (c) Band structure of pristine bilayer graphene showing gapless dispersion at the Dirac points. Applications of an external potential would open a gap at the Dirac points.

Taylor expansion of the dispersion Equation 26.6 about a Dirac point and keep only small linear terms in momentum, that is, $k = K + q$ and q is a small wave vector. This would give the energy dispersion as

$$E(\vec{q}) = \pm \hbar v_F |\vec{q}| \qquad (26.8)$$

where $v_F = 3at/2\hbar$ (~10^6 m/s), and it is approximately 1/300 of the speed of light. The complete analogue of the Dirac Hamiltonian can also be obtained by replacing the operators in Equation 26.5 with slowly varying field operators defined as

$$a_n = e^{-i\vec{K}\cdot\vec{R}_n} a_{1n} + e^{-i\vec{K}'\cdot\vec{R}_n} a_{2n} \quad \text{and} \quad b_n = e^{-i\vec{K}\cdot\vec{R}_n} b_{1n} + e^{-i\vec{K}'\cdot\vec{R}_n} b_{2n} \qquad (26.9)$$

After expanding the operators and keeping the leading order, using the prescription $\vec{q} = -i\vec{\nabla}$ and making use of the Pauli matrices $\vec{\sigma}$, we can rewrite the Hamiltonian in terms of the momentum operators as

$$H = v_F \sum_{\eta=1,2} \psi_\eta^\dagger \vec{\sigma}\cdot\vec{p}\,\psi_\eta \qquad (26.10)$$

The four component field operators $\psi_\eta^{(\dagger)}$ has the so-called valley index η (η = 1, 2) which is associated with K and K' points in the Brillouin zone. So the Hamiltonian in Equation 26.10 is a 4 × 4 matrix with each 2 × 2 block making up each valley at K and K', respectively. Every 2 × 2 block in Equation 26.10 resembles a spinor space (or pseudospin) resulting from the sublattice A and B. The term chiral is often used to describe quasiparticles in graphene because the eigenstates at K and K' associated with each valley would swap over (reverses sign) upon a rotation by 180°,

$$\psi_{\pm,K}(q) = \frac{1}{\sqrt{2}}\begin{bmatrix} e^{-i\theta_q/2} \\ \pm e^{i\theta_q/2} \end{bmatrix} \quad \text{and} \quad \psi_{\pm,K'}(q) = \frac{1}{\sqrt{2}}\begin{bmatrix} e^{i\theta_q/2} \\ \pm e^{-i\theta_q/2} \end{bmatrix}$$

$$(26.11)$$

where $\theta_q = \arctan(q_x/q_y)$. This picture resembles chiral particle behavior, but the apparent spin space implied here is not the real spin of the particles, and we shall call it pseudospin. There have been suggestions, however, to harvest the valley electronic behavior in graphene as a "valleytronics" device [9]. Nevertheless, this idea has so far remained in confines of theoretical proposals only [36]. The low-energy Hamiltonian model is valid up to a certain cut-off wave vector, above which the linear dispersion becomes inaccurate. The common method to determine the cut-off wave vector is to impose limit to the carrier energy to be less than $0.4t$ (around 1 eV), which corresponds to $k = 0.25$ nm^{-1}.

Apart from monolayer graphene, multilayer graphene is more commonly found in exfoliated graphite. In a multilayer graphene or graphite flakes in general, the most common stacking configuration is shown in Figure 26.3b, where one layer has half of the atoms directly above the one below it while keeping the same lattice orientations. The most common stacking orders like ABAB and sometimes ABCABC stacking are also found in different types of graphite. A bilayer graphene may be obtained by combining two relaxed monolayers, and for an AB-Bernal stacked bilayer graphene with A sublattice atoms directly above A' sublattice atoms below it, the tight-binding model Hamiltonian is given by

$$H = -\gamma_0 \sum_{<ij>} (a^\dagger_{mi\sigma} b_{mj\sigma} + H.c.) - \gamma_1 \sum_{<j\sigma>} (a^\dagger_{1j\sigma} a_{2j\sigma} + H.c.)$$
$$- \gamma_4 \sum_{j\sigma} (a^\dagger_{1j\sigma} b_{2j\sigma} + a^\dagger_{2j\sigma} b_{1j\sigma} + H.c.) - \gamma_3 \sum_{j\sigma} (b^\dagger_{1j\sigma} b_{2j\sigma} + H.c.)$$

$$(26.12)$$

where m is the monolayer plane index ($m = 1, 2$), and a typical set of parameters for the hopping energy are $\gamma_0 = t \approx 3$ eV, $\gamma_1 \approx 0.4$ eV, $\gamma_4 \approx 0.04$ eV, and $\gamma_3 \approx 0.3$ eV. The band structure of a bilayer graphene is shown in Figure 26.3c. It is also possible to open a gap in bilayer graphene by applying a gate potential [37]. The tight-binding model in Equation 26.12 can be written in a 4×4 matrix form as

$$H = \begin{bmatrix} -V & -\gamma_0 f(\vec{k}) & \gamma_4 f(\vec{k}) & -\gamma_3 f^*(\vec{k}) \\ -\gamma_0 f^*(\vec{k}) & -V & \gamma_1 & \gamma_4 f(\vec{k}) \\ \gamma_4 f^*(\vec{k}) & \gamma_1 & V & -\gamma_0 f(\vec{k}) \\ -\gamma_3 f(\vec{k}) & \gamma_4 f^*(\vec{k}) & -\gamma_0 f^*(\vec{k}) & V \end{bmatrix}$$

and $\quad f(\vec{k}) = \sum e^{i\vec{k}\cdot\vec{\delta}_l}$ $\quad$ (26.13)

In a bilayer graphene, the application of an extrinsic potential V causes the breaking of the AB lattice symmetry, and it enables the opening of bandgap around the Dirac points. This makes bilayer or multilayer graphene more interesting for device applications where a bandgap is needed on a large flake.

An infinite pristine graphene does not have a bandgap, thus many efforts have been attempted to create the bandgap necessary for useful applications. Among many techniques that already exist today, strain engineering of graphene has been proposed to create all-graphene electronics [13]. The first-order effects of strain on graphene electronic properties can be modeled by the perturbation of the hopping energy $t_i \to t_i + \delta t_i$. Within the low-energy Hamiltonian, the effects of this energy perturbation can be modeled as a pseudo-magnetic field by introducing a pseudo-vector potential $\vec{A}$ into the momentum operator,

$$H = v_F \int d\vec{r} \Psi^\dagger \begin{bmatrix} \vec{\sigma} \cdot \left(\vec{p} - \dfrac{1}{v_F} \vec{A} \right) & 0 \\ 0 & -\vec{\sigma} \cdot \left(\vec{p} + \dfrac{1}{v_F} \vec{A} \right) \end{bmatrix} \Psi \quad (26.14)$$

The field operators there contain electron fields at each valley and sublattice,

$$\Psi = \left[\psi^A_K(r), \psi^B_K(r), \psi^A_{K'}(r), \psi^B_{K'}(r) \right] \quad (26.15)$$

In this case, the pseudo-vector potential in Equation 26.14 arises from the perturbation energy term δt_i as a result of the strain,

$$\vec{A}(r) = A_x(r) - iA_y(r)$$
$$= \sum_i \delta t_i e^{i\vec{K}\cdot\vec{n}_i} \quad (26.16)$$

and the pseudo-vector potential appears with its sign reversed for K' valley which ensures time-reversal symmetry of the system. Strain engineering in graphene can potentially be used for generating confinements as well, which is difficult for Dirac particles due to Klein tunneling effect. It may also be used to study tunneling effects in general, beam collimations, and so on. In terms of generating bandgaps, strain engineering is more effective when it is applied on finite size nanoribbons [38,39].

26.2.2 DENSITY OF STATES AND CARRIER CONDUCTIVITY OF GRAPHENE

As the band structure of graphene is identified, its electron density of states ($D(E)$), carrier concentration ($n(E)$), and electron transport property (conductivity, $\sigma(E)$) can be theoretically investigated. Based on the fundamentals of semiconductor physics [40,41], first, the equations for the electronic properties of two-dimensional materials are given as follows:

$$D(E)v(E)p = 2\tilde{N} \quad (26.17)$$

$$\sigma(E) = g_s \frac{q^2}{h} \frac{M(E)}{W} \lambda(E) \quad (26.18)$$

$$\tilde{N} = \pi \frac{WL}{(h/p)^2} g_s g_v \quad (26.19)$$

$$M(E) = \frac{2W}{(h/p)} g_v \quad (26.20)$$

where v is the carrier velocity, p the carrier moment, $\tilde{N}$ the number of states for the electrons, M the number of modes in the system, λ the carrier mean free path, W and L are the width and the length of the system, h the Planck constant, and g_s and g_v are the degeneracy of spin and the valley in the electronic band structure, respectively. For graphene, electron behavior follows the Dirac dispersion relation, $E = \hbar v_f k = v_f p$, and g_s and g_v are 2. As a result, it can be obtained that

$$\tilde{N} = \pi \frac{WL}{(h/p)^2} g_s g_v = \frac{Ap^2}{\pi \hbar^2} = \frac{AE^2}{\pi \hbar^2 v_f^2}, \quad (26.21)$$

$$M(E) = \frac{2W}{h/p} g_v = \frac{2WE}{\pi \hbar v_f}, \quad (26.22)$$

and

$$D(E) = \frac{\tilde{N} \cdot d}{v_f \cdot p} = \frac{2}{E} \times \frac{AE^2}{\pi \hbar^2 v_f^2} = \frac{2AE}{\pi \hbar^2 v_f^2}. \quad (26.23)$$

From these results, it is clear that near the Fermi level at the Dirac point (E_f), D is linearly proportional to energy. It is consistent with DOS shown in Figure 26.2b as energy near the Dirac point. Furthermore, assuming the temperature is close to zero to simplify the calculations, carrier concentration can be derived from the above equation to obtain

$$n(E_f) = \frac{AE_f^2}{\pi \hbar^2 v_f^2}. \quad (26.24)$$

Therefore, the conductivity of electrons in graphene can be generalized as

$$\sigma(E) = g_s \frac{q^2}{h} \frac{M}{W} \lambda = \frac{2q^2}{h} \cdot \left(\frac{2E}{\pi \hbar v_f}\right) \cdot \lambda(E), \quad (26.25)$$

where $\lambda(E) = \pi/2 v_f \tau(E)$, and $\tau(E)$ is the life time of the carriers for different scattering mechanisms.

In this chapter, we will simply consider three common mechanisms that influence transport in solid-state materials as examples. The first case is ballistic transport, in which λ can be treated as a constant for the entire transport length. As a result, $\sigma(E)$ is proportional to energy, and therefore proportional to $\sqrt{n}$. The second case is carrier transport in the presence of acoustic phonon scattering. From the scattering theory, the mean-free time (τ_{AP}) is scales with energy as $(1/\tau_{AP}(E)) \propto D(E) \propto E \Rightarrow \tau_{(AP)}(E) \propto (1/E)$. This indicates that σ is not a function of energy or carrier concentration. It is a very interesting result, since the conventional treatment to enhance conductivity in normal semiconductors is to increase the carrier concentration. The Drude model, $\sigma(E) = qn\mu$, where μ is the carrier mobility, indicates that the carrier mobility of graphene decreases as the carrier concentration increases under the consideration of the acoustic phonon

scattering. Finally, we consider the case of ionic impurity scattering, whose scattering rate is inversely proportional to energy, $(1/\tau_I(E)) \propto (1/E) \Rightarrow \tau_I(E) \propto E$. Therefore, it can be found that conductivity is linearly proportional to the carrier density, $\sigma(E) \propto E^2 \propto n(E_f)$. The effective carrier conductivity should consider all effects via Matthiessen's rule,

$$\frac{1}{\sigma(E)} = \sum_{j=\text{all conditions}} \frac{1}{\sigma_j(E)}.$$

The recent experimental measurement [21] shows a near linear relation between conductivity and carrier concentration, indicating that in graphene, the backscattering caused by phonon is suppressed and the ionic impurity scattering dominates the carrier transport characteristics even at room temperature, in contrast to most of the semiconductor materials in which acoustic phonon scattering has a dominant role at room temperature.

26.2.3 GRAPHENE P–N JUNCTION

Unlike any preexistent semiconductor, the gapless Dirac energy bands in graphene have facilitated the study [42] and manipulation [43] of unconventional physics of negative refraction, a phenomenon that was previously limited to optical metamaterials. An electronic analogue of Veselago lens [44,45] effectuates at the p–n junction in graphene, because of the apparent negative refractive index that focuses the electron waves [42] traversing across the junction. In addition to the negative refraction, the reflection-less (perfect-reflection) propagation of normally incident wave through the potential barrier [46] at p–n junction for monolayer (bilayer) graphene has disinterred the examination of Klein tunneling [47–51]. Since p–n junctions are an elementary unit of electronic devices, the understanding of quantum transport through these is of prime importance. This section therefore appraises the effect of negative refraction and Klein tunneling on electron transport through graphene p–n junctions.

26.2.3.1 Negative Refraction

The Fermi level at the Dirac point in pristine graphene can be easily shifted into conduction (valence) band by small n-type (p-type) charge doping or positive (negative) gate potential to create a spatial distribution of charge along the transport direction [52]. Conversely, for a given Fermi level, the Dirac point can be shifted to engineer a slowly varying potential barrier of width D_w and height V_0, in the form of a p–n junction, as shown in Figure 26.4. An electron wave packet of Fermi energy E_F, with respect to a local Dirac point, impinging the barrier at an angle θ, with respect to normal of the boundary, must conserve the transverse component of the crystal momentum k whose magnitude is given by $|E|/\hbar v_F$, where v_F is the Fermi velocity. Furthermore, the group velocity of electron waves is parallel (antiparallel [53]) to the wavevector in conduction (valence) band. The conservation of momentum, therefore, gives an equivalent Snell's law as

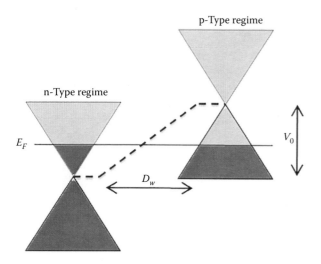

FIGURE 26.4 Schematic to illustrate a p–n junction of width D_w and barrier height V_0. The electron filled energies below the Fermi level E_F are shaded in dark gray, whereas empty band is marked in light gray.

$k_{Fn} \sin(\theta_n) = -k_{Fp} \sin(\theta_p)$, where k_{Fn} and k_{Fp} are Fermi wave-vectors on n and p side, respectively. The refractive index for wave going from n-type region to p-type becomes $\eta_{np} = \sin(\theta_n)/\sin(\theta_p) = -k_{Fp}/k_{Fn}$. This expression although describes the path of the electrons, the transmission is governed by conservation of pseudospin [54]. Therefore, below the critical angle given by Snell's law, the transmission [46] across symmetrical p–n junction is expressed as

$$T(k_F, \theta_n) = \left[\frac{\cos \theta_n \cdot \cos \theta_p}{\cos^2 \left(\frac{\theta_n + \theta_p}{2} \right)} \right] e^{-\pi \hbar v_F k_F^2 D_w \sin^2 \theta_n / V_o} \qquad (26.26)$$

In particular, the above equation implies that for a normally incident wave, that is, $\theta_n = 0$, there is a perfect transmission irrespective of the magnitude of V_0, which is a signature of Klein tunneling. This happens because for backscattering, the pseudospins must be flipped [54] that requires fast spatially varying potential to act differently on the two sublattices (A and B) of graphene. Also note that in Equation 26.26, the exponential factor is present only for a slowly varying bipolar junction, and therefore, for the unipolar junction only cosines terms from pseudospin conservation determine the transmission. For a detailed derivation, we suggest readers to refer [46]. Furthermore, for the p and n regimes, the wavefunction is formulated [55] as $|\Psi_n\rangle = (1, e^{i\theta})/\sqrt{2}$ and $|\Psi_p\rangle = (1, e^{i(\theta+\pi)})/\sqrt{2}$, where $\theta = \arctan(k_y (k_f^2 - k_y^2)^{-0.5})$. Now, the transmission probability for any given mode is $|\langle \Psi_n | \Psi_p \rangle|^2$, which becomes unity for $\theta = 0$. However, a wavefunction mismatch is observed between n-type and p-type regimes for nonzero θ. Unequal conductance is, therefore, observed for an equivalent n–n and n–p doping, which induces asymmetry in the conductance measurements [54]. Now, we briefly peruse two important effects of negative refraction on electron transport through the p–n junction. First, a p–n junction acts as an angle selective filter

[11,56]. The electron waves incident at an angle greater than $\theta_C = \arcsin(|\eta_{np}|)$ are not allowed to cross the junction, a property that can be manipulated for electro-optic-based graphene devices as discussed in Volume 6, Chapter 20. The angular bandwidth of $2\theta_C$ therefore makes a p–n junction more resistive compared to homogeneous graphene. Second, the transition width D_w reduces the angular bandwidth [55] considerably, in addition to which the obliquely incident electron waves have to quantum mechanically tunnel, as an evanescent wave [22], through the mode-dependent band gap, that further reduces the conductance of the p–n junction. The simultaneous deterrence of conductance has been verified through experiments [23]. However, for n–p–n (or p–n–p) junctions, the transmission [24] becomes a function of not only incidence angle but also of barrier width (see Equations 26.3 and 26.4 in Reference 48). Therefore, for certain electron energies and incidence angle, the resonant tunneling [48] may increase the conductance.

26.2.3.2 Klein Tunneling

Quantum mechanics permits an intraband transmission of a particle as an evanescent wave through a classically forbidden energy barrier but the probability of such phenomenon decreases exponentially with barrier height (V_0) and its spatial width (W_B). Klein paradox [47], however, suggests that an electron moving at a relativistic speed can tunnel through the barrier with transmission probability approaching unity, as the barrier height tends to infinity. The extreme conditions [48] required to verify the paradox or its promulgated resolutions [57,58] have so far eluded experimental observation. The Fermi velocity of $c/300 \sim 10^6$ m/s in Dirac bands of graphene, therefore, has provided a solid-state platform to examine the theory, as originally suggested in Reference 48 and later experimentally demonstrated in Reference 49. It is, however, imperative to understand that Klein tunneling in graphene is subtly different [22] from both quantum tunneling and Klein paradox. This is because, in Klein tunneling, an electron makes an interband transition, and there is no paradoxical charge conservation owing to natural charge-conjugate symmetry between electron and holes (condensed-matter equivalents of positrons). This effect has been comprehensively reviewed in References 22, 59, and 60; especially note that Reference 22 explains some common misconceptions about this phenomenon in graphene.

The solution of Schrodinger equation at an interface mandates the continuity of wavefunction and its differential. At the p–n junction, for monolayer graphene, the wave function for an electron (conduction band in n-type region) propagating with wavevector k matches with the corresponding wavefunction of hole (valence band in p-type region), which propagates with a real wavevector—k. For a normally incident electron wave, the conservation of pseudospin, which acts as a selection rule [22], complements the wavefunction matching to yield a reflection-less transport across the barrier. However, for the bilayer graphene, the solution results in an evanescent wave for the hole wavefunction in barrier (p-type region) with an imaginary wavevector—ik [48], which totally suppresses the electron propagation across the interface. In addition to the tunneling through transition width D_W (c.f. Figure 26.4) and Klein tunneling through barrier W_B for normally incident wave, the oblique incidence of electron wave on a square barrier (e.g., n–p–n junction) may result in Fabry–Perot resonances [22] due to multiple interferences because of oscillations between n–p and p–n interfaces. The resonance periodically modulates the transmission with incidence angle, and consequently the conductance, as theoretically elaborated in Reference 61 and experimentally observed in Reference 49 by controlling the Fabry–Perot fringes via magnetic field. It should, however, be noted that the impurity concentration [62] must be sufficiently low ($<10^{12}$/cm^2) so that the impurity limited mean-free path is much longer than the spatial dimensions for respective quantum effect to be observed, for example, W_B for Fabry–Perot resonance and D_W for Klein tunneling.

26.2.4 QHEs in Graphene

QHE is one of the most important fundamental phenomena in condensed matter physics. Under strong magnetic field, the diagonal conductivity σ_{xx} is vanishing to zero and the nondiagonal conductivity shows quantized values of $\sigma_{xy} = -\nu e^2/h$. The factor ν can be integer (IQHE) or fractional (FQHE) depending on the system. Graphene shows an anomalous quantum Hall effect (AQHE) due to its relativistic-like energy dispersion behaviors, which differ from ordinary quadratic dispersions in typical 2D electron gas. In tight-binding models, the effects of magnetic field can be studied by using Peierls substitution in the hopping energy t, and Landau levels can be deduced from the spacing of the flat bands near the Dirac points [5]. For an infinite monolayer graphene, however, the low-energy Hamiltonian sufficiently describes the physics of the first few Landau levels, and the solutions are much simpler to obtain than tight-binding models. In low-energy Hamiltonian models, we can replace the momentum operator with canonical momentum operator $\vec{p} \to (\vec{p} + e\vec{A})$, and by using the Landau gauge $\vec{A}(x) = Bx\hat{y}$, the analytical solutions for the eigenenergies and eigenstates can be obtained as [25]

$$E_{n,k_y} = \text{sgn}(n)\sqrt{2|n|}\frac{\hbar v_F}{l_c} \qquad (26.27)$$

$$\psi_{n,k_y}(\xi) = e^{ik_y y}\begin{bmatrix} \text{sgn}(n)\varphi_{|n|-1}(\xi) \\ i\varphi_n(\xi) \end{bmatrix} \qquad (26.28)$$

where $l_c = \sqrt{\hbar/eB}$, $\xi = (x + l_c^2 k_y)/l_c$ and n is integer. The function $\varphi_n(\xi)$ is the familiar harmonic oscillator eigenfunction in quantum mechanics. The analytical solutions above perfectly explain the experimentally observed unusual $\sqrt{|n|}$ dependence of discrete Landau levels in graphene, which makes it different from other ordinary IQHE of 2D electron gas due to its zero mode Hall conductivity [7]. The zero mode is shared equally by the electron and hole. The Hall conductivity of graphene can be derived directly from its energy spectrum using simple classical arguments, and at the half-filling chemical potential, it is given by

$$\sigma_{xy} = -\frac{2e^2}{h}(2n+1) \qquad (26.29)$$

The Hall conductivity shows the presence of zero mode shared by two Dirac points, which eliminates the typical Hall plateau at $n = 0$ as shown in Figure 26.5. We also see from the eigenvalues above that the cyclotron energy scales linearly with $\sqrt{nB}$ instead of nB, and this makes the energy spacing rather large for the first few Landau levels. Observation of QHE requires energy spacing to be greater than the temperature smearing, which implies that the energy scales of Dirac fermions would be observable at much higher temperatures compared to the ordinary 2D electron gas in semiconductors. This enables observations of AQHE in graphene at room temperatures [7]. Graphene is the only material whose QHE is observable at room temperature. Magnetic fields are also useful for particle confinement in graphene because relativistic particles cannot be confined with a conventional electric barrier. Manipulation of the Landau states in a patterned magnetic lattice can also be used for confining Dirac particles in graphene [26]. A simple way to systematically study a graphene system with magnetic lattice, is to use the low-energy Hamiltonian at one Dirac point (neglecting intervalley scattering) and expand the eigensolution in terms of the Bloch functions,

$$\psi(x, y) = e^{i(k_x x + k_y y)}\begin{bmatrix} \phi_1(x, y) \\ \phi_2(x, y) \end{bmatrix} \qquad (26.30)$$

where $\phi_{1,2}(x, y)$ are periodic. The discrete Fourier transform can be applied,

$$\phi_1(x, y) = \sum_{m,n} a_{mn} e^{im\omega_x x} e^{in\omega_y y} \qquad (26.31)$$

$$\phi_2(x, y) = \sum_{m,n} b_{mn} e^{im\omega_x x} e^{in\omega_y y} \qquad (26.32)$$

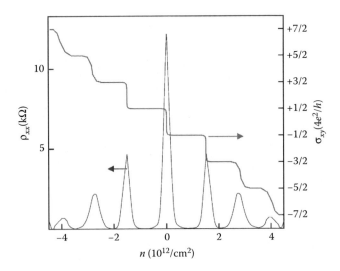

FIGURE 26.5 AQHE in graphene as function of charge-carrier concentration. There is no Hall conductivity plateau at $n = 0$, which marks the relativistic Dirac particle behavior in graphene. Carrier concentration is tunable with a simple gate. (Adapted from K.S. Novoselov et al. *Nature*, 438;2005:197–200.)

where $\omega_x = 2\pi/T_x$, $\omega_y = 2\pi/T_y$ with T_x and T_y the periodicity of magnetic lattice. The Hamiltonian can then be expanded easily with these plane-wave basis, and it enables analysis of the Landau level without resorting to Landau gauge. If we need to model the intervalley scatterings, then a 4×4 Hamiltonian for four orbital basis should be used instead.

26.2.5 Phonon and Heat Conduction for Graphene

To study the heat conduction of graphene, it is important to understand the phonon dispersions first because of its importance in determining phonon density of states, phonon scatterings, thermal conductivity, and so on. Raman measurements are used to determine the phonon dispersions experimentally, such as the phonon dispersions reported for graphite along the Γ–M–K–Γ directions measured by x-ray inelastic scattering [63,64]. Theoretically, besides *ab intio* calculations, other approaches like fourth-nearest-neighbor force constant (4NNFC) method, valence force field (VFF) model, and the Tersoff and Brenner potentials are based on fitting parameters from experimental data. Figure 26.6 shows the phonon dispersion of graphene calculated using 4NNFC method. With two atoms in one unit cell in graphene, there are six phonon branches comprising three acoustic (A) and three optical (O) phonon modes [1]. These modes are associated with out-of-plane (Z), in-plane longitudinal (L), and in-plane transverse (T) atomic motions. The phonon dispersion is the relation of phonon energy or frequency and phonon wave vector. The acoustic modes originate from Γ-point, and TA and LA modes have linear dispersions at small wave vector near Γ-point with much higher group velocities due to strong in-plane sp^2 bonds and small mass of carbon atoms; whereas ZA modes are quadratic with lowest acoustic phonon energy.

The thermal conductivity (κ) is the physical property of a material to conduct heat. It is defined in terms of Fourier's law, $Q = -\kappa\nabla T$, where Q is the heat flux and ∇T is the temperature gradient. In general, heat is carried by phonons and electrons, so $\kappa = \kappa_p + \kappa_e$, showing the lattice thermal conductivity and electron thermal conductivity. For carbon-based materials, the acoustic phonons dominate the heat conduction, and the contribution of electron can become important for doped structures. In the diffusive regime, the sample size (L) is larger than the mean free path, and the Umklapp scattering process limits thermal transport. When L is comparable or even smaller than the mean free path, the continuous transport model is broken and phonon transport is in the ballistic regime. Fourier's law is only applicable in the diffusive regime [65].

Thermal conductivities of graphene can be measured using micro-Raman spectroscopy. Balandin et al. [12] have reported that the suspended single-layer graphene has thermal conductivity in the range of $\sim(4.84 \pm 0.44) \times 10^3$ to $(5.30 \pm 0.48) \times 10^3$ W m^{-1} K^{-1} at room temperature comparable to the predicted value by molecular dynamics (MD) simulations when isolated graphene monolayer was still hypothetical [27]. The phonon mean free path of ~775 nm near room temperature is extracted [66]. Other groups have measured the in-plane thermal conductivity in the range of ~2000–5000 W m^{-1} K^{-1} for suspended graphene samples of various sizes and geometries at different temperatures [28–30]. These results show that graphene has extremely high thermal conductivity that exceeds the bulk graphite limit, among the highest of any known materials, suggesting its potential use as the future thermal management material. Besides freely suspended graphene, it is also very important to study the effect of substrate on supported graphene [67,68]. Contact with a substrate could decrease the thermal conductivity of graphene significantly, since supported graphene cannot benefit from the suspension that reduces thermal coupling and scattering on the substrate leading to the thermal conductivity of suspended graphene close to its intrinsic value. The measurement by Seol et al. has shown that the thermal conductivity of monolayer graphene exfoliated on a silicon dioxide support is ~600 W m^{-1} K^{-1} near room temperature [69]. It is much lower than the reported values of suspended graphene because of the phonon leaking across the graphene-support interface and strong interface-scattering of flexural modes.

On the theoretical side, several methods are employed to model the thermal transport of graphene, including common atomistic techniques like MD, nonequilibrium Green's function (NEGF), and Boltzmann transport equation (BTE) simulations. Using the optimized Tersoff and Brenner empirical potentials, Lindsay et al. have found that the thermal conductivity of graphene is ~3500 W m^{-1} K^{-1}, better than those obtained from the original parameter sets [70]. The optimized parameter sets provide better fitting to phonon dispersions and measured data as well. With the Green–Kubo method, Zhang et al. have demonstrated that pristine graphene has thermal conductivity of 2903 ± 93 W m^{-1} K^{-1}, and the out-of-plane phonon mode contribution is 1202 ± 32 W m^{-1} K^{-1} [71]. Other calculations have also shown that the intrinsic thermal conductivity of graphene is in the range of 2000–5000 W m^{-1} K^{-1} at room temperature depending on size, defect, and edge roughness, in good agreements with experimental measurements [72–74].

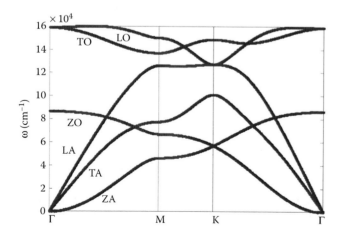

FIGURE 26.6 The phonon dispersion of graphene along high symmetric points calculated using 4NNFC method. (The results are benchmarked with R. Saito, G. Dresselhaus, M.S. Dresselhaus, *Physical Properties of Carbon Nanotubes*, World Scientific, Singapore, 1998.)

In the ballistic regime, a formula is derived by Saito et al. [75] to calculate the ballistic limit of the thermal conductance per unit length of graphene when L is smaller than the mean free path. Their formula shows that, for intrinsic graphene, the electron thermal conductance increases in proportion to T^2 with temperature, while the lattice thermal conductance increases in proportion to $T^{1.5}$ due to the quadratic dispersion relation of the out-of-plane acoustic mode.

26.2.6 GRAPHENE WITH GRAIN BOUNDARY

Structural defects appearing during growth or processing are of practical significance on the performance of graphene, such as point and line defects. Grain boundaries are line defects separating single-crystalline domains with preferably periodic structure due to minimum formation energy. After individual dislocation is imaged in free-standing graphene layer by transmission electron microscopy (TEM) [76], recently, more reports have shown grain boundaries in graphene produced by chemical vapor deposition (CVD) [14,77], opening an opportunity for studies on the structure, properties and control of grain boundaries in graphene. Theoretically, first principles calculations are used to study the electronic properties of graphene with grain boundaries [31–33]. The structure is dictated by the misorientation angle of two crystallites. For any possible misorientation angle, a grain boundary structure can be constructed in the form that all carbon atoms maintain their threefold coordination. The grain boundaries can be classified into symmetric and asymmetric. The electronic properties are different for graphene with different classes of grain boundaries from metallic to semiconducting based on the theory of transverse momentum conservation, for example, large transport gap can be introduced by asymmetric grain boundary with misorientation angle of 30° as shown in Figure 26.7. These transport gaps and electrical conductance can be modulated by strains on grain boundaries [33]. Yazyev et al. have also found that the symmetric large-angle grain boundaries are favorable due to low formation energy, and there is a strong tendency toward an out-of-plane deformation that further reduces their formation energies in small-angle regimes [32].

The effect of grain boundaries on the thermal conductivity of graphene is also studied theoretically using the NEGF approach [34,35]. Unlike the structural dependence on

electronic transport, graphene with all types of grain boundaries have excellent thermal conductivity, among which, symmetric zigzag grain boundaries show the highest thermal conductance. Graphene with grain boundaries can yield about 80% transmission of pristine graphene, and as temperature increases, the thermal conductance increases. Similar to pristine graphene, as discussed above, the out-of-plane acoustic mode is dominant in thermal conductivity of graphene with grain boundaries at low temperatures [34]. In addition, Serov et al. [35] have shown the dependence of thermal conductivity in substrate-supported graphene with grain boundaries that grain type and grain size are crucial when grain sizes become comparable to or smaller than several hundred nanometers, which is important for practical applications.

26.3 GRAPHENE NANORIBBONS

The boundary effects of different edge types and quantum confinement make GNRs very different from graphene. AGNRs are semiconducting with oscillatory bandgaps depending on width, and all ZGNRs with very small band gaps. The edge effects also have great impacts on magnetic properties, that is, magnetism in ZGNRs and magnetoresistive effect in AGNRs. Although the thermal conductivity of GNRs is greatly reduced compared with graphene, it offers an opportunity for GNRs to be potential thermoelectric materials. In this section, the fundamentals of GNRs are introduced followed by the review on magnetic field, thermal, and thermoelectric properties for GNRs of different structures.

26.3.1 GNR FUNDAMENTALS

The electronic properties of graphene beyond its bulk properties show many important features, which are vital for every device application. GNRs are the most basic geometry of interest where electron transport takes place along the GNR's length. Regular GNRs can either be cut with zigzag edges (ZGNR) or armchair edges (AGNR) as shown in Figure 26.8 [78]. We shall use theoretical models with effectively infinite length of GNR to predict their transport properties. In theory, ZGNRs have conducting states with rather peculiar magnetic properties at their edges regardless of the ribbon width [18]. Narrow AGNRs can be either semiconducting or metallic, and their bandgap oscillates with the ribbon width parameter as will be explained later. Obviously, the bandgap also diminishes with increasing ribbon width due to less confinement in the ribbon. The atomic structures of both ZGNRs and AGNRs are shown in Figure 26.8, together with their band structures from π-orbital tight-binding model. Many detailed edge effects on the electronic structure of GNRs have been extensively studied in the literature due to ubiquitous existence of edge roughness and defects in practice and their ability to change the transport properties of GNRs significantly. Structural stability of GNRs is provided by the edge atoms passivation mechanism typically by hydrogen atoms. Otherwise crumpling and other undesired chemical reconstructions of the ribbons would occur. Reconstructions of the

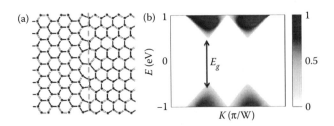

FIGURE 26.7 (a) Atomic structure of the graphene with θ = 30.0° asymmetric grain boundary. (b) Transmission map. (The results are benchmarked with S.B. Kumar, J. Guo, *Nano Letters*, 12;2012:1362–1366.)

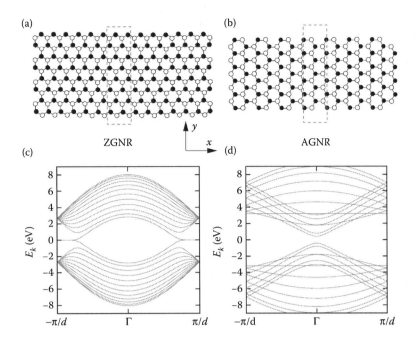

FIGURE 26.8 (a) Zigzag graphene nanoribbon (ZGNR). (b) Armchair graphene nanoribbon (AGNR). Unit cells are highlighted. Black and white atoms describe sublattice *A* and *B*. ZGNR has conducting states at its edges. (c) Band structure for ZGNR and (d) for AGNR, calculated from tight-binding model. (The results are benchmarked with G. Liang et al. *Electron Devices, IEEE Transactions on*, 54, 677–682, © 2007 IEEE.)

edge atoms by around 3% also introduce around 12% increase in the nearest-neighbor hopping energy.

Sufficiently narrow ZGNRs have been predicted to have a finite but small direct bandgap in the order of 20 meV [18], though the experimental observation of this energy gap remains elusive due to impurities in GNRs. The origin of this bandgap comes from a staggered sublattice potential due to spin ordered states at the edges. ZGNRs have ferromagnetic ordering on every edge, with opposite spin orientation between them. However, this energy gap and the antiferromagnetic coupling between the edges diminish as the width of the ribbon increases to be comparable to the spin diffusion length in the ribbon. Recently, Li et al. has suggested to improve the preservation of the edge states by terminating them with ethylene [79]. Opposite spin states on opposite edges in ZGNRs occupy different sublattice, so in a sense they resemble boron nitride (BN) nanoribbons whose bandgaps arise from ionic potential difference between B and N in different sublattices [80]. In practice, exploitation of the magnetic properties in ZGNRs remains nontrivial because the effects from the underlying substrate and intrinsic charge puddles on the entire graphene layer is significant [81].

Theoretically, AGNRs are semiconducting and their bandgap varies for three distinct families of $3p$, $3p + 1$, and $3p + 2$, where n is a positive integer that parameterizes the number of carbon atoms along the ribbon width. The bandgap oscillations are shown in Figure 26.9a, which are the results from tight-binding calculation [82]. A consistent trend shows that $3p + 1$ family has the largest bandgap while $3p + 2$ family has the smallest. *Ab initio* calculations with local spin density approximation (LSDA) for exchange correlation predict

semiconducting properties for all families in narrow AGNRs. While this phenomenon may look interesting theoretically, its observation is obscured by defects and impurities in GNRs. In fact, the existing monolayer GNRs fabricated in the lab so far show no dependence of the bandgap on the orientations of the GNRs [81].

Electronic transport properties of perfect GNRs may be derived from the band structure using NEGF formalisms. The conductance ($G(E)$) through the device will show the typical quantization in the form,

$$G(E) = \frac{2e^2}{h} \sum_n T_n(E) \qquad (26.33)$$

and the transmission probability for a given channel n is either one or zero depending on whether the channel is occupied or not at a given energy E. This would show step-like conductance across energy. Real GNRs always show significant difference in their transport behaviors due to inherent disorders such as defects, impurities, and ripples across the ribbons, which modify the transmission coefficient above. Theoretical models for disorders in graphene have been discussed in the literature [81,83–85]. Mostly, we use tight binding for the microscopic model and generally a local disorder at atomic site i may be modeled by putting a high on-site energy value at i, making electrons difficult to hop into it. A uniform disorder on graphene may be modeled using Anderson model by making the on-site energy randomly distributed across a certain energy interval $[-U, U]$ with a uniform probability distribution, for example, $P = 1/2Ut$. Edge disorders in

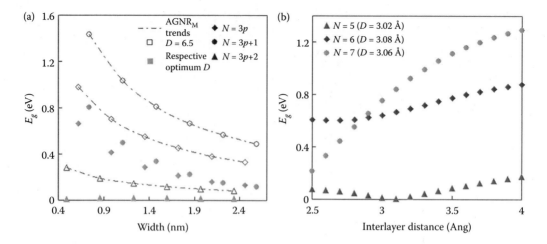

FIGURE 26.9 (a) Energy gap (E_g) as a function of the width of monolayer AGNR (dot-dash lines) and bilayer AGNR with interlayer distance (D) = 6.5 Å (hollow points) and the respective optimum D (solid points). Diamond, circle, and square points represent the three different families of $N = 3p$, $3p + 1$, and $3p + 2$, respectively. (b) Dependence of E_g on D of bilayer AGNRs for $N = 5$, 6, and 7. (Adapted from K.-T. Lam, G. Liang, *Applied Physics Letters*, 92;2008:223106.)

GNRs reduce the conductance and introduce fluctuations of the order of e^2/h, which are sensitive to the relative arrangement of the scatterers. An average over a large number of edge disorders will smear the conductance steps, which eventually makes conductance linear, and results in transport energy gap. Wider GNRs experience less sensitivity to edge disorders, and solution-phase fabrication methods are claimed to enhance the smoothness of the GNR edges [86]. Disorders may also enhance the performance of graphene device for thermoelectric applications [87].

Different disorder length scales lead to different transport regimes. The phase coherence length scale in graphene can be up to five micron long [88]. Therefore, for most nanoscale devices that we model in this book, quantum model effects are important, especially for low disorder systems. The mean free path length scale determines the ballistic-diffusive transport regimes where system size smaller than its mean free path is said to be in the ballistic regime. Another length scale is the localization length scale within which a state can be bounced back and forth between two impurities (forming a standing wave) leading to Anderson localization at certain disorder level, or else the state may also be bound to the impurity site. Shorter localization length induces greater reduction to the conductance. Graphene behaves in similar fashion to ordinary parabolic dispersion materials under uniform Anderson-type disorders [84].

On the other hand, bilayer GNRs have some qualitatively similar electronic properties compared to the monolayer GNRs, although the bilayers show some additional features. For example, the energy gap in bilayer GNRs depends on the interlayer spacing, as well as the nanoribbon width due to nonuniform charge distributions at the edges as shown in Figure 26.9 [82]. Armchair bilayer GNRs are also predicted to have three distinct families of $3p$, $3p + 1$, and $3p + 2$ despite their energy gaps being smaller than the monolayer counterparts [82]. The effects of edge doping with boron and nitrogen give the same effects on both monolayer and bilayer GNRs.

Boron doping makes p-types, while nitrogen doping makes n-types. These edge dopings show reduced energy gap compared to the undoped cases [82]. As explained previously, bilayer graphene can exhibit an energy gap upon an application of perpendicular electric field. This makes bilayer GNRs attractive for many electronic applications, such as nanoelectromechanical sensor (NEMS). A pressure acting on bilayer GNRs can change the interlayer distance, which in turn will change the energy gap significantly, useful for pressure-sensitive NEMS. A theoretical model elucidated this NEMS device concept and predicted up to 3 order of magnitude on–off current ratio under 20 meV bias [89]. This is more attractive than the bulk bilayer graphene, which requires relatively higher pressure to achieve the same effects on its energy gap.

26.3.2 MAGNETISM IN ZGNRS AND MAGNETIC-FIELD-EFFECT IN AGNRS

Both tight-binding models and first-principles calculations predict that ZGNRs are metallic with doubly degenerate flat band at the Fermi level when spins are not considered [90]. The flat band extends over one-third of the 1D Brillouin zone at $K \in ((2\pi/3a),(\pi,a))$ ($a = 0.246$ nm is the unit cell of the zigzag edge), corresponding to a highly localized electronic distribution near the zigzag edges. As a result, the on-site electron–electron Coulomb interaction leads to an instability and suggests possible edge magnetism. This edge magnetic ordering has been proved by the mean-field Hubbard-model calculation and the spin polarized DFT calculation [18,91]. The ground state of ZGNRs shows the antiferromagnetic coupling between the magnetic zigzag edge states. By applying an external magnetic field, the antiferromagnetic (AFM) ground states (GS) can be switched to ferromagnetic (FM) metastable states [91]. The band structure of ZGNRs is closely related to the magnetic states. As shown in Figure 26.10b for the GS-ZGNR, a direct bandgap and energy splitting at X point is present due to the interaction of spin polarized edge states.

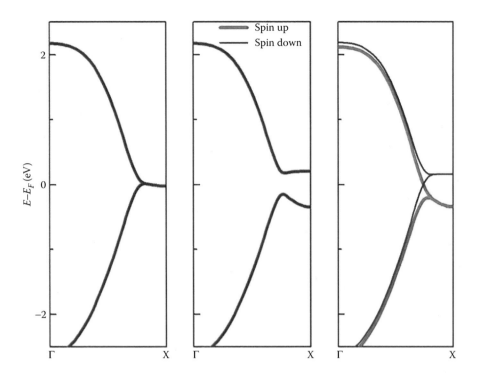

FIGURE 26.10 Band structures of ZGNRs in (a) nonmagnetic state, (b) antiferromagnetically coupled ground state, and (c) ferromagnetically coupled magnetized state. (The results are benchmarked with Y.-W. Son, M.L. Cohen, S.G. Louie, *Physical Review Letters*, 97;2006:216803; M. Zeng et al. *Physical Review B*, 83;2011:115427.)

However, FM-ZGNR shown in Figure 26.10c is metallic with two bands crossing the Fermi level at $K = 2\pi/3a$; and NM-ZGNR is a semimetallic with zero bandgap.

The existence of edge magnetic ordering at GS-ZGNR has been considered as a basis for novel spintronics devices. However, the length of this magnetic ordering, or the spin correlation length, is limited at finite temperature due to the unstable 1D spin ordering. Yazyev et al. have predicted that the spin correlation length in a clean ZGNR is inversely proportional to temperature and the value of spin correlation length at room temperature is around 1 nm [92]. To overcome this temperature-related limitation, magnetic anisotropy arising from substrate effects or edge-terminated heavy metals can be utilized to enhance the spin ordering. As a result, spin correlation length at the scale of several tens of nanometer at finite temperature may be realized for spintronic applications. Rhim and Moon have investigated the spin stiffness of the ZGNR as a function of the lateral electric field [93]. They have found that the spin stiffness of the system shows a nonmonotonic dependence on the lateral electric field and a cusp at the critical value of the lateral electric field.

Because of the antiferromagnetic coupling between the magnetic zigzag edge states, the net magnetism in the ground state of a ZGNR is zero, which agrees with Lieb's theorem of zero net magnetization for equal bipartite (A, B) sublattice structures. Several interesting methods have been proposed to introduce magnetism in ZGNRs. Topsakal et al. have proposed a repeated heterostructure of ZGNRs with different widths that can form multiple quantum well structures, in which edge states of specific spin directions can be confined,

and the magnetic ground state of whole heterostructure may change from antiferromagnetic to ferrimagnetic in specific geometries, in which the spin-down states remain distributed at the flat edge of the superlattice, while spin-up states are predominantly confined at the opposite edge of the wide segments [94]. The magnetism comes from confinement of states and absence of reflection symmetry, which breaks the symmetry between spin-up and spin-down edge states. The net magnetic moment is calculated to be $2\mu_B$, which agrees well with Lieb's theorem by counting the difference of the number of A,B sublattice atoms.

Besides, Sawada et al. have reported that the magnetic phase of the ZGNR can be controlled by injecting carriers [95]. As the carrier density increases, the magnetic coupling can change from antiferromagnetic coupling to ferromagnetic coupling. As shown in Figure 26.11, the phase transition (an antiparallel interedge spin state → a canted interedge spin state → a parallel interedge spin state) occurs when $|x|$ (x is carrier doping concentration) increases from 0 to 0.24 e/nm. The stable phase transition at a given carrier doping concentration is determined by the minimum of interedge exchange coupling $E_x(\theta)$, which is the total energy difference between the ground state and the carrier doping state.

ZGNRs show strong potential for creating spin polarized currents. Son et al. have predicted the half-metallic property of ZGNRs in the presence of a sufficiently large transverse electric field [96]. Half-metallic materials are unique because their conducting electrons are 100% spin-polarized, making them ideal for spintronic applications. Using an electric field, instead of the magnetic field, to manipulate the spin

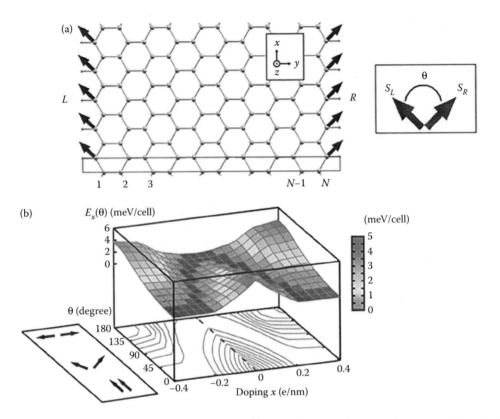

FIGURE 26.11 (a) Lattice and magnetic structures of ZGNR. The black arrows denote the magnetic moments S_L and S_R at the two edges. The relative angle between S_L and S_R is represented by θ. (b) Noncollinear magnetic phase diagram of ZGNR, which is determined by the interedge exchange coupling $E_x(θ)$ per unit cell as a function of carrier doping x (e/nm) and θ (degree). See Sawada et al. (2008) for interpreting colorscale. (Adapted from K. Sawada et al. *Nano Letters*, 9;2008:269–272.)

polarization of carriers is particularly useful for spintronic applications because the requirement of magnetic fields usually involves large pieces of components, which makes the production of small devices difficult. In Son's proposal [96], the applied transverse electric fields can induce opposite energy-level shifts for the spatially separated spin polarized edge states. As a result, the energies for localized edge states on one side are shifted upward and those on the other side downward, eventually leaving states of only one spin orientation at E_F (Figure 26.12b). The critical electric field for achieving half-metallicity in ZGNRs decreases as the width increases. For a ZGNR with width of 67.2 Å, the critical electric field is around 0.045 V/Å. The requirement of very large electrical field in Son's proposed devices is difficult to realize. An approach to reduce the critical electric field for half-metallic ZGNRs is to select the edge functional group. Hod et al. have reported that edge oxidation can effectively lower the onset electric field required to induce half-metallic behavior [19]. The oxidation groups include hydroxyl, lactone, and ketone. ZGNRs with these edge functional groups are more stable than hydrogen-terminated ones. Even though these edge functional groups are unable to change the antiferromagnetic ground state of ZGNRs, they can effectively lower the critical electric field required to induce half-metallic behavior. More interestingly, the requirement of a transverse electric field may be eliminated with a suitable choice of the edge functional group. For example, Kang et al. have

investigated an asymmetrical ZGNR in which sp^3-like C–H$_2$ group terminates at one edge while sp^2-like C–H terminates at the other [97]. Molecular dynamics simulations further show that H$_2$-ZGNRs-H is stable under room temperature. Even though H$_2$-ZGNR-H is a ferromagnetic semiconductor at its ground state, half-metallic H$_2$-ZGNR-H can be achieved by p-type or n-type doping. Boron and nitrogen are used as dopants for p- and n-type doping, respectively.

Localized spin-polarized edge states and edge magnetism are absent in AGNRs. However, quantum-Hall-like edge states can be induced by applying an external magnetic field to AGNRs. For a magnetic field $\vec{B} = (0,0,B_z)$, the magnetic flux passing through each hexagonal ring of the honeycomb carbon structure can be expressed by a dimensionless quantity $\phi = qSB_z/h$, where S is the area of the hexagon. The magnetic field induces a vector potential $\vec{A} = (-B_z y, 0, 0)$, which satisfies $\nabla \times \vec{A} = \vec{B}$. Following Peierls phase approximation, the tight-binding hopping integral between neighboring atoms acquires a phase. Therefore, a finite magnetic field can distort the Bloch wavefunction of AGNRs. Moreover, electrons in AGNRs can be constrained by the applied magnetic field to a cyclotron motion. As the B-field increases, the cyclotron radius of the electron decreases. The dispersionless Landau levels and edge states are developed while the ribbon width is wide enough to be comparable to the cyclotron radius. Wu et al. [98] have investigated the effect of the period and the strength of a spatially modulated

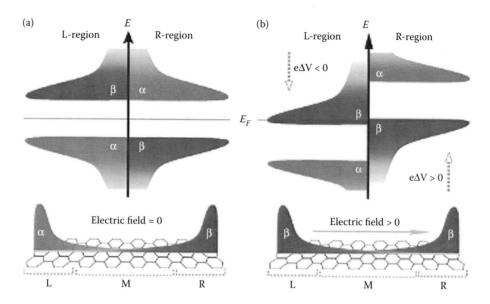

FIGURE 26.12 (a) An AFM coupling between the occupied localized spin polarized edge states in the absence of an applied electric field. (b) Applying a transverse electric field moves the energy band on both edges, but in an opposite direction. As a consequence, half-metallic ZGNRs, that is, conducting to electrons of one spin orientation, but insulating to those of the opposite orientation, can be realized. (Adapted from Y.-W. Son, M.L. Cohen, S.G. Louie, *Nature*, 444;2006:347–349.)

perpendicular magnetic field on the electronic structure of AGNRs with different widths, $N = 27$ and 26, which represent semiconducting and metallic AGNRs, respectively. The modulated magnetic field modifies energy dispersions (band curvatures or effective masses), as well as subband spacing and energy gaps. The bandgap is narrowed when the magnetic field is applied to $N = 27$ semiconducting AGNR. For $N = 26$ metallic AGNR under an applied magnetic field, the linear bands at $E_F = 0$ are separated into two parabolic bands and a bandgap is derived. As shown in Figure 26.13, Kumar et al. [99] have calculated the band structure and the conductance of AGNRs under a perpendicular B-field and found that the MR effect is originated from the narrowed bandgap. They have also revealed that the AGNRs with number of dimers, $N = 3p + 1$, show the largest bandgap variation, and hence, the most promising MR ratio (see Figure 26.14) [99]. Besides, they have also studied the spatial distributions of electrons under a moderate magnetic field to obtain the insights of the various signature Hall effects in disordered AGNRs [85].

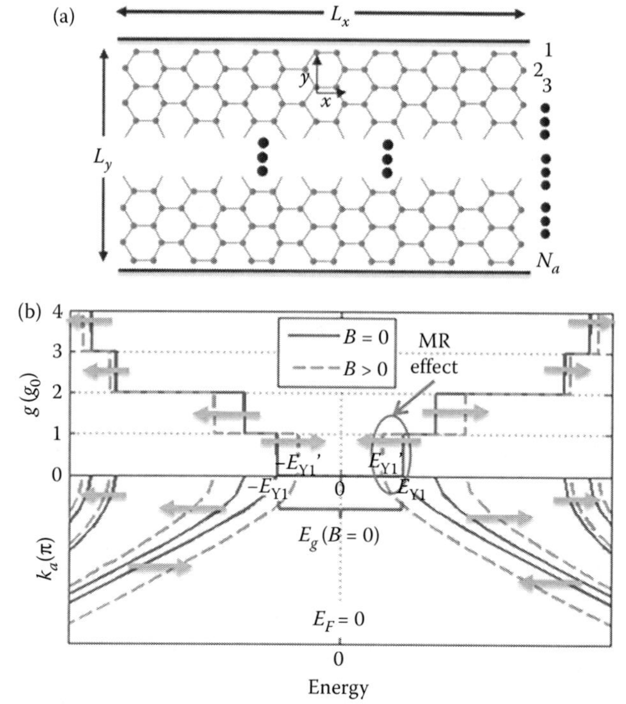

FIGURE 26.13 (a) Schematic of AGNRs. (b) The conductance and band structure of AGNRs. An applied B-field can reduce the bandgap and tune the conductance of AGNRs, indicating a MR effect. (Adapted from S.B. Kumar et al. *Journal of Applied Physics*, 108;2010:033709.)

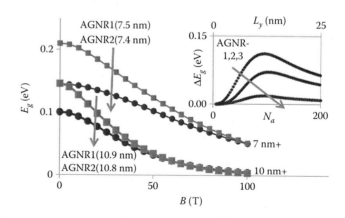

FIGURE 26.14 The bandgap variation as a function of applied B-field in two types of AGNRs. The inset shows $\Delta E_g = E_g(B = 0T) - E_g$ ($B = 50T$) as a function of the ribbon width. (Adapted from S.B. Kumar et al. *Journal of Applied Physics*, 108;2010:033709.)

26.3.3 THERMAL AND THERMOELECTRIC PROPERTIES FOR GNRs

Compared to graphene, the thermal conductivity of GNR decreases significantly due to dimensional confinement. On experimental side, Murali et al. have extracted the thermal conductivity of more than 1000 W m^{-1} K^{-1} at self-heated temperatures of 700–800°C for sub-20 nm GNRs during the measurement of the breakdown current density of GNRs using electrical self-heating method [100]. At room temperatures, much lower values of the thermal conductivity of ~80 W m^{-1} K^{-1} for ~15 nm wide GNR [101] and ~100 W m^{-1} K^{-1} for ~65 nm wide GNR [102] are reported using similar method. The experiments have demonstrated that heat flow is sensitive to the ribbon width and edge disorder for substrate-supported GNRs. There are also many theoretical works on thermal conductivity of GNRs in nanoscale, such as those using MD simulation [103–107], NEGF approach [108–112], and BTE method [113,114]. The thermal conductivities obtained in MD simulations are in the same order with experimental results at room temperature as shown in Table 26.2.

Munoz et al. [107] have presented a theoretical model to estimate the ballistic thermal conductance limit of GNRs with finite width. At low temperature, a power law is proposed for the ballistic thermal conductance of ~T^β, where $\beta = 1$ for narrow ribbons and $\beta = 1.5$ for a large graphene sheet. The analysis for graphene is in good agreement with the results previously obtained in [75] for the ballistic limit of both lattice thermal conductance and electron thermal conductance. It is found that thermal conductivity is very sensitive to edge types, and the dependence on width is different. In addition, the edge roughness and tensile/compressive uniaxial strain of GNRs reduce the thermal conductivity by a significant amount due to phonon scatterings. For example, Hu et al. [103] have investigated the effects of edge chirality and found that ZGNRs have appreciably larger thermal conductivity than AGNRs, and the thermal conductivity increases monotonically with temperature. The NEGF method is also employed by several groups to study the

thermal transport in nanoscale taking into account the quantum effect. Coupled with the Naval Research Laboratory tight-bind method to accurately describe the elastic constants and phonon dispersion of carbon systems, Lan et al. [108] have found that thermal transport properties of GNRs are strongly dependent on their widths described by the number of atoms (N) along the direction perpendicular to the ribbon axis with a threshold of $N = 12$. The quantized thermal transport is destroyed when $n < 12$, and the thermal conductance is suppressed significantly. At $N = 2$, a perfectly quantized thermal transport is restored with a zero transmission bandgap when the narrowest GNR is sandwiched between two wide GNR leads. The NEGF approaches combined with first principles are used to study the effect of edge hydrogen passivation, which modifies the atomic structure and phonon transport differently for the AGNR and ZGNR as reported in [110]. Using similar method, Jiang et al. [111] have demonstrated the dependence on defect position. The thermal conductance is suppressed greatly for inner vacancy defects due to the formation of a saddle-like surface, but it is reduced slightly for edge vacancy defects, which only create reconstructions of the edge. In addition, it is insensitive to the position of silicon substitution defect. Employing 4NNFC model combined with NEGF, the geometrical and roughness effect on AGNR is studied by Karamitaheri et al. [112], and an analytical model is developed to interpret the numerical results.

Another interesting property related to thermal transport is the thermoelectric effect, which can generate electrical energy from waste heat, which is normally lost to environment and conversely cool down heat when a voltage gradient is applied [115]. The efficiency of the conversion between energy and heat is measured by a dimensionless figure of merit, $ZT = S^2 \sigma T / (\kappa_p + \kappa_e)$, where S, σ, T, κ_p, and κ_e denote the Seebeck coefficient, electrical conductivity, absolute temperature, lattice thermal conductivity, and electron thermal conductivity, respectively [116]. The difficulty of maximizing ZT lies in the fact that the above parameters are interdependent [20], so it is not easy to alter one without affecting the others. To obtain a high ZT, the material should have high S and σ, but low κ. Due to quantum confinement, low-dimensional structure is one potential candidate for advanced thermoelectric materials, such as 1D semiconducting nanowires [117–119]. With excellent electronic properties and reduced phonon conductivity, GNR can also be considered as a promising candidate for thermoelectric material.

Many theoretical works have studied the thermoelectric properties of different GNR structures, mostly using the NEGF method. As discussed in Section 26.3.1, the bandgaps of AGNRs depend on their widths, but all ZGNRs have very small bandgaps. Hence, the width dependence on thermoelectric performance is observed for AGNRs, but not for ZGNRs. AGNRs of $3p$ and $3p + 1$ family are semiconducting; therefore, more suitable to be thermoelectric materials. Ouyang et al. [120] have shown that $N = 15$ AGNR has much better thermoelectric performance than 2D graphene even though the peak value of ZT is still below 1. The width dependence follows AGNR's family behavior, while the thermoelectric

TABLE 26.2
Summary of Thermal Conductivities Obtained by Experiments and MD Simulations for GNRs

Reference	Thermal Conductivity (W m^{-1} K^{-1})	GNR Type	Width (nm)	Length (nm)
Experiment [101]	~80	–	~15	200–700
Experiment [102]	~100	–	~65	260
Simulation [104]	218	AGNR	~2	11
Simulation [104]	472	ZGNR	~2	11
Simulation [105]	~500	Rough A/ZGNR	~2	10
Simulation [105]	~3000	Smooth A/ZGNR	~2	10
Simulation [106]	16.2	Rough ZGNR	~1	78.7
Simulation [106]	147	Smooth ZGNR	~1	78.7

TABLE 26.3
Summary of Methods to Enhance *ZT* of GNRs

Reference	Method	Type of GNR	Maximum *ZT* Attainable	Mechanism
[87]	Hydrogen vacancies	AGNR	~5.8	Suppressing κ_p
[122]	Defects	AGNR	~0.2	Suppressing κ_p
[122]	Magnetic field	AGNR	~0.3	Improving σ and S
[123]	Tensile strain	AGNR	~0.7	Improving σ
[124]	Edge disorder	ZGNR	~4	Suppressing κ_p
[125]	Line defects with impurities and roughness	ZGNR	~5	Suppressing κ_p

performance of all ZGNRs is similar and their *ZT* is smaller [121]. Other than perfect GNR structures, significant effects of defect [120,122], disorder [87], strain [123], and magnetic field [122] have also been studied for AGNRs. When introducing vacancies and edge defects, the electronic transport is degraded, but κ_p is greatly suppressed, resulting in higher *ZT* values. Ni et al. [87] have reported that *ZT* of AGNRs can be enhanced five times by randomly introducing hydrogen vacancies, and the imperfect saturation of the hydrogen–carbon bonds can remarkably reduce κ_p without degrading the electronic transport in a great amount. However, the tensile strain increases *ZT* for AGNRs of $3p$ and $3p + 2$ families mainly due to the increase in electronic transport. Similarly for ZGNRs, engineering design steps, such as extended line defects and edge roughness, can be introduced to enhance the thermoelectric performance by improving σ and S, or suppressing κ_p [124,125]. The engineering methods to enhance *ZT* for GNRs are summarized in Table 26.3.

Because of the advancements in experimental process for fabricating precise and complex shapes of GNR structures, the thermoelectric performances are also studied for GNRs with mixed structures, like junction [126–128], stub [129], resonant tunneling [130], and assembled kinked structures [131–134]. Across the GNR junction, the thermoelectric performance is mainly controlled by the narrower part of the junction [126]. Pan et al. [127] have demonstrated that the reduction in κ_p is due to the mismatched interface, and the electronic transport is very sensitive to geometrical parameters that the zigzag-edge two-side dogbone junction shows the highest *ZT* exceeding 0.9. The kinked GNRs, also called graphene nanowiggles, have better thermoelectric performance than their pristine straight GNR counterparts. Figure 26.15 [131] shows some possible kinked GNR structures connected with AGNR or ZGNR segments with different angles. Huang et al. [131] have shown that with the presence of kink structures, thermoelectric properties are less sensitive to edge geometries, which may be preferable in fabrication. In contrast to GNR bandgap behaviors, first peak of *ZT* (ZT_{max}) of kinked AA-GNRs increases as the kinked width decreases, while that of kinked ZZ-GNRs is oscillatory. In addition, the structures with smaller width of the connecting segments have better performance among various hybridized kinked GNRs with different connecting segments and angles, and structures with two ZGNR segments connected by 120° result in larger ZT_{max}, but structures with a horizontal ZGNR segment have smaller ZT_{max}. In addition, by introducing elastic strain and structural dislocation, Liang et al. [133] have found that *ZT* can be further increased to values beyond 1.

26.4 SUMMARY

In summary, in this chapter, we review the fundamental insights in graphene and GNRs including electronic band structure, phonon dispersion relation, spin polarization, and magnetism

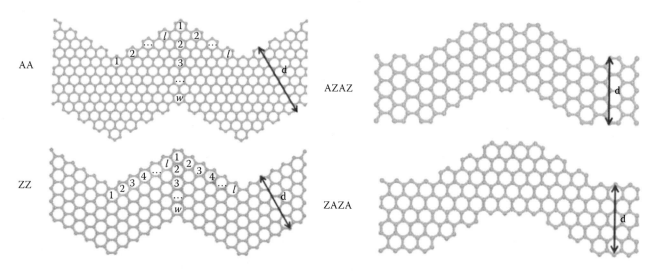

FIGURE 26.15 Structures of kinked AA/ZZ-GNRs that have two segments with the same edge shape, armchair or zigzag, connected by 120°, and kinked AZAZ/ZAZA-GNRs that have two segments with different edge shapes connected by 150° with horizontal armchair-zigzag segment, respectively. (Adapted from W. Huang, J.-S. Wang, G. Liang, *Physical Review B*, 84;2011:045410.)

properties as well as transport characteristics. Graphene has become an interesting material in various research fields because of its strong in-plane sp^2 σ-bonds with its nearest neighbors and an out-of-plane dangling π-bond at each site in its 2D hexagonal lattice. The resulting linear relation at K points in the electron structure can be described by Dirac equation with massless condition. Then, based on this linear band relation, we give detailed derivation from π orbital tight-binding model to the simple Dirac equation, and discuss several interesting and unique phenomenon in graphene, such as negative refraction, Klein tunneling, and QHE. Furthermore, the fundamental properties of graphene such as density of states and conductivity in the presence of different scattering mechanisms (ballistic, ion impurities, and lattice vibrations) are also discussed. Lastly, heat conduction of graphene and the grain boundary effects are also briefly introduced to comprehensively expound the fundamentals of graphene. Next, we also discuss the details of the electronic structure of GNRs, a 1D strip cut from graphene. Due to spatial confinement effects, GNRs can exhibit either semiconducting or metallic behavior according to their chirality. Most interestingly, an external electrical and magnetic field is able to modulate GNRs' properties, for example change them from the normal metals to ferromagnets or half metals. Besides, due to its $n = 0$ Landau level, the carrier conductivity can be modulated by the magnetic fields, a so-called magnetoresistance or magnetoconductance effect. These unique behaviors can be implemented in spintronic device applications. Finally, we also briefly discuss the phonon dispersion relations of various shapes of GNRs. The capability to change shape of GNRs begets the possibility of controlling their heat conductance and its possible application in thermal dissipation and thermoelectric devices. All of the unique characteristics of graphene and GNRs stimulate the device engineering community to look for potential applications for future devices. For a detailed discussion regarding this, interested readers may refer Chapter 19 in Volume 6 of this book series.

REFERENCES

1. R. Saito, G. Dresselhaus, M.S. Dresselhaus, *Physical Properties of Carbon Nanotubes*, World Scientific, Singapore, 1998.
2. H.W. Kroto, J.R. Heath, S.C. O'Brien, R.F. Curl, R.E. Smalley, C60: Buckminsterfullerene. *Nature*, 318;1985:162–163.
3. S. Iijima, Helical microtubules of graphitic carbon. *Nature*, 354;1991:56–58.
4. K.S. Novoselov, A.K. Geim, S.V. Morozov, D. Jiang, Y. Zhang, S.V. Dubonos, I.V. Grigorieva, A.A. Firsov, Electric field effect in atomically thin carbon films. *Science*, 306;2004:666–669.
5. A.H. Castro Neto, F. Guinea, N.M.R. Peres, K.S. Novoselov, A.K. Geim, The electronic properties of graphene. *Reviews of Modern Physics*, 81;2009:109–162.
6. J.-H. Chen, C. Jang, S. Xiao, M. Ishigami, M.S. Fuhrer, Intrinsic and extrinsic performance limits of graphene devices on SiO_2. *Nature Nano technology*, 3;2008:206–209.
7. K.S. Novoselov, A.K. Geim, S.V. Morozov, D. Jiang, M.I. Katsnelson, I.V. Grigorieva, S.V. Dubonos, A.A. Firsov, Two-dimensional gas of massless Dirac fermions in graphene. *Nature*, 438;2005:197–200.
8. K.S. Novoselov, Z. Jiang, Y. Zhang, S.V. Morozov, H.L. Stormer, U. Zeitler, J.C. Maan, G.S. Boebinger, P. Kim, A.K. Geim, Room-temperature quantum Hall effect in graphene. *Science*, 315;2007:1379.
9. A. Rycerz, J. Tworzydlo, C.W.J. Beenakker, Valley filter and valley valve in graphene. *Nature Physics*, 3;2007:172–175.
10. M. Liu, X. Yin, E. Ulin-Avila, B. Geng, T. Zentgraf, L. Ju, F. Wang, X. Zhang, A graphene-based broadband optical modulator. *Nature*, 474;2011:64–67.
11. R.N. Sajjad, A.W. Ghosh, High efficiency switching using graphene based electron "optics." *Applied Physics Letters*, 99;2011: 123101.
12. A.A. Balandin, S. Ghosh, W. Bao, I. Calizo, D. Teweldebrhan, F. Miao, C.N. Lau, Superior thermal conductivity of single-layer graphene. *Nano Letters*, 8;2008:902–907.
13. V.M. Pereira, A.H. Castro Neto, Strain engineering of graphene's electronic structure. *Physical Review Letters*, 103;2009:046801.
14. A.W. Tsen, L. Brown, M.P. Levendorf, F. Ghahari, P.Y. Huang, R.W. Havener, C.S. Ruiz-Vargas, D.A. Muller, P. Kim, J. Park, Tailoring electrical transport across grain boundaries in polycrystalline graphene. *Science*, 336;2012:1143–1146.
15. X. Wang, H. Dai, Etching and narrowing of graphene from the edges. *Nature Chemistry*, 2;2010:661–665.
16. J. Cai, P. Ruffieux, R. Jaafar, M. Bieri, T. Braun, S. Blankenburg, M. Muoth et al., Atomically precise bottom-up fabrication of graphene nanoribbons. *Nature*, 466;2010:470–473.
17. V. Barone, O. Hod, G.E. Scuseria, Electronic structure and stability of semiconducting graphene nanoribbons. *Nano Letters*, 6;2006:2748–2754.
18. Y.-W. Son, M.L. Cohen, S.G. Louie, Energy gaps in graphene nanoribbons. *Physical Review Letters*, 97;2006:216803.
19. O. Hod, V. Barone, J.E. Peralta, G.E. Scuseria, Enhanced half-metallicity in edge-oxidized zigzag graphene nanoribbons. *Nano Letters*, 7;2007:2295–2299.
20. G. Chen, M.S. Dresselhaus, G. Dresselhaus, J.P. Fleurial, T. Caillat, Recent developments in thermoelectric materials. *International Materials Reviews*, 48;2003:45–66.
21. J.H. Chen, C. Jang, S. Adam, M.S. Fuhrer, E.D. Williams, M. Ishigami, Charged-impurity scattering in graphene. *Nature Physics*, 4;2008:377–381.
22. P.E. Allain, J.N. Fuchs, Klein tunneling in graphene: Optics with massless electrons. *European Physics Journal B*, 83;2011:301–317.
23. B. Huard, J.A. Sulpizio, N. Stander, K. Todd, B. Yang, D. Goldhaber-Gordon, Transport measurements across a tunable potential barrier in graphene. *Physics Review Letters*, 98;2007:236803.
24. M.R. Setare, D. Jahani, Electronic transmission through p–n and n–p–n junctions of graphene. *Journal of Physics: Condensed Matter*, 22;2010:245503.
25. V. Lukose, R. Shankar, G. Baskaran, Novel electric field effects on landau levels in graphene. *Physical Review Letters*, 98;2007:116802.
26. S. Liu, A. Nurbawono, N. Guo, C. Zhang, Massless Dirac fermions in graphene under an external periodic magnetic field. *Journal of Physics: Condensed Matter*, 25;2013:395302.
27. S. Berber, Y.-K. Kwon, D. Tománek, Unusually high thermal conductivity of carbon nanotubes. *Physical Review Letters*, 84;2000:4613–4616.
28. W. Cai, A.L. Moore, Y. Zhu, X. Li, S. Chen, L. Shi, R.S. Ruoff, Thermal transport in suspended and supported monolayer graphene grown by chemical vapor deposition. *Nano Letters*, 10;2010:1645–1651.

29. S. Chen, A.L. Moore, W. Cai, J.W. Suk, J. An, C. Mishra, C. Amos, C.W. Magnuson, J. Kang, L. Shi, R.S. Ruoff, Raman measurements of thermal transport in suspended monolayer graphene of variable sizes in vacuum and gaseous environments. *ACS Nano*, 5;2010:321–328.

30. S. Chen, Q. Wu, C. Mishra, J. Kang, H. Zhang, K. Cho, W. Cai, A.A. Balandin, R.S. Ruoff, Thermal conductivity of isotopically modified graphene. *Nature Materials*, 11;2012:203–207.

31. O.V. Yazyev, S.G. Louie, Electronic transport in polycrystalline graphene. *Nature Materials*, 9;2010:806–809.

32. O.V. Yazyev, S.G. Louie, Topological defects in graphene: Dislocations and grain boundaries. *Physical Review B*, 81;2010:195420.

33. S.B. Kumar, J. Guo, Strain-induced conductance modulation in graphene grain boundary. *Nano Letters*, 12;2012:1362–1366.

34. Y. Lu, J. Guo, Thermal transport in grain boundary of graphene by non-equilibrium Green's function approach. *Applied Physics Letters*, 101;2012:043112.

35. A.Y. Serov, Z.-Y. Ong, E. Pop, Effect of grain boundaries on thermal transport in graphene. *Applied Physics Letters*, 102;2013:033104.

36. Y. Jiang, T. Low, K. Chang, M.I. Katsnelson, F. Guinea, Generation of pure bulk valley current in graphene. *Physical Review Letters*, 110;2013:046601.

37. E. McCann, M. Koshino, The electronic properties of bilayer graphene. *Reports on Progress in Physics*, 76;2013:056503.

38. L. Sun, Q. Li, H. Ren, H. Su, Q.W. Shi, J. Yang, Strain effect on electronic structures of graphene nanoribbons: A first-principles study. *The Journal of Chemical Physics*, 129;2008:074704.

39. O. Hod, G.E. Scuseria, Electromechanical properties of suspended graphene nanoribbons. *Nano Letters*, 9;2009:2619–2622.

40. S. Datta, *Lessons from Nanoelectronics: A New Perspective on Transport*, World Scientific, Singapore, 2012.

41. M. Lundstrom, C. Jeong, R. Kim, *Near-Equilibrium Transport: Fundamentals and Applications*, World Scientific, Singapore, 2013.

42. V.V. Cheianov, V. Fal'ko, B.L. Altshuler, The focusing of electron flow and a Veselago lens in graphene p-n junctions. *Science*, 315;2007:1252–1255.

43. J.R. Williams, T. Low, M.S. Lundstrom, C.M. Marcus, Gate-controlled guiding of electrons in graphene. *Nature Nanotechnology*, 6;2011:222–225.

44. V.G. Veselago, Electrodynamics of substances with simultaneously negative values of sigma and mu. *Soviet Physics Uspekhi*, 10;1968:509–514.

45. S. Park, H.S. Sim, π Berry phase and Veselago lens in a bilayer graphene np junction. *Physical Review B*, 84;2011:235432.

46. V.V. Cheianov, V.I. Fal'ko, Selective transmission of Dirac electrons and ballistic magnetoresistance of n-p junctions in graphene. *Physical Review B*, 74;2006: 041403(R).

47. O. Klein, Die reflexion von elektronen an einem potentialsprung nach der relativistischen dynamik von Dirac. *Zeitschrift für Physik*, 53;1929:157–165.

48. M.I. Katsnelson, K.S. Novoselov, A.K. Geim, Chiral tunnelling and the Klein paradox in graphene. *Nature Physics*, 2;2006:620–625.

49. A.F. Young, P. Kim, Quantum interference and Klein tunnelling in graphene heterojunctions. *Nature Physics*, 5;2009:222–226.

50. F. Hund, Materieerzeugung im anschaulichen und im gequantelten Wellenbild der Materie. *Zeitschrift für Physik*, 117;1941:1–17.

51. D. Dragoman, Evidence against Klein paradox in graphene. *Physica Scripta*, 79;2009:015003.

52. J.R. Williams, L. DiCarlo, C.M. Marcus, Quantum Hall effect in a gate-controlled p-n junction of graphene. *Science*, 317;2007: 638–641.

53. D.O. Guney, D.A. Meyer, Negative refraction gives rise to the Klein paradox. *Physical Review A*, 79;2009:063834.

54. R.N. Sajjad, C.A. Polanco, A.W. Ghosh, Atomistic deconstruction of current flow in graphene based hetero-junctions. *Journal of Computational Electronics*, 12;2013:232–247.

55. T. Low, S. Hong, J. Appenzeller, S. Datta, M.S. Lundstrom, Conductance asymmetry of graphene p-n junction. *IEEE Transactions on Electronic Devices*, 56;2009: 1292–1299.

56. S. Sutar, E.S. Comfort, J. Liu, T. Taniguchi, K. Watanabe, J.U. Lee, Angle-dependent carrier transmission in graphene p-n junctions. *Nano Letters*, 12;2012:4460–4464.

57. F. Sauter, Über das Verhalten eines Elektrons im homogenen elektrischen Feld nach der relativistischen Theorie Diracs. *Zeitschrift für Physik*, 69;1931:742–764.

58. A. Hansen, F. Ravndal, Klein's paradox and its resolution. *Physica Scripta*, 23;1981:1036–1042.

59. J.M. Pereira, F.M. Peeters, A. Chaves, G.A. Farias, Klein tunneling in single and multiple barriers in graphene. *Semiconductor Science and Technology*, 25;2010:033002.

60. C.W.J. Beenakker, Colloquium: Andreev reflection and Klein tunneling in graphene. *Reviews of Modern Physics*, 80;2008: 1337–1354.

61. A.V. Shytov, M.S. Rudner, L.S. Levitov, Klein backscattering and Fabry-Pérot interference in graphene heterojunctions. *Physical Review Letters*, 101;2008:156804.

62. M.M. Fogler, D.S. Novikov, L.I. Glazman, B.I. Shklovskii, Effect of disorder on a graphene p-n junction. *Physical Review B*, 77;2008:075420.

63. J. Maultzsch, S. Reich, C. Thomsen, H. Requardt, P. Ordejón, Phonon dispersion in graphite. *Physical Review Letters*, 92;2004:075501.

64. M. Mohr, J. Maultzsch, E. Dobardžiæ, S. Reich, I. Miloševiæ, M. Damnjanoviæ, A. Bosak, M. Krisch, C. Thomsen, Phonon dispersion of graphite by inelastic x-ray scattering. *Physical Review B*, 76;2007:035439.

65. A.A. Balandin, Thermal properties of graphene and nanostructured carbon materials. *Nature Materials*, 10;2011: 569–581.

66. S. Ghosh, I. Calizo, D. Teweldebrhan, E.P. Pokatilov, D.L. Nika, A.A. Balandin, W. Bao, F. Miao, C.N. Lau, Extremely high thermal conductivity of graphene: Prospects for thermal management applications in nanoelectronic circuits. *Applied Physics Letters*, 92;2008:151911.

67. W. Jang, Z. Chen, W. Bao, C.N. Lau, C. Dames, Thickness-dependent thermal conductivity of encased graphene and ultrathin graphite. *Nano Letters*, 10;2010:3909–3913.

68. Z.-Y. Ong, E. Pop, Effect of substrate modes on thermal transport in supported graphene. *Physical Review B*, 84;2011:075471.

69. J.H. Seol, I. Jo, A.L. Moore, L. Lindsay, Z.H. Aitken, M.T. Pettes, X. Li et al., Two-dimensional phonon transport in supported graphene. *Science*, 328;2010:213–216.

70. L. Lindsay, D.A. Broido, Optimized Tersoff and Brenner empirical potential parameters for lattice dynamics and phonon thermal transport in carbon nanotubes and graphene. *Physical Review B*, 81;2010:205441.

71. H. Zhang, G. Lee, K. Cho, Thermal transport in graphene and effects of vacancy defects. *Physical Review B*, 84; 2011:115460.

72. D. Nika, E. Pokatilov, A. Askerov, A. Balandin, Phonon thermal conduction in graphene: Role of Umklapp and edge roughness scattering. *Physical Review B*, 79;2009:155413.

73. B.D. Kong, S. Paul, M.B. Nardelli, K.W. Kim, First-principles analysis of lattice thermal conductivity in monolayer and bilayer graphene. *Physical Review B*, 80;2009:033406.

74. M. Park, S.-C. Lee, Y.-S. Kim, Length-dependent lattice thermal conductivity of graphene and its macroscopic limit. *Journal of Applied Physics*, 114;2013:053506.

75. K. Saito, J. Nakamura, A. Natori, Ballistic thermal conductance of a graphene sheet. *Physical Review B*, 76;2007: 115409.

76. A. Hashimoto, K. Suenaga, A. Gloter, K. Urita, S. Iijima, Direct evidence for atomic defects in graphene layers. *Nature*, 430;2004:870–873.

77. P.Y. Huang, C.S. Ruiz-Vargas, A.M. van der Zande, W.S. Whitney, M.P. Levendorf, J.W. Kevek, S. Garg et al., Grains and grain boundaries in single-layer graphene atomic patchwork quilts. *Nature*, 469;2011:389–392.

78. G. Liang, N. Neophytou, D.E. Nikonov, M.S. Lundstrom, Performance projections for ballistic graphene nanoribbon field-effect transistors. *Electron Devices, IEEE Transactions on*, 54;2007:677–682.

79. Y. Li, Z. Zhou, C.R. Cabrera, Z. Chen, Preserving the edge magnetism of zigzag graphene nanoribbons by ethylene termination: Insight by Clar's Rule. *Scientific Reports*, 3;2013:2030.

80. K.-T. Lam, Y. Lu, Y.P. Feng, G. Liang, Stability and electronic structure of two dimensional Cx(BN)y compound. *Applied Physics Letters*, 98;2011:022101.

81. S. Das Sarma, S. Adam, E.H. Hwang, E. Rossi, Electronic transport in two-dimensional graphene. *Reviews of Modern Physics*, 83;2011:407–470.

82. K.-T. Lam, G. Liang, An ab initio study on energy gap of bilayer graphene nanoribbons with armchair edges. *Applied Physics Letters*, 92;2008: 223106.

83. V.M. Pereira, J.M.B. Lopes dos Santos, A.H. Castro Neto, Modeling disorder in graphene. *Physical Review B*, 77;2008:115109.

84. P. Dietl, G. Metalidis, D. Golubev, P. San-Jose, E. Prada, H. Schomerus, G. Schön, Disorder-induced pseudodiffusive transport in graphene nanoribbons. *Physical Review B*, 79;2009:195413.

85. S.B. Kumar, M.B.A. Jalil, S.G. Tan, G. Liang, The effect of magnetic field and disorders on the electronic transport in graphene nanoribbons. *Journal of Physics: Condensed Matter*, 22;2010:375303.

86. X. Li, X. Wang, L. Zhang, S. Lee, H. Dai, Chemically derived, ultrasmooth graphene nanoribbon semiconductors. *Science*, 319;2008:1229–1232.

87. X. Ni, G. Liang, J.-S. Wang, B. Li, Disorder enhances thermoelectric figure of merit in armchair graphane nanoribbons. *Applied Physics Letters*, 95;2009:192114.

88. F. Miao, S. Wijeratne, Y. Zhang, U.C. Coskun, W. Bao, C.N. Lau, Phase-coherent transport in graphene quantum billiards. *Science*, 317;2007:1530–1533.

89. K.-T. Lam, C. Lee, G. Liang, Bilayer graphene nanoribbon nanoelectromechanical system device: A computational study. *Applied Physics Letters*, 95;2009:143107.

90. M. Fujita, K. Wakabayashi, K. Nakada, K. Kusakabe, Peculiar localized state at zigzag graphite edge. *Journal of the Physical Society of Japan*, 65;1996:1920.

91. W.Y. Kim, K.S. Kim, Prediction of very large values of magnetoresistance in a graphene nanoribbon device. *Nature Nanotechnology*, 3;2008:408–412.

92. O.V. Yazyev, M.I. Katsnelson, Magnetic correlations at graphene edges: Basis for novel spintronics devices. *Physical Review Letters*, 100;2008:047209.

93. J.-W. Rhim, K. Moon, Spin stiffness of graphene and zigzag graphene nanoribbons. *Physical Review B*, 80;2009:155441.

94. M. Topsakal, H. Sevincli, S. Ciraci, Spin confinement in the superlattices of graphene ribbons. *Applied Physics Letters*, 92;2008:173118.

95. K. Sawada, F. Ishii, M. Saito, S. Okada, T. Kawai, Phase control of graphene nanoribbon by carrier doping: Appearance of noncollinear magnetism. *Nano Letters*, 9;2008:269–272.

96. Y.-W. Son, M.L. Cohen, S.G. Louie, Half-metallic graphene nanoribbons. *Nature*, 444;2006:347–349.

97. J. Kang, F. Wu, J. Li, Doping induced spin filtering effect in zigzag graphene nanoribbons with asymmetric edge hydrogenation. *Applied Physics Letters*, 98;2011:083109.

98. J.Y. Wu, J.H. Ho, Y.H. Lai, T.S. Li, M.F. Lin, Electronic properties of 1D nanographite ribbons in modulated magnetic fields. *Physics Letters A*, 369;2007:333–338.

99. S.B. Kumar, M.B.A. Jalil, S.G. Tan, G. Liang, Magnetoresistive effect in graphene nanoribbon due to magnetic field induced band gap modulation. *Journal of Applied Physics*, 108;2010:033709.

100. R. Murali, Y. Yang, K. Brenner, T. Beck, J.D. Meindl, Breakdown current density of graphene nanoribbons. *Applied Physics Letters*, 94;2009:243114.

101. A.D. Liao, J.Z. Wu, X. Wang, K. Tahy, D. Jena, H. Dai, E. Pop, Thermally limited current carrying ability of graphene nanoribbons. *Physical Review Letters*, 106;2011:256801.

102. M.H. Bae, Z. Li, Z. Aksamija, P.N. Martin, F. Xiong, Z.Y. Ong, I. Knezevic, E. Pop, Ballistic to diffusive crossover of heat flow in graphene ribbons. *Nature Communications*, 4;2013: 1734.

103. J. Hu, X. Ruan, Y.P. Chen, Thermal conductivity and thermal rectification in graphene nanoribbons: A molecular dynamics study. *Nano Letters*, 9;2009:2730–2735.

104. Z. Guo, D. Zhang, X.-G. Gong, Thermal conductivity of graphene nanoribbons. *Applied Physics Letters*, 95;2009:163103.

105. W.J. Evans, L. Hu, P. Keblinski, Thermal conductivity of graphene ribbons from equilibrium molecular dynamics: Effect of ribbon width, edge roughness, and hydrogen termination. *Applied Physics Letters*, 96;2010:203112.

106. A.V. Savin, Y.S. Kivshar, B. Hu, Suppression of thermal conductivity in graphene nanoribbons with rough edges. *Physical Review B*, 82;2010:195422.

107. E. Munoz, J. Lu, B.I. Yakobson, Ballistic thermal conductance of graphene ribbons. *Nano Letters*, 10;2010: 1652.

108. J. Lan, J.-S. Wang, C. Gan, S. Chin, Edge effects on quantum thermal transport in graphene nanoribbons: Tight-binding calculations. *Physical Review B*, 79; 2009:115401.

109. Z. Huang, T.S. Fisher, J.Y. Murthy, Simulation of phonon transmission through graphene and graphene nanoribbons with a Green's function method. *Journal of Applied Physics*, 108;2010:094319.

110. Z.W. Tan, J.S. Wang, C.K. Gan, First-principles study of heat transport properties of graphene nanoribbons. *Nano Letters*, 11;2011:214.

111. J.-W. Jiang, B.-S. Wang, J.-S. Wang, First principle study of the thermal conductance in graphene nanoribbon with vacancy and substitutional silicon defects. *Applied Physics Letters*, 98;2011:113114.

112. H. Karamitaheri, M. Pourfath, R. Faez, H. Kosina, Atomistic study of the lattice thermal conductivity of rough graphene nanoribbons. *Electron Devices, IEEE Transactions on*, 60;2013:2142–2147.

113. Z. Aksamija, I. Knezevic, Lattice thermal conductivity of graphene nanoribbons: Anisotropy and edge roughness scattering. *Applied Physics Letters*, 98;2011: 141919.

114. Z. Wang, N. Mingo, Absence of Casimir regime in two-dimensional nanoribbon phonon conduction. *Applied Physics Letters*, 99;2011:101903.

115. G.J. Snyder, E.S. Toberer, Complex thermoelectric materials. *Nature Materials*, 7;2008:105–114.

116. H.J. Goldsmid, *Thermoelectric Refrigeration*, Plenum Press, New York, 1964.

117. A.I. Hochbaum, R. Chen, R.D. Delgado, W. Liang, E.C. Garnett, M. Najarian, A. Majumdar, P. Yang, Enhanced thermoelectric performance of rough silicon nanowires. *Nature*, 451;2008:163–167.

118. G. Liang, W. Huang, C.S. Koong, J.-S. Wang, J. Lan, Geometry effects on thermoelectric properties of silicon nanowires based on electronic band structures. *Journal of Applied Physics*, 107;2010:014317.

119. W. Huang, S.K. Chee, G. Liang, Theoretical study on thermoelectric properties of Ge nanowires based on electronic band structures. *Electron Device Letters, IEEE*, 31;2010:1026–1028.

120. Y. Ouyang, J. Guo, A theoretical study on thermoelectric properties of graphene nanoribbons. *Applied Physics Letters*, 94;2009:263107.

121. H. Zheng, H.J. Liu, X.J. Tan, H.Y. Lv, L. Pan, J. Shi, X.F. Tang, Enhanced thermoelectric performance of graphene nanoribbons. *Applied Physics Letters*, 100;2012:093104.

122. W. Zhao, Z.X. Guo, J.X. Cao, J.W. Ding, Enhanced thermoelectric properties of armchair graphene nanoribbons with defects and magnetic field. *AIP Advances*, 1;2011:042135.

123. P.S.E. Yeo, M.B. Sullivan, K.P. Loh, C.K. Gan, First-principles study of the thermoelectric properties of strained graphene nanoribbons. *Journal of Materials Chemistry A*, 1;2013:10762.

124. H. Sevinçli, G. Cuniberti, Enhanced thermoelectric figure of merit in edge-disordered zigzag graphene nanoribbons. *Physical Review B*, 81;2010:113401.

125. H. Karamitaheri, N. Neophytou, M. Pourfath, R. Faez, H. Kosina, Engineering enhanced thermoelectric properties in zigzag graphene nanoribbons. *Journal of Applied Physics*, 111;2012:054501.

126. Y. Chen, T. Jayasekera, A. Calzolari, K.W. Kim, M.B. Nardelli, Thermoelectric properties of graphene nanoribbons, junctions and superlattices. *Journal of Physics: Condensed Matter*, 22;2010: 372202.

127. C.-N. Pan, Z.-X. Xie, L.-M. Tang, K.-Q. Chen, Ballistic thermoelectric properties in graphene-nanoribbon-based heterojunctions. *Applied Physics Letters*, 101;2012:103115.

128. F. Mazzamuto, J. Saint-Martin, V. Nguyen, C. Chassat, P. Dollfus, Thermoelectric performance of disordered and nanostructured graphene ribbons using Green's function method. *Journal of Computational Electronics*, 11;2012:67–77.

129. Z.-X. Xie, L.-M. Tang, C.-N. Pan, K.-M. Li, K.-Q. Chen, W. Duan, Enhancement of thermoelectric properties in graphene nanoribbons modulated with stub structures. *Applied Physics Letters*, 100;2012:073105.

130. F. Mazzamuto, V. Hung Nguyen, Y. Apertet, C. Caër, C. Chassat, J. Saint-Martin, P. Dollfus, Enhanced thermoelectric properties in graphene nanoribbons by resonant tunneling of electrons. *Physical Review B*, 83;2011:235426.

131. W. Huang, J.-S. Wang, G. Liang, Theoretical study on thermoelectric properties of kinked graphene nanoribbons. *Physical Review B*, 84;2011: 045410.

132. L. Liang, E. Cruz-Silva, E.C. Girão, V. Meunier, Enhanced thermoelectric figure of merit in assembled graphene nanoribbons. *Physical Review B*, 86;2012:115438.

133. L. Liang, V. Meunier, Electronic and thermoelectric properties of assembled graphene nanoribbons with elastic strain and structural dislocation. *Applied Physics Letters*, 102;2013:143101.

134. H. Sevinçli, C. Sevik, T. Çağın, G. Cuniberti, A bottom-up route to enhance thermoelectric figures of merit in graphene nanoribbons. *Science Reports*, 3;2013:1228.

135. G. Liang, N. Neophytou, M.S. Lundstrom, D.E. Nikonov, Ballistic graphene nanoribbon metal-oxide-semiconductor field-effect transistors: A full real-space quantum transport simulation. *Journal of Applied Physics*, 102;2007:054307.

136. M. Zeng, L. Shen, M. Zhou, C. Zhang, Y. Feng, Graphene-based bipolar spin diode and spin transistor: Rectification and amplification of spin-polarized current. *Physical Review B*, 83;2011:115427.

27 Graphene-Enabled Heterostructures
Role in Future-Generation Carbon Electronics

Nikhil Jain and Bin Yu

CONTENTS

Abstract 423
27.1 Introduction 423
27.2 Fabrication Scheme 424
    27.2.1 Graphene Growth and Transfer 424
    27.2.2 Device Fabrication Processing 424
27.3 Characterization Methods 426
    27.3.1 Optical Microscopy 426
    27.3.2 Raman Spectroscopy 426
    27.3.3 Scanning Electron Microscopy 426
    27.3.4 Atomic Force Microscopy 427
    27.3.5 Electrical Characterization 427
27.4 Sample Cleaning Methods 427
    27.4.1 Pretesting Thermal Anneal 427
    27.4.2 Electrical Stress Annealing 427
27.5 Performance Characterization Scheme 429
    27.5.1 Electrical Performance 429
    27.5.2 Reliability Characterization 429
    27.5.3 Dielectric Behavior of *h*-BN 430
    27.5.4 Hexagonal Boron Nitride Encapsulation 431
27.6 Summary 432
References 433

ABSTRACT

Since being discovered in 2004, graphene has crossed several barriers to emerge as the frontrunner to potentially take the integrated chip fabrication technology forward in the post-silicon era. While graphene has been reported to have exceptional electrical, thermal, and electromechanical properties, significant impacts of the nonideal substrate (e.g., SiO_2) on key material properties limit the potential applications of graphene in nanoelectronics. In this chapter, hexagonal boron nitride (*h*-BN) is explored as an alternative substrate material for graphene-based electronics. Improvements in electrical behavior (carrier mobility and electrical conduction) and reliability (maximum current and power density) of devices fabricated using graphene/*h*-BN heterostructures have been demonstrated in comparison with the graphene/SiO_2 stack. It is noted that *h*-BN is not only an ideal substrate for graphene field effect transistors (GFETs)/interconnects, but also acts as a robust gate dielectric in GFETs). Moreover, owing to its superb thermal conductivity, *h*-BN is shown to act as a heat sink, thereby improving the graphene breakdown threshold. The viable graphene/*h*-BN heterostructures pave the way for realizing the true potential of graphene as both active and passive components in nanoelectronics.

27.1 INTRODUCTION

The structure of graphene shows a honeycomb lattice of sp^2-bonded carbon atoms in layered two-dimensional form [1]. From being a material that was not supposed to exist [2] to being the "rising star" [3], graphene has shown great potential in future generation electronics owing to its exceptional physical properties, namely high electrical and thermal conduction, high mechanical strength, transparency, and linear $E–k$ dispersion near the Dirac point [4]. While several reports have demonstrated transistors [5] and interconnects [6] made by monolayer and multilayer graphene, concerns exist on adverse effects of the substrate/gate dielectric (primarily SiO_2) on graphene's electrical behavior (e.g., carrier mobility and current density) [7]. Most notably, the hexagonal lattice symmetry of the graphene lattice is believed to be broken due to the presence of spatially dependent perturbations resulting in conformation of graphene to the top surface of the SiO_2 substrate [8]. Additionally, at room temperature, extrinsic scattering by surface phonons at the graphene–substrate interface imposes a limit on the carrier mobility in graphene [9]. One approach initially explored was suspending the graphene on trenches made in SiO_2, resulting in exceptional carrier transport properties, that is, low-temperature mobility as high as 200,000 cm^2 V^{-1} s^{-1} at carrier densities below 5×0^9 cm^{-2} [10]. Alternative approaches have included *in situ* growth of graphene on insulating substrates like SiC [11]. However, it leads to mechanical strain due to lattice mismatch that causes degradation of graphene's performance [12]. Additionally, SiC substrates are expensive and not practical for manufacturing uses.

Recently, hexagonal boron nitride (*h*-BN), an isomorph of graphene (lattice constant mismatch ~1.7% with graphene) has been shown as an excellent substitute to SiO_2 as the substrate and gate dielectric/passivation material for graphene-based FETs and interconnects [13,14]. Hexagonal boron nitride is a chemically inert material, and its layered crystalline structure allows for an atomically smooth surface that is free of dangling bonds. Consequently, when acting as a substrate, *h*-BN helps to suppress the rippling effect in graphene. Since graphene conduction is adversely affected because of scattering originating from the resonance of its carriers with the substrate phonons, it is critical that *h*-BN's optical phonon energy is twice that of SiO_2 that results in lesser scattering related transport degradation [13]. Additionally, *h*-BN is a wide bandgap insulator (E_G = 5.97 eV) and a medium-K dielectric ($\varepsilon \approx 4$) allowing its use as a gate dielectric material in graphene transistors [15]. High thermal conductivity of *h*-BN allows it to act as a heat sink, thereby reducing heat-induced failure and improving power dissipation at breakdown in graphene [16].

We observed excellent improvement in graphene conduction and mobility while studying graphene/*h*-BN heterostructures. The devices are also more robust with improvement in both breakdown electrical current and power density. Another issue affecting transport in graphene is the adsorption of water and oxygen molecules. The charged molecules can create large electric fields (e.g., up to ~10^9 V m^{-1}). The presence of such strong electric field on the surface of graphene can result in band-gap opening and strong scattering of the charge carriers, resulting in reduced current-carrying capability of the device [17,18]. A top layer of *h*-BN allows for encapsulation of graphene shielding it from the negative impact of environmental adsorbates [19]. A novel technique is described in this chapter for fabrication of *h*-BN/graphene/*h*-BN heterostructures. It is observed that encapsulated graphene can support much higher current density compared with uncovered graphene without any clear mobility degradation. Furthermore, the breakdown power density in encapsulated graphene is almost doubled. Improved electrical performance of graphene when using *h*-BN in its direct vicinity has opened up a new field of graphene-enabled heterostructures where pristine properties of graphene can be potentially exploited for future generation electronics. This chapter provides a basic understanding of the fabrication and characterization of graphene/*h*-BN and *h*-BN/graphene/*h*-BN heterostructures.

27.2 FABRICATION SCHEME

27.2.1 GRAPHENE GROWTH AND TRANSFER

Chemical vapor deposition (CVD) method has been adopted to obtain monolayer graphene. Alfa Aesar Cu foil (99.999% pure; 25 μm thick) is sectioned into rectangular pieces (1 cm × 5 cm) and native copper oxide is dissolved in acetic acid (CH_3COOH) for 15 min. The copper strips are loaded in a CVD tube furnace for graphene growth. The process involves raising the temperature slowly to 1000°C in an Ar (80 sccm) + H_2 (4.5 sccm) environment. Once the temperature is reached, Cu is annealed for 60 min. At this point, methane (CH_4, 30 sccm) is introduced into the chamber that acts as the carbon source because, at high temperature, CH_4 decomposes into carbon and hydrogen while Cu acts at a catalyst for graphene growth [20]. After cooling-down, the Cu strip with graphene is coated with PMMA and annealed on a hot plate. At this point, small pieces of the Cu/graphene/PMMA stack were cut and placed over the surface of iron chloride solution that etches away the Cu, leaving the graphene/PMMA bilayer floating on top of the solution. The graphene was cleaned by repeatedly washing in deionized water and then transferred onto the target substrate. The PMMA is dissolved by warm acetone (65°C).

27.2.2 DEVICE FABRICATION PROCESSING

Si/SiO_2 substrate is used as the starting point for creation of graphene/*h*-BN nanostack. While interconnects are fabricated on a Si/SiO_2 substrate, FETs are made on a Si/SiO_2 substrate with buried gates of TiN. Figure 27.1 shows a schematic of the cross-section of both configurations.

The fabrication scheme is shown in Figure 27.2a for graphene/*h*-BN interconnects and Figure 27.2b for graphene/*h*-BN FETs. The creation of buried gates allows the use of *h*-BN as not just a substrate but also as the gate dielectric avoiding any SiO_2 in the gate stack (which would otherwise be a part of the gate stack if using bottom Si as the gate). Starting point is a bare Si wafer that is covered with a uniform 100 nm thick SiO_2 layer by thermal oxidation. This results in an oxide layer on the front-side as well as the back-side of the wafer. The back-side oxide is etched away with hydrofluoric acid (HF). The top oxide is patterned using DUV lithography to create 250 nm wide lines connected with bond pads (50 × 50 μm). Once again HF is used for etching to create trenches that are 50 nm deep. In the following step, 200 nm TiN is deposited on the wafer followed by chemical mechanical planarization (CMP), which results in a smooth surface with TiN-filled trenches exposed at the top. Another step of DUV lithography results in a pattern for e-beam alignment marks (necessary for e-beam patterning later). Metal (Au) deposition in these patterns is done using an e-beam evaporator at high vacuum followed by lift-off in warm acetone (60°C). For interconnects, there is no need for buried gates, so the steps creating TiN trenches are avoided as shown in Figure 27.2b. Once the wafer is ready with e-beam alignment marks, *h*-BN is exfoliated using the

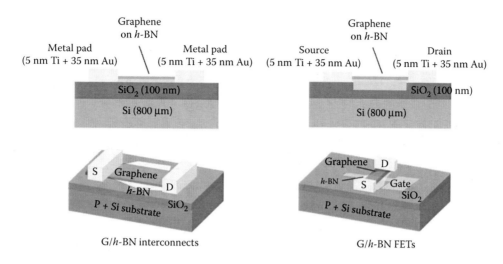

FIGURE 27.1 Schematic representation of graphene/h-BN interconnects and FETs.

scotch-tape method. Using an optical microscope, thin flakes (<20 nm) that are crossing over the metal trench are identified. Graphene is transferred over the substrate at this stage. To pattern graphene, a double layer of e-beam resists is used. PMMA A2 is spun-coated on the substrate at 2000 rpm followed by a 1 min bake at 180°C. This is followed by spin-coating of HSQ resist at 4000 rpm and 1 min bake at 150°C. The resulting bilayer resist is patterned with e-beam lithography. Since HSQ is a negative resist, after developing in CD-26 for 2 min, HSQ (with PMMA underneath) stays back

only at the exposed regions. The unexposed HSQ is washed away leaving behind PMMA on the rest of the substrate. The sample is then kept in O_2 plasma (10 sccm O_2 at 37 W) for 2 min in a micro RIE chamber. This results in the formation of a h-BN/graphene/PMMA/HSQ stack at sites that were exposed with e-beam. After keeping the sample in an acetone bath for 2 h, PMMA dissolves and lifts-off exposed HSQ leaving behind the graphene/h-BN structure. Another step of e-beam patterning and metallization results in the formation of Ti/Au contacts to graphene.

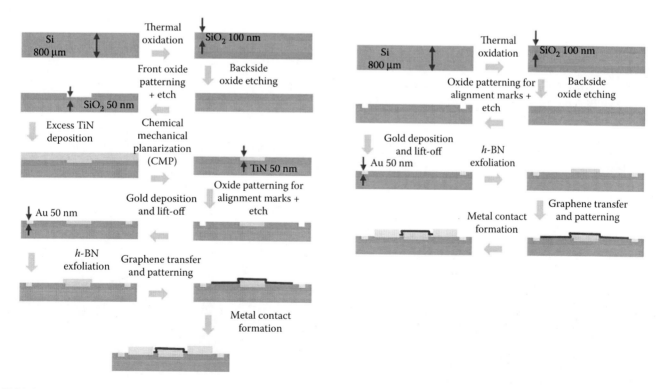

FIGURE 27.2 Fabrication process flow explaining the creation of graphene/h-BN interconnects and FETs including the formation of TiN trenches in Si/SiO2 wafers.

27.3 CHARACTERIZATION METHODS

27.3.1 OPTICAL MICROSCOPY

Two-dimensional materials are very thin (single or few layered), which causes difficulty in viewing them under the optical microscope. This makes it imperative to select the correct substrate oxide thickness for identifying thin flakes of the layered material under investigation. Since the material is essentially transparent, it adds up to the optical path of the incident light, thereby resulting in a slight contrast compared to the bare wafer. For graphene, it has been shown that the contrast is best for oxide thickness of 300 and 100 nm [21]. Simulations conducted in our lab have shown the corresponding oxide thicknesses of 70 and 250 nm for viewing h-BN on SiO_2. The details of these simulations are omitted for the purpose of brevity. For this study, 100 nm oxide thickness was used (instead of 70 nm) since the wafers were available as part of graphene study. The sample images of h-BN flakes on 100 nm SiO_2 are shown in Figure 27.3.

27.3.2 RAMAN SPECTROSCOPY

Raman spectroscopy is the single most efficient way of establishing the thickness and quality of graphene and other 2-D materials. Monolayer graphene (MLG) has a characteristic Raman spectrum and so does h-BN. The Raman spectrum was obtained with a Horiba Scientific micro-Raman system. A 532 nm laser (2.33 eV) was used as the excitation source. Laser power is kept below 0.1 mW to avoid laser-induced heating of the sample. A 1 μm laser spot size at focus was obtained with a 100× objective lens (NA = 0.95). Since graphene is sitting on a flake of h-BN, the resulting Raman signal is a superposition of the typical individual spectra of both materials. Figure 27.4 shows the Raman spectrum obtained from the graphene/h-BN sample. All the G- and D-mode features were adequately fitted with a Lorentzian component of the Voigt profiles (fitting curves not shown). The CVD-grown graphene on h-BN shows distinct G and 2D peaks and a characteristic h-BN peak along with a weak D peak. The G peak is associated with the doubly degenerate (iTO and LO) phonon modes (E_{2g} symmetry) at the Brillouin zone center. It comes

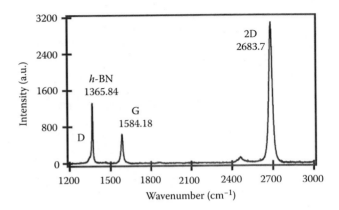

FIGURE 27.4 Raman spectrum of graphene/h-BN heterostructure showing distinct peaks for both materials. (Reprinted from *Elsevier. Carbon*, 54, Nikhil Jain et al., Monolayer graphene/hexagonal boron nitride heterostructure. 396–402, Copyright 2013, with permission from Elsevier.)

from the normal first-order Raman scattering process in graphene. The D and 2D peaks originate from the second-order process. The D peak comes from the involvement of one iTO phonon and one defect while the 2D peak involves two iTO phonons. The sharpness/height of the D peak indicates the defect level in graphene. A small/weak D peak signal indicates defect-free graphene. The h-BN peak occurs around 1366 cm^{-1}. We observed the ratio of the intensity of the 2D peak and the G-peak (I_{2D}/I_G) to be approximately equal to 5.09. This confirms the presence of monolayer graphene of good quality.

27.3.3 SCANNING ELECTRON MICROSCOPY

The response of graphene (and h-BN) to secondary electrons is very similar to the way it responds to incident light. Correspondingly, it is observed that graphene and h-BN are only faintly visible in scanning electron microscopy (SEM). Nevertheless, it is a useful metrology technique for graphene enabled heterostructures. An SEM micrograph of a graphene/h-BN nanostack on a TiN trench is shown in Figure 27.5.

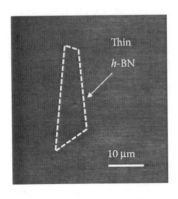

FIGURE 27.3 Optical microscope image of a thin h-BN flake.

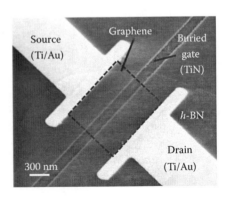

FIGURE 27.5 Scanning electron microscope image of the fabricated graphene/h-BN FET.

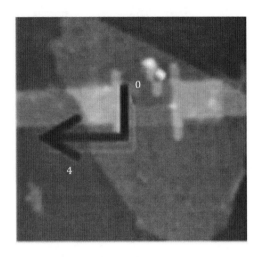
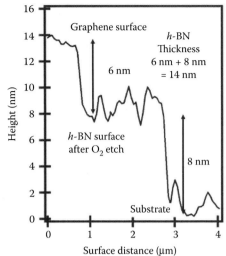

FIGURE 27.6 Atomic force microscope image showing a line scan from "0" to "4" and the corresponding data showing h-BN flake thickness for the particular device. (Reprinted from *Elsevier. Carbon*, 54, Nikhil Jain et al., Monolayer graphene/hexagonal boron nitride heterostructure. 396–402, Copyright 2013, with permission from Elsevier.)

27.3.4 ATOMIC FORCE MICROSCOPY

Thickness measurements of the thin flakes are done using atomic force microscopy (AFM) line scan. In particular, MLG as well as monolayer h-BN are 0.33 nm thick. We do not use monolayer h-BN as it is expected to be a leaky dielectric. The minimum thickness of h-BN used in this study is 14 nm as shown in Figure 27.6. Since the oxygen plasma etches h-BN as well as graphene, the resulting graphene/h-BN nanostack is surrounded by a region where h-BN is partially etched. To take h-BN thickness measurements, a line scan going across from top of graphene to bare SiO_2 is necessary. It can be observed from the AFM plot that the eroded h-BN surface is quite rough, which could be a result of oxygen damage in RIE.

27.3.5 ELECTRICAL CHARACTERIZATION

The devices were tested in an Agilent B1500A Semiconductor Parameter Analyzer using a Lakeshore probe station that is equipped with a high vacuum turbo pump. The devices were tested at room temperature.

27.4 SAMPLE CLEANING METHODS

27.4.1 PRETESTING THERMAL ANNEAL

The fabricated chip is annealed in a tube furnace at 700 mTorr in an $Ar + H_2$ environment for 5 h at 300°C. Processing steps like graphene transfer and e-beam lithography are known to leave behind resist residues (PMMA) on the surface of graphene. Moreover, oxygen and water molecules adhering to graphene surface can be removed by this process. This results in a sharp rise in current conduction capability of the graphene device as shown in Figure 27.7, where it is observed that the current has increased by almost four orders of magnitude as a direct consequence of thermal annealing.

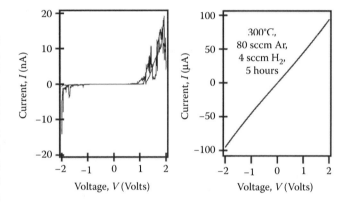

FIGURE 27.7 Improvement in device performance due to annealing in a forming gas environment. (Reprinted from *Carbon*, 54, Nikhil Jain et al., Monolayer graphene/hexagonal boron nitride heterostructure. 396–402, Copyright 2013, with permission from Elsevier.)

27.4.2 ELECTRICAL STRESS ANNEALING

This has been a reported technique to reduce contact resistance and graphene resistivity by cleaning up graphene–substrate interface [22]. Graphene is known to breakdown permanently, when the voltage applied across it increases a certain value, defined as V_{break} and is shown in Figure 27.8. For electrical-stress-induced annealing, voltage across the device is ramped up to a predetermined value (smaller than V_{break}) at a controlled rate, while the current is measured. To further illustrate the impact of this process, resistance versus gate voltage ($R - V_g$) characteristics are measured after each step of electrical stressing. The value of resistance is the total device resistance also known as R_{total}. It consists of two parts, the channel resistance (R_{ch}) and the contact resistance (R_C). The resistance of graphene is a function of gate voltage,

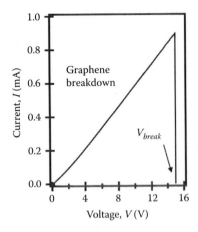

FIGURE 27.8 Ramping up the applied voltage across the graphene device results in physical breakdown of graphene as evident from the sudden drop in current. (Reprinted from *Carbon*, 54, Nikhil Jain et al., Monolayer graphene/hexagonal boron nitride heterostructure, 396–402, Copyright 2013, with permission from Elsevier.)

whereas the contact resistance is a function of the metal/graphene interface quality. For the purpose of this study, we can assume that R_C is a constant. It can be estimated by looking at the valley region of the curve ("far away" from change neutrality point), where the channel resistance is negligible and the major contribution comes from the metal/graphene contacts. When this value is removed from R_{total}, the $R_{ch} - V_g$ characteristics are obtained. Further calculation of resistivity requires the knowledge of the physical dimensions (length L and width W) of the graphene channel, assuming the thickness (t) of monolayer graphene sheet is ~0.34 nm. The resistivity (ρ) can then be given by $\rho = R_{ch}Wt/L$. Since W, t, and L are constants for a device, $R_{ch} - V_g$ can be converted to $\rho - V_g$ characteristics. This plot is shown in Figure 27.9 for several values of electrical stress annealing voltage. The Dirac peak location provides a reasonable approximation of the doping level in the graphene channel. A positive Dirac bias indicates a p-type doping behavior in graphene resulting from a positive shift in the Fermi level. This can be explained by the fact that graphene's work function (4.48 eV) is smaller than that of TiN (5.3 eV), which results in the formation of a potential difference across the thin *h*-BN dielectric. Even though *h*-BN screens a part of this potential by forming dipoles (resulting from interface traps) at the interfaces with TiN as well as with graphene [23], the screening is incomplete due to the absence of many such trap states that can help in the creation of dipoles. The remaining charge is balanced by trapping some of graphene's electrons resulting in hole-dominated carrier transport behavior in graphene (also known as p-doping).

It can also be observed that when the stressing voltage is low, the $\rho - V_g$ plot shows a slight downward shift. This can be attributed to cleaning up of the graphene channel caused by desorption of impurities from the surface due to the Joule heating effect of the current passing through graphene. At sufficiently high voltage, the channel becomes fairly clean and further increase in stressing voltage results in left shift of the Dirac peak along with the lowering of the maximum

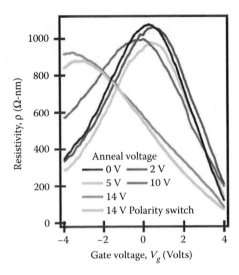

FIGURE 27.9 Electrical stress-induced annealing helps in reducing the resistivity of graphene. An additional effect is the left-shift of the charge neutrality point. (Reprinted from *Carbon*, 54, Nikhil Jain et al., Monolayer graphene/hexagonal boron nitride heterostructure, 396–402, Copyright 2013, with permission from Elsevier.)

resistivity. This observation has its origins in charge carrier trapping/release from the interfacial traps due to the impact of the stressing voltage. Additionally, with the *h*-BN flake being thin, high value of stress voltage can even pull in deep level trapped charges from underlying SiO$_2$ substrate as well as electrons from the gate metal [24]. This results in availability of more charge carriers in the channel while also making the channel n-doped. Depending on the application, this is a

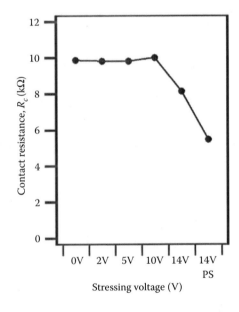

FIGURE 27.10 Electrical stress induced annealing results in cleaning up the metal–graphene contacts as evidenced from the lowering of contact resistance as the stressing voltage is increased. (Reprinted from *Carbon*, 54, Nikhil Jain et al., Monolayer graphene/hexagonal boron nitride heterostructure, 396–402, Copyright 2013, with permission from Elsevier.)

useful method of altering the graphene conductivity at zero gate voltage. For instance, when using graphene as an interconnect material, its resistivity should be as low as possible at zero gate voltage, so its Dirac point can be shifted by applying a strong stressing voltage.

Additionally, there is a significant impact of the stressing voltage on the metal–graphene contact resistance as shown in Figure 27.10. While low values of stressing voltage show a minor reduction, large values of stressing voltage can result in appreciable improvement in terms of contact conductivity. An interesting observation is that when the polarity of the stressing voltage is switched, a further reduction in contact resistance is observed in the device. This improvement could be due to preferential cleaning of the metal/graphene interface at the terminal where bias is applied. A further impact on resistivity indicates the introduction of carriers of opposite polarity being introduced from interfacial charge traps when the stress voltage polarity is switched.

27.5 PERFORMANCE CHARACTERIZATION SCHEME

27.5.1 Electrical Performance

The $\rho - V_g$ characteristics of graphene/h-BN device are compared with the exfoliated graphene and CVD graphene devices supported on SiO$_2$ as shown in Figure 27.11. It can be observed that when no gate bias is applied, the resistivity is lowest in graphene/h-BN structure as compared with the CVD/exfoliated graphene on SiO$_2$ structures. This significant improvement is attributed to the fact that both h-BN and graphene have isomorphic 2D hexagonal crystal lattices free of dangling bonds. The stack of two 2D-layered structures leads to a low density of interfacial states, which largely

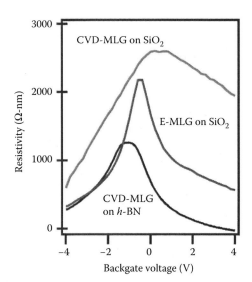

FIGURE 27.11 Graphene resistivity as a function of backgate voltage for three material systems, that is, CVD monolayer graphene (MLG) on h-BN, CVD-MLG on SiO$_2$ and exfoliated MLG on SiO$_2$. CVD-MLG on h-BN shows least resistivity as compared to the reference samples.

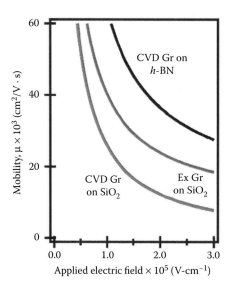

FIGURE 27.12 Mobility of charge carriers in graphene as a function of applied electric field shows a higher mobility in graphene/h-BN system. (Reprinted from *Carbon*, 54, Nikhil Jain et al., Monolayer graphene/hexagonal boron nitride heterostructure, 396–402, Copyright 2013, with permission from Elsevier.)

contribute to the degradation of carrier transport in graphene/SiO$_2$ system.

A direct impact of this behavior is observed in the enhancement of carrier mobility in CVD graphene stacked on h-BN. Effective carrier mobility was extracted by the following expression, $\mu_{eff} = g_d LG/WC_{ox}V_{BG}$, where g_d is the drain conductance given by $g_d = dI_D/dV_{DS}$ (slope of the drain current versus drain–source voltage, $I_D - V_{DS}$, curve) and C_{ox} is the gate capacitance given by, $C_{ox} = \varepsilon_r \varepsilon_0/t_{dielectric}$. Here, $t_{dielectric}$ is the thickness of the gate dielectric (SiO$_2$ + h-BN for Figure 27.1a, and h-BN for Figure 27.1b). Mobility is plotted against the effective electric field applied through the gate (Figure 27.12), which is then defined as the ratio of the applied gate voltage, V_{bg} and $t_{dielectric}$. Observing the trend in carrier mobility of graphene on h-BN and SiO$_2$ substrates, it can be established that the h-BN is clearly a superior substrate than SiO$_2$ for high-speed and high-performance graphene-based circuits.

27.5.2 Reliability Characterization

Graphene is known to undergo irreversible physical breakdown under high-level of electrical stress, resulting in a discontinuity in the sheet. This breakdown results in a drop in conduction current by several orders of magnitude, rendering device failure. Therefore, to explore the reliability limit of graphene, the I–V behavior needs to be explored in the near-breakdown region. To offset the effect of different geometries in different samples, current density (J) is plotted as a function of the applied voltage (V) across the graphene channel. The breakdown event is characterized by a sudden drop in current as seen in Figure 27.13. It is observed that the current density before breakdown is higher in CVD graphene/h-BN

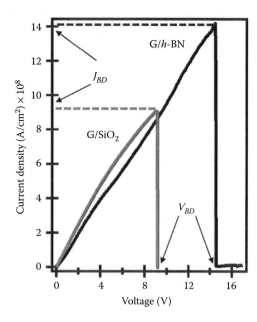

FIGURE 27.13 Improvement in breakdown current density and breakdown voltage in the graphene/h-BN system.

heterostructures as compared with that in the graphene/SiO$_2$ stack. To further investigate the breakdown characteristics of graphene, power density being dissipated at breakdown (P_{BD}) is studied for graphene on h-BN and SiO$_2$ substrates as shown in Figure 27.14. P_{BD} is defined by $P_{BD} = J_{BD} (V_{BD} - J_{BD}R_C)$, where V_{BD} and J_{BD} are the voltage and current density at which graphene breakdown occurs. We observed that P_{BD} is increased by over seven times in graphene/h-BN as compared with graphene/SiO$_2$. This difference can explained by the excellent thermal conductivity in h-BN (~600 W/mK in-plane and ~30 W/mK through-plane, compared to ~1 W/mK of amorphous SiO$_2$), which results in better heat dissipation through h-BN than that through SiO$_2$ under the 3D heat

spreading model for thermal-induced breakdown in graphene [25]. While improvement in J_{BD} provides an initial estimate of the advantages of using h-BN as a substrate for graphene, strong improvement in P_{BD} provides concrete evidence that h-BN substrate improves the performance window of graphene electronics.

27.5.3 Dielectric Behavior of h-BN

The real induction of h-BN as a substrate for graphene-based circuits depends not only on the properties of h-BN as a substrate but also as a dielectric material. In this regard, so far very little work has been done on exploring the through plane dielectric behavior of h-BN. Based on the application, h-BN can have different roles. When acting as a substrate for graphene FET, h-BN can also double up as a gate dielectric. In this configuration, h-BN is expected to avoid any leakage issues arising from the strong values of gate voltage. However, when acting as a support for graphene interconnects, it is also expected to shield the noise and prevent crosstalk. We investigate the insulating properties of the thin h-BN multilayer. In Figure 27.15, electrical current density through a 55 nm thick h-BN multilayer (in the through-plane direction) is shown as a function of electric field applied in the same direction. The first interesting observation is that even at an electric field as high as 15 MV/cm, no dielectric breakdown is seen. This is attributed to the excellent crystallinity and strong B–N covalent bonding in the h-BN lattice resulting in a very robust dielectric behavior. Since h-BN is a layered 2D lattice structure with weak van der Waals forces binding the adjacent layers, the separation between layers is much larger than a typical bond length. Such a structure with no actual bonds in the through-plane direction makes it very difficult to create a conduction filament or pathway. While reports on

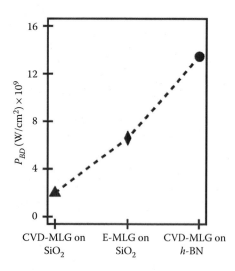

FIGURE 27.14 Comparison of the power being dissipated at breakdown for the three configurations studied in this chapter. CVD-MLG on h-BN shows a marked improvement over G/SiO$_2$ structures.

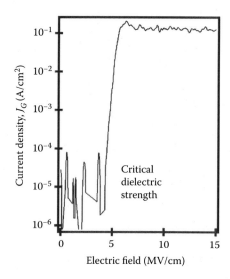

FIGURE 27.15 Study of metal/h-BN/metal heterostructure to explore the dielectric behavior of h-BN in the through-plane configuration. (Reprinted from *Carbon,* 54, Nikhil Jain et al., Monolayer graphene/hexagonal boron nitride heterostructure, 396–402, Copyright 2013, with permission from Elsevier.)

ultra-thin *h*-BN exhibiting *Direct Tunneling* (DT) at low-bias and *Fowler–Nordheim Tunneling* (FNT) at high-bias have been published recently [26], we observed neither DT nor FNT for our *h*-BN multilayer (ranging from 17–55 nm). At low bias, thin *h*-BN multilayer exhibits an excellent insulating behavior resulting in low leakage current density (~10^{-5} A/cm^2). However, as the voltage across the dielectric is ramped up, current density shows a sudden rise by approximately four orders of magnitude at a certain bias level, defined as *Transitional Voltage*, V_{trans}. This tunneling does not fall under the category of either DT or FNT as the leakage current density stays constant as the bias is increased further.

Several samples with different *h*-BN thickness were tested to observe the trend in V_{trans} in relation to dielectric thickness (using AFM line-scan to confirm the thickness). Figure 27.16 shows the observed dependence of V_{trans} on *h*-BN thickness. We define the electric field at V_{trans} as the *Critical Dielectric Strength* (CDS), given by the slope of V_{trans} versus *h*-BN thickness curve. A linear fit indicates the slope to be ~3.4 MV/cm. It should be noted that while dielectric breakdown strength is calculated in a similar manner, CDS is a different quantity as dielectric breakdown is an irreversible process making a dielectric leaky permanently whereas even after *h*-BN experiences a voltage more than its V_{trans}, it comes back to being a perfect insulator once the field is reduced below CDS. Repeating the voltage scan (0–100 V) several times does not bring about any change in dielectric behavior of *h*-BN with no occurrence of dielectric breakdown. The maximum current density remains almost constant in the sub-A/cm^2 level at high voltage bias. We present a possible mechanism to explain this volatile resistance switching behavior in *h*-BN based on

carrier hopping theory. It is well known that impurity substitution/intercalation in *h*-BN lattice introduces energy states in its wide bandgap [27–29]. At a sufficient concentration, these states can provide a pathway for carriers to get excited from the valence band and become available for carrying current. Moreover, since the number of states is constant depending on the thickness of the dielectric, the hopping current should not depend on the applied field across the dielectric. While higher electric field makes the carriers more energetic, conduction is limited by the density of available states. This is in accord with the observed dielectric behavior of *h*-BN. We believe this hopping of charge carriers through the impurity-related energy states could be the reason behind increased conduction through *h*-BN as elevated bias. At sufficiently low voltages (as in normal FET operation), the carriers do not possess sufficient energy to overcome the hopping barrier, resulting in an insulating behavior. The low-leakage behavior and the high breakdown threshold make the thin *h*-BN multilayer a robust gate dielectric.

27.5.4 Hexagonal Boron Nitride Encapsulation

Graphene devices face a further challenge when integration on an electronic chip is considered. Graphene conduction reduces with time due to adsorption of impurities from the environment on to the surface. Additionally, traditional capping/encapsulating materials for electronic circuits are expected to cause severe degradation in graphene transport since graphene is very sensitive to its dielectric environment. However, since *h*-BN lattice is very similar to that of graphene, *h*-BN was studied as an encapsulating material for graphene circuits. Figure 27.17 explains the encapsulation process. Using magnetic stirring for 1 h at 90°C, a viscous solution of

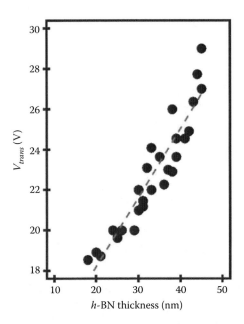

FIGURE 27.16 Voltage causing transition in *h*-BN from insulating to conducting state versus dielectric thickness shows a linear trend. (Reprinted from *Carbon*, 54, Nikhil Jain et al., Monolayer graphene/hexagonal boron nitride heterostructure, 396–402, Copyright 2013, with permission from Elsevier.)

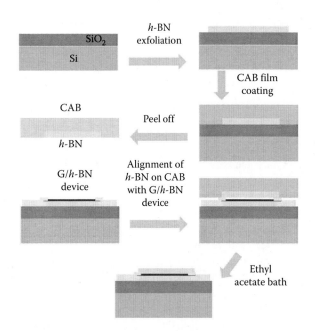

FIGURE 27.17 Process flow diagram explaining the passivation process introducing a top layer of *h*-BN over a graphene/*h*-BN device.

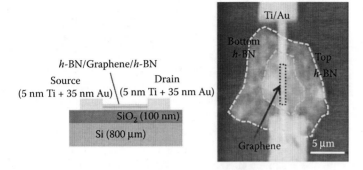

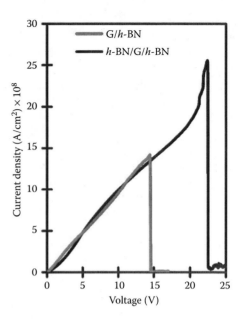

FIGURE 27.18 Schematic representation and optical microscope image of the encapsulated graphene heterostructure.

cellulose acetate butyrate (2 g) in ethyl acetate (15 mL) is created. Micromechanical exfoliation of thin flakes of multilayer h-BN onto a SiO₂/Si substrate is done followed by the identification of a useful flake with an optical microscope. Once the cellulose acetate butyrate (CAB) solution cools down, it is spin-coated on the substrate containing h-BN flakes for 60 s at 3200 rpm. After a quick 1-min anneal at 90°C, the CAB film is peeled off the substrate. This results in the film carrying many of the h-BN flakes with it. The film is placed with h-BN flakes facing down on top of the target substrate that already has a graphene/h-BN device. Using high-resolution microscope, the h-BN flake is aligned with the target substrate. Subsequently, using a drop of ethyl acetate the CAB film is dissolved leaving h-BN covering the target device. It is seen that the top h-BN layer encapsulates the entire graphene device including the metal contacts to graphene. In Figure 27.18, one such encapsulated graphene device is shown both schematically and in an optical microscope image. When compared with devices that do not have an encapsulating layer of h-BN, the encapsulated device shows a marked improvement in reliability characteristics (current density at breakdown and breakdown power density) while preserving the electrical characteristics, most notably the contact resistance and the carrier mobility which are known to suffer when graphene is contacted with other materials or left open. The current density improvement can be seen in Figure 27.19 while the power density at breakdown enhancement is shown in Figure 27.20. This improvement can again be attributed to the higher thermal conductivity of h-BN. Since the heat generated by the operating current in graphene device spreads in an isotropic manner, presence of h-BN on both sides of graphene provides better dissipation than air/graphene/h-BN system.

While the reliability characteristics show a distinct improvement, this enhancement does not come at a cost of inferior electrical performance as previously suspected. In fact, encapsulating the device passivates the metal–graphene contact making it insensitive to environmental degradation. When measuring the contact resistance in air and in vacuum, no appreciable difference was observed. This is in contrast to increase in contact resistance when measuring the open devices (no top passivating layer) in air compared to vacuum as shown in Figure 27.21. Further analysis of graphene's carrier mobility in the encapsulated system indicates very little

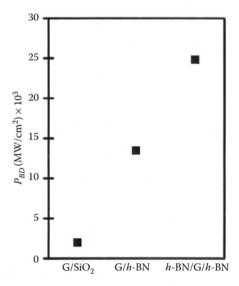

FIGURE 27.19 Effect of encapsulation on the breakdown current density of graphene/h-BN heterostructures.

FIGURE 27.20 Encapsulation of graphene in h-BN results in significant improvement in power density at breakdown, making it a more robust system.

reduction as shown in Figure 27.22, which is in contrast with previous reports on graphene encapsulation by other materials demonstrating appreciable reduction in carrier mobility [30]. Clearly, graphene passivated by h-BN on both sides performs better than any other known configuration so far.

27.6 SUMMARY

Compared with SiO₂, h-BN is an ideal substrate for graphene-based FETs and interconnects with demonstrated enhancement in electrical characteristics. Even CVD-grown graphene, known for its inferior nature to exfoliated

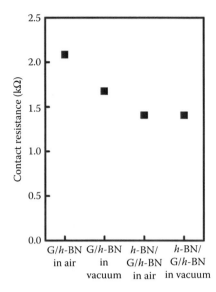

FIGURE 27.21 Graphene encapsulation in *h*-BN results in desensitization of the graphene–metal contacts to the effect of surrounding environment, resulting in uniform device behavior in either air or vacuum.

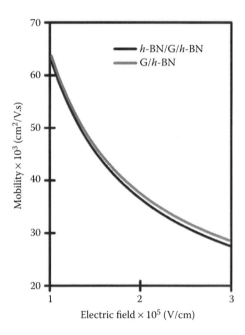

FIGURE 27.22 Carrier mobility in graphene is not adversely impacted by *h*-BN encapsulation.

graphene, performs better on *h*-BN substrate than exfoliated graphene on SiO_2. The observation paves the way for integrating graphene with standard silicon fabrication processes. We also demonstrated graphene/*h*-BN heterostructure FET in a locally buried metal-gate configuration where the thin *h*-BN multilayer acts as both gate dielectric and supporting layer for the graphene channel. In the metal/*h*-BN/graphene FET structure, graphene shows spontaneous doping due to work function difference between graphene and the gate. A volatile resistance switching behavior is observed in *h*-BN,

which does not conform to the tunneling mechanism proposed in the literature for ultra-thin *h*-BN. We hypothesize a hopping mechanism through the energy states introduced by interfacial impurity doping/intercalation into *h*-BN lattice to account for this abnormal conductive behavior under high electrical fields. No dielectric breakdown is observed even at very high electric field (15 MV/cm). Additionally, we have demonstrated that encapsulation of graphene with *h*-BN as the passivating layer improves current and power density without compromising on carrier mobility. Another advantage of graphene encapsulation is desensitization of the metal–graphene contact to the ambient effect. With the demonstrated current density in excess of 10^9 A/cm^2 in an encapsulated configuration, graphene emerges as a viable functional material in the post-Si era.

REFERENCES

1. Novoselov, K. S., A. K. Geim, S. V. Morozov, D. Jiang, M. I. Katsnelson, I. V. Grigorieva, S. V. Dubonos, and A. A. Firsov. 2005. Two-dimensional gas of massless Dirac Fermions in graphene. *Nature* 438 (7065) (November 10): 197–200. doi:10.1038/nature04233.
2. Peierls, R. 2013. Quelques propriétés typiques des corps solides. *Annales de l'institut Henri Poincaré* 5 (3): 177–222.
3. Geim, A. K. and K. S. Novoselov. 2007. The rise of graphene. *Nature Materials* 6 (3) (March): 183–191. doi:10.1038/nmat1849.
4. Castro Neto, A. H., F. Guinea, N. M. R. Peres, K. S. Novoselov, and A. K. Geim. 2009. The electronic properties of graphene. *Reviews of Modern Physics* 81 (1) (January 14): 109–162. doi:10.1103/RevModPhys.81.109.
5. Schwierz, Frank. 2010. Graphene transistors. *Nature Nanotechnology* 5 (7) (July): 487–496. doi:10.1038/nnano.2010.89.
6. Awano, Y.. 2009. Graphene for VLSI: FET and interconnect applications. In *Electron Devices Meeting (IEDM), 2009 IEEE International*, Baltimore, MD, 1–4. doi:10.1109/IEDM.2009.5424381.
7. Chen, J-H, C. Jang, S. Xiao, M. Ishigami, and M. S. Fuhrer. 2008. Intrinsic and extrinsic performance limits of graphene devices on SiO_2. *Nature Nanotechnology* 3 (4) (April): 206–209. doi:10.1038/nnano.2008.58.
8. Fratini, S. and F. Guinea. 2008. Substrate-limited electron dynamics in graphene. *Physical Review B* 77 (19) (May 13): 195415. doi:10.1103/PhysRevB.77.195415.
9. Romero, H. E., N. Shen, P. Joshi, H. R. Gutierrez, S. A. Tadigadapa, J. O. Sofo, and P. C. Eklund. 2008. n-type behavior of graphene supported on Si/SiO_2 substrates. *ACS Nano* 2 (10) (October 28): 2037–2044. doi:10.1021/nn800354m.
10. Du, X., I. Skachko, A. Barker, and E. Y. Andrei. 2008. Approaching ballistic transport in suspended graphene. *Nature Nanotechnology* 3 (8) (August): 491–495. doi:10.1038/nnano.2008.199.
11. Virojanadara, C., M. Syväjarvi, R. Yakimova, L. I. Johansson, A. A. Zakharov, and T. Balasubramanian. 2008. Homogeneous large-area graphene layer growth on 6H–SiC(0001). *Physical Review B* 78 (24) (December 1): 245403. doi:10.1103/PhysRevB.78.245403.
12. Ferralis, N., R. Maboudian, and C. Carraro. 2008. Evidence of structural strain in epitaxial graphene layers on 6H–SiC(0001). *Physical Review Letters* 101 (15) (October 6): 156801. doi:10.1103/PhysRevLett.101.156801.

13. Dean, C. R., A. F. Young, I. Meric, C. Lee, L. Wang, S. Sorgenfrei, K. Watanabe et al. 2010. Boron nitride substrates for high-quality graphene electronics. *Nature Nanotechnology* 5 (10) (October): 722–726. doi:10.1038/nnano.2010.172.

14. Xue, J., J. Sanchez-Yamagishi, D. Bulmash, P. Jacquod, A. Deshpande, K. Watanabe, T. Taniguchi, P. Jarillo-Herrero, and B. J. LeRoy. 2011. Scanning tunnelling microscopy and spectroscopy of ultra-flat graphene on hexagonal boron nitride. *Nature Materials* 10 (4) (April): 282–285. doi:10.1038/nmat2968.

15. Lipp, A., K. A. Schwetz, and K. Hunold. 1989. Hexagonal boron nitride: Fabrication, properties and applications. *Journal of the European Ceramic Society* 5 (1): 3–9. doi:10.1016/0955-2219(89)90003-4.

16. Jain, N., T. Bansal, C. Durcan, and B. Yu. 2012. Graphene-based interconnects on hexagonal boron nitride substrate. *IEEE Electron Device Letters* 33 (7): 925–927. doi:10.1109/LED.2012.2196669.

17. Moser, J., A. Verdaguer, D. Jiménez, A. Barreiro, and A. Bachtold. 2008. The environment of graphene probed by electrostatic force microscopy. *Applied Physics Letters* 92 (12): 123507. doi:10.1063/1.2898501.

18. Ito, J., J. Nakamura, and A. Natori. 2008. Semiconducting nature of the oxygen-adsorbed graphene sheet. *Journal of Applied Physics* 103 (11): 113712–113712-5. doi:10.1063/1.2939270.

19. Wang, L., Z. Chen, C. R. Dean, T. Taniguchi, K. Watanabe, L. E. Brus, and J. Hone. 2012. Negligible environmental sensitivity of graphene in a hexagonal boron nitride/graphene/h-BN sandwich structure. *ACS Nano* 6 (10) (October 23): 9314–9319. doi:10.1021/nn304004s.

20. Li, X., W. Cai, J. An, S. Kim, J. Nah, D. Yang, R. Piner et al. 2009. Large-area synthesis of high-quality and uniform graphene films on copper foils. *Science* 324 (5932) (June 5): 1312–1314. doi:10.1126/science.1171245.

21. Novoselov, K. S., A. K. Geim, S. V. Morozov, D. Jiang, Y. Zhang, S. V. Dubonos, I. V. Grigorieva, and A. A. Firsov. 2004. Electric field effect in atomically thin carbon films. *Science* 306 (5696) (October 22): 666–669. doi:10.1126/science.1102896.

22. Yu, T., C.-W. Liang, C. Kim, and B. Yu. 2011. Local electrical stress-induced doping and formation of monolayer graphene P-N junction. *Applied Physics Letters* 98 (24) (June 13): 243105–243105-3. doi:doi:10.1063/1.3593131.

23. Bokdam, M., P. A. Khomyakov, G. Brocks, Z. Zhong, and P. J. Kelly. 2011. Electrostatic doping of graphene through ultrathin hexagonal boron nitride films. *Nano Letters* 11 (11) (November 9): 4631–4635. doi:10.1021/nl202131q.

24. Giovannetti, G., P. A. Khomyakov, G. Brocks, V. M. Karpan, J. van den Brink, and P. J. Kelly. 2008. Doping graphene with metal contacts. *Physical Review Letters* 101 (2) (July 10): 026803. doi:10.1103/PhysRevLett.101.026803.

25. Liao, A. D., J. Z. Wu, X. Wang, K. Tahy, D. Jena, H. Dai, and E. Pop. 2011. Thermally limited current carrying ability of graphene nanoribbons. *Physical Review Letters*, 106 (25) (June 20): 256801. doi:10.1103/PhysRevLett.106.256801.

26. Lee, G.-H., Y.-J. Yu, C. Lee, C. Dean, K. L. Shepard, P. Kim, and J. Hone. 2011. Electron tunneling through atomically flat and ultrathin hexagonal boron nitride. *Applied Physics Letters* 99 (24): 243114–243114-3. doi:10.1063/1.3662043.

27. Park, H., A. Wadehra, J. W. Wilkins, and A. H. Castro Neto. 2012. Magnetic states and optical properties of single-layer carbon-doped hexagonal boron nitride. *Applied Physics Letters* 100 (25): 253115–253115-4. doi:10.1063/1.4730392.

28. Okada, S. and M. Otani. 2010. Stability and electronic structure of potassium-intercalated hexagonal boron nitride from density functional calculations. *Physical Review B* 81 (23) (June 9): 233401. doi:10.1103/PhysRevB.81.233401.

29. Shen, C., S. G. Mayorga, R. Biagioni, C. Piskoti, M. Ishigami, A. Zettl, and N. Bartlett. 1999. Intercalation of hexagonal boron nitride by strong oxidizers and evidence for the metallic nature of the products. *Journal of Solid State Chemistry* 147 (1) (October): 74–81. doi:10.1006/jssc.1999.8176.

30. Bolotin, K. I., K. J. Sikes, Z. Jiang, M. Klima, G. Fudenberg, J. Hone, P. Kim, and H. L. Stormer. 2008. Ultrahigh electron mobility in suspended graphene. *Solid State Communications* 146 (9–10) (June): 351–355. doi:10.1016/j.ssc.2008.02.024.

31. Nikhil Jain et al. Monolayer graphene/hexagonal boron nitride heterostructure. *Elsevier. Carbon*, 54, 396–402, 2013.

28 Recent Progresses and Understanding of Lithium Storage Behavior of Graphene Nanosheet Anode for Lithium Ion Batteries

Xifei Li and Xueliang Sun

CONTENTS

Abstract..435
28.1 Introduction 435
28.2 Synthesis of GNSs 436
28.3 GNS Anode in LIBs 438
28.4 Modification of GNS Anode..440
    28.4.1 GNSs-Based Composites 442
    28.4.2 Doped GNSs 443
28.5 Freestanding GNSs in LIBs...445
28.6 Conclusion and Perspectives...448
References..450

ABSTRACT

High-performance rechargeable lithium ion batteries (LIBs) have been one of the most important power sources in today's portable electronics and practical electric vehicle and plug-in electric vehicle applications. However, many crucial challenges could be addressed to obtain high-performance LIBs for further practical applications. Since the successful isolation of single graphene nanosheets via mechanical exfoliation in 2004 awarded the Nobel Prize in Physics in 2010, graphene nanosheets (GNSs), a new class of two-dimensional (one atom thick) carbon allotrope arranged in a hexagonal lattice with very strong sp^2-hybridized bonds, have attracted great interests in high-performance LIBs. Owing to various remarkable properties including ultra-high surface area, high electrical conductivity, and high chemical stability, GNSs show promising advantages in the applications of high-performance LIBs. This chapter highlights the recent development and understanding of GNS anodes and their applications, including (i) brief synthesis methods of GNSs; (ii) detailed modification approaches of GNSs; (iii) designs of free-standing GNS paper; and (iv) significant effects of various GNS anodes on lithium storage behavior, and thereby increased anode performance of LIBs.

28.1 INTRODUCTION

Recently, the transition from petroleum to an electrified road transportation system has been a crucial and significant social target. The researchers have focused on developing electric vehicles (EVs) and plug-in hybrid electric vehicles (PHEVs). The successful applications of EVs and PHEVs are expected to eventually replace conventional fossil fuel vehicles and help to reduce carbon dioxide emissions.[1] In comparison to other potential energy strategies (such as fuel cells and nickel metal hydride batteries), lithium ion batteries (LIBs) show a lot of obvious advantages in terms of higher working voltage, higher energy density, and longer cycle life. As a result, they have gained enormous commercial success, mainly in portable electronics, which also have been considered as one of the most important power sources for EVs and PHEVs.[2] However, LIBs fall short of satisfying needs for high power and high capacity for applications such as power tools, EVs, and PHEVs or efficient use of renewable energies.[3,4] Further performance improvement of advanced LIBs is still needed in terms of specific capacity, energy density, cost, and safety.

A typical LIB consists of cathode, anode, separator, and electrolyte. To some degree, the LIB performance mainly depends on the cathodes and the anodes. Besides the commercial graphite anode, there are some novel anodes such as high-capacity Si-based materials and Sn-based materials, and some of them have been commercialized at small scale.[5,6] However, challenge still exists to improve their cycle life and other aspects of performance. On the other hand, in addition to the traditional graphite and various natural or synthesized carbon-based anode materials, nano-carbon materials have far more potential to make contributions. Nano-carbon materials, including carbon nanotubes, fullerene (C_{60}), and diamond, have already been investigated to increase the specific capacities of anodes for LIBs.

Graphene nanosheets (GNSs), a new class of two-dimensional carbon allotrope (one atom thick), are arranged in sp^2 hybrid carbon atoms. In each carbon atom, three atomic orbitals (s, p_x, and p_y) form three strong bonds with other three surrounding atoms. In addition, the remaining p_z orbital overlaps with neighboring carbon atoms, which produces a filled band of p orbital (called the valence band), while an empty band of p* orbital is considered as the conduction band.[7,8] 2D GNSs are the basic building blocks for graphitic materials with other dimensionalities. Consequently, GNSs have been known as the "mother of all carbon forms." As shown in Figure 28.1, 2D GNSs could be wrapped up into 0D spherical bucky balls (fullerenes), rolled into 1D nanotubes (further categorized into single- or multiwalled carbon nanotubes depending on the number of graphene layers present), and stacked into 3D graphite with multilayered graphene sheets which in turn can be clustered into hard carbons. It is worth mentioning that a single GNS refers to graphene as its standard form, that is, a single layer of graphene.

GNSs possess various remarkable properties including ultra-high surface area (2630 m^2 g^{-1}), high electrical conductivity (resistivity: 10^{-6} Ω cm), and high chemical stability that are superior to those of CNTs and graphite,[7,10] which sparked widespread interests and investigations in both fundamental science and applied research around the world. Recent work has indicated that GNSs employed as anodes of LIBs have higher capacity than commercialized graphite.[11-14] All of the aforementioned properties combined with potential low manufacturing costs make GNS a promising anode candidate in LIBs. In this chapter, we highlighted the importance of GNSs and made an overview of its promising application in LIB.

28.2 SYNTHESIS OF GNSs

In 2004, the first synthesis of GNSs was obtained by mechanically separating individual graphene sheets from the strongly bonded layered structure of graphite,[15] and this important work was awarded the Nobel Prize in Physics in 2010. Since then, the researchers investigated some approaches to synthesize GNSs. GNSs were mostly synthesized through chemical, physical, and electrochemical approaches, where graphene layers were extracted from graphite, carbon fibers, or carbon nanotubes, for instance, physical/mechanical or chemical exfoliation,[16] and the unzipping of CNTs by electrochemical, chemical, or physical methods.[17] Other methods have also been proposed to obtain GNSs, for example, epitaxial growth via chemical vapor deposition,[18] microwave radiation,[19] and the reduction of sugars by some agents such as glucose and sucrose.[20]

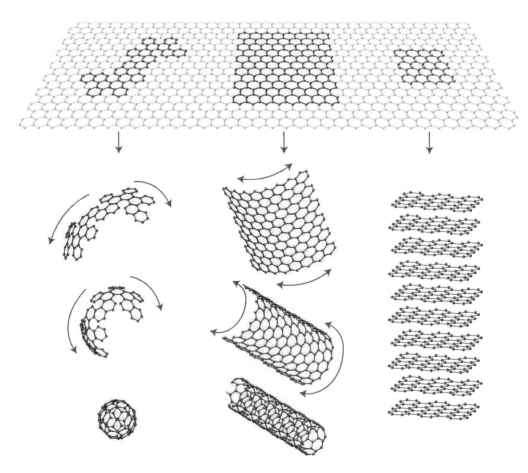

FIGURE 28.1 Mother of all graphitic forms. Graphene is a 2D building material for carbon materials of all other dimensionalities (0D buckyballs, 1D nanotubes, 3D graphite). (From A. K. Geim and K. S. Novoselov, *Nat. Mater.*, 2007, 6, 183–191.)

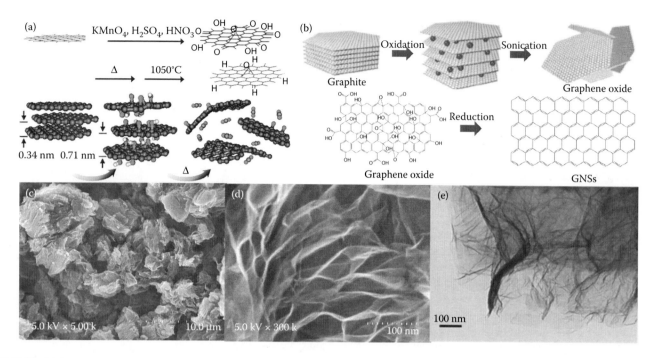

FIGURE 28.2 Synthesis mechanism of GNSs by modified Hummers of (a) thermal exfoliation (From X. Su et al. *Adv. Powder Technol.*, 2013, 24, 317–323; M. J. McAllister et al. *Chem. Mater.*, 2007, 19, 4396–4404.) and (b) sonication. (From S.-K. Lee, K. Rana and J.-H. Ahn, *The Journal of Physical Chemistry Letters*, 2013, 4, 831–841; G. Kucinskis, G. Bajars and J. Kleperis, *J. Power Sources*, 2013, 240, 66–79.) Typical (c, d) SEM and (e) TEM images of GNSs. (From X. Li et al. *Electrochem. Commun.*, 2011, 13, 822–825.)

Of the aforementioned methods, the most popular one for large-scale GNS production is Hummers' strategy followed by thermal exfoliation or sonication exfoliation due to easy operation, low cost, and easy scale-up. In the case of Hummers' strategy,[21] the precursor graphite was treated into graphite oxide (oxidized graphite) by a mixture of three types of oxidizing agents (such as H_2SO_4, $NaNO_3$ and $KMnO_4$), followed by rinsing with water and hydrogen peroxide afterward. As shown in Figure 28.2a, the graphite oxide shows similar stacked layer structure of graphite, but it contains many oxygen and hydrogen surface groups, which increase the interlayer distance. The larger distance between interlayers provides a possibility of further exfoliation into nanosheets.

The electrical conductivity of GNSs is dependent on the long-range conjugated network of graphitic lattice.[22,23] Hummers' strategy causes configuration change from a planar sp^2-hybridized to a distorted sp^3-hybrized geometry. Thus, both graphite oxide and graphene oxide show electrically insulating property resulting from the disrupted sp^2 bonding networks. The reduction of graphite oxide or graphene oxide can restore the π network, and produce a partial restoration of the electrical conductivity. Thermal exfoliation or sonication exfoliation followed by a strong reduction could meet this requirement, being two important approaches to synthesize GNSs in large scale.

To obtain GNSs using thermal exfoliation requires sufficient oxidation of the precursor (graphite) during Hummers' procedure, thereby adequate pressure obtained during the thermal treatment.[10] More oxygen-containing hydrophilic groups (such as hydroxyl, carbonyl, and epoxy) could be formed on or between the layers (Figure 28.2a). The obtained graphite oxide is suddenly heated at high temperature (such as 1050°C), where the pressure caused by the decomposition of oxygen-containing groups in the graphite oxide counteracts the van der Waals forces holding the graphene layers, and the interlayer spacing increases, therefore, the graphite oxide is split into the nanosheets. Simultaneously, the nanosheets could be reduced into GNSs.

After thermal exfoliation, the loose GNSs tend to overlap, and stick together to form fluffy agglomerates with a worm-like appearance (Figure 28.2c). An increased magnification in Figure 28.2d reveals that the "worm" consists of many ultrathin nanosheets with wavy structures. GNSs are with an ultrathin, wrinkled, and curved gossamer-like structure due to the thermodynamic stability of 2D materials; as shown in Figure 28.2e, they exhibit characteristic transparent gossamer sheets consisting of only a few graphene layers. It was found that the treatment at various temperatures exhibits some effects on GNSs. As shown in Table 28.1, specific surface areas of graphite oxide and GNSs obtained at different exfoliation temperatures were compared. The graphite oxide showed low specific surface area (59 $m^2\ g^{-1}$) due to similar structure to the graphite. After thermal exfoliation, the specific surface area of obtained GNSs significantly increased. However, the thermal exfoliation at different temperature produces various GNSs with different microstructures, resulting in different specific surface area. It can be observed that the specific surface area of GNSs produced at 300°C was the highest (559 $m^2\ g^{-1}$). With increased temperature, the specific surface area of GNSs decreased.[29]

TABLE 28.1

Specific Surface Areas of Graphite Oxide and GNSs Obtained at Various Exfoliation Temperatures

Samples	Graphite Oxide	GNSs		
		At 300°C	At 600°C	At 800°C
Specific surface area $(m^2\,g^{-1})$	60	559	412	380

Source: L. Wan et al. *Diamond Relat. Mater.*, 2011, 20, 756–761.

In addition to thermal exfoliation, the sonication can also be carried out to exfoliate the graphite oxide. This process typically involves complete exfoliation of graphite oxide into the graphene oxide sheets using sonication followed by the reduction of graphene oxide to obtain GNSs. As shown in Figure 28.2b, the sonication facilitates the intercalation of water molecules, and produces the graphene oxide suspension with an insulating property. GNSs are obtained by chemical reduction of the graphene oxide using a strong reductant. The hydrazine hydrate was reported to be the best reducing agent in synthesizing very thin GNSs.[31] Other reductants including hydroquinone and $NaBH_4$ were also reported.[26] Alternatively, some metal powders (e.g., Fe and Al) were also employed in the acidic solution to reduce graphene oxide,[32–34] and its mechanism is similar to the above-mentioned reductants.

28.3 GNS ANODE IN LIBs

Graphite is employed as the commercial anode material, but its theoretical specific capacity is limited (only 372 mAh g^{-1}) because it stores up to one Li$^+$ for every six carbon atoms between its graphene layers based on forming intercalation compounds LiC_6.[35] GNSs show vast surface-to-volume ratio and highly conductive nature, which could deliver high energy capacity. Lithium ions are stored on both sides of GNSs, where they are arranged like a "house of cards" in hard carbons, resulting in two layers of lithium for each graphene sheet.[32] Moreover, lithium ion can be stored within the cavities, covalent site, and hydrogen terminated edges of graphene fragments.[33,35] Therefore, the GNS anode was reported to show increased reversible capacities, moreover, it was observed that the different structured GNSs derived by various synthesis methods can contribute various reversible capacity (400–1100 mAh g^{-1}),[19,28,30,36–38] for example, microwave radiation derived GNSs could deliver a specific capacity of 420 mAh g^{-1} after 50 cycles.[19]

Besides high reversible capacity, in the first discharge process, some irreversible capacities can be observed due to the formation of a solid electrolyte interphase (SEI) film on the surface of the GNSs, resulting from electrolyte decomposition and formation of lithium organic compounds.[39] In addition, different from the graphite anode, the obtained large voltage hysteresis of GNS anode is ascribed to some active defects in the disordered GNSs. It was reported that the active defects of the GNS anode electrochemically react with lithium in discharge processes at lower voltages. However, in charge process, the bond break of lithium with the defects occurs at higher voltages, thereby resulting in the large voltage hysteresis.[40]

The GNS electrochemical reaction with lithium is dependent on the number of graphene layers and the defect sites on the basal plane of graphene. The divacancies, higher order defects, and few-layer GNSs assisting lithium ion diffusion through the basal plane are beneficial for increasing the specific reversible capacity in LIBs.[41] The number of graphene layers could be controlled by the intensity of oxidation of graphite. The oxidation intensity of graphite is increased with the increase of the dosage of oxidizer. As shown in Figure 28.3a, the oxidation easily causes more oxygen-containing groups on or between the layers. The GNS layers could be controlled by the different oxidation degree of graphite oxide (GO), which is recognized by the color of GO suspension. The colors of GO1, GO2, and GO3 are gradually fade, indicating controlled oxidation degree of GO. According to the statistics results and TEM images in Figure 28.3c through e, as-prepared GNSs derived from GO3, GO2, and GO1 are single-layer, triple-layer, and quintuplicate-layer graphene sheets, respectively. More importantly, the electrochemical results in Figure 28.3f demonstrated the significant influence of GNS layers on anode performance. The GNSs with fewer layers deliver higher reversible capacity.[42] The Cardema's group also demonstrated this effect.[36]

Besides the effect of GNS layers, the defects show some important influences on lithium storage of GNS anode. In Figure 28.4a, a decrease of GNS size results in an increase of available graphene edge sites, and may determine important properties of GNS. The carbon atoms at the edge of graphene display increased activity and energy in comparison to sp^2 bonded carbon atoms located within the basal plane.[15] As a result, smaller sized GNS-III anode with pronounced edge sites possesses the ability of encouraging increased lithium storage performance. Raman spectroscopy confirmed that GNSs with smaller size show higher I_D/I_G value, which indicates the presence of increased disorder and larger number of defects, encouraging additional lithium storage sites on the GNS anode. Thus, this type of GNSs produced an increased energy capacity. Interestingly, GNSs with more defects could reach a high reversible discharge capacity of 1348 mAh g^{-1}, and demonstrate an elevated value of 691 mAh g^{-1} in the 100th cycle.[43] It was reported that the GNS anode with more disordered structures, more defects, and smaller sp^2 domains showed large reversible capacity of 1013–1054 mAh g^{-1}.[40] As the researchers demonstrated in Figure 28.4b, the defects at edge sites and internal (basal-plane) defects (vacancies) of the nanodomains embedded in GNSs could involve reversible lithium storage. Some lithium ions are reversibly stored between (002) planes, but the defect-based reversible storage possibly predominate.[40] Therefore, the GNS anode with

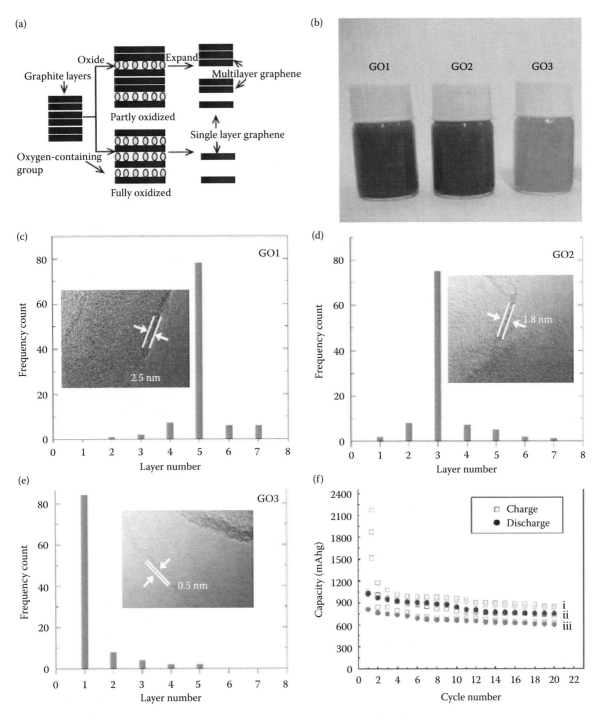

FIGURE 28.3 (a) Schematic illustration of formation process of GNSs with controlled numbers of layers; (b) The photos of different graphite oxide suspension: GO1 (partial oxidized), GO2 (inadequacy oxidized), and GO3 (fully oxidized); Statistics of controlled layer numbers of GNSs by using different oxidation degree: (c) GO1, (d) GO2, and (e) GO3; (f) capacity with cycle numbers at a current density of 300 mA g^{-1} (i) GNSs derived from GO3, (ii) GNSs derived from GO2, and (iii) GNSs derived from GO1. (From X. Tong et al. *J. Solid State Chem.*, 2011, 184, 982–989.)

more disordered structures and more defects may deliver high reversible capacity.

The aforementioned discussion is based on casting GNSs on the current collector (say copper foil). On the other hand, the direct growth of GNSs on a current collector could promote stronger adhesion without losing electrical contact;

moreover, the obtained electrodes without binder and conductive agent are beneficial for increasing energy density. Vertically aligned GNSs in Figure 28.5d were grown in a Ni substrate using a microwave plasma-enhanced chemical vapor deposition. The designed structure could decrease lithium diffusion distance, which showed some

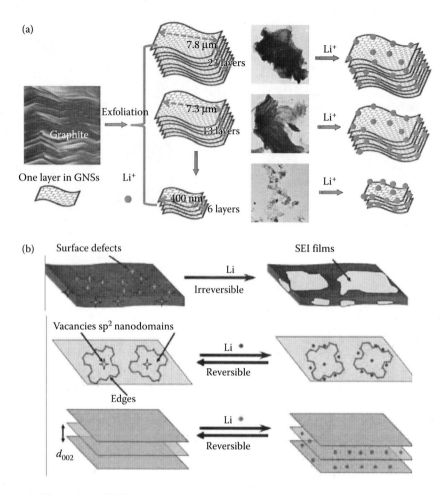

FIGURE 28.4 (a) Schematic illustration of lithium storage behavior of three types of GNSs with different layers and defects. (From X. F. Li et al. *Nanoscale*, 2013, 5, 12607–12615.) (b) Irreversible and reversible lithium storage at the interface, edge sites, and internal defects of nanodomains embedded in GNS anode. (From D. Pan et al. *Chem. Mater.*, 2009, 21, 3136–3142.)

positive effects of the improved kinetics. Moreover, the vertical alignment of GNSs with effective electronic connection to the Cu foil are beneficial for minimizing electrical resistance and improving the rate capability of GNS anode.[44] To tailor the structure of the vertical alignment, GNSs can increase battery performance. It was reported that the highly branched GNSs were directly grown on a Cu substrate, as shown in Figure 28.5a. The nanosheets were extremely small with lateral dimension down to less than 5 nm (see Figure 28.5b). These types of GNSs on Cu foil reported have some advantages compared with others, thereby showing better rate capability in Figure 28.5c.[45] Further optimization was carried out to increase the rate capability of GNSs on the current collectors. The 2D ordered mesoporous GNSs was proposed by Zhao's group. As shown in Figure 28.5e, ordered mesopore arrays with pore diameters about 9 nm were formed on the surface of the GNSs (highlighted by white arrows). Interestingly, all the mesopores were laid on one plane within a 2D mesostructure. The 2D mesopores GNSs delivered high surface area for Li+ intercalation/de-intercalation exhibiting a high reversible capacity of 1040 mAh g⁻¹ at 100 mA g⁻¹.

Even at a high current density of 5000 mA g⁻¹, the reversible capacity was delivered around 255 mAh g⁻¹, as shown in Figure 28.5f, indicating that the 2D mesoporous showed good effect in increasing rate capability and coulombic efficiency of GNSs. Therefore, the precursors and synthesis approaches obviously affect lithium storage of GNS anodes, and the comparison is summarized in Table 28.2.

28.4 MODIFICATION OF GNS ANODE

Owing to the aforementioned advantages, the GNS anode has been regarded as one of the most promising anode materials in LIBs. GNSs could exhibit high discharge capacity in the first several cycles, but the capacity rapidly drops with the increase of cycles and rates, showing the poor stability. The reason is that GNSs tend to agglomerate together to lose some merits of 2D GNSs.[47] Two approaches, that is, composites (including the combination with other carbon materials and the insertion of metal particles) and heteroatom doping, have been proposed to address this challenge, for example, by incorporation of the nanomaterials without electrochemical activity, the aggregation of GNSs could be suppressed

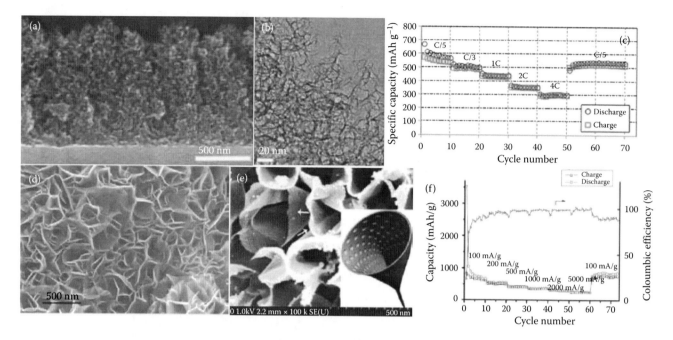

FIGURE 28.5 (a) SEM image of side view and (b) TEM image of the highly branched GNSs on a Cu foil, and (c) obtained rate capability of the highly branched GNSs at various current densities. (From H. Kim et al. *J. Mater. Chem.*, 2012, 22, 15514–15518.) (d) SEM image of GNSs grown on Ni substrate. (From X. Xiao et al. *Electrochem. Commun.*, 2011, 13, 209–212.) (e) SEM images and schematic (inset) of the mesoporous GNSs observed along the nanosheets and (f) rate capability and the coulombic efficiency of the mesoporous GNSs at various current densities. (From Y. Fang et al. *J. Am. Chem. Soc.*, 2013, 135, 1524–1530.)

TABLE 28.2
An Effect Summary of the Precursor and Synthesis Methods on GNS Anodes

Precursor	Synthesis Methods	Layers or Thickness	Coulombic Efficiency in the Initial Cycle (%)	Reversible Discharge Capacity in the 2nd Cycle (mAh g⁻¹)	Capacity Retention (%)	References
Natural graphite	Microwave radiation	\	86.21	500	84.0 (after 50 cycles)	[19]
Graphite powder	Hummers	5–13	46.51	1200	39.8 (after 100 cycles)	[30]
Graphite foils	Electrolytic exfoliation	1–10	32.72	440	70.5 (after 30 cycles)	[37]
Natural graphite	Hummers	3–8 nm	51.61	1054	74.0 (after 15 cycles)	[39]
Graphite powder	Hummers	1–5	50.34	1175	72.0 (after 20 cycles)	[42]
Graphite powder	Hummers	6–23	48.14	1348	51.3 (after 100 cycles)	[43]
Acetylene, hydrogen, and argon	Chemical vapor deposition	\	50.00	380	73.7 (after 150 cycles)	[44]
Methane, hydrogen, and argon	Chemical vapor deposition	8	32.00	550	75.0 (after 30 cycles)	[45]
Mesoporous carbon nanosheets	Solution deposition method	3	22.63	1040	80.0 (after 60 cycles)	[46]

essentially. It is believed that additional modification of GNSs could increase lithium storage performance.

28.4.1 GNSs-Based Composites

CNTs show 1D structure, different from 2D GNSs. A 3D homogeneous hybrid of GNSs wrapping CNTs, as indicated in Figure 28.6a, was demonstrated to deliver better cyclic performance and higher coulombic efficiency compared to the pure counterpart.[47] In addition to CNTs, 0D carbon nanospheres were also used to create GNSs-based composites. A series of composites of GNSs and the carbon nanospheres were designed,[48] where the carbon nanospheres (CNS) were fully covered and bridged with GNSs forming a 3D network with cavities and pores (Figure 28.6b). This structure provides many transportation pathways for electron and lithium, and

thereby delivers a high reversible capacity up to 925 mAh g^{-1} at a high rate of 5000 mA g^{-1}, which is much better than the anode without the carbon nanospheres, as shown in Figure 28.7a. The reaction of GNS anode with lithium is influenced by the layer spacing between the GNSs. A control of the intergraphene sheet distance through interacting molecules such as CNTs and fullerenes (C60) shows some effects on the lithium storage performance of the GNSs. The researchers studied the effect of different carbon materials (CNTs and C60) on GNS performance. It was found that the specific capacity of GNSs, GNSs/CNT, and GNSs/C60 significantly varied, which is related to the d-spacing of the GNS anode. In Figure 28.7c, the incorporation of macromolecules of CNTs and C60 to the GNSs can increase lithium storage performance resulting from the expansion in the d-spacing of the graphene layers with additional sites for Li$^+$ accommodation.[11]

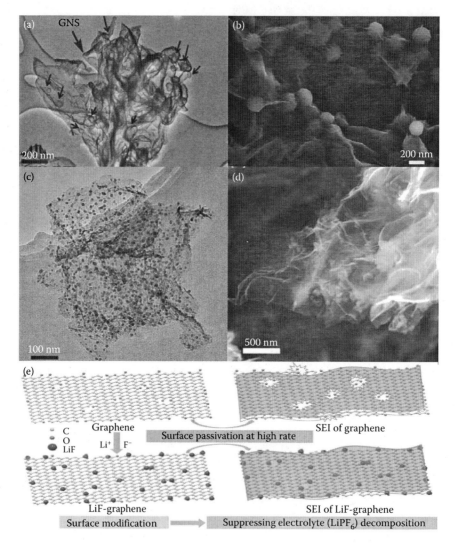

FIGURE 28.6 (a) TEM image of GNS–CNT hybrid materials. (From S. Chen et al. *Carbon*, 2012, 50, 4557–4565.) (b) SEM image of the GO/CNS composites with GO to CNS ratios (in weight) of 5/1. (From Y. Yang et al. *J. Mater. Chem.*, 2012, 22, 23194–23200.) (c) TEM image of graphene–Ni hybrid material. (From Y. J. Mai et al. *J. Power Sources*, 2012, 209, 1–6.) (d) SEM image of the LiF/graphene anode. (From Z.-S. Wu et al. *Adv. Funct. Mater.*, 2012, 22, 3290–3297.) (e) Diagram of the formation of a solid electrolyte interphase (SEI) film at a high rate in the GNSs (top) and LiF-modified GNSs (bottom). (From Z.-S. Wu et al. *Adv. Funct. Mater.*, 2012, 22, 3290–3297.)

Besides carbon materials, the nickel nanoparticles were designed to be uniformly anchored on the surface of GNSs (Figure 28.6c). According to the EIS results, Ni nanoparticles could enhance the electrical conductivity of the anode. Thus, this hybrid anode was found to increase electronic transport and lithium migration through the SEI film. As shown in Figure 28.7b, the hybrid anode is capable of showing better rate capability than the pristine GNSs. Another attempt is the GNSs/LiF composites shown in Figure 28.6d. Under the electrostatic interaction, the positively charged Li⁺ ions were easily adsorbed on the negatively charged oxygenated groups on the edges and defect sites of GNSs. In this hybrid composites, as shown in Figure 28.6e, LiF nanoparticles exhibit three functions for the GNS anode: (i) they are used as a $LiPF_6$ salt stabilizer, and thereby decrease the surface side reactions of the electrolyte decomposition and enhance the interface stability; (ii) as a SEI forming

inhibitor, they are directly a component of the SEI film on the GNS anode, and decrease the SEI film thickness; and (iii) they also increase the adhesion of the GNS surface to the SEI layer, and consequently enhance the thermodynamic stability.[50] Therefore, the GNSs/LiF composites have high rate capability in Figure 28.7d.

28.4.2 Doped GNSs

Chemical heteroatom doping into GNSs has been an effective approach to tune materials intrinsically, tailor electronic properties, manipulate surface chemistry, and produce local changes to the elemental composition of host materials.[51–53] Since nitrogen is of comparable atomic size to form strong valence bonds with carbon atoms, it is considered to be a good chemical doping of GNSs among potential dopants.[54] Nitrogen atom has one additional electron in comparison to

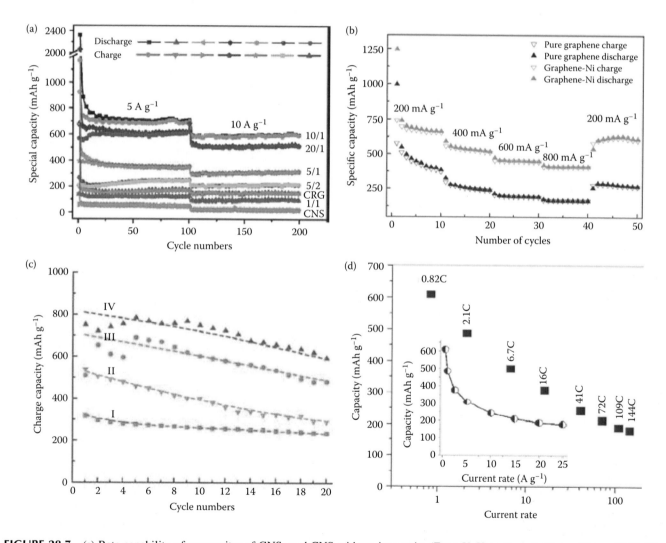

FIGURE 28.7 (a) Rate capability of composites of GNSs and CNS with various ratio. (From Y. Yang et al. *J. Mater. Chem.*, 2012, 22, 23194–23200.) (b) Rate capability of GNSs and hybrid GNSs–Ni anode. (From Y. J. Mai et al. *J. Power Sources*, 2012, 209, 1–6.) (c) Cycling performance of (I) graphite, (II) GNSs, (III) GNSs + CNT, and (IV) GNSs + C60. (From E. Yoo et al. *Nano Lett.*, 2008, 8, 2277–2282.) (d) Reversible capacity as a function of charging rates (inset: a plot of specific capacity vs. current rate). (From Z.-S. Wu et al. *Adv. Funct. Mater.*, 2012, 22, 3290–3297.)

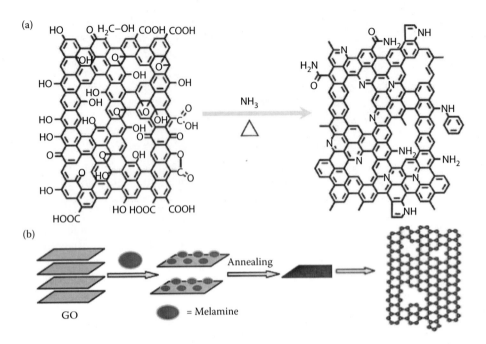

FIGURE 28.8   (a) Structure change of annealing of graphene oxide under NH₃. (From X. Li et al. *J. Am. Chem. Soc.*, 2009, 131, 15939–15944.) (b) Illustration of the nitrogen doping process of melamine into GO layers. (From Z.-H. Sheng et al. *ACS Nano*, 2011, 5, 4350–4358.)

carbon. After nitrogen is incorporated into the basal plane of graphene, it denotes electrons into graphene, which produces n-type doping of GNSs.[8] Nitrogen doped GNSs (N-GNSs) are obtained via two different approaches: (i) direct synthesis and (ii) post treatment. Direct synthesis consists of chemical vapor deposition, segregation growth, solvothermal, and arc-discharge approaches. However, the yield of direct synthesis is very limited, showing some difficulties in LIB applications. Thus, here we mainly discuss N-GNSs via post treatment.

Thermal treatment is one of the most important post treatments to obtain N-GNSs. Thermal treatment refers to treat GNSs at high temperature to produce N-GNSs. This method is simple, easy for scale-up, and, thus, it is popular in LIB application. At high temperature, to treat GNSs in the presence of various nitrogen precursors (like NH₃) can obtain N-GNSs, as previously reported.[28] In addition to GNSs, graphene oxide can also be used to produce N-GNSs using thermal treatment. In Figure 28.8a, thermal annealing graphene oxide at 500°C in an ammonia atmosphere could reduce graphene oxide and produce N-GNSs with 5 at.% nitrogen doping level.[55] During the annealing process, NH₃ reacts with oxygen groups (carboxylic, carbonyl, and lactone) in graphene oxide and forms C–N bond. More importantly, the graphene oxide is simultaneously reduced. For different graphene oxide with varying oxidation degrees, the reaction between graphene oxide and NH₃ is related to the amounts of these oxygen groups at the defect and edge sites of the graphene oxide, which gives rise to various nitrogen doping into GNSs. In addition to NH₃, other nitrogen precursors could produce N-GNSs, for instance, annealing the mixture of graphene oxide and melamine at high temperature (700–1000°C) can also obtain N-GNSs (Figure 28.8b).[52]

Normally the nitrogen atoms are substituted to incorporate into the basal plane of graphene in three forms: (i) pyridinic-N, (ii) quaternary-N, and (iii) pyrrolic-N bonding configurations, as shown in Figure 28.9a. Pyridinic-N contributes to the π-system with one p-electron, represented by a well-defined peak at the binding energy of 398.6 eV. Nitrogen atoms doped into the graphitic structure of graphene layer forms

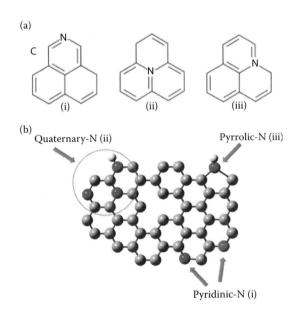

FIGURE 28.9   (a) Possible nitrogen positions in the structure; (i) pyridinic-N, (ii) quaternary-N, and (iii) pyrrolic-N. (From T. Sharifi et al. *ACS Nano*, 2012, doi: 10.1021/nn302906r.) (b) Schematic representation of N-GNSs. (From Y. Wang et al. *ACS Nano*, 2010, 4, 1790–1798.)

quaternary-N. Thus, quaternary-N can also be described as "graphitic-N." It constitutes the most complicated assignment, represented in the N_{1s} spectra by a broad peak positioned on 401.3 eV. Pyrrolic-N contributes to the π-system with two p-electrons, and hence has higher binding energy of about 400.3 eV (unnecessarily bond into the 5-membered ring). Essentially, nitrogen atoms doped in graphene could form pyridinic-N and pyrrolic-N, while nitrogen atoms substitute for carbon atoms in the hexagonal ring to form quaternary-N.[56] Among these nitrogen doping, pyridinic-N and quaternary-N are sp^2 hybridized, but pyrrolic-N is sp^3 hybridized. The difference among these three nitrogen configurations in a graphene basal plane is illustrated in Figure 28.9b. The pyridinic and pyrrolic nitrogen mainly lays at the edge or the defect sites, and the quaternary nitrogen refers to the nitrogen replacing carbon in graphene plane.

Due to the well-bonded nitrogen atom, nitrogen doping into GNSs can tailor both chemical and electronic properties of GNS anode. A lot of advantages of the disordered surface morphology, heteroatomic defects, good electrode/electrolyte wettability, increased intersheet distance, and high electrical conductivity (high electron transport) provide more active sites, rapid lithium surface absorption, high kinetics of lithium diffusion and transfer, thereby enhances the electrochemical reaction of GNSs with lithium.[58,59] As reported, the reversible discharge capacity of N-GSNs is almost double compared to pristine GNSs.[60] Therefore, it is expected that N-GNSs can increase anode performance in LIBs.

Similar to the microstructures of GNS anode (Figure 28.2c through e), N-GNSs also has a rippled and crumpled structure, as shown in Figure 28.10a. It is suggested that the nitrogen doping in the structure of GNSs could not change the nanosheet morphology. As discussed above, three types of nitrogen (pyridinc-N, quaternary-N, and pyrrolic-N) exist in N-GNSs. It is observed in Figure 28.10b that the pyridinc-N (area percentage: 55.7%) structure is more abundant than quaternary-N (13.3%) and pyrrolic-N (30.3%) structures. Nitrogen doping could increase the specific surface area of the GNS anode from 456 and 599 m^2 g^{-1}; moreover, it enhances the defect sites and vacancies as Li^+ active sites.[28] As a result, N-GNSs showed a higher discharge capacity and improved cyclic performance (Figure 28.10c). A CVD technique based on liquid precursor was employed to obtain N-GNS film on Cu foil. As demonstrated in Figure 28.10d, this film consisting of nitrogen doping showed double reversible capacity than the un-doped film. The researchers found that the improvement is ascribed to the large number of surface defects induced due to nitrogen doping.[60] Besides improvement of cyclic performance, N-GNSs doping of nitrogen could suppress the electrolyte decomposition and surface side reactions of the anodes, which significantly increase the coulombic efficiency of the anodes. For example, the coulombic efficiency of N-GNSs was increased to 49.0% in comparison to 43.8% of the pristine GNSs.[59]

Nitrogen atom with five electrons could be incorporated into the basal plane of graphene, and denote electrons into graphene. As a result, an increase of GNS electrical conductivity is obtained, which has some influences on the rate capability of the GNS anode. In Figure 28.10e, when N-GNSs was quickly charged/discharged for a very short time of 1 h to several tens of seconds, it exhibited high rate capability, for instance, a significant capacity of ~199 mAh g^{-1} at an ultra-high rate of 25,000 mA g^{-1}.[59] The improvement of rate capability was also demonstrated by Dr. Cui's group. As shown in Figure 28.10f, at each rate from C/20 to 5C, N-GNSs is capable of showing much higher reversible capacity than the pristine GNS anode.

Different from nitrogen doping, boron doping into graphene nanosheets (B-GNSs) produces an electron deficient system, thus B-GNSs could capture more lithium ions than the pristine GNSs. DFT calculations show that the specific capacity can be dramatically increased to 2271 mAh g^{-1} by boron doping, which is almost six times that of graphite anode.[61] Cheng et al. compared the anode performance of N-GNSs and B-GNSs. At a very high current density of 25 A g^{-1}, they reported that B-GNSs delivered a higher capacity of 235 mAh g^{-1} than N-GNSs.[59]

It can be seen that various modification in GNS anodes shows significant impact on lithium storage of GNSs in LIBs. The detailed comparison is summarized in Table 28.3.

28.5 FREESTANDING GNSs IN LIBs

Conventional methods for the fabrication of LIB electrodes usually involve mixing, casting, and pressing the mixed constituents, including a cathode or anode for lithium storage, a binder such as polyvinylidene fluoride to inhibit the collapse of the active materials from metal current collectors, and an electrical conductor to maintain the electrode conductivity onto the metal current collectors. It is worth noting that binders and metal current collectors do not contribute to lithium storage, and the electrical conductor exhibits minimal lithium storage performance; thus, these components decrease the energy density of LIBs. Moreover, the presence of binders in the electrodes decreases the accessible specific area of the active materials and increases the electrochemical polarization of the electrodes, undermining effective lithium ion transport.[62] From the perspective of optimal performance, the optimization of the electrode system may increase the energy density of LIBs, for example, by decreasing the dosage of the binder, the electrical conductor, and the current collector.[63] In particular, binder-free, electrical conductor-free, and current collector-free configurations can effectively meet this goal.[64,65] GNSs are with beautiful 2D structure. It is available to assemble GNSs into papers with highly freestanding and usually flexible. The freestanding graphene paper offers many advantages in LIBs, for instance, it was reported that the graphene papers are mechanically strong and electrically conductive with Young's modulus of 42 GPa, tensile strength of 293 MPa, and electrical conductivity of 351 S cm^{-1}.[66] Therefore, the development of flexible, lightweight, and environmentally benign binder-free, electrical conductor-free, and current collector-free graphene papers has drawn some research attention in LIBs.[67,68]

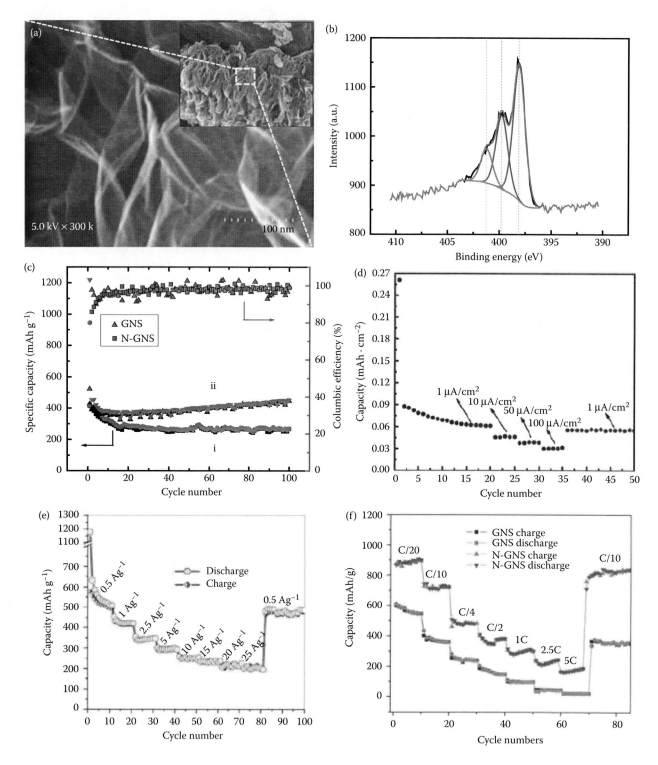

FIGURE 28.10 (a) Typical SEM image of N-GNSs. (From X. Li et al. *Electrochem. Commun.*, 2011, 13, 822–825.) (b) XPS spectra of N1s fitting for N-GNS anode. (From X. Li et al. *Electrochem. Commun.*, 2011, 13, 822–825.) (c) Reversible charge/discharge capacity versus cycle number of (i) GNS and (ii) N-GNS. (From X. Li et al. *Electrochem. Commun.*, 2011, 13, 822–825.) Rate capability of (d) N-GNS films. (From A. L. M. Reddy et al. *ACS Nano*, 2010, 4, 6337–6342.) and (e) N-GNS anode obtained. (From Z.-S. Wu et al. *ACS Nano*, 2011, 5, 5463–5471.) (f) The comparison of rate capability of GNSs and N-GNSs. (From H. Wang et al. *J. Mater. Chem.*, 2011, 21, 5430–5434.)

TABLE 28.3

A Summary of Modification Impacts on GNS Anode Performance

Material	Methods	Layers	Coulombic Efficiency (%) in the Initial Cycle	Reversible Discharge Capacity in the 2nd Cycle	Capacity Retention (%)	References
N-doped GNSs	Ammonia gas heat treatment	\	44.00	454 mAh g^{-1}	99.6 (after 100 cycles)	[28]
GNSs-CNTs	Ultrasonic treatment	2–5	34.83	618 mAh g^{-1}	78.5 (after 100 cycles)	[47]
GNSs-carbon nanospheres composite	Ultrasonication	\	50.00	810 mAh g^{-1}	86.4 (after 100 cycles)	[48]
GNSs-Ni hybrid	Freeze drying	5–10	60.00	790 mAh g^{-1}	85.0 (after 35 cycles)	[49]
LiF nanoparticle modified GNSs	Sonication	\	53.00	710 mAh g^{-1}	81.7 (after 150 cycles)	[50]
N-doped GNSs	Ammonia gas heat treatment	\	\	900 mAh g^{-1}	100 (after 10 cycles)	[58]
N-doped GNSs	Ammonia gas heat treatment	\	49.00	1000 mAh g^{-1}	83.6 (after 30 cycles)	[59]
N-doped GNSs films	Chemical vapor deposition	1–3	20.00	0.8 mAh cm^{-2}	62.5 (after 50 cycles)	[60]

Graphene paper has been successfully fabricated through vacuum-assisted filtration of graphene oxide dispersion.[68–71] Graphene paper is electrically conductive and mechanically strong, which shows promising for flexible LIB application. Wallace and coworkers prepared the graphene paper (see Figure 28.11a), and for the first time studied its LIB performance. Its initial discharge capacity was delivered to be 680 mAh g^{-1}, while decreased to only 84 mAh g^{-1} of the second discharge capacity.[68] The researchers modified the graphene paper to increase its performance, for example, graphene oxide paper was treated in supercritical ethanol, and obtained graphene paper in Figure 28.11b showed an increased capacity,[69] which is still very low. Graphene has a high in-plane Young's modulus but is easily warped in the out-of-plane direction due to its atomically 2D thin structure.[16] When randomly stacked graphene sheets are dried, the surface tension of the retreating liquid meniscus collapses the spacing between sheets and leads to intimate van der Waals contact between them.[72] It was reported that in-plane lithium diffusion coefficient is high (10^{-8} cm^2 s^{-1}), but cross-plane diffusivity of graphene paper is very limited. As a result, lithium migration into and out of a graphene stack is greatly restricted to stack edges,[72] which results in poor battery performance.

Annealing graphene oxide paper was reported to be one effective strategy to increase the performance of graphene paper, but its capacity is still unsatisfactory,[74] which is because thermally treated graphene stacks show severe intersheet aggregation limiting electrolyte permeation between the layers. Annealing graphene oxide papers with different thicknesses (1.5, 3, and 10 μm) were investigated. Three papers delivered evidently different lithium storage performances that thinner papers outperform thicker ones, as shown in Figure

28.11c.[70] The capacity decrease with the paper thickness was found to be related with the dense restacking of GNSs and a large aspect ratio of the paper, where the lithium diffusion proceeds mainly in in-plane direction, and cross-plane diffusion is restrained. To decrease current density could increase the reversible specific capacity of graphene paper in Figure 28.11d, indicating that the battery performance of graphene paper is a kinetically related process.[70] To address this challenge, it was reported that photothermally reduced graphene paper is structurally robust and displays superior rate capability (Figure 28.11e), for example, a steady capacity of 156 mAh g^{-1} at a very high rate of 40C.[71] The improved performance is due to the rapid outgassing of photothermal reduction creating microscale pores, cracks, and voids in graphene paper, which increases lithium intercalation kinetics.

Very limited cross-plane lithium diffusion causes low reversible capacity of graphene paper. Thus, via a wet chemical method that combined ultrasonic vibration and mild acid oxidation, in-plane porosity was created into the basal planes of graphene oxide, as described in Figure 28.12a. Using facile microscopic engineering, this type of holey graphene paper possesses abundant ion binding sites and enhanced ion diffusion kinetics, thereby delivers superior lithium storage capabilities at high rates. Another improved strategy is the design of graphene paper with folded structure in Figure 28.12b. Freeze-drying graphene oxide aqueous dispersion can form frozen graphene oxide dispersion and graphene aerogel. Subsequent thermal reduction followed by mechanical pressing would produce folded structured graphene paper. Obtained graphene paper used as LIB anode showed significantly improved rate performances compared to graphene paper synthesized by a flow-directed assembly method (Figure 28.11f) because the

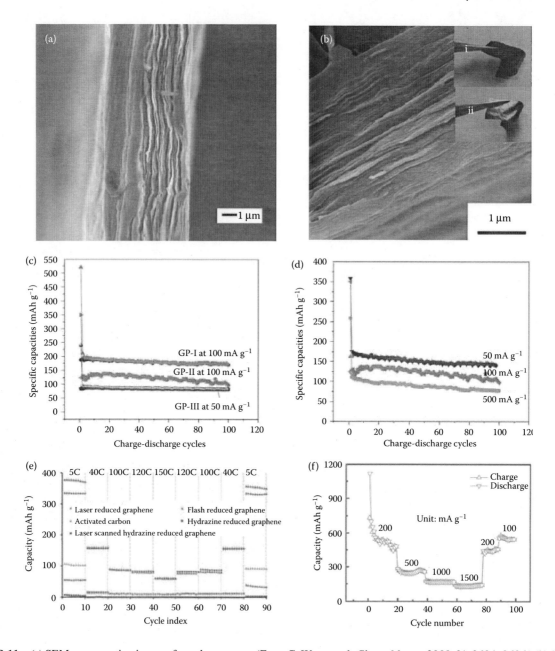

FIGURE 28.11 (a) SEM cross-section image of graphene paper. (From C. Wang et al. *Chem. Mater.*, 2009, 21, 2604–2606.) (b) SEM cross-section image of the graphene oxide paper after supercritical ethanol treatment. Inset: photo of graphene oxide paper (i) before and (ii) after supercritical ethanol treatment. (From S. Liu et al. *Appl. Surf. Sci.*, 2012, 258, 5299–5303.) (c) Cyclic performance of graphene papers with different thicknesses and (d) cyclic performance of GP-II tested at various current rates. (From Y. Hu et al. *Electrochim. Acta*, 2013, 91, 227–233.) (e) The comparison of rate capability of various graphene papers. (From R. Mukherjee et al. *ACS Nano*, 2012, 6, 7867–7878.) (f) Rate capability of the graphene paper at different current densities. (From F. Liu et al. *Adv. Mater.*, 2012, 24, 1089–1094.)

existence of graphene sheet folding increases the accessibility of lithium ions and electrolyte.[73] Furthermore, a porous structure of GNS papers was proposed to increase the discharge capacities of free-standing anodes via pressing a graphene cryogel.[75] In addition, some noble metals such as Pt and Ag were introduced to make flexible graphene composite paper. This design could increase anode performance of flexible paper, delivering a high charge capacity of 689 mAh g^{-1} at a current density of 20 mA g^{-1}.[76]

28.6 CONCLUSION AND PERSPECTIVES

In summary, the present research has shown convincing proofs that the GNS anode has been demonstrated to exhibit promising potentials in the fabrication of LIB anode materials resulting from various remarkable properties including ultra-high surface area, high electrical conductivity, and high chemical stability. More interestingly, modification (like doping and composites) of GNSs as well as free-standing GNS

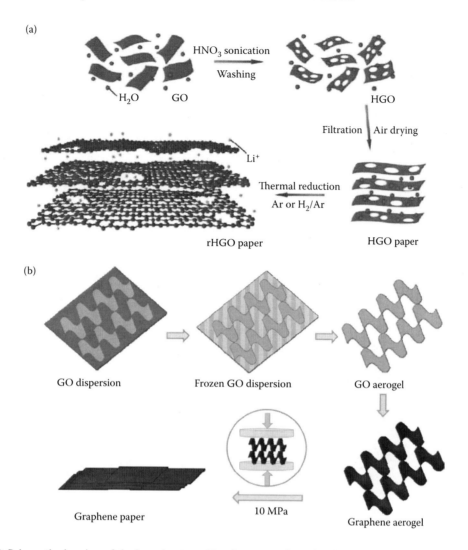

FIGURE 28.12 (a) Schematic drawing of the introduction of in-plane pores into chemically exfoliated graphene oxide and the subsequent filtration into a holey graphene oxide paper. (From X. Zhao et al. *ACS Nano*, 2011, 5, 8739–8749.) (b) Illustration of the formation process of graphene paper. (From F. Liu et al. *Adv. Mater.*, 2012, 24, 1089–1094.)

paper could further increase anode performance in LIBs. It is expected that this chapter opens the door to use GNSs as potential anodes for LIB applications. However, making GNS-based rechargeable LIBs is extremely challenging. Extensive fundamental studies on cost control, increase of coulombic efficiency, further increasing energy capacity, and so on, are needed to enable practical application of GNS anodes in rechargeable LIBs.

The promising purpose in the LIB research has been the fabrication of active materials with high energy capacity, good cyclic performance, and rate capability. Recently, various GNS anode materials have been developed rapidly; however, there remains one significant challenge to be tackled: how to further increase the energy capacity of the active materials in LIBs. To this end, the modification (doping, microstructure, and compositions, etc.) of GNSs itself is one good way to reach this purpose. On the other side, the integration of graphene with some electrochemically active materials showing high

theoretical capacity in a rational manner is another important approach to create high-performance LIBs. For example, the 2D core/shell architecture was constructed by confining the well-defined graphene-based metal oxides nanosheets (G@MO) within carbon layers (Figure 28.13a). The resulting G@MO@C hybrids exhibit outstanding reversible capacity and excellent rate performance for lithium storage.[77] GNSs, also, are having a function to increase cathode performance in the hybrid systems. For instance, GNS-wrapped cathode hybrid materials improved the cyclic stability due to promoted charge transfer reaction and lowered charge polarization (Figure 28.13b).[29] This chapter, thus, forecasts a bright future in electrode design of GNS-based active materials, which may be regarded as an emerging area of hybrid materials, but also remarks to foster their larger-scale industrial use. In benefit of the significant advantages of GNSs, it is highly expected that GNS-based hybrids could reward the new generation of high-performance LIBs in the future of energy storage.

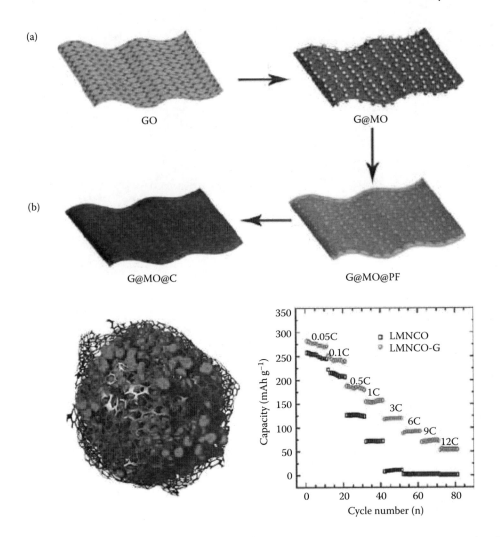

FIGURE 28.13 (a) Schematic illustration of synthesis of core-shelled GNSs@MO@C hybrids. (From Y. Su et al. *ACS Nano*, 2012, 6, 8349–8356.) (b) Schematic illustration of the hybrid cathodes and the comparison of rate performance of LMNCO and LMNCO-G. (From K.-C. Jiang et al. *ACS Applied Materials and Interfaces*, 2012, 4, 4858–4863.)

REFERENCES

1. G. Girishkumar, B. McCloskey, A. C. Luntz, S. Swanson and W. Wilcke, Lithium–air battery: Promise and challenges, *J. Phys. Chem. Lett.*, 2010, 1, 2193–2203.

2. V. Etacheri, R. Marom, R. Elazari, G. Salitra and D. Aurbach, Challenges in the development of advanced Li-ion batteries: A review, *Energy Environ. Sci.*, 2011, 4, 3243–3262.

3. P. L. Taberna, S. Mitra, P. Poizot, P. Simon and J. M. Tarascon, High rate capabilities Fe₃O₄-based Cu nano-architectured electrodes for lithium-ion battery applications, *Nat. Mater.*, 2006, 5, 567–573.

4. T. Song, J. Xia, J.-H. Lee, D. H. Lee, M.-S. Kwon, J.-M. Choi, J. Wu et al., Arrays of sealed silicon nanotubes as anodes for lithium ion batteries, *Nano Lett.*, 2010, 10, 1710–1716.

5. C. K. Chan, H. Peng, G. Liu, K. McIlwrath, X. F. Zhang, R. A. Huggins and Y. Cui, High-performance lithium battery anodes using silicon nanowires, *Nat. Nanotechnol.*, 2008, 3, 31–35.

6. Y. Idota, T. Kubota, A. Matsufuji, Y. Maekawa and T. Miyasaka, Tin-Based amorphous oxide: A high-capacity lithium-ion-storage material, *Science*, 1997, 276, 1395–1397.

7. K. S. Novoselov, D. Jiang, F. Schedin, T. J. Booth, V. V. Khotkevich, S. V. Morozov and A. K. Geim, Two-dimensional atomic crystals, *Proc. Natl. Acad. Sci. USA*, 2005, 102, 10451–10453.

8. H. Liu, Y. Liu and D. Zhu, Chemical doping of graphene, *J. Mater. Chem.*, 2011, 21, 3335–3345.

9. A. K. Geim and K. S. Novoselov, The rise of graphene, *Nat. Mater.*, 2007, 6, 183–191.

10. J. Hou, Y. Shao, M. W. Ellis, R. B. Moore and B. Yi, Graphene-based electrochemical energy conversion and storage: Fuel cells, supercapacitors and lithium ion batteries, *PCCP*, 2011, 13, 15384–15402.

11. E. Yoo, J. Kim, E. Hosono, H.-S. Zhou, T. Kudo and I. Honma, Large reversible Li storage of graphene nanosheet families for use in rechargeable lithium ion batteries, *Nano Lett.*, 2008, 8, 2277–2282.

12. A. V. Murugan, T. Muraliganth and A. Manthiram, Rapid, facile microwave-solvothermal synthesis of graphene nanosheets and their polyaniline nanocomposites for energy strorage, *Chem. Mater.*, 2009, 21, 5004–5006.

13. S.-H. Lee, S.-D. Seo, Y.-H. Jin, H.-W. Shim and D.-W. Kim, A graphite foil electrode covered with electrochemically

exfoliated graphene nanosheets, *Electrochem. Commun.*, 2010, 12, 1419–1422.

14. D. Cai, S. Wang, L. Ding, P. Lian, S. Zhang, F. Peng and H. Wang, Superior cycle stability of graphene nanosheets prepared by freeze-drying process as anodes for lithium-ion batteries, *J. Power Sources*, 2014, 254, 198–203.

15. X. Wang and H. Dai, Etching and narrowing of graphene from the edges, *Nat. Chem.*, 2010, 2, 661–665.

16. K. S. Novoselov, A. K. Geim, S. V. Morozov, D. Jiang, Y. Zhang, S. V. Dubonos, I. V. Grigorieva and A. A. Firsov, Electric field effect in atomically thin carbon films, *Science*, 2004, 306, 666–669.

17. D. V. Kosynkin, A. L. Higginbotham, A. Sinitskii, J. R. Lomeda, A. Dimiev, B. K. Price and J. M. Tour, Longitudinal unzipping of carbon nanotubes to form graphene nanoribbons, *Nature*, 2009, 458, 872–876.

18. A. Reina, X. Jia, J. Ho, D. Nezich, H. Son, V. Bulovic, M. S. Dresselhaus and J. Kong, Large area, few-layer graphene films on arbitrary substrates by chemical vapor deposition, *Nano Lett.*, 2008, 9, 30–35.

19. A. M. Shanmugharaj, W. S. Choi, C. W. Lee and S. H. Ryu, Electrochemical performances of graphene nanosheets prepared through microwave radiation, *J. Power Sources*, 2011, 196, 10249–10253.

20. C. Zhu, S. Guo, Y. Fang and S. Dong, Reducing sugar: New functional molecules for the green synthesis of graphene nanosheets, *ACS Nano*, 2010, 4, 2429–2437.

21. W. S. Hummers and R. E. Offeman, Preparation of graphitic oxide, *J. Am. Chem. Soc.*, 1958, 80, 1339–1339.

22. M. J. Allen, V. C. Tung and R. B. Kaner, Honeycomb carbon: A review of graphene, *Chem. Rev.*, 2009, 110, 132–145.

23. J. Wang, K. K. Manga, Q. Bao and K. P. Loh, High-yield synthesis of few-layer graphene flakes through electrochemical expansion of graphite in propylene carbonate electrolyte, *J. Am. Chem. Soc.*, 2011, 133, 8888–8891.

24. X. Su, G. Wang, W. Li, J. Bai and H. Wang, A simple method for preparing graphene nano-sheets at low temperature, *Adv. Powder Technol.*, 2013, 24, 317–323.

25. M. J. McAllister, J.-L. Li, D. H. Adamson, H. C. Schniepp, A. A. Abdala, J. Liu, M. Herrera-Alonso, D. L. Milius, R. Car, R. K. Prud'homme and I. A. Aksay, Single sheet functionalized graphene by oxidation and thermal expansion of graphite, *Chem. Mater.*, 2007, 19, 4396–4404.

26. S.-K. Lee, K. Rana and J.-H. Ahn, Graphene films for flexible organic and energy storage devices, *J. Phys. Chem. Lett.*, 2013, 4, 831–841.

27. G. Kucinskis, G. Bajars and J. Kleperis, Graphene in lithium ion battery cathode materials: A review, *J. Power Sources*, 2013, 240, 66–79.

28. X. Li, D. Geng, Y. Zhang, X. Meng, R. Li and X. Sun, Superior cycle stability of nitrogen-doped graphene nanosheets as anodes for lithium ion batteries, *Electrochem. Commun.*, 2011, 13, 822–825.

29. K.-C. Jiang, X.-L. Wu, Y.-X. Yin, J.-S. Lee, J. Kim and Y.-G. Guo, Superior hybrid cathode material containing lithium-excess layered material and graphene for lithium-ion batteries, *ACS Appl. Mater. Interfaces*, 2012, 4, 4858–4863.

30. L. Wan, Z. Ren, H. Wang, G. Wang, X. Tong, S. Gao and J. Bai, Graphene nanosheets based on controlled exfoliation process for enhanced lithium storage in lithium-ion battery, *Diamond Relat. Mater.*, 2011, 20, 756–761.

31. S. Stankovich, D. A. Dikin, R. D. Piner, K. A. Kohlhaas, A. Kleinhammes, Y. Jia, Y. Wu, S. T. Nguyen and R. S. Ruoff, Synthesis of graphene-based nanosheets via chemical reduction of exfoliated graphite oxide, *Carbon*, 2007, 45, 1558–1565.

32. Z.-J. Fan, W. Kai, J. Yan, T. Wei, L.-J. Zhi, J. Feng, Y.-M. Ren, L.-P. Song and F. Wei, Facile synthesis of graphene nanosheets via Fe reduction of exfoliated graphite oxide, *ACS Nano*, 2010, 5, 191–198.

33. Z. Fan, K. Wang, T. Wei, J. Yan, L. Song and B. Shao, An environmentally friendly and efficient route for the reduction of graphene oxide by aluminum powder, *Carbon*, 2010, 48, 1686–1689.

34. D. Wan, C. Yang, T. Lin, Y. Tang, M. Zhou, Y. Zhong, F. Huang and J. Lin, Low-temperature aluminum reduction of graphene oxide, electrical properties, surface wettability, and energy storage applications, *ACS Nano*, 2012, 6, 9068–9078.

35. N. Kheirabadi and A. Shafiekhani, Graphene/Li-ion battery, *J. Appl. Phys.*, 2012, 112, 124323.

36. G. Radhakrishnan, J. D. Cardema, P. M. Adams, H. I. Kim and B. Foran, Fabrication and electrochemical characterization of single and multi-layer graphene anodes for lithium-ion batteries, *J. Electrochem. Soc.*, 2012, 159, A752–A761.

37. J. M. Kim, W. G. Hong, S. M. Lee, S. J. Chang, Y. Jun, B. H. Kim and H. J. Kim, Energy storage of thermally reduced graphene oxide, *Int. J. Hydrogen Energy*, 2014, 39, 3799–3804.

38. S. Petnikota, N. Rotte, V. S. S. Srikanth, B. R. Kota, M. V. Reddy, K. Loh and B. V. R. Chowdari, Electrochemical studies of few-layered graphene as an anode material for Li ion batteries, *J. Solid State Electrochem.*, 2014, 18, 941–949.

39. S.-H. Lee, S.-D. Seo, K.-S. Park, H.-W. Shim and D.-W. Kim, Synthesis of graphene nanosheets by the electrolytic exfoliation of graphite and their direct assembly for lithium ion battery anodes, *Mater. Chem. Phys.*, 2012, 135, 309–316.

40. D. Pan, S. Wang, B. Zhao, M. Wu, H. Zhang, Y. Wang and Z. Jiao, Li storage properties of disordered graphene nanosheets, *Chem. Mater.*, 2009, 21, 3136–3142.

41. F. Yao, F. Güneş, H. Q. Ta, S. M. Lee, S. J. Chae, K. Y. Sheem, C. S. Cojocaru, S. S. Xie and Y. H. Lee, Diffusion mechanism of lithium ion through basal plane of layered graphene, *J. Am. Chem. Soc.*, 2012, 134, 8646–8654.

42. X. Tong, H. Wang, G. Wang, L. Wan, Z. Ren, J. Bai and J. Bai, Controllable synthesis of graphene sheets with different numbers of layers and effect of the number of graphene layers on the specific capacity of anode material in lithium-ion batteries, *J. Solid State Chem.*, 2011, 184, 982–989.

43. X. F. Li, Y. H. Hu, J. Liu, A. Lushington, R. Y. Li and X. L. Sun, Structurally tailored graphene nanosheets as lithium ion battery anodes: An insight to yield exceptionally high lithium storage performance, *Nanoscale*, 2013, 5, 12607–12615.

44. X. Xiao, P. Liu, J. S. Wang, M. W. Verbrugge and M. P. Balogh, Vertically aligned graphene electrode for lithium ion battery with high rate capability, *Electrochem. Commun.*, 2011, 13, 209–212.

45. H. Kim, Z. Wen, K. Yu, O. Mao and J. Chen, Straightforward fabrication of a highly branched graphene nanosheet array for a Li-ion battery anode, *J. Mater. Chem.*, 2012, 22, 15514–15518.

46. Y. Fang, Y. Lv, R. Che, H. Wu, X. Zhang, D. Gu, G. Zheng and D. Zhao, Two-dimensional mesoporous carbon nanosheets and their derived graphene nanosheets: Synthesis and efficient lithium ion storage, *J. Am. Chem. Soc.*, 2013, 135, 1524–1530.

47. S. Chen, W. Yeoh, Q. Liu and G. Wang, Chemical-free synthesis of graphene–carbon nanotube hybrid materials for reversible lithium storage in lithium-ion batteries, *Carbon*, 2012, 50, 4557–4565.

48. Y. Yang, R. Pang, X. Zhou, Y. Zhang, H. Wu and S. Guo, Composites of chemically-reduced graphene oxide sheets and carbon nanospheres with three-dimensional network structure as anode materials for lithium ion batteries, *J. Mater. Chem.*, 2012, 22, 23194–23200.

49. Y. J. Mai, J. P. Tu, C. D. Gu and X. L. Wang, Graphene anchored with nickel nanoparticles as a high-performance anode material for lithium ion batteries, *J. Power Sources*, 2012, 209, 1–6.

50. Z.-S. Wu, L. Xue, W. Ren, F. Li, L. Wen and H.-M. Cheng, A LiF nanoparticle-modified graphene electrode for high-power and high-energy lithium ion batteries, *Adv. Funct. Mater.*, 2012, 22, 3290–3297.

51. B. G. Sumpter, V. Meunier, J. M. Romo-Herrera, E. Cruz-Silva, D. A. Cullen, H. Terrones, D. J. Smith and M. Terrones, Nitrogen-mediated carbon nanotube growth: Diameter reduction, metallicity, bundle dispersability, and bamboo-like structure formation, *ACS Nano*, 2007, 1, 369–375.

52. Z.-H. Sheng, L. Shao, J.-J. Chen, W.-J. Bao, F.-B. Wang and X.-H. Xia, Catalyst-free synthesis of nitrogen-doped graphene via thermal annealing graphite oxide with melamine and its excellent electrocatalysis, *ACS Nano*, 2011, 5, 4350–4358.

53. C. Zhang, L. Fu, N. Liu, M. Liu, Y. Wang and Z. Liu, Synthesis of nitrogen-doped graphene using embedded carbon and nitrogen sources, *Adv. Mater.*, 2011, 23, 1020–1024.

54. Y. Wang, Y. Shao, D. W. Matson, J. Li and Y. Lin, Nitrogen-doped graphene and its application in electrochemical biosensing, *ACS Nano*, 2010, 4, 1790–1798.

55. X. Li, H. Wang, J. T. Robinson, H. Sanchez, G. Diankov and H. Dai, Simultaneous nitrogen doping and reduction of graphene oxide, *J. Am. Chem. Soc.*, 2009, 131, 15939–15944.

56. T. Hu, X. Sun, H. Sun, G. Xin, D. Shao, C. Liu and J. Lian, Rapid synthesis of nitrogen-doped graphene for a lithium ion battery anode with excellent rate performance and super-long cyclic stability, *PCCP*, 2014, 16, 1060–1066.

57. T. Sharifi, G. Hu, X. Jia and T. Wågberg, Formation of active sites for oxygen reduction reactions by transformation of nitrogen functionalities in nitrogen-doped carbon nanotubes, *ACS Nano*, 2012, 6, 8904–8912.

58. H. Wang, C. Zhang, Z. Liu, L. Wang, P. Han, H. Xu, K. Zhang, S. Dong, J. Yao and G. Cui, Nitrogen-doped graphene nanosheets with excellent lithium storage properties, *J. Mater. Chem.*, 2011, 21, 5430–5434.

59. Z.-S. Wu, W. Ren, L. Xu, F. Li and H.-M. Cheng, Doped graphene sheets as anode materials with superhigh rate and large capacity for lithium ion batteries, *ACS Nano*, 2011, 5, 5463–5471.

60. A. L. M. Reddy, A. Srivastava, S. R. Gowda, H. Gullapalli, M. Dubey and P. M. Ajayan, Synthesis of nitrogen-doped graphene films for lithium battery application, *ACS Nano*, 2010, 4, 6337–6342.

61. X. Wang, Z. Zeng, H. Ahn and G. Wang, First-principles study on the enhancement of lithium storage capacity in boron doped graphene, *Appl. Phys. Lett.*, 2009, 95, 183103.

62. G. Zhou, F. Li and H.-M. Cheng, Progress in flexible lithium batteries and future prospects, *Energy Environ. Sci.*, 2014, 7, 1307–1338.

63. M. S. Whittingham, Lithium batteries and cathode materials, *Chem. Rev.*, 2004, 104, 4271–4302.

64. R. A. DiLeo, S. Frisco, M. J. Ganter, R. E. Rogers, R. P. Raffaelle and B. J. Landi, Hybrid germanium nanoparticle–single-wall carbon nanotube free-standing anodes for lithium ion batteries, *J. Phys. Chem. C*, 2011, 115, 22609–22614.

65. H. Gwon, J. Hong, H. Kim, D.-H. Seo, S. Jeon and K. Kang, Recent progress on flexible lithium rechargeable batteries, *Energy Environ. Sci.*, 2014, 7, 538–551.

66. H. Chen, M. B. Müller, K. J. Gilmore, G. G. Wallace and D. Li, Mechanically strong, electrically conductive, and biocompatible graphene paper, *Adv. Mater.*, 2008, 20, 3557–3561.

67. H. Gwon, H.-S. Kim, K. U. Lee, D.-H. Seo, Y. C. Park, Y.-S. Lee, B. T. Ahn and K. Kang, Flexible energy storage devices based on graphene paper, *Energy Environ. Sci.*, 2011, 4, 1277–1283.

68. C. Wang, D. Li, C. O. Too and G. G. Wallace, Electrochemical properties of graphene paper electrodes used in lithium batteries, *Chem. Mater.*, 2009, 21, 2604–2606.

69. S. Liu, K. Chen, Y. Fu, S. Yu and Z. Bao, Reduced graphene oxide paper by supercritical ethanol treatment and its electrochemical properties, *Appl. Surf. Sci.*, 2012, 258, 5299–5303.

70. Y. Hu, X. Li, D. Geng, M. Cai, R. Li and X. Sun, Influence of paper thickness on the electrochemical performances of graphene papers as an anode for lithium ion batteries, *Electrochim. Acta*, 2013, 91, 227–233.

71. R. Mukherjee, A. V. Thomas, A. Krishnamurthy and N. Koratkar, Photothermally reduced graphene as high-power anodes for lithium-ion batteries, *ACS Nano*, 2012, 6, 7867–7878.

72. X. Zhao, C. M. Hayner, M. C. Kung and H. H. Kung, Flexible holey graphene paper electrodes with enhanced rate capability for energy storage applications, *ACS Nano*, 2011, 5, 8739–8749.

73. F. Liu, S. Song, D. Xue and H. Zhang, Folded structured graphene paper for high performance electrode materials, *Adv. Mater.*, 2012, 24, 1089–1094.

74. O. C. Compton, B. Jain, D. A. Dikin, A. Abouimrane, K. Amine and S. T. Nguyen, Chemically active reduced graphene oxide with tunable C/O ratios, *ACS Nano*, 2011, 5, 4380–4391.

75. K. Shu, C. Wang, M. Wang, C. Zhao and G. G. Wallace, Graphene cryogel papers with enhanced mechanical strength for high performance lithium battery anodes, *J. Mater. Chem. A*, 2014, 2, 1325–1331.

76. Y. Dai, S. Cai, W. Yang, L. Gao, W. Tang, J. Xie, J. Zhi and X. Ju, Graphene cryogel papers with enhanced mechanical strength for high performance lithium battery anodes, *Carbon*, 2012, 50, 4648–4654.

77. Y. Su, S. Li, D. Wu, F. Zhang, H. Liang, P. Gao, C. Cheng and X. Feng, Two-dimensional carbon-coated graphene/metal oxide hybrids for enhanced lithium storage, *ACS Nano*, 2012, 6, 8349–8356.

29 Study of Transmission, Transport, and Electronic Structure Properties of Periodic and Aperiodic Graphene-Based Structures

*Heraclio García-Cervantes, Rogelio Rodríguez-González,
José Alberto Briones-Torres, Juan Carlos Martínez-Orozco,
Jesús Madrigal-Melchor, and Isaac Rodríguez-Vargas*

CONTENTS

Abstract.. 453
29.1 Introduction 453
29.2 Electrostatic Barriers in Graphene 454
29.3 Breaking-Symmetry Substrates in Graphene 456
29.4 Theoretical Models.. 456
29.5 Periodic Structures 458
29.6 Aperiodic Structures.. 461
    29.6.1 Cantor-Like Structures 462
    29.6.2 Fibonacci Quasi-Periodic Structures.. 470
29.7 Concluding Remarks 474
29.8 Perspectives 474
Acknowledgments.. 475
References... 475

ABSTRACT

We study the transmission, transport, and electronic structure properties of periodic (Superlattices) and aperiodic (Cantor and Fibonacci) monolayer graphene-based structures. The transfer matrix method has been implemented to obtain the transmittance, linear-regime conductance, and electronic structure. In particular, we have studied two types of periodic and aperiodic graphene-based structures: (1) electrostatic graphene-based structures (EGBSs), structures formed with electrostatic potentials, and (2) substrate graphene-based structures (SGBSs), obtained alternating substrates that can open and non-open, such as SiC and SiO$_2$, an energy bandgap on graphene. We have found that the transmission properties can be modulated readily by changing the main parameters of the systems: well and barrier widths, energy and angle of incident electrons, and the number of periods as well as the degree of aperiodicity. The linear-regime conductance turns out that it diminishes various orders of magnitude increasing the barrier width for SGBSs. On the contrary, Klein tunneling sustains the conductance in EGBSs. Calculating the electronic structure or miniband-structure formation and fragmentation of periodic and aperiodic graphene-based structures, we establish a direct connection between the conductance peaks and the opening and closure of energy minibands for both EGSLs and SGSLs.

29.1 INTRODUCTION

Graphene, a two-dimensional honeycomb lattice of carbon atoms, has attracted a lot of attention since its discovery [1–3]. This renewed interest in carbon-based materials comes from: the unusual properties present in graphene, such as minimum conductivity [1–3], odd-integer-quantum-Hall-effect [3–6], Klein tunneling [7–10], zitterbewegung [11–14], and atomic collapse [15–17]; the natural connection between condensed matter physics and quantum electrodynamics giving place to new terms as relativistic condensed matter physics or relativistic solid state physics [18–20]; the outstanding intrinsic physical properties such as mobility, thermal conductivity, flexibility, strength, and stiffness, which give rise to potential device applications [21–24]. All these novel properties and possible technological breakthroughs rely on the gapless linear dispersion relation of graphene [25,26], $E = \pm \hbar v_F k$. In particular, Klein tunneling results from suppression of back-scattering due to carrier's pseudo-spin conservation [9,10].

On the other hand, it is well known that periodic and ape-riodic order appears in multiple aspects of nature as well as that both orders can be exploited technologically [27,28]. For instance, periodic modulation was the instrument to unveil quantum effects in semiconductor artificial structures [29,30], giving rise to what later on was known as semicon-ductor superlattice. Among the most important properties of semiconductor superlattices, we can find excitonic effects, miniband transport, Wannier–Stark localization, Bloch oscil-lations, resonant tunneling, and electric field domains [27]. Most of these properties rely on the miniband structure or miniband dispersion presented in semiconductor superlattices [27,29,30]. Aperiodic order turns out that it was the answer to the unexpected crystals with fivefold symmetry or quasi-crystals [31,32]. Aperiodic or quasiregular structures can be obtained if a pattern contains two elements that repeat with different periods, and the ratio of those periods is irratio-nal. In this way, such patterns can evade the prohibitions on certain rotational symmetries. It was also showed that these structures could be grown using the same techniques used for regular crystals [33]. Among the peculiar characteristics of quasi-periodic structures, we can mention the highly frag-mented electronic spectra, which give place to self-similar-ity, criticality, and fractality [34]. It is important to mention that these characteristics are more than peculiar since they can affect the optical, electronic, and transport properties of semiconductor-based devices [28]. In short, aperiodicity or quasi-periodicity can be used as an additional mechanism to modulate the fundamental properties of semiconductor-based structures, and consequently expanding their possibilities for technological applications. So, by considering the importance of the periodic and quasi-periodic modulations, from both the fundamental and technological standpoints, it seems natural that they be an extension for any novel material.

To this respect, graphene is not an exception, and shortly after its discovery, it was subjected to an intense research, especially, with regard to how its fundamental properties are modified under the influence of a periodic modulation [35–82]. There are different mechanisms or effects proposed to modulate graphene in periodical fashion. Among them, we can mention electric and magnetic fields [35–65], breaking-symmetry substrates [66–69], strain [70–74], hydrogenation [75–79], and disorder [80–82]. Graphene superlattice (GSL) is the term used to refer graphene under periodic modulation, irrespective of the mechanism used to generate the periodic pattern. The effects produced by a periodic modulation in gra-phene are remarkably different to what we see in traditional semiconductors. Firstly, the periodic pattern creates additional Dirac cones in the energy dispersion of graphene. Secondly, the propagation of charge carriers through GSL is highly anisotropic, and in some cases, results in carrier velocities that are diminished to zero in one direction and are unchanged in another. The density and the type of carrier species are also pretty sensitive to the periodic pattern considered. Thirdly, by changing the structure parameters of GSL—potential of bar-riers and wells, period of the potential and transverse wave number—the angular-averaged conductivity can be controlled

readily. On the other hand, in the last years the interest in ape-riodic or quasi-periodic modulation in graphene is rising [83–92]. All these studies focus primarily on monolayer graphene, and the preferred mechanism to create the quasi-periodic pat-tern has been the electrostatic field effect [84–94]. So far, the quasi-periodic patterns studied in graphene have been Cantor [83,87,94], Fibonacci [84–86,93], Thue-Morse [88–91,93], and Double-Period [92]. One of the most remarkable charac-teristics of quasi-periodic patterns in graphene, regardless of the quasi-periodic sequence used, is a zero-k gap associated to an unusual Dirac point. Besides, this especial gap is robust against incident angles and lattice constants, and it also has a great effect on transmission, conductance, and shot noise. Therefore, quasi-periodic patterns in graphene may enrich and facilitate the development of graphene-based electronics. As we can corroborate above, periodic and quasi-periodic pat-terns in graphene have been subjected to an intense research, see also Table 29.1; however, there are some issues that as far as we can see are not explained. Specifically, we have not found a clear and concise explanation about how periodic and quasi-periodic modulations in graphene affect the oscillatory nature of the linear-regime conductance.

Within this context, the aim of the present work is to address the main differences between periodic and aperiodic systems in graphene. Specifically, we compare the transmis-sion, transport, and electronic structure properties of the men-tioned systems when the barriers that constitute them are of electrostatic and breaking-symmetry-substrate character. We consider this kind of barriers because they are opposite, one (electrostatic case) in which Klein tunneling is presented and the other (substrate case) in which it is ruled out. Following the lines of our previous work [95], the main concern of this study is to find out how periodic and aperiodic modulations affect the intrinsic oscillatory nature of the linear-regime con-ductance in graphene as well as how these oscillations are correlated with the energy level structure, and what is the role played by Klein tunneling.

29.2 ELECTROSTATIC BARRIERS IN GRAPHENE

In semiconductor-based devices is well known that the elec-tric field effect plays a fundamental role due to its ability to modulate, in a very simple way, the optoelectronic and transport properties. This is possible due to the advance and sophistication of the growth and deposition techniques, specifically thanks to the deposition of metallic contacts on semiconductor structures or in other words, which is known as metal–semiconductor contacts [96]. Graphene is not the exception to this respect despite its intrinsic two-dimensional configuration. A graphene-based device consists of a gra-phene sheet sited on a non-breaking-symmetry substrate like SiO_2, a back gate (together with the substrate's doping) is used to control the Fermi energy of the Dirac electrons, a top gate suspended at a certain distance from the graphene sheet is used to control the width and the height of the electrostatic barrier, and finally two leads are attached to the left and right of the graphene sheet. This is a typical device configuration

TABLE 29.1

Physical Properties Stemming from the Different Types of Periodic and Aperiodic Graphene Superlattices Reported So Far

Graphene-Based System	Type of Potential Barriers	Physical Properties	Reference
Periodic	Electrostatic	Klein effect and resonant tunneling	[35]
		Extra Dirac points and anisotropic propagation of charge carriers	[37,40–44,49]
		Ballistic electron beam propagation: collimation	[38,39]
		Low-bias negative differential resistance	[45]
		Tunable transmission gaps	[46]
		Resonant valley filtering	[48]
		Spin and electron transport, magnetoresistance	[50,51,54,59,60,62–65]
		Wave-vector filtering, resonant effects, wave-vector-dependent gap	[52]
Periodic	Magnetic	Electron beam collimation	[53,57]
		Negative differential conducting regions	[54]
		Effective magnetic field	[56]
		New Dirac point and anisotropic dispersion	[58]
		Dirac points in vortex superlattices	[61]
		Dispersion relation	[66]
		Effect of gap opening on the transmission and transport properties	[67]
Periodic	Substrate	Transverse current rectification	[68]
		Extra Dirac point and normalization of the group velocities	[69]
		Corrugation leads to electronic superlattice: electron transport	[70]
Periodic	Strain	Conductivity, shot noise, and density of states	[71]
		Dichroic absorption: quantum ratchets	[72]
		Resonant modes	[73]
		Dirac points and space-dependent Fermi velocity	[74]
Periodic	Hydrogenated	Electronic structure	[75,77–79]
		Band gap opening	[76]
Periodic	Disorder	Transmission and conductance	[80–82]
	Magnetic-Cantor-like	Transmission and transport properties	[83]
	Electrostatic-Fibonacci	Resonant tunneling, spectrum of charge carriers	[84,85,94]
	Electrostatic-Fibonacci	Electronic band gap and transport	[86]
Aperiodic	Electrostatic-Fractal	Modulation of defect modes	[87]
	Electrostatic-Thue-Morse	Electronic band gaps, transport, shot-noise	[88–91,94]
	Electrostatic-Double-Periodic	Non-Bragg band gap and electronic transport	[92]
	Ferromagnetic-Fractal	Spin conductance and tunneling magnetoresistance	[93]

for electronic transport, or magnetoelectronic transport if a magnetic field perpendicularly to the graphene sheet is applied. It is important to mention that this device has been the backbone element to unveil the fundamental properties of monolayer, bilayer, and trilayer graphene [97–100]. From a theoretical perspective, a perpendicular electrostatic field can be taken into account directly because the main effect of it is a shifting of the Dirac cones proportional to the field strength. This shifting can be obtained through the massless Dirac equation:

$$\left[v_F \left(\sigma \cdot p \right) + V(x) \right] \psi(x,y) = E \psi(x,y), \quad (29.1)$$

where σ is a vector given by the Pauli matrices $\sigma = (\sigma_x, \sigma_y)$, $p = (p_x, p_y)$ is the in-plane momentum operator, $V(x) = V_0$ is

the one-dimensional potential along the x direction, and $\psi(x, y)$ represents the bispinor function. This equation can be readily solved giving the following dispersion relation:

$$E - V_0 = \pm \hbar v_F q, \quad (29.2)$$

where v_F is the Fermi velocity of Dirac electrons in graphene ($v_F = c/300$), q is the two-dimensional wave vector, and "$\pm$" states electrons and holes, respectively. The corresponding wavefunctions, normalized to the graphene sheet area, can be written as

$$\psi_{\pm}(x,y) = \frac{1}{\sqrt{2}} \begin{pmatrix} 1 \\ v_{\pm} \end{pmatrix} e^{\pm i q_x x + i q_y y}, \quad (29.3)$$

with q_x and q_y the components of the two-dimensional wave vector q given by Equation 29.2. The coefficients $v_{\pm}$ are given by

$$v_{\pm} = \frac{\hbar v_F \left(\pm q_x + i q_y \right)}{E - V_0},\qquad (29.4)$$

As a final remark of this section, we have to mention that electrostatic barriers in graphene support effects such as Klein tunneling and collimation, as a result of the preservation of the Dirac cone structure under electrostatic field effects [9,10].

29.3 BREAKING-SYMMETRY SUBSTRATES IN GRAPHENE

Like the electric field effect, substrates have turned out the cornerstone of the microelectronics and nanoelectronics era. A silicon wafer was the answer to carve out and interconnect millions and millions of transistors [101]. Substrates are also a basic element in the current growth techniques, being fundamental to obtain the quantum semiconductor structures on which are based the cutting-edge electronic and optoelectronic applications. To this respect, substrates in graphene are not the exception since they can open an energy band gap as well as change the form of the dispersion relation [102]. In particular, it is experimentally reported that graphene grown on SiC, also known as epitaxial graphene, shows a band gap due to the non-negligible interaction between graphene and SiC substrate, which turns out in a breaking of symmetry of the hexagonal honeycomb lattice of graphene. From a theoretical standpoint, this can be treated through the following Dirac-type equation:

$$\left[v_F \left(\sigma \cdot p \right) + t' \sigma_z \right] \psi(x,y) = E \psi(x,y), \qquad (29.5)$$

where $t' = m v_F^2$ is the mass term, and σ_z is the z component of the Pauli-matrix vector. As in the case of electrostatic field, this equation can be straightforwardly solved giving the parabolic dispersion relation:

$$E = \pm \sqrt{\hbar^2 v_F^2 q^2 + t'^2}, \qquad (29.6)$$

where t' is proportional to the bandgap, $E_g = 2t'$. It is important to mention that in this case the number of graphene layers can control the bandgap. In particular, a bandgap of 0.26 eV, 0.14 eV, and 0.066 eV is reported for monolayer, bilayer, and trilayer graphene, respectively. From now on, t' will be 0.13 eV since this work deals with monolayer graphene. The associated wavefunctions keep the same mathematical form as in the case of graphene subjected to electrostatic field, but now the coefficients of the bispinor come as

$$v_{\pm} = \frac{E - t'}{\hbar v_F (\pm q_x - i q_y)}. \qquad (29.7)$$

As a final comment of this section, it is important to mention that owing to the bandgap opening induced by the substrate, quantities such as pseudo-spin are not conserved and consequently Klein tunneling is ruled out [95].

29.4 THEORETICAL MODELS

As we already mentioned in the preceding sections, multibarrier structures in graphene can be obtained through different mechanisms or external effects. From a theoretical standpoint—irrespective of the external effect used to create the barriers—the transmission, transport, and level structure properties can be obtained readily through the well-known transfer matrix method [103,104]. For instance, consider an arbitrary multibarrier structure as the one depicted in Figure 29.1. Here, we have represented the external effect as vertical arrows and shadowed stripes in certain regions along the x coordinate, Figure 29.1a. Depending on the external effect, we can have barriers for electrons, holes, or both electrons and holes, Figure 29.1b.

This information will be enclosed in the dispersion relation and the coefficients of the bispinor wavefunction of the different regions that constitute our multibarrier structure. Then, taking into account the conservation of the transversal momentum, $q_y = k_y$, and imposing the continuity condition to the wavefunction in the different interfaces along the propagation direction (x coordinate), we can obtain a relation between the coefficients of the forward and backward wavefunctions of the left semi-infinite region, A_0 and B_0, and the forward wavefunction of the right semi-infinite region, A_{N+1} and $B_{N+1} = 0$ (see Figure 29.1a), through the transfer matrix as

$$\begin{pmatrix} A_0 \\ B_0 \end{pmatrix} = M \begin{pmatrix} A_{N+1} \\ 0 \end{pmatrix}, \qquad (29.8)$$

where the transfer matrix is given by

$$M = D_0^{-1} \left(\prod_{j=1}^{N} D_j P_j D_j^{-1} \right) D_0 \qquad (29.9)$$

and defined through the dynamic D_j and propagation P_j matrices,

$$D_j = \begin{pmatrix} 1 & 1 \\ v_{+,j} & v_{-,j} \end{pmatrix} \qquad (29.10)$$

and

$$P_j = \begin{pmatrix} e^{-i q_{x,j} d_j} & 0 \\ 0 & e^{i q_{x,j} d_j} \end{pmatrix}. \qquad (29.11)$$

(a)

(b)

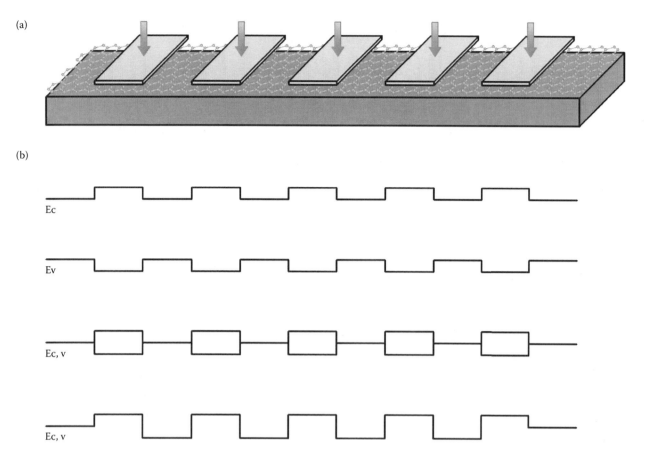

FIGURE 29.1 Graphene multibarrier structures: (a) Schematic representation of graphene subjected to an arbitrary external effect (shadowed stripes) in different regions along the propagation direction (*x* coordinate); (b) Band-edge profile possibilities depending on the external effect used, from top to bottom: barriers for electrons, barriers for holes or barriers for both electrons and holes, respectively.

Here, $j = 1, 2, \ldots, N$, q_{xj}, and d_j are the *x*-component of the wave vector and the width of the *j*-th region which can be barrier or well depending of the specific multibarrier structure, and D_0 is the dynamic matrix of the semi-infinite left and right regions, which in our model are the same. Knowing the transfer matrix, we can easily calculate the transmission probability:

$$T = \left| \frac{A_{N+1}}{A_0} \right| = \frac{1}{|M_{11}|^2}, \qquad (29.12)$$

with M_{11} the (1,1) element of the transfer matrix. The transmission probability allows us to compute the linear-regime conductance straightforwardly through the lines of the Landauer–Büttiker formalism [105]:

$$G(E_F) = \frac{2e^2}{h} \int_{-E_F/\hbar v_F}^{E_F/\hbar v_F} T(E_F, k_y) \frac{dk_y}{(2\pi/L_y)}, \qquad (29.13)$$

where L_y is the system's size in the transversal direction, and it is assumed large compared to the system's size in the propagation direction L_x, and E_F the Fermi energy of the Dirac

electrons. The conductance can be written as the angular average of the transmission probability changing k_y by θ through the relation, $k_y = \dfrac{E_F}{\hbar v_F} \sin \theta$:

$$G(E_F) = \frac{2e^2 L_y E_F}{h^2 v_F} \int_{-\pi/2}^{\pi/2} T(E_F, \theta) \cos \theta \, d\theta, \qquad (29.14)$$

and normalizing E_F to the barrier height energy $E_F^* = (E_F/E_0)$,

$$\frac{G(E_F^*)}{G_0} = E_F^* \int_{-\pi/2}^{\pi/2} T(E_F^*, \theta) \cos \theta \, d\theta, \qquad (29.15)$$

with $G_0 = 2e^2 L_y E_0/h^2 v_F$ the unit of the linear-regime conductance. Finally, the spectrum of bound states is calculated changing from open boundary conditions (Figure 29.2a) to hard-wall boundary conditions (Figure 29.2b), that is, the widths of the first and last barriers of the multiple structures are extended to infinity.

Likewise, we have to require a pure imaginary wave vector for the semi-infinite barrier regions, which turns out in a

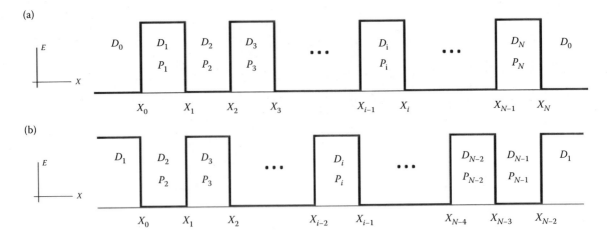

FIGURE 29.2 Schematic representation of the band-edge profile of a multibarrier structure. (a) The case of open boundaries: two semi-infinite free regions enclose the multibarrier structure. In this case free means that the semi-infinite regions are not subjected to electrostatic or breaking-symmetry-substrate effects. (b) The case of hard-wall boundaries: two semi-infinite regions subjected to external effects (electrostatic or substrate ones) enclose the multibarrier structure. In each case, we depict the dynamic (D_j) and propagation (P_j) matrices of the different regions of the system. We also represent the width of the different regions of the multibarrier structure, $d_j = x_j - x_{j-1}$.

transcendental equation between the energy and the transversal wave vector of Dirac electrons as

$$M_{11}^{BS}(E, k_y; q_x \rightarrow i\alpha_x) = 0, \qquad (29.16)$$

where q_x is the wave vector along the x-coordinate and it will change according to the specific mechanism to generate the barriers, Equations 29.2 and 29.7. M^{BS} is given by

$$M^{BS} = D_1^{-1}\left(\prod_{l=2}^{N-1} D_l P_l D_l^{-1}\right)D_1, \qquad (29.17)$$

where the superscript "BS" has been used to differentiate this transfer matrix Equation 29.17 with respect to the transfer matrix of open boundary conditions, Equation 29.9.

29.5 PERIODIC STRUCTURES

The first system that we will analyze is a graphene superlattice. Particularly, we will focus our attention in GSLs in which the periodic patterns could be created by electrostatic and breaking-symmetry-substrate means, see Figure 29.3. In the case of electrostatic GSLs (EGSLs), we will consider the same polarity (positive) for all regions subjected to a perpendicular electrostatic field, in such a way that we will have a Dirac-cone distribution as in Figure 29.3a. This Dirac-cone distribution turns out in a periodic band-edge profile for electrons as depicted in Figure 29.3b. In the case of breaking-symmetry substrate GSLs or simply substrate GSLs (SGSLs), the changes induced by the substrates, the bandgap opening, and the modification of the energy-dispersion relation, turn out in a Dirac-cone distribution or more precisely in a Dirac-paraboloid distribution as the one depicted in Figure 29.3c,

that in turn results in a band-edge profile for electrons and holes as the one depicted in Figure 29.3d. As we can see from the band-edge profiles of EGSLs and SGSLs, the main difference between these structures is the presence of barriers for both electrons and holes in the case of SGSLs, and only barriers for electrons in the case of EGSLs. It is important to mention that the transmission, transport, and electronic structure properties are quite sensitive to this difference. Specifically, it determines the presence or not of Klein tunneling, the dependence or not of the conductance on the system's size, and the onset-end-degeneration and the onset-end-degeneration-closure of the energy minibands, as we shall see shortly.

As we are interested in superlattices, we shall address the evolution of the transmission, transport, and electronic structure properties as a function of the number of barriers. So, throughout this section, we fix the other parameters of the system, specifically we consider: (1) barrier and well widths of $50a$, where a is the carbon–carbon distance in graphene; (2) a barrier height of 0.13 eV, in order to make the comparison between EGSLs and SGSLs reliable, since the bandgap generated by substrates such as SiC in monolayer graphene is of the order of 0.26 eV [102]. Within this context, we show in the first place the transmission probability or transmittance as a function of the energy of the incident electrons, see Figure 29.4.

Here, we have chosen two angles of incidence for the Dirac electrons; particularly, we have considered normal ($\theta = 0$) and oblique ($\theta = \pi/6$) incidence, solid and dashed lines, respectively. The left (Figure 29.4a, c, and e) and right (Figure 29.4b, d, and f) columns of this figure correspond to EGSLs and SGSLs, whilst the first (Figure 29.4a and b), second (Figure 29.4c and d), and third (Figure 29.4e and f) rows to the number of barriers 3, 6, and 9, respectively. In the case of EGSLs, we can see the Klein tunneling (perfect transmission) for normal incidence irrespective of the number of barriers, solid line of

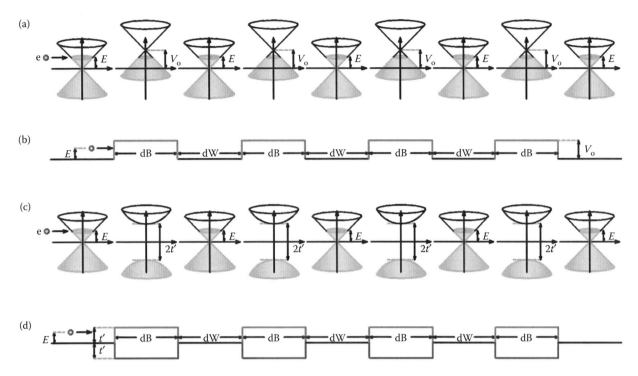

FIGURE 29.3 Dirac-cone distribution and band-edge profile of typical GSLs. (a) Dirac-cone distribution and (b) band-edge profile of Electrostatic GSLs. Dirac cones are up shift because the polarity of the electrostatic field considered is positive, and as a result, a periodic band-edge profile arises for electrons. In the case of breaking-symmetry substrate GSLs or simply substrate GSLs, the opening of the bandgap and the modification of the energy-dispersion relation turn out in a Dirac-paraboloid distribution as depicted in (c) as well as in a band-edge profile for electrons and holes as depicted in (d).

Figure 29.4a, c, and e. On the contrary, SGSLs show a quite different transmission spectrum with transmission windows at different energy ranges separated by transmission gaps, solid line of Figure 29.4b, d, and f. In general, we can see that the transmission windows are formed by resonances, and that the number of resonances corresponds to the number of well regions in the system, being 2, 5, and 8 for NB = 3, NB = 6, and NB = 9, respectively. For oblique incidence, we found a similar transmission pattern for both massless (EGSLs) and massive Dirac electrons (SGSLs) as in the case of normal incidence of SGSLs. The main difference between normal and oblique incidence for SGSLs is a shift of the transmission windows and the transmission gaps to greater energies, in absolute value, for electrons and holes, see Figure 29.4b, d, and f. As we have already mentioned, our main concern is to address the evolution of the transmission probability as a function of the number of barriers in the system. To this respect, we can see that as the number of barriers increases, the number of transmission windows and transmission gaps increases as well. We also noticed that the transmission windows and transmission gaps are better defined, that is, we have regions of perfect transmission and perfect reflection which are typical of the superlattice behavior.

In Figure 29.5a and b, we show the angular distribution of the transmission probability for EGSLs and SGSLs, respectively. The solid, dashed, and dotted lines correspond to 3, 6, and 9 barriers. The energy of the incident electrons was fixed to 0.1 eV. As we can notice for EGSLs, the Klein and

collimation effects are dominant, that is, we have non-zero transmission probability for normal incidence and small angles, see Figure 29.5a. The net effect of the change in the number of barriers is the emergence of some resonances, being 2 and 4 for NB = 6 and NB = 9, respectively. For SGSLs, the angular distribution of the transmittance is quite different, see Figure 29.5b. Firstly, the non-zero transmission range is larger with respect to EGSLs. Secondly, there is equal number of transmission resonances for positive and negative angles, being these resonances proportional to the number of well regions, NR = NW = NB − 1, where NR, NW, and NB state the number of transmission resonances, wells, and barriers, respectively. Thirdly, there is a high transmission probability for normal incidence in the case of SGSLs with six barriers, dashed line in Figure 29.5b, which comes from the overlap of transmission resonances at small angles.

In order to have a broader perspective of the transmission properties, contour plots of the transmittance are shown in Figure 29.6. In these plots, the horizontal and vertical axes correspond to the energy and the angle of the incident electrons, while the grayscale from black to gray correspond to zero and perfect transmission. As in Figure 29.4, the first and second columns correspond to EGSLs and SGSLs, whilst the first, second, and third rows to GSLs with 3, 6, and 9 barriers. As we can notice, the transmittance patterns are asymmetric with respect to the energy and symmetric with respect to the angle for EGSLs, Figure 29.6a, c, and e. For SGSLs, on the other hand, the patterns are symmetric for energy and angle,

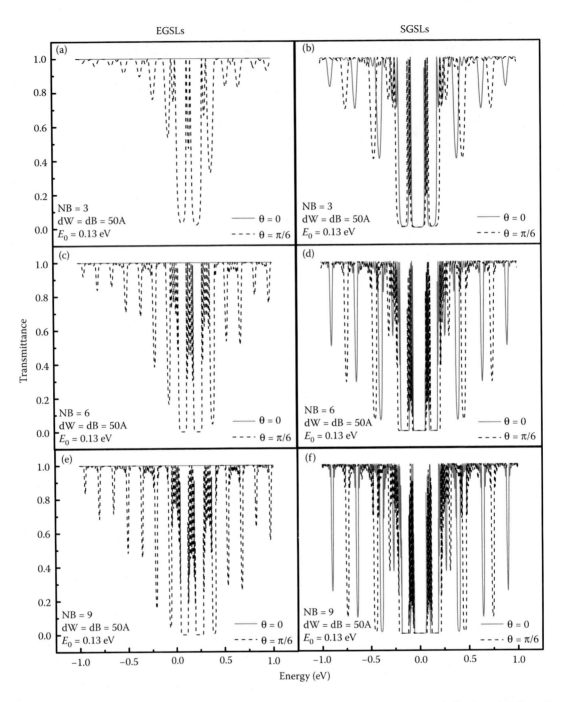

FIGURE 29.4 Transmission probability versus the energy of the incident Dirac electrons for normal ($\theta = 0$) and oblique ($\theta = \pi/6$) incidence, solid and dashed lines, respectively. The first column [(a), (c), and (e)] corresponds to EGSLs, whilst the second one [(b), (d), and (f)] to SGSLs. In terms of rows, the first one [(a) and (b)] corresponds to NB = 3, the second one [(c) and (d)] to NB = 6, and the third one [(e) and (f)] to NB = 9, respectively. In all our computations, the height of the barriers is 0.13 eV, the widths of the barriers and wells are 50a, with a the carbon–carbon distance in the honeycomb lattice of graphene.

Figure 29.6b, d, and f. We can also see that Klein and collimation effects are present, no matter the number of barriers, in EGSLs. For SGSLs, we can notice that the transmission probability is pretty sensitive to the system's size, such that, by increasing the number of barriers the transmission patterns start to be dominated by more and more regions of negligible transmission probability. For both EGSLs and SGSLs, we find a periodic pattern of regions of high- and low-transmission

probabilities along the energy axis. These regions are quite sensitive to the system's parameters; particularly, they are better defined: as the number of barriers increases, their number, width, and location depend directly on the height of the barriers, and the width of barriers and wells.

It is now time to analyze the linear-regime conductance; to do this, we show our results in Figure 29.7. The distribution of columns and rows is the same as in Figures 29.4 and 29.6. As

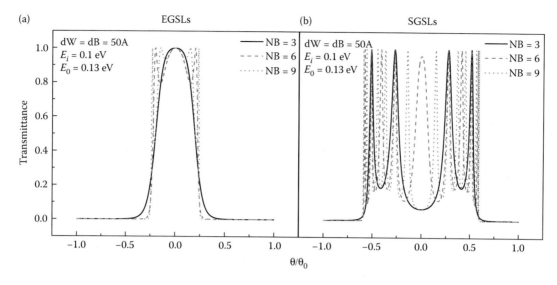

FIGURE 29.5 Angular distribution of the transmittance for (a) EGSLs and (b) SGSLs. The solid, dashed, and dotted correspond to 3, 6, and 9 barriers, respectively. The height of the barriers, the width of barriers and wells are the same as in Figure 29.4, while the energy of the incident electrons is 0.1 eV.

we can see, the transmission properties are directly reflected in the conductance, due to the simple relationship between them. Therefore, we also have a periodic pattern in the linear-regime conductance or in other words oscillations. In general, these oscillations get pronounce with the number of barriers, no matter if the barriers are of electrostatic or substrate character. On the contrary, the form and location of them are quite sensitive to the type of barriers considered. In particular, we can see peaks in the conductance at low Fermi energy, which resemble the ones found for two barriers [95]. However, there are some peculiarities that arose for EGSLs and SGSLs. For instance, the peak in the interval of zero to the half of the barrier (Figure 29.7a, c, and e) is not present for two electrostatic barriers [95], and consequently is not related to the bound states of the system. Likewise, the second peak gets steeper with the number of barriers, contrary to the smoother form of the peak in the case of two barriers. In the case of SGSLs, the envelopes of the conductance peaks (Figure 29.7b, d, and f) resemble the envelopes we found in the system of two barriers [95]; however, in this case, the periodic pattern manifests in a series of small peaks within the envelope, being these peaks proportional to the number of wells in the structure.

As we already stated in our previous work [95], the opening and opening-closure of energy subbands determine the form and location of the peaks in the conductance for Klein (electrostatic) and non-Klein (substrate) structures. So, following this line, we need to calculate the spectrum of bound states or, in the context of GSLs, the miniband structure, in order to know how the start, end, degeneration, and closure of minibands impact in the location and form of the peaks in the conductance. To this respect, Figure 29.8 depicts our results about bound states, or in other words, the allowed values of energy with respect to the transversal wave vector, $E = E(k_y)$. The distribution of columns and rows is the same as in the preceding figures; however, in this case, we refer to the

number of wells instead to the number of barriers (hard-wall conditions) with the first, second, and third row corresponding to 2, 5, and 8 wells. As a reference, we have included the results of single well, solid lines. As in the case of traditional SLs (Schrödinger electrons), the multiwell structure splits the energy subbands to form energy minibands [27]. As in the mentioned case, the energy subbands split such that the same number appears above and below the reference (single well case). Up to this point, these characteristics are equivalent for Schrödinger and Dirac electrons in a periodic potential; however, there are two new features that arose due to the relativistic character of electrons in graphene, which are degeneration and closure of energy minibands. Degeneration shows up for EGSLs, while degeneration-closure for SGSLs. These new characteristics together with the start and end of the energy minibands (arrows at the left) determine the location and form of the peaks in the conductance. Specifically, the start and the end of the energy minibands mark, qualitatively and quantitatively, the onset and fall down of the conductance oscillations, whilst the degeneration and closure the steeping of them. This trend is presented for all minibands, despite the particularities that changes in the system's parameters (barrier height, barrier, and well widths) can bring, results not shown.

29.6 APERIODIC STRUCTURES

Aperiodic or quasi-periodic structures can be divided into what are known as substitutional sequences and model fractal structures [26]. The former are generated via repeated substitution procedure and the latter in a way similar to fractal sets. Typical substitutional sequences are Fibonacci, Thue-Morse, Double-Period, Rudin-Shapiro, and Silver-Mean, while standard fractal structures are Cantor sets or Koch fractals. So far, Fibonacci and Cantor-like structures are the most studied, from both fundamental and technological standpoints

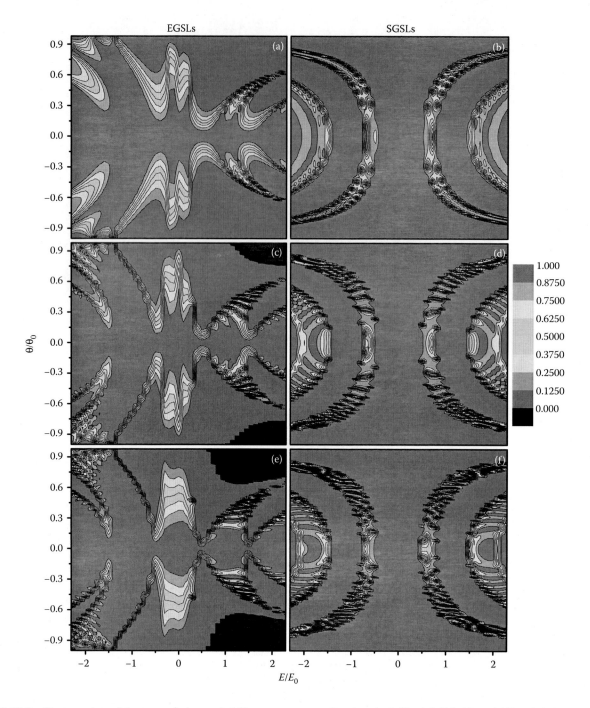

FIGURE 29.6 Contour plots of the transmission probability versus energy E and angle θ. The left [(a), (c), and (e)] and right [(b), (d), and (f)] columns correspond to EGSLs and SGSLs, whilst the first [(a) and (b)], second [(c) and (d)], and third [(e) and (f)] rows to 3, 6, and 9 barriers. The other parameters of the system, height and width of barriers as well as width of wells are the same as in Figures 29.4 and 29.5.

[26]. So, we shall focus in these systems, and, specifically, how the structural and geometrical characteristics of them are manifested in the transmission, transport, and electronic properties.

29.6.1 CANTOR-LIKE STRUCTURES

As we have already mentioned above, the Cantor-like or simply Cantor structures can be generated in a way similar

to the Cantor set construction [106]. There are two basic parameters that characterize a Cantor structure, the generator $G = 3, 5, 7, \ldots$ and the generation number $N = 1, 2, 3, \ldots$ The generator states the number of divisions of a specific region, and the generation number states the stage or generation of the system. As we are dealing with triadic Cantor structures, $G = 3$, which means that at every stage (except the first one, that for us is the single barrier case) the regions subjected to an external effect are divided by three sub-regions of equal

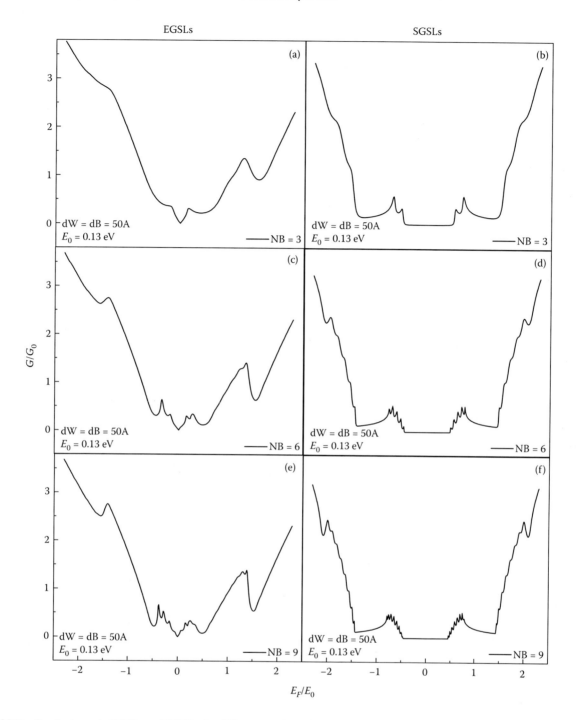

FIGURE 29.7 Conductance in EGSLs and SGSLs for different number of barriers versus the Fermi energy E_F. The distribution of columns and rows as well as the parameters of the system are the same as in Figures 29.4 and 29.6.

width, two with external effects sandwiching the remaining effect-free one. As an example, in Figure 29.9, we show the schematic representation of the Dirac-cone distribution and band-edge profile of the third generation of Electrostatic Cantor Graphene Structures (ECGSs) and Substrate Cantor Graphene Structures (SCGSs). Figure 29.9a and b correspond to ECGSs, while Figure 29.9c and d to SCGSs. It is also important to mention that the first and second generations correspond to the single and double barrier cases of GSLs.

Likewise, we want to mention that the main parameter that we shall analyze in the transmission, transport, and electronic structure is the generation number N. Therefore, we will consider an initial width of $540a$ and the same barrier height (0.13 eV) used in GSLs.

In Figure 29.10, we depict the transmission probability in ECGSs and SCGSs for different generation numbers versus the energy of the incident electrons. As we can see, in the case of ECGSs, Klein tunneling (normal incidence) is present

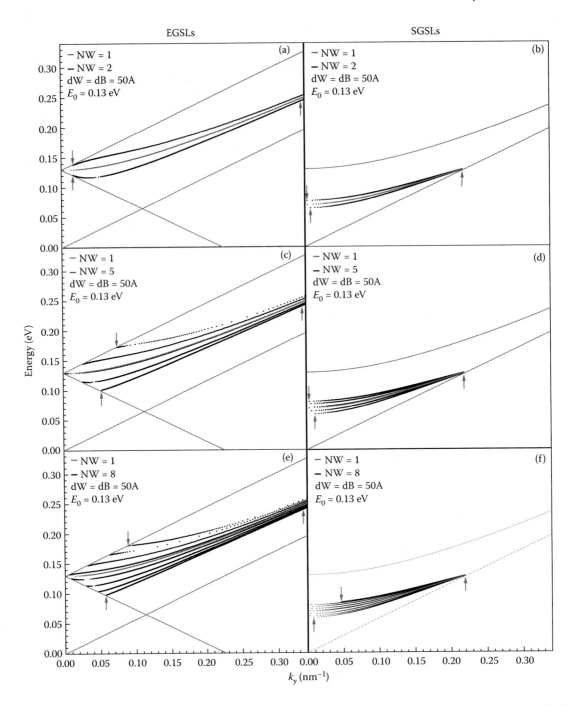

FIGURE 29.8 Spectrum of confined states in EGSLs and SGSLs for different number of wells versus the wave vector k_y. The first [(a), (c), and (e)] and second [(b), (d), and (f)] columns correspond to EGSLs and SGSLs, while the first [(a) and (b)], second [(c) and (d)], and third [(e) and (f)] rows to 2, 5, and 8 wells, respectively. The height and width of barriers and wells are the same as in Figure 29.7. The arrows in the left part state the onset and end of the miniband, whilst the arrows in the right part mark the degeneration and closure of them.

no matter the generation number considered. For oblique incidence, on the contrary, we find regions of perfect transmission and perfect reflection, in other words, windows and gaps of transmission. It is important to mention that the main difference with respect to GSLs comes in the number of resonances within a transmission window, in this case 3. We also see that when the generation number changes most resonances split and group; the splitting and grouping are proportional to the structural difference between consecutive generations, which

is three for triadic Cantor structures [106]. For SCGSs, there is no Klein tunneling, and the normal and oblique transmission spectra show a similar pattern to the oblique incidence one in ECGSs.

Now, it is turn to analyze the transmittance in ECGSs and SCGSs for different generation numbers versus the angle of incidence. To this respect, Figure 29.11 depicts the angular distribution of the transmittance, the first and second columns correspond to ECGSs and SCGSs, while the first, second,

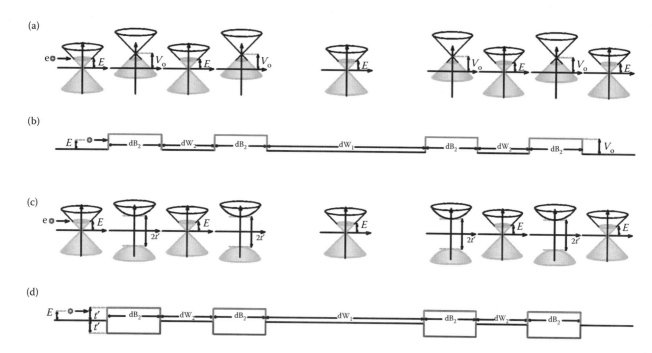

FIGURE 29.9 Dirac-cone distribution and band-edge profile of the third generation of [(a) and (b)] ECGSs and [(c) and (d)] SCGSs.

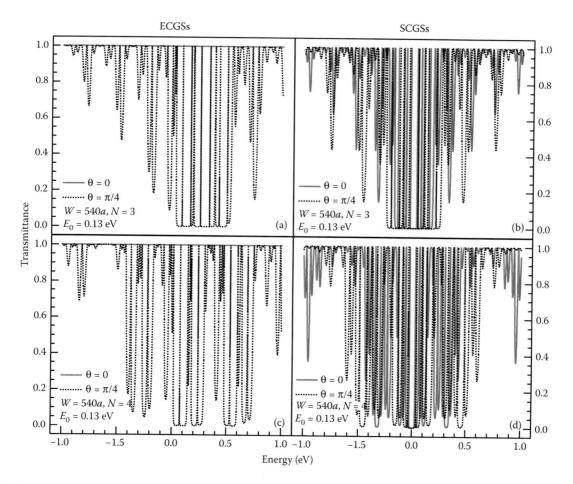

FIGURE 29.10 Transmittance in ECGSs and SCGSs for different generation numbers versus the energy E. The first [(a) and (c)] and second [(b) and (d)] columns correspond to ECGSs and SCGSs, while the first [(a) and (b)] and second [(c) and (d)] rows to generation numbers 3 and 4, respectively. The solid and dotted lines depict normal ($\theta = 0$) and oblique ($\theta = \pi/4$) incidence, respectively. In this figure, and in the rest ones of this subsection, the initial width ($540a$) and the barrier heights (0.13 eV) will be the same.

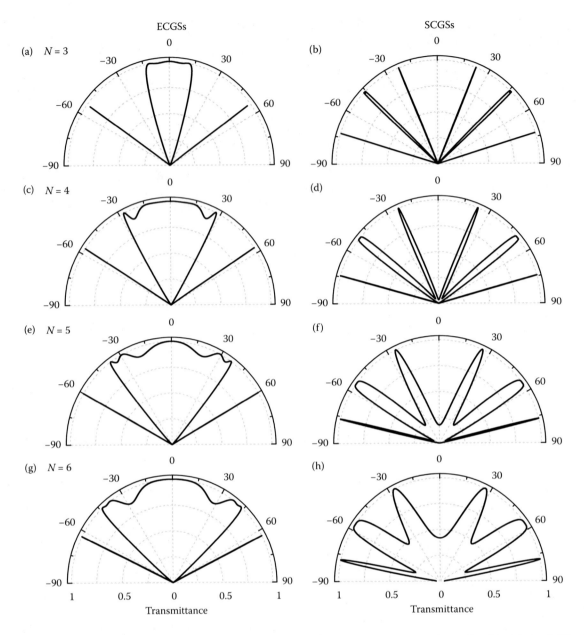

FIGURE 29.11 Transmittance in ECGSs and SCGSs for different generation numbers versus the angle of incidence θ. The first [(a), (c), (e), and (g)] and second [(b), (d), (f), and (h)] columns correspond to ECGSs and SCGSs, while the first [(a) and (b)], second [(c) and (d)], third [(e) and (f)], and fourth [(g) and (h)] rows to generation numbers 3, 4, 5, and 6, respectively. Here, the energy of the incident electrons is 0.1 eV.

third, and fourth rows to generation numbers 3, 4, 5, and 6 respectively. The energy of the incident electrons was fixed to 0.1 eV. For ECGSs, we can see collimation and Klein tunneling (preference for small angles) irrespective of the generation number. In particular, for $N = 3$, there is a perfect transmission from −15° to 15°, and a couple of resonances close to −60° and 60°. The range of perfect transmission expands with the increase in the generation number. For SCGSs, on the contrary, there is no collimation and Klein tunneling, instead, there are resonances symmetrically distributed for positive and negative angles. As we increase the generation number each resonance expands, such that the transmission patterns present an envelope with self-similar characteristics. We can

also note that the transmittance is pretty sensitive to the size's system, since it increases as the generation number rise, due to the reduction of the effective width of the system.

To have a general perspective of the transmission properties, the contour plots of the transmittance are depicted in Figure 29.12. The distribution of columns and rows is the same as that in Figure 29.11, except that, in this case, we only analyze generation numbers 3, 4, and 5. As we can notice, the contour plots for ECGSs are asymmetric with respect to the energy axis, which is quite natural since we have only barriers for electrons. However, as the generation number increases they get symmetric. We can also see that for higher generations the contours are more fragmented, for both ECGSs and

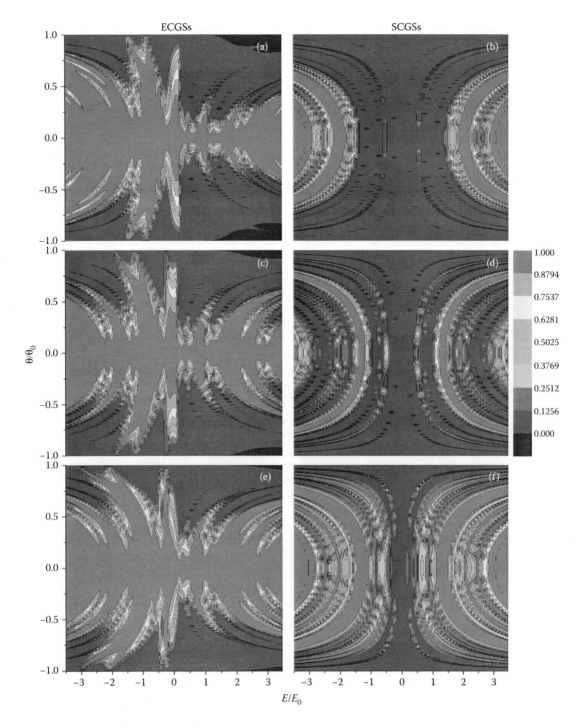

FIGURE 29.12 Contour plots of the transmission probability versus energy E and angle θ. The left [(a), (c), and (e)] and right [(b), (d), and (f)] columns correspond to ECGSs and SCGSs, whilst the first [(a) and (b)], second [(c) and (d)], and third [(e) and (f)] rows to generation numbers 3, 4, and 6, respectively.

SCGSs. This fragmentation is intricate, but we can see that some of the transmission windows are clustered according to the geometrical characteristics of our system, that is, we can find clusters with three semi-circular windows, surrounded by gaps and isolated regions of perfect transmission. Other feature clearly seen is the widening of the angular range of perfect transmission for ECGSs when the generation number increases.

The conductance in ECGSs and SCGSs for different generation numbers versus de Fermi energy, E_F, is presented in Figure 29.13. The first and second rows correspond to ECGSs and SCGSs, while the second column represents zooms of the figures in the first column. The solid, dashed, dotted, and dashed-dotted lines correspond to structures with generation numbers 3, 4, 5, and 6, respectively. For ECGSs, all conductance curves are asymmetric, contrary to the symmetric curves

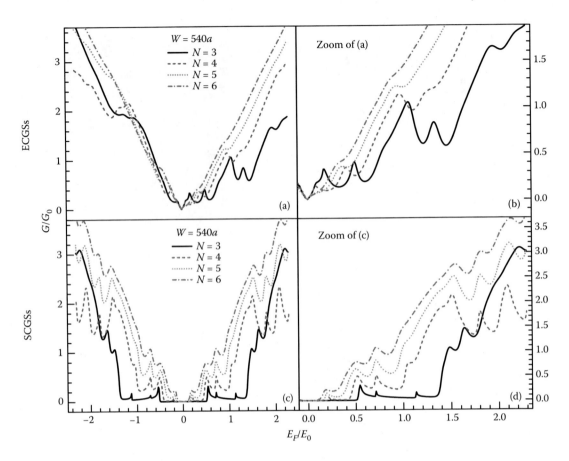

FIGURE 29.13 Conductance in ECGSs and SCGSs for different generation numbers versus the Fermi energy E_F. The first [(a) and (b)] and second [(c) and (d)] rows correspond to ECGSs and SCGs, respectively. The solid, dashed, dotted, and dashed-dotted lines correspond to generation numbers 3, 4, 5, and 6. Zooms of (a) and (c) are depicted in (b) and (d).

presented for SCGSs. In both cases, ECGSs and SCGSs, we can see a rise in the conductance with the generation number, which is a consequence of the overall increase of the transmission properties due to the reduction of the effective width of the system, see contour plots of the transmittance. Another characteristic that we can observe is the oscillating character of the conductance. However, the form and location of the peaks that give rise to the oscillations are quite different between ECGSs and SCGSs. We can also notice that the oscillations gradually disappear by increasing the generation number in ECGSs. In SCGSs, on the contrary, the oscillations are preserved in a remarkable self-similar fashion. Particularly, our outcomes for generation numbers 4, 5, and 6 are quite similar (see Figure 29.13d). For these generations, we can also notice that the geometrical characteristics of our system are manifested in the conductance, since the oscillations are clustered. Specifically, we find three oscillations at the low energy range, then a pronounced rise of the conductance, followed by another three oscillations.

To end this subsection, the spectrum of bound states in ECGSs and SCGSs for different generation numbers versus the transversal wave vector k_y is presented in Figure 29.14. The distribution of columns and rows is the same used in Figure 29.11. That is, Figure 29.14a, c, e, and g are the spectrum of bound states of ECGSs, while Figure 29.14b, d, f, and h are

our outcomes for SCGSs. Likewise, Figure 29.14a and b, c and d, e and f, and g and f depict our results for generation numbers 3, 4, 5, and 6, respectively. As we can see, the spectrum of bound states is not an exception, and like in the case of transmission and conductance the geometrical characteristics of our system are manifested. For instance, we can see clusters of subbands, and remarkably the number of subbands in each cluster is three, which is the basic factor in triadic Cantor structures. These clusters tend to degenerate and degenerate-close for ECGSs and SCGSs, respectively. It is important to mention that some of the subbands are not sensitive to the clustering effect, remaining isolated even when the generation number increases. Last but not least, is the reasonable correlation that we can see between the spectrum of bound states and the conductance in SCGSs. The basic characteristics to understand this correlation are the opening, degeneration, and closure of the subbands. It is well known that when a subband opens, a decrease in the transport properties takes place, while when the subbands degenerate or close an increase happens. By analyzing Figure 29.14d, we can identify three regions, one at low energy with three oscillations followed by another with a steep rise and finally other with three oscillations. These regions match reasonably with the three subbands clustered at low energy, the region of degeneration and closure close to 0.1 eV, and the other three subbands grouped

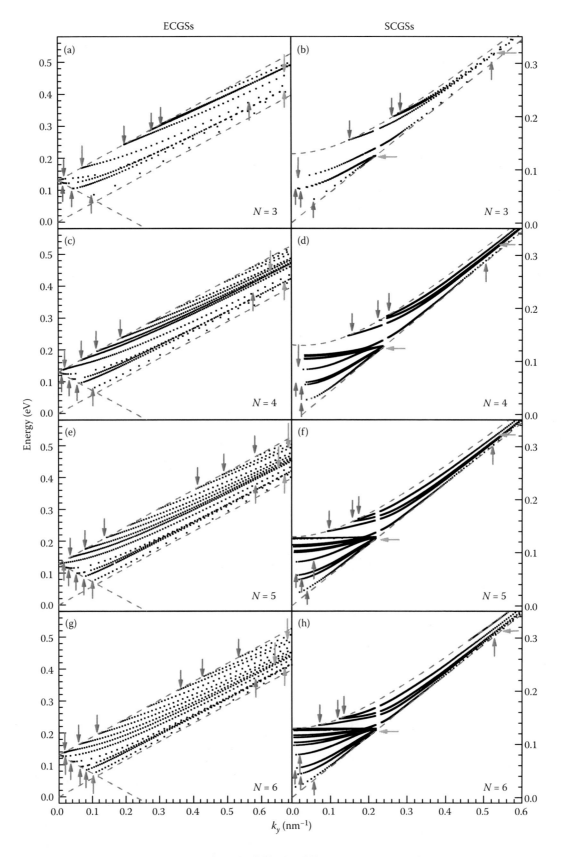

FIGURE 29.14 Spectrum of confined states in ECGSs and SCGSs for different generation numbers versus the wave vector k_y. The first [(a), (c), (e), and (g)] and second [(b), (d), (f), and (h)] columns correspond to ECGSs and SCGSs, while the first [(a) and (b)], second [(c) and (d)], third [(e) and (f)], and fourth [(g) and (h)] rows to generation numbers 3, 4, 5, and 6, respectively. The arrows state the onset and end of the minibands, the degeneration and closure of them, and those subbands that are not clustered—subbands that remain isolated even when the generation number increases.

above the degeneration-closure region. For ECGSs, this correlation is not found, possibly as a result of collimation and Klein tunneling which are present in these structures. In other words, in order to see self-similar characteristics in the conductance of Cantor graphene structures, it is crucial, in principle, the non-conservation of the pseudo-spin.

29.6.2 FIBONACCI QUASI-PERIODIC STRUCTURES

Now, it is turn to analyze our outcomes for Fibonacci graphene structures (FGSs). As in the case of GSLs and Cantor graphene structures (CGSs), we will study systems in which the barriers are generated by electrostatic and substrate means. We will call these systems as Electrostatic FGSs (EFGSs) and Substrate FGSs (SFGSs). In particular, we shall pay attention to the evolution of the transmission, transport, and spectrum of bound states for different generations. So, the widths and heights of the barriers and wells will remain fixed throughout the analysis. Specifically, we have chosen barrier and well widths of $50a$, and barrier heights of 0.13 eV. Likewise, we have included the results of the periodic case as reference.

To obtain our quasi-periodic structure, we need seeds A and B, where A represents a barrier of width and height of $50a$ and 0.13 eV respectively, while B corresponds to a well of width $50a$. After this, we apply the substitutional rules of the Fibonacci sequence, $\zeta(A) = AB$ and $\zeta(B) = A$, with A as the starting generation. Finally, the quasi-periodic structure is sandwiched with two semi-infinite free (barrier) regions for transmission–transport (bound state) calculations. As an example, the Dirac-cone distribution and the band-edge profile of the fifth generation of FGSs are depicted in Figure 29.15. Figure 29.15a and b correspond to EFGSs, whilst

Figure 29.15c and d to SFGSs. It is important to mention that the first and third generations coincide with the single and double barrier cases of GSLs, so we will focus on higher generations; particularly, in generations 5 and 7, as we shall see below.

The transmission probability in EFGSs and SFGSs for different generations versus the energy is depicted in Figure 29.16. The first column corresponds to EFGSs, while the second and third one represent our outcomes for SFGSs. In terms of rows, the first and second one correspond to generations 5 and 7, respectively. As we can see, the main effect of the quasi-periodicity (dotted lines) is to reduce and fragment the transmission windows presented in the periodic case (solid lines). With reduce and fragment, we mean that resonances are reduced and re-arranged. Here, it is important to mention that, in order to make the comparison between Periodic Graphene Structures (PGSs) and FGSs reliable, we have considered periodic systems with the same number of barriers in quasi-periodic ones. As in the cases of EGSLs and ECGSs, Klein tunneling is presented in EFGSs at normal incidence $\theta = 0$, while typical transmission curves (with transmission windows and gaps) are presented at oblique incidence $\theta = \pi/6$. In SFGSs, Klein tunneling is ruled out, and similar transmission curves for both normal and oblique incidence are obtained.

In the case of the angular distribution of the transmittance, we can see that quasi-periodicity favors collimation and Klein tunneling in EFGSs, see Figure 29.17a and c. In SFGSs, the transmission resonances are reduced and diminished with respect to PGSs; compare dotted lines with respect to solid ones in Figure 29.17b and d. In all these curves, the energy of the incident electrons was fixed to 0.1 eV. In these figures,

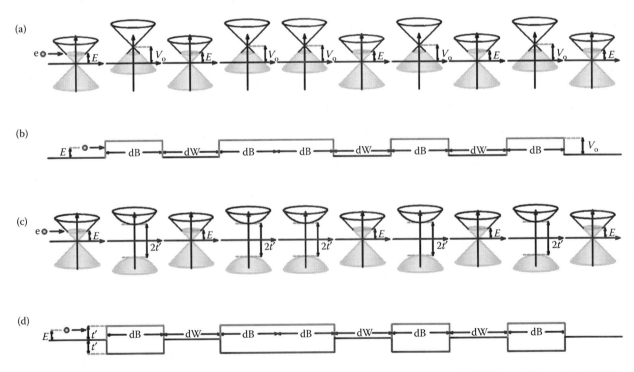

FIGURE 29.15 Dirac-cone distribution and band-edge profile of the fifth generation of [(a) and (b)] EFGSs and [(c) and (d)] SFGSs.

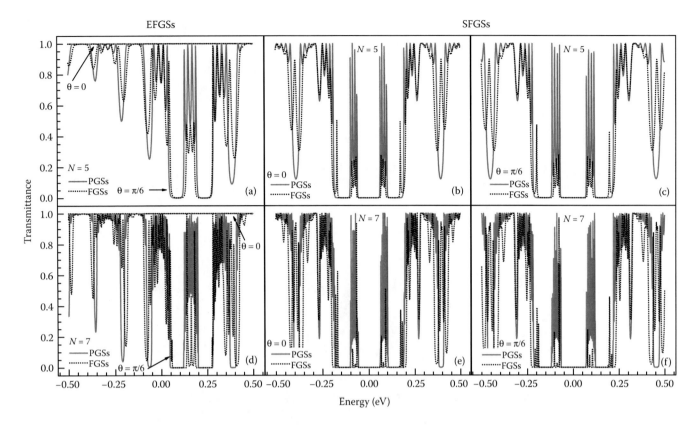

FIGURE 29.16 Transmittance in EFGSs and SFGSs for different generation numbers versus the energy E. The first column [(a) and (d)] corresponds to SFGSs, while the second [(b) and (e)] and third [(c) and (f)] columns represent EFGSs. In terms of rows, the first [(a), (b), and (c)] and second [(d), (e), and (f)] one correspond to generations 5 and 7. The solid and dotted lines depict the periodic and quasi-periodic cases, respectively.

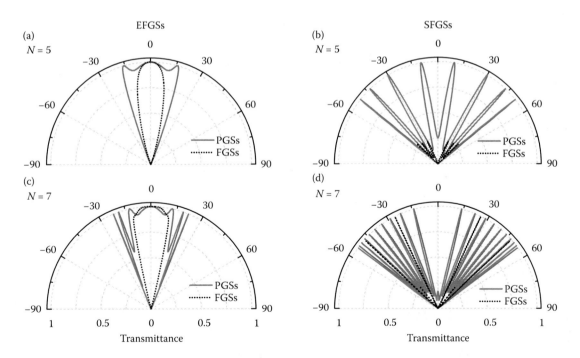

FIGURE 29.17 Transmittance in EFGSs and SFGSs for different generation numbers versus the angle of incidence θ. The first [(a) and (c)] and second [(b) and (d)] columns correspond to EFGSs and SFGSs, while the first [(a) and (b)] and second [(c) and (d)] rows to generation numbers 5 and 7, respectively. The solid and dotted lines depict the periodic and quasi-periodic cases, respectively. Here, the energy of the incident electrons is 0.1 eV.

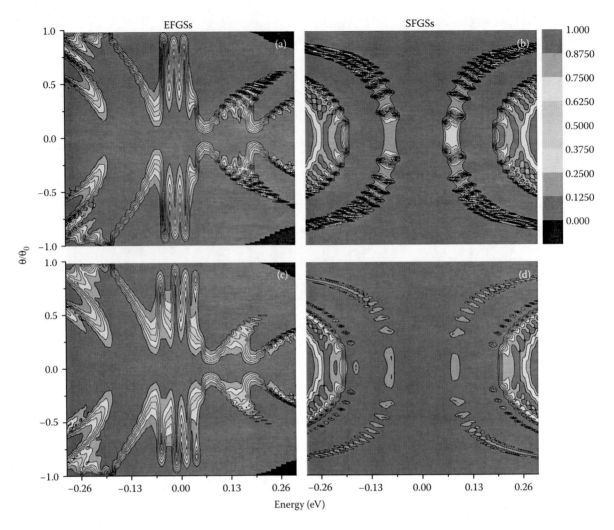

FIGURE 29.18 Contour plots of the transmission probability in FGSs for the fifth generation versus energy E and angle θ. The left [(a) and (c)] and right [(b) and (d)] columns correspond to EFGSs and SFGSs, whilst the first [(a) and (b)] and second [(c) and (d)] rows represent our outcomes for periodic and quasi-periodic structures.

the first and second columns correspond to Electrostatic and Substrate Graphene Structures, whilst the first and second rows represent PGSs and FGSs, respectively.

In order to have a broader perspective of the transmission properties, the contour plots of the transmittance of EFGSs and SFGSs, for generations 5 and 7 are depicted in Figures 29.18 and 29.19, respectively. In these graphs, the first and second columns correspond to Electrostatic and Substrate Graphene Structures, whilst the first and second rows represent PGSs and FGSs. As we can see, the typical transmission contours of GSLs, already shown and analyzed in Section 29.5, are obtained for PGSs, see Figure 29.18a and b as well as Figure 29.19a and b. That is, semi-circular regions of high transmission bounded by regions of low transmission and transmission gaps are presented for electrostatic and substrate PGSs, respectively. These regions are fragmented and diminished once the quasi-periodicity is considered, see Figure 29.18c and d as well as Figure 29.19c and d.

The conductance curves of EFGSs and SFGSs are shown in Figure 29.20 for different generations. Here, Figure 29.20a and b depict our outcomes for $N = 5$, and Figure 29.20c and d show our results for $N = 7$, first and second column, respectively. The solid lines represent the conductance of our reference system (PGSs). As we can notice, the quasi-periodicity diminishes the conductance in general, and also reduces and redistributes the number of peaks that give rise to the oscillations in the conductance. In particular, SFGSs are pretty sensitive to the generation, since the conductance decreases drastically when the generation increases. On the contrary, the collimation and Klein tunneling sustain the conductance in EFGSs at the same order of magnitude that the conductance in PGSs.

Finally, it is important to remark that the oscillations in the conductance and its changes as a result of the quasi-periodicity can be interpret through the changes in the spectrum of bound states, see Figures 29.21 and 29.22. The spectra of bound states for EFGSs are shown in Figure 29.21, and the corresponding ones to SFGSs in Figure 29.22. We have included our reference system PGSs, Figure 29.21a and c as well as Figure 29.22a and c. The results for generations 5 and

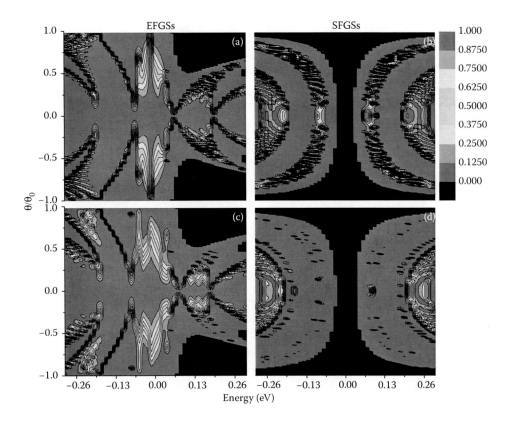

FIGURE 29.19   Contour plots of the transmission probability in FGSs for the seventh generation versus energy E and angle θ. The left [(a) and (c)] and right [(b) and (d)] columns correspond to EFGSs and SFGSs, whilst the first [(a) and (b)] and second [(c) and (d)] rows represent our outcomes for periodic and quasi-periodic structures.

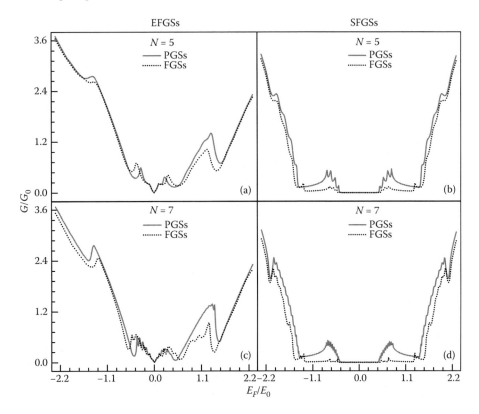

FIGURE 29.20   Conductance in EFGSs and SFGSs for different generation numbers versus the Fermi energy E_F. The first [(a) and (c)] and second [(b) and d)] columns represent EFGSs and SFGSs, while the first [(a) and (b)] and second [(c) and (d)] rows correspond to generations 5 and 7, respectively. The solid and dotted lines depict our results for the periodic and quasi-periodic structures.

EFGSs

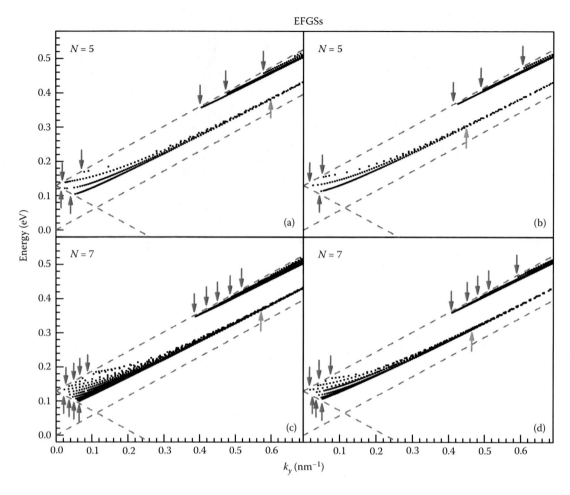

FIGURE 29.21 Spectrum of confined states in EFGSs for different generation numbers versus the wave vector k_y. The first [(a) and (c)] and second [(b) and (d)] columns correspond to the periodic and quasi-periodic structures, while the first [(a) and (b)] and second [(c) and (d)] rows to generation 5 and 7, respectively. The arrows state the onset and end of the minibands as well as degeneration and closure of them.

7 are depicted in Figure 29.21a and b and Figure 29.22a and b, respectively. As in the transmission properties, the main effect of quasi-periodicity is to decrease the number of subbands as well as the degeneration and degeneration-closure of the energy minibands, see Figure 29.21b and d and Figure 29.22b and d.

29.7 CONCLUDING REMARKS

In summary, we have studied the transmission, transport, and spectrum of bound states of periodic and aperiodic graphene-based structures. We have implemented the transfer matrix approach and the Landauer–Büttiker formula to calculate the mentioned properties for graphene superlattices, Cantor and Fibonacci graphene structures in which the barriers that form the periodic and quasi-periodic patterns are of electrostatic and breaking-symmetry-substrate character. We have found that in order to understand the conductance and its particularities, like the form and location of the peaks that give rise to oscillations, it is mandatory to calculate the spectrum of bound states. As in the case of Schrödinger electrons, the periodic pattern of barrier-well in graphene superlattices

turns out in energy minibands; however, there is a new feature for Dirac electrons, and this feature is known as degeneration. Degeneration means that an energy miniband collapses to a single subband for a specific transversal wave vector. Degeneration together with the opening and opening-closure [95] of the energy minibands determine the form and location of the conductance peaks in electrostatic and substrate graphene superlattices. In the case of Cantor and Fibonacci graphene structures, we have found that the structural and geometrical characteristics of them are manifested in practically all properties. In particular, the transmission and conductance present self-similarity, while the spectrum of bound states is fragmented and grouped according to the geometrical characteristics of the aperiodic structure considered. Finally, as in the case of graphene superlattices, the spectrum of bound states allowed us to understand the form and location of the conductance peaks in quasi-periodic graphene structures.

29.8 PERSPECTIVES

Taking into account the variety of effects that could be used to generate potential barriers in graphene, we believe that

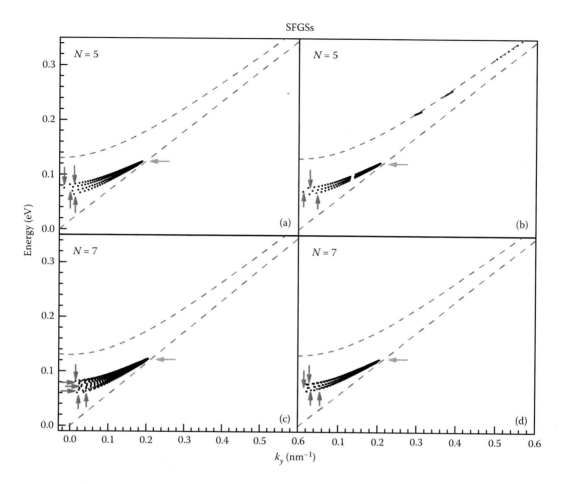

FIGURE 29.22 Spectrum of confined states in SFGSs for different generation numbers versus the wave vector k_y. The first [(a) and (c)] and second [(b) and (d)] columns correspond to the periodic and quasi-periodic structures, while the first [(a) and (b)] and second [(c) and (d)] rows to generation 5 and 7, respectively. The arrows state the onset and end of the minibands as well as degeneration and closure of them.

there is a plenty of work to unveil the transmission, transport, and electronic properties of periodic and aperiodic graphene-based structures. Particularly, we consider that the computation of the spectrum of bound states is fundamental to understand the transport properties in graphene, specifically the oscillatory behavior of the conductance, that is, the form and location of the conductance peaks.

ACKNOWLEDGMENTS

The authors acknowledge the financial support of CONACyT through Grant CB-2010-151713.

REFERENCES

1. Novoselov, K. S., A. K. Geim, S. V. Mrozov, D. Jiang, Y. Zhang, S. V. Dubonos, I. V. Grigorieva and A. A. Firsov. 2004. Electric field effect in atomically thin carbon films. *Science* 306:666–669.
2. Novoselov, K. S., A. K. Geim, S. V. Morozov, D. Jiang, M. I. Katsnelson, I. V. Grigorieva, S. V. Dubonos and A. A. Firsov. 2005. Two-dimensional gas of massless Dirac fermios in graphene. *Nature* 438:197–200.
3. Zhang, Y., Y.-W. Tan, H. L. Stormer and P. Kim. 2005. Experimental observation of the quantum Hall effect and Berry's phase in graphene. *Nature* 438:201–204.
4. Gusynin, V. P. and S. G. Sharapov. 2005. Unconventional integer quantum Hall effect in graphene. *Phys. Rev. Lett.* 95:146801.
5. McCann, E. and V. I. Falko. 2006. Landau-level degeneracy and quantum Hall effect in graphite bilayer. *Phys. Rev. Lett.* 96:086805.
6. Peres, N. M. R., F. Guinea and A. H. Castro Neto. 2006. Electronic properties of disordered two-dimensional carbon. *Phys. Rev. B* 73:125411.
7. Calogeracos, A. and N. Dombey. 1999. History and physics of Klein paradox. *Contemp. Phys.* 40:313–321.
8. Itzykson, C. and J.-B. Zuber. 2006. *Quantum Field Theory*. Dover, New York.
9. Katsnelson, M. I., K. S. Novoselov and A. K. Geim. 2006. Chiral tunneling and the Klein paradox in graphene. *Nat. Phys.* 2:620–625.
10. Allain, P. E. and J. N. Fuchs. 2011. Klein tunneling in graphene: Optics with massless electrons. *Eur. Phys. J. B* 83:301–317.
11. Trauzettel, B., Y. M. Blanter and A. F. Morpurgo. 2007. Photon-assisted electron transport in graphene: Scattering theory analysis. *Phys. Rev. B* 75:035305.

12. Rusin, T. M. and W. Zawadzki. 2007. Transient Zitterbewegung of charge carriers in mono- and bilayer graphene, and carbon nanotubes. *Phys. Rev. B* 76:195439.

13. Rusin, T. M. and W. Zawadzki. 2009. Theory of electron Zitterbewegung in graphene probed by femtosecond laser pulses. *Phys. Rev. B* 80:045416.

14. Jung, E., D. Park and C.-S. Park. 2013. Zitterbewegung in bilayer graphene: Effects of trigonal warping and electric field. *Phys. Rev. B* 87:115438.

15. Shytov, A. V., M. I. Katsnelson and L. S. Levitov. 2007. Atomic collapse and quasi-Rydberg states in graphene. *Phys. Rev. Lett.* 99:246802.

16. Shytov, A., M. Rudner, N. Gu, M. Katsnelson and L. Levitov. 2009. Atomic collapse, Lorentz boosts, Klein scattering, and other quantum-relativistic phenomena in graphene. *Solid State Commun.* 149:1087–1093.

17. Wang, Y., D. Wong, A. V. Shytov, V. W. Brar, S. Choi, Q. Wu, H.-Z. Tsai et al. 2013. Observing atomic collapse resonances in artificial nuclei on graphene. *Science* 340:734–737.

18. Castro Neto, A. H., F. Guinea and N. M. R. Peres. 2006. Drawing conclusions from graphene. *Phys. World* 19:33.

19. Katsnelson, M. I. 2006. Zitterbewegung, chirality, and minimal conductivity in graphene. *Eur. J. Phys. B* 51:157–160.

20. Katsnelson, M. I. and K. S. Novoselov. 2007. Graphene: New bridge between condensed matter physics and quantum electrodynamics. *Solid State Commun.* 143:3–13.

21. Schedin, F., A. K. Geim, S. V. Morozov, E. W. Hill, P. Blake, M. I. Katsnelson and K. S. Novoselov. 2007. Detection of individual gas molecules adsorbed on graphene. *Nature Mater.* 6:652–655.

22. Geim, A. K. and K. S. Novoselov. 2007. The rise of graphene. *Nature Mater.* 6:183–191.

23. Hill, E. W., A. K. Geim, K. Novoselov, F. Schedin and P. Blake. 2006. Graphene spin valve devices. *IEEE Trans. Magn.* 42:2694–2696.

24. Ohishi, M., M. Shiraishi, R. Nouchi, T. Nozaki, T. Shinjo and Y. Suzuki. 2007. Spin injection into a graphene thin film at room temperature. *Jpn. J. Appl. Phys. Part 2* 46:L605–L607.

25. Wallace, P. R. 1947. The band theory of graphite. *Phys. Rev.* 71:622–634.

26. Castro-Neto, A. H., F. Guinea, N. M. R. Peres, K. S. Novoselov and A. K. Geim. 2009. The electronic properties of graphene. *Rev. Mod. Phys.* 81:109–162.

27. Grahn, H. T. 1995. *Semiconductor Superlattices: Growth and Electronic Properties*. World Scientific.

28. Maciá, E. 2006. The role of aperiodic order in science and technology. *Rep. Prog. Phys.* 69:397.

29. Esaki, L. and R. Tsu. 1970. Superlattice and negative differential conductivity in semiconductors. *IBM J. Res. Develop.* 14:61–65.

30. Tsu, R. and L. Esaki. 1973. Tunneling in a finite superlattice. *Appl. Phys. Lett.* 22:562.

31. Shechtman, D., I. Blech, D. Gratias and J. W. Cahn. 1984. Metallic phase with long-range orientational order and no translational symmetry. *Phys. Rev. Lett.* 53:1951–1953.

32. Levine, D. and P. J. Steinhardt. 1984. Quasicrystals: A new class of ordered structures. *Phys. Rev. Lett.* 53:2477–2480.

33. Merlin, R., K. Bajema, R. Clarke, F.-Y. Juang and P. K. Bhattacharya. 1985. Quasiperiodic GaAs–AlAs Heterostructures. *Phys. Rev. Lett.* 55:1768–1770.

34. Pérez-Álvarez, R., F. García-Moliner and V. R. Velasco. 2001. Some elementary questions in the theory of quasiperiodic heterostructures. *J. Phys. Condens. Matter* 13:3689–3698.

35. Bai, C. and X. Zhang. 2007. Klein paradox and resonant tunneling in a graphene superlattice. *Phys. Rev. B* 76:075430.

36. Barbier, M., F. M. Peeters, P. Vasilopoulos and J. M. Pereira Jr. 2008. Dirac and Klein–Gordon particles in one-dimensional periodic potentials. *Phys. Rev. B* 77:115446.

37. Park, C.-H., L. Yang, Y.-W. Son, M. L. Cohen and S. G. Louie. 2008. Anisotropic behaviours of massless Dirac fermions in graphene under periodic potentials. *Nat. Phys.* 4:213–217.

38. Park, C.-H., Y.-W. Son, L. Yang, M. L. Cohen and S. G. Louie. 2008. Electron beam supercollimation in graphene superlattices. *Nano Lett.* 8:2920–2924.

39. Barbier, M., P. Vasilopoulos and F. M. Peeters. 2009. Dirac electrons in a Kronig–Penney potential: Dispersion relation and transmission periodic in the strength of the barriers. *Phys. Rev. B* 80:205415.

40. Barbier, M., P. Vasilopoulos and F. M. Peeters. 2010. Extra Dirac points in the energy spectrum for superlattices on single-layer graphene. *Phys. Rev. B* 81:075438.

41. Arovas, D. P., L. Brey, H. A. Fertig, E.-A. Kim and K. Ziegler. 2010. Dirac spectrum in piecewise constant one-dimensional (1D) potentials. *New J. Phys.* 12:123020.

42. Burset, P., A. Levy Yeyati, L. Brey and H. A. Fertig. 2011. Transport in superlattices on single-layer graphene. *Phys. Rev. B* 83:195434.

43. Savel'ev, S. E. and A. S. Alexandrov. 2011. Massless Dirac fermions in a laser field as a counterpart of graphene superlattices. *Phys. Rev. B* 84:035428.

44. Matthes, L., K. Hannewald, J. Furthmüller and F. Bechstedt. 2011. Screening and band structure effects on a quasi-one-dimensional transport in periodically modulated graphene. *Phys. Rev. B* 84:115427.

45. Ferreira, G. J., M. N. Leuenberger, D. Loss and J. C. Egues. 2011. Low-bias negative differential resistance in graphene nanoribbon superlattices. *Phys. Rev. B* 84:125453.

46. Lu, W.-T., S.-J. Wang, W. Li, Y.-L. Wang, H. Jiang. 2012. Tunable electronic transmission gaps in a graphene superlattice. *Physica. B* 407:918–921.

47. Silveirinha, M. G. and N. Engheta. 2012. Effective medium approach to electron waves: Graphene superlattices. *Phys. Rev. B* 85:195413.

48. Moldovan, D., M. Ramezani Masir, L. Covaci and F. M. Peeters. 2012. Resonant valley filtering of massive Dirac electrons. *Phys, Rev. B* 86:115431.

49. Ang, Y. S. and C. Zhang. 2012. Enhanced optical conductance in graphene superlattice due to anisotropic band dispersion. *J. Phys. D: Appl. Phys.* 45:395303.

50. Niu, Z. P., F. X. Li, B. G. Wang, L. Sheng and D. Y. Xing. 2008. Spin transport in magnetic graphene superlattices. *Eur. Phys. J. B* 66:245–250.

51. Wu, Q.-S., S.-N. Zhang and S.-J. Yang. Transport of the graphene electrons through a magnetic superlattice. *J. Phys. Condens. Matter* 20:485210.

52. Dell'Anna, L. and A. De Martino. 2009. Multiple magnetic barriers in graphene. *Phys. Rev. B* 79:045420.

53. Ramezani Masir, M., P. Vasilopoulos and F. M. Peeters. 2009. Magnetic Kronig-Penney model for Dirac electrons in single-layer graphene. *New J. Phys.* 11:095009.

54. Li, Y.-X. 2010. Transport in a magnetic field modulated graphene superlattice. *J. Phys. Condens. Matter* 22:015302.

55. Biswas, R., A. Biswas, N. Hui and C. Sinha. 2010. Ballistic transport through electric field modulated graphene periodic magnetic barriers. *J. Appl. Phys.* 108:043708.

56. Sun, J., H. A. Fertig and L. Brey. 2010. Effective magnetic fields in graphene superlattices. *Phys. Rev. Lett.* 105:156801.

57. Ramezani Masir, M., P. Vasilopoulos and F. M. Peeters. 2010. Kronig–Penney model of scalar and vector potentials in graphene. *J. Phys. Condens. Matter* 22:465302.

58. Dell'Anna, L. and A. De Martino. 2011. Magnetic superlattice and finite-energy Dirac points in graphene. *Phys. Rev. B* 83:155449.

59. Guo, X.-X., D. Liu and Y.-X. Li. 2011. Conductance and shot noise in graphene superlattice. *Appl. Phys. Lett.* 98:242101.

60. Ke, Q.-R., H.-F. Lü, X.-D. Chen and X.-T. Zu. 2011. Enhanced spin polarization in an asymmetric magnetic graphene superlattice. *Solid State Commun.* 151:1131–1134.

61. Kamfor, M., S. Dusuel, K. P. Schmidt and J. Vidal. 2011. Fate of Dirac points in a vortex superlattice. *Phys. Rev. B* 84:153404.

62. Liu, Z.-F., N.-H. Liu and Q.-P. Wu. 2011. Electronic spin precession in graphene-based superlattice with periodical gate and magnetic modulation. *Physica. E* 44:609–613.

63. Liu, Z.-F., Q.-P. Wu and N.-H. Liu. 2012. Electronic energy band and transport properties in monolayer graphene with periodically modulated magnetic vector potential and electrostatic potential. *Commun. Theor. Phys.* 57:315–319.

64. Lu, W.-T., Y.-L. Wang, C.-Z. Ye, H. Jiang and W. Li. 2012. Resonant peak splitting through magnetic Kronig–Penney superlattices in graphene. *Physica B* 407:4735–4737.

65. Lu, W.-T., S.-J. Wang, Y.-L. Wang, H. Jiang and W. Li. 2013. Transport properties of graphene under periodic and quasiperiodic magnetic superlattices. *Phys. Lett. A* 377:1368–1372.

66. Ratnikov, P. V. 2009. Superlattice based on graphene on a strip substrate. *JETP Lett.* 90:515–520.

67. Esmailpour, M., A. Esmailpour, R. Asgari, M. Elahi and M. R. R. Tabar. 2010. Effect of a gap opening on the conductance of graphene superlattices. *Solid State Commun.* 150:655–659.

68. Zavialov, D. V., V. I. Konchenkov and S. V. Kruchkov. 2012. Transverse current rectification in a graphene-based superlattice. *Semiconductors* 46:109–116.

69. Maksimova, G. M., E. S. Azarova, A. V. Telezhnikov and V. A. Burdov. 2012. Graphene superlattice with periodically modulated Dirac gap. *Phys. Rev. B* 86:205422.

70. Isacsson, A., L. M. Jonsson, J. M. Kinaret and M. Jonson. 2008. Electronic superlattices in corrugated graphene. *Phys. Rev. B* 77:035423.

71. Gattenlöhner, S., W. Beizig and M. Titov. 2010. Dirac–Kronig–Penney model for strain-engineered graphene. *Phys. Rev. B* 82:155417.

72. Kiselev, Y. Y. and L. E. Golub. 2011. Optical and photogalvanic properties of graphene superlattices formed by periodic strain. *Phys. Rev. B* 84:235440.

73. Pellegrino, F. M. D., G. G. N. Angilella and R. Pucci. 2012. Resonant modes in strain-induced graphene superlattices. *Phys. Rev. B* 85:195409.

74. Yan, H., Z.-D. Chu, W. Yan, M. Liu, L. Meng, M. Yang, Y. Fan et al. 2013. Superlattice Dirac points and space Fermi velocity in a corrugated graphene monolayer. *Phys. Rev. B* 87:075405.

75. Chernozatonskii, L. A., P. B. Sorokin and J. W. Brüning. 2007. Two-dimensional semiconducting nanostructures based on single graphene sheets with lines of adsorbed hydrogen atoms. *Appl. Phys. Lett.* 91:183103.

76. Balog, R., B. Jørgensen, L. Nilsson, M. Andersen, E. Rienks, M. Bianchi, M. Fanetti et al. 2010. Bandgap opening in graphene induced by patterned hydrogen adsorption. *Nat. Mater.* 9:315–319.

77. Yang, M., A. Nurbawono, C. Zhang, Y. P. Feng and Ariando. 2010. Two-dimensional graphene superlattice made with partial hydrogenation. *Appl. Phys. Lett.* 96:193115.

78. Hernández-Nieves, A. D., B. Partoens and F. M. Peeters. 2010. Electronic and magnetic properties of graphene/graphene nanoribbons with different edge hydrogenation. *Phys. Rev. B* 82:165412.

79. Huang, H., Z. Li and W. Wang. 2012. Electronic and magnetic properties of oxygen patterned graphene superlattice. *J. Appl. Phys.* 112:114332.

80. Abedpour, N., A. Esmailpour, R. Asgari and M. R. R. Tabar. 2009. Conductance of a disordered graphene superlattice. *Phys. Rev. B* 79:165412.

81. Cheraghchi, H., A. H. Irani, S. M. Fazeli and R. Asgari. 2011. Metallic phase of disordered graphene superlattices with long-range correlations. *Phys. Rev. B* 83:235430.

82. Jia, S., J. Wang, G. Yang, Y. Yang and C. Bai. 2012. Disordered effect on a graphene-based spin–orbit interactions superlattice. *Physica E* 45:146–150.

83. Sun, L., C. Fang, Y. Song and Y. Guo. 2010. Transport properties through graphene-based fractal and periodic barriers. *J. Phys.: Condens. Matter* 22:445303.

84. Sena, S. H. R., J. M. Pereira Jr., G. A. Farias, M. S. Vasconcelos and E. L. Albuquerque. 2010. Fractal spectrum of charge carriers in quasiperiodic graphene structures. *J. Phys. Condens. Matter* 22:465305.

85. Mukhopadhyay, S., R. Biswas and C. Shina. 2010. Resonant tunneling in a Fibonacci bilayer graphene superlattice. *Phys. Status Solidi B* 247:342–346.

86. Zhao, P.-L. and X. Chen. 2011. Electronic band gap and transport in Fibonacci quasi-periodic graphene superlattice. *Appl. Phys. Lett.* 99:182108.

87. Zhang, Y.-P., Y. Gao and H.-Y. Zhang. 2012. Independent modulation of defect modes in fractal potential patterned graphene superlattices with multiple defect layers. *J. Phys. D: Appl. Phys.* 45:055101.

88. Ma, T., C. Liang, L.-G. Wang and H.-Q. Lin. 2012. Electronic band gaps and transport in aperiodic graphene superlattices of Thue-Morse sequence. *Appl. Phys. Lett.* 100:252402.

89. Zhang, Z., H. Li, Z. Gong, Y. Fan, T. Zhang and H. Chen. 2012. Extend the omnidirectional electronic gap of Thue-Morse aperiodic gapped graphene superlattices. *Appl. Phys. Lett.* 101: 252104.

90. Huang, H., D. Liu, H. Zhang and X. Kong. 2013. Electronic transport and shot noise in Thue-Morse sequence graphene superlattice. *J. Appl. Phys.* 113:043702.

91. Xu, Y., J. Zou and G. Jin. 2013. Exotic electronic properties in Thue-Morse graphene superlattices. *J. Phys. Condens. Matter* 25:245301.

92. Chen, X., P.-L. Zhao and Q.-B. Zhu. 2013. Double-periodic quasi-periodic graphene superlattice: Non-Bragg band gap and electronic transport. *J. Phys. D: Appl. Phys.* 46:015306.

93. Lu, W.-T., S.-J. Wang, Y.-L. Wang, H. Jiang and W. Li. 2013. Transport properties of graphene under periodic and quasiperiodic magnetic superlattices. *Phys. Lett. A* 377:1368.

94. Liu D. and H. Zhang. 2014. Spin conductance and tunneling magnetoresistance in a fractal graphene superlattice with two ferromagnetic graphene electrodes. *J. Phys. D: Appl. Phys.* 47:185301.

95. Rodríguez-Vargas, I., J. Madrigal-Melchor and O. Oubram. 2012. Resonant tunneling through double graphene systems: A comparative study of Klein and non-Klein tunneling structures. *J. Appl. Phys.* 112:073711.

96. Rhoderick, E. H. and R. H. Williams. 1988. *Metal–Semiconductor Contacts*. Clarendon Press, Oxford, UK.

97. Stander, N., B. Huard and D. Goldhaber-Gordon. 2009. Evidence for Klein tunneling in graphene p–n junctions. *Phys. Rev. Lett.* 102:026807.

98. Young, A. F. and P. Kim. 2009. Quantum interference and Klein tunneling in graphene heterojunctions. *Nat. Phys.* 5:222–226.

99. Maher, P., C. R. Dean, A. F. Young, T. Taniguchi, K. Watanabe, K. L. Shepard, J. Hone and P. Kim. 2013. Evidence for a spin phase transition at charge neutrality in bilayer graphene. *Nat. Phys.* 9:154–158.

100. Campos, L. C., A. F. Young, K. Surakitbovorn, K. Watanabe, T. Taniguchi and P. Jarillo-Herrero. 2012. Quantum and classical confinement of resonant states in trilayer graphene Fabry-Pérot interferometer. *Nature Communications* 3:1239.

101. Goldstein, A. 1997. Tackling tyranny. *Sci. Am.* 8:80–81.

102. Zhou, S. Y., G.-H. Gweon, A. V. Fedorov. P. N. First, W. A. de Heer, D.-H. Lee, F. Guinea, A. H. Castro-Neto and A. Lanzara. 2007. Substrate-induced bandgap opening in epitaxial graphene. *Nat. Mater.* 6:770–775.

103. Yeh, P. 2005. *Optical Waves in Layered Media*. Wiley-Interscience, New Jersey.

104. Markos, P. and C. M. Soukoulis. 2008. *Wave Propagation: From Electrons to Photonic Crystals and Left-Handed Materials*. Princeton University Press, New Jersey.

105. Datta, S. 1995. *Electronic Transport in Mesoscopic Systems*. Cambridge University Press, New York.

106. Feder, J. 1988. *Fractals*. Plenum Press, New York.

30 Benefits of Few-Layer Graphene Structures for Various Applications

I. V. Antonova and V. Ya. Prinz

CONTENTS

Abstract.. 479
30.1 Introduction 479
30.2 FLG for Sensor Applications... 480
30.3 Substrates for High Carrier Mobility in Graphene.. 482
    30.3.1 Creation of Highly Resistive Substrates and Their Properties 483
    30.3.2 Cleaning of the Top Layer for High Conductivity... 483
30.4 Functionalization of FLG for Electronic Applications... 486
    30.4.1 Superlattice Hydrogenated Graphene/NMP Monolayer with High Carrier Mobility..................... 486
    30.4.2 Few-Layer Fluorographene: A Simple Approach to Creation and Potential for Applications......... 489
30.5 FLG for Memory and Electrode Applications.. 492
30.6 Summary 494
References.. 494

ABSTRACT

This chapter demonstrates new routes and extends the range of future applications for few-layer graphene (FLG). Few examples of utilization of the FLG for the creation of new advanced materials and graphene-based structures are presented. There are several distinctive properties that make the FLG eligible for such applications. Among them are (1) the high sensitivity of FLG for sensor application, (2) hybrid substrates provided the high carrier mobility in graphene, (3) superlattice hydrogenated graphene/NMP monolayer with high carrier mobility, (4) fluorination of graphene or FLG by chemical functionalization, (5) advantages of FLG for flash memory effects, and so on.

30.1 INTRODUCTION

The discovery of graphene holds the promise of unique device architectures and functionalities and new physical phenomena [1,2]. The present period of graphene study opens new routes and extended the range of future applications [3,4]. For many practical applications, a few-layer graphene (FLG) structure often demonstrates even more exiting properties than does graphene itself. The discovery of FLG provides a basis for the development of graphene-based materials for future applications. Nearly all the intrinsic properties of graphene are also inherent to FLG (including its high conductivity, high mobility, high thermal conductivity, thickness-dependent optical transparency, and good mechanical properties) except for linear energy dispersion. Moreover, the utilization of FLG provides additional possibilities for chemical functionalization: such approaches include intercalation, the modification of an individual layer or one side of a layer and the creation of a vertical superlattice. The novel properties of such modified layers and new functionalities due to interlayer interactions also look very promising. As a result, extremely thin vertical FLG-based heterostructures can be created, and new functionalities and physical phenomena in vertical transport can be expected. The creation of an FLG-based structure with different properties of the separate layers leads to prospects for the 3D architecture of future devices. Figure 30.1 presents few examples of the prospective approaches for FLG-based future devices. These properties, in combination with the possibility to fabricate large-area 2D films of FLG, extend the spectrum of potential applications and provide a strong possibility for the realization of these applications.

It is necessary to mention here that FLG films generally demonstrate interesting properties when the film thickness is in the order of a few nanometers (<10 nm). We consider such FLG films in this chapter.

We have demonstrated a set of attractive FLG properties and created some advanced FLG-based structures. First, the gas-sensing properties of FLG are extremely high in comparison to those with graphene. The resistivity response to ammonia adsorption is strongly dependent on film thickness and is higher than that of graphene by 1–8 orders of magnitude for an FLG thickness of 2–6 nm. Second, an FLG-based hybrid material with high stability and reproducible properties was fabricated by the intercalation of *N*-methylpyrrolidone (NMP) into FLG combined with heat treatment. Low-temperature annealing (100–180°C) leads to a strong increase in the resistivity (~7 orders of magnitude) of intercalated films. Recovery of the top-layer conductivity by chemical cleaning allows us to create structures of graphene with high carrier mobility

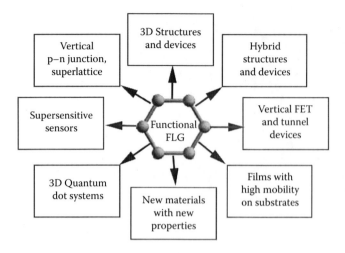

FIGURE 30.1 Examples of the prospective approaches for FLG-based future devices.

(up to 40,000 cm²/Vs) on an atomically flat, highly resistive substrate. The annealing of the intercalated structures at higher temperatures of 190–280°C was found to lead to the formation of a hydrogenated graphene/NMP superlattice with relatively high carrier mobility (3000–5000 cm²/Vs on SiO_2/Si). These and other prospective FLG-based materials will be discussed in this chapter.

30.2 FLG FOR SENSOR APPLICATIONS

Extremely high gas-sensing properties of p-type FLG flakes exfoliated from highly oriented pyrolytic graphite have been demonstrated [5]. The structures initially had p-type conductivity. Adsorption of ammonia leads to n-type doping of FLG flakes [6]. Thus, exposure to ambient ammonia increases the resistance of graphene or FLG flakes (see Figure 30.2). A relatively weak change in resistivity due to ammonia absorption

was observed for graphene. The response of graphene to ammonia adsorption is ~4% (see References 7–10), while the response of bigraphene (see Figure 30.2) was found to be ~70% [5]. An increase in graphene thickness leads to an increase in the number of gas-adsorption sites as a result of summing the defect points of both layers. For an FLG thicknesses of <2 nm, the current–voltage characteristics of the structures measured in transistor mode after long exposures to ammonia revealed n-type conductivity in graphene flakes. An increase in flake thickness leads to an increase in response of up to 7 orders of magnitude for a thickness of ~2 nm. Once the flake thickness becomes larger than the interlayer screening length, the response begins to decrease. Samples with flake thicknesses ranging from 3 to 10 nm exhibited a decrease in their response to ammonia as their thickness was increased (100%–25% change in resistivity). No conversion to n-type conductivity was found in those films, but local areas with n-type conductivity formed at the surface near defects and grain boundaries, and there was evidence of a decrease in the occurrence of these areas with depth.

The sensitivities of different graphene structures to ammonia adsorption as a function of flake thickness are illustrated in Figure 30.3. The resistivity response to ammonia adsorption is strongly dependent on film thickness and is higher than that of graphene by 1–8 orders of magnitude. The effect is attributed to the formation of multiple p–n–p junctions at the grain boundaries in the polycrystalline graphene flakes exposed to ambient ammonia-containing agents. Graphene exfoliated from graphite is known to exist on a large scale in polycrystalline form (see, e.g., Reference 11). The charge screening length is often estimated as 0.5 nm [12] in graphene and 0.4 nm in graphite for a sheet carrier density of ~1×10^{13} cm⁻² [13]. Making graphite samples thinner than approximately 50 nm will eventually lead to a measurable

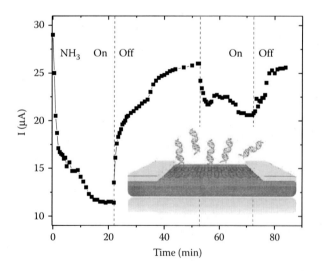

FIGURE 30.2 Change in the current passing through FLG flakes with a thickness of ~3 nm under NH_3 absorption as a function of time for a structure exposed to NH_3. The applied voltage was 0.1 V. The inset shows the configuration of the resistivity measurements.

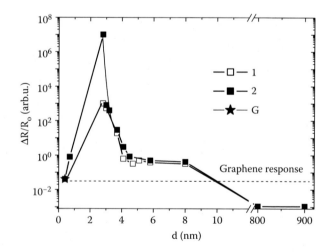

FIGURE 30.3 Change in resistivity of FLG flakes under NH_3 absorption as a function of flake thickness for two ammonia sources, 1 and 2 (the amount of ammonia in case 2 is higher than that in case 1), used to provide the gaseous air/ambient ammonia in this study. Point G and the line indicate the known 4% response of graphene to ammonia adsorption from References 2 and 5–9.

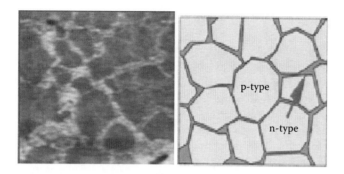

FIGURE 30.4 Atomic force microscopy image of samples with grain boundaries visualized by means of chemical functionalization. The image size is 10 μm. The inset shows a sketch of the n- (network of boundaries) and p-type (grain) areas after ammonia adsorption.

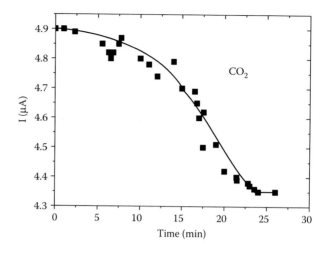

FIGURE 30.5 Change in the current passing through an FLG flake under CO_2 absorption as a function of time for a structure exposed to CO_2. The thickness of the FLG flake was 3.7 nm. The applied voltage was 0.5 V.

field effect, as the properties of the induced charge density against the background of the unaffected bulk charge density become significant [14]. The carrier density in our samples was less than or equal to approximately 1×10^{12} cm^{-2} per graphene sheet. As a result, a greater screening length in our samples could be expected. H. Miyazaki et al. experimentally obtained an inter-layer screening length of 1.2 nm in structures with a sheet carrier density of $\sim 1 \times 10^{12}$ cm^{-2} [15,16]. In our polycrystalline films (see Figure 30.4), the average sheet density of the charge carriers was $(0.5–2) \times 10^{12}$ cm^{-2}; at the grain boundaries, an even lower density of the charge carriers is generally expected. Thus, the largest inter-layer screening length is expected to occur at the grain boundaries. Therefore, n-type doping could be expected to proceed nonuniformly over graphene sheets, causing the formation of multiple p–n–p junctions in them. A similar effect is expected due to the presence of monolayer steps on the surface. A high activity of these steps leads to first-priority ammonia adsorption on these steps with the formation of additional p–n junctions.

Two main conclusions can be drawn from Figure 30.3. The first conclusion is that the number of ammonia-adsorption sites increases as the number of monolayers (<2 nm) increases. The second conclusion is that our FLG structures with thicknesses of greater than ~2 nm exhibited a decrease in response due to the shunting of conductivity by the deeper layers. Generally, we must conclude that FLG structures demonstrate a higher response to ammonia in comparison to that of graphene.

Ultrafast and sensitive room temperature NH_3 gas sensors based on chemically reduced graphene oxide were recently demonstrated by Lu et al. [17]. The high sensing performance of these sensors with resistance change as high as 2.4% and response time as fast as 1.4 s was realized when the concentration of NH_3 gas was as low as 1 ppb on relatively thick graphene oxide films.

Using different gases (for instance, CO_2), we also demonstrated that the response of FLG is also higher than that of graphene for gases other than ammonia. The change in current for FLG with a thickness of 3.7 nm due to CO_2 absorption is equal to 11% and is depicted in Figure 30.5. The response of graphene is equal to 2% [18]. Yoona et al. [19] have demonstrated a current response of up to 26% for an FLG thickness

of 1.2 nm. The above responses again demonstrate a dependence on thickness that reaches a maximum response.

The obtained high FLG sensitivity can be well understood in terms of the effect of defects in graphene or in the substrate discovered by B. Kumar et al. [20,21]: the sensitivity of pristine graphene is increased by an increase in defect concentration in graphene. The operational mode of pristine graphene sensors is dependent on defects because the sp^3-like character of defect sites in graphene sensors can form low-energy sorption sites for a broad spectrum of molecules. It was also demonstrated that the sensitivity of pristine graphene is not necessarily intrinsic to graphene, but rather it is facilitated by external defects in the insulating substrate, which can modulate the electronic properties of graphene [21]. An increase in the number of polycrystalline monolayers also leads to a collection of defect sites for the adsorption of gas molecules within a range of a few monolayers.

Graphene multilayer films (platelet network), created from graphene oxide, reduce graphene oxide or graphene flakes are one of the practical solutions to optimize gas detection and selectivity [22–25]. Theoretical studies support the fact that graphene functionalization is the best way to improve sensing performance, the gas molecules being weakly adsorbed on pristine graphene and more stronger on doped or defective graphene [26,27]. The high sensing performance of the reduced graphene oxide sensors with resistance change as high as 2.4% and response time as fast as 1.4 s when the concentration of NH_3 gas was as low as 1 ppb and ~20% for 10 ppm has recently been realized by Hu et al. [17]. The authors demonstrated ultrafast and sensitive room temperature NH_3 gas sensors by means of creation of the platelet networks between Cr/Au electrodes. Platelets with a thickness of 0.7–3.2 nm were formed from the reduction of graphene oxide platelets with their functionalization by pyrrole. These sensors, with low cost, low power, and easy fabrication, as well as scalable properties, showed great potential for ultrasensitive detection of NH₃ gas in a wide variety of fields. The sensitivity of these sensors for dimethyl methylphosphonate, methanol, dichloromethane, cyclohexane, chloroform, triethylamine vapors, CO_2 or CO, and other vapors was also measured. Graphene sponges that possess excellent conductivity coupled with high surface area have been shown to make an excellent platform for electrochemical sensors to detect certain important biological moieties at very low concentrations [28].

30.3 SUBSTRATES FOR HIGH CARRIER MOBILITY IN GRAPHENE

2D materials require a high-quality substrate to provide their functionality. For instance, polycrystalline graphene on an SiO_2/Si substrate has a carrier mobility at room temperature of approximately 2000–4000 cm²/Vs due to carrier scattering on the charges in SiO_2 [29]. The mobility in the monocrystalline domain is higher, up to 10,000 cm²/Vs [30]. Graphene on hexagonal boron-nitride (h-BN) demonstrates a carrier mobility of approximately 35,000–40,000 cm²/Vs [31] at room temperature, and graphene on CVD-grown h-BN has a carrier mobility from 1100 to 2500 cm²/Vs [32] in the case of relatively low-quality h-BN or a low-quality grown layer and up to 20,000 cm²/Vs for single-domain graphene [33]. The highest carrier-mobility values at room temperature (~100,000 cm²/Vs [34,35]) have been observed in suspended graphene (graphene on nothing). Higher values of the carrier mobility have been observed only at low temperatures [36–38]. Therefore, the problem of choosing the optimum substrate for graphene and FLG is still unresolved, and the search for new substrates is still ongoing. The main idea driving the present approach to identifying a substrate that can provide high carrier mobility in graphene is as follows. We have proposed a new approach to solving this problem by means of the chemical functionalization of FLG, which has enabled the formation of a highly resistive derivative with subsequent cleaning of the surface and recovery of the top-layer conductivity [39].

The strategy for the creation of new substrates for graphene is given below (see also Figure 30.6). Graphite intercalation compounds are a well-known class of hybrid materials (see, for instance, Reference 40). Graphene and FLG are known to

actively interact with a wide spectrum of organic and nonorganic materials [41–43]. The intercalation of nonconductive organic molecules and the interaction of these molecules with graphene plates are expected to lead to an increase in film resistivity. In fact, it was found that annealing at relatively low temperatures (100–180°C) of NMP-intercalated FLG films leads to a strong increase in film resistivity (10²–10⁴ Ω cm), leading to the creation of a stable material with interesting and reproducible properties [44,45]. Repairing the top layer by treating it with hydrofluoric acid (HF) allows us to create a highly conductive single layer with a carrier mobility of up to 16,000–42,000 cm²/Vs [44]. Moreover, the entire cycle of functionalization and cleaning treatments is capable of providing a strong gate-voltage-induced modulation (by 4–5 orders of magnitude) of the current in the top monolayer (modified graphene). The creation of highly resistive FLG film with subsequent cleaning and repair of the top-layer conductivity may be a solution to the problem of choosing the proper substrate for graphene.

Polar solvents, such as NMP and C_5H_9NO, that are perfectly matched to the graphene surface energy are very efficient chemicals for stabilizing FLG sheets in solution, and they provide a unique basis for further processing [44]. The structural formula of NMP is given in the inset of Figure 30.7. The intercalation agent was chosen based on the well-known capability of NMP to penetrate between graphene layers [44]. The ability to create chemically non-modified graphene–NMP solutions is widely used for the preparation of graphene suspensions for various applications [40]. It is very important that a suspension of graphene in NMP provides chemically non-modified graphene sheets at room temperature. We have found that annealing NMP-intercalated graphene drastically modifies the properties of FLG structures [44,45]. As a result, we gain the possibility to fabricate a highly resistive material

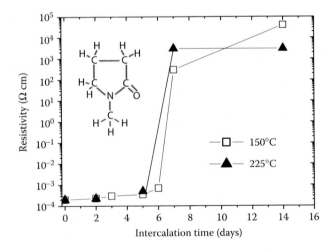

FIGURE 30.7 Step-like increase in resistivity of intercalated ribbons with a width of 5 μm as a function of the intercalation time for annealing at different temperatures (150°C and 225°C). (From Antonova, I.V., Kotin, I.A., Soots, R.A. et al. 2012. *Fullerenes, Nanotubes, and Carbon Nanostructures* 20: 543–547.) The structural formula of NMP is shown in the inset.

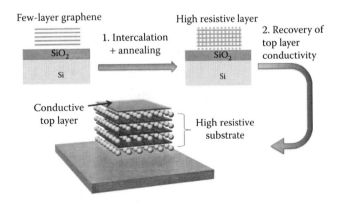

FIGURE 30.6 Sketch of the creation of a highly resistive substrate for graphene with high carrier mobility.

with stable electronic properties, offering another advanced application for the employment of NMP.

30.3.1 CREATION OF HIGHLY RESISTIVE SUBSTRATES AND THEIR PROPERTIES

Ribbons of FLG were created for NMP intercalation. These ribbons were created as follows. First, the surface of HOPG was structured using lithography and dry etching to a depth of several micrometers through a lithography mask. Ribbons with a width of 2–20 μm were then transferred onto a 300-nm SiO$_2$/Si substrate with the help of a printing technique implemented on a 6-in. nano-imprinting equipment that is commercially available from Obducat and the intermediate polymer substrate. The details of the sample preparation are given in References 39 and 44. We have evaluated the time required for NMP intercalation based on an NMP-intercalation velocity of 0.5 μm per day [45]. Annealing of the NMP-intercalated structures was performed at a set of temperatures in the range from 80°C to 180°C for a duration of 5 min in air ambient; this annealing was the final step leading to the creation of a stable highly resistive material in this study. The NMP-intercalation velocity was estimated from the data shown in Figure 30.7. Intercalation of the proper time leads to the introduction of a dense monolayer of NMP molecules and a step-like increase in resistivity after annealing due to the interaction of the NMP molecules with the nearby graphene layers.

Figure 30.8 presents the dependence between resistivity and carrier mobility for different hybrid structures created at different temperatures. The maximum carrier mobility corresponds to annealing temperatures of 140–150°C. In reference (non-intercalated) graphene ribbons annealed at different temperatures within the indicated temperature range, the resistivity showed no substantial change. For structures created at 100°C < T < 180°C, we assume that the intercalated NMP molecules interacted with the carbon atoms of the graphene layers via the formation of bonds through oxygen. The variation of the structure thicknesses in the range from 3 to 6 nm led to a slight variation in the resistivities of the structures. Repeated measurements (over a period of up to one year) have demonstrated that the properties of the hybrid structures were stable.

The influence of NMP coverage on the vibrational and electronic properties of graphene layers was studied under ambient conditions using Raman spectroscopy. Raman spectroscopy is a powerful technique used to observe the dominant vibrational modes in a system. Figure 30.9 shows the Raman spectra of several samples annealed at different temperatures. The most prominent peak G at ~1580 cm^{-1} is known to arise from in-plane vibrations of sp^2 carbon atoms. The peak D at 1350 cm^{-1} is an indicator of intrinsic defects present in graphene sheets or a basal-plane chemical reaction that breaks the π-bonds and converts the sp^2 carbon-atom hybridization into sp^3 hybridization [46,47]. It is surprising that our highly resistive FLG structures, after intercalation and annealing at 100°C < T < 180°C, demonstrated no pronounced changes in their Raman spectra. This means that interaction with NMP, as suggested above, does not lead to graphene oxidation.

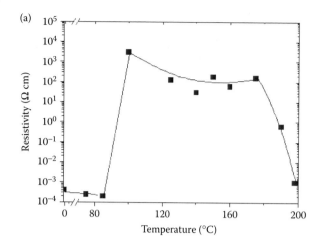

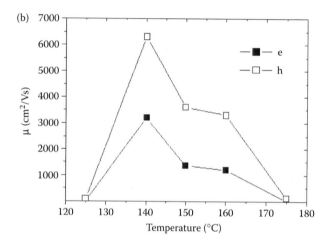

FIGURE 30.8 Resistivity (a) and carrier mobility (b) in intercalated FLG structures annealed at different temperatures. The FLG thicknesses in the various structures ranged from 4 to 6 nm. (The source of the material Kotin, I. A., Antonova, I. V., Komonov, A. I. et al. High carrier mobility in graphene on atomically flat high-resistivity layer. *Journal of Physics.* is acknowledged.)

We must stress here that the created structures were vertical superlattices of functionalized graphene and organic (dielectric) single layers. The thicknesses of these structures were increased due to intercalation by only ~15%. For every temperature used for the fabrication of our hybrid structures, we prepared several structures to evaluate the reproducibility of the resistivity of our hybrid material; as a result, a rather good reproducibility was identified. We also performed repeated measurements of the Raman scattering and *I–V* characteristics of our structures over a one-year period with the aim of testing our hybrid structures for long-term stability. In this way, a good stability of the measured properties was confirmed.

30.3.2 CLEANING OF THE TOP LAYER FOR HIGH CONDUCTIVITY

Let us consider first the hybrid structure created from single-layer graphene (NMP-treated structure annealed at 150°C). The treatment of this highly resistive structure with HF led

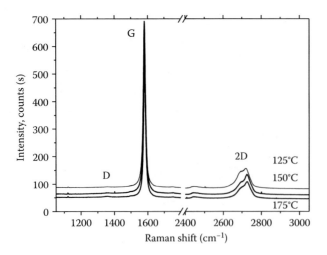

FIGURE 30.9 Raman spectra of hybrid structures created at different temperatures. The laser power incident on the samples during the registration of the Raman spectra was 3 mW. The spectra are shifted in the vertical direction. (The source of the material Kotin, I. A., Antonova, I. V., Komonov, A. I. et al. High carrier mobility in graphene on atomically flat high-resistivity layer. *Journal of Physics.* is acknowledged.)

to a strong increase (recovery) of the structure's conductivity. Figure 30.10a shows the current–voltage characteristics measured for the single-layer graphene–NMP hybrid structure. The resistivity of this structure was 3.5×10^8 $\Omega/\square$ before cleaning, and it was decreased by seven orders of magnitude after 1 min of treatment in HF vapor. Presumably, such treatments caused the removal of NMP molecules from graphene surfaces. The low resistivity of the repaired graphene is most likely related to the strong modification of graphene throughout the entire cycle of treatments (intercalation, annealing, and chemical cleaning).

Figure 30.10b shows the drain–source current as a function of gate voltage $I_{ds}(V_g)$ measured using the Si substrate as the gate. From the measured $I_{ds}(V_g)$ characteristics in the linear region, the field-effect carrier mobility (holes or electrons) could be estimated using the traditional approach for a semiconductor field-effect transistor (expression 30.1), which is commonly used for the estimation of carrier mobility in graphene and FLG (see, for instance, References 31, 48, and 49):

$$\mu = \Delta I_{DS} / \left(C_{OX} \frac{W}{L} V_{DS} \Delta V_g \right) \qquad (30.1)$$

In Equation 30.1, μ is the carrier mobility; W and L are the flake width and the flake length, respectively; $C_{ox} = \varepsilon_{ox}\varepsilon_o/t_{ox}$ is the gate oxide capacitance ($\varepsilon_{ox} = 3.9$ is the relative permittivity of silicon dioxide and $t_{ox} = 300$ nm is the initial thickness of the gate oxide layer); and ΔI_{DS} is the shift in the current I_{DS} induced by a gate-voltage change of ΔV_g. Treatment with HF leads to a decrease in thickness of the oxide underlayer t_{ox}. Therefore, the real thicknesses t_{ox} estimated from the SiO_2 color were used to determine mobility in these cases. The minimum thickness of the SiO_2 after the HF treatments used in this study was 60–80 nm. We would like to note here that the strong modulation of the current observed after HF treatment gives the recovered graphene very promising new properties. The carrier mobility determined from the $I_{ds}(V_g)$ characteristics for graphene increases after a 1-min HF treatment from 650 to 11,000 cm²/Vs for holes and from 50 to 30,000–41,000 cm²/Vs for electrons. The strong increase in carrier mobility was attributed to the cleaning and suspending of the graphene obtained in this case.

For FLG with highly resistive substrates created at different temperatures, the HF treatment led to a decrease in flake resistivity by a factor of 10^2–10^4. The changes in resistivity and carrier mobility for two FLG hybrid structures as

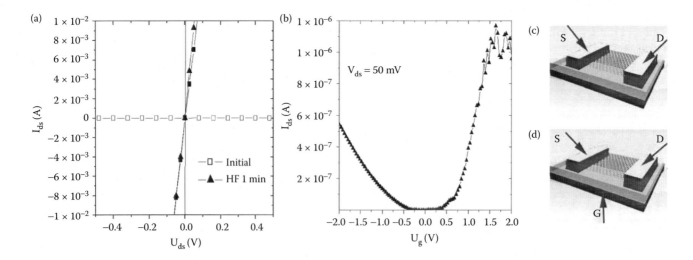

FIGURE 30.10 Current–voltage characteristics (a) of a single-layer graphene-based hybrid structure measured in diode configuration (c) after annealing and additional treatment in HF vapor for a duration of 1 min. $I_{ds}(V_g)$ characteristics of the same structure (b) measured in transistor configuration (d) after the HF treatment. S is the source contact created on the graphene and SiO_2 surface; D is the drain contact; and G is the gate (silicon substrate).

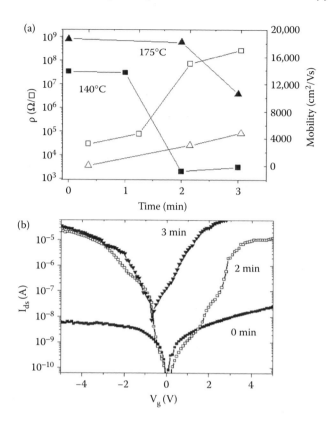

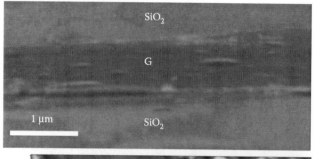

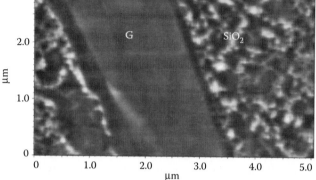

FIGURE 30.11 Resistivity and field-effect carrier mobility (a) in FLG structures created at temperatures of 140°C (square points, film thickness ~4 nm) and 175°C (triangle points, film thickness ~3 nm) as a function of the duration of their treatment in HF vapor. (b) $I_{ds}(V_g)$ characteristics of the structure created at temperatures of 175°C before and after HF treatment.

FIGURE 30.12 SEM and AFM images of HF-treated (cleaned) FLG-NMP structures of 3 nm in thickness. (The source of the material Kotin, I. A., Antonova, I. V., Komonov, A. I. et al. High carrier mobility in graphene on atomically flat high-resistivity layer. *Journal of Physics.* is acknowledged.)

a function of the duration of the HF treatment are shown in Figure 30.11. In this case, the relatively low decrease in flake resistivity is most likely because only one side of the top graphene layer was given the cleaning treatment. The observed shift of the Dirac point in the $I_{ds}(V_g)$ characteristics of the structures toward negative voltages after the HF treatment proved that, during the HF treatment, the samples were given an n-type doping.

SEM and AFM images of the surfaces of the FLG structures obtained after the treatment of the samples in HF vapor are shown in Figure 30.12. Both images illustrate the good surface quality of the treated samples. Generally, our ribbons lie flat on the oxide residue left on the sample surface after the HF treatments. The SEM and AFM images were also indicative of the fact that the oxide surface was flat, and the oxide was uniformly etched under our ribbons in the HF vapor. Thus, the thickness of the residual oxide can be used as the gate oxide thickness to calculate the carrier mobility. The Raman spectra of the structures treated in HF vapor exhibit a very weak D band. This observation provides evidence for a "soft" functionalization of the FLG hybrid structures, including the top monolayer with recovered conductivity.

With the aim of checking our hypothesis that the HF treatment causes a recovery of the conductivity in the top layer, we examined the surface conductivity with the help of a

conductive probe operating in contact mode (SpRI measurements). Figure 30.13a shows the current–voltage characteristics for a 4-nm-thick structure created by means of annealing at a temperature of 150°C before and after a 1-min treatment of the sample in HF vapor measured at approximately the same point on the sample surface. A strong increase in the current value is evident. The current maps shown in Figure 30.13b, c also lend support to the conclusion that follows from an inspection of Figure 30.13a with larger statistics. The latter provides direct evidence for the increase in conductivity of the top graphene layer during treatment in HF vapor.

The Raman spectra of the structures treated in HF vapor look like the spectra shown in Figure 30.9. It is essential to note that the structures exhibited a very weak D band in their Raman spectra. This observation provides evidence for the high quality of FLG hybrid structures, including the top monolayer with recovered conductivity. It is necessary to remember that one property of NMP is that it is perfectly matched to graphene in terms of surface energy, leading to a "soft" functionalization of the FLG without introducing any additional disorder into the functionalized material.

The most interesting phenomenon observed as a result of the HF cleaning of structures was a strong increase in the mobility of both holes and electrons in FLG structures (6000–42,000 cm²/Vs). This result is most likely related to a low charge density in the NMP single layers due to the close matching of NMP and graphene in terms of energy [50]. As a result, a partial screening of the top graphene layer from the

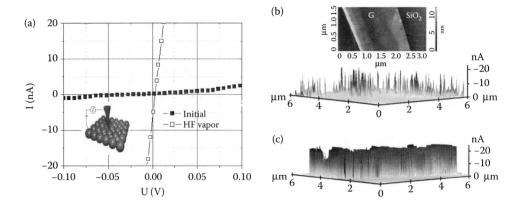

FIGURE 30.13 Current–voltage characteristics measured with the use of a conductive probe operating in contact mode on 4-nm-thick FLG structures created at 150°C before and after treatment in HF vapor for a duration of 1 min (a). Current maps measured on the same structures in SpRI mode before (b) and after (c) their treatment in HF vapor for a duration of 1 min (with an applied voltage of 0.05 V). Inset in (a) gives the sketch for measurements. Inset in (b) gives the AFM image of graphene on SiO$_2$ substrate.

charges in the SiO$_2$ substrate was observed. This effect is similar to the one observed in the case of hNB layers with thicknesses of several nanometers used as an additional substrate on SiO$_2$. Thus, treatment in HF vapor offers a very promising technological approach to controlling the properties of NMP-FLG structures with the possibility of local modification of the mobility and with opportunities for 3D design. As mentioned above, h-BN substrates provide a carrier mobility from 15,000 to 40,000 cm^2/Vs (even up to 40,000 cm^2/Vs in the case of special efforts [51]) at room temperature [32,33]; with our method, we have obtained similar values for carrier mobility. Graphene on CVD-grown h-BN layers does not demonstrate a high carrier mobility [52–54].

The main finding in this study of the structural and electrical properties of the top graphene layer on a highly resistive NMP-FLG substrate is high conductivity in combination with high carrier mobility (up to 16,000–42,000 cm^2/Vs). Gate-voltage-induced current modulation of up to 4–5 orders of magnitude opens the way to the development of a material for a graphene-based transistor. These structures are quite stable: the material properties in the top layer show no variations over time (at least over a period of one year). It seems that our highly conductive graphene layer on a highly resistive substrate is a very promising structure for electronic applications. Using lithography or other structuring methods, one can adjust the conductivity directly for particular device designs. This approach widens the range of possible applications of our structures and provides an additional tool for the control of the local properties of the material and for 2D nanodesign.

30.4 FUNCTIONALIZATION OF FLG FOR ELECTRONIC APPLICATIONS

30.4.1 SUPERLATTICE HYDROGENATED GRAPHENE/NMP MONOLAYER WITH HIGH CARRIER MOBILITY

A very interesting and promising material was created from the NMP-intercalated FLG described in 30.2.1 by annealing

it at higher temperatures (190–270°C) [48,55]. The main idea was to create a source of free hydrogen located between the graphene layers: in other words, the goal was to modify the organic molecules to release hydrogen. Annealing of the intercalated structures creates the conditions for hydrogen release by causing the formation of oligomers and/or the dehydrogenation of NMP molecules.

Annealing the NMP-intercalated FLG at temperatures higher than 190°C led to substantial transformations in the Raman spectra (see Figure 30.14). The intensity of the D line was increased. Moreover, a 1500-cm^{-1} band was revealed in the Raman spectra; this band is known to be a disorder-induced first-order scattering band similar to the D line [51]. The spectra also revealed peaks at 1625 and 2980 cm^{-1}. According to [50], the growth of the sharp D peak and the appearance of an additional peak at ~1620 cm^{-1}, called D′, as well as the

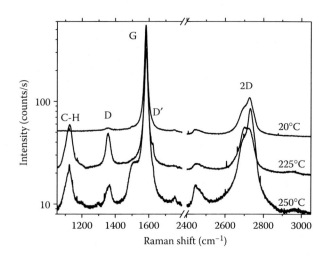

FIGURE 30.14 Raman spectra of hybrid structures created at different temperatures. The laser power incident onto the samples during the registration of the Raman spectra was 3 mW. The spectra are shifted in the vertical direction. (From Kotin, I.A., Antonova, I.V., Soots, R.A. et al. 2013. *Nanotechnologies in Russia* 9–10: 621–626.)

emergence of a peak at approximately 2950 cm⁻¹, which is a combined mode of D and D′ vibrations, are attributed to interruptions of π-electron delocalization subsequent to the formation of C–H sp³ bonds [51,56]. Moreover, the Raman spectra for these structures manifest a line that directly corresponds to the formation of C–H bonds in graphene. This is the line at 1120 cm⁻¹ [57]. This line clearly appeared in the Raman spectra only for structures annealed at temperatures greater than 190°C and disappeared once again for structures annealed at T ≥ 270°C. Therefore, it can be concluded that hydrogenated FLG (stoichiometrically hydrogenated graphane or partially hydrogenated graphone) was created only in this temperature range (190–270°C).

In the hybrid structures fabricated in the temperature range 190°C ≥ T > 270°C, a more strongly increased resistivity appeared (see Figure 30.15a). The increase in resistivity also correlated with the changes in the Raman spectra and the formation of the hydrogenated FLG. It is well known that hydrogenated graphene is a highly resistive material

[51]. Using the $I_{ds}(V_g)$ characteristics (see the inset in Figure 30.15a) measured using the Si substrate as the gate, Equation 30.1 and the parameters of our structures, we have also evaluated the field-effect mobility of electrons and holes. The obtained values are given in Figure 30.15b. With the exception of the values for the structure created at 225°C, the carrier mobility is very similar to the mobility observed in the initial non-intercalated structures, shown in Figure 30.15b as values for T = 0°C. It is necessary to stress here that in the case of the creation of hydrogenated graphene by hydrogen plasma (Figure 30.16), mobility strongly decreases even for short treatment times [58].

The surfaces of our structures were inspected with AFM, and the images are provided in Figure 30.17. Bright spots are found after annealing under these conditions (190–270°C). It was found by Balog et al. [59] that the formation of graphane-like islands is energetically favorable for adsorbed hydrogen on graphene. The spot size and concentration are maximal for structures created at T ~225°C. These structures have the maximal resistivity and minimal carrier mobility. This means that the maximal extent of hydrogenation for this temperature was achieved.

The results of STS measurements of the probe-film tunneling current, in the form of dI/dV characteristics measured for structures annealed at different temperatures, are shown in Figure 30.18. The dI/dV characteristics clearly demonstrate that a bandgap was present in our structures. According to the data measured at different points, the bandgap values ΔE_G in our structures varied up to 2.9 eV. The bandgap values extracted from the STS measurements versus the coverage of the surface with hydrogen are shown in Figure 30.19 and compared with the simulation of E_g for two types of hydrogenated structures from [59,60]. Our experimental results are more similar to the results for graphone. An estimation of the number of hydrogen atoms released from the

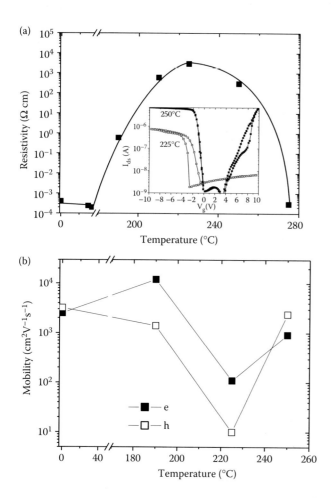

FIGURE 30.15 Resistivity (a) and carrier mobility (b) in intercalated FLG structures annealed at different temperatures. The FLG thicknesses in the various structures ranged from 4 to 6 nm. The inset shows the I_{ds} (V_g) characteristics measured at $V_{ds} = 0.2$ V for structures annealed at 225°C and 250°C. (From Kotin, I.A., Antonova, I.V., Soots, R.A. et al. 2013. *Nanotechnologies in Russia* 9–10: 621–626.)

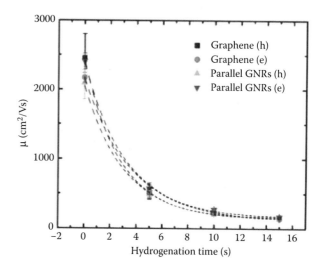

FIGURE 30.16 Carrier mobilities μe and μh versus hydrogenation time for a graphene sheet and 100-nm graphene nanoribbons (GNR). The hydrogenation time is the time of exposure to hydrogen plasma during treatment. (From Jaiswal, M., Haley, C., Lim, Y.X. et al. 2012. *ACS Nano* 5: 888–896.)

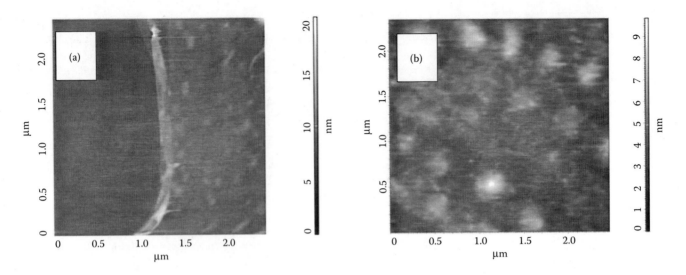

FIGURE 30.17 AFM images of FLG-NMP structures annealed at 190°C (a) and 225°C (b). (From Kotin, I.A., Antonova, I.V., Soots, R.A. et al. 2013. *Nanotechnologies in Russia* 9–10: 621–626.)

NMP during annealing due to the formation of oligomers or the dehydrogenation of organic molecules also supported this observation.

One of the main conclusions is that the studied approach allows us to obtain hydrogenated FLG with tunable electronic properties and high carrier mobility. This material has possibilities for use in electronic applications. The main disadvantage of this material is that the C–H bonds are not as stable as required for practical applications. After 1.5–2 years, the properties of the initial FLG reassert themselves.

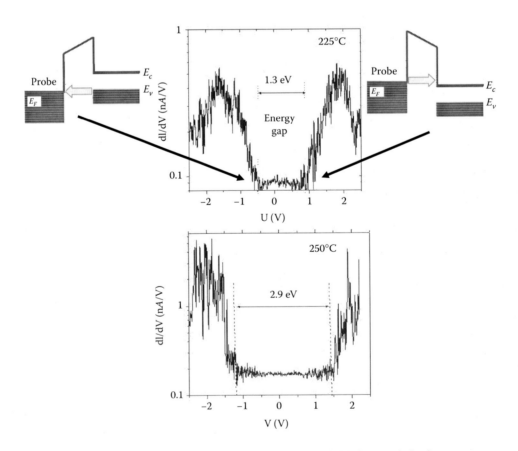

FIGURE 30.18 Determination of the bandgap by means of STS measurements: dI/dV characteristics for structures annealed at different temperatures. The FLG thicknesses ranged from 3 to 4 nm. (From Kotin, I.A., Antonova, I.V., Soots, R.A. et al. 2013. *Nanotechnologies in Russia* 9–10: 621–626.) The schematic illustrations provide a sketch of the tunneling current which appeared due to threshold electron tunneling from valence band to probe (left) or from probe to conductive band (right).

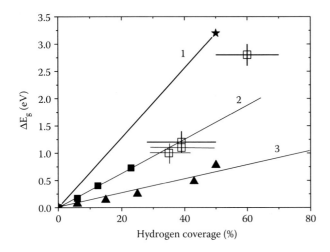

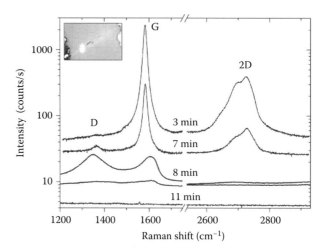

FIGURE 30.19 Bandgap in hydrogenated graphene. The points are the results of the STS study from Reference 55 and the simulation results [59]: 1—data from Reference 60 for graphane, 2, 3—data from Reference 59 (2 is for graphone, 3 is for graphane).

FIGURE 30.20 Raman spectrum of a 5-nm-thick FLG flake treated with a 5% solution of HF in water for different durations. The inset shows the position of the excitation laser beam on one of the graphene flakes. (Reprinted from *Physica E*, 52, Nebogatikova, N.A., Antonova, I.V., Volodin, V.A. et al. Functionalization of graphene and few-layer graphene with aqueous solution of hydrofluoric acid, 106–111, Copyright 2013, with permission from Elsevier.)

30.4.2 FEW-LAYER FLUOROGRAPHENE: A SIMPLE APPROACH TO CREATION AND POTENTIAL FOR APPLICATIONS

Fluorographene is a more stable material than hydrogenated graphene. The study [61,62] revealed the conditions for an efficient modification of graphene and FLG films with aqueous solutions of HF, and a method for the local protection of graphene against such modification with isopropyl alcohol (IPA) was identified. It was found that a few-minute treatment of graphene or FLG in aqueous HF solutions (~1 min for graphene and ~5 min for FLG films approximately 5 nm in thickness) led to strong changes in the structural and electrical properties of graphene involving a step-like increase in resistivity (up to 10^{11} Ω/□). Evidence for the fluorination of the FLG under HF treatment was observed. Two types of materials were obtained after different treatment durations due to the differing extent of FLG functionalization: (1) a promising material for electronic applications due to a combination of high carrier mobility, high conductivity, and strong current modulation by a gate voltage (up to four orders of magnitude) and (2) a material with insulating properties and graphene quantum dots embedded in an insulating matrix. A preliminary treatment in IPA completely suppressed the effect of HF functionalization. A combination of the two treatments may provide a key to the nanodesign of graphene-based 2D devices.

Further details on the effect of HF treatment on graphene or FLG films follow.

Raman spectroscopy shows (see Figure 30.20) changes in the Raman spectra of several samples treated in an aqueous HF solution for different durations (the FLG samples were subjected to treatments in 5% aqueous solutions of HF). The spectra of the pristine samples and the spectrum of a sample treated in HF:H$_2$O for 3 min were nearly identical. The dynamics of the Raman-spectrum changes for longer durations of

treatment include an initial increase in the D-peak intensity and a subsequent decrease of the intensity of all other peaks. The position of the excitation laser spot on the samples was checked using an optical camera. One of the images taken with the optical camera is shown in the inset to Figure 30.20. The changes in the Raman spectra are similar to the dynamics of the Raman-spectrum changes observed for fluorographene [62,63]. We therefore conclude that the HF treatments of the samples created by electrostatic exfoliation from HOPG for durations longer than 5 min led to the formation of fluorographene or few-layer fluorographene films. Variation of the HF:H$_2$O solution demonstrates that HF functionalization occurs only in a very narrow range of solution concentrations (2%–7%). Variation of the FLG thickness also leads to a variation in the time of fluorination (see details below). It is necessary to mention here that the utilization of samples created by CVD growth on Cu substrates and transference onto SiO$_2$/Si substrates requires a longer duration of treatment because of the necessity to first remove the residual organic contamination from the surface.

Dramatic changes in the Raman spectra show that the HF functionalization of graphene occurs in all monolayers of the FLG. The intercalation of HF into the FLG interlayer space during a few-minute treatment is impossible. Therefore, modification of the graphene surface morphology is assumed to be due to the functionalization of graphene with HF. AFM images of the surface of a pristine FLG sample and images after different durations of treatment in an aqueous solution of HF are shown in Figure 30.21. The HF treatment leads to the formation of an irregular surface morphology in samples treated for a relatively short time (the first stage of graphene functionalization with HF); later, this morphology transforms into a regular swell-like periodic morphology (corrugation) of the surface (the second functionalization stage). An increase

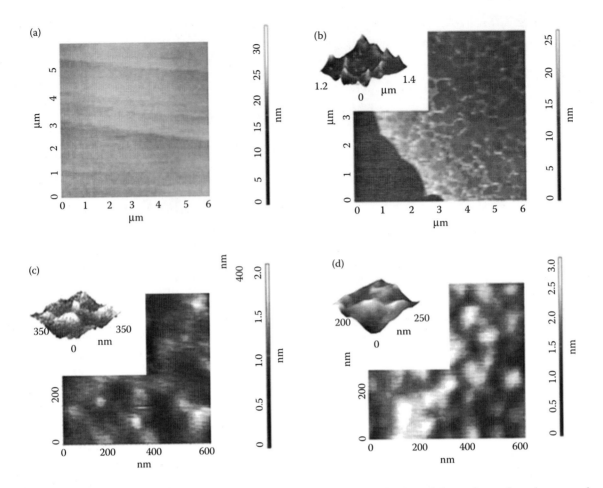

FIGURE 30.21 AFM images of the surface of a pristine (nontreated) graphene flake (a) and the surfaces of graphene samples treated in a 5% solution of HF in water for different durations of 4, 7, and 11 min (b, c, and d, respectively). (Reprinted from *Physica E, 52,* Nebogatikova, N.A., Antonova, I.V., Volodin, V.A. et al. Functionalization of graphene and few-layer graphene with aqueous solution of hydrofluoric acid 106–111, Copyright 2013, with permission from Elsevier.)

in the duration of HF treatment at the second stage induces no further changes in the size and height of the corrugation. An increase in FLG thickness of more than 15–20 nm suppresses changes in the surface morphology of FLG films.

These surface transformations are expected to manifest themselves in the electrical properties of partially and completely fluorinated FLG. The formation of an insulator network, a strong abrupt increase in the conductivity of graphene structures as a result of the completion of this network, the presence of conductive islands in the insulating matrix, and the subsequent vanishing of these islands from the graphene films are the main steps in the transformation of electrical properties of the examined graphene and FLG films revealed in the present study during the prolonged functionalization of graphene with HF.

The times of the resistivity steps for samples with different FLG-flake thicknesses show a good correlation with the decrease of the peak intensities in the Raman spectra and with the appearance of a regular swell-like periodic morphology on the surface of the HF-treated samples observed by AFM. An increase of the flake thickness in excess of 15–20 nm completely suppresses all manifestations of the HF-induced

functionalization of graphene films; in these cases, no nanoswell morphology on the sample surface and no resistivity step were observed.

The properties of the HF-treated structures were found to be quite stable. This was proved by repeated measurements over a period of 1.5 year of the electrical and structural properties of our graphene samples subjected to the HF-functionalization procedure.

The current–voltage characteristics of our samples measured in a two-terminal configuration are shown in Figure 30.22a. The resistivities of FLG samples with different thicknesses are plotted in Figure 30.22b,c as a function of the HF-treatment duration. As seen from the plot, in the first stage of HF treatment, only weak changes in I_{ds}–V_{ds} characteristics and the FLG resistivity are observed. After longer durations of HF treatment, a step-like increase in the FLG resistivity was revealed (a transition to an insulating state). The abrupt increase in the resistivity of samples was observed at longer treatment durations as the thickness increased: ~1 min for graphene, 2 min for bigraphene, and ~5 min for FLG films of approximately 5 nm in thickness. These times correspond to the HF functionalization of all monolayers in

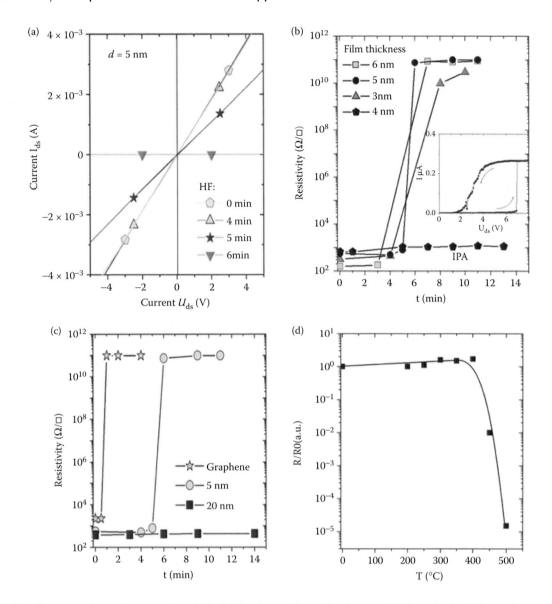

FIGURE 30.22 Current–voltage characteristics of FLG flakes treated in a 5% solution of HF in water for different durations (a) and the step-like increase in resistivity of various flakes with thicknesses of 3 to 6 nm versus the HF-treatment duration t (b). The FLG thickness is indicated in the figure as a parameter. The curve labeled IPA in (b) refers to a sample that initially (before the treatment in HF:H$_2$O) was treated for 20 min in isopropyl alcohol (IPA). The inset in (b) shows the current–voltage characteristic of a sample with 5-nm-thick FLG treated in HF:H$_2$O for 5 min. Flake resistivity of samples with widely varied FLG thicknesses as a function of the HF-treatment duration (c). Resistivity of a 4-nm-thick FLG flake treated in HF:H$_2$O (treatment duration of 8 min) versus the isochronal annealing temperature (d). The annealing time was 30 min. (Reprinted from *Physica E,* 52, Nebogatikova, N.A., Antonova, I.V., Volodin, V.A. et al. Functionalization of graphene and few-layer graphene with aqueous solution of hydrofluoric acid, 106–111, Copyright 2013, with permission from Elsevier.)

the examined structures. An increase in flake thickness in excess of 15–20 nm leads to the vanishing of the resistivity step, at least for HF treatments of more than 40 min. In the case of CVD-grown graphene transferred onto a SiO$_2$/Si substrate, the time of the resistivity step is increased from 1 to 6 min due to the residual contamination left behind after the transferring procedure.

The thermal stability and the reversible behavior of our functionalized samples were tested by subjecting them to additional annealing. It is known that hydrogen desorption from graphene occurs at temperatures of ~200–300°C [55]. On the other hand, C–F bonds are stable up to ~400°C [63,64]. We have performed an annealing of one of our FG

samples at a temperature of 300°C for 1 h. After the annealing, this structure exhibited the same high resistivity as found previously. Isochronal (30 min) annealing in ambient Ar was found to lead to a recovery of conductivity at 450°C (see Figure 30.22d). The annealing activation energy extracted from the curve of isochronal annealing in Figure 30.22d is equal to 2 eV. These observations correlate well with known parameters for the annealing of fluorographene [63,64] and provide an additional argument that HF treatment causes fluorination, not graphene hydrogenation or oxidation. More evidence of fluorination, including the direct observation of the C–F bonds obtained by x-ray photoelectron spectroscopy, is given in References 61 and 62.

Let us consider now in more detail the results of the electrical measurements that were obtained for the HF-treated samples for treatment durations that are slightly shorter than those necessary for the resistivity step (e.g., ~5 min in Figure 30.22b). The $I_{ds}(V_g)$ characteristics measured in the transistor configuration at two different temperatures (80 and 300 K) using the substrate as the gate electrode for an FLG flake treated in a 5% solution of HF in water for a duration of 4 min surprisingly show that the drain–source current I_{ds} varied over 4–5 orders of magnitude with the variation of the gate voltage V_g. Before the HF functionalization, the current modulation by the gate voltage was very weak in this structure (only a few percent). Thus, our films demonstrated a high potential for the management of their conductivity. The field-effect mobility estimated from Equation 30.1 accounting for a decrease in the thickness of SiO_2 during HF treatment has values of 2700 cm²/Vs at room temperature and 1600 cm²/Vs at T = 80 K for holes and lower values for electrons. In pristine, untreated samples, the carrier mobility had similar values (~3100 cm²/Vs at room temperature).

The observed transformations of the network surface morphology into a nanoswell relief during longer HF treatments allow us to hypothesize the presence of conductive graphene islands (quantum dots) in the functionalized insulating graphene matrix after the step-like increase of the structure's resistivity. This assertion is based on the following experimental results. It was found that immediately following the step-like increase in resistivity (treatment duration ~5 min in Figure 30.22b), the conductivity of the samples could be increased by applying higher voltages (~7 V) to the samples. The inset in Figure 30.22b shows the increase in the electric current through an HF-treated (5 min) sample that was induced by a high applied voltage, and it also shows the hysteresis demonstrated by this during the reverse voltage sweep. Moreover, the application of pulsed voltages (charge deep-level transient spectroscopy) clearly demonstrates the capture of charge carriers by and their emission from the conductive islands and quantum dots [65]. It was found [65] that it is possible to govern the carrier emission time from graphene quantum dots: an increase in quantum-dot thickness leads to a strong decrease in the carrier emission time, from milliseconds for graphene to microseconds for FLG with a thickness of 3 nm. This unique possibility is very promising for flash memory applications.

An increase in the HF-treatment duration leads to a complete vanishing of conductivity in our samples up to voltages of ±20 V. This effect is due to the increase in thickness of the HF-functionalized part of our structures and the decrease in the size of conductive graphene islands. The possibility of graphene-quantum-dot formation in a fluorinated graphene matrix was demonstrated in a theoretical study by Ribas et al. [66]. The formation of graphene quantum dots was also predicted for a corrugated graphene surface because of the huge changes in the strain at the nanometer scale [67]. The size of the graphene quantum dots estimated from the nanoswell relief observed in our samples (period ~150–200 nm) was found to be ≤80–100 nm.

To control the functionalization of graphene with HF, we have identified a new treatment capable of preventing the HF modification of graphene samples: treatment of the graphene samples in IPA prior to their HF functionalization. The resistivity data for a sample initially treated for 20 min in IPA and then subjected to HF treatment are shown in Figure 30.22b. Here, no increase in the resistivity value of the sample was observed. Moreover, the initial 20-min treatment in IPA was found to suppress the changes in AFM images and the Raman spectra during the subsequent treatment of IPA-treated samples in aqueous HF solutions.

Local treatment in IPA, in combination with the HF functionalization of graphene or FLG, can provide an approach to the nanostructuring of FLG and the design of graphene-based devices.

Thus, the chemical functionalization of FLG with organic molecules (NMP in our case) or nonorganic atoms (F) leads to the appearance of a bandgap and the modification of the electrical properties of the material. The network morphology functionalization is typical for polycrystalline FLG. Partial functionalization very often allows the creation of a material that exhibits a combination of high conductivity, high carrier mobility, and the possibility to achieve a strong modulation (up to 4 orders of magnitude) of the current by the gate voltage, which is very promising for electronic applications. It was found that the functionalization of graphene or FLG can be controlled by a preliminary treatment, providing a key to the nanodesign of graphene-based devices. Moreover, the functionalization of FLG allows the creation of materials with different monolayer properties in a common structure, which is very promising for a wide spectrum of applications.

30.5 FLG FOR MEMORY AND ELECTRODE APPLICATIONS

Flash memory, in general, is composed of a p-type silicon channel substrate, a tunnel oxide, a semiconducting highly doped n-type polysilicon or an insulating silicon nitride data storage layer, a control oxide, and a gate electrode. Writing is achieved by applying a voltage pulse on the gate electrode, which allows electrons to tunnel through the tunnel oxide from the silicon channel to the storage layer. The primary goal of most research into flash memory is to increase the density of the storage elements such that their parent devices can be miniaturized, which requires confronting several challenges that would normally jeopardize device performance at such a reduced scale. Graphene has the potential to exceed the performance of current flash memory technology by utilizing the exceptional intrinsic properties of graphene, such as its high density of states, high work function, and low dimensionality in comparison to conventional materials. Recently, it was shown that the incorporation of graphene oxide flakes as a float gate acts as an effective charge-storing layer in memory devices [68]. In Reference 69, the benefits of metallic graphene and FLG in a flash memory structure were considered along with a demonstration of such parameters as the memory window, retention time, and cell-to-cell crosstalk at low operating voltages. Large-area graphene sheets grown by chemical vapor deposition were integrated into a floating gate structure in the last case.

As reported in Reference 69, to investigate the electrical characteristics of graphene flash memory, a number of devices were fabricated with the following process: a p-Si wafer was cleaned, and a 5-nm SiO_2 tunnel oxide was grown, then graphene sheets or FLG with a thickness of 5 nm were transferred onto the tunnel oxide surface, a 35-nm control oxide was formed, and gate electrodes of Ti/Al/Au were deposited and an isolated memory device was created by plasma etching. An important parameter for flash memory is the memory window, which refers to the shift in the threshold voltage of the memory device when switching between the binary states of 0 and 1. Industry standards suggest that a minimum width of 1.5 V is necessary to produce a reasonable on/off ratio for reliable memory functionality. For standard float gate devices using polysilicon, this requires a program/erase voltage of approximately ±20 V [70]. This large voltage requirement is because of the low density of states in the degenerately doped polysilicon. Figure 30.23 demonstrates the capacitance–voltage characteristics of graphene and few-layer-graphene (thickness of 5 nm) devices and of a control sample without graphene. While the single-layer graphene devices exhibit a window width of ~2 V using a program/erase voltage of ±7 V, the FLG devices show a window width of ~6 V for the same program/erase voltage. The greater memory window of the FLG devices is directly attributable to the larger thickness of FLG compared with that of single-layer graphene [69]. FLG memory elements also show a long retention time of more than 10 years at room temperature. Additionally, simulations suggest that FLG memory elements suffer very little from cell-to-cell interference, potentially enabling scaling down far beyond current state-of-the-art flash memory devices.

The high conductivity of graphene in combination with its high transparency has attracted attention for its potential use in transparent, flexible electrodes in solar cells, light-emitting devices, and electronic touchscreens. Currently, indium tin oxide (ITO) with a sheet resistivity of 30–80 $\Omega/\square$ [71] is widely used for these applications. The relatively high cost of this material has stimulated a search for other transparent conducting films for use in electrodes. The sheet resistivity of undoped graphene is ~300 $\Omega/\square$. The utilization of FLG (usually four-layer graphene, as for practical usage in transparent electrodes, a transmittance of ~80%–90% is required) and FLG doping allows this resistivity to be decreased to 30–150 $\Omega/\square$ [72,73]. The best variants at the current time are hybrid films fabricated by using a metallic grid and graphene on a transparent substrate such as glass or polyethylene terephthalate films. The sheet resistance of such fabricated transparent electrodes is as low as 3 $\Omega/\square$ with a transmittance of ~80% or ~20 $\Omega/\square$ with a transmittance of 90% [71].

FLG electrodes with nanometer-sized gaps (1–2 nm) are now used for gateable molecular junctions [74,75]. Burzuri et al. [64] found that thicker flakes are more suitable for the fabrication of electrodes. Gateable transport through molecular contact between the electrodes demonstrates the potential

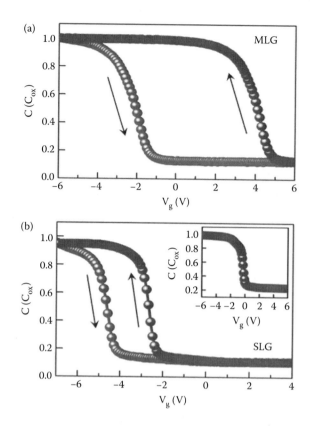

FIGURE 30.23 Capacitance–voltage (C–V) measurements on graphene and few-layer-graphene devices and a control sample without graphene. Few-layer-graphene (MLG) flash memory exhibits counterclockwise hysteresis with a 6-V memory window using a program/erase voltage of ±7 V (a). Single-layer graphene (SLG) flash memory shows a 2-V memory window at a negative threshold voltage due to hole doping in graphene (b). Inset: C–V characteristics of the control device without graphene show negligible hysteresis. All capacitance values were normalized with respect to the capacitance of the control oxide (C_{ox}). (Reprinted with permission from Hong, A.J., Song, E.B., Yu, H.S. et al. Graphene flash memory. *ACS Nano* 5: 7812–7817. Copyright 2011 American Chemical Society.)

for room temperature operation of molecular devices. By demonstrating temperature-independent nanogaps in electrodes combined with the long-term stability (weeks) of the FLG electrodes, these studies showed that few-layer-graphene nanogaps are an interesting alternative to metal electrodes for molecular electronics operating at room temperature and in ambient conditions. The authors concluded that the optimum flakes for the fabrication of nanometer-spaced graphene electrodes are those with initial resistances lower than 1 kΩ.

Practical applications are also considered the FLG-based electrode materials for biosensors [76], electrochemical sensors, absorbers for both gases and liquids, and electrode materials for devices involved in electrochemical energy storage and conversion. Several advantages of both graphene and graphene oxide sponges such as three-dimensional graphene networks, high surface area, high electro/thermo conductivities, high chemical/electrochemical stability, high flexibility and elasticity, and extremely high surface hydrophobicity are demonstrated in review [28].

Multilayered graphene membranes with different packing densities (electrodes formed from graphene sheets prepared by vacuum filtration of CCG dispersion with an average inter-sheet spacing between 1 and 10 nm) were used as electrodes for supercapacitors to probe nanoconfined electrosorption in porous carbon materials [77]. The authors highlighted the great potential of using easily available and structurally tunable graphene membranes as a model system for energy storage, membrane separation, and for nanofluidics. Use in electrochemical capacitors (or supercapacitors) is most likely one of the applications that may directly involve multilayered graphene membranes.

30.6 SUMMARY

Utilization of FLG instead of graphene widens the spectrum of possible approaches for functionalization and application. In the case of FLG materials, the various monolayers in a common structure can have different properties, which is a promising framework for a wide spectrum of applications. We have found an interesting combination of materials (FLG and NMP) for high carrier mobility in graphene or functionalized graphene. The intercalation of NMP into FLG in combination with annealing has the possibility to provide advanced films for electronic applications (highly resistive substrates for graphene, superlattice hydrogenated graphene/NMP). Fluorinated FLG with various extents of fluorination also demonstrates interesting properties for applications including memory applications. FLG (flexible and/or transparent) electrodes are already used in various structures at present.

In addition to the advanced FLG-based structures mentioned above, there are a number of other interesting prospective structures. Among them are 3D corrugated heterostructures for a wide spectrum of applications, novel materials exploiting effects such as the moire effect to provide new properties, hybrid heterostructures with tunable properties, p–n junction superlattices, vertical field-effect transistors with tunable properties, heterostructures with the individual properties of certain FLG layers, and QDs with tunable emission time.

REFERENCES

1. Novoselov, K.S., Fal'ko, V.L., Colombo, L. et al. 2012. A roadmap for graphene. *Nature* 490: 192–200.
2. Geim, A.K. and Novoselov, K.S. 2007. The rise of graphene. *Nature Materials* 6: 183–191.
3. Lee, J.-H., Lee, E.K., Joo, W.J. et al. 2014. Wafer-scale growth of single-crystal monolayer graphene on reusable hydrogen-terminated germanium. *Science* 344: 286–289.
4. Liu, C.-H., Chang, Y.S., Noris, T.B. et al. 2014. Wafer-scale growth of single-crystal monolayer graphene on reusable hydrogen-terminated germanium. *Nature Nanotechnology* 9: 273–278.
5. Antonova, I., Mutilin, S., Seleznev, V. et al. 2011. Extremely high response of electrostatically exfoliated few-layered graphene flakes to ammonia adsorption. *Nanotechnology* 22: 285502.
6. Wang, X., Li, X., Zang, L. et al. 2009. N-doping of graphene through electrothermal reactions with ammonia. *Science* 324: 768–771.
7. Shedin, F., Geim, A.K., Morozov, S.V. et al. 2007. Detection of individual gas molecules adsorbed on graphene. *Nature Materials* 6: 652–655.
8. Fowler, J.D., Alle, M.J., Tung, V.C. et al. 2009. Practical chemical sensors from chemically derived graphene. *ACS Nano* 3: 301–305.
9. Leenaerts, O., Partoens, B., and Peeters, F.M. 2008. Adsorption of H_2O, NH_3, CO, NO_2, and NO on graphene: A first-principles study. *Physical Review B* 77: 125416.
10. Dan, Y., Lu, Y., Kybert, N.J. et al. 2009. Intrinsic response of graphene vapor sensors. *Nano Letters* 9: 1472–1475.
11. Yazyev, O.V. and Louie, S.G. 2010. Electronic transport in polycrystalline graphene. *Nature Matrials* 9: 806–809.
12. Visscher, P.B. and Felicov, L.M. 1971. Dielectric screening in a layered electron gas. *Physical Review B* 3: 2541–2547.
13. Nagase, M., Hibino, H., Kageshima, H. et al. 2009. Local conductance measurements of double-layer graphene on SiC substrate. *Nanotechnology* 20: 445704.
14. Graf, D., Molitor, F., Ihn, T. et al. 2007. Phase-coherent transport measured in a side-gated mesoscopic graphite wire. *Physical Review B* 75: 245429.
15. Miyazaki, H., Odaka, S., Sato, T. et al. 2008. Inter-layer screening length to electric field in thin graphite film. *Applied Physics* 1: 034007.
16. Miyazaki, H., Li, S., Kanda, A. et al. 2010. Resistance modulation of multilayer graphene controlled by the gate electric field. *Semiconductor Science and Technology* 25: 034008.
17. Hu, N., Yang, Z., Wang, Y. et al. 2014. Ultrafast and sensitive room temperature NH_3 gas sensors based on chemically reduced graphene oxide. *Nanotechnology* 25: 025502.
18. Kritzinger, P.C. 2010. The Development of Carbon Nanostructured Sensors. Master's dissertation, Stellenbosch University.
19. Yoona, H.J., Junb, D.H., Yanga, J.H., Zhouc, Z., Yangb, S.S., and Cheng, M.M.-C. 2011. Carbon dioxide gas sensor using a graphene sheet. *Sensors and Actuators B* 157: 310–313.
20. Salehi-Khojin, A., Estrada, D., Lin, K.Y. et al. 2012. Polycrystalline graphene ribbons as chemiresistors. *Advanced Materials* 24: 53–57.
21. Kumar, B., Min, K., Bashirzadeh, M. et al. 2013. The role of external defects in chemical sensing of graphene field-effect transistors. *Nano Letters* 13: 1962–1968.
22. Yuan, W., Liu, A., Huang, L. et al. 2013. High-performance NO_2 sensors based on chemically modified graphene. *Advanced Materials* 25: 766–771.
23. Prezioso, S., Perrozzi, F., Giancaterini, L. et al. 2013. Graphene oxide as a practical solution to high sensitivity gas sensing. *Journal of Physical Chemistry C* 117: 10683–10690.
24. Salehi-Khojin, A., Estrada, D., Lin, K.Y. et al. 2012. Graphene ribbons as chemiresistors. *Advanced Materials* 24: 53–57.
25. Huang, Q., Zeng, D., Li, H., and Xie, C. 2012. Room temperature formaldehyde sensors with enhanced performance, fast response and recovery based on zinc oxide quantum dots/graphene anocomposites. *Nanoscale* 4: 5651–5658.
26. Zhang, Y.H., Chen, Y.B., Zhou, K.G., Liu, C.H., Zeng, J., Zhang, H.L., and Peng, Y. 2009. Improving gas sensing properties of graphene by introducing dopants and defects: A first-principles study. *Nanotechnology* 20: 185504.
27. Chi, M. and Zhao, Y.P. 2009. Adsorption of formaldehyde molecule on the intrinsic and Al-doped graphene: A first principle study. *Computational Materials Science* 46: 1085–1090.
28. Chabot, V., Higgins, D., Yu, A. et al. 2014. A review of graphene and graphene oxide sponge: Material synthesis and applications to energy and the environment. *Energy and Environmental Science* 7: 1564–1596.

29. Nagashio, K., Yamashita, T., Nishimura, T. et al. 2011. Electrical transport properties of graphene on SiO_2 with specific surface structures. *Journal of Applied Physics* 110: 024513.

30. Dorgan, V.E., Bae, M.-H., and Pop, E. 2010. Mobility and saturation velocity in graphene on SiO_2. *Applied Physics Letters* 97: 082112.

31. Britnell, L., Gorbachev, R.V., Jalil, R. et al. 2012. Field-effect tunneling transistor based on vertical graphene heterostructures. *Science* 335: 947–950.

32. Bresnehan, M.S., Hollander, M.J., Wetherington, M. et al. 2012. Integration of hexagonal boron nitride with quasi-free-standing epitaxial graphene: Toward wafer-scale, high-performance devices. *ASN Nano* 6: 5234–5241.

33. Zhang, Y., Zhang, L., Kim, P. et al. 2012. Vapor trapping growth of single-crystalline graphene flowers: Synthesis, morphology, and electronic properties. *Nano Letters* 12: 2810–2816.

34. Bolotin, K.I., Sikes, K.J., Jiang, Z. et al. 2008. Ultrahigh electron mobility in suspended graphene. *Solid State Communications* 146: 351–355.

35. Du, X., Skachko, I., Barker, A. et al. 2008. Suspended graphene: A bridge to the Dirac point. *Nature Nanotechnology* 3: 491–495.

36. Gannett, W., Regan, W., Watanabe, K. et al. 2011. Boron nitride substrates for high mobility chemical vapor deposit graphene. *Applied Physics Letters* 98: 242105.

37. Mayorov, A.S., Gorbachev, R.V., and Morozov, S.V. 2011. Micrometer-scale ballistic transport in encapsulated graphene at room temperature. *NANO Letters* 11: 2396–2399.

38. Petrone, N., Dean, C.R., Meric, I. et al. 2012. Chemical vapor deposition-derived graphene with electrical performance of exfoliated graphene. *NANO Letters* 12: 2751–2756.

39. Kotin, I. A., Antonova, I. V., Komonov, A. I. et al. 2013. High carrier mobility in graphene on atomically flat high-resistivity layer. *Journal of Physics D* 46: 285303.

40. Dresselhaus, M.S. and Dresselhaus, G. 2002. Intercalation compounds of graphite. *Advances in Physics* 51: 1–186.

41. Lim, H.N., Huang, N.M., Chia, Ch. H. et al. 2013. Inorganic nanostructures decorated graphene. *Advanced Topics on Crystal Growth*, ed. S.O. Ferreira, InTech Available from: http://www.intechopen.com/books/advanced-topics-on-crystal-growth/inorganic-nanostructures-decorated-graphene.

42. Georgakilas, V., Otyepka, M., Bourlinos, A.B. et al. 2012. Functionalization of graphene: Covalent and non-covalent approaches, derivatives and applications. *Chemical Reviews* 112: 6156–6214.

43. Gun'ko, M.V., Turov, V.V., Whitby, R.L.D. et al. 2013. Interactions of single and multi-layer graphene oxides with water, methane, organic solvents and HCl studied by ^1H NMR. *Carbon* 57: 191–201.

44. Antonova, I.V., Kotin, I.A., Soots, R.A. et al. 2012. Tunable properties of few layer graphene—N-methylpyrrolidone hybrid structures. *Nanotechnology* 23: 315601.

45. Antonova, I.V., Kotin, I.A., Soots, R.A. et al. 2012. Novel graphene based hybrid material with tunable electronic properties. *Fullerenes, Nanotubes, and Carbon Nanostructures* 20: 543–547.

46. Luo, Z., Yu, T., Kim, K. et al. 2009. Thickness-dependent reversible hydrogenation of graphene layers. *ACS Nano* 3: 1781–1788.

47. Elias, D.C., Nair, R.R., Mohiuddin, T.M.G. et al. 2009. Control of graphene's properties by reversible hydrogenation: Evidence for graphane. *Science* 323: 610–613.

48. Burg, B.R., Schneider, J., Maurer, S. et al. 2010. Dielectrophoretic integration of single- and few-layer graphenes. *Journal of Applied Physics* 107: 034302.

49. Han, M.E., Brant, J.C., and Kim, F. 2010. Electron transport in disordered graphene nanoribbons. *Physical Review Letters* 104: 056801.

50. Hernandez, Y., Nicolosi, V., Lotya, M. et al. 2008. High-yield production of graphene by liquid-phase exfoliation of graphite. *Nature Nanotechnology* 3: 563–568.

51. Wang, L., Meric, I., Huang, P.Y. et al. 2013. One-dimensional electrical contact to a two-dimensional material. *Science* 342: 614–617.

52. Wang, M., Jang, S.K., Jang, W.J. et al. 2013. A platform for large-scale graphene electronics—CVD growth of single-layer graphene on CVD-grown hexagonal boron nitride. *Advanced Materials* 25: 2746–2752.

53. Lee, K.H., Shin H.-J., Lee, J. et al. 2012. Large-scale synthesis of high-quality hexagonal boron nitride nanosheets for large-area graphene electronics. *Nano Letters* 12: 714–718.

54. Tang, S., Wang, H., Zhang, Y. et al. 2013. Precisely aligned graphene grown on hexagonal boron nitride by catalyst free chemical vapor deposition. *Scientific Reports* 3: 2666.

55. Kotin, I.A., Antonova, I.V., Soots, R.A. et al. 2013. Laminated hydrogenated graphene-based structures with high mobility. *Nanotechnologies in Russia* 9–10: 621–626.

56. Gupta, A., Chen, G., Joshi, P. et al. 2006. Raman scattering from high-frequency phonons in supported n-graphene layer films. *Nano Letters* 6: 2667–2673.

57. Byun, I.-S., Yoon, D., Choi, J.S. et al. 2011. Nanoscale lithography on monolayer graphene using hydrogenation and oxidation. *ACS Nano* 5: 6417–6424.

58. Jaiswal, M., Haley, C., Lim, Y.X. et al. 2012. Controlled hydrogenation of graphene sheets and nanoribbons. *ACS Nano* 5: 888–896.

59. Balog, R., Jørgensen, B., Nilsson, L. et al. 2010. Bandgap opening in graphene induced by patterned hydrogen adsorption *Nature Materials* 9: 315–319.

60. Lebegue, S., Klintenberg, M., Eriksson, O. et al. 2009. Accurate electronic band gap of pure and functionalized graphane from GW calculations. *Physical Review B* 79: 245117.

61. Nebogatikova, N.A., Antonova, I.V., Volodin, V.A. et al. 2013. Functionalization of graphene and few-layer graphene with aqueous solution of hydrofluoric acid. *Physica E* 52: 106–111.

62. Nebogatikova, N.A., Antonova, I.V., Volodin, V.A. et al. 2013. Graphene and few-layer graphene functionalization in the aqueous solution of hydrofluoric acid. *Nanotechnologies in Russia* 9: 51–59.

63. Cheng, S.-H., Zou, K., Okino, F. et al. 2010. Reversible fluorination of graphene: Towards a two-dimensional wide bandgap semiconductor. *Physical Review B* 81: 205435.

64. Nair, R.R., Ren, W., Jalil, R. et al. 2010. Fluorographene: A two-dimensional counterpart of Teflon. *Small* 6: 2877–2884.

65. Antonova, I.V., Nebogatikova, N.A., Prinz, V.Ya. 2014. Self-organized arrays of graphene and few-layer graphene quantum dots in fluorographene matrix: Formation of quantum dots and charge spectroscopy. *Applied Physics Letters* 104: 193108.

66. Ribas, M.A., Singh, A.K., Sorokin, P.B. et al. 2011. Patterning nanoroads and quantum dots on fluorinated graphene. *Nano Research* 4: 143–152.

67. Taziev, R.M. and Prinz V.Ya. 2011 Buckling of a single-layered graphene sheet on an initially strained InGaAs thin plate. *Nanotechnology* 22: 305705.

68. Wang, S., Pu, J., Chan, D.S.H. et al. 2010 Wide memory window in graphene oxide charge storage nodes. *Applied Physics Letters* 96: 143109.

69. Hong, A.J., Song, E.B., Yu, H.S. et al. 2011. Graphene flash memory. *ACS Nano* 5: 7812–7817.

70. International Technology Roadmap for Semiconductors. ITRS reports. 2009–2010 www.itrs.net/reports.html.

71. Zhu, Y., Sun, Z., Yan, Z. et al. 2011. Rational design of hybrid graphene films for high-performance transparent electrodes. *ACS Nano* 5: 6472–6479.

72. Bae, S., Kim, H., Lee, Y. et al. 2010. Roll-to-roll production of 30-inch graphene films for transparent electrodes. *Nature Nanotechnology* 5: 574–578.

73. Kim, K.K., Reina, A., Shi, Y. et al. 2010. Enhancing the conductivity of transparent graphene films via doping. *Nanotechnology* 21: 285205.

74. Burzuri, E., Prins, F., and van der Zant, H.S.J. 2012. Characterization of nanometer-spaced few-layer graphene electrodes. *Graphene* 1: 26–29.

75. Prins, F., Barreiro, A., Ruitenberg, J.W. et al. 2011. Room-temperature gating of molecular junctions using few-layer graphene nanogap electrodes. *Nano Letters* 11: 4607–4611.

76. Matsumoto, K., Maehashi, K., Ohno, Y. et al. 2014. Recent advances in functional graphene biosensors. *Journal of Physics D (Applied Physics)* 47: 094005.

77. Cheng, C., Uhe, J., and Yang, X. 2012. Multilayered graphene membrane as an experimental platform to probe nano-confined electrosorption *Progress in Natural Science: Materials International* 22: 668–672.

31 Designing Carbon-Based Thin Films from Graphene-Like Nanostructures

Cecilia Goyenola and Gueorgui K. Gueorguiev

CONTENTS

Abstract..497
31.1 Introduction 497
31.2 Synthetic Growth Concept..499
31.3 Modeling Carbon-Based Thin Films..501
    31.3.1 Carbon Nitride, CN_x: The First Compound in the FL Class 501
    31.3.2 Phospho Carbide, CP_x: The Realized Idea..503
    31.3.3 Sulfo Carbide, CS_x: The Prediction 506
    31.3.4 Carbon Fluoride, CF_x: The Structural Diversity...509
31.4 Fullerene-Like Carbon-Based Thin Films Class in a Nutshell 511
31.5 Concluding Words about the SGC...512
Acknowledgments...512
References..512

ABSTRACT

This work intends to provide an overview of our experience in predicting and guiding the synthesis of carbon-based thin films with graphene-like and fullerene-like (FL) features, and how this is made possible by studying the effects of dopant elements in graphene-like nanostructures.

In the same way as graphite can be modeled and understood as a stack of graphene layers, some nanostructured carbon-based thin films, such as FL thin films, can be modeled as assemblies of doped graphene-like low-dimensional units.

When atoms of an element different from carbon, such as fluorine or sulfur, substitute carbon atoms in a graphene-like network, important bonding and structural changes occur. Defects, such as pentagons or heptagons, become energetically feasible, inducing curvatures in an otherwise planar network, intersheet cross-linkages, disruptions, chains, and so on.

With the aim of predicting the structural patterns and their impact on the properties of such compounds, computational methods can provide invaluable insight, allowing models of graphene-like nanostructures and derived thin films to be studied. In this context, we have developed an original theoretical approach, the synthetic growth concept based on the density functional theory. It has been a powerful simulation tool, which helped to define a whole new class of nanostructured compounds: carbon-based thin films with FL and graphene-like structural features.

31.1 INTRODUCTION

Thin films can be defined as layers of materials with thicknesses that can range from a few nanometers to several micrometers and nowadays they have a crucial technological role in many industries. Thin films are formed over surfaces by depositing layers of atoms and/or molecules with the objective of improving the properties of the underlying material or giving them new functionality [1]. A large variety of thin films have been discovered and widely applied. To cite some examples, titanium nitride and titanium nitride-based thin films have been widely used as protective hard coatings for cutting tools [2,3], uniform semiconductor thin films (e.g., $SnSe_2$, SnS_2) are used as active components in thin-film transistors [4], and thin films such as amorphous silicon have been employed as the light-absorbing material in solar cells [5].

The focus of this chapter is set to the structural features, properties, and design of carbon-based thin films and, in particular, of FL carbon-based thin films. Table 31.1 summarizes previous studies on carbon-based thin films as related to graphene-like nanostructures. Details and aspects of these studies and corresponding references will be further discussed in this chapter.

Carbon-based thin films are a large family of materials that exhibit a wide variety of structural features and properties, ranging from those of graphite to those of diamond. They have a vast range of applications, from lubrication to ultradense hard coatings (e.g., protective coatings for cutting tools), electronic device applications (e.g., insulators in metal–insulator semiconductors), and medical applications (e.g., coatings for implants), among others [6,7].

The microstructure of many of the carbon-based thin films is mostly characterized by short-range atomic order (~10 Å) [8] and implies coexistence of carbon atoms with different hybridization [6,9]. In order to control their microstructure, carbon-based thin films are frequently deposited by vapor-phased deposition techniques such as chemical vapor

TABLE 31.1

Summary of Previous Studies on Carbon-Based Thin Films and Their Relation to Graphene-Like Structures

Carbon-Based Thin Films			
	Dopant	Type of Film	References
Pure C films		a-C (DLC, ta-C)	[6–10]
		FL	[39,42]
Doped C films	H	a-C	[8–11]
	N	a-C	[9,12,14,15]
		FL	[13,38,40,41,43–46,67]
	P	a-C	[16,19]
		FL	[17,18,20,43,46]
	Si	a-C	[21,22]
	S	a-C	[23,24,26]
		FL	[25,27]
	F	a-C	[28–30,32]
		FL	[31,88]

deposition or physical vapor deposition with a critical control of the growth environment [1,9].

The properties of carbon-based thin films can also be tailored by the addition of hydrogen [8–11] or small amounts of different p-elements such as nitrogen [9,12–15], phosphorus [16–20], silicon [21,22], sulfur [23–27], and fluorine [28–32], expanding the range of possible properties and applications. In addition, amorphous carbon and transition metal nanocomposites, such as titanium carbide (TiC) have also gained importance due to their high hardness, low friction coefficient, and high wear resistance [33,34].

Before focusing on the design of carbon-based thin films, it is worth to remember the bonding features of carbon atoms. Carbon can form a large variety of structures, from crystalline to amorphous, due to its ability to exist in different hybridization states. Figure 31.1 illustrates the sp^3-, sp^2-, and sp-hybridization configurations of carbon [35]. A carbon atom with sp^3 hybridization has its four valence electrons distributed in four hybridized orbitals arranged tetrahedrally (Figure 31.1a). Each of these orbitals can form strong directional σ bonds. This type of bonding is the reason for diamond being the material with largest known bulk modulus, since it comprises a tetrahedral network of sp^3-hybridized

carbon atoms [36]. The sp^2 configuration has three of the four valence electrons distributed in three hybridized orbitals that are trigonally arranged forming a plane. The fourth electron is located in a p orbital perpendicular to this plane (Figure 31.1b). While the three hybridized orbitals can form strong σ bonds, the p orbitals can participate in weaker π bonds with other p orbitals of neighboring atoms. Graphite is entirely composed of sp^2-hybridized carbon atoms forming layers. Within each layer, the atoms form strong σ bonds giving a high in-plane strength, while the bonding between layers is given by weak van der Waals interactions, allowing them to slide over each other if stress is applied [37]. Finally, in the sp configuration, two of the valence electrons are located in two hybridized orbitals that can form σ bonds along an axis (x axis), and the other two electrons lay in two p orbitals in the y and z directions that can form π bonds (Figure 31.1c).

Carbon-based thin films are mainly composed of sp^3- and sp^2- hybridized carbon atoms, while sp hybridized atoms are not expected to play any significant role [8]. Thus, the relative content of sp^3- to sp^2- hybridized carbon atoms provides the diversity in terms of structure and properties of these films, and different categories of carbon-based thin films can be identified.

Amorphous carbon (a-C) thin films are a type of carbon-based thin films with higher content of sp^3 bonds than sp^2. Within the a-C group, diamond-like carbon (DLC) and tetrahedral amorphous carbon (ta-C) are the most widely studied and also those with most applications. The classification of different a-C films derives not only from the relative content of sp^3 and sp^2, but also from their hydrogen content. For example, ta-C have the highest sp^3 content, which can be as high as 90 at.%, and the lowest hydrogen content. For comprehensive reviews discussing the a-C thin films see References 6, 8, and 9.

The carbon-based thin films category most relevant to this chapter is the FL carbon-based thin films. In contrast to a-C films, FL carbon-based thin films possess higher content of sp^2 bonding than sp^3. These films can be described as an array of parallel curved graphene fragments that are packed in different orientations [38,39]. Figure 31.2 is a cross-sectional high-resolution transmission electron microscopy (HR-TEM) image of a carbon nitride ($CN_{0.12}$) thin film with FL characteristics. It can be seen that the parallel bent graphene basal planes also intersect and cross-link within the film [40].

It has been determined that most of the carbon in the FL carbon-based thin films is incorporated into quasi-planar sp^2-hybridized graphene-like sheets. As in graphene, the directional σ sp^2 bonds provide an exceptional in-plane strength [41]. However, the fact that the graphene fragments are curved and occasionally cross-linked, extends the extraordinary strength of the planar sp^2 network in three dimensions. The reason for bending of the graphene planes is attributed to the incorporation of odd-membered ring defects, such as pentagons, in an hexagonal network [38]. At the same time, some of the incorporated carbon atoms possess sp^3 hybridization, promoting cross-linking between the graphene-like planes. In fact, the cross-linkages prevent the graphene-like sheets

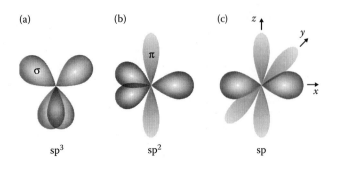

FIGURE 31.1 The sp^3-, sp^2-, and sp-hybridization configurations found in carbon.

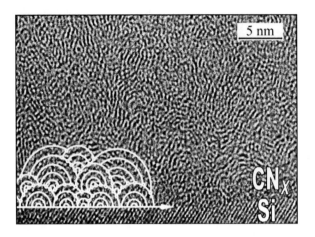

FIGURE 31.2 Cross-sectional HR-TEM micrograph from the interface region of a $CN_{0.12}$ film deposited on Si. Superimposed is the model proposed for the film formation [13]. (Reprinted with permission from L. Hultman et al. *MRS Bulletin*, 28(03):194–202, 2003.)

from gliding with respect to each other. The incorporation of defects and cross-linkage sites is promoted by the presence of a dopant element, illustrated in Figure 31.2, where nitrogen is used as a dopant to obtain a FL carbon-based thin film [13,38,40,41].

FL carbon-based thin films have not only been synthesized as carbon nitride (CN_x), but also as pure carbon [39,42] and phospho carbide (CP_x) [18] thin films. Depending on the bonding properties of the dopant, its concentration, and the growth technique, these materials exhibit diversity in physical properties: high hardness, resilience and high wear resistance [13,43], and electrical properties of interest such as increased conductivity [43]. Since FL carbon-based thin films are mainly composed of carbon, but also have small to average amounts (0–40 at.%) of a dopant element A, one can generally refer to them as FL-CA_x.

In order to obtain a better understanding of the relationship between the microstructure of a C-based thin film and its properties, as well as to help in the interpretation of experimental observations and be able to predict new FL-CA_x compounds, computer simulations of their structure arise as an essential tool. However, the application of standard methodology such as density functional theory (DFT) or molecular dynamics to noncrystalline materials or materials that lack long-range order remains a challenge. Among the usual problems is the difficulty to design appropriate models for a material system that describes it realistically and, at the same time, is not prohibitive from the point of view of consumption of computational resources.

One way to approach FL-CA_x thin films through computer simulations is to describe them as assemblies of doped graphene-like low-dimensional units. By exploiting this particular characteristic of FL carbon-based thin films, we have introduced and developed the synthetic growth concept (SGC) [44–46]. The SGC has since been established as an original approach based on DFT for modeling thin solid films formation during vapor phase deposition.

This chapter is organized in the following way: first the SGC is discussed in detail as a computational approach to the design of carbon-based thin films followed by examples of its application. The chapter finishes with a short review of the FL-CA_x thin films as a class of materials. Even though the chapter is focused on the SGC and its applications to carbon-based thin films, it is of great importance to also discuss the results on the extensive experience in carbon-based thin films synthesis and characterization achieved by our colleagues belonging to the research team at the Thin Film Physics Division at Linköping University. This experimental success is a validation for the SGC.

31.2 SYNTHETIC GROWTH CONCEPT

FL-CA_x thin films can be described as assemblies of graphene-like pieces that are arranged parallel but also cross-link and intersect. In this context, graphene-like pieces are understood as C-based two-dimensional networks based on hexagonal rings, that is, the honeycomb structure. These networks may include ring defects, such as pentagons or heptagons, directly due to incorporation of atoms of the dopant elements. Such a graphene-like nanostructure is typically obtained by vapor phase deposition techniques in which precursor species (atoms, molecules, and radicals) are deposited onto a substrate to form the desired film.

The SGC [44–46] postulates that in order to unravel the relationship between structure, properties, and synthesis, it is important to comprehend the defect formation in graphene-like model systems and their interaction with the precursor species used for carbon-based thin films deposition. The DFT methodology is adequate for both these tasks.

DFT is an accurate and efficient electronic structure method widely used in material science, physics, and chemistry. It provides a reliable description of stable geometries for condensed phases of a variety of compounds together with their properties. Thus, DFT is frequently a method of choice for studying defect formation and to account for the role of precursor species during film synthesis.

Now we focus on the SGC simulation of FL-CA_x films. The first step is to design appropriate graphene-like model systems, which realistically describe the FL-CA_x films. In Figure 31.3, such graphene-like model systems are displayed. They may be chosen in different sizes and containing defects such as pentagons, heptagons, or combinations of them. The goal is to understand how the existence and the location of these defects influences the structure and the stability of pure carbon and dopant-containing graphene-like model systems. For example, the incorporation of a pentagon defect in an inherently hexagonal network induces local curvature, an effect inherent to FL-CA_x. At the same time, the effect of the dopant element A and its concentration can be simulated by the substitution of some of the C atoms in the model systems by the adequate quantity of A atoms.

The main SGC stage is to reliably simulate the formation of the film structure during deposition. In simple terms, the vapor phase deposition of CA_x films occur in a chamber

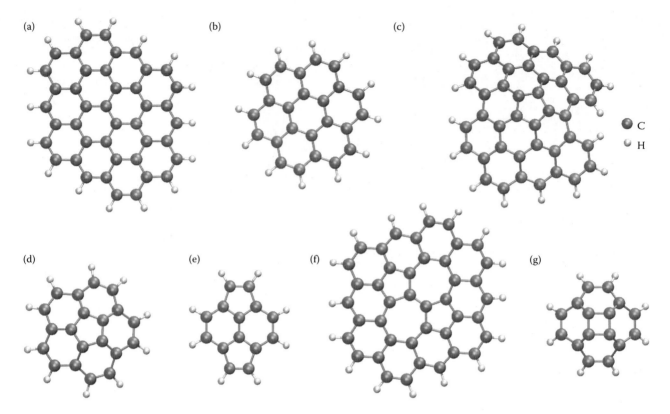

FIGURE 31.3 Graphene-like model systems adopted for the study of the influence of nonhexagonal ring defect incorporation in a graphene-like network, as well as the effect of dopant: (a,b) pure hexagonal model systems ($C_{42}H_{16}$ and $C_{24}H_{12}$, respectively); (c,d) model systems containing a single pentagon defect ($C_{42}H_{16}$ and $C_{20}H_{10}$, respectively); (e) model system containing a double pentagon defect ($C_{14}H_8$); (f) model system containing a Stone–Wales defect (combination of two pentagons and two heptagons, $C_{42}H_{16}$); (g) model system containing a tetragon defect ($C_{16}H_8$).

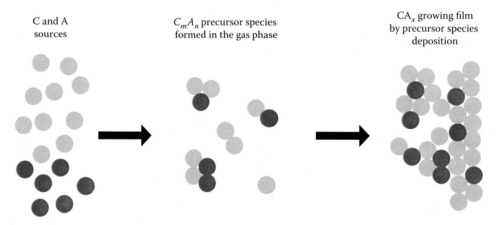

FIGURE 31.4 Sketch of vapor phase deposition of a CA_x film. Atoms are represented by filled circles, and different colors refer to different types of atoms C and A.

where C and A sources are introduced. Depending on the specific process, the sources can be solid, liquid, or in gas phase. During the deposition, C_mA_n precursor species are formed. These precursor species are consecutively deposited onto a substrate, becoming the "building blocks" of the film. Figure 31.4 illustrates such process. Predicting which precursor species, C_mA_n, are relevant to the deposition process, together with their structures and stability, followed by the study of the interaction between them and with the growing film is of

paramount importance if the structural evolution of the film is to be understood.

In particular, our colleagues at the Thin Films Physics Division at Linköping University have extensively used magnetron sputtering as the deposition method of choice for the FL-CA_x thin films. Magnetron sputtering provides small C_mA_n ($m \leq 4$ and $n \leq 4$) precursor species and also single atoms (C, A) that can be directly accounted for in the SGC simulations. The availability of small precursor species during deposition is especially important to avoid dopant atom A segregation. In addition, the deposition process, when magnetron sputtering is adopted, can occur at a relatively low substrate temperature, reducing the probability for contamination and segregation of the dopant atoms [43]. It is noteworthy that defect and dopant incorporation in graphene-like sheets occurs at an energy cost and this energy is provided by the deposition method.

Within the SGC, usually a diversity of precursor species C_mA_n are simulated: different compositions ($m \leq 4$ and $n \leq 4$) and different chemically viable isomers for each composition have to be studied. Since the number of such precursor species can become large it is useful, when selecting the viable precursors for a given compound, to account for: (i) known molecules and radicals considered in the organic chemistry books [35,47,48] and (ii) experimental data about precursor species availability in a given deposition process. By such an approach, the number of possible precursor candidates can be significantly reduced.

Once the selection of precursor species C_mA_n has been defined, the simulation of growth evolution can begin. By sequentially adding precursor species to previously relaxed model systems, the interaction between precursors and the growing film can be elucidated.

The starting point of such a simulation is a relaxed structure of one of the graphene-like nanotemplates optimized previously (Figure 31.3) with some of the C atoms substituted by atoms of the dopant element, A. A precursor species, C_mA_n, is chosen randomly from the previously established selection and it is placed in the vicinity of the nanotemplate. The ensemble is submitted to relaxation in order to obtain a new energetically and structurally stable system. This procedure is continuously repeated, resulting in an iterative simulation process where a sequence of precursor species is attached to the growing model system. The iterative process is illustrated in Figure 31.5. It is of paramount importance for the SGC that different starting points and a variety of sequences of precursor species are considered, in order to reduce the dependence of the results on the starting geometries and structural fluctuations.

To summarize, the SGC is understood as a structural evolution by sequential steps of atomic rearrangement where each step is assigned according to the previous relaxed states. The precursor and/or building blocks for nanostructured compounds are described quantitatively together with their packing rules when they form condensed phases. By SGC we also address the interplay between bonding at surfaces/interfaces and the properties of the compounds.

It is important to state that the applicability of the SGC is not limited to FL-CA$_x$ thin films. The concept is transferable and can be extended to the study of a large variety of inherently nanostructured materials. The SGC was successfully applied to explore the family of silicon transition metal cage-like molecules (MSi_{12} [49,50]) as building blocks for cluster-assembled materials, such as solids [51] and nanowires [52]. Recently, the fluorination of the corannulene molecule was approached by means of the SGC. This is a molecular system that provides structural, energetic, and electronic information on curved two-dimensional C-based compounds and shows how properties can be tuned by controlling the degree of fluorination [53]. Nitrogen-doping of carbon chains connecting graphene-like terminations were also studied along the line of the SGC; this work featuring emphasis on electronic properties [54]. Aspects of the SGC have also been instrumental in addressing the properties of low-dimensional III-nitrides [55] as well as in understanding features of the gas phase chemistry during growth of III-nitrides [56–58].

31.3 MODELING CARBON-BASED THIN FILMS

31.3.1 CARBON NITRIDE, CN$_x$: THE FIRST COMPOUND IN THE FL CLASS

When the super-hard crystalline phase of β-C_3N_4 was predicted by Liu and Cohen using computational methods [59,60], many research teams embarked on a quest of synthesizing it. A large number of different vapor phase deposition methods and techniques have been employed in this pursuit

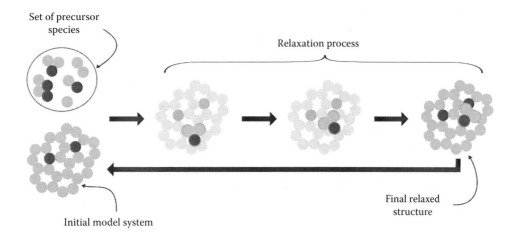

Set of precursor species

Relaxation process

Initial model system

Final relaxed structure

FIGURE 31.5 Iterative process for growth evolution simulation of FL-CA$_x$ thin films. In each cycle, a precursor species randomly selected from the previously established set is attached to the initial model system and submitted to relaxation. The final relaxed structure becomes the initial model system for the next cycle.

[12,61], but up to date, there have been no convincing reports indicating the success in its synthesis. In general, the films obtained show amorphous or short-range order CN_x structures with a N deficiency with respect to β-C_3N_4.

Trying to synthesize β-C_3N_4, Sjöström et al. at Linköping University used reactive magnetron sputtering [38]. Even though they could not achieve this compound, these experiments were, in a sense, successful since they led to the discovery of FL-CN_x thin films ($0 < x < 0.3$). These FL-CN_x films proved to have valuable mechanical properties, exhibiting an unusual combination of high hardness and high elastic recovery [38,62]. Their high elasticity and their low tendency to plastic deformation point toward a resilient nature that can be described as "superhard rubber" [13].

The mechanical properties found for these films are directly related to their microstructure. Figure 31.2 illustrates the typical microstructure of FL-CN_x films, that can be described as having the shape of a finger print. Defects due to N-incorporation lead to curved graphene-like sheets in contrast to the flat sp^2-hybridized graphite planes. The curvature of the graphene-like sheets occurs in the three dimensions, creating buckled "graphite" domains that intersect each other at their boundaries. In extreme cases, these domains may become well-defined nano-onions as shown in Figure 31.6 [13,38,63–65].

X-ray photoelectron spectroscopy demonstrated the presence of N atoms bonded to C atoms in a sp^3 configuration, raising the possibility that the graphene-like basal planes are interlocked with covalent bonds of a much shorter bond length than the van der Waals intersheet distance in graphite [38,64].

The increased elasticity attributed to FL-CN_x films originates in the buckling and bending of their basal planes, which gives the possibility of the action of restoring forces upon deformation [13,62,64]. In addition, strong C–C cross-linking prevents

the graphene-like sheets from slipping between each other and gives rise to a low tendency to plastic deformation [13].

During the last two decades, FL-CN_x films have been largely studied and, in particular, at Linköping University it resulted in a large number of research papers ([13,38,40,44,45,62,65–67] and many more) and doctoral theses [41,43,68]. Therefore, a much better understanding of the films' microstructure, their mechanical properties, and the deposition processes has been achieved [41,63,65,69]. It is noteworthy that owing to the collaborative effort between diverse characterization techniques [13,38,66] and computer simulations, which reached maturity with the SGC [44,45], a reliable structural model for the FL-CN_x compound was elaborated. In fact, this structural model corroborates the first structural guess for FL-CN_x, as proposed by Sjöström et al. [38].

Following a historical perspective, the first calculations performed on FL-CN_x were carried out using the semi-empirical Hartree–Fock-based AM1 method [70]. Two graphene-like model systems were employed (Figure 31.7), one of them containing a single pentagon defect and the other one being a pure hexagonal network. The goal was, by comparing the stability of these model systems in the cases of pure C and when N was incorporated at C sites, to evaluate the energy cost for pentagon defects in a graphene-like network [38].

The results showed that pure carbon systems favors the planar graphene structure instead of the inherently curved defect-containing one. But, when N atoms are incorporated, the energy barrier to form pentagons in graphene planes is reduced, in particular, when a N atom participates in the pentagon defect. It was also confirmed that the N atoms prefer a nonplanar surrounding, as is the case of the curved structure around a nonhexagonal ring defect [38]. These results have been found to be consistent with the microstructure observed in the films by HR-TEM [38,66] and were corroborated by DFT calculations for a larger number of model systems as those shown in Figure 31.3 [45,67].

Ring defects other than pentagons, such as heptagons and tetragons, also induce curvature in an hexagonal network, so the possibility for their role in FL-CN_x films needed to be considered. Stafström et al. showed that contrary to the fact that N lowers the cost for pentagon formation, N increases the energy

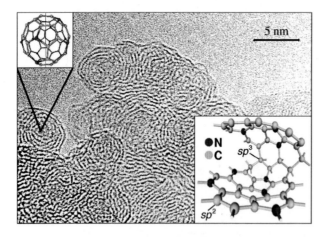

FIGURE 31.6 HR-TEM image of a FL $CN_{0.12}$ film [40]. The upper left inset shows the structure and size of a predicted $C_{48}N_{12}$ [91]. The lower right inset is a schematic illustration of the buckling of an sp^2 coordinated basal plane, caused by the incorporation of pentagons, which facilitates cross-linking through sp^3-hybridized carbon (structure proposed by Sjöström et al. [38]). (Reprinted with permission from L. Hultman et al. Sundgren. *MRS Bulletin*, 28(03):194–202, 2003.)

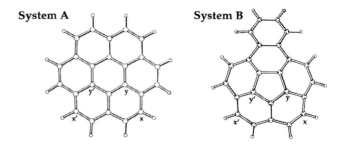

FIGURE 31.7 First two model systems used for calculation of total energies of pentagon containing systems versus pure hexagonal systems [38]. (Reprinted with permission from H. Sjöström. et al. Superhard and elastic carbon nitride thin films having fullerene like microstructure. *Physical Review Letters*, 75, 1336–1339. Copyright 1995 by the American Physical Society.)

cost for heptagon formation [67]. This indicates that the role of N in CN_x systems is to lower the energy formation of pentagons, while heptagon defects were predicted to be less likely.

Already applying the SGC to the FL-CN_x compound, Gueorguiev et al. considered the stability of different defects and combinations of defects for different N concentrations [45]. In the SGC, the stability of the model systems is evaluated by cohesive energy per atom ($E_{coh/at}$). $E_{coh/at}$ is defined as the energy required to break the system into isolated atoms normalized by the total number of C and A atoms and is calculated as

$$E_{coh/at} = \left| \frac{E_{system} - n_H \times E_H - n_C \times E_C - n_A \times E_A}{n_C + n_A} \right|, \quad (31.1)$$

where E_{system} is the total energy of the model system, n_H, n_C, and n_A are the number of H, C, and A atoms, and E_H, E_C, and E_A are the energies of the corresponding free atoms in the ground state, respectively. In the case of FL-CN_x, A ≡ N.

In Figure 31.8, the relative stability of different defects is shown as a function of the N concentration [45]. These results indicate that even though N reduces the energy barrier for pentagon formation [38,67], the pure hexagonal system is energetically more favorable for N concentrations up to ~20 at.%. Increasing the N content beyond this value makes the pentagon-containing networks more stable than a hexagonal sheet [45]. Regarding alternative defects and combinations of them, it was found that the double pentagon defect becomes more favorable than the single pentagon defect at a N concentration of 17.5 at.%, suggesting that the buckling of graphene layers in FL-CN_x may result from the combination of closely packed pentagon defects instead of single pentagons. This is especially valid for larger N concentrations. These results are in agreement with the observations that the degree of curvature in CN_x films depends on their N content as well as on the deposition conditions [13,62,64,65].

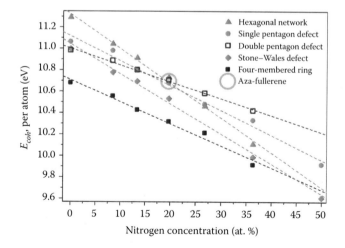

FIGURE 31.8 Cohesive energies per atom for CN_x model systems as a function of the N concentration [45]. The values were obtained using the PW91 exchange and correlation potential [90]. (Reprinted from *Chemical Physics Letters*, 410, G. K. Gueorguiev. et al. First-principles calculations on the curvature evolution and crosslinkage in carbon nitride. 228–334, Copyright 2005, with permission from Elsevier.)

Model systems containing Stone–Wales defects (a combination of a pentagon and a heptagon) are less stable than the pure hexagonal and the pentagon-containing ones, but they still may coexist with pentagon defects. On the other hand, tetragon defects have shown to be quite energetically costly over the whole range of experimentally viable nitrogen concentrations and, therefore, are not likely to appear in FL-CN_x films.

Another important aspect of the incorporation of a N atom in a C ring is that it increases the reactivity of the surrounding C atoms, promoting a change in the hybridization from sp^2 to sp^3, leading to three-dimensional cross-linkage [38,67].

The knowledge about the stability of CN_x graphene-like model systems helps the simulation of FL-CN_x film formation. SGC simulation efforts [45] combined with plasma characterization during FL-CN_x magnetron sputtering deposition [63,66] indicated that C_mN_n (m, $n \leq 2$) species emitted from the sputtered target, as well as N and C atoms, are the main films forming species by acting as building blocks during the film growth.

The exact role of the C_mN_n precursor species was revealed by SGC simulation of FL-CN_x film evolution by sequential addition of precursor species C_mN_n (m, $n \leq 2$) to CN_x graphene-like model systems (Section 31.3) [44,45]. The results showed that the incorporation of the CN dimer at the film edge during its formation is one of the most viable mechanisms for introducing pentagon defects to the graphene-like sheet bundles. Such pentagon defects, once incorporated, exhibit remarkable stability in a curved environment and are likely to prevail during the subsequent growth process [44].

Also through the SGC, the role of N incorporation as a promoter of the evolution from planar layers to curved and cross-linked graphene multilayers was understood [45]. It was observed that N incorporation at the edge of a model system produces a bond rotation that contribute to three-dimensional cross-linking. Figure 31.9 shows an evolution path represented by a chain of growth events. It is made evident how a bond rotation, originating from the presence of a N atom in the model system, can evolve to a cross-linkage site.

In agreement with experimental observations, the application of the SGC to FL-CN_x added the knowledge and understanding summarized in the following points:

- N atoms reduce the energy barrier for pentagon formation in graphene planes, producing the necessary curvature to obtain a FL-CN_x structure.
- The incorporation of N in a C ring promotes the change in hybridization (from sp^2 to sp^3) of adjacent C atoms, promoting three-dimensional cross-linking.
- N atoms can also be part of a cross-linkage by means of bond rotation.

31.3.2 PHOSPHO CARBIDE, CP_x: THE REALIZED IDEA

The experimental and theoretical studies on FL-CN_x thin films showed that their remarkable mechanical properties are a consequence of the incorporation of N atoms in the graphene

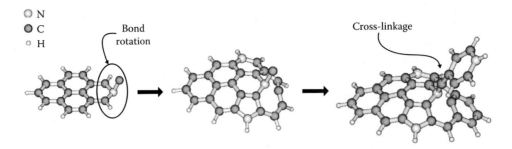

FIGURE 31.9 Chain of events for synthetic growth of CN$_x$. A bond rotation at the end of the model system evolves to a cross-linkage site [45]. Results obtained using PW91 exchange and correlation potentials [90].

planes and its peculiar structural implications. A significant research effort was invested in optimizing and controlling the mechanical properties of FL-CN$_x$ films by using different deposition techniques and adjusting the deposition parameters [41,63,64,68]. Along this line of thinking, changing the dopant element may be an efficient option in the path toward a FL material with even better mechanical properties than FL-CN$_x$.

In this context, phosphorus was the new element chosen by Furlan et al. to achieve a new member of the class of FL compounds: FL-phospho carbide (FL-CP$_x$) [43]. P follows N in group 15 of the periodic table, which is a hint of the chemical similarities between them, like their distribution of valence electrons and their low polarizabilities. On the other hand, P has lower electronegativity than N and C, and it possess a larger covalent radius. P also exhibits trends of hybridization configurations that include its d orbitals. These similarities, but also the subtle differences, between P, N, and C gave promise for a new FL material with enhanced mechanical properties compared to CN$_x$. Due to its low electronegativity and its tendency toward tetrahedral coordination, P brings new bonding characteristics, and an expectancy to incorporate a larger density of cross-linkages between the graphene sheets. In addition, the larger atomic radius of the P atoms was expected to promote a larger deformation of the graphene planes, with a tendency to enhanced local curvature and increased density of cross-linking sites.

All these features of P as a dopant motivated focused simulations within the SGC to evaluate the feasibility of synthesising FL-CP$_x$. The study comprised the stability as well as the energy cost for defects in CP$_x$ graphene-like pieces, the possibility for cross-linking, the precursor species expected to play a role during the magnetron sputtering deposition process, and the growth evolution of CP$_x$ systems [17,43,71].

The cost for P substitution and the stability of defects in finite graphene-like model systems (similar to those shown in Figure 31.3) were studied by geometry relaxations and cohesive energy calculations (Equation 31.1, A ≡ P). It was found that the incorporation of P atoms in graphene-like networks is more costly than the N case, due to the larger covalent radius and the lower electronegativity of P with respect to N ($E_{coh/at,Hex^N} = 6.53$ eV/at. $\geq E_{coh/at,Hex^P} = 6.30$ eV/at.). P incorporation results in enhanced reactivity in the atomic region close to the incorporation site of the dopant atom and also

considerably enhanced curvature in comparison to similar configurations in FL-CN$_x$ [71].

Regarding the defects' stability, the double pentagon defect was found to be the most stable one, showing energetic advantages with respect to single pentagon defects, as is also the case for CN$_x$ systems. The Stone–Wales and the tetragon defects exhibit lower stability. However, it is remarkable that in the CP$_x$ graphene-like networks, tetragons become energetically feasible with a stability close to pentagon and Stone–Wales defects. This is due to the ability of P to extend its d orbitals and promote the formation of four-membered rings [72]. The feasibility of the tetragonal defect in CP$_x$ points toward films with strongly curved and much shorter graphene-like pieces as well as higher density of cross-linking than in FL-CN$_x$ films [71].

The stability of two-dimensional CP$_x$ graphene-like model systems was also evaluated in terms of P content. It was shown that CP$_x$ thin films can accommodate up to 25 at.% of P by more regular distribution of the P atoms and the corresponding defect formation within the network, but they become increasingly unstable with a $E_{coh/at}$ lower than 1 eV/at. Regarding the geometry, CP$_x$ systems with P content higher than 20 at.% are characterized by disrupted and intersected conformations, tending to amorphousness with P segregation. Well-structured CP$_x$ thin films with elements of FL structure have been predicted by the SGC for P contents up to 10 at.% with a smooth transition to amorphous structures as the P content increases [17].

Following the SGC line of simulations, CP$_x$ film formation characteristics were also studied. A variety of precursor species with compositions C$_n$P$_m$ ($1 \leq n, m \leq 3$) as well as pure P species, P$_n$, was considered. Figure 31.10 shows the most stable ones and their $E_{coh/at}$ (calculated with Equation 31.1, A ≡ P) are

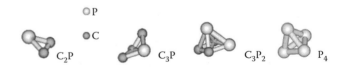

FIGURE 31.10 Most stable C$_n$P$_m$ precursor species [71]. The relaxed structures were obtained using the B3LYP exchange and correlation potential [89]. (Reprinted from *Thin Solid Films*, 515, A. Furlan et al. Fullerene-like CPx: A first-principles study of the relative stability of precursors and defect energetics during synthetic growth. 1028–1032, Copyright 2006, with permission from Elsevier.)

TABLE 31.2

$E_{coh/at}$ for the Most Stable CP_x Precursor Species Shown in Figure 31.10

Species	C_2P	C_3P	C_3P_2	P_4
$E_{coh/at}$ (eV/at.)	5.50	5.85	5.77	4.73

Source: A. Furlan et al. *Thin Solid Films*, 515(3):1028–1032, 2006.

Note: The results were obtained using the B3LYP exchange and correlation functional [89].

listed in Table 31.2. The higher stability of C_2P, C_3P, and C_3P_2 compared to the rest of the C_mP_n precursors was attributed to the fact that they have larger cohesive energies, smaller number of dangling bonds per atom, and smaller number of P–P bonds, which are energetically less favored than C–P bonds. These precursors may be formed in the deposition chamber during magnetron sputtering as a direct consequence of the sputtering process of a target consisting of a special mix of graphite and red phosphorus, as well as due to recombination processes in the deposition environment [43,71].

The precursors derived from SGC simulations together with the single atoms, C and P, and the dimer species, C_2 and P_2, were considered in the exploration of CP_x growth evolution. One of the most important results of the growth evolution simulation was the prevalence of tetragon defects when different precursor sequences are attached to the system. Furthermore, it was found that the tetragon defects behave as

nucleation centers for characteristic conformations, such as cage-like and onion-like conformations. Figure 31.11 illustrates an example of growth evolution for a CP_x system. In this example, a cage-like conformation forms around a tetragon defect after the precursor species C_3P and C_2 were added. By successive addition of large sequences of diversified precursor species, the model system evolves into an onion-like conformation typically observed in FL-CA$_x$ films [17]. Thus, due to particular dopants (in this case P), a pure graphene-like system can evolve into a much different, but to a large extent still ordered, nanostructured network.

Another new feature arising from CP_x's structural evolution is the fact that P atoms promote not only cross-linking between graphene-like sheets, but interlinking as well. Figure 31.12 shows the difference between these two structural concepts. Two graphene-like sheets are cross-linked when only one bond is formed between atomic sites belonging to the sheets (see Figure 31.12a). Of course, there may be several cross-link sites between two particular graphene-like sheets or finite model systems representing them. However, interlinking refers to the connection between two graphene layers by more than one bond that originate from the same site in one of them (see Figure 31.12b). It should be mentioned that interlinkage may lead to branched and densely intersected graphene sheets. Obviously, interlinkages favor amorphousness, and a high density of them leads inevitably to a loss of any ordered structure, that is, it tends to an amorphous compound [17].

SGC simulations of CP_x solid compounds resulting from magnetron sputtering showed that the incorporation of P

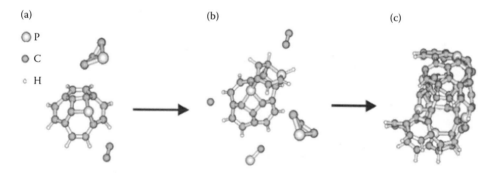

FIGURE 31.11 Growth evolution simulation path for CP_x: (a) model system containing a tetragon defect and the precursor species C_3P and C_2 added in this sequence to obtain a cage-like conformation; (b) resulting cage-like conformation formed around a tetragon defect and the precursor species C, C_2, CP, and C_3P; and (c) onion-like conformation resulting from the continuous evolution of the cage-like system after adding several sequences of precursor species [17]. (Reprinted from *Chemical Physics Letters*, 426, G. K. Gueorguiev et al. Hultman. First-principles calculations on the structural evolution of solid fullerene-like CPx. 374–379, Copyright 2006, with permission from Elsevier.)

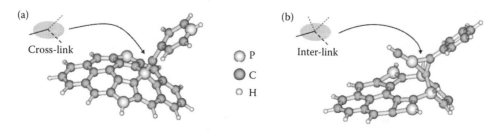

FIGURE 31.12 Representation of cross-linking and interlinking sites in CP_x resulting from growth evolution simulation [17].

atoms in graphene-like systems promotes more radical changes in the graphene-like networks than the same amounts of another dopant, such as N in CN_x. These results encouraged and guided the synthesis of CP_x thin films with P content between 2.5 and 10 at.% [18,43]. In 2008, Furlan et al. reported the first successful synthesis of FL-CP_x [18]. FL-CP_x films, as predicted by the SGC, proved to be harder than their CN_x counterparts while exhibiting similar resilience.

The main points of the SGC predictions for CP_x, later confirmed experimentally, are

- Besides pentagons, in CP_x, tetragon rings become feasible defects, resulting in strongly bent graphene planes;
- P atoms are responsible for promoting the formation of cross-linking and inter-linking sites when incorporated in a graphene-like network;
- FL-CP_x films can be synthesized by magnetron sputtering for P contents between 5 and 15 at.%.

31.3.3 SULFO CARBIDE, CS_x: THE PREDICTION

The application of the SGC to FL-CN_x and FL-CP_x added significant structural knowledge about the incorporation of dopant atoms in graphene/graphite matrices as a key element to manipulate nanostructure and tailor the properties of C-based thin films. Such structural changes are relevant not only to mechanical properties but also to the electrical and optical properties of these films. Consequently, it is natural to submit new dopant elements to simulation tests in an effort to add even more to the wealth of properties that FL C-based thin

films may offer. In this context, sulfur is another prospective dopant candidate for testing the possibility to synthesize a new FL compound, the FL sulfo carbide (FL-CS_x).

Previously available experimental [73,74] and theoretical [74–76] studies focused on doping graphene and carbon nanotubes with S reported that when S is incorporated in the sp^2 C network, the structure loses its completely planar (or cylindrical in case of nanotubes) shape and starts exhibiting a smooth local bending close to the S atom incorporation site [74,75]. At the same time, it was observed that S favors the formation of pentagons and heptagons, enhancing the curvature [73]. Regarding the electronic properties, Denis et al. [75,76] observed that S-doped graphene may become a small-gap semiconductor or may even show better metallic properties than pristine graphene depending on the concentration of S. In addition, diamond and DLC films were successfully doped with S, improving their optoelectronic properties as n-type materials [23,24,77]. These studies on S-doped DLC reported that S promotes graphitization within the mainly sp^3 carbon film. S-doped graphite has also been synthesized [78,79] and revealed to be a particularly interesting system in which superconductivity has been demonstrated at 35 K [78].

These research efforts corroborated the idea for testing the feasibility of using S in the FL family and, by application of the SGC to the CS_x system, work out if FL-CS_x may be realistically synthesized.

As in the CN_x and CP_x cases, the question of the stability for the defects in a S-doped graphene-like model system is the first task in line for the application of the SGC to this new compound. Figure 31.13 shows the most significant resulting

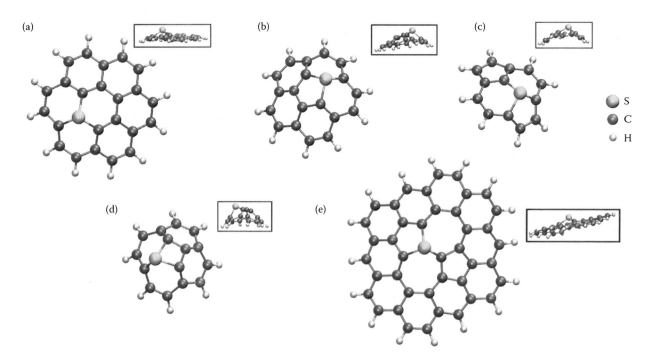

FIGURE 31.13 Relaxed structures for defect containing S-doped model systems: (a) pure hexagonal model system; (b) pentagon defect; (c) double pentagon defect; (d) tetragon defect; and (e) Stone–Wales defect [25]. The insets represent different views of the same model systems. Results obtained with the PW91 exchange and correlation potentials [90].

TABLE 31.3

$E_{coh/at}$ for CS_x Model Systems and Energy Cost for Ring-Defects Incorporation

Defect	$E_{coh/at}$ (eV/at.)	ΔE_T (eV/at.)
Single pentagon	10.58	0.53
Double pentagon	10.47	0.64
Stone–Wales	10.56	0.55
Tetragon	10.30	0.82
Hexagonal	10.73	0.38

Note: $E_{coh/at}$ for the CS_x model systems containing ring-defects (shown in Figure 39.13) are listed in the second row. The third row lists the energy cost ΔE_T for introducing a ring defect in a pure carbon hexagonal graphene-like sheet. The $E_{coh/at}$ for an hexagonal CS_x model system and the energy cost for incorporating a S atom in a pure carbon hexagonal model system are listed as Reference 25. The results were obtained using the PW91 exchange and correlation potential [90].

structures after relaxation [25]. As can be seen in Figure 31.13, most of the nonhexagonal ring defects in a graphene network are feasible and they may be incorporated without graphene disruption. In other words, the S atoms can effectively assume the role of C in a graphene-like network.

Cohesive energies per atom were calculated for all the systems as seen in Figure 31.13 by applying Equation 31.1 (A ≡ S). Table 31.3 lists the values for all these relaxed structures. Following this, the energy cost to form a S-containing ring defect from a pure C hexagonal network was calculated. The structural changes relevant to the formation of such a defect are the following:

$$Hex^C + S \rightarrow Hex^S + C \rightarrow Def^S + C, \qquad (31.2)$$

where Hex^C and Hex^S represent a pure C graphene network and a graphene network with one C atom substituted by a S atom and Def^S represents the nonhexagonal ring becoming a S-containing defect. Therefore, the cost for the formation of a S-containing nonhexagonal ring defect starting from a pure graphene network is the total change in energy (ΔE_T) for the structural sequence in Equation 31.2. The third column in Table 31.3 lists the ΔE_T values for the different types of defects. The single pentagon defect in CS_x is the most favorable energetically and is expected to be the most abundant and structure-defining defect. In energetic feasibility, it is followed by the Stone–Wales defect and the double pentagon defect. This differs from CN_x and CP_x, where the double pentagon showed to be the most stable defect.

The tetragon defect is energetically the most costly in CS_x, so its effect in a CS_x system and its structure-defining role is expected to be relatively low. In particular, tetragon defects in CS_x should be considerably less frequent than in CP_x, where the stability of the tetragon defect is comparable to the stability of the other defects [71]. These results reveal the type of CS_x film structure to expect in terms of curvature:

an intermediate state between CN_x (lower curvature) and CP_x (higher curvature).

The stability of graphene-like model systems was also considered for different concentrations of S [27]. As expected, the observed trend is a decrease in the $E_{coh/at}$ values with increasing S content, since the incorporation of S atoms into the C networks destabilizes the structure. This relation is valid for all type of defects and holds independently of the S incorporation sites. Figure 31.14 shows the dependence of $E_{coh/at}$ on the S concentration in a graphene (pure hexagonal) model system

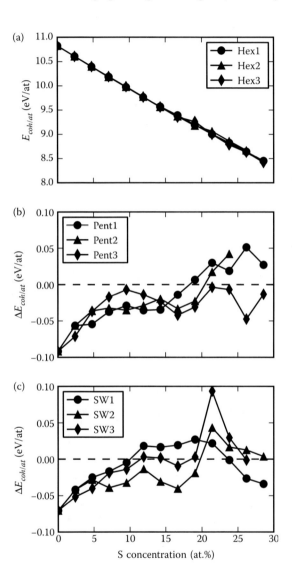

FIGURE 31.14 (a) $E_{coh/at}$ as a function of S concentration for a pure hexagonal model system for three different substitution patterns (Hex1, Hex2, and Hex3). Cohesive energy differences between (b) the pentagon-containing network and Hex1 for three different substitution patterns (Pent1, Pent2, and Pent3) and (c) the Stone–Wales defect-containing network and Hex1 for three different substitution patterns (SW1, SW2, and SW3) [27]. Results obtained with the PW91 exchange and correlation potentials [90]. (Reprinted with permission from C. Goyenola et al. Structural patterns arising during synthetic growth of fullerene-like sulfocarbide. *Journal of Physical Chemistry* C, 116, 21124–21131. Copyright 2011 American Chemical Society.)

(Figure 31.3a) for three different patterns of substitution (Hex1, Hex2, and Hex3). The behavior observed for the single pentagon defect and the Stone–Wales defect is very similar.

To compare the functional dependence of $E_{coh/at}$ on the S concentration for the pure hexagonal network and those containing odd membered rings, cohesive energy differences ($\Delta E_{coh/at}$) between a model system containing defects and a reference graphene network (Hex1) were computed for each S concentration [27]. Figure 31.14b and c show the dependence of $E_{coh/at}$ on S content calculated for the model system containing a single pentagon defect and the Stone–Wales defect, respectively (three different substitution patterns for each of them were worked out: Pent1, Pent2, and Pent3; SW1, SW2, and SW3). It is observed that a pure hexagonal network is more stable than the defect-containing ones for S contents up to 10 at.%. Above this concentration, all the systems exhibited similar cohesive energies and became equally feasible.

The fact that most nonhexagonal defects are energetically feasible under a wide range of S concentrations is an important result for predicting the CS_x structure. Regarding the structural implication of the incorporation of S atoms in the graphene-like networks: low S contents (up to 10 at.%) promote smooth bending of the graphene-like planes. At higher S concentrations, the graphene-like sheets curvature quickly increases, leading to heavy bending of planes and a higher defect concentration including a combination of different kinds of defects and new types of defects such as large N-membered rings ($N = 8$–12). Further increment of the S concentration reaches a structural limit at which the graphene-like network cannot accommodate any more S atoms without disrupting its graphene-like structure. Depending on the relative position of the incorporated S atoms in the network, the limit of well-structured FL-CS_x is between 15 and 20 at.% of S. At even higher S content, the graphene-like network is highly interlocked, resulting in a partially disrupted system. This indicates that in the case of CS_x, a S content around 15 at.% is the most appropriate for achieving a well-structured FL-CS_x film [27].

After establishing the S concentration range for FL-CS_x films, the SGC continues by exploring the growth evolution features. As discussed in Reference 25, in the case of CS_x, a large variety of C_mS_n molecules and radicals were considered possible candidates for growth precursors. Their final selection is based on their relative stability, the similarity to already known molecules and radicals [47], and previous theoretical and experimental experience with FL-CN_x and FL-CP_x

TABLE 31.4
$E_{coh/at}$ for the Most Relevant C_mS_n Precursor Species (Shown in Figure 31.15) and the Dimers C_2 and S_2

Species	SCCS	CCS	SCS	C_2S_4	C_2S	CS	C_2S_2
$E_{coh/at}$ (eV/at.)	6.52	6.21	6.17	5.97	5.77	5.73	5.54
Species	CSCS	CS_2	CSC	SSC	C_2	S_2	
$E_{coh/at}$ (eV/at.)	5.37	5.24	4.96	4.68	4.39	3.64	

Source: C. Goyenola et al. *Chemical Physics Letters*, 506(1–3):86–91, 2011.
Note: The results were obtained using the PW91 exchange and correlation potential [90].

growth by magnetron sputtering. The relaxed structures of the most relevant precursor species are shown in Figure 31.15 and their $E_{coh/at}$ are listed in Table 31.4.

The structural evolution of CS_x is addressed by considering S-doped graphene-like model systems similar to those shown in Figure 31.3 and taking into account random sequences of the precursors listed in Table 31.4. Like other dopants (e.g., N, P) in graphene-like matrices, S atoms induce a change in the hybridization of the adjacent C atoms from sp^2 to sp^3 during film growth. This results in the creation of cross-linking sites. It is a particular feature of CS_x that cross-linkages are not expected to play such an important role as they do in CN_x and CP_x, since during the CS_x film evolution some of the cross-linking sites are assimilated back into a graphene-like sheet that intersects the original one. An example of a cross-linking with subsequent assimilation is shown in Figure 31.16 [27].

Another notable CS_x feature found during the SGC growth evolution of this compound is the formation of cage-like conformations (see Figure 31.17a) [27] similar to those observed for CP_x [17]. However, in the case of CS_x, the cage-like conformations are not necessarily initiated by the presence of tetragons in the network, but they must be associated with pentagon defects. In addition, conformations in which large graphene-like pieces grow parallel (Figure 31.17b) as well as graphene-like sheets intersecting each other at angles close to 90° were also observed (Figure 31.17c) [27].

To sum up, the SGC arrived to the following conclusions with respect to the CS_x compound:

- Well-structure FL-CS_x is predicted for S contents between 10 and 20 at.%

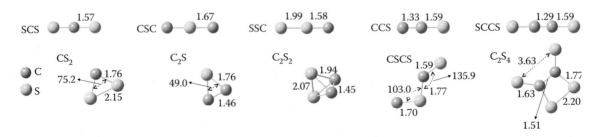

FIGURE 31.15 Relaxed structures for the most relevant C_mS_n precursor species [25]. Bond lengths are in Å and bond angles are in degrees. The structures were obtained employing the PW91 exchange and correlation potential [90].

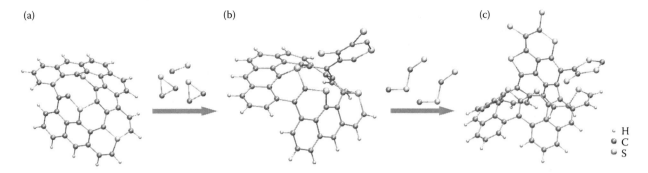

FIGURE 31.16 Chain of bonding events leading to the formation of a cross-linkage site and its subsequent assimilation into the growing planar network [27]. (Reprinted with permission from C. Goyenola et al. Structural patterns arising during synthetic growth of fullerene-like sulfocarbide. *Journal of Physical Chemistry C*, 116, 21124–21131. Copyright 2011 American Chemical Society.)

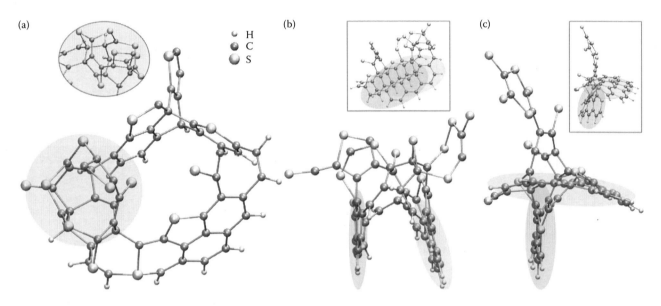

FIGURE 31.17 Structural patterns arising during growth evolution of CS_x: (a) a cage-like conformation indicated by the shadowed area; (b) parallel growing planar graphene-like sheets; and (c) intersecting planar CS_x graphene-like sheets [27]. Different views of each system are shown in the insets. (Reprinted with permission from C. Goyenola et al. Structural patterns arising during synthetic growth of fullerene-like sulfocarbide. *Journal of Physical Chemistry C*, 116, 21124–21131. Copyright 2011 American Chemical Society.)

- Structurally, and therefore in terms of mechanical properties too, CS_x occupies an intermediate position between CN_x and CP_x
- Fine structural features (e.g., cage-like systems without tetragons, parallel growing of graphene-like sheets) make the CS_x unique

The results described in this section are the reason why experimental attempts to deposit CS_x films by magnetron sputtering may be worth trying in order to obtain yet another member of the class of the FL carbon-based thin films with tunable mechanical properties.

31.3.4 CARBON FLUORIDE, CF_x: THE STRUCTURAL DIVERSITY

Enlarging the family of FL compounds by considering new dopant elements is not about mechanical properties alone. It

also provides the opportunity to synthesize thin films with new optical, electrical, and thermal properties.

Fluorine is chemically very different from both the matrix element in C-based thin films and the dopants N, P, and S. The electronic configuration of F is such that it leads to one single C–F bond in C networks. Unlike N, P, and S, F cannot take the role of C when incorporated into a graphene-like system. F is also the most electronegative element and possess low polarizability, bringing new bonding characteristics to a C-based thin solid film doped with F (CF_x).

In particular, fluorinated C-based thin films have been synthesized by different methods of vapor phase deposition [28–30,68,80–84]. The CF_x films have shown a variety of properties, such as low dielectric constant and low refractive index [80–82], moderate hardness [29,30,83], low friction coefficient [83], high wear resistance [83], chemical inertness, and biocompatibility [29]. The microstructure of the CF_x thin films reported so far is of an a-C nature (DL, polymer-like, etc.). Additionally, considerable research efforts have been dedicated to a variety of CF_x compounds [85,86], attributable to the interest in Teflon®-like materials with improved thermal resistance.

Before the SGC simulations of CF_x [30,87], the possibility of obtaining FL-CF_x has not been evaluated. This could add properties such as high hardness and high elasticity inherent to a FL compound to Teflon-like materials.

In the spirit of the SGC, the stability of typical defects inherent to FL compounds were evaluated for the case of CF_x graphene-like networks (pentagons, heptagons, tetragons, and combinations of them, see Figure 31.3) [30,68,87]. As expected, SGC simulation results revealed the CF_x compound as a very different case from the previously discussed ones (CN_x, CP_x, and CS_x). It was observed that the usual ring defects frequently found in CN_x, CP_x, and CS_x are not stable (Figure 31.3) when F atoms are incorporated in the graphene-like sheets. Instead, since F can form one single bond, it disrupts the graphene network where it bonds and new structural patterns characteristic for CF_x arise [30,87].

One of the structural defects typical for CF_x is large N-membered ($N = 7$–8) rings (see Figure 31.18a). The opening of such large rings is promoted and stabilized by the formation of C–F bonds involving a C atom that belongs to the ring. The resulting C–F bond undergoes a bond rotation and, after relaxation, the F atom ends up sticking out from the graphene-like sheet (observe the inset in Figure 31.18a). The combination of a large ring and a C–F bond rotation introduces curvature to the graphene-like plane, but this effect is less pronounced than the curvature that originates due to pentagon and tetragon defects in inherently hexagonal networks [30,87]. A particular case of this specific CF_x structural pattern is the stabilization of a pure C pentagon at the edge of a large ring opposite to the C–F bond (Figure 31.18b). The cost for the formation of the pentagon defect in CF_x is reduced by the strain and curvature provoked by the C–F bond rotation [87].

Another typical defect pattern in CF_x is the branching of a graphene sheet, where the incorporation of an F atom induces the creation of C chains at the edge of the model system (see Figure 31.18c). Generally, the branching in CF_x induces increased reactivity and a tendency to a less-ordered network. The resulting C chains resemble short polymeric chains, pointing to soft CF_x films of polymer-like nature. The film formation during vapor phase deposition is a stochastic process and, therefore, the as-formed C chains are not strictly polymers in the sense that they do not consist in rigorously repetitive C_xF_y units [30,87].

A peculiar variant of the branching defect in CF_x is revealed by the formation of hexagonal nanoribbons at the edge of the model system. Figure 31.18d shows an example where the incorporation of a F atom induces the splitting of the model system in two arms consisting of a sequence of hexagons. This type of branching illustrates the possibility of splitting of a CF_x graphene-like plane in two pieces that may grow independently with the possibility for intersection

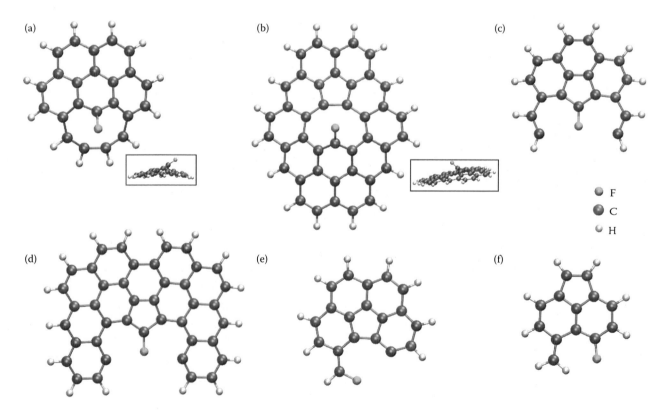

FIGURE 31.18 Structural patterns characteristic for CF_x: (a) large N-membered ring, $N = 9$; (b) large N-membered ring, $N = 9$, with stabilization of a pure C pentagon; (c) branching of the model system in two C chains; (d) branching of the model system in two ribbons; and (e,f) simple bond rotation [31].

and cross-linking. This mechanism of film formation, when predominating, would eventually lead to frequent interlocking and mostly amorphous structures [87].

Also, simple C–F bond rotation at the edge of the model system was observed and analyzed (Figure 31.18e and f). The C–F bond rotation is suggestive for cross-linking involving the C atom participating in the C–F bond. This feature is not unique to CF_x and has also been observed in CN_x [45] and CP_x [17]. The main difference between CF_x on one hand and CN_x and CP_x on the other is that while N and P can actually participate in the cross-linking by forming more than one bond, F atoms are not able to do so. This results in a preferential direction for cross-linking, opposite to the direction where the F atom is located [87].

While the formation of large rings contributes to preserving the integrity of an ordered graphene-like network and induce smooth curvature, the branching mechanism promotes amorphization of the compound by formation of polymer-like chains or intersection of graphene-like pieces. The SGC results showed that the gain in $E_{coh/at}$ is larger for the former (0.405 eV/at.) compared to the latter (0.383 eV/at.). Consequently, during film growth evolution at low F concentrations (up to 10 at.%), many F atoms will contribute to the curvature of the graphene sheets, while a lower number of them will induce structural patterns tending to amorphization. Therefore, FL types of defects can be formed with moderate curvature and relatively low degree of cross-linking. Higher F concentrations would result in the incorporation of a larger number of defects leading to less-ordered films (amorphous and polymer-like) [30,87,88].

The described relation between F concentration and the expected film microstructure is in excellent agreement with experimental observations. All characterization efforts of CF_x thin films [28–30,68,80–84,88] agree on the strong dependence of the microstructure of these films on the F content. The microstructure of CF_x films can go from graphite-like and DL to polymer-like in order of increasing F content. There are still no reports on the synthesis of FL-CF_x. This is due to the fact that during CF_x film growth by magnetron sputtering it is difficult to achieve the low F incorporation necessary for FL-CF_x [30,68].

Summarizing the results obtained by the SGC for CF_x thin films:

- The most typical defects characteristic for the other FL compounds (CN_x, CP_x, CS_x) are not feasible in CF_x
- F incorporation into graphene-like networks induces the formation of large N-membered ($N = 8–12$) rings and network disruptions by branching mechanisms
- FL-CF_x films could be achieved for F concentrations lower than 10 at.% while larger F contents lead to amorphization of the film by accumulation of defects
- Polymer-like CF_x is promising as a new Teflon-like material with possibly improved mechanical and thermal properties in comparison to Teflon

31.4 FULLERENE-LIKE CARBON-BASED THIN FILMS CLASS IN A NUTSHELL

From Sections 31.4.1 to 31.4.4 it becomes evident that due to the success of the SGC simulation approach, a new class of C-based thin films is emerging: that of the inherently nanostructured and FL C-based thin films.

FL-CN_x has shown a unique combination of high hardness and high elasticity due to its particular microstructure consisting of buckled graphene sheets that also cross-link. The SGC of CN_x showed that the incorporation of N in the graphene network leads to a completely new material, which is susceptible to tailoring during magnetron sputtering deposition. By studying precursor species and evolution, the SGC shows how to tailor the structure and properties of CN_x by changing the deposition conditions. In particular, N incorporation in a graphene network induces the formation of pentagon defects and changes in the hybridization of C atoms from sp^2 to sp^3. Pentagon defects are the mainly responsible for the curvature of the graphene planes and the sp^3-hybridized C atoms are the main responsible for cross-linking. While the elasticity originates in the curved graphene pieces, the hardness is related to the cross-linking between them and their intersection [13,41,44].

FL-CP_x was first predicted by the SGC and then successfully synthesized at the Thin Films Physics Division at Linköping University. CP_x films were a success story by their own right, since their synthesis has been entirely guided by the SGC and they can come as an improvement on CN_x films. It was shown that P atoms incorporated in the graphene network promotes not only the formation of pentagon defects, but also tetragon defects. The curvature induced by tetragon defects is more pronounced than that produced by pentagons. This means that CP_x films can be expected to have graphene pieces with larger curvature than CN_x, but shorter and more intersected. In addition, besides cross-linking, a mechanism for interlinking was also found. Both these mechanisms contribute to a more interlocked film. The synthesized FL-CP_x films showed less-pronounced FL structural features compared to FL-CN_x, with increased hardness but lower elasticity. This is in agreement with the prediction of a more interlocked film [17,43].

FL-CS_x has been predicted to exist for S concentrations between 10 and 20 at.%. Pentagons are the most relevant defects in film formation, but other defects including heptagons and tetragons are expected to coexist. The variety of defects predicted contributes to curvature and intersections of graphene pieces similar to the CP_x case, but less pronounced. In addition, cross-linking mechanisms were found to be less frequent in CS_x than in CN_x and CP_x. The combination of these structural features puts FL-CS_x films in an intermediate position between FL-CN_x and FL-CP_x regarding microstructure and mechanical properties [25,27], which makes CS_x films a very attractive proposal for future synthesis.

Finally, CF_x films have been shown by SGC simulations to exhibit large structural diversity. The structural diversity arise from new structural patterns that become feasible in CF_x: large N-members ($N = 8–12$) rings and branching of graphene-like sheets. While FL features are expected at low F concentrations (≤10 at.%), they evolve into softer polymer-like films at larger F contents. In very recent experimental work at the Thin Films Physics Division at Linköping University [30,32,68], this structural diversity was confirmed.

31.5 CONCLUDING WORDS ABOUT THE SGC

The SGC is an original approach especially adequate for modeling and designing nanostructured materials that lack periodic structure, but possess atomic order and structural patterns. The SGC approach combines small but characteristic subunits or nanotemplates that retain the basic structural characteristics of the compound. Structural evolution, careful precursor studies, and exploring different growth sequences are strong points for the SGC. In particular, the class of FL C-based thin films has been introduced by employing the SGC, demonstrating the possibility of modeling C-based thin films using low-dimensional graphene-like nanostructures.

Another highlight of the SGC and of the new class of FL C-based thin films is the close collaboration with experimentalists and the impact of the SGC predictions on experimental success. This also shows how remarkable scientific results can be obtained when experimental work is accompanied by an appropriate theoretical approach and vice versa.

ACKNOWLEDGMENTS

The authors gratefully acknowledge the long-term, close and efficient collaboration with their colleagues at the Thin Film Physics Division at Linköping University, Sweden: Andrej Furlan, Esteban Broitman, Hans Högberg, Jörg Neidhardt, Lars Hultman, Susann Schmidt, and Zsolt Czigány. Without their work and expertise, particularly in film growth and film characterization, the simulation concepts and results presented in this chapter, as well as the overall success of the novel class of nanostructured C-based thin films would not have been possible. The same gratefulness goes out to Sven Stafström, head of the Computational Physics Division at Linköping University for his long-term insightful contribution to the establishment of the FL C-based thin films as an original class of materials.

The authors acknowledge the support by Linköping Linnaeus Initiative on Novel Functionalized Materials (VR) and by FunMat (Functional Nanoscale Materials)—a VINN Excellence Centre (Swedish Agency for Innovation Systems VINNOVA). G.K.G. acknowledges further the support by the Swedish Foundation for Strategic Research (SSF) Synergy Grant #RMA11-0029 on Functional Carbides and Advanced Surface Engineering (FUNCASE), as well as the Carl Trygger Foundation for Scientific Research.

REFERENCES

1. M. Ohring. *Materials Science of Thin Films*. Academic Press, Burlington, MA, 2nd edition, 2001.
2. J.-E. Sundgren. Structure and properties of TiN coatings. *Thin Solid Films*, 128(1–2):21–44, 1985.
3. S. P. Chuan, J. A. Ghani, S. H. Tomadi, and C. Hassan. Analysis of Ti-base hard coating performance in machining process: A review. *Journal of Applied Sciences*, 12(18):1882–1890, 2012.
4. M. G. Kanatzidis. Semiconductor physics: Quick-set thin films. *Nature*, 428(6980):269–271, 2004.
5. B. Parida, S. Iniyan, and R. Goic. A review of solar photovoltaic technologies. *Renewable and Sustainable Energy Reviews*, 15(3):1625–1636, 2011.
6. J. Robertson. Diamond-like amorphous carbon. *Materials Science and Engineering: R: Reports*, 37(4–6):129–281, 2002.
7. A. H. Lettington. Applications of diamond-like carbon thin films. *Carbon*, 36(5–6):555–560, 1998.
8. J. Robertson. Amorphous carbon. *Advances in Physics*, 35(4):317–374, 1986.
9. S. R. P. Silva, editor. *Properties of Amorphous Carbon*. Institution of Engineering and Technology (IET), London, 2002.
10. L. Ilberg, H. Manis-Levy, A. Raveh, Y. Lifshitz, and M. Varenberg. Effect of structure of carbon films on their tribological properties. *Diamond and Related Materials*, 38:79–86, 2013.
11. A. Aijaz, K. Sarakinos, M. Raza, J. Jensen, and U. Helmersson. Principles for designing sputtering-based strategies for high-rate synthesis of dense and hard hydrogenated amorphous carbon thin films. *Diamond and Related Materials*, 44:117–122, 2014.
12. S. Muhl and J. M. Méndez. A review of the preparation of carbon nitride films. *Diamond and Related Materials*, 8(10):1809–1830, 1999.
13. L. Hultman, J. Neidhardt, N. Hellgren, H. Sjöström, and J.-E. Sundgren. Fullerene-like carbon nitride: A resilient coating material. *MRS Bulletin*, 28(03):194–202, 2003.
14. B. Marchon, X. Guo, B. K. Pathem, F. Rose, Q. Dai, N. Feliss, E. Schreck et al. HeadDisk interface materials issues in heat-assisted magnetic recording. *IEEE Transactions on Magnetics*, 50(3):137–143, 2014.
15. N. Tamura, M. Aono, T. Harata, H. Kishimura, N. Kitazawa, and Y. Watanabe. DC electrical conductivity study of amorphous carbon nitride films prepared by reactive RF magnetron sputtering. *Japanese Journal of Applied Physics*, 53(2S):02BC03, 2014.
16. F. Claeyssens, G. M. Fuge, N. L. Allan, P. W. May, S. R. J. Pearce, and M. N. R. Ashfold. Phosphorus carbide thin films: Experiment and theory. *Applied Physics A*, 79(4–6):1237–1241, 2004.
17. G. K. Gueorguiev, A. Furlan, H. Högberg, S. Stafström, and L. Hultman. First-principles calculations on the structural evolution of solid fullerene-like CP_x. *Chemical Physics Letters*, 426(4–6):374–379, 2006.
18. A. Furlan, G. K. Gueorguiev, Zs. Czigány, H. Högberg, S. Braun, S. Stafström, and L. Hultman. Synthesis of phosphorus-carbide thin films by magnetron sputtering. *Physica Status Solidi (RRL)—Rapid Research Letters*, 2(4):191–193, 2008.
19. J. N. Hart, P. W. May, N. L. Allan, K. R. Hallam, F. Claeyssens, G. M. Fuge, M. Ruda, and P. J. Heard. Towards new binary compounds: Synthesis of amorphous phosphorus carbide by pulsed laser deposition. *Journal of Solid State Chemistry*, 198:466–474, 2013.
20. A. Furlan, G. K. Gueorguiev, Zs. Czigány, V. Darakchieva, S. Braun, M. R. Correia, H. Högberg, and L. Hultman. Structure and properties of phosphorus-carbide thin solid films. *Thin Solid Films*, 548:247–254, 2013.

21. M. Künle, T. Kaltenbach, P. Löper, A. Hartel, S. Janz, O. Eibl, and K.-G. Nickel. Si-rich a-SiC:H thin films: Structural and optical transformations during thermal annealing. *Thin Solid Films*, 519(1):151–157, 2010.

22. X.-A. Fu, J. L. Dunning, M. Mehregany, and C. A. Zorman. Low stress polycrystalline SiC thin films suitable for MEMS applications. *Journal of The Electrochemical Society*, 158(6):H675–H680, 2011.

23. I. Sakaguchi, M. N. Gamo, Y. Kikuchi, E. Yasu, H. Haneda, T. Suzuki, and T. Ando. Sulfur: A donor dopant for n-type diamond semiconductors. *Physical Review B*, 60(4):R2139–R2141, 1999.

24. S. Wan, L. Wang, and Q. Xue. Investigation of microstructure and photo-magnetic properties of sulfur-doped DLC nanocomposite films by electrochemical method. *Applied Physics A*, 102(3):753–760, 2010.

25. C. Goyenola, G. K. Gueorguiev, S. Stafström, and L. Hultman. Fullerene-like CS$_x$: A first-principles study of synthetic growth. *Chemical Physics Letters*, 506(1–3):86–91, 2011.

26. A. Saeheng, N. Tonanon, W. Bhanthumnavin, and B. Paosawatyanyong. Sulphur doped DLC films deposited by DC magnetron sputtering. *The Canadian Journal of Chemical Engineering*, 90(4):909–914, 2012.

27. C. Goyenola, S. Stafström, L. Hultman, and G. K. Gueorguiev. Structural patterns arising during synthetic growth of fullerene-like sulfocarbide. *Journal of Physical Chemistry C*, 116(39):21124–21131, 2012.

28. H. Touhara and F. Okino. Property control of carbon materials by fluorination. *Carbon*, 38(2):241–267, 2000.

29. A. Bendavid, P. J. Martin, L. Randeniya, M. S. Amin, and R. Rohanizadeh. The properties of fluorine-containing diamond-like carbon films prepared by pulsed DC plasma-activated chemical vapour deposition. *Diamond and Related Materials*, 19(12):1466–1471, 2010.

30. S. Schmidt, G. Greczynski, C. Goyenola, G. K. Gueorguiev, Zs. Czigány, J. Jensen, I. G. Ivanov, and L. Hultman. CF$_x$ thin solid films deposited by high power impulse magnetron sputtering: Synthesis and characterization. *Surface and Coatings Technology*, 206(4):646–653, 2011.

31. G. K. Gueorguiev, Zs. Czigány, A. Furlan, S. Stafström, and L. Hultman. Intercalation of P atoms in fullerene-like CP$_x$. *Chemical Physics Letters*, 501(4–6):400–403, 2011.

32. S. Schmidt, C. Goyenola, G. K. Gueorguiev, J. Jensen, G. Greczynski, I. G. Ivanov, Zs. Czigány, and L. Hultman. Reactive high power impulse magnetron sputtering of CF$_x$ thin films in mixed Ar/CF4 and Ar/C4F8 discharges. *Thin Solid Films*, 542:21–30, 2013.

33. A. A. Voevodin and J. S. Zabinski. Load-adaptive crystalline-amorphous nanocomposites. *Journal of Materials Science*, 33(2):319–327, 1998.

34. U. Jansson, E. Lewin, M. Rå Sander, O. Eriksson, B. André, and U. Wiklund. Design of carbide-based nanocomposite thin films by selective alloying. *Surface and Coatings Technology*, 206(4):583–590, 2011.

35. F. A. Carey. *Advanced Organic Chemistry Part A, Structure and Mechanisms*. Kluwer Academic, New York, 4th edition, 2002.

36. J. C. Angus and C. C. Hayman. Low-pressure, metastable growth of diamond and "diamondlike" phases. *Science*, 241(4868):913–921, 1988.

37. P. Delhaès, editor. *Graphite and Precursors*. Gordon and Breach Science Publishers, Amsterdam, The Netherlands, 2001.

38. H. Sjöström, S. Stafström, M. Boman, and J.-E. Sundgren. Superhard and elastic carbon nitride thin films having fullerene like microstructure. *Physical Review Letters*, 75(7):1336–1339, 1995.

39. I. Alexandrou, H.-J. Scheibe, C. Kiely, A. Papworth, G. Amaratunga, and B. Schultrich. Carbon films with an sp^2 network structure. *Physical Review B*, 60(15):10903–10907, 1999.

40. Zs. Czigány, I. F. Brunell, J. Neidhardt, L. Hultman, and K. Suenaga. Growth of fullerene-like carbon nitride thin solid films consisting of cross-linked nano-onions. *Applied Physics Letters*, 79(16):2639–2641, 2001.

41. J. Neidhardt. *Fullerene-like carbon nitride thin solid films*. PhD thesis, Linköping University, 2004.

42. P. Wang, X. Wang, B. Zhang, and W. Liu. Structural, mechanical and tribological behavior of fullerene-like carbon film. *Thin Solid Films*, 518(21):5938–5943, 2010.

43. A. Furlan. *Fullerene-like CN$_x$ and CP$_x$ thin films; synthesis, modeling and applications*. PhD thesis, Linköping University, 2009.

44. G. K. Gueorguiev, J. Neidhardt, S. Stafström, and L. Hultman. First-principles calculations on the role of CN precursors for the formation of fullerene-like carbon nitride. *Chemical Physics Letters*, 401(1–3):288–295, 2005.

45. G. K. Gueorguiev, J. Neidhardt, S. Stafström, and L. Hultman. First-principles calculations on the curvature evolution and cross-linkage in carbon nitride. *Chemical Physics Letters*, 410(4–6):228–234, 2005.

46. G. K. Gueorguiev, E. Broitman, A. Furlan, S. Stafström, and L. Hultman. Dangling bond energetics in carbon nitride and phosphorus carbide thin films with fullerene-like and amorphous structure. *Chemical Physics Letters*, 482(1–3):110–113, 2009.

47. S. Oae. *Organic Sulfur Chemistry: Structure and Mechanism*. CRC Press, Boca Raton, FL, 1st edition, 1991.

48. K. B. Dillon, F. Mathey, and J. F. Nixon. *Phosphorus: The Carbon Copy*. John Wiley & Sons Ltd., New York, 1998.

49. G. K. Gueorguiev, J. M. Pacheco, S. Stafström, and L. Hultman. Siliconmetal clusters: Nano-templates for cluster assembled materials. *Thin Solid Films*, 515(3):1192–1196, 2006.

50. G. K. Gueorguiev and J. M. Pacheco. Silicon and metal nanotemplates: Size and species dependence of structural and electronic properties. *Journal of Chemical Physics*, 119(19):10313–10317, 2003.

51. J. Pacheco, G. Gueorguiev, and J. Martins. First-principles study of the possibility of condensed phases of endohedral silicon cage clusters. *Physical Review B*, 66(3):033401, 2002.

52. G. K. Gueorguiev, S. Stafström, and L. Hultman. Nano-wire formation by self-assembly of siliconmetal cage-like molecules. *Chemical Physics Letters*, 458(1–3):170–174, 2008.

53. R. B. Dos Santos, R. Rivelino, F. de B Mota, and G. K. Gueorguiev. Exploring hydrogenation and fluorination in curved 2D carbon systems: A density functional theory study on corannulene. *The Journal of Physical Chemistry. A*, 2012.

54. R. B. dos Santos, R. Rivelino, F. de Brito Mota, and G. K. Gueorguiev. Effects of N doping on the electronic properties of a small carbon atomic chain with distinct sp^{2} terminations: A first-principles study. *Physical Review B*, 84(7):075417, 2011.

55. E. F. Almeida, F. Brito Mota, C. M. C. Castilho, A. Kakanakova-Georgieva, and G. K. Gueorguiev. Defects in hexagonal-AlN sheets by first-principles calculations. *European Physical Journal B*, 85(1):48, 2012.

56. A. Kakanakova-Georgieva, G. K. Gueorguiev, S. Stafström, L. Hultman, and E. Janzén. AlGaInN metal-organic-chemical-vapor-deposition gas-phase chemistry in hydrogen and nitrogen diluents: First-principles calculations. *Chemical Physics Letters*, 431(4–6):346–351, 2006.

57. R. Yakimova, A. Kakanakova-Georgieva, G. R. Yazdi, G. K. Gueorguiev, and M. Syväjärvi. Sublimation growth of AlN crystals: Growth mode and structure evolution. *Journal of Crystal Growth*, 281(1):81–86, 2005.

58. A. Kakanakova-Georgieva, G. K. Gueorguiev, R. Yakimova, and E. Janzen. Effect of impurity incorporation on crystallization in AlN sublimation epitaxy. *Journal of Applied Physics*, 96(9):5293–5297, 2004.

59. A. Y. Liu and M. L. Cohen. Prediction of new low compressibility solids. *Science (New York, N.Y.)*, 245(4920):841–842, 1989.

60. A. Liu and M. Cohen. Structural properties and electronic structure of low-compressibility materials: β-Si3N4 and hypothetical β-C3N4. *Physical Review B*, 41(15):10727–10734, 1990.

61. E. Kroke. Novel group 14 nitrides. *Coordination Chemistry Reviews*, 248(5–6):493–532, 2004.

62. H. Sjöström, L. Hultman, J. E. Sundgren, S. V. Hainsworth, T. F. Page, and G. Theunissen. Structural and mechanical properties of carbon nitride CN_x $(0.2 \leq x \leq 0.35)$ films. *Journal of Vacuum Science and Technology A: Vacuum, Surfaces, and Films*, 14(1):56–62, 1996.

63. R. Gago, G. Abrasonis, A. Mucklich, W. Moller, Zs. Czigany, and G. Radnoczi. Fullerenelike arrangements in carbon nitride thin films grown by direct ion beam sputtering. *Applied Physics Letters*, 87(7):071901, 2005.

64. D. G. Liu, J. P. Tu, C. D. Gu, C. F. Hong, R. Chen, and W. S. Yang. Synthesis, structure and mechanical properties of fullerene-like carbon nitride films deposited by DC magnetron sputtering. *Surface and Coatings Technology*, 205(7):2474–2482, 2010.

65. S. Schmidt, Zs. Czigany, G. Greczynski, J. Jensen, and L. Hultman. Ion mass spectrometry investigations of the discharge during reactive high power pulsed and direct current magnetron sputtering of carbon in Ar and Ar/N2. *Journal of Applied Physics*, 112(1):013305, 2012.

66. J. Neidhardt, L. Hultman, B. Abendroth, R. Gago, and W. Moller. Diagnostics of a N[sub 2]/Ar direct current magnetron discharge for reactive sputter deposition of fullerene-like carbon nitride thin films. *Journal of Applied Physics*, 94(11):7059, 2003.

67. S. Stafström. Reactivity of curved and planar carbonnitride structures. *Applied Physics Letters*, 77(24):3941–3943, 2000.

68. S. Schmidt. *Carbon nitride and carbon fluoride thin films prepared by HiPIMS*. PhD thesis, Linköping University, 2013.

69. A. A. Voevodin, J. G. Jones, J. S. Zabinski, Zs. Czigány, and L. Hultman. Growth and structure of fullerene-like CN_x thin films produced by pulsed laser ablation of graphite in nitrogen. *Journal of Applied Physics*, 92(9):4980–4988, 2002.

70. M. J. S. Dewar, E. G. Zoebisch, E. F. Healy, and J. J. P. Stewart. Development and use of quantum mechanical molecular models. 76. AM1: A new general purpose quantum mechanical molecular model. *Journal of the American Chemical Society*, 107(13):3902–3909, 1985.

71. A. Furlan, G. K. Gueorguiev, H. Högberg, S. Stafström, and L. Hultman. Fullerene-like CP_x: A first-principles study of the relative stability of precursors and defect energetics during synthetic growth. *Thin Solid Films*, 515(3):1028–1032, 2006.

72. L. D. Quin. *The Heterocyclic Chemistry of Phosphorus: Systems Based on the Phosphorus-Carbon Bond*. Wiley-Interscience, New York, 1981.

73. J. M. Romo-Herrera, D. A. Cullen, E. Cruz-Silva, D. Ramírez, B. G. Sumpter, V. Meunier, H. Terrones, D. J. Smith, and M. Terrones. The role of sulfur in the synthesis of novel carbon morphologies: From covalent Y-junctions to Sea-Urchin-like structures. *Advanced Functional Materials*, 19(8):1193–1199, 2009.

74. G. Kucukayan, R. Ovali, S. Ilday, B. Baykal, H. Yurdakul, S. Turan, O. Gulseren, and E. Bengu. An experimental and theoretical examination of the effect of sulfur on the pyrolytically grown carbon nanotubes from sucrose-based solid state precursors. *Carbon*, 49(2):508–517, 2011.

75. P. A. Denis, R. Faccio, and A. W. Mombru. Is it possible to dope single-walled carbon nanotubes and graphene with sulfur? *ChemPhysChem: A European Journal of Chemical Physics and Physical Chemistry*, 10(4):715–722, 2009.

76. P. A. Denis. Band gap opening of monolayer and bilayer graphene doped with aluminium, silicon, phosphorus, and sulfur. *Chemical Physics Letters*, 492(4–6):251–257, 2010.

77. D. Saada, J. Adler, and R. Kalish. Sulfur: A potential donor in diamond. *Applied Physics Letters*, 77(6):878–879, 2000.

78. R. da Silva, J. Torres, and Y. Kopelevich. Indication of superconductivity at 35 K in graphite-sulfur composites. *Physical Review Letters*, 87(14):147001, 2001.

79. E. Kurmaev, A. Galakhov, A. Moewes, S. Moehlecke, and Y. Kopelevich. Interlayer conduction band states in graphite-sulfur composites. *Physical Review B*, 66(19):1–3, 2002.

80. K. Endo, K. Shinoda, and T. Tatsumi. Plasma deposition of low-dielectric-constant fluorinated amorphous carbon. *Journal of Applied Physics*, 86(5):2739–2745, 1999.

81. S. Agraharam, D. W. Hess, P. A. Kohl, and S. A. Bidstrup Allen. Comparison of plasma chemistries and structure-property relationships of fluorocarbon films deposited from octafluorocyclobutane and pentafluoroethane monomers. *Journal of Vacuum Science and Technology B: Microelectronics and Nanometer Structures*, 19(2):439–446, 2001.

82. H.-S. Jung and H.-H. Park. Structural and electrical properties of co-sputtered fluorinated amorphous carbon film. *Thin Solid Films*, 420–421:248–252, 2002.

83. M. Ishihara, M. Suzuki, T. Watanabe, T. Nakamura, A. Tanaka, and Y. Koga. Synthesis and characterization of fluorinated amorphous carbon films by reactive magnetron sputtering. *Diamond and Related Materials*, 14(3–7):989–993, 2005.

84. Y. Yun, E. Broitman, and A. J. Gellman. Oxidation of fluorinated amorphous carbon (a-CF_x) films. *Langmuir: The ACS Journal of Surfaces and Colloids*, 26(2):908–914, 2010.

85. A. Milella, F. Palumbo, P. Favia, G. Cicala, and R. D'Agostino. Continuous and modulated deposition of fluorocarbon films from c-C4F8 plasmas. *Plasma Processes and Polymers*, 1(2):164–170, 2004.

86. E. Sardella, F. Intranuovo, P. Rossini, M. Nardulli, R. Gristina, R. D'Agostino, and P. Favia. Plasma enhanced chemical vapour deposition of nanostructured fluorocarbon surfaces. *Plasma Processes and Polymers*, 6(S1):S57–S60, 2009.

87. G. K. Gueorguiev, C. Goyenola, S. Schmidt, and L. Hultman. CF_x: A first-principles study of structural patterns arising during synthetic growth. *Chemical Physics Letters*, 516(1–3):62–67, 2011.

88. C. Goyenola, S. Stafström, S. Schmidt, L. Hultman, and G. K. Gueorguiev. Carbon fluoride, CF_x: Structural diversity as predicted by first principles. *Journal of Physical Chemistry C*, 118(12):6514–6521, 2014.

89. A. D. Becke. Density-functional thermochemistry. III. The role of exact exchange. *Journal of Chemical Physics*, 98(7): 5648–5652, 1993.

90. J. P. Perdew, K. A. Jackson, M. R. Pederson, D. J. Singh, and C. Fiolhais. Atoms, molecules, solids, and surfaces: Applications of the generalized gradient approximation for exchange and correlation. *Physical Review B*, 46(11):6671–6687, 1992.

91. L. Hultman, S. Stafström, Zs. Czigány, J. Neidhardt, N. Hellgren, I. Brunell, K. Suenaga, and C. Colliex. Cross-linked nano-onions of carbon nitride in the solid phase: Existence of a Novel C48N12 Aza-fullerene. *Physical Review Letters*, 87(22):225503, 2001.

32 Graphene-Based Hybrid Composites

Antonio F. Ávila, Diego T. L. da Cruz,
Hermano Nascimento Jr., and Flávio A. C. Vidal

CONTENTS

Abstract...517
32.1 Carbon-Based Nanostructures: A Brief Overview...517
32.2 Hybrid Composites in Context: Adhesives Nano-Modified by Graphene.......................................520
32.3 Experimental Procedures 521
32.4 Data Analysis and Discussion 522
32.5 Closing Comments...525
Acknowledgments...526
References..526

ABSTRACT

This chapter discusses the advantages and disadvantages of hybrid composites nano-modified by carbon nanostructures. A brief review on carbon nanotubes and graphene describes the major techniques employed into the "composite's world." To be able to exemplify the graphene use, this chapter reports the effect of graphene dispersion into epoxy adhesives and its aging by UVA light exposure. To achieve this goal, two approaches were employed. The first one was direct exposure of the nano-modified AR300/AH30-150 samples to UVA light for 400 hours. After the aging process, nanoindentation tests were performed. The second approach was based on tensile tests of aged single lap joints (SLJs). The graphene nanostructures formed inside the AR300/AH30-150 nano-modified adhesive seems to block the aging process, as none of the specimens presented a decrease on stiffness. The force–displacement curves obtained by nanoindentation seem to indicate a good dispersion process, as the large majority of the curves laid down at the same path. When the SLJs were tested, the results also indicated an average increase on bearing-load capacity of 40.96% and 72.03% for 1 wt.% and 2 wt.%, respectively for 100 hours aging. When the aging reached 200 hours, there is a decrease on load capacity when compared against the 100 hours results. The average load capacity was 27.03% and 58.19% higher than the not aged AR300/AH30-150 SLJs. At 400 hours, there is another increase on load capacity, that is, 32.06% and 74.43% for 1 wt.% and 2 wt.% graphene, respectively. A finite element simulation revealed that peel and shear stresses at adhesive edge increased by 82%. The Fourier Transformed Infrared tests revealed that chemical changes on hydroxyl, carbonyl, and epoxy components could be the reason for this behavior. The graphene dispersion into the epoxy adhesives seems to have a double folded effect; in one hand, it increases the bonded joint capacity, and on the other hand, blocks the aging effect of UVA light.

32.1 CARBON-BASED NANOSTRUCTURES: A BRIEF OVERVIEW

As discussed by Mauter and Elimelech [1], carbon's exclusive hybridization properties and its structure morphing capability to perturbations in synthesis conditions allow for tailor-made manipulation to a degree not yet matched by inorganic nanostructures. Among the most important carbon-based nanostructures, carbon nanotubes (CNTs) and graphene nanosheets (GNs) are truly the most important ones. As described by Saito et al. [2], CNT is a honeycomb lattice rolled into a cylinder. CNTs have been the center of many researches due to their dimensions and remarkable electro-mechanical properties. In general, a CNT diameter has a nanometer size and its length can be more than 1 µm. Its large aspect ratio (length/diameter) is appointed as one of the reasons for the CNTs notable properties. According to Kalamkarov et al. [3], single-walled nanotubes (SWNTs) have predicted specific strength around 600 times larger than steel. As described by Gein and Novoselov [4], graphene is the designation of a single layer of carbon atoms tightly packed into a two-dimensional (2D) honeycomb lattice. This carbon atom monolayer array is the building block for graphitic materials. As commented by Lee et al. [5], the graphene effective elastic moduli follow a normal distribution with a peak close to 1.0 TPa, which is equivalent to the SWNT. This high stiffness and elevated strength (Tang et al. [6] reported a 130 GPa value) can be attributed to two factors, that is, the elevated specific surface area (≈2600 m²/g) and the strong carbon–carbon covalent bonds. Based on mechanical properties of these carbon-based nanostructures, Odegard et al. [7] postulated that they are valuable options for improving mechanical properties of composites. Furthermore, these reinforcements at micro/nanoscale can be used for creating the so-called multi-scaled composites (MSC). The MSC are multi-phase reinforced composites, that is, in addition to traditional reinforcement carbon fibers, the matrix is replaced by nanocomposites. As commented by Joshi and Dikshit [8], nanocomposites can be obtained by dispersing nanoparticles/nanostructures into the polymeric matrix. The dispersion process can create inside the polymeric matrix three different nanostructures, that is, intercalated, exfoliated, and mixed. According to Gouda et al. [9], the exfoliated nanostructured are the ones with best mechanical performance due to the largest surface area. Although CNTs have great potential for applications in a large variety of usages, for example, aerospace industry, medical, and electronic devices, there is no consensus about their exact mechanical properties.

CNT capabilities have been observed experimentally and verified by numerical analysis. Frankland et al. [10], Jin and Yuan [11], and Agrawal et al. [12] are among those researchers who employed molecular dynamics for analyzing CNTs. The atomistic simulation approach was employed by Belytschko et al. [13], Lurie et al. [14], Gates et al. [15], while the nano-mechanics modeling was described by Liu et al. [16], Ruoff and Pugno [17], Li and Chou [18], and Ávila et al. [19]. The basic difference between the two groups of modeling is the approach employed. The molecular mechanics ones are based on finite element simulations where beam elements replaced the covalent bonds while Van der Waals bonds were represented by spring elements. The atomistic modeling employed chemical potentials, for example, the Moore's one, to describe the carbon–carbon bonds. Although CNTs have tremendous potential in a large variety of applications, for example, aerospace and medical industries, there is no consensus about their exact mechanical properties. The experiments performed up to now have presented large variability due to the inherent complexity of manipulating these materials. However, their potential is unquestionable, in special for composites.

As mentioned by Ávila et al. [20], carbon-based nanostructures, that is, CNTs and GNs, can be combined to traditional composites for a multi-scale reinforcement. Moreover, the recent developments on CNT synthesis have led to dramatic decrease into its cost. As a consequence, the number of researchers using carbon-based nanostructures increased, and the results on nano-reinforcement of composites laminates are encouraging. Among those researchers are Kim et al. [21] who described no significant increase on tensile properties of the addition of CNTs to carbon fibers/epoxy laminates. Nonetheless, they noticed an enhancement on flexural modulus (≈12%) and strength (≈18%) with the addition of 0.3 wt.% of CNT to the epoxy system. This increase can be attributed to changes into flexural failure mechanisms. Following the same idea, Chou et al. [22] discussed the influence of CNTs into the failure of laminated composites. They even proposed the concept of a multi-phase inter-laminar architecture that can bridge inter-laminar cracks. Wicks et al. [23] actually produced the multi-phase nano-reinforced laminated composites proposed by Chou et al. [22]. In their laminate, CNTs were grown *in situ* in all fibers leading to a "fuzzy" fibers configuration. As mentioned by Wicks, aligned CNTs bridges the plies interfaces, which can lead to an increase on toughness, for the steady state condition, 76% higher than the conventional laminated systems. Notice that for the interlayer

nano-reinforcement some issues must be considered, that is, the interfacial bonds between CNTs, fiber/matrix system, and the length effect into this "grip condition." To understand the failure mechanism, Shokrieh and Rafiee [24] modeled the CNT length effect on reinforcement effectiveness. Moreover, they concluded that for CNTs with length less than 100 nm, the improvement on stiffness for CNT/polymeric systems is negligible. Experimental data provided by Ma et al. [25] demonstrated the limitations of using CNTs with aspect ratio smaller than 100 into polymeric systems. The "fuzzy" fibers configuration developed by Wicks et al. [23] is also limited as all plies have to be loaded with CNTs. This increase on "fiber density" due to the "CNTs loads" can lead to manufacturing limitations, for example, a severe decrease on resin flow channels into vacuum-assisted impregnation. It is clear that alternative techniques must be developed.

Different techniques have been tested for incorporating CNTs into composite materials. The CNT infusion into laminated composites and its alignment by applying an electric field after the infusion was studied by Domingues et al. [26]. The major criticism on Domingues' work is the amount of CNT dispersed which is around 0.1 wt.%. Another approach tried to link CNTs to laminated composites was implemented by Wu et al. [27]. Wu's work was based on electrochemical grafting of CNTs on carbon fibers surface. Although the technique described by Wu et al. [27] seems to be effective, it is limited to the CNT concentration into the solution. Moreover, as noticed by Wu, there were "preferential regions" for CNTs direct attachment to carbon fibers. These preferred sites were fibers' grooves and edges. This phenomenon led to a non-uniform distribution of CNT on carbon fibers surface. Another technique used for attaching CNTs to carbon fibers was studied by De Riccardis et al. [28] and Vilatela et al. [29]. In their case, the chemical vapor deposition (CVD) technique was employed for directly grown CNTs into carbon fibers. De Riccardis' work was based on deposition of nickel clusters and later on the CNTs were grown by hot filament chemical vapor deposition (HFCVD) technique. By using ferrocene as precursor and CVD as the growing process, Vilatela was also able to obtain good quality CNTs. Moreover, the CVD technique employed by Vilatela and collaborators [29] seems to be much simpler and easier to control. Although the results presented by De Riccardis et al. [28] and Vilatela et al. [29] seem to be encouraging, much work has to be done for applications to laminated composites, in special high performance carbon fiber/epoxy systems.

It is clear that CNTs can be used as potential reinforcement for nanocomposites and/or multi-scale composites, but recently GNs are emerging as another option to CNTs. According to Yasmin et al. [30], graphene-based nanocomposites can be an alternative option for engineering applications due to their outstanding specific strength and stiffness. However, the different routes to obtain GNs and their dispersion processes can lead to a large variety of mechanical properties. The first route for obtaining GNs was based on mechanical cleavage. Nevertheless, as discussed by Balandin et al. [31], this technique is time consuming and a series of defects can be introduced during this process. An alternative to mechanical cleavage is the chemical route.

As discussed by Stankovich et al. [32], a common route for graphite platelets, which can be used as GNs precursor, is based on graphite expansion. The expanded graphite (EG) is produced from graphite intercalation compounds through intercalant rapid evaporation at elevated temperature. Once the EG is obtained, techniques like ultrasonication and/or ball milling can be used to obtain graphite nanoplatelets and later on GNs. These graphite nanoplatelets consist of hundreds of stacked graphene layers. Few layers of graphene can be obtained by dispersing the graphite nanoplatelets into an aqueous solution using ultrasonication followed by ultracentrifugation. To overcome this extra procedure, Ruoff's research group [32] proposed the use of graphite oxide (GO) instead of EG. As commented by Allen et al. [33], the advantages of GO method are the low cost and massive scalability. However, the major criticism to GO method is the use of hydrazine for GO's chemical reduction. Cooper et al. [34] pointed out the hydrazine high toxicity and the environmental concerns about its use. As noted by Stankovich et al. [35], the chemical reduction has the objective of restoration of graphitic network of sp^2 bonds and consequently increases on electrical conductivity. However, coagulation can occur during the reduction of exfoliated GO nanoplatelets, which makes virtually impossible to disperse these nanostructures within polymeric matrices. Another problem was detected by Dreyer et al. [36]. According to them, the degree of oxidation caused by differences in starting graphite sources or oxidation protocol can cause substantial variations in the GO structure and properties. This was the case of the work reported by Marcano et al. [37]. By introducing a modification on Hummer's method, that is, they used a mix of H_2SO_4/H_3PO_4 at a ratio of 9:1 and by excluding the $NaNO_3$, they were able to obtain a larger amount of GO. However, the use of hydrazine for GO reduction was kept. Again, this harmful chemical component was employed, which is the major criticism of their work. A possible solution for hydrazine use was proposed by Shahil and Balandin [38]. They proposed a methodology based on natural graphite ultrasonication on an aqueous solution of sodium chlorate followed by centrifugation and mechanical exfoliation by ultrasonication and high shear mixing. This methodology is capable of obtaining multilayer graphene (MLG) consisting of 1–10 stacked atomic monolayers. A similar methodology was proposed by Ávila et al. [20]. Ávila and coworkers [20], however, employed N,N-Dimethylformamide (DMF) as a solvent. No aqueous solution of sodium chlorate was used. Ultrasonication (20 kHz for 2 hours) followed by a high shear mixing at 17,400 RPM (2 hours) lead to MLG with the number of graphene monolayers stacked between 2 and 50. After the DMF evaporation, the nanocomposite consolidation was made by dispersing the MLGs into the epoxy resin again using the high shear mix procedure under restrict temperature control (<70°C).

To be able to obtain an epoxy–MLG-based nanocomposite, Shahil and Balandin [38] simply dispersed the MLG into epoxy system by mechanical mixing. This procedure does not guarantee the homogeneity required for engineering applications. Moreover, it is a possible source of voids due to air bubbles entrapment during the mixing/cure procedure. Another dispersion procedure was employed by Yang et al. [39]. Yang's work, however, took the advantage of the hydrophilic condition of GO nanoplatelets after oxidation. Therefore, they first disperse the GO nanoplatelets into an aqueous (they employed a 1 mg/mL concentration) solution using ultrasonication and later on the resulting solution was added to the polyvinyl alcohol (PVA) solution. This procedure can only be employed due to the special nature of such nanocomposite (both PVA and GO are hydrophilic). Their mechanical results indicated an increase on tensile strength around 30% with addition of 3.5% graphene. This increase on tensile strength could be related to the decrease on crystallinity reported by Yang and coworkers [39]. It is a well-known fact that high degree of crystallinity makes polymer brittle and stiffer. By observing fracture surfaces, Yang et al. [39] noticed layered structures with uniformly dispersed GNs into PVA matrix. Moreover, the graphene dispersion into PVA matrix leads to a change into the overall macroscopic behavior from brittle to ductile. This phenomenon can be related to the decrease on crystallinity. A much higher increase on tensile strength was reported by Kuilla et al. [40], where the addition of 0.7% by weight of GO leads to an improvement on tensile strength close to 150%. This substantial increase seems to be related to the crystallinity changes and the solution blending technique employed. According to Wang et al. [41], an "extra" exfoliation is provided by this technique due to GO's swelling. Young et al. [42], however, provided another explanation for such increase. They recalled that GO's individual nanoplatelets are often wrinkled, which provides an additional superficial area. Rafiee et al. [43] went further, as they were able to link this "extra" superficial area effect into matrix toughening (in their case epoxy). It is important to point out that superficial area is not the only issue that must be considered. The total number of layers (which can be translated into total thickness), separation of graphene layers and shape are also important, as they can influence the matrix toughness by controlling the crack propagation. Moreover, according to Mukhopadhyay and Gupta [44], the surface energy is also another issue, as high surface energy help to improve dispersion into polymeric matrices. Unfortunately, a much easier dispersion provided by high surface energy also leads to a decrease on conductivity. This trade-off is the key issue for polymeric matrix/graphene nanocomposites.

A series of experiments on polymeric matrix/graphene nanocomposites were reported by Kuilla et al. [40]. Two are the major procedures that can be employed for nanocomposites' synthesis. The first one is the solution intercalation. This procedure is based on a solvent system in which graphene or MLG or GO is dispersed (in general by ultrasonication and/or high shear mixing) and allowed to swallow. Later on the intercalated solution is mixed to the polymeric matrix, and the solvent is evaporated. This procedure is particularly suitable for epoxy systems. The second major technique is related to melted intercalation. In this case, no solvent is required,

TABLE 32.1

Overall Description of Work Done on Carbon-Based Nanostructures

Work Description	References
Overall description of CNT nanostructure and bases and critical review	Mauter and Elimelech [1], Saito et al. [2]
Modeling and prediction of CNTs mechanical properties by discrete, continuum, and atomistic models	Kalamkarov et al. [3], Belytschko et al. [13], Liu et al. [16], Ruoff et al. [17]
Modeling CNTs by molecular dynamics approach	Frankland et al. [10], Jin and Yuan [11], Agrawal et al. [12]
Modeling CNTs by molecular mechanics approach	Li and Chou [18], Ávila et al. [19], Chou et al. [22]
Modeling CNTs-based composite materials	Odegard et al. [7], Lurie et al. [14], Gates et al. [15], Kim et al. [21], Chou et al. [22]
CNTs synthesis and growth	Saito et al. [2], Riccardis et al. [28], Vilatela et al. [29], Wicks et al. [23]
Experimental analysis of CNTs-based composites	Joshi et al. [8], Kim et al. [21], Wicks et al. [23], Shokrieh and Rafiee [24], Ma et al. [25], Domingues et al. [26], Wu et al. [27], De Riccardis et al. [28]
Graphene nanostructure description and applications	Geinand and Novoselev [4], Allen et al. [33], Cooper et al. [34]
Graphene synthesis	Tang et al. [6], Stankovich et al. [32], Dreyer et al. [36], Marcano et al. [37]
Measurements of graphene properties	Lee et al. [5], Balandin et al. [31]
Mechanics of graphene nanocomposites	Young et al. [42], Ávila et al. [19]
Graphene-based nanocomposites description and critical review	Mukhopadhyay and Gupta [44]
Synthesis and testing of graphene-based composite materials	Ávila et al. [20], Yasmin et al. [30], Stankovich et al. [35], Shahil and Balandin [38], Yang et al. [39], Kuilla et al. [40], Wang et al. [41], Rafiee et al. [43]
Experimental analysis of CNTs and graphene-based composites	Gouda et al. [9]

Source: Diego Thadeu Lopes da Cruz. Efeito da exposocao a luz ultravioleta em adesivo nanomodificado por grafeno. Thesis. Used with permission 2014.

but a homogeneous mixture of graphene or MLG or GO and the polymeric matrix has to be prepared. This homogenous mix is then melted by extrusion or injection molding. This practice is appropriate for thermoplastics, for example, high density polyethylene (HDPE), polypropylene (PP), and poly methyl methacrylate (PMMA). In all cases, that is, thermoplastics and thermosets, the addition of graphene or MLG or GO improves mechanical and/or electrical properties.

In summary, the use of carbon-based nanostructures such as CNTs and graphene/MLG is a viable option for composite materials. Table 32.1 shows an overall description of the work cited in this sub-section.

32.2 HYBRID COMPOSITES IN CONTEXT: ADHESIVES NANO-MODIFIED BY GRAPHENE

Adhesives have been used since the early ages of mankind, but until a century ago, the vast majority was obtained from natural products such as bones, skins, fish, milk, and plants. It was only in the beginning of the twentieth century that adhesives based on synthetic polymers have been introduced, but their usages were limited due to their high cost and poor mechanical properties. As discussed by Crocombe and Ashcroft [45], it was only in 1940s that a more scientific approach has employed to understand the adhesion phenomenon. The progresses into polymer sciences lead to the development of new adhesives and consequently an increase on load-bearing capacity. Furthermore, new designs [46–50] also contributed to the increase on bonded joints efficiency. A comprehensive study on single lap joints (SLJs) was performed by Hart-Smith [46]. The designs created by Zeng and Sun [47], and Ávila and

Bueno [48] employed the concept of changing the peel stress distribution by employing the wavy shape configuration to the bonded area. A similar idea was employed by Ashrafi et al. [49] in which the wavy shape was applied at the adherent adhesive interface. An analytical study on damage tolerance of bonded joints was described by Romilly and Clark [50]. They were able to develop a close form solution for cracked bonded structures. However, no environmental conditions were considered.

As commented by Banea and da Silva [51], bonded joints are frequently expected to sustain static or cyclic loads for considerable periods of time without any adverse effect on the load-bearing capacity of the structure. Even more important, as described by Petrova and Lukina [52], adhesive joints used in aerospace industry, for example, airplane and helicopter structures operate in the loaded state. Therefore, they must hold a high stability toward a variety of mechanical, chemical, and physical changes under service conditions. These changes are, in most cases, triggered by environmental conditions. According to Kablov et al. [53], the climate changes can be represented by hygrothermal cycling. Although moisture diffusion and temperature changes are important factors to understand the adhesive mechanical and physical chemical changes, another issue must be considered, that is, UV light exposure.

The polymeric matrices photo-degradation singularity by UV light was addressed by various researchers [54–59] in different ways. Woo et al. investigated the residual properties of epoxy/nanoclay nanocomposites exposed to UV light and moisture. As expected, the flexural modulus decreased due to plasticization effect of moisture. Another well-known effect of UV light, the embrittlement, was also detected by Woo and collaborators. However, Woo was not able to clearly identify the nanoclay effect into the nanocomposites' aging process.

Another investigation into the nanostructures effect into polymeric composite aging was performed by Allaoui et al. [55], where CNTs were dispersed into epoxy systems. They were not able to clearly assess the influence of CNTs into the aging process. Their conclusions were mostly overshadowed by the mechanical improvements provided by the CNTs. The work done by Mailhot et al. [56] focused on how the nanoparticle/nanostructures morphology affected the aging process. Again, the conclusions seem to be obvious, the nanoparticles/nanostructures' morphology did not affect the aging process. Chemical changes are driven by photo-degradation and not by the nano-fillers. The work done by Larché et al. [57] addressed this issue for the phenoxy resin. The mechanism proposed by Larché et al. [57] proposed a correlation between the molecular cross-link caused by UV radiation and the micro-cracks formed into the composite's surface. The same analogy was proposed by Dao et al. [58] for epoxy-based composites, in their case 8552/IM7, a year early. As discussed by Dao, the thermal-oxidation phenomenon leads to molecular stiffening, shrinking, and micro-cracking. Furthermore, the chemical degradation of 8552/IM7 fiber composite indicated that the degradation phenomenon was multifaceted and temperature and humidity dependent. Their conclusions were that modeling of real life degradation by thermally accelerated aging and Arrhenius extrapolations of the results were not able to produce very accurate predictions. The work done by Chang and Chow [59] went further, as they were capable of correlating the UV exposure with interface debonding for glass/epoxy/organoclay hybrid composites. Although no information about the debonding mechanism was provided by the authors, a possible explanation for this phenomenon could be the mismatch on thermal expansion coefficient between the glass fibers and the nano-modified epoxy/MMT system. Even though the literature on polymeric matrices aging by hygrothermal and UV radiation addressed different aspects of mechanical properties degradation, bonded joints have special characteristics and are treated using a different approach.

Higgins [60] discussed the durability of adhesive bonds as a function of weathering conditions. As Higgins pointed out that subsonic airplanes, bonded joints have been applied for primary structures and one of the most important issues is their shear lap strength after exposition to temperature variations (typically an airplane has to operate in temperatures from −55°C to +80°C), humidity and most of the fluids used during the plane's operation. Sugiman et al. [61] focused their attention to the humidity diffusion process through the bonded joints, in their case SLJs, and the correspondent loss of strength. In their work, the SLJs made of aluminum 2024-T3 adherents and FM73 adhesives were immersed into deionized water at 50°C for one and two years, respectively. They noticed that after a year, the FM73 reached its saturation limit. Furthermore, they also observed a decrease on bearing-load capacity of 22.1% after one year of water immersion and 24.2% after the second year. By analyzing Sugiman's results, it is possible to conclude that water diffusion is an important issue. However, for real life airplane operations (passing through clouds, rain, de-icing, etc.), the water diffusion effect plays a secondary role, as there is no enough time for water diffusion into the bonded joints. More specifically, there is no enough time for the moisture to damage the adhesive/adherent interface region. As commented by Datla et al. [62], the loss of bearing-load capacity in bonded joints is, in general, attributed to the weaknesses into this region. It is possible, however, to have damage inside the adhesive itself. The work done by Knight et al. [63] concluded that under hydrothermal conditions (moisture diffusion associated to temperature variations) lead to changes on failure mode from cohesive zone (within adhesive layer) to fiber tear (within the top layers of the adherent). These facts can be attributed to the adhesive layer porosity and the voids formation during the laminate consolidation. Baldan [64], however, pointed out another important issue, the adhesive and adherent degradation by UV radiation.

32.3 EXPERIMENTAL PROCEDURES

The SLJs were made of aluminum 2024-T3 with a thickness of 2.8 mm and a width of 25.4 mm. The SLJs geometry followed the ASTM D 5868-10 standard [65]. Figure 32.1 describes the single-lap dimensions. The adhesive employed in this research is an epoxy system AR300/AH30-150 supplied by Barracuda

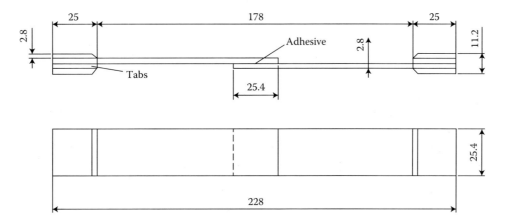

FIGURE 32.1 Single-lap joint dimensions in millimeters. (From Diego Thadeu Lopes da Cruz. Efeito da exposocao a luz ultravioleta em adesivo nanomodificado por grafeno. Thesis. Used with permission 2014.)

Composites. The adhesive was nano-modified by graphene dispersion. The graphene nanostructures dispersion process into the epoxy system followed the methodology described by Ávila et al. [20], that is, ultra-sonication at 20 KHz for 60 minutes followed by high shear mixing at 17400 RPM for 60 minutes. The graphene concentrations dispersed into the adhesives were 1 wt.% and 2 wt.% with respect to the adhesive weight. For benchmark purposes, a set of SLJs without graphene was also prepared and tested.

To be able to reproduce the day light and following the ASTM G154-12 standard [66], four UVA light of 30 watts with a wavelength of 340 nm were employed. Each set of SLJs (5 specimens) were exposed to the UVA lights in a sealed chamber. The UVA light exposure times were no aging (time zero), 100 hours, 200 hours, and 400 hours. These time intervals were selected based on similar experiment performed by Lin et al. [67]. Once the SLJs were aged, the tensile tests following the ASTM D 5868-10 standard [65] were performed. The nanoindentation tests were accomplished by Asylum

Research's Atomic Force Microscope; model MFP-3D, with a diamond Berkovich tip. Finally, to understand the adhesive's chemical changes under UVA light a Fourier Transform Infrared (FTIR) analysis was executed in all samples. The experiments were carried out on an IF66 spectrometer (Brucker Optics Inc.) at transmission mode and mid-range infrared, that is, from 400 to 4000 cm^{-1}, with a nominal resolution of 4 cm^{-1} resolution and 32 summations.

A finite element analysis using plane strain condition and linear 4 node elements [68] was performed to be capable of tracking the peel and shear stress fields. The adhesive was considered isotropic and its Young's modulus was based on the results from nanoindentation tests.

32.4   DATA ANALYSIS AND DISCUSSION

To understand how UVA light affects the nano-modified AR300/AH30-150 adhesive, a set of nanoindentation tests were performed. The nanoindentation tests were done in

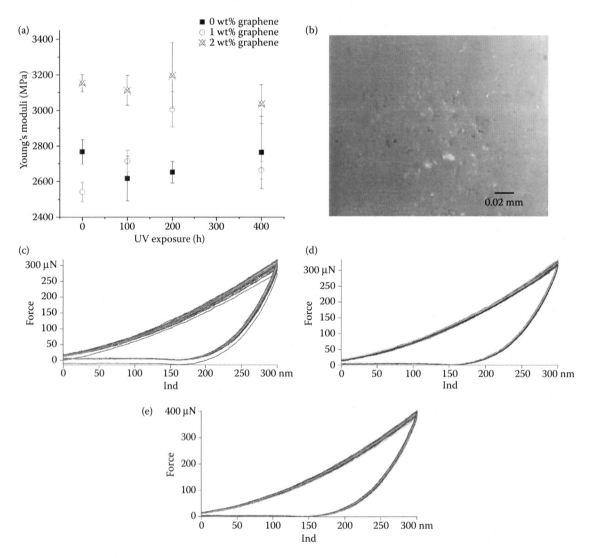

FIGURE 32.2   Nanoindentation results: (a) Young's moduli variation; (b) porous surface; (c) force–displacement for 0 wt.%; (d) force–displacement for 1 wt.%; (e) force–displacement for 2 wt.%. (From Diego Thadeu Lopes da Cruz. Efeito da exposocao a luz ultravioleta em adesivo nanomodificado por grafeno. Thesis. Used with permission 2014.)

different locations to be able to check if the graphene nano-structures were evenly distributed. As suggested by Mailhot et al. [69], the elastic moduli of aged polymeric matrices were kept unchanged for layers deeper than 40 μm from the exposed surface. Therefore, all nanoindentation tests were done at the same distance from the surface, that is, 300 nm. This depth is enough to avoid surface roughness influence into the measurements. Figure 32.2a shows the nanocomposite Young's moduli obtained by nanoindentation for specimens with 0 wt.%, 1 wt.%, and 2 wt.% of graphene, at time equals to 0, 100, 200, and 400 hours, respectively. As expected, the graphene addition led to an increase on stiffness. However, an interesting phenomenon can be spotted for the 1 wt.% at time zero and 400 hours. An apparent softening process seems to occur. The decrease on stiffness can be explained by void formation during the cure process (Figure 32.2b). Although some variations seem to be spotted, the curves (force–displacement) laid down in the same path (Figures 32.2c and e). By observing these graphs, two conclusions can be done: (i) the graphene nanostructures seem to be uniformly distributed; (ii) the UVA radiation is also uniformly spread through the entire sample area. It is important to mention that in each sample at least 36 nanoindentation tests were done in different locations.

The nanoindentation results seem to indicate that graphene nanostructures dispersed into the AR300/AH30-150 epoxy system somehow blocked the polymer degradation due to UVA radiation. This shielding effect can be explained by the 2D shape of graphene blocks dispersed into the polymeric matrix as it can be observed in Figures 32.3. To understand how the graphene nanostructures affected the AR300/AH30-150 adhesive degradation, the FTIR tests were performed for each set of specimen. Figure 32.4a, c shows the FTIR signatures at the transmission mode to identify the degradation products as described by Xiao et al. [70]. A tentative assignment of the major bands in FTIR of the studied epoxy system is listed in Table 32.2, as in Ngono and Maréchal [71].

As it can be noticed, during the aging process, the hydroxyl stretching increases, which can be an indication of water intake. The aromatic ring stretching decreased with the aging process for the no-graphene samples. A different pattern was noted for the samples with the graphene addition. The aromatic ring stretching (1612-1545-1512 cm^{-1}) increased with the aging time. This stretching associated to the increase on epoxy functional (918 cm^{-1}) and p-Phenylene groups (829 cm^{-1}) can explain the increase on stiffness observed.

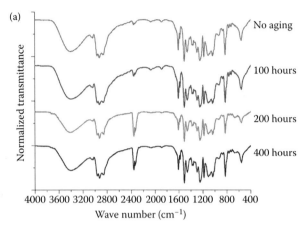

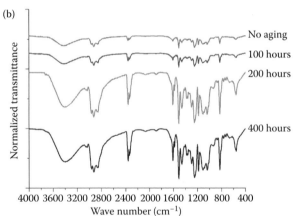

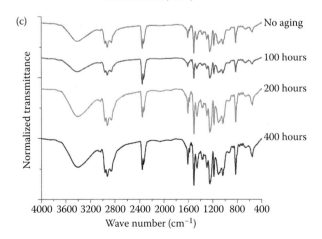

FIGURE 32.4 FTIR signature: (a) control samples; (b) 1 wt.% graphene samples; (c) 2 wt.% graphene samples. (From Diego Thadeu Lopes da Cruz. Efeito da exposocao a luz ultravioleta em adesivo nanomodificado por grafeno. Thesis. Used with permission 2014.)

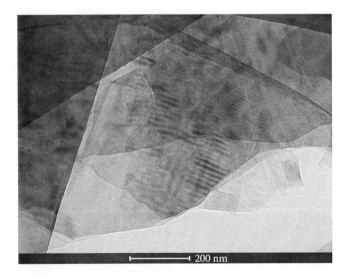

FIGURE 32.3 Graphene 2D structure by TEM. (From Diego Thadeu Lopes da Cruz. Efeito da exposocao a luz ultravioleta em adesivo nanomodificado por grafeno. Thesis. Used with permission 2014.)

TABLE 32.2
Tentative Assignment of Major Bands

Bands (cm⁻¹)	Assignment
3406	Hydroxyl (–OH) stretching
2968–2850	Alkyl units (C–H) stretching
1653	Carbonyl group (C = O)
1612–1545–1512	Aromatic ring stretching
1297	Twisting mode of –CH$_2$– groups
1255	Stretching mode for aromatic ether
1182	C–C stretching of two p-Phenylene groups
1041	Stretch of the *trans* forms of the ether linkage
918	Epoxy functional group
829	p-Phenylene groups

Source: Diego Thadeu Lopes da Cruz. Efeito da exposocao a luz ultravioleta em adesivo nanomodificado por grafeno.Thesis. Used with permission 2014.

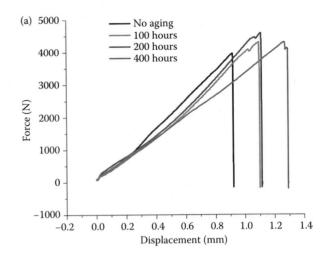
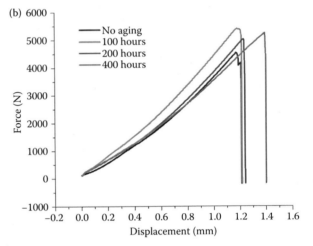
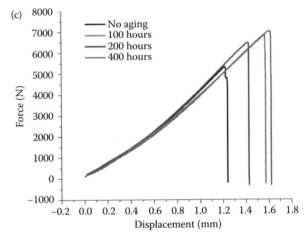

FIGURE 32.6 Single lap joint force–displacement plots: (a) control samples; (b) 1 wt.% graphene samples; (c) 2 wt.% graphene samples. (From Diego Thadeu Lopes da Cruz. Efeito da exposocao a luz ultravioleta em adesivo nanomodificado por grafeno. Thesis. Used with permission 2014.)

However, for the 400 hours aging samples with graphene, a small decrease on stiffness was observed. This decrease can be explained by the formation of micro-cracks into the samples' surface. These micro-cracks create a false sense of softening. This phenomenon is demonstrated by the hardness measurements obtained during the nanoindentation tests and is represented by Figure 32.5.

By analyzing Figure 32.6a through c, it is possible to conclude that graphene dispersion leads to an increase on SLJ load capacity. As commented by Woo et al. [54], previous studies on epoxy systems photo-degradation suggested formation of carbonyl and hydroxyl groups through chain scission and hydrogen abstraction from the polymer backbone. The graphene blocks nanostructures seem to delay the oxygen and

free radical penetration and the degradation as described by Woo et al. [54]. However, as the number of hours of UVA exposure increases, another factor takes place, that is, the formation of small micro-cracks, which could be the reason for this decrease on SLJ load capacity. Yet, a possible explanation

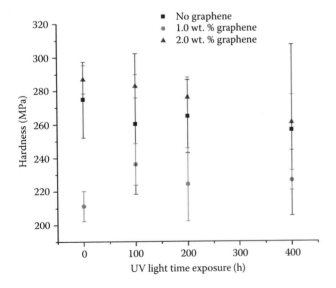

FIGURE 32.5 Nanoindentation hardness. (From Diego Thadeu Lopes da Cruz. Efeito da exposocao a luz ultravioleta em adesivo nanomodificado por grafeno. Thesis. Used with permission 2014.)

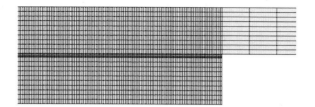

FIGURE 32.7 Finite element mesh representation. (From Diego Thadeu Lopes da Cruz. Efeito da exposocao a luz ultravioleta em adesivo nanomodificado por grafeno. Thesis. Used with permission 2014.)

for the increase in strain (large displacement), at the same time the stress (load capacity) is increased, is the competing process between the carbonyl and hydroxyl groups formation and recombination of hydrogen and carbon in the surrounded areas of graphene nanostructures. This competing process could cause a swelling that can be the reason for such increase on displacement. It is important to mention that Fernandez-Garcia and Chiang [72] detected a swelling into epoxy systems during the aging process. Unfortunately, a relationship between the swelling phenomenon and the aging process is not completely established.

To understand how the graphene dispersion and UVA light aging affected the peel and shear distribution inside the SLJs, a finite element simulation was performed. Notice that as discussed by Kumar et al. [73], the two most important stresses into SLJs are σ_{yy} (peel) and τ_{xy} (shear). The finite element model has 18,096 elements and 112,186 degrees-of-freedom. Figure 32.7 shows a representation of the mesh at the bonded line. The two stress fields (peel and shear) are summarized in Table 32.2. As discussed by Turaga and Sun [74], the stress distribution inside the adhesive calculated by the finite element method is meshing dependent. Therefore, a very refined mesh was used and this mesh was kept the same for all simulations. The adhesive was assumed to be isotropic and homogeneous with Young's modulus given by the nanoindentation tests, and Poisson's ratio of 0.34 was from Turaga and Sun [74].

By observing Table 32.3 it is possible to detect an increase on peel and shear stresses at failure. The increase on peel (σ_{yy})

and shear (τ_{xy}) stresses when compared against the reference samples (no aging and no graphene dispersion) to the 2 wt.% graphene samples aged for 400 hours was approximately of 82%. The two major factors for such increase seem to be the graphene addition and the UVA exposure. The Von Mises equivalent stress distribution, for no graphene addition at time zero, is shown in Figure 32.8. As it can be observed, the maximum value is located at the edge, as predicted by He [75].

32.5 CLOSING COMMENTS

After studying the thermo-mechanical behavior of epoxy/ graphene hybrid composite, some conclusions can be drawn. First, the force–displacement curves obtained by nanoindentation seem to indicate a good dispersion process, as the large majority of the curves laid down at the same path. Furthermore, the nanoindentation revealed an increase on stiffness when graphene was added to the epoxy system. Another interesting phenomenon is the UVA light exposure somehow seems that it favorably affected the epoxy system. The UVA exposure promoted chemical changes detected by FTIR techniques. These chemical changes on hydroxyl, carbonyl, epoxy, and p-Phenylene groups seem to allow an increase on stretching by swelling that reflected an overall increase on load capacity. When the SLJs were tested, the results also indicated an average increase on bearing-load capacity of 40.96% and 72.03% for 1 wt.% and 2 wt.%, respectively for 100 hours aging. When the aging reached 200 hours, there is a decrease on load capacity when compared against the 100 hours results. The average load capacity was 27.03% and 58.19% higher than the not aged nano-modified AR300/AH30-150 SLJs. At 400 hours, there is another increase on load capacity 32.06% and 74.43% for 1 wt.% and 2 wt.% graphene. A finite element simulation revealed that peel and shear stresses at adhesive edge increased by 82%. The Fourier Transformed Infrared tests revealed that chemical changes on hydroxyl, carbonyl, and epoxy components could be the reason for this behavior. The graphene dispersion into the epoxy adhesives seems to have a double folded effect; in one hand, it increases the bonded joint capacity, and on the other hand, blocks the aging effect of UVA light.

TABLE 32.3
Peel and Shear Stresses as Function of Time and Graphene Concentration

UV Time Exposure	No Graphene		1 wt.% Graphene		2 wt.% Graphene	
	Peel (kPa)	Shear (kPa)	Peel (kPa)	Shear (kPa)	Peel (kPa)	Shear (KPa)
0	2993.8	−315.8	3266.4	−342.8	4233.3	−450.7
100	3223.0	−338.6	4187.0	−440.9	5406.4	−574.9
200	3420.0	−359.8	3935.0	−417.5	5024.7	−535.4
400	3426.1	−361.2	3892.0	−409.4	5425.9	−575.9

Source: Diego Thadeu Lopes da Cruz. Efeito da exposição a luz ultravioleta em adesivo nanomodificado por grafeno. Thesis. Used with permission 2014.

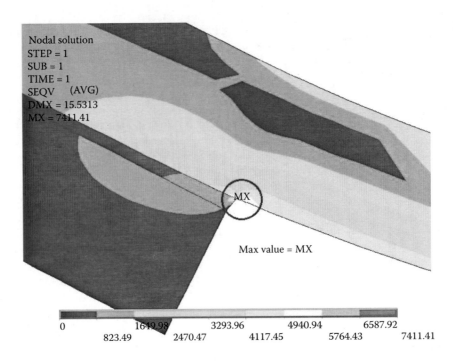

FIGURE 32.8 Von Mises stress distribution at the failure region. (From Diego Thadeu Lopes da Cruz. Efeito da exposocao a luz ultravioleta em adesivo nanomodificado por grafeno. Thesis. Used with permission 2014.)

ACKNOWLEDGMENTS

This research was supported in part by the AFOSR under contract FA9550-10-1-0050, the Brazilian Research Council (CNPq) under grants numbers 303447/2011-7, 472583/2011-5 and the Minas Gerais State Research Foundation (FAPEMIG) grant TEC-PPM00192-12. The authors are grateful to the UFMG's Center of Microscopy and Microanalysis for the technical support and the Nacional Grafite Incorporated for supplying the graphite from which the graphene blocks were obtained.

REFERENCES

1. Mauter, M.S. and Elimelech, M. 2008. Environmental applications of carbon-based nanomaterials, *Environmental Science and Technology*, 42, 5843–5859.
2. Saito, R., Dresselhaus, G., and Dresselhaus, M.S. 2005. *Physical Properties of Carbon Nanotubes*, Imperial College Press, London, UK.
3. Kalamkarov, A.L., Georgiades, A.V., Rokkam, S.K., Veedu, V.P., and Ghasemi Nejhad, M.N. 2006. Analytical and numerical techniques to predict carbon nanotubes properties, *International Journal of Solids and Structures*, 43, 6832–6854.
4. Gein, A.K. and Novoselev, K.S. 2007. The rise of graphene, *Nature Materials*, 6, 183–191.
5. Lee, C., Wei, X., Kysar, J.W., and Hone, J. 2008. Measurements of the elastic properties and intrinsic strength of monolayer graphene, *Science*, 321, 385–388.
6. Tang, H., Ehlert, G.J., Lin, Y., and Sodano, H.A. 2012. Highly efficient synthesis of graphene nanocomposites, *Nano Letters*, 12, 84–90.
7. Odegard, G.M., Gates, T.S., Wise, K.E., Park, C., and Siochi, E.J. 2003. Constitutive modeling of nanotube-reinforced polymer composites, *Composites Science and Technology*, 63, 1671–1687.
8. Joshi, S.C. and Dikshit, V. 2012. Enhancing interlaminar fracture characteristics of woven CFRP prepreg composites through CNT dispersion, *Journal of Composite Materials*, 46, 665–675.
9. Gouda, P.S.S., Kulkarni, R., Kurbet, S.N., and Jawali, D. 2013. Effects of multi walled carbon nanotubes and graphene on the mechanical properties of hybrid polymer composites, *Advanced Materials Letters*, 4, 261–270.
10. Frankland, S.J.V., Harik, V.M., Odegard, G.M., Brenner, D.W., and Gates, T.S. 2003. The stress–strain behavior of polymer–nanotube composites from molecular dynamics simulation, *Composites Science and Technology*, 63, 1655–1661.
11. Jin, Y. and Yuan, F.G. 2003. Simulation of elastic properties of single-walled carbon nanotubes, *Composites Science and Technology*, 63, 1507–1515.
12. Agrawal, P.M., Sudalayandi, B.S., Raff, L.M., and Komanduri, R. 2006. A comparison of different methods of Young's modulus determination for single-wall carbon nanotubes (SWCNT) using molecular dynamics simulations, *Computational Materials Science*, 38, 271–281.
13. Belytschko, T., Xiao, S.P., Schats, G.C., and Ruoff, R.S. 2002. Atomistic simulations of nanotube fracture, *Physics Review B*, 65, 2354301–2354308.
14. Lurie, S., Belov, P., Volkov-Bogorodsky, D., and Tuchkova, N. 2003. Nanomechanical modeling of the nanostructures and dispersed composites, *Computational Materials Science*, 28, 529–539.
15. Gates, T.S., Odegard, G.M., Frankland, S.J.V., and Clancy, T.C. 2005. Computational materials: Multi-scale modeling and simulation of nanostructured materials, *Composites Science and Technology*, 65, 2416–2434.
16. Liu, W.K., Karpov, E.G., Zhang, S., and Park, H.S. 2004. An introduction to computational nanomechanics and materials, *Computational Methods in Applied Mechanics and Engineering*, 193, 1529–1578.

17. Ruoff, R.S. and Pugno, N. 2006. *Mechanics of Nanostructures*, in: Chuang, T.S. et al. (Eds.) *Nanomechanics of Materials and Structures*, Springer, New York, pp. 199–203.

18. Li, C. and Chou, T-W. 2003. A structural mechanics approach for the analysis of carbon nanotubes, *International Journal of Solids and structures*, 40, 2487–2499.

19. Ávila, A.F., Eduardo, A.C., and Silva Neto, A. 2011. Vibrational analysis of graphene based nanostructures, *Computers and Structures*, 89, 878–892.

20. Ávila, A.F., Yoshida, M.I., Carvalho, M.G.R., Dias, E.C., and Ávila Jr., J. 2010. An investigation on post-fire behavior of hybrid nanocomposites under bending loads, *Composites Part B*, 41, 380–387.

21. Kim, M., Park, Y-B., Okoli, O.I., and Zhang, C. 2009. Processing, characterization, and modeling of carbon nanotube-reinforced multiscale composites, *Composites Science and Technology*, 69, 335–342.

22. Chou, T-W., Gao, L., Thostenson, E., Zhan, Z., and Byun, J-H. 2010. An assessment of the science and technology of carbon nanotube-based fibers and composites, *Composites Science and Technology*, 70, 1–19.

23. Wicks, S.S., Villanova, R.G., and Wardle, B.L. 2010. Interlaminar and intralaminar reinforcement of composite laminates with aligned carbon nanotubes, *Composites Science and Technology*, 70, 20–28.

24. Shokrieh, M.M. and Rafiee, R. 2010. Investigation of nanotube length effect on the reinforcement efficiency in carbon nanotube based composites, *Composite Structures*, 92, 2415–2420.

25. Ma, P-C., Siddiqui, N.A., Marom, G., and Kim, J.K. 2010. Dispersion and functionalization of carbon nanotubes for polymer-based nanocomposites: A review, *Composites Part A*, 41, 1345–1367.

26. Domingues, D., Logakis, E., and Skordos, A.A. 2012. The use of an electric field in the preparation of glass fibre/epoxy composites containing carbon nanotubes, *Carbon*, 50, 2493–2503.

27. Wu, G-P., Wang, Y.Y., Li, D-H., Lu, C-X., Shen, W-Z., Li, X.T., and Feng, Z-X. 2011. Improving the electrical conductivity of multi-walled carbon nanotube networks by H ion beam irradiation, *Carbon*, 49, 2141–2161.

28. De Riccardis, M.F., Carbone, D., Makris, Th. D., Giorgi, R., Lisi, N., and Salernitano, E. 2006. Anchorage of carbon nanotubes grown on carbon fibres, *Carbon*, 44, 671–674.

29. Vilatela, J.J., Deng, L., Kinloch, I.A., Young, R.J., and Windle, R.D. 2011. Structure of and stress transfer in fibres spun from carbon nanotubes produced by chemical vapor deposition, *Carbon*, 49, 4149–4158.

30. Yasmin, A., Luo, J-J., and Daniel, I.M. 2006. Processing of expanded graphite reinforced polymer nanocomposites, *Composites Science and Technology*, 66, 182–1189.

31. Balandin, A.A., Ghosh, S., Bao, W., Calizo, I., Teweldebrhan, D., Miao, F., and Lau, C.N. 2008. Superior thermal conductivity of single-layer graphene, *Nano Letters*, 8, 902–907.

32. Stankovich, S., Dikin, D.A., Piner, R.D., Kohlhaas, K.A., Kleinhammes, A., Jia, Y., Wu, Y., Nguyen, S.T., and Ruoff, R.S. 2007. Synthesis of graphene-based nanosheets via chemical reduction of exfoliated graphite oxide, *Carbon*, 45, 1558–1565.

33. Allen, M.J., Tung, V.C., and Kaner, R.B. 2010. Honeycomb carbon: A review of graphene, *Chemical Review*, 110, 132–145.

34. Cooper, D.R., D'Anjou, B., Ghattamaneni, N., Harack, B., Hilke, M., Horth, A., Majlis, N., Massicotte, M., Vandsburger, L., Whiteway, E., and Yu, V. 2012. Experimental review of graphene, *Condensed Matter Physics*, 1, 1–56.

35. Stankovich, S., Dikin, D., Dommett, G.H.B., Kohlhaas, K.M., Zimmey, E.J., Stach, E.A., Piner, R.D., Nguyen, S.T., and Ruoff, R.S. 2006. Graphene-based composite materials, *Nature*, 442, 282–286.

36. Dreyer, D.R., Park, S., Bielawski, C.W., and Ruoff, R.S. 2010. The chemistry of graphene oxide, *Chemistry Society Review*, 39, 228–240.

37. Marcano, D.C., Kosynkin, D.V., Berlin, J.M., Sinitskii, A., and Sun, Z. 2010. Improved synthesis of graphene oxide, *ACS Nano*, 4, 4806–4814.

38. Shahil, K.M.F. and Balandin, A.A. 2012. Graphene-multilayer graphene nanocomposites as highly efficient thermal interface materials, *Nano Letters*, 12, 861–867.

39. Yang, X., Li, L., Shang, S., and Tao, X-M. 2010. Synthesis and characterization of layer-aligned poly(vinyl alcohol)/graphene nanocomposites, *Polymer*, 51, 3431–3435.

40. Kuilla, T., Bhadra, S., Yao, D., Kim, N.H., Bose, S., and Lee, J.H. 2010. Recent advances in graphene based polymer composites, *Progress in Polymer Science*, 35, 1350–1375.

41. Wang, M., Cheng, Y., and Lin, M. 2012. Graphene nanocompostites, *Composites and their Properties*, 1, 17–36.

42. Young, R.J., Kinloch, I.A., Gong, L., and Novoselov, K.S. 2012. The mechanics of graphene nanocomposites: A review, *Composites Science and Technology*, 72, 1459–1476.

43. Rafiee, M.A., Rafiee, J., Srivastava, I., Wang, Z., Song, H., Yu, Z-Z., and Koratkar, N. 2010. Fracture and fatigue in graphene nanocomposites, *Small*, 6, 179–183.

44. Mukhopadhyay, P. and Gupta, R.K. 2011. Trends and frontiers in graphene-based polymer nanocomposites, *Plastics Engineering*, 1, 32–42.

45. Crocombe, A.C. and Ashcroft, I.A. 2008. Single lap joint geometry, in *Modeling of Adhesively Bonded Joints*, da Silva, L.C.M. and Ochsner, A. (Eds.) Springer, Berlin, pp. 3–24.

46. Hart-Smith, L.J. 1973. Adhesive bonded single lap joints, *NASA CR 112236*.

47. Zeng, Q. and Sun, C.T. 2001. Novel design of bonded lap joint, *AIAA Journal*, 39, 1991–1996.

48. Ávila, A.F. and Bueno, P.O. 2004. Stress analysis on a wavy-lap bonded joint for composites, *International Journal of Adhesives and Adhesion*, 24, 407–414.

49. Ashrafi, M., Ajdar, A., Rahba, N., Papadopoulos, J., Nayeb-Hashemi, H., and Vaziri, A. 2012. Adhesively bonded single lap joints with non-flat interfaces, *International Journal of Adhesives and Adhesion*, 32, 46–52.

50. Romilly, D.P. and Clark, R.J. 2008. Elastic analysis of hybrid bonded joints and bonded composite repairs, *Composite Structures*, 82, 563–576.

51. Banea M.D. and da Silva, L.F.M. 2009. Adhesively bonded joints in composite Materials: An overview, *Proceedings of the Institution of Mechanical Engineers, Part L: Journal of Materials Design and Applications*, 223, 1–18.

52. Petrova, A.P. and Lukina, N.F. 2007. Behavior of epoxy adhesive joints under service conditions, *Polymer Science Series. C*, 49, 99–105.

53. Kablov, E.N., Startsev, O.V., Krotov, A.S., and Kirillov, V.N. 2011. Climatic aging of composite aviation materials: I. Aging mechanisms, *Russian Metallurgy*, 2011, 993–1000.

54. Woo, R.S.C., Zhu, H., Leung, C.K.Y., and Kim, J-K. 2008. Environmental degradation of epoxy-organoclay nanocomposites due to UV exposure: Part II residual mechanical properties, *Composites Science and Technology*, 68, 2149–2155.

55. Allaoui, A., Evesque, J.P., and Bai, B. 2008. Effect of aging on the reinforcement efficiency of carbon nanotubes in epoxy matrix, *Journal of Materials Science*, 43, 5020–5022.

56. Mailhot, B., Morlat-Thérias, S., Bussière, P-O., Pluart, L.L., Duchet, J., Sautereau, H., Gérard, J-F., and Gardette, J-L. 2008. Photoageing behavior of epoxy nanocomposites: Comparison between spherical and lamellar nanofillers, *Polymer Degradation and Stability*, 93, 1786–1792.

57. Larché, J-F., Bussière, P-O., Thérias, S., and Gardette, J-L. 2012. Photooxidation of polymers: Relating material properties to chemical changes, *Polymer Degradation and Stability*, 97, 25–34.

58. Dao, B., Hodgkin, J., Kristina, J, Mardel, J., and Tian, W. 2006, Composite materials. II. Chemistry of thermal aging in a structural composite, *Journal of Applied Polymer Science*, 102, 3221–3232.

59. Chang, L.N. and Chow, W.S. 2010. Accelerated weathering on glass fiber/epoxy/organo-montmorillonite nanocomposites, *Journal of Composite Materials*, 44, 1421–1434.

60. Higgins, A. 2000. Adhesive bonding of aircraft structures, *International Journal of Adhesion and Adhesives*, 20, 367–376.

61. Sugiman, S., Crocombe, A.D., and Aschroft, I.A. 2013. Experimental and numerical investigation of static response of environmental ageing adhesively bonded joints, *International Journal of Adhesion and Adhesives*, 40, 224–237.

62. Datla, N.V., Ulicny, J., Carlson, B., Papini, M., and Spelt, J.K. 2011. Mixed-mode fatigue behavior of degraded epoxy adhesive joints, *International Journal of Adhesion and Adhesives*, 31, 88–96.

63. Knight, K.A., Hou, T.H., Belcher, M.A., Palmieri, F.L., Wohl, C.J., and Connell, W.J. 2012. Hygrothermal aging of composite single lap shear specimens comprised of AF-555M adhesive and T800H/3900-2 adherent, *International Journal of Adhesion and Adhesives*, 39, 1–7.

64. Baldan, A. 2004. Adhesively-bonded joints and repairs in metallic alloys, polymers and composite materials: Adhesives, adhesion theories and surface pre-treatment, *Journal of Materials Science*, 39, 1–39.

65. ASTM D5868-10, 2010. Test method for lap shear adhesion for fiber reinforced plastic (FRP) bonding, *ASTM Standards*, 518–529.

66. ASTM G154-12, 2012. Standard practice for operating fluorescent ultraviolet (UV) lamp apparatus for exposure of nonmetallic materials, *ASTM Standards*, 1–12.

67. Lin, Y.C., Chen, X., Zhang, H.P., and Wang, Z.P. 2006. Effects of hygrothermal aging on epoxy-based anisotropic conductive film, *Materials Letters*, 60, 2958–2963.

68. Gu, L., Kasavajhala, A.R.M., and Zhao, S. 2011. Finite element analysis of cracks in aging aircraft structures with bonded composites-patch repairs, *Composites Part B*, 42, 505–510.

69. Mailhot, B, Bussiere, P-O., Rivaton, A., Morlat-Therias, S., and Gardette, J.L. 2004. Depth profiling by AFM nanoindentations and micro-FTIR spectroscopy for the study of polymer ageing, *Macromolecules Rapid Communications*, 25, 436–440.

70. Xiao, G.Z., Delamar, J.M., and Shanahani, M.E.R. 1997. Irreversible interactions between water and DGEBA/DDA epoxy resin during hygrothermal aging, *Journal of Applied Polymer Science*, 65, 449–458.

71. Ngono, Y. and Maréchal, Y. 2000. Epoxy–amine reticulates observed by infrared spectrometry. II. Modifications of structure and of hydration abilities after irradiation in a dry atmosphere, *Journal of Polymer Science: Part B: Polymer Physics*, 38, 329–340.

72. Fernandez-Garcia, M. and Chiang, M.Y.M. 2002. Effect of hygrothermal aging history on sorption process, swelling, and glass transition temperature in a particle-filled epoxy-based adhesive, *Journal of Applied Polymer Science*, 84, 1581–1591.

73. Kumar, B., Sun, C.T., Wang, P.H., and Sterkenburg, R. 2010. Adding additional load paths in a bonded/bolted hybrid joint, *AIAA Journal*, 47, 1593–1598.

74. Turaga, U.V.R.S. and Sun, C.T. 2008. Improved design for metallic and composite single-lap joints, *Journal of Aircraft*, 45, 440–447.

75. He, X. 2011. A review of finite element analysis of adhesively bonded joints, *International Journal of Adhesion and Adhesives*, 31, 248–264.

33 Graphene Dispersion by Polymers and Hybridization with Nanoparticles

Po-Ta Shih, Kuo-Chuan Ho, and Jiang-Jen Lin

CONTENTS

Abstract 529
33.1 Introduction 529
33.2 Dispersion of Carbon Materials via Surface Modifications 530
    33.2.1 Surface Modifications of Carbon Nanotubes 530
        33.2.1.1 Via Covalent Bonding 530
        33.2.1.2 Via Non-Covalent Bonding (π–π Interactions) 531
    33.2.2 Surface Modifications of Graphite 532
        33.2.2.1 Via Graphene Oxide 532
        33.2.2.2 Via Covalent Bonding with Organic Functionalities 533
        33.2.2.3 Via Non-Covalent π–π Interactions 533
33.3 Hybridization of Graphene with Various Functional Nanoparticles 534
    33.3.1 Nanoparticles-on-Graphene 534
    33.3.2 Applications of Graphene Hybrids 534
    33.3.3 AgNP/Grpahene Hybrids and Applications 535
    33.3.4 PtNP/Grpahene Hybrids and Applications 536
33.4 Conclusions 536
References 536

ABSTRACT

Dispersion of sp^2 carbon materials such as carbon nanotubes (CNTs) and graphene in polymer matrices is considered to be essential for effectively utilizing the material in its primary structure. A great number of literatures have focused on the chemical modifications on the nanomaterials' surface through covalent bonding reactions; however, the methods are generally disadvantageous because of the possible disruption of the π-surface conjugation and the decrease of electric, optical, and thermal properties. Other methods of using polymeric dispersants for non-covalent bonding interaction with the target materials could also reach a homogeneous dispersion while maintaining the intrinsic characteristics without disrupting the covalent bonds of the materials. This review covers both methods of covalent and non-covalent interactions with the π-surface of sp^2 carbons including CNT and graphene, but emphasizing on the new developments of polymeric dispersants for achieving dispersion homogeneity and stability. In addition, the homogeneous dispersion could be the precursors for hybridizing metal nanoparticles (NPs) such as Ag and Pt to produce hybrid materials. These dispersion methods are further applied for maximizing the performances of the NP interaction with the fundamental graphene platelets. The fine dispersion of graphene in primary platelet units and their hybridizations with NPs have a myriad of applications for fabricating devices with advanced properties of electric conductivity, heat dissipation, and catalyst in dye-sensitized solar cell.

33.1 INTRODUCTION

Fullerene [1,2], carbon nanotube (CNT), and graphene are allotropes of carbon [3,4], exhibiting sp^2 hybridization within the spherical, tubular, and planar structures, respectively, and possessing the combination of excellent electrical, mechanical, and thermal properties [5–7]. Among these carbon materials, CNTs have been proven to be very effective as nanofillers [8–11], and graphene for superior mechanical, thermal, gas barrier, and electrical properties in polymer nanocomposites [12–14]. All of these advances in enhancing polymers' physical properties relied on the fine dispersion of nanomaterial units in the polymer matrix and their intensive interfacial bonding forces.

With respect to three different geometric shapes of spherical, fibril, and platelet-like shapes for all nanomateials, graphene at its extremely thin thickness has particularly intensive van der Waals and π–π stacking forces that are most difficult to be overcome for being compatible with organic polymers. The self-aggregation of these nanomaterials is often a serious problem in polymer blending process [15]. The methods of making a homogeneous dispersion of graphene primary structure have been widely reported. In general, the route of oxidization, blending/mixing, and reduction via graphene oxide (GO) intermediates was used, in which organic moieties such as hydroxyl, epoxide, ketone, carbonyl, and carboxyl groups were generated to increase their organophilic property and polymer compatibility [16]. However, one of the drawbacks in the process is the possible disruption of graphene π-surface. Even partial destruction of the conjugated π-surface may be detrimental to the property of electrical conductivity of graphene. The surface modification without disrupting π-surface is an important research topic.

In this review, an overview of surface modifications through covalent and non-covalent bonding interactions is first covered. Surface modifications via covalent bonding can largely improve the dispersion of carbon materials that subsequently enhance physical, mechanical, and thermal properties of nanocomposites. In the distinct approach, non-covalent bonding interaction required the addition of a foreign organic compound such as a surfactant or a polymeric dispersant in order to facilitate the homogeneity of dispersion. Further, the review covers the recent research efforts on hybridizing graphene with various nanoparticles including silver (AgNP) and platinum nanoparticles (PtNP). Their diversity of graphene–particle hybrid applications including antibacterial and dye-sensitized solar cells (DSSCs) is also reviewed.

33.2 DISPERSION OF CARBON MATERIALS VIA SURFACE MODIFICATIONS

CNTs and graphenes are promising nanomaterials for fabricating the devices with unique properties including electronic properties, thermal conductivity, and mechanical properties. Since the nanomaterials have the inherent tendency of aggregation into secondary structures due to strong van der Waals force, a dispersing method is required for graphene before blending into polymer matrices. The fine dispersion of primary units in polymers is the ultimate goal for achieving high performance of the nanocomposites. Hence, the approaches of functionalizing CNT and graphene are the essential steps to prevent the nanomaterial self-aggregation. These methods are summarized in Table 33.1.

33.2.1 SURFACE MODIFICATIONS OF CARBON NANOTUBES

The surface modifications of CNTs can be divided into two types: (1) adding organic functionalities by covalent bonding reaction to the conjugated and (2) non-covalent interaction of carbon material π-surface with surfactant molecules or polymeric dispersants.

33.2.1.1 Via Covalent Bonding

By attaching covalent bonding species [17], the surface structure of CNT is disrupted by transforming sp² to sp³ orbitals of carbon atoms. As a result, the functionalization of CNT rendered the solubility improvement as well as the compatibility for polymers. The covalent functionalization was commonly performed by oxidation into carboxylic acids for the second-step transformation and organic introduction on the side walls of CNTs. Generally, sulfuric acid, nitric acid, and hydrogen peroxide were used to generate the oxygenate functionalities such as –COOH

TABLE 33.1
Dispersion of Carbon Materials via Surface Modifications

Carbon Materials	Dispersing Interaction	Dispersing Method	Functional Groups	References
CNT	Covalent bonding	Acidification	–COOH and –OH	[18–21]
		Amidation	Amide	[22]
		Acidification; amidation	Amide	[25]
		Fluorination	Fluoride	[26,27]
		Cycloaddition	Carbene	[28]
		Arylation	Aryl	[29]
		Esterification	Ester	[30]
		Amidation; esterification	Amide; ester	[31]
		Organometallic reagent	Lithium	[32]
		Polymer grafting	Various polymers	[33–36]
CNT	Non-covalent bonding	Small molecule surfactants		[37–39]
		Immobilization of DNA and proteins		[40]
		Polymer wrapping		[42–48]
Graphene	Covalent bonding	Inorganic acidification	Hydroxyl, epoxy, carboxyl	[59–61]
			Amide; carbamate	[58]
		Induced by amino acids		[62]
Graphene	Non-covalent bonding	PSS		[55]
		TCNQ		[63]
		PBA		[64,65]
		POE-imide (home-made)		[66]
		SPANI		[67]

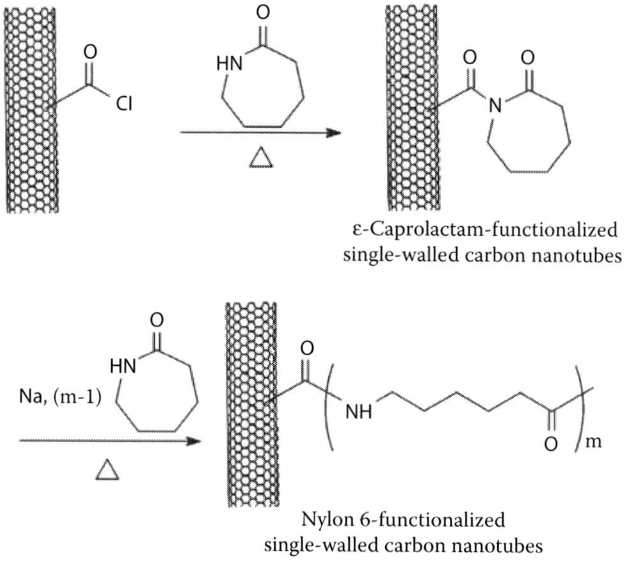

ε-Caprolactam-functionalized
single-walled carbon nanotubes

Na, (m-1)

Nylon 6-functionalized
single-walled carbon nanotubes

FIGURE 33.1 Synthesis of nylon-6-functionalized single-walled carbon nanotubes (Nylon-SWNT) (Reprinted with permission from Qu L, Veca LM, Lin Y et al., Soluble nylon-functionalized carbon nanotubes from anionic ring-opening polymerization from nanotube surface. *Macromolecules* 38:10328–31. Copyright 2005 American Chemical Society.)

or –OH on the CNTs [18]. Simultaneously, the acid treatment also destroyed some of CNT carbon–carbon double bonds including the conjugation and the diffraction pattern of lattice structure [19–21]. In the second-step, a variety of chemical reactions can be conducted [22]. Finally, the modified CNTs could be dispersed in polymer matrices [17,23]. In an example, the synthetic routes including direct thermal, acylation-mediated and DCC-coupling amidation for grafting poly(oxyalkylene)-amines onto MWNT–COOH were reported. The method allowed the tethering of different polyether backbones, including hydrophobic poly(oxypropylene) and hydrophilic poly(oxyethylene), onto CNT. The dispersion of the modified CNT either in water or in organic solvents was possible [24]. In another example, it was demonstrated that fluorine could be added onto CNTs and transformed further by using anhydrous hydrazine [25]. The single-walled CNT can be fluorinated and then

sonicated in alcohols to form "fluorotubes" [26]. The transformation of CNT by direct functionalization with carbenes [27] and arylation were reported [28]. Functionalization was also affected by attaching polymer molecules [29] via the carboxylic acid/thionyl chloride treatment to acyl chloride intermediates, and followed esterification leading to the attachment of dendritic poly(ethylene glycol). The polymer grafting to highly soluble linear copolymers such as poly(propionylethylenimine-co-ethylenimine) (PPEI-EI) via amide linkages or poly(vinyl acetate-co-vinyl alcohol) (PVA-VA) via ester linkages was reported [30]. Polymer-functionalized CNT using organometallic approach was revealed [31]. It was demonstrated that multi-walled CNTs can be effectively functionalized using organometallic reagents such as *n*-butyllithium. The introduction of lithium species led to the diversified route of organometallic reactions. *In situ* polymerization of monomers in the presence of activated CNT was successfully developed to graft different polymers such as polyamide 6 (Figure 33.1) [32], PMMA [33], PS [34], poly(*N*-isopropylacrylamide) (NIPAM) [35]. The common polymerization was suitable for applying to CNT surface via radical, cationic, anionic, ring-opening, and condensation polymerization.

33.2.1.2 Via Non-Covalent Bonding (π–π Interactions)

Non-covalent functionalization of CNT has attracted particular interests due to the advantages of the remaining pristine structure of CNT. The association of CNT by non-covalent wrapping or interaction with the hydrophobic portion of surfactants or their micelles was proposed, particularly for the presence of aromatic groups in the micelles. Among the effective surfactants of anionic, cationic, and non-ionic species, a stable aqueous colloidal dispersion of CNT was demonstrated by using sodium dodecyl sulfonate (SDS) [36] and sodium dodecylbenzene sulfonate (NaDDBS) [37]. The presence of benzene rings for promoting π–π interaction was shown to be an important non-covalent bonding force for the association [38]. In Figure 33.2, the conceptual diagram for the adsorption of different surfactants onto the nanotube surfaces is illustrated. In another example, the direct non-surfactant immobilization of protein and DNA on CNTs was reported [39]. The CNT/polymer interactions could alter the polymer morphology through the chiral

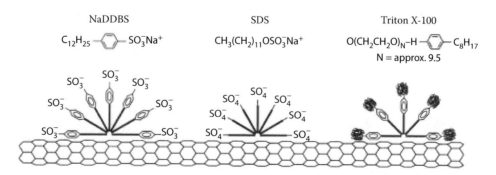

FIGURE 33.2 Adsorption of different surfactants onto the nanotube surfaces. (Reprinted with permission from Islam MF et al., High weight fraction surfactant solubilization of singlewall carbon nanotubes in water. *Nano Lett* 3:269–73. Copyright 2003 American Chemical Society.)

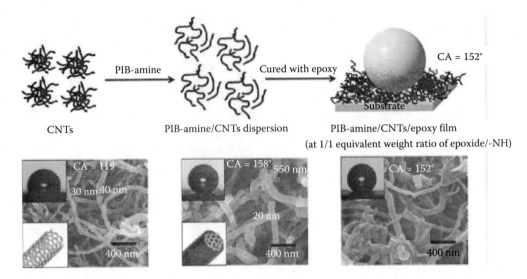

FIGURE 33.3 Manipulation of the surface roughness by dispersing CNT with a PIB-amine copolymer, and the formation of a superhydrophobic film. (Reprinted with permission from Hsu CP et al. Facile Fabrication of Robust Superhydrophobic Epoxy film with polyamine dispersed carbon nanotubes. *ACS Appl Mat Interfaces* 5:538–45. Copyright 2013 American Chemical Society.)

nature of the nanotube surface [40]. Further, CNT can also be successfully solubilized in organic [41] or aqueous [42] after non-covalent interaction. The introduction of polymer generally increased the dispersion of CNT. The polymer "wrapping" around the CNTs can be attributed to the specific interactions [43,44]. In a specific example, oil-soluble poly(isobutylene)-amines, consisting of hydrophobic alkyls and multiple amine functionalities, were synthesized and used to disperse CNT in an organic medium. The amine molecules wrapped CNTs enabled the surface roughness formation of epoxies that consequently resulted in a superhydrophobicity on glass substrate, as shown in Figure 33.3 [45]. The amphiphilic poly(styrene)-*b*-poly(acrylic acid) copolymer in forming micelles was used to encapsulate CNT [46] and render the dispersion and stabilization. The block copolymers adsorbed to the nanotubes by a non-wrapping mechanism were also suggested [47]. The strong π–π interaction and steric barrier for aggregation between polymer backbone and nanotube surface led to disperse single-walled CNT in mediums.

33.2.2 Surface Modifications of Graphite

Owing to the phase–phase stacking force, graphene is expected to have a higher tendency for self-aggregation than CNT. To overcome the inherent self-aggregating of the platelet-like nanomaterials is considered to be difficult [15]. Stable suspension of graphene in water or organic mediums is required for the fabrication of graphene-based composites [48]. To avoid the aggregation of graphene in polymer matrix [49], the process of oxidation and chemical derivation has been generally adopted similar to the CNT oxidation to carboxylic acid [50,51]. Functional groups attached to graphene included small molecules [52] or polymer chains [53,54]. The dispersion and processability can be improved through the chemical functionalization [55], such as esterification [56], isocyanate reaction [57], and polymer wrapping [53,54]. The

general approaches for the organic modification are shown in the following sections.

33.2.2.1 Via Graphene Oxide

GO is the graphite derivative with covalently attached oxygen-containing groups on the layer surface. These oxygenate groups are generated by inorganic acid oxidation [58]. The lamellar structure with randomly distributed oxygenates in aromatic regions (sp^2 carbon atoms) and some six-membered aliphatic regions (sp^3 carbon atoms) is the result of oxidation. Mainly, the exposed oxygen-containing functional groups such as hydroxyl, epoxy, and carboxyl, are embedded in the carbon layers, as shown in Figure 33.4. It has been proposed that the epoxy and hydroxyl functionalities lie above and below each carbon layer, while the carboxylic acid groups are located near the edges of the layers [59]. The GO precursor may be dispersed in water and followed by the addition of an aqueous KOH solution. A good electrical conductivity for a homogeneous aqueous suspension of chemically modified graphene was reported [60]. The addition of KOH to GO sheet could generate large negative charges of potassium salts from

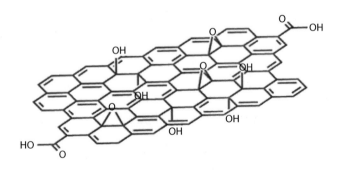

FIGURE 33.4 GO contains a high concentration of exposed oxygen-containing functional groups, like hydroxyl, epoxy, and carboxyl, embedded in its carbon layers.

the reactive hydroxyl, epoxy, and carboxylic acid groups. The further addition of hydrazine monohydrate to KOH-treated GO produces a homogeneous suspension that can be maintained for months.

33.2.2.2 Via Covalent Bonding with Organic Functionalities

GO can be reduced back to the surface with high conjugation of aromaticity by hydrazine reduction or other reagents. However, the residual functional groups are undesirable for some applications but remained. Another drawback of using a hazardous reducing agent or solvent may affect the downstream uses involving biomolecules and conjugated polymers [48]. Covalent-bond modification involved the reactions with organic amines, and isocyanates as the modifiers [57,58]. It was demonstrated that the treatment of GO with organic isocyanates resulted in a new class of iGO derivatives. In polar aprotic solvents, the materials were further exfoliated into GO nanoplatelets. These surface modifying agents were converted to the GO sheets containing amide and carbamate groups [57]. The face-to-face stacking among the GO sheets was effectively limited and easily dispersed [58].

The surface of graphene can also be derived by amines and amino acids. The combination of redox reactions and the surface modification of GO with amines led to novel reconstructed GO and graphite derivatives [61]. The covalent functionalization of graphene sheets with PVA by two synthetic strategies was reported (Figure 33.5) [53]. Direct esterification of GO and acyl chloride derivation (GOCl) routes led to the products that were soluble in dimethyl sulfoxide (DMSO) and water.

33.2.2.3 Via Non-Covalent π–π Interactions

The structural similarity between CNT and graphene is the sp^2 carbons on graphene platelets and cylinders with several nanometers in diameter. In polymer matrices, the ultrasonic treatment in water in the presence of surfactants such as SDS and Triton X-100 could render the dispersion of the

nanomaterials. In the presence of polymer/polymeric anions, stable polymer-grafted dispersion of graphene was achieved. Stable aqueous dispersions of graphene nanoplatelets were prepared by dispersing reduced graphite oxide by amphiphilic polymer, poly(sodium 4-styrenesulfonate) (PSS) [54]. An aqueous dispersion of graphene could be prepared from the expanded graphite using 7,7,8,8-tetracyanoquinodimethane (TCNQ) anion as the stabilizer. The facile route may be developed into the aqueous dispersion of graphene sheets for a number of applications [62]. It appeared that TCNQ anions enabled to adsorb on graphene sheet surface and drive the material into water. The TCNQ-stabilized graphene sheets could be re-dispersible in water, DMF, and DMSO. The preparation of stable aqueous dispersions of graphene by pyrene derivatives, 1-pyrenebutyrate (PB⁻), as the stabilizer was reported due to its strong π-surface affinity to the basal plane of graphite [63]. The process was performed by using alkaline solution of pyrene butyric acid (PBA) to react with a GO dispersion and followed by hydrazine reduction [64]. A homemade polymeric dispersant, POE-imide, was synthesized for the dispersion of graphene through non-covalent interactions; the aromatic imide functionalities of POE-imide facilitate the π–π interactions between the POE-imide and the graphene. The presence of lone pair electrons of POE-imide in its poly(oxyethylene) and imide functionalities is supposed to be capable of facilitating the polar interactions between the PtNPs and the graphene nanoplatelets [65]. A conjugative polymer, sulfonated polyaniline (SPANI), was used to produce a water-soluble and electroactive composite [66]. The successful reduction of GO was confirmed by electrical conductivity measurement of 30 S m^{-1} for the composite film.

Overall, the utilization of sp^2 carbon materials in its primary structure required the materials to be organophilic and compatible with polymers. Both of covalent and non-covalent modifications were useful for the purpose of dispersing the materials in primary units. The graphene surface may be oxidized into GO as the intermediates for the subsequent reactions. However, the π-surface conjugation is often disrupted

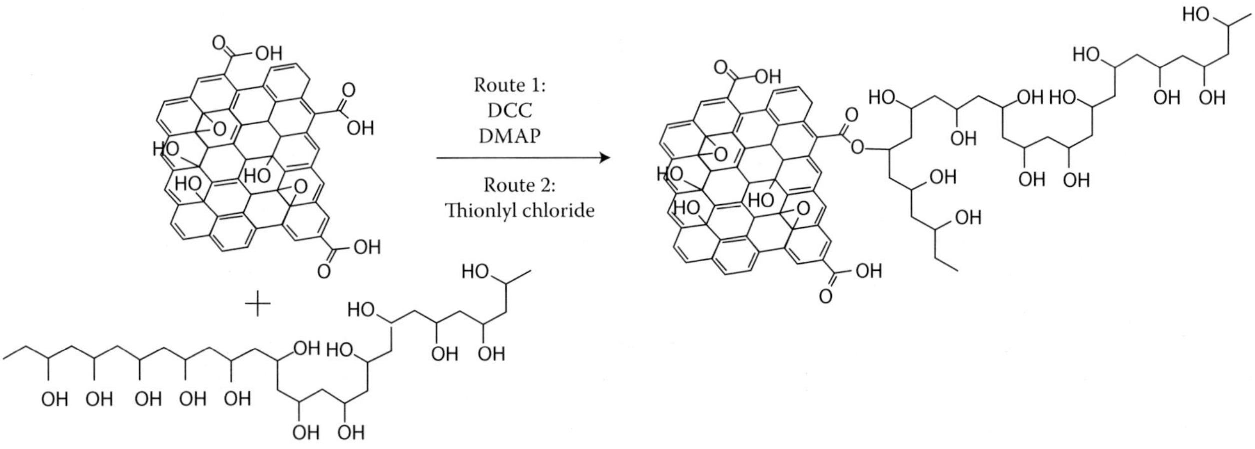

FIGURE 33.5 Two synthetic strategies of the covalent functionalization of graphene sheets with PVA. (Reprinted with permission from Salavagione HJ, Gomez MA, Martinez G. Polymeric modification of graphene through esterification of graphite oxide and poly(vinyl alcohol). *Macromolecules* 42:6331–4. Copyright 2009 American Chemical Society.)

and the original natures such as electric conductivity and thermal properties are unintentionally reduced. On the other hand, the non-covalent bonding method involving the use of polymeric dispersants can largely maintain the pristine characters of sp^2 conjugation. The ultimate goal is to make homogeneous dispersion of the nanoscale carbon materials via the π-surface chemistry in achieving the exfoliation of graphene platelets, dispersion homogeneity, and stability.

33.3 HYBRIDIZATION OF GRAPHENE WITH VARIOUS FUNCTIONAL NANOPARTICLES

GO contains oxygen-containing functional groups such as hydroxyl, epoxide, carbonyl, and carboxylic acid. The multilayered structure is possibly swelled and intercalated by incoming organics. Hence, GO is an excellent host matrix for the incorporation of aliphatic hydrocarbons, transition metal ions [67], and hydrophilic molecules and polymers [68]. It is also possible to fabricate these derivatives into thin films with smart properties [69,70]. Since the oxidative functionalities can be further removed by elimination into C–C conjugation, the graphene–metal nanocomposites are ultimately produced.

The attachments of metal nanoparticles (NPs) on graphene surface could generate new graphene–NP hybrids. Similar strategy has been reported in synthesizing CNT–AgNP hybrids. The hybrid film with surface resistance as low as

1×10^{-2} Ω/sq was obtained by melting the AgNPs at controlled annealing temperature in air. The important factors of the Ag/CNT/dispersant composition and AgNP migration kinetics were considered (Figure 33.6) [71]. It is expected that the NPs on CNT or graphene hybrids may exhibit novel catalytic, magnetic, and optoelectronic properties. The utilization of graphite oxide [72–74] for hybridizing the metal–graphene composites was performed by *in situ* reduction of metallic salts on GO sheets [75,76].

33.3.1 NANOPARTICLES-ON-GRAPHENE

Due to the unique geometric shape and sp^2 character, graphene has been used to hybridize with many different metal NPs. In order to add on new properties such as antibacterial and catalytic activity, inorganic NPs have been hybridized into the interlayer of the graphene sheets. Various metal and metal oxide NPs included Ag, Pt, Au, Pd, Cu, Ni, Co, and Fe, as listed in Table 33.2.

33.3.2 APPLICATIONS OF GRAPHENE HYBRIDS

Au nanoparticles (AuNPs) dispersed onto GO sheets exhibited excellent surface enhanced Raman scattering (SERS) and catalytic activities, as compared with the pristine AuNPs [77]. The preparation of Au–GO and Au–rGO composites by

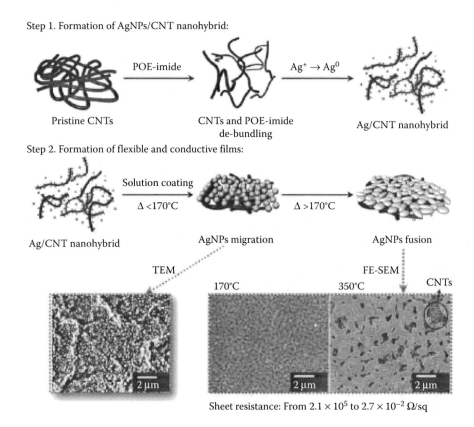

FIGURE 33.6 Illustration of the process making polymer dispersed CNT–AgNP nanohybrids. (Reprinted with permission from Dong RX. Controlling formation of silver/carbon nanotube networks for highly conductive film surface. *ACS Appl Mat Interfaces* 4:1449–55. Copyright 2012 American Chemical Society.)

TABLE 33.2

Graphene Hybridization with Nanoparticles

Particle Type	Graphene Type	Method or Process	Application	References
Au	GO; rGO	π–π interaction	SERS	[77]
Au	GO	TWEEN 20 surfactant	Hydrazine detection	[78]
Au, Pt	GO	EG	DFAFC	[79]
Co, Pt	GO	EG/microwave-assisted	Fuel cells via oxygen reduction	[80]
Cu, Pt	Ionic liquid/graphene	$NaBH_4$	DMFC	[81]
Fe_3O_4	GO	EN, EG	Biomolecule immobilization, enrichment, and separation	[82]
Fe_3O_4	GO	DEG, EG	Adsorbent of organic dyes	[83]
Fe_3O_4	GO	Sonochemical reduction	Biosensor detecting H_2O_2	[84]
Ni	Ni/graphene	Methanol	Sorbent for aromatics	[85]
Pd	GO	Graphene (Na_2PdCl_4)	DFAFCs	[86]
Ag	GO	Aniline ($AgNO_3$)	Enzymeless hydrogen peroxide detection	[90]
Ag	rGO	$scCO_2$	Antibacterial	[91]
Ag	GO		H_2O_2 detection	[92]
Ag	GO	PDDA	Antibacterial activity	[93]
Ag	GO	Sodium citrate	Antibacterial activity	[94]
Ag	GO	Hydrazine hydrate	High capacitance electrode	[95]
Pt	Graphene	PDDA	DSSC	[96]
Pt	rGO	Pulsed laser deposition	DSSC	[97]
Pt	rGO	$NaBH_4$	DSSC	[98]

a simple physisorption method could afford the composition with controlled NP size and size distribution. The hybrids demonstrated an enhanced SERS absorption and catalytic performance. The GO was functionalized by surfactants and rendered homogeneous dispersion in the presence of NPs. The synthesis of AuNPs on TWEEN-functionalized GO afforded the Au nanocomposites, which showed a remarkable catalytic performance for hydrazine oxidation [78]. A good dispersion of Pt–Au nanoparticles on graphene was prepared by EG reduction [79]. The composites were used as an anode for DFAFCs showing high performance in comparison with that of Pt/graphene. Binary Pt–Co-alloyed nanoparticles could be conveniently immobilized onto graphene by a simultaneous reduction of the mixtures of GO, Pt(IV), and Co(II) ions in ethylene glycol and followed by annealing in H_2 at 300°C [80]. The composite catalyst of Pt–Co/G thus obtained displayed promising electrocatalytic performance for oxygen reduction conducted on acidic electrolytes. The Pt–Co/G catalytic performance in fuel cells was optimized by reducing the Pt-Co size. Uniformity of CuPt nanoparticles on ionic-liquid/graphene for catalyzing methanol oxidation was reported [81]. Referenced to carbon blacks, the hybrid composite showed an enhancement in electrocatalytic activity for the oxidation of methanol. The increase in conductivity by graphene/ionic liquid was explained. Hence, a promising low-Pt content catalyst for DMFC could be potentially developed. For Fe_3O_4 nanoparticles with magnetic property, the GO–Fe_3O_4 composites with hydrophilic moieties were prepared [82] and used for immobilization of proteins. The water-dispersible rGO–Fe_3O_4 composites had the characters of high surface area for enriching peptides. A binary graphene–Fe_3O_4 hybrid [83] was prepared

and evaluated for the adsorption performances of removing organic dyes. The presence of magnetic Fe_3O_4 nanoparticles facilitated the separation and regeneration of the hybrid materials in wastewater treatments. The fabrication of Fe_3O_4/rGO [84] nanocomposite by a sonochemical method produced the NP of 30–40 nm in diameter, which uniformly dispersed on the graphene surface. The Fe_3O_4/RGO was immobilized and used as a biosensor for detecting H_2O_2. The composite of Ni@GSs–C(CH_3)$_2$COONa [85] was prepared for the purpose of absorbing aromatic compounds from waste water. For the possible replacement of Pt catalyst by Pd due to the advantage of high resistance to CO poisoning in fuel cells, the graphene-supported Pd was prepared [86]. The preparation involved the uses of Pd salt precursor in water without any extra reducing agent or stabilizer. Compared to the conventional Pd/C electrocatalyst, the Pd/graphene hybrid showed an enhancement of electrocatalytic activity for formic acid electro-oxidation in DFAFCs. More importantly, the hybrid displayed an improved durability, which is the important bottleneck to be overcome in formic acid electro-oxidation.

33.3.3 AgNP/Grpahene Hybrids and Applications

The hybrids of AgNPs on graphene are considered to have many potential applications in optics, electronics, catalysis, and electrochemistry [87–89]. The synthesis of AgNP/GO and AgNP-reduced graphene oxide (rGO) composites was reported [90]. The stable aqueous dispersion of rGOs in the presence of aniline was prepared and compounding AgNPs by *in situ* reduction from silver salts. The hybrid composite was fabricated into devices that allowed the detection of H_2O_2

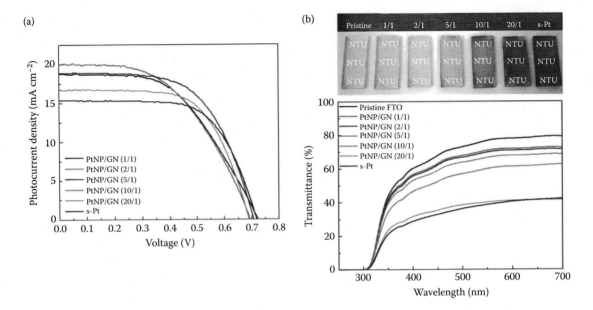

FIGURE 33.7 (a) *J–V* characteristics of the DSSCs with different weight ratios of platinum and graphene, obtained at 100 mW cm^{-2}; (b) Photographs of the counter electrodes with the PtNP/GN films formed by using PtNP and graphene at the ratios of 1/1, 2/1, 5/1, 10/1, 20/1 and with the s-Pt film and the transmittances of these counter electrodes [65].

without using enzyme in the electrode. In another application, the hybrid of AgNPs decorated on rGO sheets by supercritical CO$_2$ was synthesized [91]. The composites were demonstrated to have the antibacterial activity against *B. cereus*, *L. anguillarum*, *E. coli*, and *S. aureus* bacteria. The hydrothermal treatment of AgNO$_3$ and preformed GO solution in strong alkaline conditions were effective for preparing AgNPs–rGO nanocomposites [92]. Similar to the above synthetic methods, Ag on graphene and the modified GO were generated and stabilized by different reducing agents and stabilizers [93–95]. In general, the size of AgNPs at 20 nm diameter was produced by mild reducing agents and further demonstrated their biomedical uses.

33.3.4 PtNP/Grpahene Hybrids and Applications

In general, Pt has been applied in a number of technological areas including chemical sensors and biosensors, as catalysts for reducing pollutants and converting methane to hydrogen and electrocatalysts in DSSCs. Platinum nanoparticle/multi-wall carbon nanotube (PtNP/MWCNT) hybrids as the counter electrodes have been prepared [96,97]. The Pt/graphene composites were synthesized and used to replace the conventional sputtered Pt in DSSC [65,99]. In the example shown in Figure 33.7, the hybrid film of PtNPs and graphene nanoplatelets showed a transparency of 70% at 550 nm, indicating its suitability as a CE material for a rear-illuminated DSSC. The DSSC with the CE having the film of PtNP/GN exhibited power conversion efficiency (η) of 8.00%, superior to 7.14% of the DSSC with a conventional sputtered platinum (s-Pt) CE [65]. Furthermore, the effect of Pt loading on the PV performance of the DSSCs was evaluated [98]. The PtNPs on graphene using a pulsed laser ablation method resulted in an increase of DSSC efficiency [99].

33.4 CONCLUSIONS

Dispersion of sp^2 carbon materials, graphene, and CNT, in aqueous or organic medium is an important process for utilizing nanomaterials in various downstream applications. The performance of adding CNT and 2D platelet-like graphene as the nanoscale fillers to nanocomposites relies on the step of homogeneously dispersing the primary structure of nanomaterials. Their surface modifications by covalent bonding to organics or non-covalent π–π interaction are two effective methods for dispersing nanomaterials. The choices of using polymeric dispersants with organic functionalities and π-surface aromatic groups were found to be effective for interrupting the pristine carbon π–π stacking aggregation. This non-covalent bonding dispersion method is favored because of its mild interruption of the graphene material surface. Furthermore, the Ag and Pt nanoparticles on graphene platelets were synthesized, characterized, and used for various uses. The diversified applications of these hybrid materials include antimicrobial and catalysts for DSSCs.

REFERENCES

1. Kroto HW, Heath JR, O'Brien SC, Curl RE, Smalley RE. 1985. C60: Buckmin-sterfullerene. *Nature* 318:162–3.
2. Dresselhaus MS, Dresselhaus G, Eklund PC. 1996. *Science of Fullerenes and Carbon Nanotubes.* 15–59. San Diego: Academic Press.
3. Iijima S. 1991. Helical microtubules of graphitic carbon. *Nature* 354:56–8.
4. Geimandk AK, Novoselov KS. 2007. The rise of graphene. *Nat Mater* 6:183–91.
5. Ajayan PM. 1999. Nanotubes from carbon. *Chem Rev* 99:1787–99.

6. Mortazavi B, Pötschke M, Cuniberti G. 2014. Multiscale modeling of thermal conductivity of polycrystalline graphene sheets. *Nanoscale* 6:3344–52.

7. Treacy MM, Ebessen TW. 1996. Gibson JM. Exceptionally high Young's modulus observed for individual carbon nanotubes. *Nature* 381:678–80.

8. Ma C, Zhang W, Zhu Y, Ji L et al. 2008. Alignment and dispersion of functionalized carbon nanotubes in polymer composites induced by an electric field. *Carbon* 46:706–10.

9. Ajayan PM, Stephan O, Colliex C, Trauth D. 1994. Aligned carbon nanotube arrays formed by cutting a polymer resin–nanotube composite. *Science* 265:1212–4.

10. Jamal A, Ali R, Somayeh M. 2007. Preparation and characterization of linear low density polyethylene/carbon nanotube nanocomposites. *J Macromol Sci Part B: Phys* 46:877–89.

11. Guo H, Sreekumar TV, Liu T, Minus M, Kumar S. 2005. Structure and properties of polyacrylonitrile/single wall carbon nanotube composite films. *Polymer* 46:3001–5.

12. He F, Lau S, Chan HL, Fan J. 2009. High dielectric permittivity and low percolation threshold in nanocomposites based on poly(vinylidene fluoride) and exfoliated graphite nanoplates. *Adv Mater* 21:710–5.

13. Green M, Marom G, Li J, Kim J-K. 2008. The electrical conductivity of graphite nanoplatelet filled conjugated polyacrylonitrile. *Macromol Rapid Commun* 29:1254–8.

14. Yamaguchi H, Granstrom J, Nie W et al. 2014. Reduced graphene oxide thin films as ultrabarriers for organic electronics. *Adv Energy Mater* 4:1–6.

15. Chiu CW, Lin JJ. 2012. Self-assembly behavior of polymer-assisted clays. *Prog Polym Sci* 307:406–44.

16. Esmaeili A, Entezari MH. 2014. Facile and fast synthesis of graphene oxide nanosheets via bath ultrasonic irradiation. *J Colloid Interface Sci* 432:19–25.

17. Niyogi S, Hamon MA, Hu H et al. 2002. Chemistry of single-walled carbon nanotubes. *Acc Chem Res* 35:1105–13.

18. Zhang X, Sreekumar TV, Liu T, Kumar S. 2004. Properties and structure of nitric acid oxidized single wall carbon nanotube films. *J Phys Chem B* 108:16435–40.

19. Georgakilas V, Kordatos K, Prato M, Guldi DM, Holzingger M, Hirsch A. 2002. Organic functionalization of carbon nanotubes. *J Am Chem Soc* 124:760–1.

20. Zhang Y, Shi Z, Gu Z, Iijima S. 2000. Structure modification of single-wall carbon nanotubes. *Carbon* 38:2055–9.

21. Datsyuk V, Kalyva M, Papagelis K et al. 2008. Chemical oxidation of multiwalled carbon nanotubes. *Carbon* 46:833–40.

22. Hemon MA, Chen J, Hu H et al. 1999. Dissolution of single-walled carbon nanotubes. *Adv Mater* 11:834–40.

23. Kobashi K, Ata S, Yamada T et al. 2013. A dispersion strategy: Dendritic carbon nanotube network dispersion for advanced composites. *Chem Sci* 4:727–33.

24. Lin ST, Wei KL, Lee TM, Chiou KC, Lin JJ. 2006. Functionalizing multi-walled carbon nanotubes with poly(oxyalkylene)-amidoamines. *Nanotechnology* 17:3197–203.

25. Mickelson ET, Huffman CB, Rinzler AG, Smalley RE, Hauge RH, Mar-grave JL. 1998. Fluorination of single-wall carbon nanotubes. *Chem Phys Lett* 296:188–94.

26. Mickelson ET, Chiang IW, Zimmerman JL et al. 1999. Solvation of fluorinated single-wall carbon nanotubes in alcohol solvents. *J Phys Chem B* 103:4318–22.

27. Holzinger M, Vostrowsky O, Hirsch A et al. 2001. Sidewall functionalization of carbon nanotubes. *Angew Chem Int Ed* 40:4002–5.

28. Bahr JL, Yang J, Kosynkin DV, Bronikowski MJ, Smalley RE, Tour JM. 2001. Functionalization of carbon nanotubes by electrochemical reduction of aryl diazonium salts: A bucky paper electrode. *J Am Chem Soc* 123:6536–42.

29. Fu K, Huang W, Lin Y, Riddle LA, Carroll DL, Sun YP. 2001. Defunctionalization of functionalized carbon nanotubes. *Nano Lett* 1:439–41.

30. Lin Y, Zhou B, Fernando KAS, Liu P, Allard LF, Sun YP. 2003. Polymeric carbon nanocomposites from carbon nanotubes functionalized with matrix polymer. *Macromolecules* 36:7199–204.

31. Blake R, Gunko YK, Coleman J et al. 2004. A generic organometallic approach toward ultra-strong carbon nanotube polymer composites. *J Am Chem Soc* 126:10226–7.

32. Qu L, Veca LM, Lin Y et al. 2005. Soluble nylon-functionalized carbon nanotubes from anionic ring-opening polymerization from nanotube surface. *Macromolecules* 38:10328–31.

33. Yao Z, Braidy N, Botton GA, Adronov A. 2003. Polymerization from the surface of single-walled carbon nanotubes-preparation and characterization of nanocomposites. *J Am Chem Soc* 125:16015–24.

34. Kong H, Gao C, Yan D. 2004. Functionalization of multi-walled carbon nanotubes by atom transfer radical polymerization and defunctionalization of the products. *Macromolecules* 37:4022–30.

35. Kong H, Li W, Gao C et al. 2004. Poly(N-isopropylacrylamide)-coated carbon nanotubes: temperature-sensitive molecular nanohybrids in water. *Macromolecules* 37:6683–6.

36. Jiang L, Gao L, Sun J. 2003. Production of aqueous colloidal dispersions of carbon nanotubes. *J Coll Interf Sci* 260:89–94.

37. Sáfar GAM, Ribeiro HB, Malard LM et al. 2008. Optical study of porphyrin-doped carbon nanotubes. *Chem Phys Lett* 462:109–11.

38. Islam MF, Rojas E, Bergey DM, Johnson AT, Yodh AG. 2003. High weight fraction surfactant solubilization of single-wall carbon nanotubes in water. *Nano Lett* 3:269–73.

39. Guo Z, Sadler PJ, Tsang SC. 1998. Immobilization and visualization of DNA and proteins on carbon nanotubes. *Adv Mater* 10:701–3.

40. Lordi V, Yao N. 2000. Molecular mechanics of binding in carbon-nanotube–polymer composites. *J Mater Res* 15:2770–9.

41. O'Connell MJ, Boul P, Ericson LM et al. 2001. Reversible water-solubilization of single-walled carbon nanotubes by polymer wrapping. *Chem Phys Lett* 342:265–71.

42. Steuerman DW, Star A, Narizzano R et al. 2002. Interactions between conjugated polymers and single-walled carbon nanotubes. *J Phys Chem B* 106:3124–30.

43. Pei X, Hu L, Liu W, Hao J. 2008. Synthesis of water-soluble carbon nanotubes via surface initiated redox polymerization and their tri-bological properties as water-based lubricant additive. *Eur Polym J* 44:2458–64.

44. Cheng F, Imin P, Maunders C, Botton G, Adronov A. 2008. Soluble, discrete supramolecular complexes of single-walled carbon nanotubes with fluorene-based conjugated polymers. *Macromolecules* 41:2304–8.

45. Hsu CP, Chang LY, Chiu CW, Lee P, Lin JJ. 2013. Facile Fabrication of Robust Superhydrophobic Epoxy film with polyamine dispersed carbon nanotubes. *ACS Appl Mat Interfaces* 5:538–45.

46. Kang Y, Taton TA. 2003. Micelle-encapsulated carbon nanotubes: A route to nanotube composites. *J Am Chem Soc* 125:5650–1.

47. Nativ-Roth E, Shvartzman-Cohen R, Bounioux C et al. 2007. Physical adsorption of block copolymers to SWNT and MWNT: A nonwraping mechanism. *Macromolecules* 40:3676–85.

48. Akhavan O, Hashemi E, Shamsara M et al. 2014. Cyto- and geno-toxicities of graphene oxide and reduced graphene oxide sheets on spermatozoa. *RSC Adv* 4:27213–23.

49. Li X, McKenna GB, Miquelard-Garnier G et al. 2014. Forced assembly by multilayer coextrusion to create oriented graphene reinforced polymer nanocomposites. *Polymer* 55:248–57.

50. Thomassin JM, Trifkovic M, Alkarmo W et al. 2014. Poly(methyl methacrylate)/graphene oxide nanocomposites by a precipitation polymerization process and their dielectric and rheological characterization. *Macromolecules* 47:2149–55.

51. Wei T, Luo G, Fan Z et al. 2009. Preparation of graphene nanosheet/polymer composites using *in situ* reduction-extractive dispersion. *Carbon* 47:2290–9.

52. Ma L, Yu B, Qian X et al. 2014. Functionalized graphene/thermoplastic polyester elastomer nanocomposites by reactive extrusion based masterbatch: Preparation and properties reinforcement. *Polym Adv Technol* 25:605–12.

53. Salavagione HJ, Gomez MA, Martınez G. 2009. Polymeric modification of graphene through esterification of graphite oxide and poly(vinyl alcohol). *Macromolecules* 42:6331–4.

54. Roghani-Mamaqani H, Haddadi-Asl V, Khezri K et al. 2014. Edge-functionalized graphene nanoplatelets with polystyrene by atom transfer radical polymerization: Grafting through carboxyl groups. *Polym Int* 63:1912–23.

55. Worsley KA, Ramesh P, Mandal SK, Niyogi S, Itkis ME, Haddon RC. 2007. Soluble graphene derived from graphite fluoride. *Chem Phys Lett* 445:51–6.

56. Niyogi S, Bekyarova E, Itkis ME, McWilliams JL, Hamon MA, Haddon RC. 2006. Solution properties of graphite and graphene. *J Am Chem Soc* 128:7720–1.

57. Stankovich S, Piner RD, Nguyen ST, Ruoff RS. 2006. Synthesis and exfoliation of isocyanate-treated graphene oxide nanoplatelets. *Carbon* 44:3342–7.

58. Kim JT, Kim BK, Kim EY, Park HC, Jeong HM. 2014. Synthesis and shape memory performance of polyurethane/graphene nanocomposites. *React Funct Polym* 74:16–21.

59. Kovtyukhova NI, Ollivier PJ, Martin BR et al. 1999. Layer-by-layer assembly of ultrathin composite films from micron-sized graphite oxide sheets and polycations. *Chem. Mater.*11:771–8.

60. Park S, An J, Piner RD et al. 2008. Aqueous suspension and characterization of chemically modified graphene sheets. *Chem Mate* 20:6592–4.

61. Ahn H, Kim T, Choi H et al. 2014. Gelation of graphene oxides induced by different types of amino acids. *Carbon* 71:229–37.

62. Hao R, Qian W, Zhang L, Hou Y. 2008. Aqueous dispersions of TCNQ-anion-stabilized graphene sheets. *Chem Commun* 6576–8.

63. Xu Y, Bai H, Lu G, Li C, Shi G. 2008. Flexible graphene films via the filtration of water-soluble noncovalent functionalized graphene sheets. *J Am Chem Soc* 130:5856–7.

64. Katz E. 1994. Application of bifunctional reagents for immobilization of proteins on a carbon electrode surface: Oriented immobilization of photosynthetic reaction centers. *J Electroanal Chem* 365:157–64.

65. Shih PT, Dong RX, Shen SY, Vittal R, Lin JJ, Ho KC. 2014. Transparent graphene–platinum nanohybrid films for counter electrodes in high efficiency dye-sensitized solar cells. *J Mater Chem A* 2:8742–8.

66. Bai H, Xu Y, Zhao L, Li C, Shi G. 2009. Non-covalent functionalization of graphene sheets by sulfonated polyaniline. *Chem Commun* 1667–9.

67. Jiang L, Gu S, Ding Y, Jiang F, Zhang Z. 2014. Facile and novel electrochemical preparation of a graphene–transition metal oxide nanocomposite for ultrasensitive electrochemical sensing of acetaminophen and phenacetin. *Nanoscale* 6:207–14.

68. Liu P, Gong K, Xiao P. 1999. Preparation and characterization of poly(vinyl acetate)-intercalated graphite oxide. *Carbon* 37:2073–5.

69. Kotov A, Dékány I, Fendler H. 1996. Ultrathin graphite oxide–polyelectrolyte composites prepared by self-assembly: Transition between conductive and non-conductive states. *Adv Mater* 8:637–41.

70. Mi B. 2014. Graphene oxide membranes for ionic and molecular sieving. *Science* 343:740–2.

71. Dong RX, Liu CT, Huang KC, Chiu WY, Ho KC, Lin JJ. 2012. Controlling formation of silver/carbon nanotube networks for highly conductive film surface. *ACS Appl Mat Interfaces* 4:1449–55.

72. Stankovich S, Dikin D, Dommett G et al. 2006. Graphene-based composite materials. *Nature* 442:282–6.

73. Nossol E, Nossol AB, Guo SX et al. 2014. Synthesis, characterization and morphology of reduced graphene oxide–metal–TCNQ nanocomposites. *J Mater Chem C* 2:870–8.

74. Park S, Lee K, Bozoklu G, Cai W, Nguyen S, Ruoff R. 2008. Graphene oxide papers modified by divalent ions—enhancing mechanical properties via chemical cross-linking. *ACS Nano* 2:572–8.

75. Ocsoy I, Gulbakan B, Chen T et al. 2013. DNA guided metal nanoparticle formation on graphene oxide surface. *Adv Mater* 25:2319–25.

76. Chen FJ, Cao YL, Jia DZ. 2013. A room-temperature solid-state route for the synthesis of graphene oxide–metal sulfide composites with excellent photocatalytic activity. *Cryst Eng Comm* 15:4747–54.

77. Hunag J, Zhang L, Chen B et al. 2010. Nanocomposites of size-controlled gold nanoparticles and graphene oxide: Formation and applications in SERS and catalysis. *Nanoscale* 2:2733–8.

78. Lu W, Ning R, Qin X et al. 2011. Synthesis of Au nanoparticles decorated graphene oxide nanosheets: Noncovalent functionalization by TWEEN 20 *in situ* reduction of aqueous chloroaurate ions for hydrazine detection and catalytic reduction of 4-nitrophenol. *J Hazard Mater* 197:320–6.

79. Rao V, Cabrera C, Carlos R, Ishikawa Y. 2011. Graphene-supported Pt–Au alloy nanoparticles: A highly efficient anode for direct formic acid fuel cells. *J Phys Chem C* 115:21963–70.

80. Ma YW, Liu ZR, Wang BL et al. 2012. Preparation of graphene-supported Pt–Co nanoparticles and their use in oxygen reduction reactions. *New Carbon Mater* 27:250–7.

81. Liu Y, Huang Y, Xie Y et al. 2012. Preparation of highly dispersed CuPt nanoparticles on ionic-liquid-assisted graphene sheets for direct methanol fuel cell. *Chem Eng J* 197: 80–7.

82. Cheng G, Liu YL, Wang ZG, Zhang JL, Sun DH, Ni JZ. 2012. The GO/rGO–Fe$_3$O$_4$ composites with good water-dispersibility and fast magnetic response for effective immobilization and enrichment of biomolecules. *J Mater Chem* 22:21998–2004.

83. Fan W, Gao W, Zhang C, Tjiu WW, Pan J, Liu T. 2012. Hybridization of graphene sheets and carbon-coated Fe$_3$O$_4$ nanoparticles as a synergistic adsorbent of organic dyes. *J Mater Chem* 22:25108–15.

84. Zhu S, Guo J, Dong J et al. 2013. Sonochemical fabrication of Fe$_3$O$_4$ nanoparticles on reduced graphene oxide for biosensors. *Ultrason Sonochem* 20:872–80.

85. Li S, Niu Z, Zhong X et al. 2012. Fabrication of magnetic Ni nanoparticles functionalized water-soluble graphene sheets nanocomposites as sorbent for aromatic compounds removal. *J Hazard Mater* 229:42–7.

86. Wang S, Manthiram A. 2013. Graphene ribbon-supported Pd nanoparticles as highly durable, efficient electrocatalysts for formic acid oxidation. *Electrochim Acta* 88:565–70.

87. Sun X, Dong S, Kang E. 2004. One-step preparation and characterization of poly (propyleneimine) dendrimer-protected silver nanoclusters. *Macromolecules* 37:7105–8.

88. Lan NT, Chi DT, Dinh NX et al. 2014. Photochemical decoration of silver nanoparticles on graphene oxide nanosheets and their optical characterization. *J Alloy Compd* 615:843–8.

89. Liu S, Tian J, Wang L, Sun X. 2011. A method for the production of reduced graphene oxide using benzulamine as a reducing and stabilizing agent and its subsequent decoration with Ag nanoparticles for enzymeless hydrogen peroxide detection. *Carbon* 49:3158–64.

90. Liu S, Wang L, Tian J et al. 2011. Aniline as a dispersing and stabilizing agent for reduced graphene oxide and its subsequent decoration with Ag nanoparticles for enzymeless hydrogen peroxide detection. *J Colloid Interf Sci* 363:615–9.

91. Nguyen VH, Kim BK, Jo YL, Shim JJ. 2012. Preparation and antibacterial activity of silver nanoparticles-decorated graphene composites. *J Supercrit Fluid* 72:28–35.

92. Qin X, Luo Y, Lu W et al. 2012. One-step synthesis of Ag nanoparticles-decorated reduced graphene oxide and their application for H_2O_2 detection. *Electrochim Acta* 79:46–51.

93. Zhou Y, Cheng X, Du D et al. 2014. Graphene–silver nanohybrids for ultrasensitive surface enhanced Raman spectroscopy: Size dependence of silver nanoparticles. *J Mater Chem C* doi: 10.1039/C4TC00658E

94. Yola ML, Gupta VK, Eren T et al. 2014. A novel electro analytical nanosensor based on graphene oxide/silver nanoparticles for simultaneous determination of quercetin and morin. *Electrochim Acta* 120:204–11.

95. Palanisamy S, Karuppiah C, Chen SM. 2014. Direct electrochemistry and electrocatalysis of glucose oxidase immobilized on reduced graphene oxide and silver nanoparticles nanocomposite modified electrode. *Colloids Surface B* 114:164–9.

96. Chang LY, Lee CP, Huang KC et al. 2012. Facile fabrication of PtNP/MWCNT nanohybrid films for flexible counter electrode in dye-sensitized solar cells. *J Mater Chem* 22:3185–91.

97. Chang LY, Lee CP, Vittal R, Lin JJ, Ho KC. 2012. Control of morphology and size of platinum crystals through amphiphilic polymer assisted microemulsions and their uses in dye-sensitized solar cells. *J Mater Chem* 22:12305–12.

98. Bajai R, Roy S, Kumar P et al. 2011. Graphene supported platinum nanoparticle counter electrode for enhanced performance of dye-sensitized solar cells. *ACS Appl Mater Interfaces* 3:3884–9.

99. Yen M, Teng C, Hsiao M et al. 2011. Platinum nanoparticles/graphene composite catalyst as a novel composite counter electrode for high performance dye-snsitized solar cells. *J Mater Chem* 21:12880–8.

34 Magnetocaloric Effect of Graphenes

M. S. Reis and L. S. Paixão

CONTENTS

Abstract 541
34.1 The Magnetocaloric Effect 541
    34.1.1 Magnetic Entropy Change 543
    34.1.2 Adiabatic Temperature Change 543
34.2 Why Study Magnetocaloric Effect of Graphenes? 544
34.3 Magnetic Entropy of Nonrelativistic Diamagnets 544
    34.3.1 Grand Potential 544
    34.3.2 Three-Dimensional Nonrelativistic Diamagnets 545
    34.3.3 Two-Dimensional Nonrelativistic Diamagnets 546
34.4 Magnetocaloric Potentials of Nonrelativistic Diamagnets 547
    34.4.1 Three-Dimensional Diamagnets 547
    34.4.2 Two-Dimensional Diamagnets 549
34.5 Magnetic Entropy of Relativistic Diamagnets: Graphenes 549
34.6 Magnetocaloric Potential of Graphene 551
34.7 Conclusions 551
Appendix 34A: Density of States of a Two-Dimensional Nonrelativistic Electron Gas 552
References 553

ABSTRACT

Magnetocaloric effect (MCE) is an interesting property of materials, that are able to expel/absorb heat from a thermal reservoir, under a magnetic field change (considering an isothermal process), or even increase/decrease their temperature (considering an adiabatic process). These effects are maximum around the critical temperature of the material and therefore ferromagnetic materials are, by far, the most and intensively studied material by the scientific community. For this reason, diamagnetic materials have never been studied in this context, until the present effort, and a number of interesting features, with quantum signatures, were discovered. The fundamental model to describe a diamagnetic material is an electron gas, and a huge applied magnetic field promotes degeneracy, named Landau levels. Oscillations on the thermodynamic quantities are found when the Landau levels cross the Fermi level of the nonperturbed gas at a low-temperature regime. This contribution therefore starts presenting an oscillatory behavior found in the MCE of diamagnetic materials, which can be tuned as either inverse or normal, depending on the value of the magnetic field change. These results open doors for applications at quite low temperatures. The MCE of non-relativistic diamagnetic materials mentioned above has an oscillatory character and this effect occurs at low temperature (~1 K) and high magnetic field (~10 T). A step forward was to consider the relativistic properties of graphenes, a two-dimensional massless diamagnetic material, and those oscillations could be preserved and the effect occurs at a much higher temperature (~100 K) due to the huge Fermi velocity (106 m/s).

34.1 THE MAGNETOCALORIC EFFECT

The magnetocaloric effect (MCE), discovered in 1881 by Warburg [1], is an exciting property, intrinsic to magnetic materials. This effect is induced via coupling of the magnetic sublattice with the magnetic field, which alters the magnetic part of the total entropy due to a corresponding change of the magnetic field. We can see the effect from either an adiabatic or an isothermal process. The material is able to increase/decrease its temperature under an adiabatic process; or even expel/absorb heat from a thermal reservoir under an isothermal process, as a consequence of changes in the magnetic field. Figure 34.1 illustrates these processes.

From a quantitative point of view, the MCE is measured through the magnetic entropy change $\Delta S = \Delta Q/T$, where ΔQ is the amount of heat exchanged between the thermal reservoir and the magnetic material, when the isothermal process is considered. Analogously, the adiabatic temperature change ΔT characterizes the effect when the adiabatic process is considered. These quantities follow from the entropy when it is expressed as a function of temperature and field, $S = S(T, B)$. Figure 34.2 shows the change in entropy due to a magnetic field, and illustrates how we can obtain the quantities that characterize the MCE, that is, the magnetocaloric potentials.

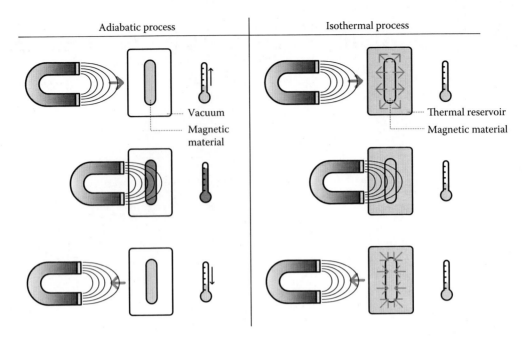

FIGURE 34.1 Fundamentals of the MCE. An applied magnetic field either changes the temperature of the magnetic material (considering an adiabatic process), or promotes a heat exchange with a thermal reservoir (considering an isothermal process). (Reprinted from *Fundamentals of Magnetism*, M. S. Reis, Copyright 2013, with permission from Elsevier.)

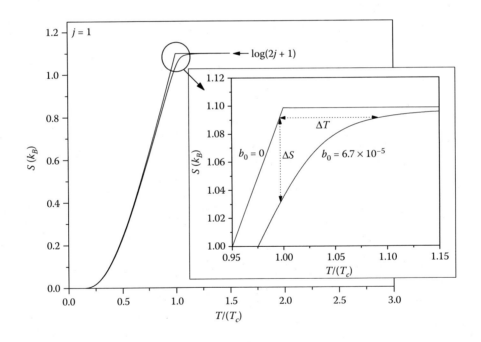

FIGURE 34.2 Numerical entropy for a localized ferromagnetic system of angular momenta $j = 1$. The reduced magnetic field is $b_0 = \mu_B B / k_B T_c$. We can see the entropy change of an isothermal process as well as the temperature change of an adiabatic process. (Reprinted from *Fundamentals of Magnetism*, M. S. Reis, Copyright 2013, with permission from Elsevier.)

It is straightforward, the idea, to produce a thermo-magnetic cycle based on the isothermal and/or adiabatic processes (like Brayton and Ericsson cycles). Figure 34.3 presents the Ericsson cycle: it is composed of two isothermal processes and two isofield processes. Indeed, the idea of magnetic refrigeration began in the late 1920s, when cooling via adiabatic demagnetization was proposed by Debye [2] and Giauque [3]. The process was later demonstrated by Giauque and MacDougall, in 1933; where they reached 250 mK [4]. The main point to the scientific community is: which kind of material optimizes the MCE, to then be used into devices? To find an answer to this question, a brief review on thermodynamics is needed; as given below.

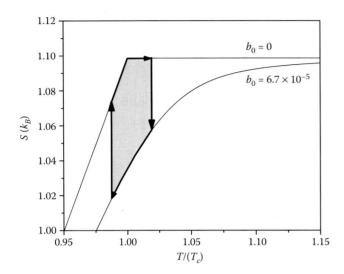

FIGURE 34.3 Ericsson cycle for the MCE. The reduced magnetic field is $b_0 = \mu_B B / k_B T_c$. This cycle is one of the proposals for magnetic refrigeration. (Reprinted from *Fundamentals of Magnetism*, M. S. Reis, Copyright 2013, with permission from Elsevier.)

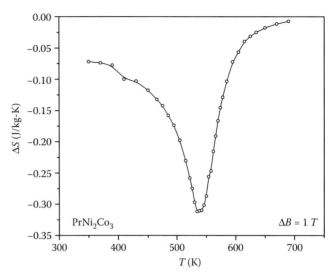

FIGURE 34.4 Experimental magnetic entropy change for a simple ferromagnetic material: $PrNi_2Co_3$. (Reprinted from *Fundamentals of Magnetism*, M. S. Reis, Copyright 2013, with permission from Elsevier.)

34.1.1 Magnetic Entropy Change

We may relate the entropy change and the magnetization of the system using the following Maxwell relation:

$$\left.\frac{\partial S}{\partial B}\right|_T = \left.\frac{\partial M}{\partial T}\right|_B. \tag{34.1}$$

After integration, the above equation reads as

$$\Delta S(T, \Delta B) = \int_{B_i}^{B_f} \left.\frac{\partial S}{\partial B}\right|_T dB = \int_{B_i}^{B_f} \left.\frac{\partial M}{\partial T}\right|_B dB. \tag{34.2}$$

Thus, one way to obtain the magnetic entropy change is via measurement of magnetization as a function of temperature and magnetic field, that is, $M(T, B)$.

There is another way to obtain the magnetic entropy change: from specific heat, which is defined by the expression

$$C_B = T \left.\frac{\partial S}{\partial T}\right|_B. \tag{34.3}$$

After a simple integration, the above equation leads to

$$\Delta S(T, \Delta B) = \int_0^T \frac{C_B - C_0}{T} dT, \tag{34.4}$$

where $C_0 = C_B = 0$.

From Equation 34.2, we see that the entropy change is a maximum when $\partial M / \partial T$ is maximum, that is, near a ferromagnetic ordering. We can also reach the same conclusion inspecting Figure 34.4, which presents the magnetic entropy change for the ferromagnetic material $PrNi_2Co_3$ [5]. The referred material

has $T_c \approx 540$ K. Therefore, ferromagnetic materials are, by far, the most and intensively studied materials by the scientific community [6] due to the high potential for application into devices. Materials with other kind of ordering, like ferrimagnets and antiferromagnets, have already been largely studied.

34.1.2 Adiabatic Temperature Change

Entropy is a function of temperature T and magnetic field B

$$S = S(T, B), \tag{34.5}$$

and then

$$dS = \left.\frac{\partial S}{\partial T}\right|_B dT + \left.\frac{\partial S}{\partial B}\right|_T dB. \tag{34.6}$$

The adiabatic condition means $dS = 0$, and therefore, considering the definition of specific heat on Equation 34.3 and the Maxwell relation on Equation 34.1, Equation 34.6 reads as

$$dT = -\frac{T}{C_B} \left.\frac{\partial M}{\partial T}\right|_B dB. \tag{34.7}$$

Finally, the adiabatic temperature change is

$$\Delta T(T, \Delta B) = -\int_{B_i}^{B_f} \frac{T}{C_B} \left.\frac{\partial M}{\partial T}\right|_B dB. \tag{34.8}$$

Thus, in conclusion, to obtain the adiabatic temperature change, we need to measure $M(T, B)$ and $C(T, B)$. However, there is another way to measure this quantity, that is, applying a magnetic field to the sample (adiabatically) and directly measuring the corresponding temperature change.

34.2 WHY STUDY MAGNETOCALORIC EFFECT OF GRAPHENES?

Owing to the large deal of attention driven to materials with long range ordering, MCE on diamagnets have been first studied very recently and it showed interesting properties [7–10]. The simplest model that describes diamagnetic materials is an electron gas; when a magnetic field is applied to this gas, highly degenerated levels (Landau levels) appear. Since the separation of these levels depend on the applied field, the Fermi energy jumps from the nth to the $(n-1)$th Landau level. Thus, since thermodynamic quantities depend on the Fermi energy, this mechanism produces the well-known magnetic oscillations. Figure 34.5 illustrates the crossing of Landau levels through the Fermi level. It is known that graphene presents oscillations on electrical conductivity [11] (Shubnikov–de Haas effect), and oscillations on magnetization [12] (de Haas–van Alphen effect), both effects caused by the jumps of the Fermi level. Therefore, it is natural to expect the oscillatory behavior to be observed in other quantities, like entropy and, consequently, the MCE.

The exotic electronic properties of graphenes are quite different from standard 3D diamagnetic materials. While for a nonrelativistic 3D diamagnetic material the Landau levels are given by a harmonic oscillator [6]:

$$E_i = \hbar\omega\left(i + \frac{1}{2}\right) \text{ with } \hbar\omega = \frac{\hbar eB}{m_e}, \quad (34.9)$$

for a 2D Dirac-like system the levels are given by [13]

$$E_j = \hbar\omega'\sqrt{j} \text{ with } \hbar\omega' = \sqrt{2\hbar eBv_F^2}, \quad (34.10)$$

where $i, j = 0, 1, 2, \ldots$ and $v_F = 10^6$ m/s stands for the Fermi velocity (300 times smaller than the speed of light). Note that the Landau levels for a nonrelativistic material are equally spaced, whereas for graphene it is not true. The reason for the difference in the behavior is that in the former case electrons obey the classical parabolic dispersion relation, while in the latter they follow the relativistic linear dispersion relation that comes from Dirac equation.

Therefore, in this chapter, we aim to connect the oscillations found in thermodynamic quantities of graphene with its magnetocaloric properties. We start the analysis with standard nonrelativistic materials and then change to the relativistic case to describe graphenes.

34.3 MAGNETIC ENTROPY OF NONRELATIVISTIC DIAMAGNETS

34.3.1 GRAND POTENTIAL

In this section, we briefly review some concepts regarding thermodynamics of electron gas that will be used later on. To discuss the MCE, we must know the entropy of the system

$$S(T,B) = -\frac{\partial\Phi(T,B)}{\partial T}, \quad (34.11)$$

which can be obtained from the grand potential

$$\Phi(T,B) = -k_BT\int_0^\infty g(\varepsilon)\ln(1 + ze^{-\beta\varepsilon})\,d\varepsilon, \quad (34.12)$$

where $z = e^{\beta\mu}$ is the fugacity, and $g(\varepsilon)$ the one-particle density of states. We considered an electron gas under an applied magnetic field following the conditions of low temperature, $\varepsilon_F \gg k_BT$, and a magnetic energy μ_BB from k_BT up to ε_F.

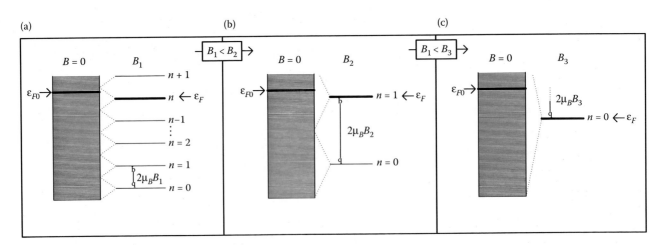

FIGURE 34.5 Electron gas under a magnetic field illustrating the crossing of Landau levels through the Fremi level, mechanism that causes oscillations on thermodynamic quantities. (Reprinted from *Fundamentals of Magnetism*, M. S. Reis, Copyright 2013, with permission from Elsevier.)

34.3.2 THREE-DIMENSIONAL NONRELATIVISTIC DIAMAGNETS

The density of states is given by [14]

$$g(\varepsilon) = \frac{3}{2} N \varepsilon_F^{-3/2} \left[\varepsilon^{1/2} + (\mu_B B)^{1/2} \sum_{l=1}^{\infty} \frac{(-1)^l}{l^{1/2}} \cos\left(\frac{l\pi}{\mu_B B} \varepsilon - \frac{\pi}{4} \right) \right]$$

$$= g_0(\varepsilon) + g_B(\varepsilon). \qquad (34.13)$$

See appendix for details about the procedure to obtain this result. From Equations 34.12 and 34.13, one easily notes that there are two contributions to the grand potential; one that does not depend on the magnetic field B and the other that does, that is,

$$\Phi(T,B) = \Phi_0(T) + \Phi_B(T,B). \qquad (34.14)$$

Thus, the total entropy of the electron gas can be written as

$$S(T,B) = S_0(T) + S_B(T,B). \qquad (34.15)$$

The first term is the well-known entropy of a free electron gas without applied magnetic field, in which, within the regime of low temperature, $k_B T \ll \varepsilon_F$, reads as

$$\frac{S_0(T)}{Nk_B} = \frac{\pi^2 t}{2}, \qquad (34.16)$$

where $t = k_B T / \varepsilon_F$. Let us now evaluate the second term that corresponds to the magnetic part of the entropy. After integration of Equation 34.12 by parts (twice), we achieve

$$\Phi_B = -\frac{1}{\beta} G_{3D}(\varepsilon) \ln\left(1 + z e^{-\beta\varepsilon}\right) \Big|_0^{\infty} - \mathcal{G}_{3D}(\varepsilon) \frac{1}{z^{-1} e^{\beta\varepsilon} + 1} \Big|_0^{\infty}$$

$$+ \int_0^{\infty} \mathcal{G}_{3D}(\varepsilon) \frac{\partial}{\partial\varepsilon} \frac{1}{z^{-1} e^{\beta\varepsilon} + 1} d\varepsilon, \qquad (34.17)$$

where

$$G_{3D}(\varepsilon) = \frac{3}{2\pi} N \left(\frac{\mu_B B}{\varepsilon_F} \right)^{3/2} \sum_{l=1}^{\infty} \frac{(-1)^l}{l^{3/2}} \sin\left(\frac{l\pi}{\mu_B B} \varepsilon - \frac{\pi}{4} \right), \qquad (34.18)$$

and

$$\mathcal{G}_{3D}(\varepsilon) = -\frac{3}{2\pi^2} N \mu_B B \left(\frac{\mu_B B}{\varepsilon_F} \right)^{3/2} \sum_{l=1}^{\infty} \frac{(-1)^l}{l^{5/2}} \cos\left(\frac{l\pi}{\mu_B B} \varepsilon - \frac{\pi}{4} \right). \qquad (34.19)$$

At low temperature, $k_B T \ll \varepsilon_F$, the chemical potential approaches the Fermi energy, and therefore $z \approx e^{\beta\varepsilon_F} \gg 1$. With this condition, Equation 34.17 resumes as

$$\Phi_B = \Phi_B^{no} + \Phi_B^o, \qquad (34.20)$$

where

$$\Phi_B^{no} = G_{3D}(0)\varepsilon_F + \mathcal{G}_{3D}(0)$$

$$= -\frac{3}{2\sqrt{2}\pi^2} N \mu_B B \left(\frac{\mu_B B}{\varepsilon_F} \right)^{3/2} \sum_{l=1}^{\infty} \frac{(-1)^l}{l^{5/2}} \left(\frac{l\pi}{\mu_B B} \varepsilon_F + 1 \right), \qquad (34.21)$$

has a non-oscillatory character and

$$\Phi_B^o = \frac{3}{2} N k_B T \left(\frac{\mu_B B}{\varepsilon_F} \right)^{3/2} \sum_{l=1}^{\infty} \frac{(-1)^l}{l^{3/2}} \cos\left(\frac{l\pi}{\mu_B B} \varepsilon_F - \frac{\pi}{4} \right) \frac{1}{\sinh(lx)}, \qquad (34.22)$$

has an oscillating behavior. In addition,

$$x = \pi^2 \frac{k_B T}{\mu_B B}. \qquad (34.23)$$

Summarizing, the grand potential (Equation 34.14) has two contributions: one that depends on B and the other that does not. The first one also has two contributions (Equation 34.20): one that depends on B as a power law (Equation 34.21), and the other that oscillates depending on the magnetic field (Equation 34.22).

From Equation 34.20, we see that the field-dependent entropy resumes as

$$S_B(T,B) = S_B^o(T,B)$$

$$= \frac{3}{2} N k_B \left(\frac{\mu_B B}{\varepsilon_F} \right)^{3/2} \sum_{l=1}^{\infty} \frac{(-1)^l}{l^{3/2}} \cos\left(\frac{l\pi}{\mu_B B} \varepsilon_F - \frac{\pi}{4} \right) \mathcal{T}(lx), \qquad (34.24)$$

where

$$\mathcal{T}(x) = \frac{xL(x)}{\sinh x}, \qquad (34.25)$$

and $L(x)$ is the Langevin function

$$L(x) = \coth x - \frac{1}{x}. \qquad (34.26)$$

Note that only $\mathcal{T}(x)$ contains information on the temperature; and the period of oscillation depends only on the magnetic field, and not on the temperature.

The hyperbolic sine in the denominator of the function $\mathcal{T}(x)$, dampens the entropy. If the thermal energy is large compared to magnetic energy $k_B T \gg \mu_B B$, the entropy becomes very small. On the other hand, if $k_B T \lesssim \mu_B B$, only the term $l = 1$ contributes notably to the summation. Therefore, we can drop out the summation and express the entropy in terms of dimensionless variables; it reads as

$$\frac{S_B(T,B)}{Nk_B} = -\frac{3}{2}\left(\frac{1}{n+1/4}\right)^{3/2}\cos(n\pi)\mathcal{T}(x), \quad (34.27)$$

where

$$n = \frac{\varepsilon_F}{\mu_B B} - \frac{1}{4}. \quad (34.28)$$

In addition, it is convenient to rewrite Equation 34.23 as

$$x = \pi^2 t\left(n+\frac{1}{4}\right). \quad (34.29)$$

34.3.3 Two-Dimensional Nonrelativistic Diamagnets

In two dimensions, the density of states of an electron gas reads (see Appendix 34A for further details)

$$g(\varepsilon) = g_0 + g_B(\varepsilon), \quad (34.30)$$

where

$$g_0 = \frac{L^2}{4\pi}\frac{2m_e}{\hbar^2}, \quad (34.31)$$

is the zero-field density of state (with no spin degeneracy), and

$$g_B(\varepsilon) = 2g_0\sum_{l=1}^{\infty}(-1)^l\cos\left(\frac{l\pi}{\mu_B B}\varepsilon\right). \quad (34.32)$$

From the density of states, it is easy again to see that the entropy can be written as a sum of two terms, as in Equation 34.15; one depends only on temperature and the other depends on both, temperature and magnetic field.

Let us now evaluate the field dependent part of the entropy, which we obtain from the grand potential using the density of states (34.32)

$$\Phi_B(T,B) = -k_B T\int_0^{\infty} 2g_0\sum_{l=1}^{\infty}(-1)^l\cos\left(\frac{l\pi}{\mu_B B}\varepsilon\right)\ln\left(1+ze^{-\beta\varepsilon}\right)d\varepsilon. \quad (34.33)$$

Analogou sly to before, after integration by parts (twice), we achieve

$$\Phi_B(T,B) = -2g_0 k_B T\sum_{l=1}^{\infty}(-1)^l\left[G_{2D}(\varepsilon)\Big|_0^{\infty} + \left(\frac{\mu_B B\beta}{l\pi}\right)\mathcal{G}_{2D}(\varepsilon)\Big|_0^{\infty}\right. \\ \left. -\frac{1}{4}\left(\frac{\mu_B B\beta}{l\pi}\right)^2\mathcal{I}(T,B)\right], \quad (34.34)$$

where

$$G_{2D}(\varepsilon) = \frac{\mu_B B}{l\pi}\sin\left(\frac{l\pi}{\mu_B B}\varepsilon\right)\ln\left(1+ze^{-\beta\varepsilon}\right), \quad (34.35)$$

$$\mathcal{G}_{2D}(\varepsilon) = -\frac{\mu_B B}{l\pi}\cos\left(\frac{l\pi}{\mu_B B}\varepsilon\right)\frac{1}{z^{-1}e^{\beta\varepsilon}+1}, \quad (34.36)$$

and

$$\mathcal{I}(T,B) = \int_0^{\infty}\frac{\cos\left(\frac{l\pi}{\mu_B B}\varepsilon\right)}{\cosh^2\left(\frac{\beta(\varepsilon-\mu)}{2}\right)}d\varepsilon \xrightarrow{\varepsilon_F \gg k_B T} \frac{4\pi^2 l}{\mu_B B\beta^2}\frac{\cos\left(\frac{l\pi}{\mu_B B}\varepsilon_F\right)}{\sinh\left(\frac{l\pi^2}{\mu_B B\beta}\right)}. \quad (34.37)$$

Above, the condition $\varepsilon_F \gg k_B T$ implies $\mu \sim \varepsilon_F$, and therefore $z \gg 1$. As a consequence, $G_{2D}(\infty)$, $G_{2D}(0)$ and $\mathcal{G}_{2D}(\infty)$ are zero; and

$$\mathcal{G}_{2D}(0) \to -\frac{\mu_B B}{l\pi}. \quad (34.38)$$

Thus, the grand potential reveals two contributions

$$\Phi_B(T,B) = \Phi_B^{no}(B) + \Phi_B^o(T,B), \quad (34.39)$$

where

$$\Phi_B^{no}(B) = \frac{1}{6}g_0(\mu_B B)^2, \quad (34.40)$$

has a non-oscillatory character and

$$\Phi_B^o(T,B) = 2g_0 k_B T\mu_B B\sum_{l=1}^{\infty}\frac{(-1)^l}{l}\cos\left(\frac{l\pi}{\mu_B B}\varepsilon_F\right)\frac{1}{\sinh(lx)}, \quad (34.41)$$

has an oscillating behavior. Above, x is the same as in Equation 34.23. From Equation 34.40, we can easily see that the non-oscillatory entropy is null. Therefore, the magnetic entropy is

$$S_B(T,B) = S_B^o(T,B)$$

$$= 2Nk_B\frac{\mu_B B}{\varepsilon_F}\sum_{l=1}^{\infty}\frac{(-1)^l}{l}\cos\left(\frac{l\pi}{\mu_B B}\varepsilon_F\right)\mathcal{T}(lx). \quad (34.42)$$

Above

$$N = \int_0^{\varepsilon_F}g_0 d\varepsilon = g_0\varepsilon_F, \quad (34.43)$$

is the number of electrons that fulfill the gas states up to the Fermi energy in the absence of magnetic field.

Following the same logic employed to 3D electron gases, we can drop the summation of the entropy and write

$$\frac{S_B(T,B)}{N k_B} = -\frac{2}{m}\cos(m\pi)\mathcal{T}(x), \qquad (34.44)$$

where

$$m = \frac{\varepsilon_F}{\mu_B B}. \qquad (34.45)$$

Equation 34.23 can be rewritten as

$$x = \pi^2 t m. \qquad (34.46)$$

34.4 MAGNETOCALORIC POTENTIALS OF NONRELATIVISTIC DIAMAGNETS

34.4.1 Three-Dimensional Diamagnets

From now on, we will investigate the magnetocaloric properties of some nonrelativistic systems, starting with a 3D diamagnetic material. The standard model for a diamagnetic material is an electron gas; for this reason, this topic was discussed in the last section, and we will apply those results below.

In an isothermal process, we define the entropy change under application of a field B as

$$\Delta S(T,\Delta B) = S(T,B) - S(T,0). \qquad (34.47)$$

As we verified in the last section, the entropy is expressed as a sum of a nonmagnetic term and an oscillating magnetic one. Consequently, the entropy change becomes

$$\Delta S(T,\Delta B) = [S_0(T) + S_B^o(T,B)] - [S_0(T) + S_B^o(T,0)]$$

$$= S_B^o(T,B), \qquad (34.48)$$

since $S_B^o(T,0) = 0$. Thus, the magnetic entropy change is given by Equation 34.27:

$$\frac{\Delta S}{N k_B} = -\frac{3}{2}\left(\frac{1}{n+1/4}\right)^{3/2}\cos(n\pi)\mathcal{T}(x). \qquad (34.49)$$

Note that the amplitude of the entropy change can be maximized and minimized by the cosine function. Thus, we can consider the condition

$$\cos(n\pi) = \pm 1, \qquad (34.50)$$

and then an interesting result arises: n even and zero implies in a negative (and minimized) ΔS; n odd implies in a positive (and maximized) ΔS; and, finally, a half-integer makes zero the magnetic entropy change (and then the magnetocaloric

potential of the material). The thermal dependence of the entropy change is ruled by the function $\mathcal{T}(x)$, and that function peaks at $x = 1.6$. Therefore, the temperature at which the entropy change is maximum lies at

$$T_{\max}[K] = 0.1\,B[T]. \qquad (34.51)$$

Let us consider a piece of gold as an example, whose Fermi energy is 5.51 eV. Consider $n = 0$ implies, thus, in $B = 380$ kT, which is absolutely out of the laboratory range; and, analogously, $n = 10^4$ implies in $B = 10$ T. Following this example, note that Figure 34.6 presents the magnetic entropy change as a function of the dimensionless temperature t. The scale of t is of the order of 10^{-5} and then, considering Au, the temperature is of the order of 1 K. Thus, for Au, this effect is comfortably visible at 10 T and 1 K. In addition, as mentioned above, the magnetic entropy change can change its sign depending on the value of n, and this behavior can also be seen in Figure 34.6.

The entropy change as a function of the reciprocal magnetic field is depicted in Figure 34.7. The effect is maximized around $t = 10^{-5}$ (see Figure 34.6); for Au it happens around 1 Kelvin as mentioned above. Note that the oscillatory behavior of this quantity is maximized for n odd, zero for n half-integer, and minimized for n even. The modulation presented is given by Equation 34.49 without the cosine.

The MCE is due to a magnetic field change $\Delta B : 0 \to B(n)$. The final value of the magnetic field $B(n)$ can tune the sign of ΔS; and the change of the magnetic field change that is able to do this inversion is (for $n \gg 1$)

$$|\Delta(\Delta B)| = |\Delta B(n+1) - \Delta B(n)| \approx \frac{\varepsilon_F}{\mu_B}\frac{1}{n^2}. \qquad (34.52)$$

Again for Au, note that around 10 T of magnetic field change, an increase of the final value of magnetic field in 10^{-3} T is enough to invert the magnetic entropy change. In other

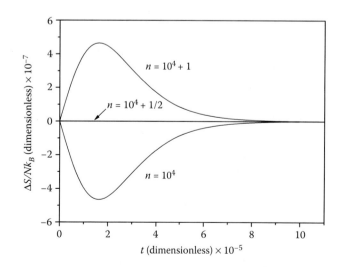

FIGURE 34.6 Magnetic entropy change for a 3D system as a function of $t = k_B T/\varepsilon_F$. (Reprinted with permission from M. S. Reis, *Appl. Phys. Lett.*, vol. 99, p. 052511. Copyright 2011, American Institute of Physics.)

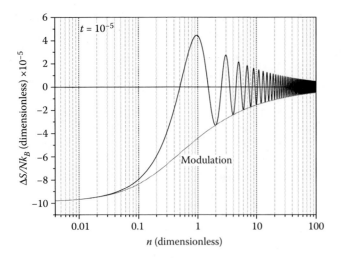

FIGURE 34.7 Magnetic entropy change for a 3D system as a function of n (a function of the inverse magnetic field B). (Reprinted with permission from M. S. Reis, *Appl. Phys. Lett.*, vol. 99, p. 052511. Copyright 2011, American Institute of Physics.)

words, this effect is sensible to a magnetic field 10,000 times smaller than the applied one.

Owing to its interesting properties this thermal system opens doors for some applications. It is possible to work as a high sensible magnetic field sensor, with a huge magnetic field on the background.

The magnetocaloric potential of the system discussed here is indeed smaller when compared to standard ferromagnetic materials [15]. However, those oscillations can be useful if improved in terms of the magnetocaloric potential; specially for magnetic cooling at quite low temperatures, since the sign of the magnetic entropy change can be ruled by the final value of the applied magnetic field. This effect can indeed be used, if improved, in adiabatic demagnetization refrigerators.

Contrary to what was discussed until now, let us focus on an adiabatic process. In such a process, the system changes its temperature from T_0 to T_B under application of a magnetic field B. The condition is imposed considering that the entropy of the system at T_0 and zero field is the same of the system at T_B and under an applied magnetic field B. Thus, from the adiabatic condition

$$S(T_0, 0) = S(T_B, B), \qquad (34.53)$$

we obtain the adiabatic temperature change

$$\Delta T = T_B - T_0. \qquad (34.54)$$

From Equations 34.16, 34.27, and the adiabatic condition stated above, we obtain

$$\frac{\pi^2}{2} t_0 = \frac{\pi^2}{2} t_B - \frac{3}{2} \left(\frac{1}{n+1/4} \right)^{3/2} \cos(n\pi) \mathcal{T}(x_B), \qquad (34.55)$$

where $t_0 = t(T_0)$, $t_B = t(T_B)$, and $x_B = x(t_B)$. This condition gives a relationship between T_0 and T_B; and then it is possible to

write the adiabatic temperature change. However, the function $T_0(T_B)$ from Equation 34.55 is not simple and an approximation is needed. As we are already considering $t \ll 1$, we can consider $x \ll 1$ (see Equation 34.29). We only need to ensure that n ranges from unity up to the order of $1/t$. Following this condition, Equation 34.25 resumes as

$$\mathcal{T}(x) \approx \frac{x}{3}. \qquad (34.56)$$

Now we can write our final result from Equation 34.55

$$\Delta T = T_0 \left[\frac{\cos(n\pi)}{\sqrt{n+1/4} - \cos(n\pi)} \right]. \qquad (34.57)$$

Note that if $\sqrt{n+1/4} = \cos(n\pi)$, the adiabatic temperature change diverges; and it occurs for $n = 1/4$, that corresponds to $B = 2\varepsilon_F/\mu_B$, that no longer fulfill the condition stated above, in which the magnetic field must be up to the order of the Fermi energy. On the other hand, for $n \gg 1$, Equation 34.57 resumes as

$$\Delta T = T_0 \left[\frac{\cos(n\pi)}{\sqrt{n}} \right]. \qquad (34.58)$$

Thus, for n even, the adiabatic temperature change is positive, that is, $T_B > T_0$, and the system warms up due to an applied magnetic field. On the other hand, for n odd, $T_B < T_0$ and the system cools down due to an applied magnetic field. Of course, for half-integer values of n, the adiabatic temperature change is zero.

Figure 34.8 presents the oscillatory behavior found for the adiabatic temperature change ΔT in Equation 34.57, as well as the approximation for large values of n, Equation 34.58. This

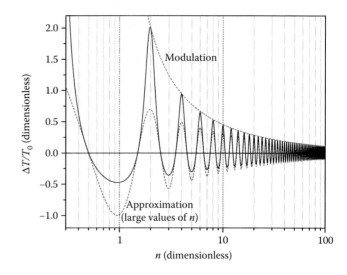

FIGURE 34.8 Reduced adiabatic temperature change $\Delta T/T_0$ for a 3D system as a function of n (proportional to the reciprocal magnetic field $1/B$). (Reprinted from *Solid State Commun.*, 152, M. S. Reis. 921, Copyright 2012, with permission from Elsevier.)

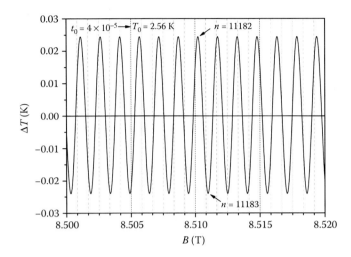

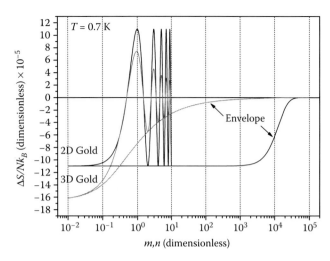

FIGURE 34.9 Adiabatic temperature change as function of the applied magnetic field. This example is for 3D Gold ($\varepsilon_F = 5.51$ eV), under accessible temperature and magnetic field ranges. (Reprinted from *Solid State Commun.*, 152, M. S. Reis. 921, Copyright 2012, with permission from Elsevier.)

FIGURE 34.10 Oscillating magnetic entropy change per electron as a function of the inverse magnetic field B (see Equations 34.28 and 34.45). The oscillatory behavior is evident and has a remarkable difference between 2D (Equation 34.60) and 3D (Equation 34.49) models. For the sake of clearness, above m, n, = 10 only the *envelope* is shown. See text for further details. (Reprinted with permission from M. S. Reis, *J. Appl. Phys.*, 113, 243901. Copyright 2013, American Institute of Physics.)

figure also contains the modulation of the oscillations found, obtained from Equation 34.57 considering the cosines equal to one.

Let us once again use gold as a real example to illustrate the effect. In this sense, Figure 34.9 presents the prediction of this effect at $T_0 = 2.56$ K and around 8.5 T of magnetic field change. Note, therefore, that this effect is comfortably visible within that range of magnetic field and temperature, reasonable for standard laboratories. The value chosen for the initial temperature T_0 corresponds to $t = 4 \times 10^{-5}$, and is a temperature range in which we can work easier.

Considering Figure 34.9, it is possible to see that Gold produces $|\Delta T(B(n = 11182)) - \Delta T(B(n = 11183))| = 50$ mK of change of the adiabatic temperature change (around 2.56 K), due to $|B(n = 11182) - B(n = 11183)| = 0.8$ mT of change of the magnetic field change (with a huge magnetic field on the background). This temperature change is easily measured with a cernox-like sensor, since its accuracy is around 5 mK for this range of temperature.

34.4.2 TWO-DIMENSIONAL DIAMAGNETS

The realization of a diamagnetic material in two dimensions is a thin film, whose magnetocaloric properties will be discussed below. As we know from the beginning of the present section, the entropy change caused by a field change $\Delta B:0 \rightarrow B$ is simply the magnetic entropy at the final field

$$\Delta S(T, \Delta B) = S_B^o(T, B). \qquad (34.59)$$

Thus, the entropy change, per electron, for a 2D diamagnetic material is given by Equation 34.44

$$\frac{\Delta S}{N k_B} = -\frac{2}{m} \cos(m\pi) \mathcal{T}(x). \qquad (34.60)$$

Let us consider a Gold thin film, with Fermi energy $\varepsilon_F = 3.62$ eV. From Equation 34.45, it is possible to see that $B(m = 1) = 6.2522 \times 10^4$ T, and therefore completely out of a laboratory range. On the other hand, $B(m = 10^4) = 6.2522$ T, and therefore this is the order of magnitude of m we must consider. Note that the magnetic field change needed to invert the magnetic entropy change (from normal/negative to inverse/positive) is $B(m = 10^4 + 1) - B(m = 10^4) = 0.7$ mT. The temperature dependence of the entropy change lies on the $\mathcal{T}(x)$ function, similarly to the 3D system. This function peaks at $x = \pi^2 t m = 1.6$, and therefore, considering $m = 10^4$, the maximum magnetic entropy change occurs at $T = 0.7$ K.

The dependence of the magnetic entropy change as a function of m is presented in Figure 34.10 for the temperature that maximizes this effect (considering $m = 10^4$), that is, 0.7 K. It has an oscillating behavior due to cosine on Equation 34.60, and the *envelope* of these oscillations comes from the same equation without the cosine function.

34.5 MAGNETIC ENTROPY OF RELATIVISTIC DIAMAGNETS: GRAPHENES

Unlike the systems we studied before, in graphene, electrons behave like relativistic particles with zero rest mass and have an effective "speed of light" equal to Fermi velocity [11]. Thus, dynamics of electrons in graphene is described by Dirac equation. Single particle Dirac Hamiltonian is

$$\mathcal{H}_D = v_F \vec{\alpha} \cdot \vec{\Pi}, \qquad (34.61)$$

where $\vec{\alpha}$ is the Pauli matrix vector and $\vec{\Pi}$ is the momentum of the particle. The eigenvalues are

$$\varepsilon_j = \text{sgn}(j)\sqrt{2\,|\,j\,|\,\hbar eB}\,v_F, \qquad (34.62)$$

where j is the Landau level index. Degeneracy of Landau levels per graphene area is

$$\tilde{g} = \frac{2eB}{\pi\hbar}, \qquad (34.63)$$

which is independent of the level. We are already taking into account a factor of 4 corresponding to spin and sublattice degeneracies [16].

To obtain the grand potential, we follow the same procedure of Zhang et al. [17]. They studied the de Haas–van Alphen effect in graphene in the presence of an additional electric field, orthogonal to the magnetic field. The first step is to obtain the total energy of the 2D gas of Dirac electrons, summing the energies of the Landau levels

$$E = \sum_{j=0}^{[N_0/\tilde{g}]} \tilde{g}\varepsilon_j + \mu_0 \tilde{g}\,\text{mod}\left[\frac{N_0}{\tilde{g}}\right]. \qquad (34.64)$$

Above $\mu_0 = \hbar v_F \sqrt{N_0\pi}$ is the zero temperature and field chemical potential, that is, the Fermi energy; and $N_0 = 10^{16}$ m^{-2} is the density of charge carriers [18]. In addition, the first term counts the energy of the completely occupied Landau levels, and the second one counts the energy of the highest level, partially occupied. Here, the notation $[z]$ means integer part of, that is, the largest integer satisfying $[z] \leq z$, and mod$[z]$ stands for the fractional part, that is, $z = [z] + \text{mod}[z]$. From the above energy, we write it as a sum of a non-oscillating term plus an oscillating one

$$E = E^{no} + E^o. \qquad (34.65)$$

where

$$E^{no} = -\frac{\zeta(3/2)v_F}{\pi^2\sqrt{2}\hbar}(eB)^{3/2} + \frac{2\mu_0^3}{3\pi(\hbar v_F^2)^2}, \qquad (34.66)$$

and

$$E^o = \frac{(eB)^{3/2}v_F}{\pi\sqrt{\hbar}}\sum_{l=1}^{\infty}\frac{1}{l\pi}J_1(lm'\pi)\cos(lm'\pi). \qquad (34.67)$$

Above J_1 is the integral

$$J_1(p) = \int_0^\infty \frac{e^{-tp}}{\sqrt{\pi t}\,(t^2+1)}dt = -\text{Im}\int_0^\infty \frac{e^{-tp}}{\sqrt{\pi t}\,(t+i)}dt, \qquad (34.68)$$

and

$$m' = N_0\frac{\phi_0}{B}, \qquad (34.69)$$

where $\phi_0 = \pi\hbar/e = 2.06\times10^{-15}$ Tm^2 is the magnetic flux quantum.

The next step is to evaluate the grand potential from the total energy above. According to the procedure reported by Sharapov et al. [12], the grand potential of electrons in graphene can be obtained evaluating

$$\Phi(T,B) = \int_{-\infty}^{\infty} P_T(\mu_0 - \mu)E(\mu_0)d\mu_0, \qquad (34.70)$$

where the distribution function $P_T(z)$ takes into account the effect of temperature

$$P_T(z) = \frac{1}{4k_BT\cosh^2\dfrac{z}{2k_BT}}. \qquad (34.71)$$

At low temperatures, we obtain for the grand potential a non-oscillatory term that depends only on the field and an oscillating term dependent on temperature and field. The non-oscillatory part of the grand potential is

$$\Phi_B^{no}(B) = -\frac{\zeta(3/2)v_F}{\pi^2\sqrt{2}\hbar}(eB)^{3/2} + \frac{2\mu_0^3}{3\pi(\hbar v_F)^2}. \qquad (34.72)$$

To obtain the oscillatory part of the grand potential, it is convenient to use the complex form of Equation 34.68 and perform the integration in the complex plane. After that we find

$$\Phi_B^o(T,B) = \frac{(eB)^2v_F^2}{\pi\mu_0}\sum_{l=1}^{\infty}\frac{\cos(lm'\pi)}{(l\pi)^2}\frac{ly}{\sinh(ly)}, \qquad (34.73)$$

where

$$y = 2\pi^2 tm'. \qquad (34.74)$$

Since the non-oscillatory term does not depend on the temperature, it follows that the non-oscillatory entropy is null. Therefore, the magnetic entropy reads as

$$S_B^o(T,B) = 2N_0k_B\frac{1}{m'}\sum_{l=1}^{\infty}\frac{\cos(lm'\pi)}{l}\mathcal{T}(ly). \qquad (34.75)$$

Note that the temperature dependence of the oscillating contribution to the magnetic entropy, which is given by

Equation 34.25, is found to be the same for both 3D and 2D nonrelativistic systems. As we argued before, the hyperbolic sine in the denominator of the function $\mathcal{T}(y)$ dampens all terms of the summation for $l \geq 2$. Thus, the magnetic entropy per electron reads as

$$\frac{S_B^o(T,B)}{N_0 k_B} = \frac{2}{m'}\cos(m'\pi)\mathcal{T}(y). \tag{34.76}$$

34.6 MAGNETOCALORIC POTENTIAL OF GRAPHENE

From the result of the previous section, it is easy to see that $S_B^o(T,0) = 0$. Therefore, the entropy change, $\Delta S(T, \Delta B) = S_B^o(T,B)$, is

$$\frac{\Delta S}{N_0 k_B} = \frac{2}{m'}\cos(m'\pi)\mathcal{T}(y). \tag{34.77}$$

The oscillating MCE for graphenes is presented as a function of m' in Figure 34.11. The change of magnetic field change required to invert the MCE (from normal to inverse) for a graphene (ca. 3.4 T) is much higher than the 3D nonrelativistic case (ca. 1 mT); therefore, it would be easier to verify the effect. In addition, the oscillation is rapidly smashed by the hyperbolic sine in the denominator of the function $\mathcal{T}(y)$ in Equation 34.77.

Figure 34.12 presents the entropy change as a function of temperature for some values of m', that is, some values of magnetic field. Note that odd (even) values of m' minimize (maximize) the magnetic entropy change. The temperature of the maximum entropy change, given by the function $\mathcal{T}(y)$, is

$$T_{\max}[K] = 5.3\, B[T]. \tag{34.78}$$

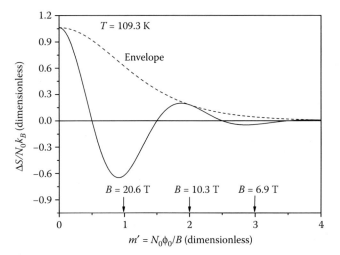

FIGURE 34.11 Oscillating magnetic entropy change as a function of m', which is inversely proportional to the magnetic field B. (Reprinted with permission from M. S. Reis, *Appl. Phys. Lett.*, 101, 222405. Copyright 2012, American Institute of Physics.)

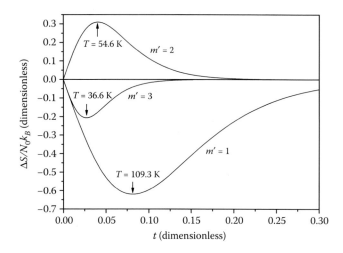

FIGURE 34.12 Oscillating magnetic entropy change as a function of temperature. We can see several curves for different values of magnetic field. (Reprinted with permission from M. S. Reis, *Appl. Phys. Lett.*, 101, 222405. Copyright 2012, American Institute of Physics.)

If we consider $m' = 3$, the corresponding magnetic field is $B = 6.9$ T. Therefore, the temperature of the maximum entropy change is 36.6 K, quite comfortable to work with. For $m' = 1$, the temperature of maximum entropy change is even higher, $T = 109.3$ K; the temperature at which Figure 34.11 was plotted.

34.7 CONCLUSIONS

The community studying the MCE sums efforts for applications considering only materials with cooperative orderings, like ferro-, antiferro-, and even ferrimagnetic materials. An interesting system with a high possibility of applications was recently presented, considering 3D [7,8] and 2D [10] standard nonrelativistic diamagnetic materials, like Gold. Their oscillating MCE is shown to work at low temperatures (ca. 1 K) considering feasible values of magnetic field change. To increase interest on this effect, the oscillating MCE of a 2D massless Dirac-like system, graphene, was explored [9]. It was shown that the temperature in which this effect appears is quite comfortable (ca. 37 K), due to the relativistic properties of the electrons in this material. In addition, the change of magnetic field change required to invert the MCE from normal to inverse is ca. 1 mT for 3D nonrelativistic material (useful for application in high sensible magnetic field sensor at low temperatures), and ca. 3.4 T for graphene. Table 34.1 summarizes some aspects of MCE of 3D and 2D diamagnets and compare with graphene. We see the applied magnetic field and the temperature of maximum entropy change for that field. Data were extracted from figures and examples discussed.

Equations 34.27, 34.44, and 34.76 suggest that the magnetic entropy has a general form that scales with the dimension d of the system as follows:

$$B^{d/2}\cos(1/B)\mathcal{T}(T/B), \tag{34.79}$$

TABLE 34.1

Summary of Maximum Entropy Change for a Feasible Applied Magnetic Field

	3D	2D	Graphene
B (T)	10	6	10
T_{max} (K)	1	0.7	53
$\Delta S/Nk_B$	4.6×10^{-7}	6.2×10^{-5}	0.3

Note: Comparison between data shown in figures and examples.

irrespective of being a relativistic material by nature (graphene) or not. This general magnetic dependence of the entropy is responsible for the unique effects found on magnetocaloric properties of 2D and 3D diamagnetic materials. However, the order of magnitude of the entropy change is different for each kind of material because of the proportionality constants involved. In the case of graphene, the proportionality involves the Fermi velocity, which is quite huge. This makes graphene special because MCE can be observed in comfortable values of temperature and field.

APPENDIX 34A: DENSITY OF STATES OF A TWO-DIMENSIONAL NONRELATIVISTIC ELECTRON GAS

We show below the procedure to obtain the density of states; we use as a simple example a 2D nonrelativistic electron gas. The one-particle density of states follows easier from the Laplace transformation of the canonical partition function $Z_1^B(\beta)$ in the Boltzmann limit ($k_B T \gg \varepsilon_F$)

$$g(\varepsilon) = \frac{1}{2\pi i} \int_{\beta-i\infty}^{\beta+i\infty} e^{\beta'\varepsilon} Z_1^B(\beta') d\beta', \quad (34A.1)$$

where, for noninteracting systems, the following relationship between canonical and grand canonical partition functions holds

$$\ln \mathcal{Z}(T,B) = z Z_1^B(\beta). \quad (34A.2)$$

Thus, at this point, we need to obtain the grand partition function in the Boltzmann limit

$$\ln \mathcal{Z}(T,B) = \sum_n z e^{-\beta\varepsilon_n}, \quad (34A.3)$$

but before, the energy spectrum ε_n for this model.

The Hamiltonian of the present model, that is, a 2D nonrelativistic electron gas with a transversal magnetic field $\vec{B} = B\hat{k}$, can be written as

$$\mathcal{H} = \frac{1}{2m_e}\left[p_x^2 + (p_y + eBx)^2\right], \quad (34A.4)$$

and then the Schrödinger equation reads as

$$\left[-\frac{\hbar^2}{2m_e}\frac{d^2}{dx^2} + \frac{1}{2}m_e\omega^2(x+x_0)^2\right]\phi_n(x) = \varepsilon_n\phi_n(x), \quad (34A.5)$$

where

$$x_0 = \frac{\hbar k_y}{eB}, \quad \omega = \frac{eB}{m_e}, \quad \text{and} \quad \varepsilon_n = \hbar\omega\left(n+\frac{1}{2}\right). \quad (34A.6)$$

Above, $k_y = 2\pi n_y/L$, where $n_y = 0, 1, 2, \ldots$, is related to the translational symmetry along the y axis; and the energy spectrum represents the Landau levels, where n is the Landau level index. Note that the harmonic oscillator of Equation 34A.5 has several centers x_0 that depend on n_y and B. Since these centers must be within the considered plane, that is, $0 \leq x_0 \leq L$ then $0 \leq n_y \leq \tilde{g}$, where $\tilde{g} = L^2 eB/h$ is the degeneracy of the Landau level n (note that for each n, there are n_y possibilities).

Thus, Equation 34A.3 can be rewritten as

$$\ln \mathcal{Z}(T,B) = \sum_{n=0}^{\infty} z\tilde{g}\exp\left[-\beta\hbar\omega\left(n+\frac{1}{2}\right)\right]$$

$$= z\frac{L^2 eB}{2h}\frac{1}{\sinh(y)}, \quad (34A.7)$$

where $y = \mu_B B\beta$. Thus, a simple comparison of Equations 34A.2 and 34A.7 leads to

$$Z_1^B(\beta) = \frac{L^2 eB}{2h}\frac{1}{\sinh(y)}. \quad (34A.8)$$

From the above and Equation 34A.1, it is possible to write the density of states we are looking for

$$g(\varepsilon) = \frac{L^2 eB}{2h}\left[\frac{1}{2\pi i}\int_{\beta-i\infty}^{\beta+i\infty}\frac{e^{\beta'\varepsilon}}{\sinh(\mu_B B\beta')}d\beta'\right], \quad (34A.9)$$

that reads as

$$g(\varepsilon) = g_0 + g_B(\varepsilon), \quad (34A.10)$$

where

$$g_0 = \frac{L^2}{4\pi}\frac{2m_e}{\hbar^2}, \quad (34A.11)$$

is the density of state (with no spin degeneracy), of the nonperturbed 2D electron gas; and

$$g_B(\varepsilon) = 2g_0 \sum_{l=1}^{\infty}(-1)^l\cos\left(\frac{l\pi}{\mu_B B}\varepsilon\right). \quad (34A.12)$$

REFERENCES

1. E. Warburg, Magnetische Untersuchungen, *Ann. Phys. (Leipzig)*, vol. 13, p. 141, 1881.

2. P. Debye, Einige bemerkungen zur magnetisierung bei tiefer temperature, *Ann. Phys. (Leipzig)*, vol. 81, p. 1154, 1926.

3. W. F. Giauque, A thermodynamic treatment of certain magnetic effects. A proposed method of producing temperatures considerably below 1° absolute, *J. Am. Chem. Soc.*, vol. 49, p. 1864, 1927.

4. W. F. Giauque and D. P. MacDougall, Attainment of temperatures below 1° absolute by demagnetization of $Gd_2(SO_4)_3 \cdot 8H_2O$, *Phys. Rev.*, vol. 43, p. 768, 1933.

5. M. S. Reis, R. M. Rubinger, N. A. Sobolev, M. A. Valente, K. Yamada, K. Sato, Y. Todate, A. Bouravleuv, P. J. von Ranke, and S. Gama, Influence of the strong magnetocrystalline anisotropy on the magnetocaloric properties of MnP single crystal, *Phys. Rev. B*, vol. 77, p. 104439, 2008.

6. M. S. Reis, *Fundamentals of Magnetism*. New York: Elsevier, 2013.

7. M. S. Reis, Oscillating magnetocaloric effect, *Appl. Phys. Lett.*, vol. 99, p. 052511, 2011.

8. M. S. Reis, Oscillating adiabatic temperature change of diamagnetic materials, *Solid State Commun.*, vol. 152, p. 921, 2012.

9. M. S. Reis, Oscillating magnetocaloric effect on graphenes, *Appl. Phys. Lett.*, vol. 101, p. 222405, 2012.

10. M. S. Reis, Oscillating magnetocaloric effect of a two dimensional non-relativistic diamagnetic material, *J. Appl. Phys.*, vol. 113, p. 243901, 2013.

11. K. S. Novoselov, A. K. Geim, S. V. Morozov, D. Jiang, M. I. Katsnelson, I. V. Grigorieva, S. V. Dubonos, and A. A. Firsov, Two-dimensional gas of massless Dirac fermions in graphene, *Nature*, vol. 438, p. 197, 2005.

12. S. G. Sharapov, V. P. Gusynin, and H. Beck, Magnetic oscillations in planar systems with the Dirac-like spectrum of quasiparticle excitations, *Phys. Rev. B*, vol. 69, p. 075104, 2004.

13. M. I. Katsnelson, *Graphene: Carbon in Two Dimensions*. Cambridge: Cambridge University Press, 2012.

14. W. Greiner, L. Neise, and H. Stocker, *Thermodynamics and Statistical Mechanics*. New York: Springer, 1995.

15. A. M. Tishin and Y. I. Spichkin, *The Magnetocaloric Effect and Its Applications*. Bristol: IoP Publishing, 2003.

16. A. H. Castro Neto, F. Guinea, N. M. R. Peres, K. S. Novoselov, and A. K. Geim, The electronic properties of graphene, *Rev. Mod. Phys.*, vol. 81, pp. 109–162, 2009.

17. S. Zhang, N. Ma, and E. Zhang, The modulation of the de Haas-van Alphen effect in graphene by electric field, *J. Phys. Condens. Matter*, vol. 22, p. 115302, 2010.

18. J. H. Chen, C. Jang, S. Xiao, M. Ishigami, and M. S. Fuhrer, Intrinsic and extrinsic performance limits of graphene devices on SiO_2, *Nat. Nanotechnol.*, vol. 3, p. 206, 2008.

35 Mode-Locked of Fiber Laser Employing Graphene-Based Saturable Absorber

Pi Ling Huang, Chao-Yung Yeh, Jiang-Jen Lin, Lain-Jong Li, and Wood-Hi Cheng

CONTENTS

Abstract 555
35.1 Introduction 555
35.2 All Fiber Ultrafast Laser: Theoretical Review 556
    35.2.1 Principle of Mode Locking 556
    35.2.2 Mode-Locking Techniques 558
    35.2.3 Graphene-Based Saturable Absorber 559
    35.2.4 Ultrafast Laser with Fast SA Mode-Locking 561
    35.2.5 Ultrafast Laser with Fast SA Mode Locking, GVD, and SPM 562
    35.2.6 Pulse Spectral Sidebands 564
35.3 Graphene Film Preparation 565
    35.3.1 Dispersant Synthetic Fluorinated MICA 565
    35.3.2 Dispersant Poly(Oxyethylene)-Segmented Imide 565
    35.3.3 Few Layer Atomic Graphene 567
35.4 Optical Properties of Graphene SA: Measurement 567
    35.4.1 Linear Optical Properties 567
        35.4.1.1 Graphene with MICA and POEM Dispersant 567
        35.4.1.2 Few Layer Atomic Graphene 567
    35.4.2 Nonlinear Optical Properties 568
        35.4.2.1 Graphene with MICA and POEM Dispersant 568
        35.4.2.2 Few Layer Atomic Graphene 569
35.5 Fiber-Laser Mode Locking with Graphene SA: Experiment 569
    35.5.1 Laser System Set Up 569
    35.5.2 Mode-Locked Laser Performance 569
        35.5.2.1 Mode-Locking Using Graphene/MICA and Graphene/POEM as SA 569
        35.5.2.2 Mode-Locking Using Few Layer Atomic Graphene as Saturable Absorber 570
35.6 Conclusion 571
References 571

ABSTRACT

Graphene is the thinnest material in the world with the lowest electrical resistance at room temperature among others. It is highly transparent and a good conductor. Graphene is not only used as new electronic materials but also recognized as novel optoelectronic devices, such as solar cells, transparent touch screens, and saturable absorber (SA). In this chapter, a stable optical pulse generation of graphene-based SA in ultrafast fiber laser system is investigated. Two different fabrication processes of graphene samples are reported, which are chemical vapor deposition (CVD) and graphene-mediated SA employing different nano-dispersants.

For monolayer graphene grown by CVD, linear and nonlinear optical properties of different stacking of atomic-layers graphene-based SA are investigated and compared. Same optical properties are also performed on dispersed few layers of graphene samples prepared by two different dispersants including fluorinated MICA clay and poly(oxyethylene)-segmented imide (POEM). Pulse duration of mode-locked fiber lasers (MLFLs) can be controlled through the modulation depth of SA by optimal selection of the number of layers of stacking of monolayer graphene SA.

MLFLs with dispersed few layers of graphene SA reveal shortened pulsewidth and enhanced modulation depth as the thickness and concentration product (TCP) of dispersed layer-graphene SAs increases.

35.1 INTRODUCTION

Graphene is the thinnest solid-state material in the world with the lowest electrical resistance at room temperature among other materials [1]. For a monolayer graphene, the optical transmission is 97.7% and the transport speed is nearly 1/300 of light. Graphene is not only used as a new electronic material such as ultra-high-speed modulator but also recognized as a novel optoelectronic device, such as solar cells, transparent touch screens, and saturable absorber (SA).

Mode-locked lasers are the important scientific tools with wide range of applications, such as optical fiber communications, ultrafast probing, nonlinear microscopy, optical coherent tomography, and frequency comb generation. Among the mode-locking techniques, passively MLFLs have received much attention due to their low cost, compact size, and long-term robustness. A passively mode-locked erbium-doped fiber laser (MLEDFL) is able to generate optical pulses in the time scale from picosecond (ps) to femtosecond (fs). Passively, MLFL is often initiated by nonlinear optical elements termed saturable absorber with intensity-dependent response to favor optical pulsation over continuous-wave lasing [2]. The SA widely used in passively mode-locked laser is semiconductor saturable absorber mirror (SESAM) or a nonlinear optical mirror [3–5]. However, the drawbacks of SESAM are cost-ineffective, time-consuming fabricated process and can only be operated in reflection mode, which is incompatibility to optical fiber structure. Recently, single-wall carbon nanotubes (SWCNTs) of 1D and graphene of 2D carbon allotrope have been noticed due to their large optical nonlinearity and low saturation intensity [6–9]. The first passive MLFL based on SWCNT SA was reported by S. Y. Set et al. in 2003 [10]. The atomic layer graphene as SA for ultrafast-pulsed lasers was first demonstrated by Q. Bao et al. [11]. Subsequent studies have indicated that graphene can be an effective alternative medium for SA [12–14]. Graphene is a one-atom-thick planar consisting of a carbon honeycomb sheet of bonded carbon atoms [15,16]. It has excellent optical properties, such as high transparency with ultra-broad flat spectra, ultrafast relaxation time, and the bandgap-limitless saturation absorption because of its point bandgap structure. Therefore, graphene can be used as fast SA with wide spectral operated range [17–20]. The graphene have been used successfully in the laser system for generation of fs-scaled pulse [21,22]. In addition, the graphene-mediated SA can be fabricated by a simple solution coating process, which allows the large-scale and reproducible production of the material. Unlike the single-wall carbon nanotubes, which required proper tube diameter for matching to the selected gain wavelength, the point bandgap structure allow graphene to operate over a wide spectral range from visible to mid-IR. Graphene-based SAs have two fast response times, $\tau_1 \sim 100$ fs due to intraband transition and $\tau_2 \sim 1$ ps due to interband transition. Due to the finite density of state graphene energy structure, the low saturation of optical absorption is easily observed.

The remaining part of this chapter is organized as follows. Section 35.2 reviews the theory of the all fiber laser system. The preparation of graphene SAs, the fabrication methods that include CVD and dispersed solution are presented in Section 35.3. Section 35.4 describes the experimental results in the linear and nonlinear optical absorption of graphene SA. Section 35.5 illustrates the performance of the fiber laser

mode-locking using graphene SA. A brief summary is provided in Section 35.6.

35.2 ALL FIBER ULTRAFAST LASER: THEORETICAL REVIEW

Laser operation can be either in continuous mode or pulse mode. Pulse mode of laser are typically classified into long pulse (ms to μs), Q-switch pulse (ns), and mode locked pulse (ps to fs). Mode-locked laser with pulse duration in the ps to fs range are critical for many scientific and industrial applications, such as materials processing, optical coherence tomography, and nonlinear effects, second-harmonic generation, and Raman shifting [23]. Since the invention of Ti:sapphire solid-state laser in 1992, it is known as the workhorse in the field of ultrafast science. Due to its nonresonant Kerr-lens mode-locking scheme, the Ti:sapphire laser suffer the drawback being not able to self-start operation. Furthermore, the Ti:sapphire requires continuous wave (CW) green laser for the optical pumping with free space optical propagation resonator geometry constraining its operation only in laboratory environment.

A compact and robust design using optical fiber waveguide resonator with all fiber optical active and passive components to form all fiber ultrafast laser can be a long-term stability alternative. In this section, we briefly introduce the basic idea of laser mode-locking in Section 35.2.1, then outline some methods for the mode locking in Section 35.2.2. In Section 35.2.3 we discuss the optical nonlinearity of the in-line fiber graphene-based SA and examine how they can be integrated into a fiber wave-guide resonator in Section 35.2.4. Finally, we explain how the generated optical pulse can be further shortened from intra-cavity optical elements control to the balancing between nonlinear self-phase modulation (SPM) and linear group velocity delay (GVD) as in Section 35.2.5.

35.2.1 PRINCIPLE OF MODE LOCKING

Ultrafast laser are generated by a technique called laser mode-locking. To construct an ultrafast laser, a pulse-shortening device (or mode-locker) is inserted inside a free-running laser system. A free-running laser is a typical CW laser including three parts, a gain medium, a pumping source, and an optical resonator. The gain medium is a material that can amplify light passing through it. Pumping source is an excited device used to provide laser gain medium with population inversion. Optical resonator provides an optical feedback to the gain medium. For sustaining laser oscillation, two conditions must be satisfied: gain condition and phase condition.

In a laser cavity, the oscillation takes place at the eigen modes [24]. The eigen modes include transverse and longitudinal modes. For a single transverse mode, the laser oscillation phase condition is when the longitudinal modes separated by the frequency $\Delta v = c/2nl$, where c/n is the speed of light in the laser material, l is the optical length of laser resonator. Each of these modes oscillates independently to the others if laser is free-running and the phase of the modes are randomly distributed between $-\pi$ and $+\pi$. This phase randomness will cause the laser intensity to fluctuate randomly and greatly reduce its usefulness for many applications where temporal coherence is important. To create streams of short intense pulses, one of the methods is mode-locking, which is a technique to control the phases of the longitudinal modes that maintains a relative value in the laser cavity. In order to mode-lock a laser, it is necessary to create periods of low loss in the cavity that synchronize with the $2nl/c$ round-trip time of the cavity.

For the longitudinal modes, the separation of these modes frequencies in the laser cavity is given by (for simplicity, the index of laser $n = 1$ is assumed)

$$\omega_q - \omega_{q-1} = \frac{\pi c}{l} \equiv \Delta\omega \qquad (35.1)$$

However, with laser gain placing inside the cavity, a laser oscillation gain condition satisfies only when the unsaturated gain is larger than cavity losses, as shown in Figure 35.1.

The total optical electric field of multimode oscillation can be expressed as

$$A(t) = \sum_n E_n e^{j[(\omega_0 + n\Delta\omega)t + \varphi_n]} \qquad (35.2)$$

where ω_0 is the reference frequency and φ_n is the phase of the nth mode [25].

After cavity round trip time $T_R \equiv 2\pi/\Delta\omega = 2l/c$, Equation 35.2 becomes

$$\begin{aligned}
A(t + T_R) &= \sum_n E_n \exp\left\{ j\left[(\omega_0 + n\Delta\omega)\left(t + \frac{2\pi}{\Delta\omega}\right) + \varphi_n \right] \right\} \\
&= \sum_n E_n \exp\{j[(\omega_0 + n\Delta\omega)t + \varphi_n]\} \\
&\quad \exp\left\{ j\left[2\pi\left(\frac{\omega_0}{\Delta\omega} + n\right) \right] \right\} \qquad (35.3)
\end{aligned}$$

Because $\omega_0/\Delta\omega$ is an integer ($\omega_0 = n\pi c/l$), therefore

$$\exp\left[2\pi j\left(\frac{\omega_0}{\Delta\omega} + n\right) \right] = 1 \qquad (35.4)$$

Hence, the property of $A(t)$ depends on the phases φ_n. As mentioned at the beginning of this paragraph, the output intensity of a typical cw laser shows a random distribution when the phase differences between the modes are random. When the phases φ_n are fixed, the optical electric field $A(t)$ shows the periodic property

$$A(t + T_R) = A(t) \qquad (35.5)$$

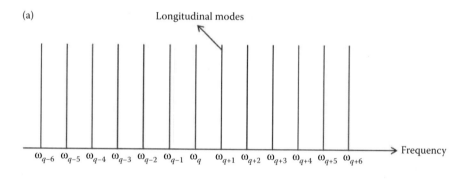

(a) Longitudinal modes

ω_{q-6} ω_{q-5} ω_{q-4} ω_{q-3} ω_{q-2} ω_{q-1} ω_q ω_{q+1} ω_{q+2} ω_{q+3} ω_{q+4} ω_{q+5} ω_{q+6} Frequency

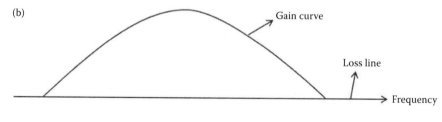

(b) Gain curve

Loss line

Frequency

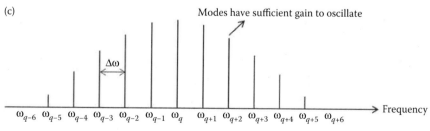

(c) Modes have sufficient gain to oscillate

$\Delta\omega$

ω_{q-6} ω_{q-5} ω_{q-4} ω_{q-3} ω_{q-2} ω_{q-1} ω_q ω_{q+1} ω_{q+2} ω_{q+3} ω_{q+4} ω_{q+5} ω_{q+6} Frequency

FIGURE 35.1 (a) Cavity longitudinal modes, (b) net gain profile, and (c) spectrum of multimode laser oscillation overlapped between (a) and (b).

The approach to force the phases φ_n to be fixed is called "mode locking."

We assume that N oscillation laser modes with equal amplitudes $E_n = E_0$ exist in a laser cavity and the phases φ_n equal to zero. Then, the optical electric field becomes

$$A(t) = E_0 \sum_{-(N-1)/2}^{(N-1)/2} e^{j(\omega_0 + n\Delta\omega)t}$$

$$= E_0 e^{j\omega_0 t} \frac{\sin(N\Delta\omega t/2)}{\sin(\Delta\omega t/2)} \qquad (35.6)$$

The resultant intensity $I(t) \propto A(t)^* A(t)$ can be expressed as

$$I(t) \propto |E_0|^2 \frac{\sin^2(N\Delta\omega t/2)}{\sin^2(\Delta\omega t/2)} \qquad (35.7)$$

Figure 35.2 shows the theoretical plot of optical intensity results in different number ($N = 5,10,20$) of longitudinal oscillation modes. It can be observed that the peak intensity I_{peak} increases as the number of modes increases ($I_{peak} \approx NI_{CW}$). The pulse width τ_p also decreases as the number of modes increases ($\tau_p \approx T_R/N$). Because the number of modes is defined as the ratio of spectral linewidth of the gain profile and round-trip frequency spacing, the pulse width and peak intensity will be affected by the spectral linewidth and the length of the cavity [26].

35.2.2 MODE-LOCKING TECHNIQUES

The methods of laser mode-locking can be achieved actively and passively. In active scheme, mode locking is performed with the aid of an externally driven loss modulator placed inside the laser cavity as shown in Figure 35.3a. The modulator is driven at a frequency corresponding to the inverse of cavity round-trip time ($\Delta\omega = 2\pi/T_R$, where T_R is the cavity round-trip time). As shown in Figure 35.3b only during the peak of the modulator transmission, the optical gain of the amplifier is high enough to overcome the losses in the cavity, resulting in positive net gain and optical pulse is generated.

However, the performance of the active mode-locking is limited due to three drawbacks: (1) an externally driven modulator is required, (2) thermal-induced cavity length change will cause the mismatch of the modulation frequency to the cavity mode spacing, and (3) pulse shortening becomes ineffective for very short pulses due to the finite bandwidth of modulator and limit the attainable pulse width.

In passive scheme, mode-locking is achieved by using a nonlinear optical element with intensity-dependent transmission to shape the intracavity pulse without external modulator as shown in Figure 35.4.

In the stationary state, the self-amplitude modulation (SAM) effect of the nonlinear optical element will lead to a net loss before and after the optical pulse. For the methods of passive mode-locking, it is typically classified as nonresonant SAM (e.g., optical Kerr effect) and resonant SAM (e.g., saturable absorber). Even though the optical Kerr-lens effect leads to an ideal SAM having a response time less than 1 fs, it suffers the problem of being incapable of self-starting the laser. In this study, we focus on the resonant-type SAM using SA to overcome the self-starting problem. In addition, the resonant type SAM, using graphene-based SA material, not only provides broad wavelength operation but also favor low mode-locking threshold operation due to its low saturation intensity.

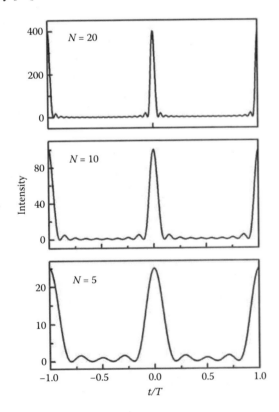

FIGURE 35.2 Theoretical plot of optical intensity results from longitudinal mode-locking with different mode numbers ($N = 5$, 10, 20).

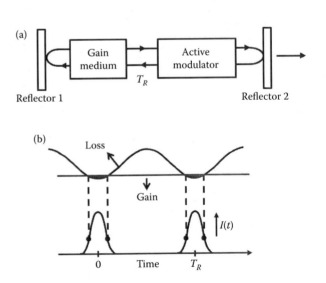

FIGURE 35.3 Schematic of (a) actively mode-locking laser with active modulator, (b) optical pulse formation and net gain in time domain.

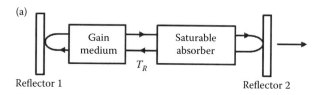

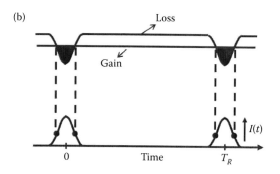

FIGURE 35.4 Schematic of (a) passively mode-locking laser with fast SA, (b) optical pulse formation and net gain in time domain.

35.2.3 Graphene-Based Saturable Absorber

Graphene is a monolayer of carbon in a two-dimensional hexagonal lattice as shown in Figure 35.5a. Many interesting electrical and optical phenomena can be attributed to its unique energy band structure, as shown in Figure 35.5b. In particular, a zero bandgap appears at the crossing of

conduction and valence band called Dirac point as shown in Figure 35.5c [19].

Single-layer graphene (SLG) exhibits electronic properties with a zero mass and move at ultrafast speed described by the relativistic Dirac's equation. From the unique energy band structure with reference near Dirac point $\vec{K}$, a dispersion relation for SLG is linear as

$$E^{\pm}(\vec{k},\vec{K}) = \pm\hbar v_F \mid \vec{k} - \vec{K} \mid \qquad (35.8)$$

where $\vec{k} = (k_x, k_y)$ is the vector in the first Brillouin zone (BZ) and v_F is the electronic group velocity ($v_F \approx 10^6$ m/s).

The optical transmission of SLG, $T \cong 1 - \pi\alpha \approx 97.7\%$ with $\alpha = 1/137$ is fine structure constant. The optical absorption of SLG, therefore, is estimated to be $A = 1 - T \approx 2.3\%$. A nearly constant flat absorption profile with a broadband wavelength range from UV (~300 nm) to mid-IR (~2500 nm) was also experimentally verified. Multilayer structure of graphene can be constructed from the stacking of monolayer and typically induced bandgap opening.

Based on the knowledge of graphene-based SA response time, one can model the SA as either a fast SA (saturation by intensity) or a slow SA (saturation by energy).

As illustrated in Figure 35.6, the electrons from the valence band are excited to the conduction band when graphene encounters light absorption. After photoexcitation, the distribution rapidly thermalizes and cools to form a hot Fermi–Dirac distribution within $\tau_1 \sim 100\, fs$. Immediately following optical excitation, the

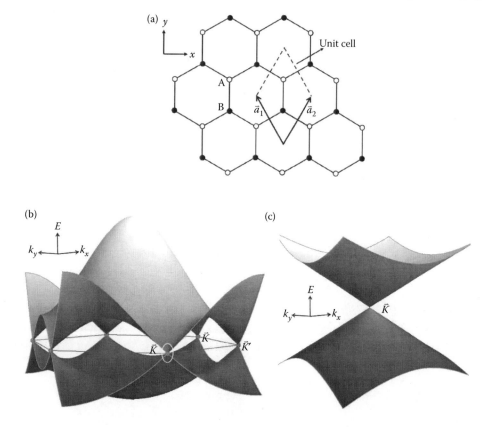

FIGURE 35.5 Graphene honeycomb lattice structure and its band structure, (a) lattice structure of graphene with cell unit vector $\vec{a}_1$ and $\vec{a}_2$, (b) graphene energy band structure, and (c) the Dirac cones are located at the $\vec{K}$ and $\vec{K}'$ points.

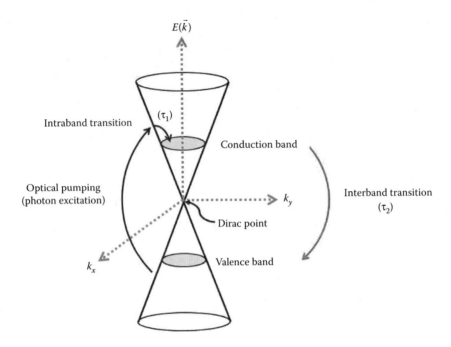

FIGURE 35.6 Carrier dynamics of graphene with two relaxation time scale: Intraband transition ($\tau_1 \sim 100$ fs) and interband transition ($\tau_2 \sim 100$ ps).

newly created electron–hole pairs could block some of the originally possible interband optical transitions around the Fermi energy and decrease the absorption of photons. The hot carriers are then cooled further due to the intraband phonon scattering with a time scale ranged between $\tau_2 \sim 1$ ps. Finally, electrons and holes recombine until the equilibrium distribution is restored [27–32]. These describe only the linear optical transition under low excitation intensity. As the light intensity increases to a higher level, the photogenerated carriers increase instantaneously and fill the energy states near the edge of the conduction and valence bands, the absorption is blocked due to the Pauli blocking process. Therefore, the additional photons with specific wavelength can transmit the graphene without absorption.

An SA is a nonlinear optical component in which the transmission increases as the incident light intensity increases. In this study, a two-level electronics model is used to describe the mechanism. At low incident light intensity, the transmission obeys the linear Beer–Lambert law and the absorption coefficient is independent of light intensity. With higher incident light intensity, electrons in the ground state are transferred to excited state and become saturated. Equally populated electrons in the excited and ground state make the SA transparent results in minimum absorption.

Since the response time of the graphene-based SA is fast, the saturation is intensity saturation instead of energy saturation, and therefore the nonlinear absorption coefficient can be expressed as

$$\alpha(t) = \alpha_0 \left(1 + \frac{I(t)}{I_s}\right)^{-1} + \alpha_{ns} \qquad (35.9)$$

where $\alpha_0 = \sigma N$ is the linear absorption coefficient, α_{ns} is the nonsaturable absorption coefficient, $I(t)$ is the intensity

incident on the absorbing medium, and $I_s = \hbar\omega_l/\sigma\tau_{rex}$ is the saturation intensity of the absorber. The σ, N, ω_l, and τ_{rex} are the absorption cross section of SA, concentration of SA, laser frequency, and energy relaxation time of SA, respectively [33,34].

From the macroscopic experimental point of view, the linear optics (such as absorption coefficient, and non-saturable loss) and nonlinear optics (such as modulation depth, and the saturation intensity) terms are commonly used for SA spectroscopy parameters. Consider an optical pulse of intensity $I(t)_{in}$ that propagates through an SA of sample thickness L and the transmitted optical pulse intensity $I(t)_{out}$. The optical transmittance $T = I(t)_{out}/I(t)_{in}$ of the SA can be expressed as [35]

$$T(t) = \exp(-\alpha(t) \times L)$$
$$= \exp\left[\frac{-\alpha_0 L}{1 + I(t)/I_s} - \alpha_{ns} L\right] \qquad (35.10)$$

Figure 35.7 is the nonlinear transmission curve of the SA; it shows the transmission behavior of the SA versus the incoming light intensity. The difference between 100% transmission and the maximum transmission of SA is defined as nonsaturable loss, which represents the linear optical loss of SA.

The modulation depth (MD) is an important optical parameter describing the strength of nonlinear coupling between optical field and the SA material. The MD is defined as the difference of maximum transparent light intensity (T_{max}) to the minimum (T_{min}). From Equations 35.9 and 35.10 it can be expressed as

$$MD = T_{max} - T_{min} = e^{-\alpha_{ns}L} \cdot (1 - e^{-\alpha_0 L}) \qquad (35.11)$$

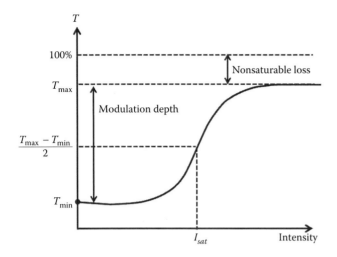

FIGURE 35.7 Nonlinear transmission versus light intensity of graphene SA.

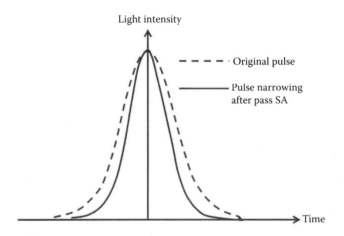

FIGURE 35.8 Pulse shaping effect due to intensity saturation of a fast SA.

The larger the MD means the better pulse shaping ability. For an idea dispersant ($\alpha_{ns} \sim 0$) the performance of MLFL correlates to the concentration and the thickness of graphene.

The pulse-shaping effect of SA can be realized as follows. Figure 35.8 sketches the pulse-shaping mechanism when a pulse passes through an SA. The output pulse is normalized to the original pulse. The light was absorbed at the tail of the pulse and remains almost unchanged at the tip of the pulse. It indicates that the pulse with low intensity encountered the higher absorption by passing the SA and the pulse with high intensity encountered the lower absorption. As a result, the pulse that passes through such an SA has been temporal narrowing.

35.2.4 ULTRAFAST LASER WITH FAST SA MODE-LOCKING

In this study, mode-locking was achieved passively by inserting graphene SA into erbium-doped fiber laser (EDFL). Fluctuations induced by spontaneous emission were enhanced by SA through many round trips. The pulse became shortened till the spectral width was comparable to the gain bandwidth. The analysis of

SA-based passive mode-locking of a homogeneously broadened laser has been presented by Haus in 1975 [36].

Here, we reviewed the theory on the modeling of ultra-fast laser system, in particular, for the mode-locked scheme based on the fast SA. Dispersion usually causes optical pulse broadening and prevents shortest pulse generation. From the following theoretical analysis, we are able to realize how one can combine negative dispersion with the nonlinear self-phase modulation to generate shortest pulse. The theory also provides the basis on the stable soliton-like pulse generation particularly useful for the experiments on the Er:doped ring fiber laser.

Under assumption of small linear and nonlinear changes in the pulse within each round-trip time in the laser cavity, the model for the pulse-shaping process was described as Haus's master equation:

$$\frac{1}{T_R}\frac{\partial}{\partial T}A(T,t) = \sum_i \Delta A_i \qquad (35.12)$$

where $A(T,t)$ is the pulse envelope, with $t \ll T \approx T_R$ and T_R is the cavity round trip time. The terms ΔA_i represents linear operator and describes the change of pulse envelope due to various optical elements (such as gain medium, SA, dispersion, and self-phase modulation) in the laser cavity. The simplest case in the passively MLFL system is with optical elements including gain medium (e.g., EDF) and fast SA (e.g., graphene-based SA). The gain medium with homogeneous broadening can be described by Lorentzian function as

$$g(\omega) = \frac{g}{1+(\Delta\omega)^2/\Omega_g^2} \qquad (35.13)$$

where $\Delta\omega = \omega - \omega_0$, g is the saturated gain coefficient, and Ω_g the gain bandwidth.

Pulse envelope after the gain medium, under small envelope change assumption as describe in the frequency domain is given by

$$\tilde{A}_{out}(\omega) = e^{g(w)}\tilde{A}_{in}(\omega) \approx \left[1 + g\frac{1-\Delta\omega^2}{\Omega_g^2}\right]\tilde{A}_{in}(\omega) \qquad (35.14)$$

Using the Fourier transform, Equation 35.14 can be written as

$$A_{out}(t) \approx \left(1 + g + \left(\frac{g}{\Omega_g^2}\right)\frac{\partial^2}{\partial t^2}\right)A_{in}(t)$$

Therefore, the change in the pulse envelope after the gain medium (ΔA_{gain}) can be modeled as

$$\Delta A_{gain} \approx \left[g + \left(\frac{g}{\Omega_g^2}\frac{\partial^2}{\partial t^2}\right)\right]A(t) \qquad (35.15)$$

where g/Ω_g^2 is the gain bandwidth filtering.

In fast SA aspect, the SA can be completely saturated and saturates linearly with intensity. While the peak intensity of pulse is much smaller than the saturation intensity, the SA is weak saturated and Equation 35.9 can be expressed as

$$\alpha(t) = \alpha_0 \left(1 + \frac{I(t)}{I_s}\right)^{-1} + \alpha_{ns} \approx (\alpha_0 + \alpha_{ns}) - \alpha_0 \frac{I(t)}{I_s} \quad (35.16)$$

Since the power in the mode indicates that the intensity is multiplied by the effective area of the mode, A_{eff}, normalize the mode amplitude and $|A(t)|^2$ equals the power (i.e., $|A(t)|^2 = A_{eff}I(t)$). The absorption coefficient can therefore be written as

$$\alpha(t) \approx (\alpha_0 + \alpha_{ns}) - \frac{\alpha_0 |A(t)|^2}{I_s A_{eff}} \equiv (\alpha_0 + \alpha_{ns}) - \gamma |A(t)|^2$$

$$(35.17)$$

where $\gamma = \alpha_0 / I_s A_{eff}$ is the self-amplitude modulation (SAM) coefficient.

Assume the unsaturated SA loss and nonsaturable loss can be incorporated into the loss coefficient, l. The pulse envelope change in passing through the fast SA is then given by

$$A_{out}(t) = e^{-\alpha(t)} A_{in}(t) \approx [1 - \alpha(t)] A_{in}(t) = [1 + \gamma |A(t)|^2] A_{in}(t)$$

$$(35.18)$$

Therefore, the change in the pulse envelop after the fast SA (ΔA_{SA}) can be modeled as

$$\Delta A_{SA} \approx \gamma |A(t)|^2 A(t) \quad (35.19)$$

At the steady state, the Haus's master equation for passive mode-locking with fast SA, according to Equations 35.15 and 35.19, becomes

$$\left[(g - l) + \left(\frac{g}{\Omega_g^2} \frac{\partial^2}{\partial t^2}\right)\right] A(t) + \gamma |A(t)|^2 A(t) = 0 \quad (35.20)$$

The analytical solution is hyperbolic secant:

$$A(t) = A_0 \sec h \frac{t}{\tau_0} \quad (35.21)$$

with pulse duration $\tau_0 = \sqrt{2g/\gamma}/A_0 \Omega_g$.

35.2.5 ULTRAFAST LASER WITH FAST SA MODE LOCKING, GVD, AND SPM

In the development of ultrafast laser, it has been realized that nonlinear self-phase modulation (SPM) effect and linear GVD plays an important role in the generation of minimum pulse temporal duration.

In GVD aspect, we examine how linear dispersion shapes the optical pulse. GVD introduces a frequency (or wavelength) dependent delay on the various spectral components of an optical pulse [24]. Under nonzero GVD, light with different frequencies has different round-trip times and it leads the broadening of pulses in the ring passive mode-locking laser. We take the frequency dependence of propagation optical pulse wave-number k to analyze the dispersion. By using Taylor series expansion, wave-number k can be written as

$$k(\omega) = k_0(\omega_0) + k'(\omega - \omega_0) + \frac{k''}{2}(\omega - \omega_0)^2 + \cdots \quad (35.22)$$

where $k' = \partial k/\partial \omega$ and $k'' = \partial^2 k/\partial \omega^2$, and the derivatives are evaluated at $\omega = \omega_0$.

For pulse propagates in a medium with thickness L, the pulse envelope output can be described in the frequency domain as

$$\tilde{A}_{out}(\omega) = \exp\{-j[k(\omega) - k(\omega_0)]L\}\tilde{A}_{in}(\omega)$$

$$\approx \left(1 - j\frac{k''}{2}(\omega - \omega_0)^2 L\right)\tilde{A}_{in}(\omega) \quad (35.23)$$

The expression above shows only the quadratic term that contributes a quadratic spectral phase variation, which describe the deformed envelope. We ignore the linear term representing the group delay of envelope that is of no consequence in the analysis. Take the Fourier transform of Equation 35.23:

$$A_{out}(t) \approx \left(1 + jD\frac{\partial^2}{\partial t^2}\right) A_{in}(t) \quad (35.24)$$

where the GVD parameter $D = k''L/2$ is defined. Therefore, the change in the pulse envelop after the GVD (ΔA_{GVD}) can be modeled as

$$\Delta A_{GVD} \approx jD\frac{\partial^2}{\partial T^2} A(t) \quad (35.25)$$

In SPM aspect, nonlinear optics effect on the ultrafast laser is particularly important. Nonlinear absorption such as self-amplitude modulation (SAM) of SA is the main process response for passively mode-locking the laser as discussed earlier. Here, we discuss how nonlinear refraction, such as self-phase modulation (SPM) effect shape the optical pulse. Phase of the optical pulse can be modulated in a nonlinear way from SPM. SPM opens a way to a spectral broadening of a pulse. As an ultra-short pulse passes through a medium, the refractive index of the medium will change as the intensity of pulse due to the optical Kerr effect:

$$n = n_2 + \frac{1}{2}n_2 I(t) \quad (35.26)$$

where n_2 is the second-order nonlinear refractive index of the medium and $I(t)$ is the intensity envelop of a light pulse.

The change of refractive index will cause a nonlinear phase shift. The nonlinear phase shift after propagating in a nonlinear medium of length L_K (Kerr effect) can be expressed as

$$\Delta\varphi(t) = \frac{-\omega}{c} n_2 I(t) L_K = \frac{-2\pi}{\lambda} n_2 I(t) L_K \quad (35.27)$$

Assuming the SPM per pass is small, the SPM effect can be shown as

$$A_{out}(t) = e^{j\Delta\varphi(t)} A_{in}(t) \approx [1 + j\Delta\varphi(t)] A_{in}(t) \quad (35.28)$$

where $A_{in}(t)$ and $A_{out}(t)$ are the pulses envelope before and after the nonlinear refractive index medium. From Equation 35.27, the nonlinear phase shift is proportional to the intensity. Set the SPM coefficient δ, then $\Delta\varphi(t) \approx -\delta|A(t)|^2$. Equation 35.28 can be expressed as

$$A_{out}(t) \approx (1 - j\delta \, |\, A_{in}(t)\,|^2) A_{in}(t) \quad (35.29)$$

where the SPM parameter $\delta = (2\pi/\lambda)n_2 L_K/A_{eff}$ is defined.

Therefore, the change in the pulse envelop from the SPM (ΔA_{SPM}) can be modeled as

$$\Delta A_{SPM} \approx -j\delta \, |\, A(t)\,|^2 \, A(t) \quad (35.30)$$

Now, we can describe the temporal evolution of a pulse envelope inside the laser cavity with gain: Equation 35.15, fast SA: Equation 35.19, SPM: Equation 35.30, and GVD: Equation 35.25, the master equation becomes

$$\frac{1}{T_R} \frac{\partial}{\partial T} A(T,t) = (g-l)A(T,t) + \left(\frac{g}{\Omega_g^2} + jD\right) \frac{\partial^2}{\partial t^2} A(T,t)$$

$$+ (\gamma - j\delta) \, |\, A(T,t)\,|^2 \, A(T,t) \quad (35.31)$$

This is the modified equation to the Haus' master equation for fast SA mode-locking as described in Equation 35.27 with new parameters: $\delta = (2\pi/\lambda)n_2 L_K/A_{eff}$ for SPM and $D = k''L/2$ for GVD.

The schematic representing the modeling of the passive mode-locking elements including gain, fast SA, SPM, and GVD was shown in Figure 35.9.

Equation 35.21 has a simple steady-state solution in the form of a chirped secant hyperbolic pulse:

$$A(t) = A_0 \left[\sec h\left(\frac{t}{\tau_0}\right)\right]^{1+j\beta} \quad (35.32)$$

where β is the chirp parameter of the optical pulse.

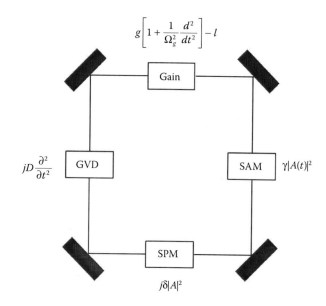

FIGURE 35.9 Passively mode-locking laser using SAM additional with intra-cavity GVD and SPM for pulse compression.

The most significant result of the Haus's analysis is relating the pulse duration and chirp as function of normalized dispersion ($D_n = (\Omega_g^2/g)D$), with normalized SPM $\delta_n = \delta(A_0^2\tau_0/g)\Omega_g$ as parameters as seen in Figures 35.10 and 35.11 [2].

Figure 35.10 shows the pulsewidth as function of normalized GVD and normalized SPM, the shortest pulse is predicted by GVD parameter $D < 0$ and SPM parameter $\delta > 0$. Figure 35.11 shows the chirp as function of normalized GVD and normalized SPM, the chirp is significantly large in the positive GVD.

The criteria of stabale solution can also be analyzed with stability parameter S satisfied:

$$S \equiv (1 - \beta^2) - 2\beta D_n > 0 \quad (35.33)$$

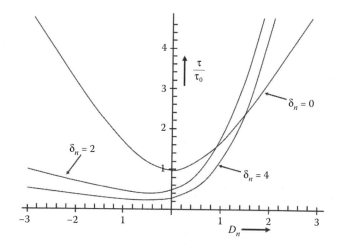

FIGURE 35.10 Pulsewidth as function of GVD and SPM. (From H. A. Haus, Mode-locking of lasers, IEEE J. Sel. Top. Quantum Electron. © 2000 IEEE.)

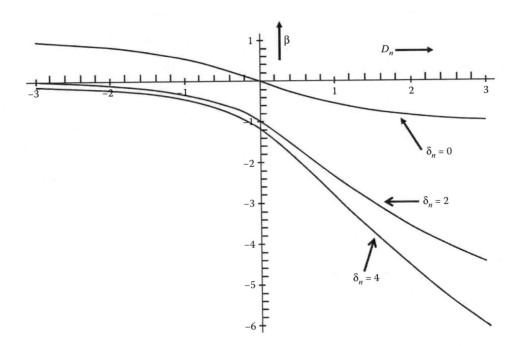

FIGURE 35.11 Chirp as function of GVD and SPM. (From H. A. Haus, Mode-locking of lasers, *IEEE J. Sel. Top. Quantum Electron.* © 2000 IEEE.)

For a chirp-free ($\beta = 0$) soliton, the pulse is always stable ($S = 1$). Instabilities in pulses are usually found for the laser with large nonlinear SPM and typically characterized with high chirping.

Furthermore, for a small SAM coefficient, weak filtering and negative D approximation, the mater equation becomes the nonlinear Schrodinger equation:

$$\frac{1}{T_R}\frac{\partial}{\partial T}A(T,t) = jD\frac{\partial^2}{\partial t^2}A(T,t) - j\delta \mid A(T,t) \mid^2 A(T,t)$$

$$(35.34)$$

Under these conditions, the stationary pulse emerges from balancing between GVD (negative) and SPM (positive) and results a chirp-free soliton-like solution:

$$A(T,t) = A_0 \left[\sec h\left(\frac{t}{\tau_0}\right) \right] e^{-j\delta|A_0|^2 T/2T_R} \qquad (35.35)$$

The duration of the chirp-free soliton-like pulse is

$$\tau = \frac{1}{|A|}\sqrt{2 \mid D \mid /\delta} \qquad (35.36)$$

The amplitude and pulsewidth product satisfied the "area theorem" $\tau \mid A \mid = \sqrt{2 \mid D \mid /\delta}$ via balance of GVD and SPM. Soliton-like pulse are found in most of the Er-doped fiber ring laser configuration where only weak SAM is required for mode-locking and stabilizing the pulse. The pulse instability can be caused by excessive SPM and positive dispersion.

35.2.6 PULSE SPECTRAL SIDEBANDS

Ultrafast all-fiber laser usually exhibits structured spectral sidebands as shown in Figure 35.12. However, those sidebands cannot be predicted from Haus' master equation. In fact, the existence of spectral sidebands in the mode-locked fiber laser limit the shorter pulsewidth generation.

A dispersive wave is formed due to the laser output coupler that provides the round-trip loss perturbation to the soliton wave and share some energy from soliton pulse. The origin of the sidebands is caused by the resonant enhancement between soliton and dispersive wave from the laser cavity periodic perturbation. The spectral sidebands were first observed by Kelly, also known as Kelly sidebands [37].

Assume the propagation constant of soliton pulse is k_S and dispersive wave is k_{lin}. As shown in Figure 35.13, the spectral sidebands occur with condition for the phase matching, $(k_S - k_{lin})Z_p = 2m\pi$, and where Z_p is the perturbation length (length of laser) and m is an integer.

Since the dispersive wave is linear and has a quadratic dependence on sideband frequency ($\Delta\omega$) on propagation constant, $k_{lin} = -\mid k'' \mid \Delta\omega^2/2$, the phase matching condition can be expressed as

$$\Delta\omega = \pm\frac{1}{\tau}\sqrt{m\left(\frac{8Z_0}{Z_p} - 1\right)} \qquad (35.37)$$

where $Z_0 = (1/2)(\pi\tau^2/\mid k'' \mid)$ is the soliton period.

The predictions of phase-matching conditions agree well with experiment results, particularly when the soliton pulses are free of chirp.

(a)

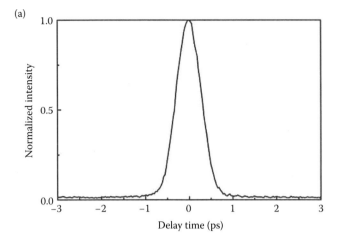

(b)

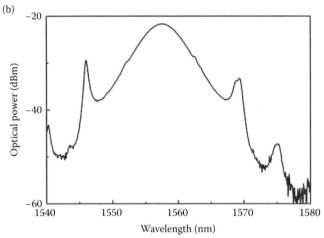

FIGURE 35.12 Typical pulse and its spectrum measurement from Er doped all fiber ring laser, (a) fs optical pulse width measure from intensity autocorrelator, and (b) the corresponding spectrum with Kelly sideband.

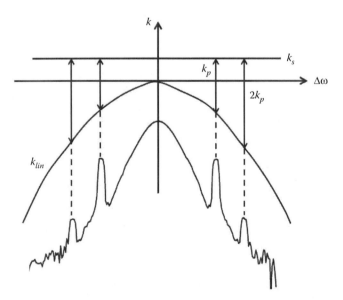

FIGURE 35.13 Phase matching condition between soliton pulse and dispersive wave showing Kelly sideband.

35.3 GRAPHENE FILM PREPARATION

The graphene-based SA can be prepared in thin solid film geometry. Various techniques, such as exfoliation, chemical vapor deposition (CVD), electrochemistry, and wet process by mixing the graphene flakes in host matrix with dispersant are viable approaches for graphene-based sample preparation. In this report, dispersant of graphene flakes from spin coating (film thickness ~ few micron) and stacking of mono-atomic layers of graphene by CVD process (film thickness ~ few nm) are fabricated.

35.3.1 DISPERSANT SYNTHETIC FLUORINATED MICA

Graphene with average thickness of 12 nm, lateral dimension of 4.5 μm, and specific surface area of 80 m/g², were supplied by UP Co. (USA). The plate-like graphene are composed of conjugated sp² orbital bonds. The materials tend to aggregate and are difficult for solvating into water or organic solvents. The self-aggregation and entanglement are mainly caused by van der Waals force attraction. However, when graphene were properly ground with plate-like clays, such as the fluorinated MICA, the pulverized powders as a physical mixture became readily dispersible in water. The fluorinated MICA (MICA), obtained from CO-OP CheMICAl Co. (Japan), was used to disperse graphene. The particular MICA of $Na_{0.66}Mg_{2.68}$ $(Si_{3.98}\ Al_{0.02})O_{10.02}F_{1.96}$, is synthesized by Na_2SiF_6 treatment of talc at high temperature. The MICA is one of anionic clays and exist in powder form of irregular aggregates with the primary units consisting of silicate platelets in stacks. The unit platelets are irregularly shaped and have ionic charges with ≡SiO⁻Na⁺. The average dimensions were estimated to be approximately $300 \times 300 \times 1$ nm³ for MICA. Due to the ≡SiO⁻Na⁺ in each unit platelet, MICA is dispersible in water [38–41].

The procedure of mixing MICA–graphene hybrid is described below. Graphene (10 mg) and MICA (30 mg) were ground adequately in an agate mortar and pestle. The sides of the mortar were occasionally scraped down with the pestle during grinding to ensure a thorough mixing. The mixture was washed from the mortar and pestle using deionized (DI) water at concentration of 10 mg of graphene in 20 g of water. The graphene/MICA hybrids were prepared at clay-to-graphene weight ratios. During the mixing, ultrasonic vibration was applied for 2 h. Ultrasonication was operated on a BRANSON 5510R-DTH (135 W, 42 kHz). A homogenous suspension of graphene was obtained in the presence of MICA. The grinding procedure for the preparation can be monitored by the transmission electron microscopy (TEM) image. The pristine graphene was observed, as shown in Figure 35.14a, and the heterogeneous mixtures were observed after grinding, as shown in Figure 35.14b.

35.3.2 DISPERSANT POLY(OXYETHYLENE)-SEGMENTED IMIDE

The oligomer dispersant and poly(oxyethylene)-segmented imide (POEM), was synthesized from the reaction of

FIGURE 35.14 TEM micrographs of (a) the pristine graphene showing serious aggregates, and (b) dispersion of graphene-MICA. (From P. L. Huang, H. H. Kuo, R. X. Dong et al. *J. Lightwave Technol.* 30;2012:3413.)

polyether-diamine and dianhydride by using the previously reported procedures [42]. The poly(oxyethylene)-diamine, specifically poly(oxypropylene-oxyethylene-oxypropylene)-block polyether of bis(2-aminopropyl ether) at averaged 2000 g/mol molecular weight (abbreviated as POE2000) was a water-soluble and crystalline poly(oxyethylene)-backboned (POE-) amine (waxy solid, mp 37–40°C, amine content of 0.95 mequiv/g with an average of 39 oxyethylene and 6 oxypropylene units in the structure), purchased from Huntsman Chemical Co. The anhydride, 4,4′-oxydiphthalic anhydride (ODPA, 97% purity), was purchased from Aldrich Chemical Co. and purified by sublimation. Tetrahydrofuran (THF) was purchased from TEDIA Company Inc.

The typical experimental procedures for synthesizing the POEM dispersant are described below for the oligomer with the composition of POE2000 and ODPA at a 6:5 equivalent ratio. To a 100-mL three-necked and round-bottomed flask equipped with a magnetic stirrer, nitrogen inlet-outlet lines, and a thermometer, POE2000 (10 g, 5 mL mmol) in THF (15 mL) was added, followed by a solution of ODPA (1.29 g, 4.2 mmol) in THF (10 mL) through an addition funnel in a drop-wise manner. During the addition, the mixture was stirred vigorously and the reaction temperature was maintained at 150°C for 3 h. During the process, samples were taken periodically and monitored using a Fourier transform infrared (FT-IR). This showed that the absorption peaks of anhydride functionality at 1780 cm^{-1} (s) and 1850 cm^{-1} (w) disappeared at the expense of the peaks at 1713 and 1773 cm^{-1} for the asymmetric stretch of imide. The characteristic absorption at 1100 cm^{-1} of oxyalkylene was observed. The product mixture was subjected to rotary evaporation under a reduced pressure and recovered as a yellowish waxy solid. The final product, used as the dispersant, can be described by the representative chemical structure, as shown in Figure 35.15.

The graphene dispersion was prepared in the presence of dispersants. Graphene (25 mg) was first dispersed in 5 mL of deionized water in a vial and sonicated under a VCX 500 Ultrasonicator at ambient temperature for 1 h. The resultant solution was dark black with some solid precipitates at the bottom of the container, indicating a low degree of dispersion.

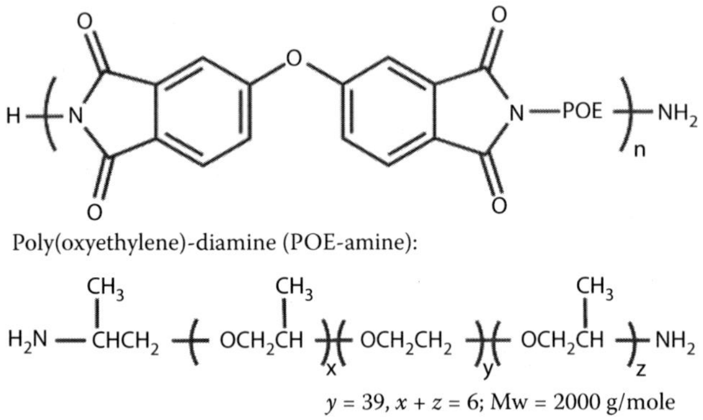

Poly(oxyethylene)-diamine (POE-amine):

$y = 39, x + z = 6$; Mw = 2000 g/mole

FIGURE 35.15 The representative structure of the poly(oxyethylene)-segmented imide dispersant (POEM). (From P. L. Huang, H. H. Kuo, R. X. Dong et al. *J. Lightwave Technol.* 30;2012:3413.)

In a separate glass container, POEM (0.5 g) was dissolved in 5 mL of deionized water (R = 18.2 MΩ/cm^2) and added to the graphene solution (5.0 mL). A homogenous suspension of graphene was obtained in the presence of POEM.

The homogeneous dispersion of graphene was resolved by compounding graphene with organic dispersants at a suitable weight ratio. It was found that the prepared POEM, dissolved in water at a concentration of 0.5 wt%, or dispersant with graphene at 2:1 weight ratio, was effective for dispersing graphene in deionized water. The fine dispersion can be recognized and differentiated by naked eyes by observing the black suspension from solid precipitates in the bottom layer, as shown in the inserted pictures of solution vials in Figure 35.16. The TEM micrographs further revealed the POEM dispersed graphene to be observed as the exfoliated state.

Mixing two precursors with an aqueous solution of polyurethane (PVA) and drop on the glass substrate to do spin coating were studied. After water evaporation, the composite film of graphene/MICA/PVA and graphene/POEM/PVA were then obtained. The thickness of thin film was varied by speed of spin coating and the concentration was controlled by the amounts of graphene [43].

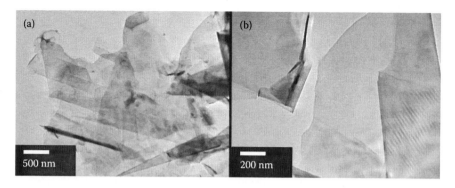

FIGURE 35.16 TEM micrographs of (a) graphene solution (sonication) and (b) graphene/POEM solution. (From P. L. Huang, H. H. Kuo, R. X. Dong et al. *J. Lightwave Technol.* 30;2012:3413.)

35.3.3 Few Layer Atomic Graphene

Different layers of graphene were produced by the CVD method [44,45]. For the CVD process of graphene on Ni substrate, the substrate structure of Ni(300 nm)/SiO$_2$(300 nm)/Si was put on a quartz plate and then loaded into the center of a tubular furnace. The chamber was evacuated to ~5 mTorr and the temperature was increased to 1000°C during the process. Prior to growth, a pretreatment step was performed under a H$_2$ atmosphere with 400 sccm flow at 2.8 Torr for 10 min. In the growth step, a gas mixture of methane and hydrogen (CH$_4$ = 80 sccm and H$_2$ = 40 sccm) was introduced for 10 min. The system was then cooled to room temperature to complete the growth. To transfer the as-grown graphene onto the substrate, the Ni substrate after the CVD growth was coated with a layer of poly (methyl methacrylate) (PMMA) by spinning-coating method, followed by baking at 90°C for 1 min. Then the PMMA-caped Ni substrate was immersed into a diluted HCl solution (HCl/Water = 1:3) for 20 min to etch away the Ni thin layer. The PMMA-caped graphene film was floated on the solution surface, and then transferred to a DI water to dilute and remove the etchant and residues. The PMMA/graphene was transferred to the receiving substrate and dried on a hot-plate. The PMMA was removed by warm acetone (90°C), and then the sample was rinsed with isopropyl alcohol and DI water. To strip off the graphene film from the quartz substrate, the graphene was covered by PVA. After water evaporation, graphene with a supporting layer of PVA film was laminated from the quartz substrate. The composite film of graphene/PVA was then obtained [21].

35.4 OPTICAL PROPERTIES OF GRAPHENE SA: MEASUREMENT

Optical transmission measurement under high intensity illumination (typically using ps or fs laser source) shows various degree of saturable absorption with low saturation intensity. The strength of MD of the SA, the key parameter for mode-locking laser, can be engineered from the process of graphene sample preparation either in the number of atomic-layers stacking of the CVD process or in the thin film concentration

and thickness control of wet spin coating process. In this section, the linear and nonlinear optical properties of graphene SA were examined. For linear optical absorption, the operating wavelength was checked by using UV–visible–NIR spectrophotometer. For nonlinear optical absorption, we examined the pulse-shaping ability of graphene SA including nonsaturable loss and MD. In the following, the graphene SA was inserted in the MLEDFL to generate the ML pulses. By examination of the oscilloscope, optical analyzer and autocorrelator, the pulse train, optical spectrum, and pulse trace are measured, respectively. Finally, we compared the characteristics of the ML pulses with different thickness and concentration of graphene SA.

35.4.1 Linear Optical Properties

35.4.1.1 Graphene with MICA and POEM Dispersant

Figure 35.17a shows the linear optical absorption spectra of graphene/MICA-SA films; the samples are prepared with 2 wt% concentration at spin rate of 800, 1200, and 1600 rpm. Figure 35.17b shows the linear optical absorption spectra of graphene/POEM-SA films; the samples are prepared with 1.65 and 3.25 wt% concentrations at spin rate of 800 and 1200 rpm. All traces showed featureless from 400 to 2000 nm as theoretically expected [46].

35.4.1.2 Few Layer Atomic Graphene

The linear absorption spectra of 7-, 11-, 14-, and 21-layers graphene is shown in Figure 35.18. The CVD-deposited graphene was well formatted close to A–A stacked structure. Figure 35.19 plots the Raman spectrum of 21 and 2 layer graphene-PVA film with two typical Raman peaks G (~1580 cm⁻¹) and 2D (~2726 cm⁻¹). The G-band was a doubly degenerate phonon mode at the Brillouin zone center that was Raman active for sp²-hybridized carbon–carbon bonds in graphene. The 2D-band was originated from a double-resonance process of crystalline graphite. The line-width broadening of the 2D-band of 21 layer graphene was due to the multilayer stacks. An increase in the number of defects among graphene would result in an increase of the D-band (~1350 cm⁻¹) intensity. In this case, the D-band was not observed in the Raman

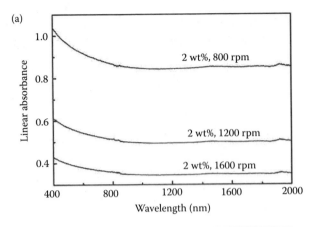

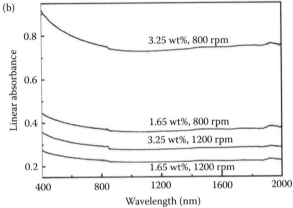

FIGURE 35.17 Linear absorption spectra of (a) graphene/MICA and (b) graphene/POEM films.

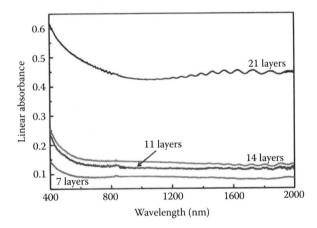

FIGURE 35.18 Linear absorption spectra of 7-, 11-, 14-, and 21-layer graphene.

spectrum, suggesting that a low defect-level of graphene was prepared [47,48].

35.4.2 NONLINEAR OPTICAL PROPERTIES

The nonlinear transmission characteristics of graphene SA were measured using a SWCNT-based SA mode-locked laser. The measurement schematic of the setup is shown in

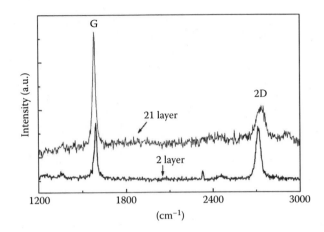

FIGURE 35.19 Raman spectrum with a 473 nm excitation laser of 2- and 21-layer graphene.

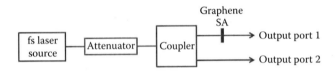

FIGURE 35.20 A system set up of nonlinear properties measurement. The laser was operated at a central wavelength of 1558.88 nm with a repetition rate of 25.51 MHz and pulse duration of 483 fs.

Figure 35.20. Through a broadband attenuator, the laser intensity was able to provide various power intensities. A coupler was connected after the attenuator so that the output power levels with and without passing the SAs could be measured simultaneously. The single-pass optical transmission can then be derived by the ratio of the intensity (output port 1) to the total input intensity [49].

35.4.2.1 Graphene with MICA and POEM Dispersant

Figure 35.21 shows the nonlinear transmission characteristics of graphene/MICA-SA and graphene/POEM-SA. The MD, nonsaturable loss and saturation intensity of graphene/MICA -SA (3 wt%, 12 µm) were measured about 2.57%, 70.92%, and 36.90 MW/cm^2, respectively, and about 1.70%, 76.94%, and 36.35 MW/cm^2 for graphene/POEM-SA (2 wt%, 19 µm). In general case, the mode-locking ability may relate to the MD and the pulsewidth shortens as the MD increases. In comparison, the SA dispersed by MICA exhibited a higher MD than that dispersed by POEM. The pulsewidth may be reduced furthermore as MD increases.

As mentioned, the difference in MD shown in Figure 35.21 can be understood from the ratio of Equation 3 with graphene/MICA-SA to graphene/POEM-SA as $MD_{MICA}/MD_{POEM} = e^{-\Delta\alpha_{ns} \cdot L}$ when the TCP of graphene are the same in both cases. The difference of mode-locking response for the MLFL using graphene/MICA and graphene/POEM may be related to the difference linear loss coefficient ($\Delta\alpha_{ns}$) from MICA and POEM dispersants in the SAs.

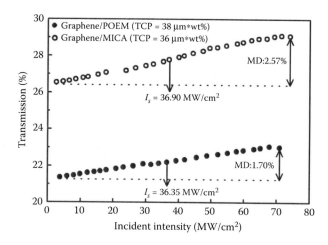

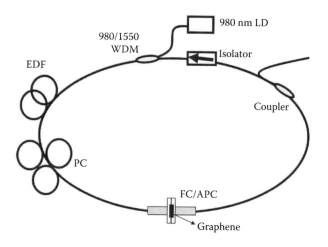

FIGURE 35.21 Nonlinear transmission of graphene/MICA and graphene/POEM films. (From P. L. Huang, H. H. Kuo, R. X. Dong et al., Performance of graphene mediated saturable absorber on stable mode-locked fiber lasers employing different nano-dispersants, *J. Lightwave Technol.* © 2012 IEEE.)

FIGURE 35.23 A graphene-based SA mode-locked fiber laser diagram.

is used as the gain medium. The 980/1550 nm WDM was used to introduce the 980 nm laser diode pumping source to excite the EDF and produce the amplified stimulated emission (ASE) with wavelength centered at 1550 nm and perform the optical feedback from the ring fiber cavity. The graphene films were inserted between two FC/APC fiber connectors as a SA to generate the mode-lock pulses. An isolator was employed to ensure the unidirectional operation; a polarization controller (PC) is utilized to distinguished two Eigen modes of state of polarization for laser operation. Total cavity length is 7.6 m with average GVD of 0.02 ps²/m. The emission light from EDF gain passed the graphene SAs then partially returned to the ring cavity and partially transmitted as cavity output from 1×2 fiber coupler. One output port was connected to power meter, autocorrelator, oscilloscope, radio-frequency spectrum analyzer, and optical spectrum analyzer to observe the pulsewidth, the pulse train, and the spectrum linewidth, respectively.

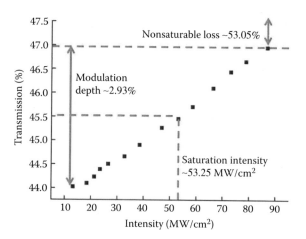

FIGURE 35.22 Nonlinear transmission of 21-layer SA. (From P. L. Huang, S. C. Lin, C. Y. Yeh et al., Stable mode-locked fiber laser based on CVD fabrication graphene saturable absorber, *Opt. Express*, 20;2012:2460. With permission of The Optical Society.)

35.4.2.2 Few Layer Atomic Graphene

The MD of the 7-, 11-, 14-, and 21-layers graphene-based SA were measured at 3.98%, 3.50%, 3.28%, and 2.93%, respectively. The nonsaturable loss of the 7-, 11-, 14-, and 21-layers graphene-based SA were also measured at 18.40%, 29.50%, 35.14%, and 53.05%, respectively. Figure 35.22 shows the nonlinear transmission characteristics of the 21-layer graphene SA.

35.5 FIBER-LASER MODE LOCKING WITH GRAPHENE SA: EXPERIMENT

35.5.1 Laser System Set Up

Figure 35.23 shows the all-fiber passive MLFL system. An 85 cm highly doped erbium fiber (LIEKKITM Er80-4/125)

35.5.2 Mode-Locked Laser Performance

35.5.2.1 Mode-Locking Using Graphene/MICA and Graphene/POEM as SA

For passive laser mode-locking using 3 wt%, and 12 μm graphene/MICA as SA, the threshold pump power in cw lasing was about 70 mW. For dispersant using MICA, the spectrum is shown in Figure 35.24. The output pulse train of MLFL had repetition rate about 25.64 MHz and the pulse width was measured as 382 fs, as shown in Figure 35.24 (inset). The average output power was measured to be about 1.16 mW. The TBP for graphene-SA was 0.322. Both optical spectra revealed the Kelly sideband indicating that soliton-like pulse was generated [50].

For passive mode-locking laser using 2 wt% and 19 μm graphene/POEM-SA; the mode-locked pulses were self-started as the pumping power increased to 78 mW. For dispersant using POEM, the optical spectrum is shown in Figure 35.25; the spectrum was centered at 1555.51 nm with 3 dB spectral bandwidth of 6.35 nm. The output pulse train of

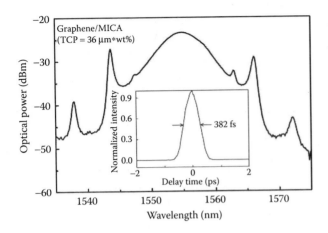

FIGURE 35.24 Optical spectrum and (inset) autocorrelator trace of mode-locked laser used 3 wt%, 12 m graphene/MICA as SA. (From P. L. Huang, H. H. Kuo, R. X. Dong et al., Performance of graphene mediated saturable absorber on stable mode-locked fiber lasers employing different nano-dispersants, *J. Lightwave Technol.* © 2012 IEEE.)

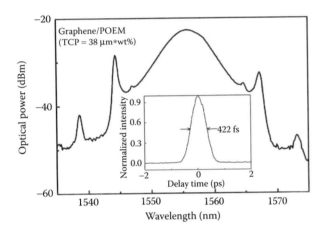

FIGURE 35.25 Optical spectrum and (inset) autocorrelator trace of mode-locked laser used 2 wt%, 19 m graphene/POEM as SA. (From P. L. Huang, H. H. Kuo, R. X. Dong et al., Performance of graphene mediated saturable absorber on stable mode-locked fiber lasers employing different nano-dispersants, *J. Lightwave Technol.* © 2012 IEEE.)

MLEDFL had exhibited repetition rate about 25.64 MHz and the pulse width measured 422 fs, the autocorrelator trace is shown in Figure 35.25 (inset). The TBP is calculated as 0.332.

35.5.2.2 Mode-Locking Using Few Layer Atomic Graphene as Saturable Absorber

The performance of MLFL using the 7-, 11-, 14-, and 21-layers graphene-based SA with different SMF fiber lengths were investigated and compared. The thinner 7-, 11-, and 14-layers of graphene-based SA were difficult to form a stable soliton-like pulse unless extra SMFs were added. The comparison of passively MLFL performance based on graphene SA is shown in Table 35.1.

For a passive MLFL using a 21-layer graphene as SA; the threshold pump power in CW lasing was about 33 mW. The mode-locked pulses were self-started as the pumping power increased to 53.30 mW. The optical spectrum of the mode-locked pulse is shown in Figure 35.26a. The spectrum was centered at 1559.12 nm with 3 dB spectral bandwidth of 6.16 nm. In Figure 35.26b, the output pulse train of MLEDF exhibited a repetition rate at about 25.51 MHz and the pulse width was measured to be 433 fs from the autocorrelator trace. Further increasing the pumping power to 73.78 mW, the harmonic mode-locking was observed, which could be confirmed by pulse train with a repetition rate about two times that of the fundamental mode-locking. The TBP was calculated to be 0.323, which was close to the bandwidth limited case. All optical spectra reveal the Kelly sideband indicating that a soliton-like pulse was generated. The laser cavity included 0.85 m of EDF (GVD:-0.02 ps^2/m), 1.35 m of corning flexcor 1060 (GVD:-0.007 ps^2/m), and 5.4 m of SMF28 (GVD:-0.023 ps^2/m). Based on the Kelly sideband location the total cavity dispersion was estimated to be 0.2124 ps/nm. The RF spectrum of ML pulses was measured by connecting a high sensitivity photo detector to a RF spectrum analyzer (HP8563E). As shown in Figure 35.27, the major peak was the cavity repetition rate of 25.67 MHz with a signal-to-noise ratio of 31 dB. In this work, the stability measurement was similar to the previous graphene-based works [11,12]. The power stability performance was monitored for 8 h a day and repeated measurements after 12 h for two weeks within 2% variation. The soliton pulse laser performance of fundamental

TABLE 35.1
Performance Comparison of MLFLs for Different Layers of Graphene SA

Number of Layers	Length of SMF (m)	Pumping Current (mA)	Spectra Width (nm)	Pulse Energy (nJ)	Pulse Duration (fs)	Pulse Stability
7	5.4 + 50	122	2.20	0.56	1147	Stable
11	5.4 + 0	108	4.48	0.01	a	Quasi-stable
11	5.4 + 10	146	3.11	0.15	715	Stable
14	5.4 + 0	109	2.86	0.03	a	Quasi-stable
14	5.4 + 5	173	3.48	0.07	563	Stable
21	5.4 + 0	142	6.16	0.05	483	Stable

a Due to the power fluctuation, the autocorrelator was not available to measure the pulsewidth.

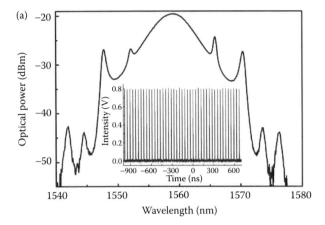

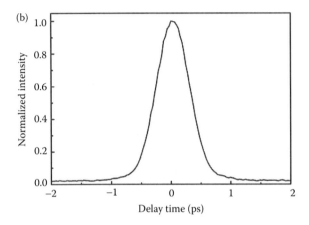

FIGURE 35.26 (a) Optical spectrum of the mode-locked laser and (b) autocorrelator trace.

mode-locking with 21-layer graphene-based SA was shown in Table 35.2. Table 35.2 indicated that the stable soliton-like operation was achieved at a pumping level from 53.30 to 63.79 mW. Second-order harmonic soliton-like pulses was achieved at a higher pumping level from 73.78 to 83.26 mW.

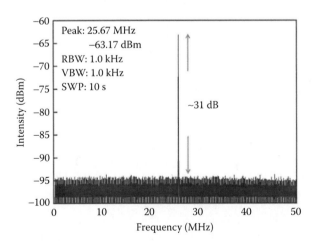

FIGURE 35.27 RF spectrum of fundamental mode-locked laser. (From P. L. Huang, S. C. Lin, C. Y. Yeh et al., Stable mode-locked fiber laser based on CVD fabrication graphene saturable absorber, *Opt. Express* 20;2012:2460. With permission of Optical Society.)

TABLE 35.2
Performance of 21-Layer Graphene-Based SA in MLFL

Pump Power (mW)	Pulse Duration (fs)	Laser Wavelength (nm)	Spectra Width (nm)	Pulse Energy (nJ)	TBP
53.30	563.64	1559.44	4.64	0.05	0.323
55.80	523.38	1559.28	5.08	0.06	0.328
58.80	483.12	1559.20	5.52	0.07	0.329
61.29	442.86	1559.28	5.92	0.07	0.323
63.79	432.47	1559.12	6.16	0.09	0.329

35.6 CONCLUSION

For the dispersed graphene-based SA, the facile process of preparing the homogeneous dispersion of graphene by a simple spin-coating technique under vacuum-free environment allows the fabrication of good quality and scalable large-area graphene film. This stable mode-locked pulse formation by employing the novel graphene-mediated SA has proven the high performance MLFLs that is potentially applicable for a myriad of low-cost nanodevices.

For few layer graphene fabricated by CVD method, the thinner layer of graphene SA had difficulty in forming a stable soliton-like pulse unless extra SMFs were added. However, a stable soliton-like mode-lock pulse train can be easily established with thicker layer of graphene SA. This might be due to the following reasons: (1) The thinner layer graphene samples exhibited relatively high MD with ISA made it difficult for a laser cavity to form a stable soliton pulse and needed additional length of SMF with sufficient abnormal dispersion to compensate nonlinear effect from graphene, for example, self-phase modulation. (2) More available density of states in the band structure of stacking layer graphene than the thinner layer favored nonlinear optics control of graphene inside the laser cavity.

In summary, graphene-polymer-based SA and the atomic-layer graphene-based SA used to achieve stable passively mode-locked laser were obtained. The results indicate that graphene can be a reliable mode-locker for stable soliton-like pulse formation of the MLFL.

REFERENCES

1. K. S. Novoselov, A. K. Geim, S. V. Morozov et al., Electric field effect in atomically thin carbon films, *Science* 306;2004:666.
2. H. A. Haus, Mode-locking of lasers, *IEEE J. Sel. Top. Quantum Electron.* 6;2000:1173.
3. M. E. Fermann, *Ultrafast Fiber Oscillators,* New York: Marcel Dekker, 2003.
4. N. J. Doran and D. Wood, Nonlinear-optical loop mirror, *Opt. Lett.* 13;1988:56.
5. E. A. De Souza, C. E. Soccolich, W. Pleibel et al., Saturable absorber modelocked polarisation maintaining erbium-doped fiber laser, *Electron. Lett.* 29;1993:447.
6. Z. Sun, T. Hasan, F. Torrisi et al., Graphene mode-locked ultrafast laser, *ACS Nano* 4;2010:803.
7. I. Hernandez-Romano, D. Mandridis, D. A. May-Arrioja, J. J. Sanchez-Mondragon, and P. J. Delfyett, Mode-locked fiber

laser using an SU8/SWCNT saturable absorber, *Opt. Lett.* 36;2011:2122.

8. J. C. Chiu, Y. F. Lan, C. M. Chang et al., Concentration effect of carbon nanotube based saturable absorber on stabilizing and shortening mode-locked pulse, *Opt. Express* 18;2010:3592.

9. Y. Hernandez, V. Nicolosi, M. Lotya et al., High-yield production of graphene by liquid-phase exfoliation of graphite, *Nat. Nanotechnol.* 3;2008:563.

10. S. Y. Set, H. Yaguchi, Y. Tanaka et al., Mode-locked fiber lasers based on a saturable absorber incorporating carbon nanotubes, in *Proc. Optical Fiber Communication Conf.* '03, Atlanta, GA, 2003, paper PD44.

11. Q. L. Bao, H. Zhang, Y. Wang et al., Atomic-layer graphene as a saturable absorber for ultrafast pulsed lasers, *Adv. Funct. Mater.* 19;2009:3077.

12. H. Zhang, D. Y. Tang, L. M. Zhao, Q. L. Bao, and K. P. Loh, Large energy mode locking of an erbium-doped fiber laser with atomic layer graphene, *Opt. Express* 17;2009:17630.

13. H. Zhang, D. Y. Tang, L. M. Zhao, Q. L. Bao, and K. P. Loh, Vector dissipative solitons in graphene mode locked fiber lasers, *Opt. Commun.* 283;2010:3334.

14. W. D. Tan, C. Y. Su, R. J. Knize et al., Mode locking of ceramic Nd:yttriu, aluminum garnet with graphene as a saturable absorber, *Appl. Phys. Lett.* 96;2010:031106-1.

15. R. Saito, G. Dresselhaus, and M. S. Dresselhaus, 1999, *Physical Properties of Carbon Nanotubes*, London, UK: Imperial College Press.

16. K. S. Novoselov, A. K. Geim, S. V. Morozov et al., Two-dimensional gas of massless Dirac fermions in graphene, *Nature* 438;2005:197.

17. F. Bonaccorso, Z. Sun, T. Hasan, and A. C. Ferrari, Graphene photonics and optoelectronics, *Nat. Photonics* 4;2010:611.

18. T. Hasan, Z. Sun, F. Wang, F. Bonaccorso, P. H. Tan, A. G. Rozhin, and A. C. Ferrari, Nanotube–polymer composites for ultrafast photonics, *Adv. Mater.* 21;2009:3874.

19. A. H. C. Neto, F. Guinea, N. M. R. Peres, K. S. Novoselov, and A. K. Geim, The electronic properties of graphene, *Rev. Mod. Phys.* 81;2009:109.

20. A. Martinez, K. Fuse, B. Xu, and S. Yamashita, Optical deposition of graphene and carbon nanotubes in a fiber ferrule for passive mode-locked lasing, *Opt. Express* 18;2010:23054.

21. P. L. Huang, S. C. Lin, C. Y. Yeh et al., Stable mode-locked fiber laser based on CVD fabricated graphene saturable absorber, *Opt. Express* 20;2012:2460.

22. G.-R. Lin and Y.-C. Lin, Directly exfoliated and imprinted graphite nano-particle saturable absorber for passive mode-locking erbium-doped fiber laser, *Laser Phys. Lett.* 8;2011:880.

23. K. J. Kuhn, 1998, *Laser Engineering*, Chapter 7, Prentice-Hall, Inc.

24. A. M. Weiner, 2009, *Ultrafast Optics*, Chapter 1, John Wiley & Sons, Inc.

25. A. Yariv, 1991, *Optical Electronics*, Chapter 13, Saunders College Publishing.

26. Jin-Chen Chiu, Study on nonlinear self-phase modulation enhancement in passive mode locked fiber laser with single-wall carbon nanotube saturable absorber, PhD dissertation, National Sun Yat-sen University, 2010.

27. G. Moos, C. Gahl, R. Fasel, M. Wolf, and T. Hertel, Anisotropy of quasiparticle lifetimes and the role of disorder in graphite from ultrafast time-resolved photoemission spectroscopy, *Phys. Rev. Lett.* 87;2001:267402.

28. T. Kampfrath, L. Perfetti, F. Schapper, C. Frischkorn, and M. Wolf, Strongly coupled optical phonons in the ultrafast dynamics of the electronic energy and current relaxation in graphite, *Phys. Rev. Lett.* 95;2005:187403.

29. D. Yoon, H. Moon, Y. W. Son et al., Strong polarization dependence of double-resonant Raman intensities in graphene, *Nano Lett.* 8;2008:4270.

30. T. Hertel and G. Moos, Electron–phonon interaction in single-wall carbon nanotubes: A time-domain study, *Phys. Rev. Lett.* 84;2000:5002.

31. J. M. Dawlaty, S. Shivaraman, M. Chandrashekhar, F. Rana, and M. G. Spencer, Measurement of ultrafast carrier dynamics in epitaxial graphene, *Appl. Phys. Lett.* 92;2008:042116.

32. P. A. George, J. Strait, J. Dawlaty et al., Ultrafast optical-pump terahertz-probe spectroscopy of the carrier relaxation and recombination dynamics in epitaxial graphene, *Nano Lett.* 8;2009:4248.

33. E. Garmire, Resonant optical nonlinearities in semiconductors, *IEEE J. Sel. Top. Quantum Electron.* 6;2000:1094.

34. R. W. Boyd, *Nonlinear Optics*, San Diego, CA: Academic Press, 2003.

35. C. Rullière, *Femtosecond Laser Pulses: Principles and Experiments*, New York: Springer, 1998.

36. H. A. Haus, Theory of mode locking with a fast saturable absorber, *J. Appl. Phys.* 46;1975:3049.

37. L. E. Nelson, D. J. Jones, K. Tamura, H. A. Haus, and E. P. Ippen, Ultrashort-pulse fiber ring lasers, *Appl. Phys. B* 65;1997:277.

38. J. J. Lin and Y. M. Chen, Amphiphilic properties of poly(oxyalkylene)amine-intercalated smectite aluminosilicates, *Langumir* 20;2004:4261.

39. Y. F. Lan and J. J. Lin, Observation of carbon nanotube and clay micellelike microstructures with dual dispersion property, *J. Phys. Chem. A* 113;2009:8654.

40. H. Tateyama, S. Nishimura, K. Tsunematsu et al., Synthesis of expandable fluorine mica from talc, *Clays Clay Miner.* 40;1992:180.

41. C. W. Chiu, C. C. Chu, S. A. Dai, and J. J. Lin, Self-piling silicate rods and dendrites from high aspect-ratio clay platelets, *J. Phys. Chem. C* 112;2008:17940.

42. R. X. Dong, C. T. Liu, K. C. Huang et al., Controlling formation of silver/carbon nanotube networks for highly conductive film surface, *ACS Appl. Mater. Interfaces* 4;2012:1449.

43. P. L. Huang, H. H. Kuo, R. X. Dong et al., Performance of graphene mediated saturable absorber on stable mode-locked fiber lasers employing different nano-dispersants, *J. Lightwave Technol.* 30;2012:3413.

44. C. Y. Su, D. Fu, A. Y. Lu et al., Transfer printing of graphene strip from the graphene grown on copper wires, *Nanotechnology* 22;2011:185309.

45. A. Reina, X. T. Jia, J. Ho et al., Large area, few-layer graphene films on arbitrary substrates by chemical vapor deposition, *Nano Lett.* 9;2009:30.

46. R. R. Nair, P. Blake, A. N. Grigorenko et al., Fine structure constant defines visual transparency of graphene, *Science*, 320;2008:1308.

47. M. A. Pimenta, G. Dresselhaus, M. S. Dresselhaus et al., Studying disorder in graphite-based systems by Raman spectroscopy, *Phys. Chem. Chem. Phys.* 9;2007:1276.

48. A. C. Ferrari, J. C. Meyer, V. Scardaci, C. Casiraghi et al., Raman spectrum of graphene and graphene layers, *Phys. Rev. Lett.* 97;2006:187401.

49. F. Wang, A. G. Rozhin, Z. Sun et al., Fabrication, characterization and mode locking application of single-walled carbon nanotube/polymer composite saturable absorbers, *Int. J. Mater. Form.* 1;2008:107.

50. M. L. Dennis and I. N. Duling, III, Experimental study of sideband generation in femtosecond fiber lasers, *IEEE J. Quant. Electron.* 30;1994:1469.

Index

A

ABC-stacking, 297
ABNR, *see* Armchair boron nitride nanoribbon (ABNR)
AC, *see* Activated carbon (AC)
a-C thin films, *see* Amorphous carbon thin films (a-C thin films)
Activated carbon (AC), 126
ADE, *see* Auxiliary differential equation (ADE)
Adhesives, 520; *see also* Graphene-based hybrid composites
Adsorption, 24
AFM, *see* Antiferromagnetic (AFM); Atomic force microscopy (AFM)
AFP, *see* Alpha-fetoprotein (AFP)
AGNR, *see* Armchair graphene nanoribbon (AGNR)
Alkyl viologens (AVs), 187
Alloy Theoretic Automated Toolkit (ATAT), 58
α-CD, *see* α-Cyclodextrin (α-CD)
α-Cyclodextrin (α-CD), 392
Alpha-fetoprotein (AFP), 101
Amorphous carbon thin films (a-C thin films), 498; *see also* Carbon-based thin films
Amorphous graphene, 254, 261; *see also* Carbon sp^2 phases
    conformational fluctuation, 257–259
    correlation between total energy per atom and magnitude of puckering, 258
    DOS of 800-atom amorphous and crystalline graphene, 255
    EDOS of, 256
    enlarged plot of puckered and smooth regions, 257
    experimental results, 254
    floppy modes, 259
    imaginary-frequency modes in flat 800 α-g model, 260
    low-frequency modes in pucker-down and-up 800 α-g models, 260
    models, 255
    molecular dynamics simulations, 256–257
    phonon calculation, 259
    pucker-up and-down 800 a-g models, 259
    random normal distortion, 256
    RDF comparison, 256
    symmetry breaking, 255–256
    time variation of two autocorrelation functions, 258
    total energy distribution functions, 258
    vibrational density of states, 259
    vibrational modes, 259–260
Amplified stimulated emission (ASE), 569
Anderson model, 27; *see also* Atomic-scale defects
Angle resolved photo emission spectroscopy (ARPES), 243, 282
Anomalous quantum Hall effect (AQHE), 408
Antiferromagnetic (AFM), 412
AQHE, *see* Anomalous quantum Hall effect (AQHE)
Armchair boron nitride nanoribbon (ABNR), 54; *see also* Atomic arrangement of graphene

Armchair graphene nanoribbon (AGNR), 50, 411; *see also* Atomic arrangement of graphene
Aromatic moleculesas dopants, 188; *see also* Doped graphene sheets
ARPES, *see* Angle resolved photo emission spectroscopy (ARPES)
ASE, *see* Amplified stimulated emission (ASE)
ATAT, *see* Alloy Theoretic Automated Toolkit (ATAT)
Atomic arrangement of graphene, 39; *see also* Tight-binding (TB)
    armchair superlattices, 56
    bands, 57, 58
    band structures of graphene, 48, 49
    brick-type lattice, 49
    dispersion relation, 50
    effects on graphene electronic structure, 55–58
    electronic structures of graphene, 48–50
    enumeration method, 58–59
    GNR electronic structures, 50–55
    Hamiltonian, 44
    Hamiltonian matrix elements, 45
    honeycomb lattice, 40, 44–48
    primitive unit cell, 44
    pristine graphene electronic structures, 44–48
    reciprocal primitive lattice vectors, 49
    unique unit cell, 58, 59
    zigzag superlattices, 55
Atomic force microscopy (AFM), 72, 277; *see also* 3D-AFM-hyperfine imaging of graphene monolayer deposits
Atomic-scale defects, 21, 34–35
    adsorption, 24
    Anderson model, 27
    appropriate Green function, 29–30
    binding and barrier energies, 34
    charge transport properties, 28–31
    chemical properties, 33–34
    defect formation and characterization, 23–24
    Dirac fermions, 29
    energetics of carbon atom vacancy, 32
    Hubbard model, 32
    magnetic properties, 31–33
    midgap states, 24–28
    π-electron cloud, 22–23
    preferential sticking mechanism, 33
    principal resonance structures, 25
    pZ vacancies, 34
    resonating valence bond model, 27
    spin-half Kondo effect, 31
    structure of carbon atom vacancy in graphene, 31
    supernumerary modes, 25
    surface reconstruction, 24, 25
    T-matrix, 29
Atom transfer radical polymerization (ATRP), 211
ATRP, *see* Atom transfer radical polymerization (ATRP)
ATS, *see* Azidotrimethylsilane (ATS)
AuNPs, *see* Gold nanoparticles (AuNPs)

Auxiliary differential equation (ADE), 367
Avrami equation, 81
AVs, *see* Alkyl viologens (AVs)
Azidotrimethylsilane (ATS), 212, 213

B

4-BBDT, *see* 4-Bromobenzene diazonium tetrafluoroborate (4-BBDT)
BCC, *see* Body-centered cube (BCC)
B-doped graphene, 184; *see also* Doped graphene sheets
Benzoyl peroxide (BPO), 211
BET method, *see* Brunauer, Emmett, and Teller method (BET method)
BHJ, *see* Bulk heterojunction (BHJ)
Bianchi identity, 328; *see also* Curved graphene nanoribbon
Bilayer graphene (BLG), 217, 344
Bipolar magnetic semiconductors (BMS), 155
BLG, *see* Bilayer graphene (BLG)
BMS, *see* Bipolar magnetic semiconductors (BMS)
BN, *see* Boron nitride (BN)
Body-centered cube (BCC), 236
Boltzmann transport equation (BTE), 409
Boron nitride (BN), 55, 411
Bovine serum albumin (BSA), 103
BPEI, *see* Branched polyethylenimine (BPEI)
BPO, *see* Benzoyl peroxide (BPO)
BPs, *see* Bucky-papers (BPs)
Branched polyethylenimine (BPEI), 104
Brillouin zone (BZ), 39
4-Bromobenzene diazonium tetrafluoroborate (4-BBDT), 211
Brunauer, Emmett, and Teller method (BET method), 111, 113, 311
BSA, *see* Bovine serum albumin (BSA)
BTE, *see* Boltzmann transport equation (BTE)
Buckyballs, 401
Bucky-papers (BPs), 376; *see also* Graphene-based 3D architectures
Bulk heterojunction (BHJ), 217
BZ, *see* Brillouin zone (BZ)

C

CAB, *see* Cellulose acetate butyrate (CAB)
Camera length (CL), 342
Camptothecin (CPT), 103
Cantor graphene structures (CGSs), 470; *see also* Graphene-based structures
Capacitive deionization (CDI), 272
Carbon, 167, 222, 249; *see also* Carbon nanotube surface modifications; Graphene (GFN); Graphene hybridization; Graphite surface modifications; Nanomedicine
    allotropes, 401, 529
    -based nanocarriers, 100
    based semiconductors, 249
    electron configurations in, 498
    material dispersion via surface modifications, 530
    self-aggregation of, 529

Carbon-based nanostructures, 517; *see also*
    Graphene-based hybrid composites
  CNT capabilities, 518
  GO method, 519
  MSC, 517
  techniques for incorporating CNTs into
    composite materials, 518
  work description, 520
Carbon-based single-phase 3D architectures,
    376; *see also* Graphene-based 3D
    architectures
  asymmetric CNT yarn structures, 384
  bucky-paper porosity, 386
  CNT alignment within CNT bucky-ribbons,
    383
  CNT BPs, 383
  CNT webs, 383
  FIB-milled cross-sections of rub-densified
    CNT yarns, 385
  FIB sections milled through CNT yarns, 384
  graphene foam, 376, 379–382
  leavening process to form GF, 381
  single-phase graphene 3D architectures, 382
  3D GF with Ni foam and without Ni
    foam, 381
Carbon-based thin films, 497
  a-C thin films, 498
  carbon fluoride, 509–511
  carbon nitride, 501–503
  chain of bonding events, 509
  chain of events for synthetic growth of
    CN_x, 504
  C_nP_m precursor species, 504
  cohesive energies per atom for CN_x model
    systems, 503
  cross-linking and interlinking sites in
    CP_x, 505
  energy required to get isolated atoms, 503,
    505, 507, 508
  fullerene-like, 499, 500, 502, 511–512
  growth simulation of FL-CA$_x$ thin films, 501
  growth simulation path for CP, 505
  HR-TEM micrograph, 499, 502
  microstructure of, 497
  modeling, 501
  phospho carbide, 503–506
  relaxed structures for defect containing
    S-doped model systems, 506
  relaxed structures for most relevant C_mS_n
    precursor species, 508
  Stone–Wales defects, 503
  structural patterns in growth of CS, 509
  structural patterns of CF, 510
  studies on, 498
  sulfo carbide, 506–509
  superhard rubber, 502
  synthetic growth concept, 499–501, 512
  tetragon defect, 507
  vapor phase deposition, 500
Carbon nanofiber (CNF), 123
Carbon nanomaterials, 105
Carbon nanospheres (CNSs), 101
Carbon nanostructures (CNSs), 123; *see also*
    Cylindrical carbon nanostructures
  classification of, 127
  and dimensionality, 126
Carbon nanosystem (CNS), 333
Carbon nanotube (CNT), 78, 231, 252–253,
    376; *see also* Carbon *sp²* phases;
    Computational modeling
  wave functions in, 365

Carbon nanotube surface modifications, 530; *see*
    also Graphite surface modifications
  via covalent bonding, 530–531
  via non-covalent bonding, 531–532
  surface roughness manipulation, 532
  surfactant adsorption, 531
  synthesis of Nylon-SWNT, 531
Carbon *sp²* phases, 249, 260–261; *see also*
    Amorphous graphene
  ab initio methods, 250–251
  carbon nanotubes, 252–253
  chiral vector in honeycomb lattice, 251
  computational methods, 250
  crystalline graphene, 251
  density functional band structure of
    crystalline graphene, 251
  density of states, 251
  empirical potentials, 250
  fullerenes, 251–252
  Schrödinger equation, 250
  schwartzite, 253
  tight-binding approximation, 250
Carcinoembryonic antigen (CEA), 101
Catalytic chemical vapor deposition (CCVD),
    123; *see also* Cylindrical carbon
    nanostructures
  set-up, 126
CBM, *see* Conduction band minimum (CBM)
C-BN HBLs, *see* Graphene-boron nitride
    heterobilayers (C-BN HBLs)
CCVD, *see* Catalytic chemical vapor deposition
    (CCVD)
CDI, *see* Capacitive deionization (CDI)
CDS, *see* Critical Dielectric Strength (CDS)
CEA, *see* Carcinoembryonic antigen (CEA)
CEG, *see* Chemically assisted exfoliated
    graphene (CEG)
Cellulose acetate butyrate (CAB), 432
Cetyl trimethylammonium bromide (CTAB),
    279, 393
CFIE, *see* Combined field integral equation
    (CFIE)
CGSs, *see* Cantor graphene structures (CGSs)
Characteristic equation, *see* Secular equation
Charge transfer, 189; *see also* Doped graphene
    sheets
Chemical doping, 215
Chemical functionalization, 147; *see also* Phase
    transformation in nanofilm
Chemically assisted exfoliated graphene
    (CEG), 211
Chemically modified graphene (CMG), 218
Chemical mechanical planarization (CMP), 424
Chemical modifications of graphene, 207, 218;
    see also Covalent functionalized
    graphene; Covalent modifications of
    graphene
Chemical vapor deposition (CVD), 63, 376, 424
Christoffel symbols, 332
CL, *see* Camera length (CL)
CMG, *see* Chemically modified graphene (CMG)
CMP, *see* Chemical mechanical planarization
    (CMP)
CNF, *see* Carbon nanofiber (CNF)
CNS, *see* Carbon nanosystem (CNS)
CNSs, *see* Carbon nanospheres (CNSs); Carbon
    nanostructures (CNSs)
CNT, *see* Carbon nanotube (CNT)
COFs, *see* Covalent organic frameworks (COFs)
Combined field integral equation (CFIE), 368;
    see also Computational modeling

Computational modeling, 359, 371
  arrangement of carbon atoms, 361
  carbon nanotubes, 365–366
  current at carbon nanotube dipole, 368
  differential-equation-based formulation,
    366–368
  electric field of two parallel graphene
    sheets, 367
  energy band structure in graphene, 361
  energy of electrons in bands, 362
  Faraday–and Ampere–Maxwell
    equations, 366
  FDTD model of cell, 367
  fermi distribution of occupied states for
    electrons, 363
  graphene, 360–365
  grapheneelectromagnetic devices, 360
  incell inductance, 368
  input impedance of CNT dipole, 370
  integral equation-based formulation,
    368–371
  interband and intraband conductivities, 369
  intraband conductivity, 365
  Kubo's formulation, 363
  Lorentz–Drude model, 370
  Maxwell's equations, 368
  quantization of transversal momentum, 365
  space–time diagram of currents in CNT
    dipole, 371
  surface conductivity of graphene, 366
  theoretical derivation of conductivity, 360
  for THz and optical regimes, 366
  unit cell and first Brillouin zone, 361
Computational simulation techniques, 113
Conduction band minimum (CBM), 152
Continuous random network (CRN), 255
Continuous wave (CW), 556
Covalent functionalized graphene, 217; *see also*
    Chemical modifications of graphene;
    Covalent modifications of graphene
  applications, 218
  biomaterials, 218
  electronic and optic devices, 217
  polymer nanocomposites, 217–218
Covalent linking, 207
Covalent modification of GO, 216–217; *see also*
    Chemical modifications of graphene;
    Covalent modifications of graphene
Covalent modifications of graphene, 208;
    see also Chemical modifications
    of graphene; Functionalization of
    graphene
  charge transfer effect, 210
  chemical doping, 215
  cyclization reaction, 209
  D-graphene/MA-POA2000 preparation, 216
  diazonium functionalization, 211
  via diazonium salt reactions, 208–211
  electronic devices, 210
  FG dispersion with different
    concentrations, 209
  FG nanoribbon synthesis routes, 209
  F-graphene/MA-POA2000 preparation, 216
  via free radical addition reactions, 208
  free radical reactions, 211
  graphene with radicals, 209
  group reaction, 209
  I–V curves of pristine and modified
    graphene, 210
  N-doped graphene, 215
  via residual groups, 216

substitutional doping, 215
via substitutional reaction and chemical
doping, 215–216
Covalent organic frameworks (COFs), 112
CPT, *see* Camptothecin (CPT)
Critical Dielectric Strength (CDS), 431
CRN, *see* Continuous random network (CRN)
Cross-linked PE (XLPE), 92
CTAB, *see* Cetyl trimethylammonium bromide
(CTAB)
Curved graphene nanoribbon, 327, 338
basic equations and spectrum of electrons,
333–335
Bianchi identity, 328
crystal lattice of graphene, 329
current–voltage characteristic of contact,
336, 337
dependence of correction to energy, 335, 338
dependence of energy correction vs. ratio of
radii of curvature, 336
Dirac equation, 333
electromagnetic field, 327
Friedmann model, 337–338
graphene and Hamiltonians, 329–330
Green's function, 330
hybridization potential, 330
invariance of Maxwell's equations, 328
long-wave approximation, 329
mathematical rules, 330–333
Mathieu equation, 337
quantum dynamics, 328
spectrum of elementary excitations, 330
spin connection, 333
tensors, 332
toroidal and helical nanoribbon, 334
tunneling characteristics, 335–336
tunneling current, 336
CV, *see* Cyclic voltammetry (CV)
CVD, *see* Chemical vapor deposition (CVD)
CW, *see* Continuous wave (CW)
Cyclic voltammetry (CV), 16, 175
Cycloaddition, 213
Cyclodextrin-functionalized graphene nanosheet
(GNS/-CD), 102
Cylindrical carbon nanostructures, 123, 140–142;
see also Carbon nanostructures
(CNSs)
basal plane lattice fringes, 130
carbon filament grown from natural gas, 140
catalytic growth model for CNS, 135
CCVD set-up, 126
classification of CNSs, 123
CNS morphology by CCVD method, 134–135
controlled defects in SWCNT, 136
dependence of melting temperature, 134
diameter variation, 140
DTA curves for nanoparticles, 138
electron diffraction pattern, 130
gas composition effecton carbon
morphology, 128
growing CNF, 139
growth of CNS, 136–140
growth of MWCNT, 138
hypothesis of melting of nanoparticles, 133
jellium model, 132
LSM model, 133
mapping of unfolded CNT, 129
melting temperature of nanoparticles, 133
metallic nanoparticles, 135
models of nanotubes, 129
morphology of CNSs, 124–125, 140

Ni catalyst nanoparticle, 138
simulated growth, 131
size-dependent melting, 134
structural classification of CNS, 127
structure of, 128–130
studies on CNS formation, 130–134
as synthesized oxide nanoparticles, 139
CYTOP, 190; *see also* Doped graphene sheets

D

Dark-field transmission electron microscopy
(DFTEM), 342
Daunorubicin (DNR), 103
DC, *see* Direct current (DC)
DDQ, *see* 2,3-Dichloro-5, 6-dicyano-1,
4-benzoquinone (DDQ)
Defects in graphene, 225; *see also* Graphene;
Hydrogen adsorption on graphene
bond length in MV, 227
elementary MV in graphene, 225
energy of vacancy formation, 227
graphene properties with MVs, 226–227
magnetic moment, 227
magnetic moment vs. MV concentration, 226
mono-vacancy defects, 225
out-of-plane displacement of dangling bond
atom, 227
Deionized water (DI water), 565
Density functional theory (DFT), 23, 53, 223,
250, 314, 499
Density of states (DOS), 22, 183
of 2D nonrelativistic electron gas, 552
Department of Energy (DOE), 229
DFT, *see* Density functional theory (DFT)
DFTEM, *see* Dark-field transmission electron
microscopy (DFTEM)
Diamagnetic material, 541; *see also*
Magnetocaloric effect (MCE)
Diamond, 401
Diamond-like carbon (DLC), 498
Diamondol, 150
Diazirines, 214
Diazonium salts, 208
2,3-Dichloro-5,6-dicyano-1,4-benzoquinone
(DDQ), 191
Diels–Alder reaction, 212
Differential scanning calorimetry (DSC), 81
Differential thermal analysis (DTA), 137
Dimethylformamide (DMF), 215, 217, 519
Dimethyl sulfoxide (DMSO), 217, 533
Dirac equation, 333
Dirac fermions, 63
Dirac point (DP), 173, 403
Direct current (DC), 63
Direct tunneling (DT), 431
DI water, *see* Deionized water (DI water)
DLC, *see* Diamond-like carbon (DLC)
DMA, *see* Dynamical mechanical analysis (DMA)
DMF, *see* Dimethylformamide (DMF)
DMSO, *see* Dimethyl sulfoxide (DMSO)
DNR, *see* Daunorubicin (DNR)
DOE, *see* Department of Energy (DOE)
Doped graphene properties, 309, 315; *see also*
Graphene doping
catalytic efficiency of Pt, 314–315
CNTs as catalyst support, 309–310
graphene as catalyst support, 310–312
graphene sheets and nanoparticle-modified
graphene sheets, 311
N-doped graphene as metal-free catalyst, 315

nitrogen doping of CNT, 310
platinum–carbon interaction strength,
312–314
proton exchange membrane fuel cell, 309
single nitrogen-doped graphene flake, 312
Doped graphene sheets, 179, 201–202; *see also*
Graphene
anode structures, 196
aromatic molecules used as dopants, 188
B-doped graphene, 184
chemical doping, 186–189
comparison of transport property of FET, 194
CYTOP, 190
De-doping phenomena, 197–201
device applications of, 191, 193
doped graphene properties, 192–193
doping by adsorbed TCNQ molecules, 190
fabricating rGO-based thin films, 181
field-effect transistors, 191
G-and G′-band in Raman spectra, 199
GaN LEDs fabrication, 198
graphene-based GaN LEDs, 196
light-emitting diodes, 195–197
methods of doping, 183
modulation of electronic structure, 187
n-and p-type doped exfoliated graphene
sheets, 189
N-doped graphene, 185
n-type doping, 188
organic photovoltaic cells, 193–195, 197
plasma processing, 185
p-type doping, 188
substitution doping, 183–186
surface charge transfer, 189–191
effect of UV/ozone-treated graphene
sheet, 198
UV photoemission spectra, 195
variation of sheet resistance, 201
viologen, 187
DOS, *see* Density of states (DOS)
Double-ζ polarized (DZP), 251
DOX, *see* Doxorubicin (DOX)
DOX-loaded GSPI-based system (GSPID), 102
Doxorubicin (DOX), 102
DP, *see* Dirac point (DP)
Drug delivery systems, 102; *see also* Graphene;
Nanomedicine; Tumor disease
diagnostics
cytostatics, 102–103
gene delivery, 103–104
human foreskin fibroblasts, 103
TRC105, 103
DSC, *see* Differential scanning calorimetry (DSC)
DSSCs, *see* Dye sensitized solar cells (DSSCs)
DT, *see* Direct tunneling (DT)
DTA, *see* Differential thermal analysis (DTA)
Dye sensitized solar cells (DSSCs), 530
Dynamical mechanical analysis (DMA), 91
DZP, *see* Double-ζ polarized (DZP)

E

EC, *see* Electro chemical (EC)
ECGSs, *see* Electrostatic Cantor Graphene
Structures (ECGSs)
ECL, *see* Electrochemiluminescence (ECL)
EDFL, *see* Erbium-doped fiber laser (EDFL)
EDLC, *see* Electrical double layer capacitors
(EDLC)
EDLs, *see* Electrical double layers (EDLs)
EDOS, *see* Electronic density of states (EDOS)

EDX, *see* Energy-dispersive x-ray (EDX)
EELS, *see* Electronic energy loss spectroscopy (EELS)
EEM, *see* Electron effective mass (EEM)
EFGSs, *see* Electrostatic FGSs (EFGSs)
EFIE, *see* Electric field integral equation (EFIE)
Efrost and Shklovskii (ES), 279
EG, *see* Expanded graphite (EG)
EGBSs, *see* Electrostatic graphene-based structures (EGBSs)
EGSLs, *see* Electrostatic GSLs (EGSLs)
Electrical double layer capacitors (EDLC), 174, 271
Electrical double layers (EDLs), 174
Electric field integral equation (EFIE), 368; *see also* Computational modeling
Electric vehicles (EVs), 435
Electro chemical (EC), 240
Electrochemiluminescence (ECL), 101
Electron effective mass (EEM), 242
Electronic density of states (EDOS), 250
Electronic energy loss spectroscopy (EELS), 239, 342
Electrostatic Cantor Graphene Structures (ECGSs), 463; *see also* Graphene-based structures
Electrostatic FGSs (EFGSs), 470; *see also* Graphene-based structures
Electrostatic graphene-based structures (EGBSs), 453; *see also* Graphene-based structures
Electrostatic GSLs (EGSLs), 458; *see also* Graphene-based structures
Electrostatic self-assembly, 389
Energy-dispersive x-ray (EDX), 267, 350
Environmental transmission electron microscope (ETEM), 123, 136
Equivalent series resistance (ESR), 177
Erbium-doped fiber laser (EDFL), 561
Ericsson cycle, 542, 543
ES, *see* Efrost and Shklovskii (ES)
ESR, *see* Equivalent series resistance (ESR)
ETEM, *see* Environmental transmission electron microscope (ETEM)
EVs, *see* Electric vehicles (EVs)
Exchange energy, 236
Expanded graphite (EG), 518
Extended tetrathiafulvalene (exTTF), 215
exTTF, *see* Extended tetrathiafulvalene (exTTF)

F

Fabry–Perot resonant cavity, 67; *see also* Graphene plasmonics
Face-centered cube (FCC), 236
Faraday–and Ampere–Maxwell equations, 366; *see also* Computational modeling
FCC, *see* Face-centered cube (FCC)
FCNTs, *see* Functionalized CNTs (FCNTs)
FDTD, *see* Finite-difference time-domain (FDTD)
FE, *see* Finite element (FE)
Ferromagnetic (FM), 412
FE-SEM, *see* Field emission scanning electron microscopy (FE-SEM)
FET, *see* Field effect transistor (FET); Future energy technology (FET)
Few-layer graphene (FG or FLG), 344, 479, 494
    bandgap determination, 488
    bandgap in hydrogenated graphene, 489
    capacitance–voltage measurements, 493
    carrier mobilities vs. hydrogenation time, 487

cleaning for high conductivity, 483–486
creation of highly resistive substrate, 482
current change in FLG flakes, 480, 481
current–voltage characteristics, 484, 486, 491
few-layer fluorographene, 489–492
FLG based future devices, 480
FLG electrodes, 493
FLG-NMP structures, 488
functionalization for electronic applications, 486
graphene multilayer films, 481
HF-treated FLG-NMP structures, 485
highly resistive substrate creation, 483
increase in resistivity of intercalated ribbons, 482
for memory and electrode applications, 492–494
properties, 479
Raman spectra of hybrid structures, 484, 486
Raman spectrum of FLG flake, 489
resistivity and carrier mobility, 483, 487
resistivity and field-effect carrier mobility, 485
resistivity change in FLG flakes, 480
samples with grain boundaries, 481
for sensor applications, 480–482
substrates for high carrier mobility, 482–483
superlattice hydrogenated graphene/NMP monolayer, 486–488
FG, *see* Few-layer graphene (FG or FLG)
FGSs, *see* Fibonacci graphene structures (FGSs)
FH, *see* Full-hydrogenation (FH)
FIB, *see* Focus ion beam (FIB)
Fiber-laser mode locking with graphene SA, 569; *see also* Graphene film preparation; Graphene SA optical properties; Ultrafast laser
    autocorrelator trace, 570, 571
    laser system set up, 569
    mode-locking using graphene, 569, 570–571
    optical spectrum, 570, 571
    performance comparison of MLFLs, 570
    RF spectrum, 571
    21-layer graphene-based SA performance, 571
Fibonacci graphene structures (FGSs), 470; *see also* Graphene-based structures
Field effect transistor (FET), 211, 232
Field emission scanning electron microscopy (FE-SEM), 267
Finite-difference time-domain (FDTD), 366; *see also* Computational modeling
    model of cell, 367
Finite element (FE), 279
Flash memory, 492
FLG, *see* Few-layer graphene (FG or FLG)
Floppy modes, 259; *see also* Amorphous graphene
Fluorine, 509
Fluorine-doped tin oxide (FTO), 240
FM, *see* Ferromagnetic (FM)
FNT, *see* Fowler–Nordheim Tunneling (FNT)
Focus ion beam (FIB), 8, 383
Förster resonance energy transfer (FRET), 101
Fourier Transform Infrared (FTIR), 81, 522
Fourth-nearest-neighbor force constant method (4NNFC method), 409
Fowler–Nordheim Tunneling (FNT), 431
FQHE, *see* Fractional quantum Hall effect (FQHE)
Fractional quantum Hall effect (FQHE), 11, 232, 408
Free-running laser, 556; *see also* Ultrafast laser

FRET, *see* Förster resonance energy transfer (FRET)
Friedmann model, 337–338; *see also* Curved graphene nanoribbon
F4-TCNQ, *see* Tetrafluorotetracyanoquinodimethane (F4-TCNQ)
FTIR, *see* Fourier Transform Infrared (FTIR)
FTO, *see* Fluorine-doped tin oxide (FTO)
Fullerenes, 251–252; *see also* Carbon sp^2 phases
Full-hydrogenation (FH), 152
Full width at half maximum (FWHM), 11, 186, 280
Functionalization of graphene, 211; *see also* Covalent modifications of graphene
    via amide-condensation reactions, 213
    with arynes, 212
    carbene insertion/addition approach, 214
    with carbenes, 214
    cycloaddition, 213
    diazirines, 214
    Diels–Alder reaction, 212
    graphene as diene–dienophile, 212
    with nitrenes, 212–214
    nucleophilic addition, 214–215
    silicon wafers, 213
    solution-phase, 213
Functionalized CNTs (FCNTs), 85
Future energy technology (FET), 306
FWHM, *see* Full width at half maximum (FWHM)

G

GB, *see* Grain boundary (GB)
GC, *see* Glassy carbon (GC)
Generalized gradient approximation (GGA), 148, 223, 236, 250
GF, *see* Glass fiber (GF); Graphene foams (GF); Graphene framework (GF)
GFETs, *see* Graphene field effect transistors (GFETs)
GFN, *see* Graphene (GFN)
GGA, *see* Generalized gradient approximation (GGA)
GGN, *see* Graphene–gold nanocomposites (GGN)
Glass fiber (GF), 85
Glassy carbon (GC), 380
GNRs, *see* Graphene nanoribbons (GNRs)
GNS/-CD, *see* Cyclodextrin-functionalized graphene nanosheet (GNS/-CD)
GNSs, *see* Graphene nanosheets (GNSs)
GO, *see* Graphene oxide (GO)
GOFs, *see* Graphene-oxide frameworks (GOFs)
GO–IONP, *see* Graphene oxide–iron oxide hybrid nanocomposite (GO–IONP)
Gold nanoparticles (AuNPs), 218
GQDs, *see* Graphene quantum dots (GQDs)
Grain boundary (GB), 241
Graph, 159, 161
Graphene (GFN), 99, 147, 159, 179, 180, 221, 403, 402, 436, 555; *see also* Computational modeling; Defects in graphene; Doped graphene properties; Doped graphene sheets; Graphene doping; Graphene nanoribbons; Graphene nanosheet anode; Hydrogen adsorption on graphene
    atomic structure of, 410
    band structure of, 404
    -based biomaterials, 104–105

characterizationmethods, 105
conductivity of electrons in, 406
conservation of pseudospin, 407
coordinates of symmetry points, 403
covalent modification and property of, 208
CVD-epitaxial graphene layer, 180
density of states and carrier conductivity of, 405
Dirac cones of, 404
Dirac points, 403
discovery of, 159
doping, 183
DOS of, 404
electronic properties, 222
electron microscopy imaging of, 320
fluorescence *in vivo* imaging, 105
with grain boundary, 410
hexagonal structure of, 222
innovative electromagnetic devices based on, 360
Klein tunneling, 407–408
large-scale graphene films, 182–183
modified graphene, 222
monolayer, 402
negative refraction, 406–407
noncovalent methods, 208
phonon and heat conduction for, 409–410
phonon dispersion of, 409
plasma-enhanced models, 307
p–n junction, 406
polyhex nanotorus, 160
pseudo-vector potential, 405
QHEs in, 408–409
reciprocal lattice vectors, 403
review articles focused on, 104
roll-based production of graphene films, 181
scotch tape method, 179, 180
structural properties, 221
structure, 306–307
symmetry, 160–162
synthesis, 180, 181, 222, 306–307
theorem, 159, 160–162, 162–164
tight-binding Hamiltonian for π orbitals, 403
tight-binding model Hamiltonian, 405
topology, 162–164
transfer processes, 182
transmission of, 404
transport properties, 402
two-dimensional structure, 249
unit lattice vectors, 403
Graphene 3D assemblies, 263
applications, 268
batteries, 270–271
bottom-up synthesis, 266–268
capacitive deionization, 272
catalysis, 273
charging induced macroscopic effects, 272–273
comparison of Raman spectra, 265
drawback of, 272
EDL model, 271
excess gravimetric density vs. BET surface area, 268
FE-SEM images, 267
graphene macroassemblies, 264
gravimetric and area-specific capacitance vs. surface area, 272
HRTEM images of nanographene, 265
hydrogen storage, 268–270
interfacial charging induced electrode response, 272

load-displacement curve, 268
loading, 268
nitrogen adsorption/desorption isotherm, 267
NMR spectra, 266
pore size distribution, 267, 268
scaffolding approach, 270
supercapacitors, 271–272
top-down synthesis, 264–266
valence and conductance band structure, 266
XRD data, 265
Graphene-based 3D architectures, 375, 396;
see also Carbon-based single-phase
3D architectures; Graphene-based 3D
composites
application, 376
bi-scrolled yarns, 394
BPs-Gold composites, 395
carbon nanotube-based 3D composites, 393–396
composite BPs, 396
fabrication, 376–378
PANI nanowires, 393
SEM images of, 391
Graphene-based 3D composites, 384; see also
Graphene-based 3D architectures
electrostatic self-assembly, 389–390
GO/multiwalled CNTs solution, 389
graphene and CNTs, 384–388
graphene and metals/metal oxide
nanoparticles, 389
graphene and polymers, 390, 392–393
graphene/Au composites, 391
graphene/CNTs composite film, 388
graphene sheets on vertically aligned
CNTs, 387
hybrid layer-by-layer films, 390
layer-by-layer self-assembly, 388–389
metal/metal oxide nanoparticle
synthesis, 389
solution mixing of CNTs and graphene, 388
thermally expanded graphene layers, 387
Graphene-based device, 454
Graphene-based hybrid composites, 517, 525;
see also Carbon-based nanostructures
data analysis, 522–525
experimental procedures, 521
finite element mesh representation, 525
FTIR signature, 523
graphene 2D structure by TEM, 523
nanoindentation, 524, 522
nano-modified adhesives, 520–521
peel and shear stresses, 525
single-lap joint dimensions, 521
single lap joint force–displacement plots, 524
tentative assignment of major bands, 524
von Mises stress distribution, 526
Graphene-based materials, 375, 401, 417–418;
see also Graphene; Graphene
nanoribbons
Graphene-based structures, 453, 474
angular distribution of transmittance, 461
aperiodic structures, 461–462
band-edge profile of multibarrier
structure, 458
breaking-symmetry substrates in
graphene, 456
cantor-like structures, 462–470
conductance, 463, 468, 473
Dirac-cone distribution and band-edge
profile, 459, 465, 470
dispersion relation, 455

electrostatic barriers in graphene, 454–456
fibonacci quasi-periodic structures,
470–474
graphene multibarrier structures, 457
graphene superlattices, 455
massless Dirac equation, 455
metal–semiconductor contacts, 454
periodic structures, 458–461
semiconductor superlattice, 454
spectrum of confined states, 464, 469,
474, 475
theoretical models, 456–458
transmission probability in FGSs, 472, 473
transmission probability vs. energy and
angle, 462, 467
transmission probability vs. energy of
incident Dirac electrons, 460
transmittance in ECGSs and SCGSs,
465, 466
transmittance in EFGSs and SFGSs, 471
Graphene-based THz metamaterial, 72–74;
see also Graphene plasmonics
Graphene-boron nitride heterobilayers
(C-BN HBLs), 242
Graphene doping, 307, 315; see also Doped
graphene properties
B-doped graphene, 308
CN$_x$ grown on carbon paper, 308
N-doped graphene, 307–308
O-doped graphene, 308
regular CNTs, 308
selective doping, 305
Graphene electronic properties modification,
231, 244
band diagram, 233
bandgap opening, 241, 242
bandgap variation of strained graphene, 238
basic methodology, 234
binding energy per unit cell, 242
boron and nitrogen doping, 239–243
change in interaction energy, 234
change in kinetic energy, 234
density gradient, 236
DFT and approximations, 235
DFT and body physics, 234–235
DFT computation, 236
doped graphene, 238–239
electronic property comparisons, 232
energy bandgap variation, 239
energy bands, 242
energy gap vs. interlayer spacing, 243
exchange energy, 236
Fermi energy deviation, 240
Fermi energy variation, 237
flow chart of DFT calculations, 235
formation energy, 241
future directions, 243–244
generalized gradient approximations, 236
ground state energy functional, 234
HK theorem, 234
local density approximations, 235–236
N-doped CNT-graphene hybrid
nanostructures, 240
N-doped reduced graphene film, 240
strained graphene, 237
tailoring electronic and electrochemical
properties, 237
total energy functional, 234
traditional wave functional methods, 234
2D graphane under uniform expansion, 238
uniform strain, 237

Graphene-enabled heterostructures, 423, 432–433
  atomic force microscopy, 427
  carrier mobility in graphene, 433
  characterization methods, 426
  device fabrication processing, 424
  dielectric behavior of h-BN, 430–431
  electrical characterization, 427
  electrical stress annealing, 427–428, 429
  fabricated graphene/h-BN FET, 426
  fabrication scheme, 424, 425
  FETs, 425
  graphene growth and transfer, 424
  graphene/h-BN interconnect, 425
  graphene heterostructure, 432, 433
  graphene resistivity vs. backgate voltage, 429
  hexagonal boron nitride encapsulation,
    431–432
  mobility of charge carriers in graphene, 429
  optical microscopy, 426
  passivation process, 431
  performance characterization scheme, 428
  pretesting thermal anneal, 427
  Raman spectroscopy, 426
  reliability characterization, 429–430
  sample cleaning methods, 427
  scanning electron microscopy, 426
  thin h-BN flake, 426
  transitional voltage, 431
Graphene field effect transistors (GFETs), 423
Graphene film preparation, 565; see also Fiber-
    laser mode locking with graphene
    SA; Graphene SA optical properties;
    Ultrafast laser
  dispersant poly(oxyethylene)-segmented
    imide, 565–567
  dispersant synthetic fluorinated MICA, 565
  dispersion of graphene-MICA, 566
  few layer atomic graphene, 567
  graphene/POEM solution, 567
  graphene solution, 567
  pristine graphene, 566
Graphene foams (GF), 376
Graphene framework (GF), 175
Graphene–gold nanocomposites (GGN), 103
Graphene heterostructures, 3, 16
  all-carbon G/CNT vertical heterostructure,
    15–16
  approaches to, 4
  atomic structures of, 6
  classes, 3
  direct growth of graphene/h-BN stacks, 14
  fabrication techniques and application
    fields of, 4
  FQHE and Hofstadter butterfly spectrum, 13
  graphene/CNTs heterostructures, 15
  impact on spatial DOS and charge density
    distribution, 12
  landscape of vertical, 10
  lateral, 3
  with patterned domains, 8
  planar h-BNC films with random
    composition, 5
  Raman spectrum and Tauc's plot, 7
  with random domains, 5–7
  surface roughness of graphene vs. h-BN, 12
  synthesis and characterizations of shape
    engineered, 9
  vertical, 8–14
Graphene hybridization, 534; see also Carbon
    nanotube surface modifications;
    Graphite surface modifications

AgNP/graphene hybrids and applications,
    535–536
  applications of graphene hybrids, 534–535
  density of states of 2D nonrelativistic
    electron gas, 552
  J–V characteristics of DSSCs, 536
  with nanoparticles, 535
  nanoparticles-on-graphene, 534
  oxygen-containing functional groups, 534
  process making polymer dispersed CNT–
    AgNP nanohybrids, 534
  PtNP/graphene hybrids and applications, 536
Graphene-like model systems, 500, 502; see also
    Carbon-based thin films
Graphene-like structures, 99, 105
Graphene multibarrier structures, 457; see also
    Graphene-based structures
Graphene nanoantenna, 72; see also Graphene
    plasmonics
Graphene nanofilms, 289, 301, 303; see also
    Graphite lattice dynamics
  ABC-stacking, 297
  formation of impurity atoms, 299
  impurities effect on phonon spectrum, 299
  localization of vibrations on point defects,
    299–300
  phonon DOS of bulk graphite and bilayer
    graphene, 296
  phonon DOS of intercalated graphite,
    302, 303
  phonon heat capacity of, 298–299
  phonon spectrum, 289, 300–301
  reconstruction and relaxation, 294–295
  spectral densities and mean-square
    amplitudes, 295–298
  temperature dependence of mean-square
    amplitudes of atoms, 297
  vibrational characteristics of, 289
Graphene nanoribbons (GNRs), 50, 410, 418;
    see also Atomic arrangement
    of graphene; Graphene; Phase
    transformation in nanofilm
  ABNR band structures, 54
  AFM coupling, 415
  AGNR, 411, 415
  armchair band structures, 52
  bandgap variation, 415
  band structures of, 149, 413
  bilayer, 412
  energy gap, 412
  fundamentals, 410–412
  Hamiltonian matrix, 53
  kinked, 417
  lattice and magnetic structures, 414
  magnetism and magnetic-field-effect, 412–415
  methods to enhance ZT of, 417
  partial charge densities for, 149
  structure of, 51
  thermal and thermoelectric properties for,
    416–417
  thermal conductivities obtained by
    experiments and MD simulations
    for, 416
  ZBNR band structures, 55
  ZGNR, 411
  zigzag band structures, 52
Graphene nanosheet anode, 435, 448–450;
    see also Graphene nanosheets (GNSs)
  doped GNSs, 443–445
  freestanding GNSs in LIBs, 445–448
  GNS–CNT hybrid materials, 442

GNS electrochemical reaction with
    lithium, 438
  GNSs-based composites, 442–443
  graphene, 436
  graphene paper, 447, 448
  highly branched GNSs on Cu foil, 441
  in-plane pores into chemically exfoliated
    graphene oxide, 449
  in LIBs, 438–440
  lithium storage behavior, 440
  modification of, 440–442, 447
  N-GNSs, 446
  precursor and synthesis methods on, 441
  rate capability of composites, 443
  specific surface areas of GO and GNSs, 438
  structure change of graphene oxide, 444
  synthesis of core-shelled GNSs@MO@C
    hybrids, 450
  thermal treatment, 444
Graphene nanosheets (GNSs), 435, 436; see also
    Graphene nanosheet anode
  synthesis of, 436–438, 439
Graphene oxide (GO), 99, 376
Graphene-oxide frameworks (GOFs), 111;
    see also Hydrogen storage inside
    graphene-based materials
Graphene oxide–iron oxide hybrid
    nanocomposite (GO–IONP), 102
Graphene paper, 447, 448; see also Graphene
    nanosheet anode
Graphene plasmonics, 63, 74
  complex phase constant of graphene
    waveguide, 70
  Fabry–Perot resonant cavity, 67
  Fermi energy, 65
  frequency dispersion of sheet
    conductivity, 65
  graphene-based THz devices and
    metamaterials, 72–74
  graphene-based THz plasmonic
    nanoantenna, 72
  graphene parallel-plate waveguide, 70
  graphene parallel-plate waveguide
    interconnect, 70–72
  graphene surface conductivity, 64–65, 66
  invisibility cloaking, 67
  nanofabricated infrared metamaterial, 74
  plasmonic cloaking, 67
  quantum-dynamical interband
    conductivity, 64
  reflectance and transmittance vs.
    frequency, 67
  reflection properties of graphene, 66
  scattering properties, 68–69
  semi-classical intraband conductivity, 64
  surface conductivity, 65
  surface wave excitation, 69–70
  terahertz wave scattering, 65
  TM-mode surface plasmon polariton wave
    propagation, 69
  transfer-matrix method, 67, 71
  transmission-line circuit model, 67, 71
  transmission properties, 67
Graphene/polymer nanocomposites, 77, 93–94
  applications, 78
  Avrami equation, 81
  Avrami plots, 81, 82
  conformational ordering and crystallization
    of PLLA, 83
  CRGO nanosheets induced PE
    crystallization, 87

crystallization behavior, 79
crystal modifications, 85
DSC thermograms, 83, 92
effects on crystallinity degree, 84–85
factors influencing mechanical response, 90–91
filler loading effects, 86–89
glass transition temperature, 91–92
graphene filler modification, 78
Halpin–Tsai model, 89, 90
hybrid crystalline structures, 86
influence of crystallinity properties, 89–90
isothermal crystallization, 81–82
mechanical properties, 86
methods for, 79, 80
methods for obtaining graphene sheets, 78
microtomed composite films, 86
non-isothermal crystallization, 82–84
nonpolar polymers used in, 79
Ozawa equation, 83
PE multilayered structure, 80
polar polymers used in, 79
polymer chain isothermal crystallization, 81
pristine graphene, 78
stabilization of γ-phase and β-phase, 85
stress–strain curves, 89
temperature vs. dynamic storage modulus, 91
TGA, 93
thermal conductivity, 93
thermal properties, 91
thermal properties of functionalized, 91
thermal stability, 92–93
thermo-mechanical properties of, 87, 88
Graphene quantum dots (GQDs), 102
Graphene SA optical properties, 567; *see also* Fiber-laser mode locking with graphene SA; Graphene film preparation; Ultrafast laser
few layer atomic graphene, 567–568, 569
graphene-based SA mode-locked fiber laser, 569
graphene with MICA and POEM dispersant, 567, 568
linear absorption spectra, 568
nonlinear properties measurement, 568
nonlinear transmission, 569
optical properties, 567, 568
Raman spectrum, 568
Graphene sheets (GSs), 263, 341; *see also* Quantitative transmission electron microscopy; Graphene 3D assemblies
electron scattering in, 353
HAADF intensities, 353
HAADF-STEM images of, 354
measuring suspended, 349
TEM measurements, 342
thickness analysis, 347
Graphene superlattice (GSL), 454; *see also* Graphene-based structures
Graphite, 401
crystal, 290
Graphite lattice dynamics, 290, 301; *see also* Graphene nanofilms
crystal lattice, 291
crystal structure, 290
dispersion curves, 293
elastic moduli, 292
force constants, 290, 293
force constants and flexural rigidity, 291–294

Graphite surface modifications, 532; *see also* Carbon nanotube surface modifications
via covalent bonding, 533
exposed oxygen-containing functional groups, 532
via graphene oxide, 532–533
via non-covalent interactions, 533–534
synthetic covalent functionalization, 533
Graphone, 148
Ground states (GS), 412
Group velocity delay (GVD), 556
GS, *see* Ground states (GS)
GSL, *see* Graphene superlattice (GSL)
GSPID, *see* DOX-loaded GSPI-based system (GSPID)
GSs, *see* Graphene sheets (GSs)
GVD, *see* Group velocity delay (GVD)
GW approximation, 236; *see also* Graphene electronic properties modification

H

HAADF, *see* High-angle annular dark-field (HAADF)
Halpin–Tsai model, 89, 90
Hartree–Fock (HF), 234
Haus's master equation, 561; *see also* Ultrafast laser
HCPT, *see* 10-Hydroxy camptothecin (HCPT)
HDPE, *see* High density polyethylene (HDPE)
Heme oxygenase 1 (HO1), 102
Hexagonal boron nitride, 424
Heyd–Scuseria–Ernzerhof (HSE), 157
HF, *see* Hartree–Fock (HF); Hydrofluoric acid (HF)
HFCVD, *see* Hot filament chemical vapor deposition (HFCVD)
High-angle annular dark-field (HAADF), 341; *see also* Quantitative transmission electron microscopy
High density polyethylene (HDPE), 520
Highest occupied molecular orbital (HOMO), 183
Highly oriented pyrolytic graphite (HOPG), 265
High-resolution (HR), 24
High-resolution transmission electron microscopy (HRTEM), 6, 129, 183, 265, 498
HK theorem, *see* Hohenberg–Kohn theorem (HK theorem)
HMM, *see* Homogeneous melting (HMM)
HO1, *see* Heme oxygenase 1 (HO1)
Hohenberg–Kohn theorem (HK theorem), 234
HOMO, *see* Highest occupied molecular orbital (HOMO)
Homogeneous melting (HMM), 133
HOPG, *see* Highly oriented pyrolytic graphite (HOPG)
Horseradish peroxidase-secondary antibodies (HRP-Ab2), 101
Hot filament chemical vapor deposition (HFCVD), 518
HPPP model, *see* Hubbard–Pariser–Parr–Pople model (HPPP model)
HR, *see* High-resolution (HR)
HRP-Ab2, *see* Horseradish peroxidase-secondary antibodies (HRP-Ab2)
HRTEM, *see* High-resolution transmission electron microscopy (HRTEM)
HSE, *see* Heyd–Scuseria–Ernzerhof (HSE)

Hubbard model, 32
Hubbard–Pariser–Parr–Pople model (HPPP model), 23
Hydrazine monohydrate, 380
Hydrofluoric acid (HF), 424, 482
Hydrogels, 392
Hydrogen adsorption on graphene, 222; *see also* Defects in graphene; Graphene
adsorption energies, 224
adsorption sites, 224
with alkali metals, 228–229
charge transfer from K atom, 228
chemisorption of hydrogen dimers, 224
chemisorptions, 223–224
dependence of binding energy vs. adsorbed H_2, 229
deviation of carbon atoms, 223
DOS for spin up and down, 223
with MVs, 227–228
physisorption, 224–225
types, 222–223
Hydrogenated wurtzite SiC nanofilm, 147; *see also* Phase transformation in nanofilm
Hydrogen storage inside graphene-based materials, 111, 120–121
adsorption characteristics, 120
atom–atom interaction, 114
basis of, 114–116
computational simulation techniques, 113
GOFs, 112
modified equations of state, 115
numerical results, 116–120
porous materials, 112
repeated unit of GOFs, 114
total gravimetric uptakes, 117, 118, 119, 120
total potential energy, 117, 118
Hydrothermal process, 172
10-Hydroxy camptothecin (HCPT), 103

I

ICVD, *see* Injection-CVD process (ICVD)
ILs, *see* Ionic liquids (ILs)
Indium tin oxide (ITO), 179
Injection-CVD process (ICVD), 281
Integer quantum Hall effect (IQHE), 408
Invisibility cloaking, 67; *see also* Graphene plasmonics
Ionic liquids (ILs), 217
IPA, *see* Isopropyl alcohol (IPA)
IQHE, *see* Integer quantum Hall effect (IQHE)
Isopropyl alcohol (IPA), 489
ITO, *see* Indium tin oxide (ITO)

J

Jellium model, 132

K

Kelly sidebands, 564; *see also* Ultrafast laser
Kerr effect, 563; *see also* Ultrafast laser
Kohn–Sham (KS), 234
KS, *see* Kohn–Sham (KS)

L

LDA, *see* Local density approximation (LDA)
LDOS, *see* Local density of states (LDOS)
LEDs, *see* Light emitting diodes (LEDs)

LEEM, *see* Low energy electron microscopy (LEEM)
LIBs, *see* Lithium ion batteries (LIBs)
Light emitting diodes (LEDs), 195
Liquid nucleation and growth (LNG), 133
Liquid skin melting (LSM), 133
Lithium ion batteries (LIBs), 435
LNG, *see* Liquid nucleation and growth (LNG)
Local density approximation (LDA), 152, 224, 235–236, 250; *see also* Graphene electronic properties modification
Local density of states (LDOS), 240
Local spin density approximation (LSDA), 411
Lorentz–Drude model, 370; *see also* Computational modeling
Low energy electron microscopy (LEEM), 319
Lowest unoccupied molecular orbital (LUMO), 183
LSDA, *see* Local spin density approximation (LSDA)
LSM, *see* Liquid skin melting (LSM)
LUMO, *see* Lowest unoccupied molecular orbital (LUMO)

M

MA, *see* Maleic anhydride (MA)
Magic bullets, 113
Magnetic entropy; *see also* Magnetocaloric effect (MCE); Magnetocaloric potentials
    change for 3D system, 547, 548
    density of states, 545
    grand potential, 544
    of nonrelativistic diamagnets, 544
    of relativistic diamagnets, 549–551
    single particle Dirac Hamiltonian, 549
    3D nonrelativistic diamagnets, 545–546
    2D nonrelativistic diamagnets, 546–547
Magnetic field integral equation (MFIE), 368; *see also* Computational modeling
Magnetocaloric effect (MCE), 541, 551–552; *see also* Magnetic entropy; Magnetocaloric potentials
    adiabatic temperature change, 543–544
    electron gas under magnetic field, 544
    Ericsson cycle, 542, 543
    fundamentals of, 542
    magnetic entropy change, 543
    MCE of diamagnets, 552
    numerical entropy for ferromagnetic system, 542
Magnetocaloric potentials; *see also* Magnetic entropy; Magnetocaloric effect (MCE)
    adiabatic temperature change, 549
    of graphene, 551
    of nonrelativistic diamagnets, 547
    oscillating magnetic entropy change, 549, 551
    oscillatory behavior for adiabatic temperature, 548
    3D diamagnets, 547–549
    2D diamagnets, 549
Maleic anhydride (MA), 216
Mathieu equation, 337
MBDT, *see* 4-Mercapto-benzenediazonium tetrafluoroborate (MBDT)
MCE, *see* Magnetocaloric effect (MCE)
MD, *see* Modulation depth (MD); Molecular dynamics (MD)
MEBI, *see* MeV electron beam irradiation (MEBI)
Mechanically exfoliated graphene (MEG), 211

MEG, *see* Mechanically exfoliated graphene (MEG)
Melamine polyphosphate (MPP), 90
MEP, *see* Minimum energy path (MEP)
4-Mercapto-benzenediazonium tetrafluoroborate (MBDT), 218
Messenger RNA (mRNA), 102
Metal–insulator–metal (MIM), 70
Metallic nanoparticles, 135
Metal-organic frameworks (MOFs), 112
Metamaterials, 72; *see also* Graphene plasmonics
MeV electron beam irradiation (MEBI), 191
MFIE, *see* Magnetic field integral equation (MFIE)
Micro-Raman spectroscopy, 277
MIM, *see* Metal–insulator–metal (MIM)
Minimum energy path (MEP), 150
    for transition processes, 151
MLEDFL, *see* Mode-locked erbium-doped fiber laser (MLEDFL)
MLFLs, *see* Mode-locked fiber lasers (MLFLs)
MLG, *see* Monolayer graphene (MLG); Multilayer graphene (MLG)
MLGNs, *see* Multilayer graphene nanosheets (MLGNs)
Mode-locked erbium-doped fiber laser (MLEDFL), 556; *see also* Ultrafast laser
Mode-locked fiber lasers (MLFLs), 555; *see also* Ultrafast laser
Mode locking, 558; *see also* Ultrafast laser
    actively mode-locking laser, 558
    drawbacks, 558
    optical pulse formation and net gain in time domain, 558, 559
    passively mode-locking laser, 559, 563
    principle of, 556–558
    ultrafast laser with fast SA, 561–564
Modulation depth (MD), 560
Modulation-doped graphene, 171; *see also* N-doped graphene (NG)
MOFs, *see* Metal-organic frameworks (MOFs)
Molecular dynamics (MD), 80, 123
Molecular graph, 159
Monolayer graphene (MLG), 402, 426; *see also* Graphene
    real space lattice structure of, 403
Mono-vacancy (MV), 225
MPP, *see* Melamine polyphosphate (MPP)
mRNA, *see* Messenger RNA (mRNA)
MSC, *see* Multi-scaled composites (MSC)
Multilayer graphene (MLG), 519
Multilayer graphene nanosheets (MLGNs), 216
Multi-scaled composites (MSC), 517
Multiwalled carbon nanotube (MWCNT), 123
MV, *see* Mono-vacancy (MV)
MWCNT, *see* Multiwalled carbon nanotube (MWCNT)

N

NaDDBS, *see* Sodium dodecylbenzene sulfonate (NaDDBS)
NADH, *see* Nicotinamide adenine dinucleotide (NADH)
Nanoelectromechanical sensor (NEMS), 412
Nanofabricated infrared metamaterial, 74; *see also* Graphene plasmonics
Nanographenes (nG), 264
Nanomedicine; *see also* Drug delivery systems; Graphene; Tumor disease diagnostics

carbon-based nanocarriers, 100
    GFN and GO in, 101
Nanoparticles (NPs), 376, 529; *see also* Graphene-based 3D architectures
Nanoribbons; *see also* Atomic arrangement of graphene
    zigzag boron nitride, 54
    zigzag graphene, 50, 411
Nanotechnology, 99; *see also* Drug delivery systems; Graphene; Nanomedicine; Tumor disease diagnostics
N-doped CNT-graphene hybrid nanostructures, 240; *see also* Graphene electronic properties modification
N-doped graphene (NG), 167, 177, 185, 215; *see also* Doped graphene sheets
    arc discharge, 170–171
    CVs. of GF electrode, 176
    deduced methods, 173–174
    denotation process, 173
    doping nitrogen into graphene, 167–170
    for energy storage in ultracapacitors, 174–177
    hydrothermal process, 172
    modulation-doped graphene, 171
    N content of, 168
    nitrogen configuration mappings, 169
    nitrogen types in, 168
    by plasma treatment, 172
    preparation procedures, 173
    redox reaction in alkaline electrolyte, 174
    solvothermal reaction, 172–173
    supercritical reaction, 173–174
    ultracapacitors' capacitances, 174
    XPS survey of synthesized, 174
Near-field scanning optical microscopes (NSOM), 67
Nearly free electron (NFE), 152
NEGF, *see* Nonequilibrium Green's function (NEGF)
NEMS, *see* Nanoelectromechanical sensor (NEMS)
NFE, *see* Nearly free electron (NFE)
nG, *see* Nanographenes (nG)
NG, *see* N-doped graphene (NG)
N-GNSs, *see* Nitrogen doped GNSs (N-GNSs)
N-GQDs, *see* Nitrogen-doped graphene quantum dots (N-GQDs)
Nicotinamide adenine dinucleotide (NADH), 197
Nitrogen-doped carbon nanomaterials, 167
Nitrogen doped GNSs (N-GNSs), 444, 446; *see also* Graphene nanosheet anode
Nitrogen-doped graphene quantum dots (N-GQDs), 102
N-Methylpyrrolidone (NMP), 479
NMP, *see* *N*-Methylpyrrolidone (NMP)
NMR, *see* Nuclear magnetic resonance (NMR)
4NNFC method, *see* Fourth-nearest-neighbor force constant method (4NNFC method)
Non-carbide forming elements, 142
Nonequilibrium Green's function (NEGF), 148, 409
NPs, *see* Nanoparticles (NPs)
NSOM, *see* Near-field scanning optical microscopes (NSOM)
Nuclear magnetic resonance (NMR), 266

O

Octadecylamine (ODA), 217
ODA, *see* Octadecylamine (ODA)

ODPA, *see* 4, 4′-Oxydiphthalic anhydride (ODPA)
OFETs, *see* Organic field-effect transistors (OFETs)
OLEDs, *see* Organic light-emitting diodes (OLEDs)
ON/OFF ratio, 16
OPVs. cells, *see* Organic photovoltaic cells (OPVs. cells)
Organic field-effect transistors (OFETs), 190
Organic light-emitting diodes (OLEDs), 189
Organic photovoltaic cells (OPVs. cells), 193–195, 197; *see also* Doped graphene sheets
Oridonin, 103
ORR, *see* Oxygen reduction reaction (ORR)
4, 4′-Oxydiphthalic anhydride (ODPA), 566
Oxygen reduction reaction (ORR), 167
Ozawa equation, 83

P

PAA, *see* Poly(acrylic acid) (PAA)
Partially reflective surface (PRS), 66
PBA, *see* Pyrene butyric acid (PBA)
PBE, *see* Perdew, Burke, and Enzerhof (PBE)
PC, *see* Polarization controller (PC)
PDDA, *see* Polydimethyldiallylammonium chloride (PDDA)
PDEA, *see* Poly[2-(diethylamino) ethyl methacrylate] (PDEA)
pDNA, *see* Plasmid DNA (pDNA)
PDOS, *see* Projected density of states (PDOS)
PE, *see* Polyethylene (PE)
PEC, *see* Perfect electric conductor (PEC)
PECVD, *see* Plasma-enhanced chemical vapor deposition (PECVD)
PEG, *see* Polyethylene glycol (PEG)
PEI, *see* Polyethyleneimine (PEI)
PEI-grafted ultrasmall graphene oxide (PEI-g-USGO), 104
PEI-g-USGO, *see* PEI-grafted ultrasmall graphene oxide (PEI-g-USGO)
PEL, *see* Potential energy landscape (PEL)
PEMFCs, *see* Proton exchange membrane fuel cells (PEMFCs)
PEO, *see* Poly(ethylene oxide) (PEO)
Perdew, Burke, and Enzerhof (PBE), 236, 250
Perfect electric conductor (PEC), 371
Perfluorophenyl azides (PFPAs), 212, 213
Periodic Graphene Structures (PGSs), 470; *see also* Graphene-based structures
PFPAs, *see* Perfluorophenyl azides (PFPAs)
P-glycoprotein (P-gp), 103
P-gp, *see* P-glycoprotein (P-gp)
PGSs, *see* Periodic Graphene Structures (PGSs)
Phase transformation in nanofilm, 147, 157
 ab initio DFT calculations, 150
 bilayer graphene chair conformations, 153
 chemical functionalized 2D SiC sheets, 153–155
 chemically induced, 147, 153
 density of states, 155
 diamondol, 150
 electronic band structure of H-(CSi)$_2$, 156
 formations of chemical bond, 150
 functionalized bilayer graphene, 150–152
 functionalized graphene/h-BN heterostructures, 152–153
 functionalized single graphene sheet, 147–149
 graphane structure in chair conformation, 148
 graphene nanoribbons, 149

graphone, 148
 hydrogenated wurtzite SiC nanofilm, 155–157
 initial and optimized geometries, 151
 MEP calculations, 150, 154
 MEP for transition processes, 151
 optimized structures of semihydrogenated SiC sheets, 154
 research on functionalized graphene sheet, 150
PHEVs, *see* Plug-in hybrid electric vehicles (PHEVs)
P3HT, *see* Poly (3-hexylthiophene) (P3HT)
Plasma-enhanced chemical vapor deposition (PECVD), 168
Plasma-enhanced models, 307
Plasma processing, 185; *see also* Doped graphene sheets
Plasmid DNA (pDNA), 103
Plasmonic cloaking, 67; *see also* Graphene plasmonics
PLED, *see* Polymer light-emitting diodes (PLED)
PLOM, *see* Polarized light optical microscopy (PLOM)
Plug-in hybrid electric vehicles (PHEVs), 435
PMDS, *see* Polydimethylsiloxane (PMDS)
PMMA, *see* Poly methyl methacrylate (PMMA)
PNIPAM, *see* Poly(*N*-isopropylacrylamide) (PNIPAM)
POA, *see* Poly (oxyalkylene) amines (POA)
POE, *see* Poly(oxyethylene) (POE)
POEM, *see* Poly(oxyethylene)-segmented imide (POEM)
Polarization controller (PC), 569
Polarized light optical microscopy (PLOM), 81
Poly(acrylic acid) (PAA), 197
Poly[2-(diethylamino) ethyl methacrylate] (PDEA), 103
Polydimethyldiallylammonium chloride (PDDA), 393
Polydimethylsiloxane (PMDS), 8
Polyethylene (PE), 80
Polyethylene glycol (PEG), 102
Poly(ethylene oxide) (PEO), 393
Polyethyleneimine (PEI), 188, 197, 393
Polyhex nanotorus, 160; *see also* Graphene (GFN)
Poly (3-hexylthiophene) (P3HT), 217
Poly(*N*-isopropylacrylamide) (PNIPAM), 211, 531
Polymer light-emitting diodes (PLED), 240
Poly methyl methacrylate (PMMA), 84, 520
 grafted GO, 85
Poly (oxyalkylene) amines (POA), 216
Poly(oxyethylene) (POE), 566
Poly(oxyethylene)-segmented imide (POEM), 555, 565, 566; *see also* Graphene SA optical properties
Poly(2-phenylpropyl) methysiloxane (PPMS), 12
Poly(propionylethylenimine-co-ethylenimine) (PPEI-EI), 531
Polypropylene (PP), 520
Poly(sodium 4-styrenesulfonate) (PSS), 390, 533
Poly(sodium acrylate) (PSA), 393
Polystyrene (PS), 12
Poly(vinyl acetate-co-vinyl alcohol) (PVA-VA), 531
Poly (vinyl alcohol) (PVA), 216, 393
Poly(vinyl pyrrolidone) (PVP), 393
Potential energy landscape (PEL), 257
1/*r* Power law, 25
PP, *see* Polypropylene (PP)
PPEI-EI, *see* Poly(propionylethylenimine-co-ethylenimine) (PPEI-EI)
PPMS, *see* Poly(2-phenylpropyl) methysiloxane (PPMS)

Preferential sticking mechanism, 33; *see also* Atomic-scale defects
Principal resonance structures, 25; *see also* Atomic-scale defects
Pristine graphene, 78
Projected density of states (PDOS), 26, 253
Prostate-specific antigen (PSA), 101
Proton exchange membrane fuel cells (PEMFCs), 309; *see also* Doped graphene properties
PRS, *see* Partially reflective surface (PRS)
PS, *see* Polystyrene (PS)
PSA, *see* Poly(sodium acrylate) (PSA); Prostate-specific antigen (PSA)
PSS, *see* Poly(sodium 4-styrenesulfonate) (PSS)
Puckering, *see* Surface reconstruction
Pulse mode of laser, 556; *see also* Ultrafast laser
PVA, *see* Poly (vinyl alcohol) (PVA)
PVA-VA, *see* Poly(vinyl acetate-co-vinyl alcohol) (PVA-VA)
PVP, *see* Poly(vinyl pyrrolidone) (PVP)
Pyrene butyric acid (PBA), 533
1,3,6,8-pyrenetetrasulfonic acid (TPA), 188
pZ vacancies, 34

Q

QDs, *see* Quantum dots (QDs)
QED, *see* Quantum electrodynamics (QED)
QFT, *see* Quantum field theory (QFT)
QHE, *see* Quantum Hall effects (QHE)
Quantitative HAADF scanning transmission microscopy, 342, 354; *see also* Quantitative transmission electron microscopy
 electron collection, 343
 graphene sheets, 345
 HAADF image, 342
 images from Au nanoparticles, 345
 images of clusters, 343, 344
 intensity analysis, 344
Quantitative transmission electron microscopy, 341, 354–355; *see also* Graphene sheets (GSs)
 calibration curves, 348, 349
 carbon nanoparticles, 350
 electron scattering in graphene sheets, 351–354
 graphene mass standard verification, 346–349
 HAADF-intensity vs. GS thickness relationship, 352
 inhomogeneous structures of carbon nanoparticles, 351
 molecular dynamic simulation, 348
 quantifying graphene sheets, 344–346
 quantitative HAADF scanning transmission microscopy, 342–344
 weighing carbon clusters, 349–351
Quantum dots (QDs), 101
Quantum dynamics, 328
Quantum electrodynamics (QED), 328
Quantum field theory (QFT), 327
Quantum Hall effects (QHE), 401

R

Radical addition fragmentation transfer (RAFT), 211
RAFT, *see* Radical addition fragmentation transfer (RAFT)
Raman breathing mode (RBM), 239
RBM, *see* Raman breathing mode (RBM)

Reaction ion etching (RIE), 8
Reactive oxygen species (ROS), 102
Reduced graphene oxide (rGO or RGO), 99, 208
Resonating valence bond model (RVB model), 23; *see also* Atomic-scale defects
Resorcinol–formaldehyde (RF), 264
Reverse osmosis (RO), 272
RF, *see* Resorcinol–formaldehyde (RF)
rGO, *see* Reduced graphene oxide (rGO or RGO)
RGO, *see* Reduced graphene oxide (rGO or RGO)
RIE, *see* Reaction ion etching (RIE)
RO, *see* Reverse osmosis (RO)
ROS, *see* Reactive oxygen species (ROS)
RVB model, *see* Resonating valence bond model (RVB model)

S

SA, *see* Saturable absorber (SA)
SAD, *see* Selected area diffraction (SAD)
SAED, *see* Selected-area electron diffraction (SAED)
SAM, *see* Self-amplitude modulation (SAM)
SAMs, *see* Self-assembled monolayers (SAMs)
Saturable absorber (SA), 555; *see also* Ultrafast laser
Scanning electron microscopy of graphene, 319, 324
    edge contrast, 323
    few-layer graphene on Ni, 322
    graphene segregation, 322, 323, 324
    graphene thickness dependence, 322
    insulator surface, 320–321
    metal surface, 321–322
    monolayer graphene, 322–323
    multilayered graphene, 323–324
    SE contrast of monolayer, 320, 321
Scattering width (SW), 68
SCGSs, *see* Substrate Cantor Graphene Structures (SCGSs)
Scherrer's formula, 280
Schrödinger equation, 250
Schwartzite, 253; *see also* Carbon sp^2 phases
Scotch tape method, 179, 180; *see also* Graphene
SDD, *see* Stuttgart—Dresden pseudopotential (SDD)
SDS, *see* Sodium dodecyl sulfonate (SDS)
SE, *see* Secondary electron (SE)
Secondary electron (SE), 320
Secular equation, 43; *see also* Atomic arrangement of graphene
SEI, *see* Solid electrolyte interphase (SEI)
Selected area diffraction (SAD), 141
Selected-area electron diffraction (SAED), 183
Self-amplitude modulation (SAM), 558, 561
Self-assembled monolayers (SAMs), 183
Self-phase modulation (SPM), 556, 562
Semiconductor saturable absorber mirror (SESAM), 556
Semiconductor superlattice, 454
Semihydrogenation (SH), 152
SERS, *see* Surface enhanced Raman scattering (SERS)
SESAM, *see* Semiconductor saturable absorber mirror (SESAM)
SFGSs, *see* Substrate FGSs (SFGSs)
SG, *see* Single layer graphene (SG or SLG)
SGBSs, *see* Substrate graphene-based structures (SGBSs)
SGC, *see* Synthetic growth concept (SGC)
SG, *see* Single layer graphene (SG or SLG)

SGs, *see* Suspended graphene sheets (SGs)
SGSLs, *see* Substrate GSLs (SGSLs)
SH, *see* Semihydrogenation (SH)
Silicon-based transistors, 249
Silicon wafers, 213
Single lap joints (SLJs), 517
Single layer graphene (SG or SLG), 217, 233, 249, 344
Single-walled carbon nanotube (SWCNT), 123
Single-ζ (SZ), 251
SL, *see* Superlattice (SL)
SLG, *see* Single layer graphene (SG or SLG)
SLJs, *see* Single lap joints (SLJs)
Sodium dodecylbenzene sulfonate (NaDDBS), 531
Sodium dodecyl sulfonate (SDS), 531
Soft x-ray emission (SXE), 265
Solid electrolyte interphase (SEI), 438
Solvent techniques, 222
Solvothermal processes, 172
SPANI, *see* Sulfonated polyaniline (SPANI)
Spin-half Kondo effect, 31; *see also* Atomic-scale defects
SPM, *see* Self-phase modulation (SPM)
SPP, *see* Surface plasmon polaritons (SPP)
Stone–Wales (SW), 232, 255; *see also* Carbon-based thin films
    defects, 503
Structural defects, 410; *see also* Graphene
Stuttgart—Dresden pseudopotential (SDD), 313
Substitutional doping, 215
Substrate Cantor Graphene Structures (SCGSs), 463; *see also* Graphene-based structures
Substrate FGSs (SFGSs), 470; *see also* Graphene-based structures
Substrate graphene-based structures (SGBSs), 453; *see also* Graphene-based structures
Substrate GSLs (SGSLs), 458; *see also* Graphene-based structures
Sulfonated polyaniline (SPANI), 533
Supercapacitors, *see* Ultracapacitors (UCs)
Superlattice (SL), 39
Surface enhanced Raman scattering (SERS), 534
Surface plasmon polaritons (SPP), 63
Surface reconstruction, 24, 25; *see also* Atomic-scale defects
Suspended graphene sheets (SGs), 78
SW, *see* Scattering width (SW); Stone–Wales (SW)
SWCNT, *see* Single-walled carbon nanotube (SWCNT)
SXE, *see* Soft x-ray emission (SXE)
Synthetic growth concept (SGC), 499
SZ, *see* Single-ζ (SZ)

T

ta-C, *see* Tetrahedral amorphous carbon (ta-C)
TB, *see* Tight-binding (TB)
TCNE, *see* Tetracyanoethylene (TCNE)
TCNQ, *see* 7,7,8,8-tetracyanoquinodimethane (TCNQ)
TCP, *see* Thickness and concentration product (TCP)
TE, *see* Transverse electric (TE)
Tensors, 332; *see also* Curved graphene nanoribbon
Terahertz (THz), 63
Tetracyanoethylene (TCNE), 239
7,7,8,8-tetracyanoquinodimethane (TCNQ), 533

Tetrafluorotetracyanoquinodimethane (F4-TCNQ), 189
Tetragon defect, 507; *see also* Carbon-based thin films
Tetrahedral amorphous carbon (ta-C), 498
Tetrahydrofuran (THF), 566
Tetramethylammonium chloride (TMAC), 393
Tetrathiafulvalene (TTF), 191, 239
TFTs, *see* Thin-film transistors (TFTs)
TGA, *see* Thermogravimetric analysis (TGA)
Thermogravimetric analysis (TGA), 92
THF, *see* Tetrahydrofuran (THF)
Thickness and concentration product (TCP), 555
Thin-film transistors (TFTs), 186
Thioredoxin reductase (TrxR), 102
3D-AFM-hyperfine imaging of graphene monolayer deposits, 277, 286
    DC-electrical conductivity measurements, 285–286
    electrons in conduction band vs. spot location, 286
    experiments and discussion, 279
    graphene characterization, 282
    graphene deposition on substrate, 281–282
    graphene film preparation, 278
    micro-Raman spectroscopy, 277
    mode frequencies of Raman spectra, 281
    monolayer tunable periodic graphene nanostructures, 279
    normal AFM-tapping mode imaging, 284
    phase identification, 280
    polypyrrole-based graphene nanostructure, 279
    Raman and nano-structural features of graphene monolayers, 282
    Raman spectrogram, 281
    Scherrer's formula, 280
    superconducting measurements, 281
    surface topology, 282–285
    3D-AFM micrograph, 283
    2D-AFM-gray scale photo-imaging, 284
    xz-graphene surface histogram, 285
    123-YBCO superconducting substrate, 279–280
THz, *see* Terahertz (THz)
THz metaferrite, 73; *see also* Graphene plasmonics
Tight-binding (TB), 39, 226; *see also* Atomic arrangement of graphene
    band structures, 40
    example of, 41–42
    Hamiltonian based on, 49
    Hamiltonian matrix elements, 43, 44
    honeycomb lattice, 40
    secular equation, 43
TM, *see* Transverse magnetic (TM)
TMAC, *see* Tetramethylammonium chloride (TMAC)
TMDs, *see* Transition metal dichalcogenides (TMDs)
TMV, *see* Tobacco mosaic virus (TMV)
Tobacco mosaic virus (TMV), 342
TPA, *see* 1,3,6,8-pyrenetetrasulfonic acid (TPA)
Traditional wave functional methods, 234; *see also* Graphene electronic properties modification
Transfer-matrix method, 67, 71; *see also* Graphene plasmonics
Transfer processes, 182; *see also* Graphene
Transitional voltage, 431
Transition metal dichalcogenides (TMDs), 3

Transmission-line circuit model, 67, 71; *see also* Graphene plasmonics
Transverse electric (TE), 69
Transverse magnetic (TM), 67, 69
TrxR, *see* Thioredoxin reductase (TrxR)
TTF, *see* Tetrathiafulvalene (TTF)
Tumor disease diagnostics, 100; *see also* Drug delivery systems; Graphene; Nanomedicine
  biosensing of tumor cells, 100–102
  imaging of tumor cells, 102
2DEG, *see* 2D electron gas (2DEG)
2D electron gas (2DEG), 232

U

UCs, *see* Ultracapacitors (UCs)
UHV, *see* Ultrahigh vacuum (UHV)
Ultracapacitors (UCs), 167, 174; *see also* N-doped graphene (NG)
  based on NG, 175
Ultrafast laser, 556; *see also* Fiber-laser mode locking with graphene SA; Graphene film preparation; Graphene SA optical properties; Mode locking
  carrier dynamics of graphene, 560
  cavity longitudinal modes, 557
  chirp as function of GVD and SPM, 564
  with fast SA mode-locking, 561–564
  free-running laser, 556
  gain medium, 556
  graphene-based saturable absorber, 559–561
  graphene honeycomb lattice structure and band structure, 559
  Haus's master equation, 561
  Kelly sidebands, 564

Kerr effect, 563
  net gain profile, 557
  nonlinear absorption coefficient, 560
  nonlinear transmission curve of SA, 560, 561
  optical electric field, 558
  passively mode-locking laser, 563
  phase matching condition, 565
  pulse and spectrum measurement, 565
  pulse mode of laser, 556
  pulse shaping effect, 561
  pulse spectral sidebands, 564–565
  pulsewidth as function of GVD and SPM, 563
  pumping source, 556
  resultant intensity, 558
  spectrum of multimode laser oscillation, 557
Ultrahigh vacuum (UHV), 6
Unique unit cell, 58, 59; *see also* Atomic arrangement of graphene

V

Valence band maximum (VBM), 152
Valence force field (VFF), 409
van der Waals (VdW), 3, 223
Variable range hopping (VRH), 279
VASP, *see* Vienna *ab initio* simulation package (VASP)
VBM, *see* Valence band maximum (VBM)
VDOS, *see* Vibrational density of states (VDOS)
vdW, *see* van der Waals (vdW)
VFF, *see* Valence force field (VFF)
Vibrational density of states (VDOS), 259
Vienna *ab initio* simulation package (VASP), 250, 314
Viologen, 187; *see also* Doped graphene sheets
VRH, *see* Variable range hopping (VRH)

W

Waterborne PU (WPU), 84
WAXD, *see* Wide-angle x-ray diffraction (WAXD)
Wide-angle x-ray diffraction (WAXD), 81
Wooten–Weaire–Winer models (WWW models), 249
WPU, *see* Waterborne PU (WPU)
WWW models, *see* Wooten–Weaire–Winer models (WWW models)

X

XAS, *see* X-ray absorption spectroscopy (XAS)
XLPE, *see* Cross-linked PE (XLPE)
XPS, *see* X-ray photoelectronic spectroscopy (XPS)
X-ray absorption spectroscopy (XAS), 265
X-ray diffraction (XRD), 265
X-ray photoelectronic spectroscopy (XPS), 11, 168
XRD, *see* X-ray diffraction (XRD)

Z

ZBNR, *see* Zigzag boron nitride nanoribbons (ZBNR)
Zeolitic imidazolate frameworks (ZIFs), 112
ZGNR, *see* Zigzag graphene nanoribbon (ZGNR)
ZIFs, *see* Zeolitic imidazolate frameworks (ZIFs)
Zigzag boron nitride nanoribbons (ZBNR), 54; *see also* Atomic arrangement of graphene
Zigzag graphene nanoribbon (ZGNR), 50, 411
Zigzag triwing graphene (ZZ-TWG), 241
ZZ-TWG, *see* Zigzag triwing graphene (ZZ-TWG)